BAYESIAN STATISTICS 3

Proceedings of the Third Valencia International Meeting

June 1–5, 1987

Edited by
J. M. Bernardo
M. H. Degroot
D. V. Lindley
and
A. F. M. Smith

CLARENDON PRESS · OXFORD
1988

Oxford University Press, Walton Street, Oxford OX2 6DP
Oxford New York Toronto
Delhi Bombay Calcutta Madras Karachi
Petaling Jaya Singapore Hong Kong Tokyo
Nairobi Dar es Salaam Cape Town
Melbourne Auckland
and associated companies in
Berlin Ibadan

Oxford is a trademark of Oxford University Press

Published in the United States
by Oxford University Press, New York

British Library Cataloguing in Publication Data
(Data available)

Library of Congress Cataloging in Publication Data
(Data available)

ISBN 0-19-852220-7

Typeset by βeta, S.A., Dr. Moliner, 2.
46010 – Valencia, Spain

Printed in Great Britain by
The Alden Press, Oxford

A MESSAGE FROM THE GOVERNOR OF THE STATE OF VALENCIA

"Como cada cuatro años, la Generalitat Valenciana tuvo la satisfaccion, el verano pasado, de patrocinar una reunión de la comunidad científica mundial interesada en los metodos estadísticos mas modernos. Con ocasion de la publicacion del volumen que recoge los resultados obtenidos en esa tercera reunión, quiero expresar mi deseo de que contribuya a potenciar el interes por nuestra comunidad entre los estadísticos de todo el mundo, y quiero adelantarles mi bienvenida para "Valencia IV" en 1991".

Joan Lerma

As is now becoming customary every four years, the Government of the State of Valencia had again last summer the satisfaction of sponsoring a meeting of the world scientific community concerned with state-of-the-art statistical methodology.
On the occasion of the publication of the Proceedings of the Third International Valencia Meeting, I wish to express my hope that it will contribute to bringing the State of Valencia to the attention of the world statistical community. I would also like to express, well in advance, a warm welcome to the 1991 Fourth Valencia Meeting.

Joan Lerma
Governor of the State of Valencia

PREFACE

The Third Valencia International Meeting on Bayesian Statistics took place from 1–5 June, 1987, at the Hotel Cap Negret, Altea, 120 kms south of Valencia, Spain.

The Scientific Programme consisted of 31 invited papers by leading experts in the field, each of which was followed by a discussion, and 3 poster sessions of contributed papers. These Proceedings contain all the invited papers, with associated discussion, together with a selection of 33 contributed papers, which were subject to a refereeing process. The selection of topics, authors and discussants ensures that those Proceedings provide a definitive up-to-date overview of current concerns and activity in Bayesian Statistics, encompassing a wide range of theoretical and applied research.

Developments arising from foundational issues are presented by **Diaconis** and by **Hill**: the former reviews recent work on exchangeability; the latter considers nonparametric data analysis. Exchangeability ideas also feature in the subjectivist approach to data analysis presented by **Goldstein**.

Issues arising in the modelling and analysis of very high-dimensional problems are considered by several authors. The important area of image-processing is reviewed by **Geman**. **H. Rubin** explores ridge-regression for high-dimensional regression problems; **Dawid** considers conjugate analysis when there are an infinite number of regressors, and **Wahba** presents recent ideas on the use of splines.

The topic of hierarchical modelling receives considerable attention: **Polasek** and **Pötzelberger** deal with robustness issues; **Pericchi** and **Nazaret** with imprecision; **Jewell** with heteroscedasticity, and **Albert** with log-linear models. Related problems with variance components are considered by **Jelenkowska**.

General robustness issues are also treated by several authors. **Berger** and **O'Hagan** consider ranges of inferences resulting from prior quantile specification; **O'Hagan** examines modelling with heavy-tailed distributions; **Bagchi** and **Guttman** also examine uses of non-normal distributions, and **Guttman** and **Peña** deal with outliers and influence.

Various mathematical aspects of inference problems are considered by **Amaral** and **Turkman**; **Cano**; **Hernández** and **Moreno**; **Cuevas** and **Sanz**; **De la Horra**; **García-Carrasco**; **Main**; **Makov**, and **Willing**.

Time series problems are considered by several authors: **Marriott** investigates reparametrisation strategies; **Pole** analyses transfer response models; **Quintana** and **West** provide an application of dynamic modelling to compositional data, and **Schnatter** uses Gaussian sum approximations in forecasting.

Decision and Design feature in the contributions of **Felsenstein**; **Owen** and **Cain**; **J. Q. Smith**, and **Spezzaferri**.

A special feature of the meeting was a panel session on computation and Bayesian software needs, with contributions from **Kass**, **Tierney** and **Kadane**; **Shaw**; **Goel**, and **A. F. M. Smith**. Related work on computational and approximation techniques is presented by **D. B. Rubin** and by **Morris**, and, in the context of finite mixture models, by **Bernardo** and **Girón**.

The important problems of multiple comparisons and testing are examined by **Berry** and **Du Mouchel.** Other aspects of inference in regression models are considered by **Consoni** and **Veronese; Mäkeläinen** and **Brown; Mendoza; Poirier** and **Sweeting**; and in surveys and observational studies by **T. M. F. Smith** and by **Sugden.**

Aspects of modelling and combining expert opinions are considered by several authors: **Barlow, Spizzichino** and **Wechsler** reexamine de Finetti's ideas on group decision-making; **Bayarri** and **De Groot** explore optimal strategies for the experts' reports; **West** considers problems of modelling, and **Huseby** deals with an application to prediction.

A variety of applications are represented in the papers by **Kadane** and **Hastorf** (paleoethnobotany); **Kim** and **Schervish** (criminінology); **Lindley** (genetics); **Spiegelhalter** and **Freedman** (clinical trials); **Deely** and **Zimmer** (quality assurance); **Dunsmore** and **Boys** (screening); **Fairley** and **Fairley** (public welfare), **Grieve** (LD50 experiments), and **Lindberg** and **Fenyo** (medical diagnosis).

Another special feature of the programme was a panel session on "The Future". **Barnard** focusses on priorities for teaching and research; **Geisser** reviews and stresses the predictivist perspective, and **Zellner** proclaims an emerging Bayesian era in statistics and econometrics.

A recent review by Michael Goldstein (in the Journal of the Royal Statistical Society) of **Bayesian Statistics 2: Proceedings of the Second Valencia International Meeting** (North-Holland) begins:

> "The meetings on Bayesian Statistics at Valencia are becoming something of an institution. Once every four years, a large number of Bayesians, quasi-Bayesians, neo-Bayesians and fellow-travellers gather at a hotel by the sea near Valencia, to exchange information, engage in friendly debate and enjoy the good weather. (As the number of participants seems to double from meeting to meeting, either the subject, or the venue, must be riding a wave.)"

In fact, participation at this third Valencia meeting was over twice that at the first meeting, with 23 countries represented. There would seem therefore to be a growing identification with both the spirit and the substance of David Blackwell's memorable rendering, at the final conference dinner, of

"*Bayes! (You're the one for me),*

to the well-known tune of

Who? (Stole my heart away)."

There is indeed a wave; and it is, of course, the subject rather than the venue which is riding it, as evidenced by the increasing Bayesian presence in conference halls and journal pages far removed from Valencian sun and sea.

We are most grateful for the financial and material support provided by: the Government of Spain; the Government of the State of Valencia; The Local Council of Altea; the University of Valencia; the US Office of Naval Research, and the British Council. In particular, we are delighted to have, once again, the promise of future support contained in the personal message from the Governor of the State of Valencia (reprinted in these Proceedings immediately before this Preface, with a translation from the original Spanish).

J. M. Bernardo
M. H. DeGroot
D. V. Lindley
A. F. M. Smith

JULY 1988

CONTENTS

I. INVITED PAPERS (with discussion)

II. CONTRIBUTED PAPERS

INVITED PAPERS

BAYESIAN STATISTICS 3, pp. 1–15
J. M. Bernardo, M. H. DeGroot, D. V. Lindley and A. F. M. Smith, (Eds.)

De Finetti's Approach to Group Decision Making

R. E. BARLOW and S. WECHSLER and F. SPIZZICHINO
Univ. of California at Berkeley, Univ. of Sao Paulo and *Univ. of Rome*

SUMMARY

A group of Bayesians must make a group decision; e.g., choose one of two final actions. De Finetti (1954a,b) considered group decision making, relative to a special sequential decision problem, when all individuals have the same loss function but different opinions. In particular, he defined the "individual horizon" and the "common horizon" relative to a given group decision rule such as a voting rule. He characterized voting rules as an "average of decisions" and argued that it is better for the group to use an "average of opinions". We generalize and extend de Finetti's ideas in his (1954a, b) papers. Finally, we present some of his ideas in his (1959) Varenna lectures.

Keywords: AVERAGE OF DECISIONS; AVERAGE OF OPINIONS; GROUP DECISIONS; INDUCTIVE RULES; SEQUENTIAL DECISION PROBLEMS; VOTING RULES.

1. INTRODUCTION

A group of individual Bayesians, $I_1, \ldots, I_N$, must make a group decision d_G: i.e., choose one of 2 final actions, a_1 or a_2, relative to an unknown quantity θ. Let $\xi_i, i = 1, 2, \ldots, N$ be the individuals' initial opinions concerning the unknown quantity θ; i.e., these are their probability distributions for θ. The group as a whole has no opinion. Only individuals have opinions. Hence the usual uni-subjective Bayesian approach to decision making does not apply, at least directly. We will only consider the case where individuals in the group have a common interest; i.e., equivalent utility functions and actually identical utility functions in Sections 2-5. Throughout the paper we make the strong assumption that the individual prior and utility functions are separable and in fact that priors do not depend on contemplated decisions [cf. Rubin (1987)]. For a recent survey of ideas related to group decision making see French (1985).

Beginning with his 1950 paper "Recent Suggestions for the Reconciliation of Probabilities", de Finetti wrote a series of papers concerning group decision making. These covered at least a ten year period and were based in part on his multisubjective interpretation of Wald's admissibility theorem. In his 1959 Varenna, Italy, lectures, he makes the point that "really new developments in the theory of inductive behaviour [i.e., decision making] arise only in so far as the decisions —unlike the opinions— are made by groups rather than individuals". [See de Finetti (1972) p. 188.]

In many situations, a group of individuals will make a group decision by voting. A vote is taken and in this way the group "decides". De Finetti interprets certain voting rules as resulting in an "average of decisions" [see Section 2]. However, as de Finetti observes, in certain circumstances this means of reaching a group decision can be improved: but still remaining in the uni-subjective Bayesian framework. De Finetti's main result is that it is better for the group to use an average of opinions rather than an "average of decisions" [de Finetti (1954a)].

The objective of this paper is to extend de Finetti's approach to group decision making, as he presented it, for a class of sequential decision problems, with special attention to the ideas in his 1954 published papers. Section 2 describes the sequential decision problem of interest and the concept of "inductive decision rules" for the group. The setup and results in this section generalize and extend the ideas of de Finetti [(1952), (1954a)].

For a specified loss function $\ell(a_h, \theta)$ $(h = 1, 2)$ and a fixed group decision rule, γ (for example a group voting rule), we characterize, in Section 3, the class of Bayes' rules which all individuals in the group would consider as good or better than γ. The "common horizon" characterizes this class of Bayes' rules. This generalizes de Finetti's results in (1954a) and (1954b). Sections 4 and 5 extend de Finetti's results relative to the example described in his (1954a) paper.

2. THE GROUP SEQUENTIAL DECISION PROBLEM

The group must choose one of two final actions (or decisions) a_1 or a_2. Any individual $I_i (i = 1, 2, \ldots, N)$ in the group has a loss function $\ell_i(a_h, \theta)$ where $\theta \in \Theta$ is the value assumed by an unobservable random quantity θ. $\xi_i^{(0)}$ denotes the probability distribution on Θ representing the initial opinion of I_i about θ and does *not* depend on the decision contemplated. The group can observe random variables $X_1, X_2, \ldots (X_j \in \mathcal{X} \subset \Re, j = 1, 2, \ldots)$. In the opinion of all individuals, $X_1, X_2, \ldots$ are conditionally i.i.d. given θ with a given conditional unidimensional density $f(x|\theta)$.

Taking an observation has a constant cost c for any individual in the group. As is usual in the "uni-subjective" framework, it is assumed that, after any observation, the group is allowed either to stop and take one of the two final actions a_1 or a_2 or to take another observation. Denote by a_0 the decision to take another observation. $\xi_i^{(n)} (n = 1, 2, \ldots)$ is the distribution on Θ for the individual I_i at the n-th stage (i. e., after n observations). Obviously $\xi_i^{(n)}$ evolves according to Bayes' formula:

$$d\xi_i^{(n)}(\theta|x_1, \ldots, x_n) \quad \propto \quad f(x_1|\theta) \ldots f(x_n|\theta) d\xi_i^{(0)}(\theta)$$

Consider the class of possible sequential decision strategies for the group. A sequential decision strategy is of course defined by a sequence of mappings

$$v^{(n)} : \chi^n \quad \longrightarrow \quad \{a_0, a_1, a_2\} \quad n = 1, 2, \ldots$$

where $v^{(n)}(x_1, \ldots, x_n)$ denotes the decision to be taken by the group at stage n if the statistical results $x_1, \ldots, x_n$ have been observed. For a fixed vector $\boldsymbol{x} \equiv (x_1, \ldots, x_n) \in \chi^n$, $v^{(n)}(x_1, \ldots, x_n)$ is defined only if $v^{(m)}(x_1, \ldots, x_m) = a_0$ for all $m < n$.

From a Bayesian point of view the group should take into account only those sequential rules which respect the "*likelihood principle*" (cf. Berger et al. [1984]); i.e.,

$$f(x_1'|\theta) \ldots f(x_n'|\theta) \propto f(x_1''|\theta) \ldots f(x_n''|\theta) \quad \forall\, \theta \in \Theta$$

$$\Longrightarrow v^{(n)}(\boldsymbol{x}') = v^{(n)}(\boldsymbol{x}''). \tag{2.1}$$

A remarkable class of such rules is formed by the "inductive rules". An inductive rule can be defined as follows: Let $\mathcal{O}$ be the space of possible probability distributions on Θ ("opinions") and $\mathcal{O}^N = \mathcal{O} \times \mathcal{O} \times \cdots \times \mathcal{O}$ (N times) be the space of possible vectors formed by the opinions of the individuals in the group. Fix a mapping

$$\gamma : \quad \mathcal{O}^N \quad \longrightarrow \{a_0, a_1, a_2\}.$$

At the n-th stage the group takes the action

$$\gamma(\xi^{(n)}) = \gamma(\xi_1^{(n)}, \xi_2^{(n)}, \ldots, \xi_N^{(n)}),$$

where γ does not depend on n. So an inductive rule is characterized by a partition (Ξ_0, Ξ_1, Ξ_2) of the space $\mathcal{O}^N$. A large class of reasonable inductive rules is formed from those which de Finetti called "average of decisions". In order to define an "average of decisions", it is necessary to fix attention on "individual Bayes' rules".

At any stage $n (n = 0, 1, \ldots)$, we denote by $a_i^{*(n)}$ the Bayes decision of individual I_i, depending on the loss function $\ell_i(a_h, \theta)$, his probability distribution $\xi_i^{(n)}$ and the continuation cost c. Obviously, $a_i^{*(n)}$ would be the decision taken by the group at stage n if I_i were a "dictator," who would choose the decision for the group.

Let

$$\varphi : \quad \{a_0, a_1, a_2\}^N \quad \longrightarrow \{a_0, a_1, a_2\}$$

be a "generalized average function" [de Finetti (1931)]; i.e.,

1) φ is symmetric (2.2)

2) if $a_{h_1} = a_{h_2} = \ldots = a_{h_N} = a_h \quad (h = 0, 1, 2)$ then

$$\varphi\,(a_{h_1}, a_{h_2}, \ldots, a_{h_N}) = a_h. \tag{2.3}$$

The group decision rule is given by an average of decisions if the group takes, at every stage n, a decision (which does not depend on n)

$$\varphi(a_1^{*(n)}, a_2^{*(n)}, \ldots, a_N^{*(n)}).$$

Particular cases of common interest are obtained by considering "voting rules; "for example de Finetti (1954a):

A) the group stops observations at stage n and chooses a_h $(h = 1, 2)$ only if a_h is the value assumed by at least M elements of the set $\{a_1^{*(n)}, \ldots, a_N^{*(n)}\}$. M is a fixed integer between $(N + 1)/2$ (simple majority) and N (unanimity).

B) the group stops observations at stage n and chooses $a_h (h = 1, 2)$ only if the number N_h of individuals in the group for which $a_j^{*(n)} = a_h$ is greater than M plus the number N_{3-h} of individuals wanting to stop and take the alternative decision $a_{3-h} (1 \leq M \leq N)$.

C) as in B) but with the additional condition that

$$N_h / (N_{3-h}) \quad > \mu$$

for some $\mu > N/(N - M)$.

We can also consider voting rules with different weights given to different individuals; obviously, in this case, we lose property (2.2). Individual Bayes' rules corresponding to utilities and initial opinions of single members in the group are very special cases of weighted voting rules. Acting according to a Bayes' rule corresponding to the utility and initial opinion of an individual chosen outside the group is still a rule respecting the likelihood principle, but it is not an average of decisions. For some individual, several Bayes' rules may exist. In this case any randomization between such rules is again a Bayes' rule for the individual. We stress this obvious fact, since randomized Bayes' rules may have a special role in the group decision problem, especially when the distribution of observations is discrete. Note that it is not necessary to specify individual utilities in order to characterize an inductive rule in general. Specification of utilities is, nevertheless, necessary, for the definition of "reasonable" sequential rules for the group.

3. THE "COMMON HORIZON" FOR THE GROUP IN THE CASE OF A COMMON LOSS FUNCTION

Now we fix attention (as de Finetti actually did in [1954a]) on the particular case when

$$\ell_i(a_h, \theta) = \ell(a_h, \theta) \quad i = 1, 2, \ldots, N.$$

Consider a fixed (completely general, in principle) inductive rule chosen by the group, determined by

$$\gamma : \quad \mathcal{O}^N \quad \longrightarrow \{a_0, a_1, a_2\}. \tag{3.1}$$

Recall that we denote by $\xi_i^{(0)}$ the i-th individual initial distribution on Θ.

Think now of a hypothetical individual with the common loss function $\ell(a_h, \theta)$. Suppose he believes that $\theta = \overline{\theta}$ but must accept, at any stage, the decision chosen by the group. This will result in a loss to him, depending on γ and $\boldsymbol{\xi}^{(0)}$, the vector of initial opinions. The expected value of such a loss (as evaluated by the individual himself) will be denoted by

$$\psi_{\overline{\theta}}(\gamma, \boldsymbol{\xi}^{(0)}) = cE(M|\theta = \overline{\theta}) + E\left[\ell(\gamma(\boldsymbol{\xi}^{(M)}), \overline{\theta})|\theta = \overline{\theta}\right] \tag{3.2}$$

where $M = \inf\{m \geq 0|\gamma\left(\boldsymbol{\xi}^{(m)}\right) \neq a_0\}$. Obviously $M = 0$, if $\gamma(\boldsymbol{\xi}^{(0)}) \neq a_0$ in which case the loss to him is deterministic. If the individual is not sure about θ and he assesses a distribution, η, for θ then he has an expected loss

$$\psi_\eta(\gamma, \boldsymbol{\xi}^{(0)}) = \int \psi_{\overline{\theta}}(\gamma, \boldsymbol{\xi}^{(0)}) d\eta(\overline{\theta}). \tag{3.3}$$

For fixed γ and $\boldsymbol{\xi}^{(0)}$, (3.3) obviously defines a linear functional of η.

Consider now an individual with the loss $\ell(a_h, \theta)$ and an initial distribution ξ on Θ. When he is supposed to choose his own strategy, he will choose a Bayes' strategy against ξ [see e.g. DeGroot, 1970]. We shall denote by Δ_ξ the set of all Bayes' strategies against ξ (including possibly also the randomized ones). For the generic $\xi \in \mathcal{O}, \Delta_\xi$ will consist of at least one strategy δ_ξ (and perhaps more if the distribution of observations is discrete). We denote, moreover, by $\rho(\xi)$ the Bayes' risk against ξ and set

$$\Delta = \bigcup_{\xi \in \mathcal{O}} \Delta_\xi.$$

For any $\delta \in \Delta$, we can consider the expected loss $\psi_\eta(\delta)$ of such a strategy as evaluated by the hypothetical individual with the same loss function $\ell(a_h, \theta)$ but with initial distribution η. This is of interest to us, since we want to compare the effects to him of following the group decision strategy in (3.1) or following the Bayes' strategy of another individual with a different opinion.

Obviously for $\hat{\delta} \in \Delta_\eta$ we have

$$\psi_\eta(\delta) \geq \psi_\eta(\hat{\delta}) = \rho(\eta) \tag{3.4}$$

$$\psi_\eta(\gamma, \boldsymbol{\xi}^{(0)}) \geq \psi_\eta(\hat{\delta}) = \rho(\eta). \tag{3.5}$$

For fixed $\delta \in \Delta, \psi_\eta(\delta)$ is a linear functional of η too. For fixed $\eta \in \mathcal{O}, \boldsymbol{\xi}^{(0)} \in \mathcal{O}^N, \gamma$ a group inductive rule, we set

$$\Delta_\eta(\gamma, \boldsymbol{\xi}^{(0)}) = \{\delta \in \Delta|\psi_\eta(\delta) \leq \psi_\eta(\gamma, \boldsymbol{\xi}^{(0)})\}.$$

$\delta_\xi \in \Delta_\eta(\gamma, \boldsymbol{\xi}^{(0)})$ means that the individual with initial opinion η prefers to act according to the Bayes' strategy of a fellow with initial opinion ξ, rather than according to the strategy γ of the group, where the i-th single individual's opinion is represented by $\xi_i^{(0)}$ $(i = 1, 2, \ldots, N)$. $\Delta_\eta(\gamma, \boldsymbol{\xi}^{(0)})$ is not empty since, obviously,

$$\Delta_\eta(\gamma, \boldsymbol{\xi}^{(0)}) \supseteq \Delta_\eta$$

by (3.5). The sets $\Delta_\eta(\gamma, \boldsymbol{\xi}^{(0)}), \eta \in \mathcal{O}$ are the Bayesian "individual horizons" with respect to γ, in the terminology introduced by de Finetti. With reference to a fixed inductive rule γ, the "common horizon for the group" is the set of (possibly randomized) individual Bayes' rules defined by the intersections of individual horizons of the members of the group:

$$\Delta_{\boldsymbol{\xi}^{(0)}}(\gamma) = \bigcap_{i=1}^{N} \Delta_{\xi_i^{(0)}}(\gamma, \boldsymbol{\xi}^{(0)}).$$

$\gamma' \in \Delta_{\boldsymbol{\xi}^{(0)}}(\gamma)$ is a decision rule which is Bayes for some individual and whose initial opinion may possibly differ from $\xi_1^{(0)}, \ldots, \xi_N^{(0)}$. For any member of the group, γ' is preferable to γ. In [1954a], de Finetti shows, by studying in detail a particular example, that in the case when Θ has only two values, it is possible to obtain explicitly individual horizons and common horizons for a group, by means of elementary geometry. We shall illustrate this in the next section.

4. THE CASE OF A PARAMETER SPACE WITH ONLY TWO VALUES

As mentioned, the Bayesian individual horizons and the Bayesian common horizon can be constructed by elementary geometric tools in the case when

$$\Theta = \{\theta_1, \theta_2\}. \tag{4.1}$$

In the following, a probability distribution on Θ will be represented by the quantity

$$\xi = P\{\theta = \theta_2\}. \tag{4.2}$$

It is no restriction to assume $\theta_1, \theta_2 \in \Re$ and $\theta_1 < \theta_2$. By (4.2) we can let $\mathcal{O}$ coincide with $[0, 1]$ and stochastic ordering among elements of $\mathcal{O}$ will coincide with the natural linear ordering for the real numbers: for $P', P'' \in \mathcal{O}$

$$P' \leq_{st} P''$$

$$\leq==\geq \xi' \equiv P'\{\theta = \theta_2\} \leq \xi'' \equiv P''\{\theta = \theta_2\}.$$

It is convenient to label individuals in the group so that

$$\xi_1^{(0)} \leq \xi_2^{(0)} \leq \ldots \leq \xi_N^{(0)}. \tag{4.3}$$

We shall also label a_1 and a_2 in such a way that

$$\ell(a_h, \theta_k) = 0 \qquad \text{if } h = k (h, k = 1, 2)$$
$$\ell(a_h, \theta_k) > 0 \qquad \text{if } h \neq k.$$

Fix now an arbitrary inductive rule γ and compute the two values

$$\psi_{\theta_1}(\gamma, \boldsymbol{\xi}^{(0)}),\ \psi_{\theta_2}(\gamma, \boldsymbol{\xi}^{(0)}).$$

According to the definition (3.2) for $\psi_{\overline{\theta}}(\gamma, \boldsymbol{\xi}^{(0)})(\overline{\theta} \in \Theta), \psi_{\theta_1}(\gamma, \boldsymbol{\xi}^{(0)})$ is the expected loss for a hypothetical individual $\hat{I}$ characterized by the following situation:

(a) $\hat{I}$ has a loss function $\ell(a_h, \theta_k)$

(b) $\hat{I}$ must accept the decision strategy γ chosen by the group of individuals $I_1, \ldots, I_N$ with initial opinions $\boldsymbol{\xi}^{(0)}$

(c) $\hat{I}$ is from the beginning sure that $\theta = \theta_1$ and

$$\eta = P\{\theta = \theta_2\} = 0$$

for him.

$\psi_{\theta_2}(\gamma, \boldsymbol{\xi}^{(0)})$ has an analogous meaning.

In the Cartesian plane, the segment joining the two points with coordinates

$$(0, \psi_{\theta_1}(\gamma, \boldsymbol{\xi}^{(0)})),\ (1, \psi_{\theta_2}(\gamma, \boldsymbol{\xi}^{(0)}))$$

is the graph of the linear function

$$W_\gamma(\eta) = \psi_\eta(\gamma, \boldsymbol{\xi}^{(0)}) \qquad (0 \leq \eta \leq 1). \tag{4.4}$$

$W_\gamma(\eta)$ is the expected loss for a hypothetical individual $\hat{I}$ with the situation described by (a), (b), and (c'), where

$(c')\hat{I}$ has initial opinion η about θ; i.e.,

$$P\{\theta = \theta_2\} = \eta.$$

As in the last section, we denote by $\psi_\eta(\delta)$, the expected loss of a Bayes' strategy $\delta \in \Delta$ for an individual with initial opinion η. The graph of $\psi_\eta(\delta)$ versus η is again a segment on a straight line. We consider the family of all such segments, together with the graph of the function $\rho(\eta)$, representing the risk associated with the Bayes' strategies as a function of the initial opinion $\eta(0 \leq \eta \leq 1)$. $\rho(\eta)$ is the curve formed by points in the plane with coordinates $(\eta, \psi_\eta(\delta_\eta))$. By (3.4) the graph of $\psi_\eta(\delta_\xi)$ is tangent to $\rho(\eta)$ at $\eta = \xi$.

Some well-known properties for $\rho(\eta)$ follow immediately. In particular

$\rho(\eta) = \eta\ell(a_1, \theta_2)$in some right neighborhood of $\eta = 0$
$\rho(\eta) = (1 - \eta)\ell(a_2, \theta_1)$ in some left neighborhood of $\eta = 1$
$\rho(\eta) \leq \min\ (\eta\ell(a_1, \theta_2),\ (1 - \eta)\ell(a_2, \theta_1))$, all $\eta \in [0, 1]$
$\rho(\eta)$ is concave.

Moreover, $\rho(\eta)$ is a continuous piecewise linear function when the statistical observations $x_1, x_2, \ldots$ have a discrete distribution since, in such a case, a Bayes' sequential decision strategy with respect to a given initial opinion will be Bayes also with respect to all initial opinions lying in some interval containing that opinion.

Now we fix an inductive rule γ and we want to determine the corresponding individual horizon for an individual with initial opinion η. Consider in the plane, the point $Q_{\eta,\gamma}$ with coordinates $(\eta, W_\gamma(\eta))$. It follows that

$$W_\gamma(\eta) \geq \rho(\eta).$$

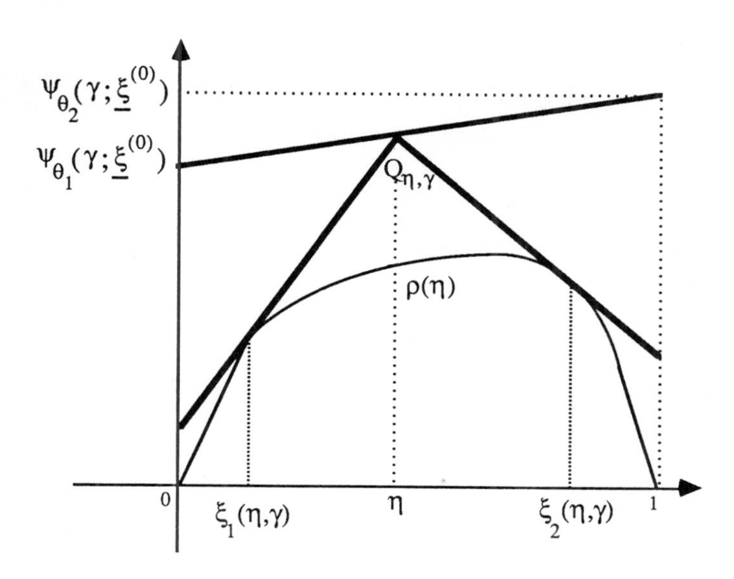

Figure 1. *The Individual Horizon*

We are interested in the case $W_\gamma(\eta) > \rho(\eta)$. From $Q_{\eta,\gamma}$ we draw the two tangent lines to the curve ρ and denote by $\xi_1(\eta,\gamma)$ and $\xi_2(\eta,\gamma)$, the abscissas of the two contact points (see Figure 4.1). From elementary considerations, it is now possible to prove the following result: The individual horizon with respect to γ is formed by the Bayes' rules corresponding to opinions ξ such that

$$\xi_1(\eta,\gamma) \le \xi \le \xi_2(\eta,\gamma).$$

If $\xi_k(\eta,\gamma)(k = 1,2)$ is a corner point for ρ then one must consider only those Bayes' strategies corresponding to lines tangent to ρ at $\xi_k(\eta,\gamma)$ and passing through points of the kind (η, y) with $y \le W_\gamma(\eta)$.

The common horizon with respect to γ for the group of individuals with initial opinions $\underline{\xi}^{(0)}$ is formed by Bayes' rules corresponding to opinions ξ such that

$$\xi_1(\xi_N^{(0)},\gamma) \le \xi \le \xi_2(\xi_1^{(0)},\gamma).$$

If $\xi_1(\xi_N^{(0)},\gamma) = \xi_2(\xi_1^{(0)},\gamma)$ and this is a corner point, then the common horizon may be formed only by randomized Bayes' rules. We remark that $\xi_1(\eta,\gamma)$ and $\xi_2(\eta,\gamma)$ are corner points $\forall\eta \in \mathcal{O}$ if the observations have a discrete distribution.

It may happen in general that the strategies in the common horizon are Bayes' only with respect to initial opinions different from those of the individuals in the group. It is also possible that they could be randomized and not correspond to even a single hypothetical opinion.

5. VOTING RULES AND GROUP SEQUENTIAL DECISION PROBLEMS

The case $\Theta = \{\theta_1, \theta_2\}$

In the uni-subjective case with $\Theta = \{\theta_1, \theta_2\}$, the Bayes' strategies are the so-called Wald sequential probability ratio tests [cf. Ferguson (1967), pp. 361-368]. Depending on the loss function, $\ell(a_h, \theta_k)$ and the cost of sampling, c, there exist numbers $\xi_L < \xi_U$ such that if

$$\xi_L < \xi_i^{(n)} < \xi_U$$

the Bayes' decision rule for individual I_i at a stage n is to continue sampling and to stop and take decision $a_1(a_2)$ if

$$\xi_i^{(n)} \leq \xi_L$$

$$(\xi_i^{(n)} \geq \xi_U).$$

It is easy to show that

$$\xi_1^{(0)} \leq \xi_2^{(0)} \leq \ldots \leq \xi_N^{(0)} \tag{5.1}$$

implies

$$\xi_1^{(n)} \leq \xi_2^{(n)} \ldots \leq \xi_N^{(n)} \tag{5.2}$$

for $n = 1, 2, \ldots$. Hence an M out of N voting group decision rule will be determined by individuals I_M and I_{N-M+1}. Obviously, if N is odd and $M = (N+1)/2$, the simple majority rule, then the median individual determines the simple majority group decision rule. In this case, the simple majority rule is also a Bayes' decision rule.

The Case When Θ *is an Interval of the Real Line*

The uni-subjective case of the sequential two terminal action decision problem for $f(x|\theta)$ a member of the exponential family of the form

$$dF(x|\theta) = \psi(\theta) e^{x\theta} d\mu(x)$$

was solved by Sobel [1953]. In this case $(y(n), n)$ is a sufficient statistic for θ where, given observations $x_1, x_2, \ldots, x_n$

$$y(n) = \sum_1^n x_i.$$

We suppose that the loss function $\ell(a_h, \theta)(h = 1, 2)$ has the following properties: $\ell(a_1, \theta)$ is increasing (that is, non-decreasing) in θ; $\ell(a_2, \theta)$ is decreasing (that is, non-increasing) in θ. The cost of another observation at stage n is $c_n > 0$.

At stage n of sequential sampling, the Bayes' rule with respect to a density (or discrete probability) $\xi(\theta)$ on Θ can be characterized by two numbers, namely $y_1(n) < y_2(n)$. The Bayes' rule is

1) Continue sampling if $y_1(n) < y(n) < y_2(n)$;
2) Stop and take action a_1 if $y(n) \leq y_1(n)$;
3) Stop and take action a_2 if $y(n) \geq y_2(n)$.

We can generalize (5.1) and (5.2) for this case as follows. Assume that $\xi_i(\theta)$ is totally positive of order 2 in $i = 1, 2, \ldots. N$ and $\theta \in \Theta$ (i.e., is TP_2), that is, there is an ordering of individuals in the group such that the following 2×2 determinant is non-negative when $i < j$ and for all $\theta_1 < \theta_2$ belonging to Θ:

$$\begin{vmatrix} \xi_i(\theta_1) & \xi_i(\theta_2) \\ \xi_j(\theta_1) & \xi_j(\theta_2) \end{vmatrix} \geq 0. \tag{5.3}$$

It is easy to see that when

$$\Theta = \{\theta_1, \theta_2\}$$

(5.3) reduces to (5.1). (Recall that $\xi_i(\theta_2) = 1 - \xi_i(\theta_1)$ in this case.)

Since $\xi_i(\theta|y(n))$ is proportional to $p(y(n)|\theta)\xi_i(\theta)$, it is obvious that $\xi_i(\theta|y(n))$ is also TP_2 in $i = 1, 2, \ldots, N$ and $\theta \in \Theta$. This generalizes (5.2).

Example. If $\xi_i(\theta) \propto \theta^{A_i}(1-\theta)^{B_i}$ and

$$0 \leq A_1 \leq A_2 \leq \ldots \leq A_N$$
$$B_1 \geq B_2 \geq \ldots \geq B_N \geq 0,$$

then (5.3) is satisfied.

Let $[y_{i1}(n), y_{i2}(n)]$ be the Bayes' rule corresponding to individual I_i at the n-th stage of sampling. By definition

$$\int_\Theta \ell(a_1, \theta)\xi_i(\theta|y_{i1}(n))d\theta < \int_\Theta \ell(a_2, \theta)\xi_i(\theta|y_{i1}(n))d\theta,$$

that is, individual I_i would prefer action a_1 to action a_2 if

$$y(n) = y_{i1}(n)$$

at stage n.

Lemma 5.1. *For $h < i$,*

$$\int_\Theta \ell(a_1, \theta)\xi_h(\theta|y_{i1}(n))d\theta < \int_\Theta \ell(a_2, \theta)\xi_h(\theta|y_{i1}(n))d\theta. \tag{5.4}$$

For $i < j$

$$\int_\Theta \ell(a_1, \theta)\xi_j(\theta|y_{i2}(n))d\theta > \int_\Theta \ell(a_2, \theta)\xi_j(\theta|y_{i2}(n))d\theta. \tag{5.5}$$

Proof. For $h < i$, it follows from the sign variation diminishing theorem [Karlin (1968)] that

$$\int_\Theta \ell(a_1, \theta)\xi_h(\theta|y_{i1}(n))d\theta \leq \int_\Theta \ell(a_1, \theta)\xi_i(\theta|y_{i1}(n))d\theta$$

$$< \int_\Theta \ell(a_2, \theta)\xi_i(\theta|y_{i1}(n))d\theta \leq \int_\Theta \ell(a_2, \theta)\xi_h(\theta|y_{i1}(n))d\theta$$

since $\ell(a_1, \theta)$ is increasing in θ while $\ell(a_2, \theta)$ is decreasing in θ. For $h < i$ and $y(n) = y_{i1}(n)$, it follows that I_h would prefer a_1 to a_2 even more than would I_i. (5.5) follows in a similar way. ◁

Theorem 5.1. *Let $\xi_i(\theta)$ be TP_2 in i and θ. If at the n-th stage $y(n) \leq y_{i1}(n)$, then individuals $h \leq i$ would prefer either to continue observation or to stop and take action a_1 rather than to stop and take action a_2.*

Similarly, if at the n-th stage $y(n) \geq y_{i2}(n)$, then individuals $j > i$ would either prefer to continue observation or to stop and take action a_2 rather than to stop and take action a_1.

Proof. From inequality (5.4) in Lemma 1 it follows that if $h < i$, then

$$y_{h2}(n) \geq y_{i1}(n).$$

Hence $y(n) \leq y_{i1}(n)$ implies that $h < i$ would prefer either to continue observation or to stop and take action a_1 rather than to stop and take action a_2.

The proof is similar for $j > i$ and $y(n) \geq y_{i2}(n)$. ◁

M-Out-Of-N Voting Rules

Following an M-out-of-N voting rule, the group stops and takes action $a_h (h = 1, 2)$ as soon as M or more members are in favor of stopping and taking action a_h. From Lemma 5.1 we can show that

$$y_{h,2}(n) \geq y_{i,1}(n) \qquad \text{when } h < i$$

and

$$y_{j,1}(n) \leq y_{i,2}(n) \qquad \text{when } i < j.$$

Hence, if at the n-th stage, $y(n) \geq y_{M,2}(n)$, then the group will either stop and take action a_2 or continue to take observations, since there will be less than M in favor of stopping and taking action a_1.

If, at the n-th stage, $y(n) \leq y_{N-M+1,1}$ then the group will either stop and take action a_1 or continue to take observations, since there will be less than M in favor of stopping and taking action a_2.

If N is odd and $M = (N + 1)/2$, then individual I_M will *not* necessarily determine the final group action to be taken as in the case $\Theta = \{\theta_1, \theta_2\}$.

6. DETERMINING THE GROUP DECISION RULE

In his summer 1959, Varenna, Italy, Lectures [cf. de Finetti (1972), pp. 196-198], de Finetti discusses the general group decision problem. In the last of 10 points, he makes the following controversial statement.

"Greater complications are encountered with more widely differing attitudes and interests of the individuals. But no new criterion is called for: one has but to apply the criterion of the maximal expected utility in different circumstances." He also presents several examples illustrating how a group of individuals, whose utilities are based on money, might reach an initial group decision rule.

Individuals With Shares in a Joint Economic Enterprise

Suppose that group losses are distributed according to share. If individual i has share w_i, decision d is taken and θ occurs, then i loses $w_i L(d, \theta)$. Individual I_i will prefer to use d_i, his best decision rule with his opinion ξ_i. The sum of the loss shares is the "group loss" $L(d, \theta)$.

There is no possible agreement between the group members as long as each considers his individual uni-subjective decision problem —even if they have the same shared loss function. Each individual has his own Bayes' rule with his opinion. The individuals *must* negotiate if the group is to make a decision. There seem to be two levels of thinking here, the individual level and the group level. The initial decision is reached by each individual considering his own decision problem. A compromise is necessary for the sake of reaching a group decision.

De Finetti suggests a compromise that leads to a group decision rule entirely acceptable to each individual. He suggests a reallocation of the group loss: individual i bears the total loss if his own Bayes' rule is actually used. Under this convention, we can consider the "randomized" group rule which selects d_i with probability w_i. The expected loss to individual i is then precisely his expected share of the group's loss were the group to agree to use his Bayes' rule with probability one (and thus minimize his true expected loss). Therefore, every individual would be satisfied "as much as if he alone were to make the decision which for him is optimal" [de Finetti (1972), p. 196].

However, the group as a whole will have even a lower expected loss, in *everyone's* opinion, if the (possibly randomized) decision rule corresponding to the parallel tangent to the risk curve is used. (Notice that the former compromised rule is not, in general, Bayesian against any possible opinion —not even of a hypothetical individual opinion). By using the rule corresponding to the parallel tangent, the amount of improvement is the same in everyone's opinion. This is the result that de Finetti considers better for all— meaning that each would consider the expected *group* loss reduced. However, in I_i's opinion, this will be *worse* for him.

ACKNOWLEDGEMENTS

R. E. Barlow's research was partially supported by the Air Force Office of Scientific Research under Grant AFOSR–81–0122 and Contract DAAG29–85–K–0208 with the University of California and the C. R. R. Visiting Professors Program with Rome University. F. Spizzichino's research was partially supported with M. P. I. funds and a C. N. R. – N. A. T. O. grant n. 217.19. S. Wechsler's research was supported by CAPES.

REFERENCES

Berger, J. O. and Wolpert, R. L. (1984). *The Likelihood Principle.* IMS Lecture Notes Monograph Series Vol 6, IMS, Hayward, CA.

de Finetti, B. (1931). Sul Concetto di Media, *Giornale dell' Istituto Italiano degli Attuari*, Anno II **3**, 3–30.

de Finetti, B. (1951). Recent Suggestions for the Reconciliation of Theories of Probability. *Proceedings 2nd Berkeley Symposium on Math. Statistics and Probability,* Univ. of California Press, 217–225.

de Finetti, B. (1952). La notion de "distribution d'opinions" comme base d'un essai d'intrepretation de la Statistique. *Publ. Inst. Stat. Univ. I* **2**, 1–19.

de Finetti, B. (1954a). Media Di Decisioni e Media Di Opinioni. *Bull. Inst. Inter. Statist.* **34**, 144–157.

de Finetti, B. (1954b). Concetti Sul "Comportamento Induttivo" Illustrati Su Di Un Esempio, *Statistica* **3**, 350–379.

de Finetti, B. (1954c). La notion de "horizon bayesien". *Colloque sur L'analyse statistique, tenue a Bruxelles,* les 15–16–17 decembre, 58–71.

de Finetti, B. (1972). *Probability, Induction and Statistics.* New York: Wiley. 188–198.

DeGroot, M. (1970). *Optimal Statistical Decisions.* New York: McGraw-Hill.

Ferguson, T. S. (1967). *Mathematical Statistics.* New York: Academic Press.

French, S. (1985). Group consensus probability distributions: a critical survey. *Bayesian Statistics* **2**. (J. M. Bernardo, M. H. DeGroot, D. V. Lindley and A. F. M. Smith, eds.). Amsterdam: North-Holland 183–201. (with discussion).

Karlin, S. (1968). *Total Positivity.* Stanford: University Press.

Rubin, H. (1987). A weak system of axioms for "rational" behavior and the non-separability of utility from prior. *Statistics & Decisions* **5**, 47–58.

Sobel, M. (1953). An Essentially Complete Class of Decision Functions for Certain Standard Sequential Problems. *Ann. Math. Statist.* **24**, 319–337.

DISCUSSION

SIMON FRENCH (*University of Manchester*)

De Finetti's writings are never easy to read; and when they are only available in French and Italian then, for me, they are quite impossible. So I am very grateful to Professors Barlow, Spizzichino and Wechsler, first, for providing me with translations of two of de Finetti's papers on group decision making (de Finetti 1952, 1954a) and second, for providing us all with so clear and lucid a summary of his work in their paper before us.

I confess that I do not find de Finetti's work on group decision making at all convincing, but I am aware that I say that with considerable hindsight. He was active in this area mainly in the early 1950's. That, as we all know, was a time when many related subjects were in ferment and I guess that the work we have before us reflects the frontiers of thought as they stood then. But frontiers have advanced and I shall comment on his his work in the light of our knowledge and understanding today. I shall begin with the relevant history of group decision theory and utility theory.

Work on voting systems and group choice dates several hundred if not thousands of years. I guess the Greek philosophers had something to say on the matter, would I but know it. However, the important dates for us are the following. In 1944 Von Neumann and Morgenstern published the *Theory of Games and Economic Behaviour*. They had 'solved' zero sum games and suggested ways forward to solve more general games. Black had been working in the late 1940's on voting structures and had found some negative results, but was optimistic of finding ways forward (Black, 1958). Generally there was a mood of optimism. The Scientific Method would, in time, prescribe fair and just solutions to problems of group choice. Then Arrow came on the scene. In 1951 he published his famous impossibility theorem. A simple result that said, twist and turn as you might, you just cannot define a system in a self consistent manner. In the 50's people did twist and turn, but, by the publication of the second edition of Arrow's book in 1963, it was clear that Arrow was right. Subject to a very minor mathematical correction, his result stood. At best it denies the existence of any straightforward democratic solution to group decision problems. At worst, and some of us accept the worst, it denies any solution, straightforward or not. In game theory things hadn't been going too well either. Completely acceptable solutions to cooperative, non-zero-sum n-person games were conspicuous by their absence. They remain so today and Arrow's Theorem predicts that they will never be found. In my Valencia II paper and my recent book, I summarise the position on the group decision problem today (French, 1985, 1986). Suffice it to say here that the waters are very, very forbidding and murky.

The history of work on utility theory is much, much happier, though it too has had its fair share of confusion. Bayesian statisticians are well familiar with the work of Bernoulli, which led eventually to the expected utility hypothesis encapsulated by the work of Von Neumann and Morgenstern (1947) and Savage (1954). They are less familiar, perhaps, with the work on preference modelling that has gone on in the economic literature. There attention has centred not just on the expected utility hypothesis but at least as much on representing strength of preference notions. In the 1920's and 30's there were many debates on exactly what needed to be represented in order to construct economic models. Unfortunately the debate did not clarify issues so much as muddle concepts. In 1954 Ellsberg wrote a paper trying to clarify matters and Chipman followed with a similar paper in 1960. Of course, matters couldn't be left clear and in the 60's work on additive utility models muddied the waters again. Today the picture is much clearer and it is recognised that there are many different aspects of preference, and only very occasionally can it be assumed that any one 'utility' function models all of these. (Dyer and Sarin, 1979; Fishburn, 1976; French, 1986). Consequently when using any numbers which model preferences, it is important to be very clear which aspects of preference are modelled.

Against this background let me comment on de Finetti's work. I should very much like to know how much de Finetti was aware of contempory work in voting systems and group decision making. Did he discuss his ideas with Arrow? There are many parallels between de Finetti's development and work elsewhere. For instance, given the structure of his example and the assumption that all the group share a common utility function, it is trivial to show that their expected utility orderings obey a condition known as "single peaked preferences".

Black had shown in the late 40's that this condition implies that simple majority voting will effectively take the median opinion: and, moreover, the simple majority rule will satisfy all the conditions that Arrow demanded of a fair and consistent voting system, including Pareto optimality as de Finetti notes. However, Black only showed this result for an odd number of members of the group. The result is false for an even number. This point which illustrates the extremely arbitrary nature of results in group decision theory: change an assumption in an intuitively irrelevant way and edifices crumble.

The fact that de Finetti did not need an assumption on the oddness of N arises because his group are all expected utility maximisers: but his result is no less spurious. Bacharach (1976) has shown that the assumption that the group members share a common utility function is vital. If either this or an alternative assumption that they share a common (prior) probability structure do not hold, then all of the problems identified by Arrow emerge.

The assumption of a common utility function is one that, I believe, de Finetti misinterprets. It means no more and no less that the group members will, if they share the same beliefs, order any two actions in the same way. It does not mean that equal changes in utility represent equal differences in strength of preference between the individuals. Firstly, that would assume that utility functions represent strength of preference information. There is no reason why they should and empirically it is know that even for coherent decision makers they often do not (Krzysztofowicz, 1983; French, 1986). Secondly, it would also assume that strength of preference were interpersonally comparable. Would that it were, for then Arrow's Impossibility might disappear; but no one has convinced me or most others of any meaningful way in which this can be achieved. De Finetti uses the assumption of interpersonal comparison in his work when he pulls the group decision back to one corresponding to a parallel tangent as in Figure 4.1 of the paper. In translation, in *Average of Decisions and Average of Opinions* he says that this choice of rule gives "the same advantage (that is, an equal decrease of the expected loss) whatever the opinion could be . . ." for all individuals.

A further point that concerns me is that de Finetti limits attention to inductive rules; i.e. those that depend in an invariant way on the current opinions of the group. Is this sensible? It ignores the varying expertise and calibration of the group members and that will affect the information content of their priors. Moreover, it ignores the possibility that the group members may update their opinions in the light of others' opinions. Such considerations may mitigate against using inductive rules if one takes certain views of group decision problems: see the discussion in Sections 2.1 and 2.5 of my Valencia II paper.

Let me finish by repeating my gratitude to the authors for making de Finetti's views available to us. De Finetti's ideas are always thoughtprovoking: but we should remember that, in this case, they may have been overtaken by later ideas. His ideas on probability will be with us for a very long time: I doubt if the same is true for his ideas on group decision making.

REPLY TO THE DISCUSSION

French has given us a lecture from his decision theory course. He has in effect discussed the question: Is it possible to define a group utility function based on the utilities of the individuals in the group alone? The answer is well known. The most recent and perhaps the best discussion of this question is Rubin (1987) [not referenced by French]. This was neither the question considered by the Finetti nor by us.

The question considered by de Finetti in his 1954 papers and by us is: *Given* an inductive group decision rule (a voting rule perhaps) and the utilities (de Finetti uses loss functions) for the individuals, are there decision rules which all would consider better and, if so, what is their characterization? In his 1954 papers de Finetti has assumed that all individuals in the group act in concert and have the *same* loss function (conditional on the "state of nature") for group decisions but not necessarily the same opinions about the "state of nature". French seems to imply that this assumption is *never* realistic! However we know that groups *do* make decisions and that they are not above giving up some of their individuality on occasion and agreeing to use a common loss function —especially in financial matters.

French makes the point that, in his opinion, de Finetti misinterprets the common utility assumption. But in our interpretation, de Finetti means that the common loss is the loss *for the group* and all agree on this as representing the strength of preference relative to group decisions. Therefore, corresponding individual unconditional utilities are comparable and Arrow's impossibility theorem does not apply.

In his 1954 papers, de Finetti was mainly dealing with admissibility and Pareto optimality for group sequential decisions. De Finetti was aware of Arrow's work in 1954 since he referenced Arrow's 1951 monograph "Social Choice and Individual Values" in his (1954c) paper which we reference. In the 1954 papers de Finetti introduced the concepts of "individual horizon" and "common horizon" with respect to a fixed group decision rule. Our present paper mainly aims to illustrate and to extend such concepts which are, in our opinion, a great part of the original and significant contribution by de Finetti to group decision theory.

De Finetti's remark number 10, which we have quoted out of context at the beginning of our Section 6, may have contributed to French's confusion. De Finetti means, in a game theory context, that a Bayesian player should keep the usual criterion of maximizing expected utility relative to his prior regarding his opponents decisions. In de Finetti's opinion, equilibrium solutions are not relevant.

French mentions that there are many parallels between de Finetti's development and work elsewhere. As far as the simple majority rule is concerned, de Finetti explicitly mentioned that it coincides with the Bayesian individual rule for the median individual *when the number of members of the group is odd* [see Media di Decisioni e Media di Opinioni, (1954), page 154]. As to de Finetti's knowledge of Black's book, *The Theory of Committees and Elections*, he reviewed this book in the journal Rivista Di Diritto Finanziario e Scienza Delle Finanze in December 1959. In the first sentence de Finetti says (in translation), "Everybody knows how hard it is to reach an agreement among the differing wishes of even a very small number of people, but it looks as if only few consider the fact that every reference to joint decisions based on the "majority will" is vague and deceptive."

French criticizes de Finetti's concept of "inductive decision rules" in the context of group decision making relative to calibration and the varying expertise of the members of the group. But who is to do the calibration? In our opinion, de Finetti assumed that the individuals' initial distributions were already updated through discussions among group members. In the examples, he also considered the possibility of weights for individuals in the group. These could be used by the group to reflect the varying expertise of the members of the group.

In summary, we and de Finetti are interested in the possible, not the impossible. De Finetti was aware of Arrow's work in 1954. Individual horizons relative to a given possible inductive group decision rule are well defined even when individuals in the group have different utilities.

The common horizon is the intersection of individual horizons. If the given rule is a voting rule and the common horizon is empty then the group may opt for the voting rule. In a personal communication, Lindley remarks that "the possibility of removing disagreement and perhaps making voting reasonable, so overcoming Arrow's result, is exciting". French is *wrong*: de Finetti's ideas on group decision making are relevant, even today!

REFERENCES IN THE DISCUSSION

Arrow, K. J. (1951–1963). *Social Choice and Individual Values.* New York: Wiley.

Bacharach, M. (1975). Group decisions in the face of differences of opinions. *Mgmt. Sci.* **22**, 186–191.

Black, R. D. (1958). *The Theory of Committees and Elections.* Cambridge: University Press.

de Finetti, B. (1959). La teoria dei comitati e delle elezioni. (A proposito di un'opera di Duncan Black) *Rivista di Diritto Finanziario e Scienza delle Finanze* **18**, 322–326.

Dyer, J. S. and Sarin, R. K. (1979). Measurable multi-attribute value functions. *Ops. Res.* **28**, 810–822.

Ellsberg, D. (1954). Classic and current notions of measurable utility. *Econ. J.* **64**, 528–556.

Fishburn, P. C. (1976). Cardinal Utility: an interpretive essay. *Revista de Scienze Economiche e Commerciali* **12**, 1102–1114.

French, S. (1986). *Decision Theory: An Introduction to the Mathematics of Rationality*. Chichester: Ellis Horwood.

Krzysztofowicz, R. (1983). Strength of preference and risk attitude in utility measurement. *Org. Behav. and Human Perform.* **31**, 88–113.

Savage, L. J. (1954). *The Foundations of Statistics*. New York: Wiley.

Von Neumann, J. and Morgenstern, O. (1944). *Theory of Games and Economic Behaviour*. Princeton: University Press.

BAYESIAN STATISTICS 3, pp. 17–24
J. M. Bernardo, M. H. DeGroot, D. V. Lindley and A. F. M. Smith, (Eds.)

The Future of Statistics: Teaching and Research

GEORGE A. BARNARD
University of Essex

SUMMARY

Attention is focussed on the need for post-introductory texts addressed to mathematics majors which incorporate "Bayesian" ideas from the beginning rather than adding them on to an essentially "classical" approach. This leads into some neglected pathways among the many directions in which future research is needed.

Keywords: HISTORY; BAYESIAN INFERENCE; ROBUSTNESS; COMPUTING.

1. SOME THOUGHTS ON TEACHING

In Box and Tiao (1973) and Berger (1985) to name only two, we already possess excellent texts for use in courses for graduate students in mathematics. And in applications fields we have books such as Zellner (1971) covering the econometric field. But the full exploitation of the insights in the foundations of statistical inference which have been reached over the past half century will require considerable reworking of old and development of new mathematical techniques. So we cannot afford to allow the cream of our mathematical talent to be completely seduced by the obvious attractions of the purely logical aspects of artificial intelligence and other mathematical areas which are experiencing vigorous development. We run the risk of allowing this to happen unless, at the undergraduate level, we can present mathematical statistics as an area calling for the exercise of all the skills a bright young mathematician possesses.

We have recently been treated to a great access of help by the publication of Stigler (1986a) on the history of statistics. The greater part of the mathematical skills which good young mathematicians take pleasure in exercising had their origin in the 18th century, and Part I of Stigler's book covers, in a thorough and witty style, the way in which these 18th century mathematical methods were used by mathematicians of that time. It provides a rich source of realistic problems, especially those arising in the analysis of astronomical data, on which junior mathematicians can cut their teeth. I have in mind especially Stigler's analysis of Laplace's early work in probability, to which I shall return below.

While historical notes form an essential part of any good teaching, this is not to say that more recent developments are to be deferred if they can be expressed in elementary terms. An outstanding example of this is provided by de Finetti's derivation of the rules of coherent guessing (see e.g. de Finetti, 1972). If we represent each of a set of propositions $a, b, c, d, e, \ldots$ by a line and on each line we intend to represent the truth or falsehood of the corresponding proposition by specifying the value, 1 or 0, of $a, b, c, \ldots$, we can express logical connections between the propositions by equations such as

$$ab = 0, \quad c = a + b, \quad d = ec, \ldots$$

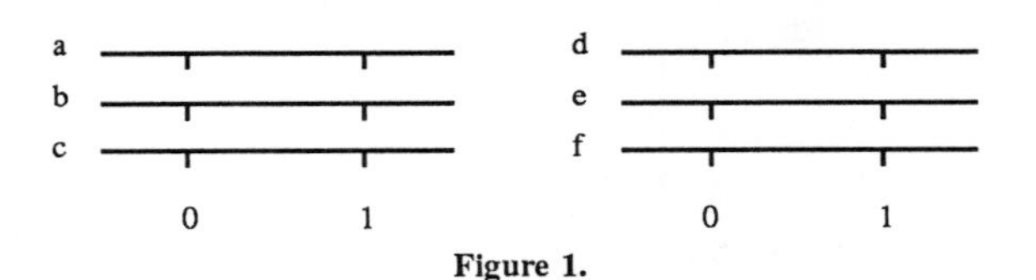

Figure 1.

and then discuss the problem of choosing points $x, y, z, \ldots$ on each line $a, b, c, \ldots$ so as to minimise a "loss"

$$L = (a-x)^2 + (b-y)^2 + (c-z)^2 + \ldots$$

and so derive the laws of personal probability in the way that de Finetti does. This is illustrated in Figure 1. Having done that we can recall the laws of "experimental probability" to which we can assume the students have already been introduced in terms of dice, coin tossing, etc., and say that, in applications, the probabilities arising will typically consist in a mixture of pure guessing probabilities with pure experimental probabilities. The fact that the laws governing them are the same means that the precise proportion of guessing to experiment in any given case need not be of immediate practical concern.

In teaching a course to psychologists some years ago I constructed an apparatus with rods representing the lines above, and bulbs at 0 and 1. The equations were represented by plug-in electrical connections and the students could "place their bets" by placing a rider on each of the lines. Mathematicians could probably dispense with such concretisation.

The availability of pocket computers with random number generators allows us to construct a large number of inferential games. The simplest is to write a programme which first generates a random θ in (0,1) and stores it unseen. Then successive (0,1) random numbers are generated, also unseen, but the computer shows, for each number, 1 if it exceeds θ and 0 if it does not. From a count of r 1's and s 0's one is required to estimate θ so as to minimise squared error loss. Elementary analysis then shows the optimum estimate to be $(s+1)/(r+s+2)$. And one can introduce Thomas Bayes' billiards table by asking for the posterior distribution of θ given r and s. Repeating the exercises with θ replaced by $\frac{1}{2}(1+\frac{1}{2}\theta)$ introduces more scope for mathematics. And a little extra complication allows, for example, for samples from a double exponential distribution with unknown location to be considered. Issues of logical principle can be raised by allowing θ to be determined in some other way, such as the fractional part of the logarithm of a fairly large number.

One of the many virtues of Stigler's book is the way in which he shows the scientific importance — for the astronomy of the time — of sound methods for the combination of observations. Indeed, if I were giving a course at the level I am considering, I would be strongly inclined to require the reading of the first three chapters of Stigler's book. This would then allow for a detailed consideration of the issues discussed in Laplace's (1774) memoir. In this, among other things, Laplace considers the problem, given 3 observations $x_i (i = 1, 2, 3)$ on the time μ of a specified event, to assess the best estimate of μ. The errors of observation are

$$p_i = (x_i - \mu)/\sigma \tag{1}$$

and Laplace gives reasons for supposing the p_i to have the joint density

$$\phi(p) = \left(\frac{1}{8}\right) \exp -(|p_1| + |p_2| + |p_3|). \tag{2}$$

He also gives reasons for thinking the best estimate of μ to be the value $\tilde{\mu}$ such that μ is as likely to exceed $\tilde{\mu}$ as to fall short of it. He first considers the case when σ, as well as μ, is unknown. In the latter case he, in effect, assumes that μ and σ are each equally likely to take any of their possible values — from $-\infty$ to $+\infty$ in the case of μ, and from 0 to $+\infty$ in the case of σ. In modern terminology Laplace sets up a model involving 5 random variables, $p_1, p_2, p_3, \mu, \sigma$, having the joint density element (j.d.e.) proportional to

$$\phi(p)\ dp d\mu d\sigma. \tag{3}$$

He is then able to condition on the known values of the x_i to obtain a posterior density for (μ, σ) and from this, by marginalisation, a density for μ. Then $\tilde{\mu}$ is found as the median of this μ distribution. Stigler (1986b) gives an interesting discussion of some mistakes made by Laplace.

Nowadays Laplace's virtually tacit assumption of independent uniform priors for both μ and σ would not be acceptable. And one would need to be clearer than Stigler shows Laplace to have been on the subjects of conditioning and the marginalisation of probabilities. Some of this clarification can come from a course in probability which should run in parallel with the course in mathematical statistics. One element of such a course, which I have found helps both to stimulate interest and to fix ideas, is to consider simplified versions of the game of "Snakes and Ladders" in which, for example, we play on a board with only 6 squares, and we read the die score modulo 3. One can then discuss where one might put a one-step "ladder" and a 3-step "snake" so as to obtain a game which, on the average, with 2 players, will last N moves for the winner. One aspect of introducing conditional probabilities in this way is that analysis of such games will make the "decision tree" analysis of some of the apparent paradoxes of conditional probability easier to appreciate.

Another way in which marginalisation and conditioning can be made simpler than they commonly are is illustrated by (3) above, where a phrase such as "the joint density element proportional to $f(u, v, w)\ dudvdw$" is to be interpreted to imply that u, v, and w are random variables, (so taking values in a measure space, with measure elements du, dv, dw), and for any subsets A, B, of the space of (u, v, w),

$$\Pr\{(u, v, w) \in A\}/\Pr\{(u, v, w) \in B\} = \int_A f(u, v, w) dudvdw \Big/ \int_B f(u, v, w) dudvdw.$$

With this notation, conditioning on $v = b$ produces the j.d.e. proportional to $f(u, b, w)dudw$, while marginalisation with respect to v gives the j.d.e. proportional to $g(u, w)dudw$ where $g(u, w) = \int f(u, v, w)dv$. The unnecessary and confusing "normalising constants" can typically be fixed at the end of the analysis, if necessary, by the condition that the total probability is 1. I have yet to be convinced that issues such a non-conglomerability have relevance to statistics.

A further advantage of this conventional notation is that a quantity such as a in the above $f(u, v, a)$ can be regarded as a fixed parameter at one stage, and if, later, it needs to be treated as a random variable with density element $h(a)da$, the joint density element of u, v, a will simply be $f(u, v, a)h(a)dudvda$. And if we transform $(u, v) 1-1$ to (u', v') with $u = U(u', v'), v = V(u', v')$, the Jacobian being $J(u', v')$, the j.d.e. of u', v', a will be proportional to $J(u', v')f(U(u', v'), V(u', v'), a)du'dv'da$. Such "deconditionings" and transformations are frequently needed in statistical reasoning, and our mathematical notations should make their formalisation easy.

Our mathematics should be "rigorisable", but we should not allow finer points of mathematical rigour to render our analysis so complex as to make us liable to strain at gnats of rigour while swallowing elephants of blunders in statistical logic. It is for debate whether what used to be called the "Annals" style, with its σ-finite measures etc., did more harm than good, in so far as it ensured the "respectability" of our subject at a time when pure mathematicians were excessively preoccupied with rigour. In any case, that time is now past.

2. AN ILLUSTRATIVE ANALYSIS OF LAPLACE'S PROBLEM

Given familiarity with conditioning, marginalisation, "deconditioning", etc, we can approach Laplace's problem by first taking p_i in (1) to have j.d.e. proportional to

$$\exp -(|p_1| + |p_2| + |p_3|).dp_1 dp_2 dp_3 \tag{4}$$

and bearing in mind that we are interested in μ, not in σ, we look for a transformation of the p_i to q_i such that q_1 involves the observables x_i and μ, only, q_2 involves the observables and σ only, while q_3 involves only the observables. Such a transformation is given by

$$p_1 = q_1 q_2,\ p_2 = q_2(q_1 + 1),\ p_3 = q_2(q_1 + q_3) \text{ with Jacobian } q_2^2,$$

because using (1) we find that

$$q_1 = (x_1 - \mu)/(x_2 - x_1),\ q_2 = (x_2 - x_1)/\sigma,\ q_3 = (x_3 - x_1)/(x_2 - x_1). \tag{5}$$

The j.d.e. of the q_i is proportional to

$$q_2^2 \exp -|q_2|(|q_1| + |q_1 + 1| + |q_1 + q_3|).dq_1 dq_2 dq_3. \tag{6}$$

We can imagine that we are first told the value m of q_3, before being told the values of x_1 and x_2. Then the joint density of q_1, q_2 is proportional to

$$q_2^2 \exp -|q_2|(|q_1| + |q_1 + 1| + |q_1 + m|).dq_1 dq_2$$

from which the marginal density of q_1 is proportional to

$$(1/\{|q_1| + |q_1 + 1| + |q_1 + m|\}^3).dq_1, \tag{7}$$

use of the principle of total probability giving the constant of proportionality as $3/(3 - z_0)$, where

$$z_0 = 1/(2 - m)^2 + 1/(1 + m)^2. \tag{8}$$

If we are told, for example, that $m = 0.6$ without being told the values of x_1 and x_2 we can evaluate the probability that q_1 lies between -1 and 0 as

$$3/(3 - z_0) \int_{-1}^{0} dq_1/\{|q_1| + |q_1 + 1| + |q_1 + m|\}^3 = (6 - 3z_0)/(6 - 2z_0),$$

which with $m = 0.6$ comes to 0.9008. But q_1 will lie between –1 and 0 if and only if μ lies between x_1 and x_2, so we can think of the interval $I(x_1, x_2)$ between x_1 and x_2 as a "net" being thrown with its landing position and its width as to some extent random. The probability that μ will be caught in this net is 0.9008. We can similarly integrate for q_1 from $-\infty$ to -1 to find the probability that μ will be found to lie to the right of the net. The interval $I(x_1, x_2)$ is then a conditional confidence, or fiducial, interval for μ, with confidence coefficient 0.9008; in the absence of any further information concerning μ itself, if we learn that $x_1 = 2$ and $x_2 = 3$, we can assess the probability that $2 < \mu < 3$ as 0.9008. We can similarly assess the probability that μ lies in any other specified interval. Of course, any further information we had concerning μ would modify these probability assessments.

We can also solve Laplace's problem, of finding the function $\tilde{\mu}(x_1, x_2, x_3)$ such that the true value μ is as likely to exceed $\tilde{\mu}$ as to fall short. If $\tilde{q}_1$ is the root of

$$3/(3 - z_0) \int_{-\infty}^{\tilde{q}_1} dq/\{|q| + |q + 1| + |q + m|\}^3 = 0.5 \tag{9}$$

then $\tilde{\mu} = x_1 - \tilde{q}_1(x_2 - x_1)$.

In fact Laplace tabulates in his 1774 paper the values of $-\tilde{q}_1$ as a function of m, for $m = 0(0.1)1$. But his values do not coincide with those given by (9). Stigler corrects an error in Laplace's values, but the values given by (9) still do not coincide with these corrected values. The reason for the discrepancy is worth noting.

In effect Laplace assumes not only a joint probability density for p_i but also σ and μ. Laplace's starting point is not simply (4) but

$$\phi(p)\ dp d\mu d\sigma, \tag{10}$$

which amounts to the assumption of a uniform prior density element for μ and for σ. If we wish to allow for an arbitrary prior density element for these two parameters we should replace (10) by

$$\phi(p)\ \pi(\mu, \sigma)\ dp d\mu d\sigma. \tag{11}$$

The effect of this is that q_3 ceases to be the only function of the five quantities p, μ, and σ which will become known when the observations x_i become known. In addition to transforming to q_i we can further transform to

$$s = \sigma q_2 = x_2 - x_1, \text{ and } x_1 = s q_1 + \mu. \tag{12}$$

If we make this transformation and then condition on the known values of q_3, s and x_1, we obtain a joint probability element for (μ, σ) from which we can marginalise σ to find the density element for μ. It turns out that to agree with the μ values derived from (9) we need to take $\pi(\mu, \sigma)$ to be proportional to $1/\sigma$ —the Jeffreys non-informative prior, rather than the uniform prior of Laplace. Given that we do this, the uniform prior for μ gives a result equivalent to that we have obtained above.

Returning to the argument we used to derive (7), the step from (6) to (7) in effect assumed the joint density of (q_1, q_2) to be given by (6), and applying this to determining μ as a function of (x_1, x_2, x_3) involves the assumption that (6) continues to give the correct density for q_2 even when $x_2 - x_1$ is known. This may perhaps not be true. If we had a fair idea that σ was not very far away from 1, say, and yet we observed $x_2 - x_1$ to have an enormous value, 1,000 say, then we surely would question our model. Strictly speaking, in passing from (6) to (7) we are assuming ignorance of σ; the relevant test of whether we really are effectively ignorant of σ is precisely to ask whether or not we feel that the density of the pivotal containing σ, namely q_2, retains the same density when we know $(x_2 - x_1)$, but nothing more about σ, as it had before we knew $x_2 - x_1$.

The analysis leading to this is both generalised and simplified if instead of taking σ as our parameter we take $\lambda = \ln \sigma$ and if instead of taking $x_2 - x_1$ as our observable we take $L = \ln |x_2 - x_1|$. We need then to add a further random variable $sgn(x_2 - x_1)$ to make our transformations retain their 1-1 character— and our model tells us that $sgn(x_2 - x_1)$ is equally likely to take the values +1 and –1, so it can easily be marginalised out. We can then replace q_2 by e^{ℓ} where $\ell = L - \lambda$. It can then easily be shown that for any pivotal such as ℓ, which is of the form $t(x) - \theta$, where t is a known function of the observations and θ is an unknown parameter, to assume that ℓ retains the same distribution when the observations become known as it had before this, is equivalent to asssuming a uniform prior for the unknown parameter λ. As is well known, a uniform prior for $\lambda = \ln \sigma$ corresponds to a prior element proportional to $d\sigma/\sigma$ for σ itself.

This approach to the problem of "non-informative priors" will not, of course, work for all kinds of parameter. As compared, however, with Bernardo's (1979) approach it can be expounded at a more elementary level. And it can be usefully extended, for example, to the correlation coefficient ρ, with its estimate r. We know that to a very good approximation, in most cases, for bivariate normality assumptions, $z-\zeta$ can be taken to have an $N(0, 1/(n-3))$ density, where $z = \tanh^{-1} r$ and $\zeta = \tanh^{-1} \rho$. This suggests we shall rarely be far wrong in taking ζ to have a uniform prior in those cases where practically nothing, apart from the data, is known about ρ. Of course I am not presenting this as a general solution to the problem of ignorance priors; only as a useful and elementary approach to those cases where it is applicable.

The relationship to Fisher's (1956) fiducial argument (including Lindley's (1958) requirement for the "consistency" of this) is easy to see. And the fact that one needs to restrict the number of unknown parameters to which a distribution over $(-\infty, +\infty)$ is assigned can be suggested at an elementary level by noticing that in any k-sphere of radius r the probability that a randomly chosen point is distant no more than $1/2r$ from the centre is only $1/2^k$.

3. NOTIONS OF ROBUSTNESS

Another line of investigation that might be pursued while still remaining within an undergraduate level introduces notions of robustness. If we define the function $M^a(u)$ by

$$M^a(u) = u^a \text{ for } u \geq 0, \qquad = (-Mu)^a \text{ for } u \leq 0,$$

then the probability density element assumed in (3) for a single p_i can be written

$$\exp -\tfrac{1}{2}M^a(p_i)dp_i, \quad \text{with } M = 1 \text{ and } a = 1.$$

Putting $M = 1$ and $a = 2$ will give us the normal case where the joint distribution of (q_1, q_2) turns out to be independent of q_3, so there is no need to know the value of m before we can condition. This, of course, relates to the fact that in the normal case we have a pair of statistics jointly sufficient for μ and σ. The same is true of the uniform $(-\frac{1}{2}, +\frac{1}{2})$ distribution for the p_i.

The effect of possible skewness can be examined by looking at $M = 2, a = 1$. In fact, for any joint density element proportional to

$$\exp -\tfrac{1}{2}M^a(p_i) \cdot d\mathbf{p} \tag{13}$$

with p_i as in (1), but with $i = 1, 2, \ldots, n$, we can extend the transformation used in Section 2 to

$$p_i = q_2(q_1 + c_i) \tag{14}$$

where, as in Section 2, $c_1 = 0, c_2 = 1$, and c_3 then replaces q_3. The Jacobian is then, as before, q_2^{n-1} and the marginal density of q_1, conditional on all the c_i becoming known, has element proportional to

$$1/M^a(q_1 + c_i) \cdot dq_1 \tag{15}$$

using the fact that $M^a(q_2(q_1 + c_i)) = q_2^a M^a(q_1 + c_i)$ to evaluate explicitly the marginal density.

4. SOME RESEARCH PROBLEMS

At this point, perhaps, the course we have been outlining verges onto the graduate research field. If n is moderately large there will be quite a lot of information as to the values of M and a embodied in the c_i, since it can be seen that c_i determine the shape of the empirical c.d.f. of the observations. If, apart from the sample, there is very little information concerning the values of M and a, it will be reasonable to choose values which appear to give a reasonable fit to the shape of sample. The formula (16) is easily programmed to take in M and a as parameters, along with the c_i values, and the sensitivity of the resultant q_1 distribution to errors of estimation of M and a therefore is easily assessed.

Another issue, at the research level, is raised by considering whether the choice of μ and σ as location and scale parameters is a good one. μ is always the mode of the x_1, and σ is the variance when $M = 1$ and $a = 2$, but for general M and a, if we define

$$K_r = \int x^r \exp -\tfrac{1}{2}M^a(x)dx$$

for $r = 1, 2, \ldots$, we find that

$$K_r = L_r(M)A_r(a),$$

where

$$L_r(M) = (M^{r+1} + (-1)^r)/M^{r+1}$$

and

$$A_r(a) = 2^{(r+1)/a}\Gamma((r+1)/a).$$

Then the r-th moment about the origin of a p_i — the mode — is

$$\mu'_r = K_r/K_o$$

the normalising constant is K_o and the mean is $\mu'_1 = K_1/K_o$. If $x_i = \mu + \sigma p_i$ then x_i has mode μ, mean $\mu + \sigma K_1/K_o$, and variance $\sigma^2(\mu'_2 - (\mu'_1)^2)$. It is to be expected that the estimate of the mean will be less sensitive to errors in M than will be the estimate of the mode; but against this must be set the fact that marginalisation of the pivotal involving σ is no longer possible.

5. CONCLUDING COMMENTS

I recall a discussion with a leading theoretical chemist, in 1946, when the possibilities being opened up by the new computers were beginning to be appreciated. He argued that we knew what the particles were — protons, neutrons, electrons, positrons and neutrinos — we could determine the Hamiltonian for any specified atom or molecule, and all (!) we had to do was to solve Schroedinger's wave equation to determine how any atom or molecule was going to behave. My referring him to William Blake's "infinity in a grain of sand and eternity in an hour", and to the fixed belief among physicists near the end of the 19th century that atoms were literally atomic which led Shuster to fail to realise he had discovered the electron, failed to move him.

I refer to this because there was a tendency in the early days of the "Bayesian revolution" to suggest that to solve any problem in statistical inference "all we had to do" was to write down the prior and the model, condition on the data, and derive the result. Whatever some enthusiasts might have thought then, I am sure no one now thinks it is as simple as that. Quite apart from the considerable difficulties, in multiparameter problems, of finding any prior which does not imply consequences at odds with the data, universal agreement on a prior will always be a rare occurrence. And we will rarely be able to find ways of expressing fully the ever present uncertainty as to the accuracy of our models. So I think the step-by step analysis of a problem which I have tried above to illustrate by reference to Laplace's problem seems likely always to be the preferred approach. In particular, there will always be some parameters like M and a above for which elimination by assuming a prior followed by marginalisation will not be a wise course. Computers now allow us to examine directly the effect of variations in such parameters, and this is what we should do. Problems remain when there are many such parameters — the "Latin hypercube" approach which has recently been advocated does not seem to help.

REFERENCES

Berger, J. O. (1985). *Statistical Decision Theory and Bayesian Analysis.* New York: Springer.

Bernardo, J. M. (1979). Reference posterior distributions for Bayesian inference. *J. Roy. Statist. Soc. B* **41**, 113–147, (with discussion).

Box, G. E. P. and Tiao, G. C. (1973). *Bayesian Inference in Statistical Analysis.* Reading, Mass.: Addison-Wesley.

de Finetti, B. (1972). *Probability, Induction and Statistics.* New York: Wiley.

Fisher, R. A. (1956). *Statistical Methods and Scientific Inference.* Edinburgh: Oliver and Boyd.

Laplace, P. S. (1774). Mémoire sur la probabilité des causes par les évènements. Reprinted in Laplace's *Oeuvres Complètes* **8**, 27–65. Paris: Gauthier-Villars. Translated into english with an introduction by Stigler (1986b).

Lindley, D. V. (1958). Fiducial distributions and Bayes' theorem. *J. Roy. Statist. Soc. B* **20**, 102–107.

Stigler, S. M. (1986a). *The History of Statistics: The Measurement od Uncertainty before 1900.* Cambridge, Mass.: Harvard University Press.

Stigler, S. M. (1986b). Laplace's 1774 memoir on inverse probability. *Statistical Science* **1**, 359–378.

Zellner, A. (1971). *An Introduction to Bayesian Inference in Econometrics.* New York: Wiley.

BAYESIAN STATISTICS 3, pp. 25–44
J. M. Bernardo, M. H. DeGroot, D. V. Lindley and A. F. M. Smith, (Eds.)

Gaining Weight: A Bayesian Approach

M. J. BAYARRI and M. H. DeGROOT
University of Valencia and *Carnegie Mellon University*

SUMMARY

Consider a problem in which the opinions of a group of experts regarding the value of an observable random variable X are to be combined. Suppose that the opinion of each expert receives a certain prior weight in the combined opinion, and that these weights are then updated based on the observed value of X. If the revised weights are going to have some future use, then it is natural for the experts to try to maximize their own weights by appropriately choosing the opinions that they report. In this paper we study optimal reporting strategies for a given expert in various situations and under various utility functions for his or her revised weight. We show how the optimal reported opinion usually differs from the expert's honest opinion, and we present the only utility functions for which it is always optimal for the expert to report his or her honest opinion.

Keywords: COMBINING OPINIONS; EXPERT OPINIONS; LINEAR OPINION POOL; OPTIMALITY; PREDICTIONS; REPORTING; STRICTLY PROPER SCORING RULES.

1. INTRODUCTION

Suppose that you and $k-1$ other experts, or advisers, must each report your predictive distribution for some random variable X whose value will be observed in the future. Let $r_1(x)$ denote the density function of your reported predictive distribution and let $r_2(x), \ldots, r_k(x)$ denote the density functions reported by the other experts. Suppose also that these k densities are to be reported to a decision maker who will combine them in some way in order to solve some particular problem. Suppose also that the rule of combination to be used by the decision maker involves the assignment of nonnegative weights $\alpha_1, \ldots, \alpha_k (\Sigma_{i=1}^k \alpha_i = 1)$ to the opinions of the different experts. Suppose, finally, that after the value of X has been observed, these weights are updated by the decision maker according to some rule that takes into account how well the observed value of X has been predicted by each expert. If these updated weights are going to be used in the future, then it is natural for the experts to try to maximize their own weights by appropriately choosing the predictive distributions that they report to the decision maker instead of necessarily reporting their honest distributions.

One model of this type that has been widely treated in the statistical literature is the linear opinion pool (see, e.g., Stone, 1961, and Genest and Zidek, 1986). In this model, the decision maker forms a convex combination of the reported densities $r_1(x), \ldots, r_k(x)$ to obtain the density

$$r_0(x) = \sum_{i=1}^{k} \alpha_i r_i(x) \quad , \tag{1.1}$$

Here, the number α_i represents the prior weight that the decision maker assigns to the report of expert i.

It is assumed in accordance with the usual development of Bayesian updating that after the value x of X has been observed, the decision maker will assign to expert i the posterior weight $\beta_i(x)$ given by

$$\beta_i(x) = \frac{\alpha_i r_i(x)}{\Sigma_{j=1}^k \alpha_j r_j(x)} \tag{1.2}$$

(see, e.g., Roberts, 1965, or McConway, 1978, who also mentions the problem of choosing an optimal report). One situation in which (1.2) arises is when a further random variable Y is to be predicted after $X = x$ has been observed. If the joint density $f(x, y)$ of the decision maker is given by the linear opinion pool

$$f(x, y) = \sum_{i=1}^k \alpha_i f_i(x, y) \quad , \tag{1.3}$$

where $f_i(x, y)$ is the reported joint density of expert i, then the marginal density for X of the decision maker is given by (1.1) and his or her conditional density for Y after $X = x$ has been observed is given by

$$r_0(y|x) = \sum_{i=1}^k \beta_i(x) r_i(y|x) \quad , \tag{1.4}$$

where $r_i(y|x)$, is the posterior predictive density of expert i and the corresponding posterior weight $\beta_i(x)$ is precisely given by (1.2).

Our focus in this paper is on your behavior as expert 1. Therefore, without loss of generality, we will regard the other $k-1$ experts as a single expert who reports the predictive density $s(x)$, where

$$s(x) = \frac{\Sigma_{j=2}^k \alpha_j r_j(x)}{\Sigma_{j=2}^k \alpha_j} \quad . \tag{1.5}$$

We can now simplify the notation by writing $r(x)$ instead of $r_1(x), \beta(x)$ instead of $\beta_1(x)$, α instead of α_1, and $\bar{\alpha}$ instead of $1 - \alpha = \Sigma_{j=2}^k \alpha_j$. With this notation, the linear opinion pool (1.1) is just

$$\alpha r(x) + \bar{\alpha} s(x) \tag{1.6}$$

and your posterior weight after $X = x$ has been observed is given by

$$\beta(x) = \frac{\alpha r(x)}{\alpha r(x) + \bar{\alpha} s(x)} \quad . \tag{1.7}$$

It should be noted that this reduction is made without loss of generality because it is consistent with the posterior linear opinion pool (1.4) if we let

$$s(y|x) = \frac{\Sigma_{j=2}^k \alpha_j r_j(x) r_j(y|x)}{\Sigma_{j=2}^k \alpha_j r_j(x)} \quad . \tag{1.8}$$

The question to be considered in this paper is the following: Suppose that your prior weight α is given and you want to maximize your posterior weight (1.7). If your honest subjective predictive density for X is $\pi(x)$, what density $r(x)$ should you report? We will discuss both problems in which you know the density $s(x)$ reported by the other expert and problems in which you are uncertain about it.

In this paper we will restrict ourselves to problems in which X can take only two values x_1 and x_2, and we will simplify the notation further by letting:

$$\begin{aligned} r(x_1) &= r, & r(x_2) &= \bar{r} = 1 - r \quad , \\ \pi(x_1) &= \pi, & \pi(x_2) &= \bar{\pi} = 1 - \pi \quad , \\ s(x_1) &= s, & s(x_2) &= \bar{s} = 1 - s \quad . \end{aligned} \tag{1.9}$$

Also, when convenient, we will simply write β instead of $\beta(x)$ to denote your posterior weight. Thus, in accordance with your honest predictive density $\pi(x)$, if you report r as your predictive probability that $X = x_1$ then your posterior weight, for any given value of s, will be

$$\beta = \begin{cases} \dfrac{\alpha r}{\alpha r + \bar{\alpha} s} & \text{with probability} \quad \pi \ , \\ \dfrac{\alpha \bar{r}}{\alpha \bar{r} + \bar{\alpha} \bar{s}} & \text{with probability} \quad \bar{\pi} \ . \end{cases} \tag{1.10}$$

Throughout the paper we will assume that $0 < \alpha < 1$, $0 < s < 1$, and $0 < \pi < 1$, unless otherwise noted, and we will allow your report r to take any value in the closed interval $0 \le r \le 1$.

In Section 2, we derive and study the optimal report r when you know the value of the other expert's report s and you are trying to maximize your expected posterior weight $E(\beta)$. A wide variety of properties are presented. In particular, we determine the conditions under which $r = 0$ and $r = 1$ are the optimal reports. It is found that for small values of α, these extreme reports are optimal over a wide range of values of s and π, and as $\alpha \longrightarrow 1$, the optimal report converges to the value $\sqrt{s\pi}/(\sqrt{s\pi} + \sqrt{\bar{s}\bar{\pi}})$, which is strictly between 0 and 1.

In Section 3, we extend the discusion in Section 2 to problems in which you are uncertain about the value of s and must choose your report r on the basis of your conditional distribution for s given $x_i (i = 1, 2)$.

In the remaining sections of the paper we consider the maximization of $E[\varphi(\beta)]$ for various nonlinear utility functions $\varphi(\beta)$ that you might have for your posterior weight β. In Section 4, we briefly consider the optimal report r when $\varphi(\beta) = \log \beta$ and its properties are contrasted with those presented in Sections 2 and 3. In particular, it is found that it always lies in the open interval $0 < r < 1$. As $\alpha \longrightarrow 0$, the optimal report converges to your honest probability π, and as $\alpha \longrightarrow 1$, it has the same limiting behavior as when $\varphi(\beta) = \beta$. Other utility functions also considered in this section are $\varphi(\beta) = \beta/(1-\beta)$ and $\varphi(\beta) = -(1-\beta)/\beta$. Under the first utility function, the optimal report can always be taken to be either $r = 0$ or $r = 1$. Under the second utility function, the optimal report does not depend on α and is shown to be identical to the common limiting value of the optimal reports for both $\varphi(\beta) = \beta$ and $\varphi(\beta) = \log \beta$ as $\alpha \longrightarrow 1$.

Finally, in Section 5, it is shown that when $\varphi(\beta) = \log[\beta/(1 - \beta)]$, the unique optimal report is always $r = \pi$, regardless of whether s is known or unknown. In other words, honesty is always the best policy under this utility function. Furthermore, $\varphi(\beta)$ is essentially the only utility function satisfying some simple regularity conditions for which this property holds. In a multiperiod problem with this utility function in which your goal is to maximize the expected utility of your final posterior weight at the end of the process, it is found that under certain conditions regarding the behavior of the other expert, the optimal strategy is again to report your honest probability at each stage.

2. MAXIMIZING $E(\beta)$

In this section we will consider problems in which you know the value of s reported by the other expert and your reported predictive probability r is to be chosen to maximize your expected posterior weight $E(\beta)$.

It follows from (1.10) that for given values of α, s, π, and r,

$$E(\beta) = \pi \frac{\alpha r}{\alpha r + \bar{\alpha} s} + \bar{\pi} \frac{\alpha \bar{r}}{\alpha \bar{r} + \bar{\alpha} \bar{s}} \ . \tag{2.1}$$

It should be noted that if you are honest and report $r = \pi$, then $E(\beta) \ge \alpha$, your prior weight, for any given value of s.

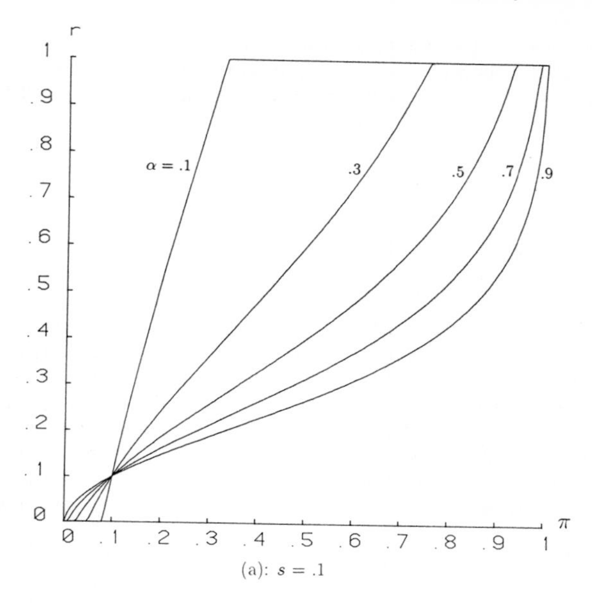

(a): $s = .1$

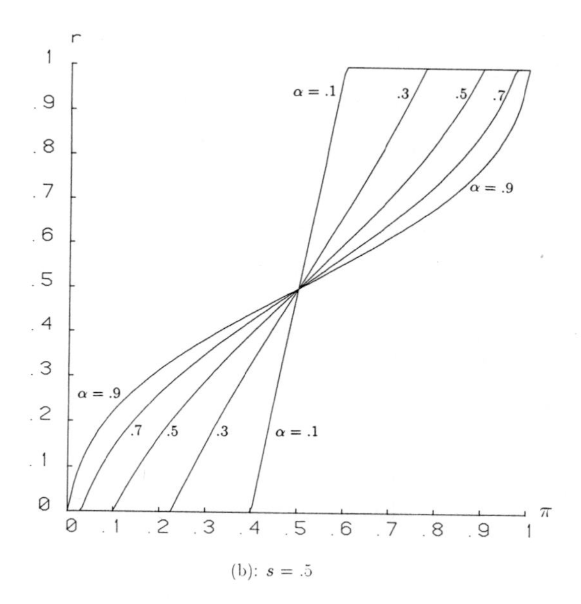

(b): $s = .5$

Figure 1. *The optimal report $r(\alpha, s, \pi)$ for maximizing $E(\beta)$ as a function of π for different values of α.*

However, for given values of α, s, and π, the choice $r = \pi$ does not necessarily maximize (2.1). In fact, it can be shown by elementary differentiation that the optimal choice of r is given by

$$r = \frac{1}{\alpha} \cdot \frac{\sqrt{s\pi}}{\sqrt{s\pi} + \sqrt{\bar{s}\bar{\pi}}} - \frac{\bar{\alpha}}{\alpha} s \tag{2.2}$$

provided that this value lies in the interval $0 \leq r \leq 1$. If the value of (2.2) is < 0, then the optimal choice is $r = 0$. If this value is > 1, then the optimal choice is $r = 1$.

We will now describe some properties of the optimal choice of r. When it is necessary to emphasize the dependence of this optimal choice on the value of α, s, and π, we will denote it by $r(\alpha, s, \pi)$. The first four properties follow easily from (2.2).

Property 2.1. If $\pi = s$, then $r(\alpha, s, \pi) = \pi$.

In words, if your honest probability π agrees with the other expert's report s, then you should be honest in your report.

Property 2.2. $r(\alpha, \bar{s}, \bar{\pi}) = 1 - r(\alpha, s, \pi)$.

Property 2.3. For any given values of α and $s, r(\alpha, s, \pi)$ as a function of π is nondecreasing, and it is strictly increasing over the interval where $0 < r(\alpha, s, \pi) < 1$.

Property 2.4. For any given values of s and $\pi, r(\alpha, s, \pi)$ as a function of α is nondecreasing when $\pi < s$ and is nonincreasing when $\pi > s$.

These properties are illustrated in Figure 1, which graphs the optimal value of the reported probability r as a function of the honest probability π for selected values of α and s. Properties 2.1, 2.3, and 2.4 are clearly displayed in each of the two parts of Figure 1. The symmetry of the functions in Figure 1(b) illustrates a special case of Property 2.2 when $s = \bar{s} = \frac{1}{2}$.

For given values of α and s, let $\pi_0(\alpha, s)$ denote the maximum value of π such that $r(\alpha, s, \pi) = 0$, and let $\pi_1(\alpha, s)$ denote the minimum value of π such that $r(\alpha, s, \pi) = 1$. Then it can be shown that $\pi_0(\alpha, s)$ and $\pi_1(\alpha, s)$ are given by

$$\pi_0(\alpha, s) = \frac{s\bar{s}\bar{\alpha}^2}{\bar{s} + \alpha^2 s} , \tag{2.3}$$

$$\pi_1(\alpha, s) = 1 - \pi_0(\alpha, \bar{s}) = 1 - \frac{s\bar{s}\bar{\alpha}^2}{s + \alpha^2 \bar{s}} . \tag{2.4}$$

Note that for given values of α and s, $\pi_0(\alpha, s)$ is the largest honest probability π for which it would still be optimal for you to report $r = 0$. For a given value of α, the largest possible value π_0^* of $\pi_0(\alpha, s)$ is attained when the other expert reports $s = s^* = 1/(1 + \alpha)$. The values $\pi_1(\alpha, s)$ and its minimum π_1^* have an analogous interpretation. Thus, for example, if $\alpha = 0.1$ and $s = 0.91$, then it is optimal for you to report $r = 0$ even when your honest probability is $\pi = 0.67$. Similarly, if $\alpha = 0.1$ and $s = 0.09$, then it is optimal for you to report $r = 1$ even when $\pi = 0.33$.

As stated in Properties 2.3 and 2.4, when $r(\alpha, s, \pi)$ is regarded as a function of π or of α, it is monotone. However, for fixed α and $\pi, r(\alpha, s, \pi)$ is not necessarily a monotone function of s. This can be seen in Figure 2 where $r(\alpha, s, \pi)$ is graphed as a function of π and the curves corresponding to different values of s cross each other, or more directly in Figure 3 where $r(\alpha, s, \pi)$ is graphed as a function of s for $\pi = .5$ and various values of α. Nevertheless, it can be shown that

$$\lim_{s \to 0} r(\alpha, s, \pi) = 0 , \tag{2.5}$$

$$\lim_{s \to 1} r(\alpha, s, \pi) = 1 . \tag{2.6}$$

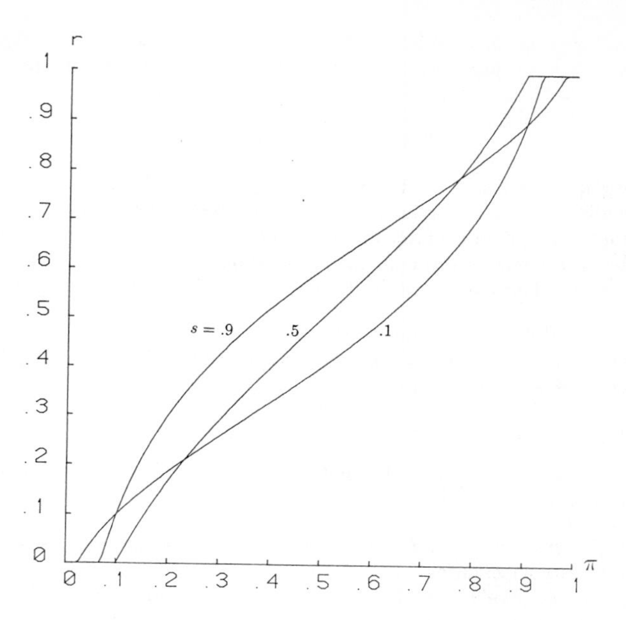

Figure 2. *The optimal report $r(\alpha, s, \pi)$ for maximizing $E(\beta)$ as a function of π for different values of s when $\alpha = .5$.*

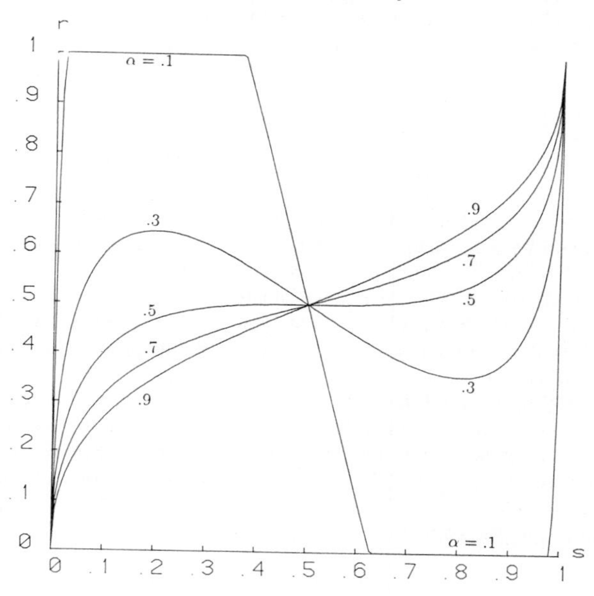

Figure 3. *The optimal report $r(\alpha, s, \pi)$ for maximizing $E(\beta)$ as a function of s when $\pi = .5$.*

The next two properties refer to special values of s and α. In Property 2.5, we will consider the values $s = 0$ and $s = 1$. Since β, as given by (1.10), is not defined for $r = s = 0$ or $r = s = 1$, and since $\beta = \alpha$ with probability 1 for $0 < r = s < 1$, we will make the natural assumption that $\beta = \alpha$ even when $r = s = 0$ or $r = s = 1$. (We continue to assume that $0 < \alpha < 1$ and $0 < \pi < 1$.)

Property 2.5. If $s = 0$, then

$$\sup_r E(\beta) = \lim_{r \to 0} E(\beta) = \alpha + \pi\bar{\alpha} \quad . \tag{2.7}$$

The supremum in (2.7) is not attained since, for $r = 0$, $E(\beta) = \alpha$. However, a value arbitrarily close to the supremum can be attained by choosing $r = \epsilon$ for an arbitrarily small $\epsilon > 0$. Similarly, if s = 1, a value of $E(\beta)$ arbitrarily close to

$$\sup_r E(\beta) = \lim_{r \to 1} E(\beta) = \alpha + \bar{\pi}\bar{\alpha} \quad . \tag{2.8}$$

can be attained by choosing $r = 1 - \epsilon$ for an arbitrarily small $\epsilon > 0$.

Property 2.6. If $\alpha = \frac{1}{2}$ and $\pi = \bar{s}$, then $r(\alpha, s, \pi) = \pi$.

In words, Property 2.6 states that when you and the other expert have equal prior weights, then it is optimal for you to report your honest probability π, not only when $\pi = s$, but also when $\pi = \bar{s}$. This property is dramatically illustrated in Figure 2 where the curves of $r(\frac{1}{2}, s, \pi)$ corresponding to $s = 0.1$ and $s = 0.9$ intersect at precisely the two values $\pi = 0.1$ and $\pi = 0.9$.

We will now consider the limiting values of $r(\alpha, s, \pi)$ for extreme values of α.

Property 2.7. For given values of s and π,

$$\lim_{\alpha \to 0} r(\alpha, s, \pi) = 0 \text{ if } \pi < s \quad , \tag{2.9}$$

$$\lim_{\alpha \to 0} r(\alpha, s, \pi) = 1 \text{ if } \pi > s \quad . \tag{2.10}$$

Proof. The optimal value of r is $r = 0$ if $\pi \leq \pi_0(\alpha, s)$ as given by (2.3). For any given values of π and s such that $\pi < s$, it can be shown that this condition will be satisfied for all sufficiently small values of α. Hence, (2.9) holds. The relation (2.10) follows from (2.9) and Property 2.2. ◁

Property 2.8. For any given values of s and π,

$$\lim_{\alpha \to 1} r(\alpha, s, \pi) = \frac{\sqrt{s\pi}}{\sqrt{s\pi} + \sqrt{\bar{s}\bar{\pi}}} \quad . \tag{2.11}$$

This property follows immediately from (2.2). If can be seen that (2.11) is always between 0 and 1. It is interesting to note that when $\pi = \bar{s}$, the optimal report is $r = \frac{1}{2}$.

The gain in your expected posterior weight $E(\beta)$ when you report the optimal value of r rather than your honest probability π is illustrated in Table 1. In that table, r denotes the optimal report as derived in this section, and $E_r(\beta)$ and $E_\pi(\beta)$ denote your expected posterior weight when you report r and π respectively. The column labeled "% Gain" gives the values of

$$100 \frac{E_r(\beta) - E_\pi(\beta)}{E_\pi(\beta)} \quad . \tag{2.12}$$

As could be expected, (2.12) is relatively large when the prior weight α is small, and it is relatively small when α is large. However, it is somewhat surprising that the smallest values of (2.12) in Table 1 over a wide range of values of π occur for $\alpha = .5$. For that value of α, it can be seen that r tends to be quite close to π. Table 1 includes just values of $s \leq .5$. A similar table for values of $s > .5$ can be easily derived by using Property 2.2.

	$\alpha = .1$				
	π	r	$E_r(\beta)$	$E_\pi(\beta)$	%Gain
$s = .1$	.2	.53	.118	.108	9.0
	.4	1	.211	.165	28.0
	.6	1	.316	.259	22.0
	.8	1	.421	.381	10.4
$s = .3$	.2	0	.110	.104	5.4
	.4	.78	.110	.104	5.9
	.6	1	.162	.133	21.9
	.8	1	.216	.189	14.4
$s = .5$	.2	0	.145	.129	12.5
	.4	0	.109	.103	5.7
	.6	1	.109	.103	5.7
	.8	1	.145	.129	12.5

	$\alpha = .5$				
	π	r	$E_r(\beta)$	$E_\pi(\beta)$	%Gai
$s = .1$	.2	.19	.510	.510	.0
	.4	.33	.563	.560	.5
	.6	.48	.643	.637	.8
	.8	.70	.750	.747	.3
$s = .3$	.2	.19	.507	.507	.00
	.4	.40	.506	.505	.00
	.6	.59	.546	.545	.00
	.8	.83	.627	.626	.07
$s = .5$	.2	.17	.550	.549	.10
	.4	.40	.505	.505	.00
	.6	.60	.505	.505	.00
	.8	.83	.550	.549	.10

	$\alpha = .3$				
	π	r	$E_r(\beta)$	$E_\pi(\beta)$	%Gain
$s = .1$	.2	.24	.314	.313	.32
	.4	.48	.388	.386	.59
	.6	.73	.500	.496	.85
	.8	1	.649	.637	1.87
$s = .3$	.2	.12	.309	.307	.62
	.4	.46	.308	.307	.34
	.6	.78	.364	.356	2.27
	.8	1	.471	.448	4.93
$s = .5$	.2	0	.369	.355	4.10
	.4	.33	.307	.306	.38
	.6	.67	.307	.306	.38
	.8	1	.369	.355	4.10

	$\alpha = .9$				
	π	r	$E_r(\beta)$	$E_\pi(\beta)$	%Gai
$s = .1$	.2	.15	.902	.901	.1
	.4	.23	.913	.903	1.0
	.6	.31	.929	.909	2.1
	.8	.43	.950	.922	2.9
$s = .3$	.2	.24	.901	.900	.0
	.4	.35	.901	.900	.0
	.6	.46	.909	.903	.6
	.8	.60	.925	.912	1.4
$s = .5$	.2	.31	.910	.905	.6
	.4	.44	.901	.900	.0
	.6	.56	.901	.900	.0
	.8	.69	.910	.905	.6

Table 1. *Gain in $E(\beta)$ from the optimal report.*

3. UNCERTAIN s

Suppose now that you are uncertain about the value of s that the other expert will report. Let ξ_1 denote your prior distribution for s if in fact $X = x_1$, and let ξ_2 denote your prior distribution for s if in fact $X = x_2$. Then for given values of α, π, and r, it follows from (1.10) that your expected posterior weight is

$$E(\beta) = \pi \int_0^1 \frac{\alpha r}{\alpha r + \bar{\alpha} s} \, d\xi_1(s) + \bar{\pi} \int_0^1 \frac{\alpha \bar{r}}{\alpha \bar{r} + \bar{\alpha} \bar{s}} \, d\xi_2(s) \quad . \tag{3.1}$$

It should be emphasized that since no relationship between the distributions ξ_1 and ξ_2 is assumed, the situation can now be drastically different from that described in Section 2. For example, when you report your honest probability π, then $E(\beta) \geq \alpha$ when you know the value of s. However, even though this relation holds for every value of s, it no longer necessarily holds when you are uncertain about s. In particular, if ξ_1 is highly concentrated near $s = 1$ and ξ_2 is highly concentrated near $s = 0$, then $E(\beta) < \alpha$ for *every* possible report r. In other words, if you think that the other expert will be almost perfect in his or her prediction, then you always expect your posterior weight to be less than your prior weight.

In this section we will consider the maximization of (3.1). The optimal value of r, that is, the value that maximizes (3.1), will depend on α, ξ_1, ξ_2, and π. When it will be helpful to exhibit this dependence explicitly, we will denote the optimal value by $r(\alpha, \xi_1, \xi_2, \pi)$. Otherwise, we will again simply write r for the optimal value.

For any distribution ξ of s on the unit interval, we shall let $\bar{\xi}$ denote the corresponding distribution of $\bar{s}$. Thus $\bar{\xi}$ is also a distribution on the unit interval, and therefore it could also serve as a distribution of s in some problem. Finally, for a given distribution $\xi_i (i = 1, 2)$ in a particular problem, we shall let E_i denote an expectation calculated with respect to the distribution ξ_i.

An explicit expression for the optimal report $r(\alpha, \xi_1, \xi_2, \pi)$ cannot be given for arbitrary distributions ξ_1 and ξ_2. However, $r(\alpha, \xi_1, \xi_2, \pi)$ exhibits some general properties that we will now discuss. The first two are generalizations of Properties 2.2 and 2.3.

Property 3.1. For any values of α and π and any distributions ξ_1 and ξ_2,

$$r(\alpha, \bar{\xi}_2, \bar{\xi}_1, \bar{\pi}) = 1 - r(\alpha, \xi_1, \xi_2, \pi) \quad . \tag{3.2}$$

Proof. By making the appropriate substitutions in (3.1), it follows that the left-hand side of (3.2) is the value of r that maximizes

$$\bar{\pi} \int_0^1 \frac{\alpha r}{\alpha r + \bar{\alpha} s} \, d\bar{\xi}_2(s) + \pi \int_0^1 \frac{\alpha \bar{r}}{\alpha \bar{r} + \bar{\alpha} \bar{s}} \, d\bar{\xi}_1(s) \quad . \tag{3.3}$$

If the distribution of s is $\bar{\xi}_i$, then the distribution of $\bar{s}$ will be ξ_i. It follows that (3.3) can be rewritten as

$$\bar{\pi} \int_0^1 \frac{\alpha r}{\alpha r + \bar{\alpha} s} \, d\xi_2(\bar{s}) + \pi \int_0^1 \frac{\alpha \bar{r}}{\alpha \bar{r} + \bar{\alpha} \bar{s}} \, d\xi_1(\bar{s}) \quad . \tag{3.4}$$

If we now change variables in each of the integrals in (3.4) by letting $u = \bar{s}$, then (3.4) becomes identical to (3.1) with r and $\bar{r}$ interchanged. Therefore, the value of r that maximizes (3.4) will be equal to the value of $\bar{r}$ that maximizes (3.1). ◁

Property 3.2. For any value of α and any distributions ξ_1 and $\xi_2, r(\alpha, \xi_1, \xi_2, \pi)$ is a nondecreasing function of π, and it is a strictly increasing over the interval where $0 < r(\alpha, \xi_1, \xi_2, \pi) < 1$.

Proof. The derivative of (3.1) with respect to r is equal to 0 if and only if

$$\frac{\int_0^1 [\bar{s}/(\alpha \bar{r} + \bar{\alpha} \bar{s})^2] \, d\xi_2(s)}{\int_0^1 [s/(\alpha r + \bar{\alpha} s)^2] \, d\xi_1(s)} = \frac{\pi}{\bar{\pi}} \quad . \tag{3.5}$$

Since the left-hand side of (3.5) is a strictly increasing function of r, there is at most one solution of (3.5) in the inverval $0 < r < 1$. When it exists, that solution must be the optimal value of r. Otherwise, the optimal value will be either $r = 0$ or $r = 1$. Property 3.2 can now easily be derived from the fact that the right-hand side of (3.5) is a strictly increasing function of π. ◁

The next property describes a special situation in which it is optimal to report your honest probability.

Property 3.3. For any value of α and any distributions ξ_1 and ξ_2 such that $\xi_1 = \bar{\xi}_2$, if $\pi = \frac{1}{2}$ then $r(\alpha, \xi_1, \xi_2, \pi) = \frac{1}{2}$.

Proof. When $\pi = \frac{1}{2}$, the right-hand side of (3.5) is 1. Also, when $\xi_1 = \bar{\xi}_2$, the two integrals on the left-hand side of (3.5) are identical except that one is a function of $\bar{r}$ and the other is the same function of r. Hence, the unique solution of (3.5) is $r = \bar{r} = \frac{1}{2}$. ◁

As in Section 2, we will now consider the limiting behavior of the optimal report as $\alpha \to 0$ and $\alpha \to 1$. These results are generalizations of Properties 2.7 and 2.8.

Property 3.4. For any value of π and any distributions ξ_1 and ξ_2 such that the expectations $E_1(1/s^2)$ and $E_2(1/\bar{s}^2)$ exist,

$$\begin{aligned} &\lim_{\alpha\to 0} r(\alpha,\xi_1,\xi_2,\pi) = 0 \quad \text{if } \pi E_1(1/s) < \bar{\pi} E_2(1/\bar{s}) \ , \\ &\lim_{\alpha\to 0} r(\alpha,\xi_1,\xi_2,\pi) = 1 \quad \text{if } \pi E_1(1/s) > \bar{\pi} E_2(1/\bar{s}) \ , \\ &\lim_{\alpha\to 0} r(\alpha,\xi_1,\xi_2,\pi) = \frac{\bar{\pi} E_2(1/\bar{s}^2)}{\pi E_1(1/s^2) + \bar{\pi} E_2(1/\bar{s}^2)} \quad \text{if } \pi E_1(1/s) = \bar{\pi} E_2(1/\bar{s}) \ . \end{aligned} \tag{3.6}$$

Proof. If $\pi E_1(1/s) < \bar{\pi} E_2(1/\bar{s})$, then for sufficiently small values of α, the left-hand side of (3.5) is larger than the right-hand side for all values of $r(0 < r < 1)$. It follows from the discussion given in the proof of Property 3.3 that $r(\alpha,\xi_1,\xi_2,\pi) = 0$ for all values of α in some interval of the form $0 < \alpha \le \alpha_0$. Similarly, if $\pi E_1(1/s) > \bar{\pi} E_2(1/\bar{s})$, then it follows that $r(\alpha,\xi_1,\xi_2,\pi) = 1$ for all values of α in some interval of the form $0 < \alpha \le \alpha_1$. Finally, suppose that $\pi E_1(1/s) = \bar{\pi} E_2(1/\bar{s})$. Then if we expand each of the integrals in (3.5) through terms order α, it is found that (3.5) reduces to the relation

$$r\pi E_1(1/s^2) = \bar{r}\bar{\pi} E_2(1/\bar{s}^2) \ , \tag{3.7}$$

from which (3.6) follows. ◁

The limiting properties of $r(\alpha,\xi_1,\xi_2,\pi)$ when the expectations in (3.6) do not exist are not considered here.

Property 3.5. For any value of π and any distributions ξ_1 and ξ_2,

$$\lim_{\alpha\to 1} r(\alpha,\xi_1,\xi_2,\pi) = \frac{\sqrt{\pi E_1(s)}}{\sqrt{\pi E_1(s)} + \sqrt{\bar{\pi} E_2(\bar{s})}} \ . \tag{3.8}$$

Proof. This property follows easily by letting $\alpha \to 1$ in (3.5). ◁

We conclude this section by considering two special cases. First, suppose that both ξ_1 and ξ_2 are uniform distributions on the inteval $0 < s < 1$. Then (3.5) becomes

$$\pi \left[\log\left(1 + \frac{\bar{\alpha}}{\alpha r}\right) - \frac{\bar{\alpha}}{\alpha r + \bar{\alpha}}\right] = \bar{\pi} \left[\log\left(1 + \frac{\bar{\alpha}}{\alpha \bar{r}}\right) - \frac{\alpha}{\alpha \bar{r} + \bar{\alpha}}\right] \ . \tag{3.9}$$

For any given values of α and π, the left-hand side of (3.9) is a decreasing function of r which becomes infinite as $r \to 0$, while the right-hand side is an increasing function of r which becomes infinite as $r \to 1$. Therefore, there must be a unique solution in the interval $0 < r < 1$ which is the optimal report. Although this optimal value of r has no expression in closed form, it is easy to find numerically from (3.9).

As a second example, suppose that you know that the other expert will report either $s = 0$ or $s = 1$. For $i = 1, 2$, let

$$\xi_i = \xi_i(s=1), \quad \bar{\xi}_i = \xi_i(s=0) = 1 - \xi_i \ . \tag{3.10}$$

Then it can be shown from (3.5) that the optimal value of r is given by

$$r = \left(1 + \frac{2\bar{\alpha}}{\alpha}\right)\left(\frac{\sqrt{\pi\xi_1}}{\sqrt{\pi\xi_1} + \sqrt{\bar{\pi}(1-\xi_2)}}\right) - \frac{\bar{\alpha}}{\alpha} \tag{3.11}$$

when (3.11) is between 0 and 1. Otherwise, $r = 0$ if (3.11) is < 0, and $r = 1$ if (3.11) is > 1.

Now suppose that $\xi_1 = \bar{\xi}_2$ in this example. Then it is seen from (3.11) that the optimal value of r does not depend on this common value. Thus, if you assign the same probability to the other expert reporting $s = 1$ when $X = x_1$ as you do to his or her reporting $s = 0$ when $X = x_2$, then your optimal report does not depend on this probability. In other words, you should give the same report regardless of whether you think that the other expert is usually right or usually wrong!

4. MAXIMIZING $E[\varphi(\beta)]$

The problems described in Sections 2 and 3 have the peculiar feature that it is often optimal for you to report either $r = 0$ or $r = 1$ for the value of $\Pr(X = x_1)$, even though your honest evaluation π of that probability is strictly between 0 and 1. Of course, if you report probability 0 for either the event that $X = x_1$ or the event that $X = x_2$, and that event then occurs, your posterior weight β will drop to 0 which you may regard as having catastrophic consequences. This discussion suggests that although you wish to make β as large as possible, the maximization of $E(\beta)$ may not be a reasonable criterion. There may be an increasing function $\varphi(\beta)$ such that the maximization of $E\,[\varphi(\beta)]$ provides a better strategy for you. In other words, $\varphi(\beta)$ is your utility function for your posterior weight β.

In this section we will just mention some results for the utility functions $\log \beta$, $\frac{\beta}{1-\beta}$ and $-\frac{\beta}{1-\beta}$. Full details as well as additional discussion of the properties of the corresponding optimal reports can be found in Bayarri and DeGroot (1987).

Consider first the utility function $\varphi(\beta) = \log \beta$, which assigns infinite negative utility to the value $\beta = 0$. Then the optimal report r is found to be

$$r(\alpha, s, \pi) = \frac{2s\pi(1 - \bar{\alpha}s)}{2\alpha s\pi + \bar{\alpha}s\bar{s} + (4\alpha s\bar{s}\pi\bar{\pi} + \bar{\alpha}^2 s^2 \bar{s}^2)^{1/2}} \quad , \tag{4.1}$$

which always lies in the interval $0 < r < 1$. It can be shown that (4.1) exhibits essentially the same properties 2.1-2.5 that were obtained when maximizing $E(\beta)$. It is also interesting to compare the limiting values of (4.1) for extreme values of α with those obtained in Section 2. It follows from (4.1) that, for given values of s and π,

$$\lim_{\alpha \to 0} r(\alpha, s, \pi) = \pi \quad . \tag{4.2}$$

It is noteworthy that with the utility function being considered, the limiting optimal report for small values of α is your honest probability π . This striking result contrasts with the degenerate limits obtained in Property 2.7 On the other hand, for any given values of s and π,

$$\lim_{\alpha \to 1} r(\alpha, s, \pi) = \frac{\sqrt{s\pi}}{\sqrt{s\pi} + \sqrt{\bar{s}\bar{\pi}}} \quad . \tag{4.3}$$

It is also noteworthy that in this case the limiting optimal report is precisely the same as that obtained in Property 2.8.

Similar limiting behavior for the optimal report is reproduced when you are uncertain about the value of s. Thus, with the notation used in Section 3, for any value of π and any distributions ξ_1 and ξ_2,

$$\lim_{\alpha \to 0} r(\alpha, \xi_1, \xi_2, \pi) = \pi \quad . \tag{4.4}$$

This property is not surprising in view of (4.2). However, it contrasts with the limiting value obtained in Property 3.4 and the integrability conditions that were needed there. On the other hand, for any value of π and any distributions ξ_1 and ξ_2,

$$\lim_{\alpha \to 1} r(\alpha, \xi_1, \xi_2, \pi) = \frac{\sqrt{\pi E_1(s)}}{\sqrt{\pi E_1(s)} + \sqrt{\bar{\pi} E_2(\bar{s})}} \quad . \tag{4.5}$$

It should be noted that this limiting value is the same as that obtained in Property 3.5 when we were maximizing $E(\beta)$.

In the remainder of this section, we will consider two other possible utility functions for this problem. Suppose first that you want to maximize $E\left(\frac{\beta}{1-\beta}\right)$, the expected ratio of your posterior weight to that of the other expert. Then it can be shown that the optimal report $r(\alpha, s, \pi)$ does not depend on α and is simply given by

$$r(\alpha, s, \pi) = \begin{cases} 0 & \text{if} \quad \pi < s \ , \\ 1 & \text{if} \quad \pi > s \ , \\ \text{arbitrary} & \text{if} \quad \pi = s \ . \end{cases} \tag{4.6}$$

If s is unknown, then in our usual notation, the optimal report is given by

$$r(\alpha, \xi_1, \xi_2, \pi) = \begin{cases} 0 & \text{if} \quad \pi E_1(1/s) < \bar{\pi} E_2(1/\bar{s}) \ , \\ 1 & \text{if} \quad \pi E_1(1/s) > \bar{\pi} E_2(1/\bar{s}) \ , \\ \text{arbitrary} & \text{if} \quad \pi E_1(1/s) = \bar{\pi} E_2(1/\bar{s}) \ . \end{cases} \tag{4.7}$$

If $E_1(1/s) = \infty$, then any report $r > 0$ will yield $E\left[\beta/(1-\beta)\right] = \infty$. Similarly, if $E_2(1/s) = \infty$, then any report $r < 1$ will yield $E\left[\beta/(1-\beta)\right] = \infty$.

It follows that, with this utility function, the optimal report can always be taken to be either 0 or 1, regardless of whether the value of s is known or uncertain.

Alternatively, suppose that instead of maximizing $E(\frac{\beta}{1-\beta})$, you want to minimize $E(\frac{1-\beta}{\beta})$ which corrresponds to using $\varphi(\beta) = -(1-\beta)/\beta$ as the utility function for β. Then it can be shown that the optimal value of r is given by:

$$r(\alpha, s, \pi) = \frac{\sqrt{s\pi}}{\sqrt{s\pi} + \sqrt{\bar{s}\bar{\pi}}} \ . \tag{4.8}$$

It should be noted that this solution always lies in the interval $0 < r < 1$ and, as before, it does not depend on α. It should also be noted that it is the same as the limiting solution when $\alpha \to 1$ for either of the utility functions $\varphi(\beta) = \beta$ or $\varphi(\beta) = \log \beta$.

When you are uncertain about s, the optimal choice of r is simply

$$r(\alpha, \xi_1, \xi_2, \pi) = \frac{\sqrt{\pi E_1(s)}}{\sqrt{\pi E_1(s)} + \sqrt{\bar{\pi} E_2(\bar{s})}} \ . \tag{4.9}$$

Again, (4.9) is the limiting solution when $\alpha \to 1$ for either of the utility functions $\varphi(\beta) = \beta$ or $\varphi(\beta) = \log \beta$.

If follows immediately from (4.9) that if $E_1(s) = E_2(\bar{s})$, then

$$r(\alpha, \xi_1, \xi_2, \pi) = \frac{\sqrt{\pi}}{\sqrt{\pi} + \sqrt{\bar{\pi}}} \ , \tag{4.10}$$

independently of the common value of $E_1(s)$ and $E_2(\bar{s})$. Of course, $E_1(s) = E_2(\bar{s})$ whenever $\xi_1 = \bar{\xi}_2$. Two special cases in which $\xi_1 = \bar{\xi}_2$ are the examples we have described in Section 3, namely (i) when both ξ_1 and $\bar{\xi}_2$ are uniform distributions on the interval $0 < s < 1$, and (ii) the example in which you know that either $s = 0$ or $s = 1$, and your probabilities are such that $\xi_1 = \bar{\xi}_2$.

5. HONESTY IS THE BEST POLICY

In the previous sections, we have encountered various special conditions under which the optimal report $r(\alpha, s, \pi)$ is your honest probability π. For example, π is the optimal report when your utility function is $\varphi(\beta) = \beta, \varphi(\beta) = \log \beta$, or $\varphi(\beta) = -(1-\beta)/\beta$, and $\pi = s$. It was also shown that π is the optimal report for all values of s when your utility function is $\varphi(\beta) = \log \beta$ and $\alpha \to 0$. It is natural to ask whether there is a utility function for which the unique optimal report is π for all values of α, s, and π. A utility function satisfying this condition might be called *strictly proper*, by analogy with the concept of a strictly proper scoring rule (see, e.g., Staël von Holstein, 1970; Savage, 1971; and DeGroot and Fienberg, 1983).

In this section we will present a strictly proper utility function and prove that it is essentially the only one that satisfies certain basic regularity conditions. Suppose that your utility function is

$$\varphi(\beta) = \log \frac{\beta}{1-\beta} \quad . \tag{5.1}$$

Then for any given values of α, s and π,

$$\begin{aligned} E\left(\log \frac{\beta}{1-\beta}\right) &= \pi \log \frac{\alpha r}{\bar{\alpha} s} + \bar{\pi} \log \frac{\alpha \bar{r}}{\bar{\alpha} \bar{s}} \\ &= \pi \log r + \bar{\pi} \log \bar{r} \\ &\quad + (\text{terms not involving } r) \quad . \end{aligned} \tag{5.2}$$

It follows from (5.2) that the optimal value of r is always

$$r(\alpha, s, \pi) = \pi \quad . \tag{5.3}$$

When s is unknown, it is again found that the unique optimal report is

$$r(\alpha, \xi_1, \xi_2, \pi) = \pi \tag{5.4}$$

for any values of α and π and distributions ξ_1 and ξ_2 such that $E_1(\log s)$ and $E_2(\log \bar{s})$ are finite.

We wil now prove that the only strictly proper utility functions satisfying certain regularity conditions are increasing linear functions of (5.1). Consider first the situation when s is known.

Theorem. Let $\psi(\beta)$ be a nondecreasing, differentiable utility function with a continuous derivative $\psi'(\beta)$ for $0 < \beta < 1$, and assume that

$$\lim_{\beta \to 0} \beta \psi'(\beta) \tag{5.5}$$

is well-defined, either as a nonnegative value or $+\infty$. Then $\psi(\beta)$ is strictly proper if and only if it is of the form

$$\psi(\beta) = c_0 \log \left(\frac{\beta}{1-\beta}\right) + c_1 \quad , \tag{5.6}$$

where c_0 and c_1 are constants with $c_0 > 0$.

Proof. For any values of α, π, and r, and any utility function ψ,

$$E[\psi(\beta)] = \pi \psi \left(\frac{\alpha r}{\alpha r + \bar{\alpha} s}\right) + \bar{\pi} \psi \left(\frac{\alpha \bar{r}}{\alpha \bar{r} + \bar{\alpha} \bar{s}}\right) \quad . \tag{5.7}$$

In order for ψ to be strictly proper, (5.7) considered as a function of r must have a unique maximum when $r = \pi$. Hence, it must be true that for any values of α, s, π, and r,

$$\pi\psi\left(\frac{\alpha r}{\alpha r + \bar{\alpha} s}\right) + \bar{\pi}\psi\left(\frac{\alpha \bar{r}}{\alpha \bar{r} + \bar{\alpha}\bar{s}}\right) \leq \pi\psi\left(\frac{\alpha \pi}{\alpha \pi + \bar{\alpha} s}\right) + \bar{\pi}\psi\left(\frac{\alpha \bar{\pi}}{\alpha \bar{\pi} + \bar{\alpha}\bar{s}}\right) \quad . \tag{5.8}$$

When each side of the inequality (5.8) is regarded as a function of π, this inequality states that the left-hand side is a line of support for the right-hand side at the point $\pi = r$. Therefore, the first derivatives of the two sides with respect to π must be equal at this point. It can be shown that this equality reduces to the relation

$$\pi\psi'\left(\frac{\alpha \pi}{\alpha \pi + \bar{\alpha} s}\right)\frac{\partial}{\partial \pi}\left(\frac{\alpha \pi}{\alpha \pi + \bar{\alpha} s}\right) = -\bar{\pi}\psi'\left(\frac{\alpha \bar{\pi}}{\alpha \bar{\pi} + \bar{\alpha}\bar{s}}\right)\frac{\partial}{\partial \pi}\left(\frac{\alpha \bar{\pi}}{\alpha \bar{\pi} + \bar{\alpha}\bar{s}}\right) \tag{5.9}$$

evaluated at $\pi = r$, which reduces to

$$\frac{sr}{(\alpha r + \bar{\alpha} s)^2}\psi'\left(\frac{\alpha r}{\alpha r + \bar{\alpha} s}\right) = \frac{\bar{s}\bar{r}}{(\alpha \bar{r} + \bar{\alpha}\bar{s})^2}\psi'\left(\frac{\alpha \bar{r}}{\alpha \bar{r} + \bar{\alpha}\bar{s}}\right) \quad . \tag{5.10}$$

Under the assumptions of the theorem, the value of the limit (5.5) must be either $0, +\infty$, or c_0, where $0 < c_0 < \infty$. Suppose first that this limit is $c_0(0 < c_0 < \infty)$. If we let $r \to 0$ in (5.10) then it follows from the continuity of ψ' that

$$\frac{c_0}{\alpha\bar{\alpha}} = \frac{\bar{s}}{(\alpha + \bar{\alpha}\bar{s})^2}\psi'\left(\frac{\alpha}{\alpha + \bar{\alpha}\bar{s}}\right) \quad . \tag{5.11}$$

If we let $x = \alpha/(\alpha + \bar{\alpha}\bar{s})$, then (5,11) becomes

$$x(1-x)\psi'(x) = c_0 \quad . \tag{5.12}$$

Since (5.12) must hold for all values of x in the interval $0 < x < 1$, it follows that ψ must have the desired form (5.6).

Note next that since the right-hand side of (5.11) is finite, the limit (5.5) cannot be infinite if ψ is to be strictly proper and satisfy the conditions of the theorem. Finally suppose that the limit (5.5) is 0. Then from (5.12), $\psi(\beta)$ must be constant, which is not a strictly proper utility function. ◁

The same theorem can be obtained when s is unknown under the additional assumption on ξ_1 and ξ_2 that the apropriate derivatives and limits can be taken inside the expectations.

The strictly proper utility function (5.1) also provides an interesting simplification in multiperiod problems. Suppose that after the value of X has been observed and your posterior weight β has been determined, β will be used as your prior weight in another similar problem in which the value of some other dichotomous random variable Y must be predicted. Suppose that you want to maximize the expected utility $E[\varphi(\gamma)]$ of your posterior weight γ at the end of the second stage. If you believe that the other expert is going to be honest in his or her report at the second stage, then it can be shown that the optimal strategy for you is to report your honest probabilities at both the first and second stages.

It follows that this result also holds for more than two stages. In other words, for this strictly proper utility function, the myopic rule is the optimal strategy.

ACKNOWLEDGEMENTS

This research was supported in part by the Spanish Ministry of Education and Science and the Fulbright Association under grant 85-07399 and by the National Science Foundation under grant DMS-8320618.

REFERENCES

DeGroot, M. H., and Fienberg, S. E. (1983). The comparison and evaluation of forecasters. *The Statistician* **32**, 12–22.

Genest, C., and Zidek, J. V. (1986). Combining probability distributions: A critique and annotated bibliography. *Statistical Science* **1**, 114–148 (with discussion).

McConway, K. J. (1978). The combination of experts' opinions in probability assessment: Some theoretical considerations. *Ph. D. Thesis*. University College London.

Roberts, H. V. (1965). Probabilistic prediction. *J. Amer. Statist. Assoc.* **60**, 50–62.

Savage, L. J. (1971). Elicitation of personal probabilities and expectations. *J. Amer. Statist. Assoc.* **66**, 783–801.

Staël von Holstein, C. -A. S. (1970). *Assessment and Evaluation of Subjective Probability Distributions*. Stockholm: Economic Research Institute, Stockholm School of Economics.

Stone, M. (1961). The opinion pool. *Ann. Math. Statist.* **32**, 1339–1342.

DISCUSSION

N. D. SINGPURWALLA (*The George Washington University, Washington, D. C.*)

This charming and well written paper addresses an interesting problem, produces elegant results and makes for enjoyable reading. In the sequel, it also advances the state of the art of Bayesian statistics. The issues raised below are intended to serve two purposes:

1. To stimulate more work along the avenue of research which Professors Bayarri and DeGroot — henceforth BAD — have initiated, and
2. To argue that the framework for "gaining weight" prescribed here can be used to show that the linear opinion pool — henceforth LOP — leads to *incoherence*.

Regarding 1, one can start off by drawing attention to the criticisms of the LOP [see Genest and Zidek (1986)] and suggest as an alternative, a consideration of the logarithmic opinion pool. The latter has the advantage of being "externally Bayesian" [which Lindley (1985) dismisses as an "adhockery"] but suffers from the disadvantage that it may lead to cumbersome results.

More substantive avenues of investigation emanate from some extensions of the formulation of this paper. To see these, focus attention on Equation (1.2) in which $\beta_i(x)$, the posterior weight given by a decision maker, the "boss" ($\mathcal{B}$) to the i-th expert δ_i is specified. Note that $\beta_i(x)$ depends on α_i, the prior weight given by $\mathcal{B}$ to δ_i and $r_i(x)$, the experts report to $\mathcal{B}$. BAD assume that δ_i's colleagues are naïve and that δ_i's reward is solely based on the magnitude of $\beta_i(x)$. Under the above set-up, δ_i aims to maximize $\beta_i(x)$, and to achieve this δ_i may resort to dishonesty. Observe that δ_i's behavior is influenced solely by the α_i's and the competence of δ_i's colleagues. A more realistic scenario would be one in which δ_i is to be rewarded based upon *both* the quality of prediction — as measured by $r_i(x)$ — and also $\beta_i(x)$. For such a generalization, the conditions under which honesty of δ_i is always the best policy needs to be investigated. A formal mechanism for addressing the above is facilitated via the decision tree of Figure 1 in which $\mathcal{D}$ denotes δ_i's decision node, $\mathcal{R}_1$, denotes a random node pertaining to $r_j(x)$ — the report of δ_i's sole colleague δ_j — and $\mathcal{R}_2$ denotes a random node pertaining to the state of nature X taking values x. If $\mathcal{U}[r_i(x), r_j(x), \beta_i(x)]$, denotes δ_i's total utility when $X = x, \delta_\bullet$ reports $r_\bullet(x), \bullet = i, j$, and the $\mathcal{B}$ assigns a posterior weight $\beta_i(x)$ to δ_i, then δ_i will choose that $r_i(x)$ which maximizes the expectation of $\mathcal{U}$ with respect to δ_i's honest distribution of X (conditional on $r_j(x)$ if $r_j(\cdot)$ is declared by δ_j or averaged over the distribution of $r_j(\cdot)$ — as perceived by δ_j).

Recall that the set-up of BAD assumes that δ_i's colleagues are naïve, in the sense that they are not interested in "gaining weight"; thus their reports to the $\mathcal{B}$ are indeed their honest opinions. If one were to consider the case of a single colleague, say δ_j, and assume that δ_j is in competition with δ_i to "gain weight", then a two-person zero sum game would result, for now δ_j also has an incentive to be dishonest. It would be interesting to see if a solution

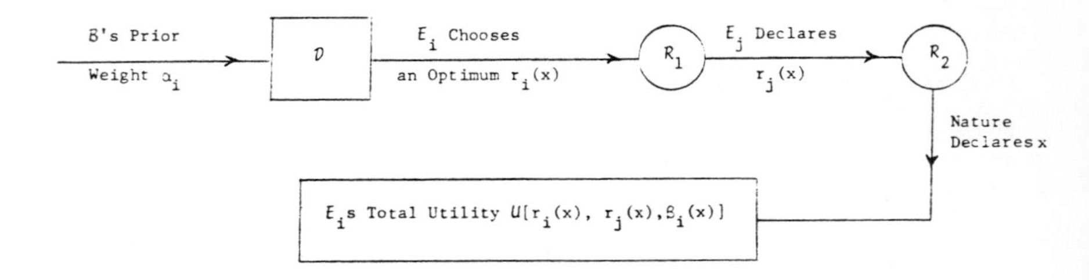

Figure 1. *δ_i's Decision Tree for an Optimal Choice of $r_i(x)$.*

to this two person zero sum game would result in the conclusion that honesty on the part of both δ_i and δ_j would be the best policy.

To see why the framework for "gaining weight" given here provides a vehicle for arguing that the LOP leads to incoherence, consider Equation (2.1) This specifies that for given values of α, s, π, and r

$$E(\beta) = \pi \frac{\alpha r}{\alpha r + \bar{\alpha} s} + \bar{\pi} \frac{\alpha \bar{r}}{\alpha \bar{r} + \bar{\alpha} \bar{s}}$$

and that the optimal choice of r, obtained by a maximization of $E(\beta)$ is

$$r = \frac{1}{\alpha} \frac{\sqrt{s\pi}}{\sqrt{s\pi} + \sqrt{\bar{s}\bar{\pi}}} - \frac{\bar{\alpha}}{\alpha} s \quad , \qquad \star$$

provided that $0 \leq r \leq 1$. The recommendation of BAD that when $\star < (>) 0(1)$, the optimal choice of r is $r = 0(1)$, is discomforting. It is an external intervention, analogous to that commonly used by frequentists who set negative estimates of variances and densities arbitrarily equal to zero. In a coherent system involving probability calculations there should be no need for interventions — the probability calculus will automatically lead to admissible answers. A detailed investigation of the behavior of $E(\beta)$ is therefore called for. Specifically, one needs to investigate conditions which ensure that the value of r which gives the global maximum of $E(\beta)$ lies between 0 and 1. Some analysis shows that there are five possible scenarios, three of which are shown in Figure 2. Of these, it is only Scenario A which results in a unique maximizing r which is between 0 and 1. Scenario D, with $0 < A^* < 1$ and $B^* > 1$, where $A^* = -(\bar{\alpha}/\alpha)s$, and $B^* = 1 + (\bar{\alpha}/\alpha)(1-s)$ is the dual of Scenario C, and is omitted. Both Scenarios C and D yield one global maximum outside $[0, 1]$ and therefore lead to inadmissibility. Omitted also from Figure 2 is Scenario E in which $E(\beta)$ attains a global maximum at A^* and B^*, both of which lie outside the interval $[0, 1]$. Scenario B yields an admissible answer but it is not unique.

Values of α and $\bar{\alpha}$ which lead to each of the above scenarios are shown in Figure 3.

It is apparent from Scenario A of Figures 2 and 3, that to obtain a unique value of r which maximizes $E(\beta)$ and which is admissible, α must equal $-\bar{\alpha}$. Furthermore, the maximizing value of r is s^+. Thus in an opinion pool with $\alpha = -\bar{\alpha}$, to maximize one's posterior weight, one must report one's colleagues answer plus a tad more.

In order to be ensured that one can obtain at least one admissible global maximizing value of r, α and $\bar{\alpha}$ must have, as a necessary (but not sufficient) condition opposite signs. This observation supports the result of Genest and Schervish (1985) who prescribe the conditions for the LOP to have a Bayesian justification. Actually Scenario A with its requirement that $\alpha = -\bar{\alpha}$ is a stronger statement than that of the above authors.

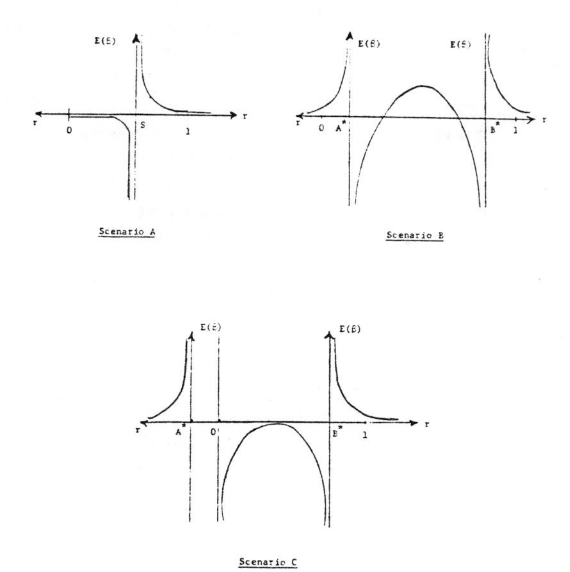

Figure 2. *Behavior of the Global Maxima of $E(\beta)$.*

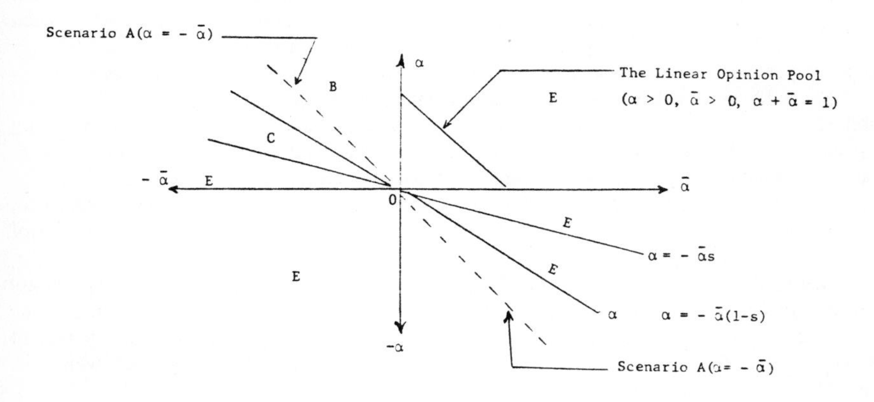

Figure 3. *Range of values of α and $\bar{\alpha}$ leading to Scenarios, A, B, C, and E.*

The LOP requires that $\alpha > 0$, $\bar{\alpha} > 0$ and $\alpha + \bar{\alpha} = 1$; however, these conditions lead us to Scenario E (see the upper right hand quadrant of Figure 3) which implies a violation of an axiom of probability.

By way of a few closing remarks, it is apparent that BAD have unveiled an avenue of research in Bayesian statistics which should generate a flood of new papers around their ideas. It is common to see research in Bayesian statistics center around established themes such as prior to posterior manipulations, approximations, sensitivities, robustification, computations, hierarchicalizations, algorithms with fancy names, and other devices which mimic the frequentists arsenal of techniques. In contrast, the problem addressed by Professors Bayarri and DeGroot is novel, stimulates thought and is fun to work upon. The author has enjoyed the opportunity to read and comment on this work and would like to thank the organizing committee for their contribution to his learning by inviting him to attend the conference and serve as a discussant. The author also acknowledges the efforts of Mr. Khalid Aboura in connection with the development of Figures 2 and 3.

M. WEST (*University of Warwick*)

Probability statements are always the result of decision processes. The authors present a framework in which the statement of a probability by an expert is modelled explicitly within a decision theoretic framework. Though concerned with linear opinion pools, a special framework leading to some nice, tractable mathematics, the underlying concept of an individual mispresenting his/her honest opinions to *gain weight* for the future is very relevant and important generally in problems of Bayesian discourse, group interactions and disclosure of personal beliefs/preferences. It is good to see the ideas crystalised in the setting of this paper.

I have one or two comments on technical aspects of the paper.

Firstly I'd like the authors to clarify for me the role played by the probability π, particularly with respect to what it is conditional upon. Suppose I am the expert interested in gaining weight as in Sections 2 and 3 of the paper. Initially with fixed s, the problem is posed as that of maximising expression (2.1), my expected posterior weight given s. Here π should represent my honest probability that $X = x_1$ *given* s; the assumption that this does not depend on s implies that learning s does not alter my true π. Now whilst this may be assumed in Section 2, it is at odds with the development in Section 3 for the case when I am initially uncertain about s. Here ξ_1 and ξ_2 provide me with a likelihood function for X when s is observed that will typically lead to $\pi = \pi(s)$ varying with s. Thus, in equation (3.1), π and $\bar{\pi}$ depend on s and should appear in the intergrands, clearly altering the problem.

Secondly what if the other expert is also interested in gaining weight as in Sections 2 and 3? In the spirit of Delphi-like techniques, suppose that the experts have the opportunity for dialogue before submitting final probabilities to the opinion pool. Each discloses his/her initial probability, $r = r_1$ and $s = s_1$, and, on seeing the other experts value, will then revise using the maximisation process, leading to new values r_2 and s_2. The process is repeated, with r and s changing at each stage. Have the authors thought of this and the possibility of convergence to stable values?

My final point concerns the user of the opinion pool ... the Boss. There is opportunity for the Boss to learn about the expert's honest probability π from the stated, optimal value r. In particular, Properties 2.3 and 3.2, concerning the monotonicity of r as a function of π, suggest that the Boss may be able to calculate π from r in certain cases. Knowing this, should the expert change his stated probability?

K. J. McCONWAY (*Open University*)

I found this paper fascinating, but I want to raise a question about the interpretation of the results. How would a decision-maker promote honesty in the assessor if the assessor's utility for weight did not take the form discussed in Section 6? Professor DeGroot in his presentation suggested that one might do this by manipulating the rewards for the assessor. But this seems a little odd. If one *can* manipulate these rewards, would one not simply reward assessors according to some strictly proper scoring rule based on the stated probability, rather than getting them to think about the weights in what may be an unnatural way? Another possibility would be to alter the method of updating the weights. Bayesian weight updating is not the only possible method (McConway 1978) and another method may promote honesty without being less effective as a method of pooling probabilities.

REPLY TO THE DISCUSSION

We are grateful to the discussants for their kind words about our work.

They have raised many interesting issues in their comments that will help to guide future research in this area. We will respond to each of the discussants separately.

To Professor Singpurwalla:

We agree with Professor Singpurwalla that the linear opinion pool is restrictive and that it would be desirable to develop other models in which there is a clear-cut definition of the weights that are assigned to the experts and a natural way to revise them. However, the logarithmic opinion pool does not seem to be a satisfactory alternative because in that model the weights never get revised.

We also agree that it would be interesting to consider models in which your reward as an expert depends on your report r and the actual outcome x in other ways than exclusively through your posterior weight β. Thus far, however, we have just considered rewards that are strictly functions of β.

It is true that in this paper we have focussed on your optimal report, without explicitly describing how the other experts choose their reports. However, it should be emphasized that we do not assume that they are necessarily honest in their reporting. In our model, if the other expert's report s is known to you, then it is irrelevant to you whether or not it is honest. If s is not known to you, then you specify a joint distribution for s and x; you do not need to quantify your beliefs about the other expert's honest opinion, although those beliefs will typically influence your specification of the distribution of s and x.

We recognize that in our model the two experts are implicitly involved in a game situation. The game will be zero-sum when both players are trying to maximize their expected posterior weights (or appropriate linear functions of them), but in general it will be non-zero-sum. We have explicitly avoided the usual complications of game-theoretic considerations by picking up the problem at the point where you must choose your report based on what you believe the other expert will report. How you arrive at your beliefs about s will be based on your beliefs about the goals, skills, expertise, and, as we have just mentioned, the honesty of the other expert.

Professor Singpurwalla raises the interesting possibility of considering models in which the weights assigned to the experts might be negative and do not necessarily sum to 1. However, two basic points should be kept in mind when exploring this possibility. First, our expression (2.1) for $E(\beta)$ is heavily based on the condition that the posterior weights are always normalized so that they sum to 1. Thus, if the prior weights do not sum to 1, expression (2.1) may not have any meaning. Second, if you have negative prior weight, then it may be optimal for you to report *small* probabilities for events that you consider *very likely* to occur. Therefore, standard rules for revising the weights assigned to the experts will be inappropriate and it will be necessary to develop new ones.

To Professor West:

It should be emphasized that we do not assume that your probability π is independent of the other expert's report s. In the situation when s is known, we assume that π is your probability at the time that you must choose your report, so that it is based on all the information available at that time, including s. Thus, our notation does not explicitly indicate this dependence because s is given and fixed at the time that π is assessed and $E(\beta)$ is maximized.

Professor West's concern about the validity of (3.1) is unfounded. When s is unknown, you have a joint distribution for s and x which in Eq. (3.1) and throughout the paper we have factored into the marginal distribution for x, represented by π and $\bar{\pi}$, and the conditional distribution for s given $x = x_i$, represented by ξ_i, for $i = 1, 2$. Of course, we could have used an alternative but equivalent factorization involving the conditional distribution of x given s which would have led to an equivalent expression for (3.1) having a somewhat different appearance.

Professor West asks what happens if the other experts are also interested in gaining weight. As we have already mentioned, our models certainly allow the other experts to have their own strategies for reporting, although we have not studied the game-theoretic details. He also raises the interesting possibility of allowing the experts to communicate with each other in a sequential process. The convergence of such a process might be achieved along lines similar to those in models of consensus (DeGroot, 1974) and of equilibrium in duopoly (Cyert and DeGroot, 1987).

With regard to Professor West's final question about whether your optimal report should be different when you know that the "Boss" can learn your honest probability π, the answer is "no" as long as the "Boss" does not change the method of revising the weights. If it is changed, then the function that you must maximize will be different.

To Professor McConway:

The use of strictly proper scoring rules to promote honesty has, of course, been widely studied in the statistical literature. However, this approach fails to capture the essence of the problems that we are discussing, namely that your reward depends not only on your own report and expertise, but on those of the other experts as well. Thus, in the final section of our paper, although your optimal report is always your honest probability, your expected reward does depend on the other expert's report s.

As Professor McConway notes, we have not considered non-Bayesian methods for updating the weights. It would be interesting to compare the optimal reports for various methods.

***

In closing this rejoinder, we trust that Professor Singpurwalla had no hidden implication when he referred to our work as BAD. We hope that in the future when statisticians consider gaining weight, they will always remember that it is a BAD idea.

REFERENCES IN THE DISCUSSION

Cyert, R. M., and DeGroot, M. H. (1987). *Bayesian Analysis and Uncertainty in Economic Theory.* Totowa, New Jersey: Rowman and Littlefield.

DeGroot, M. H. (1974). Reaching a consensus. *J. Amer. Statist. Assoc.* **69**, 118–121.

Genest, C. and Schervish, M. J. (1985). Modeling expert judgements for Bayesian updating. *Ann. Statist.* **13**, 1198–1212.

Lindley, D.V. (1985). Reconciliation of discrete probability distributions. *Bayesian Statistics* **2**. (J. M. Bernardo, M. H. DeGroot, D. V. Lindley and A. F. M. Smith, eds.). Amsterdam: North-Holland, 375–390.

BAYESIAN STATISTICS 3, pp. 45–65
J. M. Bernardo, M. H. DeGroot, D. V. Lindley and A. F. M. Smith, (Eds.)

Ranges of Posterior Probabilities for Unimodal Priors with Specified Quantiles

JAMES O. BERGER and ANTHONY O'HAGAN
Purdue University and *University of Warwick*

SUMMARY

Suppose several quantiles of the prior distribution for θ are specified or, equivalently, the prior probabilities of a partitioning collection of intervals $\{I_i\}$ are given. Suppose, in addition, that the prior distribution is assumed to be unimodal. Rather than selecting a single prior distribution to perform a Bayesian analysis, the class of all prior distributions compatible with these imputs will be considered. For this class and unimodal likelihood functions, the ranges of the posterior probabilities of the I_i, and of the posterior c.d.f. at the specified prior quantiles, are determined. Small ranges ensure robustness with respect to the exact functional form of the prior.

Keywords: PRIOR QUANTILES; UNIMODAL PRIOR; BAYESIAN ROBUSTNESS; RANGES OF POSTERIOR PROBABILITIES.

1. INTRODUCTION

In any practical Bayesian analysis a prior distribution for a continuous parameter cannot be specified in complete detail. To do so would imply infinitely many prior probability judgements. Instead, only a few judgements are actually made. The prior specification is then usually completed by assuming a reasonable, and preferably tractable, form of distribution which fits the judgements which have been made. Furthermore, the judgements which are required are often not prior probabilities but complex functions of them, such as prior means and variances. We consider here a scenario of prior specification that does not require the fitting of a specific distribution to the prior judgements. The user assigns his prior probabilities for the parameter lying in each of the intervals $I_1, I_2, \ldots, I_m$, which are contiguous and partition the real line; this amounts to asserting $m - 1$ points of the prior c.d.f. of the parameter. We then assume only that the prior density π lies in some set Π of densities, all of which agree with the stated prior probabilities over the intervals $\{I_i\}$. The posterior density π^* then lies in a corresponding set Π^*, and we consider what bounds are thereby implied for relevant posterior probabilities.

i	Interval I_i	$p_i^A = \Pr(\theta \in I_i \mid A)$	$p_i^B = \Pr(\theta \in I_i \mid B)$
1	[0, 1000)	0.01	0.15
2	[1000, 2000)	0.04	0.15
3	[2000, 3000)	0.20	0.20
4	[3000, 4000)	0.50	0.20
5	[4000, 5000)	0.15	0.15
6	[5000, ∞)	0.10	0.15

Table 1. *Specified Prior Probabilities of Intervals*

Example 1. In Martz and Waller (1982), Example 5.1 supposes that two engineers are concerned with the mean life θ of a proposed new industrial engine. The two engineers, A and B, quantify their beliefs about θ in terms of the probabilities given in Table 1 for being in specified intervals. Note that A has substantially more precise beliefs than does B.

Either of these probability specifications determines a class of prior distributions π, namely

$$\Pi_0 = \{\pi : p_i = \Pr(\theta \in I_i) = \int_{I_i} \pi(d\theta) \text{ for all } i\}. \tag{1.1}$$

Berliner and Goel (1986) determine the ranges of the posterior probabilities of the I_i when Π_0 is the assumed class of priors. Earlier, DeRobertis (1978) had considered the related class of priors with $\Pr(\theta \in I_i) \geq \gamma_i$ for all i. For both this class and Π_0 the ranges of the posterior probabilities are often quite large because the classes include discrete distributions concentrated at "least favorable" configurations. The engineers in Example 1 might well deny that such discrete priors are plausible reflections of their prior beliefs for a continuous parameter, and might insist that they actually have smooth prior densities. Indeed, it would not be uncommon to encounter the belief that Π is actually unimodal, leading to a class such as

$$\Pi_2 = \{\text{unimodal } \pi : p_i = \Pr(\theta \in I_i) \text{ for all } i\}. \tag{1.2}$$

Use of this more realistic class can sharply reduce the variability in posterior answers, and is the subject of Section 4. It is shown that, although Π_2 is a huge class of priors, maximizations and minimizations over this class can be reduced to low dimensional numerical optimization. Section 2 presents basic notation that will be needed, while Section 3 illustrates the type of answers obtained, through several examples.

Previous work on finding ranges of posterior quantities for classes of priors has mainly dealt with conjugate priors (e.g. Leamer (1978, 1982) and Polasek (1984)). Huber (1973) was the first to explicitly consider a large "non parametric" class of priors. He determined the range of the posterior probability of a set when Π is an ε-contamination class of priors having the form $\pi = (1-\varepsilon)\pi_0 + \varepsilon q$; here π_0 is a single elicited prior, ε reflects the uncertainty in π_0, and q is a "contamination". Huber considered the case where all contaminations (even discrete) are allowed. Berger and Berliner (1986), Sivaganesan (1986a, 1986b), and Sivaganesan and Berger (1986) consider a variety of generalizations, to different classes of contaminations (e.g. unimodal) and different posterior criteria (e.g. the posterior mean and variance). DeRobertis and Hartigan (1981) consider a large class of priors specified by a type of upper and lower envelope on the prior density, and also find ranges of posterior quantities of interest. Each of these classes is plausible as a model of prior uncertainty. Classes such as Π_0 and Π_2 perhaps have the advantage of being the simplest to understand and elicit. Other work dealing with similar classes of priors includes Bierlein (1967), Kudō (1967), West (1979), Manski (1981), Lambert and Duncan (1986), Cano, Hernández, and Moreno (1985), and Lehn and Rummel (1987). Related analyses for testing situations can be found in Edwards, Lindman, and Savage (1963), Berger and Sellke (1987), Berger and Delampady (1987), Casella and Berger (1987), and Delampady (1986). Other related works include Kadane and Chuang (1978), Wolfenson and Fine (1982), Berger (1984, 1985, 1987), and Walley (1986). These latter works of Berger and Walley also include general review and history of the subject.

2. NOTATION AND THE FORMAL PROBLEM

Prior information is to be stated for an unknown, continuous parameter $\theta \in [a_0, a_m]$ by giving

$$p_i = \Pr(\theta \in I_i) = \Pr(a_{i-1} \leq \theta \leq a_i)$$

for $i = 1, 2, \ldots, m$. The intervals partition the parameter space $[a_0, a_m]$ and their endpoints $a_1, a_2, \ldots, a_{m-1}$ are arbitrary, possibly even specified by the user. Infinite parameter spaces are included by $a_0 = -\infty$ and/or $a_m = \infty$. It is assumed that there is an underlying prior density $\pi(\theta)$ on $[a_0, a_m]$ constrained by

$$\int_{a_{i-1}}^{a_i} \pi(\theta)d\theta = p_i \qquad (i = 1, 2, \ldots, m). \tag{2.1}$$

Data are obtained, yielding a likelihood function $\ell(\theta)$. We will assume that $\ell(\theta)$ is unimodal, with mode θ_0. For an arbitrary prior density π, the posterior density π^* is, by Bayes theorem,

$$\pi^*(\theta) = \pi(\theta)l(\theta) / \int_{a_0}^{a_m} \pi(t)l(t)dt. \tag{2.2}$$

Of interest is some set

$$C = \bigcup_{i \in \Omega} I_i, \tag{2.3}$$

where Ω is some subset of the indices $\{1, 2, \ldots, m\}$. We will seek bounds on $\Pr^*(C)$, the posterior probability of the set C. The two cases of most common interest will be $C = I_i$ and $C = \cup_{i \leq n} I_i$; for the latter case, $\Pr^*(C)$ is the posterior c.d.f. evaluated at a_n. Sets, C, more general than (2.3) can be considered (see Section 4.2), as could quantities such as the posterior mean, but the analyses then become substantially messier.

For a given C and class of priors, Π, we seek the range of the posterior probability of C as π ranges over Π. Specifically, we will calculate

$$\overline{P^*_\Pi}(C) = \sup_{\pi \in \Pi} \Pr^*(C) \tag{2.4}$$

$$\underline{P^*_\Pi}(C) = \inf_{\pi \in \Pi} \Pr^*(C). \tag{2.5}$$

The classes Π_0 and Π_2, defined in (1.1) and (1.2), will be considered. Consideration of Π_2 is sensible, of course, only when the specified p_i are compatible with the unimodality condition (so that Π_2 is nonempty). This will be the case when the constants

$$q_i = p_i/(a_i - a_{i-1}), \qquad i = 1, 2, \ldots, m \tag{2.6}$$

(defined to be zero if the denominator is infinite) satisfy

$$q_1 \leq q_2 \leq \cdots \leq q_k \geq q_{k+1} \geq q_{k+2} \geq \cdots \geq q_m \tag{2.7}$$

for some k. Note that q_i is the uniform density on I_i which has mass p_i. We will henceforth assume that (2.7) holds, and that the prior mode is known to be in I_k.

In addition to Π_0 and Π_2, two simplified versions of Π_2 will be considered. The first (in section 4.1) is

$$\Pi_2' = \{\pi \in \Pi_2 : a_k \text{ is the mode of } \pi \text{ and } \pi(a_k) \leq h^*\}. \tag{2.8}$$

The motivation behind consideration of this class is twofold. First, it substantially simplifies the analysis, there being no uncertainty in the location of the prior mode, and no need to allow for point masses at the prior mode. The additional inputs needed for Π_2' are, however, certainly ascertainable. Indeed, elicitation processes often begin by determining the "most likely" value of θ, with the partitioning intervals I_i determined by working away from this mode. Allowing a maximum prior density of h^* is also very reasonable, and "safe" values of this upper bound are not hard to elicit. (Of course, setting h^* equal to a very large number will yield answers more or less equal to the answers with no constraint on this height.)

A reasonable value for this upper bound in many situations is

$$h^* = 3 \max_i \{q_i\}, \tag{2.9}$$

the reasoning being that it would typically be unreasonable for the prior density to exceed the largest "average" density on an interval by more than a factor of 3. In all examples, we will assume (2.9) as a "default" value of h^* for Π_2'.

While Π_2' results from applying additional restrictions to Π_2, a certain loosening of the unimodality condition leads to a class Π_1, defined in O'Hagan and Berger (1988), which involves only univariate optimizations, and is hence very easy to analyze. Furthermore, it will be seen to frequently give answers similar to those from the more complicated Π_2. (We do not repeat this analysis here but will include calculations for this class in our examples, for comparative purposes.) Note that

$$\Pi_0 \supset \Pi_1 \supset \Pi_2 \supset \Pi_2'$$

so that the posterior ranges $(\underline{P}^*_\Pi(C), \overline{P}^*_\Pi(C))$ will be nested in reverse order.

3. EXAMPLES

As the calculations are rather complex, we delay their development until after the presentation of several illustrative examples. In these examples we calculate the range of the posterior probabilities of the intervals, I_i and also of the c.d.f. evaluated at the a_i. The format used for each example is to present the range of the posterior probability of the relevant set C as the interval $(\underline{P}^*_\Pi(C), \overline{P}^*_\Pi(C))$. For each example, the classes Π_0 and Π_2 defined in Section 1, and the classes Π_2' and Π_1 defined in Section 2, will be considered.

Example 1 (continued). Data becomes available in the form of two independent life-times which are exponentially distributed with mean θ. The observed life-times are 2000 and 2500 hours, leading to a likelihood function

$$l(\theta) = \theta^{-2} \exp(-4500/\theta).$$

Tables 2 and 3 present the ranges of the posterior probabilities of the intervals I_i for engineers A and B, respectively. Tables 4 and 5 present the ranges of the posterior c.d.f.s evaluated at the a_i for A and B respectively. For engineer B we make the natural assumption that the prior mode is specified to be $a_k = 3000$ (see Table 1) in calculations with Π_1, Π_2 and Π_2'. For engineer A, the probabilities in Table 1 lead to no natural restriction on the prior mode, other than that it be in the interval [3000, 4000). The calculations with Π_1 and Π_2 allow an arbitrary mode in this interval, but Π_2' admits only the choice of 3000 or 4000 as the mode. It somewhat surprisingly makes very little difference to the answer for Π_2' whether 3000 or 4000 is chosen as the mode; the tables present the answers for the choice 3000.

Note first that Π_1, Π_2, and Π_2' yield usefully small ranges of posterior probabilities, in all cases. For instance, if engineer A is willing to assume unimodality as well as the given p_i, then he knows that his posterior probability that $\theta \in [3000, 4000)$ lies between 0.517 and 0.584, while his posterior probability that $\theta \leq 2000$ lies between 0.039 and 0.050. For engineer B, the corresponding ranges are 0.248 to 0.288, and 0.192 to 0.221. These ranges are small enough that the engineers can probably make decisions on this basis, obviating the need for more detailed prior specification.

Note also that Π_1, Π_2 and Π'_2 tend to yield similar answers, so that the particular manner in which one choses to implement unimodality does not seem to matter greatly. On the other hand, Π_0 yields substantially broader intervals (typically 2 to 4 times larger than Π_2, say), indicating that the unimodality assumption has a pronounced effect.

I_i	p_i	Π_0	Π_1	Π_2	Π'_2
[0, 1000)	0.01	(0,0.006)	(0.001,0.004)	(0.001,0.004)	(0.001,0.004)
[1000, 2000)	0.04	(0.019,0.057)	(0.037,0.049)	(0.037,0.049)	(0.038,0.049)
[2000, 3000)	0.20	(0.214,0.291)	(0.222,0.260)	(0.225,0.260)	(0.229,0.260)
[3000, 4000)	0.50	(0.476,0.613)	(0.512,0.584)	(0.517,0.584)	(0.517,0.579)
[4000, 5000)	0.15	(0.106,0.164)	(0.121,0.147)	(0.121,0.147)	(0.122,0.146)
[5000, ∞)	0.10	(0,0.083)	(0,0.071)	(0,0.071)	(0,0.071)

Table 2. *Posterior Ranges for $C = I_i$, Engineer A*

I_i	p_i	Π_0	Π_1	Π_2	Π'_2
[0, 1000)	0.15	(0,0.111)	(0.020,0.023)	(0.020,0.023)	(0.020,0.023)
[1000, 2000)	0.15	(0.088,0.255)	(0.171,0.197)	(0.172,0.197)	(0.172,0.197)
[2000, 3000)	0.20	(0.235,0.391)	(0.282,0.327)	(0.283,0.327)	(0.284,0.327)
[3000, 4000)	0.20	(0.197,0.349)	(0.247,0.288)	(0.248,0.288)	(0.248,0.288)
[4000, 5000)	0.15	(0.125,0.233)	(0.149,0.175)	(0.149,0.175)	(0.149,0.175)
[5000, ∞)	0.15	(0,0.146)	(0,0.121)	(0,0.121)	(0,0.121)

Table 3. *Posterior Ranges for $C = I_i$, Engineer B*

a_i	Π_0	Π_1	Π_2	Π'_2
1000	(0,0.006)	(0.001,0.004)	(0.001,0.004)	(0.001,0.004)
2000	(0.0194,0.062)	(0.038,0.053)	(0.039,0.050)	(0.039,0.050)
3000	(0.241,0.341)	(0.262,0.310)	(0.265,0.308)	(0.268,0.308)
4000	(0.769,0.886)	(0.794,0.871)	(0.800,0.870)	(0.801,0.869)
5000	(0.917,1)	(0.929,1.000)	(0.929,1)	(0.929,1)

Table 4. *Posterior Ranges for $C = [0, a_i]$, Engineer A*

a_i	Π_0	Π_1	Π_2	Π'_2
1000	(0,0.111)	(0.020,0.023)	(0.020,0.023)	(0.020,0.023)
2000	(0.096,0.327)	(0.191,0.221)	(0.192,0.221)	(0.192,0.221)
3000	(0.388,0.623)	(0.474,0.547)	(0.476,0.547)	(0.477,0.547)
4000	(0.659,0.860)	(0.725,0.830)	(0.728,0.830)	(0.728,0.830)
5000	(0.854,1)	(0.879,1)	(0.879,1)	(0.879,1)

Table 5. *Posterior Ranges for $C = [0, a_i]$, Engineer B*

A secondary point of interest is the very small interval of posterior probabilities that is obtained for interval I_1 of engineer B when Π_1, Π_2, or Π'_2 is used. The reason serves as a warning about casual assumption of the unimodality constraint. It is easy to see that, when two adjacent intervals have equal q_i (as do I_1 and I_2 for engineer B), then any unimodal prior must have its mode in one of the intervals or be uniform over those intervals. In Table 3 the mode could only be between 2000 and 4000, so all priors in Π_1, Π_2, and Π_2' are uniform over I_1 and I_2. Thus there may be little variation in the prior (over Π) under the unimodality assumption if certain of the adjacent q_i are nearly equal (the central intervals excepted.)

Example 2. As a second example, we illustrate the methodology on a standard type of Bayesian example. Subjective elicitation yields the following intervals and corresponding prior probabilities, p_i, for a normal mean θ:

I_i	$(-\infty,-2)$	$(-2,-1)$	$(-1,0)$	$(0,1)$	$(1,2)$	$(2,\infty)$
p_i	0.08	0.16	0.26	0.26	0.16	0.08

Table 6. *Intervals and Prior Probabilities: Normal Example*

A "textbook" Bayesian analysis would be to notice that the p_i are a good match to a $\mathcal{N}(0,2)$ (normal, with mean 0 and variance 2) prior distribution. Suppose now that $x = 1.5$ is observed from a $\mathcal{N}(\theta,1)$, experiment. Then usual conjugate prior Bayesian theory would be employed, resulting in a $\mathcal{N}(1,2/3)$ posterior distribution. The resulting posterior probabilities of the I_i are listed in Table 7 as p_i^*.

As an indication of the robustness of the p_i^* to the prior normality asumption, we can calculate the ranges of the posterior probabilities of the I_i for the various classes of priors we are considering. These results are given in Table 7.

I_i	p_i^*	Π_0	Π_1	Π_2	Π_2'
$(-\infty,-2)$	.0001	(0,0.001)	(0,0.0002)	(0,0.0002)	(0,0.0002)
$(-2,-1)$	.007	(0.001,0.029)	(0.006,0.011)	(0.006,0.011)	(0.006,0.010)
$(-1,0)$	.103	(0.024,0.272)	(0.095,0.166)	(0.095,0.166)	(0.095,0.155)
$(0,1)$	.390	(0.208,0.600)	(0.320,0.447)	(0.322,0.447)	(0.332,0.447)
$(1,2)$	.390	(0.265,0.625)	(0.355,0.475)	(0.357,0.473)	(0.360,0.467)
$(2,\infty)$	.110	(0,0.229)	(0,0.156)	(0,0.156)	(0,0.154)

Table 7. *Posterior Ranges for $C = I_i$, Normal Example*

The p_i^* are reasonably robust, except possibly for p_6^*. Also of interest is the now very dramatic difference between the Π_0 ranges and the ranges for the unimodality classes; their sizes differ by roughly a factor of 4. This provides further evidence of the value of incorporating the unimodality assumption (if subjectively warranted). Of course, these are but two examples, and situations can be constructed where there is little difference between the results for Π_0 and Π_1, but our general experience in looking at a variety of examples is that incorporation of unimodality typically has a substantial effect.

One final comment: the degree of robustness in situations such as Example 2 will typically depend strongly on the data x. In particular, as x gets extreme, so that the likelihood and the prior clash, substantially less robustness will be observed (cf. Berger and Berliner (1986) and Sivaganesan and Berger (1986)).

4. ANALYSIS UNDER STRICT UNIMODALITY CONSTRAINTS

In this section we present analyses for the classes Π_2 and Π_2' defined in (1.2) and (2.8), respectively. Since Π_2' is easier to analyze, it is discussed first. Section 4.2 considers Π_2.

4.1 Solutions for Π_2'.

Minimizations and maximizations of $\Pr^*(C)$ over Π_2' can be reduced to minimizations and maximizations over a small dimensional class of extreme points. We describe these extreme points and the algorithm for calculating $\overline{P}^*(C)$ and $\underline{P}^*(C)$ (dropping the subscript Π_2') below, in a series of steps. Proofs of most assertions are given in the appendix.

Step 1: Classification of Intervals.

There will be two relevant classifications of intervals. The first classification is as a "maximizing" interval (Max) or a "minimizing" interval (Min). Essentially, these are intervals in which it is desired to maximize or minimize the posterior probability, respectively. Table 8 presents this classification.

	Interval i	
Goal of analysis	In Ω	Not in Ω
$\overline{P}^*(C)$	Max	Min
$\underline{P}^*(C)$	Min	Max

Table 8. *Min-Max Interval Classification*

The second classification is according to the form of the "optimizing" prior in each interval. It will be shown that there is a prior distribution (possibly a subprobability distribution) at which $\overline{P}^*(C)$ or $\underline{P}^*(C)$ is actually achieved, and that in each interval this "optimizing" prior is one of the following types:

Uniform (denoted by U): on I_i,

$$\pi(\theta) = q_i; \tag{4.1}$$

Step (denoted by S): on I_i,

$$\pi(\theta) = \begin{cases} h_{i-1} & \text{if } a_{i-1} \le \theta \le s_i \\ h_i & \text{if } s_i < \theta \le a_i, \end{cases} \tag{4.2}$$

where s_i is defined in (4.8) and makes the total probability in I_i equal to p_i. Each interval is classified as a U or an S in Table 9. In the table, "likelihood form" refers to whether $l(\theta)$ is increasing, decreasing, or both (called modal) in the interval. In the modal interval (note that there can be only one), it is necessary to distinguish between three cases, depending on a comparison of the values of the likelihood at the endpoints with the "average" likelihood over the interval; in the table and in general, we define

$$L(x, y) = \int_x^y l(\theta) d\theta. \tag{4.3}$$

Step 2: Fill in Uniforms.

After classifying each interval, choose π in each "U" interval according to (4.1). These will define the actual "optimizing" π in such intervals. Note that then

$$\int_{I_i} l(\theta)\pi(\theta) d\theta = q_i L(a_{i-1}, a_i). \tag{4.4}$$

Step 3: Classify Chains.

A *chain* is a set of consecutive intervals among $\{I_1, \ldots, I_k\}$ *or* among $\{I_{k+1}, \ldots, I_m\}$ which consists only of "S" intervals and which is bordered on each side either by a "U"

Likelihood Form	Class	Intervals I_1 to I_k	Intervals I_{k+1} to I_m
Increasing	Max	S	U
	Min	U	S
Decreasing	Max	U	S
	Min	S	U
Modal			
(i) $L(a_{i-1}, a_i) \leq l(a_{i-1})(a_i - a_{i-1})$	Max	U	S
	Min	S	U
(ii) $L(a_{i-1}, a_i) \leq l(a_i)(a_i - a_{i-1})$	Max	S	U
	Min	U	S
(iii) otherwise	Max	S	S
	Min	S	S

Table 9. *U-S Interval Classification*

interval or by a_0, a_k, or a_m. Such a chain is *simple* if it consists only of "Max" or only of "Min" intervals; otherwise it is a *compound* chain.

It turns out that, within a chain, the "optimizing" π is one in which the adjoining step functions match up at the boundaries to form a stairway. This leads to the following algorithms for calculating the contributions of chains to $\overline{P}^*(C)$ or $\underline{P}^*(C)$.

Step 4: Evaluating Simple Chains.

Let $\{I_l, I_{l+1}, \ldots, I_n\}$ be a simple chain, and let $h_i = \pi(a_i)$ for $i = l-1, \ldots, n$. Again, the "optimizing" prior over a chain is essentially a stairway ascending or descending from h_{l-1} to h_n, with one (or no) steps occurring in each interval (at s_i), and h_i being the height of each step. (The special case of the chain consisting of only a single interval is separately treated, for convenience, as Case 4.) The step heights are constrained by the probability and unimodality requirements to satisfy

$$q_i \leq h_i \leq q_{i+1}, \quad i = l, l+1, \ldots, n-1. \tag{4.5}$$

(Specification of h_{l-1} and h_n will be considered later; they are defined in terms of $h_l, \ldots, h_{n-1}$.) Define

$$\Lambda = \{h = (h_l, \ldots, h_{n-1}) : \quad (4.5) \text{ holds}\}. \tag{4.6}$$

For any such "stairway" prior, the function determining the contribution of the chain to $\Pr^*(C)$ is

$$\begin{aligned}\Psi(h) = \Psi((h_l, \ldots, h_{n-1})) &= \sum_{i=l}^{n} \int_{I_i} l(\theta)\pi(\theta)d\theta \\ &= h_{l-1}L(a_{l-1}, s_l) + \sum_{i=l}^{n-1} h_i L(s_i, s_{i+1}) + h_n L(s_n, a_n).\end{aligned} \tag{4.7}$$

The step locations, s_i, are defined by the constraint that I_i have total mass p_i; (4.4) thus yields

$$s_i = [p_i + h_{i-1}a_{i-1} - h_i a_i] \,/\, [h_{i-1} - h_i] \tag{4.8}$$

(to be understood as "empty", i.e. there is no "step", if $h_{i-1} = h_i$). We will need, depending on whether this is a "Max" or a "Min" chain, either

$$\overline{\Psi} = \sup_{h\in\Lambda} \Psi(h) \quad \text{or} \quad \underline{\Psi} = \inf_{h\in\Lambda} \Psi(h). \tag{4.9}$$

Numerical maximization or minimization is generally needed to find $\overline{\Psi}$ or $\underline{\Psi}$. This is typically easy, since $\Psi(h)$ turns out to be concave on Λ if it is a "Max" chain and convex if it is a "Min" chain. Additional comments about this calculation are given in the appendix.

It remains only to determine the boundary values, h_{l-1} and h_n, for each simple chain. Defining

$$\underline{h} = \begin{cases} 0 & \text{if } l = 1 \\ h^* & \text{if } l = k+1 \\ q_{l-1} & \text{otherwise,} \end{cases} \tag{4.10}$$

$$\overline{h} = \begin{cases} h^* & \text{if } n = k \\ 0 & \text{if } n = m \\ q_{n+1} & \text{otherwise,} \end{cases} \tag{4.11}$$

three cases need to be considered.

Case 1. (No Modal Interval).

If the chain does *not* contain the modal interval, then set

$$h_{l-1} = \underline{h}, \qquad h_n = \overline{h}. \tag{4.12}$$

Case 2. (Modal; Min-Left or Max-Right).

If the chain contains the modal interval and is a "Min" chain lying to the left of a_k or a "Max" chain lying to the right of a_k, then it must be the case that the first interval in the chain, I_l is the modal interval. To solve this case, find the unique $v > \theta_0$ in I_l such that

$$(v - a_{l-1})l(v) = L(a_{l-1}, v), \tag{4.13}$$

define

$$f = [p_l - h_l(a_l - v)]/(v - a_{l-1}), \tag{4.14}$$

and set

$$h_n = \overline{h}, \text{ and } h_{\ell-1} = \max\{f, \underline{h}\}(\text{left case}) \textit{ or } h_{\ell-1} = \min\{f, \underline{h}\}(\text{right case}). \tag{4.15}$$

Note that h_{l-1} depends on h_l, one of the variables over which optimization is performed.

Case 3. (Modal; Max-Left or Min-Right).

If the chain contains the modal interval, and is a "Max" chain lying to the left of a_k or a "Min" chain lying to the right of a_k then it must be case that the last interval, I_n, is the modal interval. To solve this case, find the unique $v < \theta_0$ in I_n such that

$$(a_n - v)l(v) = L(v, a_n), \tag{4.16}$$

define

$$f = [p_n - h_{n-1}(v - a_{n-1})]/(a_n - v), \tag{4.17}$$

and set

$$h_{l-1} = \underline{h}, \text{ and } h_n = \min\{f, \overline{h}\} \text{ (left case) } \textit{or } h_n = \max\{f, \overline{h}\} \text{ (right case).} \tag{4.18}$$

Note that h_n depends on h_{n-1}, one of the variables over which optimization is performed.

Case 4. (Single Interval).

The preceding formulas are formally correct when $l = n$ (i.e., there is a single interval in the chain), but the notation might become garbled in a computer program; also no numerical optimization is needed. Thus we present the formulas for this special case separately. Labelling the single S-interval I_l, it is clear that

$$\Psi(h) = h_{l-1}L(a_{l-1}, s_l) + h_l L(s_l, a_l), \tag{4.19}$$

and that $\overline{\Psi}$ or $\underline{\Psi}$ are achieved at the following choices of h_{l-1} and h_l analogous to Cases 1 through 3. Here $\underline{h}$ is as in (4.10), and $\overline{h}$ is as in (4.11) with n replaced by l.

No Modal Interval: $h_{l-1} = \underline{h}$ and $h_l = \overline{h}$.

Modal; Min-Left or Max-Right: $h_{l-1} = \max\{f, \underline{h}\}$ and $h_l = \overline{h}$, where $v > \theta_0$ satisfies (4.13) and f is defined by $f = [p_l - \overline{h}(a_l - v)]/(v - a_{l-1})$.

Modal; Max-Left or Min-Right: $h_{l-1} = \underline{h}$ and $h_l = \min\{f, \overline{h}\}$, where $v < \theta_0$ satisfies (4.16) and f is defined by $f = [p_l - \underline{h}(v - a_{l-1})]/(a_l - v)$.

Step 5. (The Solution If No Compound Chain is Present).

The contributions of the "U" intervals to $\Pr^*(C)$ are determined by

$$K_1 = \sum_{i \in \Omega : I_i \text{ is a "U"}} q_i L(a_{i-1}, a_i),$$

$$K_2 = \sum_{i \notin \Omega : I_i \text{ is a "U"}} q_i L(a_{i-1}, a_i). \tag{4.20}$$

Next, suppose that there are r_1 simple chains consisting of "Max" intervals, and r_2 simple chains consisting of "Min" intervals. Denote the corresponding $\overline{\Psi}$ or $\underline{\Psi}$ by $\overline{\Psi}_1, \ldots, \overline{\Psi}r_1$ and $\underline{\Psi}_1, \ldots, \underline{\Psi}r_2$. Then

$$\overline{P}^*(C) = \left[1 + \frac{(K_2 + \sum_{i=1}^{r_2} \underline{\Psi}_i)}{(K_1 + \sum_{i=1}^{r_1} \overline{\Psi}_i)}\right]^{-1}, \tag{4.21}$$

$$\underline{P}^*(C) = \left[1 + \frac{(K_2 + \sum_{i=1}^{r_1} \overline{\Psi}_i)}{(K_1 + \sum_{i=1}^{r_2} \underline{\Psi}_i)}\right]^{-1}. \tag{4.22}$$

(These are *not* meant to be the same $\underline{\Psi}_i$ or $\overline{\Psi}_i$ in the two formulas; one generally has to start again at the beginning for each separate calculation.)

Step 6. (Solution If a Compound Chain Exists).

First calculate K_1, K_2, $\overline{\Psi}_1, \ldots, \overline{\Psi}r_1$, and $\underline{\Psi}_1, \ldots, \underline{\Psi}r_2$ as in Step 5. Now at most one compound chain can exist, and it must be a simple chain of "Max" intervals followed by a simple chain of "Min" intervals, or vice versa. Furthermore, the modal interval must be at the boundary of the Max chain and the Min chain. Let a_t be this boundary, and label the simple chains to the left and right by S_L and S_R, respectively. We present the solution only for the compound chain being to the left of the prior mode a_k; if it is to the right, use the obvious reflection symmetry to reduce it to the left-side case.

A compound chain on the left must have S_L being a "Max" simple chain and S_R being a "Min" simple chain. The formulas differ slightly depending on whether a_t is at the left or right boundary of the modal interval.

Case 1: a_t is the right boundary of the modal interval.

Let $v \leq \theta_0$ be the unique solution to

$$(a_t - v)l(v) = L(v, a_t), \tag{4.23}$$

and define

$$\tilde{h} = \min\left\{q_{t+1}, \quad \frac{p_t - q(v - a_{t-1})}{(a_t - v)}\right\}, \tag{4.24}$$

where q equals zero if $t = 1$ and equals q_{t-1} otherwise. For fixed h_t, $q_t \le h_t \le \tilde{h}$, let

$$\varphi_L(h_t) = \overline{\Psi},$$

where $\overline{\Psi}$ is the supremum of (4.7) for the simple chain S_L; here, of course, $n = t$. The supremum should be calculated over Λ in (4.6), with the slight modification that the range for $h_{n-1} = h_{t-1}$ be changed to

$$q_{t-1} \le h_{t-1} \le \frac{p_t - h_t(a_t - v)}{v - a_{t-1}}. \tag{4.25}$$

Also, set $l_{t-1} = \underline{h}$ (see (4.10)) and $h_n = h_t$ (the fixed value above). Note that if S_L consists only of the single interval I_t, then $\varphi_L(h_t)$ is given by

$$\varphi_L(h_t) = \underline{h}L(a_{t-1}, s_t) + h_t L(s_t, a_t). \tag{4.26}$$

Similarly, optimize over the right chain S_R for fixed h_t, finding

$$\varphi_R(h_t) = \underline{\Psi},$$

where $\underline{\Psi}$ is the infimum of (4.7) for the simple chain S_R; here, of course, $l = t + 1$. The infimum should be calculated over Λ as in (4.6), with h_{l-1} set equal to the fixed h_t and $h_n = \overline{h}$. Note that if S_R consists only of the single interval I_{t+1} then $\varphi_R(h_t)$ is given by

$$\varphi_R(h_t) = h_t L(a_t, s_{t+1}) + \overline{h}L(s_{t+1}, a_{t+1}). \tag{4.27}$$

Finally,

$$\overline{P}^*(C) = \sup_{q_t \le h_t \le \tilde{h}} \left[1 + \frac{K_2 + \sum_{i=1}^{r_2} \underline{\Psi}_i + \varphi_R(h_t)}{K_1 + \sum_{i=1}^{r_1} \underline{\Psi}_i + \varphi_L(h_t)}\right]^{-1} \tag{4.28}$$

and

$$\underline{P}^*(C) = \inf_{q_t \le h_t \le \tilde{h}} \left[1 + \frac{K_2 + \sum_{i=1}^{r_1} \underline{\Psi}_i + \varphi_L(h_t)}{K_1 + \sum_{i=1}^{r_2} \underline{\Psi}_i + \varphi_R(h_t)}\right]^{-1}. \tag{4.29}$$

Case 2: a_t is the left boundary of the modal interval.
Let $v \ge \theta_0$ be the unique solution to

$$(v - a_t)l(v) = L(a_t, v), \tag{4.30}$$

and define

$$\boldsymbol{h} = \max\left\{q_t, \frac{p_{t+1} - \tilde{q}(a_{t+1} - v)}{(v - a_t)}\right\}, \tag{4.31}$$

where $\tilde{q}$ equals h^* if $t = h + 1$ and equals q_{t+2} otherwise. For fixed $h_t, \boldsymbol{h} \le h_t \le q_{t+1}$, let $\varphi_L(h_t)$ be $\overline{\Psi}$, the supremum of (4.7) for the simple chain S_L; here $n = t$. The supremum should be calculated over Λ in (4.6), with $h_{l-1} = \underline{h}$ and h_n equal to the fixed h_t. If S_L consists only of the single interval I_t, then $\varphi_L(h_t)$ is given by (4.26).

For fixed h_t, let $\varphi_R(h_t)$ be $\underline{\Psi}$, the infimum of (4.7) for the simple chain S_R; here $l = t+1$. The infimum should be calculated over Λ as in (4.6), with the range of $h_l = h_{t+1}$ being changed to

$$\frac{p_{t+1} - h_t(v - a_t)}{(a_{t+1} - v)} \leq h_{t+1} \leq q_{t+2}, \tag{4.32}$$

and with h_{l-1} equal to the fixed h_s and $h_n = \overline{h}$. If S_R consists only of the single interval I_{t+1}, then $\varphi_R(h_t)$ is given by (4.27).

Equation (4.28) or (4.29) can be used to complete the calculation.

4.2 Solution for Π_2 *and for Arbitrary Intervals C.*

The algorithm in section 4.1 for Π_2' can be modified to solve the problem for Π_2. Let a^* be the location of the prior mode in the interval I_k. It is easy to show that a^* must be one of the endpoints of the interval, unless the interval is the modal interval (of the likelihood); then a^* could also be at the mode of the likelihood. In the former case the analysis is as in Section 4.1, while in the latter case the original I_k can be considered to be two intervals $I^* = [a_{k-1}, a^*)$ and $I^{**} = [a^*, a_k)$. Let p^* be the prior probability assigned to I^*, so that $p_k - p^*$ is that assigned to I^{**}. To ensure that the unimodality and probability constraints are satisfied, it is easy to check that p^* must satisfy

$$(a^* - a_{k-1})q_{k-1} \leq p^* \leq p_k - q_{k+1}(a_k - a^*).$$

Finally, let h^* be the maximum allowed prior density at a^* (as before).

For specified a^*, p^*, and h^*, one has a Π_2' problem with $a_k = a^*$. Hence the algorithm in the preceding section can be applied to find $\overline{P}(C^*)$ or $\underline{P}(C^*)$ for this Π_2'. One can then optimize these quantities over p^*. (Note that the optimizing h^* will be ∞, i.e. a point mass will creep in. A reasonable practical way to deal with this is just to choose a very large fixed h^*, such as $10^4 q_k$; this will produce essentially the same answer as $h^* = \infty$.)

The analysis above suggests the manner in which one can handle an arbitrary interval $C = [c_1, c_2]$. Simply create new intervals in each of the I_i where a c_j occurs, with each c_j becoming an endpoint. Of course, two new parameters, p_1^* and p_2^* (analogous to p^*), might then be introduced, and require additional optimizations.

APPENDIX

The purpose of the appendix is to outline a proof of the results in Section 4. For simplicity, we consider only the calculation of

$$\begin{aligned}\overline{P}^*(C) &= \sup \ \Pr{}^*(C) \\ &= \sup \frac{\sum_{i\in\Omega} \int_{I_i} l(\theta)\pi(\theta)d\theta}{\sum_{\text{all } i} \int_{I_i} l(\theta)\pi(\theta)d\theta} \\ &= \left[1 + \inf \frac{\sum_{i\notin\Omega} \int_{I_i} l(\theta)\pi(\theta)d\theta}{\sum_{i\in\Omega} \int_{I_i} l(\theta)\pi(\theta)d\theta}\right]^{-1}.\end{aligned} \tag{A1}$$

Verification of Step 1.

Expression (A1) makes it clear that one wants to choose π to make the $\int_{I_i} l(\theta)\pi(\theta)d\theta$ small for $i \notin \Omega$ and large for $i \in \Omega$. This is the basis for the classification of an interval as a "Max" or "Min" in Step 1.

The first key to the proof is establishing that, on each I_i, the optimizing π should be either a U or an S, as in Table 9. For this purpose, consider the interval (a_{l-1}, a_l), and consider arbitrary allowable fixed values, h_{l-1} and h_l, for $\pi(a_{l-1})$ and $\pi(a_l)$. (The allowable values must satisfy $h_{l-1} \leq q_l \leq h_l$; if this is violated it is easy to see that π cannot be unimodal.) We are done if we can show that a U or an S in the interval is optimal for any allowable h_{l-1} and h_l. (There is the technical point here that uniform segments violate unimodality; the uniform segments arise as the extreme point limits of strictly unimodal priors, however, so that we can ignore the distinction.)

Consider first the case where $h_{l-1} < h_l$ and $l(\theta)$ is increasing on I_l. By unimodality, $\pi(\theta)$ must be nondecreasing on I_l. Depending on whether I_l is a "Max" or a "Min" interval, we thus seek to maximize or minimize $\int_{I_l} l(\theta)\pi(\theta)d\theta$ over

$$\Gamma = \{\pi : \quad \pi(\theta) \text{ is nondecreasing on } I_l, \pi(a_{l-1}) = h_{l-1}, \pi(a_l) = h_l, \text{ and } \int_{I_l} \pi(\theta)d\theta = p_l\}.$$

It is trivial that the minimum is attained by a uniform segment on I_l (since $\pi(\theta)$ and $l(\theta)$ are both increasing); the probability constraint then implies that $\pi(\theta) = q_l$ on I_l. That the maximum is obtained by a step distribution follows from observing that as much of the mass, p_l, should be shifted towards high values of $l(\theta)$ as is possible. Since $l(\theta)$ is increasing, this means that as much mass should be given to large θ as is possible and, correspondingly, as little mass to small θ as is possible. This is clearly achieved, subject to being in Γ, by a step density as in (4.2). Noting that the intervals I_1 to I_k have increasing $\pi(\theta)$, while those from I_{k+1} to I_m have decreasing $\pi(\theta)$, variants of these arguments can be used to fill in the "Increasing" and "Decreasing" Likelihood Form sections of Table 9.

The analysis for the modal interval follows by breaking up the modal interval into the intervals $[a_{l-1}, \theta_0)$ and $[\theta_0, a_l)$. Since (i) we will either want to maximize or minimize the relevant integrals over both intervals jointly, (ii) $\pi(\theta)$ is either increasing or decreasing over both intervals, and (iii) $l(\theta)$ is increasing on $[a_{l-1}, \theta_0)$ but decreasing on $[\theta_0, a_l)$, the previous argument indicates that the optimizing π will be uniform on one interval and a step density on the other; for illustration, suppose it is uniform with density h on $[\theta_0, a_l)$ and is a step density on $[a_{l-1}, \theta_0)$, with $\pi(a_{l-1}) = h_{l-1}$. But the argument of the preceding paragraph applies to show that π must be of the from (on $[a_{l-1}, \theta_0)$)

$$\pi(\theta) = \begin{cases} h_{l-1} & \text{if } a_{l-1} \leq \theta < s \\ h & \text{if } s \leq \theta < \theta_0, \end{cases}$$

for some point s. Hence we have that the optimizing π is actually a single step function on I_l given by

$$\pi(\theta) = \begin{cases} h_{l-1} & \text{if } a_{l-1} \leq \theta < s \\ h & \text{if } s \leq \theta < a_l, \end{cases} \tag{A2}$$

where $a_{l-1} \leq s \leq \theta_0$.

To complete the argument for Table 9, it must be shown that, under the conditions indicated therein, the step function in the modal interval is actually a uniform density. We illustrate the argument in the case where I_l is a "Max" interval to the left of a_k (which is consistent with our earlier choice of $[\theta_0, a_l)$ as the uniform subinterval, since $\pi(\theta)$ is then decreasing on $[\theta_0, a_l)$).

Consider the condition (from Table 9)

$$L(a_{l-1}, a_l) \leq l(a_{l-1})(a_l - a_{l-1}).$$

A consequence of this condition is that

$$L(s, a_l) < l(s)(a_l - s) \tag{A3}$$

for $a_{l-1} < s \le \theta_0$ (since $L(s,a_l) - l(s)(a_l - s)$ is a decreasing function of s in this range). Now

$$\begin{aligned}\int_{I_l} l(\theta)\pi(\theta)d\theta &= h_{l-1}L(a_{l-1},s) + hL(s,a_l)\\ &= h_{l-1}L(a_{l-1},a_l) + (h - h_{l-1})L(s,a_l) \\ &= h_{l-1}L(a_{l-1},a_l) + (q_l - h_{l-1})(a_l - a_{l-1})\frac{L(s,a_l)}{(a_l - s)},\end{aligned} \tag{A4}$$

the last equality following from the facts that

$$h_{l-1}(s - a_{l-1}) + h(a_l - s) = p = q_l(a_l - a_{l-1}).$$

Consider (A4) as a function of s, for fixed h_{l-1}. Since

$$\begin{aligned}\frac{d}{ds}\frac{L(s,a_l)}{(a_l - s)} &= \frac{-l(s)}{(a_l - s)} + \frac{L(s,a_l)}{(a_l - s)^2}\\ &= (a_l - s)^{-2}[-l(s)(a_l - s) + L(s,a_l)],\end{aligned} \tag{A5}$$

which is negative for $a_{l-1} \le s \le \theta_0$ by (A3), it follows that $\int_{I_l} l(\theta)\pi(\theta)d\theta$ is decreasing in s. The conclusion is that $s = a_{l-1}$ is optimal, i.e. π is uniform on I_l. This proves the validity of the relevant entry in Table 9.

Verification of Step 2.

We demonstrated that, for any allowable h_{l-1}, h_l, the "local" optimizing π in a U interval, I_l, is $\pi(\theta) = q_l$. Since $h_{l-1} \le q_l \le h_l$, this satisfies the global umimodality condition. Furthermore, it does not depend on the particular choice of h_{l-1} and h_l, so it must define the global optimizing π on I_l.

Verification of Step 3 and 4.

The key feature which greatly simplifies the problem is that two intervals with adjoining step densities will have the steps match at the boundary. The reason for this is that at least one of the intervals must be a non-modal interval, and hence have monotonic $l(\theta)$ in the interval. As argued before, any such interval, if constrained at its boundaries by values h_{l-1} and h_l for π, will be optimized by a step function attaining these heights. Thus π must be continuous at the boundary of two step functions. This fact verifies the nature of chains, as described in Step 3. The formula (4.7) follows immediately.

The formula (4.12), for the boundary values when both end intervals of the chain are non-modal, follows from the previous observation that the step heights for a non-modal interval will always seek the extremes allowed. The possible extremes in (4.10) and (4.11) correspond to the possibilities that an end interval of the chain is either I_1, I_k, I_{k+1} or I_m (with corresponding extreme allowed boundary values of 0, h^*, h^*, and 0), or that the end interval adjoins to a U interval I_r (with the corresponding extreme allowed boundary value of h to be q_r).

Cases 2 and 3 are more complex in that, if the end of a chain is a modal interval, its outer step height need not match up with the density in the adjoining interval. As an illustration of the argument, consider the case where the modal interval is a "Max" interval to the left of a_k; this can only occur when the modal interval is I_n, the rightmost interval of the chain, and when (see Table 9)

$$L(a_{n-1},a_n) > l(a_{n-1})(a_n - a_{n-1}). \tag{A6}$$

Differentiating with respect to s shows that

$$L(s,a_n) - l(s)(a_l - s)$$

is a decreasing function of s on (a_{n-1}, θ_0) (since $l(\theta)$ is increasing there), so that (by (A6) and (4.16)),

$$L(s, a_n) - l(s)(a_l - s) > 0$$

for $a_{n-1} < s < v$, with the inequality reversed for $s > v$. Finally using (A4) and (A5) (with I_n instead of I_l), it can be concluded that $\int_{I_n} l(\theta)\pi(\theta)d\theta$ has a unique maximum at $s = v$, providing this leads to an allowable step height at the right boundary. The correct step height, corresponding to a step at v, is f in (4.17), but (as before) the extreme allowable height is $\overline{h}$. This is the reason for the restriction in (4.18). (If $f > \overline{h}$, the maximizing s will be the point in (a_{n-1}, v) which results in a step height of $\overline{h}$, but this is automatically taken care of by the formulas.)

The arguments for Case 4 are entirely analogous, though now they lead to a deterministic solution (i.e., optimization over interval step interval heights of the chain is not necessary.)

Verification of Step 5.

This is just bookkeeping, adding up the contributions of all the separately maximized or minimized components in (A1). It is important to realize that the decomposition into U intervals and chains decomposed the global problem into a set of local problems which could be analyzed separately and then combined to yield the global solution.

Verification of Step 6.

The reason only one compound chain can exist, and that it must be of the indicated form, follows from examining Table 9 and realizing that a chain on one side of a_k can have adjacent Max-S and Min-S intervals only if one corresponds to a modal interval. Furthermore, on one side of this modal interval the chain must be all "Max" intervals, and on the other side it must be all "Min" intervals. All subsequent comments about the nature of a compound chain follow from similar examination of Table 9.

The remaining analysis in Step 6 is very similar to that in Cases 2 through 4 of Step 4, and will be omitted. The essential difference here is that the modal interval cannot have a step height that is less than the adjoining step height. This leads to the possible constraints on the heights h_{t-1} in (4.25) and h_{t+1} in (4.32).

The Numerical Calculation in Step 4.

For a simple chain $\{I_l, I_{l+1}, \ldots, I_n\}$, $l \neq n$, Ψ in (4.7) turns out to be a convex function of $h_l, \ldots, h_{n-1}$ if the chain is a "Min" chain, and a concave function if the chain is a "Max" chain. To verify this, note that first and second partial derivatives of Ψ are as follows (using (4.8) to simplify the expressions):

I. *First Order Partials:* For $i = l, \ldots, n-1$,

$$\frac{\partial}{\partial h_i}\Psi = (s_i - a_i)l(s_i) + (a_i - s_{i+1})l(s_{i+1}) + L(s_i, s_{i+1}). \tag{A7}$$

II. *Second Order Partials:* For $i = l, \ldots, n-1$,

$$\frac{\partial^2}{\partial h_i^2}\Psi = \frac{(s_i - a_i)^2}{(h_{i-1} - h_i)}l'(s_i) + \frac{(a_i - s_{i+1})^2}{(h_i - h_{i+1})}l'(s_i); \tag{A8}$$

except, if Case 2 of Step 4 applies *and* $h_{l-1} = f$, then

$$\frac{\partial^2}{\partial h_l^2}\Psi = \frac{(a_l - s_{l+1})^2}{(h_l - h_{l+1})^2}l'(s_{l+1}); \tag{A9}$$

or, if Case 3 of Step 4 applies *and* $h_n = f$, then

$$\frac{\partial^2}{\partial h_{n-1}^2}\Psi = \frac{(s_{n-1} - a_{n-1})^2}{(h_{n-2} - h_{n-1})}l'(s_{n-1}); \tag{A10}$$

here $l'(s) = \frac{d}{ds}l(s)$.

III. *Second Order Mixed Partials:* For $i = l+1, \ldots, n-1$,

$$\frac{\partial^2}{\partial h_i \partial h_{i-1}} \Psi = \frac{(s_i - a_i)(a_{i-1} - s_i)}{(h_{i-1} - h_i)} l'(s_i),$$

and

$$\frac{\partial^2}{\partial h_i h_j} \Psi = 0 \text{ for } |i-j| > 1.$$

Note: It can happen that $h_i = h_{i-1}$ for some i in a chain (i.e., the interval has collapsed to a U interval), and this needs to be considered in a numerical program so that formulas such as those above are not ill-defined.

Convexity or Concavity of Ψ.

The matrix of mixed second partial derivatives of Ψ is tri-diagonal, and an induction argument shows that, in Case 1 of Step 4, the upper $(r \times r)$ determinant of this matrix is

$$D_r = (-1)^r \sum_{i=1}^{(r+1)} \left[\left(\prod_{j=1}^{r} \{c_j^2 \chi_{(r \geq i+j-1)} + d_j^2 \chi_{(r < i+j-1)}\} \right) \frac{\prod_{j=1}^{r+1} l'(s_j)/w_j}{l'(s_i)/w_i} \right],$$

where $c_j = s_j - a_j$, $d_j = a_j - s_{j+1}$, $w_j = h_j - h_{j-1}$, and χ denotes the indicator function, as usual. (Slightly different formulas hold in Cases 2 and 3 when $h_{l-1} = f$ or $h_n = f$, but the conclusions are identical.)

Consider the case where the simple chain is a "Max" chain to the left of a_k. Then each $w_j > 0$. From Table 9, it further follows that $l(\theta)$ must be increasing over the chain, so that each $l'(s_j) > 0$. Thus the sign of D_r is $(-1)^r$, establishing the concavity of Ψ in this case. All other cases are handled similarly.

The Optimization of Ψ.

Because of the convexity or concavity of Ψ, no special problems are encountered in its maximization or minimization (other than the reduction of dimension caused by possible equality of adjacent h_i). Almost any maximization program should work well; note the availability of analytic derivatives in (A7) through (A10).

The maximization (or minimization) is over an $n-l-1$ dimensional rectangle. There is a way to reduce the problem to almost a two dimensional optimization. The idea is to fix $\{h_l, h_{l+2}, \ldots\}$, and then use the deterministic relationships in Step 4 to calculate the optimal corresponding $\{h_{l+1}, h_{l+3}, \ldots\}$. Then consider these heights fixed, and recalculate the $\{h_l, h_{l+2}, \ldots\}$. Continue iterating between these two "dimensions" until convergence. Note that one has to pay careful attention to h_{l-1} and h_n in Step 4 at each stage of the iteration. Also, one must "fill in" a sequence from the outside in. Although this is, in a sense, just a two-dimensional problem, the number of iterations needed may be much larger than a good general-purpose $(n-l-1)$ dimensional algorithm, unless $(n-l-1)$ is quite large.

ACKNOWLEDGEMENTS

The authors are grateful to Mr. Kun-Liang Lu for performing the numerical work, and to the National Science Foundation (U.S.), Grant #DMS-8401996, the U.S.-Spain Joint Committee for Scientific and Technological Cooperation, Grant CC B 8409-025, and the Science and Engineering Council (U.K.) for support of the research.

REFERENCES

Berger, J. (1984). The Robust Bayesian Viewpoint, (with discussion). *Robustness of Bayesian Analyses*, (J. Kadane, ed.) Amsterdam: North-Holland.

Berger, J. (1985). *Statistical Decision Theory and Bayesian Analysis*. New York: Springer-Verlag.

Berger, J. (1987). Robust Bayesian Analysis: Sensitivity to the Prior. *Tech. Rep.* #87–10, Department of Statistics, Purdue University.

Berger, J. and Berliner, L. M. (1986). Robust Bayes and empirical Bayes analysis with ε-contaminated priors. *Ann. Statist.* **14**, 461–486.

Berger, J. and Delampady, M. (1987). Testing precise hypotheses. *Statistical Science* **2**, 317–352.

Berger, J. and Sellke, T. (1987). Testing a point null hypothesis: The irreconcilability of significance levels and evidence. *J. Amer. Statist. Assoc.* **82**, 112–122.

Berliner, L. M. and Goel, P. (1986). Incorporating Partial Prior Information: Ranges of Posterior Probabilities. *Tech. Rep.* **357**, Department of Statistics, Ohio State University.

Bierlein, D. (1967). Zur einbeziehung der erfahrung in spieltheoretische modelle. *Z. Oper. Res. Verfahren* **III**, 29–54.

Cano, J. A., Hernández, A., and Moreno, E. (1985). Posterior measure under partial prior information. *Statistica* **2**, 219–230.

Casella, G. and Berger, R. (1987). Reconciling Bayesian and frequentist evidence in the one-sided testing problem. *J. Amer. Statist. Assoc.* **82**, 106–111.

Delampady, M. (1986). Testing a precise hypothesis: interpreting p-values from a robust Bayesian perspective. *Ph. D. Thesis*, Purdue University, West Lafayette.

De Robertis, L. (1978). The use of partial prior knowledge in Bayesian inference. *Ph. D. Thesis*, Yale University, New Haven.

De Robertis, L. and Hartigan J. A. (1981). Bayesian Inference using intervals of measures. *Ann. Statist.* **1**, 235–244.

Edwards, W., Lindman, H. and Savage, L. J. (1963). Bayesian statistical inference for pychological research. *Psychological Review* **70**, 193–242.

Good, I. J. (1983). *Good Thinking: The Foundations of Probability and Its Applications*. Minneapolis: University of Minnesota Press.

Huber, P. J. (1973). The use of choquet capacities in statistics. *Bulletin of the International Statistical Institute* **45**, 181–191.

Kadane, J. B. and Chuang D. T. (1978). Stable decision problems. *Ann. Statist.* **6**, 1095–1110.

Kudō, H. (1967). On partial prior information and the property of parametric sufficiency. *Proceedings of Fifth Berkeley Symposium on Statistics and Probability* **1**, Berkeley: University of California Press.

Lambert, D. and Duncan, G. T. (1986). Single-parameter inference based on partial prior information. *Canadian J. of Statistics* **14**, 297-305.

Leamer, E. E. (1978). *Specification Searches*. New York: Wiley.

Leamer, E. E. (1982). Sets of posterior means with bounded variance prior. *Econometrica* **50**, 725–736.

Lehn, J. and Rummel, F. (1987). Gamma minimax estimation of a binomial probability under squared error loss. *Statistics and Decisions* **5**, 229–250.

Manski, C. F. (1981), Learning and decision making when subjective probabilities have subjective domains. *Ann. Statist.* **9**, 59–65.

Martz, H. F. and Waller, R. A. (1982). *Bayesian Reliability Analysis*. New York: Wiley.

O'Hagan, A. and Berger, J. (1988). Ranges of posterior probabilities for quasi-unimodal priors with specified quantiles. (To appear in *J. Amer. Statist. Assoc.*).

Polasek, W. (1985). Sensitivity analysis for general and hierarchical linear regression models. *Bayesian Inference and Decision Techniques with Applications*. (P. K. Goel and A. Zellner, eds.). Amsterdam: North-Holand.

Sivaganesan, S. (1986a). Robust Bayes Analysis with Arbitrary Contaminations. *Tech. Rep.* SMU-DS-TR-198, Department of Statistical Science, Southern Methodist University.

Sivaganesan, S. (1986b). Sensitivity of the Posterior to Unimodality Preserving Contaminations. *Tech. Rep.* SMU-DS-TR-199, Department of Statistical Science, Southern Methodist University.

Sivaganesan, S. and Berger, J. (1986). Ranges of Posterior Measures for Priors with Unimodal Contaminations. *Tech. Rep.* #86–41, Department of Statistics, Purdue University.

Walley, P. (1986). *Rationality and Vagueness*. (To appear).

West, S. (1979). Upper and lower probability inferences for the logistic function. *Ann. Statist.* **7**, 490–413.

Wolfenson, M. and Fine, T. L. (1982). Bayes-like decision making with upper and lower probabilities. *J. Amer. Statist. Assoc.* **77**, 80–88.

DISCUSSION

J. DE LA HORRA (*Universidad Autónoma de Madrid*)

Congratulations to Professors Berger and O'Hagan on an interesting and stimulating paper. I shall organize my discussion into four different points.

1. The first point deals with the class Π_0. I agree with the authors when they say that people define their prior beliefs by means of quantiles more easily than by means of moments. These are, therefore the most natural specifications we can consider on the prior distribution, and this is true not only for the continuous case. If the parameter space Θ is finite, only a finite number of judgements are needed for specifying the prior distribution in complete detail. If, however, Θ is large (although finite), it is going to be very difficult, in practice, to obtain the exact prior, and therefore, it is very natural, also in this case, to specify only some quantiles for the prior distribution.

It is necessary to say that mathematical work with fixed quantiles is more difficult (in general) than work with fixed moments; but this is nothing but a challenge.

On the other hand, the class of prior distributions with some specified quantiles have very interesting properties. For instance, García-Carrasco and de la Horra (1987) have proved that, if we consider this class of prior distributions, the reference (noninformative) prior distribution relative to any reasonable uncertainty function is always the same. This property does not necessarily hold true when constraints on moments are considered.

2. The second point is about the class Π_2. I think it is convenient to write the exact definition of Π_2, as used in the paper:

$$\Pi_2 = \{\pi \in \Pi_0 \quad : \pi \text{ is unimodal and the mode is in } I_k\}$$

where $q_1 \leq q_2 \leq \cdots \leq q_k \geq q_{k+1} \geq \cdots \geq q_m$. The idea behind Π_2 is to add unimodality to the class Π_0. So, it is not too clear to me what the reason is for demanding that "the prior mode is known to be in I_k". It is easy to prove the following result:

If $\pi \in \Pi_0$ is unimodal and $q_1 \leq q_2 \leq \cdots < q_k > q_{k+1} \geq \cdots \geq q_m$, then the mode is either in I_{k-1}, or in I_k, or in I_{k+1}.

Therefore for engineer A, the hypothesis of unimodality would be compatible with taking the mode in the interval [2000, 3000), for instance. Is there any reason for taking the mode in the interval [3000, 4000) as in the paper? Are calculations easier with this additional hypothesis?

3. The third point is on the class Π_2'. I should like to remark that the hypothesis $\pi(\theta) \leq h^*$ for all $\theta \in \Theta$ (with $h^* = 3\max_i\{q_i\}$) is very weak (by the way, is there any reason for taking $h^* = 3\max_i\{q_i\}$, instead of, say, $h^* = 4\max_i\{q_i\}$?). It is easy to prove that:

If $\pi \in \Pi_2$ with $q_1 \leq q_2 \leq \cdots \leq q_k \geq q_{k+1} \geq \cdots \geq q_m$, then $\pi(\theta) \leq q_k$ for all $\theta \in I_i (i \neq k-1, k, k+1)$.

Therefore, the constraint $\pi(\theta) \leq h^*$ for all $\theta \in \Theta$, says nothing about intervals I_i with $i \neq k-1$, k, $k+1$. It only says that the density over the intervals I_{k-1}, I_k and I_{k+1} is not very sharp and this does not seem too restrictive for the prior specification. Only if the mode is assumed to be in I_1 (or in I_m), can the class Π_2' omit possibly interesting densities).

4. And the fourth point is about a related problem. The analysis has been carried out for a single sample and we know nothing about what is going on with other samples. Of course, if the experiment has been fixed, we shall observe a single sample and the robustness for this sample is all we need to study. I think, however, it is of interest to consider the following problem: let us suppose that we have at our disposal different experiments to be performed, and we need to choose the most robust (in some sense); so, in this case, we want to know something about the robustness of each experiment (not for a single sample, but for the set

of all the possible samples). For instance, for finding the reference prior distribution, we also need to consider all the samples (Bernardo (1979)).

A study of robustness (in this last sense) under changes on the likelihood has been carried out by Cuevas (1987), by applying Hampel's concept of qualitative robustness (Hampel (1971)) to the sequence of posterior distributions.

M. WEST (*University of Warwick*)

This is a nice paper providing some useful results, the lucid presentation and examples of which belie the considerable underlying mathematical development. I particularly like the emphasis on partial prior specification using quantiles (a focus of my paper at the conference) and would like to think that these sorts of robustness studies will become routine for most of us in applied work. To this end and following the discussion meeting at the conference on the availability of algorithms and packages for Bayesian methods, it would be useful to have the (non-trivial) algorithms presented here available for wider use.

A couple of specific points arising from the paper are as follows.

Firstly the examples provide insight into the differences in ranges of posterior probabilities across prior classes Π_0, Π_1, etc. In almost every case, the likelihoods are in fairly close agreement with the priors, as can be seen from the fact that most of the ranges of posterior probabilities include the prior probabilities (see Tables 2 to 5, and 7). Can the authors comment on the qualitative effect on the posterior ranges of conflict between the data and the prior, when the likelihood favours values in the tails of the prior? In this situation, the prior tails really matter. Unimodality allows, for example, the possibilities of normal and Cauchy priors, which have radically different effects on the posterior in cases of conflict. Sometimes it may be desirable to strengthen the restrictions on the class of priors from unimodality in order to differentiate between such cases. In the normal location problem of Example 2, for instance, excluding *strongly unimodal* priors (such as the normal) from the class would leave us with priors that would have reduced effects on the posterior in cases of conflict with the data. Have the authors considered such subclasses of priors and the effects on ranges of posterior probabilities?

Secondly, in a decision analysis the focus may shift from posterior probabilities to point estimates of θ, e.g. posterior means, and expected utilities. It is natural then to consider whether variation in posterior probabilities over particular intervals matters. Can we calculate ranges of posterior expectations for simple functions to examine sensitivity of decisions?

Finally, a related question concerns the extension of the method to consider the class of unimodal priors having specified mean and variance. I am particularly interested in this in conection with my own, current use of convenient, conjugate priors in dynamic generalised linear modelling.

REPLY TO THE DISCUSSION

We thank both discussants for their valuable insights and elaborations. There is little in the discussions with which we disagree; our reply thus focuses on merely answering questions and providing explanations or additional background. We follow the format and order of the discussions in our reply.

to Professor de la Horra:

1. The last paragraph of point 1 refers to the interesting alternative idea of finding the reference (or noninformative) prior subject to the constraints imposed by Π. Reference distributions are usually inherently robust; thus this approach may well provide a reasonable adhoc method of dealing with prior uncertainty. The work of García-Carrasco and the discussant in this direction is very interesting, though it should be noted that

extension to unbounded Θ has a problem: the reference prior, constrained by a finite set of specified quantiles, does not exist!

2. We utilized the constraint "the mode is in I_k" partly for simplicity in presenting the algorithm, and partly to indicate another perhaps easily elicited feature of the prior. To analyze the problem without this constraint, one simply repeats the algorithm over each of the three possible regions for the mode, as pointed out by de la Horra, and selects the maximum or minimum value. We found that the location of the mode made vitually no difference in our examples, and thus neglected to emphasize this possibly important point.
3. The "default" choice $h^* = 3 \max \{q_i\}$ was quite arbitrary. The choice of h^* will make little difference unless the likelihood is very concentrated or, as de la Horra observes, unless the mode is in I_1 or I_m.
4. Pre-experimental robustness concerns and robustness with respect to the likelihood tend to be entirely different (though indeed important) than posterior robustness with respect to the prior.

to Professor West:

The three interesting specific questions raised here could each be best answered by an entire paper. Presumably our "authors' license" does not extend quite that far; thus we will merely indicate what these papers would be about.

Question one is concerned with the situation in which the likelihood is concentrated in the tail of the prior. Bayesian analysis is notorious for a lack of robustness in this situation and our development, if anything, exaggerates this lack. For instance, in Example 2 if the likelihood is N (3.645, 1) (which gives 95% of its mass to the "tail interval" $(2, \infty)$ of the prior), then the range of the posterior probability of the interval can be shown to be (0, 0.824). The lower bound of zero arises because the prior tail is allowed to become uniform over a huge range, and hence essentially zero. The upper bound arises from concetrating the tail mass of 0.08 near the likelihood; the upper bound is large because the likelihood gives virtually zero weight to the other intervals.

This extreme nonrobustness in the tails can arise because the classes of priors considered in the paper provide only minimal control over the prior tails. It should be mentioned, however, that to obtain robustness in this situation it is often necessary to very precisely pin down the prior tail. Global properties of the class, such as strict unimodality or even fixed mean and variance, often do not pin down the tail sufficiently for the posterior range to be small.

Question 2 concerns the possibility of finding posterior ranges for other Bayesian quantities, such as the posterior mean. This can most likely be done, and probably by methods close to those in the paper. For instance, Kun-Liang Lu has found that, under rather mild conditions on the likelihood function, the algorithms in the paper need only slight adaption to provide the range of the posterior mean. What is not clear, however, is if it is possible to produce a general algorithm that will work for a wide variety of posterior quantities of interest.

Question 3 inquires as to the possibility of investigating robustness for the class of unimodal priors with specified mean and variance. This is indeed possible, but requires methods that are completely different than those used in the paper. These methods are based on representing the class of unimodal priors as an appropriate class of mixtures of uniform distributions, and then writing the prior moment constraints as (generalized) moment constraints on the mixing distribution. Finding the range of Bayesian quantities of interest often then reduces to solving a type of generalized probabilistic moment problem. Such methods can be seen in action in Sivaganesan and Berger (1986).

An interesting side issue is the suitability of using classes of priors with specified mean and variance to investigate robustness. Ideally, one should use a class of priors that is compatible with prior beliefs and that contains a suitably wide representation of these prior beliefs. The difficulty with basing the class on specification of moments is that moments are extremely sensitive to the prior tail and the prior tail is probably the most difficult feature of a prior distribution to specify. Thus a class based on moments omits prior distributions that can be visually indistinguishable from the distributions in the class, and that are hence typically also plausible priors.

This concern with definition of Γ in terms of moments is admittedly rather "soft", not being based on a study of actual examples. Indeed, the "unimodal-moment" class may well be broad enough to obtain very reasonable indications of robustness. These is no compelling reason to consider such classes, however (unless, of course, the moments truly are known precisely). One can more easily (and with more confidence) verify the robustness of conjugate prior analyses by, say, using ε-contamination classes based on the conjugate prior, as in Sivaganesan and Berger (1986).

REFERENCES IN THE DISCUSSION

Bernardo, J. M. (1979). Reference posterior distributions for Bayesian inference. (with discussion). *J. Roy. Statist. Soc. B* **41**, 113–147.

Cuevas, A. (1988). Qualitative robustness in abstract inference. *Journal of Statistical Planning and Inference* **18**, 277–289.

García-Carrasco, P. and de la Horra, J. (1988). Maximizing uncertainty functions under constraints on quantiles. *Statistics. & Decisions.* (To appear).

Hampel, F. R. (1971). A general qualitative definition of robustness. *Ann. Math. Statist.* **42**, 1887–1896.

BAYESIAN STATISTICS 3, pp. 67–78
J. M. Bernardo, M. H. DeGroot, D. V. Lindley and A. F. M. Smith, (Eds.)

A Bayesian Analysis of Simple Mixture Problems

J. M. BERNARDO and F. J. GIRÓN
Universidad de Valencia and *Universidad de Málaga*

SUMMARY

A large number of interesting problems may be described by finite mixture models; these include outlying observations, probabilistic classification, unsupervised sequential learning and clustering. There are, however, important difficulties with the implementation of those models: (i) the combinatorial explosion of the likelihood function effectively prevents the derivation of exact posterior distributions in virtually all practical problems; (ii) the lack of general results on the joint asymptotic posterior distribution of the parameters involved precludes the use of asymptotic approximations, even if large samples are available; and (iii) although it is well known that, in complex models, the posterior distribution of the parameter(s) of interest may be very sensitive to the joint prior, there are no results on the form of sensible reference priors in the context of mixture models. In this paper, we explore the simplest mixture models, those where the mixands are totally specified, in an attempt to identify possible directions for further progress.

Keywords: APPROXIMATIONS; FINITE MIXTURES; LOGARITHMIC DIVERGENCE; PROBABILISTIC CLASSIFICATION; REFERENCE PRIORS.

1. INTRODUCTION

Mixture models are often useful to describe complex statistical problems. Indeed, identification of outlying observations, probabilistic classification, unsupervised sequential learning, and clustering are all problems which may naturally be modelled in mixture form; see Everitt and Hand (1981), Titterington, Smith and Makov (1985), and references therein. A *mixture model* is a probabilistic model described by the density

$$p(x \mid \boldsymbol{\lambda}, \boldsymbol{\theta}) = \sum_{j=1}^{k} \lambda_j \, p(x \mid \theta_j), \qquad \lambda_j > 0, \quad \sum_{j=1}^{k} \lambda_j = 1 \tag{1}$$

where $\boldsymbol{\lambda} = \{\lambda_1, \ldots, \lambda_k\}$, $\boldsymbol{\theta} = \{\theta_1, \ldots, \theta_k\}$ and k denotes the number of mixands in the mixture; in this model, $p(x \mid \theta_j)$ describes the probabilistic mechanism of generating data x within population P_j, which is completely identified by its corresponding parameter θ_j, and λ_j denotes the probability that a random observation comes from population P_j. The appropriate choice of the number of mixands, and of their functional form, depends on the particular statistical problem that the statistician intends to model. Often, the functional form of all the terms in the mixture will be the same. For instance, the mixands may all be assumed to be normal distributions with possibly different location and scale parameters. Mixture models have inherent theoretical and computational difficulties which may deter practitioners from using them. Indeed, when dealing with mixtures, two important problems typically arise: one is computational, due to the combinatorial explosion of terms in the likelihood function and, hence, in the posterior distribution; the other is more theoretical and refers to the difficulties encountered in the definition of an appropriate joint prior for the unknown parameters, $(\boldsymbol{\lambda}, \boldsymbol{\theta})$.

If $z = \{x_1, \dots, x_n\}$ is a random sample from (1), then the likelihood of z is

$$\prod_{i=1}^{n}\left[\sum_{j=1}^{k}\{\lambda_j\, p(x_i \,|\, \theta_j)\}\right]$$

which is a sum of k^n individual terms. It is well known that conjugate families, in the strict sense, do not exist for mixture models, even if each of the individual mixands does admit a conjugate family. However, in this case, a weak form of conjugacy still holds: if the prior belongs to the class of finite mixtures of the usual conjugate family, then the posterior also belongs to this class. For the case of finite mixtures of normal distributions, the corresponding extended conjugate families are described in Bernardo and Girón (1986).

Unfortunately, even if we restrict the choice of prior distribution to this extended conjugate family, the problems that mixtures typically entail do remain; in particular, the derivation of an appropriate "non-informative" reference prior is less than obvious: *reference* priors (Bernardo, 1979) depend on the asymptotic behaviour of the relevant posterior distributions, and very little is known about the asymptotic behaviour of the posterior distribution of the parameters of mixture models. Indeed, although both Kazakos (1977) and Smith and Makov (1978) have shown that certain recursive estimators are consistent and, more recently, Redner and Walker (1984) and Hathaway (1985) have stated the limiting properties of maximum likelihood estimators in mixture models, the general conditions under which consistent estimators exist for the parameters of a general mixture model are still unknown.

In this paper, we shall concentrate on mixture models where the mixand distributions $p(x \,|\, \theta_j)$, $j = 1, \dots, k$, are totally specified, so that $\{\lambda_1, \dots, \lambda_k\}$ are the only unknown parameters; this can be viewed as a conditional analysis of the posterior distribution of the weights to changes in the mixands. Section 2 discusses the learning process within these simple mixture models and, in particular, the choice of the prior distribution of the λ_j's. Section 3 presents some new approximation procedures to the corresponding posterior distributions, and compares them with those advanced by Smith and Makov (1978). Finally, Section 4 briefly outlines interesting problems for further research.

2. THE LEARNING PROCESS

2.1. The Model

In this section we consider the problem of learning from the data about the unknown parameters $\{\lambda_1, \dots, \lambda_k\}$ of the mixture model

$$p(x \,|\, \boldsymbol{\lambda}) = \sum_{j=1}^{k} \lambda_j\, p_j(x), \qquad \lambda_j > 0, \quad \sum_{j=1}^{k} \lambda_j = 1 \tag{2}$$

where the $p_j(x)$'s are totally specified densities with respect to some dominating measure, defined on a sample space X, which describe the individual populations P_j. Note that the model (2) may be regarded as a hierarchical model in two stages:

(i) the observation x has a distribution $p(x \,|\, \omega)$ where ω is a discrete hyperparameter with possible values $\{1, 2, \dots, k\}$, which identifies the population to which x belongs, so that $p(x \,|\, \omega = j) = p_j(x)$, and

(ii) the prior distribution of ω is $p(\omega = j) = \lambda_j$, $j = 1, \dots, k$.

Moreover, the likelihood for a sample $z = \{x_1, \dots, x_n\}$ of size n is given by

$$L(\boldsymbol{\lambda}; z) = \prod_{i=1}^{n}\left[\sum_{j=1}^{k}\{\lambda_j\, p(x_i \,|\, \theta_j)\}\right] = \sum_{j(1)=1}^{k} \cdots \sum_{j(n)=1}^{k} \prod_{i=1}^{n}\{\lambda_{j(i)}\} \prod_{i=1}^{n}\{p_{j(i)}(x_i)\}. \tag{3}$$

2.2. *Posterior and Predictive Distributions*

It follows from (3) and Bayes' theorem that the posterior distribution of $\boldsymbol{\lambda}$ given the data z is

$$p(\boldsymbol{\lambda} \mid z) \propto p(\boldsymbol{\lambda}) \sum_{j(1)=1}^{k} \cdots \sum_{j(n)=1}^{k} \prod_{i=1}^{n} \{\lambda_{j(i)}\} \prod_{i=1}^{n} \{p_{j(i)}(x_i)\} .$$

Note that given the special form of the likelihood given in (3), if $p(\boldsymbol{\lambda})$ is a mixture of Dirichlet distributions, $p(\boldsymbol{\lambda} \mid z)$ will also be a mixture of Dirichlet distributions, so that the extended conjugacy property mentioned above will hold.

The problem of probabilistic classification of a new observation x into one of the k populations reduces to the derivation of

$$\Pr\{x \in P_j \mid x, z\} = \frac{p(x \mid x \in P_j) \Pr\{x \in P_j \mid z\}}{\sum_{j=1}^{k} p(x \mid x \in P_j) \Pr\{x \in P_j \mid z\}}$$

$$= \frac{p_j(x) \Pr\{x \in P_j \mid z\}}{\sum_{j=1}^{k} p_j(x) \Pr\{x \in P_j \mid z\}} = \frac{p_j(x) E[\lambda_j \mid z]}{\sum_{j=1}^{k} p_j(x) E[\lambda_j \mid z]}$$

since

$$\Pr\{x \in P_j \mid z\} = \int \Pr\{x \in P_j \mid \boldsymbol{\lambda}\} p(\boldsymbol{\lambda} \mid z)\, d\boldsymbol{\lambda} = \int \lambda_j\, p(\boldsymbol{\lambda} \mid z)\, d\boldsymbol{\lambda} = E[\lambda_j \mid z].$$

2.3. *Reference Distributions*

It may be verified that if the densities $p_j(x)$ are linearly independent almost everywhere, then the model (2) satisfies sufficient regularity conditions to guarantee the asymptotic posterior normality of $\boldsymbol{\lambda}$. Hence, the reference prior for $\boldsymbol{\lambda}$ is Jeffreys' prior, that is, $\pi(\boldsymbol{\lambda}) \propto |\, H(\boldsymbol{\lambda}) \,|^{1/2}$, where $H(\boldsymbol{\lambda})$ is the matrix whose typical element is given by

$$- \int p(x \mid \boldsymbol{\lambda}) \frac{\partial^2}{\partial \lambda_i \partial \lambda_j} \log p(x \mid \boldsymbol{\lambda})\, dx.$$

Let us now specialize to the case of only two mixands, so that

$$p(x \mid \lambda) = \lambda\, p_1(x) + (1 - \lambda)\, p_2(x). \tag{4}$$

In this case, $|\, H(\lambda) \,|$ reduces to the real function $h(\lambda)$ defined by

$$h(\lambda) = - \int p(x \mid \lambda) \frac{\partial^2}{\partial \lambda^2} \log\{\lambda\, p_1(x) + (1 - \lambda)\, p_2(x)\}\, dx$$

$$= \int \frac{\{p_1(x) - p_2(x)\}^2}{\lambda\, p_1(x) + (1 - \lambda)\, p_2(x)}\, dx \tag{5}$$

and, hence, $\pi(\lambda) \propto \{h(\lambda)\}^{1/2}$ which, in general, cannot be evaluated in explicit analytic form. However,

Proposition 1. *For arbitrary density functions* $p_1(x)$ *and* $p_2(x)$ *consider the model*

$$p(x \mid \lambda) = \lambda\, p_1(x) + (1-\lambda)\, p_2(x).$$

Then, the reference prior for λ *is* $\pi(\lambda) \propto \{h(\lambda)\}^{1/2}$, *where*

$$h(\lambda) = \int \frac{\{p_1(x) - p_2(x)\}^2}{\lambda\, p_1(x) + (1-\lambda)\, p_2(x)}\, dx.$$

Moreover,

(i) $h(\lambda)$ *is a convex function of* λ.

(ii) $h(\lambda) = 0$ *iff* $p_1(x) = p_2(x)$, *almost everywhere.*

(iii) $h(\lambda) \leq \lambda^{-1}(1-\lambda)^{-1}$, *with equality iff* $p_1(x)$ *and* $p_2(x)$ *have disjoint support almost everywhere.*

Proof. The form of the reference prior has been established above; (i) is trivial: it suffices to check the sign of $\partial^2 h(\lambda)/\partial\lambda^2$. The second part is also immediate for, in this case, the model reduces to $p(x \mid \lambda) = p_1(x)$ almost everywhere.

To prove (iii), consider the following sequence of equalities and inequalities

$$(p_1 - p_2)^2 \leq (p_1 + p_2)^2 \leq p_1^2 + p_2^2 + p_1 p_2 \left[\frac{\lambda}{1-\lambda} + \frac{1-\lambda}{\lambda}\right]$$

$$= [\lambda p_1 + (1-\lambda) p_2] \left[\frac{p_1}{\lambda} + \frac{p_2}{1-\lambda}\right] \tag{6}$$

so that,

$$\frac{[p_1(x) - p_2(x)]^2}{\lambda\, p_1(x) + (1-\lambda)\, p_2(x)} \leq \frac{p_1(x)}{\lambda} + \frac{p_2(x)}{(1-\lambda)}$$

and therefore, using (5),

$$h(\lambda) \leq \frac{1}{\lambda} \int p_1(x)\, dx + \frac{1}{1-\lambda} \int p_2(x)\, dx$$

$$= \frac{1}{\lambda} + \frac{1}{1-\lambda} = \frac{1}{\lambda(1-\lambda)}.$$

If X_1 and X_2 denote almost everywhere disjoint supports of $p_1(x)$ and $p_2(x)$ respectively, then $\{p_1(x) - p_2(x)\}^2 = p_1^2(x) + p_2^2(x)$, for the product term vanishes (a.e.); thus,

$$h(\lambda) = \int_X \frac{p_1^2(x) + p_2^2(x)}{\lambda\, p_1(x) + (1-\lambda)\, p_2(x)}\, dx$$

which, adding the separate integrals in X_1 and X_2, reduces to $h(\lambda) = \lambda^{-1}(1-\lambda)^{-1}$.

Conversely, if $h(\lambda) = \lambda^{-1}(1-\lambda)^{-1}$, all the inequalities in (6) become equalities and, hence, $[p_1(x) - p_2(x)]^2 = [p_1(x) + p_2(x)]^2$ almost certainly. This, in turn, implies that $|\, p_1(x) - p_2(x)\,| = p_1(x) + p_2(x)$ and, therefore, the supports of $p_1(x)$ and $p_2(x)$ are almost everywhere disjoint. ◁

Corollary. *The reference prior* $\pi(\lambda)$ *is always proper.*

Proof. Since $\pi(\lambda) \propto \{h(\lambda)\}^{1/2}$ and, by part (iii) above, $\{h(\lambda)\}^{1/2}$ is bounded above by the integrable function $k\,\lambda^{-1/2}(1-\lambda)^{-1/2}$, for some $k > 0$, $\pi(\lambda)$ must be integrable. ◁

Proposition 1 (iii) shows that when the two probabilistic models described by $p_1(x)$ and $p_2(x)$ do not overlap, so that it is known almost surely which of the two populations each sample element belongs to, model (4) reduces to the usual Bernoulli model. Note also, in this case, that the values of the sample elements $x_1, \ldots, x_n$ are irrelevant; all we require is the number of them belonging to each population, r and $n-r$, respectively. Indeed, in this case, the likelihood is proportional to $\lambda^r(1-\lambda)^{n-r}$.

An alternative way of proving (iii) is to see the mixture model (4) as a problem with incomplete data: it is not known to which population each sample element belongs to; the hyperparameters $\omega_1, \ldots, \omega_n$ are part of the complete data. The upper bound $\lambda^{-1}(1-\lambda)^{-1}$ is Fisher's information for the complete data model while $h(\lambda)$ is the corresponding information for the incomplete data problem, i.e. the mixture model. Since Fisher's information is always smaller for the incomplete data problem, the result follows.

The theorem and its corollary suggest that a beta distribution $\text{Be}(\lambda \mid \alpha_0, \beta_0)$ with both parameters in the range $[\frac{1}{2}, 1]$ may be a good approximation to the reference prior $\pi(\lambda)$ regardless of the densities $p_1(x)$ and $p_2(x)$. Indeed, $\text{Be}(\lambda \mid \frac{1}{2}, \frac{1}{2})$ could be expected to be a good approximation to $\pi(\lambda)$ for well separated densities $p_1(x)$ and $p_2(x)$ (even if, technically, their supports overlap), while the uniform distribution $\text{Be}(\lambda \mid 1, 1)$ would approximate $\pi(\lambda)$ when $p_1(x)$ and $p_2(x)$ are very close. This, in turn, shows that the reference prior for the mixing parameter is fairly robust under changes in the specification of the individual mixture terms.

Example 1. Consider the case of a mixture of two normal densities, so that the model considered becomes

$$p(x \mid \lambda) = \lambda\, N(x \mid \mu_1, \sigma_1) + (1-\lambda)\, N(x \mid \mu_2, \sigma_2).$$

Using Proposition 1, we have numerically evaluated the exact reference priors which correspond to various combinations of $(\mu_1, \sigma_1, \mu_2, \sigma_2)$, and found that these are graphically indistinguishable from appropriately chosen Beta densities; some of those results are shown in Table 1.

Case	*Population 1*	*Population 2*	*Approximate* $\pi(\lambda)$
(i)	$N(x \mid -2, 0.25)$	$N(x \mid 2, 0.25)$	$Be(\lambda \mid 0.500, 0.500)$
(ii)	$N(x \mid 0, 1)$	$N(x \mid 0.01, 1.01)$	$Be(\lambda \mid 1.001, 0.989)$
(iii)	$N(x \mid 0, 1)$	$N(x \mid 0, 0.5)$	$Be(\lambda \mid 0.660, 0.912)$
(iv)	$N(x \mid 0, 1)$	$N(x \mid 0.5, 1)$	$Be(\lambda \mid 0.954, 0.968)$

Table 1. *Approximate reference priors for the mixture of two normals*

It may be appreciated that, as one could intuitively expect from Proposition 1, the reference prior is virtually Jeffreys' $\text{Be}(\lambda \mid \frac{1}{2}, \frac{1}{2})$ when the two normal densities are well separated, as in case (i), and it is practically uniform when the two normal densities are very close, as in case(ii). Variations in the standard deviation seem to be more important within this context than variations in the mean, as illustrated in cases (iii) and (iv). ◁

Unfortunately the extension of Proposition 1 and its corollary to the case of k mixands ($k \geq 3$) is not readily available. It may be established, however, that for the limiting case of all the mixands having pairwise disjoint supports, the reference prior approaches the usual reference prior $\pi(\lambda) \propto \prod_{j=1}^{k} \lambda_j^{-1/2}$ while, at the other extreme, when all mixands converge to the same distribution, the reference prior tends to the uniform distribution on the k-dimensional simplex.

3. APPROXIMATIONS

From the preceding results, in a mixture problem where the mixands are totally specified, it seems reasonable to approximate the, typically proper, reference prior of the unknown weights $\{\lambda_1, \ldots, \lambda_k\}$ by a Dirichlet distribution with parameters ranging in the interval $[\frac{1}{2},1]$. Of course, any proper prior can be approximated by a finite mixture of Dirichlet distributions (Diaconis and Ylvisaker, 1985; Dalal and Hall, 1983); the corresponding posterior would then also be a mixture of Dirichlet distributions. We shall now consider the case of a Dirichlet prior density; it is clear, however, that the procedures presented may be easily adapted to the case where the prior is a finite mixture of Dirichlet distributions.

The standard procedure considered in the literature to avoid the combinatorial explosion of the likelihood function is to apply Bayes theorem sequentially, with one or more observations considered at a time, followed by suitable approximations to the resulting posterior in such a way as to obtain recursive estimates of the parameters characterizing an approximate posterior within some specified class, typically the class of Dirichlet distributions. See, Makov (1980), Titterington, Smith and Makov (1985) and references therein.

We propose, at each step, to approximate the true posterior distribution $p(\boldsymbol{\lambda} \mid z)$ by the "closest" tractable distribution, defined as that $p^*(\boldsymbol{\lambda})$ which minimizes, within a given class $\mathcal{P}$, the logarithmic divergence

$$\delta(p, p^*) = \int p(\boldsymbol{\lambda} \mid z) \log \frac{p(\boldsymbol{\lambda} \mid z)}{p^*(\boldsymbol{\lambda})} \, d\boldsymbol{\lambda}, \quad p^*(.) \in \mathcal{P}.$$

This procedure has an interesting decision-theoretical justification, as that which minimizes the expected loss when the decision space consists of all available approximations and the utility function is a proper, local scoring rule (Bernardo, 1987).

Let us begin by applying Bayes theorem sequentially, one observation at a time. If $p(\boldsymbol{\lambda})$ is a Dirichlet distribution $\mathrm{Di}(\boldsymbol{\lambda} \mid \alpha_1^{(0)}, \ldots, \alpha_k^{(0)})$, then the posterior distribution after x_1 has been observed is

$$p(\boldsymbol{\lambda} \mid x_1) = \sum_{j=1}^{k} \Pr(x_1 \in P_j \mid x_1) \, \mathrm{Di}(\boldsymbol{\lambda} \mid \alpha_1^{(0)} + \delta_{1j}, \ldots, \alpha_k^{(0)} + \delta_{kj}) \tag{7}$$

where $\Pr(x_1 \in P_j \mid x_1)$ is the probability that observation x_1 belongs to population P_j, and δ_{ij} is Kronecker's delta.

It is easily verified, by differentiation, that minimization of the logarithmic divergence of $p(\boldsymbol{\lambda} \mid x_1)$ from a member of the Dirichlet family implies that the parameters of the approximating distribution $\mathrm{Di}(\boldsymbol{\lambda} \mid \alpha_1^{(1)}, \ldots, \alpha_k^{(1)})$ are the solutions to the implicit system, defined in terms of the digamma function $\psi(x) = d\{\log \Gamma(x)\}/dx$,

$$\psi(\alpha_1^{(1)} + \cdots + \alpha_k^{(1)}) - \psi(\alpha_j^{(1)})$$
$$= \psi(\alpha_1^{(1)} + \cdots + \alpha_k^{(1)} + 1) - \psi(\alpha_j^{(0)}) - \frac{1}{\alpha_j^{(0)}} \Pr(x_1 \in P_j \mid x_1), \quad j = 1, \ldots, k.$$

Note from (7) that the mixands which define $p(\boldsymbol{\lambda} \mid x_1)$ are Dirichlet densities whose parameters are such that two of them differ in precisely one component, the rest being identical.

If the approximation $\mathrm{Di}(\boldsymbol{\lambda} \mid \alpha_1^{(1)}, \ldots, \alpha_k^{(1)})$ to the true posterior $p(\boldsymbol{\lambda} \mid x_1)$ is used as a prior for the next updating, the *approximate* posterior $p^*(\boldsymbol{\lambda} \mid x_1, x_2)$ replacing $p(\boldsymbol{\lambda} \mid x_1, x_2)$ is given by

$$\sum_{j=1}^{k} \Pr{}^*(x_2 \in P_j \mid x_1, x_2) \, \mathrm{Di}(\alpha_1^{(1)} + \delta_{1j}, \ldots, \alpha_k^{(1)} + \delta_{kj})$$

where the $\Pr^*(x_2 \in P_j \,|\, x_1, x_2)$'s, the approximate posterior probabilities, given x_1 and x_2, that x_2 belongs to each of the populations, are given by

$$\Pr^*(x_2 \in P_j \,|\, x_1, x_2) \propto p_j(x_2) E[\lambda_j \,|\, x_1], \quad j = 1, \ldots, k.$$

Let us denote by $\alpha_+^{(i)} = \alpha_1^{(i)} + \cdots + \alpha_k^{(i)}$, the sum of the k parameters of a Dirichlet distribution $\mathrm{Di}(\boldsymbol{\lambda} \,|\, \alpha_1^{(i)}, \ldots, \alpha_k^{(i)})$, often referred to as its *sample size equivalent*; then,

Proposition 2. *Let $z = \{x_1, \ldots, x_n\}$ be a random sample from the mixture model*

$$p(x \,|\, \boldsymbol{\lambda}) = \sum_{j=1}^{k} \lambda_j \, p_j(x).$$

Then, the posterior distribution of $\boldsymbol{\lambda} = \{\lambda_1, \ldots, \lambda_k\}$ is approximately given by

$$p^*(\boldsymbol{\lambda} \,|\, z) = \sum_{j=1}^{k} \Pr^*(x_n \in P_j \,|\, z) \, Di(\boldsymbol{\lambda} \,|\, \alpha_1^{(n-1)} + \delta_{1j}, \ldots, \alpha_k^{(n-1)} + \delta_{kj})$$

where the $\alpha_j^{(k)}$'s are recursively obtained from the system

$$\psi(\alpha_+^{(i+1)}) - \psi(\alpha_j^{(i+1)}) = \psi(\alpha_+^{(i)}) - \psi(\alpha_j^{(i)}) - \frac{p_{ij}^*}{\alpha_j^{(i)}}, \quad j = 1, \ldots, k,$$

with

$$p_{ij}^* = \Pr^*(x_{i+1} \in P_j \,|\, x_1, \ldots, x_i, x_{i+1}) \propto p_j(x_{i+1}) \frac{\alpha_j^{(i)}}{\alpha_+^{(i)}}$$

and $\alpha_j^{(0)} \in [\frac{1}{2}, 1]$, $j = 1, \ldots, k$

Proof. This follows by induction from the preceding argument. ◁

No explicit solution to the implicit system of equations in Proposition 2 is known; some useful approximations are given in Caro, Domínguez and Girón (1986). However,

Proposition 3. *With the notation established above,*

(i) $\alpha_+^{(i+1)} \le \alpha_+^{(i)} + 1$ *with equality if, and only if, one of the classification probabilities $p_{ij}^* = Pr^*(x_{i+1} \in P_j \,|\, x_1, \ldots, x_i, x_{i+1})$, $j = 1, \ldots, k$, is equal to one.*

(ii) $\alpha_j^{(i+1)} = \alpha_j^{(i)}$ *for every $j = 1, \ldots, k$, iff, $p_{ij}^* \propto \alpha_j^{(i)}$ for every $j = 1, \ldots, k$, that is, if the classification probabilities of the i-th observation are proportional to the corresponding parameters of the current prior.*

Proof. Using the recursive property of the digamma function $\psi(x+1) = \psi(x) + (1/x)$, it is easily checked that both (i) and (ii) are verified by the solutions to the updating system of equations described in Proposition 2. ◁

The first part of Proposition 3 implies that, at each iteration, one cannot learn more about the λ_j's than in the case of perfect or error free classification; moreover, it is only in this case where the amount of information obtained is a full one unit. In fact, it can be verified that in some instances $\alpha_+^{(i+1)} < \alpha_+^{(i)}$, so that the "uncertainty" about $\boldsymbol{\lambda}$ may increase. This typically happens when there are "unexpected" observations, abnormally difficult to identify.

The second part of Proposition 3 also has an obvious intuitive appeal: if the probabilities of the i-th observation belonging to each of the populations are identical to the current expected values of the λ_j's, $\alpha_j^{(i)}/\alpha_+^{(i)}$, then no learning occurs: such a type of observation adds nothing to the learning process.

We claim that Proposition 3 contains sensible *desiderata* for any updating procedure; however the so-called Quasi-Bayesian (QB) procedures considered in the literature (see Makov, 1980, and references therein) do not satisfy them. Indeed, for QB procedures, the equality $\alpha_+^{(i+1)} = \alpha_+^{(i)} + 1$ always holds, regardless of the classification probabilities.

From the viewpoint of correctly classifying observations in the sense of giving highest probability to the true mixand, the two procedures yield similar results, as shown by extensive simulation. Yet, the approximate posterior distribution of the weights, derived using QB procedures may be very misleading when there is some overlap in the mixands, as Laird and Louis (1982) have pointed out. In fact the QB procedure is mathematically equivalent to ignoring the possible overlap of the mixands. As could be expected from (i), and can be verified by simulation, our procedure does take into account any degree of overlap.

With only two populations ($k = 2$) and assuming the prior distribution to be Beta, with parameters α_0 and β_0, the recursive equations take the form

$$\begin{cases} \psi(\alpha_{i+1} + \beta_{i+1}) - \psi(\alpha_{i+1}) = \psi(\alpha_i + \beta_i + 1) - \psi(\alpha_i) - p^*/\alpha_i \\ \psi(\alpha_{i+1} + \beta_{i+1}) - \psi(\beta_{i+1}) = \psi(\alpha_i + \beta_i + 1) - \psi(\beta_i) - (1 - p^*)/\beta_i \end{cases} \tag{8}$$

where p^* is the current approximate probability that a random observation (the i-th) belongs to the first population. Figure 1 provides, as a function of p^*, the new values of α_{i+1}, β_{i+1} and $\alpha_{i+1} + \beta_{i+1}$ which are obtained from both Equation 8 (convex lines) and QB procedures (straight lines) when the previous values are $\alpha_i = 3$ and $\beta_i = 1$.

These illustrate properties (i) and (ii) of Proposition 3, and make apparent the existing differences with the QB recursive updating rules,

$$\alpha_{i+1} = \alpha_i + p^*, \qquad \beta_{i+1} = \beta_i + (1 - p^*),$$

and those obtained above.

An obvious improvement over the procedure we have advocated is to update by taking observations in batches, small enough to ensure that the computational requirements of coherent Bayesian updating are within reasonable limits, and then making suitable approximations. This can be done in a number of different ways. For instance, one may compute the true posterior distribution of $\boldsymbol{\lambda}$ given x_1 and x_2, $p(\boldsymbol{\lambda} \mid x_1, x_2)$, i.e.,

$$\sum_{j=1}^{k} \sum_{i=1}^{k} \Pr(x_2 \in P_j, x_1 \in P_i \mid x_1, x_2)\, \mathrm{Di}(\boldsymbol{\lambda} \mid \alpha_1^{(0)} + \delta_{1i} + \delta_{1j}, \ldots, \alpha_k^{(0)} + \delta_{ki} + \delta_{kj}), \tag{9}$$

then approximate this by a mixture of k Dirichlet mixands, and finally apply Bayes' theorem sequentially replacing, at each iteration, a mixture of k^2 terms by one of k terms. The problem of finding the mixture of k Dirichlet terms which minimizes the logarithmic divergence from (9) does not lend itself to analytic treatment. Instead, the following procedure may be considered: write (9) as

$$p(\boldsymbol{\lambda} \mid x_1, x_2) = \sum_{j=1}^{k} p(\boldsymbol{\lambda} \mid x_2 \in P_j, x_1, x_2)\, \Pr(x_2 \in P_j \mid x_1, x_2)$$

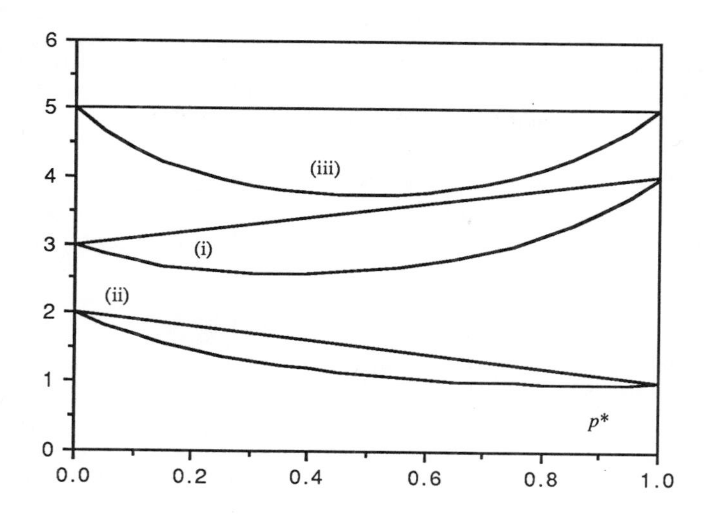

Figure 1. *Updating chart (k=2), for* (i) α_{i+1}, (ii) β_{i+1}, *and* (iii) $\alpha_{i+1}+\beta_{i+1}$, *when* $\alpha_i=3$ *and* $\beta_i=1$

where each of the conditional densities of $\boldsymbol{\lambda}$ above may be written as

$$\sum_{i=1}^{k} p(x_1 \in P_i \mid x_2 \in P_j, x_1, x_2)\,\mathrm{Di}(\boldsymbol{\lambda} \mid \alpha_1^{(0)} + \delta_{1i} + \delta_{1j}, \ldots, \alpha_k^{(0)} + \delta_{ki} + \delta_{kj}),$$

i.e., a mixture of k Dirichlet mixands which can be approximated by a single Dirichlet density by minimizing the logarithmic divergence using the procedure described before; thus, $p(\boldsymbol{\lambda} \mid x_1, x_2)$ may be approximated by

$$\sum_{j=1}^{k} \mathrm{Di}(\boldsymbol{\lambda} \mid \alpha_{1j}^{(1)}, \ldots, \alpha_{kj}^{(1)})\,\Pr(x_2 \in P_j \mid x_1, x_2).$$

Combination of this approximate posterior with the likelihood of the next observation, via Bayes theorem, produces a new mixture of k^2 terms which can be handled analogously. We want to stress, however, that the procedure just described is only one of the many possible generalizations of the method presented and no claim is made of its overall superiority.

4. CONCLUSION

The combination of a sensible reference prior for the weights, and a tractable, but appropriate, sequential approximation procedure seems to produce a pragmatic solution to the problem of making inferences on the weights of a mixture model, when the mixands are totally specified. Interesting as this particular case might be, this barely scratches the surface of the formidable problems posed by general mixture models. Even in relatively simple cases such as the mixture of two normal distributions with unknown parameters, progress is difficult; indeed, it seems clear that the joint posterior distribution of the five parameters involved in that model is not asymptotically normal without further restrictions, its precise form not being known; hence, a reference prior is not readily available. This is unfortunate, for it is known that the different marginal posterior distributions dramatically depend on the specification of the prior, as illustrated by the limiting case provided by Lindley's paradox. We hope that future work will provide further light on the issues involved.

ACKNOWLEDGEMENTS

This research has been supported by the *Comisión Asesora de Investigación Científica y Técnica*, under grant PR84–0674 and by the *Junta de Andalucía* under grant 07–CLM–MDM.

REFERENCES

Bernardo, J. M. (1979). Reference posterior distributions for Bayesian inference. *J. Roy. Statist. Soc. B* **41**, 113–147 (with discussion).

Bernardo, J. M. (1987). Approximations in statistics from a decision-theoretical viewpoint. *Probability and Bayesian Statistics.* (R. Viertl, ed.) New York: Plenum, 53–60.

Bernardo, J. M. and Girón, F. J. (1986). A Bayesian approach to cluster analysis. Invited paper at the *Second Catalan International Symposium on Statistics.* Barcelona, Spain. September 18–19, 1986.

Caro, E., Dominguez, J. I. and Girón, F. J. (1986). Métodos Bayesianos aproximados para mixturas de normales. *Actas de la XVI Reunión Nacional de la S. E. I. O.*, Málaga; (to appear).

Dalal, S. R. and Hall, W. J. (1983). Approximating priors by mixtures of natural conjugate priors. *J. Roy. Statist. Soc. B* **45**, 278–286.

Diaconis, P. and Ylvysaker, D. (1985). Quantifiying prior opinion. *Bayesian Statistics* **2**. (J. M. Bernardo, M. H. DeGroot, D. V. Lindley and A. F. M. Smith, eds.). Amsterdam: North-Holland , 133–156 (with discussion).

Everitt, B. S. and Hand, D. J. (1981). *Finite Mixture Distributions*. London: Chapman and Hall.

Hathaway, R. J. (1985). A constrained formulation of maximum-likelihood estimation for normal mixture distributions. *Ann. Statist.* **13**, 795–800.

Kazakos, D. (1977). Recursive estimation of prior probabilities using a mixture. *IEEE Trans. Inform. Theory* **IT-23**, 203–211.

Laird, N. M. and Louis, T. A. (1982). Approximate posterior distributions for incomplete data problems. *J. Roy. Statist. Soc. B* **44**, 190–200.

Makov, U. E. (1980). Approximations of unsupervised Bayes learning procedures. *Bayesian Statistics.* (J. M. Bernardo, M. H. DeGroot, D. V. Lindley and A. F. M. Smith, eds). Valencia: University Press, 69–81 (with discussion).

Redner, R. A. and Walker, H .F. (1984). Mixture densities, maximum likelihood and the EM algorithm. *SIAM Review* **26**, 195–239.

Smith, A. F. M. and Makov, U. E. (1978). A quasi-Bayes sequential procedure for mixtures. *J. Roy. Statist. Soc. B* **40**, 106–111.

Titterington, D. M., Smith, A. F. M. and Makov, U. E. (1985). *Statistical Analysis of Finite Mixture Distributions.* New York: Wiley.

DISCUSSION

U. E. MAKOV (*University of Haifa*)

Bayesian treatment of mixture models is often very complex both in terms of the mathematical analysis involved and in terms of implementation. This is clearly reflected in the small number of papers dealing with the subject. The authors are therefore to be congratulated for looking into this subject and for providing us with new results concerning a mixture model where the mixand distributions are fully specified.

The first important result is to do with reference priors for the mixing parameters. In most existing work in this area, attention is devoted to the development and study of means to curb the combinatorial explosion which is inherent in mixture models. The authors provide us with a reference prior and demonstrate that, for a two-category case, this prior can be approximated by a Beta distribution which is shown to be robust for a particular choice of hyperparameters. It will be interesting to see this study extended to the multiple-category case, both for the purpose of understanding the quality of Dirichlet priors (some doubts about their adequacy are raised in Brown, 1980) and for the purpose of finding robust priors.

The second novel result is the authors' suggestion to check the combinatorial explosion by employing an approximate posterior Dirichlet distribution with hyperparameters chosen so that the logarithmic divergence from the actual posterior distribution is minimized. In order to give this approach a wider perspective, we shall compare it to other existing methods in the context of a mixture of two distributions, $\lambda p_1(x) + (1-\lambda)p_2(x)$ (see Makov, 1980, for details).

Let the Beta prior distribution of λ be $\text{Be}(\lambda|\alpha_0, \beta_0)$ and let δ_n, the 'teacher' as it is termed in the engineering literature, be

$$\delta_n = \begin{cases} 1 & \text{if} \quad x_n \in P_1 \\ 0 & \text{if} \quad x_n \in P_2 \end{cases}$$

When the origin of each observation is known, the posterior distribution of λ, given a sample of size n, $x_1, \ldots, x_n$, is $\text{Be}(\lambda|\alpha_n, \beta_n)$, where

$$\alpha_n = \alpha_0 + \sum_{i=1}^{n} \delta_i, \qquad \beta_n = \beta_0 + n - \sum_{i=1}^{n} \delta_i$$

In the case where the δ's are unknown, the posterior distribution can be approximated by inputing them, in terms of $p_1^*(n) = \Pr(x_n \in P_1|x_1, \ldots, x_{n-1})$ by one of the following methods:

1. *Decision Directed*

$$\delta_n = \begin{cases} 1 & \text{if} \quad p_1^*(n) \geq \frac{1}{2} \\ 0 & \text{otherwise} \end{cases}$$

2. *Probabilistic Teacher*

$$\delta_n = \begin{cases} 1 & \text{with probability} \quad p_1^*(n) \\ 0 & \text{with probability} \quad 1 - p_1^*(n) \end{cases}$$

3. *Quasi Bayes*

$$\delta_n = p_1^*(n)$$

On the other hand, the updating procedure of Bernardo and Girón is an element of the more general class

$$\alpha_n = \alpha_0 + \sum_{i=1}^{n} \eta_i\{p_1^*(i)\}, \qquad \beta_n = \beta_0 + \sum_{i=1}^{n} \zeta_i\{p_2^*(i)\}$$

where, typically, $\zeta_i\{p_1^*(i)\} \neq 1 - \eta_i\{p_2^*(i)\}$.

The QB attraction lies in its cautious updating based on the strength of evidence as reflected by $p_1^*(i)$. Contrary to the authors' claim, these probabilities do reflect the degree of overlap between the two underlying distributions. The QB has, however, an obvious pitfall. While it guarantees an approximate posterior distribution with mean identical to that of the true distribution, its precision is over-estimated. In the method suggested by the authors, α and β are not updated by $p_1^*(i)$ and $1-p_1^*(i)$, respectively, but by non-linear functions of these probabilities. I am, however, not entirely happy about the fact that $\alpha_{i+1} + \beta_{i+1} \leq \alpha_i + \beta_i + 1$, Proposition 3(i); indeed, this implies that it is possible that α_i or β_i be 'incremented' by a negative number, and hence the total evidence is less than unity. An 'unexpected' observation, as the authors put it, should result in $p_1^*(i)$ close to zero or one and thus $\eta_i\{p_1^*(i)\}$ should be positive and similarly close to zero or one. Perhaps a small sample simulation study of the proposed method may reveal whether the reservations made are well founded. In such a case, a modification which truncates $\eta_i(.)$ to only positive values may be considered.

With the interesting results given in this paper in mind, I am looking forward to the authors' extension of their work to more complicated mixture models.

REPLY TO THE DISCUSSION

We are very grateful to Dr. Makov for his comments. However, we would like to point out that

(i) While Dirichlet distributions seem to provide good approximations to the exact *reference* priors in the general case, there is nothing in our argument to support the use of this particular family of distributions to describe *informative* prior beliefs; it is only a mathematical curiosity that those distributions which maximize the expected missing information about λ happen to be well approximated by elements of the Dirichlet family, at least in the case of two mixands.

(ii) The amount of *information*, in the sense of divergence between prior and posterior, is known to be positive for any model and any data. However, we find no support for Dr. Makov's assumption that the amount of 'evidence' about the mixing parameter λ provided by each observation should be positive (let alone constant!) for all observations. Indeed, if one is fairly sure that $x_i \in P_1$, so that $p_1^*(i) \simeq 1$, then $\eta_i\{p_1^*(i)\} \simeq 1$ and hence $\alpha_{i+1} \simeq \alpha_i + 1$ as he suggests; however, if x_i is a 'puzzling' observation, unexpectedly difficult to identify, with $p_1^*(i) \simeq 0.5$, our uncertainty about the true value of λ will often increase, and this is described by a flatter posterior.

(iii) The declared objective of Bayesian inference is to provide a posterior distribution of the parameter of interest. We claim that minimizing the divergence from the exact posterior gives better final solutions than any other proposed approximations.

REFERENCES IN THE DISCUSSION

Brown, P. J. (1980). Contribution to the discussion of 'Approximations of unsupervised Bayes learning procedures', by U. E. Makov. *Bayesian Statistics.* (J. M. Bernardo, M. H. DeGroot, D. V. Lindley and A. F. M. Smith, eds). Valencia: University Press, 132–134.

BAYESIAN STATISTICS 3, pp. 79–94
J. M. Bernardo, M. H. DeGroot, D. V. Lindley and A. F. M. Smith, (Eds.)

Multiple Comparisons, Multiple Tests, and Data Dredging: A Bayesian Perspective

DONALD A. BERRY
University of Minnesota

SUMMARY

Multiple inferences are discussed from a Bayesian point of view and related to the classical statistical approach. A special case is testing one of the myriad of possible hypotheses that could have been generated by "dredging" through the data. Another special case is multiple comparisons. In the case of exchangeable parameters, a multiple comparisons problem can be viewed as a Bayes empirical Bayes problem. When the parameters are not exchangeable the problem is not as amenable to analysis.

Keywords: MULTIPLE INFERENCES; MULTIPLE TESTS; INTERIM ANALYSIS; VARIABLE SELECTION IN REGRESSION; DATA DREDGING; EMPIRICAL BAYES; HIERARCHICAL PRIORS.

1. INTRODUCTION

An experiment compares a new procedure with a control on the basis of k variables. The results suggest that new is better than control for some variables but not for others. Statistical hypothesis tests find that new is "significantly better" ($\alpha = .05$) for two of the variables.

Classical analysis does not stop here. Suppose new and control are actually identical insofar as these k variables are concerned; this is the intersection of the k individual null hypotheses. Then one expects new to be statistically significantly better for 5% of the variables. For example, if $k = 40$ then observing two significant results is actually expected when all 40 null hypotheses are true.

Suppose that the k variables $X_1, \ldots, X_k$ are differences, new minus control, standardized to have unit variance. Suppose further, that each X_i is normally distributed with mean μ_i and that the X_i's are independent given the μ_i's. The ith null hypothesis is that $\mu_i = 0$. The intersection of all k null hypotheses is $H_0 : \mu_1 = \cdots = \mu_k = 0$. The null probability of $X_i > 1.645$ is .05 since $\Phi^{-1}(.95) = 1.645$, the 95th percentile of the standard normal. But, assuming independence, the null probability that at least one $X_i > 1.645$ is $1 - .95^k$. This obviously increases rapidly; for $k = 40$ it is about .87. So the null probability of at least one rejection can be much larger that the "nominal level" .05. Classical statistical wisdom dictates an adjustment. One way to make the "actual level" equal to .05 is to increase the rejection limits for each X_i from 1.645 to $\Phi^{-1}(.95^{1/k})$. For example, $.95^{1/40} = .9987$ and $\Phi^{-1}(.9987) = 3.00$. So to attain significance for the ith variable *at the .05 level* requires $X_i > 3.00$.

This seems ludicrous from a scientific point of view: How can the mere fact that bilirubin levels were measured (not what they were, just that they were measured!) affect inferences about blood pressure? An investigator who carried out only one test might find a significant difference whereas the same difference would not have been significant had the investigator tested enough other variables. (There is a strong temptation to cheat—cf. Berger and Berry (1987). Especially since such cheating is impossible to uncover—it depends on the *intentions* of the investigator rather than on the *data*, which are unadulterated. And it would be regarded as cheating at most by those few scientists who understand this statistical construction.)

Such classical statistical adjustments also seem inconsistent with a Bayesian point of view; the mere fact that variables were measued is irrelevant to the current distribution of the μ_i's. Of course, the actual observations (or partial information about these observations) can change the current distribution. The new distribution is a function of those observations and so is random—its average being the current distribution.

Still, this issue is not clearcut. For example, when observing a large number of exchangeable random variables, some will be larger than others. Selecting those that are extreme can be misleading. In both this example and the previous setting, the infinitely careful Bayesian need not worry, but the Bayesian who assumes a prior distribution without careful reflection may obtain a posterior distribution that is far from what it should be. While this statement is true generally, the issue is more critical when making multiple inferences than when making inferences in a univariate setting.

Adjustments for simultaneously testing many null hypotheses are similar to those that arise in many guises in classical statistics. Historically the most important of these has been the problem of "multiple comparisons", in which many treatments (and their interactions) are compared on the basis of measuring some variable for each treatment (see Section 3). Another is applying several different statistical tests of the same hypothesis to the same data (t-test, signed-rank test, etc.). An area that has recently been the focus of much research is "interim analysis", in which tests of an hypothesis are carried out periodically as data accumulate.

The Bayesian approach to interim analysis is straightforward (Berry 1985, 1987) and uncomplicated by some of the considerations of multiple comparisons and testing many variables (see Section 4). The distinction is that in the problem of interim analysis (and also that of several different tests for the same hypothesis) the hypothesis being tested involves a one-dimensional parameter (e.g., $\mu_1 = 0$ as opposed to $\mu_1 = \ldots = \mu_k = 0$). This might seem like a trivial distinction—one simply applies Bayes's theorem in both cases. But, much greater care seems necessary in assessing prior information when the null hypothesis is more complicated.

Still another area of some concern to classical statistics is variable selection in regression (Freedman 1983). Letting the data indicate which of many variables to use in regression can greatly exaggerate the appropriateness of such a data-indicated model from a classical statistical point of view. However, there would be no objection to the *same* model if the variables had been selected in advance, or separate from the data! Suppose a researcher uses a linear model involving a number of independent variables. If the researcher chose the variables in advance then the results are "believable"; otherwise, they are not. If you cannot tell which of the two from a report of the study, then (as I have heard classical statisticians advise) you have to write to the reasearcher to find out which! (What to do if the researcher has since died? Randomize?) It is hard for me to understand how classical statisticians can continue to adhere to a philosophy that takes them down so many roads that lead to nonsense.

A more general question of great importance to statistics and all of science is whether and how it is possible to learn from data. Sounds silly! Statistics *is* learning from data. But classical statistics qualifies this statement. A fundamental tenet of classical statistics is that one cannot test a hypothesis using the same data that generated the hypothesis. If one notices a tendency through "data dredging" then one must get *another* data set to verify that the tendency is real. The problem, classical statisticians say, is that there are so many tendencies that could dredged up, noticing one tendency is not very surprising.

A Bayesian can calculate the posterior probability that any given tendency is real, but requires a prior probability. It seems impossible to, in advance of an experiment, assess one's prior probability for each tendency that might arise. But is it possible to assess one's *prior* probability *after* seeing the data? I think it is possible, but it is very difficult: a curious failing of the human mind is the ability to rationalize and explain virtually any observation, however false! Obviously, it is wrong to incorporate the same data into a posterior probability twice. So one must be able to say with some confidence, "This is what I thought before".

Alternatively, one can simply assess one's current probability after having seen the data, eschewing the use of Bayes's theorem and relying instead on one's internal analogue. We do this all the time. Our internal Bayes's theorem may not process information as well as does the real thing, and some of us may process it incredibly badly, but we are not likely to use the data twice. To check an assessor's ability in this regard, apply Bayes's theorem in reverse, dividing the posterior by the likelihood to yield the prior. This may not be possible and, if possible, when there are zeroes in the likelihood the result is not uniquely defined. If it is not possible then the assessor can be instructed on removing this inconsistency. In any case, it is appropriate to adjust the posterior if the assessor "never could have had that prior".

Still a third possibility is to find people who haven't seen the data and have them serve as surrogate assessors. This procedure can serve to educate the assessor, and can be combined with the other two procedures.

In the next section I discuss the possibility of, and diffculties associated with, making Bayesian inferences concerning hypotheses generated by the data. In Section 3 I define multiple comparisons and multiple tests. Section 4 draws a parallel with the so-called empirical Bayes problem and suggests this as a way to view some problems in multiple inference.

My goal throughout is not to give results that are immediately useful to the practitioner, but only to elucidate the major issues. So the examples and settings I use are rather simple. For useful Bayesian results in the area of multiple comparisons the reader is referred to Cornfield (1969), Dixon and Duncan (1975), DuMouchel (1988), Louis (1984), and Waller and Duncan (1969); also, see the discussion in Spiegelhalter and Freedman (1987, Section 5). In particular, DuMouchel's approach is very appealing; still, more work needs to be done in this area. Some multiple testing questions (especially (ii) of Section 3) are very difficult to address from a Bayesian point of view and to my knowledge there are no useful results in the literature. By their nature some multiple testing problems require very careful specification of prior probabilities and using a prior because it is convenient can lead to conclusions that are obviously wrong.

2. DATA DREDGING; SIMULTANEOUS LEARNING AND TESTING

Consider this scenario. A large study was conducted by randomly selecting 30-year-old men and following them for a long period of time. The investigators measured hundreds of variables at five-year intervals, with no particular plan for testing them all. Some conclusions are boringly predictable: men who were overweight tended to die at an earlier age and were generally less healthy; similarly for men who smoked cigarettes. But there's something new and quite unexpected: men who chewed gum regularly lived six years longer on average than men who never chewed gum! This difference is "highly statistically significant": nominal P-value $< .01$. Moreover, this difference persists upon adjusting for all available covariates.

As I have indicated, classical statisticians would consider the number of tests that had been carried out and suitably adjust the P-value upwards, for example, using Bonferroni's inequality. In particular, if the number of tests is sufficiently large, the result will no longer be significant. And if this number is not available then a good classical statistician would say that correct inferences regarding this issue require another study.

What about the Bayesian point of view? Figure 1 shows the likelihood function for μ, the mean increment in length of life (in years) as a reasult of chewing gum. This is reasonably approximated by $N(6, 2^2)$. In particular, the likelihood ratio of $\mu = 6$ (obviously the extreme case) compared with $\mu = 0$ is about 89.

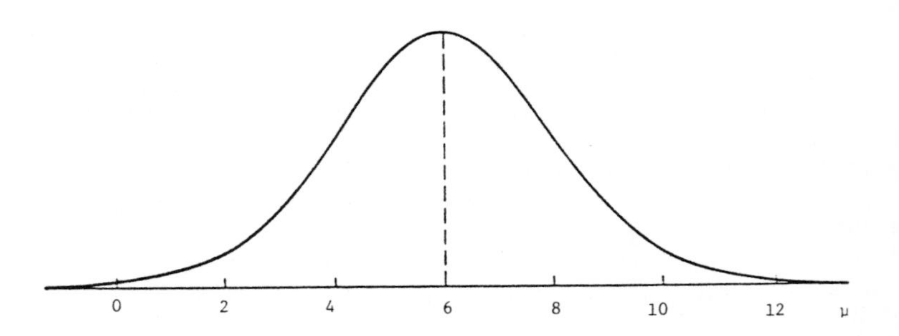

Figure 1. *Likelihood of* μ

My prior distribution on μ (which I can assess unencumbered by knowing the data because I also know that the data are fake!) is approximately $.6\delta_0 + .4N(-.1, 1.2^2)$—the probability of $\mu = 0$ is .6 and the rest of my probability is dispersed rather near 0. So I'm not convinced that chewing gum has no effect but I doubt that it has much. (The negative mean, $E(\mu) = -.04$ years, reflects my pessimism regarding the healthfulness of regular intake of sugar. This pessimism is partially balanced by the possibility that the miniscule amount of exercise one gets while chewing gum might be beneficial!)

To be somewhat more general, suppose that the prior distribution of μ is

$$p\delta_0 + \bar{p}N(\mu_0, \sigma_0^2), \tag{1}$$

where $\bar{p} = 1 - p$. Then the posterior distribution when the likelihood is $N(6, 2^2)$ is

$$(\mu \mid \text{data}) \sim p'\delta_0 + \bar{p}'N(\mu_1, \sigma_1^2) \tag{2}$$

where

$$\sigma_1^2 = 4\sigma_0^2/(4 + \sigma_0^2),$$
$$\mu_1 = \sigma_1^2\left[6/4 + \mu_0/\sigma_0^2\right],$$

and p' is defined by the posterior odds ratio:

$$\frac{p'}{\bar{p}'} = \frac{p}{\bar{p}} \cdot \frac{\sqrt{\sigma_0^2 + 4}}{2} \exp\left\{\frac{1}{2}\left[\frac{(\mu_0 - 6)^2}{\sigma_0^2 + 4} - 9\right]\right\}.$$

The mean of this distribution is $E(\mu \mid \text{data}) = \bar{p}'\mu$.

Substituting $\sigma_0 = 1.2, \mu_0 = -.1$, and $p = .6$ gives

$$(\mu|\text{data}) \sim .37\delta_0 + .63N(1.51, 1.03^2),$$

which has mean $E(\mu|\text{data}) = .95$ years. So the data has little effect on my rather strongly held opinion that chewing gum cannot increase one's life expectancy by anything like six years.

I will consider two alternative prior distributions. The first Bayesian is rather cavalier, and the second's prior has been affected by the data. These types of behavior are never good, but I want to show how bad they can be in the current setting. Though both priors are rather extreme to best make my point, I have kept the prior probability of $\mu = 0$ at $p = .6$ to facilitate comparison.

The Bayesian who goes overboard in being open-minded might choose $p = .6, \mu_0 = 0$, and $\sigma_0^2 = 10^2$ in (1). Substituting these into (2) gives

$$(\mu|\text{data}) \sim .09\delta_0 + .91N(5.77, 1.96^2),$$

which has mean $E(\mu|\text{data}) = 5.2$ years. Obviously, the data have substantially changed this prior. (Incidentally, the Bayesian who assumes $\sigma_0^2 \to \infty$ in (1) is actually being dogmatic rather than open-minded. For, $p' \to 1$ for any μ_0 and $p > 0$; and so the distribution of $(\mu \mid \text{data}) \to \delta_0$! Also, of course, the distribution of $(\mu|\text{data}) \to \delta_0$ as $\sigma_0^2 \to 0$, though the functional relation above gives $p' \to p$. (Neither of these results depends on the data.) As indicated by Edwards, Lindman, and Savage (1963), between these two extremes, p' has a minimum value which for these data and $\mu_0 = 0$ occurs at $\sigma_0^2 \doteq 5.7^2$; the minimum value of p' when $p = .6$ is .076, which is rather close to $p' = .09$ corresponding to $\sigma_0^2 = 10^2$.)

It is not an easy matter to put Figure 1 out of one's mind having seen it, or to ever be convinced of having put it out of mind! The Bayesian who forms an opinion after looking at the data and then updates via Bayes's theorem using *the same data* is obviously acting unreasonably. As an extreme case, suppose $p = .6$ and the prior mean and variance of $(\mu \mid \mu \neq 0)$ are the maximum likelihood estimates: $\mu_0 = 6$ and $\sigma_0^2 = 2^2$. Substituting these into (2) gives

$$(\mu|\text{data}) \sim .02\delta_0 + .98N(6, 2),$$

which has mean $E(\mu|\text{data}) = 5.9$ years. The prior weight of $p = .6$ on $\mu = 0$ has been essentially annihilated; such a Bayesian is pretty convinced that μ is near 6 years. The data have virtually the entire say when allowing one aspect of the prior to depend on the data.

I hope this example convinces you first that it *is* possible to test hypotheses using the data that generated them, and second that doing so is not without peril.

3. MULTIPLE COMPARISONS AND MULTIPLE TESTS

To set up some terminology and conventions for the next section, consider k populations with means $\mu_1, \ldots, \mu_k$. Observations $X_1, \ldots, X_k$ are available on these populations such that $EX_i = \mu_i$. Given $\mu_1, \ldots, \mu_k$ the X_i's are independent. Let $X_{(1)} \leq \cdots \leq X_{(k)}$ be the ordered observations and $\mu_{[i]} = EX_{(i)}$ (so $\mu_{[i]}$ is the mean of the population with the ith smallest observation—the $\mu_{[i]}$'s may not be ordered).

The following two problems involve multiple inferences:

Multiple Comparisons: Based on $X_1, \ldots, X_k$, make inferences concerning the relationships among the various μ_i's.

Multiple Tests: Based on $X_1, \ldots, X_k$, make inferences concerning the various μ_i's individually.

An example of the first is testing the hypothesis that $\mu_{[1]} = \mu_{[2]}$. An example of the second is testing the hypothesis that $\mu_{[1]} = 0$.

There are at least two types of multiple testing problems:

(i) The populations are k different treatments and the X_i's refer to the same measurement (blood pressure, e.g.), and

(ii) There is one treatment or experimental setting and the populations refer to k different measurements (blood pressure, heart rate, bilirubin levels, etc.)

Obviously, in a multiple comparisons problem, only type (i) is appropriate: one would hardly be interested in the difference between mean blood pressure and mean heart rate. Some of the discussion so far in this article deals with (ii) in the context of multiple tests. The next section focuses mainly on (i) in the context of multiple comparisons and multiple tests.

4. AN EMPIRICAL BAYES CONNECTION

Though the setup in this section is rather simple, the conclusions correspond rather closely with the way Bayesians should think about comparisons and tests when there are multiple treatments. For example, to make inferences about $\mu_{[1]}$ one may need to know all the data, not just $X_{(1)}$. Also $\mu_{[1]}$ is typically positively correlated with $X_{(i)}, i = 1, \ldots, k$.

Suppose that the treatment responses are

$$X_i \sim N(\mu_i, 1), i = 1, \ldots, k.$$

The μ_i's are unknown and so are themselves random variables, their joint distribution reflecting information about, and relationships among, the various treatment responses.

While most of this section treats problems of type (i) defined in the previous section, I want to consider type (ii) problems briefly. Suppose the X_i's are measurements on k different variables in the same experimental setting. It seems reasonable a priori to consider the possibility that the μ_i are independent. But one might also allow for the possibility that they are related. For example, a drug that decreases blood pressure is likely to affect (positively or negatively) heart rate, left ventricular ejection fraction, vascular resistance, etc. Such possibilities should be considered in multiple testing problems. It may be wrong to suppose independence but it would also be wrong to suppose that the μ_i's are positively correlated, say.

If the μ_i's are independent a priori then they will also be independent given $X_1, \ldots, X_k$. So in this case the posterior distribution of μ_i given $X_1, \ldots, X_k$ is simply the posterior distribution of μ_i given X_i, and making inferences about μ_i is a one-dimensional problem. In particular, *there is no multiple inference issue*: the distribution of $\mu_{[i]}$ depends on $X_1, \ldots, X_k$ only through $X_{(i)}$.

While it may be reasonable for someone to regard the means of k *variables* (type (ii)) as being independent, this seems less reasonable for k *treatments* (type (i)). Independence across treatments assumes very firm information about the treatments and the experimental setting. The careful probability assessor may well recognize that there is an underlying unknown effect which influences all observations similarly, irrespective of treatment. For example, consider a clinical trial involving several treatments for breast cancer. This disease continues to be diagnosed earlier and earlier in its cycle. Thus *every* treatment should be more effective now than it ever was before, though how much more effective would not be clear in advance. This is certainly true of no treatment because a patient "treated" earlier will obviously live longer after such "treatment". In addition, the entrance criteria and the way these criteria are administered by clinicians vary from one trial to the next; the differences can seem minor but still show up in the results very dramatically. Because individual trials tend to involve rather homogeneous populations, the various treatments used in the current trial may seem more like each other than a single treatment seems like itself in previous trials. (A consequence of these considerations is that there is no such thing as a "known" treatment, one whose effectiveness in an experiment can be predicted up to statistical error. Even if an experimenter is meticulous about ensuring that all aspects of the current trial duplicate those of a previous trial, time and its many covariates will be different.)

There are many ways that the μ_i can be dependent. One suggested by the previous paragraph is that $(\mu_1, \ldots, \mu_k)$ is a random sample from some distribution G, which is itself unknown. This is precisely the setup assumed in the "empirical Bayes" problem proposed by Robbins (1956). Converting one's available information about G into a probability distribution gives rise to what Deely and Lindley (1981) call a "Bayes empirical Bayes" problem; see also Berry and Christensen (1979). When I say empirical Bayes I mean Bayes empirical Bayes. The empirical Bayes objective is usually to estimate G or the various μ_i. The adaptation here is to hypothesis tests concerning the μ_i.

There is a large literature dealing with the use of "hierarchical priors" in which a prior distribution is assigned to G, the distribution of the μ_i's; see Berger (1985) for a list of references. Berger and Deely (1988) is particularly relevant for the use of hierarchical priors in multiple inference issues.

To keep things reasonably simple, suppose treatment i has either no effect ($\mu_i = 0$) or a known positive effect ($\mu_i = 1$). Distribution G has parameter p, which is the probability that any particular treatment has no effect:

$$(\mu_i \mid p) \sim p\delta_0 + \bar{p}\delta_1.$$

Conditional on p, the μ_i are independent, and so they are exchangeable. (This assumption is clearly inappropriate in type (ii) problems.) The proportion p of treatment with no effect is unknown, with a priori density uniform on $(0, 1) : \pi(p) = 1$, which then implies the initial distribution for G and for the μ_i. In particular, the marginal distribution of μ_i is

$$\mu_i \sim \frac{1}{2}\delta_0 + \frac{1}{2}\delta_1.$$

Consider X_1, the observed response to treatment 1. Its conditional distribution given p is

$$(X_1|p) \sim pN(0,1) + \bar{p}N(1,1);$$

unconditionally,

$$X_1 \sim \frac{1}{2}N(0,1) + \frac{1}{2}N(1,1).$$

The density of p on (0,1) given X_1 is

$$\begin{aligned}\pi(p|X_1) &= 2\frac{p \cdot \exp\left[-\frac{1}{2}X_1^2\right] + \bar{p} \cdot \exp\left[-\frac{1}{2}(X_1 - 1)^2\right]}{\exp\left[-\frac{1}{2}X_1^2\right] + \exp\left[-\frac{1}{2}(X_1 - 1)^2\right]} \\ &= 2\frac{p + \bar{p} \cdot \exp\left[X_1 - 1/2\right]}{1 + \exp\left[X_1 - 1/2\right]}.\end{aligned}$$

If, for example, $X_1 = 1/2$, halfway between the two candidates for μ_1, then $\pi(p|X_1 = = 1/2) = \pi(p)$, corresponding with the obvious fact that $x_1 = 1/2$ is noninformative as regards $\mu_1 = 0$ vs $\mu_1 = 1$. Also, $\pi(p|X_1) \to 2p$ or $2\bar{p}$ according as $X_1 \to -\infty$ or $X_1 \to +\infty$, equivalent to actually observing $\mu_1 = 0$ and $\mu_1 = 1$, respectively.

Now consider the distribution of μ_1 given X_1. The new probability of $\mu_1 = 0$ is

$$P(\mu_1 = 0|X_1) = \int_0^1 \frac{2pdp}{1 + \exp(X_1 - 1/2)} = \frac{1}{1 + \exp(X_1 - 1/2)}, \tag{3}$$

a decreasing function of X_1. So we have

$$(\mu_1|X_1) \sim \frac{1}{1 + \exp(X_1 - 1/2)}\delta_0 + \frac{\exp(X_1 - 1/2)}{1 + \exp(X_1 - 1/2)}\delta_1.$$

Again, $X_1 = 1/2$ leaves the prior unchanged.

Now consider responses X_1 and X_2 on treatments 1 and 2. We have

$$(X_1, X_2|p) \sim [pN(0,1) + \bar{p}N(1,1)]^2$$

and so

$$\pi(p|X_1, X_2) \propto \prod_{i=1}^{2} [p + \bar{p} \cdot \exp(X_i - 1/2)] .$$

It follows that

$$\begin{aligned} P(\mu_1 = \mu_2 = 0|X_1, X_2) &\propto 2 \\ P(\mu_1 = 0, \mu_2 = 1|X_1, X_2) &\propto \exp(X_2 - 1/2) \\ P(\mu_1 = 1, \mu_2 = 0|X_1, X_2) &\propto \exp(X_1 - 1/2) \\ P(\mu_1 = \mu_2 = 1|X_1, X_2) &\propto 2 \cdot \exp(X_1 + X_2 - 1). \end{aligned}$$

As a consequence,

$$P(\mu_1 = 0|X_1, X_2) = \frac{2 + \exp(X_2 - 1/2)}{2 + \exp(X_2 - 1/2) + \exp(X_1 - 1/2) + 2 \cdot \exp(X_1 + X_2 - 1)}. \tag{4}$$

This tends to 1 or 0 according as $X_1 \to -\infty$ or $+\infty$. And it tends to $2/[2+\exp(X_1 - 1/2)]$ or $1/[1 + 2 \cdot \exp(X_1 - 1/2)]$ according as $X_2 \to -\infty$ or $+\infty$. These latter conclusions are the same as the conditional distribution of μ_1 given X_1, having already conditioned on μ_2 (the conditional distribution of μ_1 given μ_2 being

$$\begin{aligned} (\mu_1|\mu_2) &\sim \frac{2}{3}\delta_0 + \frac{1}{3}\delta_1 \quad \text{if } \mu_2 = 0 \\ &\sim \frac{1}{3}\delta_0 + \frac{2}{3}\delta_1 \quad \text{if } \mu_2 = 1). \end{aligned}$$

Compare (3) with (4). Obviously, they are equal if $X_2 = 1/2$—there is no information about p in $X_2 = 1/2$. When X_2 is greater than 1/2 then (4) > (3): if treatment 2 gives a large response then it is more difficult to decide that treatment 1's effect is small. For example, if $X_2 = 2$ then .43 has to be subtracted from X_1 to keep the same probability of $\mu_1 = 0$. On the other hand, if $X_2 = -2$ then to keep the probability of $\mu_1 = 0$ the same requires that .58 be added to X_1.

Consider the impact that this has on the question of multiple tests. A researcher reports that $X_{(1)} = -1$. If $k = 1$ then the probability of $\mu_{[1]} = 0$ from (3) is $1/(1+\exp(-3/2)) = .82$. But if $k = 2$ then the probability of $\mu_{[1]} = 0$ from (4) is

$$P\left(\mu_{[1]} = 0|X_{(1)} = -1, X_{(2)}\right) = \frac{2 + \exp\left(X_{(2)} - 1/2\right)}{2 + \exp\left(X_{(2)} - 1/2\right) + \exp(-3/2) + 2 \cdot \exp\left(X_{(2)} - 2\right)}.$$

Figure 2 shows this as a function of $X_{(2)}$. For comparison, Figure 3 shows the posterior joint probabilities of $\mu_{[1]}$ and $\mu_{[2]}$. Evidently, if $X_{(1)}$ is much less than $X_{(2)}$ then the evidence in favor of $\mu_{[1]} = 0$ is not as strong as if $X_{(1)}$ and $X_{(2)}$ were both small.

The joint probabilities of $\mu_{[1]}$ and $\mu_{[2]}$ are especially relevant for the question of multiple comparisons. Figure 4 shows how much less $\mu_{[1]}$ is expected to be than $\mu_{[2]}$ as a function of $X_{(2)}$. Figure 4 also shows the corresponding difference ignoring the relationship between $\mu_{[1]}$ and $\mu_{[2]}$. The two curves in Figure 4 are analogous to the classical statistician considering

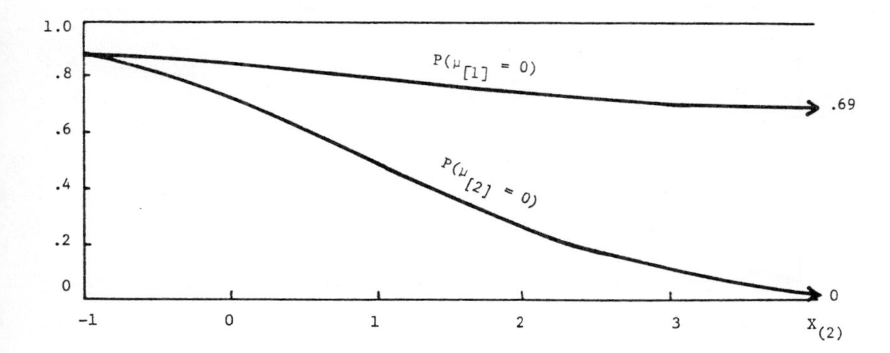

Figure 2. *Posterior probabilities of* $\mu_{[1]} = 0$ *and* $\mu_{[2]} = 0$ *given* $X_{(1)} = -1$ *as a function of* $X_{(2)}$.

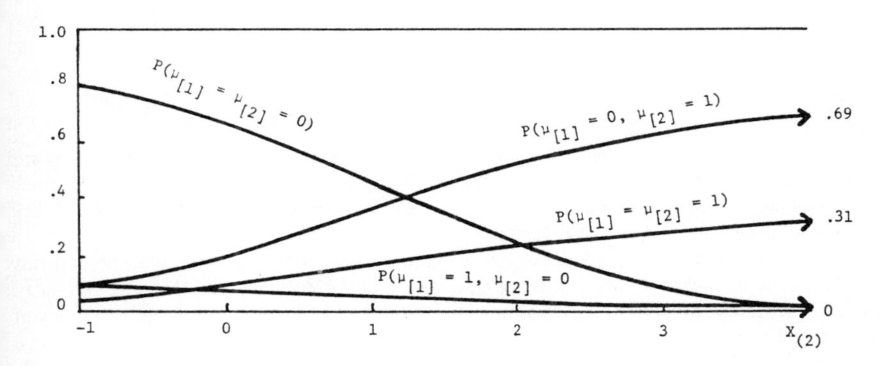

Figure 3. *Posterior probabilities of* $\mu_{[1]}, \mu_{[2]}$ *given* $X_{(1)} = -1$ *as a function of* $X_{(2)}$.

and not considering the multiple comparisons question. Obviously, ignoring the relationship exaggerates the difference between the means of $\mu_{[1]}$ and $\mu_{[2]}$.

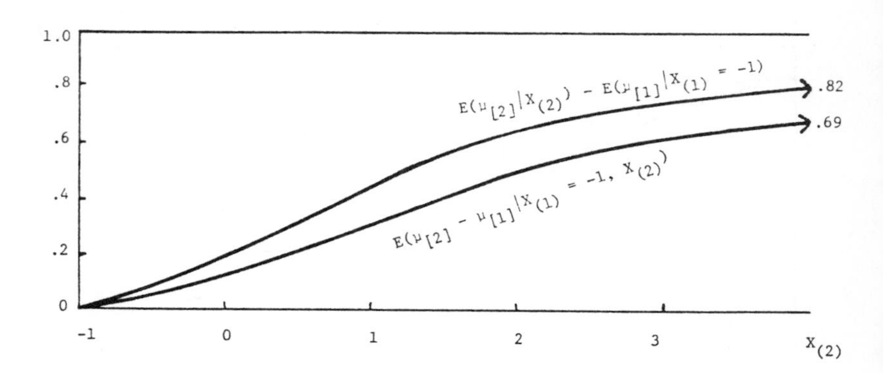

Figure 4. *Difference between posterior means of $\mu_{[2]}$ and $\mu_{[1]}$ recognizing and not recognizing the relationship between μ_1 and μ_2.*

The more general case of making inferences about the μ_i's given $X_1, \ldots, X_k$ involves more complicated calculations, but no new ideas. For larger k any inference about μ_1 (or $\mu_{[1]}$) depends less on X_1 (or $X_{(1)}$) and more on responses to the other treatments. For example,

$$P(\mu_1 = 0 | X_1, X_2 = \cdots = X_k \to -\infty) = \frac{k}{k + \exp(X_1 - 1/2)}$$

and

$$P(\mu_1 = 0 | X_1, X_2 = \cdots = X_k \to +\infty) = \frac{1}{1 + k \cdot \exp(X_1 - 1/2)}.$$

So it is easier to conclude that μ_1 is small when the responses to the other treatments are small but it is difficult (though not impossible!) to conclude that μ_1 is small when the other responses are large. In particular, given X_1 and $X_2 = \cdots = X_k \to= \infty$, for the probability of $\mu_1 = 0$ to be greater than 1/2 requires $X_1 < 1/2 - \log k$.

An empirical Bayes approach which is quite promising for the multiple comparisons problem uses mixtures of Dirichlet processes (Antoniak 1974, Berry and Christensen 1979). In this approach, estimating G and the various μ_i requires finding the posterior probabilities of all possible combinations of equality and inequality among the μ_i. For example, when $k = 3$ there are five possibilities: $\mu_1 \neq \mu_2 \neq \mu_3 \neq \mu_1, \mu_1 = \mu_2 \neq \mu_3, \mu_1 = \mu_3 \neq \mu_2, \mu_1 \neq \mu_2 = \mu_3$, and $\mu_1 = \mu_2 = \mu_3$. One then calculates $P(\mu_1 = \mu_2 | X_1, X_2, X_3)$, say, by adding the second and fifth of these. A less than attractive aspect of using mixtures of Dirichlet processes is that the number of terms in the mixture increases very fast as a function of k (Berry and Christensen 1979).

The empirical Bayes approach in settings of multiple comparisons and multiple tests gives results which incline toward the classical statistical view. For example, $X_{(1)}$ and $X_{(2)}$ can be further apart than would be expected by the naive analyst and still be consistent with the null hypothesis $\mu_{[1]} = \mu_{[2]}$. And a researcher cannot simply report $X_{(1)}$ as an estimate of $\mu_{[1]}$

or as a statistic for tests concerning $\mu_{[1]}$; the rest of the data $(X_{(2)}, \ldots, X_{(k)})$ also contains relevant information. But while classical "adjustments" depend only on k, empirical Bayes adjustments depend on the actual data. In analogy with classical adjustments, there would also be a Bayesian adjustment to inferences concerning $\mu_{[1]}$ if for some reason the Bayesian does not know all the data but only knows $X_{(1)}$ and k. I have not addressed this problem here.

5. CONCLUSIONS

From a Bayesian point of view it is possible to dredge data to generate hypotheses and then to test these hypotheses using the same data. However, the path to such inferences is perilous.

Correcting for comparisons and tests involving multiple treatments, so widely espoused by classical statisticians, has an analogue in Bayesian statistics. Namely, one assumes that the treatments are themselves sampled from an unknown distribution, as in the empirical Bayes problem. When the k treatments have means μ_i that are exchangeable a priori, the following rough interpretation is consistent with the empirical Bayes approach, and it seems reasonable in the problem at hand. The posterior distribution of μ_i is pulled toward treatment i response X_i, but also in the directions of the responses to the other treatments. For any pair of treatments (i, j), the estimated distance between μ_i and μ_j given $X_1, \ldots, X_k$ is less than that between X_i and X_j but the signs of the two differences are the same. In particular, if $X_i = X_j$ then μ_i and μ_j are exchangeable a posteriori as well as a priori.

The empirical Bayes approach is more relevant for simultaneous inferences concerning many treatments, problem (i) of Section 3, than concerning many variables (same "treatment"), problem (ii) of Section 3. Regarding the latter, I do not see that an adjustment along the lines of Section 4 would ever be appropriate because the μ_i's cannot be exchangeable a priori. An appropriate analysis must take into account one's prior information concerning the relationships among the variables.

ACKNOWLEDGEMENTS

This paper was supported by National Science Foundation Grant DMS 8505023. Professors J. O. Berger and R. Christensen made helpful comments.

REFERENCES

Antoniak, C. E. (1974). Mixtures of Dirichlet processes with applications to Bayesian nonparametric problems. *Ann. Statist.* **2**, 1152–1174.

Berger, J. O. (1985). *Statistical Decision Theory and Bayesian Analysis.* New York: Springer-Verlag.

Berger, J. O. and Berry, D. A. (1988). The relevance of stopping rules in statistical inference (with discussion). *Statistical Decision Theory and Related Topics IV,* Vol. 1, 29–72. New York: Springer-Verlag.

Berger, J. O. and Deely, J. (1988). A Bayesian approach to ranking and selection of related means with alternatives to AOV methodology. (To appear in *J. Amer. Statist. Assoc.* **83**.)

Berry, D. A. (1985). Interim analysis in clinical trials: Classical vs. Bayesian approaches. *Statistics in Medicine* **4**, 521–526.

Berry, D. A. (1987). Interim analysis in clinical trials: The role of the likelihood principle. *The American Statistician* **41**, 117–122.

Berry, D. A. and Christensen, R. (1987). Empirical Bayes estimation of a binomial parameter via mixtures of Dirichlet processes. *Ann. Statist.* **7**, 558–568.

Cornfield, J. (1969). The Bayesian outlook and its application. *Biometrics* **24**, 617–657.

Deely, J. J. and Lindley, D. V. (1981). Bayes empirical Bayes. *J. Amer. Statist. Assoc.* **76**, 833–841.

Dixon, D. O. and Duncan, D. B. (1975). Minimum Bayes risk t-intervals for multiple comparisons. *J. Amer. Statist. Assoc.* **68**, 117–130.

DuMouchel, W. (1988). A Bayesian model and a prior elicitation procedure for multiple comparisons. (in this volume).

Edwards, W., Lindman, H. and Savage, L. J. (1963). Bayesian statistical inference for psychological research. *Psychological Review* **70**, 193–242.

Freedman, D. A. (1983). A note on screening regression equations. *The American Statistician* **37**, 152–155.

Louis, T. A. (1984). Estimating a population of parameter values using Bayes and empirical Bayes methods. *J. Amer. Statist. Assoc.* **79**, 393–398.

Robbins, H. (1956). An empirical Bayes approach to statistics. *Proceedings of the Third Berkeley Symposium on Mathematical Statistics and Probability* **1**, 157–163.

Spiegelhalter, D. J., and Freedman, L. S. (1987). Bayesian approaches to clinical trials. (in this volume).

Waller, R. A. and Duncan, D. B. (1969). A Bayes rule for the symmetric multiple comparisons problem. *J. Amer. Statist. Assoc.* **64**, 1484–1503.

DISCUSSION

P. J. KEMPTHORNE* (*Sloan School of Management, M.I.T.*)

Professor Berry raises a most interesting issue in Section 1 concerning the use of prior distributions specified a posteriori. To fix notation, define: X, data to be observed; μ, an unknown parameter indexing the possible distributions of X; π, the distribution of μ prior to observing X, and π^*, the distribution of μ posterior to observing X. Valid inferences by a statistician about μ given X, including tests of data-generated hypotheses, are based on π^*. The problem for the statistician is how to obtain π^* when the prior distribution was not assessed before observing the data. Professor Berry discusses three schemes.

- Specify π, claiming "This is what I thought *before*", and deduce π^* given X.
- Specify π^* directly, claiming "This is what I believe *now*", and, if necessary, deduce π (modulo zeros in the likelihood of X).
- Specify π, using surrogate assesors and deduce π^* given X.

The logic of the last scheme is questionable. Why should the beliefs of surrogates be an adequate substitute for one's own? Even if the incorporation of surrogates' opinions were appropriate to the construction of π^*, perhaps because of their knowledge or previous experience, its implementation would be complex. The statistician would need an adequate model of surrogate behavior. This thorny issue has been the subject of considerable research and discussion. See, for example, French (1985), Lindley (1985), and Kempthorne and Mendel (1987).

The first scheme, specifying π after observing X, is arguably unrealistic. Bayesians are (or should be) experienced at specifying their current beliefs with a probability distribution. Techniques for such specification exist because current beliefs can be measured directly. Past beliefs cannot be measured directly. It is not surprising that such an indirect approach could result in observers suggesting that the statistician is "cheating".

I find most appealing the author's suggestion of assessing one's posterior distribution directly. After observing X, this simply requires the statistician to characterize his or her current beliefs. Observers may be interested in how such beliefs are a composition of sample and prior information. As the author remarks, using the likelihood of the data, the implied prior distribution can be deduced, modulo zeros in the likelihood. Also, the coherency check, provided by the statistician judging the acceptability of such prior beliefs, would be useful in practice.

Professor Berry comments that some individuals' internal processing of Bayes Theorem may not be good enough for this approach to be of any use. In contrast, I believe that for many people, their assessments of beliefs, a posteriori, are better than applying Bayes Theorem blindly, given prior distributions which only approximate their true past beliefs.

*Supported by Grant DMS-8605819 from the National Science Foundation.

An approach I find appealing for implementing this scheme of specifying the posterior distribution directly, is as follows:

1. Specify $\mathcal{C} = \{\tilde{\pi}'\text{s}\}$, a class of possible prior distributions.
2. Calculate: $\mathcal{C}^* \equiv \{\tilde{\pi}^* | X, \tilde{\pi}; \tilde{\pi} \in \mathcal{C}\}$.
3. Choose $\hat{\pi}^*$ in $\mathcal{C}^*$ (and deduce $\hat{\pi}$, if necessary).

The critical aspect in this approach is how to define $\mathcal{C}$, the class of possible priors. In linear models, hierarchical priors are often convenient and rich enough to approximate prior beliefs well. For some individuals, the choice of the distribution $\hat{\pi}^* \in \mathcal{C}$ to summarize one's posterior beliefs might coincide with existing approaches of empirical Bayes analysis; see e.g., Morris (1983). Such an approach is clearly coherent. Furthermore, when calculating posterior distributions for their own purposes, aspiring Bayesians could justifiably act in this manner.

Turning to Section 2, perhaps Professor Berry would share further his insight or experience regarding the potential "peril" and "possibility" of testing hypotheses using the data that generated them. I did not find helpful the numerical example comprising this section of the paper. The perils of a sloppy choice of the prior distribution are obvious whenever a posterior distribution is sensitive to the specification of the prior distribution.

Defining multiple comparisons and multiple testing problems was potentially the most important section of the paper. The definitions given in Section 3 are too vague. I advocate defining these problems as formal Bayesian decision problems. Historically, a number of issues motivated various approaches to problems of simultaneous inference, including controlling experiment-wise versus comparison-wise error rates in testing and providing greater protection for a "global" null hypothesis. The Bayesian decision-theoretic perspective provides a compelling framework for incorporating such objectives and might convince practitioners to use Bayesian methods.

Miller (1981) provides an excellent overview of classical approaches to multiple comparisons, including those of Duncan, Scheffe, and Tukey, the three major contributors to current theory. Interestingly, it is Duncan who advocated a Bayesian decision-theoretic approach to this problem more than twenty-five years ago. The articles by Duncan (1961), (1965), (1975), Waller and Duncan (1969), Dixon and Duncan (1975), and work by Aleong (1974), (1980), comprise much of the Bayesian research in this area.

Many of these papers assume a balanced one-way analysis of variance model for Gaussian data from k populations and use an additive bilinear loss function for the collection of all pairwise component problems $\{P_{ij} :$ deciding whether $\mu_i \le \mu_j$ or $\mu_i > \mu_j;\ i, j = 1, \ldots, k\}$. Duncan (1975) provides an excellent review of this approach as well as details for an empirical Bayes implementation. The empirical Bayes techniques for testing and interval estimation are appealing because they derive from models typically assumed for data in a one-way analysis of variance, and the hypotheses being tested are not necessarily sharp hypotheses. Furthermore, the similarity of the Bayes rule to Fisher's Least-Significant-Difference test might increase the appeal of this approach to those with a non-Bayesian background.

In contrast, I find the empirical Bayes approach suggested in Section 4 of the paper less than promising. The use of sharp null hypotheses and mixtures of Dirichlet processes is likely to be unrealistic and cumbersome in many applications.

One of the key assumptions to the Bayesian decision-theoretic approaches cited above is the additivity of the loss with respect to incurred losses for each of the pair-wise problems P_{ij}. Further research might investigate the applicability of alternate loss structures. In a Bayesian analysis of the variable-selection problem in regression, Lindley (1968) uses a loss function equal to squared error of estimation plus complexity cost depending on the subset of explanatory variables used in the fitted regression. Identifying components of the regression parameter with contrasts among population means in a one-way analysis of variance, an interesting class of loss functions results from assuming different forms for the complexity cost function; for example, additive, sub-additive or super-additive functions of subsets of contrasts. Without pursuing the details here, we note that Bayes procedures similar to the classical methods of Tukey and Scheffe would result for certain definitions of the complexity cost function.

It is important to note the relevance of past and current research in ranking and selection. Such problems are often appropriate formulations for applications involving multiple comparisons or multiple testing. Many of the results in this area derive substantially from Bayesian decision theory. Gupta and Huang (1981) review developments up to 1980 and recent work includes that of Berger and Deely (1986) and Fong (1987). Existing frequentist approaches are dissatisfying to many because they are overly conservative. Consequently, the potential is high for significant Bayesian contributions to the practice of multiple comparisons and multiple testing. I commend Professor Berry for stimulating research in this direction.

W. POLASEK (*University of Basel*)

To what extent can multiple comparisons replace multiple tests? Multiple tests often arise in econometric model checking: autocorrelation, heteroscedasticity, structural change, etc. What would be a coherent Bayesian perspective?

REPLY TO THE DISCUSSION

to Professor Kempthorne:

I want to thank Professor Kempthorne for his insightful comments. I agree with much of what he has to say, but I do have a few reservations. I shall address them in what I perceive to be the order of their importance rather than in the order they appear in Professor Kempthorne's comments.

Professor Kempthorne advocates defining multiple comparisons and multiple tests as "formal Bayesian decision problems". While I too find comfort in a formal decision-theoretic structure, most statisticians do not. We Bayesians should not build a wall of decision theory between us and statistical practitioners. The distance between Bayesians and non-Bayesian practitioners results mainly from Bayesians' use of probability to describe all uncertainty, and from practitioners' rejection of the requisite subjective interpretation of probability. We should not insist on a decision-theoretic approach, or we may never close the gap. Decision theory "provides a compelling framework" for Professor Kempthorne, but there is no chance that it will "convince practitioners to use Bayesian methods", at least not the practitioners I know. My approach in the paper has been to describe the posterior distribution of the various unknowns. The appropriate Bayes procedures can be calculated from this distribution in a straightforward way for any specified loss function.

In this same vein Professor Kempthorne suggests that my definitions of multiple comparisons and multiple testing problems are too vague. Perhaps they are. While preparing this paper I spent some time mulling over various issues in the area of multiple inference. It seems hard to believe now, but it was a revelation when I eventually realized that these two problems, while seemingly two branches on the same tree, are not even on the same planet! In multiple comparisons problems the variables in question are *comparable*, while in multiple testing problems they may not be. As I tried to make clear in my paper, this distinction gives rise to incredible differences in analysis. In particular, there is a wealth of more or less standard methods (e.g., the use of hierarchical priors) for addressing the former and no standard way for addressing the latter.

Professor Kempthorne makes two (apparently separate) comments about Section 4; I disagree with both. Bayes empirical Bayes methods and the use of hierarchical priors seems to me a natural and promising way of viewing problems of multiple comparisons. I am not sure what Professor Kempthorne means regarding "The use of sharp null hypotheses". It seems to be "in" to criticize the assumption of a sharp null hypothesis. I happen to think most such criticism is unwarranted and that there are many settings in which sharp nulls are very realistic—but none appear in Section 4. I did use a two-point distribution for values of μ, and this is clearly unrealistic. I wanted to "keep things reasonably simple" while illustrating the ideas and the calculations: "My goal throughout is not to give results which are immediately useful to the practitioner, but only to elucidate the major issues". In an actual problem the practitioner must assess his own prior distribution of the values of μ.

I agree with Professor Kempthorne that mixtures of Dirichlet processes are very cumbersome. But I continue to feel, as I said in Section 4, that they represent a natural and promising approach to problems of multiple comparisons. Professor Kempthorne suggests that I share further my insight or experience regarding the potential "peril" and "possibility" of testing hypotheses using data that generated them. I am sorry that my example was not sufficiently illustrative. Perhaps any single example is misleading in that it focuses on a tree rather than on the forest. I know of no more fundamental tenet of applied statistics than this: Because there are so many possibilities, most data sets are unusual in some unexpected regard; to test whether such an observation is real rather than artifactual requires another data set. This tenet is far from silly. I tried to show that the Bayesian approach provides a legitimate alternative to collecting more data. But a posterior has to assessed while keeping in mind that unusual things usually happen! I would have to be convinced that there exists even a single human being who can do this well. There may be some, but such inferences are—if not "perilous"—built on very shaky ground.

Regarding the use of surrogates to help assess one's prior distribution after having seen some data, I did not mean to suggest that I would simply accept a surrogate's prior. How much I weigh a surrogate's prior depends on how well-informed and able I think the surrogate is, and on my perception of the surrogate's prior in comparison with other information at my disposal. I did not have a formal analysis in mind; I thank Professor Kempthorne for the several references in this regard.

Professor Kempthorne incorrectly states that I said some individuals' internal processing a la Bayes's theorem may not be good enough for this approach to be of any use. Not only is it of use, we do this all the time—as I indicated in the paper. What I did say is that some of us process information "incredibly badly". (Research of Ward Edwards and other psychologists verifies this.) Still, I wholeheartedly agree with Professor Kempthorne that processing internally can be better than applying Bayes's theorem blindly.

to Professor Polasek:

I turn to Professor Polasek's questions. Unfortunately, the application of standard Bayesian hierarchical priors to problems of multiple comparisons do not carry over to multiple testing. In problems of multiple testing one must take pains in assessing a prior on $(\mu_1, \mu_2, \ldots, \mu_k)$. If the μ's are independent a priori then there is no multiple testing issue. So it is essential that the prior distribution adequately reflect one's knowledge of the interdependencies among the μ's. I am very interested in this issue but know of no literature that deals with it from a Bayesian perspective.

REFERENCES IN THE DISCUSSION

Aleong, J. (1980). Alternate Bayes rules for the multiple comparison problem. *1980 Statistical Computing Section Proceedings of the American Statistical Association*, 268–270.

Aleong, J. (1974). *Aspects of Simultaneous Inference*. PhD thesis, Iowa State University.

Berger, J. O. and Deely, J. (1986). Hierarchical Bayesian selection and ANOVA of means. Presented at NSF-CBMS Regional Conference on Hierarchial and Empirical Bayesian Models and Methods, Bowling Green State University.

Dixon, D. O. and Duncan, D. B. (1975). Minimum Bayes risk t-intervals for multiple comparisons. *J. Amer. Statist. Assoc.* **70**, 822–831.

Duncan, D. B. (1961). Bayes rules for a common multiple comparisons problem and related student-*t* problems. *Ann. Math. Statist.* **32**, 1013–1033.

Duncan, D. B. (1965). A Bayesian approach to multiple comparisons. *Technometrics* **7**, 171–222.

Duncan, D. B. (1975). *t*-tests and intervals for comparisons suggested by the data. *Biometrics* **31**, 339–359.

Fong, D. (1987). Ranking and estimation of exchangeable means in balanced and unbalanced models: A Bayesian approach. *Ph.D thesis*, Department of Statistics, Purdue University.

French, S. (1985). Group consensus probability distributions: a critical survey. *Bayesian Statistics* **2**. (J. M. Bernardo, M. H. DeGroot, D. V. Lindley and A. F. M. Smith, eds.). Amsterdam: North-Holland , 183–202.

Gupta, S. S. and Huang, D. Y. (1981). *Multiple Statistical Decision Theory: Recent Developments*. New York: Springer-Verlag.

Kempthorne, P. J. and Mendel, M. E. (1987). Adjusting experts' probabilistic forecasts: the use and implication of exchangeability assumptions in the calibration problem. *Tech. Rep.*; Sloan School of Management, M. I. T.

Lindley, D. V. (1968). The choice variables in multiple regression. *J. Roy. Statist. Soc. B* **30**, 31–51.

Lindley, D. V. (1985). Reconciliation of discrete probability distributions. *Bayesian Statistics* **2**. (J. M. Bernardo, M. H. DeGroot, D. V. Lindley and A. F. M. Smith, eds). Amsterdam: Nort-Holland, 375–390.

Miller, R. G. Jr. (1981). *Simultaneous Statistical Inference*. New York: Springer-Verlag.

Morris, C. (1983). Parametric empirical Bayes inference: theory and applicatons. *J. Amer. Statist. Assoc.* **78**, 47–65.

Waller, R. A. and Duncan, D. B. (1969). A Bayes rule for the symmetric multiple comparisons problem. *J. Amer. Statist. Assoc.* **64**, 1484–1503, Corrigenda: **67** (1972), 253–255.

BAYESIAN STATISTICS 3, pp. 95–110
J. M. Bernardo, M. H. DeGroot, D. V. Lindley and A. F. M. Smith, (Eds.)

The Infinite Regress and its Conjugate Analysis

A. P. DAWID
University College London

SUMMARY

We consider the problem of regression on an unlimited number of explanatory variables. This is approached through its finite regression counterparts. Prior distributions for the parameters of these distributions must be suitably compatible, and it is shown how this may be achieved through using the conjugate inverse Wishart prior for an underlying dispersion matrix. The implications of such a prior assignment are considered in detail, and shown to have some disturbing aspects. In particular they incorporate strong beliefs about the deterministic predictability of the response.

Keywords: LINEAR REGRESSION; BAYESIAN INFERENCE; CONJUGATE PRIOR; INVERSE WISHART DISTRIBUTION; PREDICTIVE DISTRIBUTION; DETERMINISM.

1. SAMPLING DISTRIBUTIONS

Consider a response variable X_0, and a potentially infinite sequence $(X_1, X_2, \ldots)$ of explanatory variables. Let $X_{(p)}$ denote $(X_1, X_2, \ldots, X_p)$. We suppose that, for each $p \geq 0$, we have a linear homoscedastic normal regression of X_0 on $X_{(p)}$:

$$X_0|X_{(p)} \sim N(X_{(p)}B_p, \Gamma_p) \tag{1.1}$$

with (for simplicity) zero intercept. In (1.1), B_p is $(p \times 1)$ and $\Gamma_p > 0$.

The different models (1.1), as p varies, are not to be considered as alternatives, only one of which can be "true". On the contrary, (1.1) is supposed to hold for every p simultaneously. This would be so if $(X_0, X_1, \ldots)$ had a joint normal distribution with zero means and unknown covariance structure, and this property may also be shown to be necessary. We can therefore replece the collection of *undermodels* given by (1.1) by a single *overmodel*:

$$(X_0, X_1, \ldots) \sim N(0^T, \Sigma) \tag{1.2}$$

where Σ is an unknown infinite matrix with (i,j) entry $\sigma_{ij} = \text{cov}(X_i, X_j)$ $(i, j \geq 0)$. In other words, for each $p \geq 0$, the distribution of $(X_0, X_1, \ldots, X_p)$ is $N(0^T, \Sigma_q)$, Σ_q being the leading $(q \times q)$ submatrix of Σ $(q = 1 + p)$. We shall assume that every Σ_q is positive definite.

To relate the parameters of (1.1) and (1.2), we first record the following familiar result.

Lemma 1. Let Y_1 $(1 \times p_1)$ and Y_2 $(1 \times p_2)$ have a joint multivariate normal distribution $(Y_1, Y_2) \sim N(0^T, \Phi)$, with Φ $(q \times q)$ positive definite $(q = p_1 + p_2)$. Partition Φ as

$$\Phi = \begin{matrix} & p_1 & p_2 \\ p_1 & \Phi_{11} & \Phi_{12} \\ p_2 & \Phi_{21} & \Phi_{22} \end{matrix}$$

and define $B = \Phi_{22}^{-1}\Phi_{21}$ and $\Gamma = \Phi_{11.2}$ (*viz.* $\Phi_{11} - \Phi_{12}\Phi_{22}^{-1}\Phi_{21}$). Then $Y_2 \sim N(0^T, \Phi_{22})$, while $Y_1|Y_2 \sim N(Y_2B, \Gamma)$. ◁

In the present context, take $Y_1 = X_0, Y_2 = X_{(p)}$, $\Phi = \Sigma_q$ $(q = 1+p)$, and write the partition of Σ_q as

$$\Sigma_q = \begin{matrix} & 1 & p \\ 1 & \Sigma_{00} & \Sigma_{0p} \\ p & \Sigma_{p0} & \Sigma_{pp} \end{matrix}.$$

Then (1.1) and (1.2) are related by:

$$B_p = \Sigma_{pp}^{-1}\Sigma_{p0} \tag{1.3}$$

$$\text{and} \quad \Gamma_p = \Sigma_{00.p} \ (\textit{viz.}\ \Sigma_{00} - \Sigma_{0p}\Sigma_{pp}^{-1}\Sigma_{p0}). \tag{1.4}$$

We now investigate the relationship between (B_p, Γ_p) and (B_{p_1}, Γ_{p_1}) $(p_1 > p)$. Write X_+ for $(X_{p+1}, \ldots, X_{p_1})$, q_1 for $1 + p_1$, and partition

$$\Sigma_q = \begin{matrix} & 1 & p & p_1 - p \\ 1 & \Sigma_{00} & \Sigma_{0p} & \Sigma_{0+} \\ p & \Sigma_{p0} & \Sigma_{pp} & \Sigma_{p+} \\ p_1 - p & \Sigma_{+0} & \Sigma_{+p} & \Sigma_{++} \end{matrix}. \tag{1.5}$$

The residual of X_0 after allowing for $X_{(p)}$ is $X_{0.p} = X_0 - X_{(p)}B_p \sim N(0, \Gamma_p)$, independently of $X_{(p)}$. Likewise we have $X_{+.p} = X_+ - X_{(p)}\Sigma_{pp}^{-1}\Sigma_{p+} \sim N(0, \Sigma_{++} - \Sigma_{+p}\Sigma_{pp}^{-1}\Sigma_{p+}) = N(0, K)$ say, independently of $X_{(p)}$. The partial covariance between X_+ and X_0, allowing for $X_{(p)}$, is $\text{cov}(X_{+.p}, X_{0.p}) = \Sigma_{+0} - \Sigma_{+p}\Sigma_{pp}^{-1}\Sigma_{p0} = \Sigma_{+0} - \Sigma_{+p}B_p$. Since Σ_{q_1} is assumed non-singular, so is K, and the regression of $X_{0.p}$ on $X_{+.p}$ therefore has the form

$$X_{0.p} - X_{+.p}K^{-1}(\Sigma_{+0} - \Sigma_{+p}B_p) \sim N(0, \Gamma_p - (\Sigma_{0+} - B_p^T\Sigma_{p+})K^{-1}(\Sigma_{+0} - \Sigma_{+p}B_p)), \tag{1.6}$$

independently of $(X_{(p)}, X_{+.p})$, and hence of $X_{(p_1)}$. Resubstituting for $X_{0.p}$ and $X_{+.p}$ yields the regression of X_0 on $X_{(p_1)} = (X_{(p)}, X_+)$:

$$\begin{aligned} X_0 \sim\ & X_{(p)}\left[B_p - \Sigma_{pp}^{-1}\Sigma_{p+}K^{-1}(\Sigma_{+0} - \Sigma_{+p}B_p)\right] \\ & + X_+K^{-1}(\Sigma_{+0} - \Sigma_{+p}B_p) + N(0, \Gamma_p - (\Sigma_{0+} - B^T\Sigma_{p+})K^{-1}(\Sigma_{+0} - \Sigma_{+p}B_p)) \end{aligned}$$

whence we have

$$B_{p_1} = \begin{matrix} p \\ \\ p_1 - p \end{matrix} \begin{pmatrix} B_p - \Sigma_{pp}^{-1}\Sigma_{p+}K^{-1}(\Sigma_{+0} - \Sigma_{+p}B_p) \\ \\ K^{-1}(\Sigma_{+0} - \Sigma_{+p}B_p) \end{pmatrix} \tag{1.7}$$

$$\text{and} \quad \Gamma_{p_1} = \Gamma_p - (\Sigma_{0+} - B_p^T\Sigma_{p+})K^{-1}(\Sigma_{+0} - \Sigma_{+p}B_p). \tag{1.8}$$

If we write B for B_p, and express B_{p_1} as $\begin{smallmatrix} p \\ p_1 - p \end{smallmatrix}\left[\begin{smallmatrix} C \\ D \end{smallmatrix}\right]$, we thus have

$$D = K^{-1}(\Sigma_{+0} - \Sigma_{+p}B) \tag{1.9}$$

$$C = B - \Sigma_{pp}^{-1}\Sigma_{p+}D \tag{1.10}$$

$$\Gamma_{p_1} = \Gamma_p - D^TKD \tag{1.11}$$

These formulae are especially useful when $p_1 = p + 1$.

We note from (1.11) that $\Gamma_{p_1} \leq \Gamma_p$, with equality if and only if $D = 0$, which is in turn equivalent, by (1.10) , to $C = B$. These equalities hold if and only if X_0 is independent of X_+ conditional on $X_{(p)}$, so that the regression of X_0 on $X_{(p_1)}$ is the same as that on $X_{(p)}$. In particular $\Gamma_{p+1} = \Gamma_p$ if and only if the last element of B_{p+1} is zero.

Suppose now that we are given $\{(B_k, \Gamma_k) : k = 0, 1, \ldots, p\}$, with $\Gamma_{k+1} > \Gamma_k > 0$ and the last element of each B_k non-zero. Then we can reconstruct a unique positive definite matrix Σ_q ($q = 1+p$) consistent with all of these. This may be performed recursively, starting with $\Sigma_1 = (\Gamma_0)$, and using equations (1.9-1.11) for $p_1 = p + 1$ to find Σ_{0+}, Σ_p and Σ_{++} in (1.5) in terms of Σ_q, $B_p = B$, $B_{p+1} = {}^{p}_{1}\left[\begin{smallmatrix} C \\ D \end{smallmatrix}\right]$, Γ_p and Γ_{p+1}, thus allowing Σ_q to be updated to Σ_{q+1}. We obtain the formulae:

$$\left.\begin{aligned} \Sigma_{p+} &= \Sigma_{pp}(B - C)J & (1.12)\\ \Sigma_{0+} &= [(\Gamma_p - \Gamma_{p+1}) + B^T \Sigma_{pp}(B - C)]J & (1.13)\\ \Sigma_{++} &= J^T[(\Gamma_p - \Gamma_{p+1}) + (B - C)^T \Sigma_{pp}(B - C)]J & (1.14) \end{aligned}\right\}$$

where J is the scalar D^{-1} in the present context.

The above procedure likewise yields the reconstruction of the unique infinite dispersion matrix Σ consistent with a collection $\{(B_k, \Gamma_k) : k \geq 0\}$ satisfying the above constraints.

Asymptotic behaviour. (In this section, all expectations and variances are taken in the sampling distribution, for fixed parameter values). Let $X_{0 \bullet p}$ denote $E(X_0 | X_{(p)})$, *viz.* $X_{(p)} B_p$. Applying the martingale convergence theorem in the overmodel (1.2), as $p \to \infty$ we must have $X_{0 \bullet p} \to E(X_0 | X_1, X_2, \ldots) = X_{0 \bullet \infty}$ say, almost surely and in mean-square. Likewise the residual $X_{0.p} = X_0 - X_{0 \bullet p} \to X_{0.\infty} = X_0 - X_{0 \bullet \infty}$. Then $\Gamma_p = \Sigma_{0.pp} = \mathrm{var}(X_{0.p}) \to \mathrm{var}(X_{0.\infty}) = \Gamma_\infty$ say, the residual prediction variance after all variation explained by the $(X_i : i = 1, 2, \ldots)$ has been removed. If $\Gamma_\infty = 0$ we shall call the model described by (1.1) or (1.2) *deterministic*, otherwise *non-deterministic*. Only in the former case would one expect to be able to predict X_0 arbitrarily closely by increasing the number of predictors —always assuming the parameters in (1.1) or (1.2) known.

Now let $\Im_p$ be the σ-field generated by $(X_p, X_{p+1}, \ldots)$, and $\Im = \cap_p \Im_p$ the tail σ-field. Then $X_{0 \bullet \Im} = E(X_0 | \Im) = \lim_{p \to \infty} E(X_0 | \Im_p)$, while in turn $E(X_0 | \Im_p) = \lim_{p_1 \to \infty} E(X_0 | X_p, \ldots, X_{p_1})$, all limits being almost sure and in mean-square. Likewise $\mathrm{var}(X | \Im), = \Gamma_\Im$ say, $= \lim_{p \to \infty} \lim_{p_1 \to \infty} \mathrm{var}(X_0 | X_p, \ldots, X_{p_1})$, which, in principle, yields an expression in terms of the infinite matrix Σ. We must have $\Gamma_\infty \leq \Gamma_\Im \leq \Gamma_0$. If $\Gamma_\Im = \Gamma_0$, then $X_{0 \bullet \Im}$ is trivial, and we call the model *tail-free*; if $\Gamma_\Im = \Gamma_\infty$, so that $X_{0 \bullet \infty} = X_{0 \bullet \Im}$, we call it *tail-dominated*; and otherwise, we call it *tail-mixed*. Any of these possibilities is consistent with determinism ($\Gamma_\infty = 0$) or non-determinism ($\Gamma_\infty > 0$).

As an example, suppose $\sigma_{00} = b + \lambda^2 a + a_0$, $\sigma_{10} = \sigma_{01} = b + \lambda a$, $\sigma_{ii} = (b + a)$ $(i > 0)$, and every other entry of Σ is equal to b. Then we calculate $X_{0 \bullet \infty} = \lambda X_1 + (1 - \lambda)\, U$, $X_{0 \bullet \Im} = U$, with $U = l.i.m.\ p^{-1} \sum_{i=1}^{p} X_i$; and $\Gamma_\infty = a_0$, $\Gamma_\Im = a_0 + \lambda^2 a$. This model is therefore deterministic when $a_0 = 0$, tail-free when $b = 0$, and tail-dominated when $\lambda = 0$. The regression $X_{0 \bullet p}$ of X_0 on $X_{(p)}$ has the form $X_{0 \bullet p} = \sum_{i=1}^{p} \beta_{pi} X_i$ in which, as $p \to \infty$, $\beta_{p1} \to \lambda$, $\beta_{pi} \to 0$ $(i > 1)$. Note that, except for the (tail-mixed) case $\lambda = 1$, the "infinite regresion" $X_{0 \bullet \infty}$ cannot be obtained by taking limits of individual regression coefficients, and this is the typical situation.

2. PRIOR DISTRIBUTIONS

Any distribution for Σ in (1.2) determines, *via* (1.3-1.4), a distribution for (B_p, Γ_p) for each p. Since $\Gamma_{p+1} \leq \Gamma_p$, these must be such that Γ_p is stochastically non-increasing in p. Conversely, suppose given a collection of such distributions for (B_p, Γ_p) where (for simplicity) Γ_p is *strictly* stochastically decreasing in p and, with probability one for each p, the last element of B_p is non-zero. Then we can construct (non-uniquely) a joint distribution for all the $\{(B_p, \Gamma_p)\}$, having the given marginals, and such that $\Gamma_{p+1} > \Gamma_p$, all p. The construction at the end of Section 1 then displays Σ as a function of the $\{(B_p, \Gamma_p)\}$, and we thus obtain a prior distribution for Σ in the overmodel (1.2), compatible with all the given prior distributions for the parameters of the undermodels (1.1).

Here we shall take the straightforward approach of assigning a distribution directly to Σ, and deriving the implied distributions for (B_p, Γ_p). The distribution of Σ may itself be specified by assigning a distribution to Σ_q for every q, subject to the compatibility condition that, for $q_1 < q$, the distribution for Σ_{q_1} induced by regarding it as a submatrix of Σ_q must agree with that assigned directly. This construction is considered in detail in Dawid (1981), which also presents the notation we shall use for matrix distributions and gives some of their properties, which we shall utilise without further comment.

In this paper we investigate the consequences of assigning to Σ the natural conjugate *inverse Wishart* prior distribution:

$$\Sigma \sim IW(\delta; Q). \tag{2.1}$$

Here $\delta > 0$ is the degrees of freedom parameter, and Q an infinite dispersion matrix. This distribution is such that, for each q,

$$\Sigma_q \sim IW(\delta; Q_q) \tag{2.2}$$

or equivalently (if Q_q is non-singular, which we henceforth assume)

$$\Sigma_q^{-1} \sim W(\delta + q - 1; Q_q^{-1}). \tag{2.3}$$

Dickey *et al.* (1985) have considered the particular "exchangeable" case in which the dispersion matrix Q is of intra-class form.

Lemma 2. With notation as in Lemma 1, if $\Phi \sim IW(\delta; G)$ then, in the induced distribution for (Φ_{22}, B, Γ),

(a) Φ_{22} is independent of (B, Γ);

(b) $\Gamma \sim W(\delta + p_2; \Lambda)$;

and (c) $B|\Gamma \sim H + \mathcal{N}(G_{22}^{-1}, \Gamma)$,
where $H = G_{22}^{-1} G_{21}$ and $\Lambda = G_{11.2}$.

A derivation of these results may be culled from Dempster (1969, Theorem 13.4.2). ◁

Using Lemma 2, it follows from (2.3) that the implied prior distribution for (B_p, Γ_p) in (1.3-1.4) is given by:

$$\left.\begin{aligned} \Gamma_p &\sim IW(\delta + p; \Lambda_p) & (2.4)\\ \text{and}\quad B_p|\Gamma_p &\sim H_p + \mathcal{N}(Q_{pp}^{-1}, \Gamma_p) & (2.5)\end{aligned}\right\}$$

where (*cf.* (1.3-1.4))

$$\left.\begin{aligned} H_p &= Q_{pp}^{-1} Q_{p0} & (2.6)\\ \text{and}\quad \Lambda_p &= Q_{00.p} = Q_{00} - Q_{0p} Q_{pp}^{-1} Q_{p0} & (2.7)\end{aligned}\right\}.$$

The implied marginal distribution for B_p is:

$$B_p \sim H_p + T(\delta + p; Q_{pp}^{-1}, \Lambda_p). \tag{2.8}$$

[In more conventional notation, these results become

$$\Gamma_p \sim \Lambda_p / \chi^2_{\delta+p} \tag{2.4$'$}$$

$$B_p|\Gamma_p \sim N(H_p, \Gamma_p Q_{pp}^{-1}) \tag{2.5$'$}$$

$$\text{and} \quad B_p \sim H_p + (\delta + p)^{-1/2} t_{\delta+p}(\Lambda_p Q_{pp}^{-1}).] \tag{2.8$'$}$$

A general natural conjugate prior for the undermodel (1.1) has the form:

$$\left.\begin{array}{l}\Gamma_p \sim IW(\delta_p^*; \Lambda_p^*) \\ B_p|\Gamma_p \sim H_p^* + \mathcal{N}(G_p^*, \Gamma_p)\end{array}\right\} \quad \begin{array}{r}(2.9)\\(2.10)\end{array}$$

where (δ_p^*, Λ_p^*, H_p^*, G_p^*) are arbitrary hyperparameters. So long as the hyperparameters vary with p in such a way that the distribution (2.9) is stochastically decreasing in p, we can find a distribution for Σ consistent with all these undermodel priors. However, we can take this to be of the form $IW(\delta; Q)$ if and only if the various hyperparameters satisfy: $\delta_p^* = \delta + p, \Lambda_p^* = Q_{00.p}$, $H_p^* = Q_{pp}^{-1} Q_{p0}$ and $G_p^* = Q_{pp}$. Note that, in this case, the hyperparameters for any value of p completely determine δ and Q_q $(q = 1+p)$, which in turn completely determine the hyperparameters for any smaller value of p. Thus the distributions for (B_p, Γ_p), as p varies, are very closely tied together. Against this limitation of the inverse Wishart overmodel should be set its property that, if we consider the regression of X_0 on *any* subset of the variables $(X_1, X_2, \ldots)$, the associated parameters will again have a natural conjugate prior distribution. The inverse Wishart also allows, as will be seen in the next Section, simple updating of the distribution (2.4-2.5) on observing data. In particular, for this case but not more generally, the posterior distribution of (B_p, Γ_p) may be found by restricting attention to the undermodel (1.1).

Determinism in the inverse Wishart. One feature of prior (2.1) is especially noteworthy. Since, in (2.4), $\Lambda_p = Q_{00.p}$ is necessarily non-increasing with p, whilst the degrees of freedom always exceed p, we must have $\Gamma_p \to 0$ in probability. But we have shown that $\Gamma_p \to \Gamma_\infty$, and hence we have $\Gamma_\infty = 0$ almost surely. Likewise we easily deduce $\Gamma_{\mathfrak{F}} = 0$ almost surely. That is to say, if one uses an inverse Wishart prior for Σ, one is automatically incorporating the judgment that, if only the parameters of the model were known, one could calculate X_0 as a *deterministic function*, with no residual error, of the infinite collection of predictor variables $(X_1, X_2, \ldots)$ —and indeed, could still do so from any infinite subcollection. In many applications such implicit determinism will be clearly inappropriate, and then so too must be the inverse Wishart prior.

3. POSTERIOR DISTRIBUTIONS

Given a prior distribution for Σ, and a set of n independent observations $(x_i = (x_{i0}, x_{i1}, \ldots) : i = 1, 2, \ldots, n)$ on the sequence $(X_0, X_1, \ldots)$ we have two routes to obtaining a posterior distribution for (B_p, Γ_p):

(i) derive the posterior distribution for Σ given all the data, and apply (1.3-1.4); or

(ii) apply (1.3-1.4) to the prior for Σ to yield a prior for (B_p, Γ_p), and derive the posterior based on this induced prior and the undermodel (1.1).

Route (i) is entirely coherent, but route (ii) involves first restricting attention to the data on $(X_0, X_1, \ldots, X_p)$ alone, and then ignoring any information in the marginal distribution of $(X_1, \ldots, X_p)$. In general, therefore, (i) and (ii) will yield different answers. However, it turns out that both methods agree in the case under consideration of an inverse Wishart prior for Σ.

Lemma 3. Let $(Y_1\ Y_2) \sim N(0^T, \Phi)$ as in Lemma 1, with Φ having prior distribution $IW(\delta; G)$ as in Lemma 2. Then

(a) the full posterior distribution of Φ_{22} is the same as that based on Y_2 alone (with model $Y_2 \sim N(0^T, \Phi_{22})$ and prior $\Phi_{22} \sim IW(\Phi; G_{22})$); and

(b) the full posterior for (B, Γ) is the same as that obtained from the model for Y_1 conditional on Y_2, *viz.* $Y_1|Y_2 \sim N(Y_2 B, \Gamma)$, and the prior for (B, Γ) given in Lemma 2.

The proof is immediate from the forms of the likelihood and prior given in Lemmas 1 and 2. It trivially extends to the case of multiple independent observations on (Y_1, Y_2). $\triangleleft$

In Lemma 3, take $Y_2 = (X_0, X_1, \ldots, X_p), Y_1 = (X_{p+1}, X_{p+2}, \ldots, X_{p_1})$ $(p_1 > p)$. Then property (a) shows that posterior distribution of Σ_q given observations on $(X_0, \ldots, X_{p_1})$ is the same as that given only the data on $(X_0, \ldots, X_p)$. Since this holds for all $p_1 > p$, the posterior distribution of Σ_q given all the data is likewise obtainable by restrincting to the data on $(X_0, \ldots, X_p)$. From (1.3-1.4) the same is true for inference about (B_p, Γ_p). Now take $Y_2 = (X_1, \ldots, X_p)$, $Y_1 = X_0$ and apply property (b) above to deduce that routes (i) and (ii) will deliver the same posterior for (B_p, Γ_p).

It is interesting to carry through the prior-to-posterior analysis by each of the above routes. Each is essentially straightforward, involving a well-known statistical model and its natural conjugate prior (although route (i) yields far the simpler method).

Define $S = s_{\alpha\beta} : \alpha, \beta \geq 0)$, with $s_{\alpha\beta} = \sum_{i=1}^{n} x_{i\alpha} x_{i\beta}$. Then under (1.2), $S \sim W(n; \Sigma)$. For any q, the posterior distribution of Σ_q, is, by the above analysis, that based on data $(x_{i\alpha} : 1 \leq i \leq n, 0 \leq \alpha \leq p)$, or equivalently on the corresponding sufficient statistic S_q with distribution $W(n; \Sigma_q)$; combined with the prior distribution $IW(\delta; Q_q)$ for Σ_q. Standard analysis then yields: $\Sigma_q|\text{data} \sim IW(\delta + n;\ Q_q + S_q)$. It thus follows that the posterior distribution of Σ, required for route (i) above, is:

$$\Sigma|\text{data} \sim IW(\delta + n; Q + S). \tag{3.1}$$

We then immediately deduce, by analogy with (2.4-2.5), the "route (i) posterior distribution":

$$\left.\begin{aligned} \Gamma_p|\text{data} &\sim IW(\delta + p + n; Q^*_{00.p}) \quad &(3.2)\\ B_p|\Gamma_p, \text{data} &\sim Q^{*-1}_{pp} Q^*_{p0} + \mathcal{N}(Q^{*-1}_{pp}, \Gamma_p) \quad &(3.3) \end{aligned}\right\}$$

where $Q^* = Q + S$.

For the alternative route (ii) analysis, we shall restrict to the case S_{pp} non-singular (so that, in particular, $n \geq p$). We first note that the sufficient statistic for undermodel (1.1), based on the data $((x_{i0}, x_{i1}, \ldots, x_{ip}) : 1 \leq i \leq n)$, comprises $\hat{B}_p = S_{pp}^{-1} S_{p0}$ and $R_p = S_{00.p}$, the sample least squares estimator of B_p and the residual sum of squares respectively (cf.(1.3-1.4)). The sampling distribution of $(\hat{B}_p, R_p)$ implied by (1.1) is given by:

$$\hat{B}_p \sim B_p + \mathcal{N}(S_{pp}^{-1}, \Gamma_p) \tag{3.4}$$

$$\text{and} \quad R_p \sim W(n - p; \Gamma_p), \tag{3.5}$$

independently.

[In more conventional notation, these are expressed as:

$$\hat{B}_p \sim N(B_p, \Gamma_p S_{pp}^{-1}) \tag{3.4$'$}$$

$$\text{and} \quad R_p \sim \Gamma_p \chi^2_{n-p}.] \tag{3.5$'$}$$

We must thus perform a prior-to-posterior-analysis based on (3.4-3.5) as model and (2.4-2.5) as prior. Application of standard results delivers the posterior distribution:

$$\left.\begin{array}{l}\Gamma_p|\text{data} \sim IW(\delta+p+n; \Lambda_p + (\hat{B}_p - H_p)^T(Q_{pp}^{-1}+S_{pp}^{-1})^{-1}(\hat{B}_p - H_p) + R_p) \\ \text{and} \\ B_p|\Gamma_p, \text{data} \sim (Q_{pp}+S_{pp})^{-1}(Q_{pp}H_p + S_{pp}\hat{B}_p) + \mathcal{N}((Q_{pp}+S_{pp})^{-1}, \Gamma_p).\end{array}\right\} \quad \begin{array}{r}(3.6)\\ \\ (3.7)\end{array}$$

[For a quick derivation of (3.7) above, we can regard Γ_p as fixed throughout, when the sufficient statistic reduces to $\hat{B}_p$ with distribution (3.4), which is combined with the conditional prior (2.5) by the standard Bayesian analysis for a multivariate normal mean. To derive (3.6) easily, we can first combine (3.4) and (2.5) to deduce that the distribution of $\hat{B}_p$ given Γ_p alone is

$$\hat{B}_p|\Gamma_p \sim H_p + \mathcal{N}(Q_{pp}^{-1} + S_{pp}^{-1}, \Gamma_p), \tag{3.8}$$

independently of R_p with distribution (3.5). The sufficient statistic for Γ_p in this joint distributon is then $R_p + (\hat{B}_p - H_p)^T(Q_{pp}^{-1}+S_{pp}^{-1})^{-1}(\hat{B}_p - H_p)$, with distribution $W(n;\Gamma_p)$ (or equivalently $\Gamma_p\chi_n^2$). Combining this with (2.4) or (2.4′) by a standard Bayesian analysis for a normal variance then yields (3.6).]

Of course, we must have (3.2) and (3.6) equivalent, and likewise (3.3) and 3.7). The latter equivalence is evident, but the former implies the useful algebraic identity:

$$(Q+S)_{00.p} = Q_{00.p} + S_{00.p} + (Q_{pp}^{-1}Q_{p0} - S_{pp}^{-1}S_{p0})^T(Q_{pp}^{-1}+S_{pp}^{-1})^{-1}(Q_{pp}^{-1}Q_{p0} - S_{pp}^{-1}S_{p0}). \tag{3.9}$$

The posterior distribution (3.1), being of the same inverse Wishart form as the prior, again incorporates an implicit determinism. Examining (3.6) it is seen that, for $p \ll n$, Γ_p will tend to be close to the sample residual variance estimate $R_p/(n-p)$. However, as p increases, this is constrained by the fact that Γ_p must decrease stochastically to 0.

One can ask whether, in light of data, one could learn enough about the unknown deterministic dependence of X_0 on $(X_1, X_2, \ldots)$ to predict X_0 arbitrarily closely. This question will be taken up in Section 5.

4. MULTIPLE RESPONSE VARIABLES

Most of the above theory applies, virtually unchanged, to the case that X_0 is a vector of response variables: $X_0 = (Y_1, Y_2, \ldots, Y_r)$. In (1.1), B_p is now $(p \times r)$ and Γ_p a $(r \times r)$ dispersion matrix. Again the collection of undermodels (1.1), for varying p, is equivalent to the overmodel (1.1). We now take $q = r + p$, and partition

$$\Sigma_q = \begin{array}{c} \\ r \\ p \end{array}\!\!\begin{array}{c} \begin{array}{cc} r & p \end{array} \\ \begin{pmatrix} \Sigma_{00} & \Sigma_{0p} \\ \Sigma_{p0} & \Sigma_{pp} \end{pmatrix} \end{array}.$$

Then (1.3-1.4) hold, as do (1.9-1.11), in which B and C are now $(p \times r)$ and D is $((p_1 - p) \times r)$. We remark on the compatibility conditions: $\Gamma_p - \Gamma_{p_1} = D^T K D$ for K a $(p_1 - p) \times (p_1 - p)$ positive definite dispersion matrix. In particular, taking $p_1 = p+1$, $\Gamma_p - \Gamma_{p+1}$ must be a positive multiple of $\omega^T\omega$, ω being the last row of B_{p+1}. It follows that $\Gamma_{p+1} = \Gamma_p$ if and only if $\omega = 0$, which in turn holds if and only if $B = C$. Subject to compatibility, and $\omega \neq 0$, Σ may be reconstructed using (1.12-1.14) where now $J = \omega^T(\omega\omega^T)^{-1}$.

The above compatibility conditions complicate any attempt to assign directly a prior distribution for each (B_p, Γ_p), compatibly across differing values of p, but are, of course, automatically satisfied if these distributions are those induced from a single distribution assigned to Σ. Taking again the conjugate prior (2.1) for Σ, this again induces the conjugate prior (2.4-2.5) for (B_p, Γ_p). Prior-to-posterior analysis may again be performed by the two routes (i) and (ii) of Section 3, which give identical answers, again expressed by (3.1-3.2) or (3.5-3.6). Both prior and posterior incorporate the determinism property $\Gamma_\infty = \Gamma_\Im = 0$ almost surely.

5. PREDICTIVE DISTRIBUTIONS

We henceforth assume multiple response variables $X_0 = (Y_1, Y_2, \ldots, Y_r)$. Our *training set* of n independent observations on $(X_0, X_1, X_2, \ldots)$ will be denoted by $(X_{t0}\ X_{t\infty})$, with $X_{t\infty}$ $(n \times \infty)$ containing the measured explanatory variables, and X_{t0} $(n \times r)$ the associated response variables. For a new *forecast set* $X_{f\infty}$ $(m \times \infty)$ of m observations on the explanatory variables $(X_1, X_2, \ldots)$, we wish to make predictive inference about the associated m response vectors X_{f0} $(m \times r)$. That is to say, we want the overall predictive conditional distribution of X_{f0}, given $(X_{t\infty}, X_{t0}, X_{f\infty})$, the parameters having been "integrated out". We can approach this through the conditional distribution of X_{f0} given $(X_{t\infty}, X_{t0}, X_{fp})$, based on the full training data but on only p explanatory variables in the forecast set, by letting $p \to \infty$. Limits as $n \to \infty$, when the number of training data is large, will also be of interest.

$$\text{Let} \quad X = \begin{matrix} \\ m \\ n \end{matrix} \begin{matrix} r \quad\ \ p \\ \begin{pmatrix} X_{f0} & X_{fp} \\ X_{t0} & X_{tp} \end{pmatrix} \end{matrix}. \tag{5.1}$$

$$(m + n = N, r + p = q).$$

The rows of X are, given Σ, independently distributed as $N(0^T, \Sigma_q)$; equivalently, $X \sim \mathcal{N}(I_N, \Sigma_q)$. Marginally $\Sigma_q \sim IW(\delta; Q_q)$. It follows (Dawid, 1981) that the overall marginal distribution of X is the matrix-t $T(\delta; I_N, Q_q)$.

Lemma 4. Let Z $(N \times q) \sim T(\delta; K, G)$ (K is $(N \times N)$, G $(q \times q)$). Partition Z, K, G according to $N = n_1 + n_2, q = p_1 + p_2$.
Then

(i) $Z_{11} | Z_{12} \sim Z_{12} G_{22}^{-1} G_{21} + T(\delta + p_2;\ K_{11} + Z_{12} G_{22}^{-1} Z_{12}^T, G_{11.2})$;
(ii) $Z_{11} | Z_{21} \sim K_{12} K_{22}^{-1} Z_{21} + T(\delta + n_2;\ K_{11.2}, G_{11} + Z_{21}^T K_{22}^{-1} Z_{21})$;
(iii) $Z_{11} \perp Z_{22} | Z_{12}$ and $Z_{11} \perp Z_{22} | Z_{21}$;
(iv) $Z_{11} | Z_{12}, Z_{21}, Z_{22} \sim A + T(\delta + p_2 + n_2; L, M)$,

where

$$A = Z_{12} G_{22}^{-1} G_{21} + (K_{12} + Z_{12} G_{22}^{-1} Z_{22}^T)(K_{22} + Z_{22} G_{22}^{-1} Z_{22}^T)^{-1}(Z_{21} - Z_{22} G_{22}^{-1} G_{21}) \tag{5.2}$$

$$= K_{12} K_{22}^{-1} Z_{21} + (Z_{12} - K_{12} K_{22}^{-1} Z_{22})(G_{22} + Z_{22}^T K_{22}^{-1} Z_{22})^{-1}(G_{21} + Z_{22}^T K_{22}^{-1} Z_{21}); \tag{5.3}$$

$$L = (K_{11} + Z_{12} G_{22}^{-1} Z_{12}^T) - (K_{12} + Z_{12} G_{22}^{-1} Z_{22}^T)(K_{22} + Z_{22} G_{22}^{-1} Z_{22}^T)^{-1} (K_{21} + Z_{22} G_{22}^{-1} Z_{12}^T) \tag{5.4}$$

$$= K_{11.2} + (Z_{12} - K_{12} K_{22}^{-1} Z_{22})(G_{22} + Z_{22}^T K_{22}^{-1} Z_{22})^{-1}(Z_{12}^T - Z_{22}^T K_{22}^{-1} K_{21}); \tag{5.5}$$

and

$$M = (G_{11} + Z_{21}^T K_{22}^{-1} Z_{21}) - (G_{12} + Z_{21}^T K_{22}^{-1} Z_{22})(G_{22} + Z_{22}^T K_{22}^{-1} Z_{22})^{-1} (G_{21} + Z_{22}^T K_{22}^{-1} Z_{21}) \tag{5.6}$$

$$= G_{11.2} + (Z_{21}^T - G_{12} G_{22}^{-1} Z_{22}^T)(K_{22} + Z_{22} G_{22}^{-1} Z_{22}^T)^{-1}(Z_{21} - Z_{22} G_{22}^{-1} G_{21}). \tag{5.7}$$

Proof. For (i): see Dickey (1967), Dawid (1981, Section 8). Then (ii) follows from (i) on using the equivalence of $Z \sim T(\delta; K, G)$ and $Z^T \sim T(\delta; G, K)$ (Dawid, 1981).
For (iii), apply (i) to the partition $N = N + 0$, $q = p_1 + p_2$ to obtain

$$\begin{bmatrix} Z_{11} \\ Z_{21} \end{bmatrix} \Big| (Z_{12}, Z_{22}) \sim \begin{bmatrix} Z_{12} \\ Z_{22} \end{bmatrix} G_{22}^{-1} G_{21} + T(\delta + p_2; K + \begin{bmatrix} Z_{12} \\ Z_{22} \end{bmatrix} G_{22}^{-1} (Z_{12}^T \; Z_{22}^T), G_{11.2}) \quad (5.8)$$

and deduce that $Z_{11} | (Z_{12}, Z_{22})$ has the same distribution as $Z_{11} | Z_{12}$ alone in (i). The second part then follows by transposition.
For (iv): Starting from the conditional distribution (5.8), apply (ii) (replacing q by p_1, partitioned as $p_1 + 0$) to $(Z_{11}^*, Z_{21}^*) = (Z_{11} - Z_{12} G_{22}^{-1} G_{21}, \; Z_{21} - Z_{22} G_{22}^{-1} G_{21})$. We obtain (iv) with the first form given for A and L and the second form given for M. The other forms follow by transposition. ◁

Now consider the joint predictive distribution of $(X_{f0} \; X_{fp})$ given $(X_{t0} \; X_{tp_1})$ $(p > p_1)$. Apply Lemma 4 (iii), taking

$$Z = \begin{array}{c} \\ m \\ n \end{array} \!\!\begin{array}{c} \begin{array}{cc} r & p_1 \end{array} \\ \begin{pmatrix} X_{f0} & X_{fp_1} \\ X_{t0} & X_{tp_1} \end{pmatrix} \end{array}, \qquad Z \sim T(\delta; I_{m+n}, Q_{q_1}),$$

but repartitioning the columns with $q = (r+p) + (p_1 - p)$. We deduce that, for any $p_1 > p$, the distribution of $(X_{f0} \; X_{fp})$ given $(X_{t0} \; X_{tp_1})$ in fact depends only on $(X_{t0} \; X_{tp})$. Consequently, the distribution of X_{f0} given $(X_{fp}, X_{t0}, X_{tp_1})$ is the same as that given only (X_{fp}, X_{t0}, X_{tp}). Letting $p_1 \longrightarrow \infty$, we deduce that, given all the training data (m observations each with infinitely many explanatory values), but only p explanatory values for the forecast set, we need only retain the same p explanatory variables from the training set when making predictive inference for X_{f0}. Our restricted data are thus X as in (5.1), with distribution $T(\delta; I_N, Q_q)$. By Lemma 4 (iv) we obtain the desired conditional distribution:

$$X_{f0} | (X_{fp}, X_{t0}, X_{t\infty}) \sim A + T(\delta + n + p; L, M), \quad (5.9)$$

where

$$\begin{aligned} A &= X_{fp} H_p + X_{fp} Q_{pp}^{-1} X_{tp}^T (I_n + X_{tp} Q_{pp}^{-1} X_{tp}^T)^{-1} (X_{t0} - X_{tp} H_p) && (5.10) \\ &= X_{fp} (Q_{pp} + S_{pp})^{-1} (Q_{p0} + S_{p0}); && (5.11) \\ L &= I_m + X_{fp} Q_{pp}^{-1} X_{fp}^T - X_{fp} Q_{pp}^{-1} X_{tp}^T (I_n + X_{tp} Q_{pp}^{-1} X_{tp}^T)^{-1} X_{tp} Q_{pp}^{-1} X_{fp}^T && (5.12) \\ &= I_m + X_{fp} (Q_{pp} + S_{pp})^{-1} X_{fp}^T; && (5.13) \end{aligned}$$

and

$$\begin{aligned} M &= \Lambda_p + (X_{t0} - X_{tp} H_p)^T (I_n + X_{tp} Q_{pp}^{-1} X_{tp}^T)^{-1} (X_{t0} - X_{tp} H_p) && (5.14) \\ &= (Q_{00} + S_{00}) - (Q_{0p} + S_{0p})(Q_{pp} + S_{pp})^{-1} (Q_{p0} + S_{p0}) && (5.15) \\ &= \Lambda_p + R_p + (\hat{B}_p - H_p)^T (Q_{pp}^{-1} + S_{pp}^{-1})^{-1} (\hat{B}_p - H_p) \quad \text{by}(3.9); && (5.16) \end{aligned}$$

with S, R_p and $\hat{B}_p$ calculated from the training data, and Λ_p, H_p from Q_q. Note that the equivalence of (5.10) and (5.11) for all X_{fp} implies the alternative formula for the mean of the posterior distribution (3.7):

$$(Q_{pp} + S_{pp})^{-1} (Q_{pp} H_p + S_{pp} \hat{B}_p) = H_p + Q_{pp}^{-1} X_{tp}^T (I_n + X_{tp} Q_{pp}^{-1} X_{tp}^T)^{-1} (X_{t0} - X_{tp} H_p). \quad (5.17)$$

6. ASYMPTOTIC PREDICTION

We now investigate the behaviour of the quantities A, L and M of (5.10-5.16), and of the associated predictive distribution (5.9), both for increasing training data ($n \to \infty$) and for increasing number of explanatory variables ($p \to \infty$). (Where "almost sure" *etc.* limits are involved, it must be understood that this is with respect to the overall marginal distribution of the data after integrating out over the inverse Wishart prior distribution. Such limits need not be meaningful if either subjective opinion or the data are not in accord with the implications of this prior.)

Consider first the case $n \to \infty$, with p and X_{fp} fixed. From (5.11) we find

$$A \sim X_{fp} S_{pp}^{-1} S_{p0} = X_{fp} \hat{B}_p. \tag{6.1}$$

Moreover, by considering the generation of the marginal distribution of the data as a mixture over Σ of independent $N(0^T, \Sigma)$ rows, we see that, almost surely, $\hat{B}_p$ converges, as $n \to \infty$, to a limit, which we may identify with $B_p = \Sigma_{pp}^{-1}\Sigma_{p0}$. Thus

$$A \to X_p^* = X_{fp} B_p \quad \text{almost surely} \quad (n \to \infty). \tag{6.2}$$

(This argument does not depend on Σ being assigned the inverse Wishart prior. Under any smooth prior, we shall have $E(X_{f0}|X_{fp}, X_{t0}, X_{tp}) \to X_{fp}B_p$, almost surely as $n \to \infty$.)

Similarly, using (5.13) and (5.15), we obtain almost surely as $n \to \infty$,

$$L \to I_m \tag{6.3}$$

and

$$n^{-1}M \to \Gamma_p, \tag{6.4}$$

while the full predictive distribution (5.9) approaches the sampling distribution:

$$X_{f0} - X_{fp}B_p|(X_{fp}, X_{t0}, X_{t\infty}) \xrightarrow{L} \mathcal{N}(I_m, \Gamma_p). \tag{6.5}$$

The above results continue to hold under an arbitrary smooth prior for Σ.

Now consider the case $p \to \infty$, n fixed. We make use of the alternative, artificial, representation of the marginal $T(\delta; I_N, Q_q)$ distribution of X:

$$X|\Phi \sim \mathcal{N}(\Phi, Q_q)$$

where $\Phi \sim IW(\delta; I_N)$.

We then have, given Φ,

$$\begin{bmatrix} X_{f0} - X_{fp}H_p \\ X_{t0} - X_{tp}H_p \end{bmatrix} \sim \mathcal{N}(\Phi, \Lambda_p)$$

and

$$\begin{bmatrix} X_{fp} \\ X_{tp} \end{bmatrix} \sim \mathcal{N}(\Phi, Q_{pp}),$$

independently.

$$\text{Partition} \quad \Phi = \begin{matrix} & \begin{matrix} m & n \end{matrix} \\ \begin{matrix} m \\ n \end{matrix} & \begin{pmatrix} \Phi_{ff} & \Phi_{ft} \\ \Phi_{tf} & \Phi_{tt} \end{pmatrix} \end{matrix},$$

and let $\Delta_n = \Phi_{ft}\Phi_{tt}^{-1}$, $\Phi_{ff.n} = \Phi_{ff} - \Phi_{ft}\Phi_{tt}^{-1}\Phi_{tf}$. Then

7. CONCLUDING COMMENTS

The above limiting properties assume that the data used for conditioning are behaving as expected under their marginal distribution, induced by mixing their normal sampling distributions with the inverse Wishart prior for Σ. Although this limiting behaviour will often be undesirable, it could be that it would be replaced by better behaviour if the conditioning data used did not look as expected under this unreasonable marginal distribution. Note however, that most of the qualitative limiting results will still hold, in general, if the inverse Wishart is replaced by an arbitrary scale-modified rotatable distribution (Dawid, 1981).

All the above analysis extends to the case that p_0 of the regressor variables are non-random and must thus be included in every regression equation. In particular, this allows a non-zero intercept. The overmodel now consists of a multivariate normal regression model for all other variables conditional on these. Its parameters are assigned a conjugate matrix normal-inverted Wishart distribution. From a purely formal point of view, all calculations can proceed as before, using (1.2) as an artificial overmodel and (2.1) as its prior, but now ony considering submodels containing the fixed regressors. We no longer require $\delta > 0$ in (2.1), but only $\delta + p_0 > 0$ in order that the true overmodel should exist. In asymptotic analysis, the values of the fixed regressors can no longer be considered as arising from a distribution, but must be held constant. However, the essential findings of our analysis with $p_0 = 0$ continue to hold when $p_0 > 0$.

REFERENCES

Dawid, A. P. (1978). Extendibility of spherical matrix distributions. *J. Mult. Anal.* **8**, 559–566.

Dawid, A. P. (1981). Some matrix-variate distribution theory: notational considerations and a Bayesian application. *Biometrika* **68**, 265–274.

Dempster, A. P. (1961). *Elements of Continuous Multivariate Analysis*. Reading, Mass.: Addison-Wesley.

Dickey, J. M. (1967). Matricvariate generalizations of the multivariate t distribution and the inverted multivariate t distribution. *Ann. Math. Statist.* **38**, 511–518.

Dickey, J. M., Lindley, D. V. and Press, S. J. (1985). Bayesian estimation of the dispersion matrix of a multivariate normal distribution. *Commun. Statist. Theor. Meth.* **14**, 1019–1034.

DISCUSSION

M. MOUCHART (*Université Catholique de Louvain*)

(i) I first want to congratulate the author for his demonstration of matrix manipulations at the service of a Bayesian analysis of a discrete time Gaussian process; note that his impressive skill was not unexpected given his previous contributions in the field of matrix-variate distribution theory. As some minor technical and notational remarks were handed to the author at the conference, these comments will concentrate on some aspects concerning possible uses of this work.

(ii) In general terms, I would have appreciated to see either some theoretical situations where such developments would provide interesting insights or (preferably ?) an actual statistical problem where the model specification and/or the data analysis would clearly benefit from these technical developments.

(iii) When analyzing a potentially infinite sequence of observables, a crucial role of statistical modelling is to look for a structurally stable parametrization or, from a slightly different point of view, to look for some "statistical regularities" which would hopefully not be affected if the data had been generated under a (not too) different environment. Such a structural stability is traditionally obtained through assumptions about the structure of the model such as exchangeability (total or, more realistically, partial), (conditional) independence, exogeneity, non-causality, etc. In this paper, no such assumptions are introduced resulting in a parametrization the dimension of which increases as we progress in the infinite sequence and the model is made conceptually manageable by ad hoc distributional assumptions. This situation raises two types of questions. First, is there any actual situations where an elicitation of such a prior distribution would be conceivable? Second, what kind of structural properties (such as exchangeability, exogeneity, etc) would actually provide a model that would be more manageable both for its manipulation and for the interpretation of its parameters and its properties. Clearly, these questions reveal a somewhat different perspective from the author's one. More specifically, suppose one is given a potentially infinite sequence of variables, suppose one is interested in the potentially infinite sequence of conditional expectations $E(x_0|x_1, \ldots, x_p)$ and suppose one agrees to use a natural conjugate prior specification along with a normal sampling specification, then one will find in this paper a host of interesting results about the properties of such a specification, a striking one being, for example, the determinism property incorporated in the inverse Wishart distribution. Suppose now that "one" is "you" and the question becomes: under which circumstances would "you" be ready for such specifications and my feeling is that both the sequential *and* the asymptotic set-up make the answer quite dubious if only because of the difficulty of understanding the meaning of the parameters and eventually eliciting their prior distribution. One aspect of particular relevance is: although the parameters of the successive regressions are functions of a some (huge !) covariance matrix, it should nevertheless be stressed that their real-life meaning will typically be different in each regression and, in general, it will be difficult to admit that all these regressions will have a valid structural interpretation: would this not imply that the real-world is conceived in a strictly static framework i.e., that *all* its characteristics never change? I rather believe that the role of statistical analysis is precisely to look for which aspects (which parameters) are structurally invariant.

(iv) A related issue regards the understanding of this determinism property. In particular, it would be interesting to know whether it may be explained by the natural-conjugacy of the inverse Wishart specification i.e., its equivalence to a fictitious prior sample; if this were true, then could we "understand" this property as a natural Bayesian use of standard results in asymptotic theory?

(v) Finally I wonder whether the author has in mind a potential use of this analysis in a prequential approach to statistical inference.

W. POLASEK (*University of Basel*)

Variable selection in Bayesian econometrics in an important issue. Leamer (1979) has pointed out than in contrast to information criteria Bayesians would maximize the squared correlation coefficient. Zellner (1978) showed that BIC can be obtained in the limit from posterior odds-ratios. I wonder if the assumptions in the paper are sensible ones: a) normal distribution for the "overmodel" b) infinitely many regressors c) zero information costs. Model choice from a Bayesian perspective will be mainly restricted to finite numbers of models and variables, else prior considerations and elicitations become practically meaningless.

REPLY TO THE DISCUSSION

Let me first clarify, in response to Polasek, that my paper does *not* deal with the problem of "model selection". I assume a single overmodel (1.2). The various undermodels (1.1) are merely different, but mutually consistent, partial facets of this—different only because they refer to different extents of conditioning. This is in contrast to the case that Polasek appears to have in mind, where the right-hand side of (1.1) represents the full regression of X_0 on all the remaining X's and thus asserts that, of all these, only the variables in $X_{(p)}$ are needed for predicting X_0. In that case, the various models (1.1), for differing p, would indeed be distinct, and we would have to choose between them. In my formulation, there is nothing to choose—except, perhaps, how many variables to observe for a specific predictive purpose. It would be easy to set this problem up in a decision-theoretic framework, incorporating non-zero information costs.

Mouchart asks, essentially, where does all this matrix theory get us, and when can we use it? To answer these questions, I should explain how I came to conduct this investigation in the first place. I was interested in the problem of allowing for overfitting when one chooses one's inference after looking at the data. The classical statistician has a whole panoply of techniques (multiple comparisons, step-wise selection, etc.) which are aimed at either reducing or allowing for the optimistic biased generated by the model-fitting process—which runs the risk of over-fitting past data, and, in consequence, of being poorly validated on future data. In particular, in a regression problem with many regressors, he would not normally seriously consider the least-squares regression on all of them, but would instead regress on a suitable subset. Similarly, if he fits the "best" of all regressions on subsets of size m, say, he would, if he were careful, appreciate that the usual estimator of residual variance is likely to be biased downwards, and try to allow for this.

Now consider the "conjugate Bayesian". If her prior is "weak", and she is not considering too many regressors simultaneously, her posterior distribution for the parameters of the regression on any subset will be numerically very similar to (although conceptually quite different from) the (unadjusted) inferences made by the classical statistician—without however any possibility of making allowance for selection or over-fitting. Does this mean that the Bayesian can ignore these problems?

I was concerned about this, since it seemed to me that the classical intuition was sound, even though classical ways of dealing with it might not be. I decided that (as Mouchart suggests) the problem must lie with the conjugate prior, and set out to investigate this in depth, thus leading to the present study. The results in my paper give some explanation for the above "paradox". To hold conjugate prior beliefs is to commit oneself to the availability of deterministic prediction, given enough regressors. This is a view which would seem implausible in most applied contexts, although there might be rare problems where it would be appropriate. If one truly believes it, then it is sensible to use as many regressors as one can get hold of, or the best subset from those available. The problem of over-fitting does not arise when there is a possibility of a perfect fit!

The paradox thus dissolves on appreciating that the classical analysis and the conjugate Bayesian analysis are trying to solve different problems. But this realisation should make us very wary of the conjugate Bayes analysis, which should not be used—or, at least, not driven to its limits—without a full appreciation of what it is assuming about the problem at hand. In this sense, all my analysis is by way of "counter-example", and leads me *not* to recommend conjugate priors for highly-parametrised problems—except in those very special circumstances where the possibility of deterministic prediction is taken seriously. Even then, I would not recommend use of a prior for which $\Lambda_n \to 0$, *viz.* the case where deterministic prediction is possible even without learning the parameter values. I can conceive of no context in which this would be realistic.

The principal virtue of the conjugate analysis is, I suggest, that one can do it! It would admittedly be more constructive to develop analyses based on non-conjugate priors, and to see whether one could sensibly incorporate indeterminism and some inbuilt sensitivity to the problem of over-fitting. However, it seems to me that any such problem that we could actually solve and understand is likely to be so special that it, too, would have problems similar to those of the conjugate case. (Of course, this can also be true, though unknowable, of a problem we can't solve!)

Mouchart asks me to tie this work up with "prequential analysis" (Dawid, 1984). A prequential approach to prediction of X_0 from $(X_1, X_2, \ldots)$ might proceed as follows (Dawid,

1986). Consider any strategy for processing a set of data on $(X_0, X_1, \ldots)$ to select a prediction formula—for example, using the linear regression of X_0 on that subset of $(X_1, X_2, \ldots)$, of size m, which minimises the residual sum of squares. The prequential assessment of the performance of such a strategy is obtained by applying the resulting formula, calculated from the first n data-vectors,to predict X_0 from $(X_1, X_2, \ldots)$ for the $(n+1)$-th; and comparing this with the corresponding realised X_0—e.g. by squared prediction error, although more sophisticated methods are available. Cumulating these comparisons for varying n permits an "unbiased" assessment of the performance of the strategy, unsullied by the usual selection effects. One can go on to construct new super-strategies by optimising over a collection of strategies (e.g. for varying m), and again assess their performance prequentially.

One can apply this method to the strategy of using the predictive mean (or, more generally, the full predictive distribution) constructed from a Bayesian analysis, as in the present paper. It can then be shown that, under the Bayesian's model and prior, this strategy is expected to perform at least as well as any other strategy. However, the data when collected may not bear this out! In this way, we can obtain empirical evidence against the Bayesian analysis. (These ideas are very close to the empirical assessment of forecasts using the calibration criterion: Dawid, 1982). We can also compare different priors by means of their implied "prequentia likelihoods" for the data.

A point to note is that, unlike the finite case, in infinitely parametrised models it matters very much what prior we use—the effect of the prior does not wear off as we collect more data, and there can be no such thing as a "vague" prior—every prior puts all its mass on some "thin" subset of the parameter space. In the finite case, the data distributions induced by two different priors are typically mutually absolutely continuous; in the infinite case, they are typically mutually singular, and so embody quite different "world-views". Thus we must choose our prior with care, and even then should be prepared to abandon it if the data suggest we have chosen unwisely. As Mouchart puts it, we cannot confine ourselves to a static view of the real world.

Mouchart makes the important point that we should look to additional structure in choosing a prior distribution, and in simplifying the elicitation problem. If possible, we should select from a family specified by only a finite number of hyper-parameters. This was the approach taken in the conjugate case by Dickey *et al.* (1985), who imposed an exchangeability assumption on the X's, thus giving Q the structure $q_{ii} = \sigma^2$, $q_{ij} = \rho\sigma^2 (i \neq j)$,and requiring the elicitation of just three hyper-parameters, $\delta > 0$, $\sigma^2 > 0$ and $\rho \in (0, 1)$. (They in fact go further, in allowing a mixing distribution over these hyper-parameters.)

Note, however, that while this may simplify the elicitation process, it does not avoid the problems associated with the conjugate prior—it still implies determinism given the parameters, a consequence which is not averted by mixing over the hyper-parameters. We have $\Lambda_p = \sigma^2(1 - p\rho^2/\{1 + (p-1)\rho\}]$, so $\Lambda\infty = \sigma^2(1-\rho) > 0$, whence we see that we are not, at least, in the worst case, in which we could predict perfectly without knowledge of the parameters. Nevertheless, the limitations of this prior, as of any other, should be clearly appreciated before it is put into service.

REFERENCES IN THE DISCUSSION

Dawid, A. P. (1982). The well-calibrated Bayesian. *J. Amer. Statist. Assoc.* **77**, 605–613. (with discussion).

Dawid, A. P. (1984). Statistical theory. The prequential approach. *J. Roy. Statist. Soc. A* **147**, 278–292. (with discussion).

Dawid, A. P. (1986). Prequential data analysis. *Tech. Rep.* **46**, Department of Statistical Science, University College London.

Leamer E. E. (1979). Information Criteria for Choice of Regression Models: A Comment. *Econometrica* **47**, 407–510.

Zellner A. (1978). Jeffreys-Bayes Posterior Odds Ratio and the Akaike Information Criteria for Discriminating between Models. *Economics Letters* **1**, 337–342.

BAYESIAN STATISTICS 3, pp. 111–125
J. M. Bernardo, M. H. DeGroot, D. V. Lindley and A. F. M. Smith, (Eds.)

Recent Progress on de Finetti's Notions of Exchangeability

P. DIACONIS
Stanford University

SUMMARY

This review covers the following topics: exchangeability, partial exchangeability with finitely many types, Markov exchangeability, random walk with reinforcement —an application of the Markov theory, mixtures of exponential families and the Küchler-Lauritzen theorem, Gibbs states, Ressell's work, and finite forms of de Finetti's theorem. It concludes by introducing some "new" results that date back to de Finetti's original paper on partial exchangeability.

Keywords: EXCHANGEABLE; PARTIALLY EXCHANGEABLE; MIXTURES; MARKOV CHAINS, DE FINETTI; CONDITIONED LIMIT THEOREMS.

1. INTRODUCTION

This paper gives a survey of work on exchangeability, between the second and third Valencia meetings. To keep things self-contained, brief reviews of previous work on exchangeability and partial exchangeability have been added. A full survey of this classical material can be found in Diaconis and Freedman (1984), Aldous (1987) or Lauritzen (1982).

Diaconis and Freedman present the material in the language of Bayesian statistics. Aldous describes developments and applications to probability. Lauritzen uses the language of extreme point models pionnered by Martin-Löf.

The review sections also present some new work —an application of de Finetti's theorem for Markov chains to a novel problem of random walk with reinforcement and the Küchler-Lauritzen theorem for characterizing mixtures of exponential families.

Sections on newer results cover Ressel's developments using semi-groups, joint work with Freedman on finite forms of the basic theorems, and a battery of recent solutions of special problems.

The final section attempts to explore some untapped streams in de Finetti's original papers on partial exchangeability using modern notation.

2. EXCHANGEABILITY

Consider a process $X_1, X_2, X_3, \ldots,$ taking two values. A probability distribution P for the processes is *exchangeable* if it is invariant under permutations:

$$P\{X_1 = e_1, \ldots, X_n = e_n\} = P\{X_{\pi(1)} = e_1, \ldots, X_{\pi(n)} = e_n\} \tag{2.1}$$

where $e_1, e_2, \ldots,$ is any sequence of possible values, and π is any permutation.

de Finetti's (1931) basic theorem supposes a potentially infinite exchangeable process. He shows that there is a unique representing measure μ on the unit interval such that for any n, and any sequence $e_1, \ldots, e_n$,

$$P\{X_1 = e_1, \ldots, X_n = e_n\} = \int x^a (1-x)^b \mu(dx), \tag{2.2}$$

with a the number of e_i of type 1 among $e_1, e_2, \ldots, e_n$ and b the number of e_i of type 2.

Expressions like the right hand side of (2.2) have been used since Bayes' original paper. In modern language μ is called the prior measure $x^a(1-x)^b$ the likelihood. This μ may be thought of as the prior opinion about the limiting proportion of type one events. Subjectivists prefer not to speak about unobservable events (like limiting frequencies). They are perfectly willing to assign prior opinions to observable events like three successes out of the next ten trials. The theorem shows these two ways of working are equivalent.

More generally, X_i can take values in any nice space $\mathcal{X}$ (such as a complete, separable metric space). Then, de Finetti (1938), Hewitt and Savage (1955), Diaconis and Freedman (1980c), show in varying degree of generality that for an infinite exchangeable process

$$P\{X_1 \in A_1, \ldots, X_n \in A_n\} = \int_{\mathcal{P}} \prod_{i=1}^{n} F(A_i)\mu(dF). \tag{2.3}$$

On the left, A_i are arbitrary Borel sets in $\mathcal{X}$. On the right $\mathcal{P}$ is the set of all probabilities on the Borel sets of $\mathcal{X}$(itself given the weak star topology) and μ is a unique probability on the sets of $\mathcal{P}$. The same μ works for any n and A_i.

Curiously, results like (2.3) require some sort of topological restriction (Dubins and Freedman (1979)). Forms for finite n, or forms involving finitely additive μ are available with only the measure structure. See Diaconis and Freedman (1980c).

Most subjectivists find (2.2) a very satisfactory theorem. It seems like a reasonable task to specify a prior distribution on $[0,1]$ if only approximately. As to (2.3), it seems like an essentially impossible task to meaningfully specify a prior on all probabilities on the real line. One searches for additional restrictions, like symmetry and smoothness to cut the problem down to manageable size. See Section 6 below.

Diaconis and Freedman (1986) show how conventional, automated methods of putting a prior on such infinite dimensional spaces can lead to silly procedures in quite practical problems such as the basic measurement error model.

Lester Dubins points out that the unit interval and the space of all probabilities on the real line have the same cardinality c (from a recursively enumerable view both are effectively countable) so it may be only a lack of experience and suitable language that makes (2.3) seem less useful than (2.2).

3. PARTIAL EXCHANGEABILITY WITH FINITELY MANY TYPES

In 1938, de Finetti broadened the concept of exchangeability. Consider first the special case with two types of observations: $X_1, X_2, \ldots; Y_1, Y_2, \ldots$. The X_i might represent binary outcomes for a group of men and the Y_i might represent binary outcomes for a group of women. *If* it were judged that the observable covariate men/women did not matter, all of the variables would be judged exchangeable. Often, the covariate is judged as potentially meaningful, the X_i's are judged exchangeable between themselves and the Y_i's are judged exchangeable between themselves. Mathematically, the joint law must be invariant under permutations within the X's and Y's:

$$\mathcal{L}(X_1, \ldots, X_n; Y_1, Y_2, \ldots, Y_m) = \mathcal{L}(X_{\pi(1)}, \ldots, X_{\pi(n)}; Y_{\sigma(1)}, \ldots, Y_{\sigma(m)}). \tag{3.1}$$

This must hold for all n and m, and permutations π and σ.

de Finetti proved that for an infinite process $\{X_i, Y_j\}$ partially exchangeable as in (3.1) implies

$$\frac{X_1 + \cdots + X_n}{n}, \frac{Y_1 + \cdots + Y_m}{m} \to (p_1, p_2) \qquad \text{almost surely,} \tag{3.2}$$

$$P\{X_1 = e_1, \ldots, X_n = e_n; Y_1 = f_1, \ldots, Y_m = f_m\} = \\ = \int\int p_1^a(1-p_1)^b p_2^c(1-p_2)^d \mu(dp_1, dp_2), \tag{3.3}$$

for a unique measure μ on the Borel sets of the unit square $[0,1]^2$. Of course, a and b are the number of ones and zeros among the e_i and c and d are the number of ones and zeros among the f_i. The same measure μ works for all n, m and all binary sequences $e_1, \ldots, e_n$; $f_1, \ldots, f_m$.

Return to the case of two types and consider situations where one is unsure if the covariate matters: e.g., if the outcome is passing a written test or not, it may well be that the covariate has only a negligible influence. de Finetti's theorem represents the joint law of the processes as a mixture over the unit square. If the X_i and Y_j were *all* exchangeable, the mixing measure μ would be supported on the diagonal $p_1 = p_2$.

de Finetti (1972, Chap. 9) has shown how to build natural prior distributions on the unit square which are supported near the diagonal as a way of allowing that a difference might show up in the data but allowing expression of the belief that most probably the covariate does not matter. The paper by Bruno translated as Chapter 10 of de Finetti (1972) works out some numerical examples with three types that are fascinating: as a constant stream of data comes in one can oscillate between the types. Diaconis and Freedman (1988c) construct examples with infinitely many types in which convergence never occurs. Alas, this seems like a model of the way things work in practical inference—as more data comes in, one admits a richer and richer variety of explanatory hypothesis. Without care, a simple message can become unrecognizable.

4. MARKOV CHAINS

de Finetti (1959) mentions the possibility of a subjective treatment of Markov chains using partial exchangeability. The idea is that the type of the i-th observable depends on the outcome of the previous observation. With binary processes, there are three types: the first observation, observations following a zero, and observations following a one. A precise mathematical formulation becomes tricky and it is simpler to proceed along the following lines developed by Freedman (1962a).

Consider two binary sequences of zeros and ones. Call them *equivalent* if they begin with the same symbol and have the same number of transitions from 0 to 0, 0 to 1, 1 to 0 and 1 to 1. Thus

$$0101101011 \quad \text{and} \quad 0110101011 \quad \text{and} \quad 0101011011$$

are all equivalent having transitions

$$\text{from} \begin{matrix} \\ 0 \\ 1 \end{matrix} \overset{\text{to}}{\begin{matrix} 0 & 1 \\ \begin{pmatrix} 0 & 4 \\ 3 & 2 \end{pmatrix} \end{matrix}} .$$

Note that the first string is obtained from the second by switching the first one and the block of 2 ones. Switching such blocks does not change the overall transitions and it is easy to show that equivalent sequences can be obtained by switching blocks.

A probability on binary sequences is called *partially exchangeable* if it assigns equal probability to equivalent strings. Freedman (1962a) showed that a stationary partially exchangeable process is a mixture of Markov chains. Diaconis and Freedman (1980a) eliminated the stationary assumption. To get a mixture of Markov chains, infinitely many returns to the starting state are needed. This is guaranteed by a recurrence assumption

$$P\{X_n = X_0 \text{ infinitely often}\} = 1. \tag{4.1}$$

Theorem. *Let* $X = (X_0, X_1, X_2, \ldots)$ *be a partially exchangeable process satisfying the recurrence condition (4.1). Then* X *is a mixture of Markov chains:* $P\{X_0 = e_0, X_1 = e_1, \ldots, X_n = e_n\} = \int \int \prod p_{ij}^{t_{ij}} \mu(dp_{ij})$ *with* t_{ij} *being the number of* i *to* j *transitions in the sequence* $e_0, e_1, \ldots, e_n, 0 \leq i, j \leq 1$. *The measure* μ *is unique and does not depend on* n.

The extension to countable state space is straightforward. On the other hand, the extension to general state spaces require sophisticated machinery, see Freedman (1963, 1984) and Kallenberg (1975, 1981). These last papers also discuss the extension to continuous time.

More generally, the extension of de Finetti's theorem to cover naturally occuring stochastic processes is a challenging research area. This was started by Freedman (1962, 1963) who characterized Poisson processes and Brownian motion with unknown parameters. Kallenberg (1976) contains more recent results.

One recently solved problem deserves special mention. Consider a pure birth process (Galton-Watson process) with one unknown parameter —the rate of births λ. This description, in terms of an unknown parameter λ, is like the right side of (2.2) in Section 1. What is the characterization in terms of observables? Roughly this: For each time t, two paths with the same number of births and area (the area under the path and above the x axis) should be assigned the same probability. The details are very tricky. See Oleg and Hamler (1988) for a careful statement and proof.

There is much further work to be done in this area.

5. RANDOM WALK WITH REINFORCEMENT – AN APPLICATION

Consider a triangle

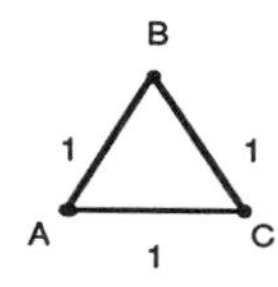

A random walk starts at A and chooses B or C with probability $\frac{1}{2}$. Each time the walk travels over an edge, 1 is added to the edge-weight. At a new vertex, the walk chooses the next vertex with probability proportional to edge-weights leading out of the present vertex.

Thus if the first choice is C, the walk moves to C and the 1 on the AC edge is changed to a 2. The walk next chooses to go back to A (probability $\frac{2}{3}$) or move to B (probability $\frac{1}{3}$). The process continues in this way.

Random walk with re-enforcement is a simple version of models for neural networks. It was introduced as a simple model of exploring a new city. At first all routes are equally unfamiliar and one chooses at random between them. As time goes on, routes that have been traveled more in the past are more likely to be traveled.

Of course, such a walk can be performed on any graph. For now, let us stick to the triangle and ask what happens as time goes on? People often guess that the walk eventually dies on an edge, or else winds up visiting each vertex about a third of the time. The answer is not so simple.

Let W_{AB}^n, W_{AC}^n, W_{BC}^n be the edge-weights at time n. These add up to $n + 3$. The theory to be described shows that the W's divided by n tend to a limit

$$\frac{1}{n}(W_{AB}^n, W_{AC}^n, W_{BC}^n) \to (L_{AB}, L_{AC}, L_{BC}) \quad \text{almost surely.}$$

Here $L_{AB} + L_{AC} + L_{BC} = 1$, and the limits are random, with an absolutely continuous distribution on the simplex $x + y + z = 1$. The limiting density is proportional to

$$(xy + xz + yz)^{1/2}(x + y)^{-1}(x + z)^{-3/2}(y + z)^{-3/2}. \tag{5.1}$$

A similar result holds for any finite graph and starting edge-weights: the edge-weights, divided by the number of steps converge almost surely. The limit is random with an absolutely continuous density which can be explicitly described in terms of the homology group of the graph.

These results are closely liked to de Finetti's theorem and indeed contribute to statistical inference for Markov chains. To describe the link, consider a simple graph, with starting edge-weights 2

2 2

$A \quad B \quad C$

A walk starts at B. At first it chooses between A and C with probability $1/2$. Say it goes to A. According to the rules, it next moves back to B and the edge-weights are

4 2

$A \quad B \quad C$

Thus the next step is twice as likely to be back to A, etc.

On reflection, one easily sees that these edge-weights evolve exactly like the balls in Polya's urn. Here one begins with an urn containing one red and one white ball. Balls are choosen from the urn at random and replaced each time along with another of the same color. Polya showed that the proportion of red balls has a uniform limit.

Starting with a star graph with d external vertices

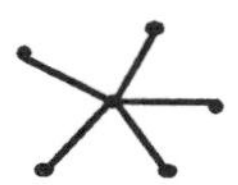

gives a Polya urn with d colors. Starting with a star having countably many extremal vertices, with the i-th edge having initial edge weight $\alpha(i)$ gives a Dirichlet random measure on the integers as limiting value.

As is well known, successive draws from a Polya urn form an exchangeable process and the limiting Dirichlet distribution is the representing measure in de Finetti's theorem.

Random walk with reinforcement on a general graph has a similar connection with the partially exchangeable prcesses discussed in Section 4 above. To describe the connection, consider a graph (V, E) with V a set of vertices and E a set of edges. There are starting weights $\alpha(e)$ for each $v \in V$, $\Sigma_{V \in e} \alpha(e) < \infty$. Start at vertix v_0 and run random walk with reinforcement. This generates a process $V : V_0 = v_0, V_1, V_2, \ldots$; the successive vertices visited.

A direct computation, similar to the proof of exchangeability for Polya's urn, show that the process V is partially exchangeable as in Section 4.

By the results of Diaconis and Freedman (1980a) a de Finetti type representation obtains: the process V can be represented as a mixture of Markov chains. Further, the empirical transition count matrix after n steps, divided by n, has an almost sure limit. This is how we know that the edge weights for the triangle converge.

Calculation of the representing measure is a far trickier business. At present, the only method involves a difficult combinatorics calculation. This is carried out by Coppersmith and Diaconis (1986). We merely state the result:

Theorem. *For random walk with reinforcement on a finite graph* (V, E) *let* $\{W_e^n\}_e \in E$ *be the edge-weights. Then*

(a) $\frac{W_e^n}{n}$ *converges almost surely to a limit on the* $|E|$ *simplex.*

(b) *the limit is random with absolutely continuous distribution having density*

$$C\Pi_e x_e^{(\alpha_e - \frac{1}{2})} \Pi_v x_v^{-(s_v+1)/2} x_{v_0}^{\frac{1}{2}} |A|^{\frac{1}{2}}. \tag{5.2}$$

In (5.2), x_e are variables on the simplex, $x_v = \sum_{v \in e} x_e, \alpha_e$ are the starting edge-weights, $s_v = \sum_{v \in e} \alpha_e, v_0$ is the starting vertix. The matrix A has dimension $|E| - |V| + 1$. This is the dimension of the first homology group(see Giblin (1977)). This is the group of "loops" c_i given with an arbitrary orientation. A has entries:

$$A_{ii} = \sum_{e \in c_i} 1/x_e$$

$$A_{ij} = \sum_{e \in c_i \cap c_j} \pm 1/x_e$$ The sign being $\pm$ as edge e has the same or opposite orientation in the i-th and j-th loop.

Finally C is a normalizing constant making the density integrate to 1 over the simplex.

As as example, the triangle only has one loop. If the starting edge-weights are taken as 1, the limiting density is proportional to

$$(x_{ab}x_{ac}x_{bc})^{1/2}(x_{ab}+x_{ac})^{-1}(x_{ab}+x_{bc})^{-3/2}(x_{ac}+x_{bc})^{-3/2}\left(\frac{1}{x_{ab}}+\frac{1}{x_{ac}}+\frac{1}{x_{bc}}\right)^{1/2}. \tag{5.3}$$

This is a density on $x_{ab} + x_{ac} + x_{bc} = 1$ with respect to the uniform density.

The theorem above gives the limiting distribution for the edge-weights. The de Finetti representation is in terms of a mixture of Markov chains for the vertix process. In probabilistic language this means there is a measure on the set of 3×3 transition matrices of form

$$\text{from} \begin{matrix} & \begin{matrix} & \text{to} & \\ A & B & C \end{matrix} \\ \begin{matrix} A \\ B \\ C \end{matrix} & \begin{pmatrix} 0 & a & 1-a \\ b & 0 & 1-b \\ c & 1-c & 0 \end{pmatrix} \end{matrix}.$$

One picks a matrix (and so a, b, and c) and then runs the Markov chain determined by this matrix from starting state A. The representation theorem says that this is an equivalent description of the process originally described by reinforcement.

The law of a, b, and c can be described as

$$a = \frac{L_{AB}}{L_{AB} + L_{AC}} \qquad b = \frac{L_{AB}}{L_{AB} + L_{BC}} \qquad c = \frac{L_{AC}}{L_{AC} + L_{BC}}$$

where the law of (L_{AB}, L_{AC}, L_{BC}) is given by (5.3). Observe that $abc = (1-a)(1-b)(1-c)$ so the mixing measure is singular, even on matrices with zeros on the diagonal.

Here is an application of the previous work to statistical inference for Markov chains. The Dirichlet prior is a basic ingredient for inference about multinomial parameters. Unfortunately, the conjugate priors for Markov chains have the rows of the unknown transition matrix as independent Dirichlet variables. See, e.g. Martin (1969). This seems to be quite a severe limitation. It seems fair to say that even in the 2×2 case

$$\begin{pmatrix} P_{00} & P_{01} \\ P_{10} & P_{11} \end{pmatrix}.$$

There is not a simple tractable family of prior distributions rich enough to capture believable prior information.

The mixing measures for random walk with reinforcement offer some variety *and* mathematical tractability. This is not evident from the density; indeed, it is not even immediately clear the density has a finite integral, and far less clear what the normalizing constant is.

The idea is to use the process representation: given the past of the process up to time n, the law of the future is given by the reinforcement description. For example, the chance of the next value being any of the vertices is proportional to the edge-weight leading to that vertix. This allows easy computation of posterior predictions and so, of posterior means of parameters. This seems to be a potentially fruitful direction for future development.

I will not take the time here to mention the interesting probability developments arising from random walk with reinforcement on infinite graphs. Pemantle(1988a, 1988b) contains much of interest here.

6. MIXTURES OF EXPONENTIAL FAMILIES

de Finetti's theorem presents a real-valued exchangeable process as a mixture over the set of all measures. Most of us find it hard to meaningfully quantify a prior distribution over so large a space. There has been a search for additional restrictions that get things down to a mixture of familiar families parametrized by low dimensional Euclidean parameter spaces.

The first result of this type was given by Freedman (1963) who characterized orthogonally invariant probabilities as scale mixtures of mean zero normals. Freedman also gave characterizations of the familiar 1 parameter exponential families such as Poisson, geometric, and gamma.

Dawid (1977) characterized covariance mixtures of multivariate normals. Diaconis, Eaton, and Lauritzen (1988) characterize the usual normal models for regression and analysis of variance in terms of symmetry or sufficiency.

By now it is clear that the correct version of these characterization results involves sufficiency: Start with a standard family on a space $P_\theta(dx)$. Let P_θ^n be product measure on $\mathcal{X}^n$. Let $T_n : \mathcal{X}^n \rightarrow \mathcal{Y}_n$ be a sufficient statistic. Let $Q_{n,t_n}(dx)$ be a regular conditional probability on $\mathcal{X}^n$ given $T_n = t_n$. By sufficiency, Q_{n,t_n} does not depend on θ.

One can define a general extension of exchangeability given T_n and Q_n by declaring a measure P on $\mathcal{X}^\infty$ to be *partially exchangeable* if for every n, Q_{n,t_n} is a regular conditional probability for the marginal law of P on $\mathcal{X}^n$ given $T_n = t_n$.

The class of partially exchangeable probabilities is a convex set and its extreme points can be identified in an indirect way as measures having a trivial partially exchangeable tail field. Then an abstract version of de Finetti's theorem follows: every partially exchangeable probability is a unique mixture of extreme points.

Myriad details have been left out of the above brief description. Careful versions have been given by Diaconis and Freedman (1984) in Bayesian language. Closely related results are given by Lauritzen (1974) in the language of extreme point models and by Dynkin (1978) or Reulle (1978) in the language of Gibbs states and statistical mechanics. The main results are essentially the same, but the language and examples vary widely between presentations.

The general theorem gives an abstract result which may require a fair amount of work to understand in specific cases. In particular, if the original family $P_\theta(dx)$ is a standard exponential family with $T_n = X_1 + \cdots + X_n$ the usual sufficient statistic, it is not at all clear if the extreme point representation says that partially exchangeable probabilities are mixtures of the exponential families that were started with. In other words, are the extreme points the P_θ^∞ and nothing else?

Following work of Martin-Löf (1970), Diaconis and Freedman (1984) solved the problem for 1-dimensional discrete exponential families. The problem for continuous families has recently been solved by Küchler and Lauritzen (1986). They assume the base measure has a continuous and everywhere positive denstity. Here is a refinement of their result from Diaconis and Freedman (1988).

Let h be a nonnegative, finite, locally integrable Borel function on $\mathcal{R}$: Let $c(\theta) = \int_{-\infty}^{\infty} e^{\theta x} h(x)dx$ and let Θ be the set of θ with $c(\theta) < \infty$. Assume Θ is non-empty. The exponential family through h is defined as $P_\theta(dx) = e^{\theta x}h(x)dx/c(\theta)$.

Let P_θ^∞ be product measure on $\Re^\infty$. Define Q_{ns} as the regular conditional probability for the coordinate functions $X_1, X_2, \ldots, X_n$ given $X_1 + \cdots + X_n = s$. This is only defined for $s \in \Re$ with $0 < h^{(n)}(s) < \infty$ with $h^{(n)}$ the n-fold convolution of h with itself. Let D_n be this set of s-values.

Define M_Q—the Q-exchangeable probabilities—as the set of probabilities P on $\mathcal{R}$ such that for every n

(a) $P\{X_1 + \cdots + X_n \in D_n\} = 1$

(b) Q_{ns} is a regular conditional P-distribution for $X_1, \ldots, X_n$ given $X_1 + \cdots + X_n = s$.

Clearly $P_\theta^\infty \in M_Q$. So is P_μ, defined as $\int_\Theta P_\theta^\infty \mu(d\theta)$ with μ a probability on the Borel sets of θ. The following version of the Küchler-Lauritzen theorem is proved in Diaconis and Freedman (1988c).

Theorem. *P is Q-exchangeable if and only if P has the unique integral representation*

$$P = \int_\Theta P_\theta^\infty \mu(d\theta).$$

For example, if $P_\theta(dx)$ is a mean zero normal scale family, then $S_n = X_1^2 + \cdots + X_n^2$ is the sufficient statistic and given $S_n = s, X_1, X_2, \ldots, X_n$ is uniformly distributed on the sphere of radius $\sqrt{s}$. This specifies $Q_{n,s}$. Now a measure is conditionally uniform if and only if it is orthogonally invariant. The theorem specializes to Freedman's original result, P on $\Re^\infty$ is orthogonally invariant if and only if it is a scale mixture of mean zero normal variables.

It is natural to try to extend this theorem to other natural sufficient statistics such as products and maxima. Vector valued versions are also natural requests. Lauritzen (1975) set things up in the language of semi-groups: one looks at the conditional law of $x_1, \ldots, x_n$ given $x_1 * x_2 * \cdots * x_n$. This idea has been put into definitive form by Ressel (1985) who showed that the extreme points of the partially exchangeable probabilities are in $1-1$ correspondence with the positive part of the dual group. de Finetti's theorem for these cases then becomes Bochner's theorem which represents a positive definite function as an integral of characters. The details make extensive use of the modern theory of Abelian semi-groups as developed in Berg, Christiansen and Ressel (1984). One nice bonus is that the duals of many semi-groups have been classified so that new theorems result. One disadvantage: the results often wind up in highly analytic form, e.g. as conditions on the characteristic function of the process instead of on observables. It is a worthwhile project to try to systematically translate these results to conditions on observables.

7. FINITE THEOREMS

The classical results on exchangeability involve infinite exchangeable processes and constructions like tail fields which are known to have no finite content. It is also known that there are finite exchangeable sequences which cannot be represented as a mixture of i.i.d. processes.

Reasonable finite versions of de Finetti theorems have been evolving.

Theorem. *(Diaconis and Freedman, 1980c). Let $X_1, X_2, \ldots, X_n$ be a binary exchangeable sequence. Then there is a μ on $[0,1]$ such that for $k \leq n$*

$$\|\mathcal{L}(X_1, \ldots, X_k) - P_\mu^k\| \leq \frac{2k}{n}. \tag{7.1}$$

In (7.1), $\mathcal{L}(X_1, \ldots, X_k)$ stands for the marginal distribution of the first k coordinates as a measure on binary k-tuples, $p_\mu^k(e_1, \ldots, e_n) = \int x^a(1-x)^{k-a}\mu(dx)$ with $a = e_1 + \cdots + e_n$. The norm is total variation distance:

$$\|P - Q\| = \sup_A |P(A) - Q(A)|.$$

The theorem says that if one is considering a binary exchangeable process of length k which can be extended to an exchangeable process of length n (so that the experiment can be repeated in principle) then, de Finetti's theorem almost holds. We showed that the k/n rate cannot be improved.

If the process takes c values instead of 2, the rate ck/n obtains. For processes taking infinitely many values, the rate $2k/\sqrt{n}$ obtains. For infinite state spaces, there are no topological restrictions (such as Polish spaces). Taking limits leads to the most general known version of the infinite form of de Finetti's theorem.

These theorems are all proved by using an exact finite form of de Finetti's theorem representing the processes as a mixture of urn processes—samples without replacement from an urn of fixed composition. Then, sharp bounds between sampling with and without replacement are derived to show that the urn processes are approximately binomial processes. Thus the original processes are approximately mixtures of binomial processes.

A crucial point: sampling with and without replacement are close in variation distance *uniformly* in the contents of the urn.

Diaconis and Freedman (1987) give similar theorems for particular versions of de Finettis' theorem like normal location, or scale parameters, mixtures of Poisson, geometric and gamma (shape or scale parameters). In each case, there is a finite theorem with rate k/n in variation distance. Alas, all of the arguments are different, and it is not at all clear where the k/n comes from.

We embarked on a systematic investigation of exponential families in (1988a) by working directly on the conditional density of $X_1, X_2, \ldots, X_k$ given $X_1 + \cdots + X_n$. This is a ratio of convolutions and Edgeworth expansions allow careful bounds to be obtained. To guarantee tht the bounds hold uniformly in the conditioning variable $X_1 + \cdots + X_n$, strong conditions are imposed on the underlying exponential family (things like uniformly bounded standardized fourth moments). These rule out some natural examples (where the uniform form of the theorem is known to hold) but they show where the k/n rate comes from (terms of form $k/\sqrt{n}$ cancel out of numerator and denominator) and allow construction of counterexamples where the k/n rate fails.

Finite versions of de Finetti's theorem for Markov chains are given by Diaconis and Freedman (1980c) and Zaman (1986). Here one of the important steps was to find an exact representation of a finite partially exchangeable process as a mixture of urn processes. Zaman's (1984) elegant form of this has been used by geneticists to simulate the law of a random finite string with fixed transition counts as a way of calibrating DNA string matching algorithms. Zaman's result has an annoying extra factor of log n. It is not clear if this is really there or just an artifact of the proof.

Lest the reader think that all the the interesting problems have been solved, I suggest two open problems: Find a finite version of Aldous' (1981) basic theorem on random binary arrays invariant under permuting rows and columns. Diaconis and Freedman (1981) apply Aldous' theorem to a problem in human perception.

A second problem: The finite version of Freedman's basic theorem on orthogonally invariant measures required a sharp version of the following: pick a point at random on a high dimension sphere; the first k-coordinates are approximately independent normal. Diaconis and Freedman (1987) prove this for $k = o(n)$. This is a very special case of the following : let Γ be a uniformly distributed random orthogonal n by n matrix. In some sense all of the entries of Γ are approximately independent normal variables. Perhaps any $o(n^2)$ can be proved to be variation distance close to independent normals. Diaconis and Shahshahani (1987) and Diaconis, Eaton and Lauritzen (1988) contain background and references.

8. ON READING DE FINETTI

de Finetti wrote about partial exchangeability in his article of 1938 (translated in 1979) and 1959 section 9.6.2 (translated in 1972). Both treatment are rich sources of ideas which take many readings to digest. His basic examples involve several types (e.g., men and women) with exchangeability within type. He derives parametric representations, with one parameter per type. He emphasizes situations of almost exchangeability where the mixing measure concentrates near the diagonal.

de Finetti (1938) briefly describes two unusual examples that I want to present here.

Example 1. Suppose we are considering families and $X_i = 1$ or 0 as the i-th family reports an accident in the coming year or not. As a covariate, we currently know the number of people in each family.

If this is all we know, it is natural to assume that families with the same size are exchangeable, so there is one type for each family size. Then, de Finetti's theorem yields one parameter per family size and a mixture over the "cube" $0 \leq P_j \leq 1; j = 1, 2, \ldots$.

Thinking further, we may, as a first approximation, judge that *all* of the people involved are exchangeable. Then, the prior on the cube will be concentrated near a curve: If θ is the proportion of people in a single person family having an accident, the proportion of i member families reporting an accident should be $1 - (1 - \theta)^i$ approximately. Such a prior is clustered about the curve $(1 - \theta, 1 - \theta^2, 1 - \theta^3, \ldots)$ $0 \leq \theta \leq 1$.

Of course, one wants to allow mass off the curve. As de Finetti suggests, priors based on a fair amount of background data will be approximately normal. A prior with the features above has density proportional to $\exp -A\{(P_1^2 - P_2)^2 + (P_1^3 - P_3) + \cdots\}$. This would have to be truncated to the cube.

In carrying out approximations to Bayes' theorem the curvature of the curve would presumably appear in the expansions of the posterior as in Efron (1975).

Example 2. This illustrates a common problem: we observe a decline in the mortality rate caused by a certain treatment for animals and expect an analogous decline for humans. A simple set-up involves four types

- untreated lab animals
- treated lab animals
- untreated humans
- treated humans.

Supposing that all observations of type i are exchangeable binary variables, de Finetti's theorem gives a representation with four parameters P_1, P_2, P_3, P_4.

There are probably real situations where the untreated survival rates P_1 and P_3 are quite different, but in which the percent improvement P_2/P_1 will be about the same for humans and animals. Then, the prior will be taken concentrated near the surface

$$\frac{P_2}{P_1} = \frac{P_4}{P_3} \quad \left(\text{or equivalently } \frac{P_2}{P_1 + P_2} = \frac{P_4}{P_3 + P_4}\right).$$

de Finetti continues: "But this can be used to explain more than the mechanism of this particular reasoning: on looking deeper, the very fact of this belief in a near-proportionality should be interpreted in the framework of these same considerations, since it turns essentially on similar observations made for other medical treatments. Let the treatments be $i = 1, 2 \ldots, c$; under an obvious interpretation of the symbols, the conclusion is established if we allow that $P_{4i+1}\colon P_{4i+2} - P_{4i+3}\colon P_{4i+4} = 0\ (i = 1, 2, \ldots, c)$ entails also $P_1 : P_2 = P_3 : P_4 = 0$. The very belief in the plausibility of extending certain conclusions concerning certain medical treatments to certain others can be explained in turn by the observation of analogies in a broader and vaguer sense, and in the same way one can explain every similar belief, which manifests itself by the formulation of a 'statistical law'".

The two examples suggests fresh avenues of research. Notice that they are stated in the language of parameters. Are there representation theorems characterizing such parametric families in terms of conditions on observables?

Consider the first example with families containing only one or two members. One can consider $2n$ single person families $X_1, X_2, \ldots, X_{2n}$ and m two person families $Y_1, Y_2, \ldots, Y_m$. Pair the X's as $(X_1, X_2), (X_3, X_4), \ldots, (X_{2n-1}, X_{2n})$ and let $Z_1 = X_1 \cdot X_2, Z_2 = X_3 \cdot X_4$, etc. Then the Z_i and Y_j are all exchangeable. This gives a characterization.

For the second example, consider $P_1/P_2 = P_3/P_4 = c$. Suppose first that c is known and equal (say) to $1/2$. Priors with $P_3 = P_4/2$ with P_i unknown can be characterized by saying the observables corresponding to P_3 "thinned" with a fair coin flip. This leads to a rather contorted characterization theorem.

The contortions above underscore the idea that parametric representations can be useful. They also point to a failing in our carrying out of de Finetti's program. Most of the extensions of de Finetti's theorem have been on the lines of taking a classical model and finding a Bayesian version.

de Finetti's alarm at statisticians introducing reams of unobservable parameters has been repeatedly justified in the modern curve fitting exercises of today's big models. These seem to lose all contact with scientific reality focusing attention on details of large programs and fitting instead of observation and understanding of basic mechanism. It is to be hoped that a fresh implementation of de Finetti's program based on observables will lead us out of this mess.

REFERENCES

Aldous, D. (1981). Partial exchangeability and $\bar{d}$- topologies. *Exchangeability Probability and Statistics*, (G. Koch and F. Spizzichino, eds.). Amsterdam: North-Holland, 23–38.

Aldous, D. (1985). Exchangeability and related topics. *Springer Lecture Notes in Mathematics* **1117**.

Berg, C., Christensen, J. P. J. and Ressel, P. (1984). *Harmonic Analysis on Semigroups*. Berlin: Springer-Verlag.

Coppersmith, D. and Diaconis, P. (1986). Random walk with reinforcement. (To appear).

Dawid, A. P. (1977). Extendibility of spherical matrix distributions. *J. Multivar. Anal.* **8**, 567–572.

de Finetti, B. (1931). Funcione caratteristica di un fenomeno aleatorio. *Atti della R. Accademia Nazionale dei Lincii Ser. 6, Memorie, classe di Scienze, Fisiche, Matamatiche e Naturali* **4**, 251–299.

de Finetti, B. (1938). Sur la condition d'equivalence partielle. *Actualites Scientifiques et Industrielles* **739**. Paris: Herman and Cii. Translated in *Studies in Inductive Logic and Probability* **II**, (R. Jeffrey, ed.). Berkeley: University of California.

de Finetti, B. (1972). *Probability, Induction and Statistics*. New York: Wiley.

Diaconis, P. Eaton, M. and Lauritzen, S. (1988). de Finetti's theorem for the linear model and analysis of variance. *Tech. Rep.* Department of Statistics, University of Minnesota.

Diaconis, P. and Freedman, D. (1980a). de Finetti's theorem for Markov chains. *Ann. Prob.* **8**, 115–130.

Diaconis, P. and Freedman, D. (1980b). de Finetti's generalizations of exchangeability. *Studies in Inductive Logic and Probabillty* **II**, (R. C. Jeffrey, ed.). Berkeley: University of California Press.

Diaconis, P. and Freedman, D. (1980c). Finite exchangeable sequences. *Ann. Prob.* **8**, 745–764.

Diaconis, P. and Freedman, D. (1981). On the statistics of vision: the Julesz conjecture. *Jour. Math. Psychol.* **24** 112, 138.

Diaconis, P. and Freedman, D. (1984). Partial exchangeability and sufficiency. *Proc. Indian Statist. Inst. Golden Jubilee Int'l Conf. on Statistics: Applications and New Directions*, (J. K. Ghosh and J. Roy, eds.). Calcutta: Indian Statistical Institute, 205–236.

Diaconis, P. and Freedman, D. (1986). On the consistency of Bayes estimates. *Ann. Statist.* **14**, 1–67.

Diaconis, P. and Freedman, D. (1987). A dozen de Finetti-style results in search of a theory. *Ann. Inst. Henri Poincaré* **23**, 394–423.

Diaconis, P. and Freedman, D. (1988a). Conditional limit theorems for exponential families and finite versions of de Finetti's theorem on exchangeability. (To appear in *Theoretical Prob.*)

Diaconis, P. and Freedman, D. (1988b). On the problem of types. *Tech. Rep.* **153**, Department of Statistics, University of California, Berkeley.

Diaconis, P. and Freedman, D. (1988c). On a theorem of Küchler and Lauritzen. *Tech. Rep.* **152**, Department of Statistics, University of California, Berkeley.

Diaconis, P. and Shahshahani, M. (1987). The subgroup algorithm for generating uniform random variables. *Prob. In Eng. and Info. Sci.* **1**, 15–32.

Dubins, L. and Freedman, D. (1979). Exchangeable processes need not to be mixtures of independent identically distributed random variables. *Z. Wahr. verw. Geb.* **48**, 115–132.

Dynkin, E. (1978). Sufficient statistics and extreme points. *Ann. Prob.* **6**, 705–730.

Efron, B. (1975). Defining the curvature of a statistical problem. *Ann. Statist.* **6**, 362–376.

Freedman, D. (1962a). Mixtures of Markov processes. *Ann. Math. Statist.* **33**, 114–118.

Freedman, D. (1962b). Invariants under mixing which generalize de Finetti's theorem. *Ann. Math. Statist.* **33**, 916–923.

Freedman, D. (1963). Invariants under mixing which generalize de Finetti's theorem: Continuous time parameter. *Ann. Math. Statist.* **34**, 1194–1216.

Freedman, D. (1984). de Finetti's theorem in continuous time. *Tech. Rep.* **36**, Department of Statistics, University of California, Berkeley.

Giblin, P. J. (1977). *Graphs, Surfaces, and Homology: An Introduction to Algebraic Topology*. London: Chapman and Hall.

Hewitt, E. and Savage, L. J. (1955). Symmetric measures on Cartesian products. *Trans. Amer. Math. Soc.* **80**, 470–501.

Kallenberg, O. (1975). Infinitely divisible processes with interchangeable increments and random measures under convolution. *Z. Wahr. verw. Geb.* **32**, 309–321.

Kallenberg, O. (1976). *Random Measures.* Berlin: Academic Press.

Kallenberg, O. (1981). Characterizations and embedding properties in exchangeability. *Tech. Rep.* **10**, Department of Mathematics, Chalmers University of Technology, Göteborg, Sweden.

Lauritzen, S. L. (1974). Sufficiency, prediction and extreme models. *Scand. J. Statist.* **2**, 128–134.

Lauritzen, S. L. (1975). General exponential models for discrete observations. *Scand. J. Statist.* **2**, 23–33.

Lauritzen, S. L. (1982). *Statistical Models as Extremal Families.* Aalborg: Aalborg University Press. (To appear in *Springer Lecture Notes in Statistics.*)

Martin, J. J. (1967). *Bayesian Decision Processes and Markov Chains.* New York: Wiley.

Martin-Löf, P. (1970). Statistika Modeller. Notes by Rolf Sundberg. Mimeographed lecture notes.

Pemantle, R. (1988). Random walk with reinforcement on trees. (To appear in *Ann. Prob.*)

Pemantle, R. (1988). Processes with reinforcement. Ph. D. thesis, Department of Mathematics, Massachusetts Institute of Technology.

Ressel, P. (1985). de Finetti-type theorems: an analytic approach. *Ann. Prob.* **13**, 898–922.

Ruelle, D. (1978). *Thermodynamic Formalism.* Reading, MA: Addison-Wesley.

Smith, A. F. M. (1981). On random sequences with centered spherical symmetry. *J. Roy. Statist. Soc. B* 208–209.

Zaman, A. (1984). Urn models for Markov exchangeability. *Ann. Prob.* **12**, 223–229.

Zaman, A. (1986). A finite form of de Finetti's theorem for stationary Markov exchangeability. *Ann. Prob.* **14**, 1418–1427.

DISCUSSION

D. BLACKWELL (*U.C. Berkeley*)

My comments on Professor Diaconis' paper are on three topics: 1) de Finetti's Theorem, 2) finite exchangeability, 3) partial exchangeability. My comments are for $0-1$ variables only.

de Finetti's Theorem

If a sequence $X = (X_1, \ldots, X_n)$ is finitely exchangeable then two sequences x, y have the same probability if they have the same number of 1s. Thus we have

$$P(X = x) = p(s) \Big/ \binom{n}{s},$$

where $s = \Sigma x_i$ and $p(s) = P(\Sigma X_i = s)$. So Diaconis and Freedman (D and F hereafter) represent the distribution P of X by

$$P = \Sigma p(s) H(n, s),$$

where H is the uniform distribution over the $\binom{n}{s}$ sequences of 0s and 1s of length n with s1s.

de Finetti's Theorem asserts that every infinite exchangeable sequences $X_1, X_2, \ldots$ is a mixture of i.i.d. Bernoulli processes. D and F's beautiful and constructive approach to this theorem shows that every finitely exchangeable sequence is *nearly* a mixture of i.i.d. Bernoulli sequences. They tell us what mixture, and how near, as follows. With $B(n, s/n)$ the distribution of n i.i.d. Bernoulli variables with parameter s/n, they show that

$$P^* = \Sigma p(s) B(n, s/n)$$

is close to P in the sense that the distributions of $X_1, \ldots, X_k$ under P and P^* are within $4k/n$ of each other. Thus the main step in their proof is a careful assessment of the difference between sampling with replacement: $B(n, s/n)$ and without: $H(n, s)$. I shall return to this assessment at the end.

This finite form of de Finetti's Theorem is very welcome. Jimmie Savage once wondered whether the Hewitt-Savage $0-1$ law and other $0-1$ laws have any finite content. Now that D and F have shown us the finite content de Finetti's Theorem, perhaps they will turn their attention to $0-1$ laws.

Finite exchangeability

Finite exchangeability is important on its own, not just as an approach to exchangeability. It already explains much about the relation between frequency and probability. For example suppose that $X_1, \ldots, X_{n+1}$ are finitely exchangeable. We observe $X_1, \ldots, X_n$ and are interested in $P(X_{n+1} = 1 | X_1, \ldots, X_n)$. We have, with $X = (X_1, \ldots, X_n)$, $s = X_1 + \cdots + X_n$, $p(t) = P(X_1 + \cdots + X_{n+1} = t)$.

$$\frac{P(X_{n+1} = 1|X)}{P(X_{n+1} = 0|X)} = \frac{p(s+1)/\binom{n+1}{s+1}}{p(s)/\binom{n+1}{s}} = \frac{p(s+1)}{p(s)} \cdot \frac{s+1}{n-s+1}.$$

Thus if, before observing $X_1, \ldots, X_n$, we considered s and $s+1$ about equally likely as values of $X_1 + \cdots + X_{n+1}$, our posterior odds for $X_{n+1} = 1$ are very nearly the frequency odds $s/(n-s)$ for 1s in the first n trials. Bayes may have made this very calculation; according to Steve Stigler [1982], Bayes used the uniform prior distribution precisely because it makes $p(s)$ independent of s.

Partial exchangeability

Partial exchangeability may turn out to be of more practical importance than full exchangeability. For instance if $X = (X_1, \ldots, X_m)$ is a sample of men and $Y = (Y_1, \ldots, Y_n)$ is a sample of women, X tells us something about Y even if the Xs and Ys are not exchangeable with each other. Say that X, Y are *partial exchangeable* if given $s = \Sigma X_i$ and $t = \Sigma Y_j$, the $\binom{m}{s}\binom{n}{t}$ possible X, Y sequences are equally likely. Thus the joint distribution of s and t describes the prior relation between X and Y.

D and F show how to define partial exchangeability in a much more general context, and obtain, for infinite partial exchangeable sequences, an extension of de Finetti's Theorem. I'm not sure, though, that they have fully captured what de Finetti had in mind. He specifically mentioned the possibility that no two variables would be exactly exchangeable, for instance with a sequence of measurements made at different temperatures (another example is estimation of a dose-response curve from responses to a sequence of different doses). Persi, does your general concept of partial exchangeability cover this case?

In the D and F formulation of partial exchangeability, sufficiency plays a central role. For instance in our X, Y example, the pair (s, t) is sufficient for X, Y. Perhaps they have revived the concept for us Bayesians. We haven't needed it up to now, since all Bayes estimates, tests, predictions,... will just naturally come out depending on sufficient statistics only.

A D and F inequality

Denote by $W(A, a)$ the chance that, in drawing a random sample of size a with replacement from a population of size A, we get a different individuals. A basic inequality in the D and F estimate of the difference between sampling with and without replacement is that, for every A, a, B, b,

$$W(A,a)W(B,b) \leq W(A+B, a+b-2).$$

(It is Lemma on p. 748 of D and F [1980], specialized to $c = 2$ and slightly rewritten).

It has the following interpretation. Suppose you have a population of size C and must draw a sample of size c. You may split the population into two subpopulations of any sizes A, B with $A + B = C$ and draw random samples with replacement of any sizes a, b with $a + b = c$ from the two subpopulations. How should you choose, A, B, a, b to maximize your chance of getting c different individuals? The inequality says that your chance does not exceed that with all C in one population, but drawing a sample of only $c - 2$. For c small compared to C, calculations seem to indicate that the D and F inequality is sharp, that how you split C into A, B doesn't matter much, but that c should be split in about the same proportions as C. For $C = 10000$, $c = 100$, here is a table showing, as a function of A, the best $a = a^*$ and the corresponding $P = W(A, a^*)W(B, b^*)$.

A	a^*	P
1	1	.61462
100	1	.61177
500	5	.61160
2000	20	.61163
4000	40	.61147
5000	50	.61162.

The D and F upper bound for P is $W(10000, 98) = .62072$.

Their proof is analytic, using for instance the concavity of $-x \log x$. It would be nice to have a combinatorial proof of their combinatorial inequality.

Thank you, Persi, for what you have taught us and for what you have given us to think about.

SIMON FRENCH (*University of Manchester*)

The Bayesian coin has two faces: probability and utility. Exchangeability ideas are a way of structuring probability distributions so that certain qualitative perceptions in the user's beliefs are represented. They imply the forms of probability models without requiring the user to provide any modelling input in the usual quantitative sense. Within utility theory similar qualitative perceptions, generally called independence conditions, have been shown to imply the form of multi-attribute preference models (Fishburn and Farquhar, 1981, 1982; French, 1986).

I have a gut feeling that there must be a close relationship between probability and exchangeability ideas on the one hand and utility and independence ideas on the other. However, exploring that relationship defeats me. Professor Diaconis said in his presentation that some of the exchangeability ideas "require a fair amount of translation". With my pidgin measure theory and group theory I cannot make that translation. Yet I am sure that relating these two areas of probability and utility is important. Apart from any theoretical cross-fertilisation of ideas that may result, there are practical implications. Decision analysis could be provided with a common approach to structuring probability models and utility functions.

REPLY TO THE DISCUSSION

Professors Blackwell and French have suggested new research projects that I find quite interesting. I do not know any finite version of the zero one laws, Hewitt-Savage, Kolmogorov, or others. I have a nice finite setting to think about what they might mean. In joint work with Freedman (1981) we used Aldous' theorem to construct counter-examples to a set of conjectures about visual perception. The results are full of trivial tail, shift, and partially exchangeable fields. Yet in the end we drew some pictures on a 100 by 100 grid which proved convincing to the experimenters involved. The connection between tail fields and reality remains beyond me, but this applied problem seems like a good place to focus.

The most exciting part of Blackwell's contribution is his suggeston about a purely probabilistic proof of the analytic fact that forms the base of my work with Freedman on finite exchangeability. It seems right to me that there is a proof along the lines that he suggests. I haven't found one.

Aldous and I have started to build a theory of sequences that are almost invariant under permutations, and so almost mixtures of i.i.d. variables. I had hoped to include details here, but there is still too much undone.

Professor French's suggestion is very welcome. There has been far too little emphasis on utility in statistical decision theory. Presumably symmetry considerations can be introduced before the decomposition into probability and utility. Perhaps the split can be made differently. I will take the suggestion to heart and report, at the next Valencia meeting. It is a pleasure to thank both discussants.

REFERENCES IN THE DISCUSSION

Diaconis, P. and Freedman, D. (1980). Finite exchangeable sequence. *Ann. of Prob.* **8**, 745–764.

Farquhar, P. H. and Fishburn, P. C. (1981). Equivalences and continuity in multivalent preference structures. *Ops. Res.* **29**, 282–293.

Fishburn, P. C. and Farquhar, P. H. (1982). Finite degree utility independence. *Math. Ops. Res.* **7**, 348–353.

French, S. (1986). *Decision Theory: An introduction to the Mathematics of Rationality.* Chichester: Ellis Horwood.

Stigler, S. M. (1982). Thomas Bayes's Bayesian inference. *J. Roy. Statist. Soc. A* **145**, 250–258.

BAYESIAN STATISTICS 3, pp. 127–145
J. M. Bernardo, M. H. DeGroot, D. V. Lindley and A. F. M. Smith, (Eds.)

A Bayesian Model and a Graphical Elicitation Procedure for Multiple Comparisons

WILLIAM DUMOUCHEL
BBN Software Products Corp. Cambridge, Massachusets.

SUMMARY

A hierarchical Bayesian model and an interactive graphical elicitation procedure are proposed for the problem of comparing several parameters measurable in the same units. Emphasis is on obtaining interval estimates for contrasts between the parameters in the presence of informative prior information. An interactive computer graphical technique for representing, describing, and eliciting the prior beliefs is proposed and demonstrated.

Keywords: HIERARCHICAL MODEL; INFORMATIVE PRIOR; MOUSE AND WINDOW INTERFACE.

1. THE BAYESIAN MODEL

1.1. Hierarchical parameter structure

Assume that there are K unknown parameters of primary interest in a statistical inference situation, represented as a column vector $\theta = (\theta_1, \theta_2, \ldots, \theta_K)'$. Suppose also that these K parameters can all be measured in the same units, and that interest centers on making comparisons among them. We consider the statistical problem of making inferences about differences between pairs of elements of θ, and of estimating other contrasts. The elements of θ are assumed to be measured on a continuous scale, and the prior probability that any two elements are exactly equal is assumed to be zero. We further assume that (perhaps after making a suitable transformation of scale) the prior distribution of θ is, conditional on two other parameters μ and σ, jointly normal with mean vector $(\mu + d_1, \ldots, \mu + d_K)'$ and covariance matrix $\sigma^2 \boldsymbol{V}$, where $\boldsymbol{d} = (d_1, \ldots, d_K)'$ is a vector of prior means and $\boldsymbol{V}$ is a $K \times K$ positive-definite prior covariance matrix. The values of $\boldsymbol{d}$ and of $\boldsymbol{V}$ are assumed known as the result of introspection or of a process of eliciting the prior distribution. One possible scenario for this elicitation process is described in Section 2. The model includes the two scalar parameters μ and σ in an effort to be more robust against errors of prior specification. The prior distribution of μ is assumed to be normal with mean 0 and standard deviation c, allowing for extra uncertainty in the overall location of the θ's. The prior distribution of σ is assumed to be centered near the value $\sigma = 1$, so that values of $\sigma > 1$ occur when the elicited elements of $\boldsymbol{V}$ are uniformly too small (an overconfident assessor), while values of $\sigma < 1$ occur when the elicited elements of $\boldsymbol{V}$ are uniformly too large (an overly conservative asessor). For convenience the form of the prior distribution for σ^{-2} is taken to be that of a gamma distribution with parameters a and b, and μ and σ are assumed to be independent.

Next, assume that a vector of estimates $\boldsymbol{y} = (y_1, y_2, \ldots, y_K)'$ is observed, and that conditional on θ and a scalar parameter τ, $\boldsymbol{y}$ has a normal distribution with mean vector θ and covariance matrix $\tau^2 \boldsymbol{C}$, where $\boldsymbol{C}$ is a known $K \times K$ positive-definite covariance matrix. Finally, assume that the avaiable information about τ^2, besides that contained in $\boldsymbol{y}$, is summarized by a prior distribution also of the gamma family, with parameters e and f, and that τ is independent of the other parameters in the model. In practice, e and f will usually be determined by the degrees of freedom and mean square of a set of residuals, while the matrix $\boldsymbol{C}$ will be determined by the design matrix of the experiment giving rise to the estimates.

The following equations restate the preceding assumptions [let $\mathbf{1} = (1, 1, \ldots, 1)'$]:

$$\mu \sim N(0, c^2) \tag{1}$$

$$\sigma^{-2} \sim \Gamma(a, b) \quad [\text{i.e. } \sigma^{-2} \text{ has density } f_{\sigma^{-2}} \propto s^{a-1}e^{-bs}] \tag{2}$$

$$\tau^{-2} \sim \Gamma(e, f) \tag{3}$$

$$\boldsymbol{\theta}|(\mu, \sigma) \sim N(\mu\mathbf{1} + \boldsymbol{d}, \sigma^2\boldsymbol{V}) \tag{4}$$

$$\boldsymbol{y}|(\boldsymbol{\theta}, \tau) \sim N(\boldsymbol{\theta}, \tau^2\boldsymbol{C}). \tag{5}$$

The problem is to make inferences about θ once the scalars a, b, c, e, and f, the vectors $\boldsymbol{d}$ and $\boldsymbol{y}$ and the matrices $\boldsymbol{V}$ and $\boldsymbol{C}$, are determined. From a practical point of view, only the determination of $\boldsymbol{d}$ and $\boldsymbol{V}$ (discussed in detail in Section 2) are difficult in this setup. The values of e, f, $\boldsymbol{y}$, and $\boldsymbol{C}$ are computed directly from the experimental design and data, using the usual theory of the general linear model. Determination of a and b depends on how much faith one wants to put on the assessment of $\boldsymbol{V}$. For example, one might choose a and b so that $P(\sigma < 1/2) = 1/10$ and $P(\sigma > 2) = 1/10$ which would be equivalent to maintaining that with 80% probability, the assessments of prior standard deviations of contrasts among the elements of θ are off by at most a factor of 2.

The exact value of c is not very important in most real situations, since c concerns the assessor's uncertainty about the grand mean of all the elements of $\boldsymbol{y}$ but does not affect uncertainty about contrasts. In everything which follows, we assume that $c \to \infty$, that is, that μ has an improper uniform prior distribution.

Once the vector of prior means $\boldsymbol{d}$ has been assessed, we can shift the scale of each element of θ and of $\boldsymbol{y}$ by subtracting $\boldsymbol{d}$ from each of them. Therefore, in the mathematical derivations which follow, without loss of generality we assume that $\boldsymbol{d} = \mathbf{0}$, since otherwise we replace θ by $\theta - \boldsymbol{d}$ and $\boldsymbol{y}$ by $\boldsymbol{y} - \boldsymbol{d}$.

1.2. The posterior distribution of θ given $\boldsymbol{y}$

Using equations (1)-(5), and Bayes theorem, the posterior distribution of θ, $P(\theta|\boldsymbol{y})$, is obtained as follows. Conditional, on μ, σ, and τ, the distribution of $\boldsymbol{y}$ is normal with mean $\mu\mathbf{1}$ and covariance matrix $\sigma^2\boldsymbol{V} + \tau^2\boldsymbol{C}$. To make the calculation, it is convenient to make the transformation $\phi = \sigma^2/\tau^2$. Standard calculations (integrate the product of the three prior densities in (1)-(3) and the normal likelihood with respect to μ, assuming that $c \to \infty$ and $\boldsymbol{d} = \mathbf{0}$) show that the joint posterior density of ϕ and τ^{-2} is

$$p(\phi, \tau^{-2}|\boldsymbol{y}) \propto \phi^{-a-1}|\boldsymbol{W}|^{.5}w^{-.5}\tau^{-2[a+e+(K-3)/2]}\exp\{-\tau^{-2}[f + b/\phi + S(\boldsymbol{y}, \phi)/2]\}, \tag{6}$$

where

$$S(\boldsymbol{y}, \phi) = (\boldsymbol{y} - m\mathbf{1})'\boldsymbol{W}(\boldsymbol{y} - m\mathbf{1}),$$

$$\boldsymbol{W} = \boldsymbol{W}(\phi) = (\phi\boldsymbol{V} + \boldsymbol{C})^{-1},$$

$$\boldsymbol{w} = \boldsymbol{W}\mathbf{1},$$

$$w = \boldsymbol{w}'\mathbf{1},$$

and where

$$m = m(\boldsymbol{y}, \phi) = \boldsymbol{w}'\boldsymbol{y}/w$$

is the posterior expectation of μ given $\boldsymbol{y}$ and ϕ. The density (6) may be integrated with respect to τ^{-2} to yield the posterior density of ϕ:

$$p(\phi|\boldsymbol{y}) \propto \phi^{-a-1}|\boldsymbol{W}|^{.5}w^{-.5}\{f + b/\phi + S(\boldsymbol{y}, \phi)/2\}^{-[a+e+(K-1)/2]}. \tag{7}$$

The posterior density of θ is then the posterior density of θ given ϕ, averaged with respect to $p(\phi|\boldsymbol{y})$:

$$p(\theta|\boldsymbol{y}) = \int p(\theta|\phi, \boldsymbol{y})p(\phi|\boldsymbol{y})d\phi, \tag{8}$$

where, for each value of ϕ, $p(\theta|\phi, \boldsymbol{y})$ is a multivariate Student's-t density with $2e+2a+K-1$ degress of freedom. However, rather than compute this mixture of t-distributions, we shall be content to derive the mean vector and covariance matrix of $p(\theta|\boldsymbol{y})$. A little manipulation shows that conditional on ϕ, τ, and $\boldsymbol{y}$ the distribution of θ is

$$\theta|(\phi, \tau, \boldsymbol{y}) \sim N((\boldsymbol{I} + \phi\boldsymbol{V}\boldsymbol{C}^{-1})^{-1}(m\boldsymbol{1} + \phi\boldsymbol{V}\boldsymbol{C}^{-1}\boldsymbol{y}), \tau^2(\phi\boldsymbol{V}\boldsymbol{W}\boldsymbol{C} + \boldsymbol{C}\boldsymbol{w}\boldsymbol{w}'\boldsymbol{C}/w)), \tag{9}$$

using the fact that $\boldsymbol{C}\boldsymbol{W} = (\boldsymbol{I} + \phi\boldsymbol{V}\boldsymbol{C}^{-1})^{-1}$. Since the mean of the normal distribution in (9) is independent of τ, the center of $p(\theta|\phi, \boldsymbol{y})$ is therefore

$$E(\theta|\phi, \boldsymbol{y}) = (\boldsymbol{I} + \phi\boldsymbol{V}\boldsymbol{C}^{-1})^{-1}(m\boldsymbol{1} + \phi\boldsymbol{V}\boldsymbol{C}^{-1}\boldsymbol{y}), \tag{10}$$

which shows that the posterior mean of θ conditional on ϕ is a matrix-weighted average of $\boldsymbol{y}$ and of the posterior mean of μ conditional on ϕ. By inspection of (6), the posterior distribution of τ^{-2} conditional on ϕ is seen to be

$$\tau^{-2}|(\phi, \boldsymbol{y}) \sim \Gamma[a + e + (K-1)/2, f + b/\phi + S(\boldsymbol{y}, \phi)/2],$$

so that the posterior expectation of τ^2 given ϕ is

$$E(\tau^2|\phi, \boldsymbol{y}) = [f + b/\phi + S(\boldsymbol{y}, \phi)/2]/[e + a + (K-3)/2]. \tag{11}$$

Therefore, by averaging over the variance in (9), and using (11),

$$\mathrm{Cov}(\theta|\phi, \boldsymbol{y}) = (\phi\boldsymbol{V}\boldsymbol{W}\boldsymbol{C} + \boldsymbol{C}\boldsymbol{w}\boldsymbol{w}'\boldsymbol{C}/w)[f + b/\phi + S(\boldsymbol{y}, \phi)/2]/[e + a + (K-3)/2]. \tag{12}$$

Finally, using (7), (10), and (12),

$$\begin{aligned} E(\theta|\boldsymbol{y}) &= \int E(\theta|\phi, \boldsymbol{y})p(\phi|\boldsymbol{y})d\phi \\ &= \int (\boldsymbol{I} + \phi\boldsymbol{V}\boldsymbol{C}^{-1})^{-1}(m\boldsymbol{1} + \phi\boldsymbol{V}\boldsymbol{C}^{-1}\boldsymbol{y})p(\phi|\boldsymbol{y})d\phi, \end{aligned} \tag{13}$$

$$\mathrm{Cov}(\theta|\boldsymbol{y}) = \int \{\mathrm{Cov}(\theta|\phi, \boldsymbol{y}) + [E(\theta|\phi, \boldsymbol{y}) - E(\theta|\boldsymbol{y})][E(\theta|\phi, \boldsymbol{y}) - E(\theta|\boldsymbol{y})]'\}p(\phi|\boldsymbol{y})d\phi. \tag{14}$$

Equations (7), (10), (12), (13) and (14) can be used as computing formulas for making inferences about θ. In practice one first computes and maximizes $p(\phi|\boldsymbol{y})$, then evaluates the integrals in (13) and (14) using at most 10 or 20 values of ϕ, or using the approximate methods of Kass, Kadane and Tierney (1987). Although these formulas give only the first two moments of θ (and from them the first two moments of any linear combination of the elements of θ), one can use a normal approximation for posterior probabilities of intervals or for graphing a posterior density. The posterior distribution is theoretically a mixture of t-distributions, each with degress of freedom $df = 2e + 2a + K - 1$, but unless df is less than 5 or so the normal approximation will be adequate for computing probabilities within the central 99% of the posterior distribution of any contrast.

1.3. Discussion of the model

The model defined by equations (1)-(5) is quite flexible. It is applicable to more situations than the classic oneway ANOVA setup. It could be useful any time differences among the elements of θ are of primary interest. For example, the elements of θ could be the individual terms associated with any factor or interaction in a multiway ANOVA. Repeated measures designs or longitudinal studies could also use this model.The elements of θ could even be regression coefficients, if the corresponding predictor variables had first been normalized to have comparable ranges, and the main interest were to compare the effects of the different predictors, as in a screening experiment. In all these cases, the matrix $\boldsymbol{C}$ is simply related to the design matrix of the experiment, For example, in the oneway ANOVA design, $\boldsymbol{C}$ is a diagonal matrix with $C_{kk} = 1/n_k$, where n_k is the sample size in the kth group, $k = 1, \ldots, K$.

Berry (1988) provides an excellent review of the different philosophies and rationales for the Bayesian and classical approach to the problem of multiple comparisons. One might add to his discussion the following remarks: Both frequentists and Bayesians agree that, when investigating a great many comparisons, it is necessary to adjust for the "surprize value" of any "significant" differences which show up. But the two schools prefer very different strategies for making such an adjustment. In the Bonferroni method, for example, the severity of the frequentist's adjustment is based solely on how many comparisons are being made, with no reference to how surprizing the value of y really was. Frequentists expect to be surprized when making a great many comparisons, and that is all they care about. In contrast, the Bayesian definition of "surprize" is directly tied to how much the data disagree with the prior distribution. The Bayesian adjustment will be practically nil if y agrees with $\boldsymbol{d}$ and $\boldsymbol{V}$, no matter how many comparisons are to be made, while the Bayesian adjustment can be severe if y disagrees with $\boldsymbol{d}$ and $\boldsymbol{V}$, even if only one comparion is made. [From equation (10) one sees that the adjusment of each component of y (actually y-$\boldsymbol{d}$) increases as its distance from m increases unless the posterior mean of ϕ increases at the same time. The posterior mean of ϕ is large when *many* of the components of y are far from m.]

Besides using radically different criteria for *when* an adjustment for surprize value is necessary, the two types of statisticians make the actual adjustment very differently as well. The usual frequentist adjustment consists solely of *increasing the stated uncertainty* of the estimate, but leaving the estimate itself unchanged at y. To the contrary, the Bayesian's use of an informative prior will actually decrease the uncertainty somewhat (compared to that in the sampling distribution of y), but the main effect of an adjustment will be to *pull in the estimate itself* from y toward the set of apriori most likely values of θ. This crucial difference in the *type* of adjustment for surprize value ensures that the frequentist's simultaneous confidence intervals are much wider than the Bayesian's posterior intervals for the same parameter when the number of comparisons being made is large. This can make the simultaneous confidence intervals look way too conservative and is no doubt one reason why many scientists and other data analysts seem reluctant to use them, even when required by strict frequentist theory.

We next discuss the rationale for the particular statistical model defined above. The traditional nonhierarchical Bayesian model for a normally distributed vector parameter, normally distributed data and all variances known would be defined by equations (4) and (5) if $\mu = 0$ and $\sigma = \tau = 1$. The three scalar hyperparameters whose distributions are defined in equations

(1)-(3) provide one degree of freedom each for uncertainty in d, V, and C. Removing any one of them would result in a procedure with undesirable properties.

First, removing the uncertainty in μ would prevent our procedure from easily handling the very common situation in which the general level of response is quite uncertain even though one might be doubtful that there is much difference between treatments. The use of equation (1), with $c \to \infty$, allows μ to be estimated by a weighted average of the elements of y, where for fixed $\phi = \sigma^2/\tau^2$, the weights are proportional to the elements of w, defined below equation (6) as the row sums of $(\phi V + C)^{-1}$. The resulting posterior expectations of the elements of θ are the elements of y, shrunken towards m, when the matrix V is diagonal. (If the user puts more structure in V, then it is possible to have different centers of shrinkage for different subsets of the parameters, as discussed in more detail in Section 2.)

Second, the distribution in equation (2) allows for uncertainty in the elicited prior covariance matrix, V. It is crucial to allow for such uncertainty, since the specification of prior covariance matrix in several dimensions is known to be difficult for most assessors of subjective probabilities. Another rationale for the use of equation (2) is to avoid the restrictive assumption of a normal prior distribution for θ. Combining equations (2) and (4) (with μ fixed), one sees that the marginal distribution of θ is multivariate Student's-t distribution with $2a$ degrees of freedom. By varying a and b, the user of this model can choose from a wide range of behaviors for the prior distributon. Of course the use of σ necessarily multiplies every value of V by the same value. The assessor is not particularly protected against the underestimation of uncertainty in merely some directions of the space of θ. In Section 2 we discuss methods of putting structure into the elicitation of V to help avoid this possibility.

Third, the uncertainty in τ described by equation (3) allows us to handle small sample sizes with unknown error variances in the usual way. In the absence of informative prior information about τ, the value of e and f can be chosen to duplicate the classical fiducial distribution of τ^2 based on the mean square and degrees of freedom of a relevant set of normally distributed residuals. Alternatively, e and f and C can be chosen in some other way to try to match any other assumed distribution of errors in the estimation of θ by y.

In summary, the model developed here seems to be a good compromise between simplicity and flexibility. It is computationally not too laborious, involving a search and a numerical integration in one dimension only. (It takes $K(K + 3)/2$ integrations to evaluate all the means and covariances of equations (13) and (14).) In return one has a fairly realistic model appropriate for a wide variety of data analysis situations, and which can handle the simultaneous estimation of many parameters if they are measured in the same units. This model is a refinement of the model developed in Hildebrandt (1983). Compared to similar models developed by Lindley and Smith (1972) and DuMouchel and Harris (1983), the current model has the simpler multiple comparisons structure relating the elements of θ to each other, but a more complex handling of the variance components, in particular the insistence that the user specify the matrix V and that σ and τ be given prior distributions and be integrated out.

A primary motivation for proposing this model, and the method of elicitation of d and V described in the next Section, is to address the current dearth of general purpose Bayesian data analysis software. If Bayesian statistics is ever to capture even one percent of all the world's statistical analyses, simple and easy-to-use Bayesian versions of the most common types of analyses, namely the various examples of the general linear model, must be developed and put into computer packages. We need to visualize and then put into place not just a mathematical model, but an entire Bayesian data analysis process, consisting of, at least, an assessment of prior belief, a graphical exploration of the data, a combining of the prior belief with the data via the statistical model, a diagnostic assessment of the way the data fit the model, and an exploration and presentation of the posterior distribution and its implications. The current proposal is not fully developed in all these aspects, but we hope it is a start in the right direction.

2. ELICITING A PRIOR DISTRIBUTION FOR SEVERAL PARAMETERS VIA INTERACTIVE COMPUTER GRAPHICS

2.1. Goals of the elicitation procedure

This Section discusses an attempt to develop an easy-to-use and effective method to help a subjective probability assessor formulate an informative multivariate prior distribution for a vector of parameters which are all measured in the same units. The parameters are required to be measured in the same units because the assessment involves making visual comparisons between pairs of the parameters on the computer screen, and interacting with a graphical display depicting the user's estimates of, and uncertainty about, differences between pairs of parameters. The procedure is intended to be useful in a wide variety of statistical situations where the data analyst has nonnegligible subjective information about a set of parameters. It is designed to be especially suitable for use with the statistical model and data analysis situations discussed in Section 1 of this paper.

The procedure merely elicits a second-order description of the multivariate distribution, assuming that such an approximation will be sufficient for its intended use. It should be appropriate for a distribution with roughly elliptical contours, but not for a joint distribution containing spikes, pronounced multimodality, or extreme skewness in one or more directions. The representation of the distribution is in terms of specified intervals containing a fixed probability, say 95%, of the prior probability for all differences between pairs of parameters. If it were assumed that the distribution is jointly normal, then these 95% intervals could be used to construct the mean and covariance matrix of the distribution. One additional subjective assessment, of the parameter σ in equation (2) of Section 1.1., allows the construction of a multivariate Student's-t distrubution, as noted in Section 1.3.

The procedure asks the assessor to describe uncertainty about the true values of the parameters directly, rather than indirectly by way of assessing a predictive distribution of future observables. Athough some authors advocate the use of predictive distributions in elicitation situations, the strategy of working directly with the parameters was adopted here for two main reasons. First, there are many inference stuations in which the parameters, and differences between parameters, are more intuitive, and are easier to form probabilities about, than are the raw data values. This is especially true in many multiple comparisons situations. For example, suppose a psychology experiment were performed to compare reaction times of subjects exposed to many different stimuli under several different conditions. The psychologist may have a theretical model which predicts certain relationships between the mean reaction times for different stimuli, and may have in mind measures of uncertainty or expected deviations from the model as well. But in order to assess 95% probability intervals for each actual reaction time, it would be necessary to also carefully assess the combined effects of variations in the experimental conditions and many characteristics of the population of subjects, even if these factors were not central to the scientific question being evaluated. Direct assessment of the stimuli effects greatly lessens, although it does not completely avoid, this potential confounding problem. This is also the reason for the use of a diffuse prior for μ in the model of Section 1.

A second reason for preferring to work with the parameters is the desire to create a computerized procedure which can be used directly in a variety of situations. The parametric setup described in Section 1 is applicable to many statistical estimation situations in which multiple comparisons are involved. An assessment procedure involving the prediction of actual observations would either have to be tailored separately for every different experimental design or have to be much more complex in order to flexibly accommodate different designs.

Finally, a primary goal of the proposed procedure is to use graphics to make the assessment process quicker and more intuitive and interesting. Earlier attempts at computer elicitation of multivariate prior distributions, like that of Kadane et al (1980), have involved long text-and-number-oriented interactions with the computer and little feedback on the implications of previous choices. The computer program was usually in control, and the assessor responded passively. The visual and icon-oriented nature of the procedure described here allows the assessor to have a feeling of being in control, and even to skip back and forth easily between the tasks required to perform the assessment.

2.2. Eliciting the marginal distributions of each parameter

The proposed elicitation procedure requires that the probability assessor interact with a computer screen using a mouse or some other pointing device. The prototype program described here was written in the computer language APL on the Apple Macintosh microcomputer. An example use of the program is used to help describe the procedure. Several figures show the computer screen at different stages of the assessment. Because the process involves motion and interaction with the computer screen, it is hard to make figures which adequately capture the dynamic flavor of the assessment. The reader is asked to try to "read between the lines" of the figures and the text to imagine this aspect of the method.

The example involves the assessment of uncertainty about the relative risk from bone cancer incurred by receiving a fixed dose of ionizing radiation from each of four radioactive isotopes. The example is inspired by a meta analysis in DuMouchel and Groër (1987) which summarized the results of 15 studies of the effects of internal deposition of alpha-emitters on bone cancer mortality in humans and other especies. The full complexity of the original study will not be discussed here. For the sake of this example, just suppose that for a particular dose to the bone of each isotope, say 100 rads, θ_1, θ_2, θ_3, and θ_4 represent the $\log_{10}$ relative risks from bone cancer to a human population exposed to the four isotopes, compared to an unexposed population. The four isotopes are Radium 226, Radium 228, Plutonium 238, and Plutonium 239. Supppose the assessor is confident that all four values of θ lie between 0 and 2 (relative risk 1 and 100).

The assessor begins the process by specifying just two quantities: the number of parameters to be assessed, in this case four, and the range over which they vary, in this case from 0 to 2. The program responds by drawing a rectangular area on the screen and marking the left side of this area with tic marks from 0 to 2. Lined up horizontally midway between the top and bottom of the rectangle are four small open circles representing the four values of θ, as shown in Figure 1. By using the mouse, whose screen position is represented by the diagonal arrowhead in the figures, the user of the program can select fields above and below the drawn rectangle and then type in labels for the units of θ and the individual parameters. This labeling process is just being finished in Figure 1.

The assessor now begins the specification process. In Figure 2 the user has just selected the parameter for Radium 226 by moving the mouse pointer over the circle representing θ_1 and depressing the mouse button, and the circle turns black to acknowledge the selection. Next, the user "drags" the circle up or down by keeping the mouse button depressed and moving the mouse at the same time.

In Figure 3, the Radium 226 parameter has been moved to a value more nearly midway between 0 and 1, in accordance with the assessor's prior expectation of θ_1. The user will now express uncertainty about this value by drawing error bars about this value. Note that, after being selected a second time, in Figure 3 the circle for Radium 226 has sprouted little "handles", one of which is now being selected by positioning the mouse on the handle just outside of the circle itself.

In Figure 4, as the handle is dragged away from the circle, a rectangular bar expands and contracts with the mouse pointer, while the assessor decides on an interval of uncertainty representing 95% of the prior probability. In Figure 5, the user has just released the mouse

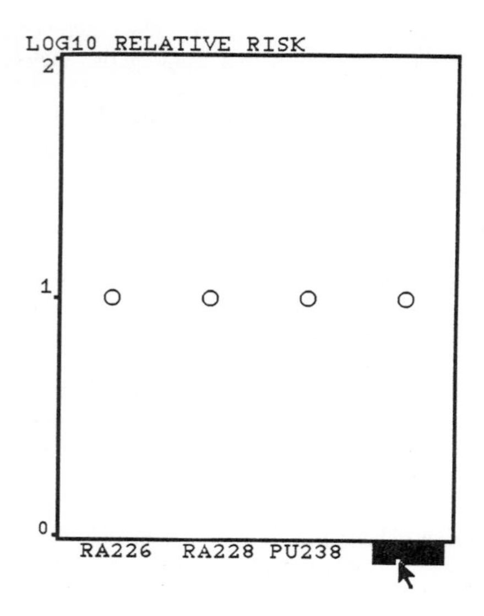

Figure 1.
Visual representation of several parameters on a common scale

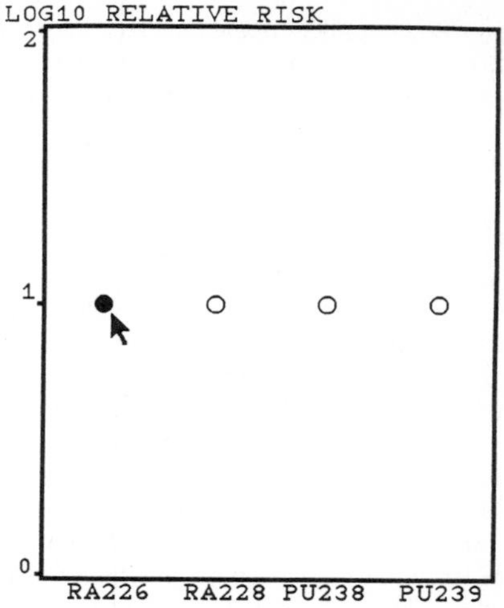

Figure 2.
Selecting a parameter

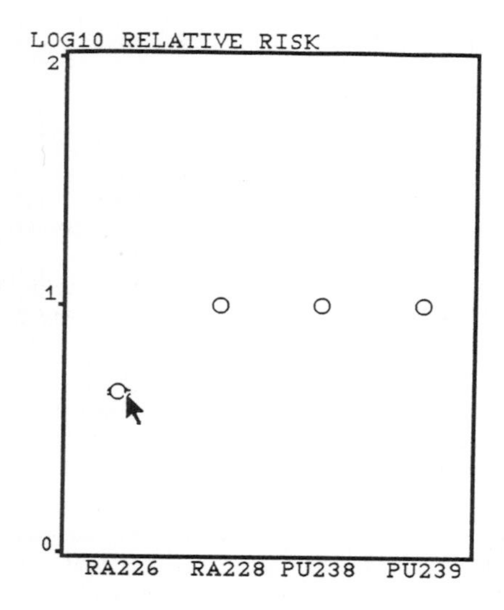

Figure 3.
Selecting an error bar

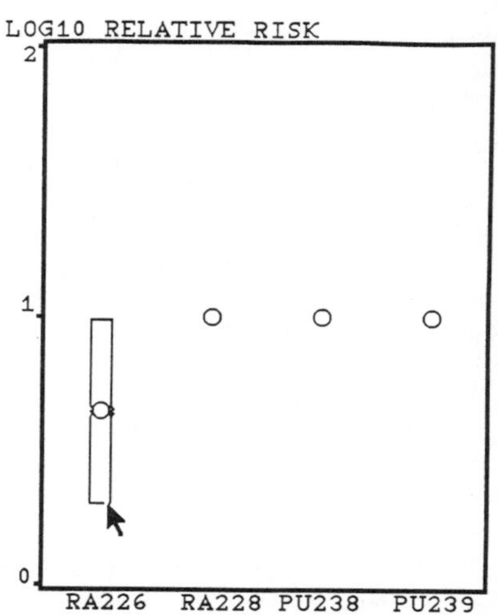

Figure 4.
Extending the error bar

button and the error bar has reverted to its conventional shape. The assessor is free to adjust the location of the circle or the length of the error bars at will.

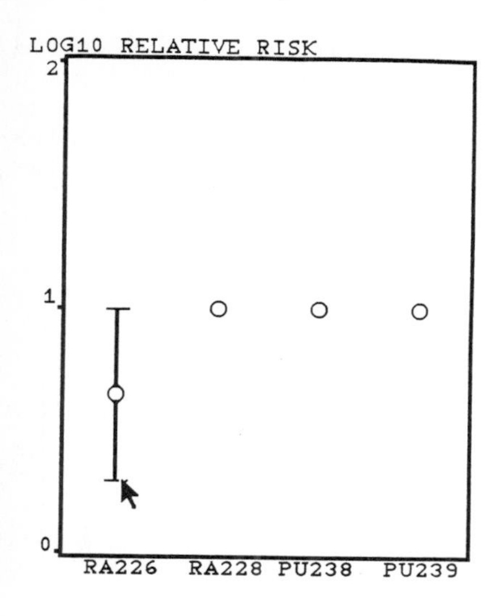

Figure 5.
The completed error bar

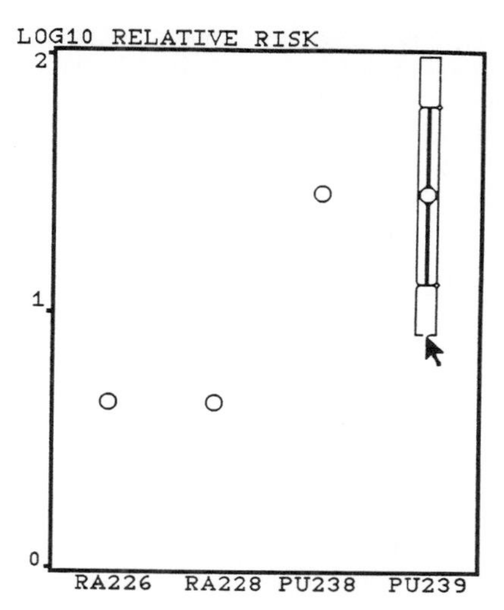

Figure 6.
Adjusting an error bar

Next, the assessor moves to consideration of one of the other isotopes and repeats this process. In Figure 6 the user is just finishing the the adjustment of the error bar for the fourth isotope, Plutonium 239. Note that previously constructed error bars temporarily disappear during consideration of the marginal distribution of any particular parameter, but that any of the bars may be redisplayed and adjusted at any time by clicking on the corresponding circle twice in a row. Note also from Figure 6 that the assessor believes that the two isotopes of plutonium are each more dangerous than the two isotopes of radium. Previous research has shown that ingested plutonium seems to concentrate more in the outer layers of bone tissue than does ingested radium. Because most bone cell division occurs in the outer layers of the bone, the assesor belives that radioactive doses to this layer are more dangerous than doses to the skeleton as a whole. However our knowledge of the mechanisms of isotope deposition and of bone cancer is so weak that this is only a conjecture.

2.3. *Eliciting uncertainty about differences between parameters*

In attempting to construct a graphical representation of the uncertainty about differences between pairs of parameters, we must overcome the fact that, when assessing K parameters, there are $K(K-1)/2$ differences to visualize. It is important to choose a representation which avoids confusion as to which parameters are compared at any one time and which also blends well with the representation of the marginal distributions. The solution proposed here involves the construction of error bars which have a different interpretation than usual, but which seem to fit the needs of the situation well.

The key idea is that when comparing two of the parameters, error bars form around *each* of the two corresponding points, and the lengths of these error bars are proportional to the *uncertainty in the difference* between the parameters, rather than to the marginal uncertainty about the individual parameters. That is, associated with each of the K points is not just one error bar, but K error bars, one representing uncertainty in the marginal distribution of that parameter, and one representing uncertainty in the distribution of each of the $(K-1)$

differences between that parameter and the other parameters. Which error bar is displayed on the screen at any one time depends on the sequence of selection of circles with the mouse. If the same circle is selected by two successive mouse clicks, then the error bar representing the uncertainty in the marginal distribution of that parameter is displayed. Otherwise, error bars will be displayed about the two most recently selected circles, and the heights of these bars will be proportional to the uncertainty in the difference between the parameters.

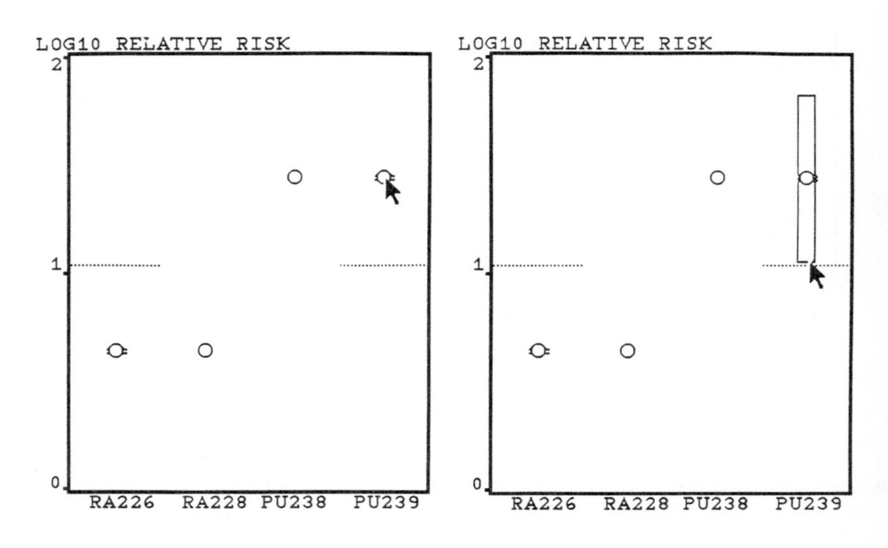

Figure 7.
Eliciting uncertainty about contrasts

Figure 8.
Extending the bar for a comparison

Figure 7 shows the assessor just beginning this process of considering differences. The circles corresponding to Radium 226 and Plutonium 239 have just been selected in succession. As a result, little handles for error bars form on both circles, and a dotted horizontal reference line appears between the two selected circles. (The vertical distance between the intermediate reference line and each of the selected circles is proportional to the uncertainty in the marginal distributions of the respective parameters. In Figure 7 the line is about halfway between the heights of the two circles because the error bars for the marginal distributions are about equally high.)

Next the user selects a handle on either of the two circles and draws out an error bar as before, and as shown in Figure 8. But, as shown in Figure 9, once the mouse button is released a corresponding error bar is also drawn about the other selected circle. (In fact, the program could be improved by having both error bars expand and contract in unison as the user adjusts one of them with the mouse.) In Figure 9, the lower tip of the one error bar and the upper tip of the other error bar just about touch the reference line. This is intended to represent the fact that a 95% error bar for the *difference* between the two parameters would almost, but not quite, intersect the origin. That is, when two parameters are selected in the elicitation procedure, each of the two corresponding vertical strips is converted into scaled error bar for the difference, with the dotted reference line serving as a temporary origin. For example, if the user chose to shrink the error bars in Figure 9 so that they reached just halfway between the reference line and the points marking the expected values of the parameters, the

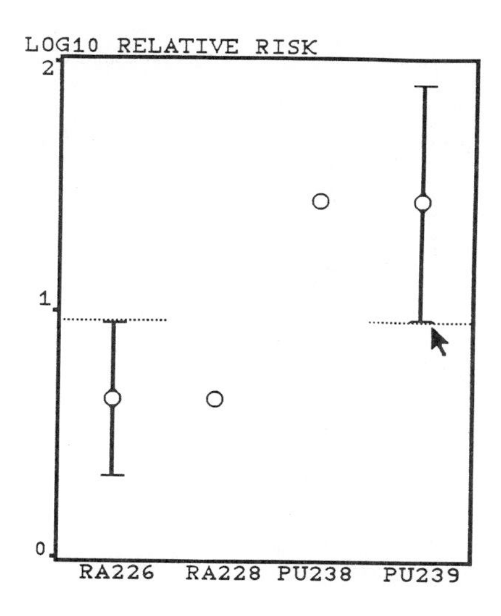

Figure 9. *Visualize the range of differences*

interpretation would be that the 95% interval for the difference between the two parameter extends only halfway from the expected difference toward the possibility of no difference (and extends an equal amount in the opposite direction, of course). In this way, the ratio of the distance of the circle from the line to the height of the error bar represents a "visual t-statistic" for the hypothesis of equality between the two parameters.

Another interpretation for these error bars focuses more on the relation of the two sets of error bars to each other than on their relation to the reference line. The 95% range of probable differences between the two parameters consists of the set of all diferences which can be formed by taking one point from each of the two bars, and subtracting their heights. In Figure 9, the extreme pairs of points are, first, the comparison of the lower limit for Plutonium 239 to the upper limit of Radium 226, where the difference is about 0, and second, the comparison of the upper limit of Plutonium 239 to the lower limit of Radium 226, where the difference is that Plutonium 239 is more potent than Radium 226 by one and one-half orders of magnitude. So the 95% probability interval for the difference extends between these extreme differences. Note that, when a difference is being assessed, *the sum of the lengths of the two bars* has the same interpretation as the length of the single bar has when a single parameter is being assessed. For visual consistency, the program keeps the ratio of the lengths of the two bars making up the sum equal to the ratio of the two corresponding single-parameter bars.

The psychological perceptual tasks necessary to make the quantitative interpretations and assessments described in the last two paragraphs take a little practice to perform smoothly and accurately, but they are reasonably natural and form a consistent whole, once the assessor has firmly in mind that there is not just one error bar for each point, but different error bars for each pair of points.

Figures 10 and 11 show the same operations being carried out for the comparison of Plutonium 238 and Plutonium 239. In Figure 10 the user is in the process of shortening the interval of uncertainty for the difference between these two isotopes, and in Figure 11 the

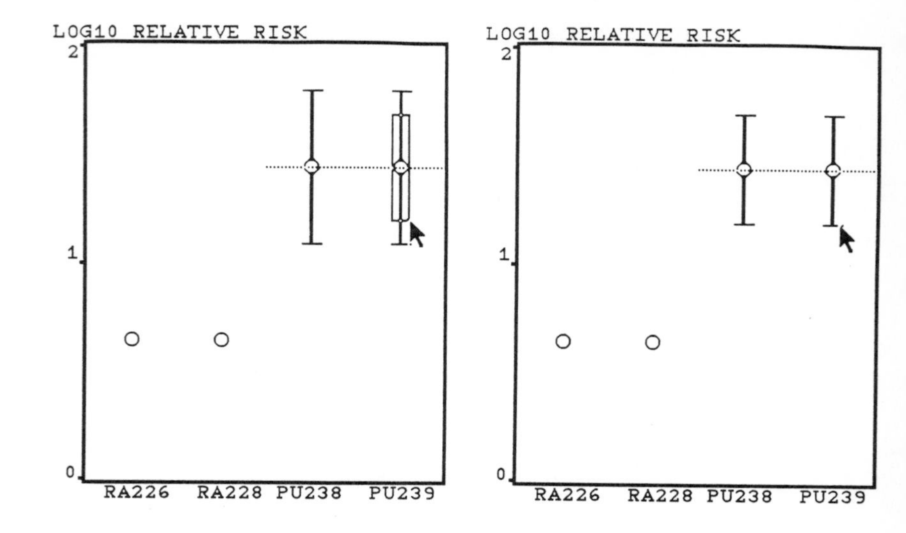

Figure 10.
Adjusting the error bar for a comparison

Figure 11.
New error bars

revised error bars are displayed. For this comparison, the distance between two prior means is very small compared to the heights of the error bars, so the reference line does not play much of a role in the visual interpretation. The principal comparison is between the upper tip of one bar and the lower tip of the other, since the assessor is claiming that differences more extreme than these have less than a 5% chance of occurring.

2.4. Relating the assessed differences to a joint distribution

Once the assessor has constructed error bars corresponding to each parameter and each pair of parameters, the question arises of how to use this assessed uncertainty in a subsequent Bayesian analysis. One simple strategy, which we adopt, is to assume that the subjective distribution is approximately a multivariate normal distribution, and to interpret the length of each 95% probability interval as 4 times the standard deviation of the quantity being estimated. That is, let L_{kk} be the length of the 95% error bar for $\theta_k, k = 1, \ldots, K$, and let L_{jk} be the sum of the lengths of the two error bars for $\theta_j - \theta_k$, $1 \le j < k \le K$. Let v_{jk} be the (j,k) element of $\boldsymbol{V}$, the approximating normal covariance matrix. Then we have the system of $K(K+1)/2$ equations:

$$L_{kk}^2 = 16 v_{kk}, \qquad k = 1, \ldots, K,$$
$$L_{jk}^2 = 16(v_{jj} + v_{kk} - 2v_{jk}) \qquad 1 \le j < k \le K.$$

This system has an easy solution, namely

$$v_{kk} = L_{kk}^2/16 \qquad k = 1, \ldots, K,$$
$$v_{jk} = (L_{kk}^2 + L_{jj}^2 - L_{jk}^2)/32 \qquad 1 \le j < k \le K.$$

It seems advisable to constrain the estimated $\boldsymbol{V}$ to be positive definite. There are several ways this might be done. First, one could modify the solution of the above system of equations whenever the solution could not be a covariance matrix. Second, one could prevent the user from assessing values of L_{jk} which would lead to an illegal $\boldsymbol{V}$ matrix. Third one could build some extra structure into $\boldsymbol{V}$ so that a positive-definite $\boldsymbol{V}$ is automatically obtained. The last two of these strategies have been implemented.

In order to keep the user from constructing a confidence bar which would lead to an illegal value of $\boldsymbol{V}$, the system converts from the L's to $\boldsymbol{V}$ after every modification of an L. Then, if the user is next about to modify L_{12}, say, the system computes the range of values of v_{12} which are compatible with the other elements of the current value of $\boldsymbol{V}$. This range of values is converted to a range of acceptable values of L_{12}. Then, if the user attempts to stretch or contract the error bar for the $\theta_1 - \theta_2$ comparison outside of the acceptable range, the program breaks the connection between the mouse and the error bar and the error bar reverts to its former size.

Although the user can be prevented from constructing a $\boldsymbol{V}$ matrix which is not positive definite by the method of the previous paragraph, that solution to the problem is not perfect. It can happen that, during the construction of a long series of error bars, the next error bar to be adjusted is constrained to be in a region which conflicts with the opinion of the assessor. The only way to adjust that error bar to a more acceptable value is to first go back and adjust other error bars to relax the constraint. But there is no easy way to tell which other error bars are most responsible for the constraint on the one which the assessor wants to change. So a trial and error process sometimes continues until the system lets the assessor make the desired change.

A more workable solution is to try to take advantage of any structure present in the substantive problem being analyzed to simplify the covariance matrix $\boldsymbol{V}$. Although some asssessment problems may indeed require all $K(K+1)/2$ different elements of $\boldsymbol{V}$, there are often subsets of the parameters whose members are exchangeable, and $\boldsymbol{V}$ can then be greatly simplified. In the radiation risk problem being used in our example, the assessor may have no opinions as to distinctions between the two isotopes of Radium, or between the two isotopes of Plutonium. It is also reasonable to suppose that θ's for two isotopes of the same element are more highly correlated than are two θ's corresponding to isotopes of different elements. This suggests that the following statistical model may well represent the assessor's opinions about the θ's. Let the θ corresponding to the jth isotope of the ith element be denoted θ_{ij} The model assumes that

$$\theta_{ij} = \mu + \alpha_i + \delta_{ij}, \tag{15}$$

where μ, the α's and the δ's are all independent, and where the variances of these components are assumed to depend only on i but not j and are given by

$$\begin{aligned} V[\mu] &= v_0, \\ V[\alpha_i] &= A_i, \\ V[\delta_{ij}] &= D_i. \end{aligned}$$

The elements of θ (actually of θ-$\boldsymbol{d}$) are thus assumed to be partially exchangeable. This model has just 5 variance parameters (v_0 plus one A_i and one D_i for each of the two elements) rather than the 10 necessary for an arbitrary 4 by 4 covariance matrix. It assumes that the variance of θ for every isotope of the ith element is $v_0 + A_i + D_i$, that the covariance between the θ's of two isotopes of the ith element is $v_0 + A_i$, and that the covariance between the θ's of all pairs of isotopes from different elements is v_0. In general, if the K parameters are divided into I groups of parameters, with the ith group having J_i members, then the $K(K+1)/2$ potentially, different v's are computed from at most $2I+1$ values of v_0, A_i and D_i. (To avoid unidentifiability, we set $D_i = 0$ whenever $J_i = 1$.)

It is easy to use the elicited lengths of error bars to estimate v_0 and the A's and the D's. If the L's are all squared and divided by 16, they estimate the variance of a θ or of a difference of θ's. From the relationships

$$V[\theta_{ij}] = v_0 + A_i + D_i, \tag{16}$$

$$V[\theta_{ij} - \theta_{ik}] = 2\, D_i, \qquad (j \neq k) \tag{17}$$

$$V[\theta_{ij} - \theta_{kl}] = A_i + D_i + A_k + D_k, \qquad (i \neq k) \tag{18}$$

one can build a linear model in which the L^2's are predicted by v_0, the A's and the D's. Solving this model by least squares or some other method leads to estimates of the parameters of $\boldsymbol{V}$. (Any estimates of v_0, A_i or D_i which turn up negative are set to 0.)

If the purpose of the elicitation is to obtain values of $\boldsymbol{d}$ and $\boldsymbol{V}$ for use in the multiple comparisons statistical model developed in Section 1 of this paper, then the estimation of v_0 becomes less important, because the use of a diffuse prior distribution for μ causes the posterior distribution to be independent of v_0, (Notice that the symbol μ has exactly the same meaning in equation (15) as in equation (4), namely a location parameter common to every element of θ.) This implies that, for use in the model of Section 1 and if $c \to \infty$, the elicitation procedure for $\boldsymbol{V}$ could be based exclusively on assessments of differences, in which case eq. (16) would not be used when building $\boldsymbol{V}$. But we prefer to use assessments of uncertainty in the marginal distributions and v_0 is estimated so as not to bias the estimates of the A's and the D's, even though its value is irrelevant to the analysis.

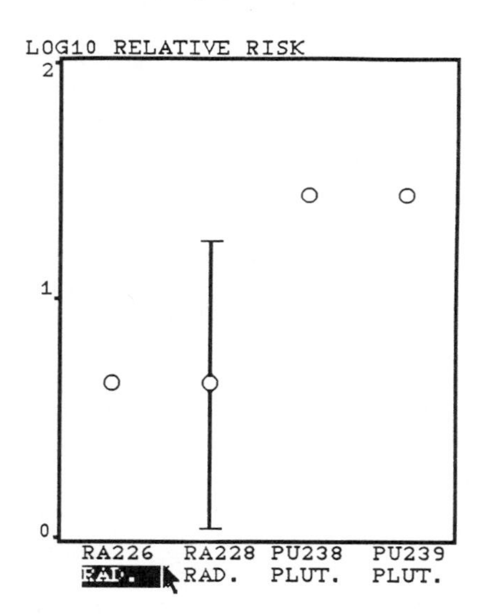

Figure 12.
Adding structure by grouping the parameters

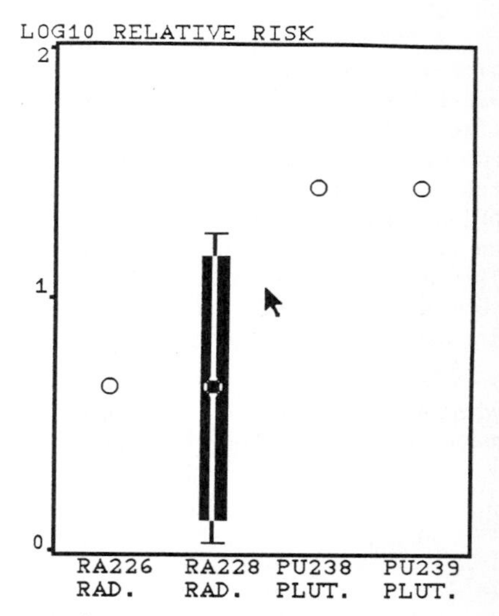

Figure 13.
Comparing fitted error bars to elicited ones

2.5. Continuation of the example elicitation

These ideas have been implemented in the prototype computer program. Figure 12 shows that the assessor has decided to force a grouping of the four parameters by typing in group names below the label for each parameter. Once groups have been formed, the system uses the model of equation (15) to estimate the covariance matrix $\boldsymbol{V}$, and from that the predicted lengths of error bars can be computed. Figure 13 shows how the system superimposes the fitted error bar on top of the user-specified error bar for easy comparison. The assessor's stated uncertainty in the potency of Radium 226 seems to be a bit greater than would be predicted by the fit of the components of variance model to all the error bars.

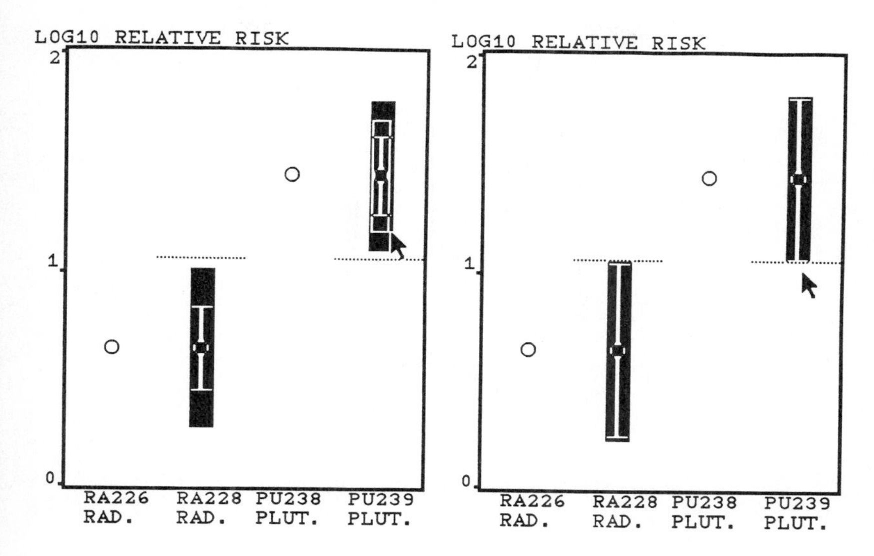

Figure 14.
Modifying an "outlier" error bar

Figure 15.
Updated fitted and elicited bars

Figure 14 shows another comparison, Radium 228 versus Plutonium 239, in which the elicited uncertainty in the difference between the two θ's was only about half of what the model fit to all the responses predicts. Upon reflection the assessor decided to increase the estimated uncertainty for that contrast, as shown in Figure 14. Figure 15 shows that after the elicited error bar is changed the fitted error bar is also updated.

Using the information that the four isotopes are from just two elements, the assessor can easily specify that differences between isotopes of the same element have a smaller prior variance than differences between isotopes of different elements. When the resulting matrix V is used to compute the posterior mean of θ, the effect will be to shrink the estimates for the two isotopes of each element toward the mean of that element, rather than merely shrink all four estimates toward the grand mean. Note that if there were, to take an extreme example, 40 isotopes of, say, 10 elements, and if the same judgement of partial exchangeability between isotopes of the same element were to apply, then a minimum of 20 assessments of differences would be required to estimate V (although making 30 or 40 might be advisable) rather than the $40(40+1)/2 = 820$ assessments required to estimate V in full generality.

2.6. Conclusion

An interactive computer graphical procedure has been proposed and implemented which makes the assessment of prior information for a particular class of multiparameter models relatively painless. The restriction to parameters all measured in the same units allows the specification of the second-order model in terms of comparisons of pairs of parameters. In addition, the program makes it easy for the assessor to specify subsets of the parameters which are exchangeable and to take advantage of this specification during the elicitation. The result is a

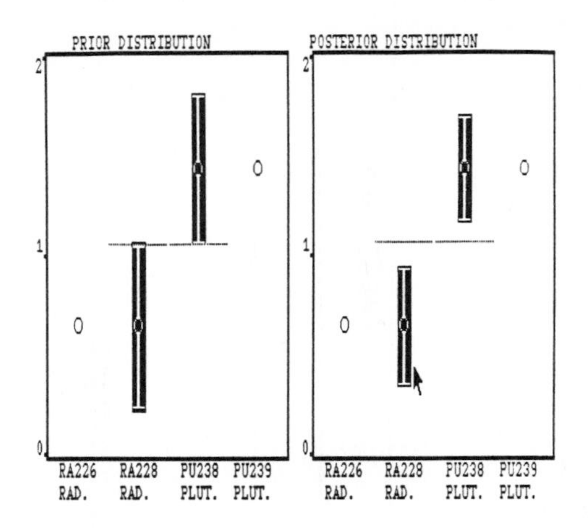

Figure 16. *Comparing prior and posterior distributions*

considerable lessening of the "curse of dimensionality" which usually accompanies attempts to work with prior distributions of many parameters.

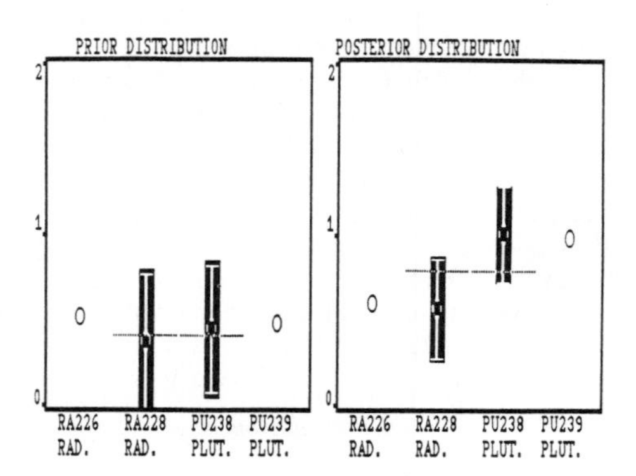

Figure 17. *"What if" the prior were changed?*

A Bayesian hierarchical model has also been developed to analyze data in conjunction with this elicitation procedure. Both the model and the elicitation procedure are especially suitable for several classes of problems involving multiple comparisons. It is especially important to develop methods for using *informative* priors in multiple comparisons problems, because otherwise the Bayesian is liable to misinterpret a surprizing result found during the course of a long investigation.

Finally, Figures 16 and 17 show one more feature of the computer program developed for this reseach. It is possible to display two distributions for θ side-by-side on the computer screen, one displaying the prior distribution and the other displaying the posterior distribution. Each distribution can be dynamically displayed and compared by selecting different parameters or pairs of parameters and viewing the corresponding estimates and error bars. In addition, if one uses the mouse to alter one more features of the prior distribution, the posterior distribution is automatically recalculated and all of its features available for investigation. Figure 16 displays a prior distribution which agrees with the (artificial) data, and we note that the posterior is very similar to the prior distribution, but with less uncertainty, at least in the contrast being displayed, Plutonium 238 versus Radium 228. Figure 17 shows what happens after the assessor changes the prior distribution to one which hardly distinguishes between the potency of Radium and Plutonium. The corresponding posterior distribution, shown in the right half of Figure 17 is now much less certain about whether Plutonium 238 is more potent than Radium 228.

REFERENCES

Berry, D. (1988). Multiple comparisons, multiple tests, and data dredging, a Bayesian perspective. (in this volume).

DuMouchel, W. and Groër, P. (1987). A Bayesian methodology for combining radiation studies. Unpublished manuscript, 21pp.

DuMouchel, W. and Harris, J. (1983). Bayes methods for combining the results of cancer studies in humans and other species (with discussion). *J. Amer. Statist. Assoc.* **78**, 293–315.

Hildebrandt, B. (1983). *A Bayesian Version of One-way ANOVA and Multiple Comparisons*, M. S. Dissertation, Massachusetts Institute of Technology Statistics Center, Cambridge, MA 02139.

Kadane, J., Dickey, J., Winkler, R., Smith, W. and Peters, S. (1980). Interactive elicitation of opinion for a normal linear model. *J. Amer. Statist. Assoc.* **75**, 845–854.

Kass, R., Kadane, J. and Tierney, L. (1987). Asymptotics in Bayesian computation. (in this volume).

Lindley, D. V. and Smith, A. F. M. (1972). Bayes estimates for the linear model (with discussion). *J. Roy. Statist. Soc. B* **34**, 1–41.

DISCUSSION

JOHN J. DEELY (*University of Canterbury*)

This paper presents a useful and necessary Bayesian model for multiple comparisons and Bill DuMouchel is to be congratulated for both his clarity of exposition and his fine judgement concerning a "compromise between flexibility and simplicity". His interactive computer package for the elicitation of personal probabilities and subsequent computation of posterior probabilities related to individual parameters and their differences in AOV type problems is a good first attempt to "address the current dearth of general purpose Bayesian data analysis software!" I think he is on the road to that million dollar pot of gold to be found somewhere in the software jungle!

My comments will be broken into two categories: one concerning the model and two concerning the elicitation procedure.

The model: It is clear that frequentists have out-distanced Bayesians by a long way in the development of software for data analysis. To wean practitioners from these crude, cumbersome and often incorrect methods, various strategies are suggested. DuMouchel argues that we must meet the practitioners on their ground; in that regard when it comes to AOV type problems they are on the multiple comparisons syndrome. Hence his effort in this paper. There is another approach suggested in Berger and Deely which also uses a hierarchical Bayesian model for AOV type problems. Their model although similar to DuMouchel's has important differences, as does their philosophy and end result. They allow a more flexible overall mean specification for $\theta|\mu, \sigma$, require a diagonal matrix V, but treat quite arbitrary distributions for σ^2. They advocate that if the practitioner is truly interested in ranking or comparing various treatment means then interest should be focused on computing for each j the posterior probability that θ_j belongs prespecified sets consistent with the overall goal of the AOV type experiment. For example if one is interested in the treatment with largest θ_j then they compute $p_j = p(\theta_j > \theta_i$ for *all* $i \neq j|$ data) for $j = 1, 2, \ldots, k$, and in fact can compute a posterior distribution of the form $p(\theta_j > \theta_i + \delta$ for *all* $i \neq j|$ data) for *any* δ. Thus the practitioner can determine quite easily which θ_j is largest and by how much. These computations require evaluating only a three dimensional integral and do not require an approximation to either the posterior distribution $p(\theta|y)$ or the subjective initial prior both of which are approximated in DuMouchel's approach. It is to be noted that no definitive errors associated with these approximations were given in DuMouchel's paper. The basic difference between the two approaches not only relates to the model for prior information but more importantly what is the most desirable output in the data analysis stage. The major feature of DuMouchel's approach is that he has computerised it and made it possible for the practitioner to easily interact with the package and obtain worthwhile analysis of the data. It remains for the same to be done with the Berger and Deely model.

The elicitation procedure: My only comment here is that the procedure seems cumbersome even though I did try to "read between the lines". Experience with practitioners will no doubt provide the appropriate comment on this procedure. Part of my complaint is that I can't see how a practitioner will be aware enough (let alone be motivated enough) to express confidence bars on *all* comparisons when k is ten or more. Partial exchangeability does help in this regard by considerably reducing the number of parameters to be elicited and hence the number of comparisons required. But what if no judgement of partial exchangeability can be made a priori. Is there a way that the computer package can look at the data and suggest possible partially exchangeable subsets for the practitioner to consider? If so, what would the effect be upon sound statistical analysis?

In conclusion I would want to strongly encourage this type of work in all areas and again to congratulate Bill DuMouchel for an excellent paper. His exposition particularly in Section 1.3, paragraphs 2 and 3 is beautifully concise and his admonition in the last paragraph of that Section speaks loudly of the Savage influence.

REPLY TO THE DISCUSSION

I would like to thank Professor Deely for his encouraging words. I will first reply to his questions about the model introduced in my paper, and then to his comments on the elicitation procedure.

My confidence in the normal approximation to the posterior distributions is based on the well-known accuracy of the normal approximation to the t-distribution. For example, a normal distribution with the same variance matches the width of 95% t-distribution intervals to within 10% whenever $df > 2$, and matches 99% intervals to the same relative accuracy whenever $df > 9$. For the practical purpose of assessing contrasts, this is almost always sufficient accuracy. Although I did not address the issue of ranking and selection of parameters, I expect that approximations assuming that the posterior distribution is normal will be adequate there also.

My proposed elicitation procedure would indeed be cumbersome if there were ten or more parameters to assess, and no assumptions of partial exchangeability. I will continue to try to improve it, and I would very much appreciate receiving suggestions for how to do so. But no less cumbersome procedure has yet been proposed in that situation. The fact is that every other multiple comparison procedure, whether Bayesian or classically inspired, makes at least implicit use of assumptions equivalent to partial exchangeability. The partial exchangeability model in Section 2.4 has several simple special cases suitable for general use. For example, if there are K parameters, setting $I = K$ and every $J_i = 1$ (uncorrelated priors) leads to a procedure requiring only $(K + 1)$ elicitations. Setting $I = 1$ and $J_1 = K$ (completely exchangeable prior) requires only 2 elicitations. And setting $I = 2$, $J_1 = K - 1$ and $J_2 = 1$ (in order to compare $K - 1$ exchangeable treatments to a single control) would require just 4 elicitations. I would not suggest that the computer estimate which parameters to group together in the context of the model proposed here. The data should not suggest the prior distribution, if these posterior distributions for contrasts are to be used. The multiple-shrinkage families of prior distributions suggested by O'Hagan (1988) are more suitable for that purpose.

REFERENCES IN THE DISCUSSION

Berger, J. O. and Deely, J. (1988). A Bayesian Approach to Ranking and Selection of Related Means with Alternatives to AOV Methodology. *J. Amer. Statist. Assoc.* **82.**

O'Hagan, A. (1988). Modeling with heavy tails. (in this volume).

BAYESIAN STATISTICS 3, pp. 147–158
J. M. Bernardo, M. H. DeGroot, D. V. Lindley and A. F. M. Smith, (Eds.)

The Future of Statistics in Retrospect

S. GEISSER
University of Minnesota

SUMMARY

We present the view that all statistical analyses with the possible exception of the measurement of certain physical entities are more suitably executed in a predictivistic manner. Also a great variety of problems can only be handled in this manner.

Keywords: BAYESIAN INFERENCE; DISCORDANCY TESTING; MODEL SELECTION; PERTURBATION ANALYSIS; PHYSICAL MODELS; PREDICTIVE INFERENCE; PREDICTIVE SAMPLE REUSE; PROBABILITIES FOR EXTREMES; SCREENING TESTS.

1. INTRODUCTION

Where are we and where are we going? This is a topic periodically discussed at conferences relating to scientific and technical disciplines. For statistics it is, in addition, the fundamental paradigm of the subject. Virtually unrecognized as such by the vast majority of statisticians for over the past 50 years, it now appears to be gaining a measure of its rightful acceptance.

A related view, implicitly shared by many over the same period is that statistical models represent a mechanistic reality in a variety of disciplines. Otherwise why would the overwhelming majority of statistical publications be replete with theoretical and methodological efforts directed towards the testing and estimation of parameters?

We shall attempt to discuss these topics in some detail in order to indicate some appropriate directions for the future of statistics.

2. THE 4 M's OF INFERENCE/DECISION

We refer to four basic and related dimensions that serve to structure statistical endeavor. They are (1) Motivation, (2) Model, (3) Mode and (4) Method. With regard to the motivation, our interest is focussed either on so-called "true entities" (often referred to as parameters) or on observables, whether potential values in the future or unobserved, lost or missing values in the past (prediction or retrodiction). For making inferences or decisions about these entities we generally employ two kinds of models, namely physical or statistical or combinations thereof. At one extreme, physical models are mechanism driven and embedded in the discipline. They may be either deterministic cum measurement error or stochastic. At the other extreme "pure" statistical models are generally method driven and off-the-shelf. They exhibit either parametric or data analytic features.

Connected to this is a third dimension, namely the method applied to the problem. Here we differentiate between Model Selection/Prediction versus Hypothesis Testing/Estimation. Classically, in a narrow frame of reference, but one almost universally applied, Hypothesis Testing/Estimation refers to inferences about the parameters assuming the truth of the model. Model Selection/Prediction refers to the activity of identifying adequate models that are suitable for prediction under particular circumstances.

Certainly not the least amongst the frenetic preoccupations of statisticians is the adherence to a mode of inference. And this bias resulted in most of the bitter polemics that have plagued or enlivened (as the case may be) the statistical enterprise. On the dimension of Mode we can traverse from a high structure approach that is Bayesian to several intermediate structures, Fiducial-Structural, Likelihood, Frequentist (parametric) to Low Structure stochastic (misnamed non-parametric) and Non-stochastic data analytic and Sample Reuse procedures.

In any event if we consider all of the possibilities on each dimension as virtually discrete, we would have roughly speaking $2^3 \times 5$ bins. However not all of them are viable combinations for inferences that are pertinent to the problems that arise.

3. WHICH BIN SHOULD WE BE IN?

Needlesss to say there are several bins that I find attractive and insufficiently utilized. In order to converge in on these I will first discuss the interplay between the dimensions of Motivation and Model. Models derived from expert knowledge of a real physical mechanism are rather scarce in statistical practice with perhaps the exception of the borderline case namely measurement error models. This measurement error convention, often assumed to be justifiable from central limit theorem considerations and/or experience, was seized upon and indiscriminately adopted for situations where its application was dubious at best and erroneous at worst. Applications of this sort, that regularly occur in technology but much more frequently in the softer sciences, constitute the bulk of practice. The variation here is often not of the measurement error variety. As a physical description its use is inappropriate if we stress hypothesis testing and estimation of the "true" entities parameters. If these and other such models are considered in their proper context, i.e., as statistical models that can yield adequate approximations for the prediction of further observables presumed to be exchangeable in some sense with those already generated from the process under scrutiny, then these statistical models are potentially very useful. Therefore on the dimension of Method, Model Selection and Prediction will be preferable over Hypothesis Testing and Estimation. Clearly Testing and Estimation assume the truth of a model and impart an inappropriate existential meaning to an index or parameter while Model Selection is an activity which consists of searching for a single model (or a mixture of several) that is adequate for the prediction of observables even though it is not likely to be the "true" one. This is particularly appropriate in those softer areas of application, which are legion, where the so-called true explanatory model is virtually so complex as to be unattainable. Lest we deceive ourselves–we need only observe the physicists' search for fundamental particles and the origin of things to realize the ultimate elusiveness of true models. Even so-called physical constants of nature appear to change their so-called true values, even if only by definition. The standard for clock time is now the average of a number of cesium beam atomic clocks, while solar time is even less constant because it is based on the earth's rotation which is more variable. All of this being said, however, should not deter one from the task of seeking better models.

But even all of these considerations are quite remote from the typical applications that statisticians deal with.

We then must realize that inferring about observables is more pertinent since they can occur and be validated to a degree which is not possible with parameters. Divesting ourselves of parameters would then be our goal if we could but model observables without recourse to parameters. Strictly speaking, in the completely observabilistic realm, modeling would require a finite number of observables whose measurements are discrete. This becomes, in most cases, such a difficult and unappealing enterprise that we tend to fall back on the infinite and the continuous and the parametric for ease in modeling. Indeed, parameters can be properly conceived as a result of passing to a limit. This needn't overly concern us as long as we focus on the appropriate aspects of this endeavor. We must acknowledge that even aside from the measurement model, e.g., $X = \theta + e$ where X is the observable measurement of θ the "true value" subject to error e, there are several other advantages to initiating the paradigm with a parametric model. Suppose our interest is in the prediction of the function $g(X_{N+1}, \ldots, X_{N+M})$ of future or unobserved values after a sample $X_1, \ldots, X_N$ is at hand. When M, the number of future values, is large so that exact computation is overwhelmingly burdensome, then going to the limit, i.e., as $M \to \infty$, may provide a simplification that is more than adequate for the purposes in mind. This limiting function, properly defined, can be considered a parameter and is a reasonable one in the context of the problem. (Something like this may have been in Fisher's mind when he defined consistency.) In fact when no particular known M is at issue the conception of a hypothetically infinite number of as yet unobserved values may be useful as a normative evaluation procedure in making comparisons among therapeutic agents, treatments, etc.

4. THE INFERENTIAL MODE

When compelling physical models are unavailable, as is usually the case in much statistical practice, we may be constrained to use statistical models. To then estimate or test certain of these parameters without embedding them in a limiting predictivistic interpretation vests them with an authority they do not possess. Hence, for statistical models, the stress should be on observables and model selection cum prediction on the Motivation and Method dimensions respectively.

The amount of structure one can reasonably infuse into a given problem or process could, very well, determine the inferential mode. Any one of them possesses a capacity for implementing the predictive approach, but only the Bayesian mode is always capable of producing probability distributions for prediction.

5. THE FIRST BAYESIAN PREDICTIVIST

There are in the folklore many candidates for what is often termed the second oldest profession. Among them are Priests, Prophets and Physicians. Priests intercede for you with the diety to ensure an unconditional prosperous future. Prophets admonish you to mend your ways lest the conditional future be worse than the present. Physicians, conditional on symptoms, retrodict (diagnose) and offer conditional predictions (prognoses) given some ailment; i.e. if you adhere to a prescribed therapy your chances are such and such of improvement.

What all of these so called second oldest professions have in common is that they are, in their own fashion, Predictivists. That such an enterprise, as old as man himself, should have virtually disappeared during one of the most productive periods of the discipline of Statistics, say 1925 to 1975, is quite remarkable. But even more astonishing is that it remained so even during the neo-Bayesian revival 1950-1975. A reason for this may be in a long standing misconception of the work of Bayes. This leads us to the following question.

Who was the first Bayesian predictivist? This, I believe, is clear on three different interpretations of Bayes' work. Although this was briefly discussed in Geisser (1985), it pays to rehearse it in somewhat more detail.

First the "Received Version":

Let $X_1, \ldots, X_N$ be independently and identically distributed binary variables with

$$\Pr[X_i = 1|\theta] = \theta = 1 - \Pr[X_i = 0|\theta].$$

Let $R = \sum_{i=1}^{N} X_i$, so that

$$\Pr[R = r|\theta] = \binom{N}{r} \theta^r (1-\theta)^{N-r}. \tag{1}$$

The "Received Version" implies that Bayes assumed that θ was subject to a prior uniform distribution in the absence of knowledge about it. Hence we can obtain, a posteriori, for the parameter θ

$$p(\theta|r) \propto \theta^r(1-\theta)^{N-r}.$$

Now it is interesting to note that Price, who communicated Bayes' posthumous essay, calculated the predictive probability of the next binary variable

$$\Pr[X_{N+1} = 1] = \frac{r+1}{N+2}.$$

It is quite possible that Bayes himself did this or discussed it with Price who merely reported it. It does indicate predictive interest which was followed up by Laplace (1774). There is a "Revised Version" due to Stigler (1982) which claims that Bayes implicitly if not explicitly assumed that a priori

$$\Pr[R = r] = \frac{1}{N+1} \text{ for } r = 0, 1, \ldots, N$$

and that from this and (1), he deduced that

$$p(\theta) = 1$$

which in the "Received Version" was assumed. In other words the uniform assumption was on the observable R rather than the unobservable θ. This is certainly an improvement, and is in line with a current view that prior distributions are more sensible when applied to observables, Geisser (1975b, 1976, 1980).

However I have put forward a "Stringent Version," Geisser (1985), which would endow Bayes with the cognomen, "The Compleat Predictivist." In the essay, Bayes imagined a ball being rolled on a unit square flat table with the horizontal coordinate of the final resting place uniformly distributed in the unit interval. Call this random variable Y. A second ball is then rolled N times and we are informed of the number of times the second ball did not exceed the horizontal coordinate of the original ball whose actual resting place y is to be inferred based on this information. Clearly then

$$p(y) = 1,$$

and assuming independent rolls of the second ball, given y we obtain

$$\Pr[R = r|y] = \binom{N}{r} y^r(1-y)^{N-r}.$$

Now by "Bayes Theorem" we have

$$p(y|r) \propto y^r(1-y)^{N-r}$$

bereft of parametric intrusions. Thus Bayes becomes the "Compleat Predictivist," stochastically speaking, as well as the first Bayesian Predictivist.

6. PROSPECT

So much for the past. What should Bayesian statisticians stress in the future in terms of their inference/decision/action approach?

The formal decision framework for parametric estimation requires an observation space X, a parameter space Θ and an action space $\mathcal{A}$. A random variable is observed at $X^{(N)} = x^{(N)} = (x_1, \ldots, x_N)$ with distribution function

$$F(x^{(N)}|\theta) \text{ for } \theta \in \Theta,$$

and prior distribution $P(\theta)$ for θ. A loss function is defined

$$L(a(x^{(N)}), \theta) \text{ for } a \in \mathcal{A}$$

for taking action a when $x^{(N)}$ is observed and θ is the true value. The average loss

$$L(a) = \int L(a, \theta)\, dP(\theta|x^{(N)}),$$

where $P(\theta|x^{(N)})$ is now the posterior distribution given $x^{(N)}$, is then minimized,

$$L(a^*) = \min_a L(a),$$

to yield the optimal decision or action a^*.

The main value of this apparatus is in those instances where θ is some true value, i.e., a physical constant or some real entity subject to measurement error, or interpreted as the limit of a pertinent function of observables.

The predictive/retrodictive approach implies that $x^{(N)}$ has been observed and inferences (decision/action) about $x_{(M)} = (x_{N+1}, \ldots, x_{N+M})$, a set of realizable observations that have not been observed, are to be made (taken). In this case we posit a joint distribution function

$$F(x^{(N)}, x_{(M)}|\theta)$$

and predictive, loss function

$$L(a(x^{(N)}), x_{(M)}),$$

which is the loss that would be incurred in taking action a upon observing $x^{(N)}$ in the event that $X_{(M)} = x_{(M)}$ will be realized.

Again the average loss

$$L(a) = \int L(a, x_{(M)})\, dF(x_{(M)}|x^{(N)})$$

is minimized with respect to a to yield the optimal action a^*, where

$$F(x_{(M)}|x^{(N)}) = \int F(x_{(M)}|x^{(N)}, \theta)\, dP(\theta|x^{(N)})$$

is the distribution of the potential observables given those whose values have been observed.

This is the formal framework for most statistical applications although in many cases it would be sufficient to report the predictive distribution itself. Of course, here a model, consisting of the joint distribution function of observables and parameters, is assumed to be "true" or, more realistically, adequate. Additionally a particular loss function needs to be posited.

The advent of the enormous computing power now available and the apparent prospect of this state of affairs increasing exponentially will undoubtedly have a prodigious impact on the statistical analysis of data. We would, of course, expect much more complex computer intensive analyses involving number crunching and graphical displays. This would facilitate the use of more complicated and realistic modeling previously precluded due to costs and the amount of computing time. Another phenomenon, perhaps less welcome to some, would be the ease with which a host of alternative (complex or otherwise) analyses could be performed. This could create havoc (or so it seems) for those who adhere to stringent versions of the Neyman-Pearson or even Bayesian approaches, where assumptions regarding error rates, models, loss functions, prior probabilities etc. presumably preceded the disclosure of the observational values. Even in the past, little attention was paid to this unrealistic view. A rationalization for circumventing this was an implicit if, then conditioning; i.e., if this were assumed then that would follow — and some choice made on a practical basis. Generally this exercise was informal and restricted to a very few alternatives, usually not more than two were even considered. In the future the capacity to produce almost instantaneously a host of analyses based on varying assumptions could become a problem for overly formal inferential philosophies.

Often there is some standard statistical model that has proven adequate in data sets similar to a current one in need of analysis. The framework for a Bayesian predictive approach which can be termed a perturbation analysis is suggested for such a situation. Let us suppose that we consider a model

$$f(x^{(N)}, x_{(M)}, \theta|\omega) = f(x_{(M)}|x^{(N)}, \theta, \omega) f(x^{(N)}|\theta, \omega) p(\theta|\omega)$$

which includes the postulated joint density of observables and parameters governed by some perturbation $\omega \varepsilon \Omega$ with $\omega = \omega_0$ representing the standard model.

Now our interest is focussed on

$$F(x_{(M)}|x^{(N)}, \omega)$$

which results from the model and whose perturbation is at issue. If we let, as before,

$$L(a, x_{(M)})$$

be the predictive loss function, then averaging it over $F(x_{(M)}|x^{(N)}, \omega)$, we obtain

$$E_{X_{(M)}}[L(a, x_{(M)})] = \bar{L}_\omega(a).$$

We note that the average loss now depends on ω.

Now for each ω, we obtain a^*_ω, say such that

$$\bar{L}_\omega(a^*_\omega) = \min_a \bar{L}_\omega(a).$$

To assess the perturbation from the standard we can examine the differential loss

$$d(w) = \bar{L}_\omega(a^*) - \bar{L}_\omega(a^*_\omega),$$

where a^* results from $\omega = \omega_0$, the standard, by varying ω over a suitable space Ω.

While this formalization often requires more information then is available, i.e., specification of an action space and a loss function it can still serve as a useful guide for a statistician's handling of a data set. In particular if a statistician's main concern is to report the predictive distribution of $X_{(M)}$, a convenient and useful assessment is the Kullback-Leibler directed divergence

$$I(\omega) = E_{X_{(M)}}[\ln f(x_{(M)}|x^{(N)}, \omega) - \ln f(x_{(M)}|x^{(N)}, \omega_0)]$$

where the expectation is with respect to the predictive distribution of $x_{(M)}$ under the perturbed model. We note that ω may be considered to be a parameter, so that the standard is embedded in a supermodel; or it may be an index; or an indicator that completely changes the model, e.g., from a normal likelihood to a Cauchy likelihood; or from a standard to one that is moderately contaminated; or to one which has just a few outliers in terms of location or variation; or even to variations in the prior assessment or likelihood of a periparametric type, by which is meant densities within a specified neighborhood of a given parametric density. Of course other loss functions can also be contemplated. If $I(\omega)$ does not vary appreciably for $\omega \varepsilon \Omega$, then the sample is robust with respect to the anticipated perturbation of the standard model. In such a case the standard analysis should adequately serve predictive needs. When this is not the case, a determination of the nature of the perturbation, the magnitude of its effect on particular predictive inferences, the possibility of the perturbed or some other model being more adequate, should be undertaken. In other words what is needed is a thorough review of the adequacy of the model, the protocols and stipulated experimental conditions and anything else one can think of that might be pertinent. When the standard appears inadequate then the Bayesian always has recourse to averaging over a prior $p(\omega)$ but this sort of activity could go on forever and requires closure at some point, Geisser (1980, p. 466). One could also report the range of effects on inferences/actions for varying ω.

For prediction the ultimate solution is to take another sample from the process and find the modeling concatenation that yields the best predictions from the old sample onto the new sample. Often this course is unavailable for one reason or another. One suggestion is predictive sample reuse assessments of models wherein a subset of the data (a single observation or more) is withheld and predicted from the remaining data and cycled through in this fashion until every subset of the required size of the original data has been predicted from non-deleted data. Comparisons on a variety of "goodness of prediction measures" are then made among alternative modeling possibilities to render a suitable choice for the entire data set at hand: Fearn, (1975), Geisser (1974, 1975a, 1981) Lee and Geisser (1972, 1975), Rao (1981, 1987), Rao and Boudreau (1985).

I believe that such informal selection methods will be used much more frequently in the future along side of more formal Bayesian model selection methods. The latter may often require more assumptions than investigators or statisticians are willing to make. What this will also demonstrate, to a large degree, is that the uncertainty in prediction is so much larger than the uncertainty in estimation, which is generally a second order effect, that the difference in results among several different, but adequate predictive methods is often negligible; Geisser (1982, p. 92), Clayton, Geissser and Jennings (1986).

Another area where we anticipate more predictive activity is in screening tests for diseases, i.e., to determine that a disease is present before the usual overt symptoms appear. For example, a test result that asserts that an individual has disease D will be denoted by S. Then the sensitivity (specificity) of the test is the probability that an individual is correctly diagnosed as belonging (not belonging) to D, so that

$$P(S|D) = \eta \text{ is the sensitivity}$$

$$P(\bar{S}|\bar{D}) = \theta \text{ is the specificity.}$$

In diseases such as AIDS, Gastwirth (1987), the prime focus is on

$$P(D|S) = \frac{\pi\eta}{\pi\eta + (1-\pi)(1-\theta)} = \tau, \text{ say,}$$

and on the false negative

$$P(D|\bar{S}) = \frac{\pi(1-\eta)}{\pi(1-\eta)+(1-\pi)\theta} = \rho, \text{ say,}$$

where $\pi = P(D)$ in the specified population.

In order to deal with these entities, n_1 individuals who are known to have the disease are tested as well as n_2 individuals known not to have the disease. Additionally n individuals are tested to estimate the fraction in the population that respond positively to the test. Hence the likelihood from such a series of samples yields

$$L(\theta,\eta,\pi) \propto \eta^{r_1}(1-\eta)^{n_1-r_1}\theta^{r_2}(1-\theta)^{n_2-r_2}\left(\frac{\pi\eta}{\tau}\right)^t\left(1-\frac{\pi\eta}{\tau}\right)^{n-t}$$

recalling that τ is a function of θ, η, and π, Geisser (1987b). Suppose a joint prior for η, θ and π, $g(\eta,\theta,\pi)$ is available, then the posterior density of θ, η, and π, is

$$p(\theta,\eta,\pi|d) \propto L(\theta,\eta,\pi)g(\eta,\theta,\pi),$$

where $d = (r_1, r_2, n_1, n_2, t, n)$.

Then for any set S on the unit square we could find

$$\Pr[(\tau,\rho)\varepsilon S] = P_S$$

or conversely for any fixed P we could find the "smallest" set S_P such that

$$\Pr[(\tau,\rho)\varepsilon S_P] = P.$$

Similar results could be obtained marginally for either ρ or τ, i.e. $P(\tau|d)$ could be computed. This would be much more informative than the usual frequentist calculation made for estimates of ρ and τ. Of course this would require a good deal of heavy calculation involving numerical integration and approximation as well as some prior knowledge about θ, η, π, which no doubt is available for many screening tests. We also note that the expected values of τ and ρ are the minimum squared error predictors of τ or ρ and the predictive probability that one of the individuals, whose disease status was unknown but tested as S, actually has the disease.

For a new individual who can be regarded as exchangeable with the previous n testees and who has been classified by the test as S, the predictive probability of being D is

$$\int \tau dP(\tau|d')$$

where $d' = (r_1, r_2, n_1, n_2, t+1, n+1)$. For other important predictive probabilities of interest in screening test situations, including calculation of the probability that K out of M future testees who tested as S have the disease, see Geisser (1987b).

More generally, in many other medical situations it is often of interest to use the results of a trial to predict the response of future subjects to the agent under study. The most informative summary is a probability distribution of the response values. For a single new individual the predictive distribution is the critical summary. For a clinician who will treat some number M (say not large) of such individuals, the interest may be focused on the fraction KM^{-1} who will be "cured" or will exceed a certain threshhold or fall in a given set. Public health authorities, for example, will often be confronted with a very large or an imprecisely known M. Here the limiting value of the fraction, i.e. as $M \to \infty$, is of value either as an approximation in a complex calculation or as a normative evaluation. This limiting value is a parameter, and at least in this sense parameters become limiting or special cases of functions of observables and hence interpretable.

A further point in stressing KM^{-1} as opposed to $\theta = \lim_{M\to\infty} KM^{-1}$ is the fact that in certain situations a comparison may be wanted in terms of $K = M$ or $K = 0$, i.e., "success" with every one of the M individuals; e.g., when the therapy does not succeed death results or when looking at adverse effects and a success is its absence. Here the best therapy may be considered as the one that has the highest probability of not incurring a death in any of the next M patients. In any model, as long as $\Pr[\theta = 1] < 1$, a priori, and one of these severe reactions occurred in the study no matter how large, the posterior $\Pr[\theta = 1] = 0$. However $P[KM^{-1} = 1|M] \neq 0$ and for moderate or even reasonably large M can be quite high and appropriately used to make comparisons of two such "life saving procedures." It is also to be noted that no procedure is used in an infinite number of cases for two reasons: the first is that there is obviously no such number of cases and the second is that every procedure is soon replaced by another.

The calculation of extremes has, among others, particular application to orphan drugs (drugs for rare diseases) and floods. For example situations exist where a drug or a therapy that is useful for a rare malady may have a serious adverse effect (lethal, disabling, or merely worse than the disease) to a small fraction of people to whom it is administered. A clinician may wish to calculate the probability that none of the next M patients subjected to the therapy, where M is not very large, will suffer an adverse effect based on previous experience with the drug which may and often has been used on patients for other purposes where the disease was worse than the adverse reaction. Clearly it is of vital interest to builders of dams, dikes and bridges to calculate the probability that a given flood level will not be exceeded in the next M years, based on the previous N years.

Significance testing, widely frowned upon by Bayesians and others, can, in certain instances, be performed in a predictivistic manner that renders them useful. In many cases of discordancy testing, situations exist such that formulating alternative hypotheses and their prior probabilities concerning an observation in question is a formidable if not impossible task. In fact, even reasonable modeling alternatives may be difficult to contemplate; for example errors in transcribing data, numbers misread, digits transposed or an incorrect sign before a number, or a stipulated experimental condition that did not actually obtain for each of the observables. Therefore the surprise engendered by a small P-value of an unconditional or reasonably conditioned predictive significance test is useful in bringing to one's attention these and other anomalies.

Predictive discordancy tests for normal and exponential sampling distributions of both a conditional and unconditional nature and possibly depending on diagnostic procedures have been initiated, Geisser (1986, 1987a). It is expected that tests of this sort will be developed for a variety of statistical paradigms as well.

The key features of this approach are to calculate the marginal probability function of the observations under model M

$$f(y^{(n)}|M) = \int f(y^{(n)}|\theta, M)g(\theta|M)d\theta$$

and partition

$$y^{(n)} = (y_i, y_{(i)}),$$

where y_i is the possibly discordant observation. We note that $f(y^{(n)}|M)$ exists only if $g(\theta|M)$ is a proper prior. If y_i were prechosen, for reasons other than diagnostic checking, we can test whether it concords with model M unconditionally, i.e. irrespective of whether $y_{(i)}$ does, by calculating the marginal

$$f(y_i|M),$$

where, e.g.,

$$\Pr[f(Y_i|M) \le f(y_i|M)] = P$$

is one way of calculating the significance level for the rejection of y_i as concording with M. A conditional predictive test, where $y_{(i)}$ is assumed to accord with M, can be calculated from the conditional predictive probability function

$$f(y_i|y_{(i)}, M),$$

whence, e.g.,

$$\Pr[f(Y_i|y_{(i)}, M) \le f(y_i|y_{(i)})] = P$$

is the significance level for rejection of y_i as concording with M assuming that $y_{(i)}$ is concordant. This will exist for certain useful reference or non-informative priors as well.

Now suppose that the suspicion cast upon an observation results solely from some diagnostic procedure D which assigns discordancy value $D(y_i)$ to y_i, with $D(y_C) = \max_i D(y_i)$, say.

Hence, to take account of this, we may calculate, from

$$f(y^{(n)}|M),$$

the predictive probability function of Y_C,

$$f(y_C|M, C);$$

i.e. taking into account the diagnostic's choice of Y_C as having the largest discordancy index. The significance level then is calculated from $f(y_C|M, C)$. As noted before, this method cannot be used with improper priors. Hence when a reference prior is used, we suggest the following procedure. Calculate

$$\Pr[D(Y_C) \ge D(y_C)|D(Y_C) \ge D(y_{C-1}); y_{(C)}, M]$$

from the standard predictive distribution

$$f(y_C|y_{(C)}, M),$$

where y_{C-1} represents the second most discordant observation. This can be considered a conditional subjective probability assessment and basically requires calculating and using the predictive distribution of an observation.

7. CONCLUDING REMARKS

We previously outlined a number of areas of application for the predictive approach, Geisser (1971, 1985). We now find that there are virtually no traditional applications that can elude this way of analysing data.

We further note that everything we have said can be summed up in the apothegm, "The future is what it used to be." Clearly the future that we have pointed towards is this relatively simple and obvious one. However, a caveat is in order, in that we recognize that the really difficult problems come about when the apothegm is negated.

ACKNOWLEDGEMENTS

Acknowledgement is made to the National Institute of General Medical Sciences GM 25271.

REFERENCES

Clayton, M. K., Geisser, S. and Jennings, D. E. (1986). A comparison of several model selection procedures. *Bayesian Inference and Decision Techniques.* (P. Goel and A. Zellner, eds.) Amsterdam: North-Holland, 425–439.

Fearn, T. (1975). A Bayesian approach to growth curves. *Biometrika* **62**, 89–100.

Gastwirth, J. L. (1987). The statistical precison of medical screening procedures: Application to polygraph and AIDS antibodies test data. *Statistical Science* **2**, 213–238.

Geisser, S. (1971). The inferential use of predictive distributions. *Foundations of Statistical Inferences* , (V. Godambe and D. Sprott, eds.) New York: Holt, Rinehart and Winston, 456–469.

Geisser, S. (1974). A predictive approach to the random effect model. *Biometrika* **61**, 101–107.

Geisser, S. (1975a). The predictive sample reuse method with applications. *J. Amer. Statist. Assoc.* **70**, 320–328.

Geisser, S. (1975b). Bayesianism, predictive sample reuse, pseudo observations and survival. *Bulletin of the International Statistical Institute* **3**, 285–289.

Geisser, S. (1976). Predictivism and sample reuse. *Proceedings of the Twenty-First Conference on the Design of Experiments.* Army Mathematics Steering Committee, 349–362.

Geisser, S. (1980a). A predictivistic primer. *Bayesian Analysis in Econometrics and Statistics: Essays in Honor of Harold Jeffreys.* 363–381 (A. Zellner, ed.) Amsterdam: North-Holland.

Geisser, S. (1980b). Predictive sample reuse techniques for censored data. *Bayesian Statistics* (J. M. Bernardo *et al.* eds.). Valencia, Spain: University Press. 430–468 (with discussion).

Geisser, S. (1981). Sample reuse procedures for predicting the unobserved portion of a partially observed vector. *Biometrika,* 243–250.

Geisser, S. (1982). Aspects of the predictive and estimative approaches in the determination of probabilities. *Biometrics* **38**, supplement, 75–93 (with discussion).

Geisser, S. (1985). On predicting observables: A selective update. *Bayesian Statistic.* **2**, (J. M. Bernardo *et al.* eds.) Amsterdam: North-Holland, 203–230. (with discussion)

Geisser, S. (1986). On predictive tests of discordance (submitted).

Geisser, S. (1987a). Influential observations, diagnostics and discordancy tests. *Journal of Applied Statistic.* **14**, 133–142.

Geisser, S. (1987b). Dicussion on The statistical precision of medical screening procedures to polygraph and AIDS antibodies test data by J. L. Gastwirth. *Statistical Science* **2**, 231–232.

Laplace, P. S. (1974). Memoir sur la probabilite des causes par les evenements. *Memoires de l'Academie Royale des Sciences* **6**, 621–656.

Lee, J. C. and Geisser, S. (1972). Growth curve prediction. *Sankhya A* **34**, 393–412.

Lee, J. C. and Geisser, S. (1975). Applications of growth curve prediction. *Sankhya A* **37**, 239–256.

Rao, C. R. (1981). Prediction of future observations in polynomial growth curve models. In *Statistics: Applications and New Directions,* 512–520. Proceedings of the Indian Statistical Institute Golden Jubilee International Conference, 1981, Calcutta.

Rao, C. R. (1987). Prediction of future observations in growth curve models. *Statistical Science* **2**, 434–471.

Rao, C. R. and Boudreau, R. (1985). Prediction of future observations in factor analytic type growth model. *Multivariate Analysis VI,* 449–466.

Stigler, S.M. (1982). Thomas Bayes's Bayesian inference. *J. Roy. Statist. Soc. A* **145**, 250–258.

DISCUSSION

H. RUBIN (*Purdue University*)

The speaker has come up with proposals which, in my opinion, make the situation much worse. We do not need *trained* statisticians; we need *educated* ones. Statistics is not a collection of procedures; it is a template on which any inference problem can be based, and the actual procedures to use in a real situation may be ones not found anywhere in the literature. I have frequently had to invent procedures for actual problems.

We also need to provide statisticians and *particulary* users of statistics with understanding. I believe that to do this requires that methods not be taught until the understanding of a decision problem in extensive form, namely, that it is necessary to simultaneously consider all the consequences of the proposed action in all states of nature, has been exemplified by a wide variety of problems. We must inform the user that the problem must be formulated without regard to a collection of standard procedures. The statistician must be prepared to inform the user of unforseen consequences of his assumptions, of the unimportance of some of the assumptions (robustness), and of the necessity to make additional assumptions (lack of robustness). A good biologist and a good statistician together are likely to be worth many typical biostatisticians.

Notice that I have emphasized the decision problem in extensive form. Once this is understood, it will be difficult to get the user to even accept most classical procedures, and such Bayesian procedures as confidence intervals, as they clearly violate the idea of decision theory. I believe that such a person will accept the need for coherence as a normative principle, and consequently become a Bayesian. I do not see how one can sell a Bayesian cookbook; without some understanding of what a statistical problem is, the superiority of Bayesian to classical procedures is by no means obvious.

For the statistician, it is easy to confuse formalism and proofs for concepts. It is easy and confusing to start with the loss function and/or prior. Whence cometh the loss function? The weight measure (loss-weighted prior) comes from the same place. That a real problem can be represented by a mathematical formalism does not mean that being able to solve the formalism enhances the understanding of the problem. Even worse is the tendency to force a problem into a standard mould when robustness arguments are not available.

For the user of statistics, the confusion of understanding with formalism and methodology is equally great. We are teaching them in disservice courses (I heard this term from Saul Blumenthal). It is much easier to find out what the problem is even if the user is largely ignorant of probability and statistics, than if the user knows methodology.

Statistics has a good future, and will probably be rational Bayesian, if we educate both users and statisticians with an understanding of the concepts and problems involved, rather than training them in formalism, methodology, and proof. These have their place for statisticians after understanding of the problems. Also, I think the implications for the proper use of computers are clear.

REPLY TO THE DISCUSSION

Professor Rubin once again imposes upon us his testy and tiresome tirade on the virtues of the Rubinesque Herman-eutics regarding statistical endeavors. His prompt but predictable pronouncements have been a consistent constituent of conferences during the last quarter century. Unfortunately what this message communicates is the necessity for treating each problem in a singularly unique manner — the extended form, thus liberating him from developing any common methodology. It does not, however relieve him of the obligation of presenting, at least once, an illustration worked out according to this loftier conception which, although correct in principle, appears to be rarely achievable in practice. The utter lack of such a delivery, over this entire period, neither advances his extensive form nor compels us to take him as seriously as perhaps we should.

BAYESIAN STATISTICS 3, pp. 159–172
J. M. Bernardo, M. H. DeGroot, D. V. Lindley and A. F. M. Smith, (Eds.)

Experiments in Bayesian Image Analysis

S. GEMAN
Brown University

SUMMARY

We propose a statistical framework for modelling and analyzing pictures. The approach is Bayesian. It is given here in brief outline, with references to various reports and papers for a full discussion and for the results of experiments with a variety of applications.

Keywords: IMAGE PROCSSING; BAYESIAN INFERENCE.

1. INTRODUCTION

Computational image analysis encompasses a variety of applications involving a sensing device, a computer, and software for restoring and possibly interpreting the sensed data. Most commonly, visible light is sensed by a video camera and converted to an array of measured light intensities, each element corresponding to a small patch in the scene (a picture element, or pixel). The image is thereby digitized, and this format is suitable for computer analysis. In some applications, the sensing mechanism responds to other forms of light, such as in infrared imaging where the camera is tuned to the invisible part of the spectrum neighboring the color red. Infrared light is emitted in proportion to temperature, and thus infrared imaging is suitable for detecting and analyzing the temperature profile of a scene. Applications include automated inspection in industrial settings, medical diagnosis, and targeting and tracking of military objects. In single photon emission tomography, as a diagnostic tool, individual photons, emitted from a radiopharmaceutical (isotope combined with a suitable pharmaceutical) are detected. The objective is to reconstruct the distribution of isotope density inside the body from the externally-collected counts. Depending on the pharmaceutical, the isotope density may correspond to local blood flow (perfusion) or local metabolic activity. Other applications of computer vision include satellite imaging for weather and crop yield prediction, radar imaging in military applications, ultrasonic imaging for industrial inspection and a host of medical applications, and there is a growing role for video imaging in robotics.

The variety of applications has yielded an equal variety of algorithms for restoration and interpretation. Unfortunately, few general principles have emerged and no common foundation has been layed. Algorithms are by and large *ad hoc*; they are typically dedicated to a single application, and often critically tuned to the particulars of the environment (lighting, weather conditions, magnification, and so-on) in which they are implemented. It is likely that a coherent theoretical framework would support more robust and more powerful algorithms. A well-studied candidate is regularization theory (see Marroquin *et al.* (1987), Poggio *et al.* (1985), and the similar "variational" approach in Blake (1983), Black and Zisserman (1986), Mumford (1986), Terzopoulos (1986)), which has been successfully applied to a variety of vision tasks. We have been exploring a related approach based upon probabilistic image models, well-defined principles of inference, and a Monte Carlo computation theory. Exploiting this framework, we have recently obtained encouraging results in several areas of application, including tomography, texture analysis, and scene segmentation.

In the following paragraphs, we lay out, briefly, our paradigm, in its general formulation. We refer the reader to various manuscripts for more complete discussions of the methodology, and for applications to texture segmentation and classification (Geman *et al.* (1987), Geman *et al.* (1988), Geman and Graffigne (1987), Graffigne (1987)), boundary detection (Chalmond (1987), Geman (1987), Geman *et al.* (1988), Geman and Geman (1984), Marroquin (1984), Marroquin *et al.* (1987)), single photon emission tomography (Geman and McClure (1987)), and complex shape modelling and recognition (Grenander (1984), Keenan and Grenander (1986), Knoerr (1988)).

2. BAYESIAN PARADIGM

In real scenes, neighboring pixels typically have similar intensities; boundaries are usually smooth and often straight; textures, although sometimes random locally, define spatially homogeneous regions; and objects, such as grass, tree trunks, branches and leaves, have preferred relations and orientations. Our approach to picture processing is to articulate such regularities mathematically, and then to exploit them in a statistical framework to make inferences. The regularities are rarely deterministic; instead, they describe correlations and likelihoods. This leads us to the Bayesian formulation, in which prior expectations are formally represented by a probability distribution. Thus we design a distribution (a prior) on relevant scene attributes to capture the tendencies and constraints that characterize the scenes of interest. Picture processing is then guided by this prior distribution, which, if properly conceived, enormously limits the plausible restorations and interpretations.

The approach involves five steps, which we shall briefly review here.

2.1. Image Model

This is a probability distribution on relevant image attributes. Both for reasons of mathematical and computational convenience, we use *Markov random fields* (MRFs) as prior probability distributions. Let us suppose that we index all of the relevant attributes by the index set S. The set S is application specific. It typically includes indices for each of the pixels (about 512×512 in the usual video digitization) and may have other indices for such attributes as boundary elements, texture labels, object labels and so-on. Associated with each $s \in S$ is a real-valued random variable X_s, representing the state of the corresponding attribute. Thus X_s may be the measured intensity at pixel s (typically, $X_s \in \{0, \ldots, 255\}$), or simply 1 or 0 as a boundary element at location s is present or absent.

The kind of knowledge we represent by the prior distribution is usually local, which is to say that we articulate regularities in terms of small local collections of variables. In the end, this leads to a distribution on $X = \{X_s\}_{s \in S}$ with a more or less local neighborhood structure. Specifically, our priors are Markov random fields: there exists a (symmetric) *neighborhood relation* $G = \{G_s\}_{s \in S}$, wherein $G_s \subseteq S$ is the set of neighbors of s, such that

$$\Pi(X_s = x_s | X_r = x_r, r \in S, r \neq s) = \Pi(X_s = x_s | X_r = x_r, r \in G_r).$$

$\Pi(a|b)$ is conditional probability, and, by convention, $s \notin G_s$. G symmetric menas $s \in G_r \Leftrightarrow r \in G_s$. (Here, we assume that the range of the random vector X is discrete; there are obvious modifications for the continuous or mixed case.)

It is well known, and very convenient, that a distribution Π defines a MRF on S with neighborhood relation G if and only if it is Gibbs with respect to the same graph, (S, G). The latter means that Π has the representation

$$\Pi(x) = \frac{1}{z} \exp\{-U(x)\}, \tag{2.1}$$

where

$$U(x) = \sum_{c \in C} V_c(x), \tag{2.2}$$

C is the collection of all cliques in (S, G) (a clique is a collection c of sites such that every two sites in c are neighbors), and $V_c(x)$ is a function depending only on $\{x_s\}_{s \in c}$. U is known as the energy, and has the intuitive property that the low energy states are the more likely states under Π. The normalizing constant, z, is known as the partition function. The Gibbs distribution arises in statistical mechanics as the equilibrium distribution of a system with energy function U.

As a simple example (too simple to be of much use for real pictures) suppose the pixel intensities are known, a priori, to be one of two levels, minus one (black) or plus one (white). Let S be the $N \times N$ square lattice, and let G be the neighborhood system that corresponds to nearest horizontal and vertical neighbors:

```
o — o — o  ...
|   |   |
o — o — o  ...
|   |   |
o — o — o  ...
⋮   ⋮   ⋮
```

For picture processing, think of N as typically 512. Suppose that the only relevant regularity is that neighboring pixels tend to have the same intensities. An energy consistent with this regularity is the "Ising potential":

$$U(x) = -\beta \sum_{[s,t]} x_s x_t, \quad \beta > 0,$$

where $\Sigma_{[s,t]}$ means summation over all neighboring pairs s, $t \in S$. The minimum of U is achieved when $x_s = x_t$, $\forall s$, $t \in S$. Under (2.1), the likely pictures are therefore the ones that respect our prior expectations; they segment into regions of constant intensities. This is called the Ising model. It models the equilibrium distribution of the spin states of the atoms in a ferromagnet. Aligned spins cooperate to produce a measurable magnetic field.

Obviously, β is an important parameter. The larger $\beta > 0$, the larger the typical regions of constant intensity. Parameters, such as β, that determine the detailed quantitative behavior of the prior, are inevitably introduced in constructing image models. Whenever possible, we *estimate* these parameters from data. This raises a host of interesting computational and theoretical issues; we refer the reader to Besag (1974), Devijver and Dekesel (1987), Geman (1984), Geman and Graffigne (1987), Geman and McClure (1987), Gidas (1985), Gidas (1988), Hinton and Sejnowski (1983), and Younes (1987) for some experiments with, discussions of, and partial solutions to the parameter estimation problem.

One very good reason for using MRF priors is their Gibbs representations. Gibbs distributions are characterized by their energy functions, and these are more convenient and intuitive for modelling than working directly with probabilities. Again, we refer the reader to the references for many more examples, and for applications.

2.2. Degradation Model

The image model is a distribution $\Pi(\cdot)$ on the vector of image attributes $X = \{X_s\}_{s \in S}$. *By design*, the components of this vector contain all of the relevant information for the image processing task at hand. Hence, the goal is to estimate X. This estimation will be based upon partial or corrupted observations, and based upon the image model, i.e., the prior distribution. In emission tomography, X represents the spatial distribution of isotope in a target region of the body. What is actually observed is a collection of photon counts whose probability law is Poisson, with a mean function that is an attenuated Radom transform of X. In the texture labelling problem, X is the pixel intensity array combined with a corresponding array of texture labels. Each label gives the texture type of the associated pixel. The observation is only partial: we observe the pixels, which are just the digitized picture, but not the labels. The purpose is then to estimate the labels from the picture.

The observations are related to the image process X by a *degradation model.* This models the relation between X and the *observation process*, say $Y = \{Y_s\}_{s \in T}$. For texture analysis, we define $X = (X^P, X^L)$, where X^P is the usual grey level pixel intensity process, and X^L is an associated array of texture labels. The observed picture is just X^P, and hence $Y = X^P$: the degradation is a projection. More typically, the degradation involves a random component, as in the tomography setting where the observations are Poisson variables whose means are related to the image process X. A simpler, and widely studied (if unrealistic), example is additive white noise. Let $X = \{X_s\}_{s \in S}$ be just the basic pixel process. In this case $T = S$, and for each $s \in S$ we observe

$$Y_s = X_s + \rho_s,$$

where, for example, $\{\rho_s\}_{s \in S}$ is Gaussian with independent components, having means 0 and variances σ^2, and $\{\rho_s\}$ is independent of the X-process.

Formally, the degradation model is a conditional probability distribution, or density, for Y given X: $\Pi(y|x)$. If the degradation is just additive white noise, as in the above example, then

$$\Pi(y|x) = \left(\frac{1}{2\pi\sigma^2}\right)^{|S|/2} \exp\left\{-\frac{1}{2\sigma^2}\prod_{s \in S}(y_s - x_s)^2\right\}.$$

For labelling textures, the degradation is deterministic; $\Pi(y|x)$ is concentrated on $y = x^P$, where $x = (x^P, x^L)$ has both pixel and label components.

2.3 Posterior Distribution

This is the conditional distribution on the image process X given the observation process Y. This posterior or *a posteriori* distribution contains the information relevant to the image restoration or image analysis task. Given an observation $Y = y$, and assuming the image model ($\Pi(x)$) and degradation model ($\Pi(y|x)$), the posterior distribution reveals the likely states of the "true" (unobserved) image X. Having constructed X to contain all relevant image attributes, such as locations of boundaries, labels of objects or textures, and so on, the posterior distribution comes to play the fundamental role in our approach to image processing.

The posterior distribution is easily derived from Bayes' rule

$$\Pi(x|y) = \frac{\Pi(y|x)\Pi(x)}{\Pi(y)}.$$

The denominator, $\Pi(y)$, is difficult to evaluate. It derives from the prior and degradation models by integration: $\Pi(y) = \int \Pi(y|x)\Pi(dx)$, but the formula is computationally intractable. Happily, our analysis of the posterior distribution will require only *ratios*, not absolute probabilities. Since y is fixed by observation, $1/\Pi(y)$ is a constant that can be ignored (see the paragraph below on Computing).

As an example we consider the simple Ising model prior, with observations corrupted by additive white noise. Then

$$\Pi(x) = \frac{1}{z} \exp\left\{ \beta \sum_{[s,t]} x_s x_t \right\}$$

and

$$\Pi(y|x) = \left(\frac{1}{2\pi\sigma^2} \right)^{|S|/2} \exp\left\{ -\frac{1}{2\sigma^2} \sum_{s \in S} (y_s - x_s)^2 \right\}.$$

The posterior distribution is then

$$\Pi(x|y) = \frac{1}{z_p} \exp\left\{ \beta \sum_{[s,t]} x_s x_t - \frac{1}{2\sigma^2} \sum_{s \in S} (y_s - x_s)^2 \right\}.$$

We denote by z_p the normalizing constant for the posterior distribution. Of course, it depends on y, but y is fixed. Notice that the posterior distribution is again a MRF. In the case of additive white noise, the neighborhood system of the posterior distribution is that of the prior, and hence local.

For a wide class of useful degradation models, including combinations of blur, additive or multiplicative colored noise, and a variety of nonlinear transformations, the posterior distribution is a MRF with a more or less local graph structure. This is convenient for our computational schemes, as we shall see shortly. We should note, however, that exceptions occur. Indeed, nonlocal graph structures that incorporate long-range interactions are useful and are completely consistent with the Bayesian paradigm. In tomography, for example, the posterior distribution is associated with a highly non-local graph. This particular situation incurs a high computational cost (see Geman and McClure (1987)) as a consequence of each site in the graph having a high degree.

2.4. *Estimating the Image*

In our framework, image processing amounts to choosing a particular image x, given an observation $Y = y$. One choice is the maximum a posteriori, or MAP estimate:

$$\text{choose } x \text{ to maximize } \Pi(x|y).$$

The MAP estimate chooses the most likely x, given the observation. In many applications, our goal is to identify the MAP estimate, or a suitable approximation. Often, though, other estimators are more appropriate (see Besag (1986) for an insightful discussion). We have found, for example, that the posterior mean ($\int x\Pi(dx|y)$) is more effective for tomography, at least in our experiments (see Geman and McClure (1987)).

The computational issues for posterior mean and MAP estimators are similar. For illustration we concentrate here on MAP estimation. In most applications we cannot hope to identify the true maximum a posteriori image vector x. To appreciate the computational difficulty, consider again the Ising model with added white noise:

$$\Pi(x|y) = \frac{1}{z_p} \exp\left\{ \beta \sum_{[s,t]} x_s x_t - \frac{1}{2\sigma^2} \sum_{s \in S} (y_s - x_s)^2 \right\}. \tag{2.3}$$

This is to be maximized over all possible vectors $x = \{x_s\}_{s \in S} \in \{-1, 1\}^S$. With $|S| \sim 10^5$, brute force approaches are intractable; instead, we will employ a Monte Carlo algorithm which gives adequate approximations. (Remarkably, this particular example can be solved exactly, using techniques from network flow theory, see Greig *et al.* (1986).)

Maximizing (2.3) amounts to minimizing

$$U_p(x) = -\beta \sum_{[s,t]} x_s x_t + \frac{1}{2\sigma^2} \sum_{s \in S} (y_s - x_s)^2,$$

which might be thought of as the posterior energy. (As with z_p, the fixed observation y is suppressed in the notation $U_p(x)$.) More generally, we write the posterior distribution as

$$\frac{1}{z_p} \exp\{-U_p(x)\} \tag{2.4}$$

and characterize the MAP estimator as the solution to the problem

$$\text{choose } x \text{ to minimize } U_p(x).$$

The utility of this point of view is that it suggests a further analogy to statistical mechanics, and a computation scheme for approximating the MAP estimate, which we shall now describe.

2.5. Computing

When exploring a specific application, it is our consistent experience that computation-intensive algorithms for image analysis can be made fast by suitable compromises and exploitation of special structure. Still, as a research tool, it has been invaluable to have available a general computational framework, which we now describe in the context of MAP estimation.

Pretend that (2.4) is the equilibrium Gibbs distribution of a real system. Recall that MAP estimation amounts to finding a minimal energy state. For many physical systems the low-energy states are the most ordered, and these often have desirable properties. The state of silicon suitable for wafer manufacturing, for example, is a low-energy state. Physical chemists achieve low-energy states by heating and then slowly cooling a substance. This procedure is called *annealing*. Cerný (1982) and Kirkpatrick *et al.* (1983) suggest searching for good minimizers of $U(\cdot)$ by *simulating* the dynamics of annealing, with U playing the role of energy for an (imagined) physical system. In our image processing experiments, we often use simulated annealing to find an approximation to the MAP estimator. (See Arts and van Laarhoven (1986) for a good review of simulated annealing.)

Dynamics are simulated by producing a Markov chain, $X(1)$, $X(2), \cdots$ with transition probabilities chosen so that the equlibrium distribution is the posterior (Gibbs) distribution (2.4). One way to do this is with the Metropolis algorithm (Metropolis *et al.* (1953)). More convenient for image processing is a variation we call *stochastic relaxation.* The full story can be found in Geman and Geman (1987), (1984), and Grenander (1984). Briefly, in stochastic relaxation we choose a sequence of sites $s(1)$, $s(2)$, $\cdots$, $\in S$ such that each site in S is visited infinitely often. If $X(t) = x$, say, then $X_r(t+1) = x_r$, $\forall r \neq s(t)$, $r \in S$, and $X_{s(t)}(t+1)$ is a sample from

$$\Pi(X_{s(t)} = \cdot | X_r = x_r, r \neq s(t)),$$

the conditional distribution on $X_{s(t)}$, given $X_r = x_r$, $\forall r \neq s(t)$. By the Markov property,

$$\Pi(X_{s(t)} = \bullet | X_r = x_r, r \neq s(t)) = \Pi(X_{s(t)} = \bullet | X_r = x_r, r \in G^p_{s(t)})$$

where $\{G^p_s\}_{s \in S}$ is the *posterior* neighborhood system, determined by the posterior energy $U_p(\cdot)$. The prior distributions that we have experimented with have mostly had local neighborhood systems, and usually the posterior neighborhood system is also more or less local as well. This means that $|G^p_{s(t)}|$ is small, and this makes it relatively easy to generate, Monte Carlo, $X(t+1)$ from $X(t)$. In fact, if Ω is the range of $X_{s(t)}$, then

$$\Pi(X_{s(t)} = \alpha | X_r = x_r, r \in G^p_{s(t)}) = \frac{\Pi(\alpha,_{s(t)} x)}{\sum_{\hat{\alpha} \in \Omega} \Pi(\hat{\alpha},_{s(t)} x)}, \tag{2.5}$$

where

$$(\alpha,_{s(t)} x)_r = \begin{cases} \alpha & r = s(t) \\ x_r & r \neq s(t) \end{cases}$$

Notice that (fortunately!) there is no need to compute the posterior partition function z_p. Also, the expression on the right-hand side of (2.5) involves only those potential terms associated with cliques containing $s(t)$, since all other terms are the same in the numerator and the denominator.

To simulate annealing, we introduce an artificial temperature into the posterior distribution:

$$\Pi_T(x) = \frac{\exp\{-U_p(x)/T\}}{Z_p(T)}.$$

As $T \to 0$, $\Pi_T(\cdot)$ concentrates on low energy states of U_p. To actually find these states, we run the stochastic relaxation algorithm while slowly lowering the temperature. Thus $T = T(t)$, and $T(t) \downarrow 0$. $\Pi_{T(t)}(\cdot)$ replaces $\Pi(\cdot)$ in computing the transition $X(t) \to X(t+1)$. In Geman and Geman (1984) we showed that, under suitable hypotheses on the sequence of site visits, $s(1)$, $s(2)$, ...:

If $T(t) > c/(1+\log(1+t))$, and $T(t) \downarrow 0$, then for all c sufficiently large $X(t)$ converges weakly to the distribution concentrating uniformly on $\{x : U(x) = \min_y U(y)\}$.

More recently, our theorem has been improved upon by many authors. In particular, the smallest constant c which guarantees convergence of the annealing algorithm to a global minimum can be specified in terms of the energy function U_p (see Chiang and Chow (1987), Gidas (1985), Hajek (1985), and Holley and Stroock (1987)). Also, see Brandt (1987) and Gidas (1988), (1989) for some ideas about faster annealing via multiresolution methods, Geman and Hwang (1986), Hwang and Sheu (1987), and Gidas (1985) for some extensions to continuous-time continuous-state annealing, and see Geman and Geman (1987) for an annealing algorithm designed for *constrained* optimization.

ACKNOWLEDGEMENTS

This is a summary of collaborative work, principally with Donald Geman, Basilis Gidas, Ulf Grenander, and Donald E. McClure.

Research partially supported by Army Research Office contract DAAL03-86-K-0171 to the Center for Intelligence Control Systems, National Science Foundation grant number DMS-8352087, and the General Motors Research Laboratories.

REFERENCES

Arts, E. and van Laarhoven, P. (1986). Simulated annealing: a pedestrian review of the theory and some applications. *NATO Advanced Study Institute on Pattern Recognition: Theory and Application.* Belgium: Spa.

Besag, J. (1974). Spatial interaction and the statistical analysis of lattice systems. *J. Roy. Statist. Soc. B* **36**, 192–236.

Besag, J. (1986). On the statistical analysis of dirty pictures. *J. Roy. Statist. Soc. B* **48**, 259-302. (with discussion).

Blake, A. (1983). The least disturbance principle and weak constraints. *Pattern Recognition Letters* **1**, 393–399.

Blake, A. and Zisserman, A. (1986). *Weak continuity constraints in computer vision.* Report CSR-197-86, Dept. of Comp. Sci., Edimburgh Univ.

Brandt, A. (1987). Multi-level approaches to large scale problems. *Proceedings of the International Congress of Mathematicians 1986*. (A. M. Gleason, ed.). Providence: American Mathematical Society.

Cerný, V. (1982). *A thermodynamical approach to the travelling salesman problem: an efficient simulation algorithm*. Bratislava: Inst. Phys. and Biophysics, Comenius Univ.

Chalmond, B. (1987). *Image restoration using an estimated Markov model*. Orsay: Dept. of Mathematics, University of Paris.

Chiang, T.-S. and Chow, Y. (1987). *On eigenvalues and optimal annealing rate*. Taipei, Taiwan: Institute of Mathematics, Academia Sinica.

Devijver, P. A. and Dekesel, M. M. (1987). Learning the parameters of a hidden Markov random field image model: a simple example. *Pattern Recognition Theory and Applications*. (P. A. Devijver and J. Kittler, eds.). Heidelberg: Springer-Verlag, 141-163.

Geman, D. (1987). A stochastic model for boundary detection. *Image and Vision Computing*.

Geman, D. (1984). Parameter estimation for Markov random fields with hidden variables and experiments with the EM algorithm. *Complex Systems, Tech. Rep. 21*. Div. of Applied Mathematics, Brown University.

Geman, D. and Geman, S. (1987). Relaxation and annealing with constraints. *Complex Systems, Tech. Rep. 35*. Div. of Applied Mathematics, Brown University.

Geman, D., Geman, S. and Graffigne, C. (1987). Locating texture and object boundaries. *Pattern Recognition Theory and Applications*. (P. A. Devijver and J. Kittler, eds.). Heidelberg: Springer-Verlag.

Geman, D., Geman, S., Graffigne, C. an Dong, P. (1987). *Boundary detection by constrained optimization*. Providence, RI: Div. of Applied Mathematics, Brown University.

Geman, S. and Geman, D. (1984). Stochastic relaxation, Gibbs distributions, and the Bayesian restoration of images. *IEEE Trans. Pattern Anal. Machine Intell.* **6**, 721–741.

Geman, S. and Graffigne, C. (1987). Markov random field image models and their applications to computer vision. *Proceedings of the International Congress of Mathematicians, 1986*. (A. M. Gleason, ed.). Providence: American Mathematical Society.

Geman, S. and Hwang, C.-R. (1986). Diffusions for global optimization. *SIAM J. Control and Optimization* **24**, 1031–1043.

Geman, S. and McClure, D. E. (1987). Statistical methods for tomographic image reconstruction. *Proceedings of the 46th Session of the International Statistical Institute*. Bulletin of the ISI, **52**.

Gidas, B. (1985). Nonstationary Markov chains and convergence of the annealing algorithm. *J. Statist. Phys.* 39, 73–131.

Gidas, B. (1985). *Global minimization via the Langevin equation*. Proceedings of 24th conference on Decision and Control, Ft. Lauderdale, Florida, 744–778.

Gidas, B. (1987). *Parameter estimation for Gibbs distributions*. Providence, RI: Div. of Applied Mathematics, Brown University.

Gidas, B. (1988). A multilevel-multiresolution technique for computer vision via renormalization group ideas. *Proceedings of SPIE-International Society for Optical Engineering* **880**, High Speed Computing, 214–218.

Gidas, B. (1988). *Consistency of maximum likelihood and pseudo-likelihood estimators for Gibbs distributions*. Proceedings of the Workshop on Stochastic Differential Systems with Applications in Electrical/Computer Engineering, Control Theory, and Operations Research, Institute for Mathematics and its Applications, University of Minnesota, Springer.

Gidas, B. (1989). A renormalization group approach to image processing problems. *IEEE Trans. Pattern Anal. Machine Intell.* (To appear).

Graffigne, C. (1987). *Experiments in texture analysis and segmentation*. Ph. D. Dissertation, Div. of Applied Mathematics, Brown University.

Greig, D. M., Porteous, B. T. and Seheult, A. H. (1986). Discussion: On the statistical analysis of dirty pictures, by J. Besag. *J. Roy. Statist. Soc. B* **48**, 259–302.

Grenader, U. (1985). *Tutorial in Pattern Theory*. Div. of Applied Mathematics, Brown University.

Hajek, B. (1985). A tutorial survey of theory and applications of stimulated annealing. *Proceedings of the 24th IEEE Conference on Decision and Control*, 755–760.

Hinton, G. E. and Sejnowski, T. J. (1983). Optimal perceptual inference. *Proc. IEEE Conf. Comput. Vision Pattern Recognition*.

Holley, R. and Stroock, D. (1987). *Simulated annealing via Sobolev inequalities*.

Hwang, C.-R. and Sheu, S.-J. (1987). *Large time behaviors of perturbed diffusion Markov processes with applications*, I, II and III. Taipei, Taiwan: Institute of Mathematics, Academia Sinica.

Keenan, D. M. and Grenander, U. (1986). *On the shape of plane images*. Reports in Pattern Analysis 145. Providence, RI: Brown University, Div. of Applied Mathematics.

Kirkpatrick, S., Gellatt, C. D., Jr. and Vecchi, M. P. (1983). Optimization by simulated annealing. *Science* **220**, 671–680.

Knoerr, A. (1988). Global models of natural boundaries: theory and applications. Ph. D. Dissertation. Providence, RI: Brown University, Div. of Applied Mathematics.

Marroquin, J. L. (1984). Surface reconstruction preserving discontinuities. *Artificial Intell. Lab. Memo 792*, M. I. T.

Marroquin, J. L., Mitter, S. and Poggio, T. (1987). Probabilistic solution of ill-posed problems in computational vision. *J. Amer. Statist. Assoc.* **82**, 76–89.

Metropolis, N., Rosenbluth, A. W., Rosenbluth, M. N., Teller, A. H. and Teller, E. (1953). Equations of state calculations by fast computing machines. *J. Chem. Phys.* **21**, 1087–1091.

Mumford, D. and Shah, J. (1986). *Boundary detection by minimizing functionals*, I.

Poggio, T., Torre, V. and Koch, C. (1985). Computational vision and regularization theory. *Nature* **317**, 314–319.

Terzopoulos, D. (1986). Regularization of inverse visual problems involving discontinuities. *IEEE Trans. Pattern Anal. Machine Intell.* **8**, 413–424.

Younes, L. (1987). Estimation and annealing for Gibbsian fields. *Tech. Rep.*, Orsay: University of Paris, Dept. of Mathematics.

DISCUSSION

S. J. PRESS (*University of California, Riverside*)

Background

There is a very large literature in which Professor Geman's interesting work is embedded.

Most of the work is frequentist; some is Bayesian.

The literature is in fields of: Pattern Recognition, Electrical Engineering, Computer Science, Statistical Physics, Applied Mathematics, Remote Sensing, Classification, Geology and Geography, and Tomography.

Work is beginning to appear in the statistics literature in sufficient abundance that we no longer have any excuse for not being aware of the interesting developments taking place, and for not contributing to this literature.

Some Bayesian statisticians and physicists have been fortunate to hear Ed Jaynes' report on the use of maximum entropy methods for image reconstruction.

Definition of Problem

The are at least two major, related, but different problems in this area:

(1) Image restoration (or reconstruction):

There is a picture that has been blurred or degraded in some way by noise, or projections, or missing variables, or whatever, and we would like to restore the original image. The picture is considered to be a lattice array made up of rows and columns of pixels (picture elements). The pixels are simultaneously classified into distinct colors (groups, or populations). The pixels are assumed to be homogeneous within a pixel (but may not be). In some problems, boundaries are sought; in others, we seek segments.

(2) Traditional Classifications

A picture (scene) is considered to be a lattice array of pixels, each of which is to be classified into one of several groups. Each pixel is represented by a vector of characteristics that is observed. There is generally assumed to be noise superimposed on the vector signals of characteristics. Pixels may be classified one at a time, or simultaneously. There are vectors of data available that are known to have come from each of the populations (these sets are sometimes called training data). Sometimes neighboring observations are used to assist in the classification procedure, and sometimes the neighboring observations are assumed to be spatially correlated). This approach is used in both. Ruben Kein and James Press reported on such classification research from a Bayesian point of view, at the spring, 1987, meeting of the NSF/NBER Bayesian Seminar on Bayesian Inference in Econometrics, at Duke University.

In the Geman paper, the author is focussed on the problem of image segmentation, i.e., the process of grouping image data into regions with similar features.

Model

The author begins with an $N \times N$ square lattice, with N perhaps 512.

There is a blurring or degrading function: $\Phi(\cdot)$.

There is a likelihood function of the observed data given the true, or original image, X. The "observed data" is recorded by a sensor. The likelihood function is assumed to be normal.

The author adopts a Gibbs distribution (using nearest neighbors) for the prior on the original image, the matrix of pixels, X. [The Gibbs distribution has maximal entropy among all distributions of image states with the same average energy.] Equivalently, the author assumes the probable configuration of the state of the neighbors of given pixels is governed by a Markov random field; i.e., the probable configuration is based upon local properties of the field only (the states of the nearest neighbors only). It can be shown that the two are equivalent (Preston, 1974).

Bayes theorem generates a posterior distribution that is also Gibbsian.

The original image is restored by finding the mode of the posterior density. This is a major computational problem. A labelling scheme is used to classify pixels from the same group (segmentation). Posterior means are used also.

Computational Solution

In a lattice 512×512, there are $262,144$ pixels to be classified. If there were only two possible states or colors (populations) for each pixel, the total number of possible states for the array would be 2 to the power $262,144$, an enormous number.

A stochastic relaxation scheme (Monte Carlo sampling) is used to iterate to the mode of the posterior density. The state of the array at time (iteration) t is constructed as a Markov chain with limiting distribution equal to the posterior.

Questions About the Paper

I would now like to raise a few questions about the paper:

1. Image segmentation can be carried out as proposed in Professor Geman's paper, or by MLE, or by use of maximum entropy, by pseudo-MLE, by Bayesian classification techniques using the predictive distribution on a pixel-by-pixel basis, and perhaps by other methods.

   How do these methods compare —in misclassification error rate; in computational ease and cost; in time required; etc? Which procedure requires the fewest number of model assumptions?

2. It has been recommended that we adopt a Gibbs family of prior distributions to express our prior belief about the original image. Are the implications of such a prior sensible in this context?

   Under what conditions are the implications of such a prior (adopted for nearest neighbors) reasonable?

3. How do we go about assessing the unknown parameters (hyperparameters) of a nearest neighbor Gibbs prior?

4. Is there a way to test the adequacy of the degrading function?

5. How do we select the parameters of the texture model, so that the model fits a particular texture?

6. How do we know, in the iterative Monte Carlo procedure, whether we have converged to the mode?
7. In our own work on Bayesian classification with spatially correlated neighbors, we found that when we assumed an observation and its neighbors all belong to the same population, we could do well (very few classification errors) in the interior of a segment, but errors increased at the boundaries of segments. Do the authors encounter the same phenomenon in their approach?
8. Besag (1986) uses an iterative approach called Iterated Conditional Modes. His approach seems not to require the assumption of Markov random field (or a Gibbs distribution prior). The approach seems to me to be very clever in that it is intuitively reasonable, appears to depend upon an algorithm that is rapidly converging, and perhaps, is computationally inexpensive. How does the work of the author compare with Besag's approach? What are the relative advantages and disadvantages of the two approaches?

REPLY TO THE DISCUSSION

Professor Press has identified eight pivotal issues involving the practical application of our methods. These issues are paraphrased and briefly discussed in the following paragraphs.

1. *How do the variety of existing segmentation and classification methods compare with respect to accuracy, computational complexity, and number of model assumptions?*

To any statistician viewing the image processing/image analysis literature, it is apparent that very little has been done in the way of systematic comparisons of existing methods. In part, this is because *meaningful* comparisons are hard to develop. For example in texture classification there are a host of variables that significantly, and idiosyncratically, influence the performance of the different algorithms. Some algorithms are unduly sensitive to lighting variations, while others (such as ours) are critically tuned to scale and orientation. Most algorithms are "feature-based" and exhibit grossly uneven performance over a wide repertoire of textures; the prefered textures are, of course, algorithm-dependent. It is our general impression, and that of many of our engineering colleagues, that our segmentation algorithms are among the best available, at least in their ability to identify visually subtle transitions between (possibly textured) regions.

On the other hand, it is indeed difficult to find approaches that are as computationally expensive as ours. In this regard, we view Monte Carlo optimization techniques as *research tools*. They are poor substitutes for the efficient dedicated algorithms that should be developed when facing applications involving a flow of data and a need for speedy analysis. It has been our consistent experience that order of magnitude speed ups can be achieved by problem-dependent compromises and reworking of the stochastic relaxation computational framework.

As for the number of model assumptions involved, there are certainly many of these behind our approach. This issue is at the heart of a time-honored Bayesian/frequentist debate, and it would be very hard to contribute something new to the general philosophical discussion. We proceed pragmatically: we recognize that our models are "wrong" in many respects, but we believe that these fictions very significantly and appropriately narrow the spectrum of reasonable image restorations and analyses. Of course, models are developed interactively, with experiments helping to shape and validate specific assumptions.

2. *Do Gibbs distributions provide a reasonable class of priors for image analysis?*

We adopt Gibbs priors for two reasons. Most important is their characterization in terms of an energy function. Energy functions are more natural and convenient vehicles for engineering prior knowledge than probabilities. Intractable consistency issues are avoided and complex relations, as between edges and pixel grey levels, are easily expressed (see, for example, Geman and Geman (1984), Marroquin *et al.* (1987), Geman *et al.* (1988)).

The second advantage of the Gibbs representation is that it suggests a *generic* computational algorithm for experiments, namely, stochastic relaxation.

Of course, esentially *every* distribution is a Gibbs distribution. For computational reasons, we work mostly with *local* Gibbs distributions, meaning that the terms in the energy functions involve small numbers of spatially near-neighbor variables. It remains to be seen what class of image processing tasks lend themselves to more-or-less local Gibbs representations. We have done well with certain kinds of restorations and reconstructions, and with segmentation and boundary placement. It is less likely that "higher level" vision tasks, such as object detection and classification, will fit into this framework without significant modification. These issues are addressed in some exciting recent work on shape recognition (see Chow *et al.* (1988), Keenan and Grenander (1988), and Knoerr(1988)).

Hierarchical Gibbs models, such as the pixel/edge model introduced in Geman and Geman (1984) (see also Marroquin *et al.* (1987)), are not as "local" as they appear. Jointly, the pixel/edge variables define a local, near-neighbor, Gibbs distribution. But the edge variables are make believe. The sensing system defines only pixel intensities; auxiliary variables such as edges are introduced for convenience. However, the *marginal* pixel intensity distribution, integrating out the "auxiliary" edge variables, has a full graph! By introducing imaginary edges, we are able to capture long range intensity effects, that arise from well-known properties of boundaries, while employing only local interactions in the Gibbs representation.

3. *How do we go about assessing the unknown parameters (hyperparameters) of a Gibbs prior?*

We have experimented with a variety of parameter estimation schemes. Some have been successful for specific problems (e.g. Geman and McClure (1987), Geman and Graffigne (1987)), but we have been unable to find a tractable general mechanism for estimating hyperparameters.

4. *Is there a way to test the adequacy of the degrading function?*

The answer is problem-specific. In tomography, for example, the degradation function depends critically on the attenuation of the particular object being imaged, and this is hard to know precisely. Although it may be possible to estimate attenuation, our current approach is to make simple approximations, similar to those used in other reconstructions algorithms. *Given* the attenuation, the degradation function can be well-approximated from elementary physics, using the detailed geometry and response characteristics of the imaging device, and the known scattering and absorption characteristics of the isotope.

Unfortunately, typically the best available assessment of the model is the accuracy or believability of the reconstructions.

5. *How do we select the parameters of the texture model, so that the models fits a particular texture?*

In our experiments (Geman and Graffigne (1987)), we have used Besag's pseudolikelihood method (Besag (1974)). The method is relatively computationally efficient, at least when compared to maximum likelihood, and is easily shown to be consistent (in the "large picture" limit) —see Geman and Graffigne (1987). Unfortunately, we must first define an appropriate parametric model, which is often quite difficult. We have developed a limited class of models, appropriate for some simple texture recognition tasks. But the problem of constructing a wide range of texture models, capable of producing credible renditions of real textures, is certainly unsolved.

6. *How do we know when the iterative Monte Carlo procedure has converged to the mode?*
Of course, we can develop artificial problems in which the mode is known, a priori, and thereby test our Monte Carlo procedures. But in real problems the mode is not known. Indeed, whether searching for a mode (as in MAP estimation) or a posterior mean (for a least squares criterion), it is difficult to know at what point stochastic relaxation type methods yield accurate approximations. Convergence theorems are reassuring, but in practice we stop at finite time, and the relevance of the asymptotic theory is unclear. Certain functionals, such as the Gibbs energy, of the Markov process produced by stochastic relaxation can be monitored, and do give some indication of convergence. Nevertheless, our practice has generally been to rework the model if we get poor answer after 100 or so iterations of stochastic relaxation.

7. *Does the author find, as we have, that enforcing high correlation between neighboring classifications increases classification errors at boundaries?*
We observe a similar phenomenon in our texture classification experiments (Geman and Graffigne (1987)). In developing the prior, we discourage very small homogenous regions. This amounts to adjusting the interaction potential in an Ising-like label model so as to promote large regions of constant label. If the potential is too large, then the posterior distribution is insensitive to the data, and boundary errors occur. All in all, though, we have not had much difficulty finding appropriate levels for the interaction potentials, at least for the limited set of experiments that have been carried out so far.

8. *How does the work of the author compare with the approach suggested in Besag (1986)?*
Besag's use of the Iterated Conditional Mode (ICM) avoids reconstructions that depend strongly on undesired long-range properties of the Markov random field priors. Furthermore, this *deterministic* algorithm is typically far more efficient than stochastic relaxation methods. We have had good luck with ICM for tomography (Geman and McClure (1987)), and with a closely related procedure for boundary placement (Geman *et al.* (1988)). On the other hand, in some applications we are interested in posterior *means*, and then ICM is not appropriate. It may be that in these situations a similar algorithm can be devised, for example, by iterating conditional *means*. In this regard, recent related results (Bilbro *et al.* (1988)) with so-called mean field theory approximations to annealing are very encouraging.

REFERENCES IN THE DISCUSSION

Bilbro, G., Mann R., Miller, T. K., Snyder, W. E., Van den Bout, D. E and White, M. (1988). Simulated annealing using the mean field approximation. (to appear).

Chow, Y., Grenander, U., and Keenan, D. (1988). Hands: a pattern theoretic study of biological shape. *Tech. Rep.*, Division of Applied Mathematics, Brown University, Providence, RI.

Knoerr, A. (1988). Global models of natural boundaries: theory and applications. Ph. D. Dissertation, Division of Applied Mathematics, Browm University, Providence, RI.

Preston, C. J. (1974). *Gibbs States on Countable Sets.* Cambridge: Cambridge University Press.

BAYESIAN STATISTICS 3, pp. 173–188
J. M. Bernardo, M. H. DeGroot, D. V. Lindley and A. F. M. Smith, (Eds.)

Software for Bayesian Analysis: Current Status and Additional Needs

P. K. GOEL
Ohio State University

SUMMARY

We make an attempt to provide comprehensive information about the existing software for data analysis within the Bayesian paradigm. The paucity of programs seems to indicate that the Bayesian software available for widespread use still in its infancy. We have a long way to go before a general purpose Bayesian Statistical Analysis Package is made available. Alternatives for reaching this goal quickly are presented in the concluding section.

Keywords: DATA ANALYSIS; STATISTICAL COMPUTING; BAYESIAN SOFWARE.

1. INTRODUCTION

The starting point for this article was the workshop on *Bayesian Statistical Computing*, which was held at The Ohio State University during May 1986 and attended by approximately 40 people involved in statistical computing, Bayesian decision analysis and AI. We organized this workshop to assess the status of Bayesian analysts software. We believe that Bayesian methodology will be used routinely by a widespread group of data analysts and scientists if a user -friendly, general purpose Bayesian Statistical Analysis Package was available which could be used for class room teaching as well as for experimental data analysis. The workshop featured eleven talks concerning various issues in computational Bayesian analysis and some applications, as well as an open forum on Bayesian computing. The two main issues were (1) desirable computing environments for Bayesian statistical analysis, and (2) potentials for a Bayesian software package. There were diverse points of view about the environment suitable for an *interactive Bayesian statistical analysis package*, although all the participants seemed to believe that wide-spread use of Bayesian methodology will not become a reality without such a package. An overwhelming majority agreed with the notion that it is too early to push for a new statistical package. Instead, we should strive for new Bayesian software to be compatible with a package like "*S*"®, in order to use its data handling and graphics capabilities. Furthermore, it was suggested that the Bayesian community should develop a "*Bayesian Bulletin Board*", which will provide news about Bayesian programs as they are developed, and a "Bayesian Software Database" which could be accessed for file transfers via popular computer networks. These task have not been initiated in any meaningful way.

The information compiled in this article was provided by the individuals listed as contributors (within the parentheses after the program name), in response to our survey questionnaire which was sent to approximately 450 Bayesian Statisticians and Econometricians in January 1987. An updated SBIE mailing list along with some other lists were used to solicit this information. We have included all the responses. Thus the information is up-to-date as of April 1987. We believe that one large group of users of Bayesian methodology, namely engineers involved in risk assessment and reliability, have developed several special purpose programs which could be easily adapted for general reliability applications. However, we did not have access to any mailing list for this group, so the reliability programs listing cannot be viewed as comprehensive.

One can view this article as an update of Press (1980) which listed Bayesian programs in existence at that time. If no update information was received for a program listed in Press (1980), we have not included it in this paper with the belief that the program has either been superseded or else is no longer available. On comparing this paper with Press (1980), it is clear that impressive gains have been made in the development of software for implementing the Bayesian analysis paradigm for realistic specifications of prior information via approximations, numerical analysis, and Monte Carlo integration. On the other hand, the small number of responses to our request for information indicates that only a few people have devoted their energy to developing general purpose Bayesian analysis software. We believe that this must change quickly if we want to see the **"Bayesian 21st Century"**.

The listing for programs in this paper follows the same general format as in Press (1980). The software has been listed according to the following categories:

(i) CADA, a general purpose data monitor (Section 2) (ii) Regression modeling, Econometric modeling and Time Series methodology (Section 3) (iii) Computation/Approximation of Posterior distribution, moments, quantiles and mode (Section 4) (iv) Elicitation of prior information (Section 5) (v) Reliability Analysis (Section 6) and (vi) Miscelleneous (Section 7).

In Section 8, we present our views on the development of a general purpose Bayesian Analysis Software Package.

2. CADA, A GENERAL PURPOSE DATA ANALYSIS MONITOR

Program Name: **CADA** [Computer Assisted Data Analysis Monitor (CADA Group)]

Function: This monitor provides a conversational Language for Bayesian analysis. It has gone through several updatings. The most recent version (1983) includes susbtantial enhancements over the (1980) version. The CADA operation has been moved to a private corporation now. CADA is a hierarchically structured system with several component groups namely: Data Management facility, Simple Parametric models, Decision theoretic models, Full-rank ANOVA models, Simultaneous estimation, Full-rank MANOVA, Exploratory Data Analysis, Psychometric methods, Probability distribution functions, and Actuarial functions.

Input: Raw data to be entered in, on-line data files to be loaded.

Output: Analysis for Beta, two-parameter normal, and multinomial models using conjugate priors; assessment of conjugate priors and utility functions; full-rank Model I ANOVA and MANOVA for multifactor designs using conjugate or noninformative priors; elementary classical statistics; graphical evaluations of various probability distributions; multiple linear regression analysis and simultaneous estimation of regression in m-groups.

Language: BASIC (Compiler or interpreter required on the machine)

Machines: DEC-PDP-11 (RSTS); DEC-VAX-11(VMS), PRIME, HP-3000. IBM PC version coming soon.

Documentation: Novick, M. L., Hamer, R. M., Libby, D. L., Chen, J. J. and Woodworth, G. G. (1983), *Manual for the Computer-Assisted Data Analysis (CADA) Monitor (1983)*, Iowa City, IA: The CADA Group, Inc.

Availability: The CADA Group, Inc., 306 Mullin Ave., Iowa City, IA 52240, Tel# (319) 351–7200

Remarks: (i) The CADA system was developed at The Iowa State University under the direction of the late Melvin R. Novick.
(ii) The software is fully supported and is available at a cost of $ 600 per copy.

3. NORMAL LINEAR REGRESSION, ECONOMETRIC MODELS AND TIME SERIES ANALYSIS

Program Name: **BRAP** [Bayesian Regression Analysis Program (Abowd/Zellner)], Version 2.0

Function: This program provides a unified package for the Bayesian analyses of the normal linear multiple regression model (MRM) with multivariate normal errors under a noninformative prior, a g-prior or a natural conjugate prior distribution. Some numerical integration capability via Simpson's rule and Monte Carlo importance sampling also provides a facility for analysis of nonstandard models. Both the prior and the posterior distributions of the regression coefficients can be analyzed. Plotting of raw data and residuals, prior and posterior marginals and bivariate contours for regression coefficients can also be done. The posterior distribution of linear functions of regression coefficients, the realized error terms, and predictive distribution of the dependent variables can also be obtained. Some transformations are already available and IMSL® could be loaded for more transformations.

Input: The control cards are in JCL format. Data Files can be easily loaded using a load command.

Output: Updates the prior parameters and plots marginal and bivariate contours of the prior and the posterior distributions of the regression coefficients; posterior distribution of the realized errors; posterior distributions of linear functions of coefficients; standard posterior information; quantiles of posterior distribution can also be obtained via numerical integration routines.

Language: FORTRAN-IV

Machines: IBM-MVS (may need some modifications for newer IBM compilers)

Documentation: Abowd, J. M., Moulton, B. R. and Zellner, A. (1985) *The Bayesian Regression Analysis Package, BRAP User's Manual Version 2.0 of Dec. 1985.*, H. G. B. Alexander Research Foundation, Graduate School Of Business, University of Chicago.

Availability: The package is available from Prof. Arnold Zellner, University of Chicago, Graduate School of Business at a very nominal cost.

Remarks: (i) A new version may be coming soon according to Prof. Zellner.
(ii) Other contributers to the development of BRAP include F. Finnegan, S. Grossman, C. Plosser, P. Rossi, A. Siow, J. Stafford, and W. Vandaele.

Program Name: **BRAP** [Bayesian Regression Analysis Package-PC version (de Alba/Rocha)]

Function: This main program and subroutine package for IBM PCs is an enhancement of BRAP in that it includes BRAP as well as subroutines for Bayesian Disaggregation and Constrained Forecasting.

Language: FORTRAN 77.

Machines: IBM PC and PC compatibles.

Availability: It is available from Prof. Enrique de Alba, Instituto Tecnológico Autónomo De México (ITAM), Río Hondo, No. 1, México, D.F. 01000, at a nominal cost of a diskette and mailing charges.

Program Name: **SEARCH** [Seeks Extreme and Average Regression Coefficient Hypothesis (Leamer/Leonard)]

Function: This is a user-oriented package for Bayesian inference and sensitivity analysis that pools prior beliefs about the regression coefficients with evidence embodied in a given data set. Prior beliefs are assumed to be equivalent to a previous, but possibly fictitious data set. SEARCH offers a study of the sensitivity of the posterior estimates to changes in features of the prior beliefs expressed in terms of a fictitious data set.

Input: Formatted or free-format Card-Image files or on-line CRT input. Input files can be prepared on SAS®, BMDP®, TSP® and SPSS®. SEARCH requires access to a double precision version of the IMSL® library.

Output: Diagnostic messages for debugging syntax errors are available. The program reports the summary of prior and data information received and computes the approximate posterior mode for the regression coefficients when the prior beliefs are modeled as if $\boldsymbol{R\beta}$ came from a normal population with a specified mean $\boldsymbol{r}$ and a covariance matrix $\boldsymbol{V}$. It also reports the sensitivity of the modal estimate to changes in the prior location $\boldsymbol{r}$ and the prior covariance matrix $\boldsymbol{V}$ in the form of of extreme bounds for any linear function of the parameters specified by the user.

Program Name: **FORTRAN IV.** The manual for Version 6 states that it is not completely available in FORTRAN source code. Several of the subroutines for performing high precision arithmetic that SEARCH calls for are object code modules (written in IBM 370 machine code) and the bulk of the SEARCH is written in FORTRAN IV that is complied at UCLA on the IBM FORTRAN IV G1 Compiler; i.e., not necessarily ANSI standard FORTRAN IV.

Machines: IBM370/3033.

Documentation: Leamer, E. E. and Leonard H. B. (1985) *User's Manual for SEARCH-A software package for Bayesian inference and sensivity analysis, SEARCH Version 6, Oct. 1985.*

Availability: The program is available from Prof. E. E. Leamer, Department of Economics UCLA, 405 Hilgard Av., Los Angeles, CA 90024, (213) 825-1011, at $100 per copy on an IBM OS standard label 9 track 1600 BPI tape containing four card-image files. Cards or a 6250 BPI tape can be made available on special request.

Remarks: (i) The program was developed by Edward E. Leamer and Herman B. Leonard. This version was programmed by Arvin Stidick and the MANUAL was extensively edited and largely rewritten by Thomas E. Wolff.

(ii) The Version 6 differs from Version 5 in its efficiency of computation and economy of imput and output.
(iii) A latest example of how SEARCH can be used is given in Leamer, E. E. and Leonard, H. B. (1983) Reporting the fragility of Regression Estimates. *The Review of Economics and Statistics.*

Program Name: **MICRO EBA** [Micro computer version of SEARCH (Fowles)]
Function: This main program is the micro computer version of Leamer and Leonard's Program SEARCH described above.
Language: GAUSS
Machines: Any personal computer with GAUSS software package Version 1.46 or higher.
Availability: The program is available free of charge from Prof. Richard Fowles, Department of Economics, Rutgers University, Newark, NJ 07102.

Program Name: **BRP** [Bayesian Regression Program (Bauwens)]
Function: The main Program executes computations necessary for Bayesian regression analysis for various standard econometric models, discussed in Drèze (1977). The prior beliefs are modeled as Poly-t densities.
Input: Raw data as card-image files. Input data are echoed as output.
Output: Posterior parameters and marginals of regression coefficients and precision and standard deviations; classical regression analysis, posterior residuals and predictive density function of the dependent variable; conditional posterior with given precision, conditional posterior of some regression coefficients given some others, marginalized over the precision.
Language: FORTRAN 77
Machines: IBM 370/158 at the University of Louvain. It is portable according to Dr. Bauwens. In the near future, a PC version is possible.
Documentation: Bauwens, L. and Tompa, H. (1977) *Bayesian Regression Program (BRP)*, CORE User's Manual Set # A-5, and Tompa, H. (1977) *Poly-t Distributions (PTD)*, CORE User's Manual Set # C-9.
Availability: It is available from Prof. Luc Bauwens, CORE, 34 Voie Du Roman Paays, B-1348 Louvain-La-Neuve, Belgian at a cost of 5,000 Belgium Francs.
Remarks: (i) BRP calls another program PTD to evaluate poly-t densities.
(ii) These programs have been developed by H. Tompa under the guidance of Profs. Jacques Drèze and Jean-Francois Richard and with assistance from Luc Bauwens, Jean-Paul Bulteau and Philippe Gille.

Program Name: **Fully Bayesian Analysis of ARMA Time Series Models** (Monahan)
Function: A collection of main program and subroutines carries out the Bayesian Analysis for ARMA time series models using natural conjugate priors as described in Monahan (1983).
Input: Information not yet available.
Output: Programs compute the posterior and predictive distribution of parameters for a given set of ARMA models using the natural conjugate prior. The graphical display can be obtained via SAS/GRAPH.

Language: FORTRAN 66
Machines: Portable
Documentation: Monahan, J. (1980) "*A Structured Bayesian Approach to ARMA time series models*, I, II, III", Technical Reports, Department of statistics, North Carolina State University, Raleigh, NC.
Availability: The programs package is available on tape from Prof. John Monahan, Department of Statistics, North Carolina State University, P. O. Box 8203, Raleigh, NC 27695 at a nominal charge.

Program Name: **Sampling the Future** (Thompson)
Function: This program simulates the predictive distribution of a set of future observations via Monte Carlo methods as discussed in Thompson (1986).
Output: The main program and several subroutines provide a Monte Carlo histogram for the predictive distribution of a future observation or a scattergram of samples from the predictive distribution of a pair of future observations. The model may contain as many as 10 ARMA parameters in up to 3 AR factors and difference factors may be used in the model. The estimation step allows either a diffuse or a conjugate normal/gamma prior distribution.
Language: FORTRAN 77 ANSI standard.
Machines: The program should run on any machine with standard FORTRAN 77 compiler and IMSL® library. Future extensions will requires a graphics terminal. The program will run on a PC with a math co-processor, but an AT type machine with a hard disk is recommended for realistic usage.
Availability: The package is available on a diskette, for a nominal charge of $10, from Prof. Patrick Thompson, Faculty of Management Sciences, The Ohio State University, 1775 S. College Road, Columbus OH 43210.
Remarks: (i) Future plans include a graphic display of predictive distributions and to add the algorithm for prediction from a set of ARMA models given in Monahan (1983)

Program Name: **Bayes & Empirical Bayes Estimation of Regression Coefficients** (Nebebe)
Function: The program computes Bayes and empirical Bayes Estimates for a multiple normal linear regression model in which the prior for the regression coefficients and the hyperparameters are assumed to have various diffuse distributions, see Nebebe, F. and Stroud, T. W. F. (1986).
Language: FORTRAN, requires access to NAG library.
Documentation: No separate documentation is available. The details are given in Nebebe, F. (1984) Ph. D. thesis, Department of Mathematics and Statistics, Queen's University, Kington, Canada.
Availability: The program is available from Prof. F. Nebebe, Dept. of Decision Sciences and MIS, Concordia University, 1455 De Maisonneuve Blvd, West, Montreal, Quebec H3G1M8, Canada.
Remarks: (i) This program provides no capability which is not available in BRAP, SEARCH or BAP. But it may be useful for individuals who are interested in giving a workout to their NAG package.

Program Name: **SHAZAM** [General Econometrics program (White)]

Function: The program provides a portable FORTRAN? program for general econometric modeling on a PC for \$250 or a main frame for \$500–\$900. The author promises that the next version will include a Bayesian Inequality regression and has not provided any other information.

Availability: Available from Prof. Kenneth J. White , Econometrics Department, University of British Columbia, Vancouver, B. C. Canada fot those curious about the name).

Program Name: **BTS** [Bayesian Time Series (Carlin/Dempster)]

Function: The program package carries out computations for Bayesian estimation of unobserved components ("seasonal"/"nonseasonal") in monthly time series under a class of Gaussian Mixed models as described in Carlin, Dempster and Jonas (1985). It uses likelihood based methods for estimation of model parameters.

Output: The program provides posterior estimates of model parameters. A nonportable version for the Apollo DN600 workstation has many graphical capabilities.

Language: FORTRAN 77 (Standard ANSI).

Documentation: Description of the program is available in Carlin, J. B. (1987) Ph. D. Thesis, Department of Stastistics, Harvard University.

Availability: Available from Prof. A. P. Dempster, Department of Statistics, Harvard University, Science Centre, 1 Oxford Street, Cambridge, MA 02138 free of charge.

Program Name: **PROC SEQ** [Sequential Scoring Algorithm (Blattenberger)]

Function: The function performs iterative computation of forecasting distribution for the dependent variable of a normal linear model with a normal-gamma prior distribution or optional g-priors. Scores for five different scoring rules are also computed.

Language: STAT80 Procedure, currently being converted to SAS® PROC MATRIX.

Availability: Available free of charge from Prof. Gail Blattenberger, Department of Economics, University of Utah, Salt Lake City, UT.

Program Name: **MAXENT** [Data Analysis by Maximum Entropy Principle Version 1.17 (Jaynes)].

Function: This beta test version of MAXENT provides fitting of an incompletely specified linear model of the form $\boldsymbol{Y} = \boldsymbol{X}\boldsymbol{F}$, where the data vector is $\boldsymbol{Y}$, the "smearing matrix" $\boldsymbol{X}$ is known but does not have full rank and the elements of the vector $\boldsymbol{F}$ are non-negative and add to 1. The Maximum Entropy Principle, see Jaynes (1983), finds the solution to the above equation which maximizes the entropy of the probability distribution $\boldsymbol{F}$.

Input: The program is interactive. One needs to decide the accuracy level for satisfying all the constraints.

Output: The optimal solution is obtained in an iterative mode, the output for each iteration can be printed.

Language: BASIC

Machines: IBM PC and compatibles. An ASCII source code file is also on the diskette for transporting the program to other micro computers.
Documentation: Help file and Manual on diskette.
Availability: Available free of charge from Prof. Ed. T. Jaynes, Department of Physics, Washington University St. Louis, MO 63130. Individuals sending comments and user experience to Prof. Jaynes will receive the Version 2.0 free.

The programs briefly discussed below have been written for applications of linear models to specific problems.

Program Name: **RECONDA** (Braithwait, Steven)
Function: This program incorporates engineering prior estimates of appliance level electricity consumption into a statistical analysis of household hourly consumption via a hierarchical linear model. The modeling details are given in Caves, Herriges, Train, Windle (1987).
Language: C
Machines: IBM PC and PC compatibles
Availability: The program will be distributed free of charge by EPRI, P. O. Box 10412, Palo Alto, CA 94303 to EPRI member utilities, government and academic institutions.

Program Name: **Statistical Cost Allocation** (Wright, Roger)
Function: Implements the indirect cost allocation methodology based on a multiple linear model as described in Wright (1983).
Language: FORTRAN 77 (Standard ANSI)
Documentation: The program description and listing are given in Wright, R. and Oberg, K. (1983). *The 1979-1980 University of Michigan Heating Plant and Utilities Cost Allocation Study*, Working Pazper #352, Graduate School of Business Administration, The University of Michigan.
Availability: Available free charge from Prof. Roger Wright, Graduate School of Business Administration, The University of Michigan, Ann Arbor, MI 48109.

4. COMPUTATION/APPROXIMATION OF POSTERIOR DISTRIBUTION FEATURES

Program Name: **Bayes Four and** ***gr*** (Smith, A. F. M.)
Function: The BAYES FOUR system consists of a library of subroutines which is primarily intended for numerical computation of multiple integrals in interactive mode. The evaluation of the posterior distribution's features for a practical implementation of the Bayesian paradigm for up to 6 paramters using numerical quadrature procedures and up to 20 parameters using Monte Carlo integration. The *GR* library consists of subroutines for an interactive color graphics system which can be used to reconstruct and display output of the BAYES FOUR system, for reference, see Smith, Skene, Shaw, Naylor, and Dransfield (1985).

Input: Solving a specific inference problem requires writing additional program code which can call BAYES FOUR and GR subroutines.

Output: The posterior moments and marginals can be evaluated by calling these menu driven subroutines. The GR package can be used to provide graphical displays of the univariate and bivariante marginal posterior densities and predictive densities from outputs of BAYES FOUR.

Language: BAYES FOUR in FORTRAN 77; GR in 68000 assembler, C and FORTRAN 77.

Machines: BAYES FOUR is portable. However GR has not yet been configured for any standard graphics system or workstation.

Documentation: Naylor J. C. and Shaw, J. E. H. (1985) *BAYES FOUR-User Guide*; Naylor J. C. and Shaw, J. E. H. (1985) *BAYES FOUR-Implementation Guide*; Shaw, J. E. H. (1985) *GR User Guide*. All these are technical reports from the Nottingham Statistics Group, Department of Mathematics; University of Nottingham.

Availability: These systems may be made available by Prof. Adrian Smith, Department of Mathematics, University of Nottingham, Nottingham, U.K.

Remarks: (i) The Nottingham Statistics Group is actively involved in developing numerical integration systems for implementing Bayesian methodology. Therefore some enhanced versions of these subroutine packages may be available soon.

(ii) For usage experience of this system on some interesting applied problems in the pharmaceutical industry, see Racine, Grieve, Fluhler, and Smith (1986).

Program Name: **Simple Importance Sampling** [Computation of Posterior moments and densities via Monte Carlo Integration (van Dijk)]

Function: The program approximates multiple integrals that arise in calculating posterior moments and marginal densities of parameters of interest in econometric and statistical modeling, via a Monte Carlo integration method known as importance sampling.

Language: FORTRAN 77

Documentation: The algorithm, the program listing and some examples are given in van Dijk, H. K., Hop, J. P. and Louter, A. S. (1986) *An algorithm for the computation of Posterior moments and densities using simple importance sampling*. Econometric Institute Report 8625/A, Erasmus University, Rotterdam.

Availability: The program is available from Prof. Herman K. Van Dijk, Econometric Institute, Erasmus University Rotterdam, P.O. Box 1738-3000 Dr. Rotterdam, The Netherlands.

Remarks: (i) Some standard programs for the method of mixed integration [see, van Dijk, Kloek and Boender (1985)] are under preparation by Prof. van Dijk.

(ii) Geweke, J. (1987) provides some interesting methods for constructing Importance Samplig density which are more flexible than the multivariable Student-t density. He has also developed a PC-AT version of a Monte Carlo integration program which uses these more flexible methods. It is available on diskettes from Prof. John

Geweke, Institute of Statistics and Decision Sciences, Duke University, Durham, NC 27706.

Program Name: **BAYES 3/3D** [Multiparameter Univariate Bayesian Analysis using Monte Carlo Integration (Stewart)]

Function: Bayesian inference for univariate response variable using Monte Carlo integration. Up to nine parameter flexibility allowed. Can handle usual random sampling data, censored data, binomial data at different stresses or times.

Input: Data and control cards as card-image files.

Output: Displays posterior means and posterior percentile curves for CDF's, hazard rate functions, or probability of failure (response) versus stress (dose) or time. (References: Stewart, L. (1979, 83, 85)).

Language: FORTRAN 77

Machines: A graphics terminal is highly desireable but not absolutely necessary. Needs DISPLA graphics software. GKS and DI-3000 versions are being written.

Documentation: Stewart, L. (1987) *User's Manual for* BAYES 3/3D, A program for multiparameter univariate Bayesian analysis using Monte Carlo integration.

Availability: The program was developed under various Federal contracts at Lockheed Palo Alto Research Laboratory, Palo Alto CA 94304. Dr. Leland Stewart, will provide the individual cases, if he can get permission from Lockheed.

Program Name: **LINDLEY. BAS** (Sloan)

Function: This BASIC subroutine performs algebraic manipulation and constructs the expanded formula for use of approximating the ratio of two integrals, required in the evaluations of posterior distribution's features, as discussed in Lindley (1980).

Input: The program prompts for the number of parameters to be estimated.

Output: The printout gives the complete algebraic equation needed to approximate the ratio of integrals.

Language: MS BASIC

Machines: IBM PC or compatibles. Special printing customized for EPSON series of printers.

Availability: Available free of charge from Prof. Jeff A. Sloan, Department of Statistics, University of Manitoba, Winipeg, Manitoba, Canada R3T 2N2.

Program Name: **SBAYES** (Tierney)

Function: The system consists of S-functions to compute approximations of posterior means, variances and marginal densities that are generally more accurate than Lindley's Method mentioned above (see for reference: Tierney and Kadane (1986)).

Language: FORTRAN 77 and C. Requires access to the S package for implementation.

Availability: Available free of charge from Prof. Luke Tierney, School of Statistics, University of Minnesota, Minneapolis, MN 55113.

5. ELICITATION OF PRIOR INFORMATION

Program Name: **Bayes** (Schervish)

Function: This program elicits priors and finds posterior and predictive distributions for samples from normal or binomial data with natural conjugate prior or mixed conjugate plus point mass priors. It also handles flat priors over bounded regions for normal data.

Language: FORTRAN IV, requires access to IMSL.

Machines: DEC-2060. Graphics are good for GIGI terminals only.

Availability: The program is not yet ready for distribution. Available on request from Prof. Mark Schervish, Department of Statistics, Carnegie Mellon University, Pittsburgh, PA 15213.

Program Name: [**B/D**] [Beliefs adjusted by Data (Goldstein/Wooff)]

Function: The program, in final development stage, provides an interactive, interpretive subjectivist analysis of general (partial specified exchangeable) beliefs as described in Goldstein (1987a, b, 1988).

Output: The program output provides summaries of as to how and why beliefs are (i) expected to change and (ii) actually change as well as system diagnostics based on comparison of (i) and (ii).

Language: PASCAL

Availability: will be available at cost of mailing and manual production from Prof. Michael Goldstein, Department of Statistics, University of Hull, Cottingham Road, Hull, U.K.

6. RELIABILITY ANALYSIS

Program Name: **BASS** [Bayesian Analysis for Series Systems (Martz)]

Function: This program performs a Bayesian reliability analysis of series systems of independent binomial subsystems and components for either prior or test data at the component, subsystem and overall system level. It uses a beta prior for the survival probabilities.

Language: FORTRAN 77

Machines: Portable, but requires DISPLA software package for graphics.

Availability: Free of charge from Dr. Harry F. Martz, Group S-1, MS F600, Los Alamos National Laboratory, Los Alamos, NM 87545.

Program Name: **BURD** [Bayesian Updating of Reliability Data (Martz)]

Function: The program performs Bayesian updating of Binomial and Poisson likelihood with a natural conjugate prior or a lognormal prior for the parameter. The updating for lognormal prior is done via Monte Carlo integration. These models are used in nuclear industry. The program is a proprietary of Babcox and Wilcox Inc.

Documentation: Ahmed, S., Metcalf, D. R., Clark, R. E. and Jacobsen J. A. (1981) BURD-*A Computer program for Bayesian updating of realiability data*, NPGD-TM-582, Babcox and Wilcox Inc., Lynchburg, VA.

Program Name: **IPRA** [An Interactive PC-Based Procedure for Reliability Assessment (Singpurwalla)]

Function: A menu driven program performs a prior assessment based on expert opinion or informed judgement for Weibull distributed life length data and the posterior analysis in a highly interactive manner. Singpurwalla (1986a). It also allows the incorporation of the analyst's opinion on the expertise of the experts.

Input: On-line data entry or use of menu option to store data in a file for a later use in the analysis.

Output: The program computes the marginal and joint posterior densities of the Weibull parameters. The prior and posterior reliability functions for a specified time interval as well as distributions of realibility for specified mission times can be computed. These quantities can be displayed in a tabular or 2-d/3-d graphics from or saved on disk.

Language: IBM BASIC

Machines: IBM PC or compatibles with math co-processor and graphics board.

Documentation: Aboura, K. N. and Soyer, R. (1986) '*A User's manualfor an Interactive PC-Based Procedure for Reliability Assessment.*, Tech. Report GWU/IRRA/ serial TR-86-14, George Washington University, Washington, D.C.

Availability: The program diskette and user's manual are available from Prof. Nozer Singpurwalla, Department of Operations Research, George Washington University, Washington, D.C. 20052 at a nominal charge.

Program Name: **IPND** [An Interactive PC-Based System for Predicting the number of defects due to fatigue in Railroad Tracks (Singurwalla)].

Function: A menu driven program performs a Bayesian analysis of a non-homogeneous Poisson process with a Weibull intensity function in which the assessment of the prior information about the parameters is induced via an engineering model based on S-N curves, Singpurwalla (1986b). The procedure is applied to the prediction of the number of defects due to fatigue in railroad tracks.

Input: On-line data entry or use of menu option to store data in a file for a later use in the analysis.

Output: The program computes the marginal and joint posterior densities of the parameters in the Weibull intensity function. The prior and posterior distribution of the number of defects due to fatigue over a time period is also computed. These quantities can be displayed in a tabular or 2-d/3-d graphics form or saved on disk.

Language: IBM BASIC

Machines: IBM PC or compatibles with math. co-processor and graphics board.

Documentation: Choksy, M. and Darynami, S. (1987) "*A interactive PC-Basic System for Predicting the Number of Defects due to Faigue in Railroad Tracks: User's manual*" Tech. Rep. GWU/IRRA/ Serial TR-87-3, George Washington University, Washington, D.C.

Availability: The program diskette and user's manual are available from Prof. Nozer Singpurwalla, Department of Operations Research, George Washington University, Washington, D.C. 20052 at a nominal charge.

Remarks: (i) Prof. Singpurwalla has communicated to us that this procedure and the program has been adopted by The Association of American.

Railroads for the analysis of fatigue defects data in railroad tracks. This is an indications that availability of appropriate software would lead to a widespread use of Bayesian methodology.

Program Name: **PREDSIM** [Prediction and Simulation for mixtures of exponentials (Sloan)]
Function: This program performs a Monte Carlo simulation of sampling from a mixtures of exponentials model using a method proposed by Marsaglia and computes Bayes estimates of the systematic parameters and reliability function and predictive intervals for future observations.
Programing languaje: PL/I
Machines: Portable but requires access to IMSL.
Availability: Available free of charge from Prof. Jeff A. Sloan, Department of Statistics, University of Manitoba, Winipeg. Manitoba, Canada R3T 2N2.

7. MISCELLENEOUS

Program Name: **DISCBDIF** (Stroud)
Function: This SAS® program classifies an input record into one of the two normal populations, based on training samples from each one. It uses either Geisser's discrimination procedure or a semi-diffuse limit of conjugate priors.
Programing languaje: Requires access to SAS® package and SAS® PROC MATRIX.
Availability: Available free of charge from Prof. Thomas W. F. Stroud, Department of Mathematics and Statistics, Queen's University, Kingston, Ontario K7L3N6.

8. CONCLUDING REMARKS

The CADA monitor was the first general purpose program for Bayesian data analysis. It has gone through several enhancements and at this time is the only general purpose package. Even though CADA was presented at several SIBE seminars and it was available in various machine versions, we believe that it has not been accepted as "*the package*" for Bayesian data analysis. This is mainly because all the analyses in CADA are carried out under a noninformative or a simplistic conjugate prior framework and it has no facility for numerical integration. Thus it precludes Bayesian analysis for some realistic prior specifications. Furthermore, the BASIC language does not provide today's state of the art computing environment. The graphics in CADA have been almost non-existent. The package was probably installed at almost all US universities with Bayesian statisticians, but was not used extensively for teaching courses. Thus the circle of CADA awareness has been quite limited. There did not seem to be any strong interest among the participants of the Bayesian Computing Workshop last year, to go with CADA as the base for the future development of a suitable Bayesian Package.

The existing enhanced version of the CADA monitor seems to be quite obsolete to us as it has not changed the basic computing environment from the earlier version. On the other hand, the package is now being marketed by a private company and depending on their future development strategy, the algorithms in CADA could become a vehicle for a state of the art system. We need to explore the future plans of the CADA group before deciding about our own strategy.

The implementation of the Bayesian paradigm for a realistic data analysis would require a variety of numerical integration and approximation routines. The growth of the methodology and software for this has been phenomenal. But we still have a long way to go for approximation and numerical integration procedures and graphical displays for high dimensional problems.

Several researchers are in the process of developing general purpose software for Bayesian analysis of data from various sampling distributions and models. We must mention the software for Bayesian dynamic linear modeling and forecasting, see West, Harrison and Migon (1985) and West and Harrison (1986), which will be an excellent addition to the Bayesian arsenal.

The strategy of writing all bayesian software in S compatible routines sounds appealing from the point of view of researchers in Statistics departments, where UNIX is slowly becoming a defacto operating system. However, S in not accesssible to a large group of statisticians and other reasearchers in Business schools, and Economics and Engineering departments. On the other hand, it is wise to develop all new Bayesian software so that it could be incorporated in an already existing and widely acceptable computing environment. Thus one does not have to worry about developing its data management and graphics capabilities and the students and data analysts do not have to learn yet another system. The only way to develop a quickly acceptable Interactive Bayesian Software Package is to adopt some of the existing main programs and subroutines as modules in some widely used statistics package which is available for mini and micro computers and add more and more modules to it as the new methodology and its software are developed. It is about time that all of us agree on one option. We believe that the most suitable package for this purpose is MINITAB, since it is very widely used for teaching and data analysis for moderate sized data sets. Furthermore, with a proper approach to Minitab Inc., we could receive suitable cooperation from them.

REFERENCES

Carlin, J. B., Dempster, A. and Jonas, J. B. (1985). Bayesian estimation of unobserved componentes in time series, *Jour. of Econometrics* **30**, 67–90.

Caves, D. W., Heeriges, J. A., Train, K. A. and Windle, R. J. (1987). A Bayesian approach to combining conditional demand and engineering models of electricity usage *Review of Economics and Statistics.* (to appear).

Geweke, John (1987). Bayesian inference in econometric models using Monte Carlo integration, **DP:87-2**, Institute of Statistic and Decision Sciences, Duke University.

Goldstein, M. (1987a). Systemic analysis of limited belief specifications. *The Statistician.* (to appear).

Goldstein, M. (1987b). Can we build a subjectivist statistical package?. *Symposium in memoriam of Bruno de Finetti.* (to appear).

Goldstein, M. (1988). The data trajectory. (in this volume).

Jaynes, E. T. (1983). Prior information and ambiguity in inverse problems. *Proc. AMS-SIAM Symposium on Inverse Problems*, New York.

Lindley, D. V. (1980). Aproximate Bayesian methods. *Bayesian Statistics* (J. M. Bernardo, M. H. DeGroot, D. V. Lindley and A. F. M. Smith, eds.), Valencia, Spain: University Press.

Monahan, J. F. (1983). Fully Bayesian analysis of ARMA time Series models, *Jour. of Econometrics* **31**, 307–331.

Nebebe, F. and Stround, T. W. F. (1986). Bayes and empirical Bayes estimation of regression coefficients, *Canadian Jour. of Statist.* **14**, 267–280.

Press, S. James (1980). Bayesian computer programs *Bayesian Analysis in Econometrics and Statistics* (A. Zellner, ed.), Amsterdam: North Holland.

Racine, A., Grieve, A. P., Fluhler, H. and Smith, A. F. M. (1986). Bayesian methods in practice: Experiences in the pharmaceutical industry (with discussion), *Applied Statistics* **35**, 93–150.

Singpurwalla, Nozer D. (1986a). An interactive PC-Based procedure for reliability assessment incorporating expert opinion and survival data. GWU/IRRA/Serial **TR-86/1**, Institute of Reliability and Risk Analysis, The George Washington University.

Singpurwalla, Nozer D. (1986b). An interactive PC-Based system for predicting the number of defects due to fatigue in railroad tracks GWU/IRRA/Serial **TR-86/7**, Institute of Reliability and Risk Analysis, The George Washington University.

Smith, A. F. M., Skene, A. M., Shaw, J. E. H., Naylor, J. C. and Dransfield, M. (1985). The implementation of the Bayesian paradigm. *Comm. Statist., Theo. & Meth.* **14 (5)**, 1079–1109.

Stewart, L. (1979). Multiparameter univariate Bayesian analysis. *J. Amer. Statist. Assoc.* **74**, 684–693.

Stewart, L. (1983). Bayesian analysis using Monte Carlo integration and a powerful methodology for handling some difficult problems. *The Statistician* **32**, 195–200.

Stewart, L. (1985). Multiparameter Bayesian inference using Monte Carlo integration-Some techniques for bivariate analysis. *Bayesian Statistics 2* (J. M. Bernardo, M. H. DeGroot, D. V. Lindley and A. F. M. Smith, eds.). Amsterdam: North Holland.

Thompson, P. and Miller, R. B. (1986). Sampling the future: A Bayesian approach to forecasting from univariate time series models. *Jour. of Business and Economic Statist.* **4**, 427–437.

Tierney, L. and Kadane, J. (1986). Accurate approximations for posterior moments and marginal densities. *J. Amer. Statist. Assoc.* **81**, 82–86.

West, M. and Harrison, P. J. (1986). Monitoring and adaptation in Bayesian forecasting models. *J. Amer. Statist. Assoc.* **81**, 741–750.

West, M., Harrison, P. J. and Migon, H. S. (1985) Dynamic generalized linear models and Bayesian forecasting (with discussion). *J. Amer. Statist. Assoc.* **80**, 73–97.

Van Dijk, H. K., Kloek, T., and Boender, C. G. E. (1985). Posterior moments computed by mixed integration, *Jour. of Econometrics* **29**, 3–18.

DISCUSSION

H. K. VAN DIJK (*Erasmus University Rotterdam and Université Catholique de Louvain*)

The fundamental mathematical operation in Bayesian analysis is that of *integration* (or, stated in even simpler terms, that of *averaging*). One may argue that the success of Bayesian studies in posterior, predictive and decision analysis is strongly dependent on the question whether one can perform the integration operation in a more or less mechanical way using modern computing equipment. The papers of this session focus in different ways on this problem. My comments are directed to the more technical aspects of the integration operation in the prior-to-posterior analysis. This does not imply that the integration operation in prediction and decision analysis is less relevant. On the contrary, but practical experience with integration using modern computers appears to be mostly in the area of posterior analysis.

Due to the great advances in computing equipment and in computational procedures there exists now an enormous flexibility with respect to performing the integration operation for many families of prior and posterior distributions. It seems, therefore, that one can discuss fruitfully the problem of *standardized* Bayesian computational procedures (or Bayesian software). But how standardized can or should Bayesian software be? A naive answer is that the software should enable a researcher to analyze any class of posterior distributions where the numerical constant is not known in terms of elementary functions as in the case of, for instance, the normal, the beta, and the gamma distribution. A more useful approach seems to me to develop a *strategy* for application of the different ways in which one can perform the integration operation. In fact, in numerical optimization one has already a collection of numerical optimization procedures and one makes use of a so-called decision tree as a tool to determine which numerical optimization method is suitable for which class of functions to be optimized. A rough attempt to classify the existing procedures for the computation of posterior moments and densities is the use of the following questions.

1. Is the posterior kernel reasonably well-behaved in the sense that it is unimodal, continuous, proper, not too skew? If the answer is yes, then one may apply the approach from Kass, Tierney and Kadane using the Laplace expansion method. A slower, but more flexible and more robust method appears to be importance sampling using a (symmetric) Student-t density as importance function (see, e.g., van Dijk, Hop and Louter (1986) who provide a diagnostic check on the approximation of the posterior by a symmetric Student-t density).

2. Is a transformation of random variables that results in a more regular shape of the posterior kernel known? If so, then one may use the approach from the University of Nottingham Group (Naylor, Shaw, Skene, Smith, *et al.*).

3. Is the posterior unimodal but is not much more known? Then one may use Monte Carlo with flexible importance functions (see, e.g., Bauwens (1984), Geweke (1987), van Dijk and Kloek (1985)). This is an area for more research. A flexible, robust, but not so efficient method of numerical integration that refers to both questions 2 and 3 is the method of mixed integration of van Dijk, Kloek and Boender (1985). This method makes use of a transformation of random variables that conditions on multivariate skewness and as a next step uses importance sampling with a symmetric normal importance function.

4. Can one generate pseudo-random numbers from the posterior? In this case may use the Monte Carlo approach described by Kass (1985), Geweke (1986) and Zellner, Bauwens and van Dijk (1987).

These questions are rather preliminary and must be developed further, but they may be of use in the classification of the algorithms that are presented by Goel. Further, this brief list of questions with respect to the choice of an integration procedure shows a particular problem, i.e., *diagnostics*, which I define as a set of (informal) checks on the shape of the surface of the posterior density. Traditional error analyses of numerical approximations are usually based on conservative error bounds or on asymptotic statistical theory. Both of these may be poor guides in the choice of a particular integration procedure. More research on diagnostics is needed.

REFERENCES IN THE DISCUSSION

Bauwens, L. (1984). Bayesian full information analysis of simultaneous equation models using integration by Monte Carlo. Berlin: Springer-Verlag.

Geweke, J. (1986). Exact inference in the inequality constrained normal linear regression model. *Journal of Applied Econometrics* **1**, 127–141.

Geweke, J. (1987). Bayesian inference in econometric models using Monte Carlo integration. *Tech. Rep.* Duke University.

Kass, R. E. (1985). Inferences about principal components and related quantities using a numerical delta method and posteriors calculated by simulation, *Tech. Rep.* **346**, Carnegie-Mellon University.

Van Dijk, H. K., Hop, J. P. and Louter, A. S. (1986). An algorithm for the computation of posterior moments and densities using simple importance sampling. *The Statistician* **35**, 83–90.

Van Dijk, H. K. and Kloek, T. (1985). Experiments with some alternatives for simple importance sampling in Monte Carlo integration. *Bayesian Statistics 2* (J. M. Bernardo, M. H. DeGroot, D. V. Lindley and A. F. M. Smith eds.). Amsterdam: North-Holland, 511–530. (With discussion).

Van Dijk, H. K., Kloek, T. and Boender, C. G. E. (1985). Posterior moments computed by mixed integration. *Journal of Econometrics* **29**, 3–18.

Zellner, A., Bauwens, L. and Van Dijk, H. K. (1987). Bayesian specification analysis and estimation of simultaneous equation models using Monte Carlo methods. To appear in *Journal of Econometrics.*

BAYESIAN STATISTICS 3, pp. 189–209
J. M. Bernardo, M. H. DeGroot, D. V. Lindley and A. F. M. Smith, (Eds.)

The Data Trajectory

M. GOLDSTEIN
University of Hull

SUMMARY

An interpretive analysis of the way in which our beliefs are changed by data has two components. Firstly, we summarise the changes over interesting subcollections of beliefs. Secondly, we identify the various aspects of such changes with different aspects of the data. This analysis proceeds by evaluation of the data trajectory (a vectorial summary of changes in belief over the subcollection). We resolve the trajectory along various paths through different portions of the data, to identify which aspects of the evidence have been most influential in modifying our beliefs, to suggest simple diagnostics for our specifications and to assess whether the various parts of the evidence are giving complementary or contradictory messages. We illustrate these ideas with the analysis of a simple regression example, and an example in which simple exchangeability judgements are used to suggest various summaries for a collection of examination data.

Keywords: BELIEF TRANSFORM; DATA SUMMARISATION; EXCHANGEABILITY; LINEAR BAYES METHODS; REGRESSION DIAGNOSTICS.

1. INTRODUCTION

"Exploratory data analysis is the manipulation, summarization, and display of data to make them more comprehensible to human minds, thus uncovering underlying structure in the data and detecting important departures from that structure". (Andrews, 1978)

In a subjectivist data analysis, such summarisation and display is explicitly related to the inferences drawn from the data. Our intention is to make the revision of belief more "comprehensible to human minds". Structures, and departures from such structures, perceived in the data are interpreted in terms of their effect upon the resulting inferences.

Many standard data-analytic methods can be formulated in this way, under diffuse prior specifications. The main interest, however, is in the way that more detailed prior specifications suggest different, and more revealing, data displays, and in the ways in which the intuitive, interpretive aspects of the subjectivist analysis are used to complement the formal, evaluative aspects of the analysis.

In this paper, we describe a few first steps towards a systematic subjectivist data analysis.

2. TRACKING THE DATA

In a formal subjectivist analysis, a collection of belief statements expressed at one time point, t_1, is modified by evidence to give a revised collection of beliefs at a second time point, t_2. A subset of the belief revisions is of primary interest, the remainder providing background inputs to facilitate the analysis. The revision of beliefs is carried out by computer analysis, leaving us the human task of unravelling, interpreting and criticising the new beliefs provided by the mechanical analysis. A natural first step is to provide a summary of the changes in belief over the "interesting" subset, expressed in a form amenable to further analysis. This summary is provided by the "bearing" of the revision, a quantity which summarises both the magnitude and the nature of the changes in belief.

2.1. Bearings

At t_1, you specify a collection of previsions $P_1(C) = [P_1(X_1), \ldots, P_1(X_k)]$ for the elements of C, a collection of random quantities $[X_1, \ldots, X_k]$. At t_2, you specify new previsions $P_2(C) = [P_2(X_1), \ldots, P_2(X_k)]$ for these quantities. (The prevision, $P(X)$, of a random quantity X can be considered to be the expectation for X, but specified directly as a primitive quantity rather than as a quantity evaluated through some intermediary probability distribution; see de Finetti (1974). For example, C represents a probability space if the X_i are the indicator functions for the elements of a partition.)

For now, we will place no constraints upon the methods by which the beliefs are revised, but simply suppose that at each time point the previsions are coherent. Thus, P_1 and P_2 are both linear, so that the comparison between the two sets of beliefs is expressed in terms of the changes in prevision over $L(C)$, the set of linear combinations, $a_1X_1 + \cdots + a_kX_k$, of elements of C. (For example, if C represents a probability space, then $L(C)$ is the space of random variables defined over the probability space.)

This summary is constructed using certain further specifications, made at t_1, namely variances and covariances between all pairs of the elements of C (again specified directly as previsions). These specifications provide an orientation for your prior beliefs in terms of which changes in these beliefs can be expressed. The first step in constructing the summary is to derive an uncorrelated "co-ordinate frame" for C, defined follows.

Definition. *A component representation of C is any collection $[E_1, \ldots, E_r]$ of r linear combinations of $X_1, \ldots, X_k$ (plus a constant), for which each E_i has prevision 0 and variance 1, each pair $E_i, E_j, i \neq j$, is uncorrelated, and each X_i in C can be uniquely expressed as a linear combination of $E_1, \ldots, E_r$ (plus a constant). (For example, the principal components of C provide a natural component representation.)*

At t_2, you may use the component representation established at t_1 to evaluate an element of $L(C)$ which expresses the position (or bearing) of your new beliefs with respect to your old beliefs. This is defined as follows.

Definition. *If $E_1, \ldots, E_r$ is a component representation of C, then the bearing of $P_2(C)$ with respect to $P_1(C)$ is the quantity Y_2 defined by*

$$Y_2 = P_2(E_1)E_1 + \cdots + P_2(E_r)E_r.$$

(The bearing, as defined above, appears to depend upon the choice of component representation, but any two different component representations for C will lead to the same bearing Y_2.)

At t_2, Y_2 is a specified element of $L(C)$, whose covariance properties (as evaluated with respect to the covariance structure declared at t_1) summarise your changes in belief over C, as follows.

Theorem 1. *With notation as above, the bearing Y_2 has the property that for every X in $L(C)$,*

$$P_2(X) - P_1(X) = \text{Cov}_1(X, Y_2).$$

(Cov_1 *is the covariance declared at* t_1.)

In particular, for any X in $L(C)$ which is a priori uncorrelated with the bearing Y_2, $P_1(X) = P_2(X)$.

Proof. Each X in $L(C)$ may be written

$$X = P_1(X) + \text{Cov}_1(X, E_1)E_1 + \cdots + \text{Cov}_1(X, E_r)E_r.$$

Thus, for each X in $L(C)$,

$$\begin{aligned} P_2(X) - P_1(X) &= \text{Cov}_1(X, E_1)P_2(E_1) + \cdots + \text{Cov}_1(X, E_r)P_2(E_r) \\ &= \text{Cov}_1(X, Y_2). \end{aligned}$$

From the theorem, we observe that all changes of belief are in the "direction" of Y_2. Further, having fixed a direction for Y_2, the extent of the change in belief over $L(C)$ is expressed by the size of Y_2 expressed as follows.

Definition. *The length, L_2, of the revision of beliefs over $L(C)$ is defined to be*

$$L_2 = SD_1(Y_2).$$

(SD_1*is the standard deviation declared at* t_1)

The length of the bearing relates to the standardised change in belief as follows.

Corollary 1.1. *The maximum value of $|P_2(X) - P_1(X)|/SD_1(X)$, over all X in $L(C)$ is L_2, this maximum being attained by Y_2.*

More generally, if $Z_2 = kY_2$, then a bearing of Z_2 would represent k times the change in prevision for every element of $L(C)$ as would a bearing of Y_2, so that the bearing expresses both the direction and the magnitude of the change in beliefs.

Note 0. To illustrate the machinery of our approach, we give a simple example in detail as an appendix. The bearing for the example is evaluated in subsection (A.1).

Note 1. The bearing is formally analogous to the likelihood in a traditional Bayes analysis. For given "sample" values, the likelihood, L, is a function of the unknown "parameter" values, i.e. the likelihood is a random quantity. For any other random quantity (i.e. function of the parameters) R, we may evaluate the change from prior to posterior expectation of R given the observed sample values by evaluating the covariance of R and (a normalised version of) L with respect to the prior measure over the parameter space.

Our approach differs in not supposing a priori the existence of a convenient, universally specified likelihood, but rather constructively evaluating the counterpart to this quantity, as restricted to the (usually small) subspace of primary interest, and expressing it in a framework in which various basic properties of interest may be directly computed. (We can rescue some of these properties for the strict Bayes counterpart to our analysis by projecting the likelihood down to the subspaces of interest, using the Hilbert space formalism of section 4.)

2.2. *The trajectory*

The changes in belief in a typical analysis arise from complicated combinations of evidence which we must unravel in order to identify which aspects of the data are giving rise to which features of the changes in belief, whether different aspects of the data are giving complementary or contradictory information and so forth. Thus we need to resolve the bearing into component parts which address these questions.

Consider the simplest case, namely that you have two pieces of "evidence" E and F. Your overall revision of belief will be based on observing both E and F, with bearing Y_2, which we write Y_{E+F}. You may view this revision as occurring in two stages. Firstly, you observe E, and revise beliefs, giving bearing Y_E. Then you also observe F, giving overall bearing Y_{E+F}, so that the change in bearing due to observing F, having already observed E, is $Y_{[F/E]} = Y_{E+F} - Y_E$. We have identified the variance of Y_{E+F} with the magnitude of the revision of beliefs over $L(C)$. (As in section (2.1), all variances and covariances are those expressed at t_1.) We may write

$$\text{Var}(Y_{E+F}) = \text{Var}(Y_E) + \text{Var}(Y_{[F/E]}) + 2\text{Cov}(Y_E, Y_{[F/E]}).$$

Thus, there are two aspects to consider in moving from Y_E to $Y_{[E+F]}$. Firstly, $\text{Var}(Y_{[F/E]})$ expresses the magnitude of the further change in belief induced by F. Secondly, $\text{Cov}(Y_E, Y_{[F/E]})$ expresses the degree of support/conflict between the two pieces of evidence in determining the revision of beliefs. If this covariance is large and positive, then the two pieces of evidence are complementary, and their combined effect is greater than the sum of the individual pieces of evidence, whereas if this covariance is large and negative then the two pieces of evidence are giving contradictory messages which in combination result in smaller changes of belief.

More generally, we may represent the revision of beliefs from P_1 to P_2 over the collection C as the result of performing a sequence of m revisions of belief $P_{[1]}, \ldots, P_{[m]}$. (For example, beliefs about C might have been informally revised at regular intervals, before the final revision. Alternately, the revision might be based upon applying some algorithm, such as Bayes theorem, to a given data set, in which case we might divide the data into m disjoint subsets and observe the cumulative changes in belief as we input portions of the data into the algorithm.)

Overall, P_1 is revised to P_2 with bearing Y_2. Construct this revision in stages as follows. From each $P_{[i]}$ construct the bearing $Y_{[i]}$. The change in the bearing, between stage $[i]$ and stage $[j]$ is $Y_{[j]} - Y_{[i]}$. Call this difference the bearing for $P_{[j]}$, *adjusted* for $P_{[i]}$, denoted $Y_{[j/i]}$. The adjusted bearing expresses the change, in both magnitude and direction, in beliefs between stage $[i]$ and stage $[j]$. As an immediate corollary of theorem 1, we have the following result.

Corollary 1.2. *For every X in $L(C)$, and each $i < j$,*

$$P_{[j]}(X) - P_{[i]}(X) = \text{Cov}_1(X, Y_{[j/i]}).$$

In particular, the maximum value of $|P_{[j]}(X) - P_{[i]}(X)|/SD_1(X)$, over all X in $L(C)$ is $SD_1(Y_{[j/i]})$.

We will be interested in the one step revisions $Y_{[i/i-1]}$, which we denote by $Y_{[i/]}$. We make the following definition.

Definition. *A trajectory over C is a sequence of adjusted bearings $Y_{[1]}, Y_{[2/]}, Y_{[3/]}, \ldots, Y_{[r/]}$.*

We have identified the variance of the bearing with the magnitude of the change in belief. We denote $V[i]$, $V[i/]$ to be the variance of $Y_{[i]}$, $Y_{[i/]}$, respectively, and we denote $C[i]$ to be the covariance between $Y_{[i-1]}$ and $Y_{[i/]}$.

For each j, we may write

$$Y_{[j]} = Y_{[1]} + Y_{[2/]} + \cdots + Y_{[j/]}.$$

We may therefore write

$$V[j] = V[1] + V[2/] + \cdots + V[j/] + 2(C[2] + C[3] + \cdots + C[j]).$$

Thus, to assess the way that the individual stages in the revision have combined to produce the overall revision, we must examine

(i) the individual adjusted bearings $Y_{[i/]}$. Those bearings with large variance identify stages at which large changes in belief occur, the effects being summarised by the direction of $Y_{[i/]}$;

(ii) the relationship between each adjusted bearing $Y_{[i/]}$ and the raw bearing for the preceding stage $Y_{[i-1]}$. The effect of the additional information upon the total information is expressed through the value of $C[i]$. Large positive/negative correlation between $Y_{[i-1]}$ and $Y_{[i/]}$ implies that the additional evidence supports/contradicts the earlier evidence and that the resulting evidence from both stages produces much stronger/weaker changes in belief than from the individual stages.

A simple summary of the trajectory is given by the pairs $(V[i/], C[i])$. For ease of interpretation, we replace each covariance, $C[i]$, by the corresponding correlation $Cr[i]$, between $Y_{[i-1]}$ and $Y_{[i/]}$. We term the collection of pairs $(V[i/], Cr[i])$ the *route* of the trajectory.

2.3. *Belief adjustment by projection*

In defining the bearing and the trajectory, we have not specified how beliefs are revised. The revision may be standard Bayes or follow some alternative formalism or be made in a purely informal manner. The general principle, however beliefs are revised, is that initial beliefs provide an efficient collection of landmarks for describing changes in beliefs.

We now describe a particular way to adjust previsions which forms the basis of our evaluations in this paper. Suppose that between t_1 and t_2 you will observe the values of the quantities $D = [D_1, \ldots, D_s]$. For a general random quantity X, let $P_D(X)$ denote the linear Bayes rule for X in $[D_0, D_1, \ldots, D_s]$ (D_0 is the unit constant), i.e. $P_D(X)$ is the linear combination minimising Var$(X - a_0 D_0 - a_1 D_1 - \cdots - a_s D_s)$ over all choices of coefficients a_i. (This choice depends only on the covariance structure: see, for example, Goldstein (1975).)

Having observed the outcomes $D_i = d_i$, $i = 1, \ldots, s$, we evaluate each adjustment of belief as

$$P_2(X) = P_d(X) = a_0 d_0 + a_1 d_1 + \cdots + a_s d_s$$

(where $a_0, a_1, \ldots, a_s$, are the coefficients of $D_0, D_1, \ldots, D_s$, in $P_D(X)$).

The bearing for this revision, i.e. the quantity Y_d for which $P_d(X) - P(X) = \text{Cov}(X, Y_d)$ for each X in $L(C)$, is termed the *data bearing*, and is calculated as in section (2.1). The type of trajectory that we will primarily be concerned with below comes from dividing the data into subsets (based on variables and/or individuals), $D[1], \ldots, D[m]$, and progressively making linear fits on $D[1], D[1]+D[2], D[1]+D[2]+D[3]$, etc. Such a trajectory is termed a *data trajectory*. (A simple example of the evaluation of the route of a data trajectory is given in subsection A.3 of the appendix.)

Note 2. Our choice of linear fitting is motivated both by theoretical and pragmatic considerations. Firstly, Bayes conditioning is the special case of linear fitting for which D represents a partition (so that $P_D(X) = P(X|D_1)D_1 + P(X|D_2)D_2 + \cdots$). The general linear fitting procedure simply lifts the restriction that $D_1, \ldots, D_k$ are the indicator functions for a partition. As argued in Goldstein (1986a), all of the inferential features usually associated with probabilistic conditioning may be viewed as simple special cases of the same properties associated with linear projection. Further, the linear fitting methodology makes the minimal demands for belief specification of any fully elaborated inferential system (see Goldstein (1987a, b)).

Further, the variances and covariances that are required in order to construct the data trajectory are precisely those which are required in order to evaluate the various linear fits that we require. Our data analytic investigations will be more straightforward if they exploit simple methods of belief adjustment reflecting actual belief specifications rather than the fitting of complicated probability measures, specified simply to allow the Bayesian machinery to perform the required computations. (Of course, an elaborate probability specification is perfectly acceptable if it expresses a genuine belief specification. Such specifications are special cases of the linear fitting methodology. We have restricted attention to finite observation structures for simplicity of exposition, but the theory follows in similar manner when the data space is given a continuous representation for which linear fitting corresponds to projection into the corresponding space of square integrable functions over the data values.)

3. EXAMPLE ONE – SIMPLE LINEAR REGRESSION

3.1. Belief specification

We have 11 pairs of values on variables Y and X, which are to be used to fit a simple linear regression

$$Y_i = a + bx_i + e_i,$$

where a, b are unknown constants and $e_1, \ldots, e_{11}$ are unobserved, random quantities, a priori uncorrelated (with each other and with a, b), each with mean zero, and a common prior variance v. As our intention is simply to illustrate the data trajectory, in a familiar context, we treat the model at face value and specify beliefs over the smallest set of quantities that might be of interest , namely the regression coefficients. i.e. $C = [a, b]$. (We may loosely identify $L(C)$ with the set of predictive values for Y given the various values of x.) Similarly, we choose the simplest functions of the data, namely the raw values, i.e. we choose $D = [Y_1, \ldots, Y_{11}]$. We specify zero prevision for both a and b, $\text{Var}(a) = 2$, $\text{Var}(b) = 1$, $\text{Cov}(a, b) = 0$, $v = 1$.

3.2. The bearing

Suppose that the data values are as follows.

Y	8.04	6.95	7.58	8.81	8.33	9.96	7.24	4.26	10.84	4.82	5.68
X	10	8	13	9	11	14	6	4	12	7	5

The adjusted previsions are $P_d(a) = 2.16$, $P_d(b) = 0.58$. The bearing is evaluated to be

$$Y_d = 1.08a + 0.58b,$$

with (prior) standard deviation 1.63. Thus, the maximum change in "predictive prevision", 1.63 standard deviations, is for values of x around 0.5. For values of x around -3.5, prior and posterior previsions for Y are about the same (namely zero).

3.3. The data trajectory

The data set given above was one of four sets of eleven pairs of readings on two variables (as given in Anscombe (1973)), constructed in such a way that each had identical summary statistics, so that a least squares fit would in each case lead to the same conclusions (i.e. the same estimates for the regression coefficients, error variance and coefficient of determination), but for which plots of the four data sets suggest very different interpretations in each case.

The four data sets are as follows (the X values are the same for sets 1 to 3).

	CASE	1	2	3	4	5	6	7	8	9	10	11
SET												
1-3)	X	10.0	8.0	13.0	9.0	11.0	14.0	6.0	4.0	12.0	7.0	5.0
1)	Y	8.04	6.95	7.58	8.81	8.33	9.96	7.24	4.26	10.84	4.82	5.68
2)	Y	9.14	8.14	8.74	8.77	9.26	8.10	6.13	3.10	9.13	7.26	4.74
3)	Y	7.46	6.77	12.74	7.11	7.81	8.84	6.08	5.39	8.15	6.42	5.73
4)	X	8.0	8.0	8.0	8.0	8.0	8.0	8.0	19.0	8.0	8.0	8.0
4)	Y	6.58	5.76	7.71	8.84	8.47	7.04	5.25	12.50	5.56	7.91	6.89

Plots of the four sets suggest that the regression may be plausible for set one, but that set two seems to follow a smooth polynomial curve. Set three suggests a much stronger linear relationship than set one, with the exception of a single case, case 3, which lies far from the fitted line. Finally in set 4, all of the information about the regression is provided by a single case, case 8. The plots are as follows.

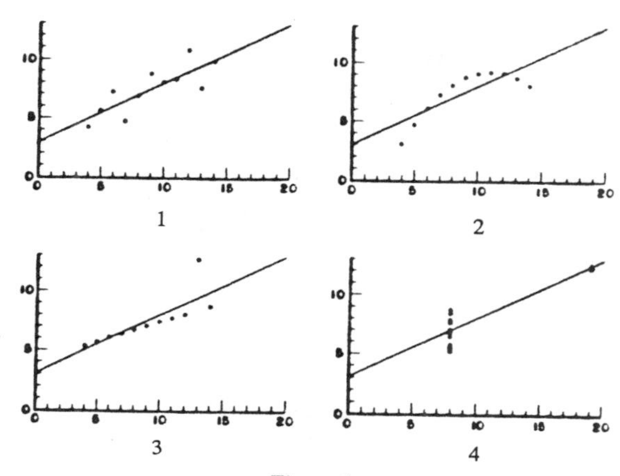

Figure 1.

In the belief specification in section (3.1), the belief structures were chosen so that the summaries of the data required for the belief adjustment were precisely the same as those used in the least squares analysis. Thus, the bearing for each of the above four data sets, under the analysis of section 3.2, would be the same. As was the case for the least squares fit, any inferential statements within any logical system are summary statements, which must be unravelled into their constituent elements, in our case by examination of the data trajectory.

For purpose of comparison, we evaluate the data trajectory for each data set as follows. We order the x values, and let $Y_{(i)}$ denote the value of Y corresponding to the ith largest value of x. (This gives the same ordering for each of the first three data sets.) To damp out the "noise" at the start of the trajectory, we set $D[1] = [Y_{(1)}, \ldots, Y_{(5)}]$. We track the data across increasing x by choosing $D[2] = [Y_{(6)}]$, $D[3] = [Y_{(7)}], \ldots, D[7] = [Y_{(11)}]$. At each step, for each set, we evaluate $V[i]$, $V[i/]$, and $Cr[i]$, as follows.

	DATA 1			DATA 2			DATA 3			DATA 4		
i	$V[i]$	$V[i/]$	$Cr[i]$	$V[i]$	$V[i/]$	$Cr[i]$	$V[i]$	$V[i/]$	$Cr[i]$	$V[i]$	$V[i/]$	$Cr[i]$
1	1.3	—	—	1.16	—	—	1.71	—	—	0.65	—	—
2	1.01	.05	–.50	1.01	.05	–.55	2.15	.06	.60	0.73	.002	1
3	1.32	.05	.36	0.94	.03	–.47	2.62	.04	.72	0.72	.000	–1
4	1.66	.05	.55	0.86	.08	–.30	3.10	.03	.80	0.76	.001	1
5	1.33	.04	–.69	1.01	.14	.02	3.54	.02	.84	0.72	.001	–1
6	2.45	.33	.60	1.56	.19	.40	1.88	.39	–.87	0.74	.000	1
7	2.66	.01	.81	2.66	.25	.69	2.66	.11	.75	2.66	1.96	0

Table 1. *Data Trajectory for linear fit*

(Thus, for example, for data set one, the variance of the bearing based on the observations at the 5 smallest x values is 1.3 —i.e. the largest change in prevision over $L(C)$ is 1.14 sd. Based on the 6 smallest x values, the variance of the bearing has reduced to 1.01. The largest change in prevision from adding the sixth point is .22 sd, and the correlation between the directions is -0.5, suggesting a degree of "conflict" between the evidence.)

The differences in the trajectories reflect the differences in the four plots. In set 1, the evidence for a line fluctuates until we add the tenth point. The curve in set 2 is reflected in the consistent pattern in the stepwise correlations. In set 3, the first 9 points show a strong and consistent pattern of evidence for the line, and the aberrant point, 10, is represented by the high negative correlation and large reduction in length of trajectory. In set 4, very little happens until the final point is added, the resulting change in beliefs being more than twice the change due to the remainder of the sample, with zero correlation as the new evidence concerns aspects of the uncertainty unrelated to the evidence provided by the first ten points.

Often, we will compare the route through the data of various different belief formulations. For example, suppose that instead of fitting a linear regression, we fit a quadratic regression

$$Y_i = a + bx_i + cx_i^2 + e_i.$$

For comparison with the above analysis, suppose that beliefs over a, b, e are as above and c is judged a priori uncorrelated with the remaining quantities, with prior prevision zero, prior variance 0.2. The trajectories corresponding to the first three of the above data sets, in the same form as the above table, are as follows.

	DATA 1			DATA 2			DATA 3		
	$V[i]$	$V[i/]$	$Cr[i]$	$V[i]$	$V[i/]$	$Cr[i]$	$V[i]$	$V[i/]$	$Cr[i]$
i									
1	1.58	—	—	0.36	—	—	1.95	—	—
2	1.12	.07	–.81	0.57	.04	.75	1.93	.00	–.77
3	1.27	.01	.56	0.84	.04	.77	1.87	.00	–.62
4	1.36	.01	.48	1.18	.04	.76	1.80	.00	–.43
5	1.17	.08	–.41	1.60	.05	.76	1.75	.01	–.27
6	1.45	.21	.07	2.12	.05	.77	2.08	.54	–.11
7	1.36	.01	–.37	2.72	.06	.79	1.72	.12	–.48

Table 2. *Data trajectory for quadratic fit*

Notice, in particular, how the correlation pattern has been straightened out for data set 2. Note also the difference in the representation of the "aberrant" point $Y_{(10)}$ in data set 3 between table 1 and table 2.

4. MAPPING THE TRAJECTORY

In the example in section 3, differences in data patterns were reflected in differences in the data trajectories. In order to interpret the trajectories (and more generally to suggest automatic approaches for generating interesting trajectories), we must develop simple descriptions of the type of trajectory that we expect. (For example, in table 1, the large jump in the trajectory for data set 4 should be expected, unlike the jump in data set 3.)

In section (2.1), we observed that any component representation $E_1, \ldots, E_r$ could be used to calculate the bearing Y_2, by the relation

$$Y_2 = P_2(E_1)E_1 + \cdots + P_2(E_r)E_r.$$

At t_1, Y_2 is a randomly selected element of $L(C)$, as we have not yet made the assignments $P_2(E_i)$. Our first step in describing beliefs about the trajectory is to identify a co-ordinate frame within which certain properties of our beliefs about Y_2, such as our prior assessment for $\text{Var}(Y_2)$, are immediate. This frame is best described by considering the belief structures which underlie the analysis.

4.1. Belief structures

All of the manipulations that we require exploit covariance structures over linear spaces. Thus, it is natural to make this structure explicit by viewing each collection of beliefs (i.e. a covariance structure declared over a collection C of random quantities) as determining an inner product space. In Goldstein (1986b), such spaces were called belief structures, and defined as follows.

Definition. *A belief structure is an inner product space constructed as follows. Begin with a set of random quantities* $C = [X_0, X_1, X_2, \ldots,]$, *including* $X_0 = 1$, *the unit constant. Denote by* L *the vector space in which each* X_i *is represented as a vector and linear combinations of vectors are the corresponding linear combinations of random quantities. Construct the inner product and norm over* L *defined for each pair of elements* U, V *in* L *by*

$$(U,V) = P(UV), \qquad ||U|| = \surd(U,U).$$

(We identify with the zero vector all elements W *for which* $P(W^2) = 0$. *We restrict* C *to elements with finite squared previsions.) The set* C *is called the base of the belief structure* A, *denoted by* $C = b(A)$.

In the adjustment of belief described in section (2.3), there are two belief structures of interest. Firstly, B, with base $[X_0, X_1, \ldots, X_t]$, the collection of quantities for which beliefs are to be revised, and secondly D, the belief structure with base $[D_0, D_1, \ldots, D_s]$, the collection of quantities which we intend to use for linear fitting. Both structures are subspaces of the combined structure $B + D$, with base $b(B + D) = b(B) + b(D)$. The linear Bayes rule $P_D(X)$ for an element X in B is the vector in D corresponding to the orthogonal projection of X into D (i.e. to the element Z in D which minimises $||X - Z||$. Note the equivalence between previsions and projections —for example conditional prevision is projection into a space of indicator functions determining a partition).

Note 3. The belief structure construction is of general interest as the minimal amount of structuring (that I know of) that incorporates all of the exchangeability, belief revision and interpretive features that we require of the subjectivist theory. For example, the existence of a bearing for any belief revision may be demonstrated in this formulation to be an immediate consequence of the Riesz representation for linear functionals. This is the approach that we would adopt if we wished to investigate existence and properties of the bearings in full generality. For example, we have restricted C to a finite number of elements to avoid various technicalities which would otherwise distract from the basic argument. However, the Riesz construction works in exactly the same way as in the finite case if C has an infinite number of quantities —for example, if C represents a continuous probability distribution in terms of the corresponding Hilbert space of square integrable functions over the space— provided that certain technical details related to compactness, finite versus countable additivity, and so forth are handled carefully.

4.2. Belief transforms

At t_1, we summarise the expected changes in belief over A by evaluating the belief transform defined as follows.

Definition. *The belief transform over B induced by D is the linear operator over B defined by*

$$T_D = P_B P_D,$$

where P_B, P_D are respectively the orthogonal projections from D to B and from B to D.

In Goldstein (1981), the following properties are demonstrated. Firstly, T_D is a self-adjoint operator on B of norm at most one. If B has a finite base, then we can select a component representation for B, consisting of orthonormal eigenvectors $E_0, E_1, \ldots, E_s$ of T_D, corresponding to the ordered eigenvalues $1 = m_0 \geq m_1 \geq \cdots \geq m_s \geq 0$, where E_0 is the unit constant (so that $[E_1, \ldots, E_s]$ is an uncorrelated collection, each of which has zero prevision, unit variance, and satisfies $P_B P_D(E_i) = m_i E_i$). The subcollection $[E_1, \ldots, E_t]$ consisting of eigenvectors with non-zero eigenvalues (but stripping out the constant), is termed a *map* over B and denoted $M(B)$. We call the values $m_1, \ldots, m_t$ the *scale* of the map. The map summarises expectations as to the revision of belief. The axes of the map corresponding to large/small eigenvalues are those for which there is large/small expected change in information. (For example, for each E_i, the ratio of residual variance from the linear fit to prior variance is $1 - m_i$.) The essential properties of the map for our purposes are as follows.

Lemma. *Suppose that $E = [E_1, \ldots, E_t]$ is a map over B induced by D. For each i let*

$$F_i = m_i^{-1/2} P_D(E_i).$$

Then $F = [F_1, \ldots, F_t]$ is a map over D induced by B, with the same scale as for the map over B induced by D. (In particular each F_i has prevision zero, unit variance, $F_1, \ldots, F_t$ are uncorrelated and $P_B(F_i) = m_i^{1/2} E_i$.)

We call the maps $M(B)$, $M(D)$ constructed in the above lemma the *twin maps* for the pair B and D, with common scale determined by the non-zero eigenvalues of $P_B P_D$ (or equivalently the non-zero eigenvalues of $P_D P_B$). We can now write the bearing in a form symmetric between the two spaces as follows.

Theorem 2. *Let $[E_1, \ldots, E_t]$, $[F_1, \ldots, F_t]$ be the twin maps for the pair of belief structures B and D, with scale $[m_1, \ldots, m_t]$. Then the bearing over B induced by D, and the bearing over D induced by B are each of the form*

$$\sqrt{m_1} E_1 F_1 + \cdots + \sqrt{m_t} E_t F_t,$$

where, for observed $D = d$, Y_d is evaluated by substituting the observed values $F_i = f_i$, whereas, for observed $B = b$, Y_b is evaluated by substituting the observed values $E_i = e_i$.

We can now informally describe the relationship between unusual data and the resulting bearings. The collection $M(D)$ forms an uncorrelated grid over the data space. Each F_i has zero prior prevision and unit variance so that peculiar data corresponds to various large observed values f_i. The representation of the data by the bearing magnifies this effect for those directions with large m_i, and discounts those directions with small m_i, thus suppressing features of the data which do not affect the belief revision.

The simplest summary of this effect is the evaluation of Var(Y_d). Large values of this expression alert us to data peculiarities. How large? One measure is given by the prior prevision for this quantity, which is evaluated as follows.

Corollary 2.1. *With notation as above,*

$$P(Var(Y_D)) = trace(T_D)$$

Proof. From the above representation, for observed $D = d$,

$$\text{Var}(Y_D) = m_1 f_1^2 + \cdots + m_t f_t^2.$$

The prevision for this value is

$$P\text{Var}(Y_D) = m_1 P(F_1^2) + \cdots + m_t P(F_t^2),$$
$$= m_1 + \cdots + m_t = \text{trace}(T_D).$$

Thus, a simple standard for the data trajectory is comparison with the cumulative trace of the belief transform. This also suggests an automatic procedure for constructing informative data trajectories, namely separating the data in ways which maximise the change in trace and we will return to this below. (In the appendix, the belief transform is evaluated in subsection A.4.)

Note 4. For simplicity, we have evaluated the expected length of the bearing for the special case in which beliefs are adjusted by linear projection. However, this result is perfectly general. For any method of belief revision, we may evaluate a corresponding belief transform (see Goldstein (1981)). The expected length of the bearing will in all cases be the trace of the belief transform. In a sense, the trace evaluation of corollary (2.1), and all the belief adjustments by projection, may be considered as linear reductions of their subjectivist counterparts in the full belief revision —see Goldstein (1986a)

4.3. Example: simple regression (continued)

We conclude the example from section (3) by evaluating the trace of the belief transform at each stage of the trajectory.

	Points	x(1)-x(5)	x(6)	x(7)	x(8)	x(9)	x(10)	x(11)
Simple linear fit								
Data set 1-3		1.33	1.42	1.50	1.57	1.62	1.67	1.70
4		0.997	0.997	0.998	0.998	0.998	0.998	1.70
Quadratic fit								
Data set 1-3		1.90	1.96	2.01	2.05	2.10	2.15	2.19

Comparing the expectations to the observed values, all of the full belief revisions seem reasonable whereas many of the paths seem to be aberrant. Note in particular that the jump in trajectory for data set 4 corresponds to the change in trace for the last data point.

4.4. Constructing a pure trajectory

The data trajectory summarises the cumulative changes in belief induced by progressively adding subsets of the data. There is another type of trajectory which is often of interest, which comes from dividing a data set into portions and evaluating the bearing for each portion of the data separately. (For example, a treatment might be given at several centres. We might evaluate the bearing for each centre individually, as a complement to the pooled bearing from the combined data.) We call such a trajectory a *raw* trajectory. A raw trajectory which is also a trajectory as previously defined is doubly useful in expressing both the cumulative information from the data and also the effect directly attributable to the various portions of the data.

We now describe a general method for constructing such trajectories. For any set of quantities $G = [G_1, \ldots, G_r]$, we denote a general partition of G into k disjoint subsets as $G[1], \ldots, G[k]$. We make the following definition.

Definition. *For any pair of structures, B and D, let the twin maps be $M(B)$, $M(D)$. Any raw trajectory based upon a partition $M[1], \ldots, M[k]$ of $M(D)$ is termed a pure trajectory $Y_{[1]}, \ldots, Y_{[k]}$.*

The basic properties of a pure trajectory follow from the representation of the trajectory given in theorem (2) and are summarised as follows.

Theorem 3. *Any pure trajectory satisfies the following properties.*

(i) *For any observed data outcomes, the data trajectory is equivalent to the raw trajectory (i.e. the bearing for data subset i, $Y_{[i]}$, is equivalent to the adjusted bearing for this subset $Y_{[i/]}$).*

(ii) *The bearings $Y_{[1]}, \ldots, Y_{[k]}$ are a collection of uncorrelated random quantities. In particular, the length of the bearing corresponding to any subcollection $M[i_1] + \cdots + M[i_r]$ is equal to the sum of the lengths of the individual bearings for $M[i_1], \ldots, M[i_r]$*

(iii) *The expected length of $Y_{[i]}$ is equal to the sum of the eigenvalues of T_D corresponding to the eigenvectors in $M[i]$.*

(The proof of (i) follows fron the mutual orthogonality of the elements of $M(D)$. The proof of (ii) follows from the mutual orthogonality of the elements of $M(B)$. The proof of (iii) follows from (ii).)

Thus, we can create interesting trajectories by examining the belief transform for the system, dividing the eigenstructure into meaningful subsets, and tracking the data across the subsets. (An example of a pure trajectory is given in appendix A.5.)

5. EXAMPLE TWO: SUMMARISING EXCHANGEABLE OBSERVATIONS

5.1. Belief structures for simple exchangeable systems

Subjectivist statistics concerns the exploitation of different aspects of exchangeability. We give a simple special case. Suppose we may make a sequence of k observations in each of a series of situations. Let Q_{ij} be an observation on quantity i for situation j. Suppose that the collections $[Q_{1j}, \ldots, Q_{kj}]$ may be considered as part of an infinite exchangeable sequence, in the sense of Goldstein (1986b). That is we suppose, for each i, r, and $j \neq h$,

(E1) $P(Q_{ij}) = m$;

(E2) $\text{Cov}(Q_{ij} Q_{rj}) = U_{ir}$ (which does not depend on j);

(E3) $\text{Cov}(Q_{ij} Q_{rh}) = V_{ir}$ (which does not depend on $j \neq h$).

As a further simplification, suppose that the quantities Q_{ij} are themselves exchangeable for each j, that is, for all $s \neq t$,

(E4) $V_{ss} = V$, $V_{st} = C$, $U_{ss} = U$, $U_{st} = B$.

From the exchangeability over situations, the representation in Goldstein (1986b) allows us to construct a sequence of (unobservable) quantities, $Q_i, i = 1, \ldots, k$ and $R_{ij}, i = 1, \ldots, k, j = 1, 2, \ldots$, where, for all i, s, t and $u \neq v$

$$P(R_{su}) = \text{Cov}(R_{su}, Q_i) = \text{Cov}(R_{su}, R_{tv}) = 0,$$

and $\text{Cov}(R_{su}, R_{tu})$ does not depend on u. We can write each Q_{ij} as

$$Q_{ij} = Q_i + R_{ij}.$$

(The specifications for the quantities follow from these properties; for example $\text{Cov}(Q_i Q_r) = V_{ir}$.)

The quantities Q_i are the subjectivist counterpart of the "population" averages for the observables, and the quantities R_{ij} express differences between individuals. (Compare the usual exchangeability representation by "parameters", given which observation vectors are iid. Here the residual belief structures are identical and uncorrelated. Our representation is weaker because it represents actual, rather than hypothetical judgments, to the minimal level required to support our inferences.)

Suppose we have, as data, a sample of size n on these quantities (i.e. we observe $D = [Q_{ij} : i = 1, \ldots, k; j = 1, \ldots, n]$). We may use the data to adjust our previsions for the belief structure B, with base $[Q_1, \ldots, Q_k]$. (This is precisely equivalent to using the data to adjust "predictive" beliefs about future observations, but computationally more straightforward.)

We can show that the belief transform over B induced by D has eigenvalues $m_1 = w = ns/(1+(n-1)s)$, $m_2 = \cdots = m_k = z = nt/(1+(n-1)t)$, where $s = (V+(k-1)C)/(U+(k-1)B), t = (V-C)/(U-B)$.

The corresponding eigenvectors of T are

$$E_1 = h(Q_1 + \cdots + Q_k),$$

where h is the normalising constant $1/\sqrt{(rV + r(r-1)C)}$, and $E_2, \ldots, E_k$ are any $k-1$ uncorrelated combinations of $Q_1, \ldots, Q_k$ which are themselves uncorrelated which E_1 (i.e. for which the coefficients of the linear combination sum to zero —a typical element would be $(Q_i - Q_j)$). Thus, we expect two types of information, the first related to the overall magnitude of the quantities Q_i, and the second equally distributed over all differences between the various Q_i. The relative amounts of information that we expect about each aspect are determined by the relative magnitudes of w and z.

The corresponding elements, $F_i = m_i^{-1/2} P_D(E_i)$, of D are

$$F_1 = h'(Q_{1.} + \cdots + Q_{k.}),$$

where $Q_{i.} = (Q_{i1} + \cdots + Q_{in})/n$ (h' is a further normalising constant), and $F_2, \ldots, F_k$ are any $(k-1)$ uncorrelated combinations of $Q_{1.}, \ldots, Q_{k.}$, with coefficients summing to zero.

(The simple structure of the belief transform reflects the exchangeability over quantities Q_i, which allows us to write each Q_i as $Q_i = Q + S_i$ where $S_1 + \cdots + S_k = 0$ and $\text{Cov}(Q, S_i) = 0$, for each i.)

An obvious separation is between "overall size" and "individual differences", which follows from setting $M[1] = [F_1]$, and $M[2] = [F_2, \ldots, F_k]$. From the above, the expected value of $\text{Var}(Y_{[1]})$ is w, and of $\text{Var}(Y_{[2]})$ is $(k-1)z$. We illustrate the analysis below.

5.2. *Example: exam data*

The data set concerns a portion of the marks of 854 candidates on a certain examination, comprising the first set of candidate returns. Each candidate had to answer six compulsory questions, each marked out of 10, with questions not attempted being scored as zero. (These were followed by certain optional questions that we shall not consider.) The immediate questions of interest to the chief examiner were firstly whether the questions were of roughly similar difficulty, secondly whether the standard was similar to previous years, and thirdly whether there were identifiable subsets of candidates who had been marked differently from other candidates.

In the notation of section (5.1), Q_{ij} is the score of candidate j on question i. Initially, the candidates were judged exchangeable and the questions were also judged exchangeable. The examiner made initial assessments, as follows.

Firstly, the quantities Q_i were identified with the final "average" mark on each question which would be calculated after all of the (very large number of) scripts had been marked. He assessed a prevision of 6.5 for each Q_i (and so, also, for each Q_{ij}), a variance of 1 for each Q_i, and a covariance of 0.25 for each pair Q_i, Q_t (that is, he felt that high or low values on a particular question would be primarily due to features of the particular question rather than, for example, variation between general standards across years). The quantities R_{ij} were interpreted as the variation in marks on the question due to candidate variation, in the operational sense of judgements as to the uncertainty about the actual score on each question of a candidate that would remain if we saw average scores over a large sample of scripts. He assessed a variance of 4 for each R_{ij}, and a covariance of 3.5 between each pair R_{ij} and R_{tj} (that is, he felt that there was high candidate variability, and high correlation between a candidate's performance on different questions, "good" students scoring consistently well, etc.) In the notation of section (5.1), this gives assessments

$$m = 6.5, \quad V = 1, \quad U = 5, \quad C = 0.25, \quad B = 3.75$$

which were validated as yielding sensible beliefs over the observable quantities Q_{ij}.

As in (5.1.) T_D is the belief transform over B, the belief structure generated by $[Q_1, \dots, Q_6]$ induced by D, the belief structure generated by all of the data values Q_{ij}, $i = 1, \dots, 6$, $j = 1, \dots, 854$. The eigenvectors of T_D are $E_1 = h(Q_1 + \cdots + Q_6)$, and $E_2, \dots, E_6$ are any 5 uncorrelated combinations with coefficients summing to zero. From E_1, we learn about the general standard of the exam, as E_1 represents the overall average mark so a high/low value suggests an easy/hard examination. The collection $E_2, \dots, E_6$ summarises all of the differences between difficulty in the various questions, and changes identify which questions candidates found relatively hard/easy.

We will evaluate the overall bearing Y_d for D and the pure trajectory $Y_{[1]}, Y_{[2]}$ based on the separation $M[1] = [F_1]$, $M[2] = [F_2, \dots, F_6]$. Thus, for example, if $Y_{[2]} = a_1 Q_1 + \cdots + a_6 Q_6$, then, for any i, j, we have

$$\begin{aligned} P_2(Q_i) - P_2(Q_j) &= \text{Cov}(Q_i - Q_j, Y_{[2]}) \\ &= (a_i - a_j)(V - C). \end{aligned}$$

so the magnitudes of the coefficients a_i are simple summaries of the relative difficulties of the questions.

5.3. Example: pure trajectory

The trajectory as evaluated on the given data is as follows.

$$\begin{aligned} Y_d &= 0.11Q_1 - 2.75Q_2 + 2.95Q_3 + 0.98Q_4 - 1.51Q_5 - 0.84Q_6 \\ Y_{[1]} &= -0.18(Q_1 + \cdots + Q_6) \\ Y_{[2]} &= 0.29Q_1 - 2.57Q_2 + 3.12Q_3 + 1.16Q_4 - 1.33Q_5 - 0.66Q_6 \end{aligned}$$

for which, denoting $V[d] = \text{Var}(Y_d)$, $V[i] = \text{Var}(Y_{[i]})$, we have

$$V[d] = 15.42, \quad V[1] = 0.42, \quad V[2] = 15.00,$$

$$P(V[d]) = 5.99, \quad P(V[1]) = 0.99, \quad P(V[2]) = 5.00.$$

From $V[d]$, $P(V[d])$ we see that overall change in beliefs is higher than expected, and from the breakdown of values, we see that the difference is due to differences between questions rather then to overall difficulty of the exam. From $Y_{[1]}$, candidates found the examination very slightly harder than expected. From $Y_{[2]}$, candidates found the questions substantially different in difficulty (to a degree which would suggest a discontinuity with previous years had the present initial judgements been formed by careful analysis of previous years performances) and, in increasing order of difficulty, to be $Q_3, Q_4, Q_1, Q_6, Q_5, Q_2$.

5.4. *Example: marker and centre differences*

One of the principal concerns of the chief examiner is to ensure that all of the examiners are marking to the same standard. This particular data set was marked by eleven different markers. A priori, markers were considered exchangeable and marker effects were thought (hoped!) to be small. The obvious next step is formal introduction of exchangeability specifications over markers. However, to illustrate the process of data summarisation and display, we will simply divide the data set into eleven portions, $D1, \ldots, D11$, corresponding to the different markers, and evaluate the raw trajectory over this division. Our immediate concern is to identify whether any of the markers appear sufficiently out of line that the chief examiner should give particularly close scrutiny to that marker's scripts.

Each bearing, Y_{di} is split into components $Y_{di[1]}$, $Y_{di[2]}$ corresponding to the division in the whole data set described in section (5.3). For each marker, we output the two values, $V[i]$, and the number in the group. The output is as follows.

Marker	1	2	3	4	5	6	7	8	9	10	11
Candidates	150	35	31	73	40	18	158	184	30	59	76
$V[1]$	0.01	1.46	0.97	2.48	0.93	2.72	0.09	0.55	0.09	0.06	0.17
$V[2]$	13.2	12.0	19.3	15.4	15.3	24.9	20.2	15.5	23.4	12.6	15.0

The expected value of $V[1]$ ranges from 0.65 to 0.95, and for $V[2]$ from 4.82 to 4.98, increasing in each case with the number in the sample. All markers find consistently more difference between questions than was expected. However, the range of values for $V[1]$ is much wider than for $V[2]$ (as the ratio of the largest to the smallest $V[1]$ value is about 250, whereas the ratio of largest to smallest $V[2]$ value is about 2). As might be expected, marker differences are more evident in differences in overall level than in differences in approaches to individual questions. We might suggest further scrutiny of markers with low $V[1]$ values (who are marking comparatively high), and in particular marker 1. From the coefficients of the bearings for each marker, marker 1 actually assigns an adjusted value for the prevision of E_1 which is higher than the prior prevision, whereas all other markers make adjusted previsions for E_1 lower than the prior prevision. (In comparision, the bearing $Y_{[2]}$ for marker 1 is

$$Y_{[2]} = -0.09Q_1 - 2.02Q_2 + 2.91Q_3 + 1.36Q_4 - 1.74Q_5 - 0.42Q_6,$$

which ranks the questions in roughly the same degree of difficultly as does the corresponding bearing for the combined data set.) We might also check markers with high $V[1]$ values (who are marking comparatively low), and in particular markers 4 and 6. (This distinction is made much sharper if we display the adjusted trajectories for those quantities, but the message is the same.)

The exam data above was collected from returns from a large number of different centres, and it is possible that marker differences are actually due to the differences between centres. (Further, it may be of general interest to investigate differences between centres —for example, if one geographic region appears to be systematically underperforming then this might suggest that further resources are required in the region.)

As an illustration, we divide the scripts of marker one by centre. Again, a detailed analysis would incorporate prior exchangeability specifications over the centres. (If we were to repeat the analysis from year to year, then beliefs in the second year would be non-exchangeable over centres, and we would produce sharper data displays.) Corresponding to the above table, we have a similar table of values for marker 1 across centres (eliminating centres with only one candidate), as follows.

Centre	1	2	3	4	5	6	7	8	9	10
Candidates	8	18	8	20	13	19	8	19	20	16
$V[1]$	+3.02	+0.07	+1.35	−0.16	−0.14	+1.08	+0.05	−3.25	−0.23	+0.10
$V[2]$	5.92	24.75	7.48	31.56	15.69	11.42	35.64	11.81	16.90	12.22

($V[1]$ is +/− according to the sign of $P_d(E_1) - P_1(E_1)$.)

The expected length of $V[1]$ rises from 0.46, on sample of 8, to 0.68 on a sample of 20. The expected length of $V[2]$ rises from 4.6 to 4.8. We observe that there is sufficient variation across centres in values of $V[1]$ to make it reasonable to anticipate that a careful analysis this year incorporating and updating initial specifications of centre effects would produce far sharper data displays for assessing marker differences for next year.

5.5. Concluding comments

Our intention is to provide a system within which whatever the level of detail of (genuine) belief specification, we may routinely generate useful data displays. Further, these displays are in a form which is amenable to analysis. (For example, a good display for relating the different centres would be to evaluate all of the pairwise correlations between the quantities $Y_{[2]}$ evaluated individually for each of the 78 centres in the total sample, and compare these evaluations with their expectations (which can be straightforwardly calculated). This should identify the varying patterns of teaching across the region. Actually, it would not be very revealing for the present data, as it concerns basic core material. However, we would expect the analysis on the optional questions to reveal very strong patterns in this plot.)

Our aim in this paper has been simply to provide an introduction to some standard summaries which can be made a routine feature of the subjectivist analysis. Systematic exploitation of such summaries is closely related to questions of "inference" and "belief validation" which require careful logical formulation. The technical aspects of such a formulation relate to higher order specifications against which data discrepancies can be judged. Such discrepancies occur in displays summarising "surprising" changes in belief. (One way to find these is by multiplying the bearing by the inverse of the belief transform —but that's another story!)

ACKNOWLEDGEMENTS

All computations in this paper were carried out using the program $[B/D]$, which acronym stands for *beliefs* adjusted by *data*. This program is currently under development by David Wooff and myself, under a grant from the U. K. Science and Engineering Research Council.

REFERENCES

Andrews, D. F. (1978). Data analysis, exploratory. *The International Encyclopaedia of Statistics*. New York: Macmillan.

Anscombe, F. J. (1973). Graphs in statistical analysis. *Amer. Statistician* **27**, 17–21.

De Finetti, B. (1974). *Theory of Probability*.**1**. London: Wiley.

Goldstein, M. (1975). Approximate Bayes solutions to some nonparametric problems. *Ann. Statist.* **3**, 512–517.

Goldstein, M. (1981). Revising previsions: a geometric interpretation. *J. Roy. Statist. Soc. B* **43**, 105–130.

Goldstein, M. (1986a). Separating beliefs. *Bayesian Inference and Decision Techniques*, (P.K. Goel and A. Zellner, eds.). Amsterdam: North-Holland.

Goldstein, M. (1986b). Exchangeable belief structures. *J. Amer. Statist. Assoc.* **81**, 971–976.

Goldstein, M. (1987a). Systematic analysis of limited belief specifications. *The Statistician* **36**, 191–199.

Goldstein, M. (1987b). Can we build a subjectivist statistical package?. *Probability and Bayesian Statistics*. (R. Viertl, ed.). New York: Plenum Press. 203–217.

APPENDIX

For illustrative purposes, we describe in detail the calculations for a simple example. Suppose $C = [X_1, X_2]$. At t_1, you assign $P_1(X_1) = P_1(X_2) = 0$, $\text{Var}(X_1) = \text{Var}(X_2) = 1$, $\text{Cov}(X_1, X_2) = 1/2$.

At t_2, you assign $P_2(X_1) = 2$, $P_2(X_2) = 1$.

A.1. The bearing

The normalised principal component representation is $E_1 = (X_1 + X_2)/\sqrt{3}, E_2 = X_1 - X_2$, so that $P_2(E_1) = \sqrt{3}$, $P_2(E_2) = 1$, and the bearing Y_2 is

$$Y_2 = \sqrt{3}(X_1 + X_2)/\sqrt{3} + (X_1 - X_2) = 2X_1.$$

As required,

$$P_2(X_1) - P_1(X_1) = \text{Cov}(X_1, Y_2) = 2,$$

$$P_2(X_2) - P_1(X_2) = \text{Cov}(X_2, Y_2) = 1,$$

and for V in $L(C)$ to be uncorrelated with Y_2 implies V must be proportional to $(X_1 - 2X_2)$, so that $P_2(V) = 0$.

The standard deviation of Y_2 is 2. This is the maximum change in prevision for any element of $L(C)$ with unit variance. The largest change is in the direction of Y_2 (in this case X_1).

A.2. Data specifications

Suppose that the revision over C was produced by linear fitting on two quantities Z_1, Z_2. At time 1, you assigned the following beliefs.

$$P(Z_1) = P(Z_2) = 0, \quad \text{Var}(Z_1) = \text{Var}(Z_2) = 10, \quad \text{Cov}(Z_1, Z_2) = 2,$$

$$\text{Cov}(Z_1, X_1) = \text{Cov}(Z_2, X_2) = 2, \quad \text{Cov}(Z_1, X_2) = \text{Cov}(Z_2, X_1) = 0.$$

Suppose that the observed values are $Z_1 = 11$, $Z_2 = 7$.

A.3. Data trajectory

The projections into D are

$$P_D(X_1) = (5Z_1 - Z_2)/24, \quad P_D(X_2) = (5Z_2 - Z_1)/24.$$

Substituting the observed values into the above formulae gives the revisions specified in (A.1), i.e. $P_d(X_1) = 2$, $P_d(X_2) = 1$, so that the bearing Y_d is as in (A.2).

Now suppose that we evaluate the trajectory in two stages. Firstly, fit on $D[1] = [Z_1]$, with bearing $Y_{[1]}$, then add $D[2] = [Z_2]$, with bearing $Y_{[2]} = Y_d$, as obtained above.

The projections of X_1, X_2 into $D[1]$ are

$$P_{[1]}(X_1) = Z_1/5, \quad P_{[1]}(X_2) = 0,$$

so that, if $Z_1 = 11$, $P_{[1]}(X_1) = 11/5$, $P_{[1]}(X_2) = 0$, and

$$P_{[1]}(E_1) = (11/5\sqrt{3}), \quad P_{[1]}(E_2) = 11/5,$$

so that the bearing $Y_{[1]}$ is given by

$$Y_{[1]} = 11(X_1 + X_2)/15 + 11(X_1 - X_2)/5 = 22(2X_1 - X_2)/15.$$

Again, we can check that if V in $L(C)$ is uncorrelated with $Y_{[1]}$, then V must be proportional to X_2, so that $P_{[1]}(V) = 0$.

Now adding Z_2 gives adjusted bearing

$$Y_{[2/1]} = Y_d - Y_{[1]} = 2X_1 - 22(2X_1 - X_2)/15,$$
$$= (22X_2 - 14X_1)/15.$$

The route of the trajectory is therefore given by

$$\text{Var}(Y_{[1]}) = 6.453, \quad \text{Var}(Y_{[2/1]}) = 1.653, \quad Cr[2] = -0.629.$$

(Note the "conflict" between the two steps in the trajectory, as indicated by the negative correlation $Cr[2]$.)

A.4. Belief transform

Denote by B, D the belief structures with base $[X_1, X_2]$, $[Z_1, Z_2]$, respectively. To evaluate the belief transform, we evaluate the projections of Z_1, Z_2 into B, giving

$$P_B(Z_1) = 4(2X_1 - X_2)/3, \quad P_B(Z_2) = 4(2X_2 - X_1)/3.$$

The belief transform T_D is determined by the values

$$T_D(X_1) = P_B P_D(X_1) = P_B(5Z_1 - Z_2)/24 = (11X_1 - 7X_2)/18,$$
$$T_D(X_2) = (11X_2 - 7X_1)/18.$$

Solving the derived matrix relations gives the eigenvalues of T_D as $m_1 = 1$, $m_2 = 0.222$, with eigenvectors $E_1 = X_1 - X_2$, $E_2 = (X_1 + X_2)/\sqrt{3}$.

The eigenvalue $m_1 = 1$ implies that the residual variance for the linear fit for E_1 is zero. (We can check that the correlation between $X_1 - X_2$ and $Z_1 - Z_2$ is one.) Thus, we obtain full information on the difference between X_1 and X_2, and reduce variance about the sum of X_1 and X_2 by about one fifth. The corresponding values $F_i = m_i^{-1/2} P_D(E_i)$ are

$$F_1 = (Z_1 - Z_2)/4, \quad F_2 = (Z_1 + Z_2)/2\sqrt{6}.$$

Since $Y_2 = \sqrt{m_1} f_1 E_1 + \sqrt{m_2} f_2 E_2$, the overall data bearing can therefore be written

$$Y_2 = (z_1 - z_2)(X_1 - X_2)/4 + (z_1 + z_2)(X_1 + X_2)/18,$$

for any observed values z_1, z_2. (Note that the bearing is very sensitive to the difference $z_1 - z_2$, and less sensitive to the value of $z_1 + z_2$.) In particular, $P(\text{Var}(Y_D))$ = trace $T_D = 1 + 0.222 = 1.222$. The observed value $\text{Var}(Y_d)$ is 4, expressing a large change relative to prior expectations. Similar calculations yield $P(\text{Var}(Y_{[1]})) = 0.533$, $P(\text{Var}(Y_{[2/1]})) = 0.689$, as compared to the observed values 6.453, 1.653 for these quantities, identifying even larger internal changes along the trajectory than for the overall revision.

A.5. Pure trajectory

The pure trajectory comes from dividing D into $M[1] = [F_1]$, $M[2] = [F_2]$. For this trajectory, for any z_1, z_2,

$$Y_{[1]} = (z_1 - z_2)(X_1 - X_2)/4, \quad Y_{[2]} = (z_1 + z_2)(X_1 + X_2)/18,$$

and

$$P(\text{Var}(Y_{[1]}) = 1, \quad P(\text{Var}(Y_{[2]})) = 0.222.$$

For the observed values $z_1 = 11$, $z_2 = 7$, we have

$$Y_{[1]} = (X_1 - X_2), \quad Y_{[2]} = (X_1 + X_2),$$

so that the observed variances are $\text{Var}(Y_{[1]}) = 1$, $\text{Var}(Y_{[2]}) = 3$. We see that the information about the difference $X_1 - X_2$ was as expected, and that the reason that we have a large overall change in belief is that the change in belief for the sum $(X_1 + X_2)$, which we expected to be small, turned out to be large (because the observed sample values were large).

DISCUSSION

G. MEEDEN (*University of Iowa*)

I wish to thank Professor Goldstein for presenting an interesting and original paper. He is to be congratulated for further advancing his program of doing Bayesian statistics without Bayes theorem. In analyzing data he prefers to work without any of the usual standard parametric models. This allows him to proceed without fully specifying a prior distribution. Even though I prefer working with models I am somewhat sympathetic to his approach, because in most problems with many parameters the conventional Bayesian approach can be quite difficult.

In the paper Goldstein has given some techniques for analyzing how our beliefs are changed by data. To this end he has defined the notion of *bearing* and identified the magnitude of its variance as a measure of the revision of our beliefs in light of the data. Then using the notion of *trajectory* he can study how various aspects of the data influence different areas of our beliefs. He then gives two intriguing examples with various sumaries of how the data influences our judgements. I am not completely comfortable with the examples, however, since sometimes it is difficult for me to interpret what the numbers really mean. Perhaps it is just my lack of practice with this approch which causes my problems and so I wish to see some more examples. In what follows I will suggest that his approach might be useful in finite population sampling.

Basu (1969) gave an elegant, and for me convincing, justification for a Bayesian approach to finite population sampling. The only problem is that, except for a few instances, it is just too difficult to carry out in practice.

Following standard notation let U denote a finite population consisting of N units labeled $1, 2, \ldots, N$. Let X_i be the value of a single characteristic attached to unit i. The vector $\boldsymbol{X} = (X_1, \ldots, X_N)$ is unknown and assumed to belong to some subset of N-dimensional Euclidean space. Let s, a proper subset of $\{1, 2, \ldots, N\}$, be the sample. For our purposes how s was chosen is not important. Then $X(s) = \{X_i : i \in s\}$ is the observed values in the sample while $X(\tilde{s}) = \{X_i : \notin s\}$ is the set of the remaining unobserved population values. The basic question of finite population sampling is "What does $X(s)$ tell about $X(\tilde{s})$?". For a Bayesian with a prior distribution over $\boldsymbol{X}$ this question is easy to answer, assuming one can specify the prior.

It seems to me that Goldstein's approach could be of some use here since he need only specify the covariance structure of X. Here is one possible scenario where such a specification seems reasonable to me. Suppose the indices have been chosen so that we believe that the X_i's tend to increase with i and pairs which are close together are more alike then pairs that are far apart. For each i we are willing to specify a mean and a variance for X_i and we choose a covariance function which, for (X_i, X_j), is a decreasing function of $|i-j|$ that eventually becomes zero. After the sample is observed we compute for any $X_j \in X(\tilde{s})$ its predicted value based on $X(s)$. These values could then be used to calculate an estimate of the population total, say. Since our final estimate depends on the specified covariance structure one would be concerned if the sample $X(s)$ does not "agree" with it. I wonder if it is possible to use the theory of this paper to help to decide when the sample contradicts our prior beliefs.

It should be noted that the approach to finite population sampling described in the previous paragraph can be considered a special case of Kriging. Journel and Huijbregts (1978) is a general introduction to this topic while Cressie (1986) is a recent article on the topic intended for statisticians.

Suppose now for each X_i we are only willing to specify a prevision m_i which is assumed to be positive. After $X(s)$ is observed it might be reasonable to assume that

$$\bar{r} = \frac{1}{n} \sum_{i \in s} X_i/m_i \tag{1}$$

is a sensible estimate of X_j/m_j for $j \notin s$, where n is the number of indicies belonging to s. That is, given the sample, $m_j\bar{r}$ is our prevision of X_j for $j \notin s$. Since no variances or covariances are specified this has less structure than assumed in Goldstein's paper, nevertheless I wonder if his theory could be adopted to say something interesting about how the sample revises our prior beliefs. The preceding argument and the resulting estimator of the finite population total is due to Basu (1971).

To see why Goldstein's theory might apply, suppose that after specifying the m_i's we degree that $E(X_i/m_i) = \mu$, $\text{Var}(X_i/m_i) = 1$ and $\text{Cov}(X_i/m_i, X_j/m_j) = a$ for $i \neq j$. The actual value of μ is not important but $\mu = 1$ would usually be a sensible choice, while a is assumed to be positive. Note that we are again in the situation of Goldstein's paper. Given the sample, a straightforward calculation, yields

$$\frac{an}{a(n-1)+1}\bar{r} + \left(1 - \frac{an}{a(n-1)+1}\right)\mu \tag{2}$$

as the predicted value of X_j/m_j when $j \notin s$. Note that if we set $a = 1$ in equation (2) it reduces to equation (1).

In closing it should be noted that estimators of the population total described above are closely related to some estimators discussed in Ericson (1969). This point is discussed in Vardeman and Meeden (1983).

W. POLASEK (*University of Vienna*)

I wonder if the name "data trajectory" is a good one: it might be even misleading, since one actually describes the change and updating of posterior beliefs. What about a graphical summary?

REPLY TO THE DISCUSSION

To Professor Meeden:

I am grateful to Professor Meeden for his comments, which I agree with, excepting for the mild caveat that my program is not "Bayesian statistics without Bayes theorem" but simply Bayesian statistics with updating appropriate to the belief inputs supplied. If you provide genuine inputs to the level of detail demanded by the formal Bayes analysis, and update with Bayes theorem, then you can use all of the methods in this paper precisely in the form described. However such updating is simply an important special case of more general updating methods, some of which I describe in the paper, and which seem to me to be more appropriate in the examples that I describe, because they depend strictly upon the simple beliefs that were elicited.

Professor Meeden observes that he does not feel completely comfortable in interpreting the output from the analysis of the trajectory. New ways of measuring changes in belief always require experience —each time I calculate a trajectory I have to think carefully about the information that it conveys. However, such trajectories are a basic part of the essential process of summarising and interpreting the analysis, and so unfamiliarity with this type of output perhaps reflects a more general carelessness in standard ways of handling the output from Bayesian analyses.

The suggested example in finite population sampling wolud provide interesting applications of the methodology. In Meeden's example, prior beliefs about members of the population are related through a covariance function which decreases monotonically with the distance between the labels of the individuals. I have considered a closely related problem, the localised regression model proposed by Tony O'Hagan, in which the prior correlation between the regression coefficients for best local linear fit at different design points decreases monotionically with distance between the design points. I treated this example as an illustration of the simplifications in the design of the experiment which follow from extracting the belief transform over carefully chosen belief structures. (Some details are provided in Goldstein (1987a).) This formulation translates directly into the problem of survey design when the population has several exploitable labels. Given the design and the sample, then a natural trajectory would follow from the sequential addition of the sample points by the ordering induced by the labelling, which should provide a detailed diagnostic of the belief specification.

Finally, Wolfgang Polasek observes a possible confusion in my choice of terminology. In a sense he is right. In the way that the concepts are introduced in the paper, a trajectory always tracks changes in beliefs (although this may be viewed as implicitly charting a route through data). Thus, I have simply contracted the somewhat clumsy expression "a trajectory across your beliefs induced purely by successive additions to the data set" down to the simpler term "data trajectory". Of course graphical summaries are an important part of the methodology. In my presentation at the conference, I graphed the changes in trajectory for the various data sets described in section (2), and each problem will suggest the appropriate display (for example, in the finite population sampling example, presumably we would plot the bearings against the labels).

REFERENCES IN THE DISCUSSION

Cressie N. (1986). Kriging nonstationary data. *J. Amer. Statist. Assoc.* **81**, 625–634.

Basu, D. (1969). Role of sufficiency and likelihood principles in sample survey. *Sankhyá A* **31**, 441–454.

Basu, D. (1971). An essay on the logical foundations of survey sampling, part one. *Foundations of Statistical Inference*, (V. P. Godambe and D. A. Sprott. eds.). Toronto: Holt, Rinehart and Winston.

Ericson, W. A. (1969). Subjective Bayesian models in sampling finite populations. *J. Roy. Statist. Soc. B* **31**, 195–233. (with discussion).

Journel , A. G. and Huijbregts, C. J. (1978). *Mining Geostatistics*. London: Academic Press.

Vardeman, S. and Meeden, G. (1983). Admissible estimators in finite population sampling employing various types of prior information. *J. Statist. Plan. and Infer.* **7**, 329–241.

BAYESIAN STATISTICS 3, pp. 211–241
J. M. Bernardo, M. H. DeGroot, D. V. Lindley and A. F. M. Smith, (Eds.)

De Finetti's Theorem, Induction, and $A_{(n)}$ or Bayesian Nonparametric Predictive Inference

BRUCE M. HILL
University of Michigan

SUMMARY

The role of the de Finetti's theorem in scientific induction is discussed for the case of nonparametric statistical inference and prediction of observables. A new subjectivistic argument is given for the nonparametric procedure of Hill (1968), Known as $A(n)$. It is suggested that this procedure, together with its generalization $H(n)$, provide a basis for induction. Connections with Zipf's Law, the species problem, and the fiducial approach of R. A. Fisher are also discussed.

Keywords: EXCHANGEABILITY; INDUCTION, BAYESIAN; NONPARAMETRIC, PREDICTION, ZIPF'S LAW.

1. INTRODUCTION

In a celebrated article, Bruno de Finetti (1937) proposed a subjective Bayesian solution to the problem of scientific induction, as formulated, for example, by the Scottish philosopher, David Hume (1748). De Finetti did so in terms of the concept of exchangeability, which is a special form of dependence that he introduced and studied extensively (1937, 1975). Other key references are Hewitt and Savage (1955), Savage (1972), Heath and Sudderth (1976), and Diaconis and Freedman (1980, 1981). In this article I shall give a somewhat personal review of the history and substance of the connection between induction and subjectivistic perceptions of symmetry with particular *attention to* $A(n)$, which I developed for the case of vague or diffuse prior knowledge as to the shape of the underlying distribution of the observables. See also Zabell (1985).

The problem of induction is the problem of drawing inference about the future based upon the past. This problem has long plagued philosophers and others, partly because there is no way to prove that induction works (apart from induction itself), and also because in the real-world it can be extremely difficult to formulate inferential or decision procedures, i.e., inductive techniques, that are appropriate in a given situation. The problem is best thought of in terms of the probabilistic prediction of potentially observable random quantities (not necessarily exchangeable), say $X(1), \ldots, X(n), \ldots$. Given the values of the first n, $X(1) = x(1), \ldots, X(n) = x(n)$, what can we say about $X(n+1)$ or any other future observations? In the Bayesian approach this is done in terms of the evaluation of a probability distribution for the future observables, given the data $X(1) = x(1), \ldots, X(n) = x(n)$. Conventional Bayesian methods, using a prior distribution for a "parameter," such as the parameter of a Bernoulli sequence, yield such a predictive posterior distribution, as well as a posterior distribution for the parameter. (Of course, using de Finetti's theorem, the conventional parameter of a Bernoulli sequence can be interpreted as the proportion of successes in a very long sequence of trials, and thus has a predictive meaning.) In such situations, once a statistical model and prior distribution have been formulated and specified, the posterior distribution of the future observations, given the past, is thereby completely determined. Such a scheme may be called inductive, since it prescribes a (coherent) mode of inference and behavior with respect to the future observables, given any set of data.

This scheme, as usually interpreted, requires that there exist "true" known probabilities that represent the conditional distributon of the data, given the parameter, i.e., a conventional statistical model. However, at the deepest level where "true" probabilities either do not exist, or even if in some as yet unknown sense they do exist, they are at least unknown, the conventional model-based Bayesian theory is incomplete, since it is difficult even to give operational meaning to the assertion that a particular model is "true", much less to find such a model. The problem that de Finetti clearly formulated and largely solved was the problem of giving meaning to Bayesian inferential procedures without relying upon the usual crutch of an assumed statistical model. Before the fundamental work of de Finetti, the assumption of such a true model was simply an unjustified act of faith. One could, of course, refer to some underlying physical theory, or to the central limit theorem, or some previous analogous data, to support belief in such a model. But deep down this remained at best a matter of delicate subjective judgment, and it was not even clear how to express what such subjective judgments concerned. For example, consider the use of the normal or Gaussian distribution. Poincaré (1912, p.171) states in connection with this distribution, "Tout le monde y croit cependant, me disait un jour M. Lippmann, car les expérimentateurs s'imaginent que c'est un théoréme de mathématiques et les mathématiciens que c'est un fait expérimental," or "everybody believes in the law of errors, the experimenters because they think it is a mathematical theorem and the mathematicians because they think it is an experimental fact." In the real world it is justified, in fact, by neither. In Hill (1969) it is shown that the use of the normal distribution can instead be based simply upon a subjective judgment of spherical symmetry for the "actual" errors in the observations. See also Borel (1914, pp.66, 90-93) and Borel (1906). (Diaconis and Freedman (1987) have done careful historical research concerning the theorem that states that under spherical symmetry the marginal distribution of a fixed number of coordinates is approximately Gaussian. This theorem is sometimes mistakenly attributed to Poincaré. Diaconis and Freedman imply that the result, in some form or other, is due to Maxwell (1878, eqn 49-55). Although there is a distant connection with Maxwell's work, the first clear statement and proof of the result that I know of is that of Borel, cited above. Borel only obtains the result for the marginal distribution of a single fixed coordinate, although his discussion suggests that he may also have derived or at least understood the asymptotics for the general case. The first clear statement and proof of the general case that I know of is my own in Hill (1969, p.95). I give the exact density for the marginal distribution of n coordinates based upon spherical symmetry, or conditional uniformity, on the N-dimensional sphere. By Scheffé's lemma that convergence of densities to a proper density implies convergence in distribution, it immediately follows thaat each fixed r-dimensional marginal distribution of the joint distribution of the n coordinates converges to the Gaussian, even as n goes to infinity as well as N. In this sense, which was perhaps Borel's intent, spherical symmetry implies approximate normality. It should be noted that my statement of the result, which is for the case of spherical symmetry without a constraint on the average of all N coordinates, agrees with that of Borel for the case of one coordinate. When there is also a constraint on the average of the N coordinates, then my exponent $N-n-2$ should be changed to $N-n-3$.)

In the theory that I proposed, spherical symmetry, or more generally, conditional uniformity on surfaces, is itself only an approximation based upon the available knowledge, and does not purport to be more than this, or to have any other objective meaning. For example, in the case of errors of measurement, one may view the usual orthogonal axes of a coordinate system as arbitrary, and therefore introduce rotational symmetry. Ultimately, it is simply a matter of judging that spherical symmetry represents a sufficiently good approximation to one's opinions in order to be useful for inference, prediction, and decision-making. In my opinion there is no hope of proving that such a judgment is either "correct" or "incorrect", other than empirically, for example, by seeing how well it works predictively.

At an even more basic level, as de Finetti first realized, induction can be based upon a direct subjective judgment of exchangeability for the sequence of observables. Once this subjective judgment is made it is a mathematical fact that one will be acting (nearly) as though some statistical model were true. Conditional upon the parameters of such a model, the data will be regarded as independent and identically distributed, if the exchangeable sequence is infinite, and approximately so if it is a sufficiently long finite sequence. See Diaconis and Freedman (1980). (It should be noted that the conventional assumption of independent, identically distributed observations, with an unknown distribution, corresponds to the subjective Bayesian assumption of exchangeability.) This is de Finetti's theorem, and its significance is that it provides a subjective justification for the use of statistical models and for conventional model-based Bayesian inference, provided that care is taken in the interpretation of such techniques. See Savage (1972) for a treatment of exchangeability from this point of view. De Finetti stressed the fact that the judgment of exchangeability is itself only a subjective judgment, and perhaps only an approximation to one's actual opinions. To the extent that one judges the sequence as exchangeable, then one is led to conventional Bayesian inferential techniques. Often, in fact, approximate, or even partial exchangeability, will suffice to justify such techniques. Questions regarding the selection of models, partial exchangeability, Bayesian data-analysis, etc., although fundamental for the future development of Bayesian statistics, will not further be addressed in this article. Except where explicitly stated otherwise, we will be considering only the case of exchangeable sequences. See Hill (1986) for a discussion of the selection of models from a subjective Bayesian viewpoint, Hill (1987a) for discussion of the deeper structures that may underlie conventional statistical models and Hill (1988) for a theory of Bayesian data analysis.

In a practical sense, based upon the subjective judgment of exchangeability, de Finetti had completely solved the problem of inductive inference for the case of Bernoulli data, or, more generally, for multinomial data with a known finite number of categories. Combined with the beautiful result of W. E. Johnson (1932), as discussed in Zabell (1982), there was little more to be said at the foundational or even practical level for these cases, other than to elaborate on the choice of prior distribution for the Bernoulli parameter p or for the parameter θ of a multinomial distribution. Thus the stable estimation argument of L. J. Savege (1961, Ch. 4; 1962, p. 20), or as presented in Degroot (1970, p. 199), deals with the case in which the prior distribution is dominated by the likelihood function, so that it is of little consequence, and the posterior density for the parameter can be approximated by the likelihood function. On the other hand, H. Jeffreys's theory of hypothesis testing covers the most important situations in which the prior is not diffuse. See Edwards, Lindman and Savage (1963), and Hill (1974a, 1982) for discussions. The problem of so-called "uninformative" priors has also been dealt with very effectively by a number of people for the case of multinomial data. See Good (1965, Ch. 4) for a review and discussion. Furthermore, it has long been recognized that many real-world problems can be adequately modelled by such finite partitions, Fisher (1959, p. 111), Savage (1961, p. 4.23), so when this can be done effectively there is available a more or less complete system (apart from details and various complications that arise in practice) for inductive inference and decision-making, including the prediction of future observations. In de Finetti's own words (1937, p. 147): "It is thus that when the subjectivistic point of view is adopted, the problem of induction receives an answer which is naturally subjective but in itself perfectly logical, while on the other hand, when one pretends to *eliminate* the subjective factors one succeeds only in *hiding* them (that is, at least, my opinion), more or less skillfully, but never in avoiding a gap in logic. It is true that in many cases —as for example on the hypothesis of exchangeability— these subjective factors never have too pronounced an influence provided that the experience be rich enough; this circumstance is very important, for it explains how in certain conditions more or less close agreement between the predictions of different individuals is produced, but it also shows that discordant opinions are always legitimate. This does not make any change in the purely subjective character of the whole theory of probability." See also de Finetti (1975, Ch. 11).

Thus in the exchangeable case the only type of situation that had not been essentially resolved was that in which no finite partition model was appropriate, or more generally, when the number of parameters required realistically to model the data is large relative to the number of observations. One can speak of this either in terms of multinomial data with an infinite number of categories, or alternatively, as I shall do, in terms of an unknown and possibly large or even infinite number of categories. Still again, to suggest the general type of problem, one can speak of Bayesian nonparametric statistics, or of Bayesian inference about an "unknown" distribution function. Whatever words we may use, what we are trying to describe is the situation in which no conventional parametric statistical model is thought to be appropriate for the exchangeable sequence of observations. In the non-Bayesian literature such situations are usually called nonparametric, and it would appear from the vast literature on such nonparametric techniques, that these situations may often arise in practice. Indeed, from my own point of view, which will be explained below, they in fact represent the great majority of statistical situations, with Gaussian and other conventional parametric models being appropriate only in very limited contexts.

What then can be said about the nonparametric case from a subjective Bayesian point of view? The first thing to observe is that de Finetti's theorem still holds, so that in the case of an infinite exchangeable sequence of observables, one will be mixing over a dummy variable that represents the "unknown" distribution, say F, in the population. De Finetti (1937, Ch. 4), had already given an insightful development of the mathematics of this situation for exchangeable random quantities. Diaconis and Freedman (1980, 1981) have presented fully rigorous and easily accessible proofs for even more general cases. Just as in the case of exchangeable events, what must be specified in order to implement the Bayesian approach, is the mixing function, or a priori distribution for F. In the case of exchangeable events, F is concentrated at only two known values, 0 and 1, while in the present case the distribution F can in principle be any distribution function on the real line. Special mixing distributions will correspond to the subset of F appropriate for a finite multinomial situation, in which the categories are coded numerically, or for sure knowledge that F is Gaussian, etc. The fully nonparametric case is that in which F cannot be restricted to such special subsets of the space of all distribution functions.

How then can a subjective Bayesian specify a prior distribution that expresses a realistic degree of vagueness about F? Since we are dealing with infinitely many parameters, it is clear that the problem is formidable even from the point of view of the mathematics involved, and of course even much more so conceptually. Furthermore, after observing a sample from the population, in order to obtain the posterior predictive distribution of future observations, one would have to integrate the conditional distribution of the future observations, given F, with respect to the posterior distribution of F. Here the "unknown" F plays the same role as the "unknown" θ of a conventional multinomial model, but the mathematics is again enormously more complicated. In addition, a basic diffulty arises here that did not appear in the case of finite multinomial models. It is no longer the case that one can rely on some form of Savage's stable estimation argument. Thus the distribution function F may have infinitely many parameters, and no matter how large a finite sample is taken from the population, the prior distribution for F may still play a crucial role. Even if more realistically, we regard F as having a large finite number of parameters, in a practical sence the same phenomenon occurs, since realistic sample sizes will be small relative to the total number of parameters. See Hill (1975b) for a discussion of this phenomenon. Typically there is no such thing as global robustness (for all possible distributions of F), or in other words, the posterior distribution may be extremely sensitive to the prior distribution F.

The problem is not, however, so hopeless of solution as may first appear. The first hint or suggestion as to the nature of a possible solution occurs in the work of R. A. Fisher (1939, 1947), who proposed a fiducial interpretation for what I later called $A_{(n)}$. (Fisher gives credit to "Student" for the underlying idea.)

Consider a conventional formulation of statistical inference, in which the observations are conditionally independent with cumulative distribution function $F(x;\phi)$, where ϕ is a conventional unknown parameter. Assume that the distribution function is continuous in x for each ϕ. Let $X_{(i)}$ denote the ascending order statistics of the data, for $i = 1,\ldots,n$. Then let $\theta_{(i)} = F(X_{(i)};\phi) - F(X_{(i-1)};\phi)$, for $i = 1,\ldots,n+1$, where by definition, $X_{(0)} = -\infty$ and $X_{(n+1)} = \infty$. *Before* the data are drawn, clearly the distribution of the $\theta_{(i)}$ is a uniform distribution on the n-dimensional simplex, i.e., a special Dirichlet distribution in which all the parameters are equal to unity. This is the fundamental frequentistic intuition with regard to $A_{(n)}$, and which Fisher presumably used to put forth his proposed fiducial solution. Thus Fisher suggested (or implied) that even when the random variables $X_{(i)}$ are replaced by their observed values $x_{(i)}$, that the uniform distribution for the $\theta_{(i)}$ would still be appropriate. It should be mentioned that the articles by Fisher (1939, 1947) only briefly and cryptically discuss the nonparametric fiducial case that we are concerned with. The first clear formal statement of something like $A_{(n)}$ is by Dempster (1963), who stated the Fisherian argument precisely, changed the name from fiducial to "direct" probability, and applied the argument for predition of future observables as well. Dempster also asserted that what I later called $A_{(n)}$ "does not appear to have a Bayesian interpretation."

In my 1968 article I showed that in fact $A_{(n)}$ does have a Bayesian interpretation. Before discussing this, however, let me note that Fisher's proposed fiducial distribution is an example of a posterior predictive distribution, since whatever the rationale, it is posterior to the data, and does partially specify a probability distribution for the future data. This predictive distribution is not completely specified, since what it does is to attach a probability of $\frac{1}{n+1}$ to each of the $n+1$ open intervals formed by consecutive order statistics of the given sample, assuming that there are no ties and, goes no further. The fiducial argument that Fisher gave for this evaluation depends upon one's willingness to persist with the pre-data evaluation of the distribution of say, $F(X_{(i)};\phi) - F(X_{(i-1)};\phi)$, *after* the $X_{(i)}$ are replaced by their observed numerical values. Such a fiducial argument, although intriguing was logically suspect. See Edwards (1972, p.207) for a totally devastating example against the logic of the fiducial argument. Also Lindley (1958) had already shown, *under certain special conditions*, that the fiducial argument leads to a genuine posterior distribution only in simple cases reducible to that of a location parameter. However Fisher, in a footnote (1959, p.51), asserted that "Probability statements derived by arguments of the fiducial type have often been called statements of "fiducial probability". This usage is a convenient one so long as it is recognized that the concept of probability involved is entirely identical with the classical probability of the early writers, such as Bayes. It is only the mode of derivation which was unknown to them. In short, the situation with regard to $A_{(n)}$ was anything but clear, and at the time it was not even known whether $A_{(n)}$ was a coherent evaluation in the sense of de Finetti. If it was, then presumably it could have been derived by means of a prior distribution for F, Bayes theorem, and an integration with respect to the posterior distribution of F, as discussed above. The problem that I addressed in my 1968 article was that of giving such a derivation of $A_{(n)}$. The first step in my formulation was to consider the case of arbitrary finite populations, in which the size of the population may be unknown, rather than infinite populations, or in other words, to deal with finite exchangeable sequences.

This step is not only more realistic, since in the real world we are not ordinarily called upon to deal with more than finite populations or sequences of observables, but it also greatly simplifies the mathematics. Indeed, the proofs of de Finetti's theorem for the infinite case by

Heath and Sudderth (1976) and by Diaconis and Freedman (1980, 1981), proceed by taking limits for the finite case. Thus for the case of events, one can condition on the sum of the indicators after N observations, and note that because of exchangeability, all paths to a given sum are equally likely. The exchangeable distribution of any N indicators (for example, for red and white balls in an urn) is then identical with one that arises from sampling without replacement from a "randomly selected" urn with N balls, i.e., the draws are made from an urn with composition (R, W), $R+W = N$, with a probability equal to the original subjective probability that the sum of the N indicators is R. More generally, one can consider the empirical distribution function that arises from "sampling" the first N coordinates from an exchangeable sequence. Conditional on this empirical distribution function, the individual coordinates are distributed uniformly over the collection of N-tuples having this empirical distribution. Because sampling without replacement is, for large N, close to sampling with replacement from this empirical distribution, and because of the convergence of this empirical distribution to some limiting distribution (with probability one, under exchangeability), one obtains de Finetti's theorem for the infinite sequence in this way. See Diaconis and Freedman (1980, p.749, 1981, p.209) for details. These authors have emphasized the importance of the finite case even for the underlying mathematics, and as I shall argue later, it is also the appropriate formulation for inferential purposes. The model for $A_{(n)}$ in Hill (1968, p.679) is actually equivalent to the specification of a diffuse prior distribution for the empirical distribution of a finite population. In the context of Heath and Sudderth, or of Diaconis and Freedman, it is equivalent to specifying the subjective distribution for the sufficient statistic based upon N trials within their models. Thus instead of specifying a diffuse prior distribution on the space of all distribution functions F, what I have done is to specify such a prior distribution for the empirical distribution function of the entire finite population from which a simple random sample has been drawn. Here the number of units in the finite populations can be unknown, as well as the number of jump-points of the empirical distribution of the population, and the points at which the jumps occur and the sizes of the jumps. The details concerning this diffuse prior distribution will be given in Section 2. Here what I want to discuss is the underlying sampling model.

The statistical model for this problem can be thought of in terms of sampling with or without replacement from a finite population of units. Imagine that each unit carries an attached tag or label, for example giving the color of the unit, or the name of the species to which the unit belongs, or a numerical value such as the mass of the unit, or the future time of death of the unit. It does no harm to visualize the population of units, with their attached labels, as sitting in an urn. A simple random sample, with or without replacement, is then drawn, and we observe the value of the label for each unit in the sample. It is assumed for simplicity here that the label or numerical value is observed without error although the theory can easily be extended to deal with errors of measurement.

The population of labels or numerical values can be described in terms of the empirical distribution of such labels or values. Indeed, because we are dealing with only finite populations, the case of colors or names can be viewed as a special case of the case of numerical values. Thus we can imagine, without loss of generality, that the finite collection of colors or names in the entire population have been encoded numerically, thus yielding values. For simplicity, we shall describe the situation for the case of such numerical values. When we return to the case of "colors," as in the species sampling problem, we shall point out the special features that arise in this case. For the time being, visualize the urn population as consisting of the numerical values attached to the units, such as their masses. This population can then be described in terms of the number of units in the population, say N, and the empirical distribution of the values in the population. Note, for example, that if sampling is with replacement from this population, and if the number of distinct values in the population is known to be say, M, then this model is a special case of a conventional multinomial model with exactly M non-empty categories. In general, of course, M need not be known, except that $M \leq N$, and sampling can be without replacement. In any case, the number of units in the population, N, and the empirical distribution of population values, completely characterize the finite population of values. In fact here the empirical distribution for the entire finite population of values plays a similar role to that of the "unknown" probabilities in a conventional statistical model. Following the spirit of de Finetti (1975, Ch. 11), I regard the fundamental problem of induction to be reducible to that which arises in sampling without replacement from an urn consisting of units that are labelled with numerical values.

The solution that I proposed for this problem, which consists of a model for a generalized version of $A_{(n)}$ in which ties can occur, will be discussed in Section 3. Historically, the sequence of events concerning $A_{(n)}$ after Dempster (1963) was as follows. I proved in my 1968 article that $A_{(n)}$ cannot hold exactly for *countably additive proper prior distributions*, in the case of exchangeable sequences in which ties have probability 0. At the same time I recommended it as an approximation for a variety of situations, that can be roughly described as situations in which the data is measured on a "rubbery scale," and gave several models in which it would be appropriate. I also proved in Hill (1968, p.686) that $A_{(n)}$ for all n implies that the posterior distribution of the $\theta_{(i)}$ defined earlier is the uniform Dirichlet distribution on the $n+1$ dimensional simplex, thus giving support to Fisher's fiducial argument. Also, Hill (1967) derived the posterior expectation of a future observation, and of the mean of the population, using $A_{(n)}$. The next historically significant development regarding $A_{(n)}$ was the proof by Lane and Sudderth (1978), using finite additivity, that $A_{(n)}$ for all n is coherent in the sense of de Finetti, i.e., it is impossible to be made a sure loser, and the further result by the same authors (1984) that it is predictively coherent. The robustness and invariance properties of $A_{(n)}$ were investigated by myself in Hill (1980a) with the general result that it is robust in the modern Bayesian sense of Berger (1985), Hill (1980b). Then my doctoral student Peter Lenk (1984) showed, along with many other things, that $A_{(n)}$ can arise as a limit of proper priors, using a log gaussian model for the prior distribution of the unknown density function of the population. Next, Berliner and Hill (1986) used $A_{(n)}$ to obtain the predictive distribution for future observations in the case of censored data, as for example in survival analysis. Finally, in Hill (1987b) I have shown that $A_{(n)}$ for all n can be obtained by means of an improper countably additive prior distribution on a conventional parameter in a very simple and natural parametric model.

In the next Section I will restate $A_{(n)}$, give my model for the basic inferential question, and suggest a new and compelling (for me) subjectivistic argument for $A_{(n)}$. In Section 3 I will review the connection of $A_{(n)}$ with the Dirichlet process prior of Ferguson and others, and in Section 4 I will return to the fiducial interpretation and other fundamental questions regarding the validity of $A_{(n)}$.

2. $A_{(n)}$ AND $H_{(n)}$

In Hill (1987) a direct specification, denoted $A_{(n)}$, for the posterior predictive distribution of future observations was proposed. $A_{(n)}$ was meant to express extremely vague subjective prior knowledge as to the form of the underlying population distribution. For the case of $n = 1$ and 2, $A_{(n)}$ follows from a conventional parametric model (Gaussian, for example) with a diffuse prior distribution on the location parameter, or on the location and scale parameters, respectively, Jeffreys (1961, p.171). For example, when $n = 1$, suppose that the parameter θ is the mean of a normal population with known standard deviation of unity. Given an observation $X_{(1)} = x$ from this population, the posterior distribution of θ is $N(x, 1)$. The predictive distribution of the next observation, is then easily seen to be $N(x, 2)$. Hence the posterior probability that the next observation is $\leq x$ is .5. Note that whenever the likelihood function is sharp relative to the prior distribution, $A_{(1)}$ will hold to a good approximation, since the posterior distribution of θ will still be approximately $N(x, 1)$. A similar analysis applies in the case $n = 2$. At the time of Hill (1968), it was not known whether $A_{(n)}$ could be obtained for conventional parametric models when $n \geq 3$. However, Hill (1987b) shows that this is the case for both $A_{(n)}$ and $H_{(n)}$.

$A_{(n)}$, for any n, is exactly appropriate for data measured on a merely ordinal scale, or with a trivial modification, for data that consists of labels (such as the name of a species, as in the species sampling problem), and can yield an extremely good approximation for data on a ratio interval scale, such as the weights in a population of penguins, as will be discussed at the end of this section. The cases where it is exactly appropriate can be described as data measured on a "rubbery" scale. Just as with other nonparametric models, it is hardly necessary for the assumptions to hold literally, in order that the conclusions be appropriate to a very good approximation.

The condition $A_{(n)}$ is defined as follows. $A_{(n)}$ asserts that conditional upon $X(1)$, $\ldots, X(n)$, the next observation $X(n+1)$ is equally likely to fall in any of the open intervals between successive order statistics of the given sample (Hill, 1968, p.677). Note that in our *definition* of $A_{(n)}$ we do not assume that the sequence is necessarily exchangeable or that ties have probability 0. Thus, we can also include cases where there is a positive probability that the next observation ties one of the previous observations, and also partially exchangeable situations that satisfy $A_{(n)}$. At the present time I wish to slightly modify this notation, use $H_{(n)}$ to denote the situation in which ties can occur, and reserve $A_{(n)}$ for the special case of $H_{(n)}$ in which there are no ties (or ties have probability 0). In this article I will also assume that the observations are exchangeable, although this will not be included in the definition of $A_{(n)}$ and $H_{(n)}$.

$A_{(n)}$ specifies a predictive distribution for one future observation. If also $A_{(n+1)}$ holds, then by conditioning upon which interval the first new observation falls in, we can obtain a predictive distribution for two new observations, and by extension for an arbitrary number of new observations. See Hill (1968, p.684) for such predictive schemes.* Furthermore, we can use this same idea to deal with censored data, again by conditioning upon which intervals the censored observations will fall in. Berliner and Hill (1986) carry through such an analysis for the case of survival data, present upper and lower bounds for the survival function, and simple algorithms with which to make the analysis. In the survival problem, for example, we assess the predictive probability for the time of death of new patients given a treatment, using as data the death times and times at which censoring occurred for a previous group of patients who were given the treatment, and with whom the new patients are regarded as exchangeable.

In addition to de Finetti (1937, 1975), other key references on exchangeability are Hewitt and Savage (1955), Savage (1972), Heath and Sudderth (1976), and Diaconis and Freedman (1980. 1981). (The article by Heath and Sudderth gives an extremely simple and yet rigorous proof of de Finetti's theorem for the case of events. The articles by Diaconis and Freedman do so for the general case.) The definition of exchangeability that we shall use is that motivated by the subjective Bayesian viewpoint, namely, in terms of a subjective judgment that the order is irrelevant. (Mathematically, this is the same as all other definitions of exchangeability but psychologically it is different, in that we do not assume the sequence is "truly" exchangeable, but merely that one regards it as exchangeable, perhaps only as an approximation to the truth.)

To be precise, let $X(1), \ldots, X(k+1)$, be $k+1$ random variables that are (finitely) exchangeable in the subjective Bayesian sense; that is, the joint distribution of any r distinct variables is the same as that for any other such r variables, $r = 1, \ldots, k$. An infinite sequence of such variables is said to be exchangeable if the above condition is true for each k. Such

---

* Note that the equation on the top of page 684 of Hill (1968) is only valid if $i \neq j$. When $i = j$ it is necessary to add another term which corresponds to the possibility that the second new observation ties the first. A similar correction is necessary in the formula for $E(\theta_{(i)} \times \theta_{(j)})$.

models arise from the following Bayesian formulation: Assume that, given some distribution, say $F, X(1), \ldots, X(n)$ are independent and identically distributed according to F. For F unknown it is natural for the Bayesian to model F itself as "random" with some apriori probability specification. This can be done either parametrically or nonparametrically. In either case, "integrating out" F leads to an exchangeable unconditional joint distribution for the $X's$. Conversely, de Finetti's theorem implies that if the exchangeable sequence is infinite, then their exists a distribution on F, called the prior distribution of F, for which the joint distribution of the observations obtained by "integrating out" F is the original exchangeable distribution. See Hewitt and Savage (1955), Heath and Sudderth (1976), and Diaconis and Freedman (1980, 1981) for proofs. The 1980 article, which emphasizes the finite exchangeable case, is particularly appropriate for my purposes. Thus the authors show that the most general exchangeable sequences arise by taking limits of the finite exchangeable sequences that arise in sampling without replacement from an urn.

I will now present my model for $A_{(n)}$, or more precisely, for my generalization of $A_{(n)}$, called $H_{(n)}$, which allows for ties.

We assume that there exists a finite population of units, with each unit having an attached value or label. For example the value might be the mass of the unit, or the label might be the name of the species to which the unit belongs. We assume that the set of values is simply ordered, or at least can be simply ordered. By a simple ordering we mean a relationship, say $\leq$, which for any two elements x, y, of the set of values, is such that either $x \leq y$ or $y \leq x$, and which is transitive. (See Jeffreys (1957, Chs. 5-6) and Luce and Narens (1987) for discussions of the concept of measurement.)

Thus masses would be on a ratio scale, and are certainly simply ordered; while labels can be simply ordered for a finite population simply by designating an ordering. (This can be done for infinite populations as well, using the well-ordering theorem, but there is no need to go into such things here.) Suppose there are N units in the population, and that the set of attached values or labels is $\{Z_{(i)}, i = 1, \ldots, N\}$. We shall now refer only to values,with it being understood that we include labels as a special case after the finite population has been simply ordered. Some of the values $Z_{(i)}$ may be equal to one another. Suppose that in fact there are only M distinct values amongst the $Z_{(i)}$, and denote these in ascending order of magnitude as $X_{(1)} < X_{(2)} < \ldots < X_{(M)}$, where of course $M \leq N$. Finally, suppose that the value $X_{(i)}$ occurs in $L_{(i)}$ units, where $L_{(i)} \geq 1$, since by assumption the value $X_{(i)}$ does in fact occur, and $\sum_{i=1}^{M} L_{(i)} = N$.

The above model consititutes our description of the finite population of values $Z_{(i)}$. Note that this determines the empirical distribution of values in the finite population, i.e., the empirical distribution has jumps occurring at $X_{(1)}, \ldots, X_{(M)}$, and the jump that occurs at $X_{(i)}$ has height $L_{(i)}/N$. Of course in general all of these quantities are unknown, i.e., N, M, the $X_{(i)}$ and the $L_{(i)}$. From the subjective Bayesian point of view one must then specify a probability distribution for all of these quantities. It should be noted that this point of view corresponds exactly to the recent probabilistic treatments of exchangeability for finite sequences, as in Diaconis and Freedman (1980), where the finite exchangeable sequence of length N is the vector $Y_{(1)}, \ldots, Y_{(N)}$, which would be generated by sampling without replacement all N elements of the finite population, so that these $Y_{(i)}$ are some permutation of the $Z_{(j)}$.

The case of an infinite exchangeable sequence may be viewed as an idealization of this scheme, and gives rise to de Finetti's theorem. But the model in terms of sampling from a finite population is simpler, avoids difficulties and paradoxes of infinity, is more realistic, and in view of the results of Diaconis and Freedman, loses no generality in any case. For example, in my model we require only a prior distribution for the composition of the finite population, i. e., for $M, N, \boldsymbol{X}, \boldsymbol{L}$, rather than a prior distribution on F, the theoretical distribution for an infinite exchangeable population. It is far simpler to specify such a prior distribution on the finite number of parameters (at most $2N + 2$) needed to describe this finite population, than to do so on the infinite dimensional space of distribution functions F. Furthermore, we shall argue that there is a natural way to represent vagueness for the finite population, which would be much more difficult to achieve for F.

Now let us consider the data that we shall be analyzing. It is assumed that a simple random sample is drawn without replacement from the finite population that we have described above. Let the sample size be n. The data will consist of the numerical values attached to the n units that are thus selected from the finite population. Let $x_{(1)} < x_{(2)} < \ldots x_{(m)}$, be the ascending order statistics of the sample, with m distinct values, $1 \leq m \leq n$, and with $n_{(i)}$ sample units having the value $x_{(i)}$. Thus $n_{(i)} \geq 1$, and $\sum_{i=1}^{m} n_{(i)} = n$. It is assumed here that the values are measured without error, so that each $x_{(i)}$ is necessarily some $X_{(j)}$ in the population, but of course we do not know with certainty which. By data we mean the set of m distinct $x_{(i)}$ values, and the $n_{(i)}$. Thus the data determines the empirical distribution of the sample, but is more informative because n and the $n_{(i)}$ are known as well. We now require only one further bit of notation. Given the data, define $J_{(i)}$ to be the rank, in the population, of the value $x_{(i)}$ in the sample, for $i = 1, \ldots, m$. The vector $\boldsymbol{J} = (J_{(1)}, \ldots, J_{(m)})$ then gives the true ranks, in the population, corresponding to the sample values $x_{(i)}$. Thus $1 \leq J_{(1)} < J_{(2)} < J_{(3)} \cdots < J_{(m)} \leq M$, because of the fact that the $x_{(i)}$ and $X_{(j)}$ are strictly ordered. We now are ready for the basic equations of the Hill model for $H_{(n)}$.

In the first equation, we condition on the true composition of the finite population, by which we mean the unknown quantities $\boldsymbol{X}, \boldsymbol{L}, M$ and N. This equation gives the probability for observing the data together with $\boldsymbol{J} = \boldsymbol{j}$, for each possible vector of ranks $\boldsymbol{j}$.

$$\Pr\{data,\ \boldsymbol{J} = \boldsymbol{j} | \boldsymbol{X}, \boldsymbol{L}, M, N\} = \binom{N}{n}^{-1} \times \prod_{i=1}^{m} \binom{l_{(j_{(i)})}}{n_{(i)}}, \tag{1}$$

if $X_{(j_{(i)})} = x_{(i)}, i = 1, \ldots, m$, and is otherwise 0.

Note that this would be the likelihood function for the population quantities, except that we have included $\boldsymbol{J} = \boldsymbol{j}$ together with the data, because this is the key to making an effective evaluation; the ordinary likelihood function will involve a mixture of (1) with respect to $\boldsymbol{j}$. For sampling with replacement, it is only necessary to replace the factor $\binom{l_{(j_{(i)})}}{n_{(i)}}$ by $(l_{j_{(i)}})^{n_{(i)}}$. We shall not further deal with the case of sampling with replacement, since sampling without replacement is the more common, more difficult to analyze, and more important form of sampling.

The next step is integrate out over the unknown $\boldsymbol{X}$ values in the population. This can be done in the conventional countably additive theory, or in the finitely additive theory assuming conglomerability as in Hill and Lane (1985), to yield

$$\Pr\{data,\ \boldsymbol{J} = \boldsymbol{j} | \boldsymbol{L} = \boldsymbol{l}, M, N\} =$$

$$\binom{N}{n}^{-1} \times \prod_{i=1}^{m} \binom{n_{(i)}}{l_{(j_{(i)})}} \times \Pr\{X_{(j_{(1)})} = x_{(1)}, \ldots X_{(j_{(m)})} = x_{(m)} | \boldsymbol{L} = \boldsymbol{l}, M, N\}. \tag{2}$$

We thus obtain the basic result,

$$\begin{aligned}
\Pr\{\boldsymbol{J} = \boldsymbol{j}, \boldsymbol{L} = \boldsymbol{l}, data | M, N\} &= \Pr\{data, \boldsymbol{J} = \boldsymbol{j} | \boldsymbol{L} = \boldsymbol{l}, M, N\} \\
&\quad \times \Pr\{\boldsymbol{L} = \boldsymbol{l} | M, N\} \\
&= \binom{N}{n}^{-1} \times \prod_{i=1}^{m} \binom{n_{(i)}}{l_{(j_{(i)})}} \\
&\quad \times \Pr\{X_{(j_{(1)})} = x_{(1)}, \ldots X_{(j_{(m)})} = x_{(m)} | \boldsymbol{L} = \boldsymbol{l}, M, N\} \\
&\quad \times \Pr\{\boldsymbol{L} = \boldsymbol{l} | M, N\}.
\end{aligned} \tag{3}$$

Clearly all that must be specified in order to make further evaluations are simply the three components of the prior distribution on the composition of the population, namely

$$\Pr\{\boldsymbol{L} = \boldsymbol{l}|M, N\}, \tag{4}$$

$$\Pr\{X_{(j_{(1)})} = x_{(1)}, \ldots X_{(j_{(m)})} = x_{(m)}|\boldsymbol{L} = \boldsymbol{l}, M, N\}, \tag{5}$$

and

$$\Pr\{M|N\} \times \Pr\{N\}. \tag{6}$$

Although our primary interest in this article is the specification of (5) in such way as to express diffuse or vague knowledge about the underlying population of values, we note that our formulation is sufficiently general so as to include conventional parametric specifications as well. For example, we may be of the opinion that the population distribution is approximately normal, in which case the distribution of $\boldsymbol{X}$ can be chosen so that the $X_{(i)}$ are order statistics of a normal population, and similarly for any other parametric distribution. We shall not pursue this idea here, however, since the most basic case is the nonparametric one.

We shall specify (4) and (5) as follows:

$$\Pr\{\boldsymbol{L} = \boldsymbol{l}|M, N\} = \binom{N-1}{M-1}^{-1}, \tag{7}$$

while, for each possible $x_{(1)}, \ldots, x_{(m)}$, and $\boldsymbol{j}$,

$$\Pr\{X_{(j_{(1)})} = x_{(1)}, \ldots, X_{(j_{(m)})} = x_{(m)}|\boldsymbol{L} = \boldsymbol{l}, M, N\}, \tag{8}$$

does not depends upon $\boldsymbol{j}$.

Any specification of (4) and (5) is equivalent to a specification of the prior distribution for the empirical distribution of the population, given M and N. Obviously this can be done in infinitely many ways, any one of which might be appropriate in a specific real-world situation. But it is of value to single out those specifications that are of special significance, such as for example corresponding to a diffuse prior distribution (as is commonly done with improper prior distributions on conventional parameters), and also those that are known to be compatible with much real-world data. The specification (7) that I originally chose was to take $\Pr\{\boldsymbol{L} = \boldsymbol{l}|M, N\} = \binom{N-1}{M-1}^{-1}$, which is the Bose-Einstein distribution for non-empty cells, as in Feller (1968, p. 40). Thus the results in Hill (1968) are based upon this choice, while those in Hill (1980a) discuss the robustness of this choice within the class of exchangeable distributions for $\boldsymbol{L}$. My doctoral student, Wen-Chen Chen, in his Ph. D. dissertation (1978) and Chen (1980) generalized this choice to include arbitrary symmetrical Dirichlet-multinomial distributions, and argued that for some data it is desirable to choose a Dirichlet prior other than the Bose-Einstein, which of course is a Dirichlet-multinomial corresponding to a uniform Dirichlet distribution. See also Lewins and Joanes (1984). My primary motivation for the Bose-Einstein distribution (which I still regard as the single most appropriate choice) is the connection with Zipf's Law. This law represents more real world data than any other known law, including the Gaussian. It is shown in my articles Hill (1970, 1974b, 1975a, 1979, 1980a, 1981), and in Hill and Woodroofe (1975) and Woodroofe and Hill (1975), that the Bose-Einstein choice yields Zipf's Law. This is why I singled it out as of special significance within the class of exchangeable prior distributions for $\boldsymbol{L}$. See also Ijiri and Simon (1975) for discussion of the Bose-Einstein distribution. Of course it is mathematically straightforward to replace the Bose-Einstein distribution by any other Dirichlet-multinomial distribution, and sometimes this may be of value in modelling the data. The logic underlying my model would at best only suggest that the distribution of $\boldsymbol{L}$ should be chosen to be exchangeable, and even this is not really necessary.

Next, one must also make some specification for the prior distribution of M and N. The most basic case for inference is simply where N is known to be large, and M has a uniform distribution, given N. This was the case considered in Hill (1968). Hill (1979) then considered the case where M has a truncated negative binomial distribution, of which the uniform is a special case. Although the specification of the distribution for M, given N, is of lesser importance here than the specification of (4) and (5), it does play a crucial role in obtaining Zipf's Law, as in the cited articles by myself and by Chen.

Even more important than the choice of the Bose-Einstein distribution for $\boldsymbol{L}$ is the choice of (8). Here we directly confront the problem of formulating a diffuse prior distribution on the empirical distribution of the population. Note that if $M = N$, so that all $L_{(i)} = 1$, then we obtain the case where ties have probability 0, and must then only express vagueness of opinion about the jump-points $X_{(i)}$ in the population. Thus the problem of expressing a diffuse distribution for the jump-points is logically independent of that of expressing one for $\boldsymbol{L}$. It was shown by Lane and Sudderth (1978, Theorem 1), defining $A_{(n)}$ for the case where ties have probability 0 and where the sequence is exchangeable, that (8) is equivalent to $A_{(n)}$.

Consider then the specification (8). What it says is that no matter what the distinct values $X_{(i)}$ may be, they contain no information whatsoever about the ranks $J_{(i)}$ of these values in the population. Clearly this is not always appropriate. For example, if one believed that the population was approximately Gaussian in form, then one would favor some $\boldsymbol{j}$ vectors over others. Or if one knew sufficiently much about the set of values in the population, then one might know, for example, that $x_{(m)}$ was in fact the largest value in the population. Or again, if the $X_{(i)}$ are necessarily integers, and if two are consecutive, then one knows that the corresponding $J_{(i)}$ are also consecutive. To understand the force of the argument for (8) however, consider the following example.

Suppose for the sake of argument that there are 100,000 adult male emperor penguins, and that their weights can be measured sufficiently precisely so that no two agree exactly. (This is assumed only to make the essential point clear. My model $H_{(n)}$, with ties, can deal with any degree of rounding.) Consider your apriori subjective opinions about the population of weights of these penguins. Suppose now that I were to give you all but one of these weights as the data $x_{(i)}$, i.e., 99999 positive numbers, no two of which are equal. The question I wish you to think about concerns your opinions about the vector $\boldsymbol{J}$, which specifies the ranks of these 99999 numbers in the population of all 100000 numbers. Condition (8) here would require that you be aposteriori indifferent as to which ranks these observations have in the population with $N = 100,000$. Note that this is meant to apply no matter what the $X_{(i)}$ values are, provided that only possible values are included, so that negative values are excluded, as well as weights that are known to be impossibly large or impossibly small. For example, if (8) holds, then you are indifferent as to which of the 100000 possible values is missing in the data. It could just as well be the largest as the smallest, or any other member of the population. Thus it would be the largest that is missing if it were the case that $\boldsymbol{J}$ consists of the ranks $1, \ldots, 99999$ in the population, and it would be the smallest that is missing if $\boldsymbol{J}$ consists of the ranks $2, \ldots, 100,000$ in the population. Are you so indifferent?

A fairly natural first reaction is to say that you might or might not be indifferent, depending upon what the numbers $x_{(i)}$ that I give you are. And you might feel that for lots of such sets of 99999 numbers you might be, and for others you might not be. But think again. Suppose, to take an extreme case that might seem to speak against $A(99999)$, that the $x_{(i)}$ that I give you are such that there is an enormous gap between the largest, $x(99999)$, and all the others. In fact suppose that $x(99999)$ is an extremely large value, say 1000 pounds, one that (although not impossible), seems highly improbable, while the other sample weights are all less than 100 pounds. You do not appreciate the full force of $A_{(n)}$ until you realize that if the largest weight in the sample were in fact 1000 pounds, then there might well be another penguin that weighs even more that this! Thus the naive reaction, which would be that no penguin weighs anything like 1000 pounds, is immediately dispelled once one fully appreciates the fact that you have already seen one such (in the scenario of the problem) and may therefore well see another one. Still another example of this type concerns human age. One might well regard it as extraordinarily improbable that any human being has lived to the age 500 years. But if one such could be demonstrated, then you might well think that another might also, and even find that your opinions were roughly in accord with $A_{(n)}$.*

What condition (8) is expressing is a completely pragmatic attitude towards the population. Such an attitude is not only a subjectively Bayesian coherent attitude but in the case at hand even seems quite compelling; and this is for the case of weights on a ratio scale, which is the worst type of example for $A_{(n)}$, as opposed to data on a merely ordinal scale, such as the Mohs scale for hardness of rocks, where the hardness values are more or less meaningless. And yet I think, after reflection, you may find it compelling even in the extreme example I have given. It would of course be even more compelling if the largest weight in the data were say, 120 pounds, rather than 1000 pounds. The general argument that I would give is that (8) with $m = M - 1$, and $N = M$ (so there are no ties), is a highly compelling subjective evaluation, and this implies $A_{(M-1)}$. Note that there is no possibility of mathematical proof that (8) is "correct," just as these is never any way of proving that one ordinary prior distribution is more appropriate than any other. All prior distributions are possible, and each is to be given "equal rights" as de Finetti says. But just as some prior distributions are sometimes regarded as more appropriate than others, for example, a uniform prior distribution on the parameter of a Bernoulli process is sometimes regarded as particularly appropriate, so too I claim that (8) is quite compelling, and I personally regard it is the most generally appropriate specification. My reasons are perhaps not entirely unrelated to the original fiducial intuitions of "Student" and Fisher.

That $A_{(n)}$ for large n should be highly compelling also agrees with certain frequentistic ideas in conventional nonparametric statistics. Very few statisticians use parametric models when dealing with large samples from some underlying population F. The reason is that one is nearly certain that the true distribution is not of any specfic parametric form, for example, Gaussian, and that with a large sample the discrepancies will almost certainly appear and be serious.This is part of the approach to hypothesis testing of J. Berkson (1938), for example, who pointed out that with a sufficiently large sample you will certainly reject most conventional null hypotheses. Thus for a sufficiently large sample one might be nearly certain that the data will allow rejection of any prespecified fixed dimensional parametric model, even using a subjective Bayesian test of the hypothesis, for which it is more difficult to reject the null hypothesis. On the other hand, if the sample from the very same population F were sufficiently small, then one might well use the Gaussian or some other parametric model. Because of the relationship of $A_{(n)}$ to the empirical distribution function (see Section 3), it is clear that the same considerations that make conventional statisticians prefer the empirical distribution function when dealing with large samples should also apply to $A_{(n)}$.

Now we come to a rather strange and interesting fact. Suppose I have managed to convince you of the appropriateness of $A_{(M-1)}$. But it is a mathematical fact, proved in Hill (1968, p. 688), that $A_{(k)}$ implies $A_{(j)}$ for $j \leq k$. Thus if you accept $A_{(M-1)}$ as appropriate exactly, then you are forced into $A_{(1)}$ as well. Of course, both $A_{(1)}$ and $A_{(2)}$ correspond to conventional Bayesian and frequentistic procedures, with a diffuse prior on location, or on location and scale parameters, respectively, and they are certainly sometimes appropriate as an approximation. But it is equally clear that they are not always appropriate. How are we to explain this? My argument for $A_{(M-1)}$, which I regard as extremely compelling when M is

* The Encyclopedia Americana, 1981, referring to penguins. states "In size they range from the gigantic emperor penguins standing about 40 inches high, and weighing up to 90 pounds, to the diminutive fairy penguin of the Australian region that attains a length of just over a foot.

large, if accepted, then implies $A_{(1)}$ as well, which is not always compelling. I believe that the explanation is as follows. In my proof that $A_{(k)}$ implies $A_{(j)}$ for $j \leq k$ there is a backwards induction. In carrying the argument backwards, it is possible that slight discrepancies from $A_{(M-1)}$ may build up, yielding a possibly much larger discrepancy for $A_{(1)}$. I should also point out that even $A_{(M-1)}$ need not hold literally. For example, suppose that you knew a great deal about the average weight in the population of penguins. Then if I gave you all but one of these weights, you would have a good idea about the missing weight. Indeed, you would know it exactly if you knew the average weight exactly. Thus one might have a discrepancy even from $A_{(M-1)}$, and this could build up even more in reaching down to $A_{(1)}$. It is considerations such as these which point out the importance of recognizing, once and for all, that we are at best only dealing with approximations. These approximations can nonetheless be very useful. It is my opinion that the nonparametric formulation, as in $H_{(n)}$, although itself only an appproximation, is ordinarily the most appropriate way to perform predictive inference, with parametric representations, such as the Gaussian, being useful primarily for inference and prediction when the sample size is small.

I remarked earlier that $A_{(n)}$ is exactly appropriate for merely ordinal data, such as hardness of rocks. The argument is as follows. Suppose one draws a simple random sample of n rocks from a population of N rocks in which no two are of the same composition or of the same hardness. (Here, as is usual, one rock is said to be harder than another if it scratches the other rock.) Before the data is taken you are surely of the opinion that $\boldsymbol{J}$ is equally likely to be any of the $\binom{N}{n}$ possible $\boldsymbol{j}$ vectors, since this is precisely what sampling without replacement means. In the present case, however even after the sample is drawn you must still be of the same opinion, since no "data" becomes available other than the relative orderings of the rocks in hardness, i.e., there are no values. (Even if some arbitrary scale is used, as the Mohs scale, it means nothing, and its "values" are totally uninformative.) Thus in this situation one is forced to make the evaluation (8). Furthermore, this provides a justification for the original fiducial intuition, which also ignores the "values" of the observations. Finally if we now consider the case of ties, as for example if we draw a simple random sample from the rocks on some mountain, then it can easily be seen that $H_{(n)}$ rather than $A_{(n)}$ applies, provided, of course, that the distribution of $\boldsymbol{L}$ is taken as Bose-Einstein (or at least exchangeable).

Finally, in the case of "colors" or "species," the natural way to proceed is as follows. Suppose that we go to a new region and, taking a sample, find n living creatures that we decide belong to m distinct species. We can number these species in any way we like, for example, we can take species 1 to be the first type caught, etc., with species m the last type caught in our sample. Suppose this system is used so that the $\theta_{(i)}$ are necessarily 0, since any creature belonging to a new species must then be given a number larger than m. However, define the quantity $\gamma_{(i)}, i = 1, \ldots, m$, to be the proportion of the unsampled population belonging to the same species as the ith sample species. Then although neither $A_{(n)}$ nor $H_{(n)}$ is exactly appropriate, it is shown in Hill (1980a) that the posterior distribution of the quantities $\gamma_{(i)}, i = 1, \ldots, m$ is exactly as under $H_{(n)}$, and that the posterior distribution of M and the posterior probability of catching a new species is as in Hill (1968, p. 681, p. 691; 1979).

3. $H_{(n)}$ AND THE DIRICHLET PROCESS PRIOR DISTRIBUTION

Because of the attention given the so-called Dirichlet process prior distribution, it is of interest to indicate the relationship between it and $H_{(n)}$. The Dirichlet process prior is a prior distribution on the space of distribution functions. It was proposed by Ferguson (1973) for Bayesian nonparametric inference. See also Blackwell and MacQueen (1973) and Kingman (1975). It yields predictions that are quite different from those of $A_{(n)}$. Thus observe that the Dirichlet process with a positive and finite parameter yields a countably additive distribution for the observations, while as proved by Hill (1968), for no n does $A_{(n)}$ hold for countably additive distributions on the space of observations. Further, when the Dirichlet parameter is identically 0, one obtains the empirical distribution as the Bayes posterior predictive distribution of a new observation, and again this is quite different from what $A_{(n)}$ yields. Of course, for large n, the empirical distribution and $A_{(n)}$ attach nearly the same weight to any interval that includes a substantial number of the $x_{(i)}$, and to this extent have asymptotically the same behaviour (so that consistency results for the empirical distribution carry over to $A_{(n)}$). For example, if I is an interval that contains exactly r of the observations, then the empirical distribution attaches weight r/n, while $A_{(n)}$ gives weight between $(r-1)/(n+1)$ and $(r+1)/(n+1)$ to I. But $A_{(n)}$ spreads all of its mass on the complement of the $x_{(i)}$ that have been observed, rather than on the $x_{(i)}$, which would appear to be more realistic for situations in which one does not anticipate ties.

Although certainly an important and interesting probabilistic development, unfortunately the Dirichlet process prior falls far short of the purpose for nonparametric Bayesian inference. The problem is not so much that a realization of this prior is discrete (with probability 1), or that in some contexts it can lead to inconsistent estimators, as in Diaconis and Freedman (1986a, 1986b), but rather that after seeing the observations $x_{(i)}$, one would regard it as highly probable that future observations will be equal to those that have already been seen. This is a terribly unrealistic mode of behavior for genuinely nonparametric problems. In Ferguson's own words (1973, p. 210): "There are disadvantages to the fact that P chosen by a Dirichlet process is discrete with probability one. These appear mainly because in sampling from a P chosen by a Dirichlet process, we expect eventually to see one observation exactly equal to another." This is precisely what $A_{(n)}$ avoids, since all the posterior predictive probability is placed on the open intervals between successive order statistics, and none at the $x_{(i)}$ that have already been seen; while $H_{(n)}$ allows for a Dirichlet-style buildup at the observed $x_{(i)}$, but with a weight that depends upon M and N in a coherent fashion. As in Hill (1968, p. 681, p. 691; 1979) one then obtains the posterior distribution of M, given N, and integrating with respect to M, this determines the posterior probability of seeing a "new species" or equivalently, of getting an observation unequal to those already observed.

$H_{(n)}$, which I believe is the most important Bayesian model for nonparametric inference, can be characterized as giving equal weight to the open intervals between the observed $x_{(i)}$ for the prediction of a new observation; while the probabilities for the new observation being equal to each of the $x_{(i)}$, given that it will tie one of them, are determined by a symmetrical Dirichlet-multinomial distribution, as in Hill (1968), Chen (1978, 1980). In fact the logic underlying $H_{(n)}$, as representing diffuse prior information, at best only requires that the distribution of $\boldsymbol{L}$ be exchangeable, although of course the Dirichlet-multinomial is especially convenient and interesting. The result of W. E. Johnson (1932) suggests the reason for the importance of the Dirichlet-multinomial prior distribution in the the case of $t \geq 3$ categories, while Hill, Lane and Sudderth (1987) show that essentially the only exchangeable urn processes are those for which the implicit prior distribution is a beta distribution, when there are only 2 categories. (Urn processes are concrete models for binary data, wherein the probability that the next observation is a red ball, given the past, is a function only of the proportion of red balls in the urn at the time.)

Thus the Hill model differs in many crucial respects from the conventional Dirichlet process model. As already remarked, $A_{(n)}$ and $H_{(n)}$ lead to consistent estimators, while those based upon the Dirichlet-process prior can be inconsistent. See Lenk (1984) for an interesting discussion of consistency in this context. It is also worthwhile to note that the predictive probabilities based on $A_{(n)}$, although derived by fiducial or Bayesian arguments, coincide with results of a distribution-free approach to tolerance analysis, as given in Wilks (1941). See Guttman (1970). The methods based upon the Dirichlet process prior distribution do not appear to have any such non-Bayesian interpretations.

Finally, both the methods based on the Dirichlet process prior and those based upon $H_{(n)}$ are coherent. Thus the choice between them in a specific situation must be based upon how appropriate the model is in the particular instance of statistical inference, i.e., upon judgement. The circumstances in which the Dirichlet process is reasonable would be in situations which are essentially finite, or to paraphrase de Finetti, in a slightly different context, "finite except for trifles". See Lane and Sudderth (1978, p. 1331) for a detailed discussion of the Hill, Ferguson, and Pólya urn models, Blackwell and MacQueen (1973) for a discussion of the Ferguson and Pólya urn models, and Kingman (1975) for a discussion of the relationship between the Ferguson model and the Poisson-Dirichlet model.

4. $A_{(n)}$: BAYES OR FIDUCIAL?

The initial intuition as regards $A_{(n)}$ seems to be due to "Student," or at least Fisher (1939) implies that this is case. Fisher then generalized the idea and interpreted it in a fiducial spirit, which Dempster (1963) crystallized and called "direct probability." Note that for all three of these authors the justification for $A_{(n)}$ seems to be purely intuitive. Thus none give anything vaguely representing a "proof" for $A_{(n)}$, or suggest a way in which its coherency or rationale can be discussed, or even indicate in what circumstances it might or might not be appropriate. While I believe that intuition (or inspiration) is what all scientific progress ultimately comes from, it is nonetheless the case that a critical attitude is necessary, and that one must ask when and why $A_{(n)}$ is sensible, and whether there are any qualifications and pitfalls associated with it. For example, one must ask immediately, when, if ever, should one be using conventional parametric models as opposed to $A_{(n)}$. It is in helping to understand such questions that I believe the subjective Bayesian approach plays a fundamental role.

Consider, for example, the case of a normal model with known standard deviation of 1 and unknown mean, θ. The fiducial argument of Fisher suggests that the pivotal quantity $(X - \theta)$ should continue to have the $N(0, 1)$ distribution even after X is replaced by its observed value x, which is a number. (The confidence argument of Neyman does not assume this, but provides no way of telling whether there is anything peculiar about the particular x for which the confidence is quoted. As is well known, there are many examples, such as the Fieller-Creasy example, in which such a procedure is patently absurd, in that the whole real line may have confidence 95 percent. More generally, such confidence procedures do not provide a way to allow one to deal with data for which the conventional confidence level is obviously inappropriate based upon prior knowledge that is generally accepted. Thus the confidence argument, as applied in practice by sensible statisticians, is instead a conditional argument, i.e., it is conditional upon not getting data that is wildly contrary to prior knowledge. As shown in Hill (1986), the "true" confidence coefficients, when adjusted to be conditional upon getting such data, are necessarily both unknown and unknowable.)

The Bayesian argument goes far beyond this. It first tells one that if one has a prior distribution for θ which is sufficiently diffuse relative to the likelihood function, then in fact Fisher's fiducial conclusion is justified. (This fact, which Harold Jeffreys had been telling Fisher for years, seems finally to have been accepted by Fisher, as the previously quoted footnote of Fisher (1959, p. 51) seems to indicate.) Next, the Bayesian argument tells you that there are many situations in which instead the prior distribution may be sharp relative to the likelihood function, in which case the appropriate conclusions are quite different; and still again, there are important cases in which the prior and the likelihood are of comparable magnitude. Thus one sees clearly in what situations the fiducial argument is relevant, and what the nature of its limitations are. See Hill (1974, p. 570) for a mathematical discussion of the behavior of a posterior distribution for various kinds of extreme data.

The situation with regard to $A_{(n)}$ is of the same general nature as that for a normal mean, except it is much more complicated. Thus the primary basis for $A_{(n)}$ or $H_{(n)}$ is (8), which really says that the observed values $x_{(i)}$ are totally uninformative about J. Once again the initial intuition comes from a form of Fisher's fiducial argument. Even if one finds his argument compelling, however, one would presumably want to put it into a broader context, including at least sampling without replacement and the case of ties. Thus the generalization from $A_{(n)}$ to $H_{(n)}$, and from sampling with replacement to sampling from a finite population without replacement, is important, since the case of ties and of sampling without replacement is both more fundamental and more realistic. The Bayesian approach does not stop here, however, for the underlying assumptions, especially (8), are themselves only approximations. They are extremely valuable, because without such approximations there is nothing that one can do in a rational and logical way. But as with all things, they too are only approximations, and the trick is to learn when they are appropriate. The model in Hill (1987b) should clarify the nature of these approximations, and further indicate in what circumstances $H_{(n)}$ is and is not appropriate. Finally, because $A_{(n)}$ is a de Finetti coherent procedure, one knows that there are no operationally meaningful ways in which one can be made a loser by using $A_{(n)}$ or $H_{(n)}$. Obviously, this property is a desirable one, but it is not sufficient to justify use of these procedures. Thus in addition to the internal coherency property, one wants also to know whether the procedure is "reasonable," that is to say, whether it corresponds to prior knowledge that is generally considered to be appropriate for the situation at hand, or for which a reasonable case can be made. In my opinion, it ordinarily is, but with a few qualifications.

Let me conclude by observing that $A_{(n)}$ is supported by all of the serious approaches to statistical inference. It is Bayesian, fiducial, and even a confidence/tolerance procedure. It is simple, coherent, and plausible. It can even be argued, I believe, that $A_{(n)}$, along with $H_{(n)}$, constitutes the fundamental solution to the problem of induction.

REFERENCES

Aitchison, J. and Dunsmore, I. R. (1975). *Statistical Prediction Analysis*. Cambridge University Press.

Berger, J. (1985). *Statistical Decision Theory and Bayesian Analysis*. 2nd. Ed. New York: Springer-Verlag.

Berger, J. and Wolpert, R. L. (1984). *The Likelihood Principle*. Institute of Mathematical Statistics Lecture Notes-Monograph Series, Vol. 6.

Berkson, J. (1938). Some difficulties of interpretation encountered in the application of the chi-square test, *J. Amer. Statist. Assoc.* **33**, 526–542.

Berliner, L. Mark. and Hill, B. M. (1986). Bayesian nonparametric survival analysis. *Tech. Rep.* Ohio State University.

Blackwell, D. and MacQueen, J. B. (1973). Ferguson distributions via Pólya urn schemes. *Ann. Statist.* **1**, 353–355.

Borel, E. (1914). *Introduction Géométrique à Quelques Théories Physiques*, Gauthier-Villars, París.

Borel, E. (1906). Sur les principles de la théorie cinétique des gaz. *Annales de l'Ecole Normale Supérieure*, 9-32.

Campbell, G. and Hollander, M. (1982). Prediction intervals with a Dirichlet-process prior distribution. *Canadian Journal of Statistics* **10**, 103–111.

Chen, Wen-Chen. (1978). *On Zipf's Law*. University of Michigan Doctoral Dissertation.

Chen, Wen-Chen. (1980). On the weak form of Zipf's law. *Journal of Applied Probability* **17**, 611–622.

Chen, Wen-Chen. (1980). On the infinite Pólya urn models. *Tech. Rep.* **167**. Department of Statistics, Carnegie-Mellon University.

Chen, Wen-Chen. Hill, B. M. Greenhouse, J. and Fayos, J. (1985). Bayesian analysis of survival curves for cancer patients following treatment. *Bayesian Statistics* **2**. (J. M. Bernardo, M. H. DeGroot, D. V. Lindley and A. F. M. Smith, eds.). Amsterdam: North-Holland pp. (with discussion).

de Finetti, B. (1937). La prévision: ses lois logiques, ses sources subjectives. *Annales de l'institut Henri Poincaré* **7**, 1–68.

de Finetti, B. (1975). Theory of Probability. **2**. London: Wiley.

Degroot, M. H. (1970). *Optimal Statistical Decisions*. New York: McGraw-Hill.

Dempster, A. P. (1963). On Direct Probabilities, *J. Roy. Statist. Soc. B* **25**, 100–114.

Diaconis, P. and Freedman, D. (1980). Finite exchangeable sequences, *Ann. Probab.* **8**, 745–764.

Diaconis, P. and Freedman, D. (1981). Partial exchangeability and sufficiency. *Proceedings of the Indian Statistical Institute Golden Jubilee International Conference on Statistics: Applications and New Directions*, 205–236.

Diaconis, P. and Freedman, D. (1986a). On the consistency of Bayes estimates, *Ann. Statist.* **14**, 1–67 (with discusion).

Diaconis, P. and Freedman, D. (1986b). On inconsistent Bayes estimates. *Ann. Statist.* **14**, 68–87.

Diaconis, P. and Freedman, D. (1987). A dozen de Finetti-style results in search of a theory. *Annales de l'Institut Henri Poincaré.* **23**, 397–423.

Edwards, A. W. F. (1972). *Likelihood.* Cambridge: University Press.

Edwards, W. Lindman, H. and Savage, L. J. (1963). Bayesian statistical inference for pychological research. *Psychological Review* **70**, 193–242.

Feller, W. (1968). *An Introduction to Probability Theory and its Applications.* 3rd Ed. New York: Wiley.

Fisher, R. A. (1939). "Student". *Annals of Eugenics* **9**, 1–9.

Fisher, R. A. (1948). Conclusions Fiduciare. *Annales de l'Institut Henri Poincaré* **10**, 191–213.

Fisher, R. A. (1959). *Statistical Methods and Scientific Inference.* 2nd Ed. New York: Hafner.

Ferguson, T. (1973). A Bayesian analysis of some nonparametric problems. *Ann. Statist.* **1**, 209–230.

Geisser, S. (1971). The inferential use of predictive distributions. *Foundations of Statistical Inference*, (V. P. Godambe and D. A. Sprott, eds.) Toronto: Holt, Rinehart and Winston, 456–469.

Geisser, S. (1982). Aspects of the predictive and estimative approaches in the determination of probabilities. *Biometrics* **38**, 75–93, (with discussion).

Geisser, S. (1985). On the prediction of observables: a selective update. *Bayesian Statistics* **2**. (J. M. Bernardo, M. H. DeGroot, D. V. Lindley and A. F. M. Smith, eds.). Amsterdam: North-Holland, 203–230, (with discussion).

Good, I. J. (1965). *The Estimation of Probabilities.* MIT Research Monograph No. 30.

Greenhouse, J. B. (1983). *Analysis of Survival Data when a Proportion of Patients are Cured: A Mixture Model.* Doctoral Dissertation, The University of Michigan.

Guttman, I. (1970). *Statistical Tolerance Regions: Classical and Bayesian.* London: Griffin.

Hartigan, J. (1983). *Bayes Theory*, New York: Springer-Verlag.

Heath, D. and Suddderth, W. (1976). De Finetti's theorem for exchangeable random variables. *Amer. Statisti.* **30**, 188–189.

Heath, D. and Suddderth, W. (1978). On finitely additive priors, coherence, and extended admissibility. *Ann. Statist.* **6**, 333–345.

Hewitt, E. and Savage, L. J. (1955). Symmetric measures on cartesian products. *The Writings of Leonard Jimmie Savage-A Memorial Selection.* Washington: ASA and IMS 1981, 244–275.

Hill, B. M. (1967). Posterior distribution of the mean of a future sample. Unpublished.

Hill, B. M. (1968). Posterior distribution of percentiles: Bayes' theorem for sampling from a finite population. *J. Amer. Statist. Assoc.* **63**, 677–691.

Hill, B. M. (1969). Foundations for the theory of least squares. *J. Roy. Statist. Soc. B* **31**, 89–97.

Hill, B. M. (1970). Zipf's law and prior distributions for the composition of a population. *J. Amer. Statist. Assoc.* **65**, 1220–1232.

Hill, B. M. (1974a). On coherence, inadmissibility, and inference about many parameters in the theory of least squares. *Studies in Bayesian Econometrics and Statistics in Honor of L. J. Savage.* (Fienberg and Zellner eds.). Amsterdam: North-Holland, 555–584.

Hill, B. M. (1974b). The rank frequency form of Zipf's law. *J. Amer. Statist. Assoc.* **69**, 1017–1026.

Hill, B. M. (1975a). A simple general approach to inference about the tail of a distribution. *Ann. Statist.* **3**, 1163–1174.

Hill, B. M. (1975b). Aberrant behavior of the likelihood function in discrete cases. *J. Amer. Statist. Assoc.* **70**,717–719.

Hill, B. M. and Woodroofe, M. (1975). Stronger forms of Zipf's law. *J. Amer. Statist. Assoc.* **70**, 212–219.

Hill, B. M. (1979). Posterior moments of the number of species in a finite population, and the posterior probability of finding a new species. *J. Amer. Statist. Assoc.* **74**, 668–673.

Hill, B. M. (1980a). Invariance and robustness of the posterior distribution of characteristics of a finite population, with reference to contingency tables and the sampling of species. *Bayesian Analysis in Econometrics and Statistics: Essays in Honor of Harold Jeffreys.* (A. Zellner ed.). Amsterdam: North-Holland, 383–395.

Hill, B. M. (1980b). Robust analysis of the random model and weighthed least squares regression. *Evaluation of Econometric Models.* (J. Kmenta and J. Ramsey, eds.). New York: Academic Press, 197–217.

Hill, B. M. (1981). A theoretical development of the Zipf (Pareto) law. *Studies on Zipf's Laws.* (H. Guiter, ed.) Bochun: Ruhr-Universitat.

Hill, B. M. (1982). Discussion of Lindley's paradox, by G. Shafer. *J. Amer. Statist. Assoc.* **77**, 344–347.

Hill, B. M. and Lane, David (1985). Conglomerability and countable additivity. *Sankhyá*, 47, *A* , 366–379.

Hill, B. M. (1986). Some subjective Bayesian considerations in the selection of models. *Econometric Reviews* **4**, No. **2**, 191–288. (with discussion).

Hill, B. M. (1987a). The validity of the likelihood principle. *The American Statistican* **41**, 95–100.

Hill, B. M. (1987b). A parametric model yielding $A_{(n)}$ and $H_{(n)}$. *Tech. Rep.* University of Michigan.

Hill, B. M. Lane, David, and Sudderth, William (1987). Exchangeable urn processes. *Ann. Probab.* **15**, 1586–1592.

Hill, B. M. (1988). A theory of Bayesian analysis. Unpublished.

Hume, David (1748). *An Enquiry Concerning Human Understanding.* London.

Ijiri, Y. and Simon, H. A. (1975). Some distributions associated with Bose-Einstein statistics. *Proc. Nat. Acad. Sci.* **72**, 1654–1657.

Jeffreys, H. (1961). *Theory of Probability.* 3rd Ed. Oxford. Clarendon Press.

Jeffreys, H. (1957). *Scientific Inference*, 2nd Ed. Cambridge University Press.

Johnson, W. E. (1932). Probability: the deductive and inductive problems. *Mind* **49**, 409-423.

Kingman, J. F. C. (1975). Random discrete distributions. *J. Roy. Statist. Soc. B* **37**, 1–22 (with discussion).

Kalbfleisch, J. O. and Prentice, R. L. (1980). *The Statistical Analysis of Failure Time Data.* New York: Wiley.

Lane, D. and Sudderth, W. (1978). Diffuse models for sampling and predictive inference. *Ann. Statist.* **6**, 1318–1336.

Lane, D. and Sudderth, W. (1984). Coherent predictive inference. *Sankhyá A* **46**, 166–185.

Lenk, P. (1984). *Bayesian Nonparametric Predictive Distributions.* Doctoral Dissertation, The University of Michigan.

Lewins, W. A. and Joanes, D. N. (1984). Bayesian estimation of the number of species. *Biometrics* **40**, 323–328.

Lindley, D. V. (1958). Fiducial distributions and Bayes theorem. *J. Roy. Statist. Soc. B* **20**, 102–107.

Luce, R. D. Narens, L. (1987). Measurement scales on the continuum. *Science* **236**, 1527–1531.

Maxwell, J. C. (1878). On Boltzmann's theorem on the average distribution of energy in a system of material points. *Trans. Cambrige Philos. Soc* **12**.

Poincaré, H. (1912). *Calcul des Probabilités*, 2ème Ed. París: Gauthier-Villars.

Savage, L. J. (1961). *The subjective Basis of Statistical Practice. Tech. Rep.* Department of Mathematics, The University of Michigan.

Savage, L. J. (1962). *The Foundations of Statistical Inference.* London: Methuen.

Savage, L. J. (1972). *The Foundations of Statistics.* 2nd Revised Edition. New York: Dover.

Wilks, S. S. (1941). Determination of Sample Sizes for Setting Tolerance Limits. *Ann. Math. Statist.* **12**, 91–96.

Woodroofe, M. and Hill, B. M. (1975). On Zipf's law. *Journal of Applied Probability* **12**, 425–434.

Zabell, S. L. (1982). W. E. Johnson's sufficientness postulate. *Ann. Math. Statist.* **10**, 1091–1099.

Zabell, S. L. (1985). Symmetry and its discontents. *Tech. Rep.* Departament of Statistics, Northwerstern University.

DISCUSSION

J. Q. SMITH (*University of Warwick*)

On learning about $A_{(n)}$ I cannot understand why Dirichlet processes have become so favoured as an expression of ignorance; $A_{(n)}$ seems based on a much more realistic modelling assumption. Indeed it is the first time I have met a useful class of models which is necessarily not countably additive. As I argued in my discussion of Racine *et al* (1986), if we find it necessary to express vagueness about a process we are far less likely to make unjustifiable probability statements if our uncertainty is expressed in terms of probability statements about observables rather than parameters. And $A_{(n)}$ does just this.

When $A_{(2)}$ is assumed to hold precisely, however, the joint distribution across observables can be quite odd. Consider first three absolutely continuous random variables X_1, X_2, X_3

where $F_3(x|x_1, x_2)$ denotes the distribution function of X_3 given $X_1 = x_1$ *and* $X_2 = x_2$. The $A_{(2)}$ hypothesis now gives us that

$$\phi(x_2) = F_3(x_1|x_1, x_2) = \begin{cases} 2/3 & x_2 < x_1 \\ 1/3 & x_1 < x_2 \end{cases}$$

It follows that $\phi(x_2)$ is a discontinuous function of x_2. Now this is a rather strange property and is certainly not shared by conditional distributions over continuous variables we usually meet (try jointly normal random variables for example).

It is not difficult to understand why you cannot have countably additive $A_{(1)}$ when X_1 and X_2 are independent. Note that $A_{(1)}$ implies

$$P(X_2 \geq X_1|X_1 = \omega) = 1/2 \qquad \text{for all } \omega. \tag{2}$$

If X_1 and X_2 were *independent* and the density of X_1 non zero on the real line then this would imply that

$$P(X_2 \geq \omega) = 1/2 \qquad \text{for all } \omega. \tag{3}$$

By a similar argument

$$P(X_2 \leq \omega) = 1/2 \qquad \text{for all } \omega.$$

It would follow that $X_2 = \begin{cases} -\infty & \omega.p.1/2 \\ \infty & \omega.p.1/2 \end{cases}$. On the other hand if X_1 only takes values in closed interval [a,b] and X_2 has the same margin as X_1 it is clear that no countably additive joint distribution could have the property (3).

Lane and Sudderth (1978) extend this type of argument to exchangeable random variables. They prove that the *only* finite additive measures consistent with $A_{(n)}$ have the following properties.

(a) When $\{X_r\}_{r \geq 1}$ are contained in a compact set

$$\text{For all } \varepsilon > 0 \qquad P(X_{(n)} - X_{(1)} < \varepsilon) = 1$$

where $X_{(i)}$ denotes the order statistic. Thus, to believe $A_{(n)}$ for bounded random variables I must be prepared to accept a bet at any odds that the distance between the smallest and largest observation is less than any number you could state. Now in most pratical problems I could state a finite interval inside which I would believe all posible observations would fall with probability 1. (Consider the weight guessing of penguins used by Bruce to illustrate $A_{(n)}$). So although I could be *coherent* in such problems. I could not be *honest* in any use of $A_{(n)}$.

(b) When $\{X_r\}_{r \geq 1}$ have infinite support assuming $A_{(n)}$ forces me to state only the slightly weaker statement that

$$\text{For all } \varepsilon > 0 \qquad P(X_{(n)} - X_{(1)} < \varepsilon \text{ or } X_{(2)} - X_{(1)} > \varepsilon^{-1}) = 1 \tag{4}$$

One practical case where Bruce (Hill, 1980) suggests $A_{(n)}$ might hold exactly is when X_n is a number measuring of the hardness of the nth rock to be classified. The hardness is measured by an "arbitrary" real index x chosen by the practitioner with the property that the nth rock will scratch all rocks tabulated so far with an index less than x. Rocks with index greater than x will scratch the nth rock. I am afraid that in all honesty I could not, in this situation, make the probability statement (4). For example my probability that $X_{(n)} - X_{(1)} > 10^{-10^{10}}$ would certainly not be zero.

However the main contention of this paper as I understand it is that $A_{(n)}$ is a good *working approximation* to beliefs I might hold, and useful for the purposes of *prediction.* I can wholeheartedly concur with this use of $A_{(n)}$. Indeed we succesfully use such approximations all the time in standard problems of inference. For example if $\{X_i\}_{i\geq 1}$ are sequence of independent random variables each with zero mean and unit variance we know that

$$S_n = (\sqrt{n})^{-1} \sum_{i=1}^{n} X_i$$

tends in distribution to a normal distribution with zero mean and unit variance. For the sake of making *predictions* I might well assume S_n is exactly normal. This assumption would, in fact, imply $X_1 \ldots X_n$ were themselves normal and this may be manifestly untrue. But provided I know my "approximation" will work well for the sort of predictions I might like to make, then this will not matter. The same argument is true for $A_{(n)}$. Although $A_{(n)}$ implies $A_{(1)}$ and I may not be prepared to accept $A_{(1)}$, this does not preclude me from using $A_{(n)}$ to predict. This is provided, of course, $A_{(n)}$ remains a reasonable approximation for my beliefs about the future functions of observables of interest.

To obtain a grasp of what I can and cannot conclude when using $A_{(n)}$ as an *approximation* to my beliefs it is helpful to consider an extreme case. So suppose I use $A_{(n)}$ to approximate my beliefs that $\{X_n\}_{n\geq 1}$ are independent with a *known* strictly increasing distribution function F. It follows that $\{Y_n\}_{n\geq 1}$ where $Y_n = F(X_n)$ are independent uniform $(0,1)$ variables. Since the property $A_{(n)}$ is invariant under any strictly increasing transformation assuming $\{X_n\}_{n\geq 1}$ are $A_{(n)}$ is equivalent to assuming $\{Y_n\}_{n\geq 1}$ are $A_{(n)}$. Well, under $A_{(n)}$,

$$P(Y_{n+1} < Y_{(r)}) = \frac{r}{n+1} \quad \text{where } Y_{(r)} \text{ is the } r^{th} \text{ order statistic}$$

and under the independence above

$$P(Y_{n+1} < y_{(r)}) = y_{(r)}$$

by the Glivenko-Cantelli Theorem, for all $\varepsilon > 0$

$$P[\sup_{0<p<1} |Y_{([np])} - p| > \varepsilon] \to 0 \qquad \text{as } n \to \infty$$

where [ ] denotes "integer part of". It follows that all statements concerning the percentiles of future observations using the two models will be approximately the same for large n. In fact if we happened to observe distances between $Y_{(1)} \ldots Y_{(n)}$ as their expected values so that $y_{(r)} = r/n+1$, independence and $A_{(n)}$ would agree exactly. If in general, we define $Y_n = F_n(Y_n)$ where F_n is the predictive distribution of X_n given $X_1 \ldots X_{n-1}$ these are independent provided $F_n \to F$ as $n \to \infty$ in an appropriate manner; for example when F_n has a fixed number of parameters all consistently estimable.

On the other hand, statements you would make about the predictive *density* of Y_{n+1} in these parametric models would not necessarily agree at all with the $A_{(n)}$ model. For example assuming $\{X_n\}$ are independent with known density f, it is easy to show that

$$P[\sup_{x} |f - f_{A_{(n)}}| < M] \to 0 \text{ as } n \to \infty$$

where $f_{A_{(n)}}$ is any density on X_{n+1} consistent with $A_{(n)}$. A similar result holds for a large class of parametric models. So statements about densities *cannot* be inferred from the $A_{(n)}$ approximation in such a cavalier fashion as statements about distribution functions.

Perhaps more seriously, by using $A_{(n)}$ I am stating that

$$P(\text{for some } 0 \leq r \leq n+1 | F(X_{n+1}) - F(x_{(r)}| \leq 2/(n+1) | x_{(1)} \ldots x_{(n)}) = 1$$

where $F(x_{(0)}) \stackrel{\text{def}}{=} 0, F(x_{(n+1)}) \stackrel{\text{def}}{=} 1$. If I believed a parametric model then this statement would be manifestly untrue. It assumes with probability 1 that there are no intervals in x space with associated non negligible probability where, by chance, no observations fall. I would expect that this discrepancy between $A_{(n)}$ and my true but intractable belief structure would, for example, tend to inflate my *certainty* about statements about percentiles given the parametric model was more appropriate. Perhaps the author would like to comment for which types of statements he believes $A_{(n)}$ is an appropriate approximation to my beliefs in this sense.

Of course, when there is disagreement between $A_{(n)}$ and an alternative model I cannot conclude that $A_{(n)}$ is a less appropriate representation of my beliefs. Indeed $A_{(n)}$ has the very pleasing property of allowing me to make good probability predictions when n is large whether or not it is exactly appropriate. And this is clearly not going to be true in general for parametric models.

I would like to congratulate Professor Hill a very thought provoking paper. I hope to see $A_{(n)}$ used more often in future. By the way, does he know of a multivariate analogue to $A_{(n)}$?

G. A. BARNARD (*University of Essex, U.K.*)

First I should note that the 1984 edition of Edwards (1972), to the first edition of which Hill refers as providing a totally devastating example against the logic of the fiducial argument, contains an Addendum in which Edwards says "On fiducial probability I now have doubts over the criticisms expressed in section 10.5. I am inclined to accept fiducial probability statements which possess the confidence property ... and which satisfy the Principle of Irrelevance...". He proceeds to withdraw some of his criticisms of Jeffreys invariant priors, and to make further interesting remarks in this vein, while insisting on the primacy of likelihood. In fact he has indicated to me broad agreement with the view put forward in Barnard (1987), in which a distinction is drawn between a quantity concerning which a probability statement can be made, and a random variable in the sense of Kolmogorof. If a maximally conditioned pivotal quantity $p = x - \theta$ is known to be $N(0,1)$, where θ is an unknown parameter and x is observable, and if x is observed =2, then if I guess that θ lies between 0.04 and 3.96, the probability that I guess right is 0.95. This is a probability statement about θ; but it does not constitute θ as a random variable in the sense of Kolmogoroff, since such a random variable has the property that any (measurable) function of it is also a random variable. But there is no function of the p above defined expressible in the form $f(p) = g(x, \theta^2)$, so that the probability statements about p which translate, given x, into statements about θ, do not serve to enable us to make probability statements about θ^2.

For further details, reference is made to Barnard (*loc. cit.*). In particular it is there pointed out that Fisher himself must, before his untimely death, have been working towards a limitation on the fiducial argument which might have been of the kind indicated above.

While Hill's reference to Edwards is an aside to his main thesis, the point I am here making seems relevant to Hill's main argument concerning $A_{(n)}$. Because it seems to me directly obvious that if we have exchangeability of any of Hill's kinds between $x_1, x_2, \ldots, x_{n+1}$, then, knowing just $x_1, \ldots, x_n$, the observable x_{n+1} must be regarded as equally likely to fall into any one of the $n+1$ intervals into which it could fall. Any preference for one interval over another would imply a violation of the exchangeability condition.

Bearing in mind de Finetti's remark (1979) that we should choose the prior-posterior *pair* which gives the subjectively most satisfactory result, could we not say, in a case in which $A_{(n)}$, after reflection, appeared "obvious", that the prior must have been such as to give this result? We might have to accept some approximation. If it seems intuitively justifiable to assign the $N(0,1)$ distribution to the pivotal $p = x - \theta$ after x has been observed, we may argue that although the prior element $1.d\theta$ which would give precisely this result is objectionable, since it is improper, nonetheless the "true", proper prior, must have been nearly uniform — in the neighbourhood of the observed value of x — so that p will be nearly $N(0,1)$ — near enough to make no difference.

It can happen that asking which prior would give us what seems to be an intuitively satisfactory result exposes an implausible assumption. In the Behrens-Fisher problem the prior element is often taken to be $d\mu_1 d\mu_2 d\sigma_1 d\sigma_2/\sigma_1\sigma_2$, or some proper approximation to it. If we transform parameters to the scale parameter for the difference of means $\bar{y}-\bar{x}$, $\sigma = (\sigma_1^2/m) + (\sigma_2^2/n)^{1/2}$, where m and n are the sample sizes, and if we take as the other parameter θ, such that $\theta\sigma^2 = \sigma_1^2/m$ and $(1-\theta)\sigma^2 = \sigma_2^2/n$, the prior element transforms to $(d\sigma/\sigma)d\theta/\theta(1-\theta)$. The second factor here is telling us that, almost certainly, the variance of $\bar{x}$ swamps that of $\bar{y}$, or vice versa. But in reality we would only be interested in comparing $\bar{x}$ with $\bar{y}$ is we thought that their variances were at least comparable. This analysis exposes the assumption, implied by $d\sigma_1 d\sigma_2/\sigma_1\sigma_2$ that, not only do we not know either σ_1 or σ_2, but, knowing σ_1 we still would have absolutely no idea of the value of σ_2.

My ignorance of emperor penguins allows me to think of several priors, any one of them plausible, which would make $A_{(n)}$ very nearly true.

S. L. ZABELL (*Northwestern University*)

Professor Hill's $A_{(n)}$ function is an interesting illustration of the use of symmetry assumptions in the probabilistic analysis of inductive inference. Such assumptions go back to Bayes, and have been a continuing thread in the work of Johnson, de Finetti, Jeffreys, Carnap, and many other statisticians and philosophers. The strength of the conclusions that flow from symmetry assumptions makes them attractive, but the obverse side of the coin is that they often have surprising and unexpected consequences. One example, closely related to $A_{(n)}$, is a little puzzle which I will call the *exchange paradox*:

> A, B, and C play the following game. C acts as referee and places an unspecified amount of money x in one envelope and amount $2x$ in another envelope. One of the two envelopes is then handed to A, the other to B.
>
> A opens his envelope and see that there is $10 in it. He then reasons as follows: "There is a 50-50 chance that B's envelope contains the lesser amount x (which would therefore be $5), and a 50-50 chance that B's envelope contains the greater amount $2x$ (which would therefore be $20). If I exchange envelopes, my expected holdings will be (1/2)$5 + (1/2)$20 = $12.50, $2.50 in excess of my present holdings. Therefore I should try to exchange envelopes."
>
> When A offers to exchange envelopes, B readily agrees, since B has already reasoned in similar fashion.

It seems unreasonable that the exchange be favorable to both, yet it appears hard to fault the logic of either. Obviously all hinges on A's apparently harmless symmetry assumption that it is equally likely that B holds the envelope with the greater or the lesser amount. In the following I will discuss what I take to be the resolution of the paradox, but since the charm of the paradox consists largely in grappling with it for oneself, it is urged that the reader first attempt to puzzle through it before proceeding further.

***

Before considering the symmetry question proper, note that in any realistic setting one would need to take into account the attitudes of the referee. 1) If you were A and felt the referee were hostile towards you, you might think it likely that he would give you the smaller amount, and it would obviously be a mistake not to take this into account. 2) The referee, however, might know that you were thinking along these lines (for example, you discussed an exchange-type paradox with him the week before) and therefore he might give you the envelope with the higher amount, banking on your attempting to exchange envelopes. 3) Realizing this, you might decide *not* to exchange envelopes. 4) In turn, of course, the referee might guess this, and give you the smaller sum. And so on. In his short story *The Purloined Letter*, Edgar Allan Poe describes an amusing case of such second-guessing: a schoolboy possessed of unusual skill at the game of successive guessing of "even or odd," due to his ability to identify with the other player, and determine the level of analysis employed by him. The celebrated Newcomb paradox (see, e.g., Jeffrey, 1983, pp. 23-25) posits, in effect, the asymptotic limit of this: the opponent can predict one's actions with unerring accuracy.

To avoid such complexities, let us assume that the envelopes are distributed at random, by tossing a coin or whatever. In that case, *before opening the envelopes*, the chances — by definition — are indeed 50-50 that your envelope contains either the greater or the lesser amount, this will remain the case if the envelopes are exchanged, and the expected amount in each is thus the same. That is, if Y_1 is the amount in your envelope, Y_2 the amount in the other envelope, and X the unknown lesser amount, then $E[Y_1] = X/2+(2X)/2 = 3X/2 = E[Y_2]$. The point is to understand how observing the specific amount $Y_1 = y$ contained in your envelope affects the amount you expect in the other envelope, i.e., $E[Y_2|Y_1 = y]$.

The Bayesian answer is straightforward. Given a continuous or discrete prior for X with density $p_{(x)}$, the posterior density for X is $P[X = y|Y_1 = y] = p_{(y)}/\{p_{(y)}+p_{(y/2)}\}$. Whether or not it is thought advantageous to exchange envelopes will then depend on the specific prior $p_{(x)}$ adopted, and the specific value of Y_1 observed. In general, it is certainly possible for both parties to wish to exchange, since they may be using different priors p_A and p_B.(But even with the same prior, both may wish to exchange; by itself, this need not be considered paradoxical since each party has different information.)

There are some obvious real-world considerations. If there is a minimum amount m or maximum amount M that the envelopes may contain, then when $Y_1 = m$ or M you know for sure that your envelope contains the lesser or greater amount, and therefore always offer or decline to exchange. To rule out edge and discreteness effects, let us make the admittedly unrealistic assumption that x may be any positive real number. Clearly, for any specific value of Y_1, there exist priors $p_{(x)}$ such that conditional on this value it is equally likely that we hold the lesser or greater amount. Indeed, it is clear there exist priors such that this is true for any finite set of values for Y_1, say $y_1, y_2, \ldots y_n$.

It is natural to ask whether it is possible to specify $p_{(x)}$ so that it is *always* equally likely, conditional on the observed value of $Y_{(1)}$, that we hold the lesser or greater amount. This is equivalent to demanding that $p_{(x)} = p_{(x/2)}$ for all $x > 0$, or equivalently $p_{(x)} = p_{(2x)}$ for all $x > 0$, and it is apparent that there is no proper prior with this property. (For in that case, either the half-open interval $[1, 2)$ would have zero probability, in which case R^+ would have zero probability, or $[1, 2)$ would have positive probability, in which case R^+ would have infinite probability mass). One might less stringently require that one's prior be such that it is always preferable to exchange envelopes, independent of the value of Y_1. A very simple argument due to I.J. Good (personal communication) shows that this is not possible: the condition translates into the functional equation $2p_{(x)} > p_{(x/2)}$ and again this can have no solution involving a proper prior p.

I have discussed this problem with several nonmathematical friends on a number of occasions and I always find the Bayesian analysis somewhat difficult to explain. Yes — exchanging the envelopes can be rational for the particular value of Y_1 you see; no — it can not be rational to exchange envelopes whatever the value of Y_1; why? — because if you were to decide beforehand on odds for all possible values of X, your behavior would in some cases be inconsistent with those odds. They always have problems with this last point. Why should I decide on the odds beforehand? For many lay people, this notion of rationality seems very artificial, especially since it seems to conflict with what appears to be a very natural conclusion.

There is a way of rescuing the popular intuition, however, which draws on the connection between the exchange paradox and $A_{(n)}$. Suppose Z_1 and Z_2 are two random variables satisfying $A_{(1)}$: i.e., Z_1 and Z_2 can assume all real values, and $P[Z_2 > Z_1|Z_1] = P[Z_2 < Z_1|Z_1] = 1/2$. As shown by Professor Hill, although there does not exist a countably additive probability space on which this can be realized, there does exist a finitely additive probability space with random variables Z_1 and Z_2 for which this is the case. Now let $Y_1 = \exp Z_1$ and $Y_2 = \exp Z_2$. Then Y_1 and Y_2 assume all positive values and $P[Y_2 > Y_1|Y_1] = P[Y_2 < Y_1|Y_1] = 1/2$, i.e., precisely the condition assumed in the exchange paradox. Thus the condition is coherent, but only if one uses finitely additive probability measures. (I have not attempted to explain this last point to my non-mathematical friends.)

The above analysis pertains to the question of whether or not you should prefer to exchange envelopes. But even if you would *prefer to exchange envelopes*, whether or not you should *actually attempt to do so* may be an entirely different matter. Suppose the maximum amount of money that may be in either envelope is \$1000. If the amount in your envelope exceeds \$500, then of course you have the higher amount and will refuse to trade. Moreover, if you find between \$250 and \$500 in your envelope, you should still avoid offering an exchange: you will only profit when your opponent has the higher amount, but in this case your opponent will undoubtedly realize this and decline to exchange. Your prior $p_{(x)}$ may be such that you think it highly likely that your opponent has the higher amount, but the harsh practical reality is that your opponent will never agree to exchange in this case. Moreover, not only should you never offer to exchange in this case, you should never accept such an offer either: your opponent would presumably never offer to exchange envelopes if he had more than \$500, so he must have less than \$250.

Curiously, it is possible to continue reasoning in similar fashion: if you find between \$125 and \$250 in your envelope, you should never offer to exchange, because in the case when you would profit (your opponent has between \$250 and \$500), he would never accept. And you should never accept: the only case where you would profit will never occur, since it would not be rational for your opponent to offer an exchange. And so on. Thus, it is argued, no matter how much is found in the envelopes, neither you nor your opponent should offer or accept an exchange.

This twist on the exchange paradox was pointed out to me by Professor H. R. Varian; as he notes, it is closely related to Aumann's discussion of common knowledge (Aumann, 1976), and the problem of drawing inferences from willingness-to-trade (Milgrom and Stokey, 1982). It also bears a close family ressemblance to the *paradox of the unexpected hanging*: a prisoner is told he will be hung some time during the next week, but that the day chosen will be a surprise; see Gardner (1969, Chapter 1) for an amusing discussion and further references, O'Beirne (1965, pp. 189-192) for an attempted resolution.

The following observations may clarify some of the issues involved. 1) The analysis presupposes the opponent to be "completely rational": if your opponent has not thought about these issues, is likely to trade, and is thought to have the higher amount, then you should certainly offer to trade. 2) It presupposes that the opponent supposes you to be "completely rational;" otherwise the chain of reasoning on his part would grind to a halt. 3) It presupposes that not only are you and the opponent "completely rational," but that each acts on his rationality: to go through such a chain of reasoning and then not act on it is as if one did not think about such matters at all.

But this last presupposition has a curious consequence. Let us return to the case where you hold betwen \$125 and \$250. As we have seen, if your opponent holds more than \$250 and is "completely rational," he will never agree to trade. But if he holds less than \$125 and is "completely rational," *he will also never agree to trade*. And the same will be true no matter what you or your opponent hold. Thus, when facing a "completely rational opponent" there is no reason not to offer to trade, merely the reality that your offer will never be accepted.

On the other hand, if your opponent is only "partially rational" (e.g., he can only be depended upon to refuse to trade when he has in excess of $500), then your analysis will depend on the degree to which he is "rational" and how far down the tree of possibilities he will reason; i.e., we are essentially back to the case of Poe's schoolboy guessing even and odd.

Another way of making the same point is to consider the case where your envelope contains y dollars. The reasoning assumes: 1) In the case where your opponent holds $2y$ dollars he *never* trades; if there is any possibility that he will trade, the logic breaks down. 2) In the case where your opponent holds $y/2$ dollars, there is a *possibility* that he will trade; otherwise, offering to trade is merely futile, not irrational. But then the reasoning continued another step contradicts this possibility. (All this presupposes, of course, that one has no further information about the behavior of the opponent. For example, if the opponent decides what do on the basis of a known randomized strategy, in some cases one can certainly profit by this knowledge.)

The exchange paradox appears to have invented by the French mathematician Maurice Kraitchik (1953, pp. 133-134). Kraitchik frames the problem in terms of two persons, each of whom claims to have the better necktie. Both agree to let a third person act as referee, the loser getting the winner's necktie as consolation. Each argues: "I know what my tie is worth. I may lose it, but I may also win a better one, so the game is to my advantage". Kraitchik notes that the problem may be framed in terms of a payoff matrix, but questions whether probabilities may be meaningfully assigned to the various possibilities.

Martin Gardner (1982, p. 106) gives a crisper version of the paradox, which he calls the "wallet game": two persons play, agreeing that whoever has the lesser amount in their wallet gets the sum total of both. In each case the argument effectively goes as follows: Suppose my wallet contains the known amount $X = x$ dollars, and the other person's wallet the unknown amount Y dollars; if $Y < x$, I lose x while if $Y > x$ I win Y, so that the expected gain $E[Y|Y > x]P[Y > x] - xP[Y < x]$ is positive if $P[Y > x] = P[Y < x] = 1/2$. For an amusing discussion of both the Kraitchik and Gardner versions of the paradox, see McGilvery (1987). (My thanks to Persi Diaconis and Martin Gardner for the above references.)

I first heard the paradox in the form discussed here from Steve Budrys of the Odesta Corporation (although it does not originate with him); arguably it has the merit of paring the elements of the paradox down to the essential minimum.

REPLY TO THE DISCUSSION

To Professor J. Q. Smith:

I would like to thank Professor Smith for an informative and interesting discussion.

As to the first point he raises, about $A_{(n)}$ being *necessarily* not countably additive, this is only true if one insists that $A_{(n)}$ hold *exactly* for cardinal data. The model of Hill (1987b) (which Jim has not yet seen), yields $A_{(n)}$ or $H_{(n)}$ to an arbitrarily good approximation, and can be obtained using a diffuse (or improper) prior distribution on the parameters of a conventional parametric model. Next, I do not find anything odd about (1) being discontinuous. Countable addivitity is itself a continuity property that is not always appropriate. One gets accustomed to certain conventions, such as continuity, on the basis of purely mathematical properties. In the real-world they may or may not be appropriate. Compare, for example, the discussion of the discontinuity of time in Whitrow (1980, p. 200). (Some physicists believe time to be discontinuous, with a quantum of time, called a chronon, that is estimated by some to be on the order of 10^{-43} seconds.)

On the question of rocks, one must recall that the usual scale hardness, called the Mohs scale, is quite arbitrary, and the "values" mean very little, if anything. For example, Whitrow (1980, p. 216) says "Consequently, Mohs" scale is purely ordinal and is not a scale of measurement. No numerical operations on the numbers of this scale have any significance." Compare the discussion in my article at the end of Section 2. Note that if sampling is simple random, and there are no ties, then one is forced into $A_{(n)}$, since one gets no information about J from the "data" that consists only of the relative ordering of the sample rocks. Similarly, for $H_{(n)}$, in the case of ties, although here one must also introduce a priori information about L, for example, exchangeability.

I think that Professor Smith and I are in substantial agreement about $A_{(n)}$ and $H_{(n)}$. Both are only approximations, but I believe they are at present the most useful ones we have for the vast number of real-world problems that require some form of nonparametric modelling. In my opinion, the most wrong-headed attitude in all of statistics is to take some approach, such as the fiducial approach, or maximum-likelihood. (both of which are sometimes sensible as an approximation), and then turn these into rigid requirements, almost as a form of idolatry. This is true even of the Bayesian approach. For example, although a diffuse prior distribution on the mean of a normal population is terribly useful as an approximation, if one thinks of the literal meaning (and incredible implications) of such a prior, then it is hard to imagine anyone taking it seriously. But, as I said, it can nontetheless be extraordinarily useful in practice, and the same goes for $A_{(n)}$ and $H_{(n)}$.

Finally, in answer to Jim's last question, there is a multivariate analogue of $A_{(n)}$, which is discussed in Hill (1987b).

To Professor G. Barnard:

I wish to thank Professor Barnard for his interesting discussion.

With regard to the example of Edwards, I am afraid that I still find it totally devastating. Edwards (1972, p. 207f) shows that the fiducial argument would provide an objective justification for the principle of insufficient reason or indifference. He says "Thus, starting with absolutely no information about H, we end up with a definite statement of probability. What is more, ' Probabilities obtained by the fiducial argument are objectively verifiable in the same sense as are the probabilities assigned in games of chance."' The principle of insufficient reason is of the greatest importance, but its basis is purely subjective, as discussed in my article in connection with spherical symmetry. By "reductio ad absurdum" one can therefore reject the fiducial argument, at least as a logical argument. (The later edition of Edwards provides no reasons to reject his example, although his own opinions seem to have changed.)

A very basic disagreement between Professor Barnard and myself concerns the question of measurability, and more generally the relevance of the Kolmogorov measure-theoretic formulation of probability. There are many quantities that are unknown that are not conventional random variables. The 100th billion digit of pi, the exact time at which Mary, Queen of Scots, was beheaded, the population of the Earth at January 1, 2000 AD, are examples. Following de Finetti, I prefer to call these random quantities rather than random variables, in order to emphasize that we know nothing whatsoever that would allow us to presume that there exists an underlying measure-theoretic structure such as that of Kolmogorov. All that we really know is that such things are not known to us. The Procrustean bed that de Finitti (1975, p. 33, p. 246) discusses is a very uncomfortable one, indeed, and leads to incredible and stifling rigidity. See also the discussion in Hill (1987a, p. 99) with regard to the likelihood principle.

Thus it does not matter to me whether θ^2 is or is not measurable. It is unknown, and we must deal with it, whether for inference, prediction, or decision-making. (Plainly the infinite precision assumed in the normal model, is totally unrealistic for real-world data. To compound such unreality with questions as to whether a function is or is not measurable, which in the finitely additive theory of de Finetti and L.J. Savage is both a useless and meaningless abstraction, and in practice is irrelevant, makes things infinitely worse.) All we are trying to do after all, in the practice of statistics is to make sensible approximations that are appropriate for the purpose at hand. Partitions can, except in very special situations, be taken as finite. Infinite partitions are sometimes useful to obtain approximations, but great care must be exercised not to lose oneself in meaningless abstractions and paradoxes of the infinite.

It is not true that "any preference for one interval over another would imply a violation of the exchangeability condition." Independent identically distributed observations are automatically exchangeable, but $A_{(n)}$ does not hold, since in this case the distribution of the next observation is independent of the past. $A_{(n)}$ is only appropriate for exchangeable distributions in which there is diffuse or vague prior knowledge about the shape of the underlying population.

Professor Barnard and I do agree, however, that it is useful to examine whether results derived from the fiducial argument can be obtained as a posterior distribution for some prior distribution. I view the fiducial argument as an original and ingenious alternative way to derive the posterior distribution in certain special cases where there is diffuse prior knowledge, much as Fisher implies in the footnote I quoted. It is, of course, delightful, when as with $A_{(n)}$ the basic result can be derived in many different ways. In summary, I think Professor Barnard and I are in substantial agreement about all the essential matters, and perhaps differ only in tactics.

To Professor Zabell:

I wish to thank Professor Zabell for his interesting presentation and discussion of the exchange paradox. I believe that this paradox is, by *reductio ad adsurdum*, implicity another demonstration that utility must be bounded, rather than calling into question the use of symmetry. The assumption that utility can be equated to monetary value is not so innocuous as might appear, when one bears in mind that Savage's axioms imply that utility must be bounded, Fishburn (1970, pp. 194, 206–207), Savage (1972, p. 80).

As Professor Zabell shows, the paradox is intimately related to $A_{(1)}$. If $A_{(1)}$ holds, and if utility is equal to monetary value, then the exchange paradox is a logical consequence. It is clear that the behavior of the participants is irrational. Thus, if I were one of the participants, then no mattter what value x is revealed to me, after observing x my posterior expected utility would be maximized by agreeing to pay a positive amount in order to exchange tickets. (In fact, using this linear utility function, it follows that for any prespecified amount C, and with $x > 4 * C$, I would be willing to pay at least C for the exchange.) On the other hand, before observing x, by symmetry I would be indifferent as to whether I receive the one ticket or the other. The tension in the example stems from the fact that under $A_{(1)}$ the value x is totally uninformative as to whether I have the smaller or larger amount, so that I have obtained no information as to which ticket is the larger from the observation, and thus should be perfectly content with the x I have, no matter what positive value it might be. (Of course, if there is a known upper bound to the value of x, then $A_{(1)}$ cannot hold when one observes x equal to this upper bound. But in many typical applications such an upper bound would itself only be vaguely known, and in any case it is desirable that the theory be able to include situations where such a bound is unknown, if only as a matter of approximation. See the discussions by Savage (1972, p. 18, p. 39, p. 81.)

Zabell shows that under appropriate real-world constraints, such as known bounds on the monetary amounts, or under a game-theory model that involves the willingness to trade of both participants, that the paradox can be resolved. I agree with this analysis, but I will argue that even without such constraints the paradox disappears if utility is restricted to be bounded; and moreover that without such boundedness of utility, the paradox would undermine the subjective Bayesian theory of maximization of expected utility. Clearly boundedness of utility is a weaker assumption than is boundedness of economic goods. The first step in my argument is to show that $A_{(1)}$ is not the source of the difficulty. I will first review some facts about $A_{(n)}$.

My proof that $A_{(1)}$ cannot hold literally, in Hill (1968, p. 687), applies not only to densities, as in the argument of Good quoted by Professor Zabell, but to any countably additive exchangeable distributions on the space of two observations. It thus includes those generated via a prior distribution on a conventional parameter of a statistical model, with the two observations being independent and identically distributed, given the parameter, provided only that ties have probability 0. (The case where ties have positive probability, is, of course, the case dealt with by $H_{(n)}$ rather than $A_{(n)}$. My original argument can be modified to show that even $H_{(1)}$ cannot hold for countably additive distributions, unless $X_{(1)} = X_{(2)}$ with probability one.) Jeffreys (1961, p. 171) had earlier discussed how $A_{(1)}$ follows from a uniform prior distribution on the mean of a normal population, and Lane and Sudderth (1978) have shown that $A_{(n)}$ is coherent in the sense of de Finetti, i.e., that there exist finitely additive distributions that yield $A_{(n)}$ for all n, so that a Dutch-book is impossible.

In addition to being coherent, I have argued in the present article and elsewhere, that $A_{(n)}$, and hence also $A_{(1)}$, is an evaluation that one is forced into, approximately, in many situations. Hence for me it is the fact that $A_{(1)}$ is approximately right, under a great many circumstances, that is most important; rather than the fact that it cannot hold literally for countably additive distributions, or the fact that it can hold literally for finitely additive distributions, although the latter fact is of course relevant for obtaining approximations. It is well known that improper prior distributions serve a useful purpose in providing approximations in Bayesian statistics, and of course, that conventional non-Bayesian statistics is implicitly (and nowadays perhaps also unconsciously) based upon such prior distributions. I believe that the finitely additive model is the best framework in which to discuss the meaning of such prior distributions. For one thing, it is fully rigorous, whereas the use of improper prior distributions has led to many so-called paradoxes and much confusion, ultimately stemming from the lack of rigor that has become common in dealing with such things. See Hill (1980) and my review of the book by Hartigan in Hill (1986), for futher discussion. (Our nonmathematical friends might actually find the finitely additive theory, which makes no presumptions about infinity, much more palatable than the host of largely irrelevant measure-theoretic considerations with which the conventional countably additive theory becomes absorbed. It should also be pointed out that only finitely additive probability has ever been demonstrated in a rational way, as for example by Savage (1972), and that there have been very serious challenges to the countably additive theory, as for example, by de Finetti (1974, p. 33, p. 246), that have never been answered.)

Be this as it may, the real-world situation is one in which the prior distribution for the mean of a finite population can be quite diffuse relative to the likelihood function, without being improper. Using the example of Jeffreys, but with a proper prior distribution on the mean of the normal population, one can simply evaluate the posterior probability that the second observation will be larger than the first, given the first, and show that for fat-tailed prior distributions this probability can be very nearly .5 for all but an apriori subjectively improbable set of x values. Indeed, this remark led me to the theorem of Hill (1974, p. 570), which characterizes all of the possible limiting distributions for extreme x and shows clearly under what circumstances one will behave very nearly in accord with $A_{(1)}$. On this basis I would argue that $A_{(1)}$ is often a very sensible approximation, in addition to being coherent and widely used, as for example is implicitly the case in non-Bayesian statistics based upon the normal distribution. It is for these reasons that I do not think the exchange paradox should be thought to stem from $A_{(1)}$.

If then one accepts the validity of the finitely additive framework of de Finetti and / or of L.J. Savage, or alternatively, if one merely regards $A_{(1)}$ as a compelling approximation for sufficiently large x, one is still led to very nearly the same behavior as in the paradox. As argued above, however, it is irrational both to view the datum as uninformative in the sense of $A_{(1)}$, and *always* to wish to trade with the other participant. Professor Zabell gives an analysis based upon both sensible real-world constraints, and also upon some game-theoretic considerations. These are very relevant for the real-world problem, and it is well-known that such constraints can dramatically change the analysis of such problems. In fact a very similar argument, using an upper bound on the data-value that can possibly be reported in finite time, was used by myself in repudiating the Monette-Fraser example against the likelihood principle in my discussion of the monograph by Berger and Wolpert in Hill (1984 p. 167-173). Here the introduction of the real-world constraint on the magnitude that can possibly be reported in a finite length of time, dramatically reversed the Monette-Fraser demomstration that the use of an improper prior led to inadmissibility in the extended sense. Similarly, the paradox of noncongiomerability of de Finetti (1974, pp. 177-178) cannot arise if one introduces finite bounds, as discussed in Hill (1980).

Nonetheless, it is interesting to ask whether if such constraints were removed, would there in fact still be a paradox. If one uses unbounded utility, and if one accepts $A_{(1)}$, the answer is yes. However, in my opinion, the weak point lies in the use of monetary value (or any other unbounded function) as equivalent to utility. I think that the excahange paradox in effect provides still another proof that utility must be bounded. Thus for a bounded utility function, it would no longer occur that one would always be willing to trade, even though one makes the $A_{(1)}$ evaluation. If $u(x)$ denotes the utility of x dollars, then one will want to trade if and only if $u(2x) - u(x) > u(x) - u(x/2)$. If we replace the inequality by an equality we have a difference equation whose solutions include $u(x)$ constant or $u(x) = \log(x)$, or linear combinations of these two solutions. The expected loss seen through trading is $u(x) - [u(2x) + u(x/2)]/2$. For a utility function that tends to a finite limit as x goes to ∞, and for sufficiently large x, this becomes negligible so that in any practical sense the paradox disappears. It is worth observing that if instead of knowing that one ticket is worth double the other, one knows, for example, that one is worth x and the other worth e^x, then one can obtain more stringent conditions that would also rule out $\log(x)$ or any other unbounded function as the utility function. In fact these considerations are reminiscent of those involved in discussions of the St. Petersburgs paradox of Daniel Bernoulli. Compare the discussion by Savage (1972, pp. 91-97).

Although my taking utility to be bounded might appear to be just another real-world constraint, such as those of Professor Zabell, I think that it is something more, since it is known that Savage's theory implies boundedness of utility, and this means that without such boundedness his axioms are inconsistent. In a practical sense the upshot is that the subjective Bayesian theory can deal with situations where economic goods are unbounded, or where one does not wish to worry about the precise value of such bounds, provided that utility is taken to be bounded. At any rate, the exchange paradox appears to cut to the core of some basic issues in the theory of subjective probability and utility, as well as of my $A_{(n)}$, and I am indebted to Zabell for bringing this interesting example to my attention.

REFERENCES IN THE DISCUSSION

Aumann, R. (1982). Agreeing to disagree. *Ann. Statist.* **4**, 1236–1239.

Barnard, G. A. (1987). *Int. Statist. Review.* **55**, 183–189.

Edwards, A. W. F. (1972). *Likelihood*, Cambridge: University Press.

de Finetti, B. (1974). *Theory of Probability*. **1**. Wiley.

de Finetti, B. (1979). *J. Roy. Statist. Soc.* **41**, 135.

Fishburn, P. C. (1970). *Utility Theory for Decision Making*, New York: Wiley.

Gardner, M. (1969). *The Unexpected Hanging and Other Mathematical Diversions*. New York: Simon and Schuster.

Gardner, M. (1982). *Aha! Gotcha: Paradoxes to Puzzle and Delight*. New York: Freeman.

Hill, B. M. (1974). On coherence, inadmissibility and inference about many parameters in the theory of least squares. *Studies in Bayesian Econometrics and Statistics in Honor of L. J. Savage*. (S. Fienberg and A. Zellner ed.) North Holland, 555–584.

Hill, B. H. (1980). Invariance & Robustness of Posterior Distribution Characteristics of a Finite Population, with reference to Contingency tables and Sampling of Species. *Bayesian Analysis in Econometrics and Statistics*. (A. Zellner. ed.) Amsterdam: North Holland 383–395.

Hill, B. M. (1980). On finite additivity, non-conglomerability, and statistical paradoxes. *Bayesian Statistics.* (J. M. Bernardo, M. H. DeGroot, D. V. Lindley and A. F. M. Smith, eds.). Valencia: University Press 39-66 (with discussion).

Hill, B. M. (1984). On the likelihood principle. *The Likehood Principle: A Review and Generalizations.* (by J. Berger and R. Wolpert) Washington: I.M.S, 161-174.

Hill, B. M. (1986). Review of Bayes Theory, by J. Hartigan. *J. Amer. Statist. Assoc.* **81**, 569–571.

Jeffreys, H. (1961). *Theory of Probability.* 3rd Ed. Oxford: Clarendon Press.

Jeffrey, R. C. (1983). *The Logic of Decision.* 2nd Ed. Chicago: University Press.

Kraitchik, M. (1953). *Mathematical Recreations.* 2nd. Ed. New York: Dover.

Lane, D. & Sudderth, W. (1978). Diffuse Models of Sampling for Predictive Inference. *Ann. Statist.* **6**, 1318–1336.

McGilvery, L. (1987). Speaking of paradoxes ... or are we? *Journal of Recreational Mathematics* **19**, 15–19.

Milgrom, P. and Stokey, N. (1982). Information, trade and common knowledge. *Journal of Economic Theory.* **26**, 17–27.

O'Beirme, T. H. (1965). *Puzzles and Paradoxes.* Oxford: University Press. Reprinted. New York: Dover, (1984).

Racine, A., Grieve, A. P., Fluhler, H. and Smith, A. F. M. (1986). Bayesian methods in practice: experiences in the Pharmaceutical industry. *Appl. Statist.* **35**, 93–150. (with discussion).

Savage, L. J. (1972). *The Foundations of Statistics.* Revised Edition. New York: Dover.

Whitrow, G. J. (1980). *The Natural Philosophy of Time.* 2nd Ed. Oxford: University Press.

BAYESIAN STATISTICS 3, *pp. 243–259*
J. M. Bernardo, M. H. DeGroot, D. V. Lindley and A. F. M. Smith, (Eds.)

Bayesian Paleoethnobotany

J. B. KADANE* and C. A. HASTORF*
Carnegie-Mellon University and *University of Minnesota*

SUMMARY

Paleoethnobotany is the use of burnt plant remains to investigate certain types of activity at archaeological sites. Ethnographic studies are used to inform opinion on the various points in the processing, storing, and cooking of grains at which the plants might come into contact with fire. Combined with opinions about the relative decay rates of different species, botanical remains are thus used to update an archaeologist's opinion of the activities carried out in various places at a prehistoric house in a large settlement. The extent of the shift from prior to posterior is a measure of the importance of botanical evidence in archaeological interpretation.

The methods are applied to excavations at a prehistoric site in Perú.

Keywords: ARCHAELOGY; BOTANY; BURNT BOTANICAL REMAINS; CROMWELL'S RULE; PATIO; PERÚ.

1. STYLES OF DATA ANALYSIS

There are many styles of data analysis and interpretation. The goal of all is to make the data more meaningful and interpretable. To illustrate how Bayesian analysis can help with meaningful interpretation, we have chosen to apply it to the archaeological problem of determining prehistoric household activities.

For the record, and so that other data analysis may compare their methods to the one used here, the full data set discussed in this paper is reported in Table 1. The columns in that table are seed-types, the rows are specific excavated locations (proveniences), and the entries are the number of burnt seeds found in soil from each place.

The method of analysis pursued here, is to set out in detail both the way the data came into being and what else is known or believed about the data by archaeologists who study this cultural area (section 2). In this case, these data are part of a larger archaeological research project and database. We then discuss the kinds of questions that paleobotanical data, such as those reported in the table, are collected to answer. For greater specificity, we choose one archaeological question to address in detail, namely, what daily life activities might have been conducted at each provenience or location within one prehistoric household (section 3). We report priors for activities based on the field notes of the archaeologists digging at each provenience (section 4). The botanical data yield a likelihood on activities (section 5), which is used to update the priors to posteriors (section 6). All these specifications must be regarded as tentative, as they represent our first attempt to quantify paleoethnobotanical beliefs in this manner. Our conclusions are reported in section 7.

---

* The authors are members of the Center for Advanced Study in the Behavioral Sciences, Stanford

OBS	FLOTNO	MZKERNEL	CUPCOBS	CHENOPOD	AMARANTH	SCIRPUS	TLEGUME	VERBENA	PLANTAGO	MALVAST	COMPOSIT	WOOD	GRSSTKS	COCA	PAPAVER	GALIUM	VITAC	SYSRINC	RELBUN	POLYGON	CYPERS	GRASS	TUBER	LEGUMES
1	275			133																				1
2	276	13		56								2										1	2	24
3	264	7		25		3	1					4									3	9	2	14
4	288	41		45		2	2					7									2	1	21	1
5	290	3		33		1						1									1			2
6	314	1	1	3		1															1	6		
7	299			20		66	2					14									66	2		9
8	191			1																				
9	302			51		1						4									1	11	6	2
10	337			14																		5		
11	284		2	17465								26	3									3	90	
12	283					1						6									1			
13	273	1	1	29	1	2	1					4									2	12		
14	175			14								7											2	
15	174	1		22	1	1						4									1	2		
16	407			7								6			47					1		1		2
17	342											1												
18	294			2								44												
19	300			1		3	1					9									3	1		
20	396			22				1				2												
21	164					1						5									1	1	3	
22	170			2		1															1	3		
23	172											7												
24	165	1		3																		1	1	
25	181			5								1										1		
26	185											4										1		
27	159			7		1						2									1	7		
28	167																					1		
29	160			1						1												4		
30	171		48	13	1		7	8	2			1	7			1	1					200	5	
31	184	1		5								2										5	2	
32	169			1								3										5		
33	182	4		26								41											15	24
34	155	3		10								37											4	6
35	157			2								3											14	
36	301		2	8		1						10									1	5	5	3
37	156		1	4								13										2	50	
38	423		1	79		63						24									63	61	989	
39	177			1																			4	
40	419			32		11	1					17			44						11	13	37	1
41	158			2								1											9	
42	279			10			1					10	7									3	25	
43	310			1																				
44	322			5		2						2									2	4	7	
45	144											5										1	5	3
46	326			3								24											1	6
47	292			48		6						31									6	8		1
48	154			1								1												
49	148	1					1																	
50	289			3								7										1		
51	291			3								9												
52	319																							
53	321											1												
54	327											3												3
55	325			13		1						3									1	1		
56	333			3								4												
57	142			2								1										1		
58	336											3												
59	340																							
60	305			1								1												
61	285		2	45								2											4	
62	149			2		1						161											1	
63	312	1										16										1		
64	139			14		1						13									1	3		1
65	168			2											42									
66	161											9											4	
67	146											2												
68	152											1												
69	188	1		2								6											4	1
70	186			12								9										5		
71	187	21		18	8	1						9									1	3	10	3
72	282			12		1						20									1	4		
73	147			1																				
74	145			4								4										2		3
75	281			10		5	1					4									5	4		
76	286	1		9							1	13										4	29	
77	287	210	15	5		1						47									1			
78	330	47	18	3		2						15									2		2	
79	339			38		1						2									1	1		
80	324			2								3										1	1	
81	303											25												
82	422			20								5										1		
83	280			1								2			26									
84	329	1	1	9								6										3		
85	313											1												
86	332	2		1								2												
87	274											2												
88	277	1		1																			4	

Table 1. *Burnt Botanical remains by Provenience as found, (1982). Paleoethnobotanical data by flotation number from J7 =2*

2. THE DATA

The archaeological data in this paper come from a region around the modern town of Jauja in the central Andes of Perú. This region is approximately 250 km east of Lima, between the two mountain ranges that run parallel to the west coast of South America. The specific region of study is the northern portion of the Wanka ethnic territory. The Wanka have lived in the area since approximately 200 B.C. In 1977 a research team, the Upper Mantaro Archaeological Research Project (UMARP, directed by T. Earle, T. D'Altroy, C. Hastorf, and C. Scott), began investigating the economy and political organization of this local group, throughout prehistory, until the arrival of the Spanish *conquistadores* in A. D. 1532. This temporal sequence has been divided into cultural periods characterized by changes in the political organization and settlement pattern. The Wanka, living between 10,000 and 12,000 feet above sea level, farmed locally adapted crops: potato (*Solanum tuberosum L.*); maize (*Zea mays L.*); quinoa (*Chenopodium quinoa* Wild.), a grain; lupine (*Lupinus mutabilis* Sweet), a legume; and a series of Andean tubers. They also herded camelids, the llama (*Llama glama* Linnaeus 1758) and the alpaca (*Llama pacos* Linnaeus 1758), and raised guinea pigs (*cavia cf. porcellus* Linnaeus 1759) in their houses (Hastorf, 1983).

This project is based on data collected during the 1982 field season, during which the excavations studied the organization of the Wanka economy during the late prehistoric periods: the Late Intermediate Period (A. D. 1350–1460), called the Wanka II Period and the Late Horizon-Inca Period (1460–1532) called the Wanka III Period (Earle et al., 1987).

During the Wanka II era the local population was organized into chiefdoms composed of groups numbering in the thousands. This organization is inferred from the settlement-pattern and artifact distribution at centers and small associated satellites. Large towns, such as Tunánmarca (labeled J7 in Figure 1), comprised habitation structures numbering in the thousands. These were associated with smaller nearby settlements. Sites occupied in this time period were located on high, rocky knolls, overlooking the small valleys and rolling countryside (Figure 1). Within stone defensive walls, these sites were filled with hundreds of household residences called patio areas (an example of which is shown in Figure 2). Each patio area was composed of one or more circular structures that opened onto an enclosed space and were joined together by stone walls. These patio areas were linked by narrow winding pathways that wove through the settlement. From the artifacts found at the site and this architectural layout, we believe that each patio area housed a family, either nuclear or extended.

The archaeological evidence suggests that local political units competed with each other for control of land and other resources. Within a polity, a large community appears to have dominated smaller nearby villages. These polities had leaders who were especially important in warfare (LeBlanc 1981).

During Late Horizon, the Wanka society was transformed through imperial conquest (D'Altroy, 1981, 1987). As the region was pacified and organized under Inca rule, the settlement locations changed radically. The Wanka population was moved into smaller, unfortified settlements at lower elevations.

To understand the overall Wanka economy, UMARP began by investigating the domestic household economy. The researchers wanted to understand the daily tasks and occupations of household groups, and consequently were interested in the activities that took place in patios. In the 1982 field season UMARP excavated patios at four different sites. In that year, six patios were excavated totally to retrieve data on a complete distribution of artifacts within individual households. In 1983 23 more patios were sampled to improve the patio sample size. Patio selection was based on a series of architectural and spatial criterial that were used to define two economic statuses within the society: elites and commoners. In each patio, excavators divided the space into units no bigger than 2 × 2 m. In each of these units, they

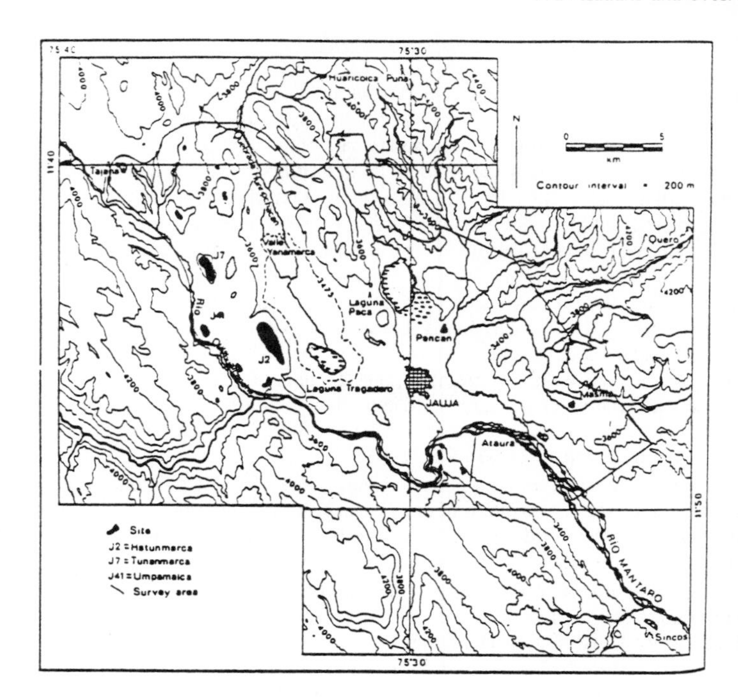

Figure 1. *Wanka II Period settlement pattern in the archaeological study area.*

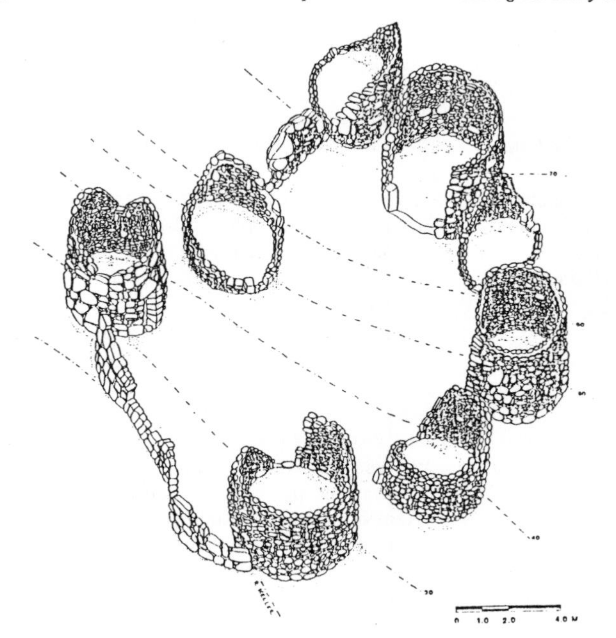

Figure 2. *A patio at Tunánmarca: An Axiometric view of J7=2 patio.*

tried to collect all cultural material: ceramic fragments (sherds), stone tools (lithics), animal bone, plant remains, metal, shell, and other miscellaneous objects that the Wanka managed to bring home. However, not all objects that were used and deposited in the past are still present in the soil. Erosion, scavenging by later people, and decomposition all affect the artifact assemblage of an archaeological site.

Of major interest to archaeologists, therefore, are (1) the relationship (spatial, temporal, or social) of the excavated objects to each other; (2) the conjectural recreation of the original deposited assemblage from the incomplete record that is actually excavated; and (3) the differential preservation of various types of artifacts. To understand the economics and politics of a group, we must begin with a reconstruction of the activities that occurred in the excavated areas. Once we know something about the activities, we can consider the causes of change.

Commonly, the excavated artifacts are divided by type and analyzed by specialists (e.g., a ceramic specialist, an osteological specialist). Each artifact type can inform us about certain aspects of the prehistoric past. For example, ceramic sherds are othen used to infer trade relations between villages and groups; lithic tools tell us about a group's complexity and type of technology.

Of special interest here are the botanical remains collected from the patios, because their information is rarely applied in archaeological research and they offer an important window into the past. Because plants and plant use in the past have a direct link with their collection and agricultural production, they should be able to give us a unique perpective on agricultural activities and the related economics (Hastorf 1988). Because of the long time between deposition and excavation (over 500 years for the Wanka II samples), and the weathering process, only charred botanical remains survive. Such remains are quite fragile, and require special procedures. When, as here, these procedures are utilized, however, the analysis results in a viable archaeological data set. In the past, botanical remains have not been considered important in archaelogy both because of the difficulty of the procedures and because of the special assumptions that need to be made, particularly concerning the differential preservation between plant taxa. We address this issue below.

3. RESEARCH PLAN

How much can archaeological data inform us about prehistoric activities? Toward this general question, we have chosen to focus on the more specific question, what were the activities in different places within a Wanka household patio? Even more specfically, what can botanical data tell us about these activities? The purpose of this paper is to see how much paleoethnobotanical data can inform us about the past in one prehistoric location. The steps in the study are:

(1) Conversion of the archaeological field notes into probabilities, here treated as prior probabilities, for the activities at each location.

(2) Development of a model for the botanical remains conditional on each activity, which plays the role of the likelihood.

(3) Computation of posterior probabilities given the actual botanical data.

(4) Comparison of prior to posterior probabilities as an indication of the extent to which botanical reamains are useful in informing the archaeologists about activities at sites.

We have chosen one household patio $J7 = 2$ (the second patio excavated on site J7), as our test case in this exploratory study. It was excavated in 1982. It was the dwelling of an elite family on the central site of Tunánmarca and included six structures and a large inner patio space (Figure 3). The excavation units divide the patio space into manageable areas. The excavation procedure included two sampling strategies. First, within each of 88 excavation sub-units (determined by location and depth) a bag of soil, 6 kg in weight, was

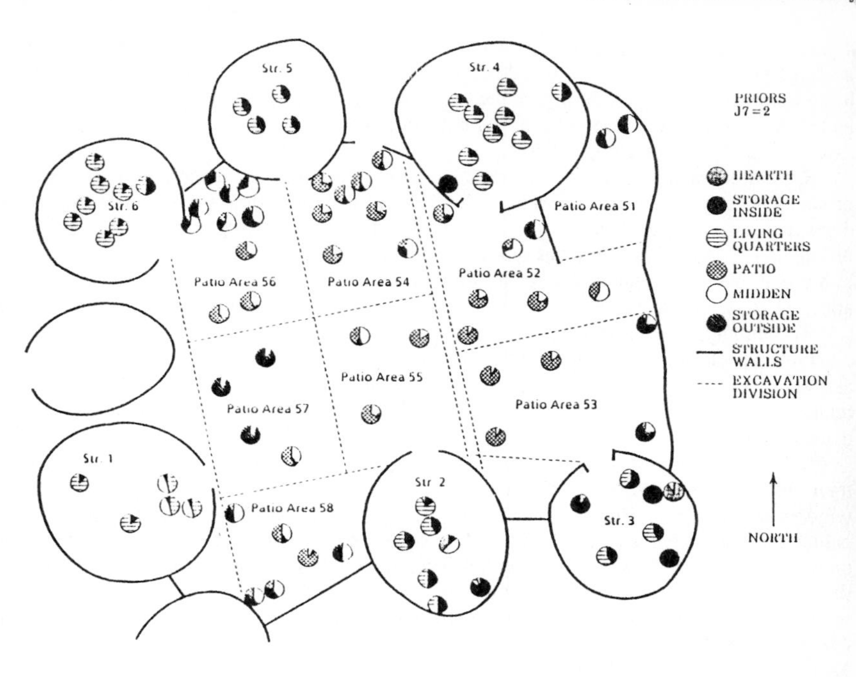

Figure 3. *Plan of patio J7=2. Structures and excavation units labeled by number. Soil sample flotation locations indicated by pie charts representing the prior probabilities.*

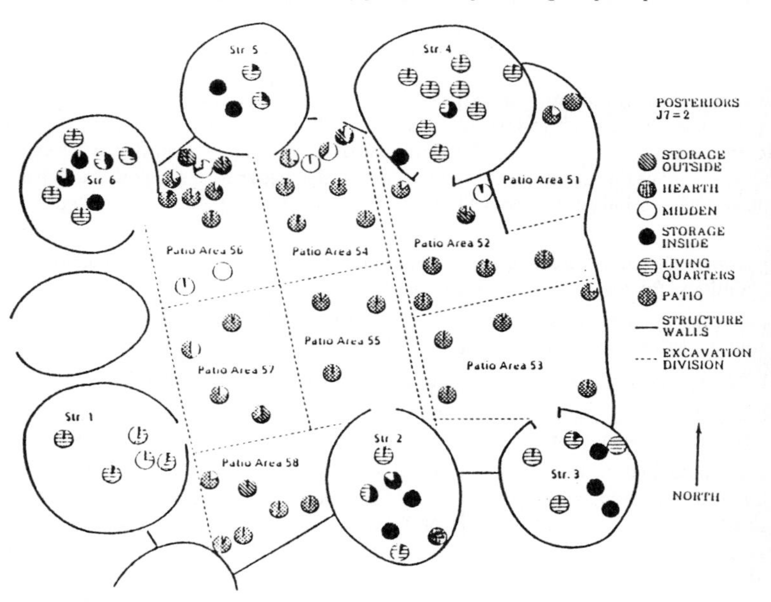

Figure 4. *Plan of patio J7=2. Structures and excavation units labeled by number. Soil sample flotation locations indicated by pie charts representing their posterior probabilities.*

collected. The specific soil collection locations are shown by pie charts in Figures 3 and 4, which are discussed in detail later. This soil then was processed by a mechanical water flotation system (Watson 1976) that gently separated the plant remains from the soil matrix. Charred or carbonized plant remains have a lighter specific gravity than water, and as the soil is lightly agitated by moving water, the plant fragments float to the surface and can be skimmed off. This procedure also collects a systematic subsample of the very small artifacts at the site.

In the second data collection strategy, all the remaining excavated soil was processed through $\frac{1}{4}$" screens.This allowed the artifacts greater than $\frac{1}{4}$" to be collected and placed in coded bags. Naturally, most seeds are too small to be recovered by this sifting procedure, although occasional wood or tuber fragments are collected. Systematically retrieved botanical remains come from the equal-sized bags of soil processed by water flotation, and are the data reported in Table 1.

The cultural description of each of provenience at J7 = 2 is based on the excavation notes, the soil changes, and the artificats that came from the screening. The observations we are comparing to the prior descriptions are the identified plants from the individual samples of bagged and floated soil. What can the botanical data tell us about the cultural activities in this prehistoric elite patio compound? Let us now turn to the construction of the prior probabilities.

4. PRIOR PROBABILITIES OF ACTIVITIES

The first step in determining the prior probabilities of prehistoric activities is to be specific about the categories used, For this paper we have chosen a rather coarse division of activities. We are treating the inside of the circular structures differently from the patio space. Within a structure, the activity areas we recognize are hearths, storage against the walls, living quarters, and midden. In the patio outside of the structures, we divide the activity areas into storage against the wall, midden, and activity center. These activities and their locations are defined by modern daily life activities, recorded from observing the Wanka who live in the region today. A hearth (A) is a place where food is cooked and heat is generated. It is a very localized spot, identified by dense carbon and *in situ* burning of the soil. Storage inside the structure (F) is located against the walls of the house structure. It is where fuel and food are stored. Living quarters (G) within the structure are where the residents eat, sleep, and socialize; the guinea pigs also reside there and eat the food scraps dropped on the floor. Midden (E) is a localized place, often near a wall, where objects and refuse are discarded. The activity of storage in the patio area (H) remains the same, with more emphasis on storing fuel, ash, and wild collectables. Midden (E) in the patio has the same use as in the structure. An activity center (I) is an area where many tasks are performed the most common are food processing, such as winnowing, sorting, drying, bagging; household goods construction or repair, such as wool carding, dyeing, weaving; tool construction, such as digging-stick making, mat weaving, pottery making, leather working etc. A particular spot may or may not have been used for only one of these categories Thus we must consider the possibility of mixed usage, where appropriate.

For each provenience, the notes of the archaeologist who excavated it provide the basis for an opinion about use of the location where the soil sample was collected. The notes incorporate evidence other than the botanical specimens recovered by flotation, of which the excavating archaeologist would have no knowledge. For each provenience, the archaeologist on our tearm (Hastorf) gave her opinion of the prior probabilities of usage, based on the notes and ethnographic knowledge. These probabilities are given in Table 2.

The probabilities given in Table 2 are incomplete, as the following example demonstrates. Consider provenience #283, a sample recovered from inside a structure. The prior probabilities

Proveniences Inside Structures

Key: A: Hearth; F: Storage (I); G: Living quarters; E: Midden

(all no Act Area, no midden, no storage (O))

FLOTATION SAMPLE NUMBER	$P(\bar{A}\bar{F}G)$	$P(\bar{A}F\bar{G})$	$P(\bar{A}FG)$
283	25	25	50
273	.33	33	34
172	25	25	50
156	25	25	.50
175	80	10	10
275	70	10	20
290	70	.10	20
174	20	30	50
314	30	20	50
302	30	20	.50
297	30	20	50
396	30	20	50
164	60	20	20
170	60	20	20
165	.60	20	.20
181	60	20	20
185	60	20	20
159	60	20	.20
167	60	.20	20
160	60	20	20
184	50	.30	.20
169	50	.30	.20
182	50	.30	.20
155	50	.30	.20
157	70	.10	.20
177	.70	.10	.20
301	70	.10	.20
158	70	.10	.20
423	70	.10	.20
419	.70	.10	.20
279	.70	.10	.20

FLOTATION SAMPLE NUMBER		FLOTATION SAMPLE NUMBER	
283, 273, 172, 288, 264, 276	$P(E\bar{F}G)$ = 50	407	$P(A\bar{F}\bar{G})$ = 0.8
	$P(EF\bar{G})$ = .05		$P(AF\bar{G})$ = 0.1
	$P(EFG)$ = .05		$P(\bar{A}\bar{F}\bar{G})$ = 0.1
	$P(E\bar{F}\bar{G})$ = .15	191	$P(A\bar{F}\bar{G})$ = .05
	$P(EFG)$ = .15		$P(A\bar{F}G)$ = .05
	$P(\bar{E}\bar{F}G)$ = .05		$P(AFG)$ = .20
	$P(\bar{E}FG)$ = .05		$P(\bar{A}\bar{F}G)$ = .40
284	$P(A\bar{F}\bar{G})$ = .05		$P(\bar{A}\bar{F}\bar{G})$ = 10
	$P(AF\bar{G})$ = 10		$P(\bar{A}FG)$ = .20
	$P(AFG)$ = 10	337	$P(A\bar{E}\bar{F})$ = .05
	$P(\bar{A}\bar{F}G)$ = .05		$P(AE\bar{F})$ = 20
	$P(\bar{A}\bar{F}\bar{G})$ = .65		$P(\bar{A}EF)$ = .10
	$P(AFG)$ = .05		$P(A\bar{E}F)$ = .10
299	$P(A\bar{F}\bar{G})$ = 1.0		$P(AEF)$ = .20
		300, 342, 171	$P(\bar{A}\bar{F}\bar{G})$ = 1.0

Proveniences Outside Structures

Key: I: Activity Area in Patio; H: Storage (O); E: Midden

(all no Hearth, no storage (O), no Living Quarters)

FLOTATION SAMPLE NUMBER	$P(I\bar{H}\bar{E})$	$P(I\bar{H}E)$	$P(IH\bar{E})$	$P(IHE)$	$P(\bar{I}\bar{H}E)$	$P(\bar{I}H\bar{E})$	$P(\bar{I}HE)$
310	.02	.02	.02	.05	.33	.33	.23
325	.05	.05	.07	.10	.15	.43	.15
285, 149	.20	.20	.80	.17	.15	.5	.15
139	.30	.35	.05	.10	.10	.05	.5
332	.03	.07	.10	.10	.30	.20	.20
292	.01	.01	.01	.9	.19	.19	.50
187, 281	.01	.01	.01	.10	.47	.20	.20
289	.50	.10	.05	.10	.10	.05	.10
291	.50	.15	.05	.10	.10	.05	.55
319	.63	.10	.05	.10	.05	.05	.02
321	.63	.10	.05	.10	.05	.05	.02
327	.60	.15	.05	.05	.10	.05	.02
142	.63	.10	.05	.10	.05	.05	.02
340	.10	.10	.10	.10	.20	.10	.30
305	.55	.15	.04	.05	.10	.01	.10
312	.55	.15	.04	.05	.10	.01	.10
161	.20	.20	.10	.15	.20	.05	.10
152	.50	.15	.05	.05	.15	.05	.05
339	.80	.10	.0	.0	.10	.0	.0
324	.80	.10	.0	.0	.10	.0	.0
329	.05	.10	.05	.10	.20	.20	.30
277	.10	.10	.10	.10	.20	.20	.20
154	.20	.35	.04	.05	.30	.02	.04
336	.30	.30	.10	.10	.10	.05	.05
147, 287	.30	.30	.05	.10	.15	.05	.05
330							
303	.30	.30	.05	.05	.20	.05	.05
144	.20	.30	.10	.10	.10	.10	.10
326	.10	.20	.05	.10	.30	.05	.20
188	.10	.10	.10	.10	.30	.10	.20
186, 286	.05	.05	.05	.10	.25	.20	.30
148	.01	.01	.04	.04	.40	.20	.30
168	.15	.20	.50	.20	.20	.05	.05
146	.55	.20	.05	.10	.10	.0	.0
282	.02	.02	.02	.05	.50	.19	.20
145	.05	.02	.05	.10	.25	.28	.30
422	.80	.10	.0	.0	.10	.0	.0
280	.20	.25	.10	.15	.15	.05	.10
274	.10	.10	.10	.10	.20	.20	.20
322	.05	.05	.05	.15	.15	.15	.40
333	.05	.05	.05	.10	.15	.20	.40
313	.60	.20	.05	.05	.05	.05	.0

Table 2. *Priors of Activities by Flotation Sample Number*

given in Table 2 are:

$$\text{P}(\bar{\text{A}}\bar{\text{E}}\bar{\text{F}}\text{G}) = 0.25$$

$$\text{P}(\bar{\text{A}}\bar{\text{E}}\text{F}\bar{\text{G}}) = 0.25$$

$$\text{P}(\bar{\text{A}}\bar{\text{E}}\text{FG}) = 0.50$$

Thus Hastorf is sure provenience 283 was *not* used as hearth (A) or midden (E). It might have been used solely as storage (F) (probability $\frac{1}{4}$), solely as living quarters (G) (probability $\frac{1}{4}$), or both (probability $\frac{1}{2}$). If it had been used as both, what is the probability that a given deposited botanical specimen came from a storage area or from living quarters?

We have chosen to treat this as a beta distribution, and here in particular a uniform distribution. In general we choose the hyperparameters α and β of this beta distribution as follows: The mean of the beta distribution is chosen to be the same as the mean of the distribution of uses conditional on the provenience being used for only one purpose. Thus if in general

$$\text{P}(\bar{\text{A}}\bar{\text{E}}\bar{\text{F}}\text{G}) = p_1$$

$$\text{P}(\bar{\text{A}}\bar{\text{E}}\text{F}\bar{\text{G}}) = p_2$$

then α and β would be chosen so that

$$\frac{\alpha}{\alpha+\beta} = \frac{p_1}{p_1+p_2}.$$

The sum, $\alpha+\beta$ is an indication of how much total informaton is available in a beta distribution. (The variance of a beta distribution is $E(p)E(1-p)(1/(\alpha+\beta+1)$.) We chose $\alpha+\beta$ to indicate considerable uncertainty about the mixture. In fact, we chose it so that it is the same as it would be for a uniform distribution, i.e., $\alpha+\beta=2$. These choices imply $\alpha = 2p_1/(p_1+p_2)$ and $\beta = 2p_2/p_1+p_2)$. This specifies the prior in all cases in which only two uses were given positive probability.

These ideas can be extended easily to cases of three or more possible activities. First, the prior probability on each combination of two activities is specified using the principles above. If k activities are possible,that they individually have probabilities, respectively, of $p_1, \ldots, p_k$. Then, analogous to the work before, a Dirichlet prior would be imposed, with parameters $\alpha_1, \ldots, \alpha_k$. These hyperparameters α would be chosen to satisfy

$$\frac{\alpha_i}{\sum_{j=1}^k \alpha_j} = \frac{p_i}{\sum_{j=1}^k p_j} \quad i = 1, \ldots, k.$$

Also, to make the uniform distribution a possibility, $(\alpha_1 = \cdots = \alpha_k = 1)$, the sum of the alphas is constrained by

$$\sum_{i=1}^k \alpha_i = k.$$

These equations have the solution

$$\alpha_i = \frac{kp_i}{\sum_{j=1}^k p_j},$$

which defines the prior on each k-dimensional space. This assumes each $p_i > 0$, $i = 1, \ldots, k$.

5. BOTANICAL REMAINS: THE LIKELIHOOD FUNCTION

For convenience, we accept a Poisson model for the number X of burnt botanical remains believed to have been at a provenience at the time of site abandonment (*ca.* A. D. 1460, when the Inca conquered and relocated the population). Suppose that X has mean λ. If the process determining whether the seed survives until 1982 is binomial with probability p, then the number of botanical remains found in 1982, Y, is Poisson with mean λp. For each kind of possible botanical specimen, then the probability p must be specified. In addition, for each activity and specimen the, mean λ (as of 1460) must be specified.

Leaving aside for the moment the issue of preservation, it is necessary to elicit Poisson rates λ for botanical remains expected by use of the provenience as, of A. D. 1460. As a first step, the group of subactivities was identified, and botanical remains were associated with each. Points were then distributed, sometimes on the basis of 100, but sometimes with other totals. The totals were chosen only for convenience, and are not regarded as having archaeological meaning. The results are listed in Table 3.

The next step was to subdivide the six activities into subactivities. These decompositions are given in Table 4. Again the totals are not regarded as meaningful, but simply as devices to get the proportions of botanical remains internal to each activity correct. Finally, we considered how many botanical remains would be expected (as of A. D. 1460) were the provenience used for each activity or combination of activities. These expectations are given in Table 5. The information in Tables 3, 4 and 5 jointly implies Poisson rates by activity in A. D. 1460, as reported in Table 6.

With respect to preservation, our first step was to rank botanical remains by how likely they are to survive. The ranking was as follows, from most to least likely to survive: small seeds, wood, *Zea mays* kernels, legume cotyledons, grass stalks, animal, dung, and tubers. After that the archaeologist chose probabilities for the survival of each kind of botanical specimen. Those are reported in column 1 of Table 7. Finally the other columns of Table 7 report the expected Poisson rates of botanical remains as found in 1982, and are the rates in Table 6 multiplied by the probabilities in column 1 of Table 7. This specifies the likelihood for the problem.

It is perhaps noteworthy that the likelihood in this problem is just as subjective as the prior. That such might be the case has been remarked many times in the Bayesian literature, and is consistent with the view of Bayarri, DeGroot, and Kadane (1987) that only the product of likelihood and prior has unique status from a Bayesian perspective. We have taken extra space to report the reasoning that leads to this specification because we think it might be interesting in its own right.

6. COMPUTATION OF POSTERIOR DISTRIBUTIONS

The problem as posed is not a convenient family to report a conjugate posterior distribution. We have chosen to report it in terms of the predictive probability of each activity. For convenience of comparison, the prior has been reformulated that way as well.

This computation distinguished each subset of possible activities. These subsets contained one to three activities each. First, each subset was checked to see if a botanical specimen, believed impossible for some of its constituents, has been found there. If so, this subset was eliminated. For the others, a computation was done: in the subsets of size one, a simple calculation of Poisson probability, for the subsets of size two, a 1-dimensional integral, and finally for subsets of size 3, a 2-dimensional integral.

The posteriors were calculated for 1-dimensional integrals using a 10-point grid, and for the 2-dimensional problems with a truncated 10×10 grid. Although these methods were adequate for such low-dimensional integrals, we would anticipate moving to more sophisticated

Subactivity / Item	
Food	
Zea mays kernels	3
Chenopodium spp.	11
tubers	4
domestic legumes	2
	20
Hearth Misc.	
Amaranthus	11
wild legumes	11
Verbena sp.	11
Plantago sp.	11
Asteraceae	11
Erthroxylum coca	1
Vitaceae	1
Sisyrinchium	11
Polygonum	11
Cyperaceae	11
grass	11
	101
Medicinals	
Verbena sp.	3
Plantago sp.	3
Sysrinchium	1
ortiga colorado	1
Vaccinium sp.	1
Valeriana sp.	1
	10
Fuel	
Zea mays cobs & cupules	5
wood	30
grass stalks	10
grass	20
animal dung	30
	95
Household Good Fabrication	
Verbena	10
Asteraceae	2.5
wood	60
grass stalks	2.5
Relbunium	2.5
Polygonum	10
Cyperaceae	10
grass	2.5
	100
Sweepings	
Chenopodium spp.	20
Amaranthus	8.3
Scirpus	2.5
wild legumes	8.3
Verbena	8.3
Plantago	2.5
Malvastrum	2.5
Asteraceae	8.3
wood	2.5
grass stalks	2.5
Papaveraceae	2.5
Galium	2.5
Vitaceae	2.5
Sysrinchium	2.5
Relbunium	0.5
Polygonum	8.3
Cyperaceae	8.3
grass	2.5
tubers	2.5
domestic legumes	2.5
animal dung	4.0
	104.3
Winnowing	
Zea mays kernels	10
Chenopodium spp.	30
Amaranthus	15
Scirpus spp.	15
Malvastrum sp.	15
Asteraceae	15
domestic legume pods	10
	110
Stored Food	
Zea mays kernels	3
Chenopodium spp.	11
Scirpus	2
Asteraceae	2
Cyperaceae	2
grass	4
tubers	4
domestic legumes	2
Minthostachys sp.	4
	34
Tool Construction	
wood	4
Scirpus	1
Polygonum	1
Cyperaceae	1
grass	1
	8

Table 3. *Botanical Remains by Subactivity (1460)*

Activity / Subactivity	
Hearth	
food	20
fuel	70
hearth misc.	5
medicinals	5
	100
Midden	
hearth misc.	70
sweepings	30
	100
Storage (inside)	
stored food	30
fuel	20
sweepings	50
	100
Storage (outside)	
fuel	1
sweepings	1
household good fabrication	1
	3
Open Area in Patio	
household good fabrication	4
tools	4
winnowing	1
	9
Living Quarters	
food	6
sweepings	6
medicinals	0.5
tools	1
	13.5

Table 4. *Activities Partitioned into Subactivities*

Hearth	200	Living Quarters	30
Midden	150	Storage (0)	50
Storage (I)	50	Open Area in Patio	10

Table 5. *Expected Total Botanical Remains (in 1460) by Activity*

	Preservation Probability	*HEARTH*	*MIDDEN*	*INDOOR STORE*	*LIVING QTR*	*OUTDOOR STORE*	*ACTIVITY CENTER*
Mzkernels	-0.50	3.00	1.58	0.66	1.00	0.00	0.05
cupcobs	0.40	2.95	1.55	0.21	0.00	0.35	0.00
Chenopodium	0.80	17.60	16.14	7.72	7.91	2.56	0.24
Amaranthus	0.80	0.87	3.32	1.59	0.85	1.06	0.12
Scirpus	0.80	0.00	0.86	1.19	0.48	1.65	0.92
wild legumes	0.80	0.87	3.32	1.59	0.85	1.06	0.00
Verbena	0.80	3.27	4.58	1.59	1.12	1.06	0.00
Plantago	0.80	3.27	2.58	0.48	0.52	0.32	0.00
Malvaistrum	0.80	0.00	0.86	0.48	0.26	0.32	0.12
Asteracae	0.80	0.87	3.32	2.30	0.85	1.39	0.21
wood	0.60	26.53	14.57	2.25	0.86	9.40	2.93
grass stalks	0.30	14.74	8.82	1.65	0.32	2.57	0.11
Coca	0.30	0.03	0.02	0.00	0.00	0.00	0.00
Papaveraceae	0.80	0.00	0.86	0.48	0.26	0.32	0.00
Galium	0.80	0.00	0.86	0.48	0.26	0.32	0.00
Vitaceae	0.80	0.08	0.90	0.48	0.26	0.32	0.00
Sisyrinchium	0.80	1.67	1.74	0.48	0.34	0.32	0.00
Relbunium	0.80	0.00	0.17	0.10	0.05	0.40	0.09
Polygonum	0.80	0.87	3.32	1.59	1.07	2.39	0.80
Cyperaceae	0.80	0.87	3.32	2.30	1.07	2.39	0.80
grass	0.80	24.45	13.70	3.58	0.48	3.46	0.53
tubers	0.10	0.80	0.53	0.24	0.30	0.04	0.00
dom. legumes	0.30	1.20	0.95	0.44	0.50	0.12	0.00
dung	0.30	13.26	7.48	1.24	0.15	1.77	0.00
Minthostachys	0.80	0.00	0.00	1.41	0.00	0.00	0.00
ortiga	0.80	0.80	0.42	0.00	0.09	0.00	0.00
Vaccinium	0.80	0.80	0.42	0.00	0.09	0.00	0.00
Valeriana	0.80	0.80	0.42	0.00	0.09	0.00	0.00
Legume pods	0.40	0.00	0.00	0.00	0.00	0.00	0.04
TOTAL		119.60	96.61	34.53	20.03	33.59	6.96

Table 6. *Expected Botanical Remains by Usage as of 1460.*

methods (Smith et al., 1985; Tierney and Kadane, 1986) as the dimensionality of the integrals grows.

Several proveniences posed special computational difficulty. In an earlier draft of this paper, numerical problems of overflow and underflow prevented calculation of posteriors in proveniences 275, 284, 171, 423, 292, 148, 149, 188, 286, 287, 330, and 329. However, careful study of these proveniences showed that multiplying the Poison probabilities by a well-chosen constant made it possible to calculate all of the posteriors. Some of the difficulty that appears to be numerical is perhaps better thought of as being a modeling problem. For example, float number 284, the observation of more than 17,000 *Chenopodium* seeds is astonishing given the likelihoods here, or anything close to them. That the posterior puts probability 1 on float number 284 being a hearth is an artifact of the huge number of *Chenopodium* seeds observed and the fact that hearths have the highest expected number of *Chenopodium* seeds among activities given positive prior weight.

A second difficulty caused one provenience (299) not to be computable. The prior puts probability one on its being a hearth. But 66 *Scirpus* seeds were found there, an impossible event because the expected number of *Scirpus* seeds found in a hearth is zero according to

	HEARTH	MIDDEN	INDOOR STORE	LIVING QTR	OUTDOOR STORE	ACTIVITY CENTER
Mzkernels -	6.00	3.15	1.32	2.00	0.00	0.10
cupcobs	7.37	3.87	0.53	0.00	0.88	0.00
Chenopodium	22.00	20.18	9.65	9.89	3.20	0.30
Amaranthus	1.09	4.15	1.99	1.06	1.33	0.15
Scirpus	0.00	1.08	1.48	0.60	2.07	1.15
wild legumes	1.09	4.15	1.99	1.06	1.33	0.00
Verbena	4.09	5.73	1.99	1.39	1.33	0.00
Plantago	4.09	3.23	0.60	0.65	0.40	0.00
Malvalstrum	0.00	1.08	0.60	0.32	0.40	0.15
Asteracae	1.09	4.15	2.87	1.06	1.74	0.26
wood	44.21	24.29	3.76	1.43	15.66	4.89
grass stalks	14.74	8.82	1.65	0.32	2.57	0.11
Coca	0.10	0.05	0.00	0.00	0.00	0.00
Papaveraceae	0.00	1.08	0.60	0.32	0.40	0.00
Galium	0.00	1.08	0.60	0.32	0.40	0.00
Vitaceae	0.10	1.13	0.60	0.32	0.40	0.00
Sisyrinchium	2.09	2.18	0.60	0.43	0.40	0.00
Relbunium	0.00	0.22	0.12	0.06	0.50	0.11
Polygonum	1.09	4.15	1.99	1.34	2.99	1.00
Cyperaceae	1.09	4.15	2.87	1.34	2.99	1.00
grass	30.56	17.12	4.47	0.60	4.32	0.67
tuber	8.00	5.28	2.36	2.99	0.40	0.00
dom. legumes	4.00	3.18	1.48	1.65	0.40	0.00
dung	44.21	29.94	4.12	0.51	5.90	0.00
Minthostachys	0.00	0.00	1.76	0.00	0.00	0.00
ortiga	1.00	0.53	0.00	0.11	0.00	0.00
Vaccinium	1.00	0.53	0.00	0.11	0.00	0.00
Valeriana	1.00	0.53	0.00	0.11	0.00	0.00
legume pods	0.00	0.00	0.00	0.00	0.00	0.10
TOTAL	200.00	150.00	50.00	30.00	50.00	10.00

Table 7. *Preservation Probabilities and Expected Botanical Remains by Usage as of 1982.*

Table 7. This problem is a straight issue of beliefs incompatible with the data. The difficulty is due to disregard of "Cromwell's Rule" (Lindley, 1985, p. 104), which says to avoid putting zero probabilities on anything. Provenience 299 is still regarded by Hastorf as a hearth, so the problem lies in the expected numbers of seeds (Tables 6 and 7), and not in the priors (Table 2). As more experience is jointly gained in this kind of analysis, we expect this problem not to recur.

Samples Inside Structures

Float Number		HEARTH Mean (SD)	MIDDEN Mean (SD)	STORAGE INSIDE Mean (SD)	LIVING QUARTERS Mean (SD)	STORAGE OUTSIDE Mean (SD)	PATIO ACTIVITY CENTER Mean (SD)	STRUCTURE
275	PRIOR			125 (331)	875 (331)			1
	POST			003 (016)	997 (016)			
276	PRIOR		459 (498)	082 (275)	459 (498)			•
	POST		424 (131)	001 (010)	574 (132)			
264	PRIOR		459 (498)	082 (275)	459 (498)			•
	POST		391 (099)	006 (032)	603 (103)			
288	PRIOR		459 (498)	082 (275)	459 (498)			•
	POST		.731 (158)	001 (009)	267 (158)			
290	PRIOR			125 (331)	875 (331)			•
	POST			.005 (022)	995 (022)			
314	PRIOR			400 (490)	600 (490)			2
	POST			502 (229)	498 (229)			
299*								•
191	PRIOR	074 (281)		.176 (301)	750 (433)			•
	POST	000 (002)		.006 (024)	993 (024)			
302	PRIOR			400 (490)	600 (490)			•
	POST			805 (206)	195 (206)			
337	PRIOR	387 (.487)	480 (.500)	.133 (340)				•
	POST	.000 (004)		1 000 (004)				
284	PRIOR	064 (244)		876 (330)	060 (238)			•
	POST	1 000 (000)		000 (000)	000 (000)			
283	PRIOR			500 (500)	500 (500)			•
	POST			069 (136)	931 (136)			
273	PRIOR			500 (500)	500 (500)			•
	POST			.973 (075)	027 (075)			
175	PRIOR			111 (314)	889 (314)			3
	POST			.006 (031)	994 (031)			
174	PRIOR			.600 (490)	400 (490)			•
	POST			.182 (221)	818 (221)			
407	PRIOR	889 (314)			111 (314)			•
	POST	000 (.004)			1 000 (004)			
312	PRIOR			1 000 (000)				•
	POST			1 000 (000)				
297	PRIOR			400 (490)	600 (490)			•
	POST			.993 (028)	007 (028)			
300	PRIOR			1 000 (000)				•
	POST			1 000 (000)				
396	PRIOR			400 (490)	600 (490)			•
	POST			016 (047)	984 (047)			
164	PRIOR			250 (433)	750 (433)			•
	POST			028 (083)	972 (083)			
170	PRIOR			250 (433)	750 (433)			•
	POST			035 (090)	965 (090)			

Samples Inside Structures (continued)

Float Number		HEARTH Mean (SD)	MIDDEN Mean (SD)	STORAGE INSIDE Mean (SD)	LIVING QUARTERS Mean (SD)	STORAGE OUTSIDE Mean (SD)	PATIO ACTIVITY CENTER Mean (SD)	STRUCTURE
72	PRIOR			500 (500)	500 (500)			4
	POST			055 (118)	945 (118)			
66	PRIOR			250 (433)	750 (433)			•
	POST			006 (027)	994 (027)			
81	PRIOR			250 (433)	750 (433)			•
	POST			008 (032)	992 (032)			
45	PRIOR			250 (433)	750 (433)			•
	POST			015 (053)	985 (053)			
59	PRIOR			250 (433)	750 (433)			•
	POST			643 (313)	357 (313)			
67	PRIOR			250 (433)	750 (433)			•
	POST			007 (028)	993 (028)			
60	PRIOR			250 (433)	750 (433)			•
	POST			055 (116)	945 (116)			
71	PRIOR			1 000 (000)				•
	POST			1 000 (000)				
84	PRIOR			375 (484)	625 (484)			5
	POST			179 (221)	821 (221)			
69	PRIOR			375 (484)	625 (484)			•
	POST			311 (294)	689 (294)			
62	PRIOR			375 (484)	625 (484)			•
	POST			988 (054)	012 (054)			
55	PRIOR			375 (484)	625 (484)			•
	POST			994 (035)	006 (035)			
57	PRIOR			125 (331)	875 (331)			6
	POST			005 (021)	995 (021)			
01	PRIOR			125 (331)	875 (331)			•
	POST			929 (162)	071 (162)			
56	PRIOR			500 (500)	500 (500)			•
	POST			295 (146)	705 (146)			
23	PRIOR			125 (331)	875 (331)			•
	POST			434 (054)	566 (054)			
77	PRIOR			125 (331)	875 (331)			•
	POST			004 (018)	996 (018)			
19	PRIOR			125 (331)	875 (331)			•
	POST			1 000 (003)	000 (003)			
58	PRIOR			125 (331)	875 (331)			•
	POST			004 (019)	996 (019)			
79	PRIOR			125 (331)	875 (331)			•
	POST			791 (255)	209 (255)			
								•

* Problem impossible due to prior contradiction

Samples Outside Structures

Float Number		HEARTH Mean (SD)	MIDDEN Mean (SD)	STORAGE INSIDE Mean (SD)	LIVING QUARTERS Mean (SD)	STORAGE OUTSIDE Mean (SD)	PATIO ACTIVITY CENTER Mean (SD)	PATIO AREA
310	PRIOR		458 (500)			488 (500)	.024 (152)	51
	POST		000 (001)			000 (005)	1 000 (005)	
322	PRIOR		452 (198)			452 (498)	096 (295)	•
	POST		209 (059)			060 (106)	731 (108)	
311	PRIOR		275 (447)			208 (406)	517 (500)	52
	POST		147 (037)			008 (035)	845 (017)	
326	PRIOR		653 (463)			106 (308)	206 (104)	•
	POST		078 (118)			766 (364)	156 (261)	
292	PRIOR		493 (500)			193 (500)	013 (115)	•
	POST		939 (090)			059 (089)	002 (013)	
154	PRIOR		576 (494)			028 (165)	396 (489)	•
	POST		000 (002)			002 (010)	998 (010)	
259	PRIOR		139 (399)			098 (294)	.706 (456)	•
	POST		027 (028)			029 (064)	914 (069)	
291	PRIOR		171 (379)			079 (270)	.717 (435)	•
	POST		024 (028)			028 (066)	948 (070)	
319	PRIOR		074 (262)			071 (256)	855 (352)	•
	POST		000 (002)			001 (006)	999 (007)	
321	PRIOR		074 (262)			071 (256)	855 (352)	53
	POST		000 (002)			001 (007)	999 (007)	
327	PRIOR		144 (351)			039 (194)	817 (386)	•
	POST		049 (012)			011 (034)	940 (030)	
325	PRIOR		250 (433)			672 (469)	078 (268)	•
	POST		172 (064)			044 (203)	.714 (173)	
333	PRIOR		250 (433)			672 (469)	078 (268)	•
	POST		003 (012)			010 (049)	988 (051)	
312	PRIOR		074 (262)			071 (256)	855 (352)	•
	POST		001 (013)			003 (017)	993 (023)	
336	PRIOR		231 (421)			032 (289)	677 (467)	54
	POST		000 (005)			003 (015)	996 (015)	
310	PRIOR		517 (500)			275 (447)	208 (406)	•
	POST		000 (001)			002 (012)	998 (012)	
305	PRIOR		222 (415)			021 (142)	.758 (428)	•
	POST		000 (005)			001 (009)	998 (010)	
255	PRIOR		112 (492)			125 (330)	463 (499)	•
	POST		620 (090)			018 (041)	362 (099)	
119	PRIOR		412 (492)			125 (330)	463 (499)	•
	POST		966 (069)			034 (069)	000 (004)	
312	PRIOR		222 (415)			024 (142)	.758 (428)	•
	POST		010 (023)			024 (069)	966 (074)	
139	PRIOR		213 (429)			085 (279)	.672 (469)	•
	POST		235 (070)			038 (117)	677 (110)	
168	PRIOR		454 (498)			122 (328)	423 (494)	•
	POST		194 (197)			719 (318)	087 (145)	

Samples Outside Structures (continued)

Float Number		HEARTH Mean (SD)	MIDDEN Mean (SD)	STORAGE INSIDE Mean (SD)	LIVING QUARTERS Mean (SD)	STORAGE OUTSIDE Mean (SD)	PATIO ACTIVITY CENTER Mean (SD)	PATIO AREA
161	PRIOR		447 (497)			107 (309)	447 (497)	55
	POST		068 (042)			029 (072)	904 (074)	
148	PRIOR		165 (371)			028 (166)	806 (395)	•
	POST		000 (003)			001 (008)	999 (009)	
152	PRIOR		233 (423)			071 (256)	697 (460)	•
	POST		000 (002)			001 (007)	999 (008)	
166	PRIOR		585 (493)			220 (414)	195 (396)	56
	POST		143 (041)			018 (049)	838 (055)	
186	PRIOR		508 (500)			413 (492)	078 (269)	•
	POST		212 (096)			240 (319)	548 (252)	
148	PRIOR		636 (481)			351 (477)	013 (112)	•
	POST		004 (014)			100 (072)	896 (063)	
282	PRIOR		699 (459)			277 (447)	024 (151)	•
	POST		050 (096)			923 (170)	027 (098)	
147	PRIOR		318 (466)			080 (271)	603 (489)	•
	POST		001 (006)			002 (011)	997 (013)	
145	PRIOR		379 (485)			550 (497)	071 (257)	•
	POST		101 (062)			165 (203)	734 (173)	
281	PRIOR		689 (463)			299 (458)	012 (110)	•
	POST		025 (065)			911 (228)	061 (171)	
286	PRIOR		508 (500)			413 (492)	078 (269)	•
	POST		677 (113)			072 (116)	250 (146)	
287	PRIOR		318 (460)			080 (271)	603 (489)	•
	POST		1 000 (000)			000 (000)	000 (000)	
330	PRIOR		318 (466)			080 (271)	603 (489)	•
	POST		985 (042)			003 (015)	011 (039)	
339	PRIOR		111 (314)				889 (314)	57
	POST		468 (075)				532 (075)	
324	PRIOR		111 (314)				889 (314)	•
	POST		051 (009)				949 (009)	
303	PRIOR		378 (485)			072 (258)	550 (497)	•
	POST		002 (011)			611 (317)	368 (315)	
322	PRIOR		111 (314)				889 (314)	•
	POST		267 (057)				733 (057)	
280	PRIOR		388 (487)			114 (318)	498 (500)	58
	POST		007 (036)			976 (099)	016 (071)	
329	PRIOR		474 (499)			434 (496)	091 (288)	•
	POST		193 (059)			049 (115)	758 (112)	
311	PRIOR		069 (253)			057 (233)	874 (332)	•
	POST		000 (003)			001 (007)	999 (007)	
332	PRIOR		540 (498)			105 (491)	055 (228)	•
	POST		000 (004)			001 (013)	998 (014)	
274	PRIOR		107 (491)			107 (491)	187 (390)	•
	POST		000 (001)			001 (010)	999 (010)	
277	PRIOR		107 (491)			107 (491)	187 (390)	•
	POST		063 (031)			008 (030)	929 (041)	

Table 8. *Prior and Posterior Means and Standard Deviations by Flotation Number and Use*

7. CONCLUSION

The most significant aspects of our study are (1) the new method by which paleoethnobotany can shed light on prehistoric behaviors, (2) the usefulness of botanical data to prehistoric interpretation generally, and (3) an application of Bayesian statistics.

Culturally, the botanical remains help us interpret the structures more explicitly. Each structure of Patio J7 = 2 now reveals a new pattern of use activity that had not been identified before our analysis. Figures 3 and 4 have presented the priors and posteriors for each flotation sample respectively. Glancing between these two figures one can begin to see the amount of information the botanical remains contribute to the priors.

In general, our statistical exercise shifted the posterior of some proveniences systematically toward one activity. Specifically, samples that produced few seeds tended to be identified with more certainty as either living quarters (inside structures) or patios-activity areas (outside). In a cultural interpretation that shift markes intuitive sense. Both in and outside the structures, the posterior pattern is quite strongly changed from the priors. The reinterpretation is notable in the proveniences against the walls in the open patio, with a dominance in patio use over midden or storage. Within the structures there is a shift toward living quarters with less storage. In the structures, the posteriors shifted the storage areas into more discrete locations, seen in five of the six structures. All six structures increased their probabilities for being living quaters with a bit of storage.

Both the priors and the posteriors define the patio use areas less clearly than the structures. The patio areas were probably used for many more amorphous and diverse activities. These activities were probably not as constrained to specific locations as those within the structures. Overall each location was used for many more activities, with only limited midden areas and minor evidence of storage than was supposed in the priors. Storage in the patio occupied two corners, areas 56 and 58. Patio midden was only along the north wall of the patio.

Both inside and outside the structures, the botanical data provided more exact information about the proveniences than we had from the field notes. Doubtless, more precision in the priors is a critcal exercise, but it is clear nevertheless that the botanical data have aided our interpretation of prehistoric activities in this specific example and are likely to do so in other investigations.

Interpreting the posteriors generated by Bayesian analysis, structure 1 was used for living quarters but shows some midden deposit near the entrance. This is supported by a disturbed human burial toward the front of structure. Structure 2 data show storage use in the center, with some daily living activities evident, supported by a hearth. Structure 3 was used mainly for living activities with some storage in the eastern half. Structure 4 was used predominantly for living activities, with a bit of storage in the western part near the entrance. Structure 5 was used for both storage and living quarters. Structure 6, like 3 and 5, also had living quarters, a hearth (not illustrated), and some storage areas. Evidently only structure 1 was used predominantly for living quarters throughout its life history, without evidence of storage. As mentioned above, this structure has additional artifactual data suggesting that it was abandoned while the patio was still occupied. This supports the lack of storage in the structure, and also its use as a garbage dump. Over all, the priors and posteriors reflect multiple usage in all structures, either sequentially or simultaneously.

In general, our analysis supports the usefulness of botanical remains in the interpretation of prehistoric dwelling patterns by a method that has not previously been tested or even applied to archaeological research. This type of analysis offers great potential for the refinement of in-field data collection and location description.

This paper suggests an approach to archaeological data interpretation not often taken. The ramifications may be several. First, our work may encourage other archaeologists to be more explicit and precise about operating assumptions in their data analysis and intepretations. Second, it may influence the way archaeological data will be analyzed. Third, it may influence the way archaeological field notes are recorded. The review of the field dig notes revealed that often they were not as informative as one might have hoped. Perhaps in a future excavation, the archaeologists will record their probabilities by provenience as part of their field recordings.

The posteriors suggest that in many proveniences the botanical data are strong enough to affect radically the archaeologist's opinion of the activities conducted there. The results of this paper thus affirm both the usefulness of botanical evidence in archaeology and the usefulness of Bayesian methods to analyze such data.

We look forward to future successful collaboration between Bayesian statisticians and paleoethnobotanists.

ACKNOWLEDGEMENTS

The authors are grateful to the Center for Advanced Study in the Behavioral Sciences which provided the opportunity for this work. The staff gave us a lot of help along the way, especially Deanna Knickerbocker and Kathleen Much. Jill Larkin provided valuable computing advice at critical points. Our research was supported in part by the office of Naval Research under Contract N00014–85-K–0539 (JBK) and by the National Science Foundation under Grants DMS–850319 (JBK), BNS 82-03723 (CAH), and BNS 84–11738 (both).

REFERENCES

Bayarri, M. J., DeGroot, M. H. and Kadane, J. B. (1987). What is the Likelihood Function?. *Statistical Decision Theory and Related Topics.* (S. S. Gupta, and J. O Berger, eds.). New York: Springer (to appear).

D'Altroy, T. N. (1981). *Empire Growth and Consolidation: The Xauxa Region of Perú under the Incas.* Ann Arbor: University Microfilms: .

D'Altroy, T. N. (1987). Transitions to power: Centralization of Wanka political organization under Inc rule, *Ethnohistory* **34**(1), 78–102.

Earle, T., D'Altroy, T., Hastorf, C., Scott, C., Costin, C., Russell, G. and Sandefur, E. (1987). *The Effect of Inca Conquest on the Wanka Domestic Economy.* Los Angeles: UCLA Institute of Archaeology.

Hastorf, C. A. (1983). *Prehistoric Agricultural Intensification and Political Development in the Jauj Region of Perú.* Ann Arbor: University Microfilms.

Hastorf, C. A. (1988). The study of paleoethnobotanical data in prehistoric crop production, processing and consumption. (Hastorf and Popper, eds.), *Current Paleoethnobotany*, (to appear), University o Chicago Press.

LeBlanc, C. J. (1981). *Late Prehispanic Settlement Patterns in the Yanamarca Valley, Perú.* Ann Arbor University Microfilms .

Lindley, D. (1985). *Making Decisions*, 2nd. Ed., New York: Wiley.

Smith, A. F. M., Skene, A. M., Shaw, J. E. H., Naylor, J. C. and Dransfield, M. (1985). The Implementation of the Bayesian Paradigm. *Commun. Statist. Theory and Meth.* **14**(5), 1079–1102.

Tierney, L. and Kadane, J. B. (1986). Accurate Approximations for Posterior Moments and Margina Densities *J. Amer. Statist. Assoc.* **81**, 82–86.

Watson, P. J. (1976). In pursuit of prehistoric subsistence: A comparative account of some contemporary flotation techniques. *Midcontinental Journal of Archaeology* **1**, 77–100.

DISCUSSION

I. POLI (*Universita di Bologna*)

This paper is to be welcomed for providing a clear illustration of the Bayesian predictive approach to a special topic such as applied archaeological research. It is original in proposing new methodology which performs well, and in investigating a type of data (burnt botanical remains) rarely considered in this area of research.

The main question that the authors consider deals with the possible activities, that might have been carried out by the Wanka people at a particular site up to 1460, to infer, subsequently, something about their economic and political system. Such problems in quantitative archaeology are often described by models that suppose a preferential distribution of archaeological items assuming that certain activities took place. Literature on this topic is mainly concerned with taxonomy procedures, factor and principal components analysis and sometimes with spatial point or lattice processes, and is mostly related to distributional patterns of artifacts (H. J. Hietala 1984, and C. R. Orton 1982).

The research described here is developed from a Bayesian point of view. The archaeologist with her own beliefs, derived from field notes, is asked to quantify such beliefs in the form of prior probabilities for a set of excavated sites. The likelihood function is thus defined, noting however that the likelihood is to be for activities occurring in 1460 while the data refers to burnt botanical remains found in 1982. A time dimension is therefore introduced into the model and subjective elements enter into both the definition of activities with respect to the plants involved and in the assessment of the survival probability of each plant. In general we can then see that the posterior probabilities of activities in each provenience show the relevance of plant remains in studying prehistorical activity patterns and the adequacy of the Bayes procedures in investigating such special areas of applied research. However it should be noticed that comments on the results are developed on aggregates of locations, namely structures and patio areas, which apparently have not been considered in the development of the research. In fact, the analysis has been conducted with respect to each single provenience, spread over all the area of interest, with no consideration of their location. Inside each provenience an hypothesis of space independence is assumed (e.g. the Poisson model) but inside the patio no hypothesis on the space distribution of data is considered. This could have undesirable consequencies.

In fact, the authors mention the serious problems which arise in evaluating posterior distributions of activities because the observation of botanical remains is sometimes in conflict with the definition of activity areas given by the archaeologist. These problems are related both to the number of remains (e.g. Chenopodia seeds, or the mixture assumed for the hearth activity) and to the type of remains (e.g. seeds extraneous to the activity area defined). Finding remains on a provenience could, of course, be the direct result of the location of human activities, but there could also be the effects of wind and water disturbance, differential erosion or simple reorganization of the sites. Actually, the authors seem to tackle this problem by assigning zero prior probability to those activities which they regard as most uncertain in a specific archaeological site. In this way, however, they prevent the data from ever influencing prior beliefs, thus denying any evidential value for the remains found in the site. This represents a violation of the well-known "Cromwell's Rule" and seems not to be a satisfactory answer to the problem. I wonder, instead, whether the archaeological mechanism which governs the preferential distributions of botanical remains in activity areas might not be more adequately described by a random process that accounts explicitly for the space dimesion of each provenience. The whole patio has to be regarded, therefore, as a random field in which the structures and the patio areas (see Figure 3) that contains proveniences, enter into the model analysis. Specifically, structure and patio areas are clusters of proveniences that seem defined by archaeological remains such as the wall around the areas or specific patterns of artifacts which receive no mention in the paper. The spatial distribution of the proveniences could then be analysed by first identifying some form of nearest neighbour structure characterizing them; afterwards, the distribution of plant remains, given the locations of sites, can be considered. In this way, we could derive a likelihood function for the problem which accounts both for the time and the spatial dimensions of data and respects Cromwell's Rule by assigning strictly positive probability to each activity. This seems to me a more flexible and useful way of learning from the data, taking into due account all of its special features.

D. A. BERRY (*University of Minnesota*)

This is a terrific application of Bayesian ideas and was a high point of the conference for me. There is, however, one small aspect of the approach that I think could be improved. An explanation being considered as part of a universe of models should seldom (if ever!) be assigned zero probability. I would have placed a positive lower bound on all prior probabilities (people are notoriously bad at assessing small probabilities). It does no harm to carry along improbable explanations whose likelihoods also turn out to be small. On the other hand, I would have been interested to see how many of these "impossible" events became the most likely of all a posteriori!

W. POLASEK (*University of Basel*)

I think the paper is a good example demonstrating the subjective nature of likelihoods and priors. Instead of listing long tables with prior-posterior probabilities, I would recommend some graphical summaries in order to facilitate the reporting process.

REPLY TO THE DISCUSSION

We welcome the comments of Poli, Berry, Polasek, and have no fundamental disagreements with any of them.

To Professor Berry:

Berry proposes that a small positive lower on prior probability be routinely used in connection with each of the models. This is an interesting suggestion. However, we doubt that it would have helped with our problem proveniences. With respect to provenience 284, the prior did put positive probability on hearth, indoor storage, and living quarters. The only conceivable addition would have been midden. But adding midden would not have saved us from the embarrassment of a truly huge number of Chenopodium seeds, far more than were to be expected under any prior. Our second kind of problem provenience was number 299, where we observed an impossible result: Scirpus seeds in a hearth. Had Berry's suggestion been followed in this instance, and we had put positive probability on storage and living quarters, the computed posterior would have eliminated hearths as a possibility. Since Hastorf is still sure that provenience 299 was a hearth, this would have led the computed posteriors to be far from the believed posterior, and would not have given the warning we got when we discovered our violation of Cromwell's rule. Perhaps extending Berry's suggestion to the likelihood as well would be a good idea. This would help with provenience 299, but not with 284.

To Professor Polasek:

We agree entirely with Polasek about the usefulness of graphical methods for displaying priors and posteriors. Figures 3 and 4 were added to the paper after the Valencia meeting to address this concern. A second method of graphical display for these priors and posteriors is given in Larkin (1989).

To Professor Poli:

Poli suggests the use of random field models to take better account of the spatial aspect of the problem. We think this would be a promissing direction for future research. One aspect that would have to be considered is elicitation, both of priors and of likelihoods.

We believe that our assumptions about preservations are the strongest and most questionable. We regard with skepticism the idea that the events of each of two burnt botanical remains from the same provenience surviving for five hundred years are independent events with a probability known to us. This would be our first priority to relax in further work on this problem.

REFERENCES IN THE DISCUSSION

Hietala, H. J. (1984). *Intersite Spatial Analysis in Archaeology.* Cambridge: University Press.

Larkin, J. (1989). Display based problem solving, (D. Klahr and K. Kotovsky eds.), *Complex Information Processing: Essays in Honor of Herbert A. Simon* Hillsdale, N. J.: Lawrence Erlbaum Associates, in press.

Orton, C. R. (1982). Stochastic Processes and Archaeological Mechanism in Spatial Analysis. *Journal of Archaeological Science* **9**, 1–23.

BAYESIAN STATISTICS 3, pp. 261–278
J. M. Bernardo, M. H. DeGroot, D. V. Lindley and A. F. M. Smith, (Eds.)

Asymptotics in Bayesian Computation

R. E. KASS, L. TIERNEY* and J. B. KADANE
Carnegie-Mellon University and *University of Minnesota**

SUMMARY

We survey approximate Bayesian methods, indicating their applicability in many inferential settings, and placing them in a context of interactive modelling, criticism, and inference. We briefly illustrate some of the approximations in two examples, indicating the benefit accrued by their rapid computation. We also discuss implementation issues, mentioning relevant advances in computing environments and an important remaining difficulty that will have to be faced in the future.

Keywords: APPROXIMATIONS; POSTERIOR EXPANSIONS; LaPLACE'S METHOD; INTERACTIVE BAYESIAN ANALYSIS.

1. INTRODUCTION

The Bayesian paradigm not only provides a philosophically sound system for articulating knowledge and making decisions, it also offers a flexible and conceptually simple framework in which to formulate problems of drawing inferences from data using parametric models. To take advantage of this flexibility and simplicity, however, obstacles of implementation must be overcome. Sometimes these require development of novel algorithmic and numerical techniques, but even when a solution seems available "in principle", there remain the deterrents associated with production, use, and running time of computer programs. For the data analyst or researcher who might employ or study Bayesian techniques, difficulties in obtaining features of posterior and predictive distributions are too often judged to overwhelm the scientific benefits they might provide (e.g., Efron 1986).

A particular task may be considered excessively time-consuming because the programming would be demanding. Then, when some relevant sofware exists, so that programming would be minimal, it may still happen that substantial time and effort is required on the part of a user in obtaining the desired output. Both of these involve the user's *working time*. Once a program is available and relatively easy to use, it may still take a long time to run. This increases the user's *waiting time*. In each case, the judgement of what constitutes a lot of time depends strongly on the context in which the particular task is to be performed. Some tasks need only be performed once, and their products may be worth waiting for, while others are useful primarily when performed repetitively.

It is worth distinguishing, roughly, between tasks that are to be accomplished interactively, and those that may be done in batch. We use these terms to describe not the running of a program, but the activity of a user. We consider an analysis to be performed in *interactive* mode when, for each task, *both* the working time and the waiting time are small. In data analysis sessions involving multiple tasks, a delay of more than a few seconds in each task may quickly become intolerable. We contrast this interactive mode of analysis with the batch mode, in which a total time delay of anywhere from several seconds to many hours (sometimes even days) might be acceptable. With this definition, in the present paper we focus primarily on some general tools for interactive Bayesian analysis.

We recognize that our dichotomy is not a clean one: it frequently happens that an analyst is willing to punctuate several interactive sessions, in our sense, with long waiting times (which may serve as "coffee breaks" or, for those who are good planners, may allow the user to swap some time on a different piece of work). In this case, techniques that produce such waiting times become acceptable, as long as they are not used too frequently. Nonetheless, it is useful for us to discuss the value of certain computational methods purely in the interactive context. We emphasize that the mode defined as "interactive" or "batch" in some operating system is not relevant here; in our terminology we refer to the user.

To elaborate further on this notion of interactive analysis, we briefly consider a hypothetical analysis of data from the Stanford Heart Transplant (SHT) study, which involved 82 patients and was described by Turnbull, Brown, and Hu (1974) and reanalyzed to illustrate Bayesian computational methods by Naylor and Smith (1982), Tierney and Kadane (1986), and Tierney, Kass, and Kadane (1988a). Considerable effort might go into the selection of a model for these data, which would be widely agreed to constitute the most important part of the data analysis project. Turnbull, Brown and Hu used three different parametric families of survival distributions in their article. It would be good to have tools that could help an analyst understand a model: the ability to rapidly plot the survival densities and display survival probabilities implied by a model makes a good start. After that, similar plots and displays for prior and predictive probabilities would be helpful. Then, the inferential consequences of a particular model and prior, and diagnostic checks on the choices, for the data at hand should be readily available. A user should be able to go back and forth between these selection, estimation, and criticism steps.

The development of the tools needed for understanding models and priors, as well as for obtaining inferential consequences and assessing possible departures from assumptions, encompasses a substantial portion of Bayesian statistics. We focus here on a few aspects of the latter part of the interactive process we have just described.

In the next section we survey results on second-order asymptotics, which provide one set of useful tools for inference and sensitivity assessment. We briefly illustrate their use in section 3. Our framework is that of making inferences about one or more real-valued functions g of an m-dimensional parameter vector θ. This is a fairly general conception of an inference problem. It includes, for example, the problem of making inferences about parameters, and also that of making inferences about probabilities of the form $P\{Y > c|\theta\}$, such as survival probabilities, the expectations of which become predictive probabilities.

In section 4 we then take up some issues of implementation for interactive analysis. We note the large savings in working time gained by an analyst working in a high-level environment, and for a user of special-purpose programs we describe the advantages of being able to call them from within the high-level environment. We also indicate some shortcomings of presently available implementation schemes, and the difficulties that must be faced in improving the situation. We close, in section 5, by tying our current efforts together with others, and returning to the goal of developing methods that will make fully interactive Bayesian analysis possible in a wide variety of settings.

2. ASYMPTOTICS

In this section we describe approximate Bayesian methods based on Laplace's method for asymptotic expansion of integrals (Laplace, 1847, Erdelyi, 1956). Assuming h is a smooth function of an m-dimensional parameter θ with $-h$ having a maximum at $\hat{\theta}$, Laplace's method approximates an integral of the form

$$I = \int f(\theta) \exp(-nh(\theta)) d\theta \tag{2.1}$$

by expanding h and f about $\hat{\theta}$. The factor $\exp(-nh(\theta))$ in the integrand is approximated by a function proportional to a Normal density determined by the second-order Taylor series approximation to h; when integrated against this Normal density, the order $O(n^{-1/2})$ terms of the expansions of f and h, which are odd functions of $\theta - \hat{\theta}$, vanish and the integral satisfies

$$I = \hat{I}\{1 + 0(n^{-1})\}, \tag{2.2a}$$

where

$$\hat{I} = f(\hat{\theta})(2\pi/n)^{m/2} \det(\Sigma)^{1/2} \exp(-nh(\hat{\theta})) \tag{2.2b}$$

and $\Sigma^{-1} = D^2h[\hat{\theta}]$ (the Hessian of h at $\hat{\theta}$). Higher-order approximations may be derived by retaining higher-order terms in the expansions of f and h. Closely related is the saddlepoint approximation, which applies Laplace's method in the complex domain. Reid (1988) reviews its (non-Bayesian) statistical applications.

Although approximation (2.2) is extremely simple, there are interesting facts associated with its application which may not be apparent from cursory inspection. In the following subsections we briefly summarize results that have been obtained by various authors.

2.1. Marginal densities

Expansion (2.2) may be applied immediately to the approximation of marginal posterior densities. Suppose $\Theta = \Theta_1 \times \Theta_2$ and we are interested in the marginal density on θ_1,

$$p(\theta_1|y) = c \cdot \int \exp[\tilde{\ell}(\theta_1, \theta_2)]d\theta_2$$

where c is the normalizing constant for the joint posterior and $\tilde{\ell} = \ell + \log \pi$ with ℓ being the loglikelihood function based on data y and π being the prior. Taking $h = h_n = n^{-1}\tilde{\ell}$ and $f = 1$ in (2.1), from (2.2) we obtain

$$\hat{p}(\theta_1|y) = c^* \det(\Sigma(\theta_1))^{1/2} \exp(\tilde{\ell}(\theta_1, \hat{\theta}_2(\theta_1))) \tag{2.3}$$

where c^* is a new normalizing constant, $\hat{\theta}_2(\theta_1)$ is the maximun of $\tilde{\ell}(\theta_1, \cdot)$, and $\Sigma(\theta_1)$ is inverse of the Hessian of $-\tilde{\ell}(\theta_1, \cdot)$ at $\hat{\theta}_2(\theta_1)$. This approximation was suggested by Leonard (1982) and Phillips (1983) and as, Tierney and Kadane (1986) noted, after the normalization constant is obtained, its relative error is of order $O(n^{-3/2})$ on neighborhoods about the mode that shrink at the rate $n^{-1/2}$. This may be compared to the approximating Normal density, which has error of order $O(n^{-1/2})$ on these shrinking neighborhoods. Perhaps more importantly, the approximation is of order $O(n^{-1})$ uniformly on any bounded neighborhood of the mode. This means that it should provide a good approximation in the tails of the distribution (and our experience shows that it often does).

The normalizing constant c may also be approximated with error of order $O(n^{-1})$ using (2.2) (as in Tierney and Kadane, 1986). This is of theoretical interest, but in practice one may as well ignore the joint normalizing constant, and then renormalize the marginal approximation after it is evaluated at suitably many points θ_1. We note that in both the Normal-Gamma and Dirichlet cases the marginal approximations, when renormalized to integrate to one, are exact. Also, any constant-order factor (such as the log prior) may be omitted from the exponent before applying Laplace's method.That is, the factor may be moved from part of h to part of f. The resulting approximation uses a different maximum $\hat{\theta}$ (such as the MLE), but it remains accurate to the same order (compare the discussion following (2.6)). Finally, we add that Box and Tiao (1973) used approximations similar to (2.3), as did Jeffreys (1967) in his treatment of odds factors.

It is often convenient to have an expression for the marginal density of a general function $g = g(\theta)$. Tierney, Kass and Kadane (1988b) note that, assuming its gradient is nonzero, g may be considered a parameter in some new (local) parameterization of the model. Writing $\gamma = g(\theta)$ for the argument of the density, we then obtain an approximation to the marginal density of g, having the properties of (2.3),

$$\hat{p}(\gamma|y) = c^*[\det(\Sigma)/((Dg)^T\Sigma(Dg))]^{1/2}\exp(\tilde{\ell}(\hat{\theta}(\gamma)))$$

where c^* is a new normalizing constant and $\hat{\theta}(\gamma)$ is the maximum of $\tilde{\ell}$ constrained by $g(\theta) = \gamma$ and Σ is the inverse of the Hessian of $-\tilde{\ell}$ at $\hat{\theta}(\gamma)$.

2.2. Second-order aproximations for expectations: Standard form

If the posterior density p on Θ is proportional to $\exp(-nh(\theta))$, the expectation

$$E(g(\theta)) = \frac{\int g(\theta)\exp(-nh(\theta))d\theta}{\int \exp(-nh(\theta))d\theta} \tag{2.4}$$

may be approximated by applying Laplace's method to both the numerator and denominator (with f respectively equal to g and 1) to yield the *first-order* expansion

$$E(g(\theta)) = g(\hat{\theta})\{1 + 0(n^{-1})\}. \tag{2.5}$$

In (2.4) note that by redefining h so that the probability density p becomes proportional to $b(\theta)\exp(-nh(\theta))$, the expectation may be written instead as

$$E(g(\theta)) = \frac{\int g(\theta)b(\theta)\exp(-nh(\theta))d\theta}{\int b(\theta)\exp(-nh(\theta))d\theta} \tag{2.6}$$

Thus, Laplace's method may be applied in any one of infinitely many diferent forms. When (2.2) is applied to the numerator and denominator of (2.6) with f respectively equal to gb and b, so that h remains the same in both the numerator and denominator, another first-order approximation $g(\hat{\theta})$ results as in (2.5) with $\hat{\theta}$ becoming the maximum of the newly-defined function $-h$.

To obtain a second-order approximation, the expansions of f and h in (2.1) must be carried out to include the order $O(n^{-1})$ and $O(n^{-3/2})$ terms with the latter dropping out on integration. The resulting formula can then be applied to the numerator and denominator of (2.6), and after simplification there is

$$\begin{aligned} E(g(\theta)) = g(\hat{\theta}) + n^{-1}\cdot\Sigma\{h^{ij}g_j[b(\hat{\theta})^{-1}b_j - (1/2)\sum_{r,s} h^{rs}h_{rsj}] \\ + (1/2)h^{ij}g_{ij}\} + O(n^{-2}) \end{aligned} \tag{2.7}$$

where subscripts indicate partial derivatives evaluated at $\hat{\theta}$ and h^{ij} is the (i,j)-component of the inverse of the Hessian matrix $(h_{ij}) = D^2h[\hat{\theta}]$. This expansion was used by Mosteller and Wallace (1964). When $p(\theta) = p(\theta|y)$ is a posterior probability proportional to $L(\theta)\pi(\theta)$, the product of likelihood function and a prior density, the exponent becomes $\ell(\theta)+\log(\ell(\theta)/b(\theta))$, where $\ell = \log L$ is the loglikelihood function. Here, n refers to the sample size, ℓ is a different function for each n, and $h(\theta) = n^{-1}[\ell(\theta) + \log(\pi(\theta)/b(\theta))]$. Choosing b equal to π and 1 lead respectively to the MLE and the posterior mode as values of $\hat{\theta}$. Lindley (1961, 1980) also used (2.7) with b equal to 1, and Johnson (1970) used it with b equal to π. A related expansion was used by Hartigan (1965).

There is also the variance approximation

$$V(g(\theta)) = (Dg[\hat{\theta}])^T (nD^2 h[\hat{\theta}])^{-1} (Dg[\hat{\theta}])\{1 + 0(n^{-1})\} \tag{2.8}$$

which may be obtained from (2.7). Note that when b is equal to π or 1, respectively, $nD^2h[\hat{\theta}]$ is the observed information matrix or the Hessian of the negative log posterior density at the posterior mode.

2.3. Fully exponential expansions for positive functions

Now suppose $g(\theta)$ is positive. If the numerator integrand in (2.4) is written in the form $\exp(-nh^*(\theta))$, where $-nh^*(\theta) = -nh(\theta) + \log g(\theta)$, and then (2.2) is used with $f(\theta) = 1$ in both numerator and denominator, there results the *second-order* expansion

$$E(g(\theta)) = \frac{\det(\Sigma^*)^{1/2} \exp(-nh^*(\theta^*))}{\det(\Sigma)^{1/2} \exp(-nh(\hat{\theta}))} \{1 + 0(n^{-2})\} \tag{2.9}$$

where θ^* is the maximun of $-h^*$ and $\Sigma = (D^2h[\hat{\theta}])^{-1}$ and $\Sigma^* = (D^2h^*[\theta^*])^{-1}$. That is, by applying Laplace's method in this alternative form a remarkable gain in accuracy is obtained. This was pointed out by Tierney and Kadane (1986).

We will call the form of approximation given by (2.9) *fully exponential,* and that given by (2.7) *standard.* The fully exponential form converts the evaluation of $m(m+1)(m+2)/6$ third derivatives (and some attendant arithmetic calculations) to a second maximization (for which a single pseudo-Newton step suffices) and an evaluation of two Hessian determinants. There is thus a modest improvement in efficiency (which increases with m) and a substantial improvement in working time when the third derivatives must be derived and correctly coded without a symbol manipulator. An additional advantage of the fully exponential method is that it yieldds a second-order approximation to the variance of a positive function,

$$V(g(\theta)) = \{\hat{E}(g(\theta)^2) - \hat{E}(g(\theta))^2\}\{1 + 0(n^{-2})\} \tag{2.10}$$

where $\hat{E}$ is the expectation approximation given by (2.9). This is an improvement on the first-order approximation (2.8). To obtain a second-order approximation form the standard form, fourth and fifth derivatives of the loglikelihood would be required.

When many different functions g are evaluated, as in Mosteller and Wallace (1964), Tsutakawa (1985), or Kass and Steffey (1986), the fully exponential method loses any computational advantage it might have for single function g. It remains very easy to code, however.

2.4. Fully exponential expansions for general functions

The fully exponential method described in the previous section applies only to positive functions g, which means that the approximations (2.9) will be useful only when the probability distribution for $g(\theta)$ is concentrated away from the origin. A simple way of obtaining the expectation of a non-positive function $g(\theta)$ is to first approximate the moment generating function $M(s) = E(\exp(sg(\theta)))$ by $\hat{M}(s)$ according to (2.9), which is applicable since $\exp(sg(\theta))$ is positive. Then the derivative of the approximation at zero $\hat{M}'(0)$ furnishes an approximation to $E(g(\theta))$. An approximation to the variance is given by $\hat{M}''(0) - \hat{M}(0)^2$. Each of these may be shown to have multiplicative error of order $O(n^{-2})$. In practice, the derivatives are computed by finite differencing.

These approximations are reexpressed and discussed by Tierney, Kass, and Kadane (1988a). Three further points made there should be mentioned here. First, the moment generating function approximation may be shown to be equivalent to adding a very large constant to g, performing the fully exponential approximation (2.9) and then subtractiong the constant from the result. Second, it turns out that the approximation $\hat{M}'(0)$ is mathematically equivalent to (2.7) when $b = 1$. Notice, however, that the moment generating function approximation is computed differently, and retains the advantages of the fully exponential method mentioned in the previous section. Third, the accuracy of any of these approximations depends on the parameterization in which it is applied.

2.5. *Influence and sensitivity*

Assessment of sensitivity of inferences to either a change in prior or the deletion of an observation is amenable to asymptotic approximation. The two alterations are related statistically in that they involve complementary components of the Bayesian specification, and they are also related mathematically, in that both involve contributions to the log posterior density that are of order $O(1)$, while the loglikelihood based on the full sample is of order $O(n)$. Asymptotic methods may be applied to both problems in analogous ways.

For a general g, either the posterior expectation or marginal mode may be approximated with error of order $O(n^{-2})$, and from this may be computed the change in either value when the prior is changed or an observation is deleted. These changes may be computed rapidly and are discussed by Kass, Tierney, and Kadane (1988). We note that when g is a probability $P\{Y > c|\theta\}$, the change in expectation becomes a change in predictive probability.

In the case of the posterior expectation of a positive function g there are two methods available. The first is to begin with some prior and the full sample, compute a second-order approximation to the expectation, and then introduce the alteration and compute a second-order approximation to the new expectation. The second method is to take $\hat{\theta}$ and θ^* as defined in (2.9) and $\hat{E}$ to be any second-order approximation to the expectation (or the exact posterior expectation) and use

$$\Delta = ((b(\theta^*)/b(\hat{\theta})) - 1) \cdot \hat{E}$$

where, in the case of deletion of an observation, $b(\theta) = 1/L_i(\theta)$ is the reciprocal of the contribution to the likelihood for the i-th observation. Here we are assuming the observations are independently distributed given θ. In the case of a change in prior b becomes the ratio of the new prior to the old prior. This second method is even faster than the first because it does not involve any evaluations of the full loglikelihood or its derivatives. In the case of a general function g, there are analogous methods based on the moment generating function approximation.

We note, in passing, that an approximate marginal mode and an approximate marginal mean, each having error of order $O(n^{-2})$, may be combined to produce an approximate Pearson skewness (mean-mode)/(standard deviation) having error of order $O(n^{-1})$. This can be useful as a fast diagnostic of marginal posterior non-Normality. See Kass, Tierney, and Kadane (1988).

2.6. *Regularity conditions*

For Laplace's method to be applicable, the function $-nh$ in the exponent must have a dominant peak and both h and f must have Taylor series expansions of the relevant order. Kass, Tierney, and Kadane (1987) formalize these notions using three conditions on a sequence of loglikelihood functions, together with simple requirements on the rest of the integrand. These ensure the validity of expansions based on Laplace's method. In that paper, conditions are also given to ensure that the sequence of loglikelihood functions is, in the appropriate sense, well-behaved with probability one.

2.7. *Further comments*

2.7.1. *Applications to hierarchical models*

Kass and Steffey (1986), following earlier work of Deely and Lindley (1981), discuss approximate Bayesian methods in what they call "conditionally independent hierarchical models", which are sometimes also called "parametric empirical Bayes models". Equation (2.5) immediately yields the well-known approximation of posterior means by parametric empirical Bayes estimates. In addition, equation (2.5) and (2.8) may be used to get a first-order approximation to the posterior variance, which provides a simple solution to the problem of accounting for estimation of a hyperparameter in using approximate ("empirical") Bayes estimates. Second-order approximations are discussed and exemplified by Deely and Lindley (1981), Tsutakawa (1985), and Kass and Steffey (1986).

2.7.2. *Distribution functions and percentiles*

Johnson (1970) derived asymptotic approximations for one-dimensional posterior distribution functions and percentiles. These involve expansions at MLE, but may be generalized to allow for expansions at other neighboring points, such as the mode (compare the discussion following (2.6) and (2.7)). Using the density approximation above together with a generalized version of Johnson's formulas, it may be possible to obtain useful approximations to distribution functions and percentiles for marginal posteriors. Since Johnson's formulas hold on shrinking neighborhoods of the MLE, however, we anticipate that direct use of the marginal density approximation above will be more accurate. We have not yet investigated this. (In a somewhat different but related context, Steven Skates discusses this problem in his University of Chicago Ph. D. thesis.)

2.7.3. *Non-interactive uses of asymptotics*

Finally, we note that asymptotic methods may be useful for purposes other than performing interactive analyses. First of all, they can sometimes produce intuitive interpretations (as in the variance approximations discussed by Kass and Steffey, 1986). Secondly, they may be used in conjunction with other methods: both Gauss-Hermite quadrature and importance-function monte carlo require initial centering and scaling which is usually done by first-order approximation. Second-order approximation should be an improvement. Similarly, methods for fitting hierarchical models using EM-like iterations sometimes use first-order approximations, and might be improved if second-order approximations were used instead.

3. EXAMPLES

We now describe, in the context of two examples, a few of the ways in which the asymptotic methods of the previous section may be used for interactive data analysis. Due to constraints on time and space, as well as inclination, we merely select illustrative portions of analysis for discussion, rather than presenting a thorough analysis in either example. Our purpose is mainly to indicate the kind of analysis that is currently possible.

Our results were obtained using Fortran programs written and added to the statistical package S (Becker and Chambers, 1984a,b) by one us (Tierney). The resulting extended version of S includes several commands that evaluate approximate posterior modes, means, variances, covariances, and one and two-dimensional marginal densities. A detailed description is given by Tierney, Kass, and Kadane (1987). For our purposes here, we need only note that there are two steps that a user must take in order to use these commands. First, Fortran subroutines must be written to evaluate interesting loglikelihoods, log priors, and functions g, and an S data structure must be written to specify various attributes of the problem. This would typically take a couple of hours. Once this initial "batch" job is accomplished, the user is able to do the desired data analysis entirely in interactive mode. For instance, marginal densities may be calculated and plotted with a single command and the waiting time (in the examples below) is small, i.e., several seconds on a VAX-750 with floating point hardware.

The first example has to do with a study of victimization, examined by Kadane (1985). The data are taken from the National Crime Survey and consist of interviews of households taken six months apart. In each interview a membrer of a household is asked whether anyone in the household has been the victim of a crime during the preceding six month period. The primary question of interest is whether there is any association between victimization in the two periods. A significant fraction of the interviews could not be completed, resulting in partially or completely missing information on some households. The data are given in the following table.

	2nd visit		
1st visit	Crime free	Victims	Non-response
Crime free	392	55	33
Victims	76	38	9
Non-response	31	7	115

Table 1. *Victimization results from the National Crime Survey.*

If there had been no missing data we would have had a 2×2 contingency table with, say, u_1 the probability that a household is crime-free in both periods, u_2 the probability that it is crime-free in period 1 and victimized in period 2, u_3 that it is victimized in period 1 and crime-free in period 2, and $u_4 = 1 - u_1 - u_2 - u_3$ that it is victimized in both periods. Kadane used several alterantive Dirichlet priors, which are conjugate to the multinomial likelihood and he made inferences about the odds ratio $\psi = u_1 u_4/(u_2 u_3)$.

The presence of missing data complicates the problem. As a first approach we might assume that the data are missing at random, or that the mechanism generating missing observations is ignorable. Under this assumption we can ignore the cell corresponding missing information in both periods and can view the two margins with information missing from only one interval as supplemental samples from two binomial populations with success probabilities $u_1 + u_3$ and $u_2 + u_4$, respectively. We might to be able to examine implications of this "missing at random" assumption. Kadane defined the probabilities $\alpha = P\{\text{non-response}|\text{victimized}\}$ and $\beta = P\{\text{non-response}|\text{crime-free}\}$, and assumed that α and β are the same for both periods and that response decisions are conditionally independent given the victimization status. The data may then be viewed as arising from a nine-cell multinomial distribution. If α and β are regarded as known then the likelihood can be shown to depend on α and β only through their ratio $\gamma = \alpha/\beta$. In particular, it is proportional to

$$u_1^{n_1} u_2^{n_2} u_3^{n_3} u_4^{n_4} (u_1 + \gamma u_2)^{n_{12}} (u_3 + \gamma u_4)^{n_{34}} (u_1 + \gamma u_3)^{n_{13}} (u_2 + \gamma u_4)^{n_{24}}$$

$$\cdot (u_1 + \gamma u_2 + \gamma u_3 + \gamma^2 u_4)^{n_{1234}},$$

with $n_1 = 392$, $n_2 = 55$, $n_3 = 76$, $n_4 = 38$, $n_{12} = 33$, $n_{34} = 9$, $n_{13} = 31$, $n_{24} = 8$ $n_{1234} = 115$. The missing-at-random model then corresponds to $\gamma = 1$.

In this problem we take the parameters to be u_1, u_2 and u_3, and the additional functions g are $u_4 = 1 - u_1 - u_2 - u_3$ and ψ. In our loglikelihood we allow γ as a "hyperparameter" so that its value may be assigned as part of an S command. Similarly, the hyperparameter values of the Dirichlet prior will be assigned within S. After entering our extended version of S we may thus perform a sensitivity analysis interactively, computing modes, moments, and marginal densities, and then manipulating results, e.g. by fitting a spline to a density or log density, and displaying them, e.g., as plots. For example, we may very quickly obtain the results based on any Dirichlet prior, for each of several values of γ. We then use the display facilities in S, notably plotting, rescaling, and multiple plotting to produce descriptions that we find informative.

One such display, shown in Figure 1, is the series of marginal posterior densities of ψ based on Kadane's informative prior and several choices of the parameter γ. The case $\gamma = 10$ corresponds essentially to $\gamma = \infty$ in which all missing observations are assumed to

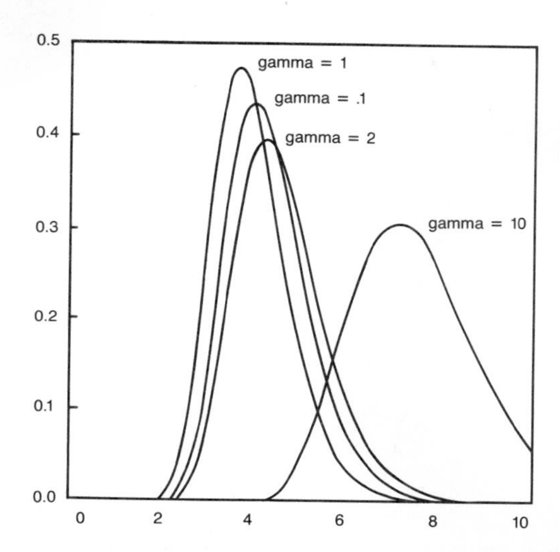

Figure 1. *Posterior density of the odds ratio*

correspond to victimizations. The case $\gamma = .1$, on the other hand, is quite close to $\gamma = 0$ in which all missing observations are assumed to be nonvictimizations. It seems plausible to us that γ should be greater than one, but not by a factor of 10; a value of two to three would seem to be more realistic. If γ were in fact in the range between 2 and 3 then this would not appear to produce conclusions that are substantially different from the conclusions under the "missing at random" assumption. A much larger value of γ would, however, shift the posterior on ψ considerably. At this point we could decide to pursue this issue further, perhaps by including α and β as parameters in our model. On the other hand, if we were primarily interested in determining whether ψ is greater than one then the results in Figure 1 would be quite conclusive: no matter what value of γ is used, all posterior densities assign essentially probability one to the event $\{\psi > 1\}$; in fact, they assign at least probability .9 to the event $\{\psi > 2\}$. (See Tierney, Kass, and Kadane, 1987, for additional details.)

As a second example we return to the SHT study and very briefly describe some analyses we performed using the Pareto model of Turnbull, Brown, and Hu (1974). The loglikelihood function on (λ, τ, p) is

$$\ell(\lambda, \tau, p) = \Sigma\{\delta_e(i) \cdot (\log(p) + \delta_t(i) \log(\tau)) \\ + p \cdot \log(\lambda) - (p + \delta_e) \log(\lambda + w_i + s_i \tau)\}$$

where $\delta_t(i)$ indicates (is one or zero) whether or not the i-th individual was in the transplant group, $\delta_e(i)$ indicates or whether not the i-th individual died, and w_i and s_i are the survival times before and after transplant.

Following Naylor and Smith (1982), Tierney and Kadane (1986), and Tierney, Kass, an Kadane (1988) we use all three parameters without integrating out p in obtaining marginal on λ and τ. We first recall the results of the analysis by Tierney and Kadane (1986). Figur 2 shows the marginal posterior on λ, which is the most skewed of the three single-paramete marginal posteriors, together with the Normal approximation based on the mode and th Hessian of the log posterior.

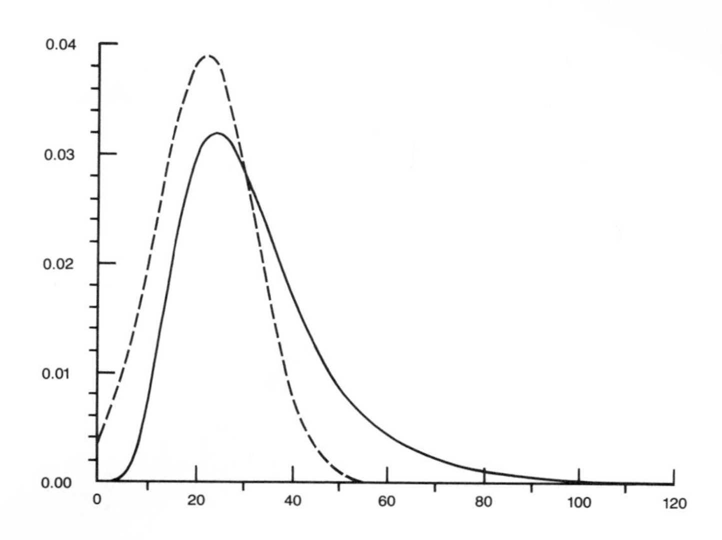

Figure 2. *Marginal posterior densities for* λ
Solid line: Laplace and 20-node adaptive Gauss-Hermite approximations.
Broken line: Asymptotic normal approximation.

To the accuracy of the figure, the appoximate density agrees with that obtained fron 20-node adaptive Gauss-Hermite quadrature, while reducing waiting time by a factor of 20 In addition, the approximate methods and those obtained by quadrature yield expectations an standard deviations that differ in all cases by less than 5%. This is an example in which th influence and sensitivity methods of section 2.5 are useful. After selecting a function g abou which we wish to assess the sensitivity of our inferences, we may proceed by recomputin the approximate marginal mode, or by using either of the two methods mentioned in sectio 2.5 to recompute the approximate mean after leaving out each successive observation. In thi example, with one of the parameters as our function g, the first method, using an S "for" function (see section 4.1) takes a little under 15 minutes on a VAX. Using a naïve applicatio of 20–node Gauss-Hermite quadrature we would expect to wait about 5 hours. (These ar rough and should not be considerd real timing comparisons since we have not tried to mak our programs efficient; intelligent use of either method, taking advantage of the parallelisn in the problem, could very greatly reduce the waiting time.) The second method of sectio 2.5 has essentially zero time. Further discussion of this method is given Kass, Tierney, an Kadane (1988).

4. IMPLEMENTATION ISSUES

4.1. Currently available facilities

The examples in the previous section showed the kind of interactive analysis that is currently available. The statistical tools were the asymptotic methods described earlier, and the computational tools were an extended version of the S system, together with some user-supplied Fortran subroutines to evaluate the loglikelihood function, the log prior, and the functions g about which inferences were to be made.

The Fortran programs that compute the approximations may be used separately, as subroutines called by a user's driver program (that is, by a Fortran program written by the user primarily to pass data to and receive results from the available subroutines). In addition, a "user-friendly" driver could have been written, which would replace the user's own driver program, thereby allowing applications to proceed more quickly. Defining a new set of functions in S serves much the same purpose, however, since the effort required of a user in applying the programs to a new problem is similar. With these functions as part of S, the user can also take advantage of the S system within the analysis. Thus, data may be entered and manipulated, analyses may be performed, and output can be displayed in prints or plots, all within S.

Furthermore, looping and control structures are available to assist with new ideas for application of methods. For instance, although we did not write a Fortran program to perform the leave-one-out calculations in S, it was easy to obtain them for the SHT example. Using the first method, the following single S expression,

```
for (i in 1:82){print(pmoment (SHT, 1:3, data = data [1:82! = i]))}
```

repeatedly calls the posterior moment function "pmoment", operating on the structure callled "SHT" and evaluating means and variances of the three parameters λ, τ, p while successively leaving out one of the 82 observations. (The symbol != means "not equal to".) In addition, using the second method, contributions to the log posterior at the mode may be computed with a single expression and stored as components of a vector. This expression has the form

$$LL := \delta_e * (\log(\hat{p}) + \delta_t * \log(\hat{\tau}) + \hat{p} * \log(\hat{\lambda}) - (\hat{p} + \delta_e) * \log(\hat{\lambda} + W + S * \hat{\tau})$$

with the i-th component of LL corresponding to the i-th components of δ_e, δ_t, W, and S, which in turn correspond to the i-th individual. (The symbol := is used here to mean "assigned the value of".) The contributions to the log posterior at (λ^*, τ^*, p^*) may be similarly computed, and the changes in the posterior mean of λ, for example, may be obtained immediately, from a single expression corresponding to Δ of section 2.5.

Finally, another advantage of adding functions to S rather than writing a front end is that S has a macro facility, which allows the user to store and subsequently execute a series of commands so that when some useful sequence of steps is discovered it may be applied to new problems with minimal typing of command lines. As one example, in adding a Gauss-Hermite quadrature function to S it is easiest to have as arguments the location vector, scaling matrix, and the number of nodes. Using this function a macro may then be written to do adaptive quadrature (Naylor and Smith, 1982), with little resulting increase in waiting time.

We mention these features of S not only because they are each important, but because they illustrate the general rule that, in terms of the user's working time, it is fastest to have as many tasks as possible executed at the highest-level environment. Shifting between environments often adds a substantial working-time overhead. Furthermore, debugging evaluation statements, like that of log posterior above, is much faster within S than within Fortran because each single line is interpreted separately. Syntactical errors will be discovered immediately and are easily corrected.

4.2 Desirable facilities

Our present implementation within S greatly decreases working time, compared with the use of the same programs as user-called subroutines, and this is achieved with a negligible increase in waiting time. With this implementation, however, the user still must write Fortran code for the loglikelihood, the log prior, and the functions g. Ideally, an environment would also allow a function definition to be used as an argument in a new function, the way FORTRAN subroutine names may be passed to other FORTRAN subroutines.

For many puposes it would be very useful to define the loglikelihood function in its symbolic form, with its arguments remaining symbols rather than being assigned current values. That is, in our high-level environment, we would like to be able to write a function definition in the form

$$\text{FUNCTION LOGLIK}(\lambda, \tau, p, \text{DATA})$$

$$\dots$$

$$\text{LOGLIK} := \cdots$$

with the function LOGLIK being defined in terms of its arguments, as in the previous section, so that we could then pass LOGLIK to another function. Thus, if we had similarly defined the functions LOG-PRIOR and G (for some function $g = g(\lambda, \tau, p)$), we might then use an approximate density routine in the form

Y: = APPROX-MARGINAL-DENSITY(LOGLIK, LOG-PRIOR, G, G-VALUES).

The point is that we would avoid leaving the high-level environment to write a new Fortran subroutine. This greatly decreases working time and, if it did not concommitently increase waiting time, it would allow us to proceed in a genuinely interactive mode, in the sense described in the introduction.

To be more concrete, currently, as soon as we think of a new likelihood or prior to use, we must stop our analysis and write a Fortran subroutine for it. In the future, we would like to be able to think of these new possibilities during our analysis, and pursue them interactively, in the high-level environment.

4.3. Difficulties

We have described some advantages of working as much as possible in a high-level environment, such as S. Meanwhile, we have noted that the ability to use one function as another function's argument is not available in the current version of S. On the one hand, there are some indications that progress may be made. Chambers (1987) is experimenting with a new environment, and a new version of S, which should allow the sort of thing we have described. Bates and Chambers (1987) discuss an application in which a user specifies a nonlinear regression model within the high-level environment; the system then passes this specification to a symbol manipulator to produce code for the derivatives and a nonlinear least squares routine to fit the nonlinear model to some data. Similary, a version of the package ISP we have used (produced by the Derpartment of Statistics, University of Washington) allows specification of a funcion to be maximized, followed by a call to a maximization program, all within ISP.

One the other hand, there are some serious difficulties associated with our desired facility. In the current implementation of our extended version of S. the user-defined subroutines may be complied before they are callled from S. Arguments (e.g. parameters of a loglikelihood are sent from S to the subroutines (which run under a separate process in UNIX), and outputs (e.g., loglikelihood evaluations) are sent back. Since the subroutines are already compiled (and communication occurs though memory, rather than disk), this is relatively rapid. In contrast, if a user-defined loglikelihood had to be interpreted each time an evaluation were needed, this could become quite slow. (In the version of ISP we have used, the function to be maximized is expanded to ISP primitive functions when the function is defined, but the addtional step of interpreting the primitives must be taken each time the functions is called.) There is, in general terms, a trade-off between the reduced working time within the high-level environment, and the reduced running time achieved by compiled code produced outside that environment. We do not yet know whether environments such as S will be able to achieve a useful balance suitable for the largely interactive analysis we have envisioned. Alternative environments (based, for instance on a LISP system allowing interactive compilation) may be preferable.

5. CONCLUSION

Computing problems in Bayesian statistics are varied. Some involve multiple integrals of high dimensionality, for which reliable answers require monte carlo methods (e.g. Geweke, 1986, Kloek and van Dyke 1978, Stewart, 1984, van Dyke and Kloek, 1984, Zellner and Rossi, 1982) or posterior simulation (Kass, 1985, Tanner and Wong, 1987, Thompson and Miller 1985). Although these methods are fairly slow, moderate or long waiting times are often tolerable. On the other hand, interactive modelling and inference requires repeated evaluation of many similar integrals. In this situation, asymptotic methods can be useful. We hasten to add that in all applications, some selection of integrals should be verified by non-asymptotic methods, such as Gauss-Hermite quadrature (Naylor and Smith, 1982). Furthermore, these can sometimes be sufficiently fast to qualify as what we have called interactive methods. Certainly Gauss-Hermite quadrature is an interactive method for one-parameter problems, and can be for somewhat higher-dimensional problems, too.

We had two aims in our presentation here. First, we tried to indicate the manner in which asymptotic approximations might be applied in a variety of inferential settings. Second, we placed asymptotic methods within a context of interactive modelling, criticism, and inference, attempting to show that they furnish one set of useful statistical tools for interactive work. In the process we mentioned both advances in and difficulties with the non-statistical component of interactive analysis. We believe that theoretical and methodological effort should continue to be spent in developing basic techniques, both asymptotic and non-asymptotic, that may become increasingly valuable as computational facilities continue to improve.

ACKNOWLEDGEMENTS

We thank Mike Meyer for many useful discussions concerning our S implementation. This research was supported by the National Science Foundation under grant number NSF/DMS-8503019.

REFERENCES

Bates, D. M. and Chambers, J. M. (1987). Statistical Models as Data Structures. *Computer Science and Statistics: Nineteenth Symposium on the Interface.* (To appear).

Becker, R. A. and Chambers, J. M. (1984a). *S - An Interactive Environment for Data Analysis and Graphics.* Belmont, Ca: Wadsworth.

Becker, R. A. and Chambers, J. M. (1984b). *Extending the S System.* Belmont, Ca: Wadsworth.

Box G. E. P. and Tiao, G. (1973). *Bayesian Inference in Statistical Analysis.* Reading, Mass: Addison-Wesley.

Chambers, J. M. (1987). Interfaces to a quantitative programming environment. *Computer Science and Statistics: Nineteenth Symposium on the Interface.* (To appear).

Deely, J. J. and Lindley, D. V. (1981). Bayes empirical Bayes. *J. Amer. Statist. Assoc.* **76**, 833–841.

Efron, B. (1986). Why isn't everyone a Bayesian? *Amer. Statist.* **40**, 1–5.

Erdelyi, A. (1956). *Asymptotic Expansions*. New York: Dover.

Geweke J. (1987). Bayesian inference in econometric models using Monte Carlo integration. *Econometrica* (to appear).

Hartigan, J. A. (1965). The asymptotically unbiased prior distribution. *Ann. Math. Statist.* **36**, 1137-1152.

Jeffreys, H. (1967). *Theory of Probability*. Oxford: University Press.

Johnson, R. A. (1970). Asymptotic expansions for posterior distributions. *Ann. Math. Statist.* **38**, 1266–1272.

Kadane, J. B. (1985). Is victimization chronic? A Bayesian analysis of multinomial missing data. *J. of Econometrics* **29**, 47–66.

Kass, R. E. (1985). Exact inferences about principal components and related quantities using posterior distributions calculated by simulation. *Tech. Rep.* **346**. Department of Statistics, Carnegie-Mellon University.

Kass, R. E. and Steffey, D. L. (1986). Approximate inference in conditionally independent hierarchical models. *Tech. Rep.* **386**. Department of Statistics, Carnegie-Mellon University.

Kass, R. E., Tierney, L. and Kadane, J. B. (1987). The validity of stochastic asymptotic expansions based on Laplace's method. *Tech. Rep.* Department of Statistics, Carnegie-Mellon University.

Kass, R. E., Tierney, L. and Kadane, J. B. (1988). Approximate methods for influence and sensitivity in Bayesian Analysis. Department of Statistics. *Tech. Rep.* Carnegie-Mellon University.

Kloek, T. and Van Dijk, H. K. (1978). Bayesian estimation of equation system parameters: an application of integration by Monte Carlo. *Econometrica* **46**, 1–19.

Laplace, P. S. De (1847). *Oeuvres 7*. Paris: Imprimerie Royale.

Leonard, T., (1982). Comment on 'A simple predictive density function'. *J. Amer. Statist. Assoc.* **77**, 657–685.

Lindley, D. V. (1961). The use of prior probability distributions in statistical inference and decisions. *Proc. 4th Berkeley Symp.* **1**, 453–468.

Lindley, D. V. (1980). Approximate Bayesian methods. *Bayesian Statistics* (J. M. Bernardo, M. H. De Groot, D. V. Lindley and A. F. M. Smith, eds.). Valencia, Spain: University Press.

Mosteller, F. and D. L. Wallence, (1984). *Applied Bayesian and classical inference: The case of the Federalist papers*. New York: Springer.

Naylor, J. C. and A. F. M. Smith, (1982). Applications of a method for the efficient computation of posterior distributions. *Applied Statistics* **31**, 214–225.

Phillips, P. C. B. (1983). Marginal densities of instrumental variable estimtors in the general single equation case. *Advances in Econometrics* **2**, 1–24.

Reid, N. (1988). Saddlepoint methods in statistical inference. *Statistical Science* (to appear).

Stewart, L. (1985). Multiparameter Bayesian inference using Monte Carlo integration: Some techniques for bivariate analysis. *Bayesian Statistics* **2**. (J. M. Bernardo, M. H. DeGroot, D. V. Lindley and A. F. M. Smith, eds.). Amsterdam: North-Holland 495–510.

Tanner, M. and Wong, W. (1987). The Calculation of posterior distributions by data argumentation. *J. Amer. Statist. Assoc.* (To appear).

Thompson P. A. and Miller, R. B. (1985). Bayesian analysis of univariate time series: forecasting via simulation. *Tech. Rep.* University of Wisconsin.

Tierney, L. and Kadane, J. B. (1986). Accurate approximations for posterior moments and marginal densities. *J. Amer. Statist. Assoc.* **81**, 82–86.

Tierney, L., Kass, R. E. and Kadane, J. B. (1987). Interactive Bayesian analysis using accurate asymptotic approximations, *Computer Science and Statistics: Nineteenth Symposium of the Interface*. (To appear).

Tierney, L., Kass, R. E. and Kadane, J. B. (1988a). Approximation of posterior expectations and variances using Laplace's method. *Tech. Rep.*, Department of Statistics, Carnegie-Mellon University.

Tierney, L., Kass, R. E., and Kadane, J. B. (1988b). Approximate marginal densities of nonlinear functions. *Tech. Rep.* Departament of Statistics, Carnegie-Mellon University.

Tsutakawa, R. K. (1985). Estimation of cancer mortality rates: a Bayes analysis of small frequencies. *Biometrics* **41**, 69–79.

Van Dijk, H. K. and Kloek, T. (1985). Experiments with some alternatives for simple important sampling in Monte Carlo integration. *Bayesian Statistics* **2**. (J. M. Bernardo, M. H. DeGroot, D. V. Lindley and A. F. M. Smith, eds.). Amsterdam: North-Holland

Zellner, A and Rosi, P. E. (1982). Bayesian analysis of dichotomous quantal response models. *Proceedings of ASA Business and Economics Section*, 15–24.

DISCUSSION

H. K. VAN DIJK (*Erasmus University*)

I would like to make some specific comments on the paper presented by Kass, Tierney and Kadane.

(i) The authors mention that they focus on interactive Bayesian statistical analysis. Any interactive statistical analysis will also involve an error-learning process, e.g., one has omitted a relevant variable in the model. I find it very hard to analyze how initial probabilities change when the model used appears to be too simple.

(ii) The Laplace expansion method works fine in the example considered in the paper. One wonders whether this is restricted to the Laplace expansion or whether other expansions like the Edgeworth, the Gram-Charlier or the Fourier expansion can also be applied in a simple way. Further, what diagnostic checks can be used to verify whether the regularity conditions hold approximately? A simple diagnostic check is to give both the first-order and second-order approximation as computer output. An other one is to use the second-order approximation to the posterior, multiply the Hessian matrix by a scalar $k(k > 1)$ and apply importance sampling with a large covariance matrix as scale matrix in the Student-t importance function.

(iii) When is the sample large enough for asymptotic approximations to hold with a reasonable degree of accuracy? In time series models one has experiments where $T \geq 50$ observations appear to give good asymptotic approximations in non-Bayesian statistics.

(iv) The authors recommend quasi-Newton procedures for numerial optimization. These methods work fast for reasonably well-behaved functions. If the posterior is flat over a large part of the parameter region one has to be careful with respect to the choice of the numerical optimization method.

(v) The proposed method of computing marginal posterior densities of a subset of parameters is related to the computation of the limited-information maximum likelihood estimator where one performs the so-called *concentration* operation with respect to a subset of parameters (see Koopmans and Hood (1953) pp. 156–157, and 163–177).

Finally, in Figure 2 of the KTK-paper it is seen that the Laplace expansion works fine compared to an asymptotic approximation. How does one know this in other cases? That is, is there a kind of general error analysis?

J. E. H. SHAW (*University of Warwick*)

I have a few comments and questions on the attactive methods developed by Kass, Kadane and Tierney.

Like French (1980), I'm worried about the accuracy of numerical differentiation in high dimensions. Estimating the 2nd (or higher) derivatives is particularly awkward: if the posterior distribution $p(\theta)$ has high correlations, then a sensible step size for estimating $\partial^2 p/\partial\theta_i^2$ and $\partial^2 p/\partial\theta_j^2$ may be too short to estimate $\partial^2 p/\partial\theta_i\partial\theta_j$ at all accurately. The estimated Hessian matrix may therefore not be positive-definite, purely because of rounding errors. Since suitable step sizes usually must be found by trial and error (for example, doubling or halving the step size in each direction as appropriate), have the authors tried any extrapolation methods to estimate derivatives? (see, for example, Johnson and Riess, 1977). Numerical partial differentiation would be more stable after an (approximately) orthogonalising transformation of θ, in the spirit of Naylor and Smith (1982), although this requires a reasonable estimate of the Hessian anyway. Alternatively, have the authors tried variable metric methods (e.g. Press *et al.* 1986), which find the mode using only 1st derivatives, but estimate the Hessian *en route*?

How should we choose the values s at which to form $\hat{M}(s)$ in Section 2.4? Instead of estimating $M'(0)$, $M''(0)$ etc. by finite differencing (possibly with extrapolation), I personally would recommend fitting a spline to $\log[M(s)]$ and differentiating the spline.

In general, how should we choose an appropriate parametrisation? This seems particularly difficult with Laplace's method since transformations to reduce tail thickness, such as

$$\psi = \log\left(\frac{\phi}{1-\phi}\right)$$

where

$$\phi = \frac{1}{2} + \frac{1}{2\pi}\tan^{-1}(a + b\theta),$$

can make the associated posterior density bimodal or give it "shoulders" (as defined in O'Hagan, 1985). Have the authors any general comments on the relative undesirability of thick tails and big shoulders?

Concerning applications to numerical integration (Section 2.7.3), importance sampling can benefit enormously from the use of initial second-order approximations, but Gauss-Hermite integration is actually quite robust to perturbations in centering and scaling. See the comments in Sections 3 and 4 of Shaw (these proceedings). However, Gauss-Hermite integration tends to be less accurate if the integrand, or one of its low-order derivatives, is discontinuous. Therefore, in contrast to Laplace's method, the performance of Gauss-Hermite integration should be much improved in the second example of Section 3 after log transformations of the parameters. Far fewer nodes would then be needed.

An alternative implemention to that in Section 4 is provided by APL, which allows functions to be defined interactively, and passed as arguments to other functions. This can be done in ways that let the APL literati modify the environment (such as the method of numerical differentiation), but without frightening off the illiterati.

Finally, I should emphasise that numerical integration can be interactive even for some high-dimensional problems. For example, the 14-parameter model in Shaw (these proceedings) was analysed in "interactive mode" (possibly excepting a 7-minute "coffee break" during the final iteration). Numerical integration and asymptotic methods complement each other nicely, and allow some valuable cross-checking of approximate Bayesian inferences.

J. C. NAYLOR (*Trent Polytechnic*)

This method is seen to provide efficient and accurate approximations especially if the derivatives required can be found analytically. If they cannot, then presumably, numerical differentiation techniques must be employed, which are generally regarded as being less reliable than numerical integration. We have then replaced a potentially easy integration problem with a potentially hazardous differentiation problem. How is this apparent contradiction to be dealt with in any automatic (as implied by suggestions of interactively defining the likelihood) implementation?

I agree with the recommendation to use numerical integration to check selected results (section 5) but am less happy with the technique demonstrated in section 3. Successful use of Gauss-Hermite quadrature rules depends more on choice of scaling than on number of points. We would expect to get good results from many fewer than 20 points with a well scaled rule and poor results even with many more than 20 points if the rule is not properly scaled. If the rules used are based simply on scaling given by the Laplace approximation, they do not really provide an independent check on the accuracy of the answers. In any case, the accuracy of the numerical integration should be checked by using more than one grid size. I do not doubt the accuracy of the results presented here but feel that considerable care is needed in checking against another (approximate) method.

REPLY TO THE DISCUSSION

Before responding in detail to some of the points raised in the discussion we would like to thank all of the discussants for their careful reading of our paper and their insightful comments.

Four central themes emerge from the discussion by Naylor, Shaw, and van Dijk. These involve (i) the use of numerical differentiation, (ii) the possibility of parameter transformation, (iii) the problem of identifying situations in which asymptotics may be relied upon, and (iv) comparison with Edgeworth expansions.

We appreciate the treachery of numerical differentiation. Kass (1987), like Shaw here, has recommended extrapolation, and has noted an efficiency in the use of many one-dimensional central differences in computing a Hessian, but the difficult problem of choosing an initial step size remains. For the time being, we might ofter the advice that failure of a careful second-differencing routine may signal poor behavior requiring more than asymptotic attention.

With regard to parameterization, the negative results about general existence of various canonical parameterization (as in Kass, 1984) make us pessimistic about finding "best" parameterizations of models. On the other hand, it may be possible to devise methods for examining quickly the approximate effects of reparameterization (as in Bates and Watts,1981). As to the specific issue of thick tails versus shoulders, we note that the the straightforward applications of Laplace's method we have described require unimodality. Thus, substantial shoulders or bimodality would probably be disastrous. Uses of Laplace's method for bimodal posterior densities are currently being explored in Ph. D. thesis work of Yves Thibodeau at Carnegie Mellon.

The adequacy of asymptotic approximations must concern anyone who uses them. Experience may become a good guide in many situations. It would be especially helpful to have even a rough idea of the way sample size (and configuration) affects approximations as the dimensionality of the problem grows. On the other hand, it would also be good to have diagnostics for use with a particular data set. One possibility is to use asymptotics to judge asymptotics in the spirit of our Pearson skewness diagnostic for non-Normality. As van Dijk suggests, one can examine departure of second-order from first-order approximations. This should work when the departure is small, but when we find substantial discrepancy we will be stuck; without third-order approximations, we are unable to get a good check on second-order asymptotics.

It seems to us that there can be no substitute for exact calculation of at least some features of the posterior. The remaining issue becomes one of choosing where to put the effort. A reminder in this and other contexts is that integration will tend to make marginal distributions better behaved than joint distributions, so we may not be able to rely on the former to tell us about the latter. It will be good for more of us to gain experience with quadrature and monte carlo methods; we hope to reach consensus about issues such as sensitivity of Gauss-Hermite quadrature to scaling, about which Naylor and Shaw appear to disagree. (Our more limited experience makes us side with Naylor.) As van Dijk suggests, part of the art of data analysis will involve the choice of integration techniques from a bayesianly-oriented mathematical and computational palette.

Finally, a technical point that we wish to emphasize in response to one of Professor van Dijk's queries is that Edgeworth series have errors that hold on shrinking neighborhoods (that is, neighborhoods of the mean or mode that have diameter of order $O(n^{-1/2})$). Laplace's method for marginal density approximation is closely related to the saddlepoint method, and shares with it errors that are uniform on bounded regions. This we believe is the theoretical explanation for the good behavior illustrated in Figure 2 of our paper. It would, of course, be interesting to compare the Edgeworth-type series of Johnson (1970).

REFERENCES IN THE DISCUSSION

Bates, D. M. and Watts, D. G. (1981). Parameter transformations for improved accurate confidence regions in nonlinear least squares. *Ann. of Statist.* **9**, 1152–1167.

French, S. (1980). in discussion of "Approximate Bayesian methods" by D. V. Lindley. *Bayesian Statistics*, (J. M. Bernardo, M. H. DeGroot, D. V. Lindley and A. F. M. Smith, eds.). Valencia: University Press, 223–237.

Johnson, L. W. and Riess, R. D. (1977). *Numerical Analysis.* Reading, Mass.: Addison-Wesley.

Kass, R. E. (1984). Canonical parameterizations and zero parameter-effects curvature. *J. Roy. Statist. Soc. B* **46**, 86–92.

Kass, R. E. (1987). Computing observed information by finite differences. *Communications in Statistics: Simulation and Camputation* **16**, 587–599.

Koopmans, T. C. and Hood, W. C. (1953). The estimation of simultaneous economic relationships, in: Hood, W.C. and Koopmans, T. C., eds., *Studies in econometric method* (Yale University Press, New Haven).

O'Hagan, A. (1985). Shoulders in hierarchical models. *Bayesian Statistics 2*, (J. M. Bernardo, M. H. DeGroot, D. V. Lindley and A. F. M. Smith, eds.). Amsterdam: North-Holland, 697–710.

Press, W. H., Flannery, B. P., Teukolsky, S. A. and Vetterling, W. T. (1986). *Numerical Recipes. The Art of Scientific Computing.* Cambridge: University Press.

BAYESIAN STATISTICS 3, pp. 279–305
J. M. Bernardo, M. H. DeGroot, D. V. Lindley and A. F. M. Smith, (Eds.)

Stochastic Models of Incarceration Careers

C. E. KIM and M. J. SCHERVISH
Carnegie-Mellon University

SUMMARY

A criminal's incarceration career consists of the sequence of times spent in and out of prison. We construct stochastic process models for the incarceration careers of a population of criminals. The models are hierarchical, allowing different criminals to have similar, but not identical career parameters. We fit the models to a data set collected in 1979 based on interviews with several thousand inmates at state correctional facilities in the United States. The problem on which we focus in this paper is that of dealing with incomplete interviews. Many of the inmates do not give complete answers to the interview questions concerning the various lengths of time they spent in or out of prison. We use as much of the interview as is available and treat the rest as missing, allowing the calculation of posterior distributions over the possible configurations of missing information. We compare the method of multiple imputation (Rubin, 1978) both numerically and by speed.

Keywords: HIERARCHICAL MODEL; ALTERNATING RENEWAL PROCESS; MISSING DATA; MULTIPLE IMPUTATION; NUMERICAL INTEGRATION.

1. INTRODUCTION

In 1979, the United States Department of Justice, Bureau of Justice Statistics prepared a survey of inmates of state correctional facilities. A total of 11,397 inmates were chosen from 215 state correctional facilities. Each inmate was asked how many times he/she had been incarcerted and, for each incarceration, was asked detailed information including type of offense and date and length of incarceration. Only 7,503 inmates provided answers which were complete for all of their incarcerations. Because so many interviews are incomplete, one must take the missing data into account in some way to avoid losing large amounts of information. This problem is considered in detail in Section 3.

Our purpose in studying the 1979 survey of inmates was to model the incarceration career. Avi-Itzhak and Shinnar (1973) introduce the notion of the criminal career process and Blumstein and Cohen (1985) attempt to estimate the duration of the criminal career. Since our data concerns only incarcerations, and not crimes, we will be concerned only with the incarceration career procees.

2. THE BASIC MODEL

2.1 The Career Process

We model the incarceration career process for each criminal as an alternating renewal process. The process alternates between states of "on the street" and "in prison". We begin by assuming that the waiting times in the states for criminal number i are independent exponential random variables conditional on parameters (Θ_i, Δ). The street time has $\exp(\Theta_i)$ distribution and the prison time has $\exp(\Delta)$ distribution. We model the parameters $\{\Theta_i, \ldots, \Theta_n, \Delta\}$ as being *a priori* independent with gamma distributions conditional on hyperparameters A, B, C, D. The Θ_i all have distribution $\Gamma(A, B)$ and Δ has distribution $\Gamma(C, D)$. We will specify the hyperparameters C and D, but we will give a prior distribution for A and$B$.

2.2 Bias Correction

2.2.1 Sampling bias

All of the data which we have to analyze was collected in a survey of inmates in state correctional facilities. Hence, every inmate on whom we have data was in prision at the time of the survey. This sampling plan means that the likelihood function from the data is the conditional density of the data given that all of the sampled inmates were in prision at a fixed time. This requires that we be able to calculate the probability of the conditioning event namely that a criminal will be in prision at a specific time. Clearly, the probability that a criminal will be in prision at a specific time depends, to some extent, on his/her age. But we do not sample inmates according to their ages, so we must find the probability of being in prison not conditional on the criminal's age.

2.2.2 Calculation of probability of being in prison

To calculate the probability that an inmate is in prison at the time of the survey, we will first condition on the age of the inmate and then integrate with respect to the distribution of the inmate's age. In order to do this, we need to specify a distribution for the ages of criminals. For convenience, we begin by assuming that the distribution of the ages of criminals is the same as the distribution of the ages of adults in the United States population at the time of the survey. Such a distribution is available for data grouped by years. Let p_j be the proportion of the United States adult population of age j in 1979. We assume all inmates are under age 85, since there are so few very old inmates.

Let IP stand for the event that a specific inmate i will be in prison at the time of the survey. Then

$$\Pr(IP|\Delta = \delta, \Theta_i = \theta) = \sum_{j=18}^{85} \Pr(IP|\Delta = \delta, \Theta_i = \theta, \text{Age} = j)p_j. \tag{2.1}$$

The term $\Pr(IP|\Delta = \delta, \Theta_i = \theta, \text{Age} = j)$ in (2.1) is calculated as follows.

$$\begin{aligned}&\Pr(IP|\Delta = \delta, \Theta_i = \theta, \text{Age} = j) = \\ &\sum_{k=1}^{\infty} \Pr(IP \text{ serving } k\text{-th incarceration } |\Delta = \delta, \Theta_i = \theta, \text{Age} = j).\end{aligned} \tag{2.2}$$

An inmate, whose age is j years, has had a career of length $t = 12(j-18)$ months. Let T denote the sum of the first $k-1$ street times, and let S denote the sum of the first $k-1$ incarceration lengths. Also, let T_k and S_k denote the k-th street time and incarceration length respectively. In order for inmate i to be in prison at time t for the k-th time it must be that $T+S+T_k$ is less than t and that $T+S+T_k+S_k$ is greater than t. In symbols

$$\Pr(IP \text{ serving } k\text{-th incarceration}|\Delta = \delta, \Theta_i = \theta, \text{Age} = j)$$

$$= \Pr(T+S+T_k < t < T+S+T_k+S_k).$$

For convenience, call this probability $P_k(t)$.

Because we model the incarceration career as an alternating renewal process,

$$P_k(t) = \int_0^t f_{S+T}(u)P_1(t-u)du, \tag{2.3}$$

where f_{S+T} denotes the conditional density $S+T$ given the parameters. We will solve the integral equation (2.3) by means of Laplace transforms. We will denote the Laplace transform of a function h by h^*. Taking Laplace transforms of both sides of (2.3) yields

$$\begin{aligned}
P_k^*(\xi) &= \int_0^\infty \int_0^t f_{S+T}(u) P_1(t-u)\exp(-\xi t)\,du\,dt \\
&= \int_0^\infty f_{S+T}(u) \int_u^\infty P_1(t-u)\exp(-\xi\{t-u\})dt\,\exp(-\xi u)du \\
&= P_1^*(\xi) f_{S+T}^*(\xi).
\end{aligned}$$

Since S and T are each the sum of $k-1$ independent exponential random variables, we easily obtain

$$f_{S+T}^*(\xi) = \left(\frac{\theta}{\theta+\xi}\right)^{k-1}\left(\frac{\delta}{\delta+\xi}\right)^{k-1}.$$

To calculate $P_1^*(\xi)$, first calculate $P_1(t)$.

$$\begin{aligned}
P_1(t) &= \Pr(T_1 < t < T_1 + S_1) \\
&= \int_0^t \int_{t-t_1}^\infty \theta\delta\exp(-t_1\theta - s_1\delta)ds_1 st_1 \\
&= \frac{\theta}{\theta-\delta}(\exp\{-t\delta\} - \exp\{-t\theta\}).
\end{aligned}$$

It easily follows that

$$P_1^*(\xi) = \frac{\theta}{(\delta+\xi)(\theta+\xi)}.$$

Combining the two Laplace transforms gives

$$P_k^*(\xi) = \frac{1}{\delta}\left(\frac{\theta}{\theta+\xi}\right)^k\left(\frac{\delta}{\delta+\xi}\right)^k.$$

It follows from (2.2) that the Laplace transform of $\Pr(IP|\Delta=\delta, \Theta_i=\theta, \text{Age}=18+t/12)$ is

$$\begin{aligned}
\sum_{k=1}^\infty P_k^*(\xi) &= \frac{\theta}{\xi(\theta+\delta+\xi)} \\
&= \frac{\theta}{\theta+\delta}\left(\frac{1}{\xi} - \frac{1}{\theta+\delta+\xi}\right),
\end{aligned}$$

which, in turn, is the Laplace transform of

$$\frac{\theta}{\theta+\delta}(1-\exp\{-t(\theta+\delta)\}). \tag{2.4}$$

Hence, (2.4) is $\Pr(IP|\Delta = \delta, \Theta_i = \theta, \text{Age} = 18 + t/12)$. Summing over the age distribution gives

$$\Pr(IP|\Delta=\delta,\Theta_1=\theta) = \sum_{j=18}^{85} p_j \frac{\theta}{\theta+\delta}(1-\exp\{-t(\theta+\delta)\}), \tag{2.5}$$

where t in (2.5) is understood to equal $12(j-18)$.

2.2.3 Sampling likelihood function

Now that we know the probability that a criminal would be in prison at the time of the survey, we can write the likelihood function for observed data. In general, if A is an event and data is observed conditional on A, the conditional likelihood is

$$\Pr(\text{observed data}|A, \text{Parameters}) = \frac{\Pr(\text{observed data}, A|\text{Parameters})}{\Pr(A|\text{Parameters})}$$

$$= \frac{\Pr(\text{observed data}, |\text{Parameters})}{\Pr(A|\text{Parameters})}$$

since the observed data already includes A. Hence, we can write the contribution to the likelihood function from the data of each inmate i by first ignoring the bias correction, and then dividing by (2.5) with $\theta = \theta_i$.

2.3 Model Variations

2.3.1 Career termination

Blumstein and Cohen (1985) give evidence to suggest that criminals terminate their careers after some time. The length of the career of one criminal may be different from that of another, so we model the career length as a random variable Y with distribution $\exp(\mu)$ conditional on parameter $M = \mu$. Assume Y is independent of all other random variables given the parameters. For an inmate who began his current incarceration at age $18 + z/12$, the likelihood function before bias correction is the same as in Section 2.2 with an extra factor of $\exp(-z\mu)$. We can calculate the probability of being in prison at the time of the survey given the parameters in a manner similar to what was done in Section 2.2.2. Equation (2.2) still holds, but now $f_{S+T}(u)$ must be multiplied by a factor of $\exp(-u\mu)$ to account for the fact that the career cannot end before the current incarceration begins. Hence, the revised Laplace transform is readily obtained as

$$f^*_{S+T}(\xi) = \left(\frac{\theta}{\theta + \mu + \xi}\right)^{k-1} \left(\frac{\delta}{\delta + \mu + \xi}\right)^{k-1}.$$

We can also calculate

$$\begin{aligned} P_1(t) &= \Pr(T_1 < t < T_1 + S_1, Y > T_1) \\ &= \int_0^t \int_{t-t_1}^{\infty} \int_{t_1}^{\infty} \theta\delta\mu \exp(-t_1\theta - s_1\delta - y\mu) dy ds_1 dt_1 \\ &= \frac{\theta}{\mu + \theta - \delta} \left(\exp\{-t\delta\} - \exp\{-t(\theta + \mu)\}\right). \end{aligned}$$

Hence,

$$P_1^*(\xi) = \frac{\theta}{(\delta + \xi)(\theta + \mu + \xi)}.$$

We now follow the same steps as we did in Section 2.2.2 to get

$$\begin{aligned} \sum_{k=1}^{\infty} P_k^*(\xi) &= \frac{\theta(\delta + \mu + \xi)}{(\mu + \xi)(\delta + \xi)(\theta + \delta + \mu + \xi)} \\ &= \frac{\theta\mu}{(\theta + \mu)(\mu - \delta)(\delta + \xi)} - \frac{\delta\theta}{(\theta + \delta)(\mu - \delta)(\mu + \xi)} \\ &\quad - \frac{\theta^2}{(\theta + \mu)(\theta + \delta)(\theta + \delta + \mu + \xi)}, \end{aligned} \tag{2.6}$$

which, in turn, is the Laplace transform of

$$\frac{\theta\mu}{(\theta+\mu)(\mu-\delta)}\exp(-t\delta) - \frac{\delta\theta}{(\theta+\delta)(\mu-\delta)}\exp(-t\mu) \\ - \frac{\theta^2}{(\theta+\mu)(\theta+\delta)}\exp\{-t(\theta+\delta+\mu)\}. \tag{2.7}$$

Notice that (2.7) reduces to (2.4) when $\mu = 0$, which corresponds to infinite career length.

2.3.2 Late starters

The assumption that every criminal begins his/her incarceration process at age 18 is not realistic. Suppose we assume that the distribution of the time until the adult incarceration career begins has a density function $f_L(t)$ which may depend on parameters. The only changes in the calculation of Section 2.3.1 which need to be made are the following. The distribution of $S+T$ is now the convolution of the old distribution with f_L, hence the Laplace transform is multiplied by f_L^*. This factor aso must be multiplied into (2.6). It follows that (2.7) must be replaced by the convolution of (2.7) with f_L.

The distribution we choose to use for f_L is a mixture of a point mass at 0 (with probability $1-p$) and an exponential $\exp(\psi)$ with probability p, conditional on parameters $P = p$ and $\Psi = \psi$. Those inmates who served time as juveniles will be assumed to start their incarceration careers immediately at age 18 or upon release from their last juvenile incarceration, whichever is later. We will then assume that $1-P$ is the probability that an inmate would have a juvenile incarceration. That is, we will make the simplifying assumption that late starting is equivalent to having had no juvenile incarcerations. In our data set there are 7,507 inmates who had no juvenile incarcerations.

2.4 The Model to Fit

2.4.1 The data model

The model which we will actually fit to the data can be described as follows. We will describe the model hierarchically. First consider a criminal numbered j. Let

$$X_j = \begin{cases} 0 & \text{if criminal } j \text{ had a juvenile incarceration.} \\ 1 & \text{if not.} \end{cases}$$

Let $T_{j,i}$ and $S_{j,i}$ denote the i-th street time and incarceration time respectivelty for criminal j. Also, let L_j denote the time from age 18 or the last release from a juvenile incarceration (whichever is later) until the start of the adult criminal career. We will assume that the time from age 18 or the last release from a juvenile incarceration until the first adult incarceration equals $L_j + T_{j,1}$. If $X_j = 1$ then $L_j = 0$. The length of the criminal career is denoted Y_j. The parameters of the model are

$$\Lambda = (A, B, \Delta, M, \Psi, P, \Theta_1, \ldots).$$

We will denote a particular value of Λ by

$$\lambda = (a, b, \delta, \mu, \psi, p, \theta_1, \ldots).$$

For convenience, the first six coordinates of the parameter vector will be denoted by

$$\Lambda_1 = (A, B, \Delta, M, \Psi, P).$$

Assume that the X_j are exchangeable and that they are independent Bernoulli random variables with probability p given $\Lambda = \lambda$. The description of the remainder of the model for criminal j proceeds conditional on the value of X_j (and on $\Lambda = \lambda$).

Given $X_j = 1$ (and $\Lambda = \lambda$), L_j has $\exp(\psi)$ distribution. All of $Y_j, L_j, T_{j,i}, S_{j,i}$, are independent. The $T_{j,i}$ have $\exp(\theta_i)$ distribution and the $S_{j,i}$ have $\exp(\delta)$ distribution. The career length Y_j has $\exp(\mu)$ distribution. The data from different criminals are independent given the parameters. Given $X_j = 0$ (and $\Lambda = \lambda$), the distributions of all of the random variables are the same as they are conditional of $X_j = 1$ except for L_j, which equals 0 with probability 1.

2.4.2 The parameter model

The Θ_j given the rest of the parameters will be independent *a priori* with distribution $\Gamma(a, b)$. We will assume that the parameters A, B, Δ, M, Ψ, and P are independent *a priori*. The distributions of Δ, B, M, and Ψ will be from the *gamma* family. The distribution of P will be from the *beta* family. We will use an exponential distribution for A to indicate that we believe A is most likely to be small, but to still allow more than negligible probability for large values. The precise prior distributions are given in Section 5.1.

3. PARTIALLY MISSING DATA

3.1 What the Problem is

Not every inmate reports a complete history of incarcerations. For example, an inmate may report having been incarcerated five times, but he only describes three of the incarcerations in detail. Possible reasons are that he answered the first question incorrectly and was only incarcerted three times, or that he did not recall sufficient information about two of the incarcerations, or perhaps that his memory is so unreliable that he should be ignored altogether. It is easiest to assume the last of these possibilities, but this is also the most naïve approach. Assuming that the first possibility is true runs the significant risk of overlooking data whose effect is partially identifiable. That is, extra incarcerations for an individual inmate make that inmate's incarceration rate seem higher. There is no possibility that the effects of several missing imprisonments being ignored for same inmate will "cancel out" in any way.

Our first approach to dealing with partially missing data is to try to find the likelihood of the parameters given the observed data. That is, we will try to integrate the missing data out of the likelihood function, when possible. To see what kinds of methods and assumptions are needed to do this, consider a simple example. We will ignore, in this example any corrections for sampling bias.

Example 3.1. An inmate claims to have been imprisoned three times (including the current incarceration) but only reports details of the current incarceration and one other. Taking the date at which the inmate became age 18 as time $t = 0$ (in months), suppose that the time of the survey is $t = 60$. Suppose that the current imprisonment began at time $t = 56$ and that the other reported incarceration began at $t = 25$ and lasted 8 months. The inmate had no juvenile incarcerations. If we were simply to assume that the missing incarceration had length zero, and we ignored late starting, then the likelihood function for the observed data would be

$$f_C(\lambda) = \delta^2\theta^3 \exp\{-(12\delta + 48\theta + 56\mu)\}. \tag{3.1}$$

If, on the other hand, we decide to treat the third incarceration as partially missing, we proceed as follows. Let the lenght of the missing incarceration be denoted s. Then, if the data were not missing, the likelihood would be

$$\delta^2\theta^3 \exp\left(-\{(12+s)\delta + (48-s)\theta + 56\mu\}\right). \tag{3.2}$$

Since we do not know what s equals, we must integrate it out of (3.2) in order to obtain the likelihood for the observed data. If we knew that the missing imprisonment occurred after the first reported one, then that interval of length 23 months would contain two free times and one imprisonment, and would therefore be the observed value of a random variable with the distribution of the convolution of two gamma distributions. The two particular gamma distributions would be $\Gamma(2, \theta)$ and $\Gamma(1, \delta)$. The density of such a random variable is

$$\begin{aligned} f(x) &= \int_0^x \theta^2 \delta u \exp\{-\theta u - \delta(x-u)\} du \\ &= \theta^2 \delta \exp(-\delta x) \left\{ \frac{-x}{\theta - \delta} \exp(-x\{\theta - \delta\}) + \frac{1}{(\theta-\delta)^2} [1 - \exp(-x\{\theta - \delta\})] \right\} \\ &= \theta^2 \exp(-\delta x) g(2, 1, \theta - \delta, x), \end{aligned} \quad (3.3)$$

where the function g is defined below and is related to the convolution of two gamma densities:

$$g(k, j, a, x) = \begin{cases} \frac{1}{(k-1)!(j-1)} \int_0^x u^{k-1}(x-\mu)^{j-1} \exp(-au) du & k > 0, j > 0 \\ \frac{1}{(k-1)!} x^{k-1} \exp(-ax) & k > 0, j = 0. \end{cases} \quad (3.4)$$

Hence, we must replace a factor of $\theta^2 \delta \exp(-23\theta)$ in (3.1) by (3.3) evaluated at $x = 23$. Notice that (3.3) evaluated at $x = 23$ has, as a factor, $\theta^2 \delta \exp(-23\delta)$. Hence, the replacement is equivalent to just multiplying (3.1) by the factor

$$\exp\{23(\theta - \delta)\} g(2, 1, \theta - \delta, 23). \quad (3.5)$$

To account for late starting, we have a similar situation in the first time interval (of length 25). This length of time is the observed value of the convolution of two random variables with $\Gamma(1, \theta)$ and $\Gamma(1, \psi)$ distributions. Hence we must replace a factor of $\theta \exp\{-25(\theta + \mu)\}$ in (3.1) by

$$p\theta\psi \int_0^{25} \exp\{-u(\theta + \mu) - (25 - u)\psi\} du = p\theta\psi \exp(-25\psi) g(1, 1, \theta + \mu - \psi, 25).$$

Had the inmate not been a late starter, we would just multiply the likelihood by $1 - p$. The effect is to produce the likelihood

$$\begin{aligned} &f_C(\lambda) \exp\{23(\theta - \delta)\} g(2, 1, \theta - \delta, 23) \\ &\quad \times [p\psi \exp\{25(\theta + \mu - \psi)\} g(1, 1, \theta + \mu - \psi, 25]^x (1 - p)^{1-x}. \end{aligned} \quad (3.6)$$

If, on the other hand, we knew that the missing imprisonment came before the first reported imprisonment, then the first interval of length 25 is the convolution of three random variables, one with $\Gamma(1, \psi)$ distribution, one with $\Gamma(2, \theta)$ distribution, and one with $\Gamma(1, \delta)$ distribution. The likelihood of such data is

$$\begin{aligned} &p \int_0^x \int_0^{x-u} \psi \delta \theta^2 u \exp\{-\psi(x - u - v) - (\theta + \mu)u - (\delta + \mu)v\} dv\, du \\ &= p \exp(-\psi x) \psi \delta \theta^2 g^*(2, 1, \theta + \mu - \psi, \delta + \mu - \psi, x), \end{aligned} \quad (3.7)$$

where the function g^* has the following general form:

$$g^*(k, j, a, b, x) = \begin{cases} \frac{1}{(k-1)!(j-1)} \int_0^x \int_0^{x-u} u^{k-1} v^{j-1} \exp(-ua - vb) dv\, du & k > 0, j > 0 \\ g(k, 1, a) & k > 0, j = 0. \end{cases} \quad (3.8)$$

Hence, the right hand side of (3.7) evaluated at $x = 25$ must replace a factor of $\theta^2\ \delta$ exp $\{-25(\ \theta + \mu)\}$ in (3.1). Since (3.7) has a factor of $\theta^2\ \delta$ exp $(-\psi x)$, the replacement is equivalent to multiplying (3.1) by

$$p \exp\{25(\theta + \mu - \psi)\}\psi g^*(2, 1, \theta + \mu - \psi, \delta + \mu - \psi, 25).$$

Had the inmate not been a late starter, the interval of length 25 would have been the observed value of convolution of a $\Gamma(2, \theta)$ random variable and a $\Gamma(1, \delta)$ random variable. This case was already handled before. The net result is a lifelihood equal to

$$\begin{aligned} &f_C(\lambda)[p\psi \exp\{25(\theta + \mu - \psi)\}g^*(2, 1, \theta + \mu - \psi, \delta + \mu - \psi, 25)]^x \\ &\times [(1-p)\exp\{25(\theta - \delta)\}g(2, 1, \theta - \delta, 25)]^{1-x}. \end{aligned} \tag{3.9}$$

Since we do not know which of the two cases is true, the data is the union of the two, and the likelihood is the sum of the two likelihoods, which equals (3.6) plus (3.9).

In the remainder of this paper, as in Example 3.1, we will assume that $\theta \neq \delta$, $\theta \neq \psi$, $\delta \neq \psi$, $\theta + \mu \neq \psi$, $\delta + \mu \neq \psi$,, and $\theta + \mu + \delta \neq \psi$, hence we will not carry along equality as special cases of the values of likelihoods.

Of course, we cannot overlook the possibility that even those inmates who reported the same numbers of incarcerations as they decribed may be mistaken about how many they actually served, when they began, or how long they lasted. We will not consider this issue in this paper, however. Similarly, there is the possibility that the missing incarcerations are not missing at random. That is, there may be a connection between the lengths of the crime types of the missing incarcerations and the fact they are missing. We will also not discuss this matter in detail in this paper.

3.2 How Big the Problem Is

To understand the magnitude of the missing data problem, we counted the numbers of inmates who had missing data of various sorts. As mentioned earlier, only 7,503 out of the 11,397 inmates provided complete information on their incarceration careers. Of the 3,894 remaining, 1,216 provided such poor or contradictory information, that we could not make any use of it whatsoever. And there were 615 who were admitted to their current incarceration before they were 18 years old. Still, it makes sense to try to use the information supplied by the other 2,063 inmates who provided partially missing incarceration information.

Missing incarcerations can come in 5 varieties. First there are those that are totally missing. For these, we know neither how long they were nor when they might have occurred. We only know how many there are. We call these *type* 1 missing incarcerations. For the remaining types of missing incarcerations, we know at least when they occurred in the sequence of imprisonments (e.g. between the first and second completely reported imprisonments). The second type of missing incarcerations is the kind for which we know only when it occurred in the sequence of incarcerations but we still do not know exactly when it began or how long it lasted (*type* 2). Thirdly, there are those for which we know when they began, but not their length (*type* 3). Fourthly, there are those for which we know how long they lasted, but not when they began (*type* 4). Finally, if we do not know when the current incarceration began, we will call it *type* 5. For convenience, we will refer to a non-missing incarceration as *type* 6.

Counts of the various types of missing incarcerations are given in Table 1.

Type Of Missing Incarceration

1	2	3	4	5	6
2214	112	81	1571	95	13961

Table 1. *Counts of Missing Incarcerations by Type*

3.3 Likelihoods For the Various Cases

We need to consider several different possible configurations of missing data. The cases of type 1 missing incarcerations provide special problems. These incarcerations could have ocurred at any time during the criminal career and could have been of any length. When they occur in conjunction with partially missing data, they require special handling. In this section, we will derive the explicit forms of the likelihoods for all of the configurations of missing data which exist in this data set.

We will consider a single inmate only, and will not use the subscript j to index any of the random variables in this section. This includes the parameter θ_j as well as indicators like x_j etc.

In all of the cases considered in this section, we use some common notation. For convenience let the time at which the inmate reaches age 18 be referred to as the end of the 0-th reported incarceration, if necessary. Inmates who served time as juveniles and who were incarcerated on their eighteenth birthday begin their adult careers upon release from the last juvenile incarceration. We define a *known event* to be any one of the following events which occurred at a reported time: the start of an incarceration, the end of an incarceration, the time of the survey. For example, the start of a type 3 incarceration is a known event, but the end is not. The start of the current incarceration is a known event unless it is a type 5 incarceration. The time of the survey is always a known event.

First, we arrange all of the known events in chronological order. The intervals of time between known events will be referred to as *intervals*. We number intervals sequentially from 1 to m, however, we will not explicitly number the interval between the start and end of a completely reported incarceration.

Q= Total length of time from age 18 until the first known event at or after the start of the current incarceration

S= Sum of lengths of all type 4 and type 6 incarcerations

I_i= The i-th interval as defined above

t_i= Length of interval I_i minus lengths of all type 4 incarcerations in I_i

$T = \sum_{i=1}^{m} t_i$= Total time unaccounted for

k= Number of incarcerations of all types, not counting current one

Because it appears a factor in the likelihood function, we define the *complete data likelihood* $f_C(\lambda)$ to be the likelihood function one would use if one were simply to treat all incarcerations of unknown length as having zero length, if one were to ignore late starting, and if one were to assume that the career lasted at least until the time of the last known event to occur at or after the start of the current incarceration. This analogous to (3.1) and has the form

$$f_C(\lambda) = \delta^k \theta^{k+1} \exp(-\{S\delta + T\theta + Q\mu\}).$$

Interval number 1 must be treated specially, because it may contain a late starting time L with distribution $\exp(\psi)$.

3.3.1 Only one missing incarceration

If the one missing incarceration is of type 1, then Example 3.1 is the prototype for calculating the likelihood. The result is

$$f_C(\lambda) \left\{ H_1(2,1,0) + \sum_{i=2}^{k} H_i(2,1,0) H_1(1,0,0) \right\}, \tag{3.10}$$

where

$$
\begin{aligned}
H_1(1,0,0) &= [p\psi\exp\{t_1(\theta+\mu-\psi)\}g(1,1,\theta+\mu-\psi,t_1)]^x(1-p)^{1-x},\\
H_1(2,1,0) &= [p\psi\exp\{t_1(\theta+\mu-\psi)\}g^*(2,1,\theta+\mu-\psi,\delta+\mu-\psi,t_1)]^x\\
&\quad\times[(1-p)\exp\{t_1(\theta-\delta)\}g(2,1,\theta-\delta,t_1)]^{1-x},\\
H_i(2,1,0) &= \exp\{t_i(\theta-\delta)\}g(2,1,\theta-\delta,t_i)\ \text{ for } i>1.
\end{aligned}
\tag{3.11}
$$

If the missing incarceration is of type 2, the likelihood has only the one term in the summation in (3.10) for i equal to the number of the reported incarceration following the missing one. (The $i=1$ term is term before the summation.)

If the missing incarceration is of type 3, then we have a case not dealt with in Example 3.1. We know $i \neq 1$ hence the interval I_i is known to contain one street time and a least part of one imprisonment. If we knew the length of the street time to be u, we would replace a factor of $\exp(-t_i\theta)$ in the complete data likelihood by $\exp\{-\delta(t_i-u)-u\theta\}$. Since we do not know u, we must integrate out u. The integral equals

$$\exp(-t,\delta)g(1,1,\theta-\delta,t_i).$$

In this case, the likelihood is

$$f_C(\lambda)H_i(1,1,0)H_1(1,0,0),$$

where

$$H_i(1,1,0)=\exp\{t_i(\theta-\delta)\}g(1,1,\theta-\delta,t_i).$$

If the one missing incarceration is of type 4, then it essentially non-missing. But, to be careful and consistent, we will write the likelihood as

$$f_C(\lambda)H_i(2,0,0)H_1(1,0,0),$$

where

$$
\begin{aligned}
H_i(2,0,0) &= \exp\{t_i(\theta-\delta)\}g(2,0,\theta-\delta,t_i)\\
&= t_i.
\end{aligned}
$$

If the missing incarceration is of type 5, but not in the first interval, then we have a situation just like type 3, except that the career may have terminated after the start of the current incarceration. In this case, we must replace a factor of $\exp\{-t_i(\theta+\mu)\}$ by

$$\int_0^{t_i}\exp\{-\delta(t_i-u)-u(\theta+\mu).\}du=\exp(-t_i\delta)g(1,1,\theta+\mu-\delta,t_i).$$

The likelihood is then

$$f_C(\lambda)H_i(1,0,1)H_1(1,0,0),$$

where

$$H_i(1,0,1)=\exp\{t_i(\theta+\mu-\delta)\}g(1,1,\theta+\mu-\delta,t_i).$$

If the missing incarceration is of type 5 and is in the first interval, we must replace a factor of $\exp\{-t_1(\theta+\mu)\}$ by

$$p\int_0^{t_1}\int_0^{t_1-u}\exp\{-u(\theta+\mu)-v\delta-(t_1-u-v)\psi\}dvdu,$$

if $x = 1$, and by

$$(1-p)\exp(-t_1\delta)g(1,1,\theta+\mu-\delta,t_1),$$

If $x = 0$. The factor to multiply by $f_C(\lambda)$ is

$$\begin{aligned} H_1(1,0,1) &= [p\psi\exp\{t_1(\theta+\mu-\psi)\}g^*(1,1,\theta+\mu-\psi,\delta-\psi,t_1)]^x \\ &\times [(1-p)\exp\{t_1(\theta+\mu-\delta)\}g(1,1,\theta+\mu-\delta,t_1)]^{1-x}. \end{aligned}$$

The important thing to notice about each of the cases considered above is that the likelihood function always takes the form

$$f_C(\lambda)\prod_{i=1}^{m} H_i(k,j,\ell)$$

(or a sum of these), where H_i is a factor corresponding to interval i and which equals 1 if there is no data missing in that interval. We will make use of this fact when dealing with two or more missing incarcerations.

3.3.2 Two or more missing incarcerations

When the data on two or more incarcerations are partially missing, we have to deal separately with those which occur in the first interval and those which occur in later intervals. If one of the missing incarcerations is of type 1, we do not know when it occurred, hence we must sum over the possibilities. If several are of type 1, this will be a multiple sum. Also, note that there can be at most one type 3 incarceration in any interval I_i and there can be at most one type 5 incarceration overall.

We will build the likelihood function one interval at a time. That is, we will construct H_i factors as in Section 3.3.1. When two or more missing incarcerations occur in the same interval, we have several cases. The cases depend both on the types of missing incarcerations and on whether or not they occur in the first interval.

Intervals other than the first. First of all, note that the lengths of type 4 missing incarcerations have already been included in the complete data likelihood, so only the street times before and after them are missing. Suppose that there are k street times of unknown length and j incarceration times of unknown length in interval i. Suppose also that there is no type 5 incarceration in interval i. Then we know only that the sum of k exp(θ) and j exp(δ) random variables equals t_i. The likelihood for such an observation is

$$\begin{aligned} &\frac{\theta^k\delta^j}{(k-1)!(j-1)!}\int_0^{t_i} u^{k-1}(t_i-u)^{j-1}\exp\{-\theta u-\delta(t_i-u)\}du \\ &\qquad = \theta^k\delta^j\exp(-t_i\delta)g(k,j,\theta-\delta,t_i). \end{aligned}$$

Hence, we must multiply f_C by

$$H_i(k,j,0) = \exp\{t_i(\theta-\delta)\}g(k,j,\theta-\delta,t_i).$$

In general, k equals one plus the number of type 1, 2, and 4 incarcerations in the interval, and j equals the number of type, 1, 2, and 3 incarcerations in the interval.

If one of the missing incarcerations is a type 5, then the career may end after the start of the current incarceration. Let j be the number of type 1, 2, and 3 incarcerations, as above. We must replace a factor of $\exp\{-t_i(\theta+\mu)\}$ in the complete data likelihood by

$$\begin{aligned} &\frac{1}{(k-1)!(j-1)!}\int_0^{t_i}\int_0^{t_i-u} u^{k-1}v^{j-1}\exp\{-u(\theta+\mu)-v(\delta+\mu)-(t_i-u-v)\delta\}dvdu \\ &\qquad = \exp(-t_i\delta)g^*(k,j,\theta+\mu-\delta,\mu,t_i). \end{aligned}$$

The factor to multiply the complete data likelihood would be

$$H_i(k,j,1) = \exp\{t_i(\theta+\mu-\delta)\}g^*(k,j,\theta+\mu-\delta,\mu,t_i).$$

The first interval. If there are no missing incarcerations in the first interval, then the factor H_1 is always $H_1(1,0,0)$ as in (3.11). Otherwise, factor H_1 depends on which types of missing incarcerations appear here. No type 3's are possible, since the inmate is always on the street at the start of the adult career. Ther are two cases which must be handled:

1. There is no type 5 in incarceration in the first interval
2. There is a type 5 incarceration in the first interval

The last case is very special. If a type 5 incarceration occurs in the first interval, that means that we do no know the starting times of any of the incarcerations. In that case there is only one interval and types 1 and 2 are the same. We discuss this case last.

No type 5 incarceration in first interval. The first case involves the same reasoning as the corresponding case in the previous section, but now, we must hold open the possibility that a random variable with $\exp(\psi)$ distribution also occurs in the first interval (if $x=1$). Another way to look at the problem is that it is equivalent to (3.9) from Example 3.1, except that there are extra missing incarcerations in the first interval. All this requires is an adjustment to the first two arguments of the functions g^* and g in H_1. If there are k unknown street times and j unknown incarceration lengths, then the H_1 factor is

$$\begin{aligned}H_1(k,j,0) = {} & [p\psi\exp\{t_1(\theta+\mu-\psi)\}g^*(k,j,\theta+\mu-\psi,\delta+\mu-\psi,t_1)]^x \\ & \times[(1-p)\exp\{t_1(\theta-\delta)\}g(k,j,\theta-\delta,t_1)]^{1-x}.\end{aligned}$$

Type 5 incarceration in the only interval. When there is a type 5 incarceration in the first interval, there is only one interval. Suppose there are k street times and j unknown completed imprisonment lengths in the first interval and one type 5 incarceration. Then, the length of time t_1 is the observed value of the sum of random variables with distributions $\exp(\psi)$ (if $x=1$), $\Gamma(k,\theta)$, $\Gamma(j,\delta)$, and censored $\exp(\delta)$. Furthermore, it is known that an $\exp(\mu)$ random variable is at least as long as the sum of the second and third of these. Hence, the likelihood is

$$\begin{aligned}
& \frac{\delta^j\theta^k}{(j-1)!(k-1)!}\Bigg[(1-p)\int_0^{t_1}\int_0^{t_1-u} u^{k-1}(t_1-u)^{j-1} \\
& \exp\{-u(\theta+\mu)-v(\delta+\mu)-\delta(t_1-u-v)\}dudv\Bigg]^{1-x} \\
& \times\Bigg[\psi p\int_0^{t_1}\int_0^{t_1-u}\int_0^{t_1-u-v} u^{k-1}(t_1-u)^{j-1} \\
& \exp\{-(\theta+\mu)u-v(\delta+\mu)-w\delta-(t_1-u-v-w)\psi\}dwdvdu\Bigg]^x \\
& = \frac{\delta^j\theta^k}{(j-1)!(k-1)!}\Bigg\{\frac{\psi p}{\delta-\psi} \\
& \Bigg[\exp(-t_1\psi)g^*(k,j,\theta+\mu-\psi,\delta+\mu-\psi,t_1)-\exp(-t_1\delta)g^*(k,j,\theta+\mu-\delta,\mu,t_1)\Bigg]\Bigg\}^x \\
& \times\Bigg\{(1-p)\exp(-t_1\delta)g^*(k,j,\theta+\mu-\delta,\mu,t_1)\Bigg\}^{1-x}.
\end{aligned}$$

This is equivalent to multiplying f_C by the factor

$$H_1(k,j,1) = \left\{ \frac{\psi p}{\delta - \psi} \left[\exp\left\{ t_1(\theta + \mu - \psi) \right\} g^*(k,j,\theta + \mu - \psi, \delta + \mu - \psi, t_1) \right.\right.$$
$$\left.\left. - \exp\left\{ t_1(\theta + \mu - \delta) \right\} g^*(k,j,\theta + \mu - \delta, \mu, t_1) \right] \right\}^x \tag{3.12}$$
$$\times \left\{ (1-p) \exp(t_1(\delta + \mu - \delta) g^*(k,j,\theta + \mu - \delta, \mu, t_1) \right\}^{1-x}.$$

This formula is also correct in the special cases in which $j = 0$.

3.3.3 The likelihood for one inmate

Let the numbers $n_{1,i}$, $n_{2,i}$, $n_{3,i}$, $n_{4,i}$, and $n_{5,i}$ denote the numbers of incarcerations of types 1, 2, 3, 4, and 5 respectively in a given interval I_i. The H_i factors involve the g and g^* functions with various different values of the integer arguments. These values, in turn, depend on how many missing incarcerations of the various types occur in interval i and on whether or not $i = 1$.

We constructed the H_i factors with three integer arguments. The first of the three arguments is always equal to the number of street times in the interval. This is equal to $n_{1,i} + n_{2,i} + n_{4,i} + 1$. Similarly, the second integer argument always equals the number of completed incarcerations of unknown length in the interval. This equals $n_{1,i} + n_{2,i} + n_{3,i}$. The third argument is always equal to $n_{5,i}$. It is convenient, then, to define

$$\begin{aligned} J_i &= n_{2,i} + n_{4,i} + 1 \\ U_i &= n_{2,i} + n_{3,i} \\ v &= \text{Number of type 1 incarcerations altogether.} \end{aligned}$$

The reason we leave $n_{1,i}$ out of these definitions is that $n_{1,i}$ changes from one configuration to the next, but J_i and U_i are functions of the inmate.

It follows that the likelihood for this configuration of missing data can be written

$$f_C(\lambda) \prod_{i=1}^{m} H_1(n_{1,i} + J_i, n_{1,i} + U_i, n_{5,i}), \tag{3.13}$$

where the general form of H_i is, for $i > 1$,

$$\begin{aligned} H_i(1,0,0) &= 1, \\ H_i(k,j,0) &= \exp\{t_i(\theta - \delta)\} g(k,j,\theta - \delta, t_i), \\ H_i(k,j,1) &= \exp\{t_i(\theta + \mu - \delta)\} g^*(k,j,\theta + \mu - \delta, \mu, t_i), \end{aligned}$$

and for $i = 1$ but $n_{5,1} = 0$,

$$H_1(k,j,0) = \left[P\psi \exp\{t_1(\theta + \mu - \psi)\} g^*(k,j,\theta + \mu - \psi, \delta + \mu - \psi, t_1) \right]^x$$
$$\times \left[(1-p) \exp\{t_1(\theta - \delta)\} g(k,j,\theta - \delta, t_1) \right]^{1-x} \tag{3.14}$$

Note that (3.14) reduces to (3.11) when $k = 1, j = 0$. When $i = 1$ and $n_{5,1} = 1$, we use (3.12).

Now, we must sum (3.12) over all possible configurations of missing data determined by the allocation of the type 1 missing incarcerations. The overall likelihood (without sampling bias correction) is

$$f_C(\lambda) \sum_{n_{1,1},\ldots,n_{1,m}=v} \binom{v}{n_{1,1}+\ldots+n_{1,m}} \prod_{i=1}^{m} H_i(n_{1,i}+J_i, n_{1,i}+U_i, n_{5,i}), \tag{3.15}$$

where the sum is m-fold over all partitions of v into m parts. For small values of v and m, calculation of this sum is a feasible task. Finally, we must divide (3.13) by the bias correction (2.1).

4. USE OF MULTIPLE IMPUTATION

4.1 Multiple Imputation in General

Rubin (1978) proposed the technique of multiple imputation for making inferences when some observations are missing. Let Λ denote the full set of parameters. The procedure is as follows:

- Approximate the posterior distribution of the parameters based on the data which has been observed. Call the approximate posterior density f_A and call the true posterior density f_T.
- Generate a large number N of values of the parameters $\lambda_1, \ldots, \lambda_N$ according to this approximate posterior.
- For each i, calculate the importance ratio

$$R_i = \frac{f_T(\lambda_i)}{f_A(\lambda_i)} \quad \text{for } i = 1, \ldots, N.$$

- Resample a small number ℓ of the existing λ_i's with probabilities proportional to R_i.
- For each of the ℓ resampled values λ_i, generate observations corresponding to the missing data using the conditional distribution of the missing data given the observed data and given that $\Lambda = \lambda_i$.
- Perform a complete data analysis for each of the ℓ sets of imputed values.
- Combine the complete data analysis in some sensible way to form an overall analysis.

The last step is often the hardest, because it must be formulated separately for each application. When comparing multiple imputation to the model for missing data discussed in this paper, we will refer to the latter as the *true posterior* approach, for lack of a better term.

4.2 Specific Implementation

We must approximate the posterior distribution of the Θ_i, A, B, Δ, M, Ψ, and P in order to implement multiple imputation. For the calculations in this paper, we will approximate the true posterior formed using the likelihood constructed in Section 3. How we do this is described in Section 4.2.1. To simulate the missing data, once we have the selected values of Λ, we need the conditional distribution of the missing data given the observed data and the parameters. We describe how we do this in Section 4.2.2.

4.2.1 Generation of parameters values

To generate the parameter values, we must first approximate the posterior distribution. The approximate distribution of the parameters A, B, Δ, $\frac{P}{1-P}$, Ψ, and M will be a multivariate log-normal distribution, with mean vector and covariance matrix of the logarithms calculated from an approximation to the true posterior distribution. This may turn out to be a dreadful approximation, but we can check it by computing the importance ratios which will be used for subsampling. The closer the importance ratios are to being equal to each other, the better the approximation is. The importance ratios would all equal 1 if the approximate and true posteriors were identical.

For the Θ_j for inmates who had partially missing data, we will approximate the conditional distribution of Θ_j, given the sampled values of the other parameters, by an exponential distribution with the same mean as the conditional posterior distribution of Θ_j. We will generate 500 sets of parameter values. In each set, there will be one Θ_j, for each inmate j who had partially missing data.

4.2.2 Generation of missing data

We will try to be more careful in generating the missing data from its exact distribution given the observed data. To generate the missing and partially missing imprisonments, we must first know what the configuration of missing data is relative to the observed data. That is, for a fixed number v of type 1 incarcerations, we must generate the numbers of type 1 incarcerations in each of the m intervals. Each term in the sum (3.13) is the joint probability-density of the observed data conditional on the parameters. Hence the conditional probability (given the parameters and the observed data) of a particular configuration of missing data, say with $n_{1,i}$ type 1 missing incarcerations in interval i for $i = 1, \ldots, m$ is

$$\frac{\binom{v}{n_{1,1},\ldots,n_{1,m}} \Pi_{i=1}^{m} H_i(n_{1,i} + J_i, n_{1,i} + U_i, n_{5,i})}{\Sigma_{n_{1,1}+\ldots+n_{1,m}=v} \binom{v}{n_{1,1}\ldots,n_{1,m}} \Pi_{i=1}^{m} H_i(n_{1,i} + J_i, n_{1,i} + U_i, n_{5,i})}. \tag{4.1}$$

Once we have used (4.1) to generate the configuration of missing data, we need the conditional distribution of the missing data given the configuration. All we really need to generate is the amount of street time in each interval and possibly the current incarceration, since the rest of the time in each interval is spent in prison.

The conditional distribution of the time spent on the street given that the sum of the street times and incarceration times equals t_i, can be obtained as follows. First, suppose that there is no type 5 incarceration in interval I_i and that either $i \neq 1$ or the inmate is not a late starter. The total time is the convolution of the street time and incarceration time. Suppose there are k street times and j missing incarceration times. Then the density of the sum at the observed value is

$$\exp(-t_i\delta)\theta^k \delta^j g(k, j, \theta - \delta, t_i).$$

Hence, the conditional density of the street time U given the sum is

$$\frac{u^{k-1}(t_i - u)^{j-1} \exp\{-u(\theta - \delta)\}}{(k-1)!(j-1)!g(k, j, \theta - \delta, t_i)} \quad \text{for } 0 < u < t_i. \tag{4.2}$$

Although this distribution is difficult to generate data from, we can use the acceptance-rejection method (cf. Law and Kelton, 1982, p. 250) by first generating $\beta(k, j)$ random variables, multiplying by t_i, and noting that the factor $\exp\{-u(\theta - \delta)\}$ is bounded between 1 and $\exp\{-t_i(\theta - \delta)\}$.

Next, suppose that there is a type 5 missing incarceration in the interval, but that either $i \neq 1$ or the inmate is not a late starter. The density of the total time is now

$$\exp(-t_i\delta)\theta^k \delta^j g^*(k, j, \theta + \mu - \delta, \mu, t_i),$$

where j is now the number of missing incarcerations that are not type 5. The conditional joint density of the street time U and the completed incarceration time V is

$$\frac{u^{k-1} v^{j-1} \exp\{-u(\theta + \mu - \delta) - v\mu\}}{(k-1)(j-1)!g^*(k, j, \theta + \mu - \delta, \mu, t_i)} \quad \text{for } 0 < u + v < t_i.$$

We can use multivariate acceptance-rejection by first generating Dirichlet random variables with parameters $(k, j, 1)$, multiplying by t_1, and noting that the exponential factor is bounded. Obtain the current incarceration length by subtracting $U + V$ from t_1. If $j = 0$, the density of the street time U is

$$\frac{u^{k-1}\exp\{-u(\theta+\mu-\delta)\}}{(k-1)!g(k,1,\theta+\mu-\delta,t_i)} \quad \text{for } 0 < u < t_i.$$

We can generate U as we did before, using Beta random variables.

Similarly, if $i = 1$ and the inmate is a late starter, but there is no type 5 incarceration in the first interval, then the density of the total time is

$$\psi\exp(-t_1\psi)\theta^k\delta^j g^*(k,j,\theta+\mu-\psi,\delta+\mu-\psi,t_1),$$

and the conditional density of U, the street time in career, and V the missing incarceration time given the total is

$$\frac{u^{j-1}v^{j-1}\exp\{-u(\theta+\mu-\psi)-v(\delta+\mu-\psi)\}}{(k-1)!(j-1)!g^*(k,j,\theta+\mu-\psi,\delta+\mu-\psi,t_1)} \quad \text{for } 0 < u+v < t_1.$$

We must generate V and obtain the street time as $t_1 - V$.

Finally, suppose that there is a type 5 missing incarceration in the only interval and the inmate is a late starter. Let j be the number of remaining missing incarcerations and k the number of street times. The conditional joint density of the street time in career U, the completed incarceration time V and the current incarceration length W is

$$\frac{u^{k-1}v^{j-1}\exp\{-(\theta+\mu)u-(\delta+\mu)v-\delta w\}}{(k-1)!(j-1)!\exp\{t,(\theta+\mu-\psi)\}g^*(k,j,\theta+\mu-\psi,\delta+\mu-\psi,t_1)-\exp\{t_1(\theta+\mu-\delta)\}g^*(k,j,\theta+\mu-\delta,\mu,t_1)},$$

for $0 < u+v+w < t_1$. Such random variables can be generated using multivariate acceptance-rejection beginning with Dirichlet random variables with parameters (k, j, 1, 1) multiplied by t_1. If $j = 0$, the joint density of U and W is

$$\frac{u^{k-1}\exp\{-u(\theta+\mu-\psi)-w(\delta-\psi)\}}{(k-1)!g^*(k,1,\theta+\mu-\psi,\delta-\psi,t_1)} \quad \text{for } 0 < u+w < t_1.$$

Acceptance-rejection. In each of the cases above, acceptance-rejection is particulary simple to implement. The ratio of the beta or Dirichlet density to the dominating function is always equal to the exponential factor divided by its maximum over the interval $[0, t_i]$. Hence, we generate a uniform X and a beta or Dirichlet Y. Plug Y into the exponential and divide by the maximum. If this ratio is larger than X, accept Y, otherwise try again.

When the exponential factor can get very large, it is very difficult to accept an observation. In these cases, the distribution is highly concentrated near the mean, so we merely generate the mean of the distribution. For example the mean of a random variable with density (4.2) is

$$\frac{g(k+1,j,\theta-\delta,t_i)}{g(k,j,\theta-\delta,t_i)}.$$

4.2.3 Full data analyses

There are many possible full data analyses which could be performed. We choose to calculate the posterior means and standard deviations of certain functions of the parameters. Those functions are the transformations used by the program of Smith *et al.* (1984) to transform the parameters to the real line. The functions are natural logarithm for the parameters, A, B, Δ, M, Ψ, and logit for P. We will also calculate the means of the original untransformed parameters.

4.2.4 Late starters

In the formulation of the model we have given, some of the inmates are late starters. For these inmates, the time from age 18 until the start of the adult incarceration career is unknown. We could treat this as missing data and use multiple imputation to generate values. We choose not to do this because such a large portion of the sample would then require at least some imputation. That would greatly increase the number of Θ_j's which need to be generated, thereby reducing the advantage gained throught multiple imputation.

However, we must still generate the first street time for those inmates who are late starters. If there is only one street time in the first interval (after generating the configuration of missing data), then the generated street time in the first interval is the length of the first street time. If there is more than one, say k street times in the first interval, with a sum of u, then we must generate one of them. The conditional distribution of one out of $k\,\exp(\theta)$ random variables given that the sum is u is $u \times \beta(1, k-1)$. Hence, we can easily generate the first street time.

5. NUMERICAL COMPUTATION

5.1 Setting Prior Distributions

To set the prior distributions for the parameters, we referred to other analyses of different criminal justice data sets. Ahn (1986) performed an analysis of an arrest data set from the state of Michigan for the years 1974-1977. The model used by Ahn was similar to the one we are using here. In his model, the incarceration process is a randomly thinned version of the arrest process, thinned by the selection of which arrestees do not go to prison. We will base our prior estimates of the parameters on the results of Ahn (1986), while making our prior precision small to reflect a great deal of uncertainty. We expect A to be slightly less than 2, so we use the prior

$$A \sim \exp(.55).$$

We expect the average incarceration rate to be around .017, so this should be about where A/B is *a priori*. Since the prior mean of A is 1/.55, we set the prior mean of B to equal $1/(.017 \times .55) = 107$. So we give B the prior distribution

$$B \sim \Gamma(.5, .00466).$$

We expect Δ to be around .042, so we give the prior

$$\Delta \sim \Gamma(.5, 11.9).$$

Both Ahn (1986) and Blumstein and Cohen (1985) suggest that the career length is around 10 years, so we give the prior

$$M \sim \Gamma(.5, 60).$$

We expect Ψ to be around .165, so we give the prior

$$\Psi \sim \Gamma(2.0, 12.12).$$

The reason the first parameter is so large is that we do not believe that it is very likely that Ψ is near zero, since this would mean that the expected time to the start of the career would be very long. We do not have any prior empirical data concerning P, so we use the prior

$$P \sim \beta(1, 1).$$

5.2 Finding Posterior Distributions

Due to the large amount of time which the likelihood function requires for calculation, we will describe results obtained from only a sample of the data. We chose a subsample of 200 inmates using simple random sampling without replacement. The result was a subsample consisting of 159 inmates with complete data and 41 with incomplete data.

To find the posterior distribution of the parameters A, B, Δ, M, Ψ, P, we need to integrate out the Θ_j parameters from the likelihood. Take the function in (3.15) based on the data from inmate j and divide it by the probability of being in prison at the time of the survey, that is, the sum (2.1). Evaluate the ratio at $\theta = \theta_j$ and call the result $f_j(\theta_j, \delta, \mu, \psi, p)$. The likelihood function for the parameters Λ_1 is then

$$L_1(\lambda_1) = \prod_{j=1}^{N} \int_0^\infty f_j(\theta_j, \delta, \mu, \psi, p) \frac{b^a}{\Gamma(a)} \theta_j^{a-1} \exp(-b\theta_j) d\theta_j, \tag{5.1}$$

where N is the number of inmates. We used Gauss-Laguerre integration with 15 points to perform the integrals in (5.1). We used the program of Smith *et al.* (1984) to find the posterior distribution of the parameters Λ_1. The (double-precision) calculations were performed in parallel on twelve Digital Equipment Corporation VAXstation II computers using the method of Eddy and Schervish (1986). Even so, it took about 9.9 seconds (elapsed time) to calculate (5.1) for the subsample of 200 inmates for each value of λ_1. The program of Smith *et al.* (1984) does Gauss-Hermite integration in six dimensions. This requires 729 evaluations of (5.1) if one wishes to use 3 point Gauss-Hermite integration. (About 2 hours per iteration.) Our calculations are all based on 2 and 3 point Gauss-Hermite integration in six dimensions for each iteration.

The posterior means and standard deviations of the transformed parameters are given in Table 2 along with the means of the original parameters. The actual procedure which produced the posterior moments in Table 2 was the following. We began by finding an approximate mode to the posterior density using the minimization program of Olsson (1974). This program calculates several dozen values of the log of the posterior density in the vicinity of the mode. These values were fit by a quadratic function in order to find an initial covariance matrix estimate. Then we performed three iterations of the program of Smith *et al.* (1984) using 2 point Gauss-Hermite integration until the answers were nearly stable. Then we formed initial posterior moments in Table 3. (These were later used to generate the multiple imputations.) We then switched to 3 point Gauss-Hermite integration and performed two more iterations of the program of Smith *et al.* (1984) until the posterior moments stabilized.

5.3 Doing Multiple Imputation

The multiple imputations were based on the initial posterior moments in Table 3.

There was very large correlation between $\log(A)$ and $\log(B)$ (not surprising, since A/B is the expected value of Θ). So we used the full covariance matrix for the parameters which was calculated along with the moments in Table 3. We imputed 500 sets of parameters for the subjects with missing data based on the posterior moments in Table 3.

To subsample the 5 imputed parameter sets, we used a probability proportional to size (pps) sampling scheme found in Brewer and Hanif (1983, p. 21) called the "ordered systematic procedure". This procedure is attributed to Madow (1949). We are not concerned with the systematic nature of the sample, since our original ordering of the sampled parameters was random. If the importance ratios vary so widely that pps sampling was impossible, we merely selected the parameters with the highest importance ratios. The logarithms of the importance ratios which we observed are given in histogram form in Figure 1. (The average has been substracted, since we did not bother to calculate the scale factor for the true posterior.) As

Original Parameter	*Transformed*		*Original Mean*
	Mean	*Standard Deviation*	
A	-0.10325	0.12915	0.90945
B	1.5047	0.19736	4.59127
Δ	-3.4789	5.8729×10^{-2}	3.0894×10^{-2}
M	-4.2946	8.6571×10^{-2}	1.3694×10^{-2}
Ψ	-4.7901	9.8052×10^{-2}	8.3515×10^{-3}
P	1.5321	0.17306	0.82091

Table 2. *Posterior Means and Standard Deviations of Transformed Parameters*

Original Parameter	*Transformed*		*Original Mean*
	Mean	*Standard Deviation*	
A	-0.10270	0.12616	0.90959
B	1.5048	0.19326	4.58772
Δ	-3.4777	6.0292×10^{-2}	3.09358×10^{-2}
M	-4.2918	8.8255×10^{-2}	1.37341×10^{-2}
Ψ	-4.7853	9.9539×10^{-2}	8.39280×10^{-3}
P	1.5246	0.14762	0.82018

Table 3. *Initial Posterior Means and Standard Deviations of Transformed Parameters*

expected, there is quite an enormous spread in these values, indicating that the posterior approximation is not very good, at least as far as particular values of the density is concerned. The selected parameters (excluding Θ_j values) are given in Table 4.

Next, we imputed the missing data, for the 41 inmates who needed it, for each of the five selected parameter sets. We then found approximate posterior modes for each imputed data set using the program of Olsson (1974). We used the posterior covariance matrix which was used for generating the imputed data as the initial covariance matrix for each imputed data set

```
MIDDLE OF   NUMBER OF
INTERVAL    OBSERVATIONS
 -120.        1    .
 -100.        1    .
  -80.        9    **
  -60.       25    *****
  -40.       43    *********
  -20.      104    *********************
    0.      118    ************************
   20.      121    *************************
   40.       57    ************
   60.       17    ****
   80.        3    +
  100.        1    .
```

Figure 1. *Histogram of Logarithms of Importance Ratios*

Imputation Number	*Original Parameter*					
	A	*B*	Δ	M	Ψ	*P*
1	0.81356	3.74296	3.56894×10^{-2}	1.60682×10^{-2}	7.78145×10^{-3}	0.82856
2	1.05498	5.76550	3.00267×10^{-2}	1.70584×10^{-2}	8.44543×10^{-3}	0.85776
3	1.24337	7.46261	3.01712×10^{-2}	1.88458×10^{-2}	1.12511×10^{-2}	0.77151
4	0.88171	4.23473	3.59601×10^{-2}	1.44294×10^{-2}	8.88867×10^{-3}	0.82025
5	1.07033	6.07386	3.34809×10^{-2}	1.58391×10^{-2}	7.91133×10^{-3}	0.84238

Table 4. *Imputed Parameter Values*

and ran three iterations of the program of Smith *et al.* (1984) using 2 point Gauss-Hermite integration and two iterations using 3 point Gauss-Hermite integration on each of five full data sets, just like we did with the original data. For some of the data sets, the posterior moments had not yet stabilized, so we ran additinal iterations with 2 point Gauss-Hermite integration. Finally, we produced means and standard deviations based on the imputed samples as follows. For the means, we took the averages of the means produced in the five full data analysis. For the standard deviations, we used the square root of

$$S^2 + \frac{5+1}{5(5-1)} \sum_{i=1}^{5} (Q_i - \hat{Q})^2,$$

where for each parameter, S^2 is the average of the squares of the five standard deviations from the full data analyses, $\hat{Q}$ is the average of the five means, and $Q_i, i = 1, \ldots, 5$ are the five means. This is the formula suggested in Rubin and Schenker (1986) for the variance of a normal mean. The results are given in Table 5.

Original Parameter	*Transformed*		*Original Mean*
	Mean	*Standard Deviation*	
A	-8.11414×10^{-2}	0.18618	0.93481
B	2.09188	0.34977	8.46975
Δ	-3.25786	6.24766×10^{-2}	3.85423×10^{-2}
M	-4.3733	8.71064×10^{-2}	1.26569×10^{-2}
Ψ	-4.65962	9.80838×10^{-2}	9.51443×10^{-3}
P	1.50326	0.17293	0.81665

Table 5. *Posterior Means and Standard Deviations From Imputed Data*

5.4 Comparisons

As far as the similarity of results is concerned, the tabled values indicate that the multiple imputation produced similar results to the true posterior approach. The similarity was due in part to the fact that we only did 2 and 3 point Gauss-Hermite integration. With only 2 or 3 point integration, there are fewer degrees of freedom for variation within a small number of iterations starting from the same initial values.

With regard to time, the imputation of the data itself took almost four hours. Most of the work was done on a single VAXstation II, while the likelihood calculations were done in parallel on 12 VAXstations. It is not surprising that the imputation took this long, given the complicated nature of the imputation process described in Section 4.2 as well as the fact that each imputation requires one evaluation of the likelihood function. For each imputed data set, finding the posterior mode took about 24 minutes on the average compared to 1 hour and 5 minutes for the true likelihood approach. Each 2 point Gauss-Hermite iteration of the program of Smith *et al.* (1984) took about 10 minutes for both approaches and each 3 point iteration took about 2 hours. It appears that the only time saved using multiple imputation is the time it took to derive the true posterior distribution theoretically. Also, the simplicity of full data analyses is partially outweighed by the complicated methods needed to impute the missing data.

6. PREVIOUS AND FUTURE WORK ON THIS DATASET

Lehoczky and Schervish (1986) (*LS*) analyzed a subset of the data described here using empirical Bayes methods. They used only those offenders who supplied complete data on all incarcerations. We will briefly summarize the results which they obtained. They found a substantial probability that inmates do drop out of their incarceration careers. The alternating renewal process model without career termination cannot account for the dramatic difference between the population age distribution and the age distribution of sampled inmates. For example, (2.4) is an increasing function of age. This would suggest that the sample of inmates, if there were no career termination, would have an age distribution with more older people and fewer young people than the population at large. But the ages of examples inmates are generally much lower than the ages in the population at large.

Also, *LS* found that there were different imprisonment lengths for violent and non-violent crimes. One thing we would like to do with our model is to extend it to treat different crime types differently. This can be done both by considering different imprisonment lengths, but also different street times before different crime types. *LS* also found substantial differences between the street times before violent and non-violent crimes. We would like to be able to extend our model to deal with different sentences and different street times for different crime types.

With regard to late starters, *LS* found that some of the inmates who did not have juvenile incarcerations still seemed to begin their adult careers much sooner than others. This suggests that late starting is not equivalent to absence of juvenile record, as we have assumed. It is a simple matter to extend the model to deal with this case, however, the resulting model has an extra parameter and it is now very difficult to fit using available computer programs.

Another possible extention of the model, to make things more realistic, is to try to model the missing data more carefully. It may be that some of so-called missing data is merely a misreported count. It may also be that missing incarcerations are shorter than the reported ones and they are missing because the inmates do not recall the details of such brief incarcerations.

Of course, the most immediate concern is to extend the analyses described above to the full data set. It would seem silly to worry about 20% of the inmates having partially missing data if one were to ignore 98% of the data due to computational considerations.

7. CONCLUSIONS

We do not claim to be experts in criminal justice, hence we will draw no conclusions as to what implications this work has for the criminal justice system in America. Our conclusions will concern the analysis of large data sets with substantial quantities of missing data in general.

The most striking, and unfortunately depressing, conclusion which we can draw concerns the computational feasibility of implementing a Bayesian analysis for such a large data set and such a complicated model. With the current state of computer technology, it is still very difficult to perform fully Bayesian analyses which require numerical integrations in multiple dimensions. Even if the data set had only been 200 observations, a tremendous amount of computation is required to fit the model. The problem, in this case is only partly due to the Bayesian paradigm. The likelihood function takes a long time to calculate. Even maximum likelihood estimation or empirical Bayes methods would face similar computational difficulties with this model. The Bayesian paradigm only compounds the problem by requiring such a large number of likelihood evaluations.

It is not easy to feel confident about results based on only 2 or 3 point Gauss-Hermite integration in 6 dimensions. It is also not easy to do 4 or higher point integration in 6 dimensions. Other methods, such as those of Tierney *et al.* (1986) may prove helpful, but they also require many likelihood evaluations. To check our results, we sampled a second set

Original Parameter	*Transformed*		*Original Mean*
	Mean	*Standard Deviation*	
A	-0.16627	0.12631	0.85360
B	1.2109	0.19433	3.42053
Δ	-3.5734	6.1554×10^{-2}	2.81142×10^{-2}
M	-4.3379	8.7978×10^{-2}	1.31140×10^{-2}
Ψ	-4.8406	9.8836×10^{-2}	7.94124×10^{-3}
P	1.6025	.17593	0.83094

Table 6. *Posterior Means and Standard Deviations From Second Subsample*

of 200 inmates. The posterior means and standard deviations are reported in Table 6. These are fairly close to those in Table 2.

With regard to multiple imputation, the most difficult decision to make is how to approximate the posterior distribution of the parameters in order to generate parameter values. We essentially did a true posterior analysis first in order to approximate the posterior for the imputation. In general, this option would not be available. In fact the main reason for doing multiple imputation is to avoid having to do a true posterior analysis.

Finally, the authors take responsibility for the correctness of the computer programs which they wrote to evaluate the various likelihood functions. Errors in any of them (and there were plenty found along the way) could greatly change the values of the estimated parameters. However, we would hope that they would not greatly change the comparisons between the true posterior approach and the approach of multiple imputation.

REFERENCES

Ahn, C. W. (1986). Hierarchical Stochastic Modelling of Arrest Careers. *Ph. D. thesis*, Departament of Statistics, Carnegie-Mellon University.

Avi-Itzhak, B. and Shinnar, R. (1973). Quantitative models in crime control. *Journal of Criminal Justice* **1**, 185–217.

Blumstein, A. and Cohen J. (1985). Estimating the duration of adult criminal careers. *Bulletin of the International Statistical Institute* **LI:4**, 29.1.

Brewer, K. R. W. and Hanif. M. (1983). *Sampling With Unequal Probabilities.* New York: Springer-Verlag.

Eddy, W. F. and Schervish, M. J. (1986). Discrete-finite inference on a network of VAXes. *Computer Science and Statistics: Proceedings of the 18th Symposium on the Interface.* (T. J. Boardman, Ed.). 30-36.

Law, A. M. and Kelton, W. D. (1982). *Simulation Modeling and Analysis.* New York: McGraw-Hill.

Lehoczky, J. P. and Shervish, M. J. (1986). Estimation of incarceration an crimininal careers using hierarchical models. Final report Bureau of Justice Statistics contract 85-BJ-CX-0004, Carnegie-Mellon University.

Madow, W. G. (1949). On the theory of systematic sampling, II. *Annals of Mathematical Statistics* **20**, 333–354.

Olsson, D. M. (1974). A sequential simplex program for solving minimization problems. *Journal of Quality Technology* **6**, 53–57.

Rubin, D. B. (1978). Multiple imputation in sample surveys - A phenomenological Bayesian approach to nonresponse. *Proceeding of the Survery Research Methods Section, American Statistical Association* 20–34.

Rubin, D. B. and Schenker, N. (1986). Multiple imputation for interval estimation from simple ramdom samples with ignorable nonresponse. *J. Amer. Statist. Assoc.* **81**, 366-374.

Smith, A. F. M., Skene, A. M., Shaw, J. E. H., Naylor, J. C. and Dransfield, M. (1984). The inmplementation of the Bayesian paradigm. *Communications in Statistics, Theory and Methods* **14**(5), 1079–1102.

Tierney, L., Kass, R. E. and Kadane, J. B. (1986). Approximation of posterior expectations and variances using Laplace's method. *Tech. Rep.***385**, Department of Statistics, Carnegie-Mellon University.

DISCUSSION

D. PEÑA (*Universidad Politécnica de Madrid*)

This is a very interesting paper and the authors should be congratulated by their deep and detailed approach to the treatment of partially missing data in alternating renewal processes.

The authors deal in the paper with four different distributions whose parameters they want to estimate:

(1) The distribution of prison time, s, that is assumed to be exponential with the same parameter $\triangle$ for all inmates and all the incarcerations.

(2) The distribution of street time, T, that is assumed to be exponential with parameter θ, that varies among inmates following a Gamma distribution.

(3) The distribution of career length, Y, that is assumed to be exponential with the same parameters μ for all criminals.

(4) The distribution of starting time, that is a mixed distribution of a point mass at 0 and an exponential distribution.

The assumptions about these distributions can and should be checked using the data set:

(1) The data on prison time can be plotted to check the shape of the distribution. Besides, exploratory data analysis could provide information about whether the distribution is the same for all incarceration (it could be argued a priori that the distribution of prison time for the first incarceration may be different from the distribution for the second or third incarceration).

(2) The distribution of street time is $f(s|\theta_i)$ where $\theta_i \sim \Gamma(A, B)$.
Therefore the predictive distribution for street time is

$$f(s) = \int f(s|\theta)g(\theta)d\theta$$

that accordingly, will be an inverted Beta distribution (Aitchinson and Dunsmore (1975)), a result that can be checked after the estimation using a goodness of fit test.

(4) The distribution of starting time for inmates who have not served as juveniles can also be checked to see if the exponential distribution is a sensible model given the data.

A crucial assumption of the model is the independence between the street and prison times. It would be interesting to do a rough check on that asumption, that, moreover, could provide some light about the effect of incarceration to prevent future crimes. If there is no relationship between the time in prison and the following length of street time we should be very worried about our social punishment system!

A check of these hypothesis given the data will also provide information about possible stratification to improve our understanding of the criminals career. For instant, as indicated in the paper the relationship between prison and street time may be different for property crimes that for violent crimes, or the distribution of street and prison time —even of starting time and drop out— may be different for different types of crimes.

I believe that this type of analysis also not only allow a more realistic model and a better understanding of the problem under study, but will provide some information about the treatment of missing data.

I fully agree that multiple imputation as developed by Rubin and others avoids the drawback of obtaining inferences that do not take into account the variability from the lack of knowledge of the missing values. However, the simulated draws from a distribution implied by multiple imputation should be done from the posterior distribution of the missing data given all the information. According to the paper, the data set includes information about the type of offence, and it seems that this information could be used to improve the estimation of the missing prison time. For instance, the paper says that it has been found elsewhere that the distribution of prison time for violent crimes will have a higher mean than for non-violent crimes, result that seems clear a priori. Why is the information about the type of offence not considered to obtain more realistic multiple imputation about the prison time?

Finally, I believe that a key point in analyzing this type of data is to use sensible approximations to simplify the problem and allow a sequential process of learning from the data. Thus, the following approach seems to be adequate in this case: First, to analyze the existing data and check the main hypothesis; second, to introduce type 1 and 4 of the missing values that, according to table 1, represent 93% of the missing incarcerations. It is doubtful that taking into account afterwards the missing data of type 2, 3 and 5 will change much the previous results.

I hope that these comments will serve to stimulate the authors to do further analysis of their data set and I want, in closing, to thank them for this interesting and thought-provoking paper.

J. C. NAYLOR (*Trent Polytechnic*)

This interesting example poses some quite extreme computational problems which I should like to comment on. Clearly the likelihood function is extremely expensive to calculate and, if alternative models and analytical approximations are not available, its computation must be examined critically. The likelihood is a product of integrals which need to be approximated and a 10 point Gauss-Laguerre rule is proposed. The form of the integrand is in fact roughly a $Ga(a + k + 1, T + b)$ density for which some rescaled generalised-Laguerre rule (see, for example, Davis and Rabinowitz page 174) would be more appropriate. An alternative, that may be easier to apply, is a Gauss-Hermite rule for log () for which a well scaled 4 point rule would probably be adequate. Two iterations to get better scaling would still be a 20% saving over the straight Laguerre rule.

The use of 6-dimensional product rules is not efficient for this problem and two points in each direction is not enough for the iterative algorithm to work properly. An alternative is provided by rules having spherical symmetry (see, for example, Stroud 1972). Substantial savings can be made using a degree 5 rule with just 76 points or degree 7 rule with 305 points, as compared with the corresponding product rules which require 729 and 4096 points respectively. These rules are available in the Nottingham BAYES 4 integration package which also includes facilities for Gauss-Hermite integrations within a likelihood function.

With such a large sample size we might expect, with the right parametrisation, the posterior density to be very nearly normal. A computationally innovative approach would be to consider the sample as a sequence of smaller samples. The results of a first analysis (on say 200 data points, and involving some investigation of alternative parametrisations) being used to define a prior for the next sub-sample and so on until all the data has been considerd. Alternative partitions of the sample could be used as a check on the acurarcy of this approach.

J. E. H. SHAW (*University of Warwick*)

Beware of 2 point Gauss-Hermite integration (Section 5.2): there is no way to disentangle the estimates of the normalising constant and the first and second moments. For example, in one dimension, even if the integration rule has been centred so that the two evaluations of the posterior density are equal, we cannot tell if we are near the mode of a diffuse density, or in the tails of a peaked density.

An alternative to Gauss-Hermite integration is provided by *spherical integration rules*; the released version of the BAYES4 package (October 1987) contains a 6-dimensional spherical rule with 272 nodes. But even this may be prohibitively expensive with a likelihood function as complicated as the authors'.

The computing time needed could probably be reduced by a two-stage analysis: first analyse only complete data, then use an approximation to the resulting posterior as the prior for analysing the incomplete data. Most of the iterations needed in the numerical integration should involve only the simpler likelihood (if several iterations are needed at the second stage, then the model for missing data is presumably inadequate). If no suitable approximation to the posterior from stage 1 can be found, then the approximate prior for the second stage might be a discretised distribution after clustering (see Shaw, these proceedings). The clustering can be defined iteratively depending on the discretised posterior from stage 2, nodes with relatively high posterior weight being replaced by the corresponding original nodes before clustering. This procedure reduces the number of evaluations of the complicated second stage likelihood.

REPLY TO THE DISCUSSION

We wish to thank the discussants for some good ideas to help make the analysis both more realistic and more computationally feasible.

In reply to Professor Peña, we will be the first to agree that exploratory data analysis should be (and, in fact, has been) done to check the assumptions. We will only report briefly on these analyses. We used only those inmates with complete data for the exploratory analyses. We drew histograms of the lengths of street times and the lengths of imprisonments for inmates with equal numbers of imprisonments to avoid confounding the effect of the number of imprisonments with the lengths. We also separated out first, second, third, etc. street times, to see if the distributions looked the same. We also disaggregated by type of crime. Being a die-hard Bayesian, one of the authors could not bring himself to perform a goodness of fit test to check the assumptions. We can report that, after disaggregating by type of crime and number of imprisonments and sequence order of imprisonment, most of the plots didn't have very many points in them. Those that did have many points showed some slight positive association between earlier and later street times. The plots of street times vs. imprisonment times showed little pattern, but there were not many points in each one, since most sampled inmates were serving their first incarceration. Other exploratory analyses included disaggregation by race, and existence of juvenile incarcerations. These suggested that blacks spent slightly less time on the street between incarcertations and slightly less time in prison per incarceration. They also suggested that those with juvenile incarcerations spent less time on the street before their first adult incarceration than did those with no juvenile record and that they also spent more time in prison during their first adult incarceration. Later adult incarcerations were of much the same length for both groups, and these were generally longer than the first incarcerations for those who had no juvenile record. These findings, and many others are reported by Lehoczky and Schervish (1986) together with many of the tables and graphs. We apologize for not saying anything about these analyses in the paper.

Professor Peña also suggests some ideas for making the model more realistic, such as differentiating certain crime types. If we ever get our hands on software that will allow us to expand the model, we intend to do precisely what Professor Peña suggests. In fact, Lehoczky and Schervish (1986) did some of that model expansion in a maximum likelihood analysis with a discrete parameter space. The idea of fitting the model to the complete data and then adding missing types 1 and 4 may also prove valuable. It will make the algorithms simpler, but the analyses we have done do not suggest that the time savings will be very great.

Both Drs. Naylor and Shaw suggest that we use spherical integration rules (as implemented in *BAYES 4*) for some of the six dimensions of integration. We agree. We await arrival of *BAYES 4* with bated breath. Spherical rules should also help us to expand the model, because we will not be subject to the exponential increase in time with the number of parameters. We intend to try to implement Dr. Naylor's suggestion of using reduced number of Laguerre points for a generalized rescaled Gauss-Laguerre integration. Dr. Shaw's suggestion of using a discrete approximation to the posterior sounds interesting, and should help cut down the computing time.

Another discussant referred us to Markov chain type models of illness and death processes as described by Chiang (1968). We believe that the model we described is a special case of such models. In fact, Ahn (1986) develops several theoretical results for general versions of such models in the context of arrest careers. The theory of these models seems to be easier to come by than good data analysis techniques.

We would also like to thank Dr. William DuMouchel for his (poolside) suggestion that we sum formula (2.3) of our paper over fewer values of t in order to save execution time. The function in (2.3) does not vary much for large values of t, and this should provide a significant time saving.

REFERENCES IN THE DISCUSSION

Chiang, C. L. (1968). *Introduction to Stochastic Processes in Biostatistics.* New York: Wiley.
Davis, P. J. and Rabinowitz, P. (1975). *Methods of Numerical Integration.* Academic Press.
Stroud, A. H. (1972). *Approximate Calculation of Multiple Integrals.* Englewood-Cliffs: Prentice-Hall.

BAYESIAN STATISTICS 3, pp. 307–326
J. M. Bernardo, M. H. DeGroot, D. V. Lindley and A. F. M. Smith, (Eds.)

Statistical Inference Concerning Hardy-Weinberg Equilibrium

D. V. LINDLEY
Somerset

SUMMARY

A probability model for the formation of genotypes from two alleles is given and expressed in terms of two parameters, α and β; $\alpha = 0$ corresponding to equilibrium. The likelihood for (α, β) given genotype data is found and an asymptotic theory provided. The posterior distribution of α is found providing estimation of α. A Bayesian test of the hypothesis that $\alpha = 0$ is also considered. The choice of prior distribution is discussed. Comparison is made with the usual tests of equilibrium proposed. An advantage of this approach is the provision, in α, of a measure of departure from equilibrium. There is a discussion of the general Bayesian procedure of investigating a scientific hypothesis.

Keywords: HARDY-WEINBERG EQUILIBRIUM; LIKELIHOOD; BAYES METHOD; POSTERIOR DISTRIBUTIONS; TRINOMIAL DISTRIBUTION; SIGNIFICANCE TESTS; SIGNIFICANCE LEVELS; INBREEDING; OUTBREEDING; NUMERICAL INTEGRATION; MAXIMUM LIKELIHOOD ESTIMATION; BOREL-KOLMOGOROV PARADOX.

1. STATISTICAL INTRODUCTION

A commonly occurring situation in science requires the testing of a hypothesis on the basis of data. The usual statistical formulation introduces a parameter α which takes the value zero, say, if the hypothesis is true and is otherwise non-zero. The statistician then describes the situation as one of testing the null hypothesis $\alpha = 0$ against the alternative $\alpha \neq 0$. It usually happens that although α is adequate to describe the hypothesis, it is inadequate to specify the probability structure of the data, so that a nuisance parameter β has to be introduced in addition to the parameter of interest.

The sampling-theory solution to the scientific problem is to construct a statistic $t = t(x)$, a function of the data x, and accept or reject the null hypothesis on the basis of t. The language sometimes differs and in place of rejection, the probability of rejection given the null value $\alpha = 0$ is quoted and described as the significance level. The only reference to the alternative hypotheses in this procedure resides in the construction of t, it being chosen so that the test has good power. But the scientist does not have to bother with this. He can rely on the statistician's wisdom in selecting the statistic: all he has to do is to calculate $t(x)$ and look up the significance level in tables. To him, all is automatic.

The presence of a nuisance parameter complicates the issue for, even when α is zero, β can still vary and the distribution of t with it, so that there is no unique significance level. Considerable ingenuity, backed by little rational argument, has been applied to this problem and good procedures have been devised that again only involve the scientist calculating the statistic and looking up the level. He need not concern himself with β or the alternatives.

The Bayesian solution to the scientific problem of testing a hypothesis is statistically simpler but scientifically harder. The coherent procedure for the elimination of the nuisance parameter is integration and the equivalent of the statistic is the posterior distribution of α. A combination of analysis and numerical methods enables the problem to be solved. However, for the scientist the task is harder. He is required to specify a joint probability distribution for α and β. The difficulties of doing this have been much exaggerated and can be overcome. However, it is necessary for the scientist to think about α and β, for even the null hypothesis $\alpha = 0$ can alternatively be described as $\phi = 0$ where ϕ is a 1-1 transformation of α, and the Borel-Kolmogorov paradox reminds us of potential difficulties. The essence of the problem is that the scientist has to think about the scientific problem, rather than adopt an automatic procedure devised by a statistician. The Bayesian argues that such thinking is to be encouraged and deplores the machine-like adoption of rote methods.

In the present paper a particular scientific hypothesis of genetical equilibrium is discussed. Special attention is paid to the choice of α and β, and to the form in which the final inference should be made. Aside from providing a solution to the genetics problems, the paper provides an example of the general, rational, Bayesian procedure for testing a scientific hypothesis. There is also some consideration of whether the value $\alpha = 0$ should be singled out for special treatment, or procedures more akin to estimation be used.

2. GENETICAL INTRODUCTION

At a single autosomal locus with two alleles, a diploid individual can be one of three possible genotypes

AA Aa aa

(Aa is indistinguishable from aA). The genotype frequencies in a population will be written

$$p_1 \quad p_2 \quad p_3$$

with $p_i \geq 0$ and $p_1 + p_2 + p_3 = 1$. Alternatively p_i may be thought of as the probability that a random individual will be of genotype i. The population is said to be in Hardy-Weinberg (HW) equilibrium if the probabilities may be written in the form

$$p^2 \quad 2p(1-p) \quad (1-p)^2$$

for some $p, 0 \leq p \leq 1$. It is an important genetical problem to test the hypothesis that a population is in HW equilibrium. To investigate this a random sample of N individuals is taken from the population and counts made of the numbers of each genotype yielding data

$$n_1 \quad n_2 \quad n_3$$

with $N = n_1 + n_2 + n_3$. The question to be addressed is whether these data support HW equilibrium.

3. STATISTICAL FORMULATION

The statistical description of the genetics is that there is a random sample (n_1, n_2, n_3) of N individuals from a trinomial distribution. The general trinomial has two parameters (p_1, p_2, p_3 with $p_1 + p_2 + p_3 = 1$) and it is desired to test the hypothesis that it really only depends on one parameter p, yielding trinomial parameters $(p^2, 2p(1-p), (1-p)^2)$. Alternatively expressed, we wish to test the hypothesis that $4p_1p_3 = p_2^2$, a relation clearly true in HW equilibrium, or equivalently that $4p_1p_3/p_2^2 = 1$. We therefore wish to test the hypothesis that a parameter takes the value 1 where, since the trinomial has two parameters, there is a single nuisance parameter, p.

The standard way to do this is due to Haldane (1954). His "exact" test is based upon the observation that if the null hypothesis of HW equilibrium is true the probability distribution of the data (n_1, n_2, n_3) given the values of $2n_1 + n_2$ and N does not depend on the nuisance parameter p. Easy calculation shows that

$$p(n_1, n_2, n_3 | 2n_1 + n_2, N) = \frac{N!}{n_1! n_2! n_3!} 2^{n_2} \frac{(2n_1 + n_2)!(n_2 + 2n_3)!}{(2N)!}$$

and the hypothesis is rejected if this probability for the data, and other data that exhibits more extreme departure from HW equilibrium yet with the same values of $2n_1 + n_2$ and N, is less than the prescribed significance level. Elston and Forthofer (1977) describe Haldane's test as "the obvious standard against which all other tests should be compared." Exact significance levels for the test are given by Vithayasai (1975). An admirable summary and comparison with other proposed tests is given by Emigh (1980). A Bayesian treatment has been provided by Pereira and Rogatko (1984).

Notice that Haldane's argument does not overtly mention α or β. What it does do is restrict the sample space to include only samples with the same values of N and $2n_1 + n_2$ as those observed. This eliminates a nuisance parameter, but only when the null hypothesis obtains. The power still depends on the nuisance parameter and is almost never mentioned by the geneticist and rarely by the statistician. In this, and similar situations, it is difficult to justify the restriction of the sample space.

As explained in section 1, the Bayesian approach requires the scientist to think about the genetical set-up with more care than the significance level procedure requires. To help in this, we formulate a model that includes both equilibrium $\alpha = 0$, and disequilibrium $\alpha \neq 0$, and introduces a natural nuisance parameter β. Giving α and β tangible meanings within the model hopefully will make it easier for the geneticist to analyze the data more sensibly.

4. A DISEQUILIBRIUM MODEL

The HW situation arises in the following way. Denote by p the proportion of A-alleles in the population. Then each genotype is produced by selecting independently an allele from each parent. The genotypes then occur in proportions

$$p^2 \quad 2p(1-p) \quad (1-p)^2 \tag{4.1}$$

as above.

Suppose now that the probability of an A-allele from the first parent remains p but that, having A, the probability of that from the other parent also being A is ξ, say. If $\xi > p$, A alleles tend to favour other A's rather than a: if $\xi < p$, the converse holds. Similarly, if the allele from the first parent is a, suppose that the probability of that from the other parent also being a is η. The genotype frequencies will then be

$$p\xi \quad p(1-\xi) + (1-p)(1-\eta) \quad (1-p)\eta. \tag{4.2}$$

(Notice that ξ and η are the probabilities that the second allele will be the *same* as the first when the first is A or a respectively.) If the proportion of A's is to remain at p in (4.2) we must have

$$p\xi + p(1-\xi)/2 + (1-p)(1-\eta)/2 = p$$

giving

$$p = \frac{1-\eta}{(1-\xi) + (1-\eta)}.$$

Inserting this value into (4.2) we have genotype frequencies

$$\frac{\xi(1-\eta)}{(1-\xi)+(1-\eta)} \quad \frac{2(1-\xi)(1-\eta)}{(1-\xi)+(1-\eta)} \quad \frac{\eta(1-\xi)}{(1-\xi)+(1-\eta)}. \tag{4.4}$$

We now have a probability mechanism that describes any mating situation. The trinomial orginally described in terms of p_1, p_2, p_3 (with $p_1 + p_2 + p_3 = 1$) is now described in terms of ξ and η, with $0 \leq \xi, \eta \leq 1$. In the null hypothesis $\xi = p$ and $\eta = 1 - p$, or eliminating p,

$$\xi = 1 - \eta. \tag{4.5}$$

Departure from HW equilibrium can, for example, be measured by the discrepancy between ξ and $1 - \eta$.

This is a parameterization in terms of ξ and η that is genetically meaningful and not just a statistical artifact. However, neither parameter describes equilibrium nor is suitable as the parameter of interest (α in the introduction). $\xi + \eta$ is possible, but then the natural nuisance parameter is $\xi - \eta$ and the pair have the awkward property that the range of one depends on the value of the other. Experience shows that log-odds are often more convenient than probabilities, so a possibility is to use

$$\alpha = \tfrac{1}{2}\left\{\log\frac{\xi}{1-\xi} + \log\frac{\eta}{1-\eta}\right\}$$

and

$$\beta = \tfrac{1}{2}\left\{\log\frac{\xi}{1-\xi} - \log\frac{\eta}{1-\eta}\right\}, \tag{4.6}$$

replacing the sums and differences of probabilities by those of log-odds. In terms of the original proportions (p_1, p_2, p_3) of the three genotypes

$$\alpha = \tfrac{1}{2}\log\frac{4p_1p_3}{p_2^2}$$

and

$$\beta = \tfrac{1}{2}\log\frac{p_1}{p_3}. \tag{4.7}$$

Equilibrium is clearly $\alpha = 0$ and we wish to test $\alpha = 0$ against $\alpha \neq 0$. $\alpha > (<)0$ corresponds to in(out)breeding. In Haldane's argument, $2n_1 + n_2$, or equally $(2n_1 + n_2)/2N$, was used. The population equivalent is $p_1 + p_2/2$, the probability of an A. Since $p_1 + p_2 + p_3 = 1$, this is $(p_1 - p_3 + 1)/2$. In place of the difference $p_1 - p_3$, (4.7) uses the ratio p_1/p_3 and half its logarithm is the nuisance parameter. (4.7) may also be written

$$\begin{aligned} 2\alpha &= \log p_1 - 2\log p_2 + \log p_3 + \log 4 \\ 2\beta &= \log p_1 - \log p_3 \end{aligned}$$

thus exhibiting orthogonality in terms of the logarithms of the original trinomial proportions. If $\alpha = 0$ and equilibrium obtains,

$$\beta = \log\frac{p}{1-p}.$$

These are the parameters that we ask the geneticist to consider. They have simple meanings within the model and, as we shall next see, lead to a rather simple likelihood function. Notice that both α and β range over the whole real line irrespective of the value of the other parameter. This is important since either parameter having a specific value possesses a meaning that does not depend on the other. The model for allele combination is due to Bernstein (1930) who uses the awkward parameter $\xi + \eta$. I am grateful to W. E. Nyquist for drawing my attention to Bernstein's work.

5. THE LIKELIHOOD FUNCTION AND ASYMPTOTIC RESULTS

The likelihood function for data (n_1, n_2, n_3) is $p^{n_1} p_2^{n_2} p_3^{n_3}$. In terms of α and β simple calculations show that this is

$$\frac{e^{\alpha(n_1+n_3)} e^{\beta(n_1-n_3)}}{(1+e^{\alpha}\cosh\beta)^N}. \tag{5.1}$$

It is required to test the hypothesis that $\alpha = 0$ against the alternatives that $\alpha \neq 0$, β being nuisance. Notice that having a parameter α enables the geneticist not merely to measure equilibrium, $\alpha = 0$, but also to describe the amount of disequilibrium through non-zero values.

According to the likelihood principle the likelihood provides all the information about α and β contained in the data, and in particular, Haldane's conditionality and the distribution of any statistic are irrelevant. We therefore concentrate on (5.1) and begin by considering its asymptotic behaviour when all of n_1, n_2 and n_3 are large.

The logarithm of the likelihood L is, from (5.1),

$$L = \alpha(n_1 + n_3) + \beta(n_1 - n_3) - N\log(1 + e^{\alpha}\cosh\beta). \tag{5.2}$$

Hence

$$\frac{\partial L}{\partial \alpha} = (n_1 + n_3) - \frac{Nc}{c + e^{-\alpha}}$$

and

$$\frac{\partial L}{\partial \beta} = (n_1 - n_3) - \frac{Ns}{c + e^{-\alpha}} \tag{5.3}$$

where $c = \cosh\beta$ and $s = \sinh\beta$. Equating these to zero and solving the resulting equations for α and β, the maximum likelihood estimates are

$$\hat{\alpha} = \tfrac{1}{2}\log\frac{4n_1n_3}{n_2^2}, \qquad \hat{\beta} = \tfrac{1}{2}\log\left[\frac{n_1}{n_3}\right]. \tag{5.4}$$

These equations should be compared with (4.7).

The second derivatives of L most conveniently depend on the data only through N, the sample size. From (5.3) they are

$$\begin{aligned} \frac{\partial^2 L}{\partial \alpha^2} &= -\frac{Nce^{-\alpha}}{(c+e^{-\alpha})^2}, \\ \frac{\partial^2 L}{\partial \alpha \partial \beta} &= -\frac{Nse^{-\alpha}}{(c+e^{-\alpha})^2}, \\ \frac{\partial^2 L}{\partial \beta^2} &= -\frac{N(1+ce^{-\alpha})}{(c+e^{-\alpha})^2}. \end{aligned} \tag{5.5}$$

By standard theory it follows that the likelihood is, for large values of N, approximately of a normal form, centered at $(\hat{\alpha}, \hat{\beta})$, equations (5.4), with dispersion matrix given by inverting the matrix with elements equal to the negatives of the second derivatives (5.5) at the values $(\hat{\alpha}, \hat{\beta})$. The dispersion matrix is therefore

$$\frac{(c+e^{-\alpha})}{N}\begin{bmatrix} c+e^{\alpha} & -s \\ -s & c \end{bmatrix} \tag{5.6}$$

evaluated at $(\hat{\alpha}, \hat{\beta})$. In particular, the variance of α is asymptotically

$$(c+e^{-\alpha})(c+e^{\alpha})/N = \sigma_{\alpha}^2$$

and therefore a reasonable range for α is $\hat{\alpha} \pm 2\sigma_\alpha$. Notice that in this range we have a set of possible hypotheses that do not conflict with the evidence, rather than just a statement about the single value $\alpha = 0$ being plausible.

The dispersion matrix (5.6) can be expressed in terms of the data. Simple calculations show that it is

$$\frac{1}{4n_1n_2n_3}\begin{bmatrix} n_2(n_1+n_3)+4n_1n_3 & n_2(n_3-n_1) \\ n_2(n_3-n_1) & n_2(n_3+n_1) \end{bmatrix}.$$

One point worth noticing is that the variance of α is larger than that of β by an amount n_2^{-1}. It is therefore harder to determine the HW parameter than β.

6. DISTRIBUTIONS FOR THE PARAMETERS

In the Bayesian view it is necessary to provide a joint probability distribution for (α, β). A major question to be settled is whether the geneticist assigns a strictly positive probability to the situation being one of HW equilibrium, or not. In symbols: is $p(\alpha = 0) > 0$? If the answer is "No" then presumably α has a density $p(\alpha)$ that does not distinguish 0 from other values of α near 0. If the genetical view is that the population is probably nearly in equilibrum, this density might have its larger values in the neighborhood of zero. Let us consider first the case where the answer is "No" and suppose there is a density $p(\alpha)$ to be assessed. There are several possibilities.

(1) Since α and β both range over the whole real line a common argument has been to let (α, β) have a uniform distribution. Various phrases, like "being ignorant of the parameters" "letting the data speak for themselves" have been used to describe this supposition.

(2) In more accord with Bayes' original idea, it could be supposed that p_1 and p_3 are uniform over the triangle of their possible values: $p_1 \geq 0, p_3 \geq 0, p_1 + p_3 \leq 1, (p_2 = 1 - p_1 - p_3)$. Simple calculation shows that the implied distribution for (α, β) has density

$$\frac{e^{2\alpha}}{(1 + e^\alpha \cosh \beta)^3}. \tag{6.1}$$

(3) A more recent idea is to take a distribution conjugate to the likelihood (5.1.). In the present case this means a density proportional to

$$\frac{e^{\alpha(m_1+m_3)} e^{\beta(m_1-m_3)}}{(1 + e^\alpha \cosh \beta)^M} \tag{6.2}$$

where m_1, m_2, and m_3 are non-negative hyperparameters, $M = m_1 + m_2 + m_3$. It is interesting to note that the distributions under (1) and (2) are of this form with $m_1 = m_2 = m_3 = 0$ and $m_1 = m_2 = m_3 = 1$ respectively .

(4) Several authors have tried to give a precise meaning to the phrase "knowing nothing about a parameter". The most successful of these is due to Bernardo (1979). He has kindly applied his results to the likelihood (5.1) where α is the parameter of interest and β nuisance. The result is a joint distribution proportional to

$$\frac{e^{\alpha/2}}{(1 + e^\alpha \cosh \beta)} \frac{1}{(\cosh \beta)^{1/2}}. \tag{6.3}$$

Notice that this has a component of conjugate form with $m_1 = m_3 = 1/4$ and $m_2 = 1/2$, modified by a multiplier $(\cosh \beta)^{-1/2}$.

(5) The best suggestion is for the geneticist to express what is known about the situation and put this into the form of a joint distribution for (α, β). As an illustrative example he might

be studying a case where A is rare, so that p_1 is small and p_3 large, implying β is appreciably negative. At the same time it might be felt that the amount of disequilibrium cannot be large, so this places limits on α. As α and β have independent ranges of variation, a possibility is to suppose them independent in a probability sense also. Then independent normals for α and β could be used with means zero and a negative value respectively, the standard deviations reflecting the uncertainties in α and β. Notice that in a conjugate distribution, α and β are not independent. Independent logistic distributions are computationally simpler.

If the distribution of (α, β) prior to seeing the data is conjugate (case (3): (1) and (2) are special cases), the distribution of (α, β) given the data is of the same form with m_i replaced by $m_i + n_i (i = 1, 2, 3)$. It will therefore suffice for us to consider the case with all $m_i = 0$. The analysis for (n_1, n_2, n_3) can be thought of as arising from the uniform case (1) with data (n_1, n_2, n_3) or from the general conjugate case (6.2) with data $(n_1 - m_1, n_2 - m_2, n_3 - m_3)$.

7. INFERENCE ABOUT EQUILIBRIUM WITH A UNIFORM PRIOR

With a uniform prior for (α, β), the posterior distribution for (α, β) is proportional to the likelihood (5.1). Standard substitutions enable the double integral with respect to α and β to be evaluated with the result that the density is

$$\frac{(N-1)!}{(n_1-1)!(n_2-1)!(n_3-1)!} \frac{1}{2^{n_1+n_3-1}} \frac{e^{\alpha(n_1+n_3)} e^{\beta(n_1-n_3)}}{(1+e^{\alpha}\cosh\beta)^N} \tag{7.1}$$

provided all $n_i > 0$. (The integral does not converge if any $n_i = 0$.) The marginal density of α is therefore

$$\frac{(N-1)! e^{\alpha(n_1+n_3)}}{(n_1-1)!(n_2-1)!(n_3-1)! 2^{n_1+n_3-1}} \int_{-\infty}^{\infty} \frac{e^{\beta(n_1-n_3)}}{(1+e^{\alpha}\cosh\beta)^N} d\beta \tag{7.2}$$

and it is this that provides a complete inference for the extent to which the population is in HW equilibrium.

The integral in (7.2) is not expressible in terms of standard functions: it is closely related to a Legendre function of the second kind. We therefore consider some numerical cases, the integration of β being performed using Simpson's rule with 21 points. Table 1 covers four situations. In each $N = 100$. All are cases carefully evaluated by Emigh so that we have his results available for comparison. It is convenient to classify cases according to the observed proportion of A alleles, $\hat{p} = (2n_1 + n_2)/2N$. In Table 1 these are $0.5, 0.17$ and 0.05, with two examples for the last. Figure 1 graphs the values $p(\alpha | n_1, n_2, n_3)$, the probability density for α, given n_1, n_2, n_3 and uniform independent priors for α and β (formula (7.2)).

Three of the cases illustrated in Figure 1 have been selected as those where the results would conventionally just be considered as indicating significant departure from HW equilibrium. In two ($\hat{p} = 0.50$ and 0.17) the exact densities of α are well-approximated by the asymptotic, normal form. For example the first case, $\hat{p} = 0.50$, exhibits only a little negative skewness. The probability of α departing by more than two asymptotic standard deviations from the mean is 0.030 in the negative direction and 0.024 in the positive: whereas normal theory would say 0.023 for both. The skewness is slightly more when $\hat{p} = 0.17$ but still not important. But by the time $\hat{p}$ has decreased to 0.05 with data $(2, 6, 92)$ the skewness is more noticeable, the tail areas are, to the left 0.041 and 0.026 to the right; also indicating a standard deviation higher than σ_α would suggest. This is the case of data with $n_1 = 2$. The last case, $(1, 8, 91)$, still with $\hat{p} = 0.05$, has $n_1 = 1$ and illustrates the skewness more sharply. The tail areas are, to the left 0.090 and to the right 0.011. Other calculations, not reported here, suggest that the asymptotic theory is a reasonable approximation provided all n_i exceed 2.

It is certainly unsatisfactory for any n_i equal to 1 (the case of any $n_i = 0$ will be discussed below).

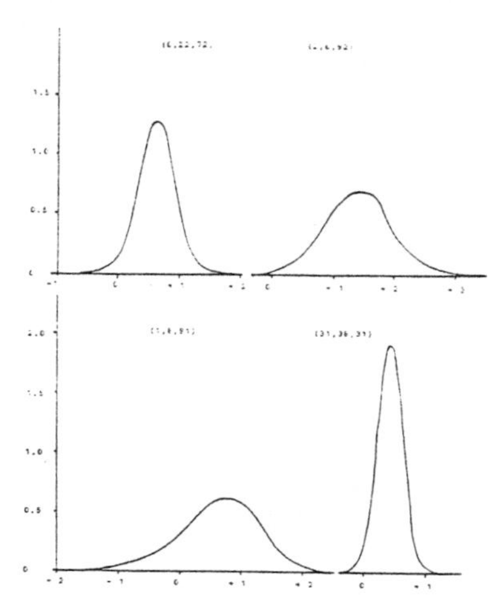

Figure 1.

The probability densities graphed in Figure 1 provide complete inference about the HW parameter α. They show what values are most probable and which are most improbable. The probability that α lies between any two selected values is the area under the curve between those values. This inference is certainly more complete and, we contend, more satisfactory than a single statement of a significance level that the usual tests produce. The question arises as to how the densities are to be compared with the levels, for they are referring to quite different things. Recall that a significance level is the probability of the observed result or more extreme *on the assumption that HW equilibrium obtains.* Our results give the probability distribution of the amount of disequilibrium α. One possibility is to look at the tail-area probabilities for α in the Bayesian approach and compare them with the tail-area significance probabilities. In all cases in Figure 1 $\hat{\alpha} > 0$ so a reasonable thing to look at is $p(\alpha < 0)$. These values are given in Table 1 under the label TA. Also included are the significance levels SL as computed by Emigh. (Actually these values depend a little on the test used: the ones quoted here are the ones given by the "exact" test and the majority of other tests.) In all cases TA is less than the significance level, SL, being between 60 and 80% of that value. This may be because Emigh has considered two-tail tests, including all possible observations with fixed N that are more extreme, in the sense of having less probability, than the actual observation. Emigh talks of them being less "consonant" with the null hypothesis $\alpha = 0$. The difficulty of deciding what is more extreme or less consonant is often present in the construction of significance levels. However, most geneticists are unlikely to be disturbed by the numerical differences between SL and TA noticed in Table 1, and we may broadly conclude that, in respect or tail areas, the two views are similar. Where the Bayesian view scores is in the extra information provided about possible departures from HW equilibrium.

The data are given in the form (n_1, n_2, n_3), the observed numbers of the three genotypes AA, Aa and aa respectively. $\hat{p}$ is the proportion of A alleles in the sample: $(2n_1+n_2)/2N$. $\hat{\alpha}$ and $\hat{\beta}$ are the maximum likelihood estimates of α and β. σ_α is the asymptotic standard error of α. SL denotes the significance level as given by Emigh for most of the tests he considers.

TA is the tail area probability

$$\int_{-\infty}^{0} p(\alpha|n_1, n_2, n_3)d\alpha$$

for an initial uniform prior on (α, β).

Data	(31,38,31)	(6,22,72)	(2,6,92)	(1,8,91)
$\hat{p}$	0.50	0.17	0.05	0.05
$\hat{\alpha}$	0.490	0.636	1.509	0.869
$\hat{\beta}$	0.000	–1.242	–1.914	–2.255
σ_α	0.206	0.301	0.543	0.614
SL	0.017	0.034	0.015	0.210
TA	0.010	0.024	0.009	0.171

Table 1.

8. INFERENCE ABOUT EQUILIBRIUM WITH NON-UNIFORM PRIORS

The asymptotic results are scarcely satisfactory if any n_i is less than 2. It therefore becomes important to study the role of the prior when this happens. This is especially true when any n_i is zero because then the posterior, with a prior uniform in (α, β), is improper. In the last section the data set $(1, 8, 91)$ was considerd. Changes in 2 of the 100 values could lead to the case $(0, 10, 90)$, still with $\hat{p} = 0.05$, with one zero value. In Table 2 summary results for this data set with 3 priors are presented with the case $(1, 8, 91)$ and uniform prior previously considered for comparison. The effect of the prior on $\hat{\alpha}$, the maximum posterior value, is dramatic. Thus the priors $(0.5, 1, 0.5)$ and $(1, 2, 1)$ — these are the m-values in (6.2) — give values of $\hat{\alpha}$ differing by a factor of over 2. The prior $(0.1, 1.8, 8.1)$, which is more in accord with the observed frequency of A-allelles, causes $\hat{\alpha}$ to change sign and its approximate standard error to more than double. It is clear that inferences when any n_i is 0 or 1 depend critically on the prior and maximum likelihood results are unsatisfactory.

Data	(1,8,91)	(0,10,90)	(0,10,90)	(0,10,90)
Prior	(0,0,0)	(0.5,1,0.5)	(1,2,1)	(0.1,1.8,8.1)
Posterior	(1,8,91)	(0.5,11,90.5)	(1,12,91)	(0.1,11.8,98.1)
$\hat{\alpha}$	0.869	0.201	0.464	-0.633
$\hat{\beta}$	-2.255	-2.599	-2.255	-3.444
σ_α	0.614	0.771	0.580	1.609

Table 2.

One possibility for a prior, case (2) in section 6, is to replace uniformity in (α, β) by uniformity in (p_1, p_3), for at least this distribution is proper. The resulting density is conjugate with $m_1 = m_2 = m_3 = 1$ and that for α is given, from (6.1), by

$$e^{2\alpha} \int_{-\infty}^{\infty} (1 + e^{\alpha} \cosh \beta)^{-3} d\beta.$$

This density is shown in Figure 2. It is doubtful if this is commonly representative of genetical experience about the amount of disequilibrium. It does favour positive α, corresponding to

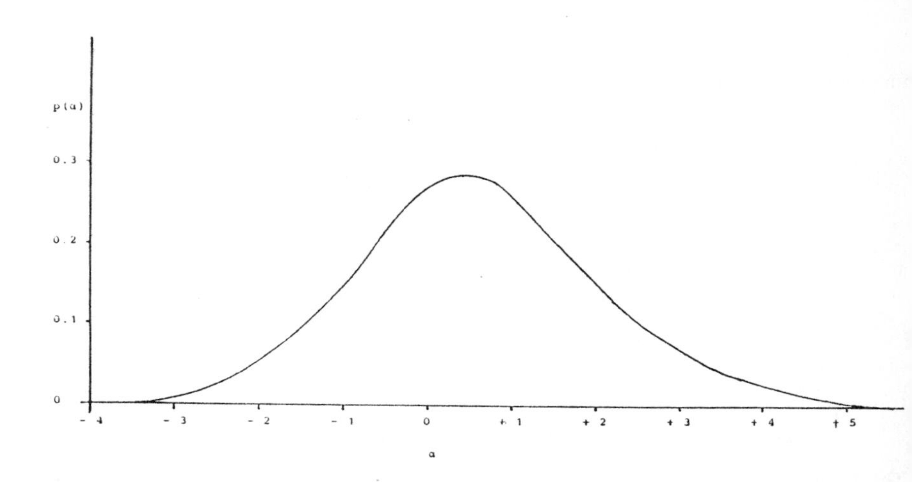

Figure 2.

inbreeding, but still seriously entertains values of α as large 4.0 when p_2 must be very small. Higher values of the m's seem to be called for. If they are used, the inference about α given the data, when that data contains a small n_i, will differ sharply form the asymptotic results.

Another possibility is to use Bernardo's suggestion, case (4). Figure 3 shows the posterior distributions of α in three cases, all for data $(1, 8, 91)$. One has the Bernardo prior (6.3), the second has the corresponding conjugate prior $(1/4, 1/2, 1/4)$ and the third uses the uniform prior as in the earlier calculations. The second has been included to show the effect of the $(\cosh\beta)^{-1/2}$ term in (6.3). The conjugate distribution gives a result somewhat to the left of the Bernardo result and with more left skewness: the uniform exaggerates both these effects. The differences are well illustrated by the values of $p(\alpha < 0)$, the three left-hand tails: these are $.078, .122$ and $.224$ respectively. The tail area, which was earlier compared with the significance level, increases by a factor of 3. The effect of the $(\cosh\beta)^{-1/2}$ term is even more dramatic for the data $(0, 10, 90)$ discussed previously. Its omission causes the posterior density (not shown) to shift to the left and become substantially more skew. The $p(\alpha < 0)$ changes from 0.56 to 0.78.

9. KNOWN β

In some genetical applications the proportion of A alleles is known fairly precisely before sampling. This is $p_1 + \frac{1}{2}p_2$, or equivalently $(p_1 - p_3 + 1)/2$, so that $p_1 - p_3$ is well-determined. In our parameterization the equivalent situation would be that $\log p_1 - \log p_3$ is known, implying knowledge of β also. If this is indeed so, the relevant posterior distribution is not the marginal distribution of α considered earlier, but the conditional distribution of α, given β. With a uniform distribution for α, easy calculations show that the distribution for α given the data is

$$\frac{e^{\alpha(n_1+n_3)}}{(1+e^{\alpha}\cosh\beta)^N}(\cosh\beta)^{n_1+n_3}\frac{(N-1)!}{(n_1+n_3-1)!(n_2-1)!}$$

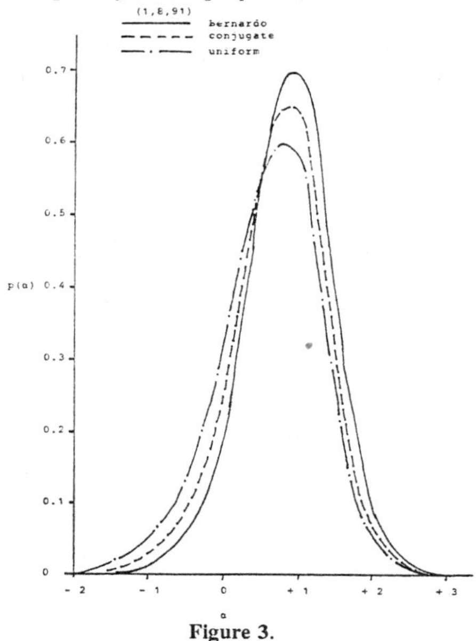

Figure 3.

This is a much simpler form than (7.2). It is a general logistic distribution, transformable to a beta-distribution by the transformation from α to $\gamma = (1 + e^{\alpha} \cosh \beta)$.

10. HYPOTHESIS TESTING

Let us now return to consideration of the null hypothesis $\alpha = 0$. The analysis of the last two sections has not distinguished $\alpha = 0$ from any other value of α, except insofar as the relationship of the posterior distribution of α to the value zero has been considered through the tail area. A possibility, extensively discussed by Jeffreys (1967), is that the scientist might orginally have a strictly positive probability for $\alpha = 0$: in our case a belief that the equilibrium *exactly* holds. In this case the posterior probability that $\alpha = 0$ will similarly be strictly positive. The odds form of Bayes theorem says

$$\frac{p(\alpha = 0|\text{data})}{p(\alpha \neq 0|\text{data})} = \frac{p(\text{data}|\alpha = 0)}{p(\text{data}|\alpha \neq 0)} \cdot \frac{p(\alpha = 0)}{p(\alpha \neq 0)},$$

so that the original odds on the null hypothesis are multiplied by

$$K = \frac{p(\text{data}|\alpha = 0)}{p(\text{data}|\alpha \neq 0)}$$

to obtain the posterior odds. K is called the Bayes factor and describes the influence of the data on the odds for the null hypothesis. Notice that K depends on the prior distribution of β and also on that of α when $\alpha \neq 0$. It does not depend on the prior odds for $\alpha = 0$.

When $\alpha \neq 0$, disequilibrium holds and the probability structure is similar to that already discussed. We shall therefore suppose that for $\alpha \neq 0$ the joint distribution is of the conjugate form given by (6.2). Denote this temporarily by $f(\alpha, \beta)$. For any α, the conditional density of β given α is, excluding $\alpha = 0$, proportional, to

$$f(\alpha, \beta) / \int f(\alpha, \beta) d\beta.$$

The assumption is now made that even on the null hypothesis $\alpha = 0$, the conditional distri bution of β given α is of the same form namely

$$f(0,\beta)/\int f(0,\beta)d\beta.$$

Effectively this is saying that the conditional distribution of the nuisance parameter β, given α is continuous, even at $\alpha = 0$. Notice that were we to change to an alternative parameterization α^*, β^*, the null still being $\alpha^* = 0$, this continuity might not persist because of the Borel Kolmogorov paradox.

This now defines a prior distribution except for the values of m_1, m_2 and m_3 in (6.2) Tedious but straightforward calculations show that $K = g(\boldsymbol{m}+\boldsymbol{n})/g(\boldsymbol{m})$ where

$$g(\boldsymbol{m}) = \frac{2^{m_2+1}(M-1)!(2m_1+m_2-1)!(m_2+2m_3-1)!}{(2M-1)!(m_1-1)!(m_2-1)!(m_3-1)!}$$

where $\boldsymbol{m} = (m_1, m_2, m_3)$ and $\boldsymbol{n} = (n_1, n_2, n_3)$, $M = m_1 + m_2 + m_3$.

Beyond this stage it is necessary to resort to numerical work. To illustrate we take the same four data sets as before (Figure 1). The values of m_1, m_2 and m_3 are chosen so tha first the mode of $f(\alpha, \beta)$ occurs at $\alpha = 0$: that is, even on the alternative hypothesis values near to the null are more likely than these further away. A second requirement is that the proportions of A-alleles in prior and data agree. Of course, this cannot be true in genera but is assumed here so that any effects observed are not confounded with the difference in the allele-proportions expected and observed. These two assumptions imply that the triple (m_1, m_2, m_3) is

$$(\hat{p}^2 M,\ 2\hat{p}(1-\hat{p})M,\ (1-\hat{p})^2 M)$$

with $\hat{p} = (n_1 + n_2/2)/N$. Only M has to be chosen.

Table 3 shows the Bayes factor for the four data sets for various values of M. Remember that for the first three the Haldane test rejects the null hypothesis of HW equilibrium. The obvious feature of all four data sets is the marked changes in the Bayes factor K with M. (Remember, a large K supports $\alpha = 0$, a small value supports $\alpha \neq 0$, $K = 1$ is neutral.) For example, with the data set $(2, 6, 92)$ $M = 1$ gives strong support for the null hypothesis (and this despite the conventional significance) whereas values around $M = 50$ support the alternative. In the first three cases K decreases as M increases to a minimum, after which it increases, the limit as $M \to \infty$ being one. In the last, the decrease of K to one is monotone. It is abundantly clear that the Bayesian and significance level approaches give different interpretations. The tendency is for the latter to favour the alternative hypothesis and too easily to reject the null. Of course, scientists like to reject the null and establish an apparent effect, but the coherent view suggests otherwise. The behaviour here is in line with the general results of Berger and Sellke (1987). Notice that there is no "natural" prior distribution. One with weak prior knowledge ($M \to 0$) gives extreme results, as does a tighter one with large M. Berger and Sellke considers the minimum of K as M varies. This is a measure but its use would be incoherent and in some cases (as with the data set $(1, 8, 91)$) would yield $K = 1$ as unrealistically M approached infinity.

For explanation see text and legend to Table 1.

The entries are Bayes factor K in favour of HW-equilibrium ($\alpha = 0$) for the conjugate priors

$$\hat{p}^2 M, 2\hat{p}(1-\hat{p})M, (1-\hat{p})^2 M).$$

For other notations see the legend to Table 1.

Data	(31,38,31)	(6,22,72)	(2,6,92)	(1,8,91)
M=1	0.99	4.52	7.51	66.0
10	.27	0.79	0.99	7.55
20	.24	.60	.64	4.30
30	.24	.54	.54	3.21
50	.27	.54	.48	2.34
100	.35	.57	.47	1.67
500	.69	.79	.68	1.14
	$\hat{p}$ =.50	$\hat{p}$ =.17	$\hat{p}$ =.05	$\hat{p}$ =.05

Table 3.

11. CONCLUDING REMARKS

The negative conclusion of this work is that the usual significance level assessment of the hypothesis of HW-equilibrium is unsatisfactory. If it is meant to be a measure of the scientist's belief in the null hypothesis, then it can be, and usually is, totally misleading. (The case $M = 1$ with data set $(2, 6, 92)$ gives 1.5% significance, yet a Bayes factor of 7.5 in favour of the null.) If the problem is regarded more as one of estimation, then the significance level, SL, is in fair agreement with the probability, TA, that α departs from its null value zero in the opposite direction to that suggested by the data, being typically a little greater than it. But this probability is a weak substitute for the full distribution of α given by the data, which shows the range of reasonable values of α and so the possible extent of disequilibrium.

Despite its advantages, it is not going to be easy to wean scientists away from the simple significance levels to the Bayesian view because the latter requires a more detailed specification of the problem. Our hope is that in this genetical situation we have been able to produce a biologically meaningful model that can be assessed by the geneticist interested in HW equilibrium with the result that data will provide more information in a readily understood form.

ACKNOWLEDGEMENTS

My interest in this problem arose through stimulating discussion with Carlos Pereira and André Rogatko at the University of Sao Paulo, Brazil. The work with a smooth prior was completed at Monash University, Australia and benefited much from discussion with W. J. Ewens. The paper was completed at Duke University, U.S.A.

REFERENCES

Berger, J. O. and Sellke, T. (1987). Testing a point null hypothesis: the irreconcilability of P values and evidence. *J. Amer. Statist. Assoc.* **82**, 112-130, (with discussion).

Bernardo, J. M. (1979). Reference posterior distributions for Bayesian inference. *J. Roy. Statist. Soc. B* **41**, 113–142. (with discussion).

Bernstein, F. (1930). Fortgesetzte Untersuchungen aus der Theorie der Blutgruppen, *Ztschr. f. Abstamm-.u. Vererbungslehre* **56**, 223–273.

Elston, R. C. and Forthofer, R. (1977). Testing for Hardy-Weinberg equilibrium in small samples. *Biometrics* **33**, 536–542.

Emigh T. H. (1980). A comparison of tests for Hardy-Weinberg equilibrium. *Biometrics* **36**, 627–642.

Haldane, J. B. S. (1954). An exact test for randomness of mating. *J. of Genetics* **52**, 631–635.

Jeffreys, H. (1967). *Theory of Probability*. Oxford: Clarendon Press.

Pereira, C. A. and Rogatko, A. (1984). The Hardy-Weinberg equilibrium under a Bayesian perspective. *Brazilian J. Genetics* **7**, 689–707.

Vithanayasai, C. (1975). Exact critical values of the Hardy-Weinberg test statistics for two alleles. *Communications in Statistics* **1**, 229–242.

DISCUSSION

I. VERDINELLI (*University of Rome*)

I enjoyed reading Prof. Lindley's paper dealing with this important genetical problem. The paper provides a complete and clear example of the use of Bayes methods to handle a practical problem.

The weakness of the classical approach to testing whether a population is in Hardy-Weinberg equilibrium is clearly demonstrated by the fact that Haldane's test power depends on the nuisance parameter $p_1 + p_2/2$.

It is important to encourage scientists to consider the use of Bayesian analysis. In order to do that it is very helpful to provide them with meaningful parametrizations. In this case the choice of α and β seems very appropriate. Both parameters refer to different features of the problem; the values of $\alpha \neq 0$ have a clear genetical meaning, since they correspond to in – or out – breeding according to whether $\alpha > 0$ or $\alpha < 0$.

The posterior distribution of α should then be useful for providing all the information the geneticist requires.

Comparisons with classical results are made when a uniform prior distribution for (α, β) is considered. In this case values of significance levels are compared with the tail area of marginal posterior distributions for α. However, as Prof. Lindley remarks, these values refer to quite different things. It seems as if this comparison is only made to make geneticists happy and more confident in using posterior distributions for α.

There are two points in this paper that I do not feel entirely happy with.

First let me mention the analysis carried out in Section 9 where the case of known β is considered. Prof. Lindley points out that in some genetical applications the proportion of A alleles is known. This means that the quantity $p_1 + \frac{1}{2}p_2 = \frac{1}{2}(p_1 - p_3 + 1)$ is known fairly precisely, so that $p_1 - p_3$ is well determined. It seems clear that this is not equivalent to the knowledge of

$$\beta = \tfrac{1}{2}\{\log p_1 - \log p_3\} = \tfrac{1}{2}\log\frac{p_1}{p_3},$$

as Prof. Lindley claims. So, if the knowledge of the proportion of A alleles was a relevant genetical hypothesis, then the parametrization in terms of (α, β) is inappropriate. A more suitable one should be looked for. As an example we might use:

$$\delta = 4p_1p_3 - p_2^2$$

$$\gamma = p_1 - p_3$$

where δ is the parameter of interest and, since γ is known, it is the only parameter involved.

In this case a different analysis could be carried out. Simple calculations show that:

$$p_1 = \frac{1}{4}\{(1+\gamma)^2 + \delta\} = \frac{1}{4}(a+\delta),$$

$$p_2 = \frac{1}{2}\{(1-\gamma^2) - \delta\} = \tfrac{1}{2}(b-\delta),$$

$$p_3 = \frac{1}{4}\{(1-\gamma)^2 + \delta\} = \frac{1}{4}(c+\delta),$$

so the likelihood for δ is:

$$\ell(\delta|\gamma, n_1, n_2, n_3) \propto (a+\delta)^{n_1}(b-\delta)^{n_2}(c+\delta)^{n_3}.$$

If for instance, Jeffreys's prior for δ was considered:

$$f(\delta) \propto \frac{(b+\delta)^{1/2}}{(a+\delta)^{1/2}(b-\delta)^{1/2}(c+\delta)^{1/2}},$$

then the posterior distributions would be

$$f(\delta|\gamma, n_1, n_2, n_3) \propto (b+\delta)^{1/2}(a+\delta)^{n_1-\frac{1}{2}}(b-\delta)^{n_2-\frac{1}{2}}(c+\delta)^{n_3-\frac{1}{2}}.$$

Figure 1 plots this posterior distribution with different values of γ, for two of the four data sets.

For data $(31, 38, 31)$, only one graph is shown, since in this case inference on δ is robust to the values of γ.

I would also like to make some remarks about the behaviour of the Bayes factor in Lindley's table 3. The values of the Bayes factor shown depend heavily on the choice of the prior distribution's parameters (m_1, m_2, m_3). Far from being just a neutral choice for illustrative purposes, the assumption that $(m_1, m_2, m_3) = (\hat{p}^2M, 2\hat{p}(1-\hat{p})M, (1-\hat{p})^2M)$ makes the prior distribution data dependent through $\hat{p} = (n_1 + \frac{1}{2}n_2)/N$. It affects the magnitude of the prior variance τ_α^2 and this has a well known effect on the Bayes factor.

Further, the choice of the value of $M = m_1 + m_2 + m_3$ is critical since τ_α^2 strongly depends on M. When M is large, the normal approximation in Lindley's section 5 can be used for the prior distribution and it is easily seen that $\tau_\alpha^2 = \{4M\hat{p}^2(1-\hat{p})^2\}^{-1}$. In this way, for increasing values of M the null hypothesis $\alpha = 0$ in no way differs from the alternative, $\alpha \neq 0$, the prior distribution for α having mode at $\alpha = 0$ and decreasing variance.

So, clearly the Bayes factor must converge to 1 for increasing values of M. On the other hand, when M is small, the normal approximation for the prior distribution is of no use, but τ_α^2 still depends on M and $\hat{p}$.

Table 1 shows the computed values of τ_α^2 and of the Bayes factor for different values of M, using three of the data sets from Lindley's table 3.

For given $\hat{p}$, τ_α^2 increases as M decreases and in fact the numerical calculations do not even converge for small values of M. Lindley points out that with the data set $(2, 6, 92)$, $M = 1$ gives strong support for the null hypothesis, but this is indeed one of the case where τ_α^2 does not converge and, as it is well known, a large prior variance favours the null hypothesis.

DATA	(31,38,31)	$\hat{p}$=.5	DATA	(6,22,72)	$\hat{p}$=.17	DATA	(2.6.92)	$\hat{p}$=.05
M	τ_α^2	B.F.	M	τ_α^2	B.F.	M	τ_α^2	B.F.
1/2	47.99	1.85						
1	13.05	.99	1	∞	4.52	1	∞	7.51
10	.466	.27	10	3.35	.79	10	∞	.99
20	.216	.24	20	1.12	.60	20	87.00	.64
30	.133	.24	30	.63	.54	30	38.70	.54
50	.082	.27	50	.33	.54	50	13.87	.48
100	.041	.35	100	.14	.57	100	4.67	.47
500	.008	.69	500	.02	.59	500	.31	.68

Table 1.

Table 1 shows how τ_α^2 increases when $\hat{p}$ decreases for any given value of M. Note for example that with the data set $(2, 6, 92)$, $\hat{p} = .05$, τ_α^2 diverges when $M = 10$, while with the data set $(31, 38, 31)$, $\hat{p} = .5, \tau_\alpha^2$, has a finite value even for $M = 1/2$.

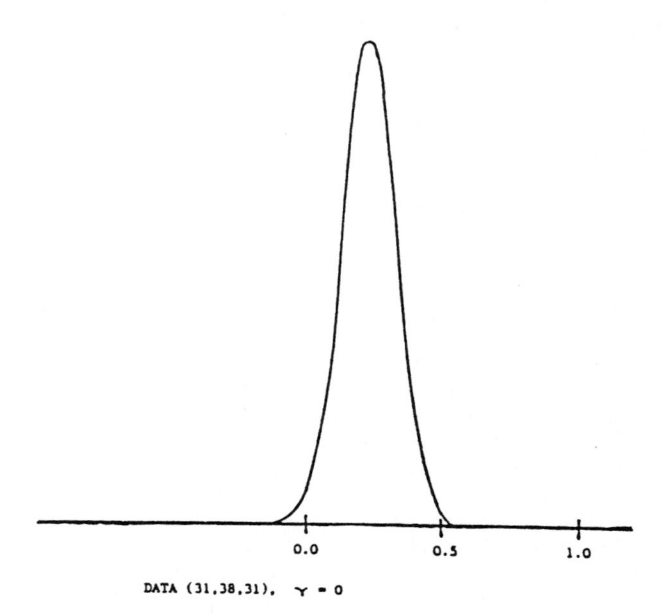

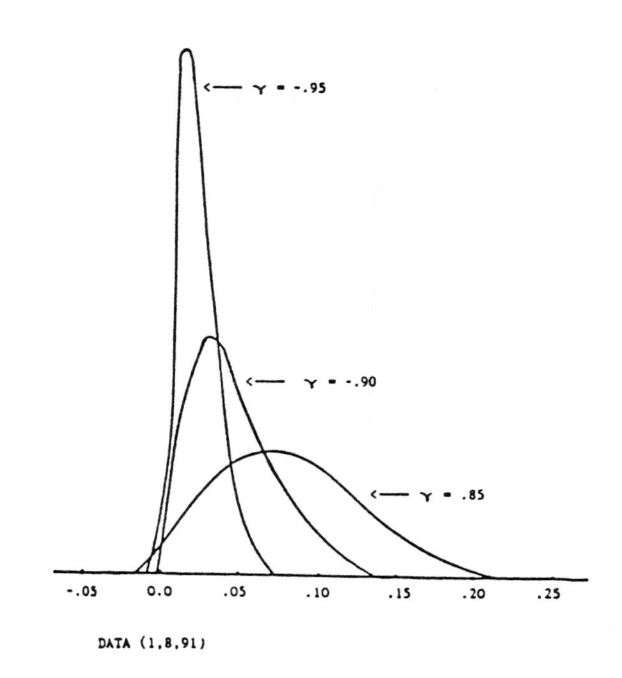

Figure 1.

This latter case shows also how a change of M from 1 to $1/2$ has a strong effect not only on the prior variance, but also on the Bayes factor, which changes from being neutral to supporting the null hypothesis.

In recent years many authors have been dealing with the difficulties arising from assigning prior distributions for Bayes hypothesis testing and various solutions have been suggested (see for example: Bernardo, 1980; Spiegelhalter and Smith, 1982; Shafer, 1982; Berger and Sellke, 1987). So I feel somewhat disappointed that this has not been taken into account here. Also if I may quote from Lindley himself (1980): "... If we have a practical problem of data analysis the quantities have a physical meaning and the scientist knows something about them. He should then be encouraged to think about them, or the parameters, and not adopt probability distributions that merely conform to... some formal model...".

Well Prof. Lindley provided us with a meaningful model; the parameters he considers are such that geneticists know something about them, so why not ask their opinions and carry out a more sensible analysis?

G. A. BARNARD (*Colchester*)

Lindley's paper is particularly interesting as affording an opportunity to contrast a Bayesian with a Fisherian approach to a scientific problem having a negligible "decision-making" aspect. That the contrasts between the two approaches can be greatly over-stressed is illustrated by comparing the "machine-like adoption of rote methods" criticised by Lindley, and widely thought to represent a Fisherian approach, with Fisher's statement: "no scientific worker has a fixed level of significance at which from year to year, and in all circumstances, he rejects hypotheses; he rather gives his mind to each particular case in the light of his evidence and his ideas."

Characteristic of the Fisherian approach here is that it focuses on the directly available data, in asking whether the number of heterozygotes falls so far below the number expected assuming Hardy-Weinberg equilibrium as to give grounds for supposing *some* disturbing factor is operating. A little calculation using tables of log-factorials and a hand-held calculator shows that on the assumption as HW equilibrium the probabilities of as few, or fewer heterozygotes for the four sets data $(31, 38, 31)$, $(6, 22, 72)$, $(2, 6, 92)$ and $(1, 8, 91)$ are, respectively, 0.0117, 0.0337, 0.0153 and 0.2105. The last probability indicatates that for the fourth data set there are no serious grounds against a supposition of equilibrium:, but for the remaining three sets, especially for the first and the third, the heterozygote count is unusually low. It is then for the scientist to consider possible reasons for this "in the light of his evidence and his ideas". He will know how the samples were collected, and perhaps something about the reprductive conditions prevailing at the relevant times. It could be, for example, that for the third set of data, the homozygote AA gave rise in the phenotype to a property which would increase the chance of such a phenotype appearing in the sample relative to the other genotypes. This might mean that the 2 specimens of AA arose from a fault in the sampling procedure, which would go a long way towards explaining the apparent defect of heterozygotes. Such an explanation might seem much less plausible in the case of the first data set; and it might be there that the population sampled was a set of plants which had recently experienced climatic conditions conducive to self-fertilisation. Anyone whose knowledge of genetics goes beyond my own schoolboy acquaintance with the subject could doubtless think of many other possibilities.

By contrast Lindley (following, he says F. Bernstein) propounds a single model for the reproductive process which involves a probability ξ for an AA allele to mate with an allele. My schoolboy knowledge makes it hard for me to form a clear conception of how such a probability would arise in the absence of special features such as are present in primroses. One would ordinarily expect such a ξ to be compounded of two conditional probabilities, one conditional on A arising in a heterozygote the other conditional on it arising in a homozygote. Whatever may be its plausibility, Lindley's model does assume knowledge going beyond the directly available data under discussion.

If Lindley's model is known to be applicable, and if there are observational grounds for supposing the priors assumed to be at least approximately true, then a Fisherian would have no objection to Lindley's analysis. When Fisher wrote, in the first edition of his "Statistical Methods" "Inferences respecting populations, from which known samples have been drawn, cannot be expressed in terms of probability, except in the trivial case when the sample is itself a sample of a super-population the specification of which is known with accuracy" he could well have had in mind Karl Pearson' objections to his treatment of the correlation coefficient on the grounds that Pearson claimed Fisher was using the wrong prior distribution. By the thirteenth edition Fisher had modified his statement to read "... cannot by this method be expressed in terms of probability, save in those cases in which there is an observational basis for making exact probability statements in advance about the population in question". This seems to be in line with Lindley's "The best suggestion is for the geneticist to express what is known about the situation and put this in the form of a joint distribution for (α, β)."

Where Lindley diverges from what *could* be a Fisherian approach (given the validity of his model and the proper expression of the geneticists *knowledge*) is in the lengthy discusion of conjugate and "ignorance" priors. Even here, of course, it could be argued that such discussion is legitimate *speculation*.

Science advances by a combination of speculation and experiment. Example abound (the double helix foundation of molecular biology is one such recent instance) to show that speculation ahead of experimential check may sometimes lead to great advances. Examples of the converse (Blackett's conjecture about magnetic fields associated with large rotating bodies is a recent instance – leading him, as it did, to fall behind with the work on polar reversal and continental drift) are, in the nature of the case, less easy to find, since their authors will not persist in advertising them; but they also are plentiful. An excessive unwillingness to speculate was a feature of the early days of the Royal Statistical Society, with its motto: "Aliis exterrendum". And down to the present statisticians have earned a reputation for reliability based on their unwillingness to depart far from what can clearly be justified by observational data.

What can ruin our reputation is any tendency to pretend that conclusions "follows" from the data, when in fact they follow only from the data *plus* further assumptions. It seems to me Lindley comes dangerously near to this in his final sentence of section 11. If we denote "biologically meaningful model" by M, and the frequency data by D, it is the *combination* M & D which provides "more information". Lindley's phrasing "the data will provide more information ..." could be taken to suggest that the additional information comes from D alone.

J. B. KADANE (*Carnegie-Mellon University*)

I wish to comment on only one aspect of Lindley's very interesting paper, the choice of parametrization. From a formal Bayesian perspective this choice does not matter, since any parametrization may be equivalently transformed to another. Ease of calculation is a possible reason for a particular parametrization, especially in a case such as this for which the likelihood is in the exponential family. In such a case, if the prior is in a conjugate family then so is the posterior, and the calculations required for the posterior are very simple.

In this case, Lindley has chosen one of the natural parametrizations, so that the log-likelihood is linear in α and β. Any linear transformation of α and β would also have this property. The chief virtue of a parametrization should be the ease of elicitation it affords. In his work with geneticists, has Lindley found this parametrization easy for them to understand? How would he propose that the elicitation be done? What questions would be asked? Is the parametrization in terms of ξ and η easier for them to work with?

I think that elicitation is a key issue in choosing a parametrization, and hope that more attention might be given to it.

J. E. H. SHAW (*University of Warwick*)

I do not see how the uniform prior used throughout Section 7 can be justified. If the prior for α in formula (4.7) is uniform, then we believe with probability 1 that either $p_1 p_3 = 0$ or that $p_2 = 0$. Similarly a uniform prior for β implies that, *a priori*, either $p_1 = 0$ or $p_3 = 0$. Therefore we have prior probability 1 that either AA or aa cannot occur, we only entertain the possibility of Hardy-Weinberg equilibrium in the degenerate case $p = 0$ or 1, and we could hardly be *less* disposed to "let the data speak for themselves"!

However, with non-uniform priors (Section 8), we avoid being infinitely surprised when the data in fact represent all three genotypes. The message from data like $(1, 8, 91)$ is then to collect more data, or to consult an expert geneticist about the prior. A simple mixture of priors each representing HW equilibrium with a different allele frequency p, could be used to avoid swamping extreme data like $(1, 8, 91)$ or $(0, 10, 90)$ while remaining analytically tractable.

Some preposterior analysis for an additional observation would give the practising geneticist a useful illustration of the sensitivity of the posterior to the data. For example, the posterior distribution for α with data $(1, 8, 91)$ decomposes into the three posteriors for α with data $(2, 8, 91)$, $(1, 9, 91)$ and $(1, 8, 92)$, scaled by the corresponding predictive probabilities. All four densities could be presented, with, say, the 1%, 5%, 25%, 50%, 75%, 95% and 99% points marked.

I would value Professor Lindley's views on when a point null hypothesis (Section 10) is appropriate in practice. I only use one when the null and alternative hypotheses represent totally different mechanisms, such as Professor Lindley's own example (in the spoken presentation) of genes on separate chromosomes (null) or on the same chromosome (alternative). But if the mechanisms are totally different, why should there be any relationship between the conditional distributions of the "nuisance" parameters under the null and alternative hypotheses? So in practice, if I need to consider the Kolmogorov paradox, am I doing the right thing?

REPLY TO THE DISCUSSION

As a result of the work reported on in this paper, I feel I understand the formal Bayesian position on Hardy-Weinberg equilibrium; and, by extrapolation, other situations where a sharp null hypothesis with nuisance parameters is being tested. Where I still feel puzzled is over the interaction between statistician and geneticist, especially over the elicitation not merely of the prior, but over the description of the null hypothesis or even whether we are testing or estimating. Kadane, especially clearly, Shaw and Verdinelli all seem to share this puzzlement with me. Frankly, I have not found geneticists interested in Bernstein's model. This could be due to the paradigmic nature of science, whereby scientists are all stuck in the Wright version. It could be due to my incompetence at presentation. What disturbs me is the majority view that alternative hypotheses are unimportant and the Haldane test, or modifications of it, adequate. It is surely true that a major problem that ought to occupy our attention in the future is the way in which scientists can think within the Bayesian paradigm. Hardy-Weinberg is merely the tip of a very large and important problem.

Shaw emphasizes one aspect of the elicitation problem in asking when a sharp null hypothesis is appropriate. The short answer is that I do not know. If null and alternative really are distinct, an unusual case, the continuity condition is inappropriate. If the hypotheses blend into one another, then perhaps it is. But, if so should we test a null? Usually I think not. Certainly in almost all linear models, parameters should be estimated. Sharpness only enters when a decision has to be made: for example, to exclude a regressor variable. But that needs a different analysis from a test of a sharp null.

Shaw's critique of the uniform distribution for (α, β) is taken. In terms of the simplex (p_1, p_2) the density is proportional to $(p_1, p_2, p_3)^{-1}$ with its emphasis on the edges. I used it because it is so popular. The unsatisfactory nature of it also helps to demonstrate the folly of methods based solely on likelihood. Shaw's idea of showing the geneticist the effect of an additional observation is good.

Verdinelli's discussion I found most valuable and I agree with it all. It is only in the neighbourhood of $\alpha = 0$ that knowledge of the proportion of A's is equivalent to knowledge of β, and her analysis is better than mine if that proportion is known. My excuse for taking a prior with the same frequency of A's as the data was merely to prevent the analysis being influenced by two factors and not just α. τ_α^2 is an excellent way of thinking about the prior effect and her analysis is most clarifying.

In reply to Profesor Barnard may I clarify the reference to "machine-like adoption of rote methods". This applied to the direct calculation of Haldane's significance level (as Barnard does) without thinking about the context for the data. This seems to me to be objectionable for at least three reasons: (i) it violates the likelihood principle, (ii) it almost ignores the alternatives to Hardy-Weinberg equilibrium, and (iii) it attempts to let data speak for themselves.

(i) Whatever model is used the likelihood principle would never involve tail areas of the type used in a conventional significance test.

(ii) The only way the alternatives enter is in the consideration of power, yet power calculations are never mentioned in the context of the significance level. If the tail area is 0.034, it may happen that there is no alternative giving this tail a probability above 0.2: or there may be one assigning a value of 0.8. Doubts concerning the null hypothesis are surely different in the two cases.

(iii) Data cannot convey a unique message that is context-free. And a scientist should not want such a message, for it is a basic idea in science that all knowledge should be considered and woven into a single pattern. So the geneticist should study the data in the light of scientific experience: this the Haldane test fails to do.

Bernstein's model deals with the formation of the individual in which egg (A or a) combines with sperm (A or a) and not with the heterozygotes themselves. Wright uses rather abtitrary parameters that do not seem to be model-based. Barnard is correct in criticizing me for claiming "more information". This is true in the sense that a probability *distribution* provides more information than a single *number* (like a significance level); but is false in that the distribution is obtained using a model where the level does not, so that part of the gain information comes from having put more information in (in the form of a model).

Others joined in the discussion at Altea but have not sent in written contributions. Many were critical of several points in the paper. Most of these criticisms I share. Hardy-Weinberg is merely an example of Bayesian thinking and if I have only demonstrated that we do not always know how to do it, then a valuable point will have been made. Statistics is not just mathematics: it involves reality as well.

REFERENCES IN THE DISCUSSION

Bernardo, J. M. (1980). A Bayesian analysis of classical hypothesis testing. *Bayesian Statistics*. (J. M. Bernardo, M. H. DeGroot, D. V. Lindey and A. F. M. Smith, eds.). Valencia: University Press. 605–618.

Lindey, D. V. (1980). Discussion to J. M. Bernardo's paper. *Bayesian Statistics*. (J. M. Bernardo, M. H. DeGroot, D. V. Lindey and A. F. M. Smith, eds.). Valencia: University Press. 637.

Shafer, G. (1982). Lindey's paradox. *J. Amer. Statist. Assoc.* **77**, 325–351.

Spiegelhalter, D. J. and Smith, A. F. M. (1982). Bayes factors for linear and log-linear models with vague prior information. *J. Roy. Statist. Soc. B* **44**, 377–387.

BAYESIAN STATISTICS 3, pp. 327–344
J. M. Bernardo, M. H. DeGroot, D. V. Lindley and A. F. M. Smith, (Eds.)

Approximating Posterior Distributions and Posterior Moments

C. N. MORRIS
University of Texas, Austin

SUMMARY

A two parameter Pearson family is fitted to univariate posterior distributions or likelihood functions by matching the first two derivatives of the log densities. This is just as easy to implement as the standard fitting of the Normal distribution. But because the Pearson families include, besides the Normal distribution a variety of skewed and bounded or semibounded distributions (e.g., Beta, Gamma, and F distributions) there is an opportunity for improved fitting. These distributional approximations also lead to estimates of integrals (moments) that generalize the method of LaPlace. When used with the Tierney-Kadane procedure, proposed to improve LaPlace's method, they provide further improvements in the estimates of posterior moments.

Keywords: PEARSON FAMILIES; LAPLACE'S METHOD; LIKELIHOOD; NATURAL EXPONENTIAL FAMILIES; TIERNEY-KADANE PROCEDURE.

1. INTRODUCTION

Maximum likelihood, or posterior mode, methods are relatively easy to apply because they require computation of just two derivatives of the log likelihood function, or the log posterior density. The first derivative locates the mode, and the second derivative, at the mode, determines the variance. With large samples, the likelihood function, and the posterior density, usually will take the form of a Normal distribution. However, this often is not the case with small samples, when the posterior distribution and the likelihood may be skewed, or limited to a semi-finite or bounded interval.

We consider here improvements in approximations to univariate probability densities (which may be either posterior densities or likelihood functions, but this will not always be mentioned) which are as easy to obtain as standard methods because they involve the use of only two derivatives. In turns out that the Pearson family of distributions (Jeffreys, 1961 or Elderton and Johnson, 1969) is very convenient to work with in this context, and provides approximating distributions with a range of skewnesses, bounded intervals, and long tails. Many of these distributions are well tabled (Gamma, or Chi-square, Reciprocal Gamma, Beta, F, t, and others, in addition to the Normal distribution), which facilitates their use. The Pearson family may be characterized as the family of conjugate prior distributions for the means of natural exponential families (NEF) with quadratic variance functions (QVF) (Morris, 1983a; Jackson, O'Donovan, Zimmer, and Deely, 1970), which are the Normal, Poisson, Gamma, Binomial, Negative Binomial, and NEF generated by the Generalized Hyperbolic Secant distributions (NEF-GHS), and all location and scale transformations of these. The approximations considered in this paper are exact in these NEF-QVF cases, but otherwise only can provide improvements on the usual method of approximation by a Normal distribution.

2. PEARSON FAMILIES AND CONJUGATE PRIOR DISTRIBUTIONS IN NATURAL EXPONENTIAL FAMILIES

A one parameter natural exponential family (NEF) with natural parameter θ and natural observation y is characterized by a cumulant generating function $\Psi(\theta)$, with density of y as

$$\exp\left(n\left(y\theta - \Psi(\theta)\right)\right) h_n(y) \tag{2.1}$$

with respect to some measure independent of n and θ. The mean of y is $\mu = \Psi'(\theta)$, the variance $\text{Var}(y) = \Psi''(\theta)/n$. The variance function is $V(\mu) = \Psi''(\theta)$ so $\text{Var}(y) = V(\mu)/n$. We allow an arbitrary convolution parameter $n > 0$ here so that y can be, for example, the average of n observations (but n need not necessarily be an integer, as happens when the family of distributions is infinitely divisible) The conjugate prior distributions for this family, for θ, are densities of the form

$$K \exp\left(m\left(\mu_0\theta - \Psi(\theta)\right)\right) d\theta \tag{2.2}$$

with $K = K(m, \mu_0)$, and we assume that (2.2) has unit integral over the natural parameter space Θ. With $\mu = \Psi'(\theta)$, (2.2) is equivalent to the density on $\mu = Ey$:

$$K \exp\left(-m \int \frac{(\mu - \mu_0)d\mu}{V(\mu)}\right) \frac{d\mu}{V(\mu)} \tag{2.3}$$

for $\mu \in \Omega$, the mean parameter space. It is easy to check in (2.3) that the mean $E(\mu)$, is μ_0, provided the density vanishes at the endpoints of Ω —see Diaconis and Ylvisaker (1979).

In the case of quadratic variance functions, $V(\mu) = v_2\mu^2 + v_1\mu + v_0$, with v_2, v_1, v_0 all known, higher moments of μ also can be found for the density (2.3). This includes the cases of the Normal (constant variance), Poisson (linear variance function), Gamma, Binomial, Negative Binomial, and the NEF-GHS. In these cases, the prior mean and variance, of μ, are μ_0 and $V(\mu_0)/(m - v_2)$. The posterior distribution of μ given y also satisfies relations of the same form, because the posterior distribution is in the same family as the prior: replace m by $N = m + n$, and μ_0 by the posterior mean $\hat{\mu} = (m\mu_0 + ny)/(m + n)$ in the formulae. See Morris (1983a, Sec. 5).

When $V(\mu)$ is quadratic, (2.3) is, by definition, the density of a Pearson distribution. Thus, the conjugate prior distribution on the mean μ of the natural observation y of a NEF-QVF family is a Pearson family, the pairs being: Normal-Normal; Poisson-Gamma; Gamma-Reciprocal Gamma (distributed as one over a Gamma variable); Binomial-Beta; Negative Binomial-F distributed; and NEF-GHS with t and other conjugate priors.

In general, the Pearson family, with respect to the quadratic function $Q(x) = q_2x^2 + q_1x + q_0 > 0$ has density

$$f(x) = K_Q(m, \mu_0) \exp\left(-m \int \frac{x - \mu_0}{Q(x)} dx\right) \frac{1}{Q(x)} \tag{2.4}$$

with respect to dx, x varying over an interval with $0 < Q(x) < \infty$. There appear to be five parameters here, $(m, \mu_0, q_2, q_1, q_0)$, but one is redundant. This representation is possible only when the mean μ_0 exists, although that unnecessarily excludes some cases, like the Cauchy distribution, which corresponds to $m = 0$. The variance $\text{Var}(x) = Q(\mu_0)/(m - q_2)$ is finite if $m > q_2$. For fixed Q, we shall think of (2.4) as a two parameter distribution, denoted by

$$\text{Pearson}(m, \mu_0; Q) = \text{Pearson}\left[\mu_0, \frac{Q(\mu_0)}{m - q_2}\right] \tag{2.5}$$

the latter indicating the mean and variance. Sometimes it will be useful to consider the exponential part of (2.4) as the density with respect to "Pearson measure", $dx/Q(x)$. Higher moments are also available from known recursions for Pearson distribution central moments. Let $M_0 = 1, M_1 = 0$, and $M_r \equiv E(X - \mu_0)^r$. Then for $r \geq 1$

$$M_{r+1} = \frac{r}{m - rq_2}\{Q'(\mu_0)M_r + Q(\mu_0)M_{r-1}\} \tag{2.6}$$

if $m > rq_2$, see e.g., (Morris, 1983, p. 522). Note that M_r is a polynomial of degree r in μ_0.

Characteristics of the most familiar Pearson families, all the NEF conjugate prior distributions with finite mean μ_0 and representation (2.4), except for NEF-GHS priors, appear in Table 1.

Distribution	Density $p(x)$	Range (x)	$Q(x)$	q_2	μ_0	m	K^{-1}
$N(\mu, \sigma^2)$	$\exp[-\frac{1}{2}(\frac{x-\mu}{\sigma})^2]$	$(-\infty, \infty)$	1	0	μ	σ^{-2}	$\sigma(2\pi)^{.5}$
Gam(a, b)	$x^a e^{-bx}$	$(0, \infty)$	x	0	a/b	b	$\Gamma(a)b^{-a}$
RGam(a, b)	$x^{-b+1}e^{-a/x}$	$(0, \infty)$	x^2	1	$a/(b-1)$	$b-1$	$\Gamma(b)a^{-b}$
Beta(a, b)	$x^a(1-x)^b$	$(0, 1)$	$x(1-x)$	-1	$a/(a+b)$	$a+b$	$\beta(a, b)$
$F^*(a, b)$	$\frac{x^a}{(1+x)^{a+b-1}}$	$(0, \infty)$	$x(1+x)$	1	$a/(b-1)$	$b-1$	$\beta(a, b)$
t_n	$\left(1+\frac{x^2}{n}\right)^{-\frac{n-1}{2}}$	$(-\infty, \infty)$	$n + x^2$	1	0	$n-1$	$\frac{\beta(.5, n/2)}{\sqrt{n}}$.

Table 1. *Some Pearson Families*

Take $a, b > 0$ in all cases with $b > 1$ for RGam and F^* and $n > 1$ for t_n so that EX is finite. RGam ("Reciprocal Gamma") is the distribution of $X = a/G_b$ and $F^*(a, b)$ is the distribution of $X = G_a/G_b$ with G_a, G_b independent Gamma distributed variates. If $m > q_1 \equiv$ coefficient of x^2 in $Q(x)$ then $\text{Var}(X) = Q(\mu_0)/(m - q_2)$, and if $m > 2q_2$ then $E(X - \mu_0)^3 = 2\text{Var}(X)Q'(\mu_0)/(m - 2q_2)$. Densities $p(x)$ satisfy $p(x) \equiv \exp(-m \int (x - \mu_0)Q^{-1}(x)dx)$, and are relative to Pearson measure. $K = K(m, \mu_0)$ satisfies $K \int p(x)Q^{-1}(x)dx = 1$.

3. APPROXIMATION BY PEARSON DENSITIES

Suppose the unimodal density $f(x) > 0$ is given, not necessarily with unit integral, and is to be approximated by a Pearson $(m, \mu_0; Q)$ density for specified Q, perhaps chosen because its range agrees with that of f. Let $l(x) \equiv \log(f(x)Q(x))$. Then, with respect to Pearson measure $dx/Q(x)$, $f(x)Q(x)$ is a density and

$$f(x)Q(x) = \exp(l(x)). \tag{3.1}$$

We also express $f(x)Q(x)$ as

$$\exp\left(-m \int \frac{x - \mu_0}{Q(x)} dx\right) \tag{3.2}$$

by matching two derivatives of the logarithms at the modal value. Letting $\dot{\ell}(x_0) = 0$, with x_0 the root of this derivative, then $\mu_0 = x_0$ and $-\ddot{\ell}(x_0) = m/Q(x_0)$, because the logarithm of (3.2) has first and second derivatives $-m(x - \mu_0)/Q(x)$ and $-m/Q(x) + m(x - \mu_0)Q'(x)/Q^2(x)$. Given $f(x)$ and $Q(x)$, one then chooses the Pearson $(m, \mu_0; Q)$ distribution with

$$\mu_0 = x_0 \quad \text{and} \quad m = -\ddot{\ell}(x_0)Q(x_0) \tag{3.3}$$

where $\dot{\ell}(x_0) = 0$. If $f(x)$ is a density function, estimates of EX and $\mathrm{Var}(X)$ are immediately determined, without further integration, as

$$EX = \mu_0, \qquad \mathrm{Var}(X) = \frac{Q(\mu_0)}{m - q_2}. \tag{3.4}$$

This is due directly to requiring the density to be defined with respect to $dx/Q(x)$, and implies that x_0 estimates the mean, not merely the mode of the distribution.

The method here permits approximation of univariate posterior densities by distributions other than the Normal distribution, and therefore more accurate approximations than the Normal for small sample sizes. Posterior probabilities of intervals can be computed approximately by reference to the tabled Pearson distributions, or general algorithms can be used for the percentage points of Pearson distributions, see *e.g.*, Davis and Stephens (1983).

4. INTEGRATION: ALTERNATIVES TO LAPLACE'S METHOD

LaPlace's method is to approximate the integral of $\exp(n\ell(x))$ by

$$\int \exp(n\ell(x))dx \doteq \sigma(2\pi/n)^{1/2}\exp(n\ell(x_0)) \tag{4.1}$$

where x_0 maximizes $\ell(x)$ and $\sigma^{-2} \equiv -\ddot{\ell}(x_0)$, see de Bruijn (1961), or Tierney and Kadane (1986) for a more recent discussion. The relative error of this approximation is $O(n^{-1})$ as $n \to \infty$. In the context of this paper, we think of LaPlace's method (4.1) as approximating $\exp(n\ell(x))$ by a Normal density, and we ask what would be obtained if a different Pearson density were used.

Theorem 1. *Let $f(x)$ be a given positive unimodal function and let $Q(x)$ be the characterizing quadratic term for a Pearson distribution chosen to approximate f. Define $\ell(x) \equiv \log(f(x)Q(x))$ and let x_0 maximize $\ell(x)$. Define $m \equiv -\ddot{\ell}(x_0)Q(x_0)$. Then*

$$\int f(x)dx \doteq f(x_0)/C_m(x_0) \tag{4.2}$$

with

$$C_m(x_0) \equiv K(m, x_0)p(x_0)/Q(x_0). \tag{4.3}$$

Note that the function $C_m(x_0)$ is just the Pearson density evaluated at the mean $x_0 = \mu_0$, since the Pearson density with parameters m, μ_0 is $p(x) \equiv p^{m,\mu_0}(x)$ with respect to the measure $K(m, \mu_0)/Q(x)$. Thus (4.2) is just the ratio of the function $f(x)$ to the matched Pearson density at its mean value.

The justification of this result will be given momentarily, but no error bound in (4.2) is provided.

Values of K^{-1}, Q, p were provided in Table 1. Values of $p(x_0)$ and $C_m(x_0)$ are given in Table 2 for the familiar Pearson distributions, C_m being defined in terms of the function $s(a) \equiv a!\exp(a)/a^a$ (which by Stirling's formula is approximately $(2\pi a)^{.5}$ for large a).

Distribution	$x_0 = \mu_0$	$p(x_0)$	$C_m^{-1}(x_0)$
$N(\mu, \sigma^2)$	μ	1	$(2\pi/m)^{.5}$
Gam(a, b)	$a/b = a/m$	$a^a m^{-a} e^{-a}$	$s(a)m^{-1}$
RGam(a, b)	$a/(b-1) = a/m$	$m^m a^{-m} e^{-m}$	$s(m)am^{-2}$
Beta(a, b)	$a/(a+b) = a/m$	$a^a(m-a)^{m-a}m^{-m}$	$\frac{s(a)s(m-a)}{s(m)m}$
$F^*(a, b)$	$a/(b-1) = a/m$	$a^a m^m/(a+m)^{a+m}$	$\frac{s(a)s(m)(m+a)}{s(m+a)m^2}$

Table 2. *Values of* μ_0, p, C_m

Here $s(a) \equiv a!a^{-a}e^a$. Refer to Table 1 for notation. $p(x_0) = p(\mu_0)$ where $\mu_0 = \mu$ for Normal, $\mu_0 = a/m$ in four other cases. $C_m \equiv Kp(x_0)/Q(x_0)$.

Justification of Theorem 1. We hope that for some constant c and Pearson family determined by $Q(x)$ and $p(x)$ that m, μ_0 exist so that

$$f(x) \doteq cp(x)/Q(x).$$

If so, then $\ell(x) \doteq \log(c) + \log(p(x))$, and so considering two derivatives, $\dot{\ell}(x_0) = 0$ requires that $\mu_0 = x_0$, and $\ddot{\ell}(x_0)$ then must be $-m/Q(x_0)$. Now

$$\int f(x)dx \doteq c\int p(x)/Q(x)dx = cK^{-1}.$$

If c satisfies, at x_0, $f(x_0) = cp(x_0)/Q(x_0)$ then

$$\int f(x)dx \doteq f(x_0)K^{-1}Q(x_0)/p(x_0) \equiv f(x_0)/C_m(x_0) \qquad \triangleleft$$

Theorem 2. *If Stirling's approximation is accurate for the values* $C_m(x_0)$ *in Table 2, then for the Pearson families of Table 2,*

$$\int f(x)dx \doteq f(x_0)\left(2\pi/(-\ddot{\ell}(x_0))\right)^{1/2} \tag{4.4}$$

with ℓ *and* x_0 *as in Theorem 1.*

Proof. The result for the Normal case is immediate, and is LaPlace's approximation. From Table 2, $x_0 = a/m$, and by Stirling's aproximation $s(a) \doteq \sqrt{2\pi a}$ when a is not small. Using this approximation in each of the last four cases of Table 2, $C_m^{-1}(x_0) \doteq (2\pi)^{1/2}m^{-1/2}Q^{1/2}(x_0)$. (This takes a little work to see, but is not difficult in any of the cases.) Because $m \equiv -\ddot{\ell}(x_0)Q(x_0)$,

$$C_m^{-1}(x_0) \doteq \sqrt{2\pi/(-\ddot{\ell}(x_0))}, \tag{4.5}$$

which gives (4.4). $\triangleleft$

Theorem 2 says that the LaPlace approximation (4.1) can be used not just for the Normal distribution, but also for the other Pearson distributions, if the arguments of $s(\cdot)$ are large enough to justify Stirling's approximation. However, the approximation (4.4) is not the same as LaPlace's, because $\ell(x)$ is not $\log(f(x))$, but instead is $\log(f(x)Q(x))$, and x_0 maximizes $f(x)Q(x)$, not $f(x)$. Thus, both x_0 and the function $\ddot{\ell}(x)$ differ from LaPlace's case.

5. EXAMPLES

If we used Section 2 examples, this would be exact. Thus, we consider the use of Theorems 1 and 2 with two skewed distributions for positive random variables which are not in the Pearson family: the Lognormal and the Inverse Gaussian distributions. The Lognormal density $f(x)$ used here is the density of $X = \exp(Z)$, with $Z \sim N(0,1)$. Thus $\int_0^\infty f(x)dx = 1$ and $\int_0^\infty xf(x)dx = \exp(.5) = 1.649$ is known. The Inverse Gaussian density is

$$g(x) = (2\pi)^{-.5}x^{-1.5}\exp(-(x-1)^2/2x), \quad x > 0 \tag{5.1}$$

and satisfies $\int_0^\infty g(x)dx = \int_0^\infty xg(x)dx = 1$ (the variance also is unity). The values of $\dot{\ell}(x)$ for the Normal, Gamma, Reciprocal Gamma, and F^* distributions, respectively, are $N: -(1+\log(x))/x; G: -\log(x)/x; R: -(\log(x)-1)/x$; and $F: (1+x)^{-1} - \log(x)/x$.

Table 3 contains approximate values for the integrals of the two densities provided by formulas (4.2) and (4.4) where the Normal (LaPlace integral method), Gamma, Reciprocal Gamma, and F^* distributions are all employed as approximating distributions. Only the latter three have the same range of integration, however, as the Lognormal and Inverse Gaussian.

	$f(x)$ = Lognormal			$g(x)$ = Inverse Gaussian		
Method	(4.2)	(4.4)	x_0	(4.2)	(4.4)	x_0
True	1.000	1.000	1.649	1.000	1.000	1.000
Approximation						
Normal (=LaPlace)	.606	.606	.368	.606	.606	.303
Gamma	1.084	1.000	1.000	1.150	1.069	.618
RGam	1.788	1.649	2.718	.711	.661	1.618
F^*	.992	.914	1.933	1.253	1.155	1.000
(5.3) Normal			1.649			1.089
(5.3) Gamma			1.649			1.000

Table 3. *Approximations to* $\int_0^\infty f(x)dx$, $\int_0^\infty g(x)dx$, $\int_0^\infty xf(x)dx$ *and* $\int_0^\infty xg(x)dx$.

True values are $1, 1.649 = \sqrt{e}, 1, 1$ respectively. Approximations (4.2) and (4.4) are provided to $\int_0^\infty f(x)dx$,, and x_0 to $\int_0^\infty xf(x)dx$, and similarly for $g(x)$. The non-Normal choices work better than the Normal (LaPlace's method). The final two rows are based on the Tierney-Kadane approach. They are described later as alternatives to the use of x_0, see (5.3).

The calculations for Table 3 are now demonstrated for the Gamma approximation to the Lognormal density, but where $\log(X)$ has mean μ, variance σ^2 (not only $\mu = 1$ and $\sigma^2 = 1$ of Table 3). Then

$$f(x) = (2\pi\sigma^2)^{-.5}x^{-1}\exp\left(-.5\sigma^{-2}(\log(x)-\mu)^2\right),$$

with $\ell(x) = \log(xf(x)), \dot{\ell}(x) = -\sigma^{-2}(\log(x)-\mu)x^{-1}$, so $x_0 = \exp(\mu)$, and $\ddot{\ell}(x_0) = \sigma^{-2}x_0^{-2}$. It follows from (4.4) that we approximate

$$\int f(x)dx \doteq f(x_0)(2\pi)^{.5}\left(-\ddot{\ell}(x_0)\right)^{-.5} = 1$$

for all μ, σ^2. This is *exact* for all μ, σ^2. LaPlace's approximation is similarly seen to be $\exp(-\sigma^2/2)$ in general, for the Lognormal. However, $\int xf(x)dx$ is approximated as $x_0 = \exp(\mu)$. The correct answer is $\exp(\mu + \sigma/2)$, and so the error ratio is $\exp(-\sigma^2/2)$. Note that since $Q(x) = x$ for the Gamma, $m = -\ddot{\ell}(x_0)Q(x_0) = \sigma^{-2}x_0^{-2}x_0 = \sigma^{-2}\exp(-\mu)$ and so $a = mx_0 = \sigma^{-2}$ in this case is the Gamma parameter. Larger values of a (smaller σ^2)

lead to better approximations in this case. We could have used (4.2) instead of (4.4), but note that (4.2) and (4.4) also become closer as a increases. This same condition, large a, improves LaPlace's approximation $\exp(-\sigma^2/2)$ to $\int f(x)dx$. It seems reasonable to calculate a when fitting Gamma approximations and to expect approximations to improve when a is large. Table 3 shows that how well these approximations work depends on one's ability to choose a good matching Pearson distribution. In this case, using the fact that $(0, \infty)$ is the known region of integration provides improvements in five of the six cases relative to the LaPlace method (use of a Normal approximation). We will consider briefly in section 7 how third derivatives can be used to help make a good choice of a particular Pearson alternative. At this point, there is no clear evidence that (4.2) generally provides better answers than (4.4), except that (4.2) is exact if $f(x)$ follows precisely the assumed Pearson distribution. Since (4.4) is generally easier to apply, it will be used whenever convenient.

Also listed in Table 3 are values x_0 satisfying $\dot{\ell}(x_0) = 0$. These values serve as estimates of then mean $\mu = \int xf(x)dx / \int f(x)dx$, for if $f(x)$ is approximately $cp(x)/Q(x)$, wiht p, Q from Table 1, then

$$\mu \doteq \frac{c\int xp(x)/Q(x)dx}{c\int p(x)/Q(x)dx} = \frac{cK^{-1}\mu_0}{cK^{-1}} = \mu_0 = x_0 \tag{5.2}$$

because $p(x)$ and $Q(x)$ are chosen so $\mu_0 = x_0$ is the mean. The values in Table 3 do not encourage the indiscriminant use of this method, although again, these approximations are better for the three non-Normal Pearson families than for the normal family. However, these examples were not chosen to be favorable to the proposed methods of integration, and even so, if one knew to use the F^* approximation, then the results would be acceptable. A more favorable example is considered in Section 7.

6. APPROACH OF TIERNEY AND KADANE

Tierney and Kadane (1986) recommend an alternative procedure for estimating μ, by applying LaPlace's approximation separately to the numerator and denominator integrals. This procedure is shown in their paper to work exceptionally well. It also can be applied using other Pearson distributions, with even better results, if the Pearson family is carefully chosen. Thus, having chosen Q, we apply (4.4) twice to

$$\mu = \frac{\int xf(x)dx}{\int f(x)dx} \doteq \frac{x_1 f(x_1)}{f(x_0)} \left(\frac{\ddot{\ell}_0(x_0)}{\ddot{\ell}_1(x_1)} \right)^{.5} \tag{6.1}$$

where $\ell_0(x) \equiv \log(Q(x)f(x)), \ell_1(x) \equiv \log(xQ(x)f(x))$ and x_0, x_1 maximize these functions. We assume here that $x > 0$ so that ℓ_1 is defined; however a variant of this procedure has been derived to obviate the need for this assumption. This provides exactly the right answer if the Gamma approximation is used for both the Lognormal and Inverse Gaussian means. The Tierney-Kadane method also works perfectly for the Lognormal, but not perfectly, for the Inverse Gaussian. See the final two rows of Table 3.

Tierney and Kadane note that if the function $f(x)$ is of the form $\exp(-nr(x))$, then the relative error, using their method for estimating the mean, is of order n^{-2} for large n, whereas the direct LaPlace technique only gives relative errors of order n^{-1}. The same result, errors of order n^{-2}, can be demonstrated when using their approach and if the Normal distribution is replaced by one of the other Pearson distributions. The details will be given elsewhere.

7. DIAGNOSTIC METHODS

One may wonder which Pearson approximation to use. Computation of a third derivative $\dddot{\ell}(x_0)$ can be useful. Define D by

$$D(x_0) \equiv \dddot{\ell}(x_0) / \left(-\ddot{\ell}(x_0)\right) \tag{7.1}$$

as a diagnostic measure, where we wish to approximate $f(x)$ by a Pearson family with quadratic function $Q(x)$. Thus $\ell(x) = \log(Q(x)f(x))$ is maximized by x_0 with $\dot{\ell}(x_0) = 0$. We note that for any Pearson family:

$$\dddot{\ell}(x_0) = 2mQ'(x_0)/Q^2(x_0) \tag{7.2}$$

and

$$-\ddot{\ell}(x_0) = m/Q(x_0) \tag{7.3}$$

so that

$$D(x_0) = 2Q'(x_0)/Q(x_0) \tag{7.4}$$

Also define the skewness diagnostic

$$\gamma(x_0) \equiv \dddot{\ell}(x_0)/(-\ddot{\ell}(x_0))^{1.5} \tag{7.5}$$

Then for the Pearson family

$$\gamma(x_0) = 2Q'(x_0)/(mQ(x_0))^{.5} = D(x_0)\left(\frac{Q(x_0)}{m}\right)^{.5}. \tag{7.6}$$

Note that $\gamma(x_0)$ is very close to the skewness of the Pearson distribution, which is, from Table 1 or (2.6), $2Q'(\mu_0)Q^{-.5}(\mu_0)(m0q_2)^{.5}/(m-2q_2)$. The recommendation is that with $f(x)$ given, one calculates D and γ in (7.1), (7.5) for each Q considered, and compares these values with the ideal values (7.4), (7.6). We illustrate for the Lognormal example of Table 3.

Values of $\dot{x}_0$ for the four Pearson families, abbreviated N, G, R, F, appear in Table 3. Values of $\dot{\ell}(x)$ are $N : -(1+\log(x))/x; G : -\log(x)/x; R : -(\log(x)-1)/x$; and $F : (1+x)^{-1} - \log(x)/x$ Values of D, γ appear in Table 4.

Distribution	$D(7.4)$	$\gamma(7.6)$
$N(\mu, \sigma^2)$	0	0
Gam	$2\mu_0^{-1}$	$2(m\mu_0)^{-.5}$
RGam	$4\mu_0^{-1}$	$4m^{-.5}$
Beta	$2\mu_0^{-1} - 2(1-\mu_0)^{-1}$	$\frac{2-4\mu_0}{\sqrt{m\mu_0(1-\mu_0)}}$
F^*	$2\mu_0^{-1} + 2(1+\mu_0)^{-1}$	$\frac{2+4\mu_0}{\sqrt{m\mu_0(1+\mu_0)}}$

Table 4. *Diagnostic values for Pearson families*

Theoretical and computed values for D, γ appear in Table 5. The best distribution is F^* because its ratios (ratios of D values and γ values are the same for each distribution) are 1.14, closer to 1.00 than the others. Note the poor ratio for the Normal, and hence one expects the LaPlace approximation to behave badly. In fact, the F^* is generally best, considering both $\int f(x)dx$ and $\int xf(x)dx$, see Table 3.

	$D(7.4)$	$D(7.1)$	Ratio	$\gamma(7.6)$	$\gamma(7.5)$	Ratio
Normal	0.000	8.155	.00	0.000	3.000	.00
Gamma	2.000	3.000	.67	2.000	3.000	.67
RGam	1.472	1.104	1.33	4.000	3.000	1.33
F^*	1.717	1.504	1.14	3.768	3.302	1.14

Table 5.

Theoretical (7.4), (7,6) and computed (7.1), (7.5) values of D, γ, assuming Lognormal density. Approximation by F^* is best because its ratio is closest to 1.00.

8. AN APPLICATION TO HIERARCHICAL BAYESIAN ESTIMATION

In a hierarchical setting, consider the problem of estimating k Normal means from observations

$$y_i|\theta_i \sim N(\theta_i, 1) \qquad \text{independently}, i = 1, \ldots, k. \tag{8.1}$$

Assume

$$\theta_i|A \sim N(0, A) \qquad \text{independently}, i = 1, \ldots, k \tag{8.2}$$

and let A have a flat distribution dA on $(0, \infty)$ (this leads for $k \geq 5$ to admissible alternatives to Stein's estimator). Letting $S = \sum y_i^2 \equiv 2T$ and $B = 1/(1 + A)$, the posterior density on B, given the data, is proportional to

$$B^{\frac{k-4}{2}} \exp(-TB) \tag{8.3}$$

for $0 \leq B \leq 1$. The details are omitted, but they are contained essentially in Morris (1983b). The Bayes estimator (posterior mean) of θ_i is $(1 - E(B \mid T))y_i$, and hence we need to compute $E(B \mid T)$. Assume henceforth that $k = 6$, a value chosen (small)) so that standard approximations are not good; substantial improvement would be available for larger k values. Then it is easily shown that

$$\begin{aligned} B^* \equiv E(B \mid T) &= \int_0^1 B^2 \exp(-BT)dB \bigg/ \int_0^2 B \exp(-BT)dB \\ &= 2T^{-1}\left(1 - \frac{T^2/2}{\exp(T) - 1 - T}\right). \end{aligned} \tag{8.4}$$

The density $B \exp(-9B/2)$ of B when $S = 9$ ($T = 9/2$) is graphed in Figure 1 as "True". The approximating Normal distribution having the same mode $= T^{-1} = .222$ and the same second derivative at the mode is the $N(T^{-1}, T^{-2}) = N(2/9, 4/81)$ density. It is plotted in Figure 1 as "Normal", and fits poorly, especially because it ignores the restriction $0 \leq B \leq 1$. The Beta distribution satisfies this restriction, and so it is chosen to be fit by the method of Section 3, as "New". The fit is reasonable, except near $B = 1$ where no Beta density can be shaped as the true density is.

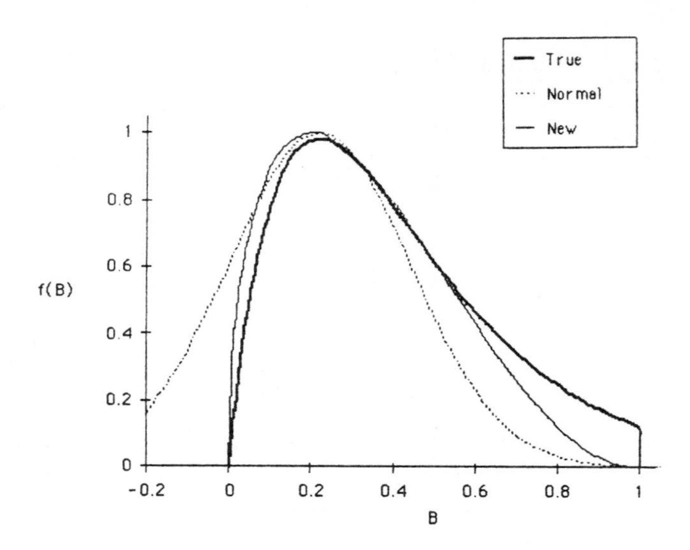

Figure 1. *Graph of "true" posterior density for* $B(8.3)$ *for* $k = 6, S = 9$ *together with "Normal" and Beta* $(1.5, 3.0)$ *("New") approximations, as obtained by method of Section 3. See calculations in Section 8.*

The fit for the Beta distributions uses the two derivatives of $\ell(B) = \log(Q(B)f(B))$ with $Q(B) = B(1-B)$ and $f(B) = B\exp(-TB)$. The maximizing value is, for general k,

$$\hat{B} = \frac{2(k-2)}{S+k+\sqrt{(S-k)^2+8S}}, \tag{8.5}$$

$$= \frac{4}{T+3+\sqrt{T^2-2T+9}} \tag{8.6}$$

when $k = 6$. This estimates $E(B \mid S)$. The approximating Beta density then has $m = a+b = -\ddot{\ell}(\hat{B})Q(\hat{B}) = (2\hat{B}^{-2} + (1-\hat{B})^{-2})\hat{B}(1-\hat{B}) = 2\hat{A} + 1/\hat{A}$, $\hat{A} \equiv (1-\hat{B})/\hat{B}$. For $S = 9$ this gives $\hat{B} = 1/3$ and $m = 4.5$. Thus, the approximating density is that of a Beta $(1.5, 3.0)$, graphed in Figure 1 as "new".

Figure 2 graphs $\hat{B}$ and B^* against S. The agreement is reasonably good (perfect as $S \to 0$ or $S \to \infty$), especially by comparison to the estimate, labeled "mode", which arises if one assumes the Normal distribution (but retains the restriction $B \leq 1$). The posterior mode estimate is $\min(1, 2/S)$, which is less than the James-Stein estimator, $\min(1, 4/S)$.

This is a case for which the Tierney-Kadane procedure, based on Laplace integration, also works badly. Their method produces estimates of $E(B \mid T)$ that are much too large (about 50% too large for $4 < S < 9$), and which can exceed unity, even though it is known in advance that $B \leq 1$. However, using their method with the Beta distribution gives an estimate significantly more accurate than $\hat{B}$, typically in these examples to within 1% of the right value. Thus, the Tierney-Kadane approach is on target, but is greatly enhanced by substitution of the Beta density for LaPlace's Normal density.

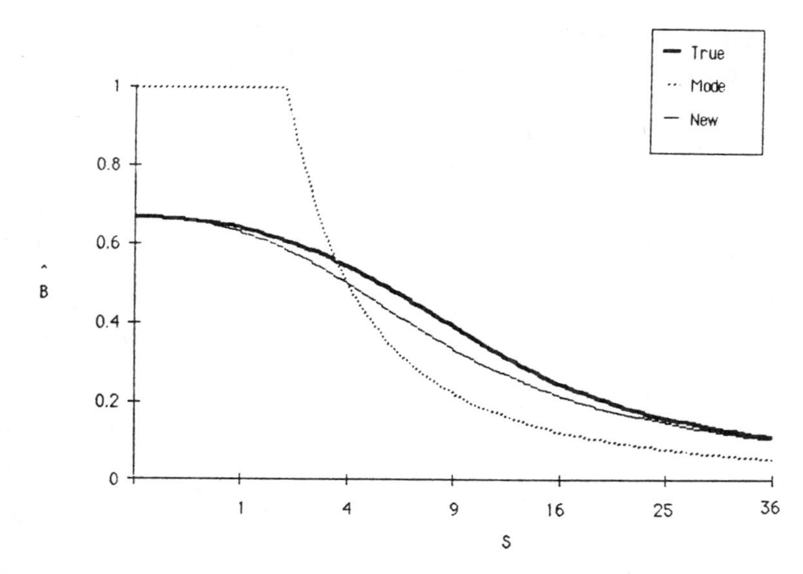

Figure 2. *Graph of B^*, the "true" posterior mean (8.4), the posterior "mode" = $min(1, 2/S)$ (= mean if (8,3) were a Normal density), and $\hat{B}$ (8.6), as obtained by the method of Section 3 using an approximating Beta density. Here $k = 6$, and the horizontal axis is S on a square root scale.*

9. CONCLUSIONS

The methods here are applied to simple problems where the correct answers can be determined easily. The methods actually were developed to deal with more complicated problems, particularly for estimating the shrinkage factor in Bayesian and empirical Bayesian models with substantially more complicated structure than that of Section 8. Some of this is being developed in another paper (Morris, 1987).

The Pearson families are one dimensional and are quite limited in that regard, except for the case of Normal distributions. Some extensions of the techniques of this paper to multivariate situations are possible, however, and are under consideration.

The simple methods here have two other advantages, even when exact numerical answers are available from computing machines.

(1) It is useful to have closed form approximations, which can be studied in their own right. For example, in the application of Section 8, $\hat{B}$ in (8.5) directly estimates the mean. Note that $\hat{B}$ actually is expressed as a function of two other variables k and S and therefore can be studied mathematically. This would not be possible using numerical computation.

(2) Approximations to density functions and also to likelihood functions are provided by the approach of this paper. This is more general than the LaPlace technique, which only approximates integrals.

Even though the methods developed here are limited dimensionally when Normal distributions are not used, they promise to add usefully to the approach of Laplace, and as improved by Tierney-Kadane. By utilizing information such as the range of integration or the suspected of direction of skewness, the methods described here can provide better approximations in small samples to likelihood functions, to posterior distributions, and to posterior moments.

ACKNOWLEDGEMENTS

Assistance provided by Alfred Stanley with the two figures is greatfully acknowledged.

REFERENCES

Davis, C. S. and Stephens, M. A. (1983). Approximate percentage points using Pearson curves. *Applied Statist.* **32**, Algorithm AS 192, 322–327.

de Bruijn, N. G. (1961) *Asymptotic Methods in Analysis*. Amsterdam: North Holland.

Diaconis, P. and Ylvisaker, D. (1979). Conjugate priors for exponential families. *Ann.Statist.* **7**, 269–281

Elderton, W. P. and Johnson, N. L. (1969). *Systems of Frequency Curves*. Cambridge: University Press

Jackson, D. A., O'Donovan, T. M., Zimmer, W. J., and Deely, J. J. (1970). Minimax estimators in the exponential family. *Biometrika* **57**, 439–443.

Jeffreys, H. (1961). *Theory of Probability*. 3rd. Ed., Oxford: University Press.

Morris, C. N. (1982). Natural exponential families with quadratic variance functions. *Ann. Statist.* **10** 65–80.

Morris, C. N. (1983a). Natural exponential families with quadratic variance functions: statistical theory *Ann Statist.* **11**, 515–529.

Morris, C. N. (1983b). Parametric empirical Bayes confidence intervals. *Scientific Inference, Data Analysis and Robustness*, 25–50, New York: Academic Press, 25–50.

Morris, C. N. (1987). Empirical Bayes Interval Estimation. Special Invited Paper. *Inst. Math. Stat. Tech Rep.*, Center Statistical Sciences, Univ. Texas (forthcoming).

Tierney, L. and Kadane, J. B. (1986). Accurate approximations for posterior moments and marginal densities. *J. Amer. Statist. Assoc.* **81**, 82–86.

DISCUSSION

G. E. KOKOLAKIS (*National Technical University of Athens*)

Professor C. N. Morris, following his excellent work on the natural exponential families of distributions with quadratic variance function (NEF-QVF), now proposes the members of the family of conjugate priors for the means of these distributions as alternatives to the usual Normal approximation of the Laplace method. The family of conjugate priors for the means of NEF-QVF enjoys a variety of skewnesses, behaviour of tails, and ranges of x that make them useful. These priors can particularly apply when the function to be approximated (likelihood function or posterior density) is far away from the Normal density.

The approximating densities that Professor Morris proposes are of the form:

$$f(x) = Kp(x)/Q(x), \quad x \in \mathcal{X} \subset \mathcal{R} \tag{1}$$

with

$$p(x) = \exp\left\{-m \int \frac{x-\mu}{Q(x)} dx\right\} \tag{2}$$

with parameters $\mu \in \mathcal{X}$ and $m > 0$, and $Q(x) = q_0 + q_1 x + q_2 x^2$ required to be positive for every $x \in \mathcal{X}$. The particular type of density is specified by the quadratic $Q(x)$. When the density $f(x)$ has well behaved tails (exponentially bounded e.g.) then $E(X) = \mu$ and $\mathrm{Var}(X) = Q(\mu)/(m - q_2)$. Clearly, densities of the above form belong to the Pearson's family of distributions, since

$$\frac{d}{dx} \log f(x) = -\frac{c(x-\alpha)}{Q(x)},$$

where $\alpha = (m\mu - q_1)/c$ and $c = m + 2q_2$, which is the defining property of a Pearson density provided $Q(x)$ is quadratic.

It must be noticed that the proposed approximating densities have a very useful property. Let $g(x) = f(x)\{Q(x)\}^r$ and $l(x) = \log\{g(x)\}$. Then

$$\frac{d}{dx}\ell(x) = -\frac{c^*(x-\alpha^*)}{Q(x)} \tag{3}$$

with $\alpha^* = \{m\mu + (r-1)q_1\}/c^*$ and $c^* = m - 2(r-1)q_2$. Consequently the function $g(x)$ can take the form (1) with parameters $\mu^* = (m\mu + rq_1)/m^*$ and $m^* = m - 2rq_2$ while $Q(x)$ remains the same. Thus $g(x) = f(x)\{Q(x)\}^r$ is a p.d. of the same type as $f(x)$ for all r's provided that $\mu^* \in \mathcal{X}$ and $m^* > 0$. It might be noticed that the t-distribution does not have this property. This is because the t-distribution is a restricted case of the Pearson's type VII distribution with density of the form $f(x) \propto \{\beta^2 + (x-\lambda)^2\}^{-b}$, $x \in \mathcal{R}$, $b > 1/2$. (c.f. Jeffreys (1961), p. 75). This type has $Q(x) = \beta^2 + (x-\lambda)^2$ and, as can be easily verified, enjoys the above property.

Another remark has to be made also. If we evaluate the derivative of (3) at the mode we get $c^* = -\ddot{\ell}(\alpha^*)Q(\alpha^*)$. In the special case where $r = 1$ we have $\alpha^* = \mu$, that is as Professor Morris mentions, the mode of $f(x)Q(x)$ is the mean of the p.d. $f(x)$ and $c^* = m = -\ddot{\ell}(\mu)Q(\mu)$.

To approximate an integral, $\int_{\mathcal{X}} f(x)dx$ say, of an arbitrary positive unimodal function $f(x)$, Professor Morris suggests two main things. Choose first a Pearson density with quadratic $Q(x)$ on the base of the shape of $f(x)$ or after applying a diagnostic test. Then fit the exponential part of the Pearson density, namely $p(x) \equiv p(x|m,\mu)$ in (2), to $f(x)Q(x)$ by equating modes and curvatures at modes. We get therefore the approximation

$$f(x)Q(x) \doteq \hat{c}p(x|\hat{m},\hat{\mu}) \tag{4}$$

with $\hat{c} = f(\hat{\mu})Q(\hat{\mu})/p(\hat{\mu}|\hat{m},\hat{\mu})$, and consequently it is expected that

$$f(x) \doteq \frac{f(\hat{\mu})Q(\hat{\mu})}{p(\hat{\mu}|\hat{m},\hat{\mu})} \times \frac{p(x|\hat{m},\hat{\mu})}{Q(x)}.$$

Thus we have

$$\int_{\mathcal{X}} f(x)dx \doteq \frac{f(\hat{\mu})}{C(\hat{m},\hat{\mu})} = f(\hat{\mu}) \times J \tag{6}$$

with $J^{-1} \equiv C(\hat{m},\hat{\mu}) = K(\hat{m},\hat{\mu})p(\hat{\mu}|\hat{m},\hat{\mu})/Q(\hat{\mu})$, where $K(\hat{m},\hat{\mu})$ is the normalizing constant of $p(x|\hat{m},\hat{\mu})/Q(x)$.

When the function $f(x)$ is a probability density then the r.h.s. of (6) has to be close to unity and thus $E(X) \doteq \hat{\mu}$ and $\text{Var}(X) \doteq Q(\hat{\mu})/(\hat{\mu} - q_2)$.

The following coments can now be made.

1) Morris' method is very conveinent since it gives direct estimates of the mean and variance without further integration. But the question is whether these estimates are robust enough outside the range of approximating densities.

2) Everything depends entirely on the choice of Q. Perhaps it would be more appropriate to fit a Pearson density to the function $g(x) = f(x)\{Q(x)\}^r$ with r chosen to make $g(x)$ sufficiently regular. For example we might require Pearson's measure of skewness of $g(x)$, namely Sk = (mean-mode)/(std. dev.), to be close to zero, for some initial rough estimates of the involved quantities.

3) Fitting $p(x|m,\mu)$ to $f(x)Q(x)$ might result in an entirely differently shaped approximating density $p(x|\hat{m},\hat{\mu})/Q(x)$. For example when we approximate a Log$N(\mu,\sigma^2)$ by a $G(a,b)$ we get $a = \sigma^{-2}$ and $b = \sigma^{-2}e^{-\mu}$. The latter is quite different from the former when $\sigma > 1$ and thus we cannot expect good estimates of the mean and variance unless $\sigma \ll 1$.

4) To overcome the above problem Professor Morris suggests the use of the Tierney-Kadan type of approximation, i.e.

$$E(X) \doteq \frac{x_1 f(x_1)}{f(x_0)} \left\{ \frac{\ddot{\ell}_0(x_0)}{\ddot{\ell}_1(x_1)} \right\}^{1/2}, \qquad (7$$

where $x_k(k = 0,1)$ maximize respectively the functions $\ell_k(x) = \log\{x^k f(x)Q(x)\}$. A similar approximation holds for the second moment, that is

$$E(X^2) \doteq \frac{x_2^2 f(x_2)}{f(x_0)} \left\{ \frac{\ddot{\ell}_0(x_0)}{\ddot{\ell}_2(x_2)} \right\}^{1/2}, \qquad (8$$

where x_2 maximizes $\ell_2(x) = \log\{x^2 f(x)Q(x)\}$. The difference from the usual Tierney Kadane approximation is that their x_k's maximize the functions $\log\{x^k f(x)\}$. It is not clea to what extent (7) and (8) improve the Tierney-Kadane approximation. It is tedious bu straightforward to see that if $x_k^{(r)}(k = 0,1,2)$ maximize respectively the functions $\ell_{kr}(x) = \log[x^k f(x)\{Q(x)\}^r]$, then the approximations (7) and (8) give exact results for the mear and variance of the Log$N(\mu, \sigma^2)$ for any μ, σ^2 and r, provided r is in the permisible regior $(r < 1+\sigma^{-2})$. Thus, at least in this case, there is no particular advantage for choosing $r = 1$
5) It seems more appropriate to use the approximation (6) for both terms of the ratio

$$E(X^k) = \frac{\int_{\mathcal{X}} x^k f(x)dx}{\int_{\mathcal{X}} f(x)dx} \quad (k = 1,2).$$

We then get the following result, which might well be called "Lindley's approximating ratio" in recognition of his original ideas on the subject,

$$E(X^k) \doteq \frac{x_k^k f(x_k)/C(\hat{m}_k, x_k)}{f(x_0)/C(\hat{m}_0, x_0)} \equiv \frac{x_k^k f(x_k) \times J_k}{f(x_0) \times J_0} \quad (k = 1,2), \qquad (9)$$

with an obvious intrepretation of the involved quantities. This approximation is exact withir the Pearson family. It is also exact within this family for any r when the functions $\ell_{kr}(x) = \log[x^k f(x)\{Q(x)\}^r](k = 0,1,2)$ are used in place of $\ell_{k1}(x) \equiv \ell_k(x) = \log\{x^k f(x)Q(x)\}$. Table 1 gives J as a function of a and b in the particular case where $r = 0$ (direct Pearsor approximation). It can be also used for any r provided we modify accordingly the parameters a and b and evaluate $p(x)/Q(x)$ at $x_k^{(r)}$ the mode of $\ell_{kr}(x)$. Thus it can be seen that the approximation (9) is exact for any $r(r < 1 + \sigma^{-2})$ when the Gamma approximation tc Log$N(\mu, \sigma^2)$ is used.
6) It would be very desirable to know which is the class of densities (outside the Pearsor family) that permit exact approximations under each of the Pearson approximating densities In this direction the paper by Daniels (1980) might be very helpful.

Concluding I must confess that I found this paper by Professor Morris very stimulating. Without doubt, a lot of us, inspired by this work, will start looking for mathematically tractable and flexible enough multivariate Pearson approximating densities. I thank him and the organizing committee for giving me the opportunity to discuss this excellent paper.

Distribution	$f(x) = p(x)/Q(x)$	Range	$Q(x)$	m	mean (μ)	mode (x_0)	$\sigma_0^{-2} = -\ddot{l}(x_0)$
$N(a,b)$	$\exp\left\{-\frac{(x-a)^2}{2b}\right\}$	$\mathbb{R}$	1	b^{-1}	a	a	$1/b$
$G(a,b)$	$x^{a-1}e^{-bx}$	$\mathbb{R}_+$	x	b	a/b	$(a-1)/b$	$b^2/(a-1)$
$RG(a,b)$	$x^{-b-1}e^{-a/x}$	$\mathbb{R}_+$	x^2	$b-1$	$a/(b-1)$	$a/(b+1)$	$(b+1)^3/a^2$
$B(a,b)$	$x^{a-1}(1-x)^{b-1}$	$(0,1)$	$x(1-x)$	$a+b$	$a/(a+b)$	$(a-1)/(a+b-2)$	$\frac{(a+b-2)^3}{(a-1)(b-1)}$
$F^*(a,b)$	$x^{a-1}(1+x)^{-(a+b)}$	$\mathbb{R}_+$	$x(1+x)$	$b-1$	$a/(b-1)$	$(a-1)/(b+1)$	$\frac{(1+b)^3}{(a+b)(a-1)}$
t_n	$\left(1+\frac{x^2}{n}\right)^{-\frac{n+1}{2}}$	$\mathbb{R}$	$1+x^2/n$	$n-1$	0	0	$(n+1)/n$

Distribution	a	n	b	K^{-1}	$\{p(x_0)/Q(x_0)\}^{-1}$	$J = \{Kp(x_0)/Q(x_0)\}^{-1}$	J^*
$N(a,b)$	x_0		σ_0^2	$\{2\pi b\}^{\frac{1}{2}}$	1	$\{2\pi b\}^{\frac{1}{2}}$	L
$G(a,b)$	$\frac{x_0^2+\sigma_0^2}{\sigma_0^2}$		$\frac{x_0}{\sigma_0^2}$	$\Gamma(a)b^{-a}$	$\frac{b^{a-1}e^{a-1}}{(a-1)^{a-1}}$	$\frac{\Gamma(a)e^{a-1}}{b(a-1)^{a-1}}$	L
$RG(a,b)$	$\frac{x_0^3}{\sigma_0^2}$		$\frac{x_0^2-\sigma_0^2}{\sigma_0^2}$	$\Gamma(b)a^{-b}$	$\frac{a^{b+1}e^{b+1}}{(b+1)^{b+1}}$	$\frac{a\Gamma(b)e^{b+1}}{(b+1)^{b+1}}$	L
$B(a,b)$	$\frac{x_0^2(1-x_0)+\sigma^2}{\sigma_0^2}$		$\frac{x_0(1-x_0)^2+\sigma_0^2}{\sigma_0^2}$	$\beta(a,b)$	$\frac{(a+b-2)^{a+b-2}}{(a-1)^{a-1}(b-1)^{b-1}}$	$\frac{\beta(a,b)(a+b-2)^{a+b-2}}{(a-1)^{a-1}(b-1)^{b-1}}$	$\frac{a+b-2}{a+b-1}\times L$
$F^*(a,b)$	$\frac{x_0^2(1+x_0)+\sigma^2}{\sigma_0^2}$		$\frac{x_0(1+x_0)-\sigma_0^2}{\sigma_0^2}$	$\beta(a,b)$	$\frac{(a+b)^{a+b}}{(a-1)^{a-1}(b+1)^{b+1}}$	$\frac{\beta(a,b)(a+b)^{a+b}}{(a-1)^{a-1}(b+1)^{b+1}}$	$\frac{b+1}{b}\times L$
t_n		$\frac{\sigma_0^2}{1-\sigma_0^2}$		$\sqrt{n}\beta(\frac{1}{2},\frac{n}{2})$	1	$\sqrt{n}\beta(\frac{1}{2},\frac{n}{2})$	$\frac{n+1}{n}\times L$

Table 1. *Characteristics of some Pearson densities. J^* is derived from J using Stirling's approximation $\Gamma(x) = \sqrt{2\pi}e^{-x}x^{x-0.5}$, $L = \{-2\pi/\ddot{\ell}(x_0)\}^{0.5}$.*

R. KASS (*Carnegie-Mellon University*)

It is clearly worthwhile to investigate alternatives to the Normal kernel appearing in Laplace's method. Carl Morris has indicated that Pearson family distributions may provide good ones. Although I have not looked at the details (which were omitted), I find the results very interesting. I would like to comment first on the three essential developments in statistical applications of Laplace's method, then on some more technical matters.

Laplace's method is a technique of classical applied mathematics (going back at least to Laplace), but in many statistical problems it is applied to a ratio of integrals. In computing a second-order approximation to a posterior expectation, Lindley (1961) noticed that the fourth derivatives of the loglikelihood (or log posterior) that appear in the numerator and denominator disappear from the ratio. This is a nice simplification. There are, however, as we have tried to emphasize in our paper in this volume, infinitely many forms in which Laplace's method may be applied (still based on a Normal kernel). This was noted by Mosteller and Wallace (1964). Tierney and Kadane (1986) observed that, in the terminology of our paper, when the fully exponential method is used, the third derivatives disappear from the ratio as well (and are replaced by an additional Hessian determinant). This is an even greater simplification. Furthermore, as we also indicated, there are other forms that can be useful, as well. Second-order expectation (and variance) approximations thus involve three things: Laplace's method, its application to a ratio of integrals, and consideration of the form in which Laplace's method is applied.

The Normal kernel is not the only one used classically with Laplace's method: the Gamma kernel is used as well, when the maximum of the exponentiated function (the log posterior) occurs on the boundary of the domain (the parameter space) rather than in its interior. It turns out that the fully exponential method again produces a simplification in this case. This is mentioned in Kass and Fu (1987), though no details are given there. Briefly, we write

$$I = \int_a^{\infty} b(\theta)\exp[-nh(\theta)]d\theta$$

where $-nh(\theta)$ could be proportional to the loglikehood or log posterior, as in our paper. Assuming $-nh$ has its maximum at a, it and b may be expanded linearly to get

$$I = \hat{I}\{1 + 0(n^{-1})\}$$
$$\hat{I} = b(a)\exp[-nh(a)][nh''(a)]^{-1}.$$

Assuming g is positive, the ratio

$$E(g(\theta)) = \frac{\int_a^{\infty} g(\theta)L(\theta)\pi(\theta)d\theta}{\int_a^{\infty} L(\theta)\pi(\theta)d\theta}$$

may be approximated by using $-nh(\theta) = \tilde{\ell}(\theta) = \log(L(\theta)\pi(\theta))$ in the denominator and $-nh^*(\theta) = \ell^*(\theta) = -nh(\theta) + \log(g(\theta))$ in the numerator and then applying the first-order approximation above to each. There is cancellation in the remainder and this produces the second-order approximation

$$E(g(\theta)) = \frac{\tilde{\ell}'(a)}{\ell^{*\prime}(a)} \cdot \exp[\ell^*(a) - \tilde{\ell}(a)]\{1 + 0(n^{-2})\}.$$

Like the calculations Morris carried out, the one above does not generalize in an obvious way to the multidimensional case. There may well be a way to push them through. In data-analytical applications, as opposed to theoretical ones, the one-dimensional case is not very interesting since Gaussian quadrature may be used instead.

In Kass, Tierney, and Kadane (1988), we summarized extensive work that had to do with (i) properties, (ii) applications and implementation, and (iii) theoretical justification of approximations based on Laplace's method. Analogous research on approximations using Pearson-family kernels would be valuable.

REPLY TO THE DISCUSSION

I thank both discussants for their encouragement and thoughtful commentary. In particular, they have provided a very useful clarification of the history of related approximation methods. I particularly welcome learning about the relationship of the pioneering work of Lindley and of Mosteller and Wallace.

The original reason for my interest in this topic was to approximate the mean of a non-linear function, i.e., the shrinkage factor in the example of Section 8. Several considerations, besides accuracy , were important, including simplicity and the ability to approximate the exact integral functionally, not just numerically. These also would be key concerns in many other problems in statistics, and certainly in Bayesian statistics. "Simplicity" here means that only two derivatives are required, because competing likelihood methods based on normal approximations use two derivatives. By widening the range of distributions useable by these techniques, one should only be able to improve on normal approximations, because the normal distribution is a particular Pearson family, besides being reached asymptotically as $m \to \infty$ by any of the other Pearson family distributions. Thus, my goal was to make improvements on standard methods, without raising the complexity cost. It was not necessarily to find the optimum fit.

In this spirit Kokolakis's idea to allow $r \neq 1$ seems to be helpful and deserves further attention because it adds flexibility to the fitting process without forfeiting other computational advantages. I preferred $r = 1$ for one main reason, however, which was to get the simple approximation $x_0 \equiv \hat{\mu}$ to the mean; $r \neq 1$ appears to forfeit that.

Both discussants raise the important question of the approximation's accuracy. While much more research is needed, two comments are in order. First, it should take no great skill to recognize some common situation in which a non-normal Pearson distribution will approximate a distribution or likelihood function better than a normal distribution. Second, approximations of the relative order $1 + O(n^{-1})$ for single integrals and $1 + O(n^{-2})$ for integral ratios (n = sample size) appear to hold quite generally, which extends the results for the normal approximations mentioned by Kass. These results are developed by making use of the general properties of the Pearson distributions, and so differ somewhat from earlier derivations that apply specifically for the normal distribution. However, there is a certain slipperiness of asymptotic calculations when applied to a data set with a particular n because $O(n^{-1})$ and $O(n^{-2})$ provide no specific bound, and therefore no specific minimum sample size n for a satisfactory approximation. Hodges (1985, Sec. 21) elaborates on this general point about approximations. Thus, the accuracy question will be best answered by more numerical study.

Kass's assertion, that the one-dimensional case is uninteresting in data-analytic cases because Gaussian quadrature can be used, overlooks the point that an entire distribution is being fitted by a Pearson family, not just the moments. Thus probability approximations are conveniently provided via standard t, chi-square, F, and other tables, but not by quadrature methods.

Both Kokolakis and Kass underscore the importance of extending these results to the multivariate case if the method is to be useful for data-analytic work, and I agree. Of course, for multivariate densities, the methods of this paper may be useful to approximate directly one "bad" (*e.g.*, highly skewed) dimension, coupled with use of the normal distribution for the remaining dimensions. Beyond that, the most important advance required for practical application seems to involve extension of the Pearson approximation methods to tractable continuous multivariate families of distributions exhibiting a variety of correlation, skewness, and support space combinations. Since such distributions are not in abundance, this search may prove to be minimally rewarding, but I think it worth a try.

REFERENCES IN THE DISCUSSION

Daniels, H. E. (1980). Exact saddlepoint approximations. *Biometrika* **67**, 59–63.

Hodges, J. S. (1985). Assessing the Accuracy of Normal Approximations. *Tech. Rep.* **449**, School o Statistics, University of Minnesota.

Jeffreys, H. (1961). *Theory of Probability*. Oxford: University Press.

Kass, R. E. and Fu, J. C. (1987). Posterior Large Deviations. *Tech. Rep.* Department of Statistics Carnegie-Mellon University.

Kass, R. E., Tierney, L, and Kadane, J. B. (1988). Asymptotics in Bayesian Computation. (In thi volume).

Lindley, D. V. (1961). The use of prior probability distributions in statistical inference and decisions *Proc. 4th Berkeley Symp.* **1**, 453–468.

Lindley, D. V. (1980). Approximate Bayesian methods. *Bayesian Statistics*, (J. M. Bernardo, M. H DeGroot, D. V. Lindley and A. F. M. Smith, eds.). Valencia: University Press.

Mosteller, F. and Wallace, D. L. (1964). *Inference and Disputed Authorship: The Federalist Papers* Reading, Mass: Addison-Wesley.

Tierney, L, and Kadane, J. B. (1986). Accurate approximations for posterior moments and margina densities. *J. Amer. Statist. Assoc.* **81**, 82–86.

BAYESIAN STATISTICS 3, pp. 345–359
J. M. Bernardo, M. H. DeGroot, D. V. Lindley and A. F. M. Smith, (Eds.)

Modelling with Heavy Tails

A. O'HAGAN
University of Warwick

SUMMARY

Heavy-tailed models for data processes permit automatic outlier accommodation and rejection. Work on this topic is reviewed before attention is directed to the use of heavy-tailed distributions in modelling prior beliefs. "Rejection" of the prior when it is "outlying" from the data is a desirable property, and is achieved through obvious heavy-tailed modelling. Less obvious are models which allow some of a group of exchangeable parameters to be "rejected" by the others, when data suggest they are dissimilar. Such models allow the data to over-rule the general shrinkage which results from normal-theory modelling. Arbitrary clumping, or "multiple shrinkage", of the parameters can be achieved.

Keywords: HEAVY-TAILED DISTRIBUTIONS; MULTIPLE SHRINKAGE; OUTLIERS; STUDENT *t* DISTRIBUTION; CREDENCE.

1. ONE PARAMETER

1.1. *Datum/Prior Conflict*

Our applications of heavy-tailed distributions will all be concerned with conflict between sources of information. The simplest and earliest useful applications are to outliers in data, where the information being presented by the outliers conflicts with the bulk of the observations. Work in this area proceeded from an even simpler case, which was analysed by Dawid (1973). We have a single observation x with location parameter θ, and a prior distribution for θ. The prior and the datum are the two information sources. They conflict if the difference between x and the prior mean is large relative to both standard deviations. If both likelihood and prior are normal the conflict is not recognised or properly resolved. The posterior mean, for example, is a strict weighted average of x and the prior mean. The posterior variance is independent of x. No matter how great the conflict between likelihood and prior, the normal model compromises.

Dawid showed that other resolutions of the conflict are possible. If, for instance, the prior distribution is normal but x has a t distribution then as the difference between x and the prior mean tends to infinity the posterior distribution tends to the prior distribution. In the limit, therefore, the observation is completely ignored; it has been rejected. Conversely if x is normally distributed but the prior is a t distribution, then in the limit the posterior density is the normalised likelihood. It is as if a uniform prior had been used, so that in this case the prior information has been ignored. In general, whichever source of information has the heavier tail will be rejected in favour of the lighter-tailed source.

A t distribution with 20 degrees of freedom has lighter tails than one with $5df$, and so will be preferred if their means conflict. The family of t distributions with decreasing df provide distributions with steadily heavier tails. The use of t distributions adds a new dimension to the combination of information sources. For, using only normal distributions the important factor is variance. Sources with lower variance, i.e. higher precision, are given more weight.

We have learnt to think of precision as representing the *strength* of an information source. This is still true with t distributions provided the sources do not conflict. When there is conflict, however, tail weight becomes important. If two sources have the same tail weight, as is the case with two normals, then neither will dominate. Otherwise, the one with the lighter tail dominates and the other is progressively ignored as the conflict increases. We can think of lighter tails (more df for t distributions) as representing greater *credibility*. When information sources do not conflict, credibility is not brought into question, and information strength (precision) is relevant. Otherwise credibility (df) matters, and becomes the key factor when conflict is great.

1.2. Outliers

Our use of words like "conflict" and "credibility" in §1.1 becomes clearer and more firmly based when we study more complex models, with many information sources. Suppose that instead of a single observation, we have a sample $x_1, x_2, \ldots, x_n$. They are identically and independently distributed conditional on the common location parameter θ. This is the problem analysed by O'Hagan (1979). If the x_is have t distributions and we let one of them become steadily more separated from the others, then in the limit it is ignored. The posterior distribution from the whole sample tends to the posterior which would be obtained using only the other $n-1$ observations; the outlier is rejected.

This happens regardless of the prior distribution for θ. Of course, if $n = 2$ and the prior distribution is uniform then neither observation can dominate: which is the outlier? But outlier rejection does occur if $n = 2$ and the prior is proper. If we hold x_1 fixed and let $x_2 \to \infty$ then x_2 will be rejected. In this case the prior is lending credibility to x_1. As $x_2 \to \infty$ it conflicts with x_1 and the prior, and so is rejected.

Information sources which do not conflict gain credibility from each other. For $n > 2$ a single outlier is always rejected, even with a uniform prior, because its credibility is only half the combined credibility of the other two observations, Formally, credibility is defined as follows. Suppose that for large x the density $f(x)$ is of order x^{-c}, then f has credibility c. The credibility of a t distribution with d df is $d+1$. Under this definition, credibility is additive. Let observations x_i have densities f_i with credibilities c_i. Thus, $p(x_i|\theta) = f_i(x_i - \theta)$ and the likelihood is their joint density

$$\Pi_i f_i(x_i - \theta),$$

which as a function of θ is of order θ^{-c}, where $c = \Sigma_i c_i$. Therefore, if two groups of observations exist, such that the observtions within a group are reasonably close but the two groups are very widely separated, then whichever group has the greater total credibility will dominate.

Three more points about credibility should be noted. First, we must qualify the implication that the information source with greater credibility dominates when there is conflict. Convergence of the posterior density does not imply convergence of posterior moments. It is clear from O'Hagan (1979) and Dawid (1973) that convergence of moments requires greater differences in tail weights between information sources. For t distributions a credibility difference of more than $2r$ allows convergence of the r-th moment.

Next, although distributions with credibilities in $(0, 1]$ are improper they can be useful. Imagine a sample of three observations from a t distribution with "$-\frac{1}{2}df$", credibility $\frac{1}{2}$. Then the tails of the likelihood are of order $\theta^{-3/2}$,and even with a uniform prior (credibility zero) the posterior will be proper.

Finally, asymmetric distributions may have tails of differing weights, and therefore we should in general distinguish between left and right credibility. Since we will only be using t distributions in this paper, the distinction is not relevant.

1.3. Linear Models, Example

Similar considerations apply to more general linear models if we replace the usual normal errors by t distributed errors. If there are no outliers, the analysis is similar to the conventional one. The notion of strength of an observation changes, of course, but could be captured by one of the many definitions of *influence*. The simple concept of credibility needs no modification. If all errors have the same *df* then all observations have the same credibility. A single outlier will eventually be rejected, regardless of how influential it might be. Consider the following example.

Five observations are taken from the "regression through the origin" model

$$Y_i = \beta x_i + E_i,$$

$i = 1, 2, \ldots, 5$. The first four observations are $(y_i, x_i) = (1,1), (1,2), (4,3)$, and $(4,4)$. The fifth is $(y, 10)$ and we let y vary. As $y \to \infty$ this observation clearly becomes an outlier. We assume a uniform prior for β. If the E_is are $N(0,1)$ errors then the posterior for β is normal with mean equal to the least squares estimate

$$\hat{\beta} = (31 + 10y)/130$$

and standard deviation $130^{-\frac{1}{2}} = 0.0877$. If we were to ignore (reject) the fifth observation, β would have mean $\hat{\beta}_0 = 31/30 = 1.033$ and standard deviation $30^{-\frac{1}{2}} = 0.1826$.

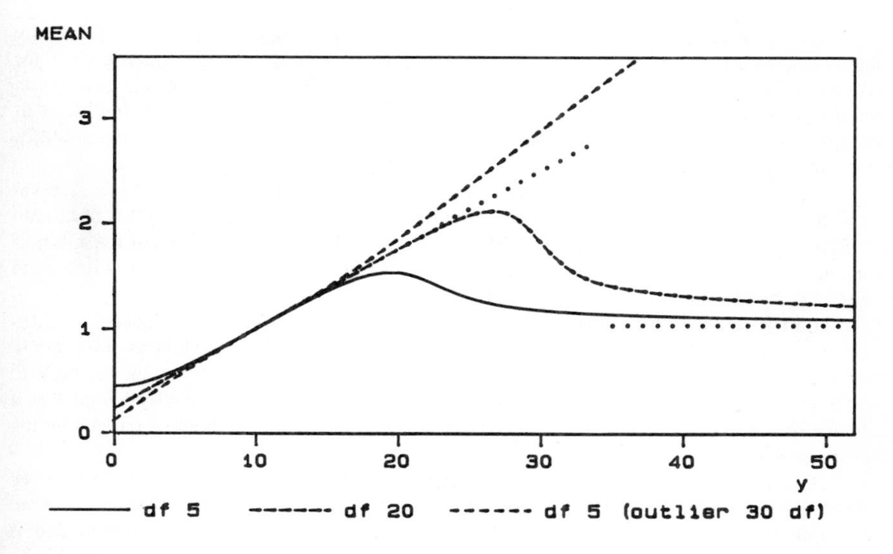

Figure 1. *Posterior means for the regression example*

Now let the E_is instead have t distributions with zero mean and unit scale parameter. Figures 1, 2 and 3 display aspects of the posterior distribution for various degrees of freedom. Figure 1 plots posterior means against y. When all observations have $5df$ we see that the posterior mean is close to the line $\beta = \hat{\beta}$ until y reaches about 17. It then curves away and eventually tends to $\hat{\beta}_0$, showing rejection of the outlier. The same happens for any *df*,

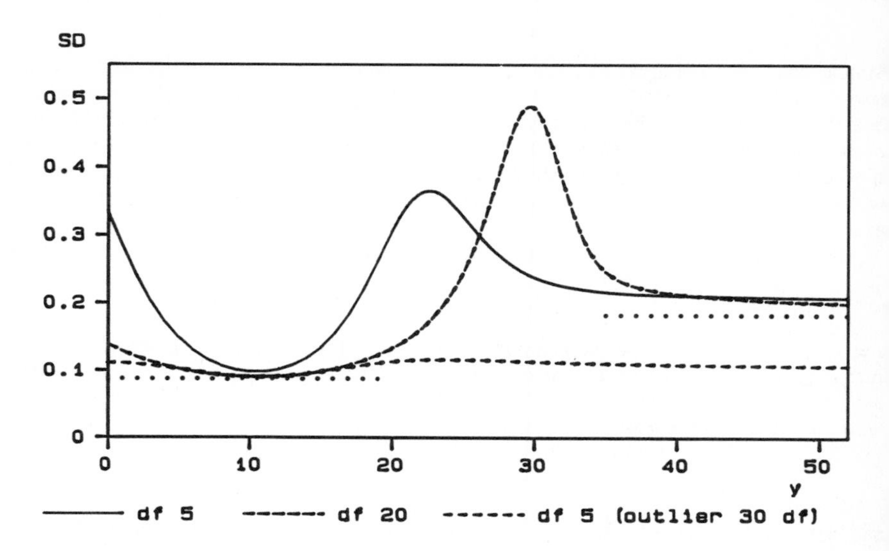

Figure 2. *Posterior standard deviations for the regression example*

provided all observations have the same *df*. We see this in Figure 1 for $20df$. The only difference from the $5df$ line is that the posterior mean stays close to the normal value $\hat{\beta}$ for longer, and only breaks away around $y = 27$. The third plot in Figure 1 is for the case where the first four observations have $5df$ but the outlier is given $30df$. The combined credibility of the group of four is $4(1+5) = 24$, whereas the outlier alone has credibility 31. Accordingly, it is the group of four that are rejected, and the posterior mean asymptotes to $y/10$.

Figure 2 shows the posterior standard deviations for the same models. When all observations have the same *df* we see the standard deviation moves from the normal value of around 0.0877 to the "reject" value of around 0.1826. Notice that at y values where the mean is turning away from $\hat{\beta}$, the posterior standard deviation peaks. This phenomenon is described in O'Hagan (1981).

The transition is shown more fully in Figure 3, which shows selected posterior distributions for the case of $20df$. In Figure 3(a) the posterior moves to the right as y increases, but at $y = 25$ it is becoming noticeably skew. In Figure 3(b) the posterior moves back to settle around $\hat{\beta}_0 = 1.033$. The transition between the two phases is marked by bimodality, a change of skewness and high variance caused by trying to encompass both of the conflicting information sources.

This analysis is easily applied in general linear models with more than one parameter. "Rejection" of outliers is automatic, and there is no need for them to be recognised or tested for explicitly. This is a useful property because identification of outliers in linear models is not a trival problem. For one Bayesian discussion if this problem see Smith and Pettit (1983), and for other Bayesian approaches to outliers generally see Freeman (1979) (and references therein) and Goldstein (1982).

A more interesting use of heavy-tailed modelling when there are several parameters is to make the prior distributions heavy-tailed. Two different heavy-tailed prior distributions are analysed in the remainder of this paper.

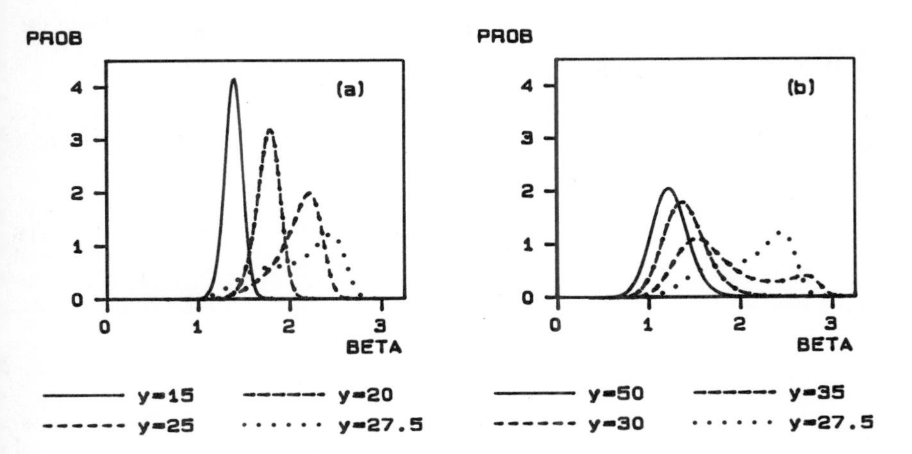

Figure 3. *Posterior densities (20df) for the regression example*

2. OUTLYING PARAMETERS

2.1. Modelling

Consider the standard hierarchical model for the "k means" or "one-way analysis of variance" problem. For $i = 1, 2, \ldots, k$,

$$y_i|\theta_i \sim N(\theta_i, 1),$$
$$\theta_i|\xi \sim N(\xi, \nu),$$

and we assume a uniform prior density for ξ. Then the posterior distribution of the θ_is shows the familiar shrinkage phenomenon:

$$E(\theta_i|y) = ay_i + (1-a)\overline{y},$$

where $a = \nu/(1+\nu)$. The shrinkage occurs because each y_i provides information about each θ_j. The information on θ_1 is of two types. First, y_1 is suggesting that θ_1 should be close to y_1. Then the other observations, indirectly through the prior exchangeability of the θ_is, suggest less strongly that θ_1 should be close to $y_2, y_3, \ldots, y_k$. Because normal distributions are used throughout, all these other observations are given equal weight, causing $E(\theta_1|y)$ to be drawn towards $\overline{y}$.

Now imagine that there is conflict between these data sources. For instance, suppose that y_k is very distant from the other y_is. With the normal model, $E(\theta_k|y)$ will be shrunk substantially, and will take a compromise value which is not actually supported by any information source. We can obtain radically different behaviour by using heavy-tailed distributions at the middle level of the hierarchy. We retain for the moment the first stage, $y_i|\theta_i \sim N(\theta_i, 1)$, but we now propose that the θ_is have t distributions given ξ with means ξ. Specifically,

$$p(\theta_i|\xi) \propto \{c + s(\theta_i - \xi)^2\}^{-\frac{1}{2}c}.$$

The parameter c is the credibility, equal to degrees of freedom plus one. We could allow c to be less than one without any impropriety occurring in the posterior distribution. The parameter s respresents strength. It is not actually defined as equal to precision, since the precision

$$\{\mathrm{Var}(\theta_i|\xi)\}^{-1} = s(c-3)/c$$

only exists if $c > 3$.

Therefore the prior joint density is

$$p(\boldsymbol{\theta}, \xi) \propto \prod_{i=1}^{k} \{c + s(\theta_i - \xi)^2\}^{-\frac{1}{2}c} \tag{1}$$

and the posterior is

$$p(\boldsymbol{\theta}, \xi|\boldsymbol{y}) \propto p(\boldsymbol{\theta}, \xi) \exp\{-\frac{1}{2}\sum_{i=1}^{k}(y_i - \theta_i)^2\}.$$

As with all the heavy-tailed models, this is rather intractable, and to understand the posterior distribution requires careful summarisation. Obtaining moments will demand numerical integration. Because of the uniform prior on ξ, the dimensionality can be decreased by one as follows. Make an orthogonal transformation from $\boldsymbol{\theta}$ to $\boldsymbol{\phi}$ such that $\phi_1 = \sqrt{k}\,\overline{\theta}$, and $\phi_2, \phi_3, \ldots, \phi_k$ therefore represent contrasts between the θ_is. Then it is easy to show that the posterior distribution of ϕ_1 is $N(\sqrt{k}\overline{y}, 1)$ independent of $\xi, \phi_2, \phi_3, \ldots, \phi_k$.

The behaviour of this model under conflict is best seen through an example.

2.2. *Example, Three Means*

Let $k = 3$ and consider the data $y_1 = 0$, $y_2 = 1$ and let y_3 increase so that it conflicts with the other two data. The prior parameters are $c = 6$, $s = 4$. Figures 4 and 5 show posterior means and standard deviations for θ_1, θ_2, and θ_3, as functions of y_3. In Figure 4 we see that for low y_3 all three means increase linearly. Around $y_3 = 5$ their behaviour changes. For θ_1 and θ_2 the information from y_3 is now progressively rejected. Their posterior means move back towards the values they would have when $k = 2$, i.e. shrunk towards the sample mean 0.5.

In contrast, the distribution of θ_3 resolves the conflict by rejecting y_1 and y_2. The information from y_3 is directly related to θ_3 through the normal likelihood, and therefore has infinite credibility. The other two observations provide information on θ_3 through the heavy-tailed prior. No matter how large k is, if we let $y_k \to \infty$ the posterior mean of θ_k will asymptote to y_k, rejecting all the other $k-1$ sources.

Figure 5 shows the usual variance behaviour. $Sd(\theta_3|y)$ rises to a peak then falls back to a new level. The limiting value is unity. For θ_3, all the prior information, which served only to tie the θ_is together, has been rejected. Its standard deviation therefore tends to that of y_3. The standard deviations of θ_1 and θ_2 settle at a lower level, since each is deriving information from both y_1 and y_2.

It is also interesting to examine posterior covariances. The posterior correlations when y_3 is small are all moderately large and positive, reflecting the general shrinkage: if one θ_i is below its mean then others are also expected to be lower. At $y_3 = 15$ the posterior correlation matrix is

$$\begin{pmatrix} 1 & .54 & -.02 \\ .54 & 1 & -.02 \\ -.02 & -.02 & 1 \end{pmatrix}$$

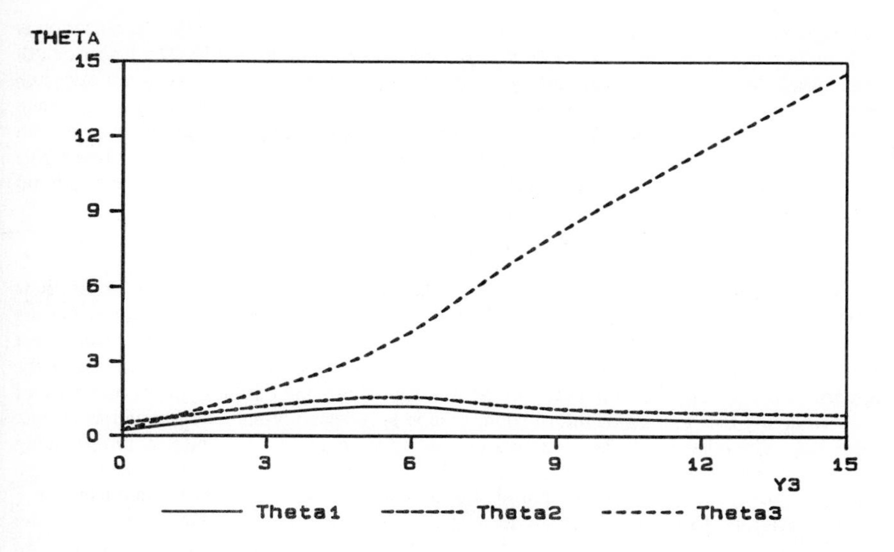

Figure 4. *Posterior means for the 3-means example*

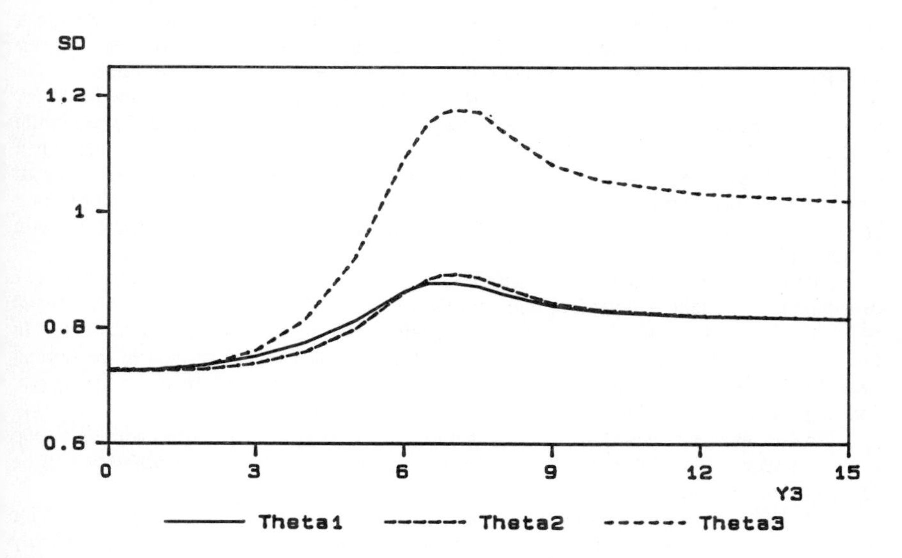

Figure 5. *Posterior standard deviations for the 3-means example*

θ_1 and θ_2 are still positively correlated (and shrunk together) but they are now essestially independent of θ_3. With the rejection of prior linkage of θ_3 with ξ, θ_3 no longer shares an information source with either θ_1 or θ_2.

The same phenomena would be seen for any other choices of prior parameters s and c.

For the example we chose high strength and low credibility partly because this combination is realistic in practice, and partly because it shows the behaviour quite clearly. The high strength (four times the data strength) results in strong shrinkage of the θ_i when y_3 is not outlying. The fact that the y_is are close together suggests that the θ_is have similar values, so strong shrinkage is realistic. Low credibility means that y_3 does not have to move far from y_1 and y_2 before credibility of the prior is brought into question and rejection begins. Again this is realistic. In an experiment, three group means $y = (0, 1, 6)$, with unit variances, should suggest that the third group is unlike the first two.

2.3. The General Case

For $k > 3$, and possibly several outlying y_is, the same general behaviour will occur. It is helpful to think of y_i giving information about θ_j via information about ξ. Integrating out the θ_is, the y_is are a sample from a heavy-tailed distribution with mean ξ. So inference about ξ reduces to the case of Section 1.2. Sufficiently outlying y_is will all be rejected and the posterior distribution of ξ will settle on the average of the non-outlying y_is. Then think of θ_i as receiving two sources of information. One is y_i, whose normal distribution has infinite credibility. The other is ξ, whose distribution has finite credibility. For the θ_is corresponding to non-outlying y_is these sources do not conflict, so these θ_is are shrunk towards their average. For a θ_i corresponding to an outlying y_i, the y_i source conflicts with and domimates the ξ source, therefore $E(\theta_i|y)$ asymptotes to y_i and its variance to unity. Outlying θ_is are not shrunk. They become independent of the other θ_is. In fact the posterior distribution is the same as would arise from the prior (1) with the product being only over "non-outlying" i.

Further generalisations are possible. The one-way analysis of variance properly has more than one observation for each i. We can think of y_i as representing the mean of n_i observations $x_{i1}, \ldots, x_{in_i}$. Then from within-group variation we can estimate the data variance. In our examples in this paper, variances/strengths are assumed known, which may not be realistic in practice. Integrating out data variance (from a vague or conjugate prior), the likelihood for the θ_is changes from multivariate normal to multivariate t. Although this gives the data finite credibility it does not produce any interesting change in behaviour. The reason is that when conflict arises between a multivariate t information source and another source, then either all or none of the multivariate t components will be rejected. In the present context, introducing an unknown data variance would in principle allow all of the data to be rejected, but not with our assumption of a uniform prior for ξ.

Allowing different data variances within each group results in the y_is having independent t distributions (assuming independent priors for the variances). The credibility of y_i will depend on n_i and it will become possible for y_i to be rejected as an information source for θ_i. In this case a θ_i corresponding to an outlying y_i will be shrunk much more than in the normal case. In the limit its mean will be shrunk completely to the average of the non-outlying y_is, and its posterior distribution as a whole will tend to that of ξ.

Finally, we can give each x_{ij} a t distribution. Then outlying observations within a group can be rejected. y_i will become the average of non-rejected x_{ij}s and its credibility will be related to the number of non-rejected x_{ij}s.

It is important to notice, however, that we cannot generally handle an unknown prior variance, i.e. unknown s. This is because k needs to be very large before we can reliably distinguish between, on the one hand, a case of outliers separated from a group with relatively low variance and, on the other hand, a single group with relatively large variance. Outlier rejection behaviour would not arise. This is not unrealistic. If in a small sample the eye discerns outliers, then either it is deluded or else there is a strong prior notion of what variance should be expected. Wherever we apply heavy-tailed distributions it is important to think about variances and to give realistic values. If uncertainty about variances is to be expressed it should be very strong and highly credible.

3. OUTLYING PARAMETER GROUPS

3.1. Multiple Shrinkage

The model of Section 2 allows individual outlying θ_is to be separated from, and become independent of, the main group of θ_is. They also become independent of each other. Suppose that $k = 5$ and we have one group of three y_is well separated from a group of two. The two would be treated as outliers. This may be reasonable in some applications, where if two outlying y_is were close together it would be regarded as fortuitous. In other applications, however, we would suspect that there are two distinct groups of θ_is. We would then want to shrink within each group. This kind of behaviour is termed "multiple shrinkage" by George (1986a, b). He proposes ad hoc estimation procedures which are capable of multiple-shrinkage. They may be derivable as Bayes estimators from mixture priors (analogous to the approach of Box and co-workers to outliers, see Freeman (1983)).

The prior distribution (1) does not admit multiple shrinkage. The only focus for shrinkage is ξ, and the θ_is are only connected through ξ.

3.2. An Alternative Prior Distribution

Consider the prior density

$$p(\boldsymbol{\theta}) \propto \prod_{i<j} \{c + s(\theta_i - \theta_j)^2\}^{-c/(2k)}. \tag{2}$$

There is no hyperparameter ξ, so there is no single focus. Instead every θ_i is linked directly to every other θ_j by a component of (2) which says that their difference is likely to be small. However, all this information has limited credibility.

The choice of power, $-c/(2k)$, for each component of (2), means that as a function of each θ_i the product is of order θ_i to the power $-c(k-1)/k$. Therefore the prior information about θ_i has total credibility $-c(k-1)/k$, or approximately c. The reason for the factor $(k-1)/k$ is as follows.

Consider the original normal prior distribution in Section 2. After integrating out ξ we have

$$p(\boldsymbol{\theta}) \propto \exp\left\{-\frac{1}{2\nu}\sum_{i=1}^{k}(\theta_i - \overline{\theta})^2\right\}, \tag{3}$$

which is of course improper because of the original impropriety of ξ. If we let the credibility parameter c in (1) tend to infinity then it is easy to see that we obtain (3) with $\nu = s^{-1}$. Now let c tend to infinity in (2). We have the limit

$$p(\boldsymbol{\theta}) \propto \exp\left\{-\frac{1}{2}\frac{s}{k}\sum_{i<1}(\theta_i - \theta_j)^2\right\}, \tag{3}$$

which is easily found to be again identical to (3). Therefore increasing credibility makes the two priors asymptotically equivalent to the same normal model. In both cases the strength s identifies with the normal precision ν^{-1}.

The prior (2) is also improper. Exactly the same trick which was used to reduce the dimensionality of the posterior distribution from the prior (1) (through an orthogonal transformation) may be used for (2). Since we do not have ξ the posterior from (2) is actually in one less dimension, and hence easier to summarise. For large k there are also computational advantages over George's approach since, although our summarisation problem increases exponentially with k. the number of possible "targets" for multiple shrinkage increases even faster.

3.3. Example, Five Means

To compare the prior models consider $k = 5$ and $y = (0, 0.5, 1, y, y+1)$ and let y tend to infinity. We then have two conflicting groups. Posterior means and standard deviations from the two priors are presented in Tables 1 and 2. For both priors we chose $c = 6$ and $s = 4$.

y	θ_1	θ_2	θ_3	θ_4	θ_5
3	1.15 (.70)	1.33 (.68)	1.48 (.67)	2.07 (.69)	2.47 (.76)
5	1.48 (.78)	1.70 (.75)	1.89 (.76)	3.31 (.83)	4.12 (1.01)
8	1.01 (.78)	1.19 (.78)	1.34 (.78)	6.88 (1.14)	8.08 (1.09)
19	.57 (.73)	.72 (.73)	.86 (.73)	18.66 (1.01)	19.68 (1.01)

Table 1. *Posterior means (standard deviations) for the five means example, prior (1)*

Table 1 shows the expected behaviour from prior (1). For small y there is general shrinkage. Then the two outliers break away, with attendant increase in variance. In the limit the group of three are shrunk but the group of two are not. Their posterior means asymptote to y and $y + 1$, with a unit difference between them. The correlation matrix at $y = 19$ is

$$\begin{pmatrix} 1 & .45 & .45 & -.01 & -.01 \\ .45 & 1 & .46 & -.01 & -.01 \\ .45 & .46 & 1 & -.01 & -.01 \\ -.01 & -.01 & -.01 & 1 & .00 \\ -.01 & -.01 & -.01 & .00 & 1 \end{pmatrix}$$

showing that θ_4 and θ_5 become independent of the other θ_is and of each other.

y	θ_1	θ_2	θ_3	θ_4	θ_5
3	.93 (.79)	1.17 (.77)	1.39 (.75)	2.24 (.79)	2.77 (.86)
5	.85 (.87)	1.14 (.86)	1.42 (.86)	4.16 (.99)	4.93 (1.00)
8	.56 (.84)	.84 (.84)	1.11 (.85)	7.64 (.96)	8.35 (.95)
19	.36 (.83)	.63 (.82)	.89 (.83)	18.97 (.92)	19.65 (.92)

Table 2. *Posterior means (standard deviations) for the five means example, prior (2)*

Table 2 shows results from prior (2) for the same y values. Separation begins somewhat earlier, but the main difference from Table 1 is in the asymptotic means of θ_4 and θ_5. In this cases they are shrunk together. The difference in their means is less than 0.7 at $y = 19$. There is a corresponding difference in the correlation matrix. At $y = 19$ it is

$$\begin{pmatrix} 1 & .24 & .22 & -.00 & -.00 \\ .24 & 1 & .24 & -.00 & -.00 \\ .22 & .24 & 1 & -.00 & -.00 \\ -.00 & -.00 & -.00 & 1 & .19 \\ -.00 & -.00 & -.00 & .19 & 1 \end{pmatrix}$$

Thus θ_4 and θ_5 have again become independent of θ_1, θ_2 and θ_3, but not of each other.

3.4. General Behaviour

In general, prior (2) allows arbitrary grouping of the y_is. Asymptotically, as between group distances increase, between group shrinkage is eliminated, but we retain shrinkage within each group. Notice, however, another feature of Tables 1 and 2. In the limit, prior (2) shrinks the large group $(\theta_1, \theta_2, \theta_3)$ less than prior (1). The reason is as follows. The limiting posterior is the same as would be achieved from the prior in which all between-group components of (2) are elimnated. So each group is independent of the others, and its posterior derived from the prior (2) where the product is only over i and j indexing the group members. There is now a mismatch between the k appearing in the power $-c/(2k)$, and the size of group, say k^*. This prior is therefore seen to have credibility

$$c^* = ck^*/k$$

and hence it has strength

$$s^* = sk^*/k.$$

The earlier group separation in Table 2 is because of the reduced credibility of a group, and the ultimate lower shrinkage of the group of three is because of the reduced strength.

It is quite realistic to have shrinkage dependent on group size in this way. The more members a group has, the more sure we can be of identifying it as a group of similar individuals.

3.5. Conclusion

Heavy-tailed prior distributions like (1) and (2), in conjunction with heavy-tailed data distributions where appropriate, have great potential for providing novel answers to standard problems. This paper has attempted to show this potential, and also to show the modelling considerations which lead to the appropriate uses of heavy-tailed distributions.

ACKNOWLEDGEMENTS

A large part of this work was done during the Bayesian Statistics Study Year at the University of Warwick. The idea behind the prior (2) arose from discussion with Bill DuMouchel. I also benefitted from discussions with others too numerous to mention, and am grateful for the sponsorship of the Science and Engineering Research Council which brought these people to Warwick.

REFERENCES

Dawid, A. P. (1973). Posterior expectations for large observations. *Biometrika* **60**, 664–667.

Freeman, P. R. (1979). On the number of outliers in data from a linear model. *Bayesian Statistics*, (J. M. Bernardo et al., eds.). Valencia, Spain: University Press, 349–365.

George, E. I. (1986a). Minimax multiple shrinkage estimators. *Ann. Statist.* **14**, 188–205.

George, E. I. (1986b). Combining minimax shrinkage estimators. *J. Amer. Statist. Assoc.* **81**, 437–445.

Goldstein, M. (1982). Contamination distributions. *Ann. Statist.* **10**, 174–183.

O'Hagan A. (1979). On outlier rejection phenomena in Bayes inference. *J. Roy. Statist. Soc. B* **41**, 358–367.

O'Hagan A. (1981). A moment of indecision. *Biometrika* **68**, 329–330.

Smith, A. F. M. and Pettit, L. I. (1983). Outliers and influential observations in linear models. *Bayesian Statistics 2*, (J. M. Bernardo et al. eds.). Amsterdam: North-Holland, 473–490.

DISCUSSION

I. GUTTMAN (*University of Toronto*)

First, congratulations to Prof. O'Hagan for a very stimulating (and may I say, credible) paper. He has given us all some very important questions to think and ponder about. The new definition of the credibility of a distribution is very pretty, and I think helps clarify ideas on the very important topic of heavy tails. I would however, like to raise the following points.

(i) A bothersome point to me that emerged on reading Prof. O'Hagan's paper is that modelling the heavy tailed aspect kind of cuts out modelling other aspects, some of which I feel are crucial when concerned about questions in this area. To be a bit specific, statements such as "suppose y_j is very distant from the other y_i's" covers a multitude of sins. What, as an opening question, could cause y_j to be distant from the other y_i's? Some needed modelling should be done here, surely. For example, based on experience we might suppose that, given parameters that include a location, say θ_i, that one of the y_i's say, y_j is spurious with

$$E(y_j|\theta_j;\text{other parameters}) = \theta_j + a_j \tag{1}$$

where a_j is a slippage parameter and j is unknown. Now y_j has "some" information re θ_j in it. If y_j has a heavier tailed distribution then $p(\theta_j)$, y_j would be rejected under O'Hagan's scheme and this I find hard to take. I would rather find the posterior of θ_j, integrating out a_j and the other parameters. What happens typically (see, for an example, Guttman, Dutter and Freeman (1978)) is that

$$p(\theta_j|\boldsymbol{y}) = \sum_1^n c_i p(\theta_j|\boldsymbol{y}_{oi}) \tag{2}$$

where, interestingly,

$$c_i = \text{posterior probability that } y_i \text{ is spurious}$$
$$p(\theta_j|\boldsymbol{y}_{oi}) = \text{posterior probability of } \theta_j,$$

given the observations, except for y_i ("oi" means omit y_i). Hence, outlying observations are not rejected, but are incorporated as in (2) in supplying information re θ_j.

Note that, in principle, we could, letting experience decide, assign heavy or light tailed distributions for $p(\boldsymbol{\theta})$, $f_i(y_i|\boldsymbol{\theta})$, etc., a point that is indirectly touched on below in point (vi).

(ii) Continuing point (i) above, in a slightly different direction, another source of heavy and/or light-tailed distributions is the family discussed in the bible - oops, book - of Box and Tiao (1973, chap. 3), viz

$$p(y|\mu,\sigma;\beta) = \omega \exp -\frac{1}{2}c(\beta)\left|\frac{y-\mu}{\sigma}\right|^{-2/(1+\beta)}$$

Here, ω is a normalizing constant and $c(\beta)$ is a standardization yielding Var$(y) = \sigma^2$, with $-\infty < \mu < \infty$, $\sigma^2 > 0$ and $-1 < \beta \leq 1$ and where we could have μ constant or $\mu = \sum_1^m \gamma_j x_j$, etc. Hence, in a routine way, we could assess, using Bayes, what the data tells us about β. This leads to interesting qustions about criterion and inference robustness. Now in the same way in principle, we could assess, if modelling $y|\mu,\sigma$ by t distributions with c d.f., what the data tells us about c. The inevitable question then to Prof. O'Hagan is has he thought about this point and its "touching down" with robustness questions as discussed in Box and Tiao's book, for example.

(iii) I have a question at this point. It seems to me that Tony's paper has only to do with symmetric distributions that are heavy-tailed, so that the modelling that underlies this paper, is, ultimately in terms of K_4, the measure of kurtosis, along with the tacit assumption that $K_3 = 0$. If so, why? In my "data snooping" experience, such as it is, I find this unrealistic, and I would now model spuriousity, for example, by a gamma of moderate order. Of course, this presents other problems—integrations cannot be performed in closed form, and numerical integration is the order of the day, but, in principle, fast, modern day computer technology allows us in this era to do just that.

(iv) Further to rejecting (almost) my own point (ii) above, and in view of (iii), I also have trouble with a situation not alluded to in this paper, and probably for the very good reason that it's difficult when approached by the methods of paper—what do you do if you have groups which are such that an inner group is flanked on both ends by an outer group, as in the case in the famous set of Darwin's data, that looks like the following:

.	.	.	..	..	..	..	.	.	.	.	.	
-67	-48	0					41	49	56	60	75	
1	2	3	..	..	..	..	..	..	..	14	15	←observation number

Figure 1.

The observations are differences in heights of self and crossed fertilized plants, measured, in one-eighth's of an inch.

Now an analysis of the type in Guttman, Dutter and Freeman (1978), based on modelling the spuriosity that might give rise to the outlying observations in Darwin's data, indicates that observations 1, 2 and 15 are suspect, if spuriosity is modelled by mean shifts. And so the question is what would heavy-tailed modelling of the O'Hagan type do in this case. One thing I suspect is that correlation patterns as described in the example of Section 3.3 will not change (as they do in the example) as outlying left and outlying right observations are allowed to proceed to $-\infty$ and $+\infty$, respectively, at the same rate.

(v) I hope that the author will reassure me that he holds the principle of coherence sacred, and that in no way does he subcribe to picking a lighter tailed source or a heavier tailed source after noticing that the data analysis described in this paper will lead to a favoured answer, favoured a priori.

After all, if we are in a loop dictated by the scientific method, "the prior is the prior is the prior" and "chosen" before seeing the data, and if for example, the observed y is disturbingly unexpected then subsequent modelling so that the distribution of y has a heavy-tailed distribution, heavier tailed than the prior, so that y is virtually ignored, is not only incoherent, it's downright unethical and in some situations, immoral.

Now I hasten to say that the sense of the paper is not quite that, but this point deserves an affirmation in no uncertain terms, so that I think this paper would be the right place for this affirmation.

(vi) This brings up the point that, for myself, I think of all modelling as a prior stage. If I worry about "the model", it's because I have concerns that alternative models hold, and as a Bayesian, I would then compute posterior probabilities of the various candidate models holding. This has been discussed in a paper with Norman Draper (1987). So point (v), coupled with this point, is of cardinal concern; at least it is to me.

(vii) Finally, referring again to the example of Section 3.3 (the 5 means example), does $\rho(\theta_i, \theta_j | y)$ depend critically on c, and/or y, and/or $(y_5 - y_4)$? [$y_5 - y_4$ is taken to be 1 in the worked example of Section 3.3].

Once again, this is a very stimulating paper, and I congratulate Prof. O'Hagan. Based on this sample of one paper, I wish that I could have participated in the Bayesian Study Year at Warwick.

W. POLASEK (*University of Basel*)

Since exchangeability is nonetheless a strong assumption, I welcome any new ideas for checking pooling priors. In particular I have suggested "hierarchical diagnostics" in Polasek (1985) for checking the question "to pool or not pool". Using a hierarchical extreme bound analysis (HEBA), I found that 2 groups would have fitted the Grunfeld data in Smith (1973) better than just one.

REPLY TO THE DISCUSSION

I am pleased to have this opportunity to thank Professor Guttman, Dr. Polasek and all those who contributed to the verbal discussion for their constructive comments.

Irwin Guttman raises many important questions which deserve a careful response. I will deal with them in the same order as his discussion.

(i) I apologise for mentioning only very obliquely, in my reference to Freeman (1979), the work on mixtures. Indeed, it was impolite of me not to refer directly to the work of my invited discussant! As a Bayesian approach to outliers, the mixture work has some advantages. In particular, it provides posterior probabilities for observations being outliers, which are valuable diagnostics. Clearly, such models have an important part to play. Equally, heavy-tailed modelling has its place. There are many data processes which are inherently heavy-tailed, with no suggestion of outliers being spurious; they are just less informative than central observations. My interest in presenting the present paper is to work towards a general scheme for routine processing of data. Often the sheer bulk of data to be processed precludes examining each suspected spurious observation. One requires an automatic robustness.

Incidentally, one can also obtain diagnostics of spuriousity within heavy-tailed models. Differentiating the posterior mean with respect to each information source provides a set of "weights" which sum to one. Sources which are in the process of being rejected as outliers will have low or even negative weight.

(ii) I discussed the Box-Tiao family in my 1979 paper. These distributions do not have sufficiently heavy tails to achieve rejection for any β in $(-1, 1]$. They do reject outliers when $\beta > 1$, but then the density is quite unrealistic because it comes to a sharp point at the centre. I much prefer the t family. In general, information provided by data about shape parameters is often extremely weak. For the kinds of models I consider in Sections 2 and 3, the number of parameters will rarely be enough for us to learn anything useful about the shape parameter of their first-stage prior.

(iii) There is no difficulty in using asymmetric distributions. One can cobble together t distributions with different tail thickness and smooth over the joint to produce a realistic density. Their properties are touched on in my 1979 paper. Whether outliers are rejected may then depend on whether they outlie on the right or left. Gamma distributions only allow rejection of outliers on the right, and are only appropriate for positive data. (Inverse gamma distributions provide left-only rejection.)

(iv) My answer here is implied by the last paragraph: outliers on both sides are rejected as easily as on one side if we use t distributions. I have analysed the Darwin data. The posterior distribution settles over the central twelve observations. Numbers 1, 2 and 15 are identified as outliers by the diagnostic referred to in (1) above. The model here is as in Section 1.2. If we saw the same data using the model of Section 2 or 3 then the correlation patterns would not show any new behaviour.

(v) This question goes right back to general principles in applying the Bayesian paradigm. I believe that one has to leave open the possibility of revising the model in the light of the data. I further believe that this is perfectly sound and accords with Bayesian principles. This is not the place to go into details: my argument is simply a development of the approach I put forward in O'Hagan (1988a).

In the context of heavy-tailed modelling, I have said in (ii) above that we cannot expect to learn about degrees of freedom from the data, and so the prior choice must be made carefully. This is possible by using what is essentially the device of imaginary data. Simply ask yourself what you would believe if certain conflicts arose between information sources. (I cannot think that you would ever want the kind of compromise engendered by normal distributions.) For instance, would you believe a group of three observations if it conflicted with your prior belief? If so, your prior credibility must be less than three times the data credibility. And so on. Not only do I believe this works, I also believe it is by far the best way of identifying tail weights in models.

(vi) I agree. In principle this is always the proper way of comparing models, and has to be done carefully when the data suggest a hitherto unconsidered model.

(vii) The correlations depend only very slightly on the data patterns within groups. The critical factor is how far the groups are separated, relative to both variances and credibilities.

I thank Wolfgang Polasek for his comments, and I certainly find his work interesting.

One comment which arose in the verbal discussion should be mentioned here. My use of the word "credibility" was objected to, on the grounds that it has an established, and different meaning in risk theory. Not surprisingly, I find my own usage more apt, but I bow to precedent. Henceforth, I shall use the term "credence" instead. This usage is established in O'Hagan (1988b), which follows up and provides some of the theoretical underpinning for the present paper. The definition of credence in O'Hagan (1988b) is as a somewhat more general measure of tail weight than as one-plus-degress-of-freedom for t distributions, and the paper presents a series of results which I hope will be the start of a "credence algebra" or "credence theory".

REFERENCES IN THE DISCUSSION

Box, G. E. P. and Tiao, G. C. (1973). *Bayesian Inference in Statistical Analysis*, Addison-Wesley.

Draper, A. R. and Guttman, I. (1987). A Common Model Selection Criterion; Proceedings of the Symposium on Probability and Bayesian Statistics, held in honour of the memory of Prof. B. De Finetti. Austria: Innerbruck. (To appear).

Guttman, I., Dutter, R. and Freeman, P. R. (1978). Care and handling of univariate outliers in the general lineal model to detect spuriosity - A Bayesian approach. *Technometrics* **20**, 187–193.

O'Hagan A. (1988a). *Probability: Methods and Measurement*. London: Chapman and Hall.

O'Hagan A. (1988b). On outliers, tail weights and credence. Statistics Departement Research Report No.131, University of Warwick.

Polasek, W. (1985). An Application of the Hierarchical Extreme Bound Analysis: The Grunfeld Data, Inst. f. Statistics, Vienna, Mimeo.

Smith, A. F. M. (1983). A general Bayesian linear model. *J. Roy. Statist. Soc. B* **35**, 67–75.

BAYESIAN STATISTICS 3, pp. 361–375
J. M. Bernardo, M. H. DeGroot, D. V. Lindley and A. F. M. Smith, (Eds.)

On Being Imprecise at the Higher Levels of a Hierarchical Linear Model

L. R. PERICCHI and W. A. NAZARET
Simon Bolivar University, Caracas

SUMMARY

Prior knowledge can be roughly classified as: structural (relationships) and parametric (assessments), the latter being potentially more controversial than the former. The hierarchical linear model has proven to be successful in modelling structural knowledge. However, the higher the stage in the hierarchy, the greater the prior imprecision on the assessments about the hyperparameters. We attempt to characterize the consequences of being imprecise in the higher levels of the hierarchy, conditional on the data. In particular, we investigate the stability of inferences about the means under the usual device of letting the covariance matrix in the last stage of the hierarchy become infinity.

Keywords: HIERARCHICAL LINEAR MODEL, POSTERIOR BOUNDS, PRIOR IMPRECISION, POSTERIOR IMPRECISION, SENSITIVITY.

1. INTRODUCTION

A crucial point in the historical development of Bayesian methods was the recognition that the statisticians's knowledge about the parameters in a model could also be subject to modelling. Good (1965) in his important work on the estimation of proportions, introduces the idea of specifying the prior distribution in stages. Lindley and Smith (1972) in their fundamental paper on the Bayesian Linear Model gave definite meaning to the term "hierarchical prior specification" and provided the basis for what has proven to be most fruitful area of development within the Bayesian paradigm.

We believe that the key concept underlying the hierarchical specification of prior beliefs is the existence of two kinds of prior knowledge: *structural and parametric.* Structural prior knowledge refers to the relationships among the parameters in the model iteself. These relationships are usually dictated by knowledge about the "physical" constraints of the problem. In Good's context, for instance, the smoothness of the proportions is a constraint induced by the belief that they arise from an underlying continuous density. Parametric prior knowledge, on the other hand, is usually related to the assessment or tuning of the relationships addressed by the structural knowledge.

It has become obvious throughout the years that this latter kind of prior knowledge is more controversial than the former. This is due in part to the analyst's traditional reluctance to quote results which are tied to a particular value of certain parameters. However, it is also apparent that many authors have attempted to magnify the virtues of hierarchical forms of prior specifications by arguing that a "loose" assessment of hyperparamenters in this context is less damaging to the inference than an improper prior specification in a non-hierachical context.

In our view what makes the difference between the two situations is the presence of structural knowledge in the hierarchical case. The effect of this prior knowledge on the inference does not vanish even if some (or all) of the hyperparameters are handled through "improper" prior specifications. It is not obvious to us, however, that one can get away, in general, with an improper prior distribution on the last stages of a hierarchical specification.

Motivated by this we attempt in this paper to characterize the impact on the inference of some "standard" ways of being imprecise at the higher levels of the hierarchy. While most authors attach the term imprecise to improper prior specifications in which one or some of the parameters take on limiting values, we use the term *imprecise* to denote a kind of prior specification in which the hyperparameters are only known to belong to a set, in contrast to *precise* priors in which the parameters take on exact values. We then define the *imprecision* as a measure of the size or volume of the set of values for the hyperparameters. Obviously for a sharp prior the imprecision is zero. With this framework in mind, we find the locus in the p-dimensional space of the posterior mean vector in a Normal Hierarchical Linear Model (NHLM) for which the location parameter at the last stage is known, but the variance is only assumed to be larger than a certain value. We show that for a class of NHLM's the locus is a straight line segment parametrized by a suitable function of the variance at the last stage. We also suggest that for the general case this locus is "monotonic" and therefore one can measure the posterior imprecision as the distance between the two ends of the line segment.

Our main conclusion is that it is not possible a priori to pick a large enough value of the variance at the last stage for which the imprecise formulation (as described above) will yield arbitrarily low imprecision and therefore similar inferences, regardless of the data and the value chosen a priori for the mean vector. Indeed we show that the imprecision is a function of the "conflict" between our prior assessment for the mean and a sample based estimate, and that therefore it is not possible to eliminate it without looking at the data first. In Section 2 we present an overview of the NHLM and set up the scenario for our work. Section 3 discusses some of the previous approaches to the handling of imprecision in the NHLM and contrasts them with our own. We also develop here our main general results within the scenario presented in Section 2. In Section 4 we particularize our results to two important cases: One-way ANOVA and Two-way ANOVA with one observation per cell. Section 4 presents the analysis of a data set, and finally Section 5 presents some conclusions and possible extensions.

2. HIERARCHICAL LINEAR MODEL (HLM): A GENERAL OVERVIEW

We are concerned with the problem of making inferences about the parameters in the linear model:

$$Y \sim N(A_1\theta_1, C_1) \tag{2.1}$$

were A_1 is an n by p matrix, θ_1 is a p-dimensional vector of parameters and C_1 is the n by n sample covariance matrix.

Proceeding along the lines of Lindley & Smith (1972) we represent our prior knowledge about θ_1 in the following hierarchical manner: given a p by q matrix A_2, a q-dimensional vector of hyperparameters θ_2 and a covariance matrix C_2:

$$\theta_1 \sim N(A_2\theta_2, C_2). \tag{2.2}$$

Finally, we assume that given a q by r matrix A_2, an r-dimensional vector θ_3, and a covariance matrix C_3 we have:

$$\theta_2 \sim N(A_3\theta_3, C_3). \tag{2.3}$$

The model (2.1) with this particular kind of prior specification is known as the Normal Hierarchical linear model (NHLM). In most cases the first stage of the prior (2.2) is used to introduce some structure into the vector of parameters of the model θ_1. This can be achieved by a suitable choice of A_2 and C_2. The second stage (2.3), on the other hand is used to express our knowledge (or the lack of it!) about the defining parameter of the previous stage. It is because of this that we say that the NHLM induces two kinds of prior knowledge: structural prior knowledge and parametric prior knowledge.

One particular example will help to clarify the situation. In One-way Analysis of Variance one case of the NHLM as described by Lindley & Smith (1972) implies that the main effects θ_1 are exchangeable. This is a way of introducing into the model our belief that the effects are similar and therefore qualifies as structural knowledge. The mean and variance at the last level represent the center and size of similarity and conform to our notion of parametric knowledge, since they merely tune the location and dispersion of the effects.

Even though both types of prior knowledge can be subject to criticism, it has become obvious throughout the years that the assessment of the parameters at the last level is the more controversial of the two. Based on this fact we focus our attention on the specification of the parametric knowledge rather than the structural one. This, in the case of the HNLM, is equivalent to assuming that C_1 and C_2 are known and the only unknown parameters are θ_3 and C_3. We would like to emphasize that this is just a working assumption which allows us to make our point without getting cluttered by the analytical complications resulting from having to specify priors for C_1 and C_2. More importantly, the practical consequences of our findings for this simpler setup are not likely to disappear for the more general one.

For the sake of simplicity and clarity of argument we will assume that C_3 is of the form $\sigma_3^2 H_3$ where H_3 is a known positive definite symmetric matrix.

3. PRIOR IMPRECISION, POSTERIOR IMPRECISION AND INFERENCE SENSITIVITY

In the NHLM as described in section 2, once θ_3 and σ_3 are specified, a Bayes estimate of θ_1 can be obtained. It is given by:

$$\theta_1^* = D[A_1^t C_1^{-1} A_1 b + \{C_2 + \sigma_3^2 A_2 H_3 A_2^t\}^{-1} A_2 A_3 \theta_3] \tag{3.1}$$

where D is the p by p matrix

$$D = [A_1 C_1^{-1} A_1^t + \{C_2 + \sigma_3^2 A_2 H_3 A_2^t\}^{-1}]^{-1} \tag{3.2}$$

and b is the least squares estimate of θ_1.

The Bayes estimator θ_1^* is a compromise between the least squares estimator of θ_1 and the prior estimate $A_2 A_3 \theta_3$, with σ_3 influencing the weight given to the estimate. Because of this, estimators of the form (3.1) are called matrix weighted averages.

Many authors have tackled the problem of how to specify the hyperparameters θ_3 and σ_3^2 when the analyst has very little information about them, or equivalently, when it is desired to minimize a priori their influence on the inference. Lindley and Smith (1972) and Smith (1973a) argue in favor of using the Bayes estimator resulting form (3.1) when σ_3^2 tends to infinity. This is mathematically sensible since it can be proven that as $\sigma_3^{-2} \to 0$

$$\theta_1^* \to D_0[A_1^t C_1^{-1} A_1 b] \tag{3.3}$$

where the D_0 is the matrix

$$D_0 = [A_1^t C_1^{-1} A_1 + C_2^{-1} - C_2^{-1} A_2 (A_2^t C_2^{-1} A_2)^{-1} A_2^t C_2^{-1}]^{-1}.$$

The above approach has become very popular because it, in fact, gets rid of θ_3 and σ_3^2 altogether and the analysis can proceed without regard for them. This, however, is not achieved with impunity. The estimator θ_1^* is not a proper or true Bayes estimator in the sense that no proper prior distribution of the form (2.2) and (2.3) will produce θ_1^*. One could obviate this seemingly technical point if it were possible to choose *a priori* a large enough value of σ_3^2 such that no matter which value of θ_3 was used, the difference between the Bayes estimate and θ_1^* were negligible. This, as we will show later, is not possible in general since θ_3 is "compared" in the final estimator with a sample statistic whose value can not be known in advance. Therefore, in practice approximating θ_1^* by a proper Bayes estimator can be "much easier said than done".

Our own approach to this problem is to legitimize what is really behind the above approach. This is, to perform inference under the assumption that we are unable to specify θ_3 and σ_3^2 precisely and that all we know is that they belong in certain ranges. For the particular case of σ_3^2 according to our discussion above, a very appropriate choice for range is an open ended interval of the form $[\sigma_3^2, \infty)$, which reflects our intention of checking what happens to the Bayes estimator if σ_3^2 is large enough. A prior specification like this is called *imprecise* because it does give a unique prior distribution for θ_1 but rather a class of prior distributions indexed by a parameter whose values are in a set.

The important point is what happens to the Bayes estimator under imprecision in the prior specification. If the Bayes estimator does not change considerably for the class of prior distributions specified, then there is really no point in choosing a particular one of them. To formalize this point we introduce the notion of "posterior imprecision" to denote a measure of how big is the set of possible answers obtained under the class of priors being considered. The idea is that if for a given data set, prior imprecision does not lead to significant posterior imprecision, then we have posterior inference stability and we do not need to be more specific in the elicitation of our prior information.

We would like to identify the locus of θ_1^* as σ_3^2 varies freely within its range. This can be done by studying the behavior of (3.1) as a function of σ_3^2. Empirical studies suggested that for several important models which fit within the NHLM context, the curve is a line whose natural parametrization depends on a function of σ_3^2 and not on σ_3^2 itself. In appendix II we prove this conjecture for a wide class of models which fit within the NHLM framework. Upon reflection we realized that we actually needed a lot less than that. What we really needed was to show that the curve was regular enough to allow us to measure the posterior imprecision by taking the distance between the two ends of the segment determined by the range of σ_3^2. A class of curves with this characteristic is the class of "monotone" curves. A monotone curve is one for which its coordinates move monotonically in the space as the parameter is changed monotonically. A line is obviously a monotone curve.

For the class of monotone curves it is meaningful to look at the derivative as a global indicator of change. The derivative of θ_1^* with respect to σ is given by:

$$B^{-1}M^{-1}A_2H_3A_2^tM^{-1}[\theta_1^* - A_2A_3\theta_3] \tag{3.4}$$

where

$$B = A_1^tC_1^{-1}A_1 + M^{-1} \tag{3.5}$$

and

$$M = C_2 + \sigma_3^2A_2H_3A_2^t \tag{3.6}$$

Notice that in (3.4) the last factor is a measure of "conflict" between the prior and the posterior Bayes estimate of θ_1. To understand the effect of σ_3^2 on the derivative it is useful to rewrite (3.4) in the following equivalent form:

$$B^{-1}M^{-1}A_2H_3A_2^tM^{-1}B^{-1}\ A_1C_1^{-1}A_1^t[b - A_2A_3\theta_3] \tag{3.7}$$

This is the contrast we alluded to previously between a sample statistic and the prior estimate. The sample statistic b happens to be the least squares estimator in this case. The advantage of (3.7) over (3.4) is that the influence of σ_3^2 is confined to the matrix product on the left. If we wanted to prove that the locus of θ_1^* is a line for a particular NHLM all we have to do is to show that the matrix product on the left of (3.7) is of the form $f(\sigma_3^2)K$ where $f(.)$ is a function and K is a matrix which does not depend on σ_3^2. This is essentially what we do in Appendix II. On the other hand, if all we wanted to prove is that the locus is a monotone curve, it is enough to check that the coordinates of the derivative vector (3.7) do not change in sign.

We would like to come back now to the analysis of the usual strategy of letting σ_3^2 be infinity. Looking at (3.7) it is now clear why we said before that it is not possible in general to choose *a priori* a large enough value of σ_3^2 such that the imprecision is negligible. The size of the derivative (3.7) depends on a contrast between the prior estimate $A_2A_3\theta_3$ and the least squares estimate b. In some situations it may be possible to bound the difference between these two quantities a prori, in which case if we choose the lower bound for σ_3^2 big enough the posterior imprecision can be made arbitrarily small. The situation is not as straightforward in general since there is always the possibility that the sample statistic exceeds the expected bounds and the imprecision becomes relevant. In any case, it seems to us that the blind use of the limiting approach is inappropriate because it sweeps the role of θ_3 and σ_3^2 under the carpet. It is preferable to examine the conflict and determine if the data follows the expected pattern. Our point is that (3.7) offers the Bayesian analyst a diagnostic tool to check how well the prior assumptions stand up to the data.

4. SOME PARTICULAR CASES

When considering imprecise prior specifications, we have already stressed the importance of entertaining classes of priors that preserve the structure of the problem as given by the model. That is, classes which while preserving the structural knowledge, are imprecise with respect to the parametric knowledge. In this section, a related fact will emerge. Namely, that the parametrization of the model is of crucial importance, not only for the interpretation of the parameters and therefore for an easier assessment, but also to isolate the influence of the imprecision on a subset of the parameters. In this way, the interpretation of the posterior imprecision becomes much simpler and natural. As an illustration of the above we first look at the One-way ANOVA.

4.1. One Way Anova

We consider here the balanced situation where

$$Y_{ij} = \beta_i + e_{ij} \quad i = 1, \ldots, a : \; j = 1, \ldots, n \tag{4.1.1}$$

and $\{e_{ij}\}$ are independent $N(0, \sigma_1^2)$. This model corresponds to NHLM as in (2.1, 2.2, 2.3), for which A_2 is an $a \times 1$ vector of ones, θ_2 is the mean of treatments' effects, and θ_3 is our prior guess for the center of the effects. It follows, that σ_3^2 measures our confidence in the chosen value for θ_3, while σ_2^2 measures the dispersion of the treatment's effects.

As was seen earlier, the two matrices that influence the means are $A_1^t C_1^{-1} A_1$ and $C_2 + A_2 C_3 A_2^t$. In the present case, since $C_2 = \sigma_2^2 I_a$ and $C_3 = \sigma_3^2$, we have

$$A_1^t C_1^{-1} A_1 = n\sigma_1^{-2} I_a \tag{4.1.2}$$

and

$$C_2 + A_2 C_3 A_2^t = \sigma_2^2 I_a + \sigma_3^2 J_a, \tag{4.1.3}$$

where the $a \times a$ matrices I_a and J_a are the identity matrix and a matrix of ones, respectively.

The parameter σ_3^2 should be regarded as by far the most imprecise of all, and following our line of thought it will only be assumed to lie in an interval of the form $[\underline{\sigma}_3^2, \infty)$. However, if we were willing to entertain an imprecise specification for σ_2^2, such as $\underline{\sigma}_2^2 \leq \sigma_2^2 \leq \overline{\sigma}_2^2$, we would have to face the unpleasant fact that in (4.1.3) all the diagonal elements remained unbounded. This is due to the particular parametrization of the problem in (4.1.1) which confounds the imprecision in σ_2^2 and σ_3^2. A more convenient parametrization is given by:

$$Y_{ij} = \mu + \alpha_i + e_{ij}, \quad i = 1, \ldots, a; \quad j = 1, \ldots, n \quad \Sigma\alpha_i = \alpha_+ = 0$$

where the e_{ij} are independent $N(0, \sigma_1^2)$. Under this new parametrization the sample covariance matrix is not invertible. Instead of setting up a general expression of the expected values in terms of generalized inverses as in Smith (1973a), we work out the particular solution which corresponds to setting $\theta_1^t = (\mu, \alpha_1, \ldots, \alpha_{a-1})$. In such a case it follows that:

$$A_1^t C_1^{-1} A_1 = n\sigma_1^{-2} \begin{pmatrix} a & 0^t \\ 0 & (I_{a-1} + J_{a-1}) \end{pmatrix} = B. \tag{4.1.4}$$

We now assume that $\mu \sim N(w, \sigma_{\mu^2})$ and that the distribution of the α's is exchangeable with $\alpha_+ = 0$. Following along the lines of Smith (1973a) we see that the NHLM under this set of assumptions corresponds to $\theta_1 \sim N(A_2\theta_2, C_2)$, with $\theta_2 = w$ and $A_2 = (1, 0, \ldots 0)^t$. The matrix C_2 becomes

$$C_2 = \begin{pmatrix} \sigma_{\mu^2} & 0^t \\ 0 & \sigma_\alpha^2(I_{a-1} - a^{-1}J_{a-1}) \end{pmatrix},$$

where $\sigma_\alpha^2 = \text{Var}(\alpha_i - \alpha_j)/2, i$ different from j.

Using the above it can be shown that

$$\{C_2 + A_2 C_3 A_2^t\}^{-1} = \begin{pmatrix} (\sigma_\mu^2 + \sigma_3^2)^{-1} & 0^t \\ 0 & \sigma_\alpha^{-2}(I_{a-1} + J_{a-1}) \end{pmatrix} = D \tag{4.1.5}$$

Hence, from (4.1.4) and (4.1.5) we obtain $[A_1^t C_1^{-1} A_1 + \{C_2 + A_2 C_3 A_2^t\}^{-1}]^{-1}$ as

$$\begin{pmatrix} \frac{\sigma_1^2(\sigma_\mu^2 + \sigma_3^2)}{na(\sigma_\mu^2 + \sigma_3^2) + \sigma_1^2} & 0^t \\ 0 & \frac{\sigma_1^2\sigma_\alpha^2}{n\sigma_\alpha^2 + \sigma_1^2}(I_{a-1} - a^{-1}J_{a-1}) \end{pmatrix} = F. \tag{4.1.6}$$

Collecting the previous expressions together we arrive at an expression for the Bayes estimator of θ_1,

$$E(\theta_1|Y) = \beta^* = F\{Bb + D[\theta_3, 0]^t\}, \tag{4.1.7}$$

which, after using (4.1.5) and (4.1.6), can be rewritten in the following equivalent form

$$\begin{pmatrix} \frac{\sigma_1^2(\sigma_\mu^2+\sigma_3^2)}{na(\sigma_\mu^2+\sigma_3^2)+\sigma_1^2} \left(\frac{naY_{..}}{\sigma_1^2} + \frac{\theta_3}{\sigma_\mu^2+\sigma_3^2}\right) \\ 0 \\ 0 \\ \cdot \\ \cdot \\ 0 \end{pmatrix} + \frac{n\sigma_\alpha^2}{n\sigma_\alpha^2 + \sigma_1^2} \begin{pmatrix} 0 \\ (Y_{1\cdot} - Y_{..}) \\ (Y_{2\cdot} - Y_{..}) \\ \cdot \\ \cdot \\ (Y_{a-1\cdot} - Y_{..}) \end{pmatrix} \tag{4.1.7}$$

Denoting by $\mu^*(\sigma_3^2)$ the Bayes estimator of μ as a function of σ_3^2 and solving from (4.1.7) we obtain:

$$\mu^*(\sigma_3^2) = \frac{\sigma_1^2(\sigma_\mu^2 + \sigma_3^2)}{na(\sigma_\mu^2 + \sigma_3^2) + \sigma_1^2} \left(\frac{naY_{..}}{\sigma_1^2} + \frac{\theta_3}{(\sigma_\mu^2 + \sigma_3^2)}\right). \tag{4.1.8}$$

It turns out that the derivative of (4.1.8) with respect to σ_3^2 does not change in sign, so that μ^* is a monotone function of σ_3^2. Hence, under an imprecise specification of σ_3^2 of the form $\sigma_3^2 \geq \underline{\sigma}_3^2$ the posterior imprecision in μ is given by:

$$(\mu^*(\underline{\sigma}_3^2) - \mu^*(\infty))^2 = (\theta_3 - Y_{..})^2[na\sigma_1^{-2}(\sigma_\mu^2 + \underline{\sigma}_3^2) + 1]^{-2}. \tag{4.1.9}$$

Notice that the first term in (4.1.9) is an expression of the conflict between θ_3 and the least squares estimate of the overall mean. The second term is just a function of the ratio of prior variances over the variance of the observations, multiplied by the sample size. A comment is now in order. It has been assumed almost universally in the literature that σ_3^2 is actually infinity. Somehow, explicitly or implicitly, it is alleged that there exists a huge σ_3^2, such that the posterior density of μ is closely approximated by the one obtained under $\sigma_3^2 = \infty$. However, the goodness of the approximation actually depends upon the particular sample obtained. In other words, it is outcome dependent. As is seen in (4.1.9) as $Y_{..}$ gets away from θ_3, the approximation becomes worse and the posterior means $\mu^*(\sigma_3^2)$ and $\mu^*(\infty)$ get further apart. In short there does not exist a finite σ_3^2, such that $\mu^*(\sigma_3^2)$ approximates $\mu(\infty)$ uniformly in $Y_{..}$.. A similar phenomenom, but in a different context, was pointed out by Walley (1988). We will touch upon this point again in section 5.

Looking back to (4.1.7), we see that if we had also introduced an imprecise specification for σ_α^2, the posterior imprecision due to σ_α^2 and σ_3^2 would not be confounded. Furthermore, σ_3^2 affects only the first component of the mean vector, and σ_α^2 affects only the remaining components. As a consequence, the imprecision involved in considering $\sigma_3^2 \geq \underline{\sigma}_3^2$ yields a range for μ only, leaving the other components of the mean vector precise. Therefore, if instead of the fixed point, a range for σ_α^2 is entertained it would not add any further imprecision to μ.

A last point concerning the role of σ_3^2 in the posterior imprecision of the Bayes estimator μ^* is to compute its derivative with respect to σ_3^2. It is given by the following expression:

$$\frac{\partial \mu^*}{\partial \sigma_3^2} = \left[\frac{\sigma_1^2 na}{\{na(\sigma_\mu^2 + \sigma_3^2\}^2}(Y_{..} - \theta_3), 0, \ldots, 0\right]^t. \tag{4.1.10}$$

As was pointed out in the previous section, for the general case the rate of change with respect to σ_3^2 is a function of the prior to sample conflict $(Y_{..} - \theta_3)$, but is zero for the last components of the vector. This fact is a convenient feature of the parametrization employed.

So far we have been working under the assumption that σ_α^2 is known exactly. We now relax this assumption by considering imprecise specifications for σ_α^2 also. In looking back to (4.1.6), we realize that the factor which involves σ_α^2, multiplies the matrix $(I_{a-1} - a^{-1}J_{a-1})$, whose structure arose from that of the problem and therefore we wish to retain. The parameter σ_α^2 represents half the variance of the contrast $\alpha_i - \alpha_j$, and so it has a "physical" meaning. Thus it has a different standing from σ_3^2. In particular, unbounded values for σ_α^2 might be considered unwise, since most people would be unwilling to believe that the treatments can have "infinitely" different effects. Under these circumstances, an interval region of the form $[\underline{\sigma}_\alpha^2, \bar{\sigma}_\alpha^2]$ seems sensible. The lower point could even be set to zero, which corresponds to the "no treatment effect" situation.

The posterior mean of α_i as a fuction of σ_α^2, for $i = 1, \ldots, a$ is

$$\alpha_i^*(\sigma_\alpha^2) = \left(\frac{n}{\sigma_1^2} + \frac{1}{\sigma_\alpha^2}\right)^{-1} \frac{n}{\sigma_1^2}(Y_{i.} - Y_{..}) \tag{4.1.11}$$

The above expression is monotone in σ_α^2, so if we entertain a finite interval for this parameter as indicated previously, it make sense to measure the posterior imprecision in the α's by:

$$\Sigma_i[\alpha_i^*(\bar{\sigma}_\alpha^2) - \alpha_i^*(\underline{\sigma}_\alpha^2)]^2.$$

This last expression is equivalent in the current context to:

$$\left(\frac{1}{n\sigma_1^2}\right)^2 \times \frac{(\bar{\sigma}_\alpha^2 - \underline{\sigma}_\alpha^2)^2}{\left(\frac{\bar{\sigma}_\alpha^2}{\sigma_1^2}+\frac{1}{n}\right)^2\left(\frac{\underline{\sigma}_\alpha^2}{\sigma_1^2}+\frac{1}{n}\right)^2}\Sigma_i(Y_i. - Y..)^2. \tag{4.1.12}$$

Notice that (4.1.12) is directly proportional to the size of the interval for σ_α^2 and also to a sampling measure of dispersion for the α's.
If $\underline{\sigma}_\alpha^2 = 0$ then (4.1.12) reduces to

$$\left(\frac{1}{\sigma_1^2} \times \frac{\bar{\sigma}_\alpha^2}{\left(\frac{\bar{\sigma}_\alpha^2}{\sigma_1^2}+\frac{1}{n}\right)}\right)^2 \Sigma_i(Y_i. - Y..)^2. \tag{4.1.13}$$

It is important to note the completely different assymptotic behaviour of (4.1.12) and (4.1.13) as n goes to infinity. The sampling term on the right is $0_p(1)$. Hence (4.1.12) is $0_p(n^{-2})$. On the other hand, (4.1.13) is $0_p(1)$ and the imprecision does not vanish as the information grows. This is due to the fact that whenever $\sigma_\alpha^2 = 0$ is included in the range, $\alpha^*(0)$ is zero regardless of the data. This behavior is typical of "dogmatic" priors.

4.2 Two Way Anova With One Observation Per Cell

We now turn to the two-way analysis of variance case. For the sake of simplicity we look to the one observation per-cell situation.

The model will be parametrized in the following fashion:

$Y_{ij} = \mu + \tau_i + \alpha_j + e_{ij}$, $i = 1, \ldots, a$; $j = 1, \ldots, b$, where the errors are independent $N(0, \sigma_1^2)$ and $\tau_+ = \alpha_+ = 0$.

Proceeding in the same manner as in the one-way case, we set $\theta_1 = (\mu, \tau_1, \tau_2, \ldots, \tau_{a-1}, \alpha_1, \ldots, \alpha_{b-1})^t$. Then it turns out that $A_1^t C_1^{-1} A_1$ results in the following matrix:

$$B' = \sigma_1^{-2}\begin{pmatrix} ab & 0^t & 0^t \\ 0 & b(I_{a-1}+J_{a-1}) & 0 \\ 0 & 0^t & a(I_{b-1}+J_{b-1}) \end{pmatrix}. \tag{4.2.1}$$

We now assume that, $\mu \sim N(w, \sigma_\mu^2)$, and that the distributions of the r's and α's are exchangeable and normal with $r_+ = \alpha_+ = 0$. From this it follows that $\theta_1 \sim N(A_2\theta_2, C_2)$, with $\theta_2 = w$, $A_2 = [1, 0^t, 0^t]^t$ and

$$C_2 = \begin{pmatrix} \sigma_\mu^2 & 0^t & 0^t \\ 0 & \sigma_\tau^2(I_{a-1} - a^{-1}J_{a-1}) & 0 \\ 0 & 0^t & \sigma_\alpha^2(I_{b-1} - b^{-1}J_{b-1}) \end{pmatrix}, \tag{4.2.2}$$

where $\sigma_\tau^2 = .5\text{Var}(\tau_i - \tau_j)$ and $\sigma_\alpha^2 = .5\text{Var}(\alpha_k - \alpha_l)$.

Note that $A_2C_3A_2^t$ is a $(a+b-1)$ by $(a+b-1)$ matrix with zeroes in all entries except the upper left corner, where the entry is σ_3^2, since $C_3 = \sigma_3^2$. This allows to write the matrix $\{C_2 + A_2C_3A_2^t\}^{-1}$ in the following fashion

$$D' = \begin{pmatrix} (\sigma_\mu^2 + \sigma_3^2)^{-1} & 0^t & 0^t \\ 0 & \sigma_\tau^{-2}(I_{a-1}+J_{a-1}) & 0 \\ 0 & 0^t & \sigma_\alpha^{-2}(I_{b-1}+J_{b-1}) \end{pmatrix} \tag{4.2.3}$$

Collecting (4.2.2) and (4.2.3), we obtain the last matrix of interest which is $[A_1^t C_1^{-1} A_1 + \{C_2 + A_2 C_3 A_2^t\}^{-1}]^{-1}$. It can be written in the form:

$$F' = \begin{pmatrix} \left(\frac{ab}{\sigma_1^2} + \frac{1}{(\sigma_\mu^2+\sigma_3^2)}\right)^{-1} & 0^t & 0^t \\ 0 & f_\tau(I_{a-1} - a^{-1}J_{a-1}) & 0^t \\ 0 & 0 & f_\alpha(I_{b-1} - b^{-1}J_{b-1}) \end{pmatrix}, \qquad (4.2.4)$$

where $f_\tau = \sigma_1^2\sigma_\tau^2/(b\sigma_\tau^2 + \sigma_1^2)$ and $f_\alpha = \sigma_1^2\sigma_\alpha^2/(a\sigma_\alpha^2 + \sigma_1^2)$.

Finally, the Bayes estimator of θ_1 can be calculated by the expression $E(\theta_1|Y) = F'[B'b + D'(\theta_3, 0^t, 0^t)^t]$, where the matrices F', B', and D' are given in (4.2.2), (4.2.3),(4.2.4) respectively, and b represents the least squares estimator

$$(Y_{..}, Y_{1\cdot} - Y_{..}, \ldots, Y_{a-1\cdot} - Y_{..}, Y_{\cdot 1} - Y_{..}, \ldots, Y_{\cdot b-1} - Y_{..})^t.$$

The resulting posterior means component by component are:

$$\mu^*(\sigma_3^2) = \frac{\sigma_1^2(\sigma_\mu^2 + \sigma_3^2)}{ba(\sigma_\mu^2 + \sigma_3^2) + \sigma_1^2}\left(\frac{baY_{..}}{\sigma_1^2} + \frac{\theta_3}{(\sigma_\mu^2 + \sigma_3^2)}\right) \qquad (4.2.5)$$

$$\tau_i^* = \frac{b\sigma_\tau^2}{b\sigma_\tau^2 + \sigma_1^2}(Y_{i\cdot} - Y_{..}), \quad i = 1, \ldots, a \qquad (4.2.6)$$

$$\alpha_j^* = \frac{a\sigma_\alpha^2}{a\sigma_\alpha^2 + \sigma_1^2}(Y_{\cdot j} - Y_{..}), \quad j = 1, \ldots, b. \qquad (4.2.7)$$

Observe that the only component which involves θ_3 and σ_3^2 is μ^*. As in the previous section, the derivative of this term with respect to σ_3^2 does not change in sign, therefore μ^* is monotone. Recalling that we are entertaining a region for σ_3 of the form $[\underline{\sigma}_3^2, \infty)$ posterior imprecision in μ can be measured by

$$(\mu^*(\underline{\sigma}_3^2) - \mu^*(\infty))^2 = (\theta_3 - Y_{..})^2[ba\sigma_1^{-2}(\sigma_\mu^2 + \underline{\sigma}_3^2) + 1]^{-2}., \qquad (4.2.8)$$

which is again a function of the conflict $(\theta_3 - Y_{..})^2$ and whose interpretation parallels that of (4.1.9).

Looking now to (4.2.6) and (4.2.7), it can be seen that these estimators as functions of σ_τ^2 and σ_α^2 respectively, are monotone functions. If we wanted to consider an imprecise prior specification for the corresponding parameters in the form of bounded intervals $[\underline{\sigma}_\tau^2, \bar{\sigma}_\tau^2]$ and $[\underline{\sigma}_\alpha^2, \bar{\sigma}_\alpha^2]$, the posterior imprecisions would come out to be

$$\left(\frac{1}{b\sigma_1^2}\right)^2 \frac{(\bar{\sigma}_\tau^2 - \underline{\sigma}_\tau^2)^2}{\left(\frac{\bar{\sigma}_\tau^2}{\sigma_1^2} + \frac{1}{b}\right)^2\left(\frac{\underline{\sigma}_\tau^2}{\sigma_1^2} + \frac{1}{b}\right)^2} \Sigma_i \hat{\tau}_i^2 \qquad (4.2.9)$$

$$\left(\frac{1}{a\sigma_1^2}\right)^2 \frac{(\bar{\sigma}_\alpha^2 - \underline{\sigma}_\alpha^2)^2}{\left(\frac{\bar{\sigma}_\alpha^2}{\sigma_1^2} + \frac{1}{a}\right)^2\left(\frac{\underline{\sigma}_\alpha^2}{\sigma_1^2} + \frac{1}{a}\right)^2} \Sigma_j \hat{\alpha}_j^2, \qquad (4.2.10)$$

where $\hat{\tau}_i$ and $\hat{\alpha}_j$ denote the least squares estimators of τ_i and α_j, respectively. The interpretation of (4.2.9) and (4.2.10) is similar to that of (4.1.12) in the previous section.

5. ANALYSIS OF A DATA SET

Smith (1973a) analyses a data set using a two-way ANOVA model with one observation per cell. The data, which are displayed in Table 1 of Appendix I, represents the yield of jute for four levels of fertilizer and five types of jute.

First of all, let us look at the posterior imprecision due to σ_3^2. Using (4.2.8) and taking square roots in order to have the same units as the overall mean we get:

$$\delta_\mu^* = |\mu^*(\underline{\sigma}_3^2) - \mu^*(\infty)| = \frac{|\theta_3 - Y..|}{ab\sigma_1^{-2}(\sigma_\mu^2 + \underline{\sigma}_3^2) + 1}. \tag{5.1}$$

To understand (5.1) above it is useful to return to a point we already made at the end of Section 3. In principle, the choice of a suitable value for $\underline{\sigma}_3^2$, such that the posterior imprecision of μ is negligible, has to be outcome dependent. This is due to the presence of the conflict factor in the numerator of (5.1). However, it might be argued that a choice of value for $\underline{\sigma}_3^2$ can be made independently of the outcome if we have some knowledge about the problem itself. For instance, if we were pretty confident about an interval in which the observations must lie, then the grand mean must fall into that interval. This information could be used to choose $\underline{\sigma}_3^2$ which bounds (5.1) a priori. But even in this fortunate situation (5.1) provides a useful diagnostic tool for the Bayesian analyst to check how well parametric prior assumptions compare to the data.

Let us illustrate the point with the data in Appendix I. There, $a = 4$, $b = 5$, and let us suppose that σ_1^2 known to be 0.2. After studying the problem it was assessed that $\theta_3 = 1$ and $\sigma_3^2 + \sigma_\mu^2 \geq 0.5$. The value of $Y.. = 1.44$ as computed from the data seems to be in good agreement with the foresesen value and $\partial\mu^* = 0.009$ is negligible with respect to Y... But, if we had assessed instead a value of 10 for θ_3, then $\partial\mu^* = 0.168$ which might be considered too large, since it represents more than 10% of relative error with respect to Y... Under this conflicting circumstance, we have several alternatives: Either our assessments were grossly wrong, or the data have a gross measurement error, or both. We can then try to reconcile prior and sample, or live with the conflict, taking into account the imprecision of μ^*, as well as its uncertainty (standard error), for the inference.

We now turn to the analysis of how sensitive inference would be to imprecision in the specification of σ_τ^2 and σ_α^2. To this end, let us suppose that all we can say about them is that: $0.06 \leq \sigma_\tau^2 \leq 0.18$ and $0.02 \leq \sigma_\alpha^2 \leq 0.06$. In Table 2 of Appendix I we show the least squares estimates of the treatment effects along with their Bayes counterparts as given in Smith (1973a). We also display the upper and lower values of the Bayes estimator for the values of the sigmas within the ranges described above. As expected the Bayes estimates display a substantial amount of shrinkage. More importantly, the posterior imprecision in our Bayes estimates is not negligible. This can checked by two different means: either by looking at the range in each coordinate, or by comparing the values of (4.2.9) and (4.2.10) with the lengths of the corresponding Bayes estimates vectors. Namely, $\partial\tau^* = 0.15$ and $\partial\alpha^* = 0.16$ as compared to 0.49 and 0.253, respectively. Finally, notice that the upper bound of the Bayes estimates computed under imprecision, compare in size and pattern to the modal Bayes estimates reported by Smith, under the assumption that σ_τ^2 and σ_α^2 are inverse Chi-square distributed.

We believe that a more reasonable strategy to deal with σ_τ^2 and σ_α^2 is to treat them as structural knowledge parameters and proceed to specify distributions on them as we did for the effects themselves. Once this is done, we are in a position to be imprecise about the parameters in these distributions, since they now represent truly parametric knowledge. We are actively pursuing this idea but it is beyond the scope of this paper.

6. CONCLUSIONS

We begin the conclusions by stressing again the usefulness of the conceptual separation between structural and parametric prior knowledge. Once this distinction is understood in a given problem, it is much easier to establish which features of the posterior inferences are affected by an imprecise specification of the hyperparameters. Typically, in a first pass analysis of the data, the structural part of our prior knowledge will be held fixed and the sensitivity of the inference to imprecision in the specification of the hyperparameters will be investigated. This is not to say that the structural part should not be put into question. It merely establishes a hierarchy of questions.

A previous development which deals with partial or imprecise prior information is that of Leamer (1982). It approaches the problem of estimation in the two stages linear model by considering ranges for the covariance matrices as opposed to fixed values. For instance in the analysis of a two level hierarchical model, he assumes that the prior covariance matrix C is only known to be larger than a lower bound positive definite matrix C_*. In his context $C \geq C_*$ means that $C - C_*$ is positive definite. Within this framework he shows that the posterior expected values lie in a particular ellipsoid. This result would seen to be useful in our case, once adapted to the three stage model. Unfortunately, an order relation between covariance matrices like the one used by Leamer yields mathematically elegant result whose practical relevance to our case is at the very least questionable. Leamer (1982, p.726) himself warns that there are better ordering schemes for ellicitation purposes, but he takes the one described above for mathematical convenience. An example would help to clarify this point.

In the context of one-way ANOVA, we would be considering the class of matrices C such that if $C \in \Gamma$, $C \geq C_2 + A_2 C_3 A_2^t = D^{-1}$ with D given by (4.1.5). It can be shown that

$$D^{-1} = \begin{pmatrix} \sigma_\mu^2 + \sigma_3^2 & 0^t \\ 0 & \sigma_\alpha^2 (I_{a-1} - a^{-1} J_{a-1}) \end{pmatrix} \tag{6.1}$$

One of the problems with this approach is that there are many matrices C in Γ such that $C - D^{-1}$ is positive definite, but which do not have the structure of a covariance matrix arking from a N.H.L.M. That is, they do not have the form (6.1) for any choice of σ_1^2, σ_μ^2 and σ_α^2.

The fact is that the N.H.L.M. under exchangeability, is a highly patterned situation, with substantial amount of structural prior knowledge built into it. Any approach which fails to take this into account and treats the covariance matrice as just an other parameter, is bound to yield meaningless results for this class of models.

Having found Leamer's approach not relevant for our purposes, we tried a direct approach, being precise with regard to the structural aspects of the prior information, but allowing imprecision with regard to the specification of the parametric knowledgde (hyperparameters). This approach yields useful diagnostic tools which help in checking "a posteriori" the sensitivity of Bayesian inference to imprecision in the prior. A general coherent setup which incorporates imprecise probabilities is described in Walley (1988).

Possible further developments include extending our approach to the case where the covariance matrix in the second level is not considered either as fixed or belonging to a range, but rather using distributions to express our uncertainty about it and then allowing imprecision in the hyperparameters of such distributions. As long as the analysis is done under the same parametrization as the one in this paper, the imprecision due to $\sigma_3^2 \geq \underline{\sigma}_3^2$ remains the same. Finally, a much harder problem is to extend the analysis of this paper to the case of σ_1^2 unknown.

REFERENCES

Good, I. J. (1965). *The Estimation of Probabilities: an Essay on Modern Bayesian Methods.* Cambridge: M.I.T. Press.

Leamer, E. E. (1982). Sets of posterior means with bounded variance priors. *Econometrica* **5**, 725–736
Lindley, D. V. and Smith, A. F. M. (1972). Bayes estimates for the linear model (with discussion). *J. Roy. Statist. Soc. B* **34**, 1–41.
Smith, A. F. M. (1973a). Bayes estimates in one-way two-way models. *Biometrika* **60**, 319–330.
Smith A. F. M. (1973b). A General Bayesian linear model *J. Roy. Statist. Soc. B* **35**, 67–75.
Walley, P. (1988). *Statistical Reasoning with Imprecise Probabilities.* London: Chapman and Hall.

APPENDIX I

Table I: Yields of jute for four different levels of fertilizer (a = 4) and five types of jute (b = 5); Smith (1973a).

	Levels of B				
Levels of A	1	2	3	4	5
1	0.9	1.5	1.4	2.3	1.5
2	2.0	1.9	2.3	1.4	1.2
3	2.4	1.3	2.2	1.2	1.0
4	1.0	1.2	1.1	0.6	0.4

In Table 2 we display the least squares estimates, Bayes estimates as in Smith (1973a) and the ranges for the Bayes estimates under imprecision.

Table 2: Least squares, Bayes and Bayes under imprecision estimates for the effects.

	α_1	α_2	α_3	α_4	α_5	τ_1	τ_2	τ_3	τ_4
L. S	.185	.085	.360	-.265	-.365	-.070	.370	.230	-.530
S. D	.182	.182	.182	.182	.182	.158	.158	.158	.158
Bayes	.121	.056	.235	-.173	-.239	-.061	.325	.202	-.466
S. D.	.118	.118	.118	-.118	-.118	-.155	.155	.155	.155
Upper	.101	.046	.197	-.075	-.102	-.042	.300	.188	-.318
Lower	.053	.024	.103	-.145	-.197	-.057	.219	.138	-.434

APPENDIX II

Here we prove that the locus of θ_1^* as σ_3^2 moves in $[\underline{\sigma}_3^2, \infty]$, is a straight line for a quite general class of models. To this end we use the matrix results 2, 3 and 4 from Smith (1973b). Using the matrix result 4 from this reference, we rewrite (3.1) as:

$$\theta_1^* = A_2A_3\theta_3 + DA_1^tC_1^{-1}A_1(b - A_2A_3\theta_3) \qquad \text{(II.1)}$$

and the only factor which depends upon σ_3^2 is D.

Denote by G the matrix $A_1^tC_1^{-1}A_1$. Now using the matrix result 3 in Smith (1973b) we obtain:

$$D = G^{-1} - G^{-1}(G^{-1} + C_2 + \sigma_3^2A_2H_3A_2^t)^{-1}G^{-1}. \qquad \text{(II.2)}$$

Let us look at the reciprocal of the factor within parenthesis in (II.2). Applying matrix results 2 in Smith (1973b) this factor can be written as: $P^{-1} - P^{-1}A_2F^{-1}A_2^tP^{-1}$ where

$P = G^{-1} + C_2$ and $F = A_2^t P^{-1} A_2 + H_3^{-1}\sigma_3^{-2}$.We have now isolated the influence of σ_3^2 in F. If we now assume that $\dim(\theta_3) = 1$, then F is a real number and it permutes with all the matrices it multiplies. It follows immediately that the locus of θ_1^* is a straight line, since (II.1) can now be written as a scalar function of σ_3^2 times a constant vector. Furthermore, it is also apparent from F, that θ_1^* is monotone with respect to σ_3^2.

DISCUSSION

A. M. SKENE (*University of Nottingham*)

This paper offers a further contribution to that collection of ideas and results concerned with the robustness of Bayesian procedures. Indeed, issues of sensitivity and robustness have received considerable attention already in earlier papers at this conference and the topic has already established itself as one of the principal themes of the meeting.

In the present paper one can identify three key ideas. First, the paper illustrates how, within the normal hierarchical linear model framework, one can describe the locus of the posterior mean of the first stage parameters, θ_1^*, as a function of third stage hyperparameters. The form of the locus and certain summaries of the locus, such as its length, provide a means of describing the robustness of the posterior mean to misspecification of third stage hyperparameters. The one-and two-way analysis of variance models are considered by way of illustration. It is pointed out that the relationship which exists between θ_1^* and, say, σ_3^2 can be made more transparent if some care is given to the choice of parametrisation at the first stage. Secondly, the paper argues that understanding the dependence is pertinent to the problem of choosing a proper third stage density which is an acceptable approximation to the more usual improper non-informative prior. Finally, it is argued that the results presented should be seen in a more general setting. Prior knowledge is classified as being either structural or parametric. The assessment of parametric prior knowledge attracts more controversy and thus sensitivity to misspecification of prior parameters needs greater attention.

In addressing these issues I find my perspective influenced by recent work at Nottingham. For the past three years I have been able to use the Nottingham numerical integration software in applications and this tool allows one to challenge the validity of many of the conventional assumptions of both classical models and the hierarchical linear model. For a rather longer period I have been involved in collaborative studies arising out of medical research or industrial settings. In none of this work has the specification of a prior for a general level parameter been a problem. Experimentalists know which scale of measurement they are going to use and can make a good guess at the range of their observations. This same work has also highlighted an aspect of sensitivity analysis which has received surprisingly little attention. The primary issue is usually the sensitivity of a *decision* to particular choices of model and prior given the data actually to hand. The robustness of an inference to varying data is largely irrelevant. Decision sensitivity is, in addition, usually more concerned with the sensitivity of tail areas or interval estimates than point estimates. These issues are discussed more fully elsewhere (Skene, Shaw and Lee (1986)).

Turning, then, to the ideas of the paper, I welcome the insight the paper gives as to the behaviour of θ_1^* as a function of hyperparameters, and the idea of a direction for θ_1^* of greatest sensitivity. Such information can, for example, be used to guide numerical investigations of sensitivity in particular problems. The authors indicate that their investigations can be taken much further. I hope they do consider the case with an unknown variance at the first stage and investigate the sensitivity of h.p.d. intervals for θ_1^* to imprecision in the hyperparameters of the prior for both location and dispersion parameters. The importance of this work to the choice of a non-informative prior for a general level parameter I find less convincing. Of much greater importance are any implications a prior or model may have for the marginal density associated with a primary treatment contrast. Finally, while I accept the structural versus parametric distinction for prior knowledge, it must be emphasized that structural prio knowledge is simply model choice by another name. In applications which use Bayesia methods, choice of model —location structure, dispersion structure and error distribution — appears to have much greater impact on inferences and subsequent decisions that choic of prior. One illustration is given by Lee(1987) who examines procedures given in Britis Standard, BS5497, for the estimation of inter- and intra-laboratory variability prior to th establishment of standard test methods for product quality control. The model proposed by th Standard is essentially the one-way analysis of variance model given in this paper. Replacin the assumption of a normal distribution at the first and second stages by a multivariate distribution has a dramatic impact on the posterior means and the posterior precision of thos functions of the inter- and intra-laboratory variances, estimates of which define the subsequen testing procedure.

K. PÖTZELBERGER (*University of Basel*)

The authors have argued that Leamer's approach to handle partial prior information cannot b relevant in the context of one-way ANOVA. Leamer deals with the problem when the prio covariance matrix C is only known to be larger than a fixed matrix C_*. In this case the set c posterior means forms a "feasible" ellipsoid. The set of covariance matrices C with $C \geq C$ is too large for the one-way ANOVA model, however, since the prior covariance matrices ar known to have a special structure.

Instead of concluding that Leamer's approach is not relevant, one should include th additional information into the analysis by considering the set of posterior means that com from prior covariance matrices C that have the desired structure and satisfy $C \geq C_*$. Thi set of posterior means does not have to be an ellipsoid any more, but its boundary can b reported and its size gives information about the robustness of the analysis. Whether the se is an ellipsoid or not, is not relevant at all.

REPLY TO THE DISCUSSION

We thank Dr. Skene and Dr. Pötzelberger for their contribution to the discussion of ou paper.

Dr Skene points out that in his experience experimentalists have very little difficult in specifying a prior for a general level parameter. As he argues: "Experimentalists kno which scale of measurement they are going to use and can make a good guess of the range o their observations." Our practical experience, being perhaps less extensive than Dr. Skenes's does not indicate that experimenters can always make a prior specification with such a level o certainty. Indeed, when people feel confortable about a particular choice of prior most probabl they are using an "ignorant" or "improper" prior distribution. In this paper we argue agains the automatic use of prior distributions, specially if they are improper. We do it choosing purposedly simple example (ANOVA with known variances). In this context we show ho the sensitivity of inferences respect to the choice of hyperparameters of the prior distributio is very much data dependent. Being so, it is important that the experimenter keeps in min that no matter how sure he/she is about the prior distribution employed, no one will mak an important decision based on his/her analysis without subjecting all assumptions to hones challenge. This should not be confounded with the traditional frequentist robustness approac of measuring the influence of varying data on the final inferences. Under the Bayesia framework there are three aspects of the robustness issue: sensitivity respect to the data sensitivity respect to the likelihood (model) and sensitivity respect to the prior information We are focusing on the third of these which happens to be the most Bayesian specific one and in fact develop some diagnostic statistics to gauge the influence that a given choice o parameters at the third level of a NHLM may have on inferences about the mean.

Dr. Skene's emphasizes that sensitivity is question dependent. We certainly agree with this. However he seems to suggest that our approach focuses on the first of the three aspects of robustness mentioned above. Namely, robustness of inference to varying data. We obviously disagree with him, since throughout the paper we state that all our conclusions are conditional to the data at hand. Equation (4.1.9), for instance, tells us that the posterior imprecision is a function of the conflict between the prior mean and the sample mean. Although this can be used to study the effect of varying data on inferences about the mean, our point is that the influence of a particular choice of $\underline{\sigma}_3^2$ on the posterior mean can not be guessed a priori since it depends on a sample statistic not known beforehand.

Since Dr. Skene is interested specially in interval estimates and tail probabilities we would like to remark that for the examples used in our paper all the posterior distributions are normal and therefore the mean and covariances govern the behavior of any tail area or interval estimates. It is then straightforward to compute their sensitivity with respect to the prior distribution. Let us denote by I the 95 percent interval of μ obtained by letting σ_3^2 go to infinity. For the data set analyzed in Section 5 we found that if the value of θ_3 is chosen as 1 then the posterior probabilities of I range from .9514 to .9500 when σ_3^2 varies in the manner explained in Section 5. However, if the value of θ_3 is chosen as 10 say, the posterior probability of I ranges now from .6118 to .9500. Walley and Pericchi (1988) discuss the sensitivity of credible intervals for more general classes of priors.

Finally our distinction between structural and parametric knowledge is not intended to place more or less importance on the different aspects of sensitivity analysis. Rather it tries to suggest a hierarchy of questions to be asked. Within the framework of a model we must worry about the assessment of prior parameters in order to obtain sensible results. One can always go one level up in the hierarchy and raise questions about the model itself. We feel that much more work is needed in the former case, since it is at the level of assessment of parameters where most of the arbitrariness is confined and paradoxically, where we have been more permissive (improper priors).

In reference to Dr. Pötzelberger's comments we just want to point out that our work shows that once you start including additional information about the form of the covariance matrix, an order relation as general as the one used by Leamer in his 1982 paper is unnecessary.

All in all if we see the Bayesian approach as a coherent mapping from prior to posterior knowledge, it does not seem judicious to evaluate that mapping on just one point, the chosen precise prior. We prefer to find directions of large sensitivity and calculate the imprecision along that path within the feasible region according with our limited initial knowledge. In such a way, we would be able to explore the structure of the map. We should start, in complex situations, with numerical aproximations, and the Nottingham software in conjunction with suitable approximations, might prove very valuable in exploring a sort of a "response surface analysis" of the Bayesian mapping.

REFERENCES IN THE DISCUSSION

Lee, T. D. (1987). Assessment of inter- and intra-laboratory variances: a Bayesian alternative to BS 5497. *The Statistician* **36** 161–170.

Skene, A. M., Shaw, J. E. H. and Lee, T. D. (1986). Bayesian modelling and sensitivity analysis. *The Statistician* **35** 281–288.

Walley, P. and Pericchi, L. R. (1988). Credible Intervals: How Credible are they? *Tech. Rep.* Universidad Simón Bolívar, Caracas, Venezuela.

BAYESIAN STATISTICS 3, pp. 377–394
J. M. Bernardo, M. H. DeGroot, D. V. Lindley and A. F. M. Smith, (Eds.)

Robust Bayesian Analysis in Hierarchical Models

W. POLASEK and K. PÖTZELBERGER
University of Basel

SUMMARY

A robust Bayesian analysis is carried out for a full 3-stage hierarchical model which was analysed by Lindley and Smith (1972). By specifying a full 3rd stage prior but leaving the covariance matrix for the second stage unknown, we derive sets for the posterior distributions of the parameter and hyperparameter in a linear regression system. In particular we show that posterior means are contained in smaller hierarchical (feasible) ellipsoids and smaller robust HPD-regions. An example is given involving seasonal hierarchical time series models of Austrian economic data.

Keywords: ROBUST BAYESIAN ANALYSIS; HIERARCHICAL MODELS; HIFI-REGIONS; SEASONAL TIME SERIES.

1. INTRODUCTION

A general linear Bayesian regression model, which has also become known as a "hierarchical" model, because the prior information is formulated in recursive stages, has been found useful in many areas of applied statistics. The model was analysed by Lindley and Smith (1972) from a very Bayesian perspective, but, surprisingly, no full Bayesian analysis was carried out. Instead, a noninformative last stage prior distribution was assumed and the proposed estimation procedure was essentially one with an empirical Bayes character. The prior coincides with the second stage distribution and is estimated in an m-group regression model. Since a second stage covariance matrix in a hierarchical model might be difficult to estimate by observed data (e.g. singularity), Polasek (1984) has suggested a robust extreme bound analysis along the lines of Leamer (1982) with symmetrical restricted empirical covariance matrices.

A full Bayesian last stage prior distribution allows two types of extreme bound analysis in a 3-stage model, if the second stage covariance matrix is not known. Within the so-called feasible ellipsoid (Leamer 1978) there exist smaller ellipsoids: one for the first stage parameters which lies closer to the ML-location, and one for the second stage parameters which lies closer to the prior location. This robust Bayesian analysis is demonstrated with a hierarchical time series model and some Austrian economic data. Furthermore, it is shown how the extreme bound analysis for the posterior mean can be extended to so-called HiFi-regions for quantiles of the posterior distribution.

In section 2 we review the general Bayesian or hierarchical model and in section 3 we give a description of our approach to Bayesian robustness. In particular we derive robust HPD-regions, also called HiFi-regions, for the hierarchical case with partial known priors. In section 4 we demonstrate our approach with an economic example and we conclude with some final remarks.

2. HIERARCHICAL MODELS

Lindley and Smith (1972) developed a general Bayesian linear model, which also become known as a (linear) hierarchical model. The hierarchy refers to the prior information of the regression coefficients, which can be extended to several stages. The usual Bayesian model can be viewed as a (informative) 2-stage model

$$Y \sim N(X\beta, \omega^2 I_N), \beta \sim N(b^*, \Sigma^{-1}), \tag{2.1}$$

and leads to the posterior mean

$$b^{**} = (\omega^{-2}X'X + \Sigma)^{-1}(\omega^{-2}X'X\hat{b} + \Sigma b^*), \tag{2.2}$$

which is a matrix weighted average between the OLS-estimate $\hat{b} = (X'X)^{-1}X'y$ and the prior location b^*. In the following we denote by a $*$-symbol the known part of a partially known prior distribution.

Theorem 2.1. *The posterior mean $b^{**}(\Sigma) = b^{**}$ in (2.2) with the partial known prior information: b^* known, Σ any positive definite (p.d.) matrix, is constrained to lie in the ellipsoid*

$$ELL(b^*, b, X'X) = \text{closure}\left\{b^{**}(\Sigma)|\Sigma \in \mathcal{M}^+\right\}, \tag{2.3}$$

where $\mathcal{M}^+$ is the set of all p.d. and symmetric matrices. $ELL(b^*, \hat{b}, X'X)$ denotes the ellipse with main diameter from the prior location b^* to the ML-location $\hat{b}$, and the metric $X'X$. Algebraically the ellipse is defined by the inequality

$$(b^{**} - f)'X'X(b^{**} - f) \leq c \tag{2.4}$$

with parameters $f = (b^* + b)/2$, the midpoint between the ML- and the prior location, and c, the boundary constant given by

$$c = (b - b^*)'X'X(b - b^*)/4. \tag{2.5}$$

Proof. This can be found in Polasek (1984).

2.1. The 3-Stage Bayesian Linear Model

$$y \sim N(A_1\theta_1, C_1), \theta_1 \sim N(A_2\theta_2, C_2), \theta_2 \sim N(A_3 b^*, H^*) \tag{2.6}$$

was analysed in Smith (1973) and the posterior means for the first and second stage parameters are matrix weighted averages (given in appendix A) between the known last stage prior location b^* and the GLS-estimates of the first and second stage, respectively. The second stage GLS-estimates is given by

$$\hat{\theta}_2 = [A_2'A_1'(C_1 + A_1C_2A_1')^{-1}A_1A_2]^{-1}A_2'A_1'(C_1 + A_1C_2A_1')^{-1}y. \tag{2.7}$$

In the case of a diffuse 3rd stage prior, denoted symbolically by $N(O, \infty)$, the 3rd stage model reduces to an "empirical Bayes" 2-stage model with estimated prior parameters θ_2 and Σ. The posterior mean of this model has the matrix average form (2.2) but with (b^*, Σ) replaced by $(\hat{\theta}_2, \hat{C}_2)$, estimated by (2.7) and $\hat{C}_2 = (\hat{b} - A_2\hat{\theta}_2)(\hat{b} - A_2\hat{\theta}_2)'$.

Unfortunately, the empirical estimation of the prior covariance matrix is usually rank deficient, and invertibility is achieved by adding a ridge type diagonal matrix. But such a procedure is equivalent to more prior information on the second stage covariance matrix, and can be avoided by going directly to a 3-stage model with known last stage. Certainly, a 3-stage prior can be sometimes difficult to assess, but the slightest amount of prior information seems to be more desirable from a Bayesian point of view than an arbitrary type ridge matrix. Hierarchical modeling can be used for (econometric) seasonal time series. Fitting a linear (or autoregressive) model for each season $s = 1, \ldots, S$, we combine the first stage estimates in a common hyper-distribution by assuming exchangeability between seasons. The second stage parameters are interpreted as aggregate coefficients of the whole non-seasonal time series, and hence it is possible to specify prior information about these coefficients. Prior information about the hyperparameters θ_2 in the seasonal model requires the same effort as an ordinary (2-stage) prior in a non-seasonal model. The seasonal full 3-stage model has the form:

$$y \sim N(X_s\beta_s, \omega_s I_T), \beta_s \sim N(\xi, \Sigma), \xi \sim N(\xi^*, H^*), s = 1, \ldots, S. \tag{2.8}$$

It is convenient to arrange the coefficients in matrix form by $B = (\beta_1, \ldots, \beta_S)$ and to use the (column by column) vectorisation operator vec if necessary. The posterior mean of the first stage coefficients is given by

$$\text{vec}B^{**} = [\text{diag}\tilde{H} + G]^{-1}[\text{diag}\tilde{H} \text{ vec } \hat{B} + G(1_S \otimes \xi^*)] \tag{2.9}$$

with $\tilde{H} = (H_1, \ldots, H_S)', H_s = X_s'X_s/\omega_s, s = 1, \ldots, S$, and $\text{diag}\tilde{H} = \text{diag}(H_S)$ being the block-diagonal matrix with blocks H_s. $\hat{B}$ is the matrix of OLS-estimates, where every column is given by $\hat{b}_s = (X_s'X_s)^{-1}X_s'y_s$. Collapsing stages 2 and 3, the prior variance G has the form of a block-intraclass correlation matrix

$$G = (1_S 1_S' \otimes H^* + I_S \otimes \Sigma)^{-1} \tag{2.10}$$

where 1_S is a $S \times 1$ vector of ones and $\otimes$ is the usual Kronecker product.

Lemma 2.2. *For unknown second and third stage covariance matrix, but known ξ^* and $D_\omega = diag(\omega_1, \ldots, \omega_S)$ the posterior mean (2.9) lies in the "surrounding first stage" ellipsoid*

$$ELL(1_S \otimes \xi^*, \text{vec}\hat{B}, \text{diag}\tilde{H}) = \text{closure}\left\{B^{**}(H^*, \Sigma|\xi^*, D_\omega)|H^* \in \mathcal{M}^+, \Sigma \in \mathcal{M}^+\right\} \tag{2.11}$$

Proof. This is just an application of theorem 2.1.

The posterior mean for the second stage hyperparameter is

$$\xi^{**} = (H^{*-1} + H_0)^{-1}(H^{*-1}\xi^* + H_0\hat{\xi}) \tag{2.12}$$

with $H_0 = \sum_{s=1}^{S}(H_s^{-1} + \Sigma)^{-1}$ and the second stage GLS-estimate

$$\hat{\xi} = \left[\sum_{s=1}^{S}(H_s^{-1} + \Sigma)^{-1}\right]^{-1} \sum_{s=1}^{S}(H_s + \Sigma)^{-1}\hat{b}_s. \tag{2.13}$$

For the hyperparameter we have the possibility of 2 types of sensitivity analysis.

Lemma 2.3.

a) *The posterior mean of the hyperparameter for known 3rd stage lies in the (surroundin second stage) ellipsoid*

$$ELL(\xi^*, \hat{\xi}, H^{*-1}) = \text{closure}\left\{\xi^{**}(H_0 | H^*, \xi^*, D_\omega) | H_0 \in \mathcal{M}^+\right\}. \quad (2.14$$

b) *The posterior mean (2.1) lies for unknown 3rd stage matrix H^* in the ellipsoid*

$$ELL(\xi^*, \hat{\xi}, H_0) = \text{closure}\left\{\xi^{**}(H^* | \xi^*, \Sigma, D_\omega) | H^* \in \mathcal{M}^+\right\}. \quad (2.15$$

Proof. By theorem 2.1.

Since we are concerned in this paper with sensitivity analysis in the 2nd stage, we wi use only the first ellipsoid (2.14). The term "surrounding" refers to the fact that we can d a better sensitivity analysis if we explore the unknown covariance structure in the posterio means more carefully. The improved sensitivity analysis results in smaller ellipsoids whic will be called in the next section "embedded hierarchical ellipsoids".

2.2. *Hierarchical Sensitivity Analysis*

In a 3-stage hierarchical model we have 3 covariance matrices to specify for every stag (D_ω, Σ, H^*) and one ultimate prior location ξ^*. In the following, we concentrate on th partially known prior structure where we specify everything except the second stage covarianc matrix Σ.

A more sophisticated sensitivity analysis makes use of the fact that the prior covarianc structure for the posterior means vecB^{**} and ξ^{**} also enters together with other matrices. I is easily seen (see Polasek (1984)) that the prior variance in (2.9) is bounded from below

$$G^{-1} = 1_S 1_S' \otimes H^* + I_S \otimes \Sigma > 1_S 1_S' \otimes H^*, \quad (2.16$$

while the variance of the GLS-estimator $\hat{\xi}$ is bounded from above

$$\text{Var}(\hat{\xi}) = \sum_{s=1}^{S} (H_s^{-1} + \Sigma)^{-1} < \sum_{s=1}^{S} H_s. \quad (2.17$$

These bounds imply a robust hierarchical analysis, which is discussed in the next section.

3. ROBUST BAYESIAN INFERENCE

By robust Bayesian inference we understand the description of the set of posterior distributions when the prior density is known only partially. For a seasonal time series model we describe the posterior mean and HPD-regions of a full 3-stage hierarchical model when the second stage covariance matrix is not known.

Other approaches to Bayesian robustness can be found in Berger (1984). Also, a different hierarchical robustness analysis for the posterior mean based on the diffuse 3-stage model (2.6) can be found in Polasek (1984). This approach involves an empirical Bayes estimation of a 2-stage model and a hierarchical EBA can be carried out for symmetric variance restrictions around the empirical estimated $\hat{\Sigma}$ based on (2.7).

In this paper we develop a robust Bayesian approach on a full 3-stage model. Since we have shown that the variances in a full 3-stage model are bounded, the set of the posterior means can be described as follows:

Theorem 3.1. *The posterior mean (2.9) for any $\Sigma > 0$ is constrained to lie in the ellipsoid*

$$ELL(1_S \otimes \xi^*, \text{vec}B_\Pi, H_L) = \text{closure}\left\{B^{**}(\Sigma|\xi^*, H^*, D_\omega)|\Sigma \in \mathcal{M}^+\right\} \tag{3.1}$$

where

$$\begin{aligned}\text{vec}B_\Pi &= \lim_{\Sigma\to 0} B^{**}(\Sigma) \\ &= [I_{hs} + (1_S 1_S' \otimes H^*)\text{diag}\tilde{H}]^{-1}[1_S \otimes \xi^* + (1_S 1_S' \otimes H^*)\text{diag}\tilde{H}\,\text{vec}\hat{B},\end{aligned} \tag{3.2}$$

is the limiting location parameter of the first stage posterior mean as $\Sigma \to 0$ (since Σ is p.d.), and the metric

$$H_L = \text{diag}\tilde{H} + \tilde{H}H^*\tilde{H}'. \tag{3.3}$$

Proof. Insert into theorem A1 of Appendix A.

Lemma 3.2. *The limiting first stage location parameter is given by*

$$\text{vec}B_\Pi + 1_S \otimes \xi_\Pi, \tag{3.4}$$

where ξ_Π is the limiting second stage parameter

$$\xi_\Pi = (H_+ + H^{*-1})^{-1}(H_+\hat{\xi}_\Pi + H^{*-1}\xi^*), \tag{3.5}$$

where $\hat{\xi}_\Pi = H_+ \sum_{s=1}^{S} X_s'Y_s/\omega_s$ is obtained by letting $\Sigma \to 0$ in the GLS-estimate (2.12) and $H_+ = \sum_{s=1}^{s} H_s$.

Proof. Apply the binomial matrix inversion lemma to (3.3). (Lemma 3.2 replaces misprints in Polasek (1984), p. 299).

Interestingly, the limiting location parameter ξ_Π is constant over seasons and appears in stage 1 and 2. Thus, the robustness analyses become easier and more comparable.

Theorem 3.3. *The posterior mean of the second stage (2.11) for free $\Sigma > 0$ is constrained to lie in the ellipsoid*

$$ELL(\xi^*, \xi_\Pi, H_u) = \text{closure}\left\{\xi^{**}(\Sigma|D_\omega, H^*, \xi^*)|\Sigma \in \mathcal{M}^+\right\} \tag{3.6}$$

with metric $H_u(H_+ + H^*)^{-1}$.

Proof. Insert into theorem A2 of Appendix A.

Note that the advantage for using ellipsoid (3.6) instead of (2.11) lies in a smaller sized ellipsoid, fully imbedded in the surrounding ellipsoid (2.11). While ξ^{**} in (2.11) lies between $\hat{\xi}$ and ξ^*, ξ^{**} in (3.6) lies between ξ^* and the limiting hyperparameter location $\xi_\Pi = \lim_{\Sigma\to 0} \xi^{**}(\Sigma)$ defined in (3.5).

The Bayesian robustness analysis can be extended in describing the set of posterior distributions not only by posterior means, but by any posterior quantiles. Such robust HPD- (or HiFi-) regions are derived in Polasek and Pötzelberger (1987).

3.1. Simplified Calculation of the Posterior Mean

Despite the fact that computer speed is improving every generation and software for statistics and programming matrix algebra becomes more and more straightforward, the inversion of matrices is still very time consuming. Any analytical simplification is therefore a computational improvement, especially for small computers. The posterior mean (2.9) of model (2.8) requires two matrix inversions which can be done analytically using the matrix binomial inversion lemma $(D + EFE')^{-1} = D^{-1} - D^{-1}E(E'D^{-1}E + F^{-1})^{-1}E'D^{-1}$. The matrix G then becomes

$$G = [1_S \otimes \xi^* + (1_S \otimes I_k)H^*(1_S' \otimes I_k)]^{-1} = I_S \otimes \Sigma^{-1} - 1_S 1_S' \otimes W \tag{3.8}$$

with $W = (S\Sigma + \Sigma H^{*-1}\Sigma)^{-1}$. Therefore the prior part in (2.9) reduces to

$$G(1_S \otimes \xi^*) = 1_S \otimes W^*\xi^*, W^* = (\Sigma + SH^*)^{-1}. \tag{3.9}$$

The inversion of the covariance matrix yields

$$\begin{aligned}[\text{diag}\tilde{H} + G]^{-1} &= [\text{diag}(H_s + \Sigma^{-1}) - 1_S 1_S' \otimes W]^{-1} \\ &= \text{diag}\tilde{F} - \tilde{F}[F_+ - W^{-1}]^{-1}F'\end{aligned} \tag{3.10}$$

with

$$\tilde{F}' = (F_1, \ldots, F_S) = \tilde{H} + 1_S \otimes \Sigma^{-1} \quad \text{and } F_+ = \sum_{s=1}^{S} F_s = H_+ + S\Sigma^{-1}. \tag{3.11}$$

The posterior mean can now be expressed as

$$\text{vec}B^{**} = (\text{diag}\tilde{F} - \tilde{F}[F_+ - W^{-1}]^{-1}\tilde{F}')(\text{diag}\tilde{H}\,\text{vec}\hat{B} + 1_S \otimes W^*\xi^*), \tag{3.12}$$

or simpler, for every season $s = 1, \ldots, S$,

$$b_s^{**} = F_s R_s - F_s[F_+ - W^{-1}]^{-1}\tilde{F}'\tilde{R}, \tag{3.13}$$

with $\tilde{R} = (R_1, \ldots, R_S)'$, $F_s = H_s + \Sigma^{-1}$, and $R_s = H_s\hat{b}_s + W^*\xi^*$.

Note that the limit $\lim_{\Sigma\to\infty} \tilde{F}'\tilde{R} = S\overline{b^{**}}$, i.e. in the diffuse case, reduces to S times the average of the posterior means across seasons.

4. EXAMPLE

A hierarchical robustness analysis is carried out for a simple 4×2-dimensional example for quarterly data of the Austrian consumption function 1964–84: $C = a + bY$. C is real consumption, Y is real income, and the data are transformed to the first differences of the logarithms (growth rates). The coefficient b is also called marginal propensity for consumption (m.p.g.). The data are plotted for all 4 quarters in Figure 4.1.

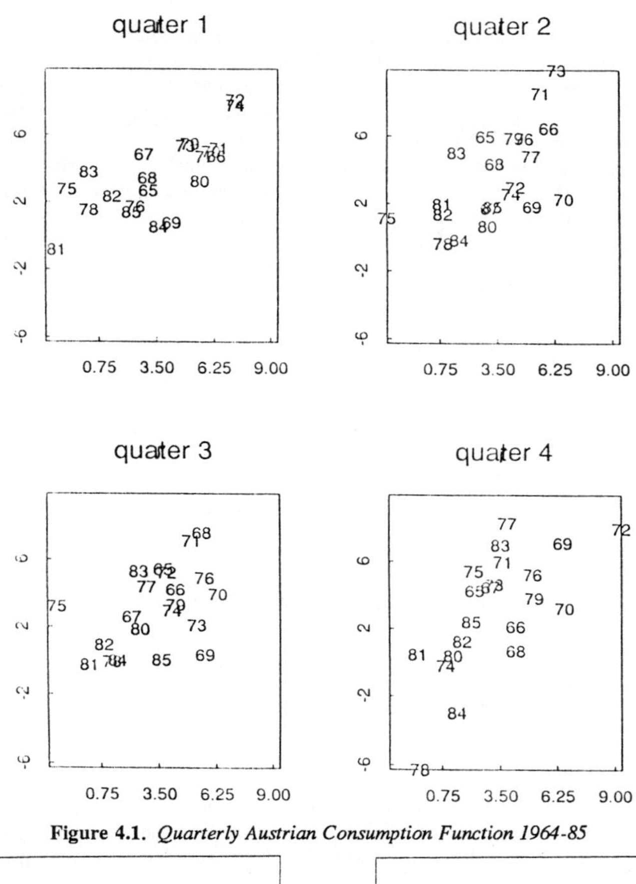

Figure 4.1. *Quarterly Austrian Consumption Function 1964-85*

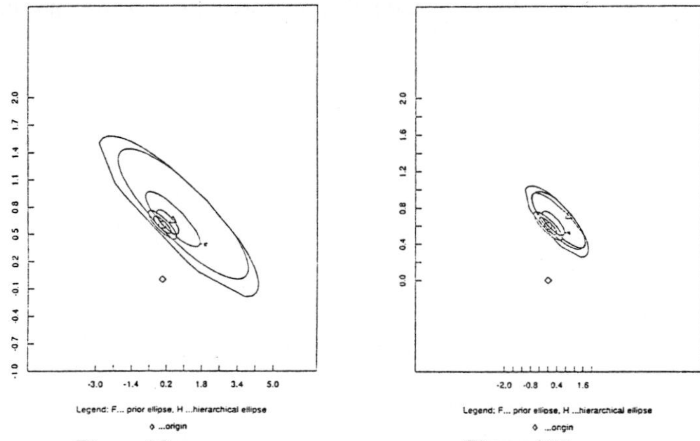

Figure 4.2.a
Hyperparameters 0.5-HiFi-region

Figure 4.2.b
Hierarchical Hyperparameters 0.5-HiFi-region

Next, we have to assess the prior distribution for a seasonal model. We expect some variation between seasons in the consumption pattern, like more expenditure in the 4th quarter because of Christmas, but the assumption of a common hyperdistribution seems quite reasonable. As 3rd stage prior mean for the hyperparameters we would expect about the same values as for a non-seasonal yearly 2-stage model, i.e. $b^* = (0, .6)$. The confidence in this prior information is expressed by the (3rd stage) prior covariance matrix $C^* = \mathrm{diag}(1, .0)$: standard deviations of more than .5 for the intercept and .05 for the m.p.c. seem to be highly unlikely (What would we expect as ML-variances in a yearly model?) For the present case we have chosen the covariance matrix of the ML-coefficients of the yearly model. The maximum likelihood estimates for the quarterly models (with standard deviations St. E. and residual variances $\hat{\omega}_s$ are listed in Table 4.1:

coeff./season	1	2	3	4
INT	1.24	.89	1.33	−.35
(St. E.)	.62	.97	.94	1.05
m.p.c.	.64	.77	.54	1.04
(St. E.)	.14	.24	.24	.26
Res Var $\hat{\omega}_s$	2.64	5.39	4.74	7.55

Table 4.1

Estimates for the hyperparameters ξ are given in Table 4.2:

	$\bar{b}$	$\hat{\xi}$	ξ^{**}	$\xi(II)$	ξ^*
INT	.776	.925	.929	.476	0
m.p.c.	.748	.718	.708	.661	.6

Table 4.2

The first column is the simple average of the 1st stage OLS-coefficients. The second contains the GLS-estimate $\hat{\xi}$ of (2.12), while the third lists the limiting posterior mean of (3.7). The pre-last column is the $\xi(II)$ estimate (3.5). Except for $\xi(II)$ they do not differ too much.

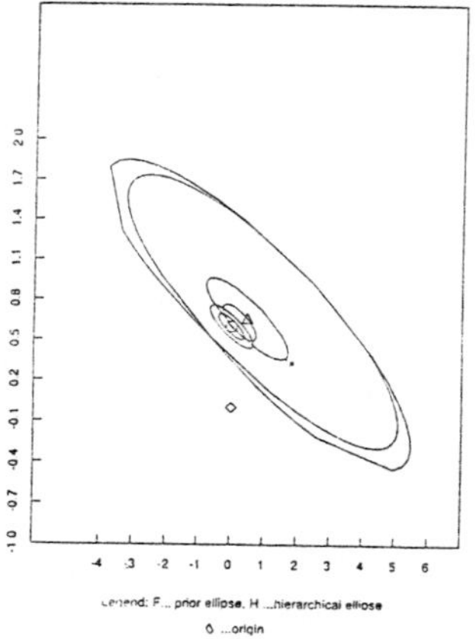

Figure 4.2.c
Hyperparameters 0.75-HiFi-region

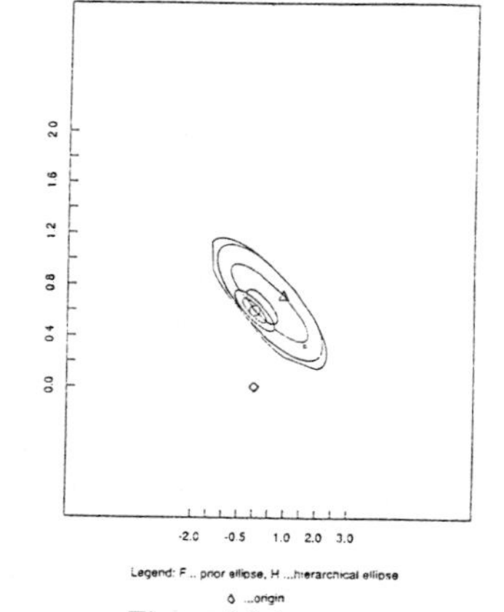

Figure 4.2.d
Hierarchical Hyperparameters 0.75-HiFi-region

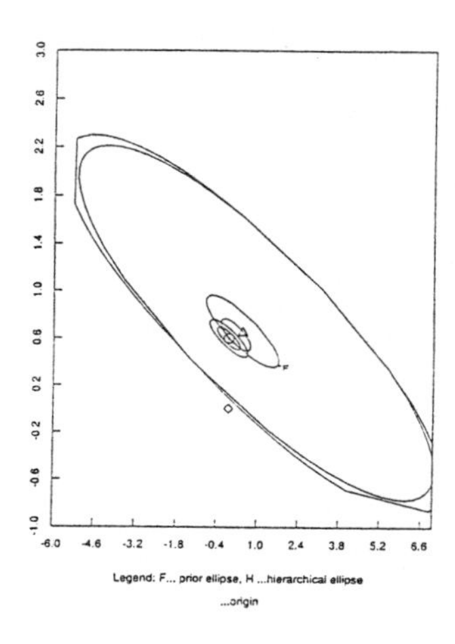

Figure 4.2.f
Hyperparameters 0.95-HiFi-region

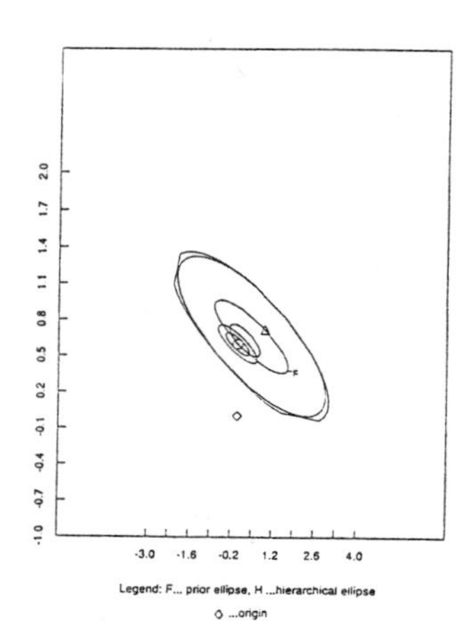

Figure 4.2.g
Hierarchical Hyperparameters 0.95-HiFi-region

A robust Bayesian analysis for the hyperparameter ξ with partial prior information $\xi^* = (0, .6)$ but any prior covariance matrix, yields the following extreme bounds:

$$\text{EBA (a)} = [-.791.7], \quad \text{EBA (b)} = [.36, .96]$$

$$\text{HEBA (a)} = [-.27.75], \quad \text{HEBA (b)} = [.49, .77]$$

The second line lists the hierarchical extreme bound analysis (HEBA) obtained from (3.6). Because the intercept-prior is 0, the associated extreme lower bounds are negative. Figure 4.2 gives the graphical results together with 50%, 75% and 80% prior HPD- and hierarchcial HiFi-regions.

Next, we sumarize the EBA for all the 1st stage coefficients in Table 4.3:

	EBA		seas. EBA		HEBA	
1.Q: a	−1.25	2.49	−.60	1.83	.26	1.45
b	.20	1.04	.35	.90	.52	.78
2.Q: a	−2.48	3.37	−1.02	1.91	−.11	1.47
b	−.04	1.41	.32	1.05	.52	.91
3.Q: a	−2.16	3.49	−.44	1.77	.40	1.41
b	−0.16	1.29	.28	.85	.47	.73
4.Q: a	−3.34	2.99	−1.49	1.14	−.79	.92
b	.04	1.60	.49	1.15	.64	1.06

Table 4.3

The first column reports the EBA for the "system estimate" of the 3-stage hierarchical model with partial prior given in (2.9). It shows the largest values, because the EBA takes into account all the variation of the covariance structure G in (2.10) for given b^*. The second column contains the separated EBA for every season or $ELL(\xi^*, b_s, X_s' X_s)$, $s = 1, \ldots, S$, which does not take into account the system variation of the 3-stage hierarchical model. The third column shows the hierarchical EBA obtained by Theorem 3.1. Clearly, the smallest bounds are obtained by HEBA for every coefficient.

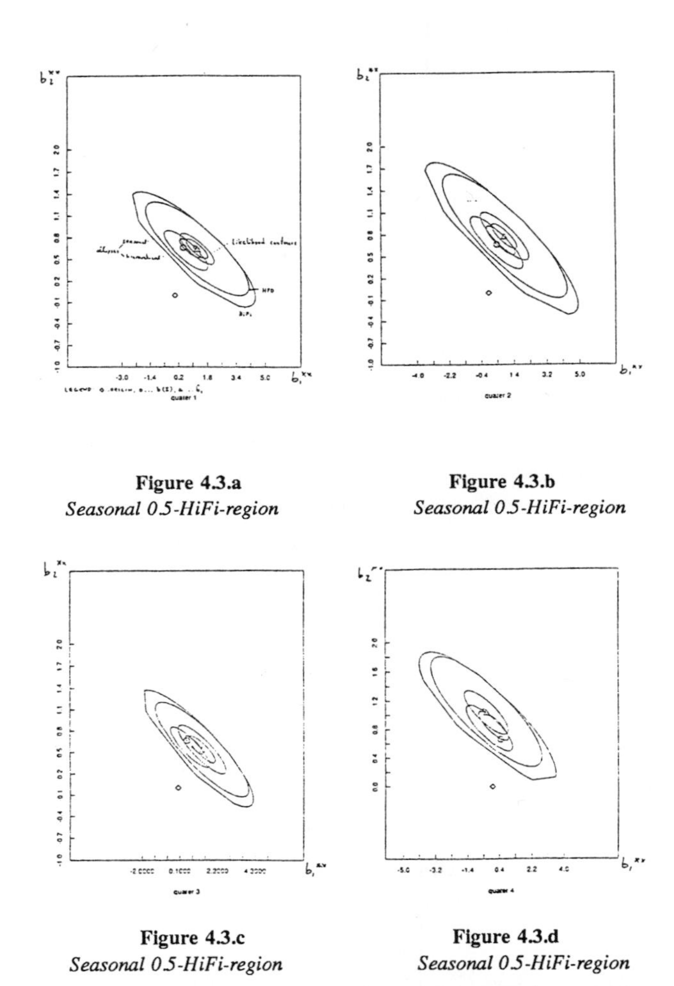

Figure 4.3.a
Seasonal 0.5-HiFi-region

Figure 4.3.b
Seasonal 0.5-HiFi-region

Figure 4.3.c
Seasonal 0.5-HiFi-region

Figure 4.3.d
Seasonal 0.5-HiFi-region

This reveals that the m.p.c. allows robust Bayesian inference for a positive coefficient for the given prior location b^* in all 4 quarters. The associated intercepts are "robustly" positive in quarters 1 and 3, but have negative lower bounds in the second and fourth quarter. The second column with "seas. EBA" are pseudo-extreme bounds. They would be obtained if only two coefficients are singled out of the system-posterior mean (2.9). If we use the correct projection technique listed in Appendix B, we get the first EBA-column again. Therefore the hierarchical column HEBA reports the extreme bounds for known first and last stage parameters. They provide sensible economic interpretation with the range of bounds between $.47$ and 1.06. furthermore, they allow a better inference-summary than the ML-results of Table 4.1, where a 2-σ range covers the whole interval from 0 to 1. Even then the high m.p.c.-estimate of 1.04 in the $4th$ quarter is explainable as lying almost on the robust upper boundary of 1.06.

The feasible ellipsoid for the hyperparameter is given in Figure 4.2 together with the prior contours for $N(b^*, H^*)$. (All the ellipses in figures 4.2-4.6 are drawn on the same scale but with different tick-marks.) They show the surrounding ellipsoids (2.14) and the hierarchical ellipsoid (3.6) with 50%, 75% and 95% HiFi-regions. The information gain for the hierarchical case is quite spectacular.

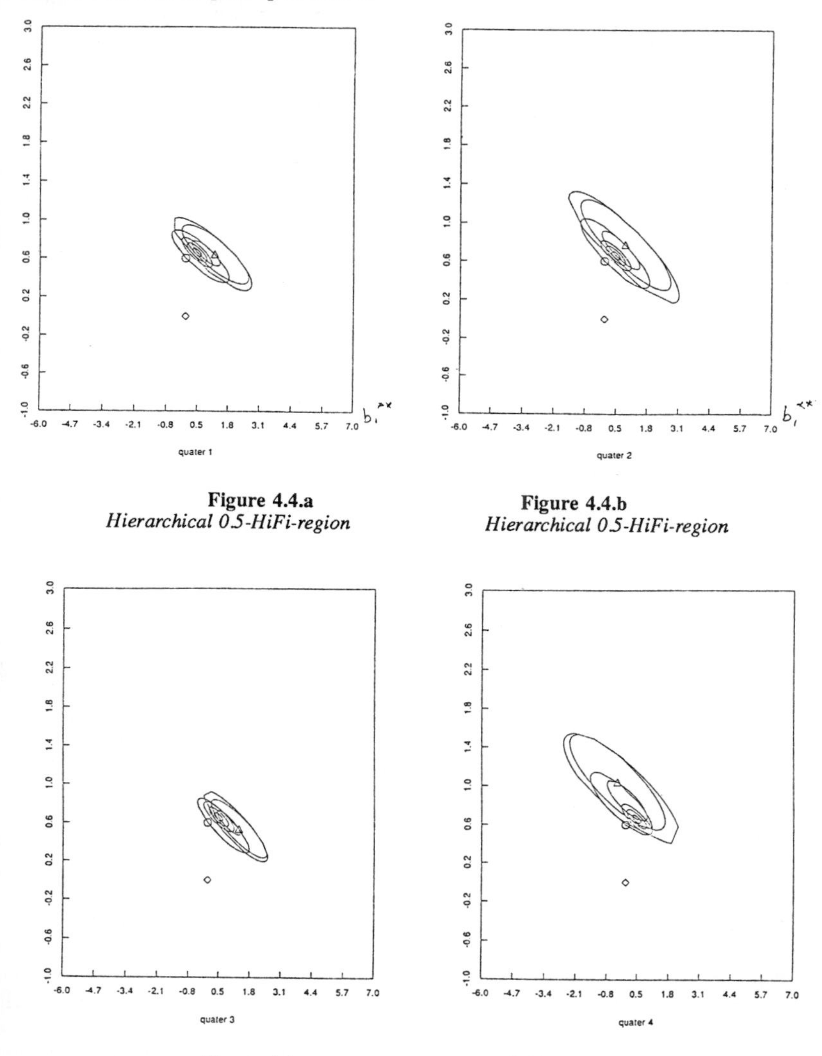

Figure 4.4.a
Hierarchical 0.5-HiFi-region

Figure 4.4.b
Hierarchical 0.5-HiFi-region

Figure 4.4.c
Hierarchical 0.5-HiFi-region

Figure 4.4.d
Hierarchical 0.5-HiFi-region

The Bayesian robustness analysis for the 1st stage coefficients are summarized in Figures 4.3-4.6. For all quarters 50%, 95% HiFi-regions, and diffuse HPD-regions are plotted for so-called "seasonal" (unrestricted) posterior means (2.9) and the "hierarchical" case (3.1). Because of the strong squeezing factor the "peeled egg shape" of the HiFi-regions can hardly be

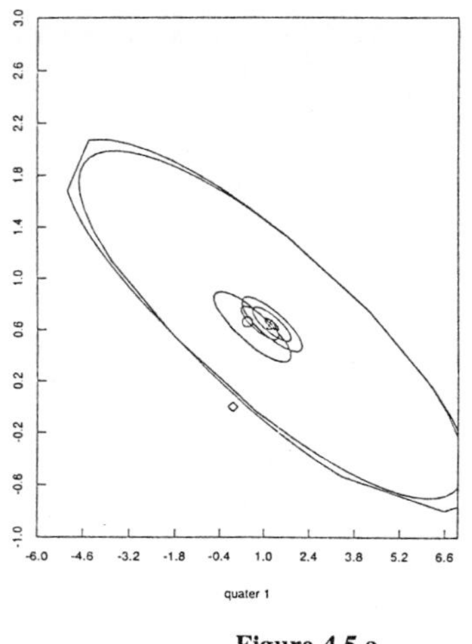

Figure 4.5.a
Seasonal 0.95-HiFi-region

Figure 4.5.b
Seasonal 0.95-HiFi-region

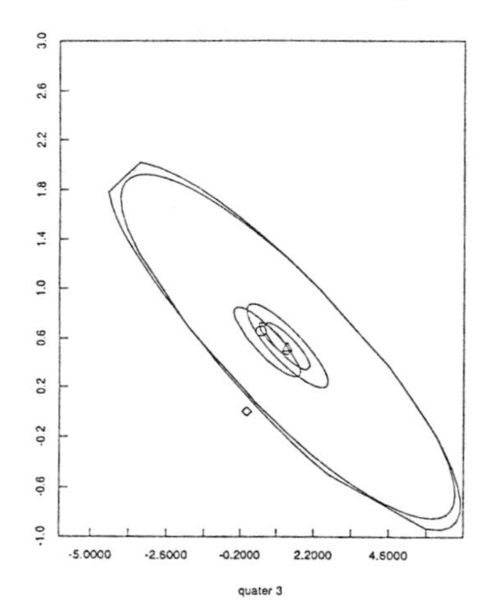

Figure 4.5.c
Seasonal 0.95-HiFi-region

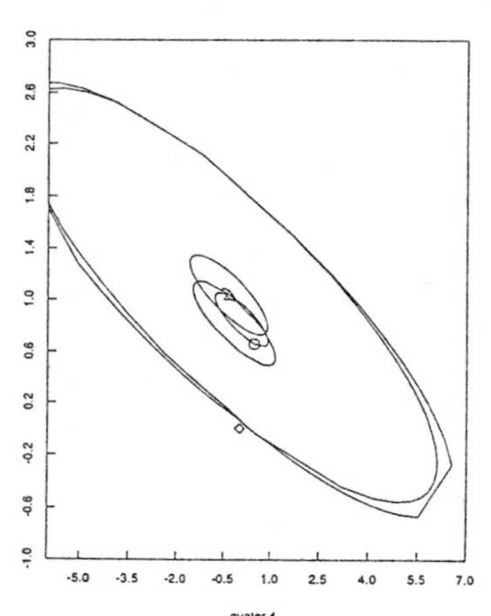

Figure 4.5.d
Seasonal 0.95-HiFi-region

seen. The eccentric ellipsoids arise from a $-.8$ correlation between the regressors. All figures show the seasonal and the imbedded hierarchical ellipsoids. Figure 4.4 shows the 50%-HPD and HiFi-regions for the hierarchical ellipsoids of theorem 3.1. (Because of the chosen coordinate system and computational constraints, the HiFi-region could not be smoothed enough.) They are much smaller and allow sensible interpretation of the coefficients. Figure 4.5 and Figure 4.6 contain the same analysis for 95%-regions. Interestingly, the HiFi-regions are only slightly larger than the diffuse (classical) HPD region, based on the ML-estimates. Note that smaller hierarchical ellipsoids require full acceptance of the 3rd stage prior: This should be quite reasonable even for cautious Bayesians (and non-Bayesians), if the prior is close to the (diffuse) estimates of the nonseasonal yearly model.

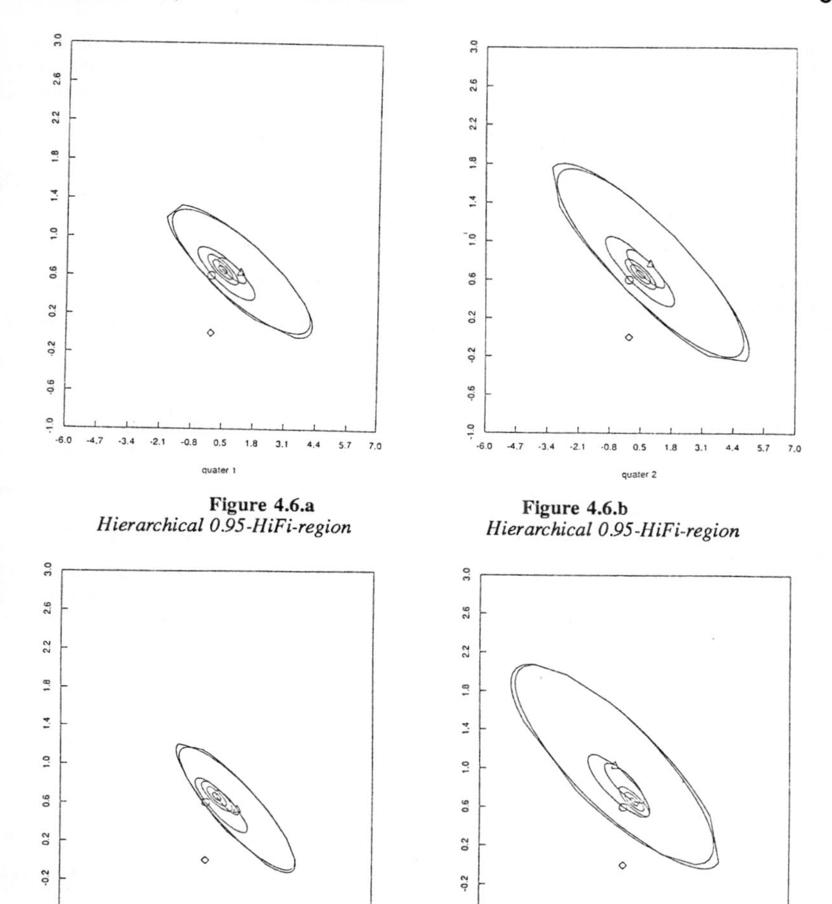

Figure 4.6.a
Hierarchical 0.95-HiFi-region

Figure 4.6.b
Hierarchical 0.95-HiFi-region

Figure 4.6.c
Hierarchical 0.95-HiFi-region

Figure 4.6.d
Hierarchical 0.95-HiFi-region

We hope to have demonstrated that hierarchical models are a good tool for analysing complex structured multivariate data, like seasonal time series. We have tried to remove to some extent the mystical "wooliness" of higher stage prior distribution and use Bayesian robustness techniques like EBA as a tool for reporting inferences.

5. CONCLUSIONS

Bayesian robustness analysis in a 3-stage hierarchical model is shown to be a reasonable alternative to the usual empirical Bayes analysis of hierarchical models (such as that of Smith 1983). While a robustness analysis for unknown 3rd stage can be carried out also for empirical

Bayes models (see Polasek 1984), we think that a 3-stage robustness analysis is closer to the Bayesian modeling philosophy, allows a better sensitivity analysis, gives more insight into multivariate structures, and is a better inference summary than non-hierarchical OLS-analyses.

ACKNOWLEDGEMENTS

A macro was written using the statistical package S (Becker *et al.* (1985)) to carry out the calculations and graphics and is available upon request.

REFERENCES

Becker, R. A. and Chambers, J. M. (1984). *S: An Interactive Environment for Data Analysis and Graphics.* Belmont CA.: Wadsworth.

Berger, J. O. (1984). The robust Bayesian viewpoint. *Robustness in Bayesian Statistics,* (J. Kadane, ed.). Amsterdam: North-Holland, 63–144.

Chamberlain, G. and Leamer, E. E. (1976). Matrix weighted averages and posterior bounds. *J. Roy. Statist. Soc. B* **38**, 73–84.

Leamer, E. E. (1978). *Specification Searches.* New York: Wiley.

Leamer, E. E. (1982). Sets of posterior means with bounded variance priors, *Econometrica* **50**, 725–36.

Lindley, D. V. and Smith, A. F. M. (1972). Bayes estimates for the linear model, (with discussion). *J. Roy. Statist. Soc. B* **34**, 1–41.

Hui, S. L. and Berger, J. O. (1983). Empirical Bayes estimation of rates in longitudinal studies. *J. Amer. Statist. Assoc.* **78**, 753–760.

Polasek, W. and Pötzelberger, K. (1987). *Robust HPD-Regions in Bayesian Regression Models,* mimeo, submitted for publication.

Polasek, W. (1984). Multivariate regression systems: Estimation and sensitivity analysis for two-dimensional data. *Robustness in Bayesian Statistics,* (J. Kadane, ed.). Amsterdam: North-Holland, 229–309.

Smith, A. F. M. (1983). Bayesian approaches to outliers and robustness. *Specifying Statistical Models,* (Florens et al. eds.). New York: Springer Verlag, 13–35.

APPENDIX A: HIERARCHICAL BOUNDS

Theorem A.1. *Lower Bound*

The posterior mean

$$\theta_1^{**} = [H + (V^* + C_2)^{-1}]^{-1}[H\hat{\theta}_1 + (V^* + C_2)^{-1}A_2A_3b^*] \tag{A1}$$

of the 3-stage hierarchical model (2.5) lies in the ellipsoid $ELL(\hat{\theta}_1, \theta_\Pi, H_L)$, *where* $H = A_1^T C_1^{-1} A_1$; $\hat{\theta} = H^{-1}A_1^T C_1^{-1} y$ *and* $H_L = diag(\tilde{H}) + \tilde{H}H^*\tilde{H}$.

Proof. The posterior mean (A1) can be transformed by the inversion lemma into

$$\begin{aligned} \theta,^{**} &= H^{-1}[H^{-1} + V^* + C_2]^{-1}[(V^* + C_2)H\hat{\theta}_1 + A_2A_3b^*] \\ &= H^{-1}[W + C_2]^{-1}[WH\theta_\Pi + C_2H\hat{\theta}_1], \end{aligned} \tag{A2}$$

where we have singled out C_2 and simplified the expression by setting $W = V^* + H^{-1}$, and θ_Π is the limiting location parameter

$$\begin{aligned} \theta_\Pi &= H^{-1}(H^{-1} + V^*)^{-1}(V^*H\hat{\theta}_1 + A_2A_3b^*) \\ &= (I + V^*H)^{-1}(V^*H\hat{\theta}_1 + A_2A_3b^*). \end{aligned} \tag{A3}$$

Note that θ_Π can be written as a matrix weighted average between the first stage ML-location $\hat{\theta}_1$ and the prior location $A_2A_3b^*$. Inserting $H^{-1}H$ between the brackets of (A2) leads to

$$\theta_1^{**} = [HWH + HC_2H]^{-1}[HWH\theta_\Pi + HC_2H\hat{\theta}_1]. \tag{A4}$$

This is the required matrix weighted average between θ_Π and $\hat{\theta}_1$ leading to the ellipse (A1) with the metric $H_L = HWH$. ◁

Theorem A.2. *Upper Bound*
The posterior mean for the hyperparameters ($A = A_2$)

$$\theta_2^{**} = [A'(H^{-1}+\Sigma)^{-1}A + H^{*-1}]^{-1}[A'(H^{-1}+\Sigma)^{-1}AA^{-1}\hat{\theta}_1 + H^{*-1}\mu] \tag{A5}$$

with $H = A_1'C_1^{-1}A_1$ *and* $\mu = A_2A_3b^*$*, lies in ellipsoid* $ELL(\theta_\Pi, \mu, H_u)$.

Proof. We show that any posterior mean (A5) lies in the smaller hierarchical ellipsoid with the parameters

$$H_u = H^{*-1} + (H^*A'HAH^*)^{-1} = H_*^{-1}((H^* + (A'HA)^{-1})H_*^{-1} \tag{A6}$$

$$\theta_\Pi = H_u^{-1}(H^{*-1}\hat{\theta} + (H^*A'HAH^*)^{-1}\mu). \tag{A7}$$

By theorem 2.1 we have to show that the following inequality is valid

$$(\theta_2^{**} - f)'H_u(\theta_2^{**} - f) \leq (\mu - f)'H_u(\mu - f) = c, \tag{A8}$$

with $f = (\mu + \theta_\Pi)/2$. Insertion and transformations lead to

$$\begin{aligned}
&(\theta_2^{**} - \mu + \mu - f)'H_u(\theta_2^{**} - \mu + \mu - f) \leq (\mu - f)'H_u(\mu - f)\\
&\Leftrightarrow (\theta_2^{**} - \mu)'H_u(\theta_2^{**} - \mu) + 2(\theta_2^{**} - \mu)'H_u(\theta_2^{**} - f) \leq 0\\
&\Leftrightarrow (\hat{\theta}_1 - \mu)'F(F + H^{*-1})^{-1}H_u(F + (F + H^{*-1})^{-1}F(\hat{\theta}_1 - \mu)\\
&\leq \hat{\theta} - \mu)'F(F + H^{*-1})^{-1}H^{*-1}(\hat{\theta}_1 - \mu).
\end{aligned} \tag{A9}$$

Now we continue the inequality by concentrating on matrices. Binomial inversion and some algebra show that

$$\begin{aligned}
&(F^{-1} + H^*)^{-1}H^*H_uH^*(F^{-1} + H^*)^{-1} \leq (F^{-1} + H^*)^{-1}\\
&\Leftrightarrow H^*H_uH^* \leq F^{-1} + H^*\\
&\Leftrightarrow H^* + (A'HA)^{-1} \leq (A'(H^{-1} + \Sigma)^{-1}A)^{-1} + H^*\\
&\Leftrightarrow AHA \geq A'(H^{-1} + \Sigma)^{-1}A.
\end{aligned} \tag{A10}$$

Since Σ and H are p.d. the last inequality is true. ◁

Note that the upper bound hierarchical ellipsoid for (A5) does not allow an easy derivation as a matrix weighted average between the limiting location parameter and μ. We can find the relationship by equating both expressions for the posterior means of the hyperparameter:

$$(F + H^{*-1})^{-1}(F\hat{\theta}_1 + H^{*-1}\mu) = (H_u + \tilde{\Sigma})^{-1}(H_u\theta_\Pi + \tilde{\Sigma}\mu), \tag{A11}$$

and solving (A11) for $\tilde{\Sigma}$. For simplicity we set $\mu = 0$, which also implies $\theta_\Pi = (H_u - H^{*-1})^{-1}H^{*-1}\mu$. Then it follows that

$$\begin{aligned}
&(F + H^{*-1})^{-1}H^{*-1}\mu = (H_u + \tilde{\Sigma})^{-1}[(H_u - H^{*-1})\mu + \tilde{\Sigma}\mu]\\
&\Leftrightarrow (F + H^{*-1})^{-1} = (H_u + \tilde{\Sigma})^{-1}[(H_u + \tilde{\Sigma})H^* - I]\\
&\Leftrightarrow (H_u + \tilde{\Sigma}) = [(H_u + \tilde{\Sigma})H^* - I](F + H^{*-1})\\
&\Leftrightarrow 0 = H_uH^*F + \tilde{\Sigma}H^*F - F - H^{*-1}\\
&\Leftrightarrow \tilde{\Sigma} = H^{*-1} - H_u + H^{*-1}F^{-1}H^{*-1} = (H^*FH^*)^{-1} - A'HA
\end{aligned} \tag{A12}$$

This shows the nonlinear dependence on $\tilde{\Sigma}$.

APPENDIX B

Projection of ellipsoids onto planes.

Let $Q = x'Ax \leq c$ a p-dimensional ellipsoid which we want to project onto a 2-dimensional plane. Let R be a $p \times (p-2)$ selection matrix $R = I_P[ij]$ selecting the i-th and j-th coefficient. $I_P[ij]$ stands for an identity matrix without the i-th and j-th column $(i \neq j)$. We write now the ellipsoid partitioned as

$$Q = (y + Rz)'A(y + Rz) \leq c, \tag{B1}$$

with $y = Tx$ and T is a $2 \times p$ 0/1-matrix consisting of the i-th and j-th unity vector. Thus, y is the vector for the selected coefficients. Now we try to find those z in (B1) which will satisfy the inequality. It is enough to show it for the minimum and since (B1) is a (p.d.) quadratic form we can set the derivative $\partial Q/\partial z$ to zero:

$$R'Ay + R'ARz = 0. \tag{B2}$$

If A has full rank then the minimum is given by

$$z = (R'AR)^{-1}R'Ay \tag{B3}$$

Now we insert (B3) into (B1) and get

$$\begin{aligned} y'(I - R(R'AR)^{-1}R'A)'A(I - R(R'AR)^{-1}R'A) \leq c \\ y'(A - AR(R'AR)^{-1}R'A)y \leq c. \end{aligned} \tag{B4}$$

The metric A is decreased by a p.d. matrix and this implies a larger ellipsoid.

DISCUSSION

L. D. BROEMELING (*University of Texas, Medical Branch*)

Polasek and Pötzelberger (1987) have developed a robust Bayesian analysis for the parameter and hyperparameter of a three-stage nested model. By leaving the second stage covariance matrix unspecified all possible posterior distributions were found. The emphasis of their paper is on the regression parameters, but I would like to direct the discussion toward estimating the second and third stage covariance matrices.

Consider a special case of a so-called random model given in three stages as

A. $\quad Y|b, \tau \sim N(Ub, \tau^{-1}I_n)$

$\quad b|\tau_1 \sim N(0, \tau_1^{-1}I_m)$

B.

$\quad \tau \sim G(\alpha, \beta)$

C. $\quad \tau_1 \sim G(\alpha_1, \beta_1),$

where Y is a $n \times l$ observation vector, and G denotes a gamma distribution. The objective is to estimate the two precision parameters τ and τ_1.

The basic questions are: (a) How do we determine the marginal posterior densities of the precision components? and (b) How do we choose values for the hyperparameters α, α_1, β and β_1? A complete solution to these problems has been difficult and Broemeling (1985) gives a short history.

Generally speaking, closed-form expressions for the marginal posterior densities of the precision components are unknown, but as will be shown, if certain approximations are made, one can determine these distributions and show how they depend on prior information. All inferences will be based on the joint posterior density

$$g(b, \tau, \tau_1|Y) = g_1(b, \tau|Y)g_2(b, \tau_1), \tag{1}$$

where $b \in R^m$, $\tau > 0$, $\tau_1 > 0$

$$g_1(b, \tau_1|Y) \propto \tau^{\frac{n+2\alpha}{2}-1} \exp -\frac{\tau}{2}[2\beta + (y - ub)'(y - ub)]$$

and

$$g_2(b, \tau_1) \propto \tau_1^{\frac{m+2\alpha_1}{2}-1} \exp -\frac{\tau_1}{2}[2\beta_1 + b'b].$$

If we want to estimate τ, the other parameters τ_1 and b must be integrated from (1). Now, after (1) is integrated with respect to τ_1, the resulting marginal posterior density of b and τ cannot be integrated (in closed form) with respect to b, however if the resulting t density on b is approximated by a normal with the same mean and covariance matrix, one may show that the approximate marginal density of τ is

$$g(\tau|Y) \propto \frac{\tau^{\frac{n+2\alpha}{2}-1}}{|\tau u'u + C_1^{-1} I_m|^{1/2}} \exp -\frac{\tau}{2}[2\beta + y'(I_n + uu'\tau C_1)^{-1} y], \tag{2}$$

where $\tau > 0$ and

$$C_1 = \beta_1(\alpha_1 - 1)^{-1}$$

is the prior mean of τ_1^{-1}. We see how posterior inferences for τ depend on prior information, namely on essentially three things: the prior mean of the parameter and the α and β parameters of τ. In a similar way, the marginal posterior density of τ_1 is derived as

$$g(\tau_1|Y) \propto \frac{\tau_1^{\frac{m+2\alpha_1}{2}-1}}{|u'u + C\tau_1 I_m|^{1/2}} \exp -\frac{1}{2}[2\beta_1\tau_1 + Cy'\left(I_n + \frac{uu'}{C\tau_1}\right) y], \tag{3}$$

where $C = \beta(\alpha - 1)^{-1}$ is the prior mean of the error variance τ^{-1}. As with (2), the dependence of (3) on prior information is elucidated. We now have a solution to the first question (a) but (b) remains. How do we choose values for the hyperparameters?

The marginal distribution of the observation y is normal with mean vector zero and covariance matrix

$$CI_n + uu'C_1$$

and C and C_1 could be chosen by say the principal of maximum likelihood. However, we would need to assign values to the four hyperparameters separately in order to implement (2) and (3). This way of assigning values to the hyperparameters is advocated by Winkler (1980) and is an empirical Bayes method.

The above gives a partial solution to making inferences about the parameters of a random model and can be extended to the more general mixed model, but some important questions remain.

(i) What is a vague prior distribution for τ and τ_1?

(ii) Can the hyperparameters be empirically estimated?

(iii) Should a fourth stage prior density for the hyperparameters be introduced?

(iv) How should robustness be approached?

I wish to thank the authors for their excellent paper.

REPLY TO THE DISCUSSION

We would like to thank the discussant for his noteworthy comments. The hierarchical model in our paper deals only with a hierarchy of (prior) means, whereas Dr. Broemeling's comments concentrate on the univariate first and second stage precision components.

Apart from the difficulties for calculations that come from the posterior distribution, there do not seem to exist natural conjugate priors for Dr. Broemeling's hierarchical models of precisions. But even if these difficulties could be overcome, the difficulty of specifying the parameters of the prior distribution (not its functional form) remains. It seems to be considerably easier to specify only the prior of a location parameter like the mean than the whole distribution and hyperparameters of the precision matrix.

This is one of the reasons why we propose in general the use of partially specified priors like the one in our paper, with the mean fixed and any prior precision matrix. While this approach seems to be a little bit outside of the traditional focus of a Bayesian analysis where uncertainty about parameters is modeled by a distribution, we justify our approach by the ease of computation, the analytical tractability at least in a conjugate framework, and the saving of elicitation costs in a multivariate problem. As long as we produce practical useful results in a robust Bayesian analysis we can avoid the burden of numerical integration or imputation.

REFERENCES IN THE DISCUSSION

Broemeling, L. D. (1985). *Bayesian Analysis of Linear Models.* New York: Marcel-Dekker Inc.

Polasek, W. and Pötzelberger, K. (1987). Robust Bayesian analyses in hierarchical models. (In this volume).

Winkler, R. L. (1980). Prior distributions model building in regression analysis. *New Developments in the Application of Bayesian Methods,* (Aykac and Brumot, eds.). Amsterdam: North-Holland.

BAYESIAN STATISTICS 3, pp. 395–402
J. M. Bernardo, M. H. DeGroot, D. V. Lindley and A. F. M. Smith, (Eds.)

Using the SIR Algorithm to Simulate Posterior Distributions

D. B. RUBIN
Harvard University

SUMMARY

The SIR (Sampling/Importance Resampling) algorithm is an ubiquitously applicable noniterative algorithm for obtaining draws from an awkward distribution: M draws from an initial approximation are made, and then $m < M$ draws are made from these with probability approximately proportional to their importance ratios. The method may have broad applicability in applied Bayesian inference, and is illustrated here using two examples. The first example is hypothetical and involves the posterior distribution of parameters and missing data in a normal/conditionally normal bivariate model. The second example involves real data from a U.S. Census Bureau project and concerns the multiple imputation of industry codes obtained by drawing logistic regression coefficients from their small sample posterior distribution.

Keywords: EM ALGORITHM; MISSING DATA; MULTIPLE IMPUTATION; PREDICTIVE DISTRIBUTIONS.

1. INTRODUCTION

Taking draws from posterior distributions can be an important component of many Bayesian data analyses. For example, in some cases the posterior distribution of the parameter, θ, $p(\theta|Y_{\rm obs})$, where $Y_{\rm obs}$ is the observed data, is analytically neither tractable nor easily approximated, and so a simulated posterior distribution based on a large number of draws will be used to represent the actual posterior distribution. The data augmentation algorithm (Tanner and Wong, (1987)) is designed for such cases. In other cases, relatively simple analytic expressions for the mean and variance of an approximately normal posterior distribution are unavailable because of the existence of a modest fraction of missing values, and inferences are to be drawn from a "multiply-imputed" data set created by drawing several, say m, repetitions of the missing values, $Y_{\rm mis}$, from their posterior distribution, $p(Y_{\rm mis}|Y_{\rm obs})$. Such procedures were proposed by Rubin (1978) and are discussed in detail in Rubin (1987a).

In either case, even the required simulation itself may be a complicated proposition; the posterior distribution may not be easy to draw from because of its difficult form (e.g., consider the problems of drawing logistic regression parameters in small samples, or missing data in a general multivariate normal). Thus it can be very useful to have an easily implemented algorithm for drawing from an awkward posterior distribution.

The method presented here, the SIR algorithm, is ubiquitously applicable in that it can be easily applied to any problem, but may not be very efficient in particular cases. Nevertheless, like many general but possibly computationally inefficient devices in statistics (e.g., the EM algorithm), it can be exceeding helpful in statistical practice.

To the best of my knowledge the SIR algorithm was first proposed in a small sample Bayesian logistic regression application (Rubin, (1983), here briefly summarized in Section 4. The algorithm was proposed as a possible general alternative to the data augmentation algorithm in Rubin (1987b), which gives the illustrative example used here in Section 3.

The SIR (*S*ampling/*I*mportance *R*esampling) algorithm is most useful when a good approximation exists to the posterior distribution being simulated from which it is easy to take draws, and only a limited number of draws are desired. The basic idea is to sample more draws than needed from this appoximation, and then resample from this finite sample with probability proportional to the importance ratios to obtain the final draws.

Some of the practical utility of SIR is related to the situation where given both observed data Y_{obs} and missing values Y_{mis}, the posterior distribution of the parameter, θ, is tractable. The essential idea of SIR, however, does not require this bifurcation of the unobserved random variable (Y_{mis}, θ). Rubin (1987a,b) describes the SIR algorithm for this case where the objective is to draw from the posterior distribution of Y_{mis} given Y_{obs} (or θ given Y_{obs}). The description here, in Section 2, focuses on the notationally simpler situation with a random variable ϕ having intractable distribution $p(\phi)$. The example in Section 3 obtains draws of Y_{mis} from $p(Y_{\text{mis}}|Y_{\text{obs}})$ in a case where $p(\theta|Y_{\text{mis}}, Y_{\text{obs}})$ is simple but both $p(\theta|Y_{\text{obs}})$ and $p(Y_{\text{mis}}|Y_{\text{obs}}, \theta)$ are intractable. The example in Section 4 obtains draws of θ from $p(\theta|Y_{\text{obs}})$ for logistic regression parameter θ, for the purpose of obtaining draws from the tractable $p(Y_{\text{mis}}|\theta, Y_{\text{obs}})$.

2. THE GENERAL SAMPLING/IMPORTANCE RESAMPLING ALGORITHM

The objective is to obtain m draws of ϕ from intractable $p(\phi)$.

Step 1. Obtain a decent first pass approximation to $p(\phi)$, say $h(\phi) > 0$ for all possible ϕ.

Step 2. Draw M values of ϕ at random from $h(\phi)$ where M is large relative to $m =$ the final number of draws desired for ϕ. Call these ϕ_j, $j = 1, \ldots, M$.

Step 3. Calculate the importance ratios for each ϕ_j:

$$r(\phi_j) \propto p(\phi_j)/h(\phi_j).$$

In practice, $p(\phi_j)$ will be constructed to be easy to evaluate up to a multiplicative constant.

Step 4. Draw m values of ϕ from $\phi_1, \ldots, \phi_M$ with probability proportional to $r_j = r(\phi_j)$, thereby creating m values of ϕ, $\phi_1^*, \ldots, \phi_m^*$. Methods for such drawing appear in the literature on p.p.s. sampling (e.g., Cochran, (1977), Chapter 9).

An important feature of the SIR algorithm is that it is noniterative: only one set of M values of ϕ need be drawn, and this drawing can be designed to be relatively inexpensive by the choice of $h(\phi)$.

The rationale for the SIR algorithm is based on the fact that as $M/m \to \infty$, the m values $\phi_1^*, \ldots, \phi_m^*$ are drawn with probabilities given by

$$h(\phi)\frac{r(\phi)}{\int h(\phi)r(\phi)d\phi} = \frac{p(\phi)}{\int p(\phi)d\phi} = p(\phi),$$

which implies that the $\phi_1^*, \ldots, \phi_m^*$ are independent draws from $p(\phi)$ as desired.

The choice of a practical ratio M/m to make the approximation adequate depends on the adequacy of the approximation $h(\phi)$: if $h(\phi)$ is perfect (i.e., if $h(\phi) = p(\phi)$), then $M/m = 1$ is fine; as $h(\phi)$ gets poorer M/m must increase. A sensible procedure in practice might be to use an adaptive version of SIR. For example, take $2m$ draws from $h(\phi)$ in Step 2, find the corresponding $2m$ importance ratios in Step 3, then calculate the variance of the log of the importance ratios, v, and select M as a monotone increasing function of v (if $v = 0$, no more draws are taken). The selection of an appropriate function seems to be a matter for future research, however.

The question of whether to sample with or without replacement deserves study. Some intuition suggests that if $h(\phi)$ is close to $p(\phi)$, sampling without replacement is preferable (e.g., consider the case $M = m$), whereas if $h(\phi)$ is a poor approximation to $p(\phi)$, sampling with replacement is preferable because it avoids the possible forced choosing of absurd values of ϕ (e.g., see the example in Section 4).

3. ILLUSTRATIVE EXAMPLE: MISSING VALUES IN A NORMAL/CONDITIONALLY NORMAL BIVARIATE SAMPLE

The case of a bivariate normal sample with missing values on both variables is a classic example (e.g., Wilks (1932)) of a missing-data problem without a general closed-form solution. This case, however, is easily handled by the EM, data-augmentation, and SIR algorithms. A slightly modified situation, which has no closed-form solution and cannot be directly handled by either the EM or data-augmentation algorithms, will be used to illustrate the SIR algorithm.

Specifically, let (y_1, y_2) be an i.i.d. sample from

$$y_1|\theta \sim N(\mu, \sigma^2), \tag{1}$$

$$y_2|\theta, y_1 \sim N(\alpha + \beta y_1 + \gamma y_1^2, \tau^2), \tag{2}$$

where

$$\theta = (\mu, \log\sigma, \alpha, \beta, \gamma, \log\tau) \tag{3}$$

and

$$p(\theta) \propto \text{constant.} \tag{4}$$

Suppose that a sample of n units is taken, where n_1 units have only y_1 observed, n_2 units have only y_2 observed, and n_{12} units have both y_1 and y_2 observed, where $N = n_1 + n_2 + n_{12} > n_1 + n_2 > n_2 > n_1$, and we assume that the missing data are missing at random (Rubin (1976)).

In our notation, Y_{obs} consists of the n_1 observations of y_1, the n_2 observations of y_2, and the n_{12} observations of (y_1, y_2), and Y_{mis} consists of the n_1 missing values of y_2 and the n_2 missing values of y_1. In the notation of Section 2, $\phi = (\theta, Y_{\text{mis}})$, $p(\phi) = p(\theta, Y_{\text{mis}}|Y_{\text{obs}})$,and $h(\phi) = h(\theta, Y_{\text{mis}}|Y_{\text{obs}})$. For convenience, we write the posterior distribution of ϕ as the posterior distribution of θ times the posterior distribution of Y_{mis} given θ:

$$p(\theta, Y_{\text{mis}}|Y_{\text{obs}}) = p(\theta|Y_{\text{obs}})p(Y_{\text{mis}}|Y_{\text{obs}}, \theta).$$

Step 1 of SIR then involves finding corresponding approximations to $p(\theta|Y_{\text{obs}})$ and $p(Y_{\text{mis}}| Y_{\text{obs}}, \theta)$

$$h(\theta, Y_{\text{mis}}|Y_{\text{obs}}) = h(\theta|Y_{\text{obs}})h(Y_{\text{mis}}|Y_{\text{obs}}, \theta).$$

Regarding $h(\theta|Y_{\text{obs}})$, it can be easily approximated by independent normal densities:

$$\log\sigma|Y_{\text{obs}} \sim N(\log s_1, [2(n_1 + n_{12}^{-1}]^{-1}), \tag{5}$$

$$\mu|Y_{\text{obs}} \sim N(\bar{y}_1, s_1^2/(n_1 + n_{12})), \tag{6}$$

$$\log\tau|Y_{\text{obs}} \sim N(\log\hat{\tau}, [2(n_{12} - 3)]^{-1}), \tag{7}$$

and

$$(\alpha, \beta, \gamma)|Y_{\text{obs}} \sim ((\hat{\alpha}, \hat{\beta}, \hat{\gamma}), \hat{\tau}^2 C), \tag{8}$$

where $\bar{y}_1$ and s_1^2 are the mean and variance of the $(n_1 + n_{12})$ observations of y_1, and $(\hat{\alpha}, \hat{\beta}, \hat{\gamma}, \hat{\tau}^2, C)$ are the standard least squares summaries obtained by regressing y_2 on $(1, y_1, y_1^2)$ using the n_{12} observations of (y_1, y_2). Similarly, $h(Y_{\text{mis}}|\theta, Y_{\text{obs}})$ can be easily approximated by $n_1 + n_2$ independent normal densities:

$$n_1 \text{ missing } y_2|\theta, Y_{\text{obs}} \overset{\text{ind}}{\sim} N(\hat{\alpha} + \hat{\beta} y_1 + \hat{\gamma} y_1^2, \hat{\tau}^2), \tag{9}$$

$$n_2 \text{ missing } y_1|\theta, Y_{\text{obs}} \overset{\text{ind}}{\sim} N(a + b y_2 + c y_2^2, s_e^2), \tag{10}$$

where a, b, c, s_e^2 are the standard least squares summaries obtained by regressing y_1 on $(1, y_2, y_2^2)$ using the n_{12} observations of (y_1, y_2). Thus $h(\theta, Y_{\text{mis}}|Y_{\text{obs}})$ is the product of the $(n_1 + n_2 + 3)$ univariate normal densities specified by (5), (6), (7), (9), and (10), and the trivariate normal density specified by (8), and is therefore easy to draw from at Step 2, and easy to evaluate as the denominator of the importance ratios, $r(\phi) = r(\theta, Y_{\text{mis}})$, in Step 3. Better approximations are available, especially for small n_{12}, but it is not clear whether they are worth the effort to develop in the context of SIR relative to increasing the ratio M/m.

The numerator of the importance ratios, $p(\phi) = p(\theta, Y_{\text{mis}}|Y_{\text{obs}})$, is also easy to evaluate up to a multiplicative constant, which is all that is needed. First note that

$$p(\theta, Y_{\text{mis}}|Y_{\text{obs}}) = p(Y_{\text{mis}}, Y_{\text{obs}}|\theta)p(\theta)/p(Y_{\text{obs}}),$$

which is proportional to $p(Y_{\text{mis}}, Y_{\text{obs}}|\theta)$ since (a) $p(\theta)$ is constant by (3) and (4), and (b) $p(Y_{\text{obs}})$ is constant since Y_{obs} is fully observed. But $p(Y_{\text{mis}}, Y_{\text{obs}}|\theta)$ is simply the product over the n units of the two normal densities implied by (1) and (2). That is, the numerator of the importance ratios is simply these products of (1) and (2) evaluated for the observed value of Y_{obs} and the drawn values of (Y_{mis}, θ).

4. CENSUS PUBLIC-USE TAPES AND MISSING OCCUPATIONAL CODES

In 1980, the U.S. Census Bureau substantially modified its coding of the descriptions of the occupations held by individuals. An important consequence of this change is that public-use tapes from the 1980 Census do not have occupational codings directly comparable to those used on public-use tapes from previous censuses, in particular, 1970 public-use tapes.

The lack of a common occupational code is considered by many economists and sociologists to be a very serious problem. If pre-1980 public-use tapes have one coding and the 1980 tapes have another coding, it will be very difficult to study such topics as occupational mobility and labor force shifts by demographic characteristics. Specific questions that would be difficult to address without 1980 codes on 1970 public-use tapes concern 1970–1980 shifts in occupational status of jobs held by males and females or by whites and nonwhites. Such questions are important because they address the issue of "equal opportunity" employment. Consequently, there is a need to supplement the 1970 codes on 1970 public-use tapes with 1980 codes.

Estimated costs for double-coding 1970 public-use tapes (i.e., for recoding all units on 1970 public-use tapes according to the 1980 new occupational coding system) are in the millions of dollars. There is available, however, an existing double-coded sample of 120,000 units from the 1970 Census, that is, with both 1970 and 1980 codes for all 120,000 units. Consequently, we can think of the lack of both codes for all units on 1970 public-use tapes as an enormous missing data problem. For 120,000 units in the 1970 Census public-use tapes, both 1970 and 1980 occupational codes are observed, whereas for all remaining units, only the 1970 code is observed. Since the public-use tapes are more than 10 times the size of the double-coded-sample, there is, at least in a naïve sense, 90% of the data missing.

Logistic regression is being used by the Census Bureau to multiply-impute the missing 1980 occupational codes. Specifically, first a logistic regression model is built from the double-coded sample to predict 1980 occupational code from 1970 occupational code and other predictors such as gender and age. Then this model is applied to the units with predictors observed, but 1980 occupational code missing in order to impute five possible occupations. The imputation is done by first drawing five sets of logistic regression coefficients from their posterior distributions and then independently drawing 1980 occupation codes for the units from their conditional posterior distributions given the drawn values of the regression coefficients.

In fact, independent logistic regressions are performed for each 1970 occupation, using a sequence of independent dichotomous logistic regressions within each 1970 occupation. Each such dichotomous logistic regression uses a convenience prior distribution on the regression coefficients, θ, that was proposed in Rubin (1983) and is studied in a very simple case in Rubin and Schenker (1987). Specifically, suppose θ is p-dimensional and the predictor variables form a contingency table with c cells. Then p/c prior observations are added to each cell, divided between the two 1980 codes according to the marginal frequency of these codes in the sample. The logistic regression are then fitted by standard maximum likelihood techniques to the data supplemented with the prior observations. Using this prior distribution guarantees that the posterior distribution of θ is unimodal and pulls the non-intercept logistic regression coefficients toward zero. It is shown in Rubin and Schenker (1987) that the p/c prior assigns the same average prior variance to the cell logits regardless of the design and model.

For many of the 1970 codes, the samples are small and the distributions into 1980 codes quite skew, resulting in very nonnormal posterior distributions for the logistic regression coefficients. In the worst example, 1970 code "317", there are 1073 units: 1071 have 1980 code "142" and 2 have 1980 code "132". Since the objective of the modelling effort is to impute five values of 1980 code for units with no 1980 code, a reasonable check on the accuracy of the implemented procedure is to impute five 1980 codes for each unit on the double-coded sample, as if the 1980 codes were missing, and compare the results to the actual 1980 codes.

Counts for males

	actual	Imp. 1	Imp. 2	Imp. 3	Imp. 4	Imp. 5
Industry						
142	584	556	581	81	201	561
132	1	29	4	504	384	24

Counts for females

	actual	Imp. 1	Imp. 2	Imp. 3	Imp. 4	Imp. 5
Industry						
142	457	229	449	171	208	364
132	1	128	9	287	250	94

Table 1. *Number of units with imputed 1980 code =142 vs. 132 using standard large sample normal approximation for the posterior distribution of the logistic regression coefficient.*

Table 1 gives the results of five imputations drawn from the posterior distribution of the 1980 codes using the standard "mode/second derivative of log posterior at the mode" normal approximation to the posterior distribution of θ. Notice the absurd results of most of the imputations. Some frequentist statisticians involved with the project at the Census Bureau were convinced that these results were due to the interjection of subjective Bayesian methods and pushed for "tried and true" methods such as step-wise techniques. In fact, however, the bizarre results are due solely to the gross nonnormality of the posterior distribution of θ.

The SIR algorithm was proposed to help correct this situation. The Census Bureau was willing to give SIR a try before retreating to step-wise-defined zero components of θ (a very poor idea). SIR worked extremely well in all but the worst (i.e., code 317) case, and even then did *much* better than the standard normal approximation using a modest 20 to 1 ratio of draws. In particular, $M = 100$ values of θ were drawn from the normal approximation to yield $M/m = 20$, a modest ratio indeed for such an exceedingly bad initial approximation. In order to demonstrate how bad this approximation is, the 10 largest of the 100 importance ratios (relative to the largest such ratio) were in one typical run:

$$1.0,\ 0.05,\ 0.2 \times 10^{-5},\ 0.7 \times 10^{-8},\ 0.1 \times 10^{-9},\ 0.4 \times 10^{-10},$$

$$0.8 \times 10^{-13},\ 0.4 \times 10^{-13},\ 0.6 \times 10^{-14}, 0.2 \times 10^{-14}, \ldots$$

When SIR was employed with approximate p.p.s. sampling without replacement, the values of θ that were drawn had the five largest importance ratios. The associated multiple imputations are summarized in Table 2. Clearly the imputations summarized in Table 2 provide a much more realistic data base than those of Table 1. When SIR was employed using approximate p.p.s. sampling with replacement, four of five times the value of θ with the largest importance ratio was chosen, and once the value of θ with the second largest importance ratio was chosen. The resultant imputations are summarized in Table 3. These results appear to be slightly preferable to those in Table 2.

Counts for males

Industry	actual	Imp. 1	Imp. 2	Imp. 3	Imp. 4	Imp. 5
142	584	581	580	584	584	578
132	1	4	5	1	1	7

Counts for females

Industry	actual	Imp. 1	Imp. 2	Imp. 3	Imp. 4	Imp. 5
142	457	457	456	458	448	456
132	1	1	2	0	10	2

Table 2. *Number of units with imputed 1980 code =142 vs. 132 using SIR with 20 to 1 ratio and p.p.s. sampling without replacement.*

Counts for males

Industry	actual	Imp. 1	Imp. 2	Imp. 3	Imp. 4	Imp. 5
142	584	584	580	579	580	582
132	1	1	5	6	5	3

Counts for females

Industry	actual	Imp. 1	Imp. 2	Imp. 3	Imp. 4	Imp. 5
142	457	457	456	458	455	456
132	1	1	2	0	3	2

Table 3. *Number of units with imputed 1980 code =142 vs. 132 using SIR with 20 to 1 ratio and p.p.s. sampling with replacement.*

Of course, in situations with as bad an initial approximation as this one, SIR by itself is not the complete answer. Several avenues are being pursued for obtaining a decent initial approximation to be used in conjunction with SIR in this problem. For example, it appears that an approximation that carefully fits the "mean" coefficient (i.e., the linear function of θ corresponding to the mean logit in the sample) as nonnormal and then the other coefficients as multivariate normal given the mean coefficient has real promise —a small skew sample implies a skew posterior distribution for the marginal proportion and relatively prior-like posterior distributions conditionally given this marginal count.

REFERENCES

Cochran, W. G. (1977). *Sampling Techniques.* New York: John Wiley.

Rubin, D. B. (1976). Inference and missing data. *Biometrika* **63**, 581–592.

Rubin, D. B. (1978). Multiple imputations in sample surveys - a phenomenological Bayesian approach to nonresponse. *Proceedings of the Survey Research Methods Section of the American Statistical Association,* 20–30. *Imputation and Editing of Faulty or Missing Survey Data.* U.S.: Dept. of Commerce, Bureau of the Census, 1–23.

Rubin, D. B. (1983). Progress report on project for multiple imputation of 1980 codes. U.S.: Census Bureau, NSF and SSRC.

Rubin, D. B. (1987a). *Multiple Imputation for Nonresponse in Surveys.* New York: John Wiley.

Rubin, D. B. (1987b). The SIR Algorithm - A discussion of Tanner and Wong's: The Calculation of Posterior Distributions by Data Augmentation. *J. Amer. Statist. Assoc.* **82**, 543–546.

Rubin, D. B. and Schenker, N. (1987). Logit-based interval estimation for binomial data using the Jeffreys prior. *Sociological Methodology,* 131–144.

Tanner, M. A. and Wong, W. W. (1987). The calculation of posterior distributions by data augmentation. *J. Amer. Statist. Assoc.* **82**, 528–540.

Wilks, S. S. (1932). Moments and distributions of estimates of population parameters from fragmentary samples. *Ann. Math. Statist.* 163–196.

DISCUSSION

I. R. DUNSMORE (*University of Sheffield, UK*)

A discussant usually does two or three things: (i) congratulates the author on a good and interesting paper, (ii) perhaps says some harsh things about the paper, and (iii) then talks about his own work (which may hopefully be related).

I can undertake the first task, and do so willingly. With regard to the third task I have not done anything at all related to this work. So what harsh things can I say about the paper from a position of total ignorance?

When I read the paper the answer presented itself. The paper claims to provide a method of finding things out about intractable situations. We have a $p(\theta, Y_{\text{mis}}|Y_{\text{obs}})$ in which we have an unknown parameter θ (what the paper is all about); some observable data Y_{obs} (in my case the paper only); and a lot of missing data Y_{mis} (papers to appear or in inaccessible sources, plus today's presentation). Of course $p(\theta|Y_{\text{mis}}, Y_{\text{obs}})$ would be perfectly tractable. So can SIR help? What does SIR mean?

I was somewhat alarmed by the extensive claims for the method. Could it really be ubiquitously applicable? Is it sensible/possible to impute 90% missing data? My initial reaction has to be that SIR is a touch of Self Indulgence, Rubin.

Consider Step 1. We need to obtain a decent first pass approximation to $p(\phi)$ —say $h(\phi)$. How do we know if it is decent? The reason for the method is that $p(\phi)$ is intractable; and presumably numerical integration is out of the question. Later we find that the choice of M/m depends on the goodness of the approximation. Will we always need $M \gg m$? It seems that SIR (Subsequent Investigative Research) is necessary here.

In Step 3 we require that $p(\phi)$ be easy to evaluate up to a multiplicative constant. My initial reaction here was that this was a considerable restriction. However it clearly works well for straight posteriors $p(\theta|Y_{\text{obs}})$ —in fact SIR is a Sticky Integration Routine— and also for the $p(\theta, Y_{\text{mis}}|Y_{\text{obs}})$ case in which the conditioning argument is very neat. Presumably, though, the assumption that Y_{mis} are missing at random is crucial.

Turning to Example 1, I was somewhat concerned with some of the assumptions made in the approximations $h(\theta, Y_{\text{mis}}|Y_{\text{obs}}) = h(\theta|Y_{\text{obs}})\ h(Y_{\text{mis}}|Y_{\text{obs}}, \theta)$. For example $h(\theta|Y_{\text{obs}})$ ignores the n_2 values with missing y_1. What if n_2 is relatively large? (The paper assumes $n_2 > n_1$.) More worrying is the estimation of the contributions to $h(Y_{\text{mis}}|Y_{\text{obs}}, \theta)$ for the n_2 missing y_1 values by i.i.d. $N(a + by_2 + cy_2^2, s_e^2)$ distributions. Firstly these cannot be normal, secondly the means are surely not quadratic, and thirdly is independence valid? The author says that "Better approximations are available ... but it is not clear whether they are worth the effort to develop in the context of SIR relative to increasing the ratio M/m?" Is SIR a recipe to Simply Input Rubbish?

The thought of imputing Y_{mis} values led me naturally to consider $p(Y_{\text{mis}}|Y_{\text{obs}})$ in the prediction context and apply SIR (Seymour's Instant Remedy). This involves considering $\phi = (\theta, Y)$, where Y now refers to a future observation and θ is a nuisance parameter. Interest lies in the marginal

$$p(y|\text{data}) = \int p(y|\theta)p(\theta|\text{data})d\theta.$$

Many are the situations where $p(y|\text{data})$ is awkward or intractable. If we are hoping to use SIR to get some picture of $p(y|\text{data})$ rather than imputing a few isolated values, disappointingly we reach a stumbling block at Step 3, since $p(y|\text{data})$ is not known up to a multiplicative constant. I have not yet been able to work out a way forward. I wonder if it would be possible to incorporate two levels of approximation in SIR —a rough first one and then a better second one— and somehow base the important ratios on these.

So how has SIR helped? My Y_{obs} now includes the presentation today. This has removed some of my worries and explained some of my difficulties. Perhaps after all SIR is Subtle in Retrospect, and that one concludes that SIR is a Sensible Investigation, Rubin.

REPLY TO THE DISCUSSION

First, I thank Ian Dunsmore for his highly refreshing general comments.

Second, regarding his specific comment concerning the possible difficulty of imputing predictions via SIR, there is no need for any concern whenever the likelihood of all data (missing and observed) can be evaluated. This fact was illustrated in Rubin (1987b, Section 3) and here in Section 3, but the generality of the idea may be difficult to see from the example.

In general, first draw M repetitions of missing (e.g., future) observations, Y_{mis}, and parameters, θ, from a convenient approximation, $h(\theta, Y_{\text{mis}}|Y_{\text{obs}})$. Second, evaluate the M importance ratios, which are proportional to the likelihood of all data, missing and observed, $p(Y_{\text{mis}}|Y_{\text{obs}}\theta)$, times the prior on θ, $p(\theta)$, divided by $h(\theta, Y_{\text{mis}}|Y_{\text{obs}})$. And third, take m draws of (θ, Y_{mis}) from the M with probability proportional to these importance ratios. Saving the m drawn values of θ and discarding the associated values of Y_{mis} (as in Section 3) yields m approximate draws from the posterior distribution of θ, whereas saving the m drawn values of Y_{mis} and discarding the associated values of θ yields m approximate draws from the posterior predictive distribution of Y_{mis} (e.g., future observations).

AYESIAN STATISTICS 3, pp. 403–410
M. Bernardo, M. H. DeGroot, D. V. Lindley and A. F. M. Smith, (Eds.)
Oxford University Press, 1988

Robustness in Generalized Ridge Regression and Related Topics

H. RUBIN
Purdue University

SUMMARY

We start out by considering the formal Bayesian version of generalized ridge regression, namely, we wish to make inferences concerning the mean of a multivariate normal distribution with normal prior, or to decide between normal priors. For estimation, if we restrict ourselves to linear procedures, the assumption of normality becomes unimportant, and in many situations the normality assumption will not be too important in practice. We will also discuss to some extent the importance of the normality assumption, both in the prior and in the distribution of the "errors". Our purpose in this paper is to attack the problem of the sensitivity to assumptions, mainly the assumptions made about the prior covariance matrix. In addition, we present some empirical Bayes type procedures, which are good in some cases. We also present some problems which are not of a regression type for which the general results apply.

eywords: ESTIMATION; ROBUST; REGRESSION; NON-PARAMETRIC; INFINITE DIMENSIONAL.

1. INTRODUCTION

Ve consider the problem of estimating the regression parameter from the Bayesian viewpoint. 'his problem was explicitly discussed by many authors, for example, by the present author ı Rubin (1974). (At the same time, the notion of ridge regression was introduced in Hoerl nd Kennard (1970). The formal similarity between ridge regression and a special case of ıe problem under discussion was easily seen, for example, in Goldstein and Smith (1974). ince ridge regression was applied to low dimensional problems where the major problem ⁄as multicollinearity, the results are not particularly sensitive to the form of the prior, but are ensitive to its scale, the "ridge prior", a multiple of the identity matrix for the appropriate caling, may be appropriate. This is, in general, *not quite* a Bayes procedure, and does ot work in the problems we are considering.) We will be concerned with the problem of ıference, *not* with the problem of prediction.

We wish to study robustness for large samples for problems with a large or even infinite umber of dimensions. The asymptotics for infinite dimensional problems can differ greatly rom that of the usual fixed dimensional versions. There are two good reasons for taking the ıfinite dimensional viewpoint; large finite dimensional problems are more likely to look like ıfinite dimensional problems than low dimensional problems for reasonable sample sizes, and ıere are many problems, including most of those for which the misnamed "nonparametric" rocedures are used, which are truly infinite dimensional.

The basic regression problem is inference on the location parameter of a distribution. ıfter the usual reduction, this means that we have a vector θ of parameters, and we observe vector $y = \theta + x$. We further need that x is independent of, or at least uncorrelated with, to avoid identification problems. Of course, other assumptions can be made, but would ou call inference on the sample size of a binomial distribution a regression problem? In the lassical regression problem, the Gauss-Markoff Theorem is the basic robustness theorem. The rguments made in that theorem require modification for Bayesian or even decision-theoretic purposes. If we keep linearity but drop unbiasedness, which we must in high dimension cases to get anything remotely reasonable, we obtain the usual Bayes procedures describe below. Accordingly, let us act in this manner.

Let us therefore make the assumption that θ is normal $(0, \mathrm{T})$, x is normal $(0, \Sigma)$, and and x are independent. We wish to investigate the effect of misspecification of the probler on the Bayes risk of the resulting procedures for estimation and testing. This is the gener problem of robustness. Other than observing that the central limit theorem can be used t justify normal errors, we consider only the problem of misspecification of the prior. This pape should be considered only as a first step in the investigation; the problem is quite complex.

There are some questions on the relevance of this paper. First, we are operating under th assumption of normality, and it is well known (?) that nothing is normal. If we look only the estimation problem under quadratic loss and consider only linear estimators, the question i irrelevant; there are other reasons, which we will point out later, which indicate that it migl not be too important in many situations. However, if we wish to use an "Occam's razor prior, or something similar, and there is a complexity cost for the estimate, i.e, estimating coordinate of θ to be non-zero increases the cost, we suspect that the situation may be muc worse. This problem is also discussed in Rubin (1974).

Second, a Bayesian specifies his prior and loss, so where is the problem? Even if w could assume that the "error" x is exactly normal, only a rash Bayesian would be that sure c his prior. The "ordinary" ridge regression prior can find justification only from a non-Bayesia viewpoint, and this does not work in the high dimensional cases, as we will demonstrate. Th prior chosen will certainly be wrong; which errors are the most important, and which ca essentially be neglected?

Third, let us consider the axiomatic Bayesian (rational decision theorist). He know that coherence requires a Bayesian procedure, but also requires an infinitely fast compute operating at zero cost. Not having one available, approximate solutions are needed; this lead to the same problem as before.

Finally, take the user of statistical procedures. We are discussing problems where Rubin' first commandment, "Thou shalt know that thou must make assumptions," cannot be ignorec We must inform our clients about which are the important assumptions.

Since the prior assumptions matter, one may ask whether non-linear techniques, such a "empirical Bayes" type procedures, can improve the situation. The answer here is decidedl yes, but we have been able to obtain such procedures only in special cases. One thing t avoid is the reckless use of the hyperparameter approach. A discussion of a special problen of Bayesian testing in the infinite dimensional normal case can be found in Cohen (1972). Th results can be extendend to other situations, such as hyperparameter inference. It is possible and indeed rather likely, for a few coordinates for which the prior variance is large to dominat a Bayesian hyperparameter inference; however, these coordinates do not appreciably affec the procedure. For the one class of problems for which we believe we have a reasonabl empirical Bayes procedure, we do not take that approach.

2. ROBUSTNESS OF BAYES PROCEDURES

Let us suppose, as in the Introduction, that the random variable y is normally distribute with mean θ and covariance matrix Σ, and that the parameter θ is normal with mean 0 an covariance matrix T. Then it is well known that if the loss of the estimate t of θ is $(t-\theta)'A(t-\theta)$ the Bayes estimator $\hat{\theta}$ of θ is $\mathrm{T}(\Sigma + \mathrm{T})^{-1}x$, and its Bayes risk is $\mathrm{tr}(A\Sigma(\Sigma + \mathrm{T})^{-1}\mathrm{T})$ It is an easily proved theorem that the matrix multiplying A is unchanged when Σ and T are interchanged, and hence is symmetric. (The problem is essentially unchanged if A i not present, but many problems are more easily formulated with it, and it adds no essentia complications.)

Now suppose that the situation is as above, but that we incorrectly use Υ instead of T in computing the estimator. Using the matrix identity $R^{-1} - S^{-1} = R^{-1}(S - R)S^{-1}$, we find that the risk is now increased by

$$\text{tr}A\Sigma(\Sigma + \Upsilon)^{-1}(\mathrm{T} - \Upsilon)(\Sigma + \mathrm{T})^{-1}(\mathrm{T} - \Upsilon)(\Sigma + \Upsilon)^{-1}\Sigma.$$

Note that this expression is *not* symmetric in T and Υ. This is not surprising. This asymmetry is apparent even in the one-dimensional case. In that case, the Bayes procedure corresponds to multiplying the observation y by $\lambda = \tau^2/(\sigma^2 + \tau^2)$. If we wish to allow the risk to increase by a factor of c, we may increase or decrease λ by $\sqrt{(c-1)\lambda(1-\lambda)}$. For λ near 0, this interval includes 0, and for λ near 1, it includes 1. The graph in the appendix shows the range by which the multiplier can vary for a given risk factor. In fact, if we set $c = 2$, we can use 0 for $\lambda \leq 0.5$ and we can use 1 for $\lambda \geq 0.5$. That is, we at most double the Bayes risk if we ignore the less concentrated of the data and the prior. This fact was used by the author to obtain preliminary results quickly.

The asymmetry is quite pronounced. In the case of large variance of the prior, even for one coordinate using too low a variance can be catastrophic. For example, if $\sigma = 1$ and we assume $\tau = 10$, the risk is 99.01 for squared error loss if in fact $\tau = 1000$, while the Bayes risk is 0.999999. However, if we assume $\tau = \infty$, we obtain a risk of 1, while the Bayes risk for $\tau = 10$ is 0.990099. Now this problem can be avoided (see Rubin (1977)) by using a non-linear estimate, and several non-normal priors are considered. However, if the prior is concentrated and does not have a variance, not much can be done about robustness. In the case of a small variance, if we assume $\tau = 0$, the error committed is small. Having a moderate error in one or a few coordinates is not too important in the case of low prior variance, but for some priors considered here, the contribution to the total Bayes risk by these coordinates is an appreciable portion of the total, and using too large prior values for τ can be important. Using too small values has little effect on the risk. Certainly little is lost by using 0 if τ is small and using ∞ if it is large.

Since using non-normal priors gives good results for large values of τ even if τ is underestimated, can this method be useful for small values of τ? Unfortunately, the answer is no. In the infinite dimensional problem, there may be millions of coordinates with $10^{-6} < \tau < 10^{-3}$. For these coordinates, the observations will not be much different than if τ were 0, and any attempt to guard against tails in the prior founders badly if there are moderate tails, such as t with 20 degrees of freedom or even logistic, in the distribution of the "errors" x. This problem also certainly occurs if the error variance must be estimated.

3. SPECIFIC RESULTS

One of the problems leading to this work is the attempt to obtain an approximate Bayes estimator for densities, suggested by the model in Chen and Rubin (1986). The problem is somewhat more complicated than this, but it suggests the problem of estimating the mean θ of an infinite dimensional vector in the special case in which the covariance matrix for a single observation is the identity and the sum of the squares of the elements of θ is 1 should be a good starting point for the solution. The method for inference on the "tail" of θ will be close to what should be done in the real problem. Now in this case the prior for θ certainly cannot be normal with mean 0, and the observations certainly are not normal; however, the central limit theorem tells us that the "errors" are approximately normal, and estimates of the early coordinates of θ will not be particularly affected by the normality assumptions. We have seen that the late ones will not be particularly affected either, since we may as well estimate them to be 0. Thus the effect is mainly in the middle. However, unless there is a considerable amount of non-normal dependency in the middle coordinates, either there are few coordinates, so the contribution is not large, or there are many, and again a central limit type effect occurs.

Now what happens if we use the classical ridge approach? The ridge approach does nc work at all! If we have an n-dimensional problem, the ridge approach says to estimate th vector θ by maximum likelihood given the length. If we have sample size N, the method ca give reasonable results only if N is considerably greater than n, since it uses a prior covarianc matrix proportional to the identity. In other words, in a problem with many dimensions, w *must* have non-trivial prior input.

For simplicity in the following, we have assumed that the covariance matrix of x fc each observation is the identity and that the covariance matrix of θ is diagonal. Furthermor let us assume that the loss is the sum of squares of the errors of the estimate. In some case for the density estimation problem, this corresponds to the Hellinger-Kakutani distance. W will discuss both the good news and the bad news.

First, the bad news. The first problem we have looked at is that for which the k-t diagonal element of T, τ_{kk}, is either k^{-2} or 2^{-k}. Now if k^{-2} is the case, the Bayes risk i on the order of $N^{-1/2}$, but if we use instead 2^{-k} it is only on the order of $1/\log N$; while i the true prior is 2^{-k} the risk is only on the order of $(\log N)/N$, whereas assuming k^{-2} pu the risk up to the order of $N^{-1/2}$. We include a table of the comparison in the appendix.

Knowing the set of values of the diagonal elements of T is also not enough. For le $k = (2r-1)2^{s-1}, r, s = 1, \ldots$. and suppose we use, as before, k^{-2}. If we interchange r an s, we get the same diagonal elements permuted. The effect on the risk is easily computed t be enormous.

Now the good news. Suppose we know the ordering of the diagonal elements. We kno that, for a sample of size N, the expected value of y_k^2 is $\tau_{kk} + \frac{1}{N}$. While we cannot expec to estimate τ_{kk} very well from the value of y_k, we can use the well-known isotonic estimatc to obtain better results. There are places where the results will not be too good. These occu for k small and where there is a rapid transition in τ_{kk}. A less critical problem occurs for large. The small k problem is unlikely to be important, especially if we robustify by usin ∞ for large values of the ratio of τ to σ. The rapid transitions can only be a real probler if the risk is greatly affected; since this is important only where the ratios are of the sam order of magnitude as 1, not very many terms can be affected. Furthermore, this can only b a major problem if a large coordinate of θ is classified as small. For this to happen, severa previous coordinates of y have to be small, which means that it is unlikely that we do not d better overall by shrinkage. The other problem is the long tail problem for k large. Agair if we robustify by using 0 for small values of the ratio, this problem disappears. Simulatio studies are being undertaken to ascertain quantitatively the penalty due to ignorance of th actual covariance matrix.

We can handle cases like $k = (2r-1)2^{s-1}, r, s = 1, \ldots$ by using bivariate isotoni estimation if we know the structure. Clearly the results will not be as good. Again, if th structure becomes so complicated that the isotonic estimates are poor, the news is bad. If w have a somewhat incorrect idea of the structure, the results are likely to be at most fair.

The observation that using only the data or the prior, whichever variance is smalle increases the risk by at most a factor of 2, suggests that in the monotone case the procedure t use the classical estimate for those coordinates of θ for which we estimate the data variance t be smaller than the prior variance, and to estimate the remaining coordinates to be 0, shoul not be too bad. In cases of smoothly dropping prior variances like k^{-2}, a moderate relativ increase in the risk can be expected, since if we know the prior variances, the risk is increase by a factor strictly greater than 1. If the prior variances decrease like 2^{-k}, the increase in th risk will correspond to a fixed number of coordinates. Some of the procedures in the literature such as some spectral density estimators do this; however, estimators considered better becaus they use "smoother" weight functions do not always have this robustness property!

4. CONCLUSIONS

The high or infinite dimensional regression problem in the Bayesian setting has some easily established robustness properties, but also can provide problems from the standpoint of robustness which we have not been able to treat. This does not mean that they can not be handled, merely that this author has not been able to find the right approach. Some of them may be very difficult.

We have only considered sums of squares of the errors. Other quadratic forms can be much easier; for example, if Σ is the infinite dimensional identity matrix, the loss is $(t-\theta)'A(t-\theta)$, and $\mathrm{tr}A < \infty$, we always have robustness for any reasonable prior, provided reasonable care is used to avoid estimating a coordinate as too small. Here the use of non-normal priors may be good, as the occasional extra-large estimate of a small coordinate of θ will be offset by its small effect on the risk.

Another thing to keep in mind is that we want to consider what happens for large but not enormous sample sizes. Most of what we have done falls into this category. Some of these procedures are reasonably good even for relatively small samples.

Note that, in some situations, relatively simple procedures have good properties, and that procedures which one would think to be better are shown to be worse. Many statistical problems are high or even infinite dimensional; for such problems prior assumptions are very important, and careful mathematical reasoning is needed to establish which assumptions are important and which are not.

ACKNOWLEDGMENTS

I would like to thank Prem Goel and Mark Berliner for their comments and discussion while the research was being done; Mark Berliner's questions were especially valuable.

APPENDIX

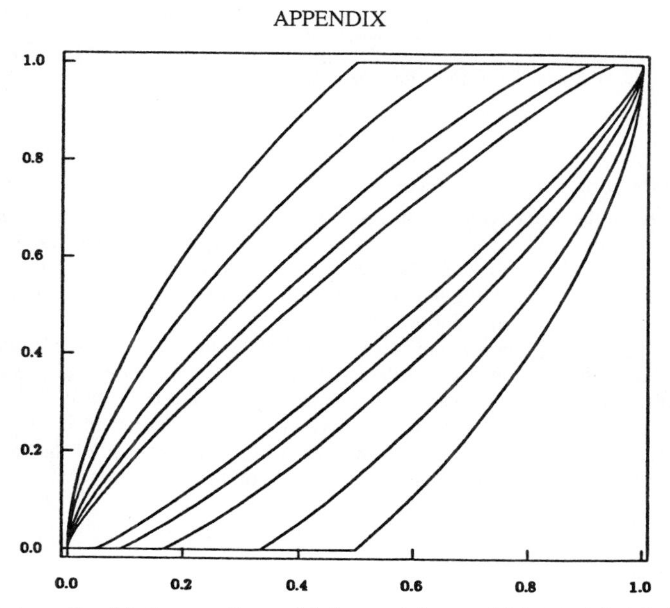

Contours for risk (incorrect) = c risk (correct) for c = 1.05, 1.1, 1.2, 1.5, 2.

Table of the risk in standard units for (a) using the smaller of the error variance and the prior variance for the correct distribution of the parameter; (b) the correct Bayes risk; (c) the risk if the wrong prior is used; (d) the risk if the procedure using the smaller of the error variance and the wrong prior variance is used. The approach to the asymptotic expression is apparent from these numbers.

Sample size	true state	true 0 or 1	true Bayes	wrong Bayes	wrong 0 or 1
1000	2^{-k}	10.953	9.468	24.701	31.000
1000	k^{-2}	62.746	49.174	99.003	114.169
10000	2^{-k}	14.221	12.788	78.748	100.000
10000	k^{-2}	199.502	156.580	705.093	753.404
100000	2^{-k}	17.526	16.110	250.108	316.000
100000	k^{-2}	631.956	496.229	5633.969	6074.754
1000000	2^{-k}	20.907	19.432	792.001	1000.000
1000000	k^{-2}	1999.500	1570.296	47281.971	51289.823

REFERENCES

Chen, J. and Rubin, H. (1986). Drawing a random sample from a density selected at random. *Computational Statistics and Data Analysis* **4**, 219–27.

Cohen, P. L. (1972). Bayes risk for the test of location – the infinite dimensional case. Unpublished dissertation, Purdue University.

Goldstein, M. and Smith, A. F. M. (1974). Ridge-type estimators for regression analysis. *J. Roy. Statist. Soc. B* **36**, 284–291.

Hoerl, A. E. and Kennard, R. W. (1970). Ridge regression: biased estimation for non orthogonal problems. *Technometrics* **12**, 55–67.

Rubin, H. (1974). Decision-theoretic approach to some multivariate problems. *Multivariate Analysis II*, (P. R. Krishnaiah, ed.). New York: Academic Press.

Rubin, H. (1977). Robust Bayesian estimation. *Statistical Decision Theory and Related Topics II*, (S. S. Gupta and D. H. Moore, eds.). New York: Academic Press.

DISCUSSION

P. J. BROWN (*University of Liverpool*)

When I received the written version of this paper I was reminded of the great mathematician E. Schroedinger describing his introduction to quantum theory. The first time he read about it he didn't understand very much. The second time he understood most of it, but there were still one or two things that didn't quite fit. The third time he read it, he decided he understood nothing! But he realised that it did not matter!

Professor Rubin's paper is mainly concerned with the simple normal model with a normal prior distribution for the mean parameter. Arguments revolve around "misspecification" of the prior distribution. The one parameter problem encapsulates the essence of the argument. Suppose Y is normal with mean θ and known variance σ^2 conditional on θ. Then θ is normal with mean zero and variance τ^2. The question posed is what if instead of the "true" variance τ^2 the investigator assumes the variance γ^2. The person using the "true" τ^2 would apply the Bayes estimator

$$[\tau^2/(\sigma^2+\tau^2)]y \tag{1.1}$$

and the Bayes risk with quadratic loss is

$$1/[(1/\sigma^2)+(1/\tau^2)] \tag{1.2}$$

whereas if γ^2 had been assumed $\tau^2 \to \gamma^2$ in (1.1), the estimator is

$$[\gamma^2/(\sigma^2+\gamma^2)]y \tag{1.3}$$

and the risk, formed by averaging the quadratic with respect to the true distribution is

$$[\gamma^2/(\sigma^2+\gamma^2)]^2\sigma^2+[\sigma^2/(\sigma^2+\gamma^2)]^2\tau^2 \tag{1.4}$$

The difference between (1.4) and (1.2) is, for $A = 1$ and many parameters given by the author's useful formula on his third page. We can see quite clearly from (1.1) and (1.3) that if γ^2 is a lot smaller than τ^2, there is too much shrinkage. Thus if $\sigma^2 = 1$, $\tau = 1000$, the Bayes risk (1.2) is 0.999 whereas if $\gamma = 10$ is assumed the risk (1.4) is 99.01.

In another example the author considers $\tau^2 = k^{-2}$, $\gamma^2 = 2^{-k}$. Here again for large $k, \gamma^2 \ll \tau^2$ since $\tau^2/\gamma^2 = 2^k k^{-2} \to \infty$ as $k \to \infty$. The risk differences accumulate when there are many parameters, $\theta_k, k = 1, \ldots, \infty$. It is here that my difficulties mount. With an infinite number of parameters in a linear model context is it reasonable to impose

(i) equal data variance?

(ii) an equal loss status on the parameters?

The author's title includes reference to regression. Let me accordingly supply the missing regression model

$$Y = X\theta + \in \tag{1.5}$$

with Y an $N \times 1$ response vector, X an $(N \times n)$ matrix of explanatory variables, $\theta(n \times 1)$ vector of parameters and $\in$ an $N \times 1$ error vector. In saying we have this regression model, we are subsuming rich strategies of model formulation. At the very least we are able to rescale the individual explanatory variables so as, for example, to induce equal variance on the components of θ. A subsequent canonical reduction, $X = Q\Lambda^{\frac{1}{2}}P$ where Q, P are orthogonal induces

$$Z_i = \sqrt{\lambda_i}\alpha_i + \in_i \qquad i = 1, \ldots n$$

with $\lambda_1 \geq \lambda_2 \geq \cdots \geq \lambda_n$ the eigenvalues of $X'X$, $Z = Q'Y$, $\alpha = P\theta$. The weight matrix, A, of quadratic loss, transforms as a consequence into PAP'. The point now is that inevitably for finite N, later eigenvalues must be smaller than earlier ones. Some will be zero when $n > N$. The lack of symmetry is crucial and is where ridge regression may score. Indeed, it will approximate to Professor Rubin's $0 - 1$ rule by shrinking to zero those components whose data variance is larger than the prior variance. To criticise ridge regression in its many variations in an entirely symmetric context is to misjudge it.

The second point mentioned above refers to the unrealistic equal loss status of the parameters. In a linear model, with possibly an infinite number of parameters, interest cannot focus on all the parameters in their own right. Of paramount importance must be prediction of the response Y given the explanatory variables. The loss function and its matrix, A, should reflect this. Concentration on prediction in the presence of an infinity of explanatory variables is explicit in Dawid (1988) and also in a particular class of ordered prior assumptions given in Mäkeläinen and Brown (1988), both published in these Proceedings. Here the motivation for the prior distributions is also in terms of observables and structural assumptions and competing prior distributions are within a coherent framework.

W. POLASEK (*University of Basel*)

Ridge regression estimates can be described in a robust Bayesian framework by the ridge curve = information contract curve (Leamer (1978)). Tracing out the risk function as function of the (unknown) ridge parameter in a multivariate ridge regression set-up (Brown and Zidek (1980)) is only a one dimensional summary. For a multivariate sensitivity analysis this seems to be a little bit thin; a combination with other sensitivity (or robust Bayesian) approaches seem to be advisable.

REPLY TO THE DISCUSSION

As I remarked in the introduction, the classical ridge approach is not exactly a Bayes procedure, and while it may be a reasonably robust approximation to a Bayes procedure in the low dimensional case where the major problem is that caused by multicollinearity, I am not completely convinced of that. I find it difficult to believe that the prior for the parameter has a covariance matrix of the type usually used; this is crucial for the Bayesian-ness of the ridge trace.

I was careful to state that this paper is only a first step in the treatment of infinite dimensional problems. Most of what I have done needs little modification if the various matrices can diagonalized simultaneously, and the trace of the loss coefficient matrix is infinite when the error variances are normalized to be one, which can be done without loss of generality. The robustness results and the counterexamples are similar.

In the standard regression model, one can, of course, use the canonical reduction. I can see no conceivable reason to rescale to induce equal variances for the components of the classical estimte of θ; doing this completely obscures the prior input about their variances. Of course in the straight finite regression model the covariance matrix of the observations is finite rank. If the observations are in principle infinte dimensional, this need not be true. I would not consider it unreasonable for there to be in actuality an infinite dimensional problem (such as Fourier series regression) for which $N = 5000$, the *effective* n used is 50000, and the estimator, when coordinates with small prior variance are made 0, only has 100 non-zero terms. There are also non-normal infinite-dimensional problems, such as arise in density estimation or functional estimation, where the structure I have used is, while not quite correct, close enough so that the results are applicable.

I have attempted to make the results reasonably resistant to those assumptions, such as normality, which cannot be correct. In some cases the establishment of robustness is trivial; in others it can be seen to hold in a wide variety of cases; in still others, only numerical work can hope to establish it. I think the approach wil turn out to be better in most cases than using such things as data-based high-dimensional conjugate priors.

The problem of prediction is different from the problem of inference. For prediction, it is certainly necessary to consider computational costs. To repeat what has been said many times before, a prior cannot be data-based. It is only reasonable to use a data based "prior" if there is at least a good reason to expect robustness.

REFERENCES IN THE DISCUSSION

Brown, P. and Zidek, J. (1980). Adaptive multivariate ridge regression. *Ann. Statist.* **8**, 64–74.
Dawid, A. P. (1988). The infinite regress and its conjugate analysis. (In this volume).
Leamer, E. (1978). *Specification Searches.* New York: Wiley.
Mäkeläinen, T. and Brown, P.J. (1988). Coherent priors for ordered regressions. (In this volume).

BAYESIAN STATISTICS 3, pp. 411–428
J. M. Bernardo, M. H. DeGroot, D. V. Lindley and A. F. M. Smith, (Eds.)

Aspects of Numerical Integration and Summarisation

J. E. H. SHAW
University of Warwick

SUMMARY

Numerical integration in Bayesian inference is equivalent to using a discrete appoximation to the posterior distribution. I discuss how this can be done efficiently, emphasising the importance of adaptive methods to find a suitable weight function. In high-dimensional problems, quasirandom sequences are recommended for producing a discretised posterior distribution, which can then be used for many exploratory purposes as well as in numerical integration. The methods are illustrated on an analytically intractable 14 parameter finite mixture model.

Keywords: DISCRETISED POSTERIOR DISTRIBUTIONS; FINITE MIXTURES; NUMERICAL BAYESIAN INFERENCE; NUMERICAL INTEGRATION; QUASIRANDOM SEQUENCES; STANDARDISED POSTERIOR DISTRIBUTIONS; SUMMARISATION; WEIGHT FUNCTIONS.

1. INTRODUCTION

The paper gives a necessarily brief, selective and personal view of some approaches to numerical integration and summarisation in practical Bayesian statistics. Section 2 discusses some desirable properties of numerical integration, and shows how numerical integration can be interpreted in terms of a discretised posterior distribution. The importance of finding a suitable weight function suggests using methods like adaptive interpolatory integration rules for low-dimensional integrands (Section 3), and adaptive importance sampling for high-dimensional integrands (Section 4). Section 4 also outlines the use of quasirandom sequences, which are usually preferable to the more familiar Monte Carlo methods. Section 5 describes how a discretised posterior distribution can be used for purposes other than numerical integration. Section 6 illustrates the use of numerical Bayesian methods on an analytically intractable 14 parameter posterior distribution.

The remainder of this Section describes the sort of integrands to be considered in the rest of the paper. Suppose that we have data y, together with a family of probability models for the mechanism giving rise to y. The individual models are indexed by a parameter $\xi = (\xi_1, \ldots, \xi_d)^T$ with prior $p(\xi)$. Suppose also that the resulting posterior is an analytically intractable high-dimensional density, so that suitable numerical ways to summarise it must be found before we can hope to use Bayesian methods routinely. A useful first step is to transform the parameter space to $\mathbf{R}^d$ by applying a one-to-one transformation $t_1 : \xi \to \theta$, examples of such transformations are given in Section 6. The posterior density of θ is then

$$\begin{aligned} p(\theta|\mathbf{y}) &= cp(\mathbf{y}|t_1^{-1}(\theta))p(t_1^{-1}(\theta))|t_1^{-1}(\theta)|, \\ &= cm(\theta), \quad \text{say} \end{aligned} \tag{1.1}$$

where c is the normalising constant that makes the right hand side integrate to 1, $|\cdot|$ represents the Jacobian of the transformation, and m is the unnormalised posterior density. I shall assume that the transformation t_1 is such that $p(\theta|\mathbf{y})$ has finite first and second moments.

In a similar manner to that described in Naylor and Smith (1982), and elsewhere, many quantities and functions that are useful in summarising and interpreting $p(\theta|\mathbf{y})$ take the form

$$S(q; m) = \int q(\theta) m(\theta) d\theta \tag{1.2}$$

where q is some function (possibly vector- or matrix-valued), and the region of integration is usually obvious from the context and may be omitted.

For example, the normalising constant c in (1.1) is given by

$$c = 1/S(1; m) \tag{1.3}$$

and the posterior mean and variance-covariance matrix of θ are

$$E(\theta) = cS(\theta; m) \tag{1.4}$$

and

$$V(\theta) = cS(\theta\theta^T; m) - c^2 S(\theta; m) S(\theta; m)^T, \tag{1.5}$$

respectively. Similarly, the posterior mean of ξ is

$$E(\xi) = cS(t^{-1}(\theta); m), \tag{1.6}$$

and other useful summaries are described in Smith *et al.* (1985) and van Dijk *et al.* (1987).

2. NUMERICAL INTEGRATION: GENERAL CONSIDERATIONS

Efficient numerical integration methods begin by expressing the unnormalised posterior density m as the product of two functions w and f, where w is called the *weight function*:

$$m(\theta) = w(\theta) f(\theta). \tag{2.1}$$

The weight function is non-negative and integrates to 1, and is chosen to have similar properties to m. For example, w should be relatively large wherever m is relatively large.

Most numerical integration methods then implicity replace the function $m(\theta)$ by a discrete approximation of the form

$$\hat{m}(\theta) = \begin{cases} w_i f(\theta) & (\theta = \theta_i, i = 1, 2, \ldots, n) \\ 0 & \text{elsewhere} \end{cases} \tag{2.2}$$

so that the integral $S(q, m)$ may be estimated by

$$\hat{S}(q; m) = \sum_{i=1}^{n} w_i f(\theta_i) q(\theta_i). \tag{2.3}$$

The points θ_i are called ***nodes***, the w_i are the ***weights***, and the nodes and weights together consititute an ***integration rule***. The integration error Δ is then a linear functional of q:

$$\Delta(q; m) = S(q; m) - \hat{S}(q; m). \tag{2.4}$$

Replacing S by $\hat{S}$ in (1-3) gives an estimate $\hat{c}$ of the normalising constant c. Similar substitutions in expressions such as (1.4) to (1.6) give estimates $\hat{E}(\theta)$, $\hat{V}(\theta)$, $\hat{E}(\xi)$,etc. Following Naylor and Smith (1982) and Naylor and Shaw (1985), when this method is used to estimate the posterior expectations of functions of θ other than the first two moments, it will be called *special function analysis*.

There are many properties we would like an integration rule to have, as discussed in Stroud (1971) and Davis and Rabinowitz (1984). The following properties are particularly desirable:

(1) The number n of nodes should be small.

(2) The nodes and weights should be easily found, or preferably calculated one and for all.

(3) The nodes should all lie in the region of integration R.

(4) The weights should all be positive.

(5) The same nodes and weights should give small integration errors $\Delta(q;m)$ for a wide range of functions q of interest. Two possible specifications are:

(5a) The error in integrating $qm = wfq = wg$, say, should be small for all functions g in some chosen function space H. For example, we could define a norm on Δ by

$$|| \Delta || = \sup_{g \in H} \frac{|| \Delta\,(1; wg)||}{||wg||},$$

and then look for minimum norm rules in which, for some prespecified n, the θ_i and w_i are chosen to minimise $|| \Delta ||$.

(5b) All integrands qm corresponding to functions $g = fq$ in some chosen space F should be integrated exactly, i.e.

$$\Delta(1; wg) = S(1; wg) - \hat{S}(1; wg) = 0 \quad \forall g \in F.$$

Properties (1) and (2) are clearly desirable for computational efficienty . Property (3) is important since m and therefore $\hat{m}$ may not be defined outside the region of integration R; this is true even if $R = \mathbf{R}^d$ because some integration rules have complex nodes. Integration rules for which (3) holds also tend to be more accurate than similar rules for which (3) doesn't hold, in the same way as interpolation tends to be more accurate than extrapolation.

Property (4) is also very important. If some of the weights are negative then there is the embarrassing possibility of estimating the normalising constant c in (1.3) to be negative; more generally there may be severe round-off errors in calculating sums of the from (2.3), which will in turn make iterative procedures to find a suitable weight function (see below) extremely unreliable. Also, theoretical bounds on the error Δ often involve the expression $\Sigma|w_i|$, which can be unduly large if the w_i are not all the same sign.

Note that if properties (3) and (4) both hold, then numerical integration over the parameter space has the natural interpretation of weighted mixing of the n "representative" models indexed by $\theta_1, \ldots, \theta_n$. Equivalently, $p(\theta|\mathbf{y})$ has been replaced by a *discretised posterior distribution* that takes values $p_i = w_i f(\theta_i)/\Sigma w_i f(\theta_i)$ at the points $\theta_1, \ldots, \theta_n$ and is zero elsewhere.

Property (5) is also necessary for computational efficiency, and implies that several numerical integrations can be carried out in parallel using just the one set of nodes and weights.

Property (5a) is a natural aim of numerical integration. Unfortunately, it is difficult to construct integration rules that optimise this sort of property, particularly with integration regions $\mathbf{R}^d$ for large d and with usefully large spaces H. The choice of H is itself unclear. For these and other reasons, as Davis and Rabinowitz (1984, page 331) state,

> "... minimum norm rules (of all varieties) are not used in computing practice and remain, after almost three decades of investigation, a plaything of the theoretician".

More Bayesian approaches, incorporating priors over function spaces, are suggested by O'Hagan (1986) and Upsdell (1986), but these also appear difficult to implement in high dimensions.

Property (5b) has so far been more useful in practice than (5a), and is intimately connected with *interpolatory integration rules*: rules that implicitly fit some interpolating function $\phi(\theta)$ to the set of points $\{(\theta_i, f(\theta_i))|i = 1, \ldots, n\}$ and then find $S(q; w\phi)$ exactly. In the important case where ϕ is a polynomial with coefficients in $\mathbf{R}$, the corresponding interpolatory integration rule will be of the form (2.3). Note that if $1 \in F$, as when F is a space of polynomials up to given degree, then the weights w_i will sum to 1.

Suitable weight functions usually have to be found iteratively, information from one iteration being used to choose the weight function for the next iteration. This gives *adaptive* numerical integration rules, in which the position of the nodes depends on observed properties of the integrand. The choice of weight function is greatly simplified by employing an affine transformation $t^2 : \boldsymbol{\theta} \to \boldsymbol{\theta}'$ such that $\boldsymbol{\theta}'$ has, at least approximately, zero mean and diagonal variance-covariance matrix. This can easily be done at each iteration via the current estimates $\hat{E}(\boldsymbol{\theta})$ and $\hat{V}(\boldsymbol{\theta})$, see Naylor and Smith (1982). A final rescaling transformation $t_3 : \boldsymbol{\theta}' \to \mathbf{x}$ makes all variances approximately 1. This idea of iterative orthogonalisation frees us to concentrate on finding weight functions and integration rules that are efficient in analysing *standardised* posterior densities $p(\mathbf{x}|\mathbf{y})$ that have roughly zero mean and identity variance-covariance matrix.

Note that a highly curved posterior density $p(\boldsymbol{\theta}|\mathbf{y})$ can similarly be made better behaved by employing a nonlinear transformation $t_2' : \boldsymbol{\theta} \to \boldsymbol{\theta}'$ such as

$$t_2' : \theta_j \to \theta_j' = \theta_j - \sum_{k=1}^{j-1} t_{2jk}'(\theta_k), \quad j = 1, \ldots, d, \tag{2.5}$$

but this procedure seems difficult to automate.

3. INTERPOLATORY INTEGRATION METHODS

Bayesian analogues of the central limit theorem indicate that the "standard Normal" weight function $w(\mathbf{x}) = (2\pi)^{-d/2} \exp\left(-\frac{1}{2}\mathbf{x}^T\mathbf{x}\right)$ is suitable for numerical integration with respect to the standardised posterior density $p(\mathbf{x}|\mathbf{y})$. Naylor and Smith (1982) show how Cartesian products of rescaled *Gauss-Hermite* rules produce efficient interpolatory rules for this weight function. If the ith Gauss-Hermite rule has $n(i)$ nodes, then the Cartesian product integration rule has $n = n(1) \times n(2) \times \cdots \times n(d)$ nodes, and is exact for all functions of the form $w(\mathbf{x})g(\mathbf{x})$ where g is a polynomial containing terms up to $x_1^{2n(1)-1} \times x_2^{2n(2)-1} \times \cdots \times x_d^{2n(d)-1}$. Thus it is *not* assumed that $p(\mathbf{x}|\mathbf{y})$ resembles a standard Normal density, but only that $p(\mathbf{x}|\mathbf{y})$ can be well approximated by the product of a standard Normal "kernel" and a low-order polynomial.

These Cartesian product have proved highly efficient in analysing a wide range of posterior distributions with up to 5 or 6 parameters. Unfortunately, the number of nodes increases exponentially with the the number of parameters d, so Cartesian product rules are impracticable for use on high-dimensional posterior distributions. Alternative interpolatory integration rules with the standard Normal weight function include the *spherical* integration rules described in Stroud (1967). These place the nodes on concentric spherical shells, and have been found useful in up to 8 dimensions (above which some of the weights become negative). Experience with spherical integration rules is described in Naylor and Shaw (1987).

Another important class of interpolatory integration rules is that of *fully symmetric rules*, see Genz (1986) and the references therein. Unfortunately, many of the weights of such rules are negative, rendering them inappropriate for use in Bayesian statistics.

4. MONTE CARLO AND QUASIRANDOM METHODS

Interpolatory integration methods use the weight function w as the kernel of as approximation to the integrand; an alternative interpretation of w is obtained by writing

$$S(q;m) = \int q(\boldsymbol{\theta})m(\boldsymbol{\theta})d\boldsymbol{\theta} = \int q(\boldsymbol{\theta})w(\boldsymbol{\theta})f(\boldsymbol{\theta})d\boldsymbol{\theta} = \int q(\boldsymbol{\theta})f(\boldsymbol{\theta})dW(\boldsymbol{\theta}) = Ew(q(\boldsymbol{\theta})f(\boldsymbol{\theta})). \tag{4.1}$$

This suggests a *Monte Carlo* approach to numerical integration: generate nodes $\theta_1, \ldots, \theta_n$ independently from the distribution W, and estimate $S(q;m)$ by $\hat{S}(q;m)$ in expression (2.3) with $w_i = 1/n$. If $q(\theta)f(\theta)$ is constant then $\hat{S}(q;m)$ will be exact; more generally $\hat{S}(q;m)$ is unbiased and its variance will be small if $w(\theta)$ has a similar shape to $|q(\theta)m(\theta)|$. The above procedure is known as *importance sampling*, since W is chosen (perhaps iteratively) to generate more nodes in "important" regions of the parameter space, i.e. regions where $|q(\theta)m(\theta)|$ is large. In practice, to allow parallel computation of $\hat{S}(q;m)$ for many function q, the weight function $w(\theta)$ may be chosen to have a similar shape to $m(\theta)$, but with heavier tails. The efficiency of importance sampling therefore depends heavily on finding a suitable weight function that not only resembles $p(\theta|\mathbf{y})$ (or, equivalently, one that resembles $p(\mathbf{x}|\mathbf{y})$), but that also allows points to be generated easily from the distribution W. Hammersley and Handscomb (1964) remains one of the best references to importance sampling and other Monte Carlo methods.

Monte Carlo integration has been used with much success in the analysis of high-dimensional posterior distributions, see, for example, Kloek and van Dijk (1978), van Dijk and Kloek (1983, 1985) van Dijk *et al.* (1987), Stewart (1979, 1983, 1985, 1987), and Stewart and Davis (1986). However, objections can be made to Monte Carlo methods on the grounds that they deliberately add random variation, and then deliberately ignore valuable information such as the positions of the generated nodes. See Bacon-Shone's comments in the discussion to van Dijk and Kloek (1985), and O'Hagan (1987). We therefore seek high-dimensional numerical integration methods that take account of the positions of the nodes; either to obtain more information about the integrand, or to place subsequent nodes more efficiently. This brings us on to so-called *quasirandom* (or *number-theoretic*) methods.

A *quasirandom sequence* is any of points $\mathbf{u}_i$ in the unit d-dimensional cube $C_d = [0,1)^d$, such that $\mathbf{u}_{i+1}$ depends on the values $\{\mathbf{u}_1, \mathbf{u}_2, \ldots, \mathbf{u}_i\}$. As in Shaw (1986a), a quasirandom sequence will be called *regular* if

$$\mathbf{u}_{i+1} = \mathbf{u}_i + \boldsymbol{\alpha} \pmod 1 \tag{4.2}$$

(i.e. $\{\alpha_1, \ldots, \alpha_d\}$ is a set of increments, and the jth coordinate of $\mathbf{u}_{i+1}$ the fractional part of the jth coordinate of $\mathbf{u}_i + \boldsymbol{\alpha}$). Any other form of algorithm generates an *irregular* sequence. Regular quasirarandom sequences will be called *rational* if each α_j is rational, *irrational* if each α_j is irrational, and *mixed* othewise.

The practical importance of quasirandom sequences is that various ideas from number theory suggest particular sequences that "fill" C_d in a uniform way (i.e. have *low discrepancy*), and the first n points of such a sequence can then be used as nodes for numerical integration. The weights w_i will usually be $1/n$, although methods using non-uniform weights may be appropriate for irrational sequences, see Haselgrove (1961). The *discrepancy* of a sequence gives theoretical upper bounds on integration errors; its precise definition may be found in Niederreiter (1978), which contains an excellent overview of quasirandom sequences. Other useful references include Zaremba (1972), Kuipers and Niederreiter (1974), and Hua Loo-Keng and Wang Yuan (1981). Quasirandom methods for numerical integration in Bayesian inference appear to have been first explicitly suggested in the "reply to the discussion" by Stewart (1985).

Note that a simple Cartesian product of points on each axis does *not* fill C_d uniformly. Consider, for example the set of points in C_{16} whose individual coordinates are all the possible combinations of $\{.1, .3, .5, .7, .9\}$. Then the distance from $(.2, .2, \ldots, .2)$ to the nearest point in the set is $\sqrt{16 \times .1^2} = .4$, even though the set contains 5^{16} points! This again demonstrates that Cartesian product methods are impracticable for high-dimensional integrands.

Much better high-dimensional generalisations of the trapezium rule are provided by rational quasirandom sequences. Suitable increments $\boldsymbol{\alpha}$ in expression (4.2) can be tabulated, but heavy computation is needed to find them in the first place. Therefore it is often easier to use other sequences. Several rational, irrational and irregular quasirandom sequences are compared in Shaw (1986a), a few aspects of irregular sequences are considered below.

Many low-discrepancy irregular sequences are based on the *radical inverse function* $\phi_b(i)$, in which the integer i is written to base b and "reflected about the decimal point". Thus 15 (base 10) is 120 (base 3), so $\phi_3(15) = .021 = \frac{2}{9} + \frac{1}{27} = \frac{7}{27}$. For example, the *Halton sequence* is give by

$$\mathbf{u}_{i+1} = (\phi_{b_1}(i), \phi_{b_2}(i), \ldots, \phi_{b_d}(i))^T,$$

where b_i is the ith prime (more generally, the b_i just need to be mutually coprime). Other sequences exist that are more efficient than the Halton sequence for the sort of numerical integrations needed in Bayesian statistics. However, the Halton sequences is very convenient for routine use: it is easy to define, it is appropriate with any number of nodes, the Halton sequence with n nodes is simply the initial segment of the Halton sequence with $n' > n$ nodes, and Halton the sequence in d dimensions is simply a projection of the Halton sequence in $d' > d$ dimensions.

The most familiar irregular sequences are probably those produced by *pseudorandom number generators*, see Niederreiter (1978) and Ripley (1983). These give rise to Monte Carlo methods as described above. However, the whole idea of a pseudorandom number sequence is that the point $\mathbf{u}_{i+1}$ should be uniformly distributed on C_d *independently* of $\{\mathbf{u}_1, \ldots, \mathbf{u}_i\}$, rather than be more likely to appear in a "gap". Therefore pseudorandom sequences are wilfully inefficient, and with very high probability have larger discrepancies than other quasirandom sequences.

Once a numerical integration rule has been defined on C_d it must be smoothly transformed into a rule that is efficient for integrating with respect to a standardised posterior density. One class of such transformations, using importance sampling ideas, is described in Shaw (1986b), and examples are given in Shaw (1986a) showing that a quasirandom integration rule with 1000 nodes can easily be as efficient as a Monte Carlo rule with 20000 nodes. Another transformation yielding *quasirandom spherical rules*, with nodes on the surfaces of concentric spherical shells, is described briefly in Shaw (1986a) and in more detail in Naylor and Shaw (1987). Similar spherical integration rules were suggested by Bacon-Shone in the discussion to Stewart (1985), but his suggestion is difficult to implement efficiently because of the need to adjust for the Jacobian of the transformation.

5. SUMMARISATION USING DISCRETISED POSTERIOR DISTRIBUTIONS

Many important summaries of the posterior distribution can be obtained by special function analysis. However, the discretised posterior distribution can be used for many other purposes, including (weighted) analogues of classical exploratory data analysis methods such as cluster analysis and projection pursuit.

For example, displays of marginal posterior distributions often give useful insights into the nature of the whole joint posterior. One approach to producing marginal distributions numerically is described in Shaw (1985) and less formally in Smith *et al.* (1987). Some examples are given in Smith *et al.* (1985), Skene *et al.* (1986), and Shaw (1987). This approach can give very accurate estimates of marginal distributions, but requires several numerical integrations, and usually has to be repeated with a different orthogonalising transformation for each marginal required.

Some simpler ways to form approximate marginal posterior distributions are described below. The resulting marginals are less accurate but may all be obtained from a single discretised posterior distribution, so are suitable for exploratory analysis. Suppose we have a discretised posterior distribution with values (weights) p_i at $\theta_i, i = 1, \ldots, n$, and that we wish to estimate the joint posterior marginal density $p(\phi|\mathbf{y})$ of $\phi = (\phi_1, \ldots, \phi_r)^T$. The vector ϕ may be a subvector of θ, or may more generally be given by $\phi^t = \mathbf{P}\theta$ where $\mathbf{P}$ is a $r \times d$ matrix. Then the discretised approximation to $p(\phi|\mathbf{y})$ is simply the projection of the discretised approximation to $p(\theta|\mathbf{y})$:

$$\hat{p}(\phi|\mathbf{y}) = \begin{cases} p_i & (\phi = \phi_i = \mathbf{P}\theta_i, i = 1, \ldots, n) \\ 0 & \text{elsewhere} \end{cases} \tag{5.1}$$

(any coincident points ϕ_i, can be combined and their weights p_i added). If $r = 2$, for example, then this approximation may be represented as a set of circles with centres ϕ_i and radii $\sqrt{p_i}$.

The resulting picture often appears messy, but a clearer picture is produced if some of the nodes are combined using, for example, *centroid cluster analysis* as described in Anderberg (1973). Possible definitions of the distance between the points ϕ_i and ϕ_j include $D_1 = \sqrt{(\phi_i - \phi_j)^T(\phi_i - \phi_j)}$, $D_2 = \sqrt{(\theta_i - \theta_j)^T(\theta_i - \theta_j)}$, $D_1/\sqrt{p_i p_j}$ and $D_2/\sqrt{p_i p_j}$. The closest 2 points ϕ_i and ϕ_j are then replaced by a single point $(p_i\phi_i + p_j\phi_j)/(p_i + p_j)$ with weight $p_i + p_j$, and the clustering process repeated. Note that this clustering method does not change the mean of the discretised marginal distribution.

An alternative and easily assimilated representation of the approximate marginal distribution is obtained by superimposing a rectangular grid on ϕ-space, and redistributing each weight p_i to the corners of the rectangular "box" in which the corresponding point $\phi_i = (\phi_{i1}, \ldots, \phi_{id})$ falls. Specifically, if $a_{j,k(j)} \le \phi_{ij} < a_{j,k(j)+1}$, then ϕ_i contributes a proportion

$$\prod_{j=1}^{r} \frac{b_{j,l(j)}}{a_{j,k(j)+1} - a_{j,k(j)}}$$

of its weight to the grid point $(a_{1,k(1)+l(1)}, a_{2,k(2)+l(2)}, \ldots, a_{r,k(r)+l(r)})^T$ where,

$$b_{j,l(j)} = \begin{cases} (a_{j,k(j)+1} - \phi_{ij} & (l(j) = 0) \\ \phi_{ij} - a_{j,k(j)} & (l(j) = 1). \end{cases}$$

It is (fairly!) easy to see that this leaves the mean and covariances of the discretised distribution unchanged, but increases the variances. Thus the picture of the marginal posterior distribution appears to have slightly smaller correlations than it should, but the effect is hardly noticeable. A three-dimensional example of this redistribution technique is given in Section 6.

6. EXAMPLE: A MIXTURE MODEL

Many alternative ways have been suggested to fit finite mixture models; see Titterington *et al.* (1985) for a comprehensive review. This Section illustrates how a fully Bayesian model can be fitted, and interpreted, using the methods of Sections 1 to 5 as implemented in Naylor and Shaw (1985).

Table 6.1, taken from Bhattacharya (1967), summarises the distribution of the lengths of 13963 porgies (a kind of fish).

The distribution of fish lengths is clearly multimodal; this is to be expected since fish tend to spawn at fixed times of year, and the distribution of fish lengths at any given time is a mixture of lengths of fish whose possible ages are roughly 1 year apart. Suppose, for the moment, that the data comprise a mixture of k Normally distributed components where the ith component has mean μ_i, standard deviation σ_i, and forms a proportion p_i of the total mixture. Since $\Sigma p_i = 1$, the number of parameters is $d = 3k - 1$:

$$\xi = (\mu_1, \sigma_1, p_1, \ldots, \mu_{k-1}, \sigma_{k-1}, p_{k-1}, \mu_k, \sigma_k)^T \tag{6.1}$$

Class range	Observed frequency	Class range	Observed frequency	Class range	Observed frequency
9–10	509	16–17	921	23–24	310
10–11	2240	17–18	448	24–25	228
11–12	2341	18–19	512	25–26	168
12–13	623	19–20	719	26–27	140
13–14	476	20–21	673	27–28	114
14–15	1230	21–22	445	28–29	64
15–16	1439	22–23	341	29–30	22

Table 6.1. *Length of porgies*

To avoid nonidentifiability, I shall assume that $\mu_1 < \mu_2 < \cdots < \mu_k$, and to hel interpretation I shall also make the natural assumption that the $(j+1)$th component of th mixture corresponds to fish approximately one year older than those represented by the jt component. The jth class limit will be denoted by x_j cm, i.e. $x_1 = 9$, $x_2 = 10$, etc. Th number of lengths falling in the jth class will be denoted N_j, the number of classes by n and the total number of lengths by N. Then the fitted proportion of fish shorter than x_j is

$$\hat{F}_j = \sum_{i=1}^{k} p_i \Phi((x_j - \mu_i)/\sigma_i), \tag{6.2}$$

where Φ is the standard Normal distribution function, and the fitted proportion of fish whos lengths lie in the interval $[x_j, x_{j+1})$ is

$$\hat{f}_j = \hat{F}_{j+1} - \hat{F}_j. \tag{6.3}$$

Table 6.1 excludes lengths outside the range $[x_1, x_{n+1})$, implying that the fitted proportion o *recorded* lengths in the interval is $\hat{f}_j = (\hat{F}_{n+1} - \hat{F}_1)$. The log-likelihood L is therefore give by

$$\begin{aligned} L(\xi) &= \sum_{j=1}^{n} N_j \log(\hat{f}_j/(\hat{F}_{n+1} - \hat{F}_1)) \\ &= \sum_{j=1}^{n} N_j \log \hat{f}_j - N \log(\hat{F}_{n+1} - \hat{F}_1) \end{aligned} \tag{6.4}$$

The parameter space is determined by the facts that all parameters are positive, tha $\mu_1 < \mu_2 < \cdots < \mu_k$, and that $p_1 + \cdots + p_{k-1} < 1$. There are several points that should be borne in mind when formulating the prior distribution:

(1) The p_i should decrease smoothly as i increases, since p_i represents the number of fish that have survived i years.

(2) The σ_i should increase smoothly as i increases, since larger fish probably have greater variability in size.

(3) The μ_i should be fairly close together. In particular, the prior must not be uniform over the μ_is (which would say that with probability 1 all fish are infinitely long).

(4) The μ_i should be locally evenly spaced, since growth is probably fairly constant from one year to the next.

(5) The σ_i should not be large (otherwise there would be an appreciable probability of negative lengths, making the Normality approximation unreasonable).

I therefore chose the prior

$$\begin{aligned} p(\xi) = \exp\left(-\frac{1}{2}\sum_{i=1}^{k-1}(\log p_{i+1} - \log p_i + 0.5)^2\right) \times \\ \exp\left(-\frac{1}{2}\sum_{i=1}^{k-1}(\log \sigma_{i+1} - \log \sigma_i + 0.2)^2\right) \times \\ \frac{1}{(\mu_k - \mu_1)^4} \times \\ \prod_{i=2}^{k-1} \frac{\mu_{i+1} - \mu_i}{\mu_{i+1} - \mu_{i-1}} \frac{\mu_i - \mu_{i-1}}{\mu_{i+1} - \mu_{i-1}} \times \\ \prod_{i=1}^{k} \frac{1}{\sigma_i}, \end{aligned} \tag{6.5}$$

where each of the each of the five components in the prior corresponds to one of the points listed above.

To emphasise the importance of point (3), note that if μ_1, σ_1 and p_1 are held fixed, then the likelihood is bounded below. Therefore a uniform prior over the μ_i implies that the posterior distribution is non-integrable.

For the numerical integration I chose the transformation

$$\begin{aligned} \theta_1 &= \mu_1 - 11 && \text{(for numerical stability)}, \\ \theta_{3i-2} &= \log(\mu_i - \mu_{i-1}) && (i = 2, \ldots, k), \\ \theta_{3i-1} &= \log \sigma_i && (i = 1, \ldots, k), \\ \theta_{3i} &= \log(p_i/(1 - \Sigma_{j=1}^{i} p_j)) && (i = 1, \ldots, k-1). \end{aligned} \tag{6.6}$$

The transformed parameter space is thus $\mathbf{R}^d$.

For compatibility with Bhattacharya's analysis, I took the number of components k to be 5. Initial guesses for the posterior means and standard deviations of θ were made by inspecting a histogram of Table 6.1, and all correlations were initialised to 0. Even starting from those crude guesses, the estimated posterior moments calculated by a quasirandom spherical rule, with iterative orthogonalisation, quickly stabilised. The iterative algorithm was halted when fewer than 50% of the correlation estimates changed by over .03 between iterations.

Table 6.2 gives the posterior moment summaries of θ produced on the final iteration, which had a total of 10000 nodes distributed on 2 shells according to a transformed Halton sequence. This final iteration took 4 minutes c.p.u. time (7 minutes connect time) on a Vax/780, which is negligible compared to the resources and patience that must have been needed to collect the original data!

Note from Table 6.2 that the standard deviations of the parameters corresponding to the ith component (i.e. $\theta_{3i-2}, \theta_{3i-1}$ and θ_{3i}) tend to increase with i as would be expected since there are fewer data on longer fish, and no distinct modes. As would also be expected, there is very little correlation between parameters corresponding to components with widely separated μ_i values, and the largest correlations (in absolute value) are between parameters corresponding to the 3rd, 4th, and 5th components, which overlap considerably.

Quantities like the posterior moments of ξ are easily estimated by special function analysis (see Section 2); the means and standard deviations are given in Table 6.3.

The methods of Section 5 can be used to explore features other than simple moment summaries. For example, Figure 6.1 represents the joint marginal posterior distribution of the parameters that define the 4th component of the mixture.

A. Means and Standard Deviations

	mean	s.d.		mean	s.d.		mean	s.d.
θ_1	.017	.012	θ_2	–.262	.014	θ_3	–.385	.018
θ_4	1.4511	.0073	θ_5	.121	.026	θ_6	.094	.037
θ_7	1.487	.027	θ_8	.300	.083	θ_9	.48	.25
θ_{10}	1.201	.094	θ_{11}	.32	.21	θ_{12}	.51	.55
θ_{13}	1.19	.16	θ_{14}	.51	.18			

B. Correlations

	θ_2	θ_3	θ_4	θ_5	θ_6	θ_7	θ_8	θ_9	θ_{10}	θ_{11}	θ_{12}	θ_{13}	θ_{14}
θ_1	.15	.13	–.38	–.23	–.15	–.03	.08	.05	.04	.06	.05	.06	–.07
θ_2		.19	–.03	–.29	–.17	–.03	.10	.04	.03	.01	–.02	–.05	.01
θ_3			–.07	–.22	–.13	.01	.09	.05	.05	–.01	.02	.01	–.03
θ_4				.51	.56	–.28	–.56	–.38	–.36	.14	.05	.12	.13
θ_5					.65	–.12	–.53	–.33	-.31	.12	.06	.14	.07
θ_6						–.24	–.65	–.43	–.39	.19	.13	.19	.06
θ_7							.72	.90	.64	–.74	–.45	–.58	–.10
θ_8								.85	.68	–.49	–.31	–.45	–.14
θ_9									.76	–.77	–.48	–.62	–.12
θ_{10}										–.35	.13	–.10	–.59
θ_{11}											.81	.83	–.27
θ_{12}												.94	–.72
θ_{13}													–.61

Table 6.2. *Internal parameter moment summaries for porgy length data*

		Component				
		1	2	3	4	5
μ_i	mean	11.017	15.284	19.71	23.09	26.46
	s.d.	.012	.029	.12	.37	.52
σ_i	mean	.769	1.129	1.35	1.40	1.69
	s.d.	.011	.030	.11	.29	.30
p_i	mean	.4051	.3112	.175	.070	.039
	s.d.	.0043	.0062	.019	.020	.011

Table 6.3. *Final "parameter estimates" for porgy length data*

Producing Figure 6.1 entailed the following steps:

(1) A $6 \times 5 \times 5$ grid of $(\xi_{10}, \xi_{11}, \xi_{12})$ values was chosen based on the estimated posterior means and standard deviation given in Table 6.3.

(2) The posterior distribution was discretised using an 800 node 2-shell spherical rule and a 1200 node 3-shell spherical rule combined, the nodes being generated in each case from a transformed Halton sequence.

(3) The nodes were transformed from $\boldsymbol{\theta}$-space to $\boldsymbol{\xi}$-space, and then projected to give the corresponding $(\xi_{10}, \xi_{11}, \xi_{12})$ values.

(4) The weights were apportioned to the points of the chosen grid, as described in Section 5.

(5) Steps (2) to (4) were repeated on a second iteration of the integration algorithm (i.e. with slightly different estimates of the posterior moments), and the average of the two sets of weights was taken.

(6) The posterior distribution was represented as a grid of circles with radii proportional to the cube root of the corresponding weight. "Hidden lines" (where the boundary of a circle fell partly within another circle lying in front) were eliminated to give a better impression of depth.

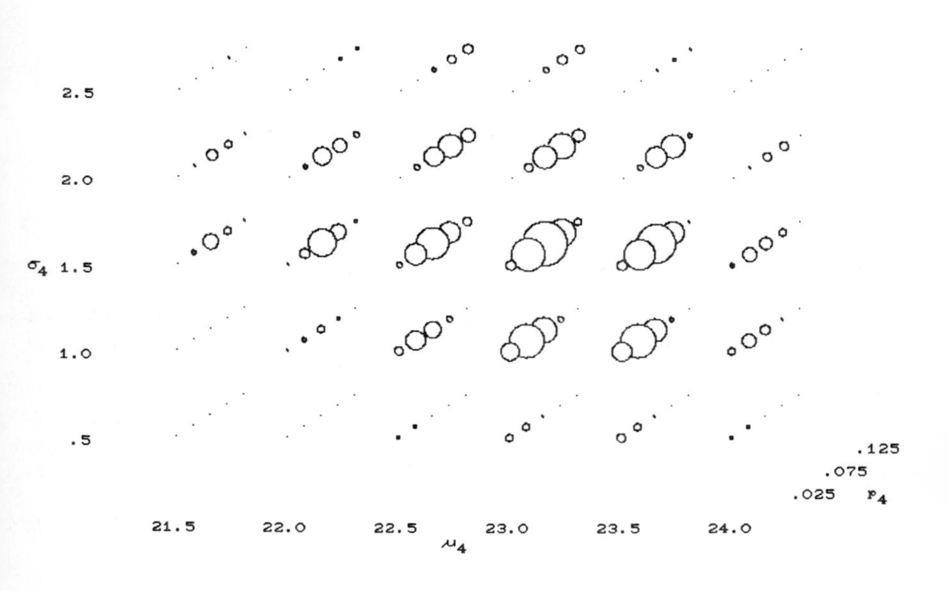

Figure 6.1. *Approximate joint marginal posterior distribution of μ_4, σ_4 and p_4.*

Figure 6.1 is actually a direct copy of a terminal screen, and indicates the sort of interactive exploratory analysis that is possible using discretised distributions. The following (tentative) conclusions can be drawn from Figure 6-1.

(1) Looking at each horizontal plane separately, μ_4 and p_4 appear virtually independent conditional on σ_4.

(2) The variance of μ_4 increases as σ_4 increases (compare the spread of weights in the plane $\sigma_4 = 2.0$ with that in the plane $\sigma_4 = 1.0$).

(3) The distribution of μ_4 seems negatively skewed, for all values of σ_4 and p_4.

(4) The vertical planes show that σ_4 and p_4 are strongly correlated for all values of μ_4.

(5) There is some negative correlation between μ_4 and σ_4, particularly at $p_4 = .05$ and $p_4 = .075$. This would be easier to see on a colour graphics display, by plotting different colours at different values of p_4.

It is noteworthy that producing Figure 6.1 is roughly equivalent to numerically estimating ($6 \times 5 \times 5 = 150$) 11-dimensional integrals, yet only 4000 posterior density evaluations were required. Of course, many other evaluations had previously been made in order to obtain reasonable estimates of the posterior moments of θ. The points in Figure 6.1 can be projected and agglomerated to give bivariate and univariate marginals; these have been omitted to save space.

As the simplest possible summary of the posterior distribution, the posterior mean of ξ can be taken as a "numerical Bayesian" point estimate. The χ^2 goodness of fit statistics corresponding to this point estimate of χ, and to four other point estimates given in Bhattacharya (1967), are shown in Table 6.4. The numerical Bayesian point estimate of χ gives by far the best fit despite the fact that it incorporates prior information shrinking the posterior parameter values towards what is biologically reasonable rather than what is suggested by the data alone.

Method	χ^2 statistic
Buchanan-Wollaston	61.3
Cassie	14.3
Tanaka	32.3
Bhattacharya	29.9
Numerical Bayesian	5.8

Table 6.4. *Goodness of fit for porgy length data*

The parameters p_i are of particular interest, since they indicate the age composition of the fish population, which needs careful monitoring to avoid overfishing. Therefore we should investigate the sensitivity of the posterior joint distribution of $(p_1, \ldots, p_k)$ to the model assumptions, such as Normality.

For example, a mixture of log-logistic distributions seems physically realistic for fish lengths; the distribution function of the ith component of the mixture is then most simply written

$$F(x_j) = \frac{1}{1 + (C_i/x_j)^{D_i}}, \tag{6.7}$$

which replaces $\Phi((x_j - \mu_i)/\sigma_i)$ in (6.2). Provided the coefficient of variation of the ith component is small (as is clearly going to be true from Table 6.3), its distribution will be close to lognormal, and we can use the prior (6.5) again if we define μ_i and σ_i by

$$\begin{aligned} W_i &= \exp\left(\frac{1}{3}(\pi/D_i)^2\right), \\ \mu_i &= C_i\sqrt{W_i}, \\ \sigma_i &= \mu_i\sqrt{W_i - 1}. \end{aligned}$$

see Johnson and Kotz (1970a) chapter 14 and (1970b) chapter 22.

A suitable transformed parameter θ for the numerical integration has the form (6.6), but with C_i and D_i replacing μ_i and σ_i respectively. With this parametrisation, the integration algorithm again converged rapidly, and the final "parameter estimates", calculated analogously to the Normal mixture ones in Table 6.3, are shown in Table 6.5. The only difference between the two sets of p_i values is that the posterior estimate of p_5 is smaller for the log-logistic mixture than for the Normal mixture, and $p_1, \ldots, p_4$ are correspondingly bigger. This is to be expected since the log-logistic distribution has a heavier upper tail than the Normal distribution, so components 1 to 4 leave less probability mass to be fitted by component 5.

Perhaps surprisingly, the log-logistic mixture model using the point estimates in Table 6.5 gives a considerably worse fit to the data than does the Normal mixture ($\chi^2 = 29.6$ compared to $\chi^2 = 5.8$).

		Component				
		1	2	3	4	5
C_i	mean	11.049	15.393	19.958	23.55	26.98
	s.d.	.013	.029	.094	.27	.33
D_i	mean	23.68	22.10	22.78	23.6	31.9
	s.d.	.41	.63	1.76	4.8	7.3
p_i	mean	.4106	.3136	.1781	.0722	.0255
	s.d.	.0046	.0068	.0156	.0174	.0085

Table 6.5. *Final "parameter estimates" assuming a log-logistic mixture*

REFERENCES

Anderberg, M. R. (1973). *Cluster Analysis for Applications*. New York: Academic Press.

Bhattacharya, C. G. (1967). A simple method of resolution of a distribution into Gaussian components. *Biometrics* **23**, 115–135.

Davis, P. J. and Rabinowitz, P. (1984). *Methods of Numerical Integration (2nd ed.)*. Orlando, Florida: Academic Press.

Genz, A. (1986). Fully symmetric interpolatory rules for multiple integrals. *SLAM J. Numer. Anal.* **23**, 1273–1283.

Hammersley, J. M. and Handscomb, D. C. (1964). *Monte Carlo Methods*. London: Methuen.

Haselgrove, C. B. (1961). A method for numerical integration. *Math. Comput.* **15**, 323–337.

Hua Loo-Keng and Wang Yuan (1981). *Applications of Number Theory to Numerical Analysis*. Berlin: Springer-Verlag.

Johnson, N. L. and Kotz, S. (1970a). *Continuous Distributions-I*. New York: John Wiley and Sons.

Johnson, N. L. and Kotz, S. (1970b). *Continuous Distributions-II*. New York: John Wiley and Sons.

Kloek, T. and Van Dijk, H. K. (1978). Bayesian estimates of equation system parameters. An application of integration by Monte Carlo. *Econometrika* **46**, 1–19.

Kuipers, L. and Niederreiten, H. (1974). *Uniform Distribution of Sequences*. New York: John Wiley and Sons.

Naylor, J. C. and Smith, A. F. M. (1982). Applications of a method for the efficient computation of posterior distributions. *Appl. Statist.* 31, 214–225.

Naylor, J. C. and Shaw, J. E. H. (1985). Bayes Four User Guide. *Tech Rep.* **09–85**, Nottingham Statistics Group, Department of Mathematics, University of Nottingham, Nottingham, U. K.

Naylor, J. C. and Shaw, J. E. H. (1987). Spherical integration rules in Bayesian statistics. (To appear).

Niederreiter, H. (1978). Quasi-Monte Carlo methods and pseudo-random numbers. *Bull. Amer. Math. Soc.* **84**, 957–1042.

O'Hagan, A. (1986). Bayesian Quadrature. *Tech Rep.* **82**, Department of Statistics, University of Warwick, Coventry, U. K.

O'Hagan, A. (1987). Monte Carlo is fundamentally unsound. *Statistician* **36**, 247–249.

Ripley, B. D. (1983). Computer generation of random variables - a tutorial. *Int.Statist. Rev.* **51**, 301–319.

Shaw, J. E. H. (1985). A strategy for reconstructing multivariate probability distributions. *Tech. Rep.* **05–85**, Nottingham Statistics Group, Department of Mathematics, Unversity of Nottingham, Nottingham U.K.

Shaw, J. E. H. (1986a). A quasirandom approach to integration in Bayesian statistics. *Bayesian Satistics Study Year Report* **14**, Department of Statistics, University of Warwick, Coventry, U.K. (To appear in *Ann. Statist.*).

Shaw, J. E. H. (1986b). A class of univariate distributions for use in Monte Carlo studies. *Tech. Rep.* **95**, Department of Statistics, University of Warwick, Coventry, U.K. (To appear).

Shaw, J. E. H. (1987). Numerical Bayesian analysis of some flexible regression models. *Statistician* **36**, 147–153.

Skene, A. M., Shaw, J. E. H. and Lee, T. D. (1986). Bayesian modeling and sensitivity analysis using interactive computing. *Statistician 35*, 281–288.

Smith, A. F. M., Skene, A. M., Shaw, J. E. H., Naylor, J. C. and Dransfield, M. (1985). The implemention of the Bayesian paradigm. *Commun. Statist. A* **14**, 1079–1102.

Smith, A. F. M., Skene, A. M., Shaw, J. E. H. and Naylor, J. C. (1987). Progress with numerical and graphical methods for practical Bayesian statistics. *Statistician* **36**, 75–82.

Stewart, L. T. (1979). Multiparameter univariate Bayesian analysis. *J. Amer. Statist. Assoc.* **74**, 684–693.

Stewart, L. T. (1983). Bayesian analysis using Monte Carlo integration - a powerful methodology for handing some difficult problems. *Statistician* **32**, 195–200.

Stewart, L. (1985). Multiparameter Bayesian inference using Monte Carlo integration - some techniques for bivariate analysis. *Bayesian Statistics* **2**. (J. M. Bernardo, M. H. DeGroot, D. V. Lindley and A. F. M. Smith, eds.). Amsterdam: North-Holland , 495–510.

Stewart, L. and Davis, W. W. (1986). Bayesian posterior distributions over sets of possible models with inferences computed by Monte Carlo integration. *Statistician* **35**, 175–182.

Stewart, L. (1987). Hierarchical Bayesian analysis using Monte Carlo integration: computing posterior distributions when there are many possible models. *Statistician* **36**, 211–219.

Stroud, A. H. (1967). Some seventh degree integration formulas for symmetric regions. *SIAM J. Numer. Anal.* **4**, 37–44.

Stroud, A. H. (1971). *Approximate Calculation of Multiple Integrals.* New Jersey: Prentice-Hall, Inc., Englewood Cliffs.

Titterington, D. M., Smith, A. F. M. and Makov, U. E. (1985). *Statistical Analysis of Finite Mixture Distributions.* New York: Wiley.

Upsdell, M. P. (1986). Bayesian inference for functions. *Tech. Rep.* **07–86**. Nottingham Statistics Group, Department of Mathematics, University of Nottingham, Nottingham, U. K.

Van Dijk, H. K. and Kloek, T. (1983). Monte Carlo analysis of skew posterior distributions: an illustrative econometric example. *Statistician* **32**, 216–223.

Van Dijk, H. K. and Kloek, T. (1985). Experiments with some alternatives for simple importance sampling in Monte Carlo integration. *Bayesian Statistics* **2**. (J. M. Bernardo, M. H. DeGroot, D. V. Lindley and A. F. M. Smith, eds.). Amsterdam: North-Holland , 511–530.

Van Dijk, H. K., Hop, J. P. and Louter, A. S. (1987). An algorithm for the computation of posterior moments and densities using simple importance sampling. *Statistician* **37**, 83–90.

Zaremba, S. K. (1972). *Applications of Number Theory to Numerical Analysis.* New York: Academic Press.

DISCUSSION

H. K. VAN DIJK (*Erasmus University*)

The paper by Ewart Shaw has two main topics:

(a) A discussion of possible transformations of random variables so that the posterior kernel has a more regular shape and efficient numerical integration routines for symmetric distributions can be applied.

(b) A discussion of the advantages of using *quasi*-random uniform numbers (where dependence between the numbers is explicitly used) compared with *pseudo*-random uniform numbers (which are constructed with the aim of making the generated numbers independently and identically distributed).

I would like to know whether in the author's experience topic (a) is more important than topic (b) in terms of numerical efficiency. Incidentally, Monte Carlo methods aim in most cases at changing the original pseudo-random uniform numbers in such a way that the resulting sequence is more efficient. Antithetic random numbers, common random and importance sampling are all techniques that aim for a more efficient use of the original sequence of uniform pseudo-random numbers. An advantage of pseudo-random numbers is that the error analysis may be based on relatively straightforward large sample statistical theory. How is error analysis done when one makes use of quasi-random numbers? Can one make use of central limit theorems for dependent observations?

With respect to the transformations one wonders whether there exists a decision strategy on the use of different transformations. In the example presented it appears that the transformation is problem-oriented. How restrictive is in this context, the condition that the first and second-order moments of the transformed parameters have to exist?

Another point is that Shaw imposes the condition that the weights of the numerical integration routines are positive. This is not a requirement for all integration routines and does such a requirement limit the author in the choice of the numerical integration method? Further, it is also my experience that the summarizations (or projections) that are discussed by Shaw are useful (van Dijk and Kloek (1977)), but given the speed of modern computers I prefer now to compute marginal posterior densities instead of projections as suggested by Shaw. With respect to the example studied, one wonders how sensitive the posterior results are for the choice of the prior. The log-logistic seems to fit poorly. It may be illuminating to show the posterior residuals between the fitted model and the data, so that one obtains an insight where the fit is poor: in the center or in the tails of the distribution.

J. C. NAYLOR (*Trent Polytechnic*)

The proposed list of desirable properties for an integration rule (Section 2) is interesting and potentially useful. As stated most rules in common use are constructed using the less natural aim described by (5b). They are exact for some well defined space, F, of functions but their properties outside F are not well defined. Experience suggests that the real motivation for properties (3) and (4) is to provide rules which we have some confidence in for integrands outside F. For the same reasons, I would suggest an additional guideline in selecting a rule: namely, that the weights should not be too different (moderate dimension rules exist with weights varying by factors greater than 10^6).

Even spacing of the integration nodes is perhaps a desirable property, but not the most important. The distance calculation given for a Cartesian product rule (Section 4) is I think a little misleading. The root mean squared distance of a uniformly distributed point to the nearest node is a little more than 0.23. The value of 0.4 is the maximum distance to a nearest node and the point given has exactly 2^{16} nodes at this distance! Are points on concentric hyper-spheres actually more evenly distributed than this? Surely the strongest argument against such a Cartesian product is the 5^{16} nodes.

The routine presentation of marginal densities is surely a primary aim of software for Bayesian statistics and so the redistribution of weights to arbitrary grid points as described in Section 5 seems especially useful. I have a few questions. How should this approach be compared with using marginal sums of a rule which includes such a rectangular grid? How should the points be chosen? It appears that it is essential that all the points fall within the chosen grid. Is this so? Might it be possible to use this technique to estimate posterior density values on a different discretisation (eg the next grid in the iterative scheme)? This could be useful for problems with extremely expensive likelihood (see, for example, the paper by Kim and Schervish in these proceedings), or prior functions.

REPLY TO THE DISCUSSION

I should like to thank the discussants for their valuable comments, and everyone else with whom I had interesting discussions at the Conference.

Beginning with Professor van Dijk's specific questions, I consider identifying suitable parameter transformations to be the single most important part of a general numerical integration strategy. The posterior density is often much more manageable after a simple transformation of the parameter space, see Smith *et al.* (1987) for some examples. The use of quasirandom sequences rather than pseudorandom numbers does, however, provide considerable extra numerical efficiency, and will be most effective after a suitable parametrisation has been found. Note that a quasirandom sequence with pseudorandom offset can be thought of as a set of mutually antithetic variates. For example, a (compound) trapezium rule gives the antithetic variates considered in formula (5.6.9) of Hammersley and Handscomb (1964), and a rational sequence gives the natural high-dimensional generalisation.

Error analysis with an integration rule whose nodes come from a quasirandom sequence can be simply performed as in Section 6.1 of Shaw (1986a). Roughly speaking, split the sequence up into several subsequences, obtain estimates of the integral(s) from each subsequence separately, and use the "between subsequence" variation as a (conservative) estimate of error variance, analogously to the error analysis for straightforward Monte Carlo.

The decision strategy I have used for choosing parameter transformations (for example in Section 6 above) has been very crude; "take logs of positive quantities; if $\theta \in (a, b)$ then transform to $\log(\theta - a) - \log(b - \theta)$, etc." More sophisticated choices of parameter transformations could use ideas from differential geometry, as in Bates and Watts (1981), Kass (1984), Amari (1985), and Cox and Reid (1987).

The existence of the first and second moments of the posterior is not really restrictive since any proper posterior can (in principle) be parametrised so that they exist. Various transformations to lighten the tails of a distribution are suggested in David and Rabinowitz (1984), for example in Section 3.4.5.

Requiring that the weights be positive does of course limit the choice of the numerical integration method, but there is surely nothing wrong with limiting the choice to more reliable methods! Integration rules with negative weights are highly unreliable if the particular conditions under which the rules are "optimal" are not satisfied.

Several plots and further sensitivity analysis on the example had to be omitted to save space. Briefly, the results are less sensitive to changing the prior (6.5) than to changing the log-likelihood (6.4), and the log-logistic model fits most poorly in the centre of the distribution around the second mode.

Professor van Dijk also makes some useful general comments on choosing an integration method and on the value of producing diagnostics of posterior shape. The projection and redistribution technique of Section 5 can be very helpful here.

As Dr. Naylor indicates, it may be valuable to find integration rules that are exact for some function space F, and still have desirable properties on a large function space H containing F. Recent work on this problem is reported in Finney and Price (1986). The additional guideline suggested by Dr. Naylor (i.e. that the weights not be too different) is certainly sensible. Indeed, a major practical difficulty in importance sampling for integration over an unbounded region is that the weights are necessarily unbounded in the tails. This again emphasises the need to choose an importance sampling density with sufficiently heavy tails, so that, in regions of high posterior probability, the weights will not be too different.

Several participants at the Conference had asked why they should bother with anything more sophisticated than a Cartesian product rule. The distance calculation for C_{16} in Section 4 was included just to emphasise the "nonuniformity" of Cartesian product rules in high dimensions. Dr. Naylor suggests that a fairer measure of nonuniformity of a sequence s of points may be r.m.s.d. (s), the root mean square distance of a uniformly distributed point to the nearest node. However, as shown below, this criterion implies that even crude Monte Carlo is better than a high-dimensional Cartesian product rule!

Consider (a) a realisation of a Poisson process with unit density and (b) the d-dimensional rectangular grid of points with integer coordinates, where both (a) and (b) are defined in a suitably large region of $\mathcal{R}^d$.

If X denotes the distance from a random point to the nearest node, then in case (a) we have

$$\Pr(X < r) = 1 - \exp(-Vr^d),$$

where

$$V = \frac{2\pi^{d/2}}{d\Gamma(d/2)}$$

is the volume of the d-dimensional unit sphere. Therefore

$$\begin{aligned}\text{r.m.s.d.}(a)^2 = E(X^2) &= \int_0^\infty V d r^{d+1} \exp(-Vr^d) dr \\ &= \int_0^\infty \left(\frac{x}{V}\right)^{2/d} \exp(-x) dx \\ &= \frac{\Gamma(1+c)(\Gamma(1+1/c))^c}{\pi}\end{aligned}$$

(after some elementary manipulation), where $c = 2/d$.

In case (b), the r.m.s.d. is given by

$$\text{r.m.s.d.}(b)^2 = \frac{d}{12}.$$

In $d = 16$ dimensions, we find that r.m.s.d.$(a) = 1.0623$ and r.m.s.d.$(b) = 1.1547$. Therefore, for the same r.m.s.d., 16-dimensional Cartesian product rule will need approximately $(1.1547/1.0623)^{16} = 3.8$ times as many points as crude Monte Carlo (and more if the coordinates of the Cartesian product rule are not equally spaced). Similarly, with the root mean square distance criterion, pseudorandom sequences on C^d are seen to be more efficient than Cartesian product rules in all dimensions $d \geq 7$, and the relative efficiency increases with d. The corresponding calculations for quasirandom sequences are intractable in general, but many well-known quasirandom sequences will undoubtedly (i.e. with personal probability $> .99$) have much higher efficiency than either Monte Carlo or Cartesian product rules.

The redistribution technique of Section 5 is very simple, efficient and general, but for accurate pictures of a given marginal density it is better to fit a surface to a grid of estimated marginal density values. As Dr. Naylor says, the redistribution grid for forming the approximate marginal density should be chosen so that all the projected nodes ϕ_i lie in the interior of the grid, other than that the choice is up to the user. The grid is probably best located and scaled based on the estimated first and second moments of the posterior, as in the example of Section 6. Preliminary cluster analysis could be used to eliminate outlying nodes, thus reducing the size of grid needed to enclose all the projected nodes. The method can be used to estimate posterior density values on a grid of points forming the nodes of an integration rule: simply divide the redistributed weight of the discretised distribution at a grid point by the corresponding weight of the integration rule at that grid point.

In a verbal contribution, Professor H. Rubin raised the issue of performing decision-oriented calculations. In fact, with the package of Naylor and Shaw (1985), Bayes risks can easily be calculated by special function analysis at the same time as other numerical computations. However, I prefer *whenever possible* to consider utility and probability separately: "the primary end-point of much, if not all, data analysis is the estimation of probabilities of uncertain events which are then used to guide a subsequent decision". (Skene *et al.* (1986), Section 3). This is partly because there may be many different decisions to be made, or many different utilities to be considered with the same posterior. For example: surgeon, patient, patient's family, patient's employer, assorted insurance companies and the Government can have completely different utilities (and even different preference orderings) for the possible outcomes of a proposed major operation, but everyone's probabilities are likely to be close to those of the surgeon, i.e. the expert. Separating the assumptions into model, prior and loss function also makes it easier to highlight which assumptions are irrelevant in determining the inferences/decisions to be made, as in the example of Section 6.

REFERENCES IN THE DISCUSSION

Amari, S. I. (1985). *Differential-Geometric Methods in Statistics. Lecture Notes in Statistics* **28**. Berlin: Springer-Verlag.

Bates, D. M. and Watts, D. G. (1971). Parameter transformation for improved approximate confidence regions in nonlinear estimation. *Ann. Statist.* **9**, 1152–1167.

Cox, D. R. and Reid, N. (1987). Parameter orthogonality and approximate conditional inference, (with discussion). *J. Roy. Statist. Soc. B* **49**, 1–39.

Finney, P. H. and Price. T. E. (1986). Minimum norm quadratures which satisfy nonstandard interpolatory conditions. *SIAM J. Numer. Anal.* **23**, 210–216.

Kass, R. E. (1984). Canonical parameterization and zero parameter-effects curvature. *J. Roy. Statist. Soc B* **46**, 86–92.

Van Dijk, H. K. and Kloek, T. (1977). Likelihood diagnostics and posterior analysis of Klein's model I, Working Paper (Econometric Institute, Erasmus University Rotterdam), presented at the fifteenth NBER-NSF seminar on Bayesian inference in econometrics, October 28–29, 1977 (Madison, Wisconsin).

BAYESIAN STATISTICS 3, pp. 429–435
J. M. Bernardo, M. H. DeGroot, D. V. Lindley and A. F. M. Smith, (Eds.)

What Should Be Bayesian about Bayesian Software?

A. F. M. SMITH
University of Nottingham

SUMMARY

Recent developments in methods of numerical integration and approximation, in conjunction with hardware trends towards the widespread availability of single-user workstations which combine floating-point arithmetic power with sophisticated graphics facilities in an integrated interactive environment, would seem to have removed whatever excuses were hitherto historically available for the lack of any generally available form of Bayesian software. This paper is an attempt to focus the growing debate about the form (or forms) that such software should take, by providing a broad-brush overview of the philosophical, theoretical and pragmatic issues which seem to be raised.

Keywords: BAYESIAN THINKING; ANALYSIS AND REPORTING; SOFTWARE NEEDS.

1. INTRODUCTION

There are a number of reasons why the time seems ripe for a serious review of issues relating to the development and dissemination of Bayesian software.

It is now some ten years since a small group of us decided to organize, here in Spain, an International Meeting on Bayesian Statistics, one consequence of which is that we are now assembled for this Third Valencia Meeting, with two successful conferences and associated proceedings behind us.

At that time, we were clearly encouraged in our enterprise by a perception that Bayesian ideas and methods were becoming firmly established and attracting the attention of an expanding community of both theoretical and applied statisticians. During this past decade, the growth in attendance at these and other Bayesian meetings, as well as in publications directly or indirectly expounding Bayesian ideas and methods, has done nothing to diminish our conviction that Bayesian thinking has an important, illuminating and fundamental contribution to make to much of the activity which takes place under the umbrella heading of statistical science.

However, if the ultimate raison d'être of the statistical sciences lies in the provision of routinely implementable methods for quantifying uncertainties of interest in actual practical problems, what price such conviction about the Bayesian approach if, still at the end of the 1980's, enabling software is simply not available to the would-be user? For better or for worse, the vast majority of statisticians are more influenced by positive, preferably hands-on, experience with applications of methods to concrete practical problems than they ever will be by philosophical victories attained through the (empirically) bloodless means of axiomatics and stylised counter-examples. It is therefore surely no longer acceptable, neither from an intellectual nor a public relations perspective, simply to proclaim and demonstrate, in the theoretical domain, the inevitability of the Bayesian position, without following the enterprise through to provide the appropriate tools in the practical domain.

But what are the appropriate tools? What software? For whom? For what kinds of problems and purposes?

The programme of talks scheduled for this meeting is itself sufficient to underline the fact that Bayesian thinking often leads to rather different perspectives on a number of aspects of the activities which statisticians are involved in, including: the assessments underlying the formulation of models; the forms of numerical algorithms required for applications; the forms of sensitivity analysis and diagnostics required for model elaboration and checking, and the ways in which inferences are reported or translated into practical actions. These different perspectives and techniques mean that it is not at all obvious that the requirements for Bayesian software simply consist in mimicking the general structures and facilities of current, popular, non-Bayesian software, but with appropriate substitution of various of the algorithmic elements.

It seems appropriate, therefore, to begin a discussion of what is required of Bayesian software by reviewing, in summary form, what is distinctive about Bayesian thinking. For convenience of exposition, various aspects of the approach are reviewed under the subheadings of modelling, analysis and reporting. In practice, of course, things are not so neatly subdivided and there is a dynamic interaction of these elements.

2. BAYESIAN THINKING

2.1. General Context

It is convenient to identify three stages in the scientific learning cycle within which various aspects of statistical activity take place:

(i) creating a formal frame of discourse for the systematic study of the problem situation;

(ii) reporting inferences and making decisions within the currently adopted formal framework;

(iii) probing and elaborating that framework.

In the first phase of this cycle, the major inputs will typically be provided by subject matter experts in demarcating the kinds of relationships and structures to be considered and explored. Data analytic techniques, particularly those directed at revealing relatively noise-free structure in multivariate contexts by computer intensive and graphical methods, also have an important role to play in attempting to arrive at an enlightening representation of the form

$$\textit{(transformed) data} = \textit{structure} * \textit{noise},$$

where * denotes the operation whereby the "noise" disturbs the "structure". Formal Bayesian procedures, in contrast to what one might think of as general Bayesian awareness, are typically irrelevant to this more exploratory phase of activity and the discussion which follows should therefore be understood as referring mainly to contexts in which a more or less formal representation (or, more usually, a series of alternative possible representations) is already being entertained, and where the second and third phases of the above cycle are in operation.

2.2. Modelling

From the subjectivistic perspective, there is increasing emphasis on the fundamental tenet that meaningful quantifications of uncertainty are those made about observables. In conventional terminology, this corresponds to a trend towards the predictivistic paradigm and away from the parametric paradigm, the latter being acknowledged as often essential for orientation, but of no intrinsic interest as an end in itself. However, in many situations parameters can be interpreted as large sample limits of observables, so that parametric inference becomes a limiting case of predictive inference, and specifications of prior distributions for parameters acquire a tangible, operational meaning.

The de Finetti injunction to "think about things" makes the routine adoption of stylised model forms —for example, one-parameter exponential family likelihoods together with conjugate priors, or linear regression likelihood together with conventional vague priors for regression coefficients and variances— rather unpalatable to anyone concerned with Bayesian thinking rather than mechanical Bayesian procedures. One response might be to move towards so-called nonparametric modelling strategies, although at the current time the available techniques themselves seem rather stylised. Another possibility might be to confine attention to Bayesian analogues of low-order moment based frameworks, but at the current time practical experience with such approaches is extremely limited. Perhaps the most widely adopted strategy at present is to continue working within a parametric framework, but with a much more creative approach to the choices of likelihood and prior forms, making extensive use of robustifying elaborations, hierarchical forms, spline functions, etc, in a systematic attempt to provide flexible inferential environments transcending the constraints of linearity, normality and immediate analytic tractability. For the purposes of this discussion, the latter general approach will be assumed.

2.3. Analysis and Reporting

From a formal perspective, Bayes' theorem and other standard results in probability theory provide the mechanisms whereby the modelling elements of likelihoods and parameter distributions can be manipulated to provide overall descriptions of uncertainly about parameters or future observables, as well as specific descriptions and summaries of uncertainty for marginal, joint or conditional features of interest. The key manipulative operation throughout is that of integration.

However, Bayesian thinking leads one to emphasise that Bayes' theorem, like any theorem in probability, is simply a form of disciplined "uncertainty accounting". For coherence, the left-hand side has to equal the right-hand side. Thus:

for given data;

> specification of a likelihood constrains the compatible pairs of prior and posterior uncertaintly descriptions;
>
> specification of a prior form defines a mapping from possible likelihoods to corresponding posterior forms;

for a given likelihood form;

> specification of a prior form defines a mapping from possible data sets to corresponding posterior forms.

There is no implied chronology in the relating of aspects of the model and descriptions or uncertainty. Viewed in this light, Bayes' theorem simply provides a powerful, flexible tool for examining both actual and potential descriptions of uncertainty arising when one or more individuals seek to interpret data in the light of possible assumptions and uncertainties about those assumptions.

The framework is ideally suited to sensitivity analysis: that is to say, to exploring the extent and nature of the dependence of the outputs of an analysis on the inputs. Here, inputs could refer to any or all of the modelling ingredients, which might be varied to reflect within or between individual hesitations or disagreements, and outputs could be any of the descriptions of uncertainties of interest, or actions based on those uncertainties. The paradigm is equally well suited to private or public analysis, and to inference or decision-making. Above all, there is a recognition that there are no magic, "objective" or "correct" specifications of ingredients in any given problem. Thus, in terms of the general representation referred to in Section 2.1, there is scope for open-minded and creative sensitivity analysis regarding any or all of the elements *data transformations; structure; form of *; noise distribution.*

In place of the traditional process

single model $\rightarrow$ *data* $\rightarrow$ *conclusion,*

Bayesian thinking leads inexorably to the routine and systematic adoption of the process

range of models $\rightarrow$ *data (with variations)* $\rightarrow$ *range of conclusions.*

3. SOFTWARE IMPLICATIONS

3.1. Modelling Facilities

For convenience of exposition, I shall talk in terms of a basic dichotomy between *likelihood management* and *prior management* facilities, with the following further subdivisions:

Likelihood Management

data :	selection, transformation
structure :	choice of variables, functional forms
noise :	link to structure, forms of distribution

Prior Management

actual :	elicitation and representation
stylised :	conjugate and limiting forms,

where the need for prior specification could be generated by unknown parameters appearing in any or all of data selection or transformation, structure description and noise characteristics.

So far as likelihood management is concerned, the facilities required would seem to be rather similar to those required by non-Bayesian statisticians. Ideally, we would like a flexible form of menu-driven interface whereby a series of selections provides sufficient instructions for generating the likelihood.

It is, of course, in the provision of facilities for prior specification that the requirements of a Bayesian software system are very different. Here, I think we need to aim at providing facilities for two rather distinct types of input. On the one hand, there will be situations where the elicitation and encoding (in some form) of one or more instances of actual individual prior beliefs is of interest. On the other hand, there will be situations where the analysis and reporting problem requires the generation of a range of prior to posterior analyses, corresponding to a suitably chosen range of stylised prior forms, typically including, as "extreme points", a representation of overall ignorance together with various forms of dogmatic beliefs about different aspects of the model. We might, of course, want to provide both types of input for the same problem.

3.2. Analysis and Reporting Facilities

In both the forms in which final inference summaries are reported, and also in the methods by which they are obtained, the Bayesian approach can differ radically from other approaches.

Whether interest is focussed on predictive or posterior distributions, descriptions of uncertainty typically take the form of density curves or surfaces (usually in the form of contour or perspective representations), together with judicious choices of presentation of univariate and bivariate marginals and conditional versions thereof, selected in an attempt to build up a "picture" of the overall and specific uncertainties of interest. Often, these will need to be supplemented by decision-related summaries, such as credible intervals for future predictions as a function of a covariate appearing in the model structure. For most such special forms of inference report, the full posterior distribution for all unknown parameters will typically be an essential intermediate calculation, whether or not inference about the individual parameters is specifically requested as an output. The key algorithmic requirement is therefore, above all, that of the efficient description (in some appropriately encoded form) of key aspects of the full joint posterior distribution for model parameters.

Clearly, we can separate the notions of *algorithm management* and *reporting management* to some extent, but in practice there is likely to be a close interaction. The output required will drive the form of analysis to be undertaken and hence the algorithms to be employed; in turn, an algorithm exploring a high-dimensional posterior density will often reveal unsuspected features of interest and suggest further interesting forms of inference report. That said, in broad-brush terms the most important subdivisions within these management strategies would appear to be as follows.

Algorithm Management

analytic :	conjugate forms, approximations (maximization, asymptotic expansion)
numerical :	integration (grids, spherical, importance sampling)
graphical :	curves, contours (splines, etc), 3^+-dimensional views

Reporting Management

graphical :	densities, contours, views required
tabular :	moment and interval summaries
diagnostics :	model fits, sensitivity analysis

Since the Bayesian approach typically has little regard for inference reporting via point estimates and standard errors, it is clear that Reporting Management will be relatively complex in Bayesian software and potentially more highly interactive (with Algorithm and Model Management) than is typically the case with non-Bayesian approaches. Similarly, since the examination of an uncertainty surface via maximization and calculation of local curvature is regarded as a poor substitute for an overall view of the shape of the whole posterior distribution, the algorithmic aspects of the Bayesian approach are also correspondingly more complex, the level of complexity required being, in part, a response to the kinds of inference reports requested.

3.3. The user

For whom should we be providing software? Here is a possible taxonomy of users, with brief notes on their requirements and the implications for the various management facilities we have been considering.

The Bayesian research statistician. Such a person should presumably be regarded as an expert user, for whom the full range of facilities should be available, with sufficient background documentation to permit critical and informed choices.

The non-Bayesian research statistician. Someone in this category might well want Bayesian output in order to undertake comparative methodological studies. He or she is unlikely to want to have to become a Bayesian algorithmic expert, but wants to be able to ask for rather sophisticated and possibly non-standard outputs. If informed choices are to be made, this user would probably like flow-chart direction of the type familiar from scientific subroutine libraries.

The broadly-focussed applied statistician. This is a user who encounters a wide range of different types of statistical problem, calling forth a correspondingly wide repertoire of models and variety of types of inference summary. Such a user is willing to provide sophisticated inputs into the comparatively few cases of non-standard problems he or she encounters, but wants extremely well-oiled user-friendly machinery for quickly dispatching the more standard cases.

The narrowly-focussed applied statistician. This is either a statistician handling a very narrow range of problems, invoking a small limited set of modelling possibilities, or even a non-statistical specialist (for example, toxicologist or geneticist) who handles all his or her own data in a very specific modelling domain (for example, using probit analysis or highly structured contingency tables). Such a user is typically looking for a rather automated system, having just a few options and only requiring manual intervention for "difficult" cases (which the system should somehow flag).

The student. This user also requires a system that is very largely automated, but which covers a rather substantial range of modelling and output options, including all those relevant to the various courses he or she is studying. Clearly, the range of options required to provide a suitable software environment for a statistics student would be very much wider than, say, for a psychology student exposed only to simple one-sample and two-sample problems and a few small-scale experimental designs.

4. SOLUTIONS?

4.1. The need for Software

I have sketched out, in a highly superficial fashion, some of what would seem to be the inescapable features and user-specific desiderata of would-be disseminated Bayesian software.

The overwhelming argument in favour of developing such software is that the environments within which most routine applied statistical practice and much statistical education take place are largely software determined. Specific novel software development for each individual problem encountered is usually beyond the resources, inclinations or abilities of hard-pressed practitioners or students with more conceptual preoccupations. If we seek the wider appreciation of Bayesian ideas and the wider use of Bayesian methods, we *must* provide appropriate user-friendly software, packaged and documented to the kind of standards the user has now become accustomed to expect from the currently widely used packages.

I used to be attracted to the purist doctrine that automated, packaged methods are so totally contrary to the spirit of Bayesian thinking that striving towards a Bayesian package would be misguided. Such an attitude has the merit of warning one that great care is required in this enterprise, but, if taken seriously, would encourage a perpetual continuation of the current dominant view of the Bayesian position as the statistical equivalent of a hard-core pornographic movie: something that the young should certainly be warned against, and —although occasionally complimented, *sotto voce*, for its suggestive insights— ultimately dismissed with lofty disdain for not being like "the real thing".

4.2. What? And for Whom?

In the light of the taxonomy of users put forward in Section 3.3, it seems to me useful to plan future developmental activity having in mind three broad-brush types of potential user: researcher, practitioner and student. My own suggestions as to how we might proceed on each of these fronts are as follows.

For researchers, we should encourage the development and dissemination of technical modules for analysis and reporting, typically in the form of libraries of subroutines to be called from within a general environment such as "S", which offers already packaged data management and display facilities. The need here is for an organisational focus for publicising, distributing and disseminating the software and documentation produced by individuals and groups.

For practitioners, in addition to the need for technical modules to provide the basis for writing one-off problem-specific software, there is a need for application-focussed and model-focussed modules, ideally as facilities within a widely available system such as SAS, or packaged separately for running on a variety of widely available PC's.

For students, there is a need for software linked to appropriate expository texts. Again, this could take the form of material packaged separately for running on PC's, or there may be advantages in aiming at an environment already widely used for statistics teaching, such as MINITAB.

4.3. When?

Clearly, there is an enormous amount of work to be done if the goal of providing readily available user-friendly software for all categories of potential user is to be realised. I sense, however, that an increasing number of those committed to the dissemination of Bayesian ideas are similarly convinced of the urgency and priority of this task. I believe, therefore, that dramatic changes will have taken place before we reassemble for the session on computation at the Fourth Valencia meeting in 1991.

DISCUSSION

H. K. VAN DIJK (*Erasmus University*)

The broad analysis of the paper presented by Adrian Smith is what I recommend as a starting point for a discussion on a Bayesian software package. In this context one has to be aware of the relation between Bayesian software and Bayesian statistical modelling. So far, techniques have been developed that give insight into the posterior kernels of certain models given a finite sample. Whether these models are inadequate or adequate representations is not discussed in great detail. The issue of computational ease and ease of modelling has several aspects of which I mention only two. On the one hand, researchers should be able to explain the implications of an ill-designed model, but it may be difficult to construct an algorithm for numerical integration for such cases. On the other hand, numerical procedures for simple models appear to be easy, but one does not know whether the model is too simple. Bayesian statistical analysis should provide some procedures that are robust with respect to a class of models and that can serve as guidelines to improve upon the design of the model. There is much that can be done in the area of Bayesian statistical modelling.

BAYESIAN STATISTICS 3, pp. 437–451
J. M. Bernardo, M. H. DeGroot, D. V. Lindley and A. F. M. Smith, (Eds.)

To Weight or not to Weight, That is the Question

(Whether 'tis nobler in the mind to suffer
The slings and arrows of outrageous fortune,
Or to take arms against a sea of troubles,
And by opposing end them?) Hamlet, Act 3, Scene 1.

T. M. F. SMITH
University of Southampton

SUMMARY

Weighting by the inverse unit selection probabilities is the basis of randomization inference. In a model-based framework probability designs are ignorable and so probability weights have no obvious role. This issue of whether to weight or not is examined by following Rubin (1983) and conditioning on the selection probabilities. Using results from size biased sampling it is shown that randomization estimators can be justified.

Keywords: RANDOMIZATION; WEIGHTING; CONDITIONAL INFERENCE; IGNORABLE DESIGNS, SIZE-BIASED SAMPLING; REGRESSION; ROBUST ESTIMATION.

1. INTRODUCTION

Statisticians frequently seek to protect themselves against outrageous fortune by an act of randomization. In sample surveys this may involve the use of different selection probabilities for different population units and the inverse selection probabilities may then be used as weights in forming estimates of population totals. These weights are basic to randomization inference and any method of estimation which fails to use them is treated with great suspicion. An alternative to randomization inference is to assume that the distribution of population values can be represented by a probability model. A sample selection mechanism using randomization can be ignored for model-based inferences and then there is no apparent role for probability weights. The problem addressed in this paper is whether probability weights have a role in model-based inference for sample surveys.

2. RANDOMIZATION INFERENCE

Let I_i be an indicator variable for unit $i, i = 1, \ldots, N$, in a finite population, such that

$$I_i = \begin{cases} 1 & \text{if } i \in s \\ 0 & \text{otherwise,} \end{cases}$$

where $s = (i_1, \ldots, i_n)$ is the set of labels selected by a sampling mechanism. For samples of fixed size n we have $\sum_{i=1}^{N} I_i = n$ and

$$\Pr(I_i = 1) = \pi_i, \tag{2.1}$$

which is the inclusion probability for the ith unit when randomization is employed. We assume that $0 < \pi_i < 1$ for all i.

A sampling mechanism is a rule for selecting s, a subset of the population units. Let $\boldsymbol{X}$ denote the prior knowledge available to a statistician before drawing the sample and let $\boldsymbol{Y}$ denote the $N \times p$ matrix of values of the survey variables of interest. A sampling mechanism of the form

$$p(s|\boldsymbol{X}) \tag{2.2}$$

for which $0 < \pi_i = \sum_{s \ni i} p(s|\boldsymbol{X}) < 1$ is called strongly ignorable, Rosenbaum and Rubin (1983). Random sampling schemes satisfy this condition, but quota sampling schemes may not, see Smith (1983). In practice the observed sample may also be determined by a non-response selection mechanism which is not under the statistician's control and may depend on the survey variables $\boldsymbol{Y}$. Such a mechanism would not be ignorable, see Little (1982). In this paper we assume throughout that the selection mechanism is strongly ignorable.

Let $\mathcal{S}$ denote the $\binom{N}{n}$ possible samples which might be drawn. The probability distribution on $\mathcal{S}$ determined by $p(s|\boldsymbol{X})$ is the randomization distribution. From the statistical point of view it has the interesting property of being completely known; it is not indexed by any unknown parameters, nor is it directly related to the survey variables $\boldsymbol{Y}$. If T is some function of $\boldsymbol{Y}$ of interest and $\hat{T}_s$ is an estimator of T then the only statistical operation of any content is to take expectations with respect to $p(s|\boldsymbol{X})$, that is to form

$$E_p(\hat{T}_s) = \sum_{s \in \mathcal{S}} \hat{T}_s p(s|\boldsymbol{X}). \tag{2.3}$$

Since $\boldsymbol{Y}$ can take any values the only useful general constraint is to require that

$$E_p(\hat{T}_s) = T \qquad \text{for all possible } \boldsymbol{Y}, \tag{2.4}$$

that is to require that estimators be p-unbiased. When T is a total and the estimators are linear in the indicator variables I_i, so that

$$\hat{T}_s = \sum_{i=1}^{N} w_i g(\boldsymbol{Y}_i) I_i, \quad \text{and} \quad T = \sum_{i=1}^{N} g(\boldsymbol{Y}_i), \tag{2.5}$$

p-unbiasedness leads to

$$E_p(\hat{T}_s) = \sum_{i=1}^{N} w_i g(\boldsymbol{Y}_i) \pi_i = \sum_{i=1}^{N} g(\boldsymbol{Y}_i) \qquad \text{for all } \boldsymbol{Y},$$

so that

$$w_i = \pi_i^{-1},$$

the inverse probability weight.

Example 1. The population mean

Let $T = \bar{Y} = \Sigma_1 Y_i / N$. An unbiased estimator is

$$\hat{T}_{1s} = \frac{1}{N} \sum_{i \in s} w_i Y_i, \quad \text{with } w_i = \pi_i^{-1}. \tag{2.6}$$

Which is the well known Horvitz-Thompson estimator. If a computer package is used for data analysis then the weighted estimator will be

$$\hat{T}_{2s} = \sum_{i \in s} w_i Y_i / \sum_{i \in s} w_i, \tag{2.7}$$

which is now a ratio and is not unbiased. However, $\hat{T}_{2s}$ is component-wise unbiased in the sense that $\Sigma_{i\in s} w_i Y_i$ is an unbiased estimator of $\Sigma_1^N Y_i$ and $\Sigma_{i\in s} w_i$ is an unbiased estimator of N. $\hat{T}_{2s}$ was suggested by Hajek (1971) as a possible solution to the Basu elephant problem and has the desirable property of being location invariant, which is not true for $\hat{T}_{1s}$.

Example 2. A regression coefficient.
Let

$$B = \sum_{i=1}^{N} (Y_{1i} - \bar{Y}_1) Y_{2i} - \bar{Y}_2) / \sum_{i=1}^{N} (Y_{2i} - \bar{Y}_2)^2 \tag{2.8}$$

be the finite population regression coefficient between Y_1 and Y_2. Apparently this is sometimes of interest. There are many alternative estimators of B, all of which are biased. Applying the weights w_i to each unit $i \in s$ gives

$$\hat{B}_w = \frac{\sum_s w_i \sum_s Y_{1i} Y_{2i} w_i - \sum_s Y_{1i} w_i \sum_s Y_{2i} w_i}{\sum_s w_i \sum_s Y_{2i}^2 w_i - \left(\sum_s Y_{2i} w_i\right)^2}, \tag{2.9}$$

which is the analogue of (2.7). In (2.9) a term like $\Sigma_s Y_{1i} Y_{2i} w_i$ is the unbiased estimator of $\Sigma_{i=1}^N Y_{1i} Y_{2i}$, and so (2.9) can be viewed as a function of unbiased estimators of totals T_j. So if $h(\boldsymbol{T})$ is the function of interest $h(\hat{\boldsymbol{T}}_s)$ is the component-wise unbiased estimator. All standard sample survey estimators are in this class so randomization inference for sample surveys is closely tied to p-unbiasedness. Taylor series expansions give the conditions under which this is a reasonable approach.

For a general multiple regression problem with Y_{1i} regressed on $\boldsymbol{Y}_{2i}, i = l, \ldots, N$, then the weighted estimator is

$$\hat{\boldsymbol{B}}_w = \left(\boldsymbol{Y}_{2s}^T \boldsymbol{w}_s \boldsymbol{Y}_{2s}\right)^{-1} \boldsymbol{Y}_{2s}^T \boldsymbol{w}_s \boldsymbol{Y}_{1s}, \tag{2.10}$$

where $\boldsymbol{Y}_{1s}$, is the $n \times 1$ vector of dependent variables in s, $\boldsymbol{Y}_{2s}$ is the $n \times p$ matrix of explanatory variables in s, and $\boldsymbol{w}_s = \text{diag}(w_i, i \in s)$ is the $n \times n$ matrix with sample weights down the diagonal. This is the solution obtained by using the weighting option in a standard package of statistical programs. It should be noted however, that the variance associated with (2.10) in packages is usually the weighted least squares variance

$$\hat{V}(\hat{\boldsymbol{B}}_w) = \left(\boldsymbol{Y}_{2s}^T \boldsymbol{w}_s \boldsymbol{Y}_{2s}\right)^{-1} \hat{\sigma}^2, \tag{2.11}$$

and this is not the randomization variance derived from $p(s|\boldsymbol{X})$, see Rao (1975).

3. ALTERNATIVES TO RANDOMIZATION INFERENCE

3.1. Ordinary Least Squares (OLS)

If Rubin (1976), Rosenbaum and Rubin (1983), Sugden and Smith (1984) etc., say that random sampling is ignorable for inference then why not ignore it? Ignoring sample weights surely implies using equal weights which in turn implies ordinary least squares as a criterion for regression analysis.

The OLS estimator is

$$\hat{\boldsymbol{B}}_0 = \left(\boldsymbol{Y}_{2s}^T \boldsymbol{Y}_{2s}\right)^{-1} \boldsymbol{Y}_{2s} \boldsymbol{Y}_{1s}, \tag{3.1}$$

which is an unweighted alternative to $\hat{\boldsymbol{B}}_w$ in (2.10). This is frequently chosen for the analysis of data from a complex survey as a default option. As we shall see in Section 3.2 this approach takes the word ignorable at face value and fails to read the small print.

Social surveys are usually designed to be self-weighting, in which case the OLS estimator is also the component-wise unbiased estimator. However, as stated above, the least squares variance is not the correct p-variance and clustering in the design can lead to considerable inflation of the true p-variance relative to the OLS variance, see Kish and Frankel (1974).

3.2. Adjusted Least Squares (ALS)

A full model of a survey requires the joint distribution of the survey variables $\boldsymbol{Y}$, the prior variables $\boldsymbol{X}$ and the sample selection variable s. Formally we can write

$$f(s, \boldsymbol{Y}, \boldsymbol{X}; \lambda) = p(s|\boldsymbol{X}) f(\boldsymbol{Y}|\boldsymbol{X}; \theta) g(\boldsymbol{X}; \phi), \tag{3.2}$$

where $\boldsymbol{\lambda} = (\boldsymbol{\theta}, \boldsymbol{\phi})$ is a vector of parameters. The sample data comprise $d_s = (s, \boldsymbol{Y}_s, \boldsymbol{X})$, and then

$$f(d_s; \lambda) = p(s|\boldsymbol{X}) g(\boldsymbol{X}; \phi) f_s(\boldsymbol{Y}_s|\boldsymbol{X}; \theta), \tag{3.3}$$

where $f_s(\boldsymbol{Y}_s|\boldsymbol{X};\theta) = \int f(\boldsymbol{Y}|\boldsymbol{X};\theta) dy_{\bar{s}}$, and $Y_{\bar{s}}$ denotes $\{Y_i : i \notin s\}$. If $\boldsymbol{X}$ is known then predictive inferences about $Y_i; i \in \bar{s}$, can be made via the conditional distribution $f(\boldsymbol{Y}|\boldsymbol{X};\theta)$ ignoring the design $p(s|\boldsymbol{X})$. If $\boldsymbol{X}_i$, $i \in \bar{s}$ is not known then the design $p(s|\boldsymbol{X})$ contains potentially usefull information that will help the satistician to predict $\boldsymbol{X}_i$, $i \in \bar{s}$ and hence $\boldsymbol{Y}_i$, $i \in \bar{s}$, see Scott (1977), Sugden and Smith (1984).

From the sample data the parameters θ can be estimated from the conditional distribution $f_s(\boldsymbol{Y}_s|\boldsymbol{X};\theta)$, and the parameters ϕ from the marginal distribution $g(\boldsymbol{X};\phi)$. In the regression problem in Section 2 the parameter of interest was a regression coefficient between Y variables and is thus defined in the marginal distribution of $\boldsymbol{Y}$, which is not directly observable from the sample data. The problem is how to use the sample data to estimate parameters in the marginal distribution of $\boldsymbol{Y}$?

If $(\boldsymbol{X}, \boldsymbol{Y})$ has a multivariate normal distribution, or equivalently if $E(\boldsymbol{Y}|\boldsymbol{X})$ is linear in $\boldsymbol{X}$ with a constant covariance matrix, where $E(\cdot)$ denotes expectation with respect to the model, then the adjusted least squares estimators of μ_y, Σ_{yy}, the mean vector and covariance matrix of the marginal distribution of $\boldsymbol{Y}$ are

$$\hat{\mu}_Y = \boldsymbol{m}_y + \boldsymbol{b}_{yx}(\boldsymbol{M}_x - \boldsymbol{m}_x),$$

and

$$\hat{\Sigma}_{YY} = \boldsymbol{s}_{yy} + \boldsymbol{b}_{yx}\left(\boldsymbol{S}_{xx} - \boldsymbol{s}_{xx}\right)\boldsymbol{b}_{yx}^T, \tag{3.4}$$

where $S = \begin{pmatrix} s_{xx} & s_{xy} \\ s_{yx} & s_{yy} \end{pmatrix}$ is the *unweighted* sample covariance matrix of (X_s, Y_s), $b_{yx} = s_{yx}^{-1} s_{xx}$, S_{xx} is the finite population (known) covariance matrix of X, and m_y, m_x are the unweighted sample mean vectors of (Y, X), and M_x is the finite population mean vector of X. If $\hat{\Sigma}_{YY}$ is partitioned according to (Y_1, Y_2), then the adjusted regression coefficient of Y_1 on Y_2 becomes

$$\hat{B}_A = \hat{\Sigma}_{Y_1Y_2} \hat{\Sigma}_{Y_2Y_2}^{-1} \tag{3.5}$$

The properties of (3.5), (3.1) and (2.10) have been compared in a simulation study by Holt, Smith and Winter (1980), Smith (1981), under various sampling schemes. For unequal probability selection schemes the OLS estimator B_0, is badly biased, while both $\hat{B}_w$ and $\hat{B}_A$ remain approximately unbiased provided the population satisfies the linearity and homoscedasticity assumption. Under this assumption $\hat{B}_A$ is generally more efficient than $\hat{B}_w$. These results are given in Table 1 for a multivariate normal model.

Design	Means $\hat{B}_0$	$\hat{B}_W$	$\hat{B}_A$	$\hat{B}_{AW}$	S.D. $\hat{B}_0$	$\hat{B}_W$	$\hat{B}_A$	$\hat{B}_{AW}$
D_1	.721	.721	.721	.721	.041	.041	.041	.041
D_2	.721	.721	.721	.721	.041	.041	.041	.041
D_3	.725	.719	.722	.719	.041	.043	.041	.043
D_4	.735	.722	.725	.722	.041	.054	.041	.054
D_5	.737	.722	.724	.722	.042	.063	.042	.062
D_6	.746	.720	.721	.719	.041	.010	.044	.109
D_7	.702	.723	.719	.723	.039	.043	.039	.043
D_8	.677	.716	.711	.716	.036	.085	.036	.085
D_9	.673	.719	.710	.719	.035	.123	.037	.123

Table 1. *Biases and standard deviations of estimated regressions.*
Simulated population, $N = 7,027$.
$Y_1 = \log(\text{expenditure on food})$
$Y_2 = \log(\text{total expenditure})$
$X = \log(\text{expenditure in housing})$
Mean and covariance matrix from Family Expenditure Survey.
Population regression $Y_1 = 1.74 + 0.71Y_2$
Finite population regression $Y_1 = 1.63 + 0.72Y_2$

What happens if the regressions are not linear or the variances are hecteroscedastic? Pfeffermann and Holmes (1985), Holmes (1987) show that $\hat{B}_A$ is not robust to these changes, whereas B_w remains approximately unbiased. In Table 2 some results are shown for repeated samples from a real finite population, the data being the U. K. Family Expenditure Survey for 1977.

3.3. A Compromise Estimator

Nathan and Holt (1980) show that the OLS estimator $\hat{B}_0$ is biased in the conditional distributions given (X, s) while the adjusted estimator $\hat{B}_A$ is approximately unbiased provided that the model is true. The empirical results in Table 2 suggest that $\hat{B}_A$ is not robust to departures from the model assumption but that the p-weighted estimator does have robustness properties. Can we get the best of both worlds by using a p-weighted version of $\hat{B}_A$? Nathan and Holt propose such an estimator and this is the estimator $\hat{B}_{AW}$ in Tables 1 and 2.

From the simulations it appears that $\hat{B}_{AW}$ shares the robustness properties of $\hat{B}_0$. When the simulation results are plotted in bands according to the value of X then it appears that

	Means				S.D.			
Design	$\hat{B}_0$	$\hat{B}_W$	$\hat{B}_A$	$\hat{B}_{AW}$	$\hat{B}_0$	$\hat{B}_W$	$\hat{B}_A$	$\hat{B}_{AW}$
D_1	.714	.714	.713	.713	.047	.047	.047	.047
D_2	.714	.714	.713	.713	.048	.048	.047	.047
D_3	.701	.712	.694	.711	.051	.049	.050	.049
D_4	.693	.711	.668	.708	.056	.063	.055	.063
D_5	.669	.706	.645	.703	.058	.065	.056	.066
D_6	.656	.699	.608	.691	.063	.111	.063	.116
D_7	.660	.701	.698	.700	.042	.088	.044	.087
D_8	.677	.716	.711	.716	.036	.085	.036	.085
D_9	.658	.712	.701	.712	.040	.123	.043	.0123

Table 2. *Biases and standard deviations of estimated regressions.*
Real population, $N = 7,027$.
Details as above
Finite population regression $Y_1 = 1.74 + 0.71Y_2$

$\hat{\boldsymbol{B}}_{AW}$ has better properties than $\hat{\boldsymbol{B}}_0$ in the conditional distribution given $(\boldsymbol{X}, s)$. It really does seem to benefit from both approaches!

Faced with these empirical results which favour p-weighting how should somebody who believes in models proceed? Brewer (1979), Little (1983), both advocate estimators based on models which are then protected against model misspecification by choosing the sampling scheme and estimator to make the estimator approximately p-unbised. The estimator $\hat{\boldsymbol{B}}_{AW}$ is chosen in this spirit. DuMouchel and Duncan (1983) have examined the issue of weighting for regression analysis in a wider context. They have considered cases where weighting should not be used, for example when certain models are strongly believed to be true, and have suggested that weighting might be used when $\boldsymbol{B}$ in (2.8) is the parameter of interest. In this latter case they advocate testing the difference between $\hat{\boldsymbol{B}}_0$ and $\hat{\boldsymbol{B}}_W$ and if there is no difference using $\hat{\boldsymbol{B}}_0$. If a difference is found then they adovocate introducing extra variables into the model to explain the difference. In our context they widen the regression to include variables in the design set $\boldsymbol{X}$. In their example this strategy works and conditional on the $\boldsymbol{X}$ variables an unweighted regression explains the data adequately.

4. A MODEL-BASED JUSTIFICATION FOR WEIGHTING

The proposals in the previous section for including probability weights into estimation are to some extent ad hoc. In Brewer's approach the design must be shown to be consistent with the model while in Little's approach the selection probabilities are stratified after selection to make the model consistent with the desing. DuMouchel and Duncan's final proposal is to condition on the $\boldsymbol{X}$ variable, thus extending the model beyond the marginal distribution of $\boldsymbol{Y}$.

In this section we show that probability weights can feature naturally in model-based inference by following Rubin (1983) and conditioning on the vector $\boldsymbol{\pi} = (\pi_1(\boldsymbol{x}), \ldots, \pi_N(\boldsymbol{x}))$ of inclusion probabilities rather than the whole design set $\boldsymbol{X}$. The target for inference is still some property of the marginal distribution of $\boldsymbol{Y}$ such as a predictive inference about $\bar{Y}$, the finite population mean. Rubin showed that the vector $\boldsymbol{\pi}$, the propensity score, is frequently an adequate summary of the prior information $\boldsymbol{X}$ in the sense that

$$p(s|\boldsymbol{X}) = p(s|\boldsymbol{\pi}), \tag{4.1}$$

and that this still enables $p(s|\boldsymbol{X})$, or $p(s|\boldsymbol{\pi})$, to be ignored for model-based inference. He then suggests using the joint distribution of $(\boldsymbol{Y}, \boldsymbol{\pi})$ rather than that of $(\boldsymbol{Y}, \boldsymbol{X})$ for constructing the model. Let the data be $d_s = (s, \boldsymbol{Y}_s, \boldsymbol{\pi})$ then

$$f(d_s;\lambda) = p(s|\boldsymbol{X})g(\boldsymbol{\pi};\phi)f_s(\boldsymbol{Y}_s|\boldsymbol{\pi};\theta) \\ = p(s|\boldsymbol{\pi})g(\boldsymbol{\pi};\phi)f_s(\boldsymbol{Y}_s|\boldsymbol{\pi};\theta); \tag{4.2}$$

and this has the same form as (3.3) and so $p(s|\boldsymbol{\pi})$ can be ignored for inference about $\boldsymbol{\theta}, \boldsymbol{\phi}$, or $\boldsymbol{Y}$. Rubin argues further that frequently it will be simpler to construct $f(\boldsymbol{Y}|\boldsymbol{\pi};\theta)$ than $f(\boldsymbol{Y}|\boldsymbol{X};\boldsymbol{\theta})$.

Unfortunately in social surveys most designs are self-weighting which means that $\pi_i(\boldsymbol{x})$ is constant for all $i = 1, \ldots, N$. In this case $\boldsymbol{\pi}$ contains no useful information. However, by expanding $\boldsymbol{\pi}$ to $\boldsymbol{\pi}^* = \left(\frac{\pi}{\boldsymbol{L}}\right)$, where $\boldsymbol{L}$ is the set of higher level labels denoting the stratification and clustering in the design, and then conditioning on $\boldsymbol{\pi}^*$ leads to stratification models and multi-level models (components of variance) which are adequate for modelling $\boldsymbol{Y}$.

When the weights in $\boldsymbol{\pi}$ are not all equal then we can distinguish two cases:

(i) stratification, with $\pi_h = \frac{n_h}{N_h}$ in stratum h, and not all π_h equal;

(ii) variable probability sampling with $\pi_i \neq \pi_j$, for some $j \neq i$.

With stratification a predictive inference about $N\bar{Y} = \Sigma_h N_h \bar{Y}_h$ leads naturally to weights involving N_h/n_h where n_h is the sample size in stratum h. But inferences about $\bar{Y}_h$ or S_h^2 do not require these weights, nor do inferences about linear combinations $\Sigma_h w_h \bar{Y}_h$, when W_h are known. These latter inferences are made when data from a survey in one area are used to predict the results for a different population in either time or space or both. For example surveys on the annoyance due to aircraft noise around London Heathrow have been used to predict possible annoyance at potential sites for a third London airport and the probability weights N_h/n_h for Heathrow have no role for such inferences.

We consider the case where the weights are a measure of size of a sampling unit. The prior data $\boldsymbol{X}$ may contain many variables and the resulting summary into $\boldsymbol{\pi}$ is at best very crude. However, for inferences about $\boldsymbol{Y}$ all that is required is $\boldsymbol{\pi}$. Before sampling, the joint distribution of $(\boldsymbol{Y}, \boldsymbol{\pi})$ is

$$f(\boldsymbol{Y},\boldsymbol{\pi}) = f(\boldsymbol{Y}|\boldsymbol{\pi})g(\boldsymbol{\pi}), \tag{4.3}$$

and after sampling on $\boldsymbol{\pi}$ the superpopulation distribution is modified to

$$f_s(\boldsymbol{Y},\boldsymbol{\pi}) = f(\boldsymbol{Y}|\boldsymbol{\pi})g_s(\boldsymbol{\pi}), \tag{4.4}$$

where strong ignorability, with $0 < \pi_i < 1$, imples that all units have a chance of inclusion the sample, so that $f(\boldsymbol{Y}|\boldsymbol{\pi})$ can be estimated from the sample data for all $\boldsymbol{Y}$. Now since unit i is selected with probability proportional to size $\pi_i, g_s(\pi_i)$ is the size biased distribution

$$g_s(\pi_i) = \frac{\pi_i g(\pi_i)}{\mu_\pi}, \tag{4.5}$$

where $\mu_\pi = \int \pi g(\pi) d\pi$. In a finite population

$$\mu_\pi = \frac{1}{N}\sum_{i=1}^{N} \pi_i = \frac{n}{N}, \tag{4.6}$$

for a fixed sample size design and then

$$f_s(\boldsymbol{Y}_i,\pi_i) = f(\boldsymbol{Y}_i|\pi_i)N\pi_i g(\pi_i)/n. \tag{4.7}$$

We assume that the sample data comprise independent observations from the size biased distribution (4.5) or (4.7).

Size-biased samples have been studied by many authors, for example, Cox (1969), Patil and Rao (1978). The moments of the sampled distribution are simply related to those of the original distribution, for example,

$$\begin{aligned} E_s\left(\frac{\mu_\pi}{\pi}Y^r\right) &= \int Y^r \frac{\mu_\pi}{\pi} f(Y|\pi) g_s(\pi) dY d\pi \\ &= \int Y^r f(Y|\pi) g(\pi) dY d\pi \\ &= E(Y^r). \end{aligned} \tag{4.8}$$

Now since the sample data can be considered as a random sample from $f_s(Y, \pi)$

$$m_s(r) = \frac{1}{n}\sum_{i\in s}\frac{\mu_\pi}{\pi_i}Y_i^r = \frac{1}{N}\sum_{i\in s}\frac{Y_i^r}{\pi_i} \tag{4.9}$$

is an unbiased estimator of $E(Y^r)$. In particular

$$m_s(1) = \frac{1}{N}\sum_{i\in s}\frac{Y_i}{\pi_i}$$

is an unbiased estimator of $\mu_y = E(Y)$. This is the well known Horvitz-Thompson estimator given by (2.6).

For more complex functions of moments such as ratios or regression coefficients component-wise unbiased estimation leads to probability weighted estimators similar to (2.9). Thus conditioning on π and using results from size-biased sampling leads to distribution free methods of moments estimators identical to the classical p-weighted estimators. Clearly if the distributions in (4.3) can be specified accurately then more efficient methods estimation can be employed. In sample surveys the populations are very complex and highly multivariate and can rarely be specified accurately. In such cases a robust estimation procedure is highly desirable and the method of moments estimator leading to the p-weighted estimator for size-biased sampling must be a serious contender.

5. THE ADJUSTED LEAST SQUARES ESTIMATOR

In Section 3.2 the ALS estimator was introduced. From the form of the estimator (3.4) it can be seen that it adjusts the unweighted estimator $\boldsymbol{m}_y$ or $\boldsymbol{s}_{yy}$ for lack of balance in the sample on the prior variables $\boldsymbol{X}$. Thus $\hat{\mu}_y$ is adjusted for the difference between $\boldsymbol{M}_x$ and $\boldsymbol{m}_x$ and $\hat{\Sigma}_{yy}$ for the difference between $\boldsymbol{S}_{xx}$ and $\boldsymbol{s}_{xx}$. These adjustments are exact if the regressions are linear and homoscedastic, but as we saw in the simulation study in Table 2 the results do not appear to be robust to departures from these assumptions. How can the model-based estimators be adjusted to take into account lack of balance in the sample on the known auxiliary variables $\boldsymbol{X}$?

If the sample selection probabilities $\pi(\boldsymbol{x})$ are related to the size of a particular variable X_1, say, then the sample points will mainly occur for large values of X_1. In Figure 1 we show a non-linear regression between Y and X_1 and the OLS regression and p-weighted regression fitted to a *pps* sample. The OLS regression line fits the data points and gives a good approximation to the true regression curve $E(Y|X_1)$ for large values of X_1. The ALS curve gives large weight to the points with small values of X_1 and gives a regression line which approximates the entire curve of $E(Y|X_1)$. Clearly it is the latter regression which is required if $E(Y|X_1)$ is to be approximated by a linear regression in the sense of Mouchart and Simar (1980). This approximate population regression can then be used for predictive inferences about unobserved Y_i.

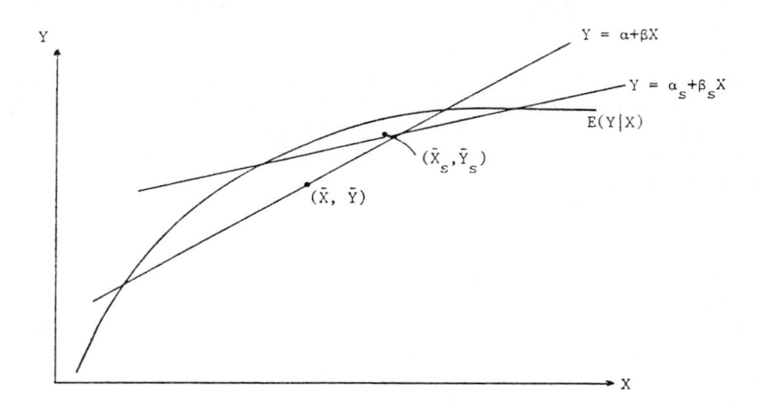

Figure 1. *Population and sample regressions*

$Y = \alpha + \beta X$ is the population linear regression.

$Y = \alpha_s + \beta_s$ is the sample linear regression.

$Y = E(Y|X)$ is the true (non-linear) population regression.

The relevant population model for regression adjustment is the joint distribution of Y and X_1, given by

$$f(Y, X_1) = f(Y|X_1)g(X_1). \tag{5.1}$$

The sampling scheme $p(s|\boldsymbol{X})$ is based in principle on the complete set of prior variables $\boldsymbol{X}$, but if in fact the size measure component is a function only of X_1 then after sampling we have

$$f_s(Y, X_1) = f(Y|X_1)\pi(X_1)g(X_1)/\mu_\pi^*, \tag{5.2}$$

where $\mu_\pi^* = \int \pi(x_1)g(X_1)dx_1$. The linear ALS estimator of μ_Y is then

$$\hat{\mu}_Y = \boldsymbol{m}_y + \boldsymbol{b}_{yx_1}^*(\boldsymbol{M}_{x_1} - \boldsymbol{m}_{x_1}), \tag{5.3}$$

where $\boldsymbol{b}_{yx_1}^* = \hat{\Sigma}_{yx_1}\hat{\Sigma}_{x_1x_1}^{-1}$.

Now the components in $\hat{\Sigma}_{yx_1}$ are the component-wise estimators of Σ_{yx_1} and using the results for size-biased sampling

$$\begin{aligned} E_s\left(YX_1\frac{\mu_\pi^*}{\pi(X_1)}\right) &= \int yx_1\frac{\mu_\pi^*}{\pi(X_1)}\frac{\pi(X_1)}{\mu_\pi^*}f(y|x_1)g(x_1)dydx, \\ &= E(YX_1). \end{aligned}$$

Thus as before $\frac{1}{N}\Sigma_{i \notin s} \frac{y_i x_{1i}}{\pi_i}$ is an unbiased estimator of $E(YX_1)$, with similar expressions for the other components.

These results suggest that the adjusted least squares estimator is not the compromise estimator $\hat{B}_{AW}$ proposed by Nathan and Holt (1980) but the modified version given by (5.3) in which only the slope is subject to p-weighting. The properties of $\hat{\mu}_Y$ in (5.3) are currently under investigation.

The overall conclusion is to agree with Rubin (1983) that the selection probabilities, π, can play a useful role in a model-based approach to finite population inference and moreover if a robust approach to inference is employed then the p-weighted estimators which are so fundamental in randomization inference appear as natural moment estimators using the ideas of size-biased sampling.

REFERENCES

Brewer, K. R. W. (1979). A class of robust sampling designs for large-scale surveys. *J. Amer. Statist. Assoc.* **74**, 911–914.

Cox, D. R. (1969). Some sampling problems in technology. *New Developments in Survey Sampling.* (N. L. Johnson and H. Smith Jr., eds.). New York: Wiley.

DuMouchel, W. H. and Duncan, G. J. (1983). Using sample weights in multiple regression analysis of stratified samples. *J. Amer. Statist. Assoc.* **78**, 535–543.

Hajek, J. (1973). Discussion of Basu, D.: "An essay on the logical foundations of survey sampling", Part I. *Foundations of Statistical Inference.* Holt, Rinehart and Winston of Canada Ltd.

Holmes D. J. (1987). *Ph. D. Thesis.* Southampton, U.K.: University of Southampton.

Holt, D. and Smith, T. M. F. (1976). The design of surveys for planning purposes. *The Australian J. of Statistics.* **18**, 37–44.

Holt, D., Smith, T. M. F. and Winter, P. D. (1980). Regression analysis of data from complex surveys. *J. Roy. Statist. Soc. A* **143**, 474–87.

Kish, L. and Frankel, M. R. (1974). Inference from complex samples (with Discussion). *J. Roy. Statist. Soc. B* **36**, 1–17.

Mouchart, M. and Simar, L. (1980). Least squares approximation in Bayesian analysis. *Bayesian Statistics,* Proceedings of the First Int'l. Meeting in Valencia, University Press, Valencia.

Little, R. J. A. (1982). Models for non-response in sample surveys. *J. Amer. Statist. Assoc.* **77**, 237–250.

Little, R. J. A. (1983). Estimating a finite population mean from unequal probability samples. *J. Amer. Statist. Assoc.* **78**, 596–604.

Nathan, G. and Holt, D. (1978). The effect of survey design on regression analysis. *J. Roy. Statist. Soc. B* **42**, 377–386.

Patil, G. P. and Rao, C. R. (1978). Weighted distributions and size biased sampling with applications, etc. *Biometrics* **34**, 179–190.

Pfeffermann, D. and Holmes, D. J. (1985). Robustness considerations in the choice of a method of inference for regression analysis of survey data. *J. Roy. Statist. Soc. A* **148**, 268–278.

Rao, J. N. K. (1975). Analytic studies of sample survey data. *Survey Methodology* **1**, supplementary issue, Statistics Canada.

Rosenbaum, P. R. and Rubin, D. R. (1983). The central role of the propensity score in observational studies for causal effects. *Biometrika* **70**, 41–55.

Rubin, D. B. (1985). The use of propensity scores in applied Bayesian inference. *Bayesian Statistics* **2**. (J. M. Bernardo, M. H. DeGroot, D. V. Lindley and A. F. M. Smith, eds). Amsterdam: North-Holland.

Scott, A. J. (1977). On the problem of randomization in survey sampling. *Sankhya C* **39**, 1–9.

Smith, T. M. F. (1981). Regression analysis for complex surveys. *Current Topics in Survey Sampling.* Academic Press, 267–92.

Smith, T. M. F. (1983). On the validity of inferences from non-random samples. *J. Roy. Statist. Soc. A* **146**, 394–403.

Sugden, R. and Smith, T. M. F. (1984). Ignorable and informative designs in survey sampling inference. *Biometrika* **71**, 495–506.

DISCUSSION

M. J. BAYARRI (*University of Valencia*)

For a long time, it has been commonly argued that the inclusion probabilities, π, had no role to play in the Bayesian approach to sample surveys. After the last two Valencia meetings the situation seems to be changing. As a matter of fact, Professor Rubin in Valencia 2 (Rubin, 1985) showed how the π's can be useful as a coarse summary of the information provided by the covariates, easing the task of modelling as well. Now, in Valencia 3, Prof. T. M. F. Smith uses Rubin's proposal for modelling and carries the argument one step further: He shows how the largely condemned (by Bayesian audiences) classical weighted estimators can also arise from a model-based approach to sample surveys. Those applied Bayesian statisticians who quietly use randomization estimators in their applications owe a debt of gratitude to both of them.

The question raised in the title of the paper, however, is not whether to use or not to use weighted estimators, but whether to weight or not to weight. This question got me interested in whether the π's could be not just useful or justifiable but even interesting. It might very well turn out that Bayesians would ask for the units to be selected with probability proportional to size, for instance, if that selection provided greater information than simple random sampling.

The following discussion is restricted to the "weighted" part of the model, as presented by Professor Smith, that is, to the size-biased version of $g(\pi)$,

$$g^b(\pi) = \frac{\pi g(\pi)}{\mu_\pi}. \tag{1}$$

This size-biased distribution is just a particular case of what Rao (1965) called *weighted distributions*, in which the original density is multiplied by some general weight function and renormalized. Professor DeGroot and myself are currently working on this topic and have already obtained some preliminary results showing that, in some situations, the experiment that selects a random sample from the weighted distribution is *sufficient*, in the Blacwkell sense (Blackwell, 1951,1953), for the experiment selecting a random sample from the original or unweighted distribution. Then, in these situations, given the choice, a Bayesian would always select the "weighted" experiment because for every decision problem involving the parameter indexing the distribution, and every prior distribution for it, the expected Bayes risk would be smaller with the weighted experiment than with the unweighted one.

Size-biased distributions being particular cases of weighted distributions, it is natural to ask what would be the case in this scenario. Needless to say, different answers will be obtained depending on the model $g(\pi)$ we have in mind. In what follows, we will study Fisher information for different models $g(\pi)$ and their size-biased versions $g^b(\pi)$. We will denote by $\mathcal{E}_{\text{original}}$ and $\mathcal{E}_{\text{size-biased}}$ the experiments in which a random sample is obtained from the original density $g(\pi)$ and its size-biased version $g^b(\pi)$, respectively. Also, $\mathcal{E}_1 \succ_F \mathcal{E}_2$ will mean that $\mathcal{E}_1$ provides greater Fisher information than $\mathcal{E}_2$ for every value of the parameter considered ($\approx_F$ will mean equal Fisher information). $I(\cdot)$ and $I_b(\cdot)$ will denote Fisher information in one observation from $g(\pi)$ and $g^b(\pi)$ respectively.

One difficulty with both Rubin's paper and Smith's paper is that they provide no hints about what a sensible model $g(\pi)$ could be, but π being a probability, the natural guess would be a beta distribution. Also, we don't expect π to be too big, particularly if N (the size of the finite population) is large, so that, as a first simple model we will consider

$$g_1(\pi|\theta) = Be(\theta, \theta + k) \tag{2}$$

where k is a constant (presumably related to N). It is easily found that the size-biased version of (2) is

$$g_1^b(\pi|\theta) = Be(\theta + 1, \theta + k) \tag{3}$$

and also that one observation from (3) provides greater Fisher information than one observation from (2) so that, in this case,

$$\mathcal{E}_{\text{size-biased}} \succ_F \mathcal{E}_{\text{original}}; \tag{4}$$

that is, it would be convenient for us to select the π's with probability proportional to size.

When thinking about selection probabilities π, we somehow feel that the value $1/N$ has some special meaning that should be reflected in $g(\pi)$. The model we will consider next is a mixture of Pareto related distributions and has the advantage over (2) of being more spiked around $1/N$ and of being far more easy to handle from a Bayesian point of view. Thus, let's consider now

$$\begin{aligned} g_2(\pi|\theta) &= \theta\pi^{\theta-1}N^{\theta-1} & \text{for } 0 \le \pi \le \frac{1}{N} \\ &= \theta(1-\pi)^{\theta-1}\left(\frac{N}{N-1}\right)^{\theta-1} & \text{for } \frac{1}{N} \le \pi \le 1 \end{aligned} \tag{5}$$

which is a density for $\theta > 0$, but values of $\theta \geq 1$ seem more sensible in this context (when $\theta < 1$, (5) is U shaped). Figure 1 shows the shape adopted by $g_2(\pi|\theta)$ for selected values of θ. Notice that the mode of this distribution (for $\theta > 1$) is precisely $1/N$ and that greater values of θ correspond to distributions which are more and more spiked around their modes. More general mixtures of this type of Pareto related distribution were studied in Bayarri (1984), and a particular mixture, which is a two parameter generalization of (5), was used in Bayarri (1985) in a Bayesian goodness-of-fit context; there it was called the alpha distribution, a name due to Bernardo (1982) who apparently first introduced it.

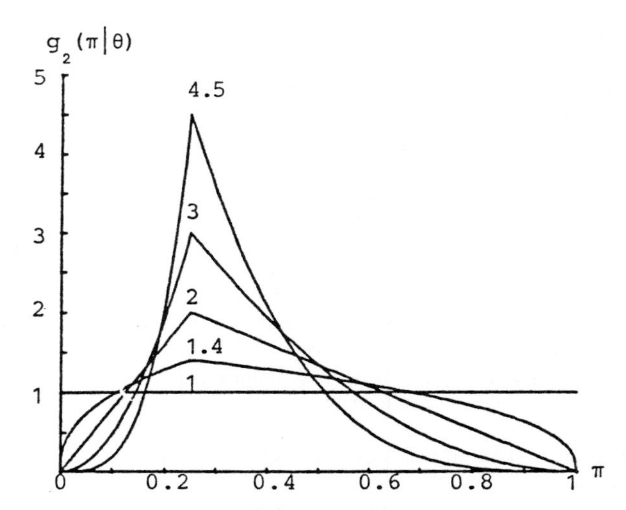

Figure 1. *The density $g_2(\pi|\theta)$ for $N = 4$ and $\theta = 1, 1.4, 2, 3, 4.5$*

The size-biased version of (5) is found to be

$$g_2^b = \frac{1+\theta}{N+\theta-1}(N\pi)^\theta \qquad \text{for } 0 \leq \pi \leq \frac{1}{N}$$
$$= \frac{1+\theta}{N+\theta-1}N^\theta \pi \left(\frac{1-\pi}{N-1}\right)^{\theta-1} \qquad \text{for } \frac{1}{N} \leq \pi \leq 1. \tag{6}$$

It can be shown that, in this case, $I(\theta) \geq I_b(\theta)$ for all values of θ, so that

$$\mathcal{E}_{\text{original}} \succ_F \mathcal{E}_{\text{size-biased}}$$

and the situation is just the opposite to the one encountered before.

In the two examples just presented, we have selected some distributions $g(\pi)$ to explain the behavior of π and assume that the data is a random sample from their size biased versions $g^b(\pi)$. But really we are not very used to thinking about models for the probabilities of selection π, so that we will deduce the last model to be studied directly from the distribution of the covariates X.

Assume X is a univariate positive random variable with density $f_X(x)$. As usual, we consider $X_1, \ldots, X_N$, the finite population, to be a random sample from f_X. We are assuming that the data we have is a sample from $X_1, \ldots, X_N$ selected with probability proportional to size.Thus, associated with $X_1, \ldots, X_N$ there is the corresponding finite population of π's : $\pi_1, \ldots, \pi_N$, where $\pi_i = X_i/(\sum_{i=1,\ldots,N} X_i)$ and we select X_i with probability π_i (notice that there is an slight variation here with respect to the paper: these π_i's add to one, while the ones in the paper add to n).

If we want to model the behavior of π instead of the behavior of X, then we assume that a sample is going to be drawn from $\pi_1, \ldots, \pi_N$ with probability proportional to size, that is, π_i is selected with probability π_i. In this process, the distribution of data, $g_s(\pi)$ in the paper, is given by:

$$g_s(\pi) = N\pi \int t f_X(\pi t) g_Y[(1-\pi)t]dt, \tag{7}$$

where Y represents the sum of $N-1$ i.i.d. random variables from f_X, and g_Y its density.

Let's take an example. Again, for a positive random variable it would be natural to try a gamma distribution, that is, $f_X(x) = Ga(\alpha, \beta)$. Then it is found that

$$g_s(\pi) = Be\{\alpha+1, (N-1)\alpha\}. \tag{8}$$

One interesting fact about (8) is that this beta distribution is just the size-biased version of the $Be\{\alpha, (N-1)\alpha\}$ distribution. So that in general, let's assume that

$$g_3(\pi) = Be\{\alpha, k\alpha\}, \tag{9}$$

where k is a constant. Notice that, if $k = N-1$ as in the gamma example, $E(\pi) = 1/N$ so that $1/N$ is again regarded as a special value in the distribution of π. As we have already said, $g_3^b(\pi) = Be(\alpha+1, \alpha k)$ and in this case it is found that $I(\alpha) = I_b(\alpha)$ for all α, so that

$$\mathcal{E}_{\text{size-biased}} \approx_F \mathcal{E}_{\text{original}}$$

and both experiments are totally equivalent with regard to this criterion.

We have thus encountered three situations in which the behaviour of Fisher information is different. Of course, all the examples refer solely to the X part of the model, without referring to the possible effects of weighting in the marginal distribution of Y. However, in our opinion the general conclusion to be drawn is that, even if Professor Smith has shown us how the π's can enter the picture, there is not yet a clear answer to the question posed.

Indeed, it is not surprising for a Bayesian to conclude that whether to weight or not to weight will depend on the particular decision problem at hand.

I wouldn't like to finish without asking Professor Smith a couple of questions: We have just seen the weighted estimators appearing as a result of both modelling π (instead of X) and using the method of moments for estimation. Rubin (1985) already cautioned us that π can be "too coarse a summary" of the information provided by X, so that my first question refers to whether this type of modelling is the only way to make the π's play a role in a model-based approach to sample surveys. Also, it is well known that the method of moments can exhibit a number of undesirable features; Has Professor Smith tried a Bayesian or at least a likelihood approach to estimation?. I am looking forward to seeing some results in this direction, but in turn, it would imply selecting suitable $f(Y|\pi)$ and $g(\pi)$. How would Professor Smith select these models?.

The second question relates to robustness: Has Professor Smith studied the issue of whether modelling $f(Y|\pi)$ and $g(\pi)$ produces more robust inferential results than modelling $f(Y|X)$ and $g(X)$?. If so, it could be another reason for using the former alternative and for weighting, thus helping to answer the question raised in the title of the paper.

R. A. SUGDEN (*Goldsmiths' College, London*)

Smith gives some answers and suggests new approaches to the vexed question for a Bayesian: What is the role of the design in survey sampling inference?

Contrary to the impression possibly given in Section 3.1., it is important to realise that inferences can depend on the design even in the ignorable case. For example under a normal error regression through the origin on a single "size" variable with error variance proportional to squared size and a probability proportional to size design, it is easy to show, through the likelihood (4.1), that the Bayes posterior mean of the population total is just the Horvitz-Thompson design-unbiased estimator (4.9) but with an additional term representing a sum of "residuals".

As shown in Sugden and Smith (1984), the design is no longer ignorable when not all the inclusion probabilities are observed. However, some aspects of ignorable inference may be preserved e.g. in the above the posterior mean depends only on the inclusion probabilities of sample units so is unaltered.

All statements about ignorability (or not) have been made by authors assuming the model is correct. A Bayesian who lacks confidence in his model must seek model elaboration —see Royall and Pfeffermann (1982)— or adopt a distribution free approach such as the method of moments that Smith suggests. In the former case ignorability may no longer hold unless sufficient design information is available. An alternative to the latter is some form of non-parametric maximum likelihood estimation of the finite population distribution function, see Vardi (1982). A problem with the method of moments here is that it essentially amounts to imposing (component-wise) design unbiasedness.

REPLY TO THE DISCUSSION

Dr. Bayarri makes many interesting points which take the discussion far beyond my modest aims. I was concerned only with the problem of inference *after* a sample has been selected using a randomized design with unequal selection probabilities. The dilemma for a Bayesian is that if the design variable $\boldsymbol{X}$ is known for all units in the population then any design of the form $p(s|\boldsymbol{x})$ contains less information than $\boldsymbol{X}$ itself and so can be ignored for inference. But sample designs are constructed by knowledgeable statisticians so surely they must contain useful information and as such should not be ignored. The resolution of the dilemma is found by constructing an appropriate conditional inference. Rubin's contribution is to show that frequently there will exist a reduction of the design (prior) information $\boldsymbol{X}$ which is an adequate summary of $\boldsymbol{X}$ for inference on Y. He shows further that under certain conditions the vector $\boldsymbol{\pi}$ of inclusion probabilities will provide such an adequate summary of $\boldsymbol{X}$. Inferences can then be made conditional on $\boldsymbol{\pi}$ and as such they will depend on $\boldsymbol{\pi}$. Dr. Sugden makes this point in his discussion but his phrasing is misleading since the inference does not depend on the sampling mechanism $p(s|\boldsymbol{x})$ but only on the units in the sample and their inclusion probabilities $\boldsymbol{\pi}$.

My contribution was to consider how the information in $\boldsymbol{\pi}$ might be used for inference about $\boldsymbol{Y}$. The complexity of most survey populations means that precise models are difficult to specify and even harder to justify and so I did not attempt to model $g(\pi)$ along the lines suggested by Dr. Bayarri. Instead I adopted one form of model-free estimation, namely the methods of moments estimators, because it gave the traditional π-weighted estimators.

As both Dr. Sugden and Dr. Bayarri point out other methods of estimation could have been considered. These would lead to different estimators and to comparisons of efficiency. Which one to choose will depend on the strenght of one's prior belief about the underlying data structure. In some further sets of simulation results based on real data we have found cases where the π-weighted estimator is inefficient unconditionally (over all samples) and is appalling conditionally (given the sample). Thus as Dr. Bayarri concludes from her models there are some cases when π-weights are good and some when they are not. There is still no simple answer to the question of whether to use π-weights or not. So the Bayesian who says that using π-weights is always wrong is wrong and the traditional statistician who says that π-weights should always be used is equally wrong.

In the absence of precise models we still need a robust procedure. My own belief is that stratification after selection on $\boldsymbol{X}$(or $\boldsymbol{\pi}$) is the best general purpose robust procedure for survey inference. This employs the π-weights indirectly through the stratification rather than directly through the weighting.

REFERENCES IN THE DISCUSSION

Bayarri, M. J. (1985). A Bayesian test for goodness-of-fit. *Tech. Rep.* Departamento de Estadística e I.O. University of Valencia. Presented at the 1985 Joint Statistical Meetings (ASA, Biometrics, IMS).

Bayarri, M. J. (1984). *Contraste Bayesiano de Modelos Probabilísticos*. Ph. D. Thesis. University of Valencia.

Bernardo, J. M. (1982). Contraste de modelos probabilísticos desde una perspectiva Bayesiana. *Trabajos de Estadística* **32**, 16–30.

Blackwell, D. (1951). Comparison of experiments. *Proceedings of the Second Berkeley Symposium on Mathematical Statistics and Probability*, 93–102. Berkeley, CA: University of California Press.

Blackwell, D. (1953). Equivalent comparison of experiments. *Ann. Math. Statist.* **24**, 265–272.

Rao, C. R. (1965). On discrete distributions arising out of methods of ascertainment. *Classical and Contagious Discrete Distributions*, (G. P. Patil, ed.), 320–333. Calcutta: Statistical Publishing Society.

Royall, R. M. and Pfeffermann, D. (1982). Balanced samples and robust Bayesian inference in finite population sampling. *Biometrika* **69**, 401–410.

Vardi, Y. (1982). Nonparametric estimation in the presence of length bias. *Ann. Statist.* **10**, 616–620.

BAYESIAN STATISTICS 3, pp. 453–477
J. M. Bernardo, M. H. DeGroot, D. V. Lindley and A. F. M. Smith, (Eds.)

Bayesian Approaches to Clinical Trials

D. J. SPIEGELHALTER and L. S. FREEDMAN
Medical Research Council, Cambridge

SUMMARY

We summarise current statistical practice in clinical trials, and review Bayesian influence over the past 25 years. It is argued that insufficient attention has been paid to the dynamic context in which development of therapeutic innovations take place, in which *experimenters, reviewers* and *consumers* form different interest groups, and may well process the same evidence in different ways. We illustrate the elicitation of quantitative prior opinion in trial design and show how graphical expression of current belief can be related to regions of possible benefit with different clinical implications. Such displays may be used both for ethical monitoring of trials and to predict the consequences of further sampling.

Keywords: DECISION THEORY; SEQUENTIAL ANALYSIS; MULTIPLICITY; PREDICTION; SUBJECTIVE PROBABILITY.

1. INTRODUCTION

The routine design and analysis of randomised clinical trials is a constant source of amazement and irritation to Bayesians interested in medical statistics. They find a combination of Neyman-Pearson theoretic use of Type I and Type II errors, point alternative hypotheses and power curves; Fisherian use of tail-areas as standardised mesures of evidence, and Waldian stopping boundaries for sequential analysis, carefully preserving Type 1 error. Aspects of Bayesian thinking, to differing degrees of purity, have been forcibly argued for a quarter of a century, and yet their impact has been negligible. In a Medline search of the Index Medicus database covering the last 20 years, the terms "Bayes" or "Bayesian" and "Clinical Trial" generated only 24 references, almost all in medically-oriented statistics journals. ("Bayes/Bayesian" alone generated 515 entries, mostly in diagnostic testing and computer-aided diagnosis).

This paper assumes that, in order to have any impact on the entrenched methodology, it is important to understand the current practice and the organisational context in which it occurs; in particular, it needs to be acknowledged that the process of development and eventual implementation of a novel therapy involves a dynamic interaction between differently motivated groups of individuals. At a basic level, three distinct bodies can be identified: the *experimenters*, who may be a pharmaceutical company or a research organisation developing the therapy; the *reviewers*, who regulate the dissemination of information and availability of the innovation and may be journal editors, funding bodies and regulatory agencies; and the *consumers*, who are the clinicians who take eventual responsibility for prescribing the therapy. This is, of course, a simplistic description of a complex process, but seems to provide a reasonable structure for discussion.

From a Bayesian and decision-theoretic perspective, it is essential to identify the body whose opinions and values are relevant in the design, analysis and reporting of any study. As Racine *et al.* (1986) and Berry (1987) have well emphasised, experimenters often have strong *private* opinions concerning new therapies, and seek to influence the consumers, but can do so only through the mediating influence of the reviewers who attempt to set standards of "objectivity". Thus it is reasonable that private opinion of experimenters is used for design and for "in-house" decision-making, but any public results are "depersonalised" to the reviewers' satisfaction. Consumers then assimilate this information according to their own private context. As we emphasise later, this diffuse process makes it somewhat difficult to adequately model the eventual consequences of an experiment.

After a brief description of current statistical practice, we use the structure described above in reviewing attempts to place "public" clinical trials in a decision-theoretic context. We then consider Bayesian assaults on sequential analysis, multiplicity, and presentation of trial results, and review the use of decision-theory in "private" experimental work. In Section 8, we argue that much of the past Bayesian effort has not been sufficiently pragmatic, and echo Racine *et al.* (1986) in claiming that the incontrovertible value and availability of prior information, and the attractiveness of graphical display of conclusions, has not been adequately exploited.

Sections 9 to 11 illustrate, with real examples, a strategy for a staged introduction of Bayesian methods, using the inadequacies of standard practice to justify the elicitation of prior opinion, and monitoring trials by means of private and public displays of belief in clinically relevant regions of efficacy. Finally, in our conclusions, we consider the consumers' viewpoint and how they are better served by a Bayesian approach.

We should emphasise in advance that we do not claim that unique "objective" Bayesian analyses exist that can be presented for public consumption, and agree with the discussion of Racine *et al.* (1986) that ideally a range of prior to posterior transformations should be presented. Nevertheless, we do refer to "reference" posterior distributions that, in general are simply normalised likelihoods, and do not necessarily correspond to the specific proposals of Bernardo (1979). It is clearly vital to establish reasonable means by which, in common problems, evidence can pass from experimenters to consumers in a manner acceptable to reviewers, preferably in a form which displays how this evidence should influence the beliefs of the consumers.

2. THE "CLASSICAL" FRAMEWORK

The randomised controlled clinical trial has become an integral part of the development and dissemination of therapeutic drugs, and —to a considerably lesser extent— surgical and technological interventions. Regulatory authorities such as the US Food and Drug Administration (FDA) have played a large part in influencing the experimental and statistical methods of pharmaceutical companies, and it is important to understand the developmental process of a new therapy in order to place the statistical methodology in context.

Within the pharmaceutical industry and research organisations, four main phases of experimentation may be generally delineated; see, for example, Pocock (1983). Phase I trials are primarily concerned with drug safety to determine dosages that do not cause toxicity in volunteers. Phase II studies are generally uncontrolled studies on small numbers of patients to obtain a rough idea of treatment effect and possibly to suggest effective dosage —as we shall see later, this may be thought of as a screening process to determine which of many contending drugs or drug regimes should go on to more extensive, and expensive, testing.

Classic randomised controlled trials are termed "phase III" and are our primary concern in this paper. The essential experimental tenets are basically unchanged from Bradford Hill's pioneering work in streptomycin (Medical Research Council, 1948): comparison of a therapy with a control group given a standard or placebo therapy, a defined protocol for selection of patients and administration of treatment, doctor and patient blinded to the treatment given (where appropriate), carefully defined outcomes measured blindly where possible and, of course, random assignment of patients to treatment.

The results of phase I to III studies are presented to regulatory authorities, and the drug may then be given a licence to be marketed.

Phase IV studies concern the long-term monitoring of the drug for effectiveness and possible adverse reactions. However, the company or research organisation may carry out further controlled trials of licenced drugs in order to assess benefit in specific groups of patients, and we shall include these as phase III studies.

The major development over the last 40 years has been in the statistical methodology used for designing and analysing phase II and III studies, and it is in this area that fundamental differences in statistical ideology are perhaps most exposed. In its archetypal form, the classical trial is concerned with a null hypothesis $H_0 : \delta = 0$ on a primary outcome parameter δ. A test statistic T is to be calculated at the end of a trial of $2n$ patients, n on each treatment arm, and an appropriate critical region C of size α defined such that $P(T \in C|H_0) = \alpha$. If the observed T_0 lies in C, we may reject H_0 at the $100\alpha\%$ level. The sample size is supposedly selected by specifying an alternative hypothesis δ_A and a power $1 - \beta$, then choosing n such that $P(T \in C|\delta = \delta_A) = 1 - \beta$.

This Neyman-Pearson formulation, with Type I error α and Type II error β, is generally adhered to in designing trials, although —as we shall discuss later— the status of the alternative δ_A is often obscure. In reporting results, a more Fisherian view is taken, with evidence against the null hypothesis summarised by a "P-value" which is the minimum α at which H_0 would be rejected, and constitutes the ubiquitous "significance test" in clinical reporting. Confidence intervals, consisting of values of δ which would not be rejected at the 95% or 90% level, are being increasingly encouraged in medical journals (Gardner and Altman, (1986)). We shall not discuss additional sophistications concerning the recognition of important background covariates influencing outcome; this has led to schemes to encourage balance in the treatment groups, such as stratified randomisation, methods which change probabilities of allocation to treatment depending on covariate imbalance, but do *not* depend on treatment response (reviewed, for example, by Kalish and Begg, (1985)), and allowance for covariates in analysis.

From a Bayesian perspective, the above description of the most basic procedure already contains many inadequacies: the existence of strong prior information being apparently ignored in design and analysis, the tenuous relationship of the null and alternative hypothesis to clinical practice, and the concentration on summary statistics rather than full graphical expressions of belief. These issues will be our major concern, although the bulk of criticism concerning classical methods has concentrated on the issue of "multiplicity", when a number of hypothesis tests and related estimates are performed on the same set of data.

As emphasised by Cornfield (1969) and Tukey (1977), multiplicity may occur through analysing many drugs simultaneously, through concern about subgroups of patients, through measuring many clinical endpoints and through repeatedly analysing the data as it accumulates. In each case, the critical region for each test should, according to Neyman-Pearson theory, be adjusted to give total significance level α to conclusions concerning hypotheses specified at the start of the trial, where each pre-specified hypothesis may be a conjunction of a number of the hypotheses tested at their own "nominal" level.

Sequential monitoring of results is increasingly popular in order to avoid continuing a trial beyond a point where firm conclusions can be drawn and hence make randomisation unethical, and the "dangers" of repeated significance testing in this context have been pointed out by many authors. For example, Armitage, McPherson and Rowe (1969) show that in sequential testing of a true hypothesis that a binomial parameter is $\frac{1}{2}$, with each test at a nominal 5% level, there is in fact 22% chance of incorrect rejection of H_0 after 100 observations. This Type I error approaches unity asymptotically ("sampling to a foregone conclusion") but rather slowly. The first edition of Armitage (1975) brought Wald's sequential analysis into clinical trials, and a boom in sequential methodology has arisen (see, for example, Whitehead (1983) and a recent issue of *Communications in Statistics* (1984), **13**, 2315–2417).

From a practical viewpoint, many trials are reviewed at, say six-month intervals by a monitoring committee. "Group sequential designs" have therefore been developed in place of continuous monitoring, as exemplified by the recent Beta-Blocker Heart Attack Trial (BHAT) reported by DeMets *et al.* (1984). Here 3837 patients after myocardial infarctions were randomised double-blind into placebo and beta-blocker groups. Mortality was to be summarised as a Z-value (standardised normal deviate under H_0: "equal mortality") based on a log-rank statistic, at seven meetings approximately 6 months apart. Observed Z-values for the first six meetings were 1.68, 2.24, 2.37, 2.30, 2.34, 2.82. At a nominal 5% level, $|Z|$ exceeded 1.96 at 1 year and the results could have been declared "significant" then. However, in view of the declared design, monitoring boundaries for the seven prospective analyses with overall Type I error of 5% had been specified early on in the trial. DeMets *et al.* (1984) describe three established strategies: Pocock's (1977) suggestion of fixed nominal P-value ($Z = 2.49, P = .012$), Peto *et al.*'s (1976) proposal of extreme interim boundaries ($Z = 3.0, P = 002$) which allow a final boundary essentially indistinguishable from that appropriate were no interim analysis intended ($Z = 1.96$) and O'Brien and Fleming's (1979) compromise boundaries that are extreme at first and then decrease (5.4, 3.82, 3.12, 2.70, 2.41, 2.20, 2.04); the investigators selected the O'Brien-Fleming boundaries. From a simplistic point of view, after 6 analyses $Z = 2.82 > 2.20$ and so the trial was stopped and the results declared "significant" at the 5% level.

DeMets (1984) and DeMets *et al.* (1984) make clear, however, that the decision to terminate was *not* a direct consequence of crossing what may seem to many a somewhat arbitrary boundary. Baseline balance, other response measures, consistency across subgroups, possible side-effects and the consequences of continuing in terms of increased precision and possible reversal of results, are among the factors considered as components of the final decision. DeMets (1984) uses a number of examples to illustrate the point emphasised by Cornfield (1976) that stopping may be appropriate for unforeseen circumstances that cannot be taken into account in the original trial design. (A notorious example is the decision to withdraw Tolbutamide, an existing profitable prescription in diabetes, from a major trial due to unforeseen excess cardiovascular mortality (UGDP, 1970). This gave rise to a decade of strong controversy, briefly reviewed by Diamond and Forrester (1983)). DeMets (1984) therefore rejects the strict Neyman-Pearson-Wald position by stating that "while they are not stopping rules, such methods can be useful guides in the decision-making process".

Closely associated with this rejection of strict stopping rules is the development of repeated confidence intervals by Jennison and Turnbull (1984), which have guaranteed coverage probabilities independent of the actual "stopping rule" and hypothesis of interest, but whose width does still depend on a particular choice of monotoring bundaries, such as Pocock's or O'Brien and Fleming's.

A classical phase III trial with fixed α and β relative to a specified alternative may still involve design choices and it is often informally acknowledged that clinical opinion as to the size of the treatment effect should be influential. So, for example, Armitage (1985) points out that a strong belief in a large effect would make Pocock's monitoring boundaries preferable to O'Brien and Fleming's, in that expected sample size would be reduced. Some authors have suggested formal use of prior opinion in design: Herson (1979) describes an application to selecting sequential designs in phase II studies, while McPherson (1982) and Freedman and Spiegelhalter (1983) consider subjective priors in selecting numbers of interim analyses in classical sequential phase III trials.

In conclusion, the emphasis has moved steadily away from a classical accept-reject formulation towards providing reasonable summaries of evidence to be used both in the decision to terminate or continue a trial, and the eventual decision of a clinician to prescribe the therapy. The complexity of the multiplicity issue has led to the frequent practice of quoting Z-values with a warning about multiple testing, but without formal adjustment. However, there has been pressure to acknowledge that a trial should lead to explicit treatment recommendations and this is reviewed in the next section.

3. INFERENCE OR DECISION IN PHASE III TRIALS?

An important contribution to understanding the varied, and possibly conflicting, aims of clinical trials was provided by Schwartz, Flamant and Lellouch (1980), who made the distinction between *explanatory* trials, designed to answer scientific questions concerning precisely defined treatments on tightly-controlled groups of patients, and *pragmatic* trials, intended to provide explicit guidance as to which therapy is most appropriate to use in routine clinical practice. They state that the analysis of a pragmatic trial should consist solely of observing whether the benefit T of therapy A over therapy B lies above a "handicap" δ_W which explicitly trades off benefit against possible side effects. If $T > \delta_W$, then A is recommended, otherwise B. Designs are evaluated by estimating the accompanying Type III error —the chance of recommending a treatment that is actually inferior.

The explanatory/pragmatic formulation illuminates many aspects of clinical trials and the value of a shifted null hypothesis to reflect secondary disadvantages has been strongly advocated by Meier (1975, 1979) and Freedman *et al.* (1984), with Meier (1975) claiming that in this context "the routine practice of formulating the proceedings in the language of significance testing has become a mischievous ritual".

Nevertheless, significance tests still remain the primary statistical tool for reporting even pragmatic trials. However, alternative strategies have been studied in considerable detail from a Bayesian perpective, starting with Colton (1963), Anscombe (1963) and reviewed in Iglewicz (1983), where it is assumed for design and stopping purposes that, after termination of the period of randomisation, the treatment with the best overall results will be given to all remaining patients in the manner originally recommended for pragmatic trials. Allowing randomisation probabilities to depend on previous results (data-dependent allocation) provides a full two-armed bandit formulation (discussed in Armitage, (1985)). The resulting procedures can be attractively simple. Anscombe considers Gaussian responses with n pairs of patients randomised equally to two groups, a total patient horizon of N, a uniform prior on treatment benefit, and loss function proportional to the number of patients given the inferior treatment times the size of the inferiority; he suggests to stop and give the "best to the rest" when the classic one-sided P-value is less than n/N —half the proportion of patients already randomised. Chernoff and Petkau (1981) have described the near optimality of this procedure. Bather (1985) considers sequential binomial experiments based on controlling the expected proportions of future patients not responding to treatment (thus avoiding the problem of stating an explicit patient horizon) and develops simple approximate minimax schemes for both symmetric and data-dependent allocation. Hilden *et al.* (1987) present a strong argument for the use of decision theory in a trial of a severely disfiguring procedure, which leads naturally to an adaptive design.

The ethical and practical issues in data-dependent allocation have been considered at length (see Armitage (1985), and following discussion), and we shall not discuss this issue: see Lindley (1975) and Kadane (1986) for Bayesian discussion on ethical randomisation. It is more fundamental to examine the appropriateness of the two-decision "best for the rest" strategy. Certainly established organisers of large trials are forthright in claiming this is an unrealistic formulation: "Bather, however, merely assumes ... 'it is implicit that the preferred treatment will then be used for all the remaining patients' and gives the problem no further attention! This is utterly unrealistic, and leads to potentially misleading mathematical conclusions." (Peto, 1985). "I believe that the promise seen by proponents of this approach is based upon their not fully appreciating that a clinical trial is a scientific experiment and not just a two-armed bandit" (Simon, 1978). Considerable support for these views comes from considering the structure for experimentation outlined in the previous section: it is clear that the publication of a phase III trial is a watershed in that the experimenters have limited control over the impact of their results on the consumers, in contrast to the transition from phase II to phase III in which explicit consideration of future patients is possible (see Section 7 fo more on this).

We therefore agree that in general the consequences of a phase III trial are difficult t formalise, and that a "medical trial is not, in any clear-cut fashion, a decision procedure" (Anscombe, 1963). Essentially, although a "private" decision may exist concerning termina tion, the consumers require *conclusions* in order to make their own decision (Meier, 1975) the "experimenter pays the piper and calls the tune he likes best, but the music is broadcas so that others may listen" (Anscombe, 1963). This does not avoid the problem of specifyin a trial objective in order to predict the consequences of a particular design, and to have som indicator when to stop, and Peto (1985) suggests that this could be based on a realistic as sessment of the impact of publication (see Section 10). But the emphasis is shifted toward developing reasonable means of displaying conclusions that may be drawn from the data whatever the private decisions of the experimenters, and this leads naturally into discussin the likelihood principle in sequential trials.

4. LIKELIHOOD PRINCIPLE AND SEQUENTIAL MONITORING

As illustrated by the BHAT example above, a classical approach requires the expression o evidence against the null hypothesis to depend on conclusions that may have been drawn had some other data been observed. The opposing view is that the evidence in the data, a expressed in the likelihood, should not depend on the experimenters intentions: Anscomb (1963) stated that "'Sequential analysis' is a hoax. The correct statistical analysis of th observations consists primarily of quoting the likelihoood function", while Meier (1975) say that "provided the investigator has faithfully presented his methods and all of his results, i seems hard indeed to accept the notion that *I* should be influenced in my judgement by how frequently *he* peeked at the data while he was collecting it". See Berger and Wolpert (1984) Berry (1985, 1987) and Berger and Berry (1987) for further recent arguments and illuminatin examples.

Cornfield (1966a, b, 1969, 1976) strongly attacks deviations from the likelihood principl in clinical trials, and provides a defence against the criticism of Armitage, McPherson an Rowe (1969) that without adjustment a true hypothesis will always be rejected at any level i one samples long enough. Cornfield points out that simple monitoring of whether a likelihoo ratio $p(T|H_0 : \delta = 0)/p(T|H_1 : \delta = \delta_A)$ crosses a particular lower boundary does not lea to certain rejection of a true null hypothesis, and terms this the "relative betting odds" (RBO since it is the ratio of posterior to prior odds on the null against the alternative (more familiarl known as the Bayes factor). In the diabetes study mentioned earlier (UGDP, 1970) an RBO o .20 is quoted for the hypothesis of "no effect on cardiovascular deaths in Tolbutamide group" against a pre-assigned alternative of 25% change, with a minimun achievable RBO of .0 under any alternative. (Fixed-sample p-values and sequential boundaries were also quoted.)

The direction of the alternative in the UGDP study was specified in the light of th observed difference, and Cornfield (1966b, 1969) avoids this rather unsatisfactory aspect b placing a symmetric Gaussian prior $p(\delta|H_1)$ centred on $\delta = 0$, to provide a "weighted" likelihood as the denominator of the RBO. The variance σ^2 of the prior is set as follows if δ_A is a plausible benefit based on past experience, then set $\sigma = (\pi/2)^{1/2}\delta_A$, so that th expected change, given it is in a particular direction, is δ_A. Subjectively assessed values fo δ_A for 11 outcome measures are reported in the Urokinase Pulmonary Embolism Trial (1973 togerther with the resulting RBOs, and the procedure is also used by the Coronary Dru Project Research Group (CDPRG) (1970). (Cornfield's influence on these major studies wa strong, with the tabulated results refreshing in the absence of P-values; however, classica analyses were also reported and in the final report of the (CDPRG) (1975) the reporting o RBOs had vanished.)

Lachin (1981a) extends the model to include interval null hypotheses $H_0 : \delta \in \Delta_0$ versus an alternative $H_1 : \delta \notin \Delta_0$, with general priors $p(\delta|H_0), p(\delta|H_1)$ and monitoring using the weighted likelihood ratio (Bayes factor) $\int p(T|\delta)p(\delta|H_0)d\delta / \int p(T|\delta)p(\delta|H_1)d\delta$.

Cornfield's approach thus followed Jeffreys (1961) in separating hypothesis testing from estimation by placing a "lump" of prior on the null hypothesis as an approximation to a tight continuous prior: "if one is seriously concerned about the probability that a stopping rule will certainly result in the rejection of a true hypothesis, it must be because some possibility of the truth of the hypothesis is being entertained" (Cornfield 1966a). Diamond and Forrester (1983) directly apply Jeffreys' suggested Bayes factor for replacing a standard t-test. Our feeling is that such an approach is only appropriate if a mixture prior genuinely reflects realistic opinion (Spiegelhalter, 1986a) and, if not, then it is better to be unconcerned with sampling behaviour conditional on an event with negligible probability. However, Lachin's idea of monitoring by the relative likelihood on ranges reflecting different clinical implications is further explored in Section 9.

Cornfield's approach was not found convincing, and —in response to a recent (rather disappointing) attack on adjusted p-values by Dupont (1983)— Royall (1983) remarked that "here we are, sixteen years after Cornfield's paper, still tyrannised by the significance test. Statisticians should have taken the initiative in overthrowing the despot, but we have instead continued to serve as his lackeys, dutifully attaching his medals (one star or two stars) to results which he finds worthy of recognition".

5. MULTIPLICITY

Sequential montoring of trial results is only one example of the "data-dredging" that frightens users of P-values when there are many outcome measures, many subgroups or many treatments, and multiple comparison adjustments to analysis of variance procedures are generally recommended. Given the complexity of large trials, these will result in very little power, and ignore the strong prior opinion that certain treatments or subgroups are "similar" to some extent. A Bayesian approach is to treat the individual response parameters as themselves being drawn from some populations, and this prior judgement should adjust the information contained in the likelihood through Bayes theorem. Thus we do not question the misgiving felt when many contrasts have been done and the most extreme noted, but interpret this as meaning that the observer is not *a priori* indifferent to all possible values of each contrast, and hence has a proper prior expressing reasonable degree of scepticism concerning large effects. Essentially, the classical approach satisfies intuition by adjusting, but does so for the wrong reasons.

Cornfield (1969) explores analysis of variance hypothesis testing assuming proper priors under the alternative hypothesis, whiile Waller and Duncan's (1969) Bayesian multiple comparisons approach has been applied in a psychiatric setting by Arnold *et al.* (1978). Donner (1982) suggests assuming exchangeable subgroup effects (identical, unknown priors) leading to a degree of pooling of results across subgroups, while Louis (1984) applies a modified hierarchical Bayesian model to a clinical trial. Tukey (1977) criticises hierarchical Bayesian approaches as requiring hypothetical well-behaved populations from which subgroup parameters are drawn, and robustness of results to assumptions about prior distributional shape is clearly important; Louis (1984) suggests a device for lessening sensitivity. Tukey also claims Bayes methods are unsatisfactory for dealing with selective reporting of extreme results. However, the Bayesian approach seems clear. If the priors for the individual subgroup effects are completely specified, the selectivity is irrelevant and the inference will adjust the likelihood with the prior for the relevant subgroup effect. If the priors depend on some common unknown parameter, say through the subgroup effects being considered exchangeable, the inference based on the selected report is affected by the knowledge that it is extreme but an adjustment may be made; see Berry (1988) for a full discussion of this. If the priors are related and the experimenter dishonestly does not reveal that his results are selected, then any procedure must be misleading. Hence a bias will still occur through publication only of extreme results, if the consumer is unaware of the selectivity.

6. EXPRESSION OF BELIEF AS A POSTERIOR DISTRIBUTION

Apart from the issues concerning repeated significance testing, we briefly describe some areas in clinical trials in which Bayesian methods have been recommended as leading to more attractive and efficient analysis.

Bioequivalence studies are intended to confirm the *similarity* of, say, two formulations of the same drug. "Similiar" is generally taken as a 20% margin in mean response, and classical approaches are based on confidence intervals lying within that region (Westlake 1979). However, it is natural to think in terms of the probability content of the relevant interval hypothesis, and Bayesian approaches with "non-informative" priors have been considered by Fluehler *et al.* (1981, 1983), Mandallaz and Mau (1981), Selwyn *et al.* (1981), Selwyn and Hall (1984) and Racine *et al.* (1986).

Some of the above studies involve two-period crossover designs, and Grieve (1985) and Racine *et al.* (1986) consider the particular problems concerning the sensitivity of the conclusions concerning the treatment effect to assumptions about the existence of a carryover effect. Different prior formulations are explored, and a mixture prior with a lump on "no carryover" leads to a Bayes factor approach.

Racine *et al.* (1986) adopt a similar procedure for assessing the degree to which historical information may be incorporated into a study, while Pocock (1976) derives a smoother pooling mechanism by not specifying a mixture prior. Van Ryzin (1980) considers using historical controls to help design and be incorporated into a non-parametric Bayesian survival analysis.

Mehta and Cain (1984) consider phase II trials which are monitored by the posterior probability that the success rate is above some level, say 20%. Charts of 90% credible intervals are provided, and the sampling properties of the procedure simulated.

7. USE OF DECISION THEORY IN PHASE II TRIALS

We note, as mentioned in Sections 1 and 3, that it may on occasions be quite reasonable to consider a phase II trial as a decision problem when it is part of an internal strategy for future experimentation. For example, Sylvester and Staquet (1980) describe a practical method for assessing the expected value of a phase II trial based on a two-point prior and a loss function incorporating the number of patients in a subsequent phase III study.

When concerned with the allocation of patients to a panel of drugs being screened in phase II trials, Whitehead (1985) considers designs to maximise the expected success rate of the single drug selected to proceed for phase III testing, using a beta distribution on the response rates. Whitehead (1986) presents an elegant use of decision theory, in which a proportion of a fixed pool of N patients need to be distributed evenly among treatments in phase II studies and the remainder to be randomised between the best of these screened therapies and the standard. The outcome measure to be optimised is the unconditional probability of obtaining a significant result, which we shall consider further in Section 9. He shows, for example that if the standard success probability is .2, and the prior on the candidate drugs is a Beta distribution with mean .2 and standard deviation .1, then with a pool of 300 patients it is optimal to screen 15 drugs with only 8 patients each, select the most effective and randomise the remaining 180 patients giving an overall 48% chance of a significant result at the 5% level.

8. A STRATEGY FOR CHANGE?

From the preceding review we draw the following conclusions concerning the limited impact of Bayesian methodology. Firstly, the two-decision formulation for phase III trials is in general unrealistic, since the implications of such studies are more complex. Secondly, Cornfield's answer to sequential analysis is not too attractive since the mixture prior is not in general realistic, and the RBO depends on the prior under the alternative and hence is not just a function of the likelihood: however, the idea of monitoring trials by relative belief in parameter intervals is appealing. Thirdly, quantitative prior knowledge *can* be obtained both from previous trials and from clinical opinion, and is finding an increasing rôle in phase II experimentation. Fourthly, Bayesian methods are attractive in their handling of nuisance parameters and the graphical display of belief. Fifthly, much more work is needed in realistic prior modelling of treatment effects and subgroup parameters in order to tackle multiplicity in a practical way.

We are currently attempting a staged introduction of Bayesian ideas into our work in clinical trials. The first steps in our strategy are as follows:

(a) Point out the absurdities in the prescribed manner of specifying an alternative hypothesis, and illustrate how prior opinion should be obtained and used to derive realistic expectation of the consequences of starting a clinical trial, initially expressed in terms of the chance of obtaining a "significant" result.

(b) Graphically superimpose prior, likelihood and posterior distributions on a parameter range divided up into areas with differing clinical implications, and recommend this as a means for ethical and efficient monitoring. The proposals of Mehta and Cain (1984) are particularly attractive.

(c) Use current belief to make predictions about the consequences of continuing a trial, possibly relative to monitoring boundaries derived from (b).

We shall illustrate these first steps in the next three sections.

9. ALTERNATIVE HYPOTHESES, PRIOR DISTRIBUTIONS, AND STRENGTHS OF TRIALS

Guidance for selecting the alternative hypothesis δ_A on which to base power calculations is varied and confusing. Spiegelhalter and Freedman (1986) found traditional suggestions that it was the difference "worth detecting" (Fleiss, 1976) or that "you would not wish to overlook" (Armitage, 1971), or the "smallest clinically worthwhile difference" (Lachin, 1981b). Others realise that δ_A should at least be plausible, and so recommend a difference that is "possible and important" (Brown, 1980) or "realistic and worthwhile" (Gore, 1981). In good-quality trials it is generally taken as the expected improvement based on previous studies, but in spoken and written work there is continual confusion between *expected* differences and differences that are *desired* in order to be willing to change treatment.

Any expected improvement must, however, carry some uncertainty, but this is universally ignored in power calculations. Consider, for example, the Beta-Blocker Heart Attack trial described in Section 2. This was designed to detect a 3.7% drop in absolute mortality "based on the European beta-blocker studies" (BHAT Research Group, 1981), and with 4020 patients there would be 90% power under this alternative to reject the null hypothesis at the 5% level. (We note in passing that there is no mention of "clinically worthwhile improvement", that the response measure used in the power calculation is not that used for monitoring the trial, and no allowance is made for the sequential intentions of the organisers.)

However, the size of the European Beta Blocker studies is vital. A repeated trial will usually mean previous results were equivocal, so suppose only 300 patients had been randomized in each group, giving an approximate 95% interval of (–2.2%, 9.6%) for the expected absolute reduction. The powers at the extremes of this interval are 0% and 100%, suggesting that the stated power of 90% may be somewhat misleading.

The *unconditional* probability of obtaining a significant result was considered as an outcome measure by Whitehead (1986), and called the "average power" by Spiegelhalter and Freedman (1986), the "predictive power" by Spiegelhalter *et al.* (1986) and the "expected power" by Brown *et al.* (1987). We shall, however, now adopt Crook and Good's (1982) use of the term *strength* to represent β_u (*unconditional* power) as defined by

$$\beta_u = \int p(T \in C|\delta)p(\delta)d\delta,$$

where $p(T \in C|\delta)$ is the conditional probability of the test statistic T falling in the critical region C at a specified δ and the chosen sample size per group. A useful alternative mean for obtaining β_u is by reversing the order of integration to give

$$\beta_u = \int_C \int p(T|\delta)p(\delta)d\delta \cdot \delta T$$

which involves integrating the predictive distribution of T over the critical region. This approach is illustrated in the next section.

When we assume a test statistic $T_n \sim N(\delta, \sigma^2/n)$, a null hypothesis $H_0 : \delta = 0$, and a prior distribution $\delta \sim N(\delta_0, \sigma^2/n_0)$, the strength to conclude an improvement (i.e. observe $T_n > z_\alpha \sigma/n^{1/2}$) is

$$\begin{aligned}\beta_u &= \int \{1 - \Phi(z_\alpha - \delta n^{1/2}/\sigma)\}p(\delta)d\delta \\ &= 1 - \Phi\{(z_\alpha - \delta_0 n^{1/2}/\sigma)/(1 + n/n_0)^{1/2}\}.\end{aligned}$$

For the BHAT trial, $n = 2010, z_\alpha = 1.96, \sigma = .52, \delta_0 = .037$, and we have assumed $n_0 = 300$ (the equivalent sample size of the prior), and hence $\beta_u = 1 - \Phi(-.44) = .67$.

The foregoing analysis shows trials which use a standard 90% power to detect an expected improvement may to some extent be deluding themselves. Nor does it help to choose some other aspect of the prior distribution, such as a particular percentile, since it is straightforward to show that the strength of a trial is equivalent to conditioning on a particular alternative hypothesis δ_u where

$$\delta_u = \delta_{1/2}\{(1 - (1 + n/n_0)^{-1/2}\} + \delta_0(1 + n/n_0)^{-1/2} :$$

a weighted average of the prior mean and the point of 50% power; i.e. $\delta_{1/2} = z_\alpha \sigma/n^{1/2}$ is the point which, if observed, would just lead to rejection of the null hypothesis. As $n_0 \to 0, \delta_u \to \delta_{1/2}$, so one should only ever expect 50% power if there is little prior information. In contrast, as $n_0 \to \infty$, $\delta_u \to \delta_0$ so the practice of conditioning just on expected benefit is only reasonable under strong prior information. (In fact, from Peto (1982) we can infer that $n_0 \approx 5500$, giving a strength of 85% for the BHAT trial.)

If no previous trials have been conducted one may fall back on observational data or even to subjective clinical opinion. A joint EORTC (European Organization for Reseach and Treatment of Cancer) / MRC multi-centre trial in osteosarcoma has recently begun, comparing a standard chemotherapy (adriamycin and cysplatinum) with a new "multidrug" regime which was suggested to prolong 5-year survival, but was envisaged to be considerably more toxic. Minimal published information was available concerning the new therapy. The baseline 5-year mortality under the standard treatment was about 50%, and a trial of $n = 200$ patients in each treatment arm was planned. Figure 1 shows the power curve of the trial assuming a final

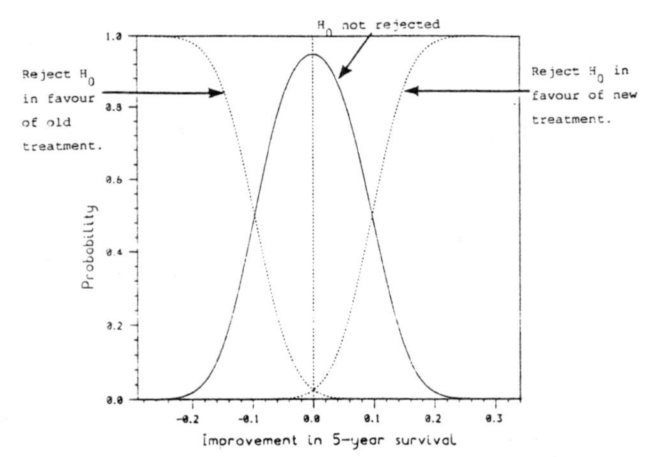

Figure 1. *Probability of rejecting null hypothesis at 5% level conditional on true improvement in 5-year survival*

2-sided test at the 5% level (assuming a normal approximation to observed absolute benefit with standard error $.7/n^{1/2}$).

At a meeting of the trial participants, 7 consultant oncologists were interviewed concerning their subjective opinion on the efficacy and disadvantages of the new regime. Using 20 minute individual structured interviews described in Spiegelhalter and Freedman (1986) the seven subjective priors shown in Figure 2 were elicited. We emphasise that no attempt is made to elicit conjugate priors.

We note the consistency in judgement among "naïve" subejects who had never undergone a similiar exercise nor discussed it amongst themselves. After noting there were no extreme opinions, a simple average was taken to provide the overall prior shown in Figure 3.

The distribution in Figure 3 is, from our experience, typical of the opinions of appropriately sceptical clinicians. There is expectation of a small benefit but some possibility of an adverse effect is entertained, with a small degree of optimism that large gains will result.

Averaging the power curve with respect to this step-function prior using the results in Spiegelhalter and Freedman (1986) gives a strength of 23% (equivalent to an alternative hypothesis of 6%). This low value seems, to us, considerably more realistic than a stated power of say, 80%, obtained by conditioning on an overly optimistic alternative of $\delta_A = 15\%$.

We feel the acknowledgement that alternative hypotheses should be based on a realistic distribution of possible benefits would be an important step in the rational decision as to whether a trial is worthwhile, and in particular may help funding bodies and pharmaceutical companies to rank competing claims on resources. Certainly we have found participants extremely enthusiastic about reporting their opinions. However, Gilbert, McPeek and Mosteller (1977) claim that "innovations brought to the stage of randomised trials are usually expected by the innovators to be sure winners" and other statisticians also warn us against naïve trust in the opinions of clinicians (especially surgeons!). For general acceptability it is better if past trials can provide the prior, and a funding body should certainly require convincing evidence as to the reasonableness of the quantitative judgement. However, it is possible to criticise the probability assessments of clinicians using measures of predictive ability described in Spiegelhalter (1986b).

It could be argued that by averaging over areas of the prior that include a *disadvantage* of the new treatment, we are distorting the idea of power to detect an improvement. One

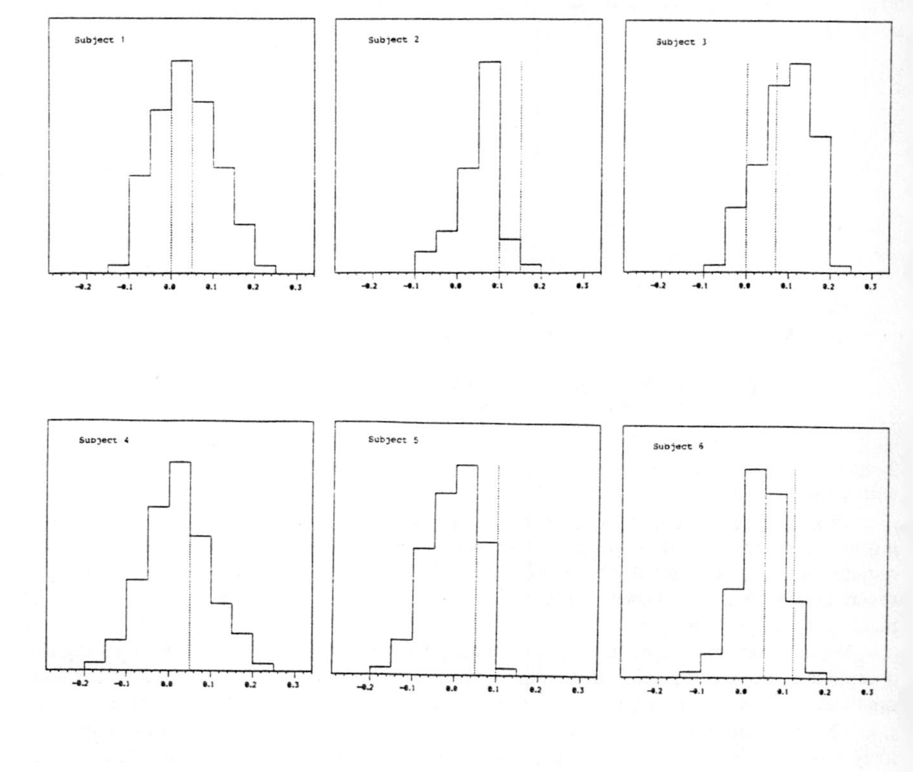

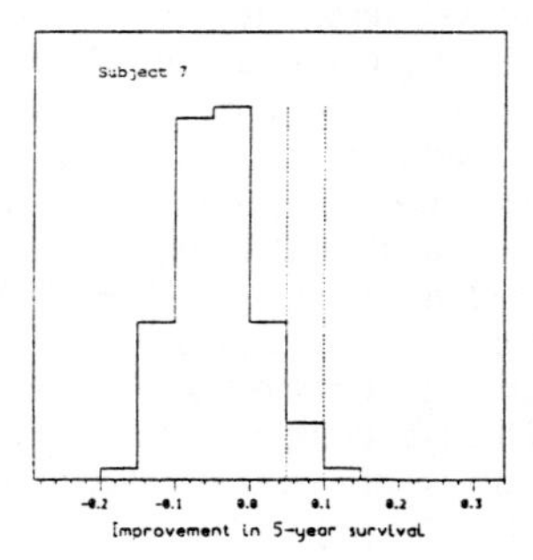

Figure 2. *Subjective priors of seven oncologists concerning true benefit, with ranges of clinical equivalence*

ossibility is to quote the strength conditional on there being a benefit, i.e. average the onditional power over the prior for $\delta > 0$, normalised to add to 1. In this instance we obtain 5% —not a substantial improvement. Perhaps the best summary measure is to quote the nconditional chance of not rejecting the null hypothesis: 69%.

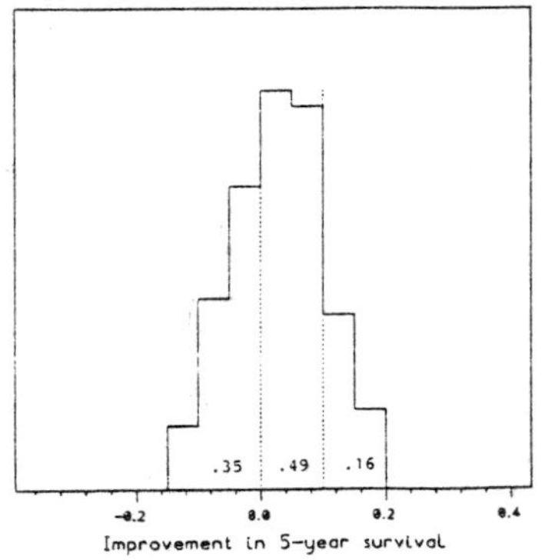

Figure 3. *"Consensus" prior opinion*

10. GRAPHICALLY RELATING BELIEFS TO VALUES

n our interviews with the oncologists in the osteosarcoma trial we also asked what improvenent they would demand before they would adopt the more toxic treatment. In general, 'ranges of equivalence" were elicited and are displayed in Figure 2: for example, subject would be willing to adopt the new regime were a 10% improvement to be convincingly hown, but would not change if the proven benefit were 5%. In between, he would be inlifferent, in the sense that he would still be prepared to randomise, even if an improvement n that range were unequivocally proven. It could be argued that it is ethical for a doctor to andomise patients provided he does not have low belief, say $< 1\%$, lying on one or other ide of his personal range of equivalence. Indifference on the primary outcome measure, as orced by Cornfield's symmetric prior under the alternative hypothesis, does not, contrary to opular opinion, seem the appropriate criteria for ethical randomisation. Our subjects appear easonable although it is perhaps surprising that subject 7 is willing to randomise his patients.

Usually one of the treatments does have secondary disadvantages; we have often found) to 10% a consensus range of equivalence, and this is superimposed on Figure 3. As the rial progresses such a range may change as more is discovered, and perhaps a separate dverse-reaction committee unaware of the primary results should adjust the monitoring range ccordingly. In any case, we feel the superimposition of belief on clinically relevant intervals nay have a number of functions: the *prior* relationship indicates the ethics of proceeding vith the trial in the first place; the *personal posterior* relationship indicates the personal thics of continuing the trial; while superimposing the reference posterior on ranges of clinical ntention provides an idea of what may be the clinical impact of the trial results when presented 'publicly", stripped of the prior opinion. In particular, provided personal ethics are satisfied, t would seem reasonable to proceed with the trial until one could present a small reference osterior probability (say $< .025$) on one or other side of the range of equivalence, and hence ould give a firm treatment recommendation based on the data alone. Using the consensus rior, the strength to obtain such a conclusive result is 62%.

Suppose, then, after 100 patients in each arm of the osteosarcoma trial that the data uggested an improvement under the new regime of 8%. Figure 4a shows the private posterior btained using the consensus prior in Figure 3, while Figure 4b shows the reference posterior assuming a uniform prior on δ, this being equivalent to the normalized likelihood. We note that the personal beliefs of the participants may in some circumstances indicate termination of the trial, but they must trade this off against the acknowledgement that unless the reference posterior shows a convincing effect, their publication is unlikely to alter clinical practice (and appallingly, may not even be accepted by the relevant journal).

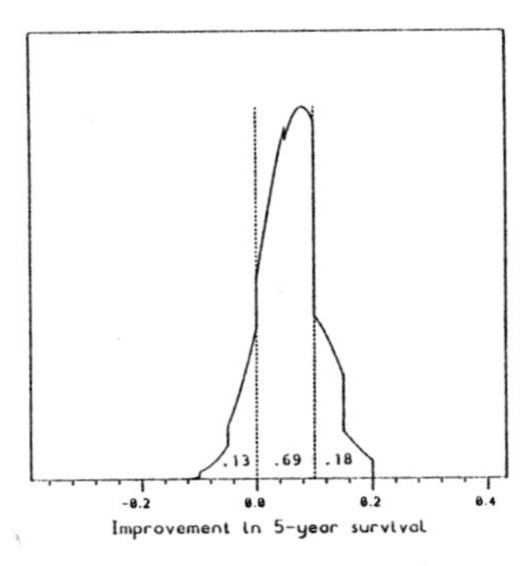

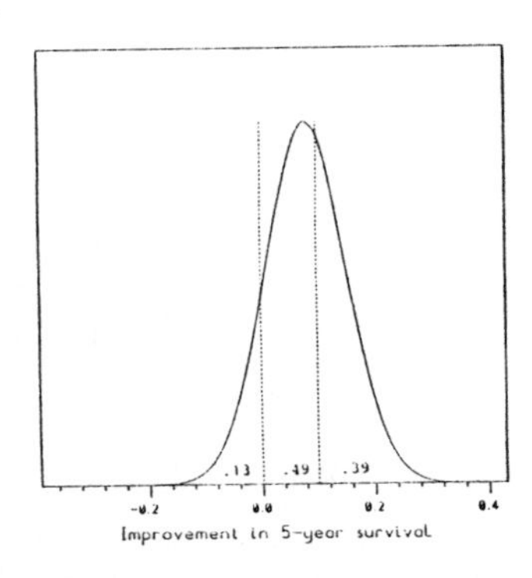

Figure 4a. **Figure 4b.**
Belief after data equivalent to 42/100 "treatment" and 50/100 "control" deaths, assuming the consensus prior (a) and uniform prior (b).

11. PLANNING OBJECTIVES IN A TRIAL

At this interim stage the participating clinicians may ask: "If we do go on and enrol another 100 patients in each group, what is the chance we'll get a "significant" result at the end?". Halperin *et al.* (1982) consider such interim predictions in the context of "stochastically curtailed testing", in which the probability of eventual rejection is calculated conditional on the original null and alternative hypothesis; this was implemented in the BHAT trial (Demets *et al.*, 1984). However, Spiegelhalter *et al.* (1986) criticise this process of making predictions conditional on circumstances that may already have been shown to be implausible, and instead suggest an unconditional predictive probability—essentially an interim calculation of the strength of the trial. Choi *et al.* (1985) make an identical sugesstion as a basis for stopping trials that appear negative. Racine *et at.* (1986) develop two-stage designs in bioequivalence studies on the basis of the predictive probability of establishing bioequivalence on the basis of the first stage sample, and Berry (1987) suggests a procedure very similar to that given below.

Our current data has classical 2-sided P-value .32, and suppose we wish to assess the chance this will drop below 5% were another 100 patients to be enrolled on each treatment. Our program, based on assuming independent Beta priors on the two success rates, produces the contours of the current joint posterior distribution, and the contours of the predictive distribution over future deaths were the trial to continue to its planned finish. Superimposed on the latter plot are the regions corresponding to the conclusions of a formal 2-sided significance test of $H_0 : \delta = 0$, and the interim strength is calculated by integrating the predictive density over these regions. The calculations are described in full in Spiegelhalter *et al.* (1986). Figures 5 and 6 show the results assuming an initial Jeffreys' prior on the success rates to resemble predictions made on the basis of the data alone, although the choice of reference prior makes little difference in this instance.

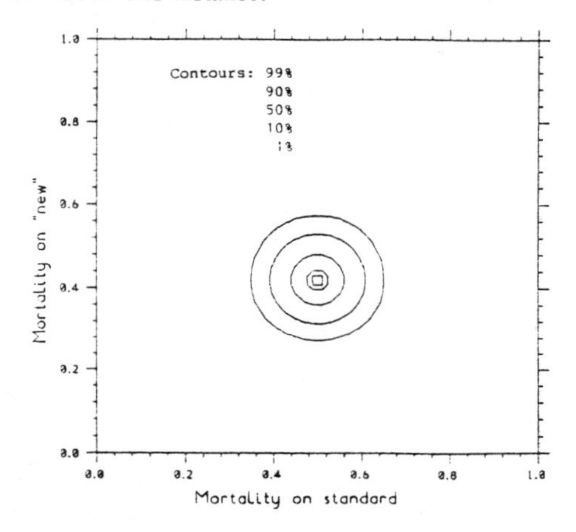

Figure 5. *Contours of joint posterior distribution of mortality rates*

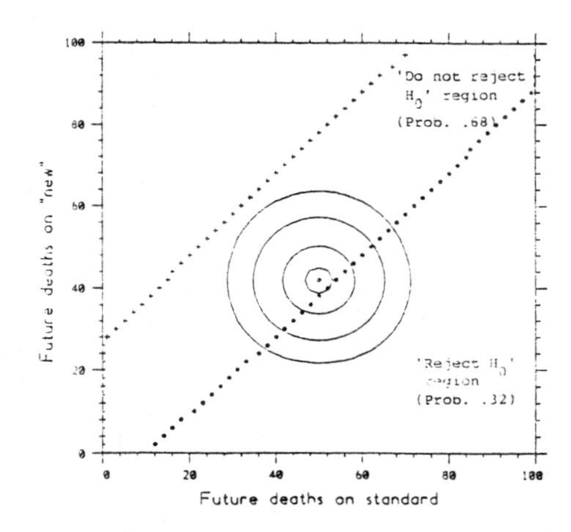

Figure 6. *Contours of predictive distribution of future mortality in 100 patients in each treatment*

The interim strength is 32% for another 100 patients in each arm, indicating that it may be reasonable to continue. The assumption of a reference posterior is particularly important if the continuation is a public decision, as when trying to convince funding bodies or other collaborators to give their support. Ranges of equivalence can also be incorporated: for example, using the reference posterior in Figure 4b to predict the consequences of continuing, we obtain a 57% chance of not obtaining a "conclusive" result, in the sense that $> 2.5\%$ tail area would still lie on each side of the range of equivalence.

Up to now we have argued that ethical monitoring may be conducted by contrasting belief with clinical intention, and predictions of the consequences of starting or continuing a trial, expressed in traditional objectives of rejecting a null hypothesis, may be expressed by prior or interim strength. We feel there is some hope for acceptance of these ideas, but are rather unhappy at adopting a "significant" result as being the sole objective. Some tentative thoughts on this are given next.

One approach has been taken by Detsky (1985), who, as suggested by Peto (1985), attempts to model the impact of a trial. Thus the annual number of patients whose lives will be saved is estimated as a function of the proportion who will be treated as a result of the trial, which depends on the observed benefit *and* its statistical significance. The marginal benefit of extra funding for a trial may then be calculated, balancing the cost against the expected benefit from increasing the numbers of patients and hence the strength of the trial to obtain an influential significant result.

A more restricted possibility is to assume that the trial will stand a good chance of making an impact if the observed difference is above the top of the range of equivalence *and* the lower end of the interval is above the lower end (usually 0). Thus the result is both clinically and statistically significant. The unconditional predictive probability of achieving this could be used as a measure for ranking candidate trials in a single clinical area.

A more Bayesian approach is to remove the criterion of statistical significance and simply predict some aspect of the beliefs one may have in the future as a result of the trial. One possibility is the probability content of a particular parameter interal. Suppose in the osteosarcoma study that we wished to predict how small our belief P in $\delta < 0$ may become as a result of further sampling. From Figure 4b our current "reference" belief is $p_0 = .13$, and conditional on that belief the relevant probability P is a martingale, as pointed out by Berry (1985). Specifically, let m patients have been treated in each group with an observed estimate t_m and $\delta \sim N(t_m, \sigma^2/m)$ under the reference prior. Then $p_0 = p(\delta < 0|t_m) = \Phi(-t_m m^{1/2}/\sigma)$. If a further n patients are studied in each group then it is reasonably straightforward to show that the distribution function of the tail-area P is

$$F_n(P) = \Phi\left[\left\{\Phi^{-1}(P) + t_m(n+m)^{1/2}/\sigma\right\}(m/n)^{1/2}\right].$$

The expected value of P is p_0 for all n —in other words we always *expect* our belief that the new drug is worse than the old to be the same as it is now, but our distribution about that expectation will change radically, although, as Dawid (1986) points out, by the Markov inequality, we always have $F_n(P) \leq (1-p_0)/(1-P)$ for $P \leq p_0$. When $n = 0$, our belief in P is a degenerate distribution on p_0; for $n = \infty$ our belief is a lump of size $1 - p_0$ on 0 and p_0 on 1. However, as the plot in Figure 7 shows, progress towards this asymptote is slow. (See Spiegelhalter (1986b) for another example of such a plot.)

Indeed, for a moderate extension of the experiment ($n = 100$), we have 33% chance that our tail area will be *greater* than that observed currently. This illustrates the dangers of extending an experiment a moderate amount in order to "clinch" the significance —there is a good chance a reversal will take place. Note that the strength calculation from the exact binomial program described above may be read off Figure 7 by considering $n = 100$, $P = .025$. Berry (1986) suggests calculating the predictive probability that the sign of the estimate would change if the experiment continued —this is simply obtained by considering $P = .50$.

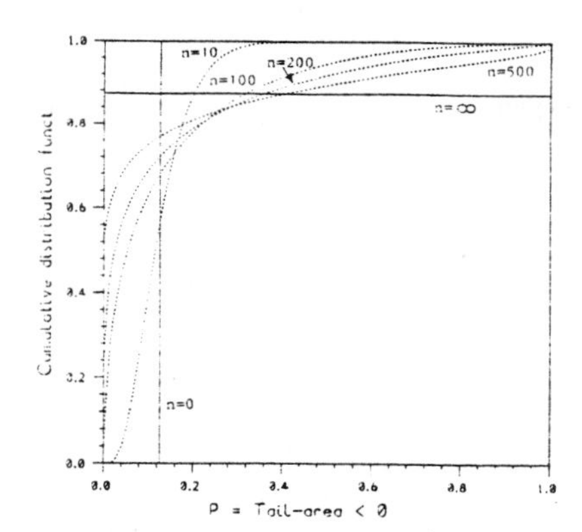

Figure 7. *Distribution function of future belief that* $\delta < 0$

12. CONCLUSION - THE CONSUMER'S VIEWPOINT

So far we have concentrated on the relationship between the experimenters' private opinions and the requirements of "objectivity" placed on them by reviewing bodies. We now consider the consumer's role in processing this depersonalised evidence.

A number of authors have pointed out that fixed Type I and Type II errors only measure the specificity and sensitivity of the trial, and that, analogously with diagnostic testing, the prior probability of a beneficial treatment needs to be taken into account in order to assess the predictive value of a trial's results. Staquet *et al.* (1979), Ciampi and Till (1980), Diamond and Forrester (1983) and Pocock (1983) all point out that with a low prior probability that a treatment is of benefit (20% is quoted) then the majority of "significant" results will, in fact, be false-positives. Hence, from the consumer's viewpoint, the classical approach is misleading. Moreover, it reveals the existence of prior beliefs that could serve to adjust reported results, and Gilbert *et al.* (1977) present an empirical distribution of observed treatment effects over a range of clinical trials.

Of course, the provision of confidence intervals is intended to take attention away from significance testing but these are almost universally interpreted as expressions of belief: even in a supportive editorial, we find the comment that "we also say using the confidence interval that there is only a 2.5% chance that the true difference in the population at large is greater than 10.mm Hg" (Langman, 1986). Full graphical expression of belief is so attractive to clinicians, compared with complex summaries by mean, standard error, confidence interval and P-value (Gardner and Altman,1986), that we feel it inevitable that an impact will be made on this area.

However, over the likelihood principle we foresee a long debate, since there is disquiet about a trial which was stopped at an opportune moment. We have argued that this discomfort is reasonable if a survey of trials is being conducted and it is felt that negative studies have not been published. Otherwise it reveals the existence of a prior scepticism that should be explicitly acknowledged and used to adjust the likelihood, reiterating that the classical urge to adjust may be appropriate but is done for the wrong reason.

Naturally there may be strong reaction from classical statisticians: "A statistical consultant who proposes a Bayesian analysis should therefore be expected to obtain a suitably informed consent from the clinical client whose data are to be subjected to the experiment" (Feinstein, 1977). However, an oncologist said recently: "We have a degree of belief in both the conventional and the new treatment we propose to test. Yet there is information we have acquired that expresses a certain degree of belief that the new treatment may be better. It seems that the trial in which we are about to engage ought to incorporate some of that prior information, and the results of the trial ought to either add or subtract from that degree of belief" (Decosse, 1983). Perhaps the greater hope for change lies in the common sense of reasonable clinicians.

REFERENCES

Anscombe, F. J. (1963). Sequential medical trials. *J. Amer. Statist. Assoc.* **58**, 365–383.

Armitage, P. (1971). *Statistical Methods in Medical Research.* Oxford: Blackwell.

Armitage, P. (1975). *Sequential Medical Trials (2nd Ed).* Oxford: Blackwell.

Armitage, P. (1985). The search for optimality in clinical trials. *Int. Statist. Rev.* **53**, 15–24.

Armitage, P., McPherson, C. K. and Rowe, B. C. (1969). Repeated significance tests on accumulating data. *J. Roy. Statist. Soc. A* **132**, 235–244.

Arnold, L. E., Christopher, J., Huestis, R. and Smeltzer, D. J. (1978). Methylphenidate vs Dextroamphetamine vs caffeine in minimal brain dysfunction. Controlled comparison by placebo washout design with Bayes' analysis. *Arch. Gen. Psychiatry* **35**, 463–473.

BHAT Research Group (1981). Beta-Blocker Heart Attack Trial design features. *Controlled Clinical Trials* **2**, 275–285.

Bather, J. A. (1985). On the allocation of treatments in sequential medical trials. *Int. Statist. Rev.* **53**, 1–13.

Berger, J. and Wolpert (1984). *The Likelihood Principle.* Hayward, CA: Institute of Mathematical Statistics.

Berger, J. and Berry, D. (1985). The relevance of stopping rules in statistical inference. *Statistical Decision Theory and Related Topics* **4**. New York: Springer-Verlag.

Bernardo, J. M. (1979). Reference posterior distributions for Bayesian inference (with discussion). *J. Roy. Statist. Soc. B* **41**, 113–147.

Berry D. A. (1985). Interim analyses in clinical trials: classical vs Bayesian approaches. *Statistics in Medicine* **4**, 521–526.

Berry, D. A. (1987). Interim analyses in clinical trials: the role of the likelihood principle. *Amer. Statist.* **41**, 117–122.

Berry, D. A. (1988). Multiple comparisons, multiple tests and data dredging: a Bayesian perspective. (In this volume).

Brown, B. W. (1980). Statistical controversies in the design of clinical trials - some personal views. *Controlled Clinical Trials* **1**, 13–28.

Brown, B. W., Herson, J., Atkinson, E. N. and Rozell, M. E. (1987). Projection from previous studies: a Bayesian and frequentist compromise. *Controlled Clinical Trials* **8**, 29–44.

Chernoff, H. and Petkau, A. J. (1981), Sequential medical trials involving paired data. *Biometrika* **68**, 119–132.

Choi, S. C., Smith, P. J. and Becker, D. P. (1985). Early decision in clinical trials when treatment differences are small. *Controlled Clinical Trials* **6**, 280–288.

Ciampi, A. and Till, J. E. (1980). Null results in clinical trials: the need for a decision-theory approach. *Br. J. Cancer* **41**, 618–629.

Colton, J. (1963). A model for selecting one of two medical treatment. *J. Amer. Statist. Assoc.* **58**, 388–400.

Cornfield, J. (1966a). Sequential trials, sequential analysis and the likelihood principle. *Amer. Statis.* **20**, 18–23.

Cornfield, J. (1966b). A Bayesian test of some classical hypotheses - with applications to sequential clinical trials. *J. Amer. Statist. Assoc.* **61**, 577–594.

Cornfield, J. (1969). The Bayesian outlook and its applications. *Biometrics* **24**, 617–657.

Cornfield, J. (1976). Recent methodological contributions to clinical trials. *Am. J. Epidemiology* **104**, 408–421.

Coronary Drug Project Research Group (1970). The Coronary Drug Project. Initial findings leading to a modification of its research protocol. *J. Amer. Med. Assoc.* **214**, 1301–1313.

Coronary Drug Project Research Group (1975). Clofibrate and Niacin in coronary heart disease. *J. Amer. Med. Assoc.* **231**, 360–381.

Crook, J. F. and Good, I. J. (1982). The powers and strengths of tests for multinomials and contingency tables. *J. Amer. Statist. Assoc.* **77**, 793–802.

Dawid, A. P. (1986). In discussion of Racine *et al.* (1986).
DeMets, D. L. (1984). Stopping guidelines vs stopping rules: a practitioner's point of view. *Commun Statistics - Theory Meth.* **13**, 2395–2417.
DeMets, D. L., Hardy, R., Friedman, L. M. and Lan, K. K. G. (1984). Statistical aspects of early termination in the Beta-Blocker Heart Attack trial. *Controlled Clinical Trials* **5**, 362–372.
Decosse J. J. (1982). Views of a surgical oncologist. *Statistics in Medical Research* (V. Mike and K. F. Stanley, eds.). New York: Wiley, 202–205.
Detsky, A. S. (1985). Using economic analysis to determine the resource consequences of choices made in planning clinical trials. *J. Chronic Diseases* **38**, 753–765.
Diamond, G. A. and Forrester, J. S. (1983). Clinical trials and statistical verdicts: probable cause for appeal. *Annals of Intern Medicine* **98**, 385–394.
Donner, A. (1982). A Bayesian approach to the interpretation of subgroup results in clinical trials. *Journal of Chronic Diseases* **35**, 429–435.
Dupont, W. D. (1983). Sequential stopping rules and sequentially adjusted P-values; does one require the other? *Controlled Clinical Trials* **4**, 3–10.
Feinstein, A. R. (1977). Clinical biostatistics XXXIX. The haze of Bayes, the aerial palaces of decision analysis, and the computerised Ouija board. *Clinical Pharmacology and Therapeutics* **21**, 482–496.
Fleiss, J. L. (1981). *Statistical Methods for Rates and Proportions* (2nd Ed.). New York: Wiley.
Fluehler, H. Grieve, A.P., Mandallaz, D., Mau, J. and Moser, H. A. (1983). Bayesian approach to bioequivalence assessment: an example. *Journal of Pharmaceutical Sciences* **72**, 1178–1181.
Fluehler, H., Hirtz, J. and Moser, H. A. (1981). An aid to decision-making in bioequivalence assessment. *Journal of Pharmacokinetics and Biopharmaceutics* **9**, 235–243.
Freedman, L.S. and Spiegelhalter, D. J. (1983). The assessment of subjective opinion and its use in relation to stopping rules for clinical trials. *The Statistician* **33**, 153–160.
Freedman, L. S., Lowe, D. and Macaskill, P. (1984). Stopping rules for clinical trials. *Biometrics* **40**, 575–586.
Gardner, M. J. and Altman, D. G. (1986). Confidence intervals rather than P values; estimation rather than hypothesis testing. *British Medical Journal* **292**, 746–750.
Gilbert, J. P., McPeek, B. and Mosteller, F. (1977). Statistics and ethics in surgery and anesthesia. *Science* **198**, 684–689.
Gore, S. M. 1981. Assessing clinical trials - trial size. *British Medical Journal* **282**, 1687–1689.
Grieve, A. P. (1985). A Bayesian analysis of the two-period crossover design for clinical trials. *Biometrics* **41**.
Halperin, M., Lan, K. K. G., Ware, J. H., Johnson, N. J. and DeMets, D. L. (1982). An aid to data monitoring in long-term clinical trials. *Controlled Clinical Trials* **3**, 311–323.
Healy, M. J. R. (1978). New methodology in clinical trials. *Biometrics* **34**, 709–712.
Herson J. (1979). Predictive probability early termination plans for phase II clinical trials. *Biometrics* **35**, 775–783.
Hilden, J., Bock, J. E. Andreasson, B. and Visfeldt, J. (1987). Ethics and decision theory in a clinical trial involving severe disfigurement. *Theoretical Surgery* **1**, 183–189.
Iglewicz, B. (1983). Alternative designs: sequential, multi-stage, decision theory and adaptive designs. *Cancer Clinical Trials: Methods and Practica*, (M. E. Buyse *et al.* eds.). Oxford: University Press, 312–324.
Jeffreys, H. (1961). *Theory of Probability* (3rd Ed). Oxford: University Press.
Jennison, C. and Turnbull, B. W. (1984). Repeated confidence intervals for group sequential clinical trials. *Controlled Clinical Trials* **5**, 33–46.
Kadane, J. B. (1986). Progress toward a more ethical method for clinical trials. *Journal of Medicine and Philosophy* **11**, 385–404.
Kalish, L. A. and Begg, C. B. (1985). Treament allocation methods in clinical trials - a review. *Statistics in Medicine* **4**, 129–144.
Lachin, J. M. (1981a). Sequential clinical trials for normal variates using interval composite hypotheses. *Biometrics* **37**, 87–101.
Lachin, J. M. (1981b). Introduction to sample size determination and power analysis for clinical trials. *Controlled Clinical Trials* **2**, 93–114.
Langman, M. J. S. (1986). Towards estimation and confidence intervals (editorial). *British Medical Journal* **292**, 716.
Lindley, D. V. (1975). The effect of ethical design considerations on statistical analysis. *Applied Statistics* **24**, 218–228.
Louis, T. A. (1984). Estimating a population of parameter values using Bayes and empirical Bayes methods. *J. Amer. Statist. Assoc.* **79**, 393–398.

Mandallaz, D. and Mau, J. (1981). Comparison of different methods for decision-making bioequivalence assessment. *Biometrics* **37**, 213–222.

McPherson, K. (1982). On choosing the number of interim analyses in clinical trials. *Statistics in Medicine* **1**, 25–36.

Medical Research Council (1948). Streptomycin treatment of pulmonary tuberculosis. *British Medical Journal* **ii**, 769–782.

Mehta, C. R. and Cain, K. C. (1984). Charts for the early stopping of pilot studies. *Journal of Clinical Oncology* **2**, 676–682.

Meier, P. (1975). Statistics and medical experimentation. *Biometrics* **31**, 5–1529.

Meier, P. (1979). Terminating a trial - the ethical problem. *Clinical Pharmacology and Therapeutics* **25**, 633–640.

O'Brien, P. C. and Fleming, T. R. (1979). A multiple testing procedure for clinical trials. *Biometrics* **35**, 549–556.

Peto, R. (1982). Long-term and short-term beta-blockade after myocardial infarction. *Lancet* **i**, 1159–1161.

Peto, R. (1985). In discussion of Bather (1985).

Peto, R., Pike, M. C., Armitage, P., Breslow, N. E., Cox D. R., Howards, S. V., Mantel, N., McPherson, K., Peto, J. and Smith, P. G. (1976). Design and analysis of randomised clinical trials requiring prolonged observation of each patient. I. Introduction and design. *British Journal Cancer* **34**, 585–612.

Pocock, S. J. (1976). The combination of randomised and historial controls in clinical trials. *Journal of Chronic Diseases* **29**, 175–188.

Pocock, S. J. (1977). Group sequential methods in the design and analysis of clinical trials. *Biometrika* **64**, 191–199.

Pocock, S. J. (1983). *Clinical Trials: a Practical Approach*. Chichester: Wiley.

Racine, A., Grieve, A.P., Fluehler, H. and Smith, A. F. M. (1986). Bayesian methods in practice: experiences in the pharmaceutical industry; (with discussion). *Applied Statistics* **35**, 93–150.

Royall, R. M. (1983). In discussion of Dupont (1983).

Schwartz D., Flamant, R. and Lellouch, (1980). *Clinical Trials*. London: Academic Press.

Selwyn, M. R. and Hall, N. R. (1984). On Bayesian methods for bioequivalence. *Biometrics* **40**, 1103–1108.

Selwyn, M. R., Dempster, A. P. and Hall, N. R. (1981). A Bayesian approach to bioequivalence for the 2×2 changeover design. *Biometrics* **37**, 11–21.

Simon, R. (1978). In response to Healy (1978).

Spiegelhalter, D. J. (1986a). In discussion of Racine *et al.* (1986).

Spiegelhalter, D. J. (1986b). Probabilistic prediction in patient management and clinical trials. *Statistics in Medicine* **5**, 421–434.

Spiegelhalter, D. J. and Freedman, L.S. (1986). A predictive approach to selecting the size of a clinical trial, based on subjective opinion. *Statistics in Medicine* **5**, 1–13.

Spiegelhalter, D. J., Freedman, L.S. and Blackburn, P. R. (1986). Monitoring clinical trials: conditional or predictive power? *Controlled Clinical Trials* **7**, 8–17.

Staquet, M. J. Rozencweig, M., Von Hoff, D. D. and Muggia, F. M. (1979). The delta and epsilon errors in the assessment of cancer clinical trials. *Cancer Treatment Reports* **63**, 1917–1921.

Sylvester, R. J. and Staquet, M. M. (1980). Design of phase II clinical trials in cancer using decision theory. *Cancer Treatment Reports* **64**, 519–524.

Tukey, J. (1977). Some thoughts on clinical trials, especially problems of multiplicity. *Science* **198**, 679–684.

University Group Diabetes Program (1970). A study of the effects of hypoglycemic agents on vascular complications in patients with adult onset diabetes. *Diabetes* **19**, Supplement 2, 747–830.

Urokinase Pulmonary Embolism Trial (1973). A National Cooperative Study. (A. A. Sasahara, T. M. Cole, F. Ederer, J. A. Murray, N. K. Wenger, S. Sherry and J. M. Stengle, eds.). *Circulation* **47**, Supplement 2, 1–108.

Van Ryzin, J. (1980). Designing for nonparametric Bayesian survival analysis using historical controls. *Cancer Treatment Reports* **64**, 503–506.

Waller, R. A. and Duncan, D. B. (1969). A Bayes rule for the symmetric multiple comparison problem. *J. Amer. Statist. Assoc.* **64**, 1484–1503.

Westlake, W. J. (1979). Statistical aspects of comparative bio-availability trials. *Biometrics* **35**, 273–280.

Whitehead, J. (1983). *The Design and Analysis of Sequential Clinical Trials*. Chichester: Ellis Horwood.

Whitehead, J. (1985). Designing phase II studies in the context of a programme of clinical research. *Biometrics* **41**, 373–383.

Whitehead, J. (1986). Sample sizes for phase II and phase III clinical trials: an integrated approach. *Statistics in Medicine* **5**, 459–464.

DISCUSSION

A. P. GRIEVE (*Ciba-Geigy Pharmaceuticals, Horsham*)

I very much enjoyed the paper by Spiegelhalter and Freedman, combining as it does a critical review of current statistical practice in clinical trials and a prescription for a staged introduction of Bayesian methods. In medical terms the authors have diagnosed the illness "lack of impact of Bayesian methods" and proposed a treatment. Two questions remain: who are the patients? what is the prognosis? The answer to the former is experimenters, reviewers and some statisticians; to the latter the answer is not clear, and we will need to consider progress made in a further twenty years, but I suspect that the prognosis for the experimenters is far better than that for the reviewers, and of the statisticians perhaps the less said the better.

In my experience it is not difficult to persuade experimenters that the correct analysis in any particular application is Bayesian. This may be because, as Rubin (1984) points out, most experimenters intuitively interpret certain classical concepts, for example confidence intervals, in a Bayesian way, or because when first confronted with graphical representations of uncertainty they realise how much information can be lost by considering only point or interval estimates. The authors give a good example of Rubin's observation in Section 12; a striking example of loss of information is given by Mandallaz and Mau (1981). Problems only arise when experimenters wish to communicate their beliefs to the reviewers. We commented in Racine *et al.* (1986):

> "experimentalists tend to draw a sharp distinction between providing their opinions and assessments for the purposes of experimental design and in-house discussion, and having them incorporated into any form of externally of disseminated report".

In practice, therefore, analyses presented to reviewers adopt some form of "non-informative", in the authors' terms "reference", prior, sometimes chosen to mirror a classical analysis. Why is this necessary? The authors' answer is that although the experimenters' objective is to influence the consumers (clinicians) they can only do so through the offices of the reviewers whose demands for "objectivity" need to be satisfied. This is a pragmatic reason and one is strongly reminded of Cox's remark:

> "one does feel that statistical techniques both of design and analysis are sometimes adopted rather as rituals designed to assuage the last holders of absolute power (editors of journals) and perhaps also regulatory agencies, and not because the techniques are appreciated to be scientifically important". (Cox, 1983)

As is well known, such rituals are often not even correctly carried out (Gore *et al.*, 1977; Altman, 1982; Gardner *et al.*, 1983).

New drugs are not discovered by serendipity, but result from a complex development process. At each stage of this process information is obtained on a chemical compound and used either to plan the next stage of experimentation, for instance to progress from Phase I to Phase II trials, or to cease investigation of the compound and possibly start investigation of a further compound discovered in animal screening studies. At the end of this process the experimenter is hopefully in a position to plan and carry out a classical Phase III trial and will have collected considerable information concerning the compound's probable efficacy and tolerability. To require that the analysis of a specific Phase III trial be analysed "objectively", or in isolation, forces the experimenter to ignore considerable amounts of information, which is surely not right, as is forcibly pointed out by Newman,

> "Each clinical trial should start with fairly strong prior information about efficacy and safety yet this is ignored as the data are subjected to techniques related to non-informative priors. The process of approval must in some way balance the benefits against the risk of serious and rare side-effects. Without consideration of the prior probabilities and the losses involved the statistics end up as numbers floating in a whirlwind of prejudice and intuition". (Newman, 1983)

and supported by Healy,

> "... is it fair to regard the results of a phase III trial in total isolation? The cost of such a trial will not be small, but it can only be one stage in a long period of development whose overall cost will usually be very large indeed. Certainly the company's prior belief in the efficacy of the new product will be fairly high, and it will be able to back this up with animal results and those from Phases I and II." (Healy, 1983)

This view is not unanimously supported as the authors' quote from Feinstein admirably demonstrates. Some people would however go even further than Feinstein,

> "... I have yet to find a scientist who would be convinced by a posterior distribution on the methotrexate and colon cancer question if the prior has been supplied by a pharmaceutical company". (LeCam, 1985)

and, in an apparent search for ultimate "objectivity", Pocock (1983) recommends that any clinician taking part in a trial sponsored by a pharmaceutical company should have the data analysed by an independent statistician. A reaction to this extreme point of view from a pharmaceutical statistician is given by Seldrup (1985).

The ideal solution, as the authors rightly say, is to provide a whole series of prior to posterior analyses rather than to rely on "magical poses and pretences" (Dickey, 1986), provided by reference priors.

An area where I feel that Bayesian methods can have an impact is in the use of historical control information. In this area the work of Pocock (1976), mentioned by the authors, seems to as yet have had little impact, but more recently there has been a considerable research interest in the use of historial control information in carcinogenicity studies (Tarone, 1982a; Tarone, 1982b; Dempster *et al.*, 1983; Hoel, 1983; Margolin, 1984; Krewski *et al.*, 1985; Tamura and Young, 1986, Hoel and Yanagawa, 1986). Most of this work is based on hierarchical models for the control data, within which structure Pocock's model may also be formulated. The interest in using historical control data in carcinogenicity studies is understandable since many tumour types are rare in control populations, and so the occurrence of such rare tumours in treated animals may strongly indicate to the toxicologist an adverse effect, but the sample sizes involved are inadequate to reach a definite conclusion. It is not so clear that such an argument has relevance in clinical studies, but it should be realised that in the context of testing a new drug therapy, say an anti-rheumatic, regulatory authorities will almost certainly not permit the use of the sorts of numbers of patients which the authors may expect in multi-centre cancer trials. Because of this, the use of historical control data to increase the precision of treatment comparison is appealing. However this appeal must be weighed against the possible dangers of incorporating data from incompatible studies. To this end, Pocock (1976) proposes a set of criteria for checking the acceptability of historical control information. Dempster *et al.* (1983) characterize these criteria as implying a judgement of exchangeability of historical and current trials; this judgement needs to be made by "the knowledgeable subject matter experts". Of the articles mentioned above, that of Dempster *et al.* is the only "pure" Bayesian method; the others may be classified as so-called "empirical Bayes". In a more recent article, Meng and Dempster (1985) generalize the Dempster *et al.* approach to multiple tumour types, that is to the problem of multiplicity. It is not clear that this generalization has direct applications to clinical studies, except perhaps to the analysis of side-effects. These two areas, together with work of Donner (1982) and Louis (1984) in subgroup analyses, the papers of Don Berry and Bill DuMouchel at this conference, and the increased interest in "meta-analyses", suggest that further research into the use of hierarchical models in clinical trials may prove fruitful. It has been a recurring theme at this conference that one should think hard about problems, and this is certainly the case with hierarchical models and assumptions of exchangeability. We should not expect, nor attempt to provide, a cook-book for the application of such methods; each application should be viewed on its own merits.

I would like to thank the authors for their stimulating paper which gave me, and I am sure many others, much food for thought.

D. A. BERRY (*University of Minnesota*)

The authors have provided a nice review of many important issues in designing and analyzing clinical trials. Many of my own thoughts are very similar to theirs but I think it is worth mentioning a few differences.

The authors take the view that the route to converting biostatisticians, medical researchers and other interested parties to Bayesian thinking should be a gradual one. This is the correct attitude, I think, and Bayesians who take the "all or nothing" attitude should pay heed to their advice. However, I think that the hardest Bayesian pill for non-Bayesians to swallow is the use of subjective probabilities, and this is precisely where the authors start; if they sell this then the remainder of the enterprise will be easy for them and for the rest of us. While I have no disagreement with the objectives they state in Section 1, I think they sometimes go overboard later in the paper in praising classical statistical ideas (for example, in Section 7 they call work of Whitehead "elegant").

The authors work in the context of randomized clinical trials and do not devote much space to alternative designs, such as adaptive treatment allocations. This is quite appropriate for them because they are restricted to eventually making classical inferences such as significance tests, in which adaptive designs are impossible. But I would like to see all designers of clinical trials keep at least one decision-theoretic principle in mind, if only qualitatively. If the trial involves a condition that is very common (coronary artery disease, say) then a large trial makes some sense. But if the trial involves a very rare condition then (making the same power calculations and) conducting a large trial seems to me unethical: anything learned from the trial will not be put to much use and some patients in the trial may have been ineffectively treated for no good reason. In the case of a rare disease, assigning the currently favored treatment or using a bandit strategy (Berry and Fristedt (1985)) seems much more appropriate.

In Section 2 the authors discuss the statistical methods used by pharmaceutical companies in the development process of a new therapy. I have suggested that a pharmaceutical company should use its prior distribution in designing a clinical trial and drug development programme, and then report the results of the trial in the usual way —sans prior (Berry (1987), Berry and Ho (1988)). This is the spirit of a remark made by Dennis Lindley during the oral discussion of Spiegelhalter-Freedman paper.

The authors mention "sampling to a foregone conclusion" in Sections 2 and 4. They are certainly aware that this is not possible in any Bayesian setting, but the issue is so important to non-Bayesians that I think statisticians should be constantly reminded of it. Posterior probabilities are martingales. If the current probability of any hypothesis H is p then the expected value of the probability of H using any bounded stopping rule is p. And if one plans to continue sampling until the probability of H is $p + \varepsilon$ (or $p - \varepsilon$) for any $\varepsilon > 0$, the expected stopping time is infinite!

Regarding Section 12 and the consumer's viewpoint, I have no concern that consumers will be unduly influenced by partial results or by small trials. At least in the United States, medical studies that contradict earlier studies are reported in the press perhaps once a month (coffee and its potential influence on heart disease may hold the record for most flip-flops!). My model of the consumer is that of a (Bayesian) statistician observing Bernoulli trials (they should perhaps be weighted somehow) and altering his or her habits when the currently available information sufficiently increases the expected utility of doing so.

S. FRENCH (*University of Manchester*)

In the presentation, Dr. Spiegelhalter said that the introduction of utility functions into the administration of clinical trials was not high on his agenda for introducing Bayesian ideas. While I appreciate the problems involved, I do think that utility function ideas should be included in the strategy for change... if only in the long term. The introduction of a utility function into the analysis of decision problems needs no further theoretical justifications amongst us surely. But perhaps a moral justification is in order. If a decision problem is analysed without a utility function, then you cannot see where the "buck stops". Who really has the responsibility? Given the awesome consequences of taking the "wrong" decision in medical trials, I can understand the desire of the various parties to try to avoid explicit value judgements. But should we let them?

R. KASS (*Carnegie-Mellon University*)

In discussing alternative hypotheses, Spiegelhalter notes the frequent confusion between what is "expected" and what is "desired". As a parenthetical remark I would add that this is much the same problem that infects the use of the word "normal" in many contexts, notably in "normal ranges". With "normal", things are even worse because of the possibility of using the Normal distribution to define what is normal.

My question, though, concerns what seems to me to be an inconsistency in Spiegelhalter's point of view. Spiegelhalter apparently feels that the Jeffreys-Cornfield approach of putting point mass on the null hypothesis is rarely realistic. Perhaps this is so, but then why should "expected" improvement be used in planning trials rather than "desired" improvement? It seems to me that one would design a trial using some amount of expected improvement based on previous studies (or other information) only if that amount were otherwise arbitrary; that is, only if essentially any amount of improvement would be of interest; and, how could this be the case unless there were substantial belief in an essentially zero effect? If it were known that an essentially zero effect had negligible probability, while any amount of improvement would be of interest, why would one wish to carry out the trial?

On the one hand, I assume that clinical trials are designed to have the ability of convincing a reasonable skeptic. But then, shouldn't a skeptic's distribution be considered? On the other hand, while I do not intend to confuse probability with utility, I assume that genuine sharp null hypotheses become decreasingly common as we move away from basic science, in which we can often ignore utility, toward policy, in which we can not. Clinical trials often seem to be caught in a fuzzy intermediate area, trying to relate a policy issue (whether a drug should be recommended), to a physiological one (whether the drug has an effect on humans). The more clearly a planning committee acknowledges policy implications, the more urgent will utility evaluations become.

REPLY TO THE DISCUSSION

We welcome the positive nature of the discussion, and the fact that Dr Grieve represents a pharmaceutical company must augur well for the future. His discussion of carcinogenicity studies is a very useful adjunct to the paper. We agree with Professor Berry's observation, and we also used to think that subjective probabilities should be the last aspect to introduce. But our experience echoes that of Dr Grieve, in that for *internal* decision-making clinical opinion is readily available (provided that the term 'gambling' is avoided in the elicitation procedure), and should be immediately exploited.

Dr French and Professor Kass raise similar points concerning our apparent avoidance of explicit consideration of utilities. Perhaps we should have discussed at greater length the role of the range of clinical equivalence which we routinely elicit. Our procedure explicitly recommends monitoring studies by means of the current probability that the treatment displays a clinically worthwhile improvement, and the predictive probability of achieving a certain confidence that this is the case. Essentially one would not even start a trial if the predictive chance of confidently asserting "clinical superiority" were either very high or very low. If it is very high, it means that there is already substantial prior belief in a worthwhile improvement, and hence it is not ethical to randomise. This would follow if there were negligible probability of zero effect and any improvement were worthwhile, and reinforces Professor Kass's comments that one would not wish to carry out the trial in these circumstances. Alternatively, the predictive probability of a firm conclusion is very *low* if, for example, there is substantial belief in zero effect and any improvement were worthwhile, or if there is prior evidence of benefit but clinical demand for a substantial benefit. In either of these cases, it may again be unreasonable to start the trial, though perhaps more on the grounds of efficiency than ethics.

Thus the monitoring and predictive statements do explicitly consider utility (expressed in the form of clinical demands), and decisions are based on a juxtaposition of expectation and demand. We feel this can provide a rational and acceptable basis for decision-making.

REFERENCES IN THE DISCUSSION

Altman, D. G. (1982). Statistics in medical journals. *Statistics in Medicine* **1**, 59–71.

Armitage, P. (1983). Trials and errors: the emergence of clinical statistics (with discussion). *J. Roy. Statist. Soc. A* **146**, 321–334.

Berry, D. A. (1987). Statistical inference, clinical trials, and pharmaceutical company decisions. *The Statistician* **36**, 181–189.

Berry, D. A. and Fristedt, B. (1985). *Bandit Problems: Sequential Allocation of Experiments*. London: Chapman-Hall.

Berry, D. A. and Ho, C.-H. (1988). One-sided sequential stopping boundaries for clinical trials: A decision-theoretic approach. *Biometrics* **44**, 219–227.

Cox D. R.(1983). In discussion of Armitage (1983).

Dempster, A. P., Selwyn, M. R. and Weeks, B. J. (1983). Combining historical and randomized controls for assessing trends in proportions. *J. Amer. Statist. Assoc.* **78**, 221–227.

Gardner, M. J., Altman, D. G., Jones, D. R. and Machin, D. (1983). Is the statistical assessment of papers submitted to the "British Medical Journal" effective?. *British Medical Journal* **286**, 1485–1488.

Gore, S. M., Jones, I. G. and Rytter, E. C. (1977). Misuse of statistical methods: critical assessment of articles in BMJ from January to March 1976. *British Medical Journal* **280**, 85–87.

Healy, M. J. R. (1983). In discussion of Lewis (1983).

Hoel, D. G. (1983). Conditional two sample tests with historical controls. *Contributions to Statistics: Essays in Honor of Norman L. Johnson.* (P. K. Sen, ed.). Amsterdam: North-Holland.

Hoel, D. G. and Yanagawa, T. (1986). Incorporating historical controls in testing for a trend in proportions. *J. Amer. Statist. Assoc.* **81**, 1095–1099.

Krewski, D., Smythe, R. T. and Burnett, R. T. (1985). The use of historical control information in testing for trend in quantal response carcinogenicity data. *Proceedings of Symposium on Long-Term Animal Carcinogenicity Studies: A Statistical Perspective.* Biopharmaceutical Section of the American Statistical Association.

Lecam, L. (1985). In discussion of Berger and Wolpert (1985).

Lewis, J. A. (1983). Clinical trials: statistical developments of practical benefit to the pharmaceutical industry (with discuussion). *J. Roy. Statist. Soc. A* **146**, 362–393.

Margolin, B. H. and Risko, K. J. (1984). The use of historical data in laboratory studies. *Proceeding International Biometrics Conference*, Japan: Tokyo.

Meng, C. Y. K. and Dempster, A. P. (1985). A Bayesian approach to the multiplicity problem for significance testing with binomial data. *Proceedings of Symposium on Long-Term Animal Carcinogenicity Studies: A Statistical Perspective.* Biopharmaceutical Section of the American Statistical Association.

Newman, G. R. (1983). In discussion of Lewis (1983).

Rubin, D. (1984). Bayesianly justifiable and relevant frequency calculations for the applied statistican. *Ann. Statist.* **12**, 1151–1172.

Seldrup, J. (1985). A review of "Clinical Trials: A Practical Approach" by S. J. Pocock. *The Statistician* **34**, 337–338.

Tamura, R. N. and Young, S. S. (1986). The incorporation of historical control information in tests of proportions: simulation study of Tarone's procedure. *Biometrics* **42**, 343–349.

Tarone, R. E. (1982a). The use of historical control information in testing for a trend in proportions. *Biometrics* **38**, 215–220.

Tarone, R. E. (1982b). The use of historical control information in testing for testing for a trend in Poisson means. *Biometrics* **38**, 457–462.

BAYESIAN STATISTICS 3, pp. 479–491
J. M. Bernardo, M. H. DeGroot, D. V. Lindley and A. F. M. Smith, (Eds.)

Partial and Interaction Spline Models

G. WAHBA
University of Wisconsin-Madison

SUMMARY

A partial spline model is a model for a response as a function of variables, which is the sum of a "smooth" function of several variables and a parametric function of the same plus possibly some other variables. Partial spline models in one and several variables, with direct and indirect data, with Gaussian errors and as an extension of Glim to partially penalized Glim models are described. Application to the modelling of change of regime in several variables is described. Interaction splines are introduced and described and their potential use for modelling nonlinear interactions between variables by semiparametric methods is noted. Reference is made to recent work in efficient computational methods.

Keywords: PARTIAL SPLINES, INTERACTION SPLINES, SEMIPARAMETRIC REGRESSION.

1. INTRODUCTION

Partial spline models have proved to be interesting both from a practical and a theoretical point of view, partly because of their dual nature both as solutions to certain intuitively reasonable variational problems, and as Bayes estimates with certain parsimonious priors. In these proceedings we will attempt to give a quick rundown concerning some of their more interesting manifestations, and to report briefly on several new developments, in particular, results on partial and interactions splines.

2. PARTIAL SPLINE MODELS–ONE SPLINED VARIABLE

A response as a function of the variables x, $z_1, \ldots, z_k$ is modelled as

$$y_i = f(x(i)) + \sum_{j=1}^{p} \theta_j \Psi_j(x(i); z(i)) + \varepsilon_i, \tag{2.1a}$$

where

$$z(i) = z_1(i), \ldots, z_k(i)), \tag{2.1b}$$

the Ψ_j's are given parametric functions and the ε_i's are independent, zero mean Gaussian random variables with common (unknown) variance. The estimate $(f_\lambda, \theta_\lambda)$, where $\theta_\lambda = (\theta_{1\lambda}, \ldots, \theta_{p\lambda})$, is found as the minimizer, in an appropriate space, of

$$\frac{1}{n}\sum_{i=1}^{n}(y_i - f(x(i)) - \sum_{j=1}^{p} \theta_j \Psi_j(x(i); z(i)))^2 + \lambda J_m(f), \tag{2.2a}$$

where

$$J_m(f) = \int_0^1 (f^{(m)}(x))^2 dx. \tag{2.2b}$$

We have the following

Theorem. *(Kimeldorf and Wahba (1971) - KW) Let $\Phi_1, \ldots, \Phi_m$ span the null space of J_m. If the design matrix for least squares regression on span $\Phi_1, \ldots, \Phi_m$, $\Psi_1, \ldots, \Psi_p$ is of full column rank, then there existis a unique minimizer $(f_\lambda, \theta_\lambda)$ for any $\lambda > 0$, and f_λ is a polynomial spline function.*

The parameter λ as well as m can be choosen by generalized cross validation (GCV).

The appropriate function space here is the Sobolev space W_2^m; however, J_m (and W_2^m) can be replaced by any seminorm in a reproducing kernel (r. k.) Hilbert space of real valued functions on $[0, 1]$ provided that least squares regression onto the null space of the seminorm is well defined —you get a Bayes estimate with the r. k. related to the prior covariance. Details can be found in KW and Wahba (1978) but we will not discuss the Bayesian aspect any further, other than to note that the prior behind J_m is the most parsimonious member of a large equivalence class of priors.

Partial spline models with one splined variable were introduced by sevaral authors in different contexts, with some interesting applications, see Anderson and Senthilselvan (1982), Engle *et al.* (1986), Green, Jennison, and Seheult (1983), Shiller (1984).

3. PARTIAL SPLINE MODELS–SEVERAL SPLINED VARIABLES

Now, let the model be

$$y_i = f(x(i)) + \sum_{j=1}^{p} \theta_j \Psi_j(x(i); z(i)) + \varepsilon_i, \tag{3.1a}$$

where

$$x = (x_1, \ldots, x_d), x(i) = (x_1(i), \ldots, x_d(i)). \tag{3.1b}$$

Again, we find f in an appropriate space to minimize

$$\frac{1}{n} \sum_{i=1}^{n} (y_i - f(x(i)) - \sum_{j=1}^{p} \theta_j \Psi_j(x(i); z(i)))^2 + \lambda J_m(f), \tag{3.2}$$

where now, we can use the "thin plate spline" penalty functional. For $d = 2$, $m = 2$, it is

$$J_m(f) = \int_{-\infty}^{\infty} \int_{-\infty}^{\infty} f_{x_1 x_1}^2 + 2 f_{x_1 x_2}^2 + f_{x_2 x_2}^2, dx_1 dx_2 \tag{3.3}$$

and for arbitrary d it is

$$J_m(f) = \sum_{\alpha_1 + \cdots + \alpha_d = m} \frac{m!}{\alpha_1! \cdots \alpha_d!} \times \int \cdots \int \left(\frac{\partial^m f}{\partial x_1^{\alpha_1} \cdots \partial x_d^{\alpha_d}} \right)^2 dx_1 \cdots dx_d, \tag{3.4}$$

provided $2m > d$. The null space of J_m is the span of the $M = \binom{m+d-1}{d}$ monomials of total degree less than m, call them $\Phi_1, \ldots, \Phi_M$. Again there will be unique minimizer $(f_\lambda, \theta_\lambda)$ for every nonnegative λ if the design matrix for least squares regression on $\Phi_1, \ldots, \Phi_M$; $\Psi_1, \ldots, \Psi_p$ is of full column rank, and f_λ is a thin plate spline function.

Partial splines with several splined variables were introduced in Wahba (1984a), Wahba (1984b), Wahba (1985), and a discrete version has been proposed by Green, Jennison, and Seheult (1985). Transportable code (GCVPACK, Bates *et al.* (October 1986)) is available for fitting the partial spline models of (3.1)-(3.4) and computing the GCV estimate of λ. This code does well with up to around 400 data points on the VAX 11/750 in the Statistics Department at Madison. The computational work primarily depends on n, and not d, but, of course good estimates with large d will require large n. Diagnostics for splines (without the "partial" part) have been developed by Eubank (1986) and it can be anticipated that this work will extend to partial spline models. Results on properties of estimates of θ may be found in Heckman (1986), Rice (1986), Shiau (1987). Hypothesis tests for the null model $f = 0$ may be found in Cox *et al.* (1988) and Wahba (in preparation).

The partial spline model may be generalized to the nonlinear partial spline model

$$y_i = f(x(i)) + \Psi(\theta, x(i), z(i)) + \varepsilon_i$$

as in (3.1a) except that Ψ may be nonlinear in the parameter vector θ, see Wahba (in preparation), for estimation and GCV for this model.

4. INDIRECT MEASUREMENTS

Let

$$g(x; z) = f(x) + \Sigma\theta_j\Psi_j(x; z), \tag{4.1}$$

and now let

$$y_i = L_i g + \varepsilon_i \tag{4.2}$$

where L_i is a bounded linear functional, for example:

$$L_i f = \int w_i(x; z) g(x; z) \pi dx \pi dz. \tag{4.3}$$

This kind of data comes up in X-ray tomography, satellite tomography, stereology, and in other remote or indirect sensing problems in the physical and biological sciences. One finds f and θ to minimize:

$$\frac{1}{n}\Sigma(y_i - L_i f - \Sigma\theta_j L_i\Psi_j)^2 + \lambda J_m(f). \tag{4.4}$$

For a recent overview of indirect sensing problems, see O'Sullivan (1986). The use of variants of (4.3), and (4.4) may also provide a good way to deal with heterogeneous aggregated economic data. For an application in stereology, see Nychka *et al.* (1984).

Data involving mildly nonlinear functions can be accommodated

$$y_i = N_i g + \varepsilon_i, \tag{4.5a}$$

where

$$N_i g = \int\int w_i(x, z, g(x; z)) \pi dx \pi dz. \tag{4.5b}$$

One then finds f and θ to minimize

$$\frac{1}{n}\Sigma(y_i - N_i(f + \Sigma\theta_j\Psi_j))^2 + \lambda J_m(f). \tag{4.6}$$

The minimization can be performed using basis functions and a Gauss-Newton iteration and λ chosen by GCV for nonlinear problems, see O'Sullivan and Wahba (1985).

5. NON GAUSSIAN ERRORS (SEMIPARAMETRIC PENALIZED GLIM MODELS)

Here,

$$g(x, z) = f(x) + \Sigma\theta_j\Psi_j(x; z) \tag{5.1}$$

but

$$y_i \sim F_g.$$

For example:

$$y_i \sim \textit{Poisson with } \Lambda_i = e^{g(x(i);z(i))},$$

$$y_i \sim \textit{Binomial with } p_i/(1 - p_i) = e^{g(x(i);z(i))},$$

etc. and one finds $f_\lambda, \theta_\lambda$ to minimize

$$L(f, \theta) + \lambda J_m(f), \tag{5.2}$$

where L is the log likelihood. O'Sullivan (1983) and O'Sullivan, Yandell, and Raynor (1986) proposed numerical methods and a GCV for penalized GLIM models. See also Green and Yandell (1985), Silverman (1982), Cox and O'Sullivan (October, 1985), Leonard (1982).

6. USE OF PARTIAL SPLINES TO MODEL FUNCTIONS WHICH ARE SMOOTH EXCEPT FOR SPECIFIED DISCONTINUITIES

Let $d = 1$ and let

$$g(x; z) = f(x) + \theta|x - x^*|, \tag{6.1}$$

that is, $\Psi_1(x; z) = |x - x^*|$. Then the partial spline estimate of g will have a jump in the first derivative at x^* of size 2θ. In two dimensions we may use a partial spline to model a jump in the first derivative with respect to x_2 along a given curve $x_2^*(x_1)$. Let

$$\gamma(x) = \gamma(x_1, x_2) = |x_2 - x_2^*(x_1)|, \tag{6.2}$$

$$g(x; z) = f(x) + \theta(x_1)\gamma(x), \tag{6.3}$$

where θ may depend on x_1. Then

$$\left.\frac{\partial g}{\partial x_2}\right]_{x_2 = x_2^*(x_1)-} - \left.\frac{\partial g}{\partial x_2}\right]_{x_2 = x_2^*(x_1)+} = 2\theta(x_1).$$

If, for example

$$\theta(x_1) = \sum_{j=1}^{p} \theta_j q_j(x_1), \tag{6.4}$$

where the q_j's are given, then

$$\Psi_j(x; z) = q_j(x_1)\gamma(x). \tag{6.5}$$

This fits right into the partial spline setup, and GCVPACK may be used to compute the estimate. A generalization to $d = 3$ with a jump in the first derivative with respect to x_3 along a surface $x_3^*(x_1, x_2)$ is straightforward. For details, and a description of an application to the three dimensional modelling of the tropopause in the atmosphere and the thermocline in the ocean, see Shiau, Wahba, and Johnson (1986).

7. LINEAR INEQUALITY CONSTRAINTS

Expressions (2.2), (3.2), (4.4), etc can be minimized subject to finite families of linear inequality constraints. See Villalobos and Wahba (1987).

8. MAIN EFFECTS AND INTERACTION SPLINES

The thin plane spline is defined on Euclidean d space for any d with $2m - d > 0$, provided there are enough data points for the mth degree polynomial regression, but unless there are very large data sets, in many applications will be desirable to reduce the amount of structure involved. Several authors have suggested modelling f as a linear combination of functions of one variable, that is,

$$f(x) = f_0 + \sum_{al=1}^{d} f_\alpha(x_\alpha),$$

where $x = (x_1, \ldots, x_d)$, and $\int_0^1 f_\alpha(x_\alpha)dx_\alpha = 0$. (Note the switch to the unit cube.) See Friedman, Grosse, and Stuetzle (1983), Stone (1985), Burman (June, 1985). We have been working on generalizations of this idea, whereby f is modelled successively as linear combinations of functions of one variable, functions of one and two variables, functions of one, two and three variables, etc. The resulting estimates may be called main effects splines, two factor interaction splines, three factor interaction splines, etc., by analogy with analysis of variance. This idea is due to Barry (1983), Barry (1986). Most of the suggestions below concerning H_{TEPR} are equivalent to proposals made by Barry, devived here from a slightly different point of view. We consider here two quite different but interesting penalty functionals which we will refer to as TEPR (for "tensor product"), and THPL (for "thin plate"). We will briefly sketch some early results of some work in progress, by describing the simplest examples.

The main ideas are most easily explained by first considering only spaces of periodic functions on the unit d-dimensional hypercube, that satisfy certain linear equality or boundary conditions, and then removing these conditions. Let $\phi_v(x_j)$ =cos$2\pi v x_j$ or sin$2\pi v x_j$ (with some abuse of notation), and let $\theta_0 = 1, \theta_v = 2\pi v, v > 0$, and let H^{per}_{TEPR} and H^{per}_{THPL} be, respectively, the collections of all functions f of the form

$$f(x_1, \ldots, x_d) = \sum_{v_1,\ldots,v_d=0}^{\infty} c_{v_1,\ldots,v_d}\phi_{v_1}(x_1)\cdots\phi_{v_d}(x_d), \tag{8.1}$$

with

$$\sum_{v_1,\ldots,v_d=0}^{\infty} [\theta_{v_1}\cdots\theta_{v_d}]^{2m}c^2_{v_1,\ldots,v_d} < \infty, \qquad H^{per}_{TEPR}, \tag{8.2}$$

or

$$\sum_{v_1,\ldots,v_d=0}^{\infty} [\theta^2_{v_1} + \cdots + \theta^2_{v_d}]^m c^2_{v_1,\ldots,v_d} < \infty, \qquad H^{per}_{THPL}. \tag{8.2}$$

It can be shown that H^{per}_{TEPR} will be a reproducing kernel hilbert space with (8.2) as squared norm for any $m > 1/2$, and H^{per}_{THPL} will be a reproducing kernel space with the squared norm (8.3) for any $m > d/2$. These spaces are not equivalent, and reflect different ideas of what is "smooth". However, each can be written as the direct sum of 2^d orthogonal subspaces, namely, H_0, the $\binom{d}{1}$ "main effects" subspaces of the form

$$H_\alpha = span\{\phi_{v_\alpha}(x_\alpha), v_\alpha = 1, 2, \ldots\}, \qquad \alpha = 1, \ldots, d,$$

the $\binom{d}{2}$ first order interaction spaces of the form

$$H_{\alpha\beta} = span\{\phi_{v_\alpha}(x_\alpha) \cdot \phi_{v_\beta}(x_\beta), v_\alpha, v_\beta > 0\}, \qquad 1 \le \alpha < \beta \le d,$$

and so on.

Letting

$$J_0(f) = \left[\int_0^1 \cdots \int_0^1 f(x_1, \ldots, x_d) \prod_\alpha dx_\alpha\right]^2, \tag{8.4}$$

the squared norm (8.2) on H_{TEPR}^{per} can be shown to be equal (in H_{TEPR}^{per}) to

$$J_0(f) + J^{THPL}(f), \tag{8.5}$$

where

$$J^{THPL}(f) = \sum_{\alpha_1 + \cdots + \alpha_d} \frac{m!}{\alpha_1! \cdots \alpha_d!} \times \int_0^1 \cdots \int_0^1 \left(\frac{\partial^m f}{\partial x_1^{\alpha_1} \cdots \partial x_d^{\alpha_d}}\right)^2 dx_1 \cdots dx_d \tag{8.6}$$

is the thin plate penalty functional.

For lack of space we will not discuss the thin plate spaces further, but analyses similar to but slightly more complicated than those below can be carried out. It what follows, we will only consider the tensor product case and sub or superscripts TEPR are to be understood.

Let

$$J_\alpha(f) = \int_0^1 dx_\alpha \left[\int_0^1 \cdots \int_0^1 \frac{\partial^m f}{\partial x_\alpha^m} \prod_{\beta \ne \alpha} dx_\beta\right]^2, \tag{8.7a}$$

$$J_{\alpha\beta}(f) = \int_0^1 \int_0^1 dx_\alpha dx_\beta \left[\int_0^1 \cdots \int_0^1 \frac{\partial^{2m} f}{\partial x_\alpha^m \partial x_\beta^m} \prod_{\gamma \ne \alpha, \beta} dx_\gamma\right]^2, \tag{8.7b}$$

$$J_{1,\ldots,d}(f) = \int_0^1 \cdots \int_0^1 \left[\frac{\partial^{2md} f}{\partial x_1^m \cdots \partial x_d^m}\right]^2 dx_1 \cdots dx_d. \tag{8.7c}$$

Then the squared norm (8.5) on H_{TEPR}^{per} can be shown to be equal to

$$J_0(f) + \sum_{\alpha=1}^d J_\alpha(f) + \sum_{\alpha<\beta} J_{\alpha\beta}(f) + \cdots + J_{1\ldots d}(f). \tag{8.8}$$

As an example, we will consider below $f \varepsilon H_{TEPR}^{per}$ which consists only of a mean, all d main effects and the two factor interaction between x_1 and x_2. Thus f is of the form

$$f(x_1, \ldots, x_d) = f_0 + \sum_{\alpha=1}^d f_\alpha(x_\alpha) + f_{12}(x_1, x_2), \tag{8.9}$$

where f_0 is a constant, $f_\alpha \varepsilon H_\alpha$, and $f_{12} \varepsilon H_{12}$. We can now define the periodic interaction smoothing spline as that function f_λ of the form (8.9) which minimizes

$$\frac{1}{n} \sum_{i=1}^n (y_i - f(x(i))^2 + \lambda \left[\sum_{\alpha=1}^d J_\alpha(f_\alpha) + J_{12}(f_{12})\right], \tag{8.10}$$

where $x(i) = (x_1(i), \ldots, x_d(i))$.

Using Lemma 5.1 in KW it can be shown that there is a unique minimizer of (8.10) in $H_0 \oplus \sum_\alpha H_\alpha \oplus H_{12}$. An explicit representation for it may be found using this lemma and the fact that the reproducing kernel $K(x,z)$ for $\sum_\alpha H_\alpha \oplus H_{12}$ is given by

$$K(x,z) = \sum_\alpha B_m(x_\alpha, z_\alpha) + B_m(x_1, z_1)B_m(x_2, z_2). \tag{8.11a}$$

where

$$B_m(s,t) = \sum_{v=1}^{\infty} \theta_v^{-2m} \left[\cos 2\pi v s \cos 2\pi v t + \sin 2\pi v s \sin 2\pi v t\right]. \tag{8.11b}$$

A closed form expression for B_m may be found in Craven and Wahba (1979). GCVPACK may be used to compute f_λ. In principle, $\sum_\alpha J_\alpha(f_\lambda)$ can be replaced by $\sum_\alpha w_\alpha J_\alpha(f_\lambda)$, where the w_α are positive weights, but problems concerning their estimation from the data have not been studied to date.

We will now sketch how to remove the rather restrictive periodicity conditions from H^{per}_{TEPR}. For g a function of one variable, let

$$L_0 g = \int_0^1 g(u)du, \tag{8.12a}$$

$$L_v g = \int_0^1 g^{(v)}(u)du = g^{(v-1)}(1) - g^{(v-1)}(0), \tag{8.12b}$$

and let $L_{v(x_\alpha)}f$ mean L_v applied to f as a function of x_α. Then $L_{v(x_\alpha)}f = 0$ for $v = 0, 1, 2, \ldots, m$, $\alpha = 1, 2, \ldots, d$, and any f in H^{per}_{TEPR}. Now, it can be shown that H_α is that subspace of the Sobolev space

$$W_2^m[0,1] = \{g : g, g', \ldots, g^{(m-1)} abs.cont., g^{(m)} \varepsilon L_2\}$$

of co-dimension $m+1$ which satisfies the $m+1$ conditions $l_v g = 0, v = 0, 1, \ldots, m$. Let $k_v = \frac{b_v}{v!}, v = 0, 1, \ldots, m$, where the b_v are the Bernoulli polynomials, we have $L_v k_\mu = 0, \mu \neq v, L_v k_v = 1, \mu, v = 0, 1, \ldots, m$, and thus k_v is not in H_α. Let $W^0 = span\{k_0, \ldots, k_{m-1}\}$ and let W^1 be isomorphic to $H_\alpha \oplus \{k_m\}$. Then it can be shown that W_2^m endowed with the inner product

$$< g, h > w_2^m = \sum_{v=0}^{v=m-1} L_v g L_v h + \int_0^1 g^{(m)}(u) h^{(m)}(u) du \tag{8.13}$$

satisfies

$$W_2^m = W^0 \oplus W^1. \tag{8.14}$$

Letting $g \varepsilon W_2^m$ with $g = g_0 + g_1, g_0 \varepsilon W_0, g_1 \varepsilon W_1$, we can call g_0 the polynomial part of g, and g_1 the "smooth" part. Now let

$$H_{TEPR} = W_2^m \otimes \cdots \otimes W_2^m d\ times = (W^0 \oplus W^1) \otimes \cdots \otimes (W^0 \oplus W^1) \tag{8.15}$$

$$= \left(\sum_{\alpha=1}^{d} W_\alpha^0\right) \oplus \left(\sum_{\alpha=1}^{d} W_\alpha^1 \otimes \prod_{\substack{\beta=1 \\ \beta \neq \alpha}} W_\beta^0\right) \oplus \left(\sum_{\alpha<\beta} W_\alpha^1 \otimes W_\beta^1 \otimes \prod_{\gamma \neq \alpha,\beta} W_\alpha^0\right) \oplus \cdots \oplus \left(\prod_{\alpha=l}^{d} W_\alpha^1\right),$$

where the Greek subscripts make explicit which variables are involved. We can now identify the "polynomial" subspace

$$H_0 = \prod_{\alpha=1}^{d} W_\alpha^0,$$

the main effects subspaces

$$H_\alpha = W_\alpha^1 \otimes \prod_{\substack{\beta=1 \\ \beta\neq\alpha}}^{d} W_\beta^0, \alpha = 1, \ldots, d,$$

the two factor interaction spaces

$$H_{\alpha\beta} = W_\alpha^1 \otimes W_\beta^1 \otimes \prod_{\gamma\neq\alpha,\beta} W_\alpha^0,$$

etc.

The induced tensor product inner product in H_{TEPR} is a natural extension of the inner product of (8.7) and (8.8). Letting J_α be the induced norm on H_α, etc., we can now seek f_λ in the new, non periodic version of, for example, $H_0 \oplus \sum_\alpha \oplus H_{12}$ to minimize

$$\frac{1}{n}\sum_{i=1}^{n}(y_i - f_\lambda x(i))^2 + \lambda\left[\sum_\alpha J_\alpha(f_\alpha) + J_{12}(f_{12})\right]. \tag{8.16}$$

Existence and uniqueness for any $\lambda > 0$ can be shown via Lemma 5.1 in KW provided the design points $x(i), i = 1, \ldots, n$ are such that least squares regression in H_0 is unique. The reproducing kernels for the various subspaces then follow. The r.k.'s R_0 and R_1 for W^0 and W^1 can be shown to be

$$R_0(u, v) = \sum_{v=0}^{m-1} k_v(u)k_v(v),$$

$$R_1(u, v) = k_m(u)k_m(v) + B_m(u, v),$$

and the r.k. for H_{TEPR} with the inner product induced by (8.3) is

$$\prod_{\alpha=1}^{d}(R_0(x_\alpha, z_\alpha) + R_1(x_\alpha, z_\alpha)),$$

so that, for example, the r.k. for $\sum_\alpha H_\alpha \oplus H_{12}$ is now

$$Q(x, z) = \sum_\alpha R_1(x_\alpha, z_\alpha)\prod_{\beta\neq\alpha} R_0(x_\beta, z_\beta) + R_1(x_1, z_1)R_1(x_2, z_2)\prod_{\beta\neq 1,2} R_0(x_\beta, z_\beta).$$

Given the r.k. an explicit representation for f_λ can be given, and, again GCVPACK can be used to calculate f_λ. For $m = 1, R_0(x_\alpha, z_\alpha) = 1$, H_0 is one dimensional as before, and we only replace B_m in the discussion of periodic spaces by R_1 and the same expression hold. For $m > 1$, a typical element of H_α with, say $\alpha = 1$ is now of the form

$$f(x_1, \ldots x_d) = \sum_{v_1,\ldots,v_d=0}^{m-1} f_{v_2,\ldots,v_d}(x_1)k_{v_2}(x_2)\cdots k_{v_d}(x_d). \tag{8.18}$$

The $v_2 = \cdots = v_d = 0$ term depends only on x_1 but the other terms do depend on other variables albeit in a parametric (i.e. polynomial) way. The case $m = 2$ is probably of special interest, then x_β with $\beta \neq \alpha$ enters at most linearly in functions in H_α.

There are now many interesting questions. Some of the major ones are —the development of good methods for choosing which interactions to include (see Chen (1987) for early results on this), numerical methods for very large data sets, methods for interpreting the results, development of confidence intervals, and so on.

REFERENCES

Anderson, J. A. and Senthilselvan, A. (1982). A two-step regression model for hazard functions. *Appl. Statist.* **31**, 44–51.

Barry, D. (1983). Nonparametric Bayesian regression. Thesis, Yale University.

Barry, D. (1985). Nonparametric Bayesian regression. *Ann. Statist.* **14**, 934–954.

Bates, D. M., Lindstrom, M. J., Wahba, G. and Yandell, B. (1987). GCVPACK - Routines for generalized crosa validation. *Commun. Statist.* **16**, 263–297.

Burman, P. (1985). *Estimation of Generalized Additive Models* Rutgers University. (manuscript).

Chen, Z. (1987). A stepwise approach for the purely periodic interaction spline model. *Commun. Statist. Theory and Methods* **16**, 877–895.

Cox, D. D. and O'Sullivan, F. (1985). *Analysis of Penalized Likelihood-Type Estimators with Application to Generalized Smoothing in Sobolev Spaces.* (manuscript).

Cox, D., Koh, E., Wahba, G. and Yandell, B. (1988). Testing the (parametric) null model hypothesis in (semiparametric) partial and generalized spline models. *Ann. Statist.* **16**, 113-119.

Craven, P. and Wahba, G. (1979). Smoothing noisy data with spline functions: estimating the correct degree of smoothing by the method of generalized cross-validation. *Numer. Math.* **31**, 377–403.

Engle, R., Granger, C., Rice, J. and Weiss, A. (1986). Semiparametric estimates of the relation between weather and electricity sales. *J. Amer. Statist. Assoc.* **81**, 310–320.

Eubank, R. L. (1986). Diagnostics for smoothing splines. *J. Roy. Statist. Soc. B* **47**. (To appear).

Friedman, J. H., Grosse, E. and Stuetzle, W. (1983). Multidimensional additive spline approximation. *SIAM J. Sci. Stat. Comput.* **4**, 291–301.

Green, P.J., Jennison, C. and Seheult, A. (1983). Comments to nearest neighbour (NN) analysis of field experiments by Wilkinson *et al.*. *J. Roy. Statist. Soc. B* **45**, 193–195.

Green, P.J., Jennison, C. and Seheult, A. (1985). Analysis of field experiments by least squares smoothing. *J. Roy. Statist. Soc. B* **47**, 299–315.

Green, P. J. and Yandell, B. S. (1985). Semi-parametric generalized linear models. *Tech. Rep.* **2847**. Math. Research Center, University of Wisconsin.

Heckman, N. E. (1986). Spline smoothing in a partly linear model, *J. Roy. Statist. Soc. B* **48**, 244–248.

Kimeldorf, G. and Wahba, G. (1971). Some results on tchebycheffian spline functions, *J. Math. Anal. Applic.* **33**, 82–95.

Leonard, T. (1982). An empirical Bayesian approach to the smooth estimation of unknown functions. *Tech. Rep.* **2339**. Math. Research Center, University of Wisconsin.

Nychka, D., Wahba, G., Goldfarb, S. and Pugh, T. (1984). Cross-validated spline methods for the estimation of three dimensional tumor size distributions from observations on two dimensional cross sections. *J. Amer. Statist. Assoc.* **79**, 832–846.

O'Sullivan, F. (1983). The analysis of some penalized likelihood estimation schemes. *Tech. Rep.* **76**. University of Wisconsin-Madison, Statistics Dept.

O'Sullivan, F. and Wahba, G. (1985). A cross validated Bayesian retrieval algorithm for non-linear remote sensing experiments. *J. Comput. Physics* **59**, 441–455.

O'Sullivan, F., Yandell, B. and Raynor, W. (1986). Automatic smoothing of regression functions in generalized linear models. *J. Roy. Statist. Soc. B* **81**, 96–103.

O'Sullivan, F. (1986). A statistics perspective on ill-posed inverse problems. *Statistical Science* **1**, 503–523.

Rice, J. (1986). Convergence rates for partially splined models. *Statistics & Probability Letters* **4**, 203–208.

Shiau, J., Wahba, G. and Johnson, D. R. (1986). Partial spline models for the inclusion of tropause and frontal boundary information in otherwise smooth two and three dimensional objective analysis. *J. Atmos. Ocean Tech.* **3**, 714–725.

Shiau, J. J. (1987). *A Note on Mse Coverage Intervals in a Partial Spline Model.* (manuscript).

Shiller, R. J. (1984). Smoothness priors and nonlinear regression. *J. Roy. Statist. Soc. B* **79**, 609–615.

Silverman, B. (1982). On the estimation of a probability density function by the maximum penalized likelihood method. *Ann. Statist.* **10**, 795–810.

Stone, C. J. (1985). Additive regression and other nonparametric models. *Ann. Statist.* **13**, 689–705.

Villalobos, M. and Wahba, G. (1987). Inequality constrained multivariate smoothing splines with applications to the estimation of posterior probabilities. *J. Amer. Statist. Assoc.* **82**, 239–248.

Wahba, G. (1978). Improper priors, spline smoothing and the problem of guarding against model errors in regression. *J. Roy. Statist. Soc. B* **40**, 364–372.

Wahba, G. (1984a). Cross validated spline methods for the estimation of multivariate functions from data functionals. *Statistics: An Appraisal, Proceedings 50th Anniversary Conference Iowa State Statistical Laboratory*, 205–235. (H. A. David, and H. T. David, eds.), Iowa. University Press.

Wahba, G. (1984b). Partial spline models for the semiparametric estimation of functions of several variables. *Statistical Analysis of Time Series*, 319-329. Tokyo: Institute of Statistical Mathematics (Proceeding of the Japan U.S. Joint Seminar).

Wahba, G. (1985). Comments to Peter J. Huber, Projection Pursuit. *Ann. Statist.* **13**, 518–521.

DISCUSSION

T. SWEETING (*University of Surrey*)

Professor Wahba's paper presents an overview of the many recent interesting avenues of research relating to spline estimation of unknown regression functions. Judging from previous reported results, I have no doubt that this work will prove to be successful, in a broad sense, in many applied problems. I find the work on interaction splines particularly interesting; one could apply similar decompositions of course whenever an unknown function occurs in a model. For example, using a multivariate version of the logistic density transform of Leonard (1978), this would be a natural approach for investigating dependence between variables.

This being a conference on Bayesian Statistics, it is perhaps a pity that in her paper Professor Wahba did not spend a little time also reviewing the Bayesian underpinning of the spline theory. In her presentation, however, she did just that, for which I am most grateful; nevertheless I still await that vital flash of insight to make (Bayesian) sense of it all. There are two main points I would like to discuss on the theme "Is it Bayesian?". The first relates to the derivation of the spline estimate as the posterior mean under a suitable partly vague Gaussian prior distribution over function space, while the second concerns the estimation of the parameter λ.

Many authors have considered the problem of estimating an unknown function $g(t)$ in a semiparametric regression model from a Bayesian point of view. A standard approach (for example, in Blight and Ott (1975), O'Hagan (1978) and Professor Wahba's work) is to assume a model of the form

$$g(t) = \sum_{j=1}^{m} \theta_j \Psi_j(t) + \beta^{1/2} X(t) \tag{1}$$

where the first term is a structured component and the second term represents a departure from this component. The prior for θ is taken as multivariate normal $MN(\theta_0, \xi D_0)$ say, while prior belief about X is modelled by a Gaussian process. This all leads to a Gaussian posterior process for g, and the case of vague prior knowledge about θ treated by letting $\xi \to \infty$. Supposing for example that g is a function of one varible, and the Ψ_j are polynomials of degree j, then the spline estimate of g satisfying (2.2a) arises in this way if the prior for $X(t)$ is an m-fold integrated Wiener process. From the standpoint of simply thinking about suitable prior specifications for X, I find such a prior difficult to comprehend. In particular, a stationary process for $X(t)$ (as is often assumed in the literature) would seem a natural starting point; nonstationarity implies to me the existence of "real" prior knowledge. Should I believe in Professor Wahba's prior because it leads to the intuitively appealing spline method? A second point here is that the parameter β in (1) controls the departure from the structured component of g, but the degree of *smoothness* appears to be prescribed. This could be controlled by replacing $X(t)$ in (1) by $X(\alpha t)$, where α is a smoothness parameter.

One drawback of using a stationary process, such as the Ornstein-Uhlenbeck process, is that usually the sample paths are nondifferentiable. One possibility is to model instead some suitable derivative of $X(t)$ by such a process, as Leonard (1978) does, but then of course the stationarity of $X(t)$ itself is lost. Another possibility was suggested in my discussion of Stewart (1985); for given β, α approximate the desired stationary Gaussian process by a process with a discrete spectrum (albeit a real version of the formula given there!). This will always provide infinitely differentiable functions, and may offer computational advantages when n is large.

My second general comment relates to the estimation of $\lambda = (n\beta)^{-1}\sigma^2$, using Professor Wahba's notation, for which she uses "generalised cross-validation" (GCV). If the purpose of estimating λ is to find the best predictor of the form $g_{n,\lambda} = Eg|\lambda, y$ (formally taking $\xi = \infty$), a Bayesian solution to this decision problem would be to choose λ to minimise the posterior expectation of $n^{-1}\sum_{i=1}^{n}(z_i - g_{n,\lambda}(t_i))^2$, where $z_i \sim N(g(t_i), \sigma^2)$. This comes down to minimizing $\sum_{i=1}^{n}(g_{n,\lambda}(t_i) - \bar{g}(t_i))^2$, when $\bar{g} = Eg|y$. Such a Bayesian approach requires the specification of a prior distribution for λ, of course. On the other hand, the arguments leading to the GCV solution involve an averaging over the sample space, and it would be interesting to know how close the two approaches really are.

Assuming σ is unknown, and taking for illustration a vague prior of the form $\pi(\sigma)\alpha\sigma^{-1}$, both g and σ can be integrated out, yielding an integrated likelihood of λ, which has the form

$$p(y|\lambda)\alpha|I - A|^{1/2}\left[y^T(I - A)y\right]^{-\frac{n}{2}} \qquad (2)$$

where $A = A(\lambda)$ is the matrix defined in Wahba (1978), for example. The solution to this decision problem involves evaluating $EA(\lambda)|y$, which appears too hard. On the other hand provided the tails of the prior $p(\lambda)$ for λ are sufficiently thin, but $p(\lambda)$ is otherwise relatively diffuse, if would be reasonable to use the maximum $\hat{\lambda}$ of (2) as a first approximation. (In practice of course it would be a good idea to plot $p(y|\lambda)$ over its entire range in order to assess the impact of $p(\lambda)$.) After some computation, I arrive at the equation

$$\left[y^T(I - A)y\right]^{-1}\left[y^T(I - A)Ay\right] = n^{-1}tr(A). \qquad (3)$$

for $\hat{\lambda}$. This may be compared with the equation satisfied by the GCV estimate $\tilde{\lambda}$, which I find to be

$$\left[y^T(I - A)^2y\right]^{-1}\left[y^T(I - A)^2Ay\right] = \left[tr(I - A)\right]^{-1}tr\left[(I - A)A\right]. \qquad (4)$$

Equations (3) and (4) are of a strikingly similar form, and I wonder how close $\hat{\lambda}$ and $\tilde{\lambda}$ would be in practice.

J. E. H. SHAW (*University of Warwick*)

If there are few data, or if I wish to be more Bayesian, I may want to treat λ as a parameter with an associated prior distribution, rather than use GCV to choose λ. I therefore need to "understand' λ. For example, a simple interpretation of λ in the one-dimensional case with penalty functional $J_2(f)$ is given in Shaw (1987), and is applied to the Binomial model of Section 5. Can Professor Wahba suggest how to think about other penalty functionals (for example, J^{THPL}), to help elicit priors for λ in more complicated cases?

REPLY TO THE DISCUSSION

In response to Prof. Sweeting's remarks, I'll try to give a little more of the Bayesian underpinning of univariate splines, and also squeeze in a few remarks related to what I actually said at the meeting. I will use the notation of my text, the reader should not confuse my $\theta_0 = 1, \theta_\nu = 2\pi\nu$ with Prof. Sweetings θ's which are random variables.

Let

$$U(t) = \sum_{\nu=1}^{\infty} \alpha_\nu \cos 2\pi\nu st + \beta_\nu \sin 2\pi\nu t,$$

with α_ν, β_ν independent, zero mean Gaussian random variables with $E\alpha_\nu^2 = E\beta_\nu^2 = \theta_\nu^{-2m}$. Then

$$EU(s)U(t) = B_m(s,t),$$

where $B_m(s,t)$ is given in (8.11b). U is obviously stationary, as can be seen by substituting

$$\cos 2\pi\nu(s-t) = \cos 2\pi\nu s \cos 2\pi\nu t + \sin 2\pi\nu s \sin 2\pi\nu t$$

into (8.11b), and it can be shown to have $m-1$ quadratic mean derivatives. Now, let

$$X(t) = \alpha_0 k_m(t) + U(t) \tag{1}$$

with $\alpha_0 \sim N(0,1)$. Formally, it can be shown that

$$X^{(m)}(t) = dW(t), \tag{2}$$

where $dW(t)$ is "white noise". "White noise" in the continuous time index case does not exist in the usual sense, so by $dW(t)$ of (2) we mean, that if f and g are any two elements of $L_2[0,1]$, then $\int_0^1 f(t)dW(t)$ and $\int_0^1 g(t)dW(t)$ are well defined as Riemann-Stieltjes integrals (see Cramer and Leadbetter (1967)) and

$$E\int_0^1 f(t)dW(t)\int_0^1 g(t)dW(t) = \int_0^1 f(t)g(t)dt. \tag{3}$$

To see that $dW(t)$ of (2) has the right properties, one only has to observe, that, formally

$$X^{(m)}(t) = a_0 + \sum_{\nu=1}^{\infty} a_\nu \cos 2\pi\nu t + \sum_{\nu=1}^{\infty} b_\nu \sin 2\pi\nu t, \tag{4}$$

where the a_ν, b_ν are i.i.d. $N(0,1)$. Again (4) does not exist but it is easy to check that (3) holds using Parseval's Theorem.

Now, let

$$g(t) = \sum_{\nu=0}^{m-1} \delta_\nu k_\nu(t) + \beta^{1/2} X(t)$$

where the δ_ν's are $N(0,\xi)$ with $\xi \to \infty$. It can be shown that this is the model for which the Bayes estimate of g given data from

$$y_i = g(t_i) + \varepsilon_i$$

is a polynomial spline of degree $2m-1$. g as well as its derivatives to order $m-1$ differ from a stationary process by a random polynomial of degree at most m.

In the talk and in my monograph in preparation based on the CBMS conference of March 1987 at Ohio State University, I expand on the fact that the prior for splines is the most parsimonious member of a large equivalence class of priors — and, given that one is inside a certain equivalence class of priors one might as well use the most parsimonious member — since one cannot estimate from the data which member of the class one has. This argument is related to recent work of Stein (1987), one consequence of which is that the thin plate spline estimate is the most parsimonious member of an equivalence class of priors including kriging estimates. In the univariate case, equivalence classes are distinguished by m and λ and these can be selected by GCV. A brief mention of the equivalence class of priors similar to the equivalence class of $U(t)$ is mentioned in Wahba (1981). Just to be concrete about it, let $\sigma_\nu^2(1)$ be the variance of α_ν and β_ν and consider a new process h constructed to be the same as g except that now the variances of α_ν and β_ν are $\sigma_\nu^2(2)$, and suppose that

$$\sum_{\nu=0}^{\infty} \left| \frac{\sigma_\nu^2(1)}{\sigma_\nu^2(2)} - 1 \right| \leq \infty.$$

Then h and g are equivalent processes. Suppose, for example, that $\sigma_\nu^2(2) = ((2\pi\nu)^m + \rho(2\pi\nu)^{m-1})^2$ then h and g are equivalent, and ρ cannot be estimated perfectly, even given h for all t in $[0, 1]$.

Prof. Sweeting and Prof. Shaw both mention the use of priors for estimating λ. I have always thought of λ as a bandwidth parameter (which means its optimum value can depend on the sample size), rather than a parameter with an associated prior. The thing about spline estimates is, that while they have a Bayesian interpretation, they also make sense, and have assorted nice properties if g is some fixed, but unknown element of the associated r.k. space. The only difference in these differing points of view (they are different, since sample functions are w. pr. 1 not in the associated r.k. space if it is infinite dimensional) is in the meaning of λ. If g is in the r.k. space, then the bandwidth parameter point of view is the correct one (the optimal λ from a predictive mean square error point of view does depend on the sample size), but if, in practice, the "sample function" point of view is the correct one, no major harm has been done since the GCV estimate gives a not unreasonable estimate for the "true" value of λ. (In that case, a maximum likelihood estimate for λ could also be used. See Wahba (1985).) If, of course one has a problem where the "sample function" point of view is the correct one, and one has valid prior information concerning λ, then certainly one should use it. I look forward to seeing Prof. Shaw's work along these lines. Prof. Sweeting's equations (2) and (3) are certainly intriguing, but I don't know how close $\hat{\lambda}$ and $\tilde{\lambda}$ would be in practice.

On another tack, there seems to be a lot of recent interest in interaction splines, so, since my contribution was written over a year ago, I'm going to take the liberty of mentioning two reports which have appeared since the text was written, they are Gu *et al.* (June 1988) and Gu (June 1988).

This looks like a good place to remark that it was a terrific meeting (and Bayesians do have more fun!!).

This research was supported by the AFOSR under Grant 87-0171.

REFERENCES IN THE DISCUSSION

Blight, B. J. N. and Ott, L. (1975). A Bayesian approach to model inadequacy for polynomial regression. *Biometrika* **62**, 79–88.

Cramer, H., and Leadbetter, M. R. (1967). *Stationary and Related Stochastic Processes*, New York: Wiley.

Gu, C. (June 1988) RKPACK - a general purpose minipackage for spline modeling, *Tech. Rep.* 832, Madison, WI: University of Wisconsin-Madison, Statistics Dept.

Gu, C., Bates, D. M., Chen, Z and Wahba, G. (June 1988). The computation of GCV functions through Householder tridiagonalization with application to the fitting of interaction spline models, *Tech. Rep* 823, Madison, WI: University of Wiscosin-Madison Statistics Dep.

Leonard, T. (1978). Density estimation, stochastic processes and prior information. *J. Roy. Statist. Soc. B* **40**, 113–146.

O'Hagan, A. (1978). Curve fitting and optimal design for prediction. *J. Roy. Statist. Soc. B* **40**, 1–12.

Shaw, J. E. H. (1987). Numerical Bayesian analysis of some flexible regression models. *Statistician* **36**. 147–153.

Stein, M. L. Minimum norm quadratic estimation of spatial variograms, *J. Amer. Statist. Assoc.* **82**, 765–772. (1987).

Stewart, L. (1985). Multiparameter Bayesian inference using Monte Carlo integration - some techniques for bivariate analysis. *Bayesian Statistics* **2**. North Holland: Elsevier Science Publishers B. V.

Wahba, G. (1981). Data-based optimal smoothing of orthogonal series density estimates. *Ann. Statist.* **9**, 146-156.

Wahba, G. (1985). A comparison of GCV and GML for choosing the smoothing parameter in the generalized spline smoothing problem. *Ann. Statist.* **13**, 1378-1402.

BAYESIAN STATISTICS 3, pp. 493–508
J. M. Bernardo, M. H. DeGroot, D. V. Lindley and A. F. M. Smith, (Eds.)

Modelling Expert Opinion

MIKE WEST
*University of Warwick**

SUMMARY

I consider problems of Bayesian information processing in which data consist of forecasts from individuals, "experts" or models. The basic concepts have been developed and extensively used by Lindley in his works on reconciliation of probabilities. Here a new class of models is introduced to provide methods for the processing of forecast information in terms of full, continuous distributions or densities, and partial information in terms of collections of quantiles. The latter use of such models is appropriate in contexts where forecasts are given in terms of simple point forecasts, with or without uncertainty measures, or histograms. The models are illustrated in special, practically useful cases.

Keywords: EXPERT OPINION; DIRICHLET DISTRIBUTION; QUANTILES.

1. INTRODUCTION

In a series of papers, most recently Lindley (1988) and references therein, Lindley has identified and developed the basic ingredients of the Bayesian approach to information processing when the information obtained consists of the statements of individuals. As a simple example, suppose I am considering the purchase of pesetas for a trip to Spain and my decision to buy or not today depends primarily on what the exchange rate is likely to be at some future time, say the day before I leave. Denote this uncertain quantity by Y. I have a view about Y and also consult a colleague who provides me with his forecast distribution for Y. This provides me with additional information that I should treat as data, processing it in more or less standard ways, to obtain my revised beliefs about the exchange rate. In order to do this I require a probability model for the stated distribution of my colleague, the "expert" providing his opinion in this example, conditional on each possible future value of Y. This probability model provides the likelihood for Y used in updating my prior opinion, via Bayes' Theorem, to process the expert information. The development of appropriate models is the central, technical problem in this area, and the focus of this paper. Note that the same principles apply to a variety of problems involving the assessment and use of information from forecasting models and bureaux, and other sources.

Lindley's models provide for cases in which the random quantity of interest, Y, is discrete. In this paper, the focus is on the wider class of problems in which expert, or other, opinion may be obtained about continuous random quantities in terms of:

(a) fully specified distribution or density functions;

(b) point estimates, such as medians, alone;

(c) collections of percentiles, such as median and quantiles, or deciles;

(d) histograms as discrete approximations to continuous distributions.

---

* Now at Duke University, North Carolina, USA.

Relative to full information on the expert distribution, cases (b), (c) and (d) represen partial knowledge. It is clearly vital in practice that such cases be considered. It is commo practice in some areas of forecasting, for example, for simple point forecasts, with or withou uncertainty measures, to be quoted with no reference to a global forecast distribution. I addition, it is often (or rather, always) difficult to elicit a full distribution with which an exper is totally comfortable, whereas a small collection of quantiles may be perfectly acceptable a a partial description.

In Section 2, the case of an event indicator Y is considered, and concepts underlying th basic approach developed by Lindley outlined. Even in that case, there is a need for partia information processing, such as with upper and lower bounds on expert probabilities. Sectio 3 develops the fundamental model for predictive distributions. A key ingredient is the focu on the quantile function of the expert as data, rather than on the distribution function directly This model is shown to provide simple, interpretable likelihoods for the quantity Y based o expert information provided in any of the forms above. Cases (b), (c) and (d) are considere together in Section 4, that of full information (a) in Section 5. Some final discussion an example are given in Section 6.

2. BASIC CONCEPTS IN THE EVENT CASE

To fix ideas suppose Y is binary and that my prior probability that $Y = 1$ is p. My consulte expert provides the probability f which comprises my only additional piece of information $H = \{f\}$. My problem is to construct the model defining the density or mass functio $p(H|Y)$, for each possible value $Y = 0$ or 1. Lindley's models (Lindley, 1988) suppos that $p(H|Y) = p(f|Y)$ is the density of the random quantity f following a logistic norma distribution. Another possibility is Beta, such as

$$(f|Y) \sim B\left[\delta\alpha_Y, \delta(1-\alpha_Y)\right], \qquad (Y = 0, 1),$$

where $\delta > 0$ is a precision parameter and for each Y, $\alpha_Y = E[f|Y]$ is my expectation of th expert's forecast. Clearly I view the expert to positively accord with reality if $\alpha_1 > \alpha_0$, an that expertise increases with $\alpha_1 - \alpha_0$. This model provides densities

$$p(f|Y) = \frac{\Gamma(\delta)}{\Gamma(\delta\alpha_Y)\Gamma(\delta[1-\alpha_Y])} f^{\delta\alpha_Y - 1}(1-f)^{\delta(1-\alpha_Y)-1},$$

for $0 < f < 1$, that form the likelihood function for updating to my posterior probabilit $p^* = P(Y = 1|H) = P(Y = 1|f)$. On the log-odds scale, routine calculations lead to

$$\log\left(\frac{p^*}{1-p^*}\right) = \log\left(\frac{p}{1-p}\right) + k + \delta(\alpha_1 - \alpha_0)\log\left(\frac{f}{1-f}\right),$$

where k involves δ, α_1 and α_0 via gamma functions.

This result is analogous to those in Lindley's normal models; my posterior log-odds ar obtained by adding a linear function of the expert's log-odds to my prior log-odds. If the exper is vague in the sense that $f = 0.5$, there will still typically be a correction due to the constan k. Only in very special cases is k zero, namely those in which $\alpha_1 + \alpha_0 = 1$. Specializing eve further, the multiplier $\delta(\alpha_1 - \alpha_0)$ being unity leads to $p^* \propto pf$ and $1 - p^* \propto (1-p)(1-f)$ so that the expert's forecast is itself the likelihood. An example, discussed below, is the case $\alpha_1 = 2/3$, $\alpha_0 = 1/3$ and $\delta = 3$. In such cases, if I am vague initially with $p = 0.5$, the $p^* = f$ and I adopt the expert's opinion.

Such models allow the processing of expert opinion in terms of bounds on f. This can b viewed as partial information, or censoring of the data f, and is particularly appropriate if th expert feels unhappy with further refinement of his statement beyond $H = \{f_l \leq f \leq f_u\}$, for lower and upper bounds f_l and f_u, respectively. The connections with the "robust Bayesian" viewpoint (Berger, 1984) are evident. Based on the Beta, or any other, model for f, the relevant likelihood components are now given by

$$p(H|Y) = \int_{f_l}^{f_u} p(f|Y)df, \qquad (Y = 0, 1),$$

providing a general solution to the censoring problem. In the very special case above, with $\alpha_1 = 2/3$, $\alpha_0 = 1/3$ and $\delta = 3$, this leads to posterior $p^* \propto p\bar{f}$ and $1 - p^* \propto (1-p)(1-\bar{f})$, where $\bar{f} = (f_l + f_u)/2$, although such simplistic cases are likely to be rare in practice. It should be stressed that this form of $p(H|Y)$ is only appropriate when it is not known in advance that f is to be censored in this way, and when the censoring mechanisms (ie. the reasons why the expert provides only bounds) provide no information about Y. Otherwise, alternative models would consider the bivariate distribution of $(f_l, f_u|Y)$, for each Y, if it were known in advance that the expert were to provide only bounds, or if the expert took an upper and lower probabilistic view of Y, such as that proposed by Walley (1988).

3. MODELLING EXPERT DISTRIBUTIONS

The ideas above extend to cases in which Y takes values in a discrete set. Lindley (1985) proposes what are essentially multivariate logistic normal models for discrete probabilities within a general framework. The case of continuous Y has also been considered by Lindley (1983, 1988) when it is assumed that the expert distribution lies in a parametrised family. In the former reference, for example, it is assumed that the expert states $Y \sim N[\mu, \sigma^2]$ and then the model defines a joint distribution for the parameters μ and σ^2, conditional on possible outcomes Y. The assumption of a particular, parametric form is, however, rather restrictive; it would be nice to have models catering for relatively general forms and partial information about the distribution. A first step towards this goal is explored here.

Suppose Y to be real-valued and that the expert is to provide information about his distribution function for Y, namely $F(\cdot)$. For clarity of notation in what follows, use X as the argument of F. Assume that $F(X)$ is monotonically increasing over the real line and differentiable with density $f(X)$. Rather than considering $F(X)$ directly, the focus is switched to the inverse of $F(X)$, namely the *quantile* function

$$Q(U) = F^{-1}(U), \qquad (0 \leq U \leq 1).$$

From the assumptions about $F(X)$ it follows that $Q(U)$ is monotonically increasing on $[0, 1]$, tending to $\pm\infty$ at the end-points, and also differentiable. Knowledge of the distribution function is equivalent to that of the quantile function and, in view of the earlier comments about partial information in terms of quantiles, it is natural to consider the latter directly. To develop a model for the distribution, individual points on the quantile function are considered, starting with the median of $F(X)$, $\mu = Q(0.5)$.

The first step is to specify the distribution for μ conditional on each possible value of Y, via the density $p(\mu|Y)$. Again the basic ideas in Lindley (1988, Section 17) are helpful; there Lindley considers the mean (assumed to exist) rather than the median of $F(X)$ and uses a normal model for the mean for each Y. The focus on the median as a point forecast links more directly to the quantile function and, in this context, μ always exists and is unique. Choice of the distribution for $(\mu|Y)$ will depend on prior information about the relationship between median point forecasts from the expert/model, and actual outcomes. Clearly in some contexts there will exist substantial relevant information on which to base this model and the parameters defining it, although these considerations are not explored further here.

For each Y, let

$$M_Y(X), \qquad (-\infty < X < \infty),$$

be a monotone, continuous distribution function over the real line. $M_Y(X)$ is chosen to model the anticipated form of the distribution of the median μ conditional on true value Y. This plays a role analogous to that of the mean α_Y in the event case model of Section 2, catering for biases and lack of calibration of the expert median forecast.

Example 3.1. A particular example is the normal model in which $M_Y(X)$ is the normal distribution $(X|Y) \sim N[c+Y, W]$. Often it may be suitable to assume that median point forecasts are related to outcomes via such a normal model, possibly after transformation; ie that forecast errors $\mu - Y$ (with Y observed therefore fixed) are normally distributed with mean (bias) c units and variance W. The empirical study of Pratt and Schlaifer (1985) provides a case study in which this form of relationship is suggested. A possibly useful refinement is to consider also a dependence of the variance W on Y in some cases. Note that the expert is viewed as *median unbiased* if $c = 0$ since then the anticipated location of $p(\mu|Y)$ is the true value Y.

If the median μ is assumed to follow the distribution $M_Y(\mu)$ conditional on Y, it follows that

$$\mu = M_Y^{-1}(\pi),$$

where π is a uniform random quantity on the unit interval, $\pi \sim U[0,1]$. In Example 3.1 $\mu = c + Y + W^{1/2}\Phi^{-1}(\pi)$ where Φ is the standard normal c.d.f. This model is completely analogous to that for the forecast mean in Lindley (1988, Section 17), and as such it suffers from the sensitivity to the assumption of the particular parametric form of M_Y. To alleviate this, the distribution of μ can be modelled as coming from a neighbourhood of the chosen M_Y by allowing the distribution of π above to be something other than uniform. Thus the following assumption is made.

Assumption 1. *Conditional on Y and the specified distribution function $M_Y(\cdot)$,*

$$\mu = M_Y^{-1}(\pi),$$

where π is a Beta random quantity,

$$\pi \sim B[\delta/2, \delta/2],$$

for some precision parameter $\delta > 0$, not depending on Y.

Under this assumption, the Beta random quantity can be written as $\pi = M_Y(\mu) = M_Y[Q(0.5)]$. Thus π is the value at $U = 0.5$ of the *compound* distribution function $M_Y[Q(U)]$ over the unit interval. Note that this is a random distribution function (for me) since $Q(U)$ is a random quantity for all U. Note that $E[\pi] = 0.5$ for all δ so that the location of the distribution of μ is essentially that determined by M_Y. If $\delta = 2$ then the distribution of μ is just the original M_Y. Otherwise $p(\mu|Y)$ is more or less diffuse than the density of M_Y according to whether δ is less than or greater than 2. In the normal case of Example 3.1, μ has a symmetric distribution with median $c+Y$ for all δ.

Consider now extending the model to a general point on the forecast quantile function,

$$q = Q(U),$$

for given U in $[0,1]$. Take $U = 0.75$ as an example so that q is the upper quantile. Clearly $q > \mu$ and therefore can be expressed as

$$q = M_Y^{-1}(\theta),$$

for some θ such that $\pi < \theta \leq 1$. Following the reasoning for the median, the obvious extension is to take θ as distribution over the unit interval with location near 0.75. Referring back to any quantile $q = Q(U)$, the immediate extension of the model for the median is given by assuming $q = Q(U) = M_Y^{-1}(\theta)$ where

$$\theta \sim B[\delta U, \delta(1-U)],$$

for each U. Considering U to vary over the unit interval lead to a Dirichlet model as the natural extension. The following definition formalises this discussion. From here on, the function $M_Y(\cdot)$ is referred to as the *target* distribution of the model for the quantiles, conditional on U.

Definition 1.

(a) *For integer $n > 1$, let $\underline{U}_n = (U_1, \ldots, U_{n-1})$ be any fixed values defining the partition of the unit interval*

$$0 = U_0 < U_1 < \cdots < U_{n-1} < U_n = 1.$$

(b) *Define corresponding quantiles of the expert distribution by $q_t = Q(U_t)$, for $t = 0, \ldots, n$, so that*

$$-\infty = q_0 < q_1 < \cdots < q_{n-1} < q_n = \infty.$$

Let $\underline{q}_n = (q_1, \ldots, q_{n-1})$.

(c) *For any fixed Y, define the probabilities $\underline{\pi}_n = (\pi_1, \ldots, \pi_{n-1})$ via*

$$\pi_t = M_Y(q_t) - M_Y(q_{t-1}), \qquad t = 1, \ldots, n-1,$$

and let $\pi_n = 1 - (\pi_1 + \cdots + \pi_{n-1})$. Note that $\underline{\pi}_n$ depends on Y although this is not made explicit in the notation. These probabilities are random, giving the masses allocated to the intervals (U_{t-1}, U_t) by the random distribution $M_Y[Q(U)]$ over the unit interval.

Assumption 2. *Let $\underline{a}_n$ be the n-vector $(1/n, \ldots, 1/n)$. Then $\underline{\pi}_n$ follows a Dirichlet distribution with mean $\underline{a}_n$ and precision parameter δ, having density*

$$p(\underline{\pi}_n|Y) = p(\underline{\pi}_n) = \Gamma(\delta) \prod_{t=1}^{n} \frac{\pi_t^{\delta/n-1}}{\Gamma(\delta/n)},$$

over the $(n-1)$-dimensional simplex.

Note again that $\underline{\pi}_n$ depends on Y; this Dirchlet model is defined conditional on Y. However, since neither δ nor $\underline{a}_n$ depends on Y, the distribution is independent of Y. Transforming the quantiles to $\underline{\pi}_n$ involving the target distribution $M_Y(\cdot)$ is essentially a *pivotal* device to obtain a distribution that does not involve Y. Since the assumption holds for all n and any partition $\underline{U}_n$ of the unit interval, then $M_Y[Q(U)]$ is a Dirichlet process. Some comments on this are in order. Firstly, the fact that a Dirichlet process is discrete with probability one implies that the model involves a discrete approximation, of essentially indeterminable accuracy, to a continuous problem. In the model analysis below, this feature is of little consequence due essentially to the use of a discretisation of the quantile function Q from the outset. The likelihood for Y given Q is constructed as the limiting from of a sequence of likehoods from discrete approximations to Q i.e. histograms. A second, related feature is the implied negative correlation between probabilities $\underline{\pi}_n$ that precludes the incorporation of smoothness assumptions. The implied distribution for Q has, however, qualitatively the right form of dependence structure between quantiles due to the ordering. Thus any two quantiles $q_t = Q(U_t)$ and $q_s = Q(U_s)$ are positively correlated, the correlation decreases as $|U_t - U_s|$ increases and tends to unity as $|U_t - U_s|$ tends to zero.

4. EXPERT OPINION: COLLECTIONS OF QUANTILES

Forecast statements are often in terms of summaries of distributions, such as point forecasts with simple uncertainty measures. As in Section 2, this can be viewed as a form of censoring, providing only a partial specification of $F(X)$. The model developed above provides a relatively easily calculated likelihood in cases when this partial information consists of selected percentage points, or quantiles, of $F(X)$. Suppose the expert provides quantiles $\underline{q}_n$ as in Definition 2. Let $m_Y(X)$ be the density of the target distribution $M_Y(X)$, for each Y. The following result now holds.

Theorem 1. *Under Assumption 2, the density function for the random quantiles $\underline{q}_n$, conditional on Y, is given by*

$$p(\underline{q}_n|Y) = c\,[1 - M_Y(q_{n-1})]^{\delta/n-1} \prod_{t=1}^{n-1} [M_Y(q_t) - M_Y(q_{t-1})]^{\delta/n-1}\, m_Y(q_t)$$

for $-\infty < q_1 < \cdots < q_{n-1} < \infty$, where c is the constant $c = \Gamma(\delta)/\prod_{t=1}^{n}\Gamma(\delta/n)$.

Proof. Directly by transformation from $\underline{\pi}_n$ to $(\underline{q}_n|Y)$ using the defining relationships in Assumption 1. Note that the Jacobian is simply given by

$$\left|\frac{d\underline{\pi}_n}{d\underline{q}'_n}\right| = \prod_{t=1}^{n-1} m_Y(q_t).$$

Theorem 1 provides the joint density for any collection of expert quantiles. Some insight into the form of this density and the implied relationships amongs quantiles can be obtained in the context of Example 3.1.

Example 4.1. Take target distributions $(X|Y) \sim N[Y, 1]$, so that the expert median μ has a distribution symmetric about Y, unimodal if $\delta > 2$. Thus the expert is viewed as median unbiased. Consider the particular case of $Y = 0$, so that the target is standard normal. Suppose in addition that $\delta = 5$. Consider two expert quantiles, $\mu = q_1 = Q(0.5)$, the median, and $q = q_2 = Q(0.75)$, the upper quantile. The marginal distributions of each and their bivariate distribution follow from Theorem 1. The marginal for μ is symmetric and unimodal at zero, the true value. That of q has mode approximately 0.73, close to the value 0.67, the upper quantile of the standard normal target. To explore the joint structure, Figure 1 displays the conditional density of $(q|\mu)$ for $\mu = -2, 0$ and 2. Clearly this density is zero for $q < \mu$. As μ decreases to negative values, the conditional distribution of q flattens out with mode tending quickly to zero. As μ takes larger, positive values, the conditional distribution for q becomes highly skewed, concentrating near the value of μ conditioned on as μ moves away from the target value of zero. Also displayed is the marginal density of q.

As a likelihood for Y given $\underline{q}_n$ the form in Theorem 1 has two components: one from the Dirichlet involving the product of the probabities $\underline{\pi}_n$; the other given by the product of densities $m_Y(q_t)$. The latter is just what would be obtained if the quantiles were treated as a random sample from the target distribution given Y. The former provides correction for the positioning of the quantiles under the targer distribution determined by the $\underline{U}_n$, and the implied dependence.

A generalisation of some practical importance (to be explored later), involves taking the same Dirichlet based model but allowing for different means of the π_t probabilities. This extension is described as follows.

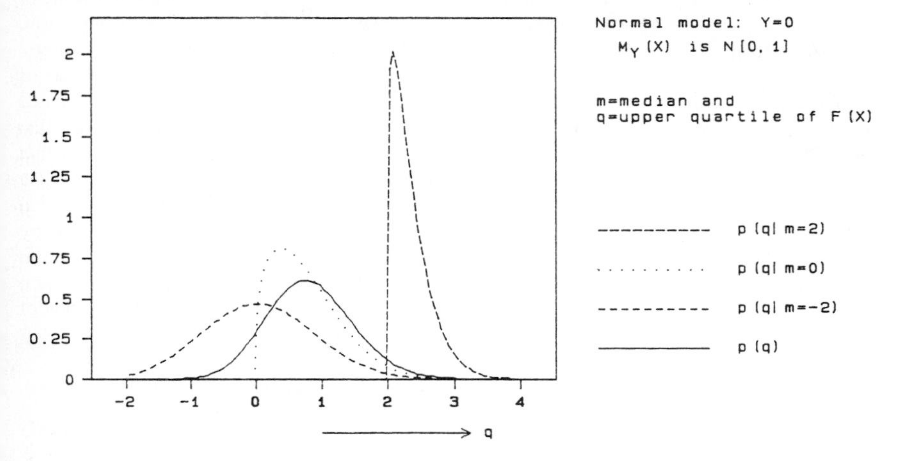

Figure 1. *Marginal density $p(q)$ and conditional densities $p(q|m)$.*

Assumption 2a.

(a) *Let $A(U)$ be a known, continous and monotone distribution function over the unit interval, having a density function $\alpha(U)$.*

(b) *In the framework of Definition 1, redefine the n-vector $\underline{a}_n$ as $\underline{a}_n = (a_1, \ldots, a_n)$ where*

$$a_t = A(U_t) - A(U_{t-1}), \qquad (t = 1, \ldots, n).$$

Then, under Assumption 2 with this modification, it follows that $\underline{\pi}_n$ has a Dirichlet distribution with mean $\underline{a}_n$ and precision parameter δ, having density

$$p(\underline{\pi}_n|Y) = p(\underline{\pi}_n) = \Gamma(\delta) \prod_{t=1}^{n} \frac{\pi_t^{\delta a_t - 1}}{\Gamma(\delta a_t)},$$

over the $(n-1)$-dimensional simplex.

Theorem 1a. *Under Assumption 2a, the density function for the random quantiles $\underline{q}_n$, conditional on Y, is given by*

$$p(\underline{q}_n|Y) = c\,[1 - M_Y(q_{n-1}]^{\delta a_n - 1} \prod_{t=1}^{n-1} [M_Y(q_t) - M_Y(q_{t-1})]^{\delta a_t - 1} m_Y(q_t)$$

for $-\infty < q_1 < \cdots < q_{n-1} < \infty$, where c is the constant $c = \Gamma(\delta)/\prod_{t=1}^{n} \Gamma(\delta a_t)$.

Proof. Directly by transformation as in Theorem 1.

Why is this generalisation of interest? Note first that the original model in Assumption 2 and Theorem 1 is the special case when $a_t = 1/n$ for all t, corresponding to the uniform distribution $A(U) = U$. The choice of target M_Y is assumed to cater for all dependence on Y in the distribution of the median in Assumption 1, and the other quantiles in Assumption 2a.

Stochastic variation away from the target is modelled and controlled by the precision parameter δ, whilst $A(U)$ may be used to cater for minor systematic departures from anticipated form. A uniform mean $A(U) = U$ will often be appropriate, implying satisfaction with the target as capturing the relevant features anticipated. An example serves to illustrate the use of alternative forms. Suppose, as in Example 4.1, that the target distribution is unit variance normal with mean Y, but that it is recognised that there is some small chance that the median may be more adequately modelled using a heavier tailed distribution, such as Cauchy. If that happens then the compound distribution $M_Y[Q(U)]$ will be lighter tailed than uniform. Use of a mean function $A(U)$ that is essentially uniform across the central part of the unit interval but that has lighter tails will lead to a discounting of the contribution made to the likelihood by quantiles in the tails of F. This feature stems directly from the focus on the quantile function of the expert rather than the distribution directly; the positioning of the tails of F is unknown whilst those of Q lie near 0 and 1. Further discussion of this appears in the illustrations of Section 6.

5. EXPERT OPINION: FULL DISTRIBUTION

Consider the generation of the expert's quantiles $\underline{q}_n$ in Definition 1. Letting n tend to infinity with the grid points U_t remaining distinct leads to $Q(U)$ being evaluated almost everywhere. Assuming continuity implies that $Q(U)$ is fully observed hence so is the inverse $F(X)$. Thus the likelihood for Y based on the full expert distribution is obtainable as the limiting form, if it exists, of the likelihood from a discrete approximation as in Theorem 1. An easy way to do this is simply to take $U_t = t/n$ and this is done here.

Let H_n denote the information set $H_n = \{\underline{q}_n\}$ where the quantiles are as in Definition 1 but now with $U_t = t/n$ for each t. Denote full information by H, so that

$$H = \lim_{n\to\infty} \underline{H}_n = \{Q(U); 0 < U < 1\}.$$

Recall that $F(X)$ has density $f(X)$, $M_Y(X)$ has density $m_Y(X)$ and $A(U)$ of Assumption 2a has density $\alpha(U)$. The following result now holds.

Theorem 2. *Suppose the model is as defined in Assumption 2a. Then, as $n \to \infty$, $H_n \to H$ and $p(\underline{q}_n|Y) = p(H_n|Y) \to p(H|Y)$ where the limiting likelihood has the form*

$$p(H|Y) \propto \exp\{-\delta D(Y)\}$$

as a function of Y, with

$$D(Y) = \int_{-\infty}^{\infty} \alpha\,[F(X)]\, f(X) \log\left[\frac{f(X)}{m_Y(X)}\right] dX,$$

whenever the integral exists for all Y.

Proof. The density of $M_Y[Q(U)]$ over $0 \leq U \leq 1$ is just the derivative which is easily seen to be given by $m_Y[Q(U)]/f[Q(U)]$. Hence

$$M_Y(q_t) - M_Y(q_{t-1}) = \frac{m_Y(q_t^*)}{f(q_t^*)}(U_t - U_{t-1})$$

where $q_t^* = Q(U_t^*)$ for some U_t^* between U_{t-1} and U_t. As n tends to infinity, the contribution to the likelihood from the last interval $(U_{n-1}, 1]$ is negligible compared to the rest of the likelihood, and so, for large n,

$$p(\underline{q}_n|Y) \simeq c \prod_{t=1}^{n-1} \left[\frac{m_Y(q_t^*)}{f(q_t^*)}\right]^{\delta a_t - 1} m_Y(q_t)$$

as a function of Y. Thus, for some constant k,

$$\log[p(\underline{q}_n|Y)] \simeq \sum_{t=1}^{n-1} \left\{ \log[m_Y(q_t)] + (\delta a_t - 1) \log \left[\frac{m_Y(q_t^*)}{f(q_t^*)} \right] \right\} + k.$$

Now $a_t = \alpha(\hat{U}_t)(U_t - U_{t-1})$ where $\hat{U}_t$ lies between U_{t-1} and U_t, and so, since $U_t - U_{t-1} = 1/n$,

$$\log[p(\underline{q}_n|Y)] - k - \sum_{t=1}^{n-1} \left\{ \log[m_Y(q_t)] + \left[\frac{\delta\alpha(U_t)}{n} - 1 \right] \log \left[\frac{m_Y(q_t)}{f(q_t)} \right] \right\}$$

tends to zero as n tends to infinity. The sum in this expression may be written as

$$\frac{\delta}{n} \sum_{t=1}^{n-1} \alpha(U_t) \log \left[\frac{m_Y(q_t)}{f(q_t)} \right] + \text{terms not involving } Y,$$

the first term of which has the limiting value

$$\delta \int_0^1 \alpha(U) \log \left[\frac{m_Y[Q(U)]}{f[Q(U)]} \right] dU$$

if this integral exists. Then, transforming to $X = Q(U)$ so that $U = F(X)$, this integral is given by $-\delta D(Y)$ where

$$D(Y) = \int_{-\infty}^{\infty} \alpha[F(X)] f(X) \log \left[\frac{f(X)}{m_Y(X)} \right] dX.$$

Thus $\log[p(\underline{q}_n|Y)] \to -\delta D(Y) +$ constant as $n \to \infty$ and so, asymptotically, $\{\underline{q}_n\} \to H$ and $p(H|Y) \propto \exp\{-\delta D(Y)\}$ as stated.

Corollary. *If my prior distribution for* Y *has density* $p(Y)$, *then fully observing the expert distribution as stated leads to posterior*

$$p(Y|H) \propto p(Y) \exp\{-\delta D(Y)\}.$$

The function $D(Y)$ determining the likelihood is a generalized *divergence* measure; it measures the discrepancy between the target density $m_Y(X)$ and the stated expert density $f(X)$, for each Y. $D(Y)$ is always non-negative, being zero for all Y if and only if $f(X) = m_Y(X)$ for all X. Thus, as the value of Y varies, a large divergence leads to a small likelihood $p(H|Y)$; conversely, if $f(X)$ and $m_Y(X)$ are close in the sense of small divergence, then $p(H|Y)$ is large. Some special cases and examples appear in Section 6 below. Here note special case in which $A(U) = U$, $0 \leq U \leq 1$, so that $\alpha(U) = 1$. This implies that, given the finite data $\underline{q}_n$, $E[\pi_t] = a_t = 1/n$ as in the original Dirichlet distribution of Assumption 2. The implication for $p(H|Y)$ is that $D(Y)$ is the well-known Kullback-Leibler directed divergence

$$\int_{-\infty}^{\infty} f(X) \log \left[\frac{f(X)}{m_Y(X)} \right] dX.$$

Note that, as mentioned in the proof, $D(Y)$ is assumed to exist for all Y. Some discussion of this appears in Section 6 below.

6. DISCUSSION AND EXAMPLES

Some general comments are in order before proceeding to examples. Sections 3, 4 and 5 detail the model that allows a variety of forms of expert opinion to be processed. If the full distribution is made available, Section 5 shows how a generalized divergence measure between the stated density and the target, $m_Y(X)$ for each Y, determines the likelihood. Given only collections of percentage points as in Section 4, the likelihood clearly shows that the global form of expert distribution is irrelevant; only the values of the chosen quantiles appear there naturally weighted with the Beta form for probabilities under the target model, the values $M_Y(q_t)$, and the target density $m_Y(q_t)$.

The choice of the Dirichlet precision and the distributions $M_Y(X)$ will typically depend on previous experience with the expert, and may be estimated based on such experience although this is not considered here. The distribution $A(U)$ must also be specified; often $A(U) = U$ will be suitable. It leads, in particular, to the Kullback-Leibler based likelihood from the full distribution. This choice is consistent with a view that the random quantile q_t obtained in Section 3 are to be treated equally; that is, the expert's assessment of his quantile function/distribution function is as sound in the tails as it is in the centre. To model the commonly held view that tail behavior is generally difficult to determine and so q_1 and q_{n-1} for example, are more likely to be subject to assessment error than, say, q_7 and q_8 alternative forms for $A(U)$ can be specified. It is clear from the form of $D(Y)$ in Theorem 1 that assessments in the tails will be discounted if the density $\alpha(U)$ decays rapidly as U tends to 0 or 1. As an extreme example, a "trimmed" assessment, ignoring the expert distribution below 5% and above 95% probabilities whilst treating the rest of the range consistently, can be modelled with

$$\alpha(U) \propto \begin{cases} 1, & 0.05 < u < 0.95; \\ 0, & \text{otherwise.} \end{cases}$$

Finally note that these features, and the forms of likelihood, derive directly from the initial focus on $Q(U)$ rather than $F(Y)$ as providing the data. This parallels experiences with elicitation where it has often been found that quantiles are more easily understood and elicited from subjects than probability distributions directly.

And now for some examples. In each of the examples, $A(U) = U$ so that $D(Y)$ is the usual, Kullback-Leibler divergence measure.

Example 6.1. The target distribution $M_Y(X)$ is that of

$$(X|Y) \sim N[c+Y, W],$$

a normal distribution with mean $c+Y$ and variance W. The constant c is an expected bias in the median as a point forecast; if $c = 0$ then the expert is viewed as essentially unbiased. Three information sets are considered: the median of $F(Y)$ alone; the median plus quantiles; and the full distribution. In the first two, the global form of $F(Y)$ is irrelevant. In the third case, suppose the expert actually states a distribution, of *any* form, with mean f and variance V. It is easily shown that, using the Kullback-Leibler divergence in Section 4, the likelihood is

$$p(H|Y) \propto \exp\left\{-\frac{\delta}{2W}(f-c-Y)^2\right\}.$$

The likelihood is the same as would be obtained from an ad-hoc model in which the point forecast f is viewed directly as a random quantity to be modelled, having a normal distribution $(f|Y) \sim N[c+Y, W/\delta]$. Such methods are used in Lindley (1983); the current approach provides further foundation. Note, however, that had the expert provided a non-normal distribution with infinite variance then the results would be rather different. In fact in such a case with $F(Y)$ Cauchy, for example, the Kullback-Leibler divergence does not exist. This highights the need for discounting of the tails of $F(Y)$ using $\alpha(U) \neq 1$, decaying to zero as U tends to 0 and 1. Generally the Kullback-Leibler divergence will exist only when the tails of $M_Y(X)$ are heavier than those of $F(X)$, for all Y. Since this cannot typically be ensured before observing the expert distribution, a weighting function $\alpha(U)$ decaying in the tails is essential if the likelihood is to exist. It is always possible, for example, to ensure a finite divergence using $\alpha(U)$ constant over most of the unit interval but zero for $U < \epsilon$ and $U > 1 - \epsilon$ where ϵ is a very small, positive quantity. It is also clear, however, that with $\alpha(U) = 1$ the likelihood based on any finite collection of quantiles from the Cauchy distribution is perfectly well-defined and appropriate, so that a discrete approximation to Q may be used in such cases.

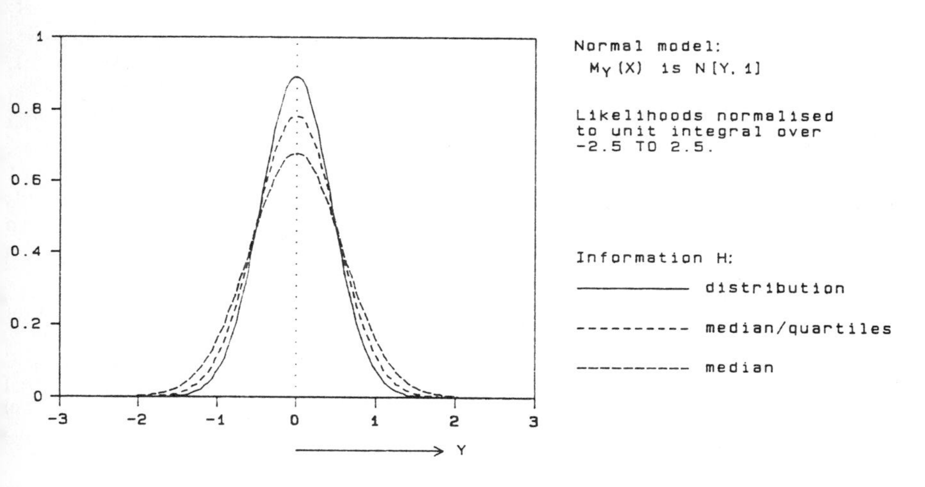

Figure 2. *Likelihoods* $p(H|Y)$.

Figure 2 shows the likelihoods for the three forms of information: median alone, $f = 0$ median $f = 0$ and quantiles ± 0.67 (coinciding with those of a unit normal distribution for illustration); and full information with forecast distribution having mean $f = 0$ and finite variance. The likelihoods have each been normalised to integrate to unity over the interval ± 2.5 for each comparison. They thus coincide with posterior densities relative to a prior $p(Y)$ being uniform over that interval. In this special case, the three forms of information can be viewed as an increasingly informative hierarchy. The model assumes $c = 0$, $W = 1$ and $\delta = 5$. The effects of increasing information are in Figure 2.

That the likelihood does not depend on the spread of the forecast distribution is a rather surprising feature of the model, suggesting a need for refinement. One refinement that is clearly necessary in practice is to allow for uncertainty about the target distribution generally, and W in particular in this example, and also about δ. Uncertainty about W represented in terms of a prior distribution would allow for learning. Although this is not pursued further here, studying past predictions from the expert/model will lead to an informed prior for W and the other components. Sequential forecasting of time series is a natural candidate application area in which such information about predictive ability is sequentially obtained.

An alternative refinement stems from the recognition that the strong (indeed, almost, sufficient) dependence of the likelihood on the location of $F(X)$ alone derives from the use of a neighbourhood of the *normal* target distribution. Use of a target more disposed to recognising spread in the quantiles, i.e. a heavier-tailed form (not strongly unimodal), provides a simple and appropriate alternative. As an example, compare the above normal model with that in which the target is replaced by the Cauchy with unit scale parameter, $M_Y(X) = 0.5 + \arctan(X - Y)$, denoted by $C[Y, 1]$. Suppose that data H consists of the quantiles $Q(0.05), Q(0.10), \ldots, Q(0.95)$ taken form the expert distribution $F(X) = \Phi(X/2)$, a normal forecast distribution with zero mean and variance 4. The likelihoods for Y from the normal and Cauchy models are displayed in Figure 3. Note that these are similiar to the limiting forms that would be obtained from Theorem 2a given the full information about F. Clearly the Cauchy based model accounts for the unanticipated spread of F much more than the normal based model; unilike the latter model, the limiting form of the likelihood in the former will be dependent on the spread. An illustration of the respose of these two models to spread amongst number of quantiles appears in Figure 4. Here the likelihoods from normal and Cauchy models are based on data H comprising the median $\mu = 0$ and the upper quantile $q = 3$ of F. As with the example illustrated in Figure 3, the data are consistent with $Y = 0$ but also with a much more diffuse expert distribution than anticipated by the targets. The more appropriate likelihood from the Cauchy model accounts for the surprise value of the data in a way that the normal model cannot.

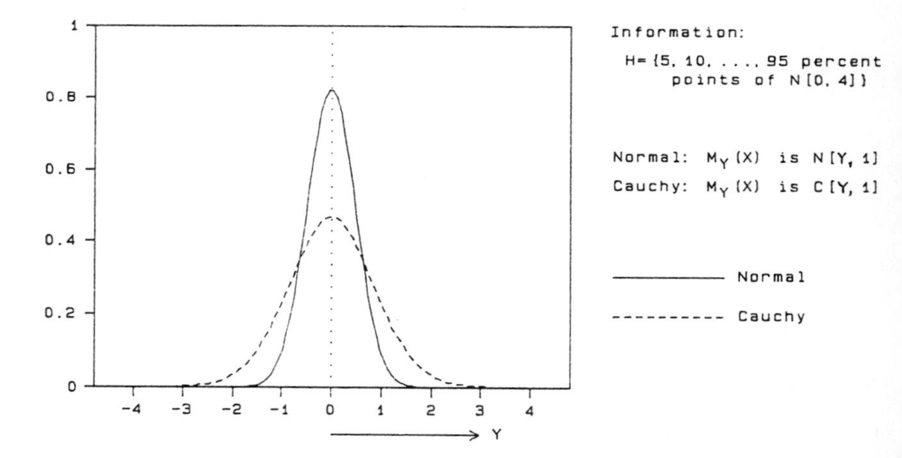

Figure 3. *Comparison of normal and Cauchy models.*

Example 6.2. Suppose $Y > 0$ is the survival or failure time of a patient or test component and that the distribution $M_Y(X)$ is gamma,

$$(X|Y) \sim G[b, bc/Y],$$

with density

$$m_Y(X) \propto Y^{-b} X^{b-1} \exp\{-bcX/Y\}, \qquad (X > 0),$$

as a function of both X and Y. Under $M_Y(X)$, $E[X|Y] = Y/c$ so that c is a multiplicative bias; $c = 1$ implies an unbiased target analogous to that in the previous example where the bias was additive. Suppose that the expert actually states a distribution, of any form such that

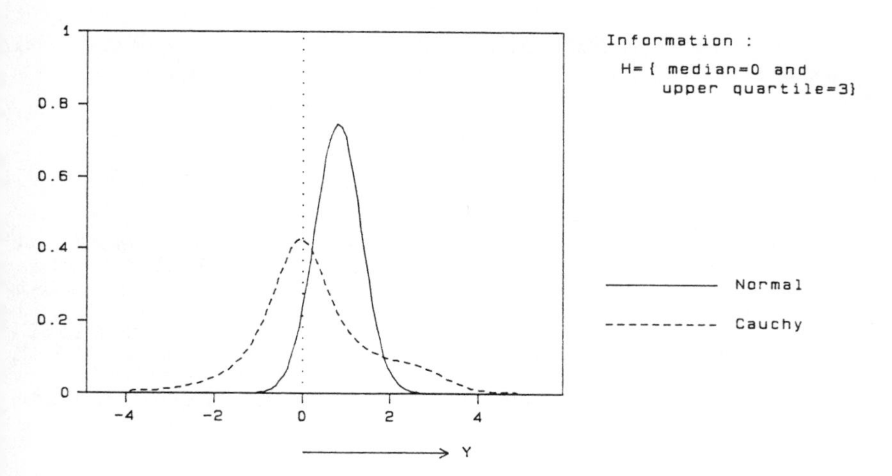

Figure 4. *Further comparison of normal and Cauchy models.*

$E[\log(X)] < \infty$, having mean f. Then the Kullback-Leibler based likelihood is easily seen to be given by

$$p(H|Y) \propto Y^{-\delta b} \exp\{-\delta bcf/Y\}.$$

This is a form that is analogous to that provided by an ad-hoc model in which the point forecast f is directly modelled as $(f|Y) \sim G[\delta b, \delta bc/Y]$; i.e. with the bias correction c and an extra scaling δ.

Example 6.3. The above examples are each special cases of the following, exponential family class models. Suppose that the target distribution has density

$$m_Y(X) = h(X, \phi) \exp\{\phi[X\mu_Y - a(\mu_Y)]\}$$

for some location parameter μ_Y (for each Y), precision $\phi > 0$, and known functions $a(\cdot)$ and $h(\cdot, \cdot)$. Note that this distribution has mean $E[X|Y] = a'(\mu_Y)$. Suppose that the expert distribution is such that $E[X] = f$ for some f, and that $E[\log\{h(X, \phi)\}] < \infty$. The Kullback-Leibler based likelihood is easily derived as

$$p(H|Y) \propto \exp\{\delta\phi[f\mu_Y - a(\mu_Y)]\}.$$

This is a form analogous to that provided by an ad-hoc model in which the point forecast $E[X] = f$ of the expert is directly modelled as coming from a distribution of the form $M_Y(\cdot)$, but with precision $\delta\phi$.

Note that in each case the expert distribution appears only through the mean. Thus, again, partial expert opinion is processed, now in terms of the mean rather than quantiles. The extensive discussion of this feature, and the types of refinement possible to alleviate it, in the normal case of Example 6.1 are clearly relevant more generally. The models require exploration to identify the effects of refinements such as: (i) the use of non-uniform functions $A(U)$; (ii) the use of alternative target distributions (as with the Cauchy replacement for normality) to those in the exponential family that lead to this strong dependence on location parameters of F and (iii) prior distributions for parameters of the target distribution to provide learning.

ACKNOWLEDGEMENTS

I am grateful to Jim Berger, Mark Schervish and Dennis Lindley for discussion of the worl reported here. An original draft was written whilst visiting the Department of Statistics a Purdue University.

REFERENCES

Berger, J. O. (1984). The robust Bayes viewpoint. *Robustness of Bayesian analysis*, (J. B. Kadane, ed.) Elsevier Science.

Lindley, D. V. (1983). Reconciliation of probability distributions. *Oper. Res.* **31**, 806–886.

Lindley, D. V. (1985). Reconciliation of discrete probability distributions. *Bayesian Statistics* **2**. (J. M Bernardo, M. H. DeGroot, D. V. Lindley and A. F. M. Smith, eds.). Amsterdam: North-Holland.

Lindley, D. V. (1988). The use of probability statements. *Accelerated Life Tests and Experts' Opinion i Reliability*, (C. A. Clarotti and D. V. Lindley, eds.). (To appear).

Pratt, J. W. and Schlaifer, R. (1985). Repetitive assessment of judgemental probability distributions: case study. *Bayesian Statistics* **2**. (J. M. Bernardo, M. H. DeGroot, D. V. Lindley and A. F. M. Smith eds.). Amsterdam: North-Holland.

Walley, P. (1988). *Statistical Reasoning with Imprecise Probabilities*. London: Chapman and Hall. (T appear).

DISCUSSION

R. J. OWEN (*University College of Wales, Aberystwyth*)

The question of how to deal formally with the opinions of persons is an important area fo Bayesians. A central feature is the apparent inability of people to say what they really know Their statements may be incomplete, biased or even "in negative accord with reality". Dr West has made here a valuable contribution to this problem area in providing a way of using all the information stated for prediction of a future variable Y. This new development both allows Y to be continuous and copes with an expert stating his distribution either completely or incompletely in quantile form. Elicitation of quantiles is of course often a more realistic proposition.

The key to the new development is the Dirichlet model for 'my' distribution of the expert's statement conditional on Y. This gives, for fixed 'statement' and generic Y, the likelihood function in Y.

Lindley (1983, 1988) has considered the case of discrete Y in detail and continuous Y (Lindley, 1988) when attention is restricted to stating location and scale measures.

It is perhaps worth stating two related problems which are not covered by this paper. Firstly, there is the given problem but without the implicit assumption that the expert is passively stating what he believes and is not contriving to influence the conclusion beyond this. In such a situation a game theoretic element would enter the analysis. Secondly, there is the problem of determining a consensus where 'I' am no longer the executive and instead the expert and 'I' must together reach a conclusion.

Some interesting features may be illustrated by the event case (binary Y) where the probability f is stated by the expert. The constant k, which depends only on the model for the likelihood function and has the same sign as $1 - \alpha_0 - \alpha_1$, corrects for possible assessed bias of the expert. If $\alpha_0 + \alpha_1 \leq 1$ and $\alpha_1 > 1$ (representing an expert viewed as being in positive accord with reality though biased towards $Y = 0$) then for large δ and any p, the resultant p^* is near 1 or 0 according as $f >$ or $< \frac{1}{2}$. Thus, in this case, the pooling of the probabilities p and f is very far from linear. To see that this is sensible, observe that when δ is large the expert is virtually expected to specify Y (though he would be unaware of this) by a value of f near α_0 or α_1. The normative approach has taken this extreme case in its stride!

Also in the binary case with large δ, a value of f near neither α_0 nor α_1 could lead to my questioning the assumed model for the likelihood function. The question of responding formally to extremely surprising utterances of the expert is considered in Example 6.1 where a Cauchy alternative to the normal target distribution is seen to alleviate the difficulty. This is a 'smooth' model allowing for surprising statements and I wonder whether a tractable alternative would be a 'discontinuous' model giving a different functional form for the likelihood function when there is a sufficiently surprising statement by the expert.

Generally the modelling and elicitation of both the target distribution and expert's statement is delicate and the sensitivity of the subsequent pooling with respect to these choices deserves scrutiny. However, sensivity requirements would typically be softened when the ultimate purpose is the choice of a decision.

The examples in the final section show that, when the target distribution belongs to a subset of the exponential family, the pooling depends on a (mildly restricted) expert distribution only through its mean. Consideration of the Cauchy alternative target distribution in Example 6.1 indicates that the property is less than universal. It would be useful to know how nearly the property holds when the target distribution is 'near' the exponential class; for instance in Example 6.1 a Student's t target distribution with large degrees of freedom would be illuminating. Generally there is a related set of questions concerning when (for which class of target distributions?) a particular partial specification of the expert distribution is sufficient. This could influence both modelling the target distribution and eliciting the expert distribution.

McConway (1981) focussed on a marginalization property to be satisfied in this context. The procedures here do not satisfy that property; however, Lindley (1985) has argued convincingly aginst marginalization being satisfied.

A possible refinement would be desirable in cases where 'my' knowledge and that of the expert overlap. This has been considered by French (1980) and Lindley (1988) with other models. In this paper consider, for example, the case where Y is binary and f is stated, then one kind of dependence of knowledge may be incorporated by allowing α_Y to depend on p.

Further possible future developments are mentioned by Dr. West. Perhaps of most interest would be the one which would allow for uncertainty about the target distribution so that learning about the expert could be incorporated when he is consulted repeatedly.

Another potential generalization would be to the case of several experts. Here the possibility of knowledge overlap would also arise.

REPLY TO THE DISCUSSION

First let me thank Dr. Owen for his comments and discussion. He raises several issues concerning features of the particular models in my paper, and others concerned with the area more generally. Suggestions for further development, including mulit-expert problems and learning about model parameters, are accepted without comment and will, hopefully, lead to further work in the area.

Several points relate essentially to the choice of target distribution $M_Y(\cdot)$. The problem of "suprising" data H is a common one in parametric models, with no particularly new features here. Standard considerations of model robustness, outlier accommodation and the use of predictive distributions (either $p(H)$ or $p(H|Y)$ when Y is observed) apply. What is certainly needed first, however, is an appreciation of the nature and extent of the dependence of the likelihood on features of H for any chosen model; Section 6, and notably Example 6.1, is devoted to this sort of study, comparing models based on normal and Cauchy targets. Figure 4 highlights the differences for a data set H "suprising" under the normal model. The effects here are, of course, to be anticipated in view of the well-known, related features of normal vs. Cauchy based models for sampling distributions and/or priors in standard applications (see, eg, O'Hagan, in these Proceedings). As Dr. Owen suggests, there is a need for wider theoretical study of the likelihoods, and apparently the most promising approach is via the limiting case and exploration of divergences (although this, limiting case is only part the story).

The question of whether a partial specification of $F(\cdot)$ suffices to determine the divergenc is intriguing; my guess is that this is only the case in particular, exponential family base models, as in Example 6.3, although I have no theoretical results here. Let me reaffirm that find this particular feature of exponential family models to be rather unsatisfactory as a genera rule, preferring models that reflect some features of the spread of uncertainty in $F(\cdot)$ rathe than just the location. From a practical viewpoint, however, the appropriateness of particula models must be addressed in the context of application. In the applied study of Pratt an Schlaifer (1985), for example, the authors conclude (from a data analytic viewpoint) that th median forecasts are essentially sufficient for the outcomes, further specification of $F(\cdot)$ vi quantiles providing relatively little additional information.

Dr. Owen raises three issues drawn from the wider field of aggregation of probabili ties/decisions. What about the concensus issue, when "the expert and I must together reac a conclusion", and game theoretic considerations? Well, my paper concerns only how I, o any other individual, may view the statements of others. The general ideas may apply, o course, within a group environment. Each individual models what he sees, having scope t adjust for perceived or anticipated biases in the statements of others, and therefore allowin for and countering possible attempts to "influence the conclusion". On the aggregation issue the Bayesian framework concerns individuals; group probabilities do not exist. For furthe discussion of this, see West (1984) and French (1985). On the issue of marginalisation, I hav nothing to add to Dr. Owen's support for the conclusions of Lindley (1988).

A final noteworthy point raised by Dr. Owen concerns "cases where my knowledg and that of the expert overlap". Clearly the models in my paper allow for dependence o $p(H|Y)$ on $p(Y)$, my prior, although this is not specifically mentioned nor made explicit. I connection with this, let me note a slightly alternative approach that obviates this problem, an the need for explicit specification of $p(Y)$ itself. Begin by switching the focus in modellin from the forecast median μ to the outcome Y. To be specific, take the example in whic $(\mu|Y) \sim N[Y, W]$ (or any other location model). This can be re-written as $Y - \mu \sim N[0, W]$. If we interpret this as a specification of $p(Y|\mu)$ rather than, as originally intended, of $p(\mu|Y)$, then, given $H = \{\mu\}$, the posterior has been specified directly: $p(Y|H) = p(Y|\mu)$ wher $(Y|\mu) \sim N[\mu, W]$. Thus $p(Y)$ is not required. Refining specifications of $F(\cdot)$ to includ further quantiles $\underline{q}$, giving data $H^* = \{\mu, \underline{q}\}$, then

$$p(Y|H^*) \propto p(Y|\mu)p(\underline{q}|\mu, Y),$$

with the second term following directly from the paper. See Pratt and Schlaifer (1985, an in reply to discussion by Lindley) for related ideas. This is simply an alternative approach to specifying (some features of) the joint distribution for (Y, H) which is, after all, the centra objective of work in this area.

REFERENCES IN THE DISCUSSION

French, S. (1980). Updating of belief in the light of someone else's opinion. *J. Roy. Statist. Soc. A* **143**, 43–48.

McConway, K.J. (1981). Marginalization and linear opinion pools. *J. Amer. Statist. Assoc.* **76**, 410–414.

West, M. (1984). Bayesian aggregation. *J. Roy. Statist. Soc. A* **147**, 600–607.

BAYESIAN STATISTICS 3, pp. 509–516
J. M. Bernardo, M. H. DeGroot, D. V. Lindley and A. F. M. Smith, (Eds.)

A Bayesian Era

A. ZELLNER
University of Chicago

SUMMARY

After discussing important aspects of the Bayesian approach and their relation to scientific methodology, it is pointed out that most problems in econometrics and statistics have been solved using Bayesian methods. Also, some recent developments in Bayesian analysis and considerations bearing on the teaching of econometrics and statistics are described and discussed. From these considerations and information on the publication of Bayesian works it is concluded that a Bayesian era in econometrics and statistics has emerged.

Keywords: BAYESIAN ERA; BAYESIAN INFERENCE AND APPLICATIONS; BAYES/NON-BAYES ISSUES; TEACHING STATISTICS AND ECONOMETRICS.

1. INTRODUCTION

At the tenth anniversary meeting of the NBER-NSF Seminar on Bayesian Inference in Econometrics, Good (1980) remarked in his prepared address:

> "In about 1946 I had a long argument about Bayesian Statistics with Maurice Bartlett... At the end of our argument Bartlett proposed the bet or test of waiting for a hundred years to see which was dominant, a classical approach or a neo-Bayesian one. After a third of a century I think Bartlett is losing the bet".

In Zellner (1981), I stated that Good's conclusion also applies to the future state of Bayesian analysis in econometrics. However, now I believe that the movement toward Bayesian statistics and econometrics has been more rapid than I anticipated in 1981, or in 1962 when I initiated my research on Bayesian econometrics and statistics.

In the present paper, I shall review what I perceive to be the major factors involved in the recent emergence of a Bayesian era in statistics and econometrics. In Section 2 issues in scientific method and the Bayesian approach will be discussed. Attention will be given to some technical aspects of the Bayesian approach in Section 3. Section 4 includes some observations on the teaching of statistics and econometrics and in Section 5 concluding remarks are presented.

2. SCIENTIFIC METHOD AND BAYESIANISM

In his ASA Presidential Address, Marquardt (1987) emphasized the important role of "scientific method" in statistics and stated, "The fundamental role of statisticians is to be *purveyors of the scientific method*" (p.6). I agree with Marquardt's views and believe that there is no better place to learn about scientific method that in the writings of Sir Harold Jeffreys, in particular his books *Scientific Inference* (1973) and *Theory of Probability* (1967). In his review of the 1961 edition of the latter book, Good (1962) wrote that Jeffreys's book on probability "is of greater importance for the philosophy of science, and obviously of greater immediate practical importance, than nearly all the books on probability written by professional philosophers *lumped together*". Also, Jaynes (1983, 1984) regards Jeffreys's work to be a direct continuation and extension of Laplace's work on Bayesian philosophy, inference and applications. That such eminent physical scientists as Laplace, Jeffreys and Jaynes find a Bayesian approach to the analysis of scientific problems to be appropriate and useful is important and has come to be more widely appreciated —see, e.g. Berger and Sellke (1987), where Jeffreys's testing procedures are explained and contrasted with non-Bayesian procedures, and the articles by Geisser, Good and Lindley in Zellner (1980) summarizing Jeffreys's contributions.

In his *Theory of Probability*, Jeffreys sets forth an axiom system for probability theory that accommodates the probabilistic reasoning of working scientists. Bayes's Theorem is central in Jeffreys's system which permits scientists to learn from their data in a coherent manner. See also Cox (1961) for a fundamental analysis of Bayes's Theorem as an appropriate model for inductive reasoning and Zellner (1986a) for a demonstration that Bayes's Theorem is a 100% efficient information-processing procedure. Also, the information-processing framework utilized in Zellner (1986a) is capable of extension to provide new and perhaps improved information-processing rules for a wide range of problems. Results such as these have put thinking non-Bayesians on the defensive and challenged them to think deeply about their informal learning procedures. As an example, Freedman (1986) has written in response to Hill (1986):

> "When drawing inferences from data, even the most hard-bitten objectivist usually has to introduce assumptions and use prior information. The serious question is how to integrate that information into the inferential process and how to test the assumptions underlying the analysis". (p. 127)

Thus Freedman recognizes the fundamental problem but does not offer any solution to it. Without a solution, non-Bayesian statistics cannot be said to be "objective".

As the Freedman quotation indicates, non-Bayesians in statistics and econometrics have become increasingly aware of the fact that prior information plays a central role in statistics and econometrics. Such information which, for example, is used to identify parameters in many statistical and econometric models is rarely, if ever, "objective" information. Indeed even in the simplest models, the introduction of unobserved errors with assumed properties is hardly an "objective" matter. Thus non-Bayesians have come to realize that they are far from being "hard-bitten objectivists" and some have come to recognize that Bayesian methods are useful in dealing with prior information in a formal, operational manner as the following words of Tukey (1978) indicate:

> "It is my impression that rather generally, not just in econometrics, it is considered decent to use judgment in choosing a functional form, but indecent to use judgment in choosing a coefficient. If judgment about important things is quite all right, why should it not be used for less important ones as well? Perhaps the real purpose of Bayesian techniques is to let us do the indecent thing while modestly concealed behind a formal apparatus. If so, this would not be a precedent. When Fisher introduced the formalities of the analysis of variance in the early 1920's, its most important function was to conceal the fact that the data was being adjusted for block means, an important step forward which if openly visible would have been considered by too many wiseacres of the time to be "cooking the data". If so, let us hope that day will soon come when the role of decent concealment can be freely admitted. . . The coefficient may be better estimated from one source or another, or, even best, estimated by economic judgment.
>
> It seems to me a breach of the statistician's trust not to use judgment when that appears to be enough better than using data." (p. 52)

Clearly Tukey is recommending that statisticians and econometricians let others know what they are sweeping under the rug, to use Good's phrase, and to consider the Bayesian approach seriously.

The following quotation from Lehmann (1959) provides further evidence that non-Bayesians are hardly "objective" in testing hypotheses:

> "Another consideration that frequently enters into the specification of a significance level is the attitude toward the hypothesis before the experiment is performed. If one firmly believes the hypothesis to be true, extremely convincing evidence will be required before one is willing to give up this belief, and the significance level will accordingly be set very low". (p. 62)

While Lehmann's advice may be sound, it is certainly not a "hard-bitten objectivist" approach to testing hypotheses.

The quotations presented above indicate that may non-Bayesian do not pursue an "objective" approach in their work. Also, the quotations indicate that these non-Bayesians are not using a frequency-based concept of probability, thus giving support to Jeffreys's (1967) observation, "In practice no statistician ever uses a frequency definition [of probability], but... all use the notion of degree of reasonable belief, usually without even noticing that they are using it and that by using it they are contradicting the principles they have laid down at the outset" (p.369). In Jeffreys (1967, Ch. 7) devastating critiques of the classical or axiomatic, the Venn limiting frequency and the Fisher hypothetical infinite population definitions of probability are presented, which have never been rebutted by non-Bayesians to the best of the author's knowledge.

Jeffreys (1967) views a numerical probability as measuring "the degree of confidence that we may reasonably have in a proposition, even though we may not be able to give either a deductive proof or disproof of it" (p.15). Savage (1962) writes, "Personal probability is a certain kind of numerical measure of the opinions of somebody about something" (p.163). These are subjective concepts of probability that Bayesians find most useful in inference and decision analysis. Non-Bayesians usually express shock that a subjective concept of probability is to be used in science even though, as pointed out above, most of them knowingly or unknowingly use subjective probability in their work. What many of these non-Bayesians fail to realize is that *subjective probabilities are used along with data to generate predictions which can be checked against actual outcomes.* For example, Jeffreys (1967) defines induction as follows, "... part [of our knowledge] consists of making inferences from past experience to predict future experience. This part may be called generalization or induction" (p.8). Further, in a review of philosophical concepts of causality, the philosopher Feigl (1953, p.408) summarizes philosophers' views by defining causality to be confirmed predictability based on a law or set of laws. Also, Friedman (1953) in economics and Geisser (1980) in statistics have emphasized that prediction of observable outcomes is a central activity in science and that subjective probability plays an important role in generating predictions. Of course, non-Bayesians use a good deal of prior subjective information in generating predictions. However, in contrast to Bayesians, they generally use their subjective information informally in a framework which fails to accommodate it in a logically satisfactory manner and thus "are contradicting the principles they have laid down at the outset", to use Jeffreys's words.

Many non-Bayesian predictions and forecasts cannot be reproduced because they have been constructed using unreported, informal subjective prior information —see McNees (1986) and Litterman (1986) for further discussion of this important issue. If predictions and forecasts are not reproducible, they are hardly scientific. Using explicit, prior distributions to represent subjective prior information, Bayesians such as West, Harrison and Mignon (1985), Litterman (1986), Highfield (1986), García-Ferrer *et al.* (1987), Zellner and Hong (1987), and others have generated reproducible forecasts which have compared favorably with those generated by non-Bayesian forecasters —see also Zellner (1986c). Note that Litterman's Bayesian forecasting models have been used by the Federal Reserve Bank of Minneapolis since about 1981. Now instead of arguing about whether or not to use prior distributions in forecasting, forecasters have begun to argue about which prior distributions to employ, a fundamental change of great importance.

In summary, it has come to be recognized that prior, subjective information is widely used in science and that the Bayesian paradigm is currently the only one capable of accommodating such prior information in an operational, formal manner. Also it has been shown that almost all sensible non-Bayesian estimation, prediction and testing results in econometrics can be produced by application of the Bayesian approach and that important new results can be produced by Bayesian methods that are difficult, if not impossible, to produce by non-

Bayesian methods —see Zellner (1971, 1986b) for references and analyses illustrating thes points. In addition, Jaynes (1984) described the situation in statistics as follows: "Indee the variety of [Bayesian] applications demonstrated by Jeffreys [1967] included all those fo which "Student", Fisher, Neyman and Pearson had developed "orthodox" methods" (p.44).

Thus it appears accurate to state that the Bayesian approach in econometrics and statistic has practically encompassed non-Bayesian approaches. It can produce sensible non-Bayesia results as well as results which are difficult, if not imposible, to produce by non-Bayesiar methods. Many have come to recognize these facts and this recognition is the opening of Bayesian era in statistics and econometrics.

3. TECHNICAL ISSUES

In this section, several technical issues will be considered that relate to the recent acceleratio of progress in Bayesian econometrics and statistics.

First, enhanced computing capabilities in recent years, particularly in the area of numerica integration have permitted Bayesians not only to solve problems more readily and accuratel but also to enrich their analyses by permitting use of a wider range of likelihood function and prior distributions —see e.g. Stewart (1977, 1979), Kloek and van Dijk (1978), Monaha (1983), Bauwens (1984), van Dijk (1984), Geweke (1986a, b), Drèze and Richard (1983) and Zellner and Rossi (1984). Further, a simple or direct Monte Carlo simulation approac has emerged that is of great importance from a practical point of view, see e.g. Kass (1985) Thompson and Miller (1986), Geweke (1986a), Zellner (1985), and Zellner, van Dijk and Bauwens (1987). To illustrate, if $p(\theta|D)$ is a posterior probability density function (*pdf*) for a vector of parameters θ, where D denotes *given* sample and prior information, and it is of interest to obtain the posterior distribution of the elements of η, given by $\eta = f(\theta)$, a one-to-one transformation, one makes independent draws from $p(\theta|D)$ and evaluates η for each draw. In this way, the complete posterior *pdfs* for the elements of η are obtained. For example θ may be elements of the inverse of a covariance matrix and η the roots of the covariance matrix, as in Kass (1985). Or θ might be a vector of autoregressive coefficients and the elements of η the roots of the time series process, as in Geweke (1986a). Or θ might be reduced form coefficients and η structural coefficients, as in Zellner (1985) and Zellner, van Dijk and Bauwens (1987). Further, in some cases $p(\theta|D)$ may have a complicated form. However, with $\theta' = (\theta_1', \theta_2')$, it is often the case that $p(\theta_1|\theta_2, D)$ and $p(\theta_2|D)$ have well known forms from which draws can be made conveniently. That is draw from $p(\theta_2|D)$ and insert the outcome in $p(\theta_1|\theta_2, D)$ and then draw from this conditional *pdf*. Such a procedure was used by Thompson and Miller (1986) to compute the joint predictive *pdf* of the elements of a vector of future observations from a time series process. These computational developments are of great importance since they are readily applicable and yield practical results that are very difficult to obtain using alternative approaches.

Second, it has come to be recognized that reference diffuse and informative prior distributions are of great value. The early, negative reactions to Jeffreys's important contributions in this area are fortunately being brushed aside. Leading younger Bayesians such as Berger (1985), Berger and Sellke (1987), Draper (1986), Morris (1986), and others have recognized the importance of having reasonable reference priors "on the shelf" to be used when appropriate, just as reasonable reference models, e.g. regression, autoregression, ARMA, and other models are "on the shelf" and employed when they are thought to be appropriate. This is not recommending that reference priors *or* models are to be used mechanically and without thought or that a "necessarist" point of view be adopted. It is just saying that in inference, as in life, conventions are often useful. In Jeffreys (1967), Jaynes (1968), Zellner (1971, 1977), Box and Tiao (1973) and other works, procedures for generating prior pdfs have appeared. Many reasonable diffuse and informative prior pdfs in use have been produced in a formal, explicit fashion by these procedures and thus some of the "fuzziness" regarding priors has been reduced which is making the Bayesian approach more palatable to non-Bayesians.

Third, several technical issues which received substantial publicity in past years and which were interpreted as negative aspects of the Bayesian approach have now been resolved. First, the Dawid-Stone-Zidek marginalization paradox has been resolved by Jaynes (1980) who concluded that, "The real cause of the paradox is not ...use of improper priors... it [the paradox] appears only when [there is a violation of] elementary Bayesian principles" (p. 44).

Second, much attention has been given to unbounded measures, improper priors etc. Fortunately, appropriate limiting operations have been indicated in Rényi (1970, p.57ff.) and Jaynes (1980, p.49ff.). Rényi (1970, p.55) writes,

> "... in the course of the development [of probability theory] there arose certain problems which could not be fitted into the frames of the theory or probability spaces of Kolmogoroff. The common feature of these problems is that unbounded measures occur in them, while in Kolmogoroff's theory, probability is a bounded measure. Unbounded measures were used to compute probabilities in statistical, mechanics, quantum mechanics, and also in mathematical statistics (in connection with the method of Bayes), in integral geometry, in probabilistic number theory, etc. ... observing attentively how unbounded measures were used in the fields mentioned, ... it turns out that such measures were used only to compute conditional probabilities as ratios of the values of an unbounded measure for two sets, the first being a subset of the second and in this way one always gets reasonable values, i.e., values between 0 and 1 for conditional probabilities". (pp.55-56)

He goes on to state that in his generalization of Kolmogoroff's theory of probability, "... probabilities obtained from unbounded measures receive full "civic rights"." (p.56) Thus the clamor about unbounded measures and improper priors is usually not of great practical relevance and has been resolved theoretically in Rényi's theory of probability in which the basic concept is conditional probability, not unconditional probability as in Kolmogoroff's theory.

Third, as regards asymptotic properties, Heyde and Johnstone (1979) conclude that conditions needed for the asymptotic normality of posterior distributions are simpler and more robust than those needed for the asymptotic normality of maximum likelihood estimators in the important case of stochastically dependent observations. This finding is just coming to be appreciated by non-Bayesian econometricians and statisticians and will probably be very influential. However, wiith good numerical integration capabilities, there is usually no need to rely on asymptotic approximations which are often not very good in small to moderate sized samples; see Zellner and Rossi (1984) for illustrations of this point.

Fourth, many non-Bayesians who are interested in estimators and predictors have come to be impressed with the ease which optimal estimators and predictors can be produced by use of the Bayesian approach; see George (1986) and Zellner (1985, 1986b, d) for some recent examples illustrating this point.

Fifth, the relationship of various Bayesian approaches, in particular the Laplace-Jeffreys and the de Finetti-Savage approaches is becoming better understood. For an early consideration of the issues involved, see Jeffreys (1967, pp.30-33). Further clarification of interrelations of various Bayesian approaches will be very helpful.

Sixth, many of those who have become better informed in the area of scientific method have become skeptical of those who advocate "objective procedures" without defining them carefully. Similarly, they have become wary of those who state that models with an infinite number of parameters are realistic, or that simple models are unrealistic, or that assuming that a parameter's value is exactly equal to zero or any other precise value is unrealistic, or that statistics should just deal with observables, etc. These and other similar statements are not in accord with Jeffreys' (1967) rules for a theory of induction, in particular his rule 5: "The theory [of induction] must not deny any empirical proposition a priori; any precisely stated empirical hypothesis must be formally capable of being accepted ... given a moderate amount of evidence". (p. 9) Thus, scientific workers can formulate their hypotheses in terms of simple or complicated models, models with parameters or in terms of observables alone, or with certain parameters assumed to have specific values say zero, one, two, etc. Finally, procedures for computing posterior probabilities associated with alternative hypotheses that automatically put an upper or lower bound, different from one and zero, respectively on probabilities associated with hypotheses are unsatisfactory. This is because such procedures place unjustified restrictions on the inductive process. The hypothesis that the sun will rise tomorrow certainly has a probability close to one and a theory of testing that restricts it a priori to a value less than or equal to .5 is absurd. That many Bayesians are in the forefront of linking econometrics and statistics to the theory of scientific method and of clearing these fields of absurd methodological views and procedures is of great importance and has given impetus to the progress of Bayesianism.

4. TEACHING ECONOMETRICS AND STATISTICS

How to teach econometrics and statistics and what should be taught are controversial issues; see Sowey (1983) for a useful discussion of these issues as they relate to econometrics. While more systematic studies of alternative teaching approaches, for example Bayesian versus non-Bayesian, are needed, casual observation indicates that the experience of many students and instructors with courses in statistics and econometrics is rather unsatisfactory. On this matter, Hey (1983) writes:

> "For more years than I care to recall, I have been teaching introductory statistics and econometrics to economics students. As many teachers and students are all too aware, this can be a painful experience for all concerned ... The fundamental malaise with most statistics and econometrics courses is that they use the *Classical* approach to inference. Students find this unnatural and contorted. It is not intuitively acceptable and does not accord with the way people assimilate information (statistical or otherwise) in their everyday life ... I first taught a Bayesian course two years ago ... To my delight, I discovered that the "fundamental problem" had disappeared ... I am now more than convinced, in the light of the experience of these two years, that the Bayesian approach is the "correct" one to adopt". (p. xi)

Whether Hey's experience is unique to him and his students or more generally applicable, is certainly worthy of systematic study. Note in this connection that Neyman-Pearson and P-value theories of testing which are featured in many current textbooks are unsound as shown in Jeffreys (1967), Jaynes (1984) and others' works. Further, as remarked above, inappropriate concepts of probability are foisted on students in most textbooks. With these and other inadequacies of current textbooks, it is hardly surprising that courses based on them are "painful experiences", as Hey states. Fortunately, several Bayesian textbooks are currently available and more will be published in the next few years. In addition, many young statisticians and econometricians have been exposed to Bayesian ideas and methods. Thus, it is not very risky to predict that the Bayesian component of teaching will grow rapidly in the next five to ten years and will provide students with a more satisfactory framework for analyzing problems. They will understand inference procedures and know how to use them effectively to solve a wide range of problems in a "natural" way. These changes will constitute a revolution in the teaching of econometrics and statistics.

5. CONCLUDING REMARKS

The main conclusion of this paper is that a Bayesian Era in statistics and econometrics has emerged, perhaps somewhat earlier in statistics than in econometrics and that this development is a major step forward in the development of econometrics and statistics. Bayesian are now very well represented in most major departments of statistics in the U.S. and several other countries. The 1985 ASA/IMS *Current Index to Statistics* lists over 300 articles and books dealing with Bayesian topics. Poirier (1986) found that Bayesian articles accounted for about 15% of the pages of leading statistics and econometrics journals in recent years. These observations, along with those presented in earlier sections of this paper, are evidence that a Bayesian Era in econometrics and statistics has arrived, a development of major importance.

REFERENCES

Bauwens, L. (1984). *Bayesian Full Information Analysis of Simultaneous Equation Models Using Integration by Monte Carlo.* Berlin: Springer-Verlag.

Berger, J. O. (1985). *Statistical Decision Theory and Bayesian Analysis* (2nd ed.), New York: Springer-Verlag.

Berger, J. O. and Sellke, T. (1987). Testing a point null hypothesis: the irreconcilability of *P* values and evidence. *J. Amer. Statist. Assoc.* **82**, 112–122 (with invited discussion).

Box, G. E. P. and Tiao, G. C. (1973). *Bayesian Inference in Statistical Analysis.* Reading, MA: Addison-Wesley.

Cox, R. T. (1961). *The Algebra of Probable Inference.* Baltimore: Johns Hopkins University Press.

Draper, D. (1986). Propagation of model uncertainty, presented at the 33rd Meeting of the NBER-NSF Seminar on Bayesian Inference in Econometrics, U. of California at Riverside and to the Dept. of Statistics Seminar U. of Chicago.

Drèze, J. H. and Richard, J. F. (1983). Bayesian analysis of simultaneous equation systems. *Handbook of Econometrics* **1**, 517–598. (Z. Griliches and M. D. Intriligator, eds.). Amsterdam: North-Holland.

Feigl, H. (1953). Notes on causality. *Readings in the Philosophy of Science*, 408–418. (H. Feigl and M. Brodbeck, eds.). New York: Appelton-Century-Crofts.

Freedman, D. A. (1986). Reply. *Journal of Business and Economic Statistics* **4**, 126–127.

Friedman, M. (1953). *Essays in Positive Economics.* Chicago: U. of Chicago Press.

García-Ferrer, A. Highfield, R. A., Palm, F., and Zellner, A. (1987). Macroeconomic forecasting using pooled international data. *Journal of Business and Economic Statistics* **5**, 53–67.

Geisser, S. (1980). A predictivistic primer. *Bayesian Analysis in Econometrics and Statistics: Essays in Honor of Harold Jeffreys*, 363–382. (A. Zellner, ed.). Amsterdam: North-Holland.

George, E. I. (1986). Combining minimax shrinkage estimators. *J. Amer. Statist. Assoc.* **81**, 437–445.

Geweke, J. (1986a). Bayesian inference in econometric models using Monte Carlo integration. Manuscript. Dept. of Economics, Duke U.

Geweke, J. (1986b). Exact inference in dynamic econometric modelling. Manuscript. Dept. of Econometrics, Duke U.

Good, I. J. (1962). Review of Harold Jeffreys' *Theory of Probability*, (3rd ed.). *Geophysical Journal of the Royal Astronomical Society* **6** (1961), 555–558; *Journal of the Royal Statistical Society A* **125**, (1962), 487–489.

Good, I. J. (1980). Prepared opening presentation. *Report of the Twenty-First NBER-NSF Seminar on Bayesian Inference in Econometrics*, prepared by Peter E. Rossi.

Hey, J. D. (1983). *Data in Doubt: An Introduction to Bayesian Statistical Inference for economists.* First published by Martin Robertson; reprinted 1985, Oxford, England: Basil Blackwell Ltd.

Heyde, C. C. and Johnstone, I. M. (1979). On asymptotic posterior normality for stochastic processes. *J. Roy. Statist. Soc. B* **41**, 184–189.

Highfield, R. A. (1986). Forecasting with Bayesian state space models, unpublished doctoral dissertation, Graduate School of Business, U. of Chicago.

Hill, B. M. (1986). Comment. *Journal of Business and Economic Statistics* **4**, 125–126.

Jaynes, E. T. (1968). Prior probabilities. IEEE Trans. Syst. Sci. and Cybern. SSC-4, Sept. 1968, 227–241, reprinted in V. M. Rao Tummala and R. C. Henshaw, eds., *Concepts and Applications of Modern Decision Models.* Michigan State U. Business Studies Series, 1976.

Jaynes, E. T. (1980). Marginalization and prior probabilities, in Zellner (1980), 43–78 and reprinted in E. T. Jaynes, *Papers on Probability, Statistics and Statistical Physics,* (R. D. Rosenkrantz, ed.). Dordrecht, Holland: D. Reidel.

Jaynes, E. T. (1983). *Papers on Probability, Statistics and Statistical Physics,* (R. D. Rosenkrantz, ed.). Dordrecht, Holland: D. Reidel.

Jaynes, E. T. (1984). The intuitive inadequacy of classical statistics. *Epistemologia* **VII**. Special Issue on Probability, Statistics and Inductive Logic, 43–74.

Jeffreys, H. (1967). *Theory of Probability.* London: Oxford University Press. 3rd ed. (1st ed., 1939).

Jeffreys, H. (1973). *Scientific Inference.* Cambridge: Cambridge University Press. 3rd ed. (1st ed., 1931).

Kass, R. E. (1985). Exact inferences about principal components and related quantities using posterior distributions calculated by simulation. Paper presented at the Thirtieth NBER-NSF Seminar on Bayesian Inference in Econometrics, May 17–18, 1985, Minneapolis, MN.

Kloek, T. and van Dijk, H. K. (1978). Bayesian estimates of equation systems parameters: an application of integration by Monte Carlo. *Econometrica* **46**, 1–19.

Lehmann, E. (1959). *Testing Statistical Hypotheses.* New York: Wiley.

Litterman, R. (1986). Forecasting with Bayesian vector autoregressions–five years of experience. *Journal of Business and Economic Statistics* **4**, 25–38.

Marquardt, D. W. (1987). The importance of statisticians. *J. Amer. Statist. Assoc.* **82**, 1–7.

McNees, S. K. (1986). Forecasting accuracy of alternative techniques: a comparison of U. S. macroeconomic forecasts (with discussion). *Journal of Business and Economic Statistics* **4**, 5–23.

Monahan, J. F. (1983). Fully Bayesian analysis of ARMA time series models. *Journal of Econometrics* **21**, 307–331.

Morris, C. N. (1986). When is Jeffreys' prior natural conjugate?, presented at the 33rd Meeting of the NBER-NSF Seminar on Bayesian Inference in Econometrics, U. of California at Riverside.

Poirier, D. J. (1986). A report from the Battlefront, *Tech. Rep.* Dept. of Economics, U. of Toronto.

Rényi, A. (1970). *Foundations of probability*. San Francisco: Holden-Day.

Savage, L. J. (1962). Bayesian statistics. *Recent Developments in Information Theory*, 161–194, (R. F. Machol and P. Gray, eds.). New York: Macmillan Co. Reprinted in *The Writings of Leonard Jimmie Savage–A Memorial Selection*, Washington, DC: American Statistical Association and Institute of Mathematical Statistics, 1981.

Sowey, E. R. (1983). University teaching of econometrics: a personal view. *Econometric Reviews* **2**, 255–289 (with invited discussion).

Stewart, L. (1977). Bayesian analysis using Monte Carlo integration. Working paper LMSC–D560641, Lockheed Palo Alto Research Lab.

Stewart, L. (1979). Multiparameter univariate Bayesian analysis. *J. Amer. Statist. Assoc.* **74**, 684–693.

Thompson, P. A. and Miller, R. B. (1986). Sampling the future: a Bayesian approach to forecasting from univariate time series models. *Journal of Business and Economic Statistics* **4**, 427–436.

Tukey, J. W. (1978). Discussion of Granger on seasonality. *Seasonal Analysis of Economic Time Series*, 50–53, (A. Zellner, ed.). Washington, DC: U. S. Government Printing Office.

van Dijk, H. K. (1984). Posterior analysis of econometric models using Monte Carlo integration, doctoral dissertation, Erasmus Universiteit Rotterdam.

West, M., Harrison, P. J. and Mignon, H. S. (1985). Dynamic generalized linear models and Bayesian forecasting. *J. Amer. Statist. Assoc.* **80**, 73–83.

Zellner, A. (1971). *An Introduction to Bayesian Inference in Econometrics*, New York: Wiley, reprinted by Krieger, 1987.

Zellner, A. (1977). Maximal data information prior distributions. *New Developments in the Applications of Bayesian Methods*, 211–232, (A. Aykac and C. Brumat, eds.). Amsterdam: North-Holland.

Zellner, A. (1980). *Bayesian Analysis in Econometrics and Statistics: Essays in Honor of Harold Jeffreys*, (A, Zellner, ed.). Amsterdam: North-Holland.

Zellner, A. (1981). The current state of Bayesian econometrics, invited address presented at the Canadian Conference on Applied Statistics, Concordia U., Montreal, Apr. 29-May 1, 1981. *Topics in Applied Statistics*, (T. D. Dwivedi, ed.). New York: Marcel Decker, 1983; reprinted in A. Zellner, *Basic Issues in Econometrics*, Chicago: U. of Chicago Press, 1984.

Zellner, A. (1985). Further results on Bayesian minimum expected loss (MELO) estimates and posterior distributions for structural coefficients. *Advances in Econometrics* **5**, 171–182.

Zellner, A. (1986a). Optimal information-processing and Bayes' theorem. *Tech. Rep.* H. G. B. Alexander Research Foundation, Graduate School of Business, U. of Chicago. (To appear in *The American Statistician*, Nov., 1988, with discussion).

Zellner, A. (1986b). Bayesian analysis in econometrics. invited paper for the International Conference on the Foundations of Statistical Inference, Israel, Dec. 16–19, 1985. *Journal of Econometrics* **37**, 27–50.

Zellner, A. (1986c). A tale of forecasting 1001 series: the Bayesian knight strikes again. *Journal of Forecasting* **2**, 491–494.

Zellner, A. (1986d). Bayesian estimation and prediction using asymmetric loss functions. *J. Amer. Statist. Assoc.* **81**, 446–451.

Zellner, A. and Hong, C. (1987). Forecasting international growth rates using Bayesian shrinkage and other procedures. *Tech. Rep.* H. G. B. Alexander Research Foundation, Graduate School of Business, U. of Chicago. To appear in the *Journal of Econometrics, Annals of Applied Econometrics*.

Zellner, A. and Rossi, P. E. (1984). Bayesian analysis of dichotomous quantal response models. *Journal of Econometrics* **25**, 365–393.

Zellner, A., van Dijk, H. K. and Bauwens, L. (1987). Bayesian specification analysis and estimation of simultaneous equation models using Monte Carlo methods. *Journal of Econometrics* **38**, 39–72.

CONTRIBUTED PAPERS

BAYESIAN STATISTICS 3, pp. 519–531
J. M. Bernardo, M. H. DeGroot, D. V. Lindley and A. F. M. Smith, (Eds.)

Bayesian Estimation of Poisson Means Using a Hierarchical Log-Linear Model

J. H. ALBERT
University of Southampton and Bowling Green State University

SUMMARY

Consider the problem of estimating means from independent Poisson distributions or linear combinations of the log means when the means are believed a priori to satisfy a log-linear model. Using a hierarchical prior distribution, approximate methods are proposed for the computation of posterior densities which use standard generalized linear model calculations. The methods are used in the estimation of linear combinations of the log means in a one-way classification and a two-way contingency table.

Keywords: CONTINGENCY TABLES; GENERALIZED LINEAR MODELS; ODDS RATIOS; SHRINKAGE.

1. INTRODUCTION

Suppose independent random variables $Y_1, \ldots, Y_N$ are observed, where Y_i is distributed Poisson with mean θ_i, $i = 1, \ldots, N$. The general problem of interest is to estimate $\boldsymbol{\theta} = (\theta_1, \ldots, \theta_N)$ or some real valued function of $\boldsymbol{\theta}$ when the user believes a priori that $\boldsymbol{\theta}$ satisfies the p-dimensional log-linear model

$$\log \theta_i = \boldsymbol{x}_i^T \boldsymbol{\beta}, \qquad i = 1, \ldots, N. \tag{1.1}$$

In (1.1), $\boldsymbol{\beta} = (\beta_1, \ldots, \beta_p)^t$ is an unknown vector of regression coefficients and $\boldsymbol{x}_i = (x_{i1}, \ldots, x_{ip})^T$ is a known vector of constants.

Classification	Frequency
1. Lower class	72
2. Working class	714
3. Middle class	655
4. Upper class	41

Table 1. *Social Class Membership Data from Haberman (1978).*

Typically it is of interest to estimate parameters that are expressible as sums of logarithms of the means θ_i. To illustrate, Table 1 gives data, from Haberman (1978), which summarizes social class membership in the 1975 general Social Survey of the National Opinion Research Center. Let Y_i denote the observed count for social class i and assume $Y_i \sim$ Poisson (θ_i), $i = 1, 2, 3, 4$. Then it may be of interest to estimate

$$\log \theta_1 - \log \theta_4 = \log(\theta_1/\theta_4) \tag{1.2}$$

the logarithm of the ratio of membership in the lower and upper classes. Haberman suggested the following model for this data:

$$\log \theta_i = \alpha + \beta_1 x_{i1} + \beta_2 x_{i2}, \tag{1.3}$$

where $x_{i1} = 2(i - 2.5)$ and $x_{i2} = (i - 2.5)^2 - 1.25$. The problem is to estimate the parameter (1.2) or other linear functions of the log means under the prior belief that model (1.3) fits the data.

As a second example, Table 2 gives a two-way contingency table from Goodman (1979) cross-classifying subjects with respect to mental health and socioeconomic status. Let Y_{ij} denote the count in the i-th row and j-th column of the table and assume Y_{ij} is distributed Poisson (θ_{ij}), $i = 1, \ldots, 4$, $j = 1, \ldots, 6$ Goodman suggested the use of the following "constant association" log-linear model for the data:

$$\log \theta_{ij} = u + u_{1(i)} + u_{2(j)} + \beta(i - 2.5)(j - 3.5). \tag{1.4}$$

In the usual log-linear notation, (Bishop, Fienberg and Holland (1975)), u is a grand mean and $u_{1(i)}$ and $u_{2(j)}$ are the additive effects corresponding to the i-th row and j-th column, respectively, in the table. The interaction in the table is represented in (1.4) by the parameter β, which equals the logarithm of the odds ratio from the 2×2 subtable formed from adjacent rows and adjacent columns; i. e.

$$\beta = \log[(\theta_{ij}\theta_{i+1,j+1})/(\theta_{i,j+1}\theta_{i+1,j})]. \tag{1.5}$$

If the user believes a priori in the model (1.4), the problem is to estimate various linear combinations of logarithms of the cell means. Logarithms of odds ratios such as the right hand side of (1.5) are of special interest, since they measure the association in the two-way contingency table.

	Parents' Socioeconomic status					
Mental Health Status	A	B	C	D	E	F
Well	64	57	57	72	36	21
Mild Sympto Formation	94	94	105	141	97	71
Moderate Sympto Formation	58	54	65	77	54	54
Impaired	46	40	60	94	78	71

Table 2. *Mental Health Data from Goodman (1979).*

The classical pretesting approach to these estimation problems (see Haberman (1978) and Goodman (1979)) is a two step approach. First, a test is performed to assess the goodness of the hypothesized model (1.1). The two statistics commonly used are the Pearson chi-squared statistic $X^2 = \Sigma(y_i - e_i)^2/e_i$ and the likelihood ratio statistic $G^2 = 2\Sigma y_i \log(y_i/e_i)$, where e_i is the MLE of θ_i under model. Then, if the result of the test is to accept the model, estimation procedures are used which assume that the θ_i actually satisfy the model (1.1). As an example, suppose that the model (1.4) is judged to be a good fit to the data based on the computation of X^2 or G^2. Then, if the parameter of interest is the "localized" log odds-ratio $\eta = \log[(\theta_{11}\theta_{22})/(\theta_{12}\theta_{21})]$, then a point estimate of η is given by the MLE of η under the model (1.4). By the restriction (1.5), any log odds-ratio formed from adjacent rows and adjacent columns will have the same MLE under this model.

One obvious criticism of this approach is that it is not clear how to set up the rejection region of the test if the primary goal is to obtain accurate estimates of the θ_i. A second criticism, which is the focus of this paper, is that the hypothesized model (1.1) is not exactly true and, by assuming it is true in this pretesting approach, one can significantly underestimate the variability of the derived estimates. In our example of estimating a localized odds ratio, the MLE of the model parameter β combines information from counts from the entire 4×6 table. The standard error of this estimate of β is relatively small and may unrealistically overestimate the precision of the estimate of one particular log odds-ratio such as η. As an alternative to this classical pretesting approach, posterior estimates will be derived using a hierarchical prior distribution for θ which allows for uncertainty in the prior belief in the model (1.1). The complete model can be written as follows:

I. Conditional on $\boldsymbol{\theta}$, $Y_i, \ldots, Y_N$ are independent and

$$Y_i|\theta_i \sim \text{Poisson } (\theta_i). \tag{1.6}$$

II. Conditional on β and α, $\theta_1, \ldots, \theta_N$ are independent and $\theta_i|\beta, \alpha \sim$ Gamma $(\xi_i(\beta)\alpha,\ \alpha)$ where $\log \xi_i(\beta) = \boldsymbol{x}_i^T \beta$.

III. $(\beta, \alpha) \sim \pi_3(\alpha, \beta)\alpha(1+\alpha)^{-2}$.

The prior distribution is described in stages II and III of (1.6). At stage II, $\boldsymbol{\theta}$ is assigned the prior distribution

$$\pi_1(\theta|\alpha, \beta) = \prod \pi_G(\theta_i|\xi_i(\beta)\alpha, \alpha),$$

where $\pi_G(u|a, b)$ denotes the gamma density proportional to $u^{a-1}\exp(-bu)$. Note that the prior mean of $\theta_i, \xi_i(\beta)$, is chosen in (1.6) to reflect the belief in the log-linear model (1.1). The hyperparameter α is a precision parameter which reflects the sureness of the prior belief in the model (1.1). At stage III of (1.6), the unknown hyperparameters (β, α) are assigned a noninformative prior π_3. At this stage, β and α are assumed independent with β assigned a flat uniform prior and α a prior density proportional to $(1+\alpha)^{-2}$. An alternative selection would be to assign to the hyperparameter α the Jeffreys' noninformative prior proportional to α^{-1}. Our experience is that the resulting posterior inferences for $\boldsymbol{\theta}$ are not very sensitive in small changes in the hyperparameters at Stage III of (1.6). This robustness with respect to the Stage III hyperparameters will be illustrated in Section 5.

A hierarchical model of the from (1.6) has been used recently by Leonard and Novick (1986) in the analysis of two-way contingency tables. Albert (1985) used the same model in the special case of Poisson sampling from two-way tables where the prior model is independence between the two attributes.

In Sections 2-4, an approximate posterior analysis is described for the vector of cell means $\boldsymbol{\theta}$ and the hyperparameters (β, α).The difficulties in exact calculations lie mainly with the posterior density of (β, α). In Leonard and Novick (1986), approximations are proposed for this posterior density and these approximations are used in the construction of posterior densities for the cell means θ_i and the "parametric residuals" $\rho_i = \log \theta_i - \log \xi_i(\beta)$. In Section 2, new approximate methods are described which use standard generalized linear model calculations and give easily computable posterior densities of α and β. The focus of the paper is on the posterior density of the hyperparameter α, which measures the goodness of fit of the model (1.1), and on the posterior density of a linear combination of the logs of the cell means.

In Section 5, the methods developed in Section 2-4 are used to analyze the data in Tables 1 and 2. In particular, Bayesian posterior densities from the hierarchical model (1.6) are compared with standard classical likelihood functions for various parameters of interest.

2. POSTERIOR DISTRIBUTIONS AND APPROXIMATIONS

The posterior density for θ is representable as

$$\pi(\theta|y) = \int \pi_1(\theta|y, \alpha, \beta)\pi_2(\alpha, \beta|y)d\alpha d\beta, \tag{2.1}$$

where π_1 is the posterior density of θ given the hyperparameters β and α and π_2 is the posterior density of β and α.

First consider the density π_1. If one combines stages I and II of the model (1.6), then an easy calculation shows that

$$\pi_1(\theta|y, \alpha\beta) = \prod_{i=1}^{N} \pi_G(\theta_i; y_i + \alpha\xi_i(\beta), 1 + \alpha), \tag{2.2}$$

a product of gamma distributions. The posterior mean of θ_i, conditional on α and β, is given by

$$\begin{aligned} E(\theta_i|y, \alpha, \beta) &= (y_i + \alpha\xi_i(\beta))/(1 + \alpha) \\ &= (1 - \gamma)y_i + \gamma\xi_i(\beta), \end{aligned} \tag{2.3}$$

where $\gamma = \alpha/(\alpha + 1)$. The posterior mean (2.3) "shrinks" the MLE y_i towards the prior mean $\xi_i(\beta)$. The reparameterized hyperparameter $\gamma (0 < \gamma < 1)$ equals the size of the proportionate shrinkage, i. e.

$$\gamma = |E(\theta_i|y, \alpha, \beta) - y_i|/|\xi_i(\beta) - y_i|. \tag{2.4}$$

Next consider the posterior density of the hyperparameters β and α. This density can be written as

$$\pi_2(\alpha, \beta|y) = C_1 m(y|\alpha, \beta)\pi_3(\alpha, \beta), \tag{2.5}$$

where m is the marginal density of y (conditional on the hyperparameters α and β), π_3 is the prior at stage III of (1.6) and C_1 is a proportionality constant. If $f(y|\theta)$ denotes the density of Y at stage I of (1.6), then the marginal density can be calculated as

$$\begin{aligned} m(y|\alpha, \beta) &= \pi_1(\theta|\alpha, \beta)f(y|\theta)/\pi_1(\theta|\alpha, \beta, y) \\ &= \exp\left[\Sigma \log \Gamma(y_1 + \alpha\xi_i(\beta)) - \Sigma \log \Gamma(\alpha\xi_i(\beta))\right] \\ &\times \exp\left[-(\Sigma(y_i + \alpha\xi_i(\beta)) \log(1 + \alpha) + \alpha\Sigma\xi_i(\beta) \log \alpha\right] \\ &\times \exp\left[-\Sigma \log \Gamma(y_i + 1)\right] \end{aligned} \tag{2.6}$$

It is difficult in practice to evaluate the posterior density of θ due to the complicated form of the posterior density of α, β. One convenient approximation to the posterior density is obtainable by an approximation to the marginal density (2.6). If we define the random variable $W_i = \gamma Y_i$, then it can be shown that $E(W_i|\alpha, \beta) = \text{Var}(W_i|\alpha, \beta) = \gamma\xi_i(\beta)$. This observation motivates the approximation of the marginal density of W_i by a Poisson density with mean $\gamma\xi_i(\beta)$. The density π_2 is then approximated by

$$\begin{aligned} \pi_2(\alpha, \beta|y) &\approx \pi_2^*(\alpha, \beta|y) \\ &= C_1\gamma^N \prod \exp[-\gamma\xi_i(\beta)][\gamma\xi_i(\beta)]^{y_i\gamma}/\Gamma(y_i\gamma + 1)\pi_3(\alpha, \beta). \end{aligned} \tag{2.7}$$

The approximation π_2^* can be further simplified by applying a Stirling's formula approximation to the log gamma functions in (2.7). If $\log \Gamma(y_i\gamma + 1)$ is replaced by the approximation $(y_i\gamma + 1/2)\log(y_i\gamma) - y_i\gamma + .5\log(2\pi)$, then, after some simplification, one obtains

$$\pi_2^*(\alpha, \beta|y) \approx C_1\gamma^{N/2} \exp\{-\gamma/2D(y; \beta)\}\pi_3(\alpha, \beta), \tag{2.8}$$

where $D(y;\beta) = -2\Sigma[y_i(\log \xi_i(\beta) - \log y_i) - (\xi_i(\beta) - y_i)]$ is the deviance between the observed data and the model $\log \theta_i = \xi_i(\beta), i = 1,\ldots,N$. The likelihood $\gamma^{N/2} \exp\{-\gamma/2\, D(y;\beta)\}$, called a quasi-likelihood in McCullagh and Nelder (1983), has been shown useful in modelling data with higher dispersion than predicted under the Poisson model. West (1985) considers Bayesian methods for estimating γ and β in (2.8). In this situation, however, the log gamma functions in (2.7) are easy to calculate and there is no need to approximate them by the crude Stirling's formula. Thus, in the remainder of this paper, the approximation (2.7) will be used.

To obtain the marginal posterior density of α, it is necessary to integrate out the p dimensional vector β. First write the log joint density as

$$\begin{aligned}\log \pi_2^*(\alpha,\beta|y) &= \gamma\ell(\beta,y) + N\log\gamma + \gamma\log\gamma\Sigma y_1\\ &\quad - \Sigma\log\Gamma(y_i\gamma+1) + \log C_1,\end{aligned} \tag{2.9}$$

where $\ell(\beta,y) = \Sigma[-\xi_i(\beta) + y_i \log \xi_i(\beta)]$. Note that $\ell(\beta,y)$ is the usual log likelihood for N independent Poisson observations with means $\{\xi_i(\beta)\}$. Let $\tilde\beta$ denote the value of β which maximizes $\ell(\beta,y)$. If we expand $\ell(\beta,y)$ in a Taylor's series about $\beta = \tilde\beta$, we obtain

$$\ell(\beta,y) \approx \ell(\tilde\beta,y) - 1/2(\beta-\tilde\beta)^T \boldsymbol{H}(\beta-\tilde\beta), \tag{2.10}$$

where H is the negative of the second derivative matrix

$$\boldsymbol{H} = -\left[\frac{\partial^2 \ell(\beta,y)}{\partial\beta_j\partial\beta_k}\right]$$

evaluated at $\beta = \tilde\beta$. For the log-linear model $\log \xi_i(\beta) = \boldsymbol{x}_i^T\beta$, it can be shown that $\boldsymbol{H} = \boldsymbol{X}^T\boldsymbol{W}\boldsymbol{X}$, where $\boldsymbol{X} = (\boldsymbol{x}_1,\ldots,\boldsymbol{x}_N)^T$ and $\boldsymbol{W}$ is the diagonal matrix with elements $(\xi_1(\beta),\ldots,\xi_N(\beta))$ evaluated at $\beta = \tilde\beta$. By substituting (2.10) into (2.9), we obtain

$$\begin{aligned}\log \pi_2^*(\alpha,\beta|y) &\approx \log \pi_2^A(\alpha,\beta|y)\\ &= \gamma\ell(\tilde\beta,y) - \gamma/2(\beta-\tilde\beta)^T\boldsymbol{H}(\beta-\tilde\beta)\\ &\quad + N\log\gamma + \gamma\log\gamma\Sigma y_i - \Sigma\log\Gamma(y_i\gamma+1) + \log C.\end{aligned} \tag{2.11}$$

Note that, since $\tilde\beta$ and the associated matrix H are found by maximizing the Poisson log-likelihood $\ell(\beta,y)$, they are easily obtainable using a generalized linear models computer program such as GLIM. In addition, since $\tilde\beta$ and H do not depend on α, they need only be computed once in the computation of the posterior density (2.11). This is in contrast to the method of Leonard and Novick (1986), which requires the computation of a maximum value $\tilde\beta_\alpha$ for each value of α.

To summarize, it is seen from (2.11) that the posterior density of β conditional on y and α is approximately $N(\tilde\beta,(\gamma\boldsymbol{H})^{-1})$. By integrating out β the marginal posterior density of α is given approximately by

$$\begin{aligned}\pi_2^A(\alpha|y) &= \exp\{\gamma\ell(\tilde\beta,y) + N\log\gamma + \gamma\log\gamma\Sigma y_i\\ &\quad - \Sigma\log\Gamma(y_i\gamma+1)\}\cdot|\gamma\boldsymbol{H}|^{-1/2}.\end{aligned} \tag{2.12}$$

Let $g_N(\cdot;\mu,B)$ denote the multivariate normal density with mean vector μ and variance-covariance matrix B. Then, by substitution of (2.11) into (2.1), one obtains the approximate posterior for $\boldsymbol{\theta}$:

$$\begin{aligned}\pi(\boldsymbol{\theta}|y) &\approx \pi^A(\boldsymbol{\theta}|y)\\ &= \int \pi_1(\boldsymbol{\theta}|y,\alpha,\beta) g_N(\beta;\tilde\beta,(\gamma\boldsymbol{H})^{-1})\pi_2^A(\alpha|y)d\beta d\alpha.\end{aligned} \tag{2.13}$$

3. POSTERIOR INFERENCE ABOUT THE HYPERPARAMETERS

The scale hyperparameter α in the hierarchical model (1.6) reflects the precision of the prior belief in the log-linear model $\log \xi_i(\beta) = x_i^T \beta$. The reparameterized hyperparameter $\gamma = \alpha/(1+\alpha)$, from (2.4), has the attractive interpretation as the size of the proportionate shrinkage of the observed counts y_i towards the prior means $\xi_i(\beta)$. Through inspection of the approximate marginal posterior density of the hyperparameter γ, one can assess the agreement of the data with the model.

Once a model has been accepted, it may be of interest to estimate or make inferences about the regression parameter β. The posterior density of β is approximately given by

$$\pi_2^A(\beta|y) = \int g_N(\beta|\tilde{\beta}, (\gamma \boldsymbol{H})^{-1} \pi_2^A(\alpha|y) d\alpha. \tag{3.1}$$

Posterior moments for β are simply obtained from (3.1). The mean and variance-covariance matrix for β are given by

$$E[\beta|y] = E^{\alpha|y}\{E[\beta|y,\alpha]\} = \tilde{\beta} \tag{3.2}$$

and

$$\begin{aligned} \text{Var}[\beta|y] &= E^{\alpha|y}\{\text{Var}[\beta|y,\alpha]\} + \text{Var}^{\alpha|y}\{E[\beta|y,\alpha]\} \\ &= \boldsymbol{H}^{-1} E[\gamma^{-1}|y]. \end{aligned} \tag{3.3}$$

(Superscripts in the expectations in (3.2) and (3.3) refer to random variables over which the expectation is taken.) Since $\tilde{\beta}$ and $\boldsymbol{H}^{-1}$ are given as output from a generalized linear models computer package, it is only necessary to compute the expectation $E[\gamma^{-1}|y]$ to find approximate posterior credible regions for β.

4. POSTERIOR INFERENCE ABOUT THE CELL MEANS

Suppose that it is of interest to estimate the individual cell means θ_i. By using the representation (2.1) and the approximate posterior density of (α, β) (2.11), one can compute posterior moments of individual θ_i. (See equation (3.10) in Leonard and Novick (1986) for an expression for the posterior mean of θ_i.) In this paper, we will focus on the computation of the posterior density of θ_i. Posterior moments are easily obtainable from the numerically computed density (along with other quantities of interest such as HPD credible sets) and the procedure described here generalizes nicely to the computation of the posterior density of a function of the cell means.

From (2.13), the approximate posterior density of θ_i is given by

$$\begin{aligned} \pi^A(\theta_i|y) = \int & \pi_G(\theta_i; y_i + \alpha\xi_i(\beta), 1+\alpha) \\ & \cdot g_N(\beta; \tilde{\beta}, (\gamma \boldsymbol{H})^{-1}) \pi_2^A(\alpha|y) d\beta d\alpha. \end{aligned} \tag{4.1}$$

First we obtain the posterior density of θ_i given α. To integrate out β, we use a Taylor's series expansion method suggested by Tierney and Kadane (1986). Expand the logarithm of the integrand of (4.1)

$$\ell_\alpha(\beta, y) = \log[\pi_G(\theta_i; y_i + \alpha\xi(\beta), 1+\alpha) g_N(\beta; \tilde{\beta}, (\gamma \boldsymbol{H})^{-1})]$$

in a second order Taylor's series about the maximum value $\tilde{\beta}_\alpha$. Then we obtain the approximation

$$\begin{aligned} \pi^A(\theta_i|y,\alpha) = & \pi_G(\theta_i; y_i + \alpha\xi_i(\tilde{\beta}_\alpha), 1+\alpha) \\ & \cdot g_N(\tilde{\beta}_\alpha|\tilde{\beta}, (\gamma \boldsymbol{H})^{-1}) |\boldsymbol{H}_\alpha|^{-1/2}, \end{aligned} \tag{4.2}$$

where $\boldsymbol{H}_\alpha$ is the negative of the Hessian matrix

$$\boldsymbol{H}_\alpha = -\left[\frac{\partial^2 \ell_\alpha(\beta, y)}{\partial\beta_l \partial\beta_k}\right]$$

evaluated at $\beta = \tilde{\beta}_\alpha$. The approximate posterior density of θ_i is then given by

$$\pi^A(\theta_i|y) \approx \int \pi^A(\theta_i|y,\alpha)\pi_2^A(\alpha|y)d\alpha. \tag{4.3}$$

The above procedure can be generalized to obtain the posterior density of an arbitrary linear combination of the logarithms of the cell means, say $\eta = \Sigma a_i \log\theta_i$. First consider the posterior density of η conditional on the hyperparameters α and β. The conditional density of θ_i is gamma $(y_i + \alpha\xi_i(\beta), 1+\alpha)$ and if the density of log θ_i is expanded in a second order Taylor's series about $\log[(y_i + \alpha\xi_i(\beta))/(1+\alpha)]$ (the log of the mean of θ_i), then one obtains the approximation

$$\log\theta_i|y,\alpha,\beta \sim N(\log[y_i + \alpha\xi_i(\beta))/(1+\alpha)], (y_i + \alpha\xi_i(\beta))^{-1}). \tag{4.4}$$

Numerical work suggests that the approximation (4.4) is adequate for large values of the observed counts say $y_i > 5$, for all i. Then by the posterior independence of $\theta_1, \ldots, \theta_N$,

$$\eta|y,\alpha,\beta \sim N(\mu(\alpha,\beta), \nu(\alpha,\beta)), \tag{4.5}$$

where

$$\mu(\alpha,\beta) = \Sigma a_i \log[y_i + \alpha\xi_i(\beta))/(1+\alpha)],$$
$$\nu(\alpha,\beta) = \Sigma a_i^2 (y_i + \alpha\xi_i(\beta))^{-1}.$$

Using the approximation (4.5), the posterior density of η is given by

$$\pi^A(\eta|y) = \int g_N(\eta; \mu(\alpha,\beta), \nu(\alpha,\beta)), \cdot g_N(\beta; \tilde{\beta}, (\gamma\boldsymbol{H})^{-1})\pi_2^A(\alpha|y)d\beta d\alpha. \tag{4.6}$$

By analogous arguments used in the derivation of (4.3),

$$\pi^A(\eta|y) \approx \int \pi^A(\eta|y,\alpha)\pi_2^A(\alpha,y)d\alpha, \tag{4.7}$$

where

$$\pi^A(\eta|y,\alpha) = g_N(\eta; \mu(\alpha,\tilde{\beta}_\alpha), \nu(\alpha,\tilde{\beta}_\alpha)) \cdot g_N(\tilde{\beta}_\alpha; \tilde{\beta}, (\gamma\boldsymbol{H})^{-1})|\boldsymbol{H}_\alpha|^{-1/2}, \tag{4.8}$$

$\tilde{\beta}_\alpha$ maximizes (for fixed α and η) the integrand of (4.6) and $\boldsymbol{H}_\alpha$ is the negative of the Hessian matrix of the log integrand of (4.6) evaluated at $\beta = \tilde{\beta}_\alpha$.

Several comments should be made about the computation of the posterior density (4.7). In practice, the density is evaluated at twenty values of η and the integration in (4.7) is performed by a summation over ten values of α. Thus it is required to compute the maximum value $\tilde{\beta}_\alpha$ and the matrix $\boldsymbol{H}_\alpha$ $20 \times 10 = 200$ times. As in Leonard and Novick (1986), the Newton-Raphson algorithm is used here to find $\tilde{\beta}_\alpha$ and, in our experience, this algorithm can be slow to converge. Thus it is desirable from the viewpoint of computational speed that the integrand of (4.6) and successive derivatives be easy to compute. The use of the approximate density $g_N(\beta; \tilde{\beta}, (\gamma\boldsymbol{H})^{-1})$, in (4.6) is far preferable to the use of the exact density (2.5) in this calculation. From our experience, the above computational method can be done quickly on a microcomputer for hierarchical models (1.6) when the number of regression parameters $p \leq 10$.

5. EXAMPLES

5.1. Bayesian Estimates Under Saturated and Unsaturated Models

In the examples which follow, it will be of interest to estimate a linear combination of the log cell means of the form $\eta = \Sigma a_i \log \theta_i$ where $\Sigma a_i = 0$. The hierarchical Bayesian estimate of η compromises between an estimate with no prior information and an estimate that assumes that the cell means satisfy the model (1.1).

First consider the situation where no prior information exists about the location of θ. In this situation (called the saturated model using the terminology of Bishop, Fienberg and Holland (1975)), the MLE $\hat{\eta} = \Sigma a_i \log y_i$, for large counts y, is approximately normally distributed with mean η and estimated variance $\Sigma a_i^2 / y_i$ (Haberman (1978)). If a flat prior is assumed for η, then approximately,

$$\eta | y \sim N(\hat{\eta}, \Sigma a_i^2 / y_i). \tag{5.1}$$

At the other extreme, suppose that $\boldsymbol{\theta}$ satisfies the "unsaturated" model (1.1), where $p < N$. Using a well-known asymptotic result (McCullagh and Nelder (1983)), the MLE of the regression vector $\tilde{\beta}$ is approximately normal with mean β and variance-covariance matrix $(\boldsymbol{X}^T \boldsymbol{V} \boldsymbol{X})^{-1}$, where $\boldsymbol{V}$ is the diagonal matrix with elements $\theta_1(\beta), \ldots, \theta_N(\beta)$ evaluated at $\beta = \tilde{\beta}$. In the examples considered here, the parameter of interest is a linear combination of the regression parameters. That is, $\eta = \boldsymbol{a}^T \beta$ where $\boldsymbol{a}^T = (a_1, \ldots, a_p)$. It then follows that approximately,

$$\tilde{\eta} = \boldsymbol{a}^T \tilde{\beta} \sim N(\eta, \boldsymbol{a}^T (\boldsymbol{X}^T \boldsymbol{V} \boldsymbol{X})^{-1} \boldsymbol{a}). \tag{5.2}$$

If (5.2) is combined with a flat prior for η, then under the model

$$\eta | y \sim N(\tilde{\eta}, \boldsymbol{a}^T (\boldsymbol{X}^T \boldsymbol{V} \boldsymbol{X})^{-1} \boldsymbol{a}). \tag{5.3}$$

5.2. Haberman's Social Class Data

First consider the analysis of the data of Table 1 when the means are believed a priori to satisfy the model (1.3). To evaluate the goodness of this model, Figure 1 gives the posterior density of the hyperparameter $\gamma = \alpha/(1+\alpha)$, where α has the posterior density (2.12). Note that this density is flat over the interval $(0.5, 1.0)$ and only gives small probabilities for small values of γ. Thus this graph indicates only modest support for the model. This conclusion is in constrast with the conclusion of the classical test which would compute $G^2 = 1.45$ with 1 d.f. and then accept the model.

Figure 2 plots the hierarchical posterior of $\eta_1 = \log(\theta_1/\theta_4)$ as a solid line. For comparison purposes, the mean and standard deviation are computed from this numerically computed density and presented in Figure 2. In addition, the figure plots the saturated model posterior of η_1 (5.1) (gray line) and the unsaturated model posterior of η_1 (5.3) (dashed line). The computation of (5.3) follows from the fact that log $(\theta_1/\theta_4) = -6\beta_1$ under the model (1.3). Using GLIM, we find that under (1.3), $\tilde{\beta} = -0.06409$ with a standard error of 0.020656. Using (5.3), this implies that η is approximately normal with mean $\tilde{\eta}_1 = -6\tilde{\beta}_1 = .3845$ and standard deviation $\sqrt{\{\boldsymbol{a}^T (\boldsymbol{X}^T \boldsymbol{V} \boldsymbol{X})^{-1} \boldsymbol{a})\}} = 6(0.020656) = 0.1239$.

Observe from Figure 2 that the hierarchical posterior is a compromise between the saturated and unsaturated model posterior both in terms of location and spread. The means and standard deviations of all three densities are given in Figure 2. Note that the hierarchical posterior standard deviation is $0.1649/0.1239 \times 100 = 33$ per cent longer that the unsaturated model posterior. This is a reflection of Figure 1 which indicated only modest support for the model.

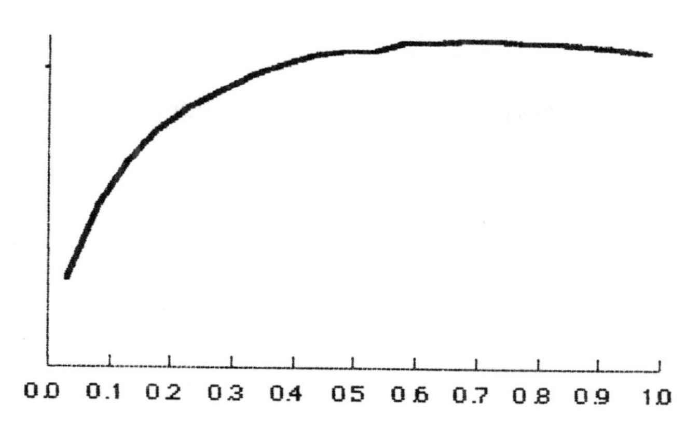

Figure 1. *Posterior density of γ for Haberman's data.*

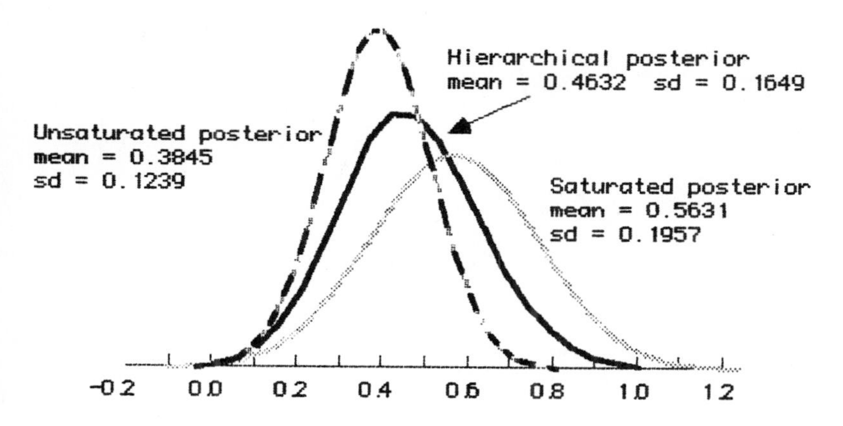

Figure 2. *Posterior densities for* $\eta_1 = \log[\theta_1/\theta_4]$.

Next, suppose the parameter of interest is

$$\eta_2 = \log(\theta_1/\theta_4) - 3\log(\theta_2/\theta_3).$$

Under model (1.3), it is easily shown that $\eta = 0$. Thus the unsaturated model posterior is degenerate at zero. Figure 3 plots the hierarchical posterior together with the saturated model posterior (5.1). Note that the hierarchical posterior is skewed with a mode close to zero. This posterior behavior is similar to the behavior of posterior densities in Albert (1987), where the usual saturated model posterior is shrunk towards a point.

Figure 4 investigates the sensitivity of the hierarchical posterior densities of η_1 and η_2 with respect to the noninformative prior chosen for the hyperparameters α and β. In the hierarchical model (1.6), the proper prior $(1+\alpha)^{-2}$ was given for α. This density implies that the shrinkage hyperparameter γ has a uniform prior density. Suppose instead that the improper prior α^{-1} (recommended by Jeffreys) is chosen for α. Figure 4 plots the posterior

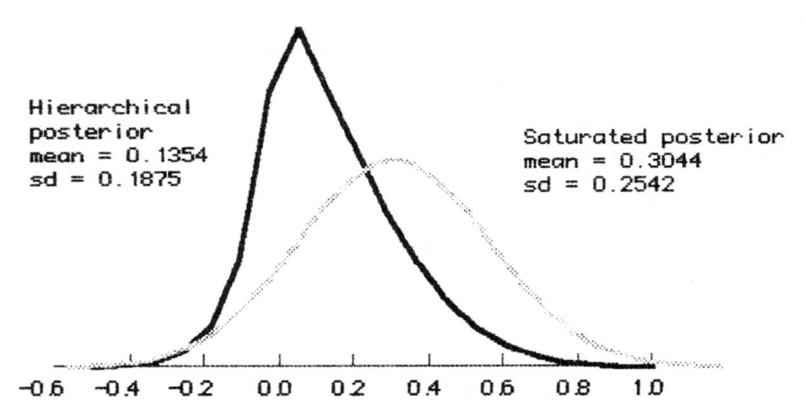

Figure 3. *Posterior densities for* $\eta_2 = \log[\theta_1/\theta_4] - 3\log[\theta_2/\theta_3]$.

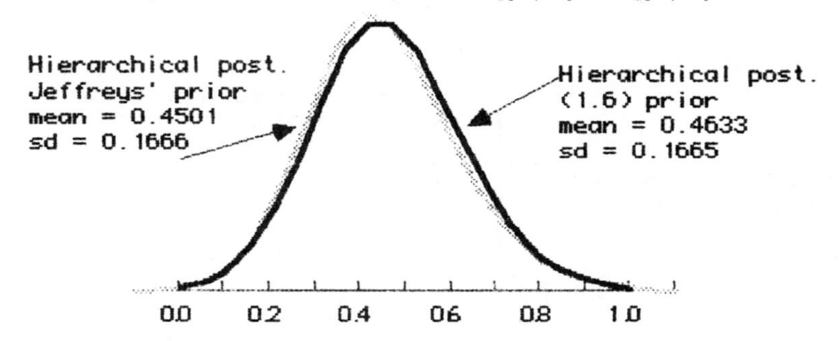

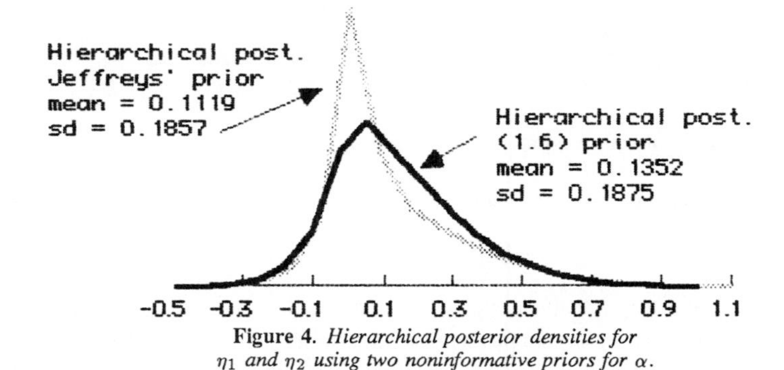

Figure 4. *Hierarchical posterior densities for* η_1 *and* η_2 *using two noninformative priors for* α.

densities for η_1 and η_2 using the two noninformative densities. Observe that the two densities for η_1 are very similar; however, the Jeffreys' hierarchical posterior density for η_2 is more peaked about the point $\eta = 0$. (From experience, it appears that there will be differences in shape when the hierarchical posterior shrinks towards a point.) Although there are significant differences in shape, note that the posterior modes and standard deviations are very similar for the two densities for η_2. Based on observations of many other examples, we would expect the usual posterior summary measures, such as measures of center and HPD credible intervals to be typically robust to the choice of the noninformative prior for α.

5.3. Goodman's Mental Health Data

Suppose we with to draw inferences about Goodman's 4×6 contingency table with the prior belief that the uniform association model (1.4) is a good fit. Figure 5 plots the posterior density of the hyperparameter γ. Unlike the first example, this graph is concentrated about large values of γ, indicating that the model is a good fit. This conclusion is consistent with the classical conclusion after calculating $G^2 = 9.89$ with 14 d.f.

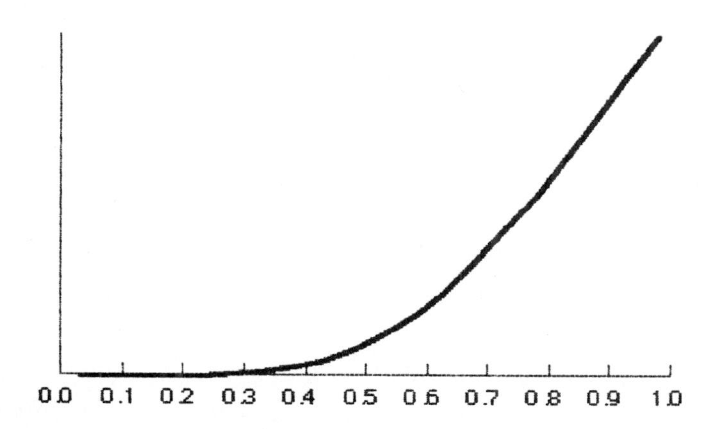

Figure 5. *Posterior density of γ for Goodman's data.*

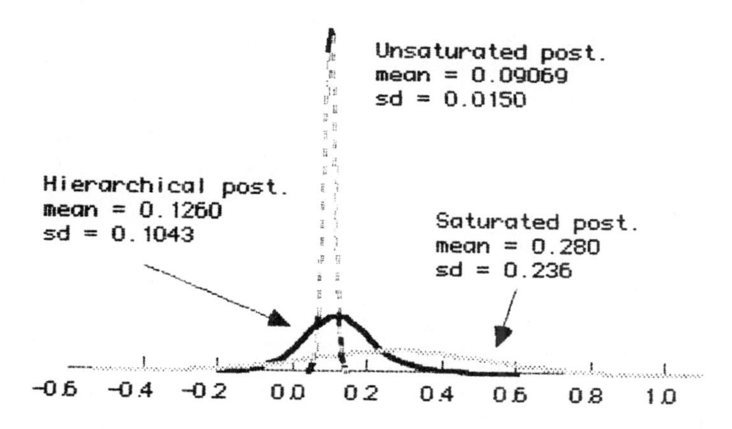

Figure 6. *Posterior densities for* $\eta_1 = \log[(\theta_{33}/\theta_{44})/(\theta_{34}\theta_{43})]$.

Although the model describes the observed data very well, it doesn't necessary imply that we should assume that the cell means satisfy the model (1.4). Figures 6 and 7 plot posterior

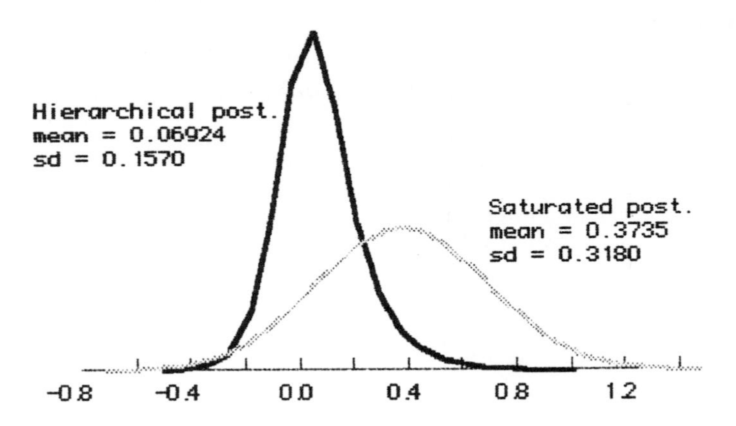

Figure 7. *Posterior densities for* $\eta_2 = \log[(\theta_{33}/\theta_{44})/(\theta_{34}\theta_{43})] - \log[(\theta_{35}/\theta_{46})/(\theta_{45}\theta_{36})]$.

densities for two parameters of interest:

$$\eta_1 = \log[(\theta_{33}\theta_{44})/(\theta_{34}\theta_{43})]$$

and

$$\eta_2 = \log[(\theta_{33}\theta_{44})/(\theta_{34}\theta_{43})] - \log[(\theta_{35}\theta_{46})/(\theta_{45}\theta_{36})].$$

The parameter η_1 is the log odds-ratio of the 2×2 subtable corresponding to the third and fourth rows and columns in the table. The parameter η_2 is the difference between η_1 and the log odds-ratio of the 2×2 subtable immediately to the right of first subtable.

As in Figures 3 and 4, the saturated model posteriors are represented by gray lines and the unsaturated model posteriors by dashed lines. Note from Figure 6 that the hierarchical posterior is much wider than that of the unsaturated posterior. This suggests that the standard error of η_1 is badly underestimated by assuming the model. The hierarchical posterior is significantly narrower that the saturated posterior, indicating the gain in posterior precision by the use of the hierarchical model.

As in Haberman's example, it is of interest to estimate the difference in log odds-ratios η_2 since this parameter is equal to zero under the constant association model. Note from Figure 7 that the standard deviation of η_2 from the hierarchical posterior is equal to 0.1570. Thus much uncertainty remains about the location of differences of the individual log odds-ratios even though the data are well explained by the constant association model.

REFERENCES

Albert, J. H. (1985). Simultaneous estimation of Poisson means under exchangeable and independence models. *J. Statist. Comput. and Simul.* **23**, 1–14.

Albert, J. H. (1987). Bayesian estimation of odds ratios under prior hypotheses of independence and exchangeability. *J. Statist. Comput. and Simul.* **27**, 251–268.

Bishop, Y., Fienberg, S. and Holland, P. (1975). *Discrete Multivariate Analysis: Theory and Practice.* Cambridge: MIT Press.

Goodman, L. (1979). Simple models for the analysis of association in cross-classifications having ordered categories.*J. Amer. Statist. Assoc.* **74**, 537–552.

Haberman, S. (1978). Analysis of Qualitative Data **1**. New York: Academic Press.

Leonard, T. and Novick, M. R. (1986). Bayesian full marginalization for two-way contingence tables. *J. Educat. Statist.* **11**, 33–56.

McCullagh, P. and Nelder, J. (1983). *Generalized Linear Models*. London: Chapman and Hall.

Tierney, L. and Kadane, J. (1986). Accurate approximations for posterior moments and marginals. *J. Amer. Statist. Assoc.* **81**, 82–86.

West, M. (1985). Generalized linear models: scale parameters, outlier accommodation and prior distributions. *Bayesian Statistics* **2**, (J. M. Bernardo, *et al.*, eds.), 531–557. Amsterdam: North Holland.

BAYESIAN STATISTICS 3, pp. 533–541
J. M. Bernardo, M. H. DeGroot, D. V. Lindley and A. F. M. Smith, (Eds.)

Bayesian Analysis of a Pure Birth Process with Linear Birth Rate

M. A. AMARAL TURKMAN and K. F. TURKMAN
Ceaul, University of Lisbon

SUMMARY

Let $\{N(t), t \geq 0\}$ be a pure birth process with linear birth rate $\lambda_n = n\lambda, n = 0, 1, \ldots,$ where $N(t)$ is the size of a population at time t.

Admitting that the process is observed either continuously or at fixed equidistant points during a certain period of time $[0, t]$, the predictive distribution of the size of the population at $t + u, u > 0$, is obtained.

These results are then applied to the study of a simple epidemic model and are compared with those obtained using an estimative method.

Keywords: PURE BIRTH PROCESSES; PREDICTIVE DISTRIBUTIONS; POPULATION SIZE.

1. INTRODUCTION

Let $\{N(t), t \geq 0\}$ be a pure birth process, that is, the Markov process in which

$$P(N(t + \delta t) = n + 1 | N(t) = n) = \lambda_n \delta t + o(\delta t)$$

and

$$P(N(t + \delta t) = n | N(t) = n) = 1 - \lambda_n \delta t + o(\delta t). \tag{1.1}$$

In this paper we will consider two special cases of pure birth processes.

As a first example, we consider a population in which each member acts independently giving birth at an exponential rate λ and no one dies. Then if $N(t)$ represents the population size at time $t, \{N(t), t \geq 0\}$ is a pure birth process with linear birth rate $\lambda_n = n\lambda, n \geq 0$. This process was first studied by Yule in connection with the mathematical theory of evolution and a similar one was used by Furry in connection with the cosmic ray phenomena.

As a second example, we consider the following simple epidemic model studied by Bailey (1963). Suppose that we have a population with m individuals such that at time 0 it consists of one infected and $m-1$ susceptibles. Once infected, an individual remains in that state forever and in any time interval δt, any given infected person will cause with probability $\lambda\delta t + o(\delta t)$, a susceptible to be infected. If we denote by $N(t)$ the number of infected individuals in the population at time t, then $\{N(t), t \geq 0\}$, is a pure birth process with

$$\lambda_n = \begin{cases} (m-n)n\lambda & n = 1, \ldots, m-1 \\ 0 & \text{otherwise.} \end{cases} \tag{1.2}$$

In such situations, one may be interested in predicting the number of individuals in the population at an instant $t + u$ given the observation of the process in a finite time interval $[0, t]$. In a Bayesian context, the key to prediction problems is the predictive distribution (Geisser (1971), Aitchison and Dunsmore (1975)) of the outcome of the future experiment of interest.

In this paper, for the two processes considered, we shall study the predictive distribution of the size of the population at some instant $t + u$ given the observation of the process in a finite time interval $[0, t]$. Two sampling schemes will be considered:

A. Continuous observations in $[0, t]$

B. Equidistant discrete observations at time points $0, \tau, \ldots, k\tau = t$.

2. PREDICTION FOR THE YULE PROCESS

Let $\{N(t), t \geq 0\}$ be a pure birth process with linear birth rate $\lambda_n = n\lambda, \lambda > 0$. Let

$$P_{ij}(u) = P(N(t+u) = j|N(t) = i) \tag{2.1}$$

represent the transition probability that the process presently in state i will be in state $j, j > i$ at time u later. Suppose that births occur at times $t_1 < t_2 \cdots$ and that $N(0) = 1$ with probability one. It is well known that the intervals between births are independently distributed as negative exponential with parameters $\lambda_i = i\lambda, i = 1, 2, \ldots$ and that the transition probabilities are given by

$$P_{ij}(u) = \begin{cases} \binom{j-1}{j-i} e^{-i\lambda u}(1 - e^{-\lambda u})^{j-i} & j \geq i \\ 0 & \text{otherwise} \end{cases} \tag{2.2}$$

where $(i, j) \in S = \{1, 2, \ldots\}$.

2.1. Continuous Observation Scheme

Let t be a fixed (total) time of observation. Suppose that we observe the process $\{N(t), t \geq 0\}$ (with $N(0) = 1$) continuously over $[0, t]$ and suppose further that $n - 1$ births occur during $[0, t]$ at time points $t_1, t_2, \ldots, t_{n-1} \leq t$. Note that $t_n > t$, where t_n is the instant of the occurence of the nth birth. Then the likelihood function is given by (e.g. Basawa and Rao (1980))

$$\begin{aligned} L(\lambda|n, t_1, \ldots, t_{n-1}) &= \prod_{k=1}^{n-1} k\lambda \exp(-k\lambda T_k) \exp(-n\lambda(t - t_{n-1})) \\ &= (n-1)!\lambda^{n-1} \exp\left(-\lambda\left(\sum_{k=1}^{n-1} kT_k + n(t - t_{n-1})\right)\right). \end{aligned} \tag{2.3}$$

Thus the outcome of the informative experiment $\mathcal{D} = \{n, t_1, \ldots, t_{n-1}\}$ can be summarized by (n, s), the realization of the minimial sufficient satistic $(N(t), S_t)$ where $T_K = t_K - t_{K-1}$ and

$$S_t = \sum_{k=1}^{N(t)-1} kT_k + N(t)(t - t_{N(t)-1}). \tag{2.4}$$

Note that S_t can be written as

$$S_t = \int_O^t N(x)dx = tN(t) - \sum_{i=1}^{N(t)-1} t_i. \tag{2.5}$$

Suppose now that the future experiment of interest is the number of individuals in the process at some instant after t, say $t + u$. Hence, for a given prior probability density function $p(\lambda)$, the predictive probability function for $N(t + u)$ given the outcome (n, s) of the informative experiment is by definition (e. g. Geisser (1985))

$$P(N(t+u) = y|(n, s)) = \int P(N(t+u) = y|N(t) = n, S_t = s, \lambda)p(\lambda|n, s)d\lambda. \tag{2.6}$$

But

$$P(N(t+u)=y|N(t)=n,S_t=s,\lambda)=P(N(t+u)=y|N(t)=n,\lambda) \\ = P_{ny}(u), \tag{2.7}$$

by the Markov property and the fact that S_t is measurable with respect to the σ-algebra spanned by $\{N(x)|x<t\}$, which follows from the representation S_t given in (2.5).

Assume that $p(\lambda)$ is $G(g,h)$, that is

$$p(\lambda)=\frac{h^g\lambda^{g-1}e^{-\lambda h}}{\Gamma(g)}, h>0, g>0.$$

Then the posterior density function $p(\lambda|n,s)$ is $Ga(n+g-1,h+s)$.

Substituting $p(\lambda|(n,s))$ and $p_{ny}(u)$ given by (2.2) in (2.6), we get

$$\begin{aligned} P(N(t+u)=y|(n,s)) &= \frac{(h+s)^{n+g-1}}{\Gamma(n+g-1)}\binom{y-1}{y-n}\int_0^\infty \lambda^{n+g-2}e^{-\lambda(h+s+nu)} \\ &\quad \times (1-e^{-\lambda u})^{y-n}\,d\lambda \\ &= \binom{y-1}{y-n}\sum_{k=0}^{y-n}\binom{y-n}{k}(-1)^k\frac{1}{\left(1+u\frac{n+k}{h+s}\right)^{n+g-1}}, \end{aligned} \tag{2.8}$$

for $g>0, h>0, u>0, n+g>1, y\geq n$ and

$$P(N(t+u)=y|(n,s))=0, \text{for } y<n.$$

Notice that a similar equation appears in Geisser (1982).

The mean and the variance of this predictive distribution are easily found to be

$$E(N(t+u)|(n,s))=n\left(\frac{h+s}{h+s-u}\right)^{n+g-1}$$

for $u<s+h$, and

$$V(N(t+u)|(n+s))=n(h+s)^{n+g-1}\left[\frac{1+n}{(h+s-2u)^{n+g-1}}-\frac{1+n\left(\frac{h+s}{h+s-u}\right)^{n+g-1}}{(h+s-u)^{n+g-1}}\right], \tag{2.9}$$

for $u<\frac{s+h}{2}$. These quantities are useful if one is interested in a point predictor for the future number of individuals in $t+u$. Indeed, for a quadratic utility structure, the optimal prediction, in the sense of maximizing expected utility, is the mean value of the predictive distribution.

2.2. *Equidistant Discrete Sampling*

Let t be a fixed total time of observation. Divide t into k equal parts and let $\tau=\frac{t}{k}$. Suppose we observe $\{N(t), t\geq 0\}$ with $N(0)=1$ at the points $0,\tau,2\tau,\ldots,k\tau=t$ and let $n_{j\tau}$ be the number of individuals in the population at instant $j\tau$, that is, $N(j\tau)=n_{j\tau}(j=0,1,2,\ldots,k)$. The likelihood is then given as

$$L(\lambda|n_{1\tau},n_{2\tau},\ldots,n_{k\tau})=\prod_{j=1}^{k}\binom{n_{j\tau}-1}{n_{j\tau}-n_{(j-1)\tau}}e^{-\lambda\tau\sum_{j=1}^{k}n_{(j-1)\tau}}(1-e^{-\lambda\tau})^{n_{k\tau}-n_0}, \tag{2.10}$$

where $n_{k\tau} = n_t$ and $n_0 = 1$.

The informative experiment can now be summarized by $(n_t, \sum_{j=1}^{k} n_{(j-1)\tau})$. Using the same prior distribution as before, we obtain for the posterior distribution

$$p(\lambda|(n,z)) \propto \lambda^{g-1} e^{-\lambda(\tau z+h)}(1-e^{-\lambda\tau})^{n-1}, \tag{2.11}$$

where $n = n_t$ and $z = \sum_{j-1}^{k} n_{(j-1)\tau}$.

Now, due to the Markov property, we have

$$P(N(t+u)=y|N(t)=n, \sum_{j=1}^{k} N((j-1)\tau)=z, \lambda) = P(N(t+u)=y|N(t)=n,\lambda)$$

and so the predictive distribution of $N(t+u)$ given the outcome of the informative experiment is:

If $u \neq \tau$,

$$\begin{aligned} P(N(t+u)=y|(n,z)) &= C\int_0^\infty \binom{y-1}{y-n} e^{-\lambda nu}(1-e^{-\lambda u})^{y-n} e^{-\lambda(\tau z+h)} \\ &\quad \times (1-e^{-\tau\lambda})^{n-1}\lambda^{g-1}d\lambda \\ &= C\sum_{i=0}^{y-n}(-1)^i\binom{y-n}{i}\sum_{j=0}^{n-1}\binom{n-1}{j}(-1)^j\frac{\Gamma(g)}{(A+iu+j\tau)^g}, \end{aligned} \tag{2.12}$$

where $A = nu + \tau z + h, n > 1, y \geq n, g > 0$ and

$$C^{-1} = \sum_{j=0}^{n-1}\binom{n-1}{j}(-1)^j\Gamma(g)((z+j)\tau+h)^{-g};$$

if $u = \tau$,

$$\begin{aligned} P(N(t+u)=y|(n,z)) &= C\int_0^\infty \binom{y-1}{y-n}(1-e^{-\lambda u})^{y-1}\lambda^{g-1}e^{-\lambda(\tau z+h)}d\lambda \\ &= C\binom{y-1}{y-n}\sum_{j=0}^{y-1}\binom{y-1}{j}(-1)^j\frac{\Gamma(g)}{((n+z+j)\tau+h)^g}. \end{aligned} \tag{2.13}$$

It can also be shown that

$$E(N(t+u)|(n,z)) = Cn\sum_{j=0}^{n-1}\binom{n-1}{j}(-1)^j\Gamma(g)(\tau(j+z)+h-u)^{-g},$$

for $\tau z + h > u$.

3. PREDICTION FOR THE SIMPLE EPIDEMIC MODEL

Consider now the simple epidemic model described in the introduction. The transition probabilities can be obtained by solving recursively the set of equations (Hill and Severo (1969))

$$\begin{cases} P_{ii}(u) &= e^{-\lambda_i u} \\ P_{ij}(u) &= \lambda_{j-1} e^{-\lambda_j u} \int_0^u e^{\lambda_j s} P_{i,j-1}(s) ds, \quad j = i+1, \ldots, m, \end{cases} \tag{3.1}$$

where λ_n is defined in (1.2), to give

$$P_{ij}(u) = c_{ij} \sum_{k=i}^{j} d_{jk} e^{-\lambda k(m-k)u}, \qquad j = 1, \ldots, m, \tag{3.2}$$

where $c_{ij} = d_{jk} = 1$ for $i = j$ and $c_{ij} = \prod_{r=i}^{j-1}(m-r)r$ for $j > i$.

d_{jk} has a simple expression when (i,j) belongs to the noncritical region i.e., when $i \geq \frac{m+\delta(m)}{2}$ or $j \leq \frac{m+\delta(m)}{2}$, where $\delta(m)$ is 0 when m is even and 1 when m is odd. In this case we have

$$d_{jk} = \frac{1}{b(j,k)},$$

where

$$b(j,k) = \prod_{\substack{l=i \\ l \neq k, m-k}}^{j} (m-(l+k))(l-k).$$

If (i,j) is in the critical region, difficulties arise since some of the λ_k are equal. (Indeed $\lambda_k = \lambda_{m-k}$.) In that case we have

$$d_{jk} = \begin{cases} \frac{1}{b(j,k)} & \text{if } i \leq k \leq m-j-1 \text{ and } k = \frac{m}{2} \text{ if } m \text{ even,} \\ \frac{\lambda u}{b(j,k)} & \text{if } \frac{m-\delta(m)}{2} + 1 \leq k \leq j, \\ c(j,k) & \text{if } m-j \leq k \leq \frac{m+\delta(m)}{2} - 1, \text{ where} \end{cases}$$

$$c(j,k) = \frac{1}{(m-(k+j))(k-j)} \left[\frac{1}{b(j,k)} - c(j-1,k) \right] \qquad m-j < k \leq \frac{m+\delta(m)}{2} - 1,$$

and

$$c(j, m-j) = \sum_{k=i}^{m-j-1} \frac{1}{b(j,k)} + (1-\delta(m)) \frac{1}{b(j,\frac{m}{2})} - \sum_{k=m-j+1}^{\frac{m+\delta(m)}{2}-1} c(j,k).$$

Continuous Observation Scheme

Suppose that we observe the process during a fixed period of time $[0,t]$ and that initially there was one infected, that is $N(0) = 1$. Suppose further that $n - 1 (n < m)$ infections occur at instants $t_1, \ldots, t_{n-1}$. Then $t_i - t_{i-1}, i = 1, \ldots, n)$, (where $t_i - t_{i-1}$ is the time for the infective population to grow from i to $i+1$) are exponential with respective rates $\lambda_1 = (m-i)i\lambda$. Since $N(t) = n, t_n > t$ and hence the likelihood function is

$$L(\lambda | t_1, \ldots, t_{n-1}, n) = (\prod_{k=1}^{n-1} (m-k)k) \lambda^{n-1} e^{-\lambda \left[\sum_{k=1}^{n-1} (m-k)k(t_k - t_{k-1}) + (m-n)n(t-t_{n-1})\right]}. \tag{3.3}$$

The informative experiment is now summarized by (n, s) where

$$s = \sum_{k=1}^{n-1} (m-k)k(t_k - t_{k-1}) + (m-n)n(t - t_{n-1}).$$

Suppose now that one is interested in predicting the number of infected individuals at some instant $t+u$, assuming a gamma prior $Ga(g,h)$ for λ. Then using similar arguments to those in Section 2, the predictive distribution of $N(t+u)$ given (n,s) can be shown to be

$$P(N(t+u) = y|(n,s)) = \begin{cases} c_{ny} \sum_{k=n}^{y} d^*_{yk} \left[\frac{1}{1+u\frac{k(m-k)}{h+s}}\right]^{n+g-1} & n \le y \le m \\ 0 & \text{otherwise} \end{cases} \tag{3.4}$$

for $n+g>1, u>0, h>0$, where $d^*_{yk} = d_{yk}$ if (n,y) belongs to the noncritical region, otherwise

$$d^*_{yk} = (1-\delta_1(k))d_{yk} + \delta_1(k)\frac{(n+g-1)u}{b(y,k)[(m-k)ku + (h+s)]},$$

where $\delta_1(k) = 0$ if $k = n, \ldots, \frac{m-\delta(m)}{2}$ and equal to 1 otherwise.

An explicit expression for $E(N(t+u)|(n,s))$ can be obtained using the fact that

$$E(N(t+u)|(n,s)) = \int E(N(t+u)|n(t) = n, \lambda)p(\lambda|n,s)d\lambda.$$

However, this would be cumbersome due to the explicit form of $E(N(t+u)|N(t) = n, \lambda)$.

Another problem of interest for this epidemic model is the prediction of the interval of time T starting from t till the total population is infected. If at instant t, there are n infected individuals and $m-n$ susceptibles, then T is given by

$$T = (t_n - t) + \sum_{j=n+1}^{m-1} (t_j - t_{j-1}),$$

where t_j represents, as before, the time the jth infection has ocurred. (At this time, there are $j+1$ infectives.) Then T is the sum of $m-n$ independent exponential random variables with parameters $\lambda_j = \lambda_j(m-j)$ for $j = n, \ldots, m-1$.

The predictive distribution function of T given the outcome (n,s) of the informative experiment is given as

$$P(T \le u|(n,s)) = \int_0^\infty P(T \le u|N(t) = n, S_t = s, \lambda)p(\lambda, n, s)d\lambda.$$

But

$$\begin{aligned} P(T|lequ/N(t) = n, S_t = s, \lambda) &= P(N(t+u) \ge m|N(t) = n, \lambda) \\ &= P(N(t+u) = m|N(t) = n, \lambda) \\ &= P_{nm}(u). \end{aligned} \tag{3.5}$$

Hence

$$P(T \le u|(n,s)) = c_{nm} \sum_{k=n}^{m} d^*_{mk} \frac{1}{\left(1 + u\frac{k(m-k)}{h+s}\right)^{n+g-1}}, \tag{3.6}$$

for $0 < u < \infty$.

Also, since

$$E(T|N(t)=n,\lambda)=\frac{1}{\lambda}\sum_{i=n}^{m-1}\frac{1}{i(m-i)} \tag{3.7}$$

and

$$V(T|N(t)=n,\lambda)=\frac{1}{\lambda^2}\sum_{i=n}^{m-1}\left(\frac{1}{i(m-i)}\right)^2, \tag{3.8}$$

it follows that the mean value and the variance of the predictive distribution of T are given as

$$E(T|(n,s))=\left(\frac{h+s}{n+g-2}\right)\sum_{i=n}^{m-1}\frac{1}{i(m-i)}, \tag{3.9}$$

for $n+g>2$ and

$$V(T|(n,s))=\frac{(h+s)^2}{(n+g-2)(n+g-3)} \left[\sum_{i=n}^{m-1}\left(\frac{1}{i(m-i)}\right)^2+\frac{1}{n+g-2}\left(\sum_{i=n}^{m-1}\frac{1}{i(m-i)}\right)^2\right]. \tag{3.10}$$

Since, for large m compared to n we have (Ross (1983))

$$\begin{aligned}\sum_{i=n}^{m-1}\frac{1}{i(m-i)}&\simeq\frac{1}{m}\int_n^{m-1}\left(\frac{1}{t}+\frac{1}{m-t}\right)dt\\ &=\frac{1}{m}\log\frac{(m-1)(m-n)}{n},\end{aligned}$$

an approximation for $E(T|(n,s))$ can be given as

$$E(T|(n,s))\simeq\left(\frac{h+s}{n+g-2}\right)\frac{1}{m}\log\frac{(m-1)(m-n)}{n}. \tag{3.11}$$

Equidistant Sampling

When the epidemic model under consideration is observed at equidistant points $0,\tau,\ldots,k\tau=t$ during the fixed interval $(0,t)$ the likelihood is

$$\begin{aligned}L(\lambda|n_\tau,n_{2\tau},\ldots,n_{k\tau})&=\prod_{j=1}^{k}P(N(j\tau)=n_{j\tau}|N((j-1)\tau)=n_{(j-1)\tau},\lambda)\\ &=\prod_{j=1}^{k}\left[c_{n_{j\tau},n_{(j-1)\tau}}\sum_{i=n_{(j-1)\tau}}^{n_{j\tau}}d_{n_{j\tau},i}e^{-\lambda i(m-i)\tau}\right],\end{aligned}$$

where $n_{j\tau}=N(j\tau)$ represents the number of infectives at time $j\tau$.

Due to the complexity of the likelihood function in this situation, we could not work out a useful formula for the predictive distribution of $N(t+u)$.

4. EXAMPLE: A COMPARATIVE STUDY WITH CLASSICAL METHODS

For illustration, we simulated an epidemic model as described with $m = 30$, $t = 3.00$, and $\lambda = 0.05$. In Table 1, we provide the observed values of n and s as well as the predictive and the estimative probabilities for the size of the population at instant $t + u$ for several values of u. (Note that $n > \frac{m}{2}$.)

	$u = 0.5$		$u = 1.00$		$u = 1.50$	
y	PM	EM	PM	EM	PM	EM
19	.006940	.003265	.000142	.000011	.000006	.0
20	.033333	.021195	.001169	.000158	.000063	.0
21	.083152	.067130	.005105	.001164	.000364	.000012
22	.141909	.136616	.015595	.005648	.001505	.000108
23	.183873	.198085	.037128	.019984	.004948	.000663
24	.189835	.214463	.072629	.054094	.013700	.003426
25	.159396	.176625	.119804	.114183	.032971	.014235
26	.108961	.110482	.167960	.187818	.070065	.047545
27	.059539	.051370	.198193	.235790	.131602	.125208
28	.024837	.016887	.190103	.215068	.214277	.247837
29	.135906	.128229	.135906	.128229	.283837	.332221
30	.001080	.000355	.056267	.037852	.246664	.228752

PM - Predictive Method (for vague prior)
EM - Estimative Method

Table 1. *Epidemic model with* $m = 30$ *observed till* $t = 3.00$
$(n = 19,\ s = 328.584)$ $\hat{\lambda} = 0.0584$

The "estimative" probabilities were computed by substituting λ in (3.2) by the maximum likelihood estimate which is given by $\hat{\lambda} = \frac{n-1}{s}$. It can be observed that the predictive method, when vague prior knowledge is considered, gives similar results to those obtained by the estimative method. This is to be expected, since the estimative and predictive probabilities can be written respectively as:

$$P(N(t+u) = y|n, s, \hat{\lambda}) = c_{ny} \sum_{k=n}^{y} d_{yk} e^{-(\frac{n-1}{s})k(m-k)u}$$

and

$$P(N(t+u) = y/|(n,s)) = c_{ny} \sum_{k=n}^{y} d_{yk} e^{-(n+g-1)\log\left(1+u\frac{k(m-k)}{s+h}\right)}.$$

If we set $g = 0$, $h = 0$ (possible provided that $n > 1$), which corresponds to vague prior knowledge about the parameter λ, and if

$$\frac{u(k(m-k))}{s} < 1,$$

for every k, then

$$\log\left(1 + u\frac{k(m-k)}{s}\right) \simeq \frac{uk(m-k)}{s}$$

and hence the similarity of the results follows.

The advantage of the predictive method over the estimative one is that the former allows the incorporation of prior knowledge when it is available. This can be particularly important when the number of infectives in $[0, t]$ is small compared to the total size of the population.

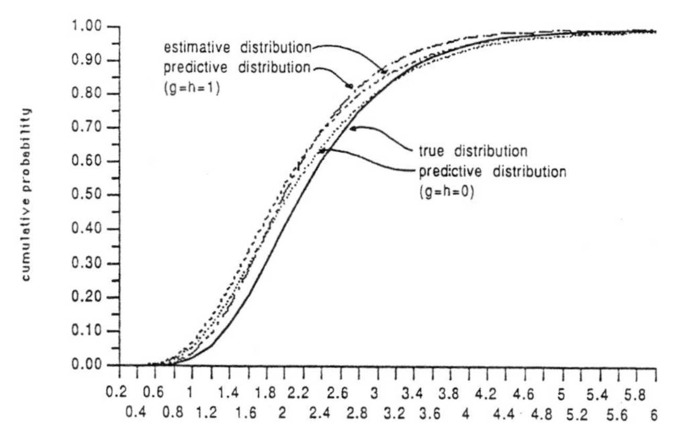

Figure 1. *Distribution of the interval of time T, after $t = 3.00$, till the total population is infected.*

In Figure 1 we plotted the predictive, estimative and the true distribution function of the random variable T, the interval of time starting from $t = 3.00$ till the total population is infected. The estimative distribution function of T was obtained by substituting λ by $\hat{\lambda} = \frac{n-1}{s}$ in (3.5) and the "true" distribution function was obtained with the value of $\lambda = 0.05$ used for the simulation of the model. For the predictive distribution function of T we considered two situations: vague prior knowledge ($g = h = 0$) and exponential prior ($g = h = 1$). Overall it seems that the predictive method, when vague prior knowledge is considered, does better than the estimative one which gives a more pessimistic picture of the process.

If interest lies in the minimum square error predictor, we have from (3.7) and (3.9) that

$$\frac{E(T|N(t) = n, \hat{\lambda})}{E(T|n, s)} = \frac{s(n + g - 2)}{(s + h)(n - 1)},$$

which is less than 1 when $g = h = 0$.

For the example considered, we have $E(T|N(t) = n, \lambda) = 2.3243$, $E(T|N(t) = n, \hat{\lambda}) = 2.1215$, $E(T|n, s) = 2.2462$ for $g = h = 0$ and $E(T|n, s) = 2.1279$ for $g = h = 1$.

REFERENCES

Aitchison, J. and Dunsmore, I. R. (1985). *Statistical Prediction Analysis.* Cambridge: University Press.

Bailey, N. T. J. (1963). The simple stochastic epidemic curve: a complete solution in terms of known functions. *Biometrika* **50**, 235–40.

Basawa, I. V. and Prakasa Rao, B. L. S. (1980). *Statistical Inference for Stochastic Processes.* New York: Academic Press.

Geisser, S. (1971). The inferential use of predictive distributions. *Foundations of Statistical Inference*, 456–469, (V. P. Godambe and D. A. Sprott, Eds.). Toronto: Holt, Rinehart and Winston.

Geisser, S. (1982). Aspects of the predictive and estimative approaches in the determination of probabilities. *Biometrics supplement: Current Topics in Biostatistics and Epidemiology* **38**, 75–93.

Geisser, S. (1985). On the prediction of observables: a selective update. *Bayesian Statistics 2*, (J. M. Bernardo, M. H. DeGroot, D. V. Lindley and A. F. M. Smith, eds.). North Holland, Valencia University Press.

Ross, S. M. (1983) *Stochastic Processes.* New York: John Wiley & Sons.

BAYESIAN STATISTICS 3, *pp. 543–552*
J. M. Bernardo, M. H. DeGroot, D. V. Lindley and A. F. M. Smith, (Eds.)

Bayesian Regression Analysis Under Non-Normal Errors

P. BAGCHI and I. GUTTMAN
University of Toronto

SUMMARY

Various assumptions involved in the analysis of linear models are studied simultaneously in the Bayesian context. We first assume i.i.d. Student-t errors and find the posterior of the regression parameters involved and the auto-correlation parameter ρ. The result for i.i.d. Student- t errors, turns out to be insensitive to the value assigned for the degrees of freedom of the Student-t distribution. A second assumption about the error distribution is also explored, namely, the errors follow a multivariate Student-t distribution with v_0 degrees of freedom, and the variance-covance matrix $\frac{v_0}{v_0-2}\sigma^2 I$. An interesting result for this case then obtains, namely, the joint posterior of the regression coefficients and ρ is independent of v_0, and has the form obtained by Zellner (1971) using i.i.d. $N(0, \sigma^2)$ errors. However, the shape parameter σ^2 has a posterior which does depend on v_0, and further, the conditional distribution of σ^2, given ρ, coincides with a result obtained by Zellner (1976), who used multivariate Student-t errors in an analysis of the linear model ($\rho = 0$). We also discuss the choice of an informative prior for ρ. An illustrative example using Haavelmo's data (Zellner (1971), page 64) is supplied.

Keywords: AUTOREGRESSIVE PROCESS; STUDENT DISTRIBUTION; BIVARIATE POSTERIOR DISTRIBUTION; MARGINAL POSTERIOR DISTRIBUTION.

1. INTRODUCTION

Most applications of univariate linear regression are considered under the assumption that the errors are independent and identically distributed normal variables with zero mean and variance $\sigma^2 > 0$. The maximum likelihood estimators are then the usual Least Squares estimators, and inferences about the regression coefficients are made on this basis. But departures from these "standard" assumptions often occur in practice and hence it is important to deal with them. Zellner (1976) has considered the problem in a multivariate setting using the broader assumption that the errors have a joint multivariate Student-t distribution with zero location vector; see the references concerning use of non-normal errors in Zellner (1976). The analysis of the assumption of independence or lack thereof has been made by Zellner and Tiao (1964); see also Zellner (1971). An excellent account of a Bayesian Analysis of Linear Models under departures from the above assumptions has been given by Broemeling (1985). We note here that under the assumptions of Zellner (1976) the errors are no longer independent but are uncorrelated. A study of this assumption has been attempted here; i.e. specifically we discuss in this paper an autocorrelated model with multivariate t errors.

The paper has been structured in the following manner. Section 2 and 3 deal with a specification of the various models when a single predictor variable is involved in the generating process, and discusses the various posterior distributions and how they are obtained. In particular, the bivariate posterior distribution of the regression coefficient and the autocorrelation parameter is obtained and discussed. In Section 4, the marginal distributions are derived, and certain ramifications of the conditional distribution of the regression parameter given the autocorrelation parameter are discussed. Section 5 gives a detailed illustrative example for the results of Section 2-4.

A generalization for the case of more than one predictor variable is given in Bagchi and Guttman (1987), along with the case of an informative prior used for the autocorrelation parameter.

2. THE MODEL

Let the model for n observations $\mathbf{y}' = (y_1, \ldots, y_n)$ be

$$\begin{aligned} y_t &= \beta x_t + u_t \\ u_t &= \rho u_{t-1} + e_t \qquad t = 1, 2, \ldots, n. \end{aligned} \tag{1.1}$$

so that the errors u_t are generated by a first order autoregressive process. Suppose too, that the e_t's are independent and identically distributed as Student-t with the p.d.f. given by,

$$f(e_t) = \frac{\Gamma\left(\frac{\nu+1}{2}\right)}{\Gamma(1/2)\Gamma\left(\frac{\nu}{2}\right)} \left(\sqrt{\nu\sigma^2}\right)^{-1} \left[1 + \frac{e_t^2}{\nu\sigma^2}\right]^{-\frac{\nu+1}{2}} \tag{1.2}$$

$$-\infty < e_t < \infty, \nu > 0, \sigma^2 > 0.$$

From (1.1) it easily follows that

$$y_t - \rho y_{t-1} = \beta(x_t - \rho x_{t-1}) + e_t \qquad t = 1, 2, \ldots, n. \tag{1.3}$$

Note that in the above equation the initial condition y_0 appears and therefore something has to be said about it. Suppose it can reasonably be assumed that the system originated in the past and hence has been operating at say $t = -T_0, \ldots, -2, -1, 0$. Then from (1.3) we obtain,

$$y_0 - \beta x_0 = \rho(y_{-1} - \beta x_{-1}) + e_0. \tag{1.4}$$

Here y_0 is observed and x_0 is known whereas y_{-1} and x_{-1} have not been observed. In this case let the unobserved (random) quantity m be defined by

$$m = \rho(y_{-1} - \beta x_{-1}) \tag{1.5}$$

so that

$$y_0 - \beta x_0 = m + e_0 \tag{1.5a}$$

or,

$$y_0 = \beta x_0 + m + e_0. \tag{1.5b}$$

Further assume that e_0 has a Student-t distribution with ν degres of freedom (df) and is independent of all e_t, $t > 0$.

Returning to (1.2) we assume initially that ν, the *df* is a known value say, ν_0. In this case the likelihood is given by,

$$\begin{aligned} l(\beta, \rho, \sigma^2, m|\mathbf{y}, y_0) &= p(y_0|\beta, \sigma^2, m)p(\mathbf{y}|y_0, \beta, \rho, \sigma^2, m) \\ &\propto (\sigma^2)^{-\frac{1}{2}} \left(1 + \frac{[y_0 - (\beta x_0 + m)]^2}{\nu_0\sigma^2}\right)^{-\frac{\nu_0+1}{2}} \times \\ &\prod_{t=1}^{n} (\sigma^2)^{-1/2} \left(1 + \frac{Q_t(\beta, \rho)}{\nu_0\sigma^2}\right)^{-\frac{\nu_0+1}{2}} \end{aligned} \tag{1.6}$$

where,

$$Q_t(\beta, \rho) = [(y_t - \beta x_t) - \rho(y_{t-1} - \beta x_{t-1})]^2. \tag{1.7}$$

If only vague information is available regarding the parameters, we assume that the non-informative prior is appropiate, that is we use the prior

$$p(\beta, \rho, \sigma^2, m) \propto \frac{1}{\sigma^2}. \tag{1.8}$$

Hence the posterior distribution of the parameters is given by,

$$\begin{aligned} p(\beta, \rho, \sigma^2, m|\mathbf{y}, y_0; \nu_0) \propto (\sigma^2)^{-\frac{1}{2}} \left(1 + \frac{[y_0 - (\beta x_0 + m)]^2}{\nu_0 \sigma^2}\right)^{-\frac{\nu_0+1}{2}} \times \\ (\sigma^2)^{-(\frac{n}{2}+1)} \prod \left(1 + \frac{Q_t(\beta, \rho)}{\nu_0 \sigma^2}\right)^{-\frac{\nu_0+1}{2}} \end{aligned} \tag{1.9}$$

where Q_t is defined in (1.7).

Suppose that the interest lies in investigating m, the initial level of the process. Then, in principle, all the variables can be integrated out from (1.9) to obtain the marginal posterior distribution of m given $\mathbf{y}$, y_0, ν_0. However, if the interest is not focussed on m, then m can be integrated out to obtain,

$$p(\beta, \rho, \sigma^2|\mathbf{y}, y_0; \nu_0) \propto (\sigma^2)^{-(\frac{n}{2}+1)} \prod \left(1 + \frac{Q_t(\beta, \rho)}{\nu_0 \sigma^2}\right)^{-\frac{\nu_0+1}{2}} \tag{1.10}$$

Hence the joint distribution of β and ρ given $\mathbf{y}$,y_o, ν_0 is obtained by integrating out σ^2 to give,

$$p(\beta, \rho|\mathbf{y}, y_0; \nu_0) \propto \int_0^\infty (\sigma^2)^{-(\frac{n}{2}+1)} \prod \left(1 + \frac{Q_t(\beta, \rho)}{\nu_0 \sigma^2}\right)^{-\frac{\nu_0+1}{2}} d\sigma^2. \tag{1.11}$$

It does not seem possible to integrate the above function in a closed form, so that we have to resort to numerical integration techniques to obtain the bivariate contours of the posterior marginal distribution of β and ρ. However if this distribution is not sensitive to a particular choice of ν_0, then it would be reasonable to take a large value for ν_0, thereby enabling us to approximate the t-errors by normal errors. From (1.11), we have that for large v_0, to the first order of approximation

$$\begin{aligned} p(\beta, \rho|\mathbf{y}, y_0; \nu_0) &\propto \int_0^\infty (\sigma^2)^{-(\frac{n}{2}+1)} \exp\left[-\frac{1}{2\sigma^2} Q(\beta, \rho)\right] d\sigma^2 \\ &\propto [Q(\beta, \rho)]^{-\frac{n}{2}}. \end{aligned} \tag{1.12}$$

where,

$$Q(\beta, \rho) = \sum_{t=1}^{n} [(y_t - \rho y_{t-1}) - \beta(x_t - \rho x_{t-1})]^2. \tag{1.12a}$$

Then the marginal posterior distributions of ρ and β can easily be obtained as shown in Section 4. In Section 5 we analyse Haavelmo's data, as reported in Zellner (1971, p64), using (1.11) and (1.12). The bivariate posterior marginal distribution of β and ρ is obtained numerically and contour & perspective plots are drawn for $\nu_0 = 1, 3, 5, 10, 15$ and 20, and are reproduced in Bagchi and Guttman (1987). For illustrative purposes, the case $\nu_0 = 10$ is reproduced in Figure 4.1 of this paper. It turns out that the posteriors, as ν_0 changes, are remarkably insensitive to the values of ν_0 (details in Bagchi and Guttman (1987)).

3. A DIFFERENT ERROR STRUCTURE

There is a considerable simplification in the calculation involved when the error structure is changed to accomodate an uncorrelated error structure. To that extent let the model for n observations $\mathbf{y}' = (y_1, y_2, \ldots, y_n)$ be

$$\begin{aligned} y_t &= \beta x_t + u_t \\ u_t &= \rho u_{t-1} + e_t \quad t = 1, 2, \ldots, n. \end{aligned} \tag{2.1}$$

$-\infty < \beta, \rho < \infty$, x_t is the nonstochastic input variable, and $\mathbf{e}' = (e_1, e_2, \ldots, e_n)$ has a multivariate Student-t distribution of order n and d.f. $= \nu_0$, so that its p.d.f. is given by,

$$p(\mathbf{e}|\nu_0, \sigma) = \frac{g(\nu_0)}{(\sigma^2)^{n/2}} \left(\nu_0 + \frac{e'e}{\sigma^2}\right)^{-\frac{n+\nu_0}{2}} \tag{2.2}$$

$$\sigma, \nu_0 > 0, -\infty < e_t < \infty, t = 1, 2, \ldots, n$$

where

$$g(\nu_0) = \frac{\nu_0^{\nu_0/2}\Gamma(\nu_0/2 + n/2)}{\pi^{n/2}\Gamma(\nu_0/2)}. \tag{2.2a}$$

From (2.1) we have that

$$y_t - \rho y_{t-1} = \beta(x_t - \rho x_{t-1}) + e_t, t = 1, 2, \ldots, n, \tag{2.3}$$

with, for $t = 0$,

$$y_0 - \rho y_{-1} = \beta(x_0 - \rho x_{-1}) + e_0 \tag{2.4}$$

so that,

$$y_0 = \beta x_0 + m + e_0 \tag{2.4a}$$

where $m = \rho(y_{-1} - \beta x_{-1})$ is an unobserved quantity and hence a random variable. We further assume that y_0 has a normal distribution with mean $(\beta x_0 + m)$ and variance σ^2, with y_0(or e_0) independent of $\mathbf{e}$. We note that the definition of our model is such that it is broad enough to include the explosive case ($|\rho| \geq 1$) as well as the nonexplosive one ($|\rho| < 1$).

Under the above assumptions the joint p.d.f. of y_0 and y is given by

$$p(\mathbf{y}, y_0|\beta, \sigma, \rho, m, \nu_0) = p(y|y_0, \beta, \sigma, \rho, m, \nu_0)p(y_0|\beta, \sigma, m, \rho, \nu_0) \tag{2.5}$$

$$\propto (\sigma^2)^{-(n+1)/2} \exp\left\{\frac{-1}{2\sigma^2}(y_0 - \beta x_0 - m)^2\right\} \left(\nu_0 + \frac{Q(\beta, \rho)}{\sigma^2}\right)^{-\frac{n+\nu_0}{2}} \tag{2.5a}$$

where,

$$Q(\beta, \rho) = \sum \{(y_t - \rho y_{t-1}) - \beta(x_t - \rho x_{t-1})\}^2.$$

With respect to prior information, we assume that β, ρ, m and $\log\sigma$ are uniformly and independently distributed so that

$$p(\beta, \rho, m, \sigma^2) \propto \frac{1}{\sigma^2} \tag{2.6}$$

It will be further assumed that ν_0 is given. On combining the prior and the likelihood we obtain the posterior distribution as

$$\begin{aligned} p(\beta, \rho, m, \sigma^2|\mathbf{y}, y_0, \nu_0) &\propto (\sigma^2)^{-((n+1)/2+1)} \\ &\times \exp\left[\frac{-1}{2\sigma^2}\{m - (y_0 - \beta x_0)\}^2\right] \left[\nu_0 + \frac{Q(\beta, \rho)}{\sigma^2}\right]^{-(n+\nu_0)/2} \end{aligned} \tag{2.7}$$

If we are interested in investigating the initial level of the process the posterior marginal distribution of m can be obtained by integrating with respect to the other parameters. This would enable us to make inferences regarding the initial level of the process. But if we wish to study the robustness of our inferential procedures with regards the distributional assumptions then we would be more interested in studying the marginal distribution of β. Hence integrating with respect to m, we have,

$$p(\beta,\rho,\sigma^2|\mathbf{y};y_0,\nu_0) \propto (\sigma^2)^{-(n+2)/2}\left[\nu_0 + \frac{\sum((y_t-\rho y_{i-1})-\beta(x_t-\rho x_{t-1}))^2}{\sigma^2}\right]^{-(n+\nu_0)/2} \tag{2.8}$$

To get the joint posterior distribution of β, ρ we now interate with respect to σ^2; i.e.,

$$p(\beta,\rho|\mathbf{y};y_0,\nu_0) \propto \int_0^\infty (\sigma^2)^{-(n+2)/2}\left[\nu_0 + \frac{Q(\beta,\rho)}{\sigma^2}\right]^{-(n+\nu_2)/2} d\sigma^2. \tag{2.9}$$

For convenience let $\sigma^2 = \theta$. In this case we have

$$p(\beta,\rho|\mathbf{y};y_0,\nu_0) \propto \int_0^\infty \theta^{-(n+2)/2}\left[\nu_0 + \frac{\sum((y_t-\rho y_{t-1})-\beta(x_t-\rho x_{t-1}))^2}{\theta}\right]^{-(n+\nu_0)/2} d\theta. \tag{2.10}$$

To evaluate this, recall that the density function of an F variable with μ_1 and μ_2 degrees of freedom is given by

$$p(F|\mu_1,\mu_2) = \frac{\left(\frac{\mu_1}{\mu_2}\right)^{\frac{\mu_1}{2}}}{B\left(\frac{\mu_1}{2},\frac{\mu_2}{2}\right)} F^{\frac{\mu_1}{2}-1}\left(1+\frac{\mu_1 F}{\mu_2}\right)^{-\frac{\mu_1+\mu_2}{2}} \tag{2.11}$$

By setting $\theta = \frac{Q}{nF}$ in the integrand of (2.10), we find

$$\begin{aligned} p(\beta,\rho|\mathbf{y},y_0,\nu_0) &\propto [Q(\beta,\rho)]^{(-n/2)} \int_0^\infty F^{(n/2)-1}\left(1+\frac{nF}{\nu_0}\right)^{-(n+\nu_0)/2} d(F) \\ &\propto [Q(\beta,\rho)]^{-(n/2)}. \end{aligned} \tag{2.12}$$

To obtain the constant of proportionality we have to appeal to numerical integration techniques. Further the contours of the bivariate p.d.f. (2.12) can be computed to provide information about its shape. Note that this p.d.f. does not depend on ν_0.

Remark. (i) It is interesting to note that exactly the same inferences would have resulted if we had assumed that y_0 has a Student-t distribution with ν_0 degrees of freedom and with location $\beta x_0 + m$ and scale σ. Thus we can say that inferential procedures regarding β are robust with respect to the distributional assumptions made about the initial level of the process. This insensitivity of the posterior to the choice of the distribution of y_0, when m is regarded as a nuisance parameter, has been noticed before —see for example, Zellner (1971)

(ii) We note, interestingly, that the result (2.12), obtained when $\mathbf{e}$ has the joint multivariate-t distribution with d.f. $=\nu_0$, is the same result as (1.12), obtained when the e_t are i.i.d. Student-t with infinite degrees of freedom.

4. INFERENCE ABOUT THE PARAMETERS

In this section we compute the marginal distributions of the parameters in question. Our starting point is (2.7)-(2.12), which of course, was obtained using the assumption for **e** implicit in (2.2). However, we note that the result (2.12) is exactly the result (1.12), which was obtained using the i.i.d. assumption for e_t's, with common distribution a Student-t, with its d.f. ν_0 large. We proceed first to investigate the autocorrelation parameter ρ, and then proceed to compute the marginal distribution of ρ given (y, y_0), from the joint posterior distribution of (ρ, β), given in (2.12).

Define

$$A_x(\rho) = \sum(x_t - \rho x_{t-1})^2 \tag{3.1a}$$

$$B(\rho) = \sum(x_t - \rho x_{t-1})(y_t - \rho y_{t-1}) \tag{3.1b}$$

$$C(\rho) = \sum(y_t - \rho y_{t-1})^2 \tag{3.1c}$$

$$H(\rho) = C - \frac{B^2}{A_x}. \tag{3.1d}$$

Then $Q(\beta, \rho)$ can be expressed as a quadratic in β and hence completing the square and assuming that $A_x(\rho) > 0$ (which is not very restrictive; see Zellner (1971)), we have that,

$$p(\beta, \rho|\mathbf{y}, y_0,) \propto \left[A_x(\rho)\left(\beta - \frac{B(\rho)}{A_x(\rho)}\right)^2 + H(\rho)\right]^{-n/2} \tag{3.2}$$

Integrating with respect to β after a suitable change of variable we have,

$$p(\rho|\mathbf{y}, y_0, \nu_0) \propto [H(\rho)]^{-(n-1)/2} A_x(\rho)^{-1/2}. \tag{3.3}$$

We note that the constant of proportionality has to be obtained by numerical integration.

In a similar manner to obtain the marginal density of β given (y, y_0) we can write down $Q(\beta, \rho)$ in terms of a quadratic in ρ and integrate. Hence it can easily be shown that,

$$p(\beta|\mathbf{y}, y_0, \nu_0) \propto [A(\beta)]^{-1/2} [D(\beta)]^{-(n-1)/2} \tag{3.4}$$

where,

$$D(\beta) = \sum(y_t - \beta x_t)^2 - \frac{[\sum(y_t - \beta x_t)(y_t - \beta x_{t-1})]^2}{\sum(y_{t-1} - \beta x_{t-1})^2} \tag{3.5}$$

and,

$$A(\beta) = \sum(y_{t-1} - \beta x_{t-1})^2. \tag{3.6}$$

As before the constant of proportionality is computed numerically.

This posterior p.d.f. of β enables us to make inferences about β by incorporating information about the assumptions. To see this note that the posterior p.d.f. of β can also be written as

$$p(\beta|\mathbf{y}, y_0) = \int_{-\infty}^{\infty} p(\beta|y, y_0, \rho)p(\rho|y, y_0)d\rho \tag{3.7}$$

and hence can be regaded as a suitably weighted average of $p(\beta|\mathbf{y}, y_0, \nu_0, \rho)$ with $p(\rho|, y_0, \nu_0)$ serving as a weighting function. When ρ is given it is relatively easy to compute the posterior conditional distribution of β given ρ, since in this case,

$$p(\beta|\mathbf{y}, y_0, \nu_0, \rho) \propto \left[1 + \frac{A_x(\rho)(\beta - \hat{\beta}(\rho))^2}{\gamma H(\rho)}\right]^{-(\gamma+1)/2} \tag{3.8}$$

where $\gamma = n - 1$ and

$$\hat{\beta}(\rho) = \frac{B}{A_x} = \frac{\sum(y_t - \rho y_{t-1})(x_t - \rho x_{t-1})}{\sum(x_t - \rho x_{t-1})^2}. \tag{3.9}$$

Thus we see that

$$\left[\frac{A_x(\rho)}{H(\rho)}\right]^{1/2} (\beta - \hat{\beta}(\rho)) \sim t_\gamma, \tag{3.10}$$

a univariate t-distribution with $\gamma = n - 1$ df. Thus the posterior p.d.f. of β given ρ can be computed and plotted for various values of ρ. It is interesting to note that (3.10) may be obtained when assuming e_t's are i.i.d. $N(0), \sigma^2)$. We have obtained (3.10) under the assumptions (2.2), and as (3.10) does not depend on ν_0, we can see that the inferential procedures are insensitive to the assumptions regarding normality.

5. AN ILLUSTRATIVE EXAMPLE

In this section, we illustrate the previous results of this paper using a data set due to Haavelmo, quoted in Zellner (1971, p64), and reproduced here in Table 4.1. Using (1.1)-(1.2) with x_t =Investment in the tth year (t = year-1921), and y_t = income in the year t, we compute the joint posterior of β and ρ for various ν_0 given in (1.11). In fact, Bagchi and Guttman (1987), did this for $\nu_0 = 1, 3, 5, 10, 15, 20$. The contours and 3-dimensional graphs of each of these posteriors is also given in Bagchi and Guttman (1987), and we reproduce these in Figures 4.1, 4.2 for $\nu_0 = 10$ by way of illustration. What is remarkable is the insensitivity found as ν_0 varies from 1 to 20. The posteriors are virtually one and the same, and the interested readers may verify this by consulting the Bagchi and Guttman (1987) paper.

HAAVELMO'S DATA ON INCOME & INVESTMENT

Year	Income	Investment
1922	433	39
3	483	60
4	479	42
5	486	52
6	494	47
7	498	51
8	511	45
9	534	60
1930	478	39
1	440	41
2	372	22
3	381	17
4	419	27
5	449	33
6	511	48
7	520	51
8	477	33
9	517	46
1940	548	54
1	629	100

*(printed from Zellner(1971))

Table 4.1.

Because the posterior (1.11) turns out to be insensitive to changes in ν_0, we assume that ν_0 is large, so that as discussed in Section 2, the posterior can be taken, to good approximation, to be as in (1.12), and so independent of ν_0. Here, the marginal posterior of ρ and the marginal posterior of β are approximated by (3.3) and (3.4) respectively. Using the data of Table 4.1, we plot these in Figure 4.3 and 4.4.

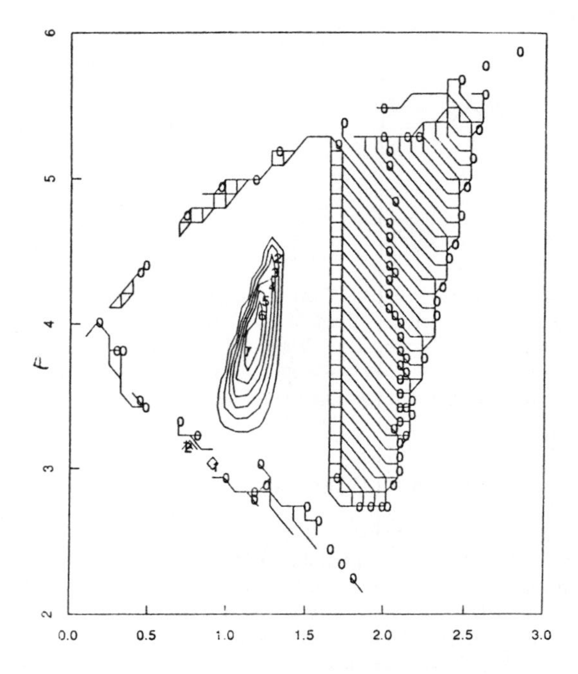

Figure 4.1. *Contours of the Bivariate Distribution of ρ and β for the Haavelmo Data*

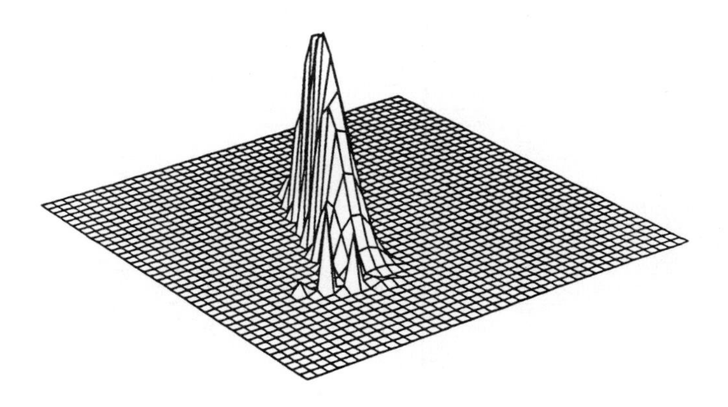

Figure 4.2. *Perspective of the Bivariate Distribution of ρ and β for the Haavelmo Data*

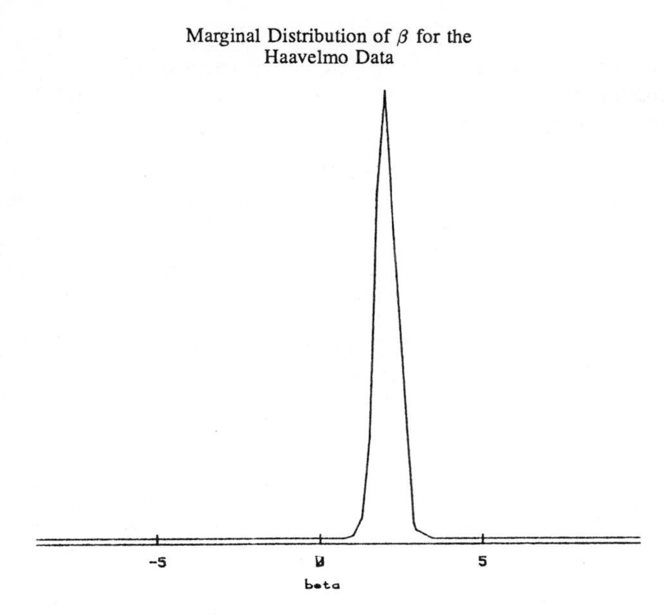

Figure 4.3. *Based on the model (2.1)-(2.2).*

Marginal Distribution of ρ for the
Haavelmo Data

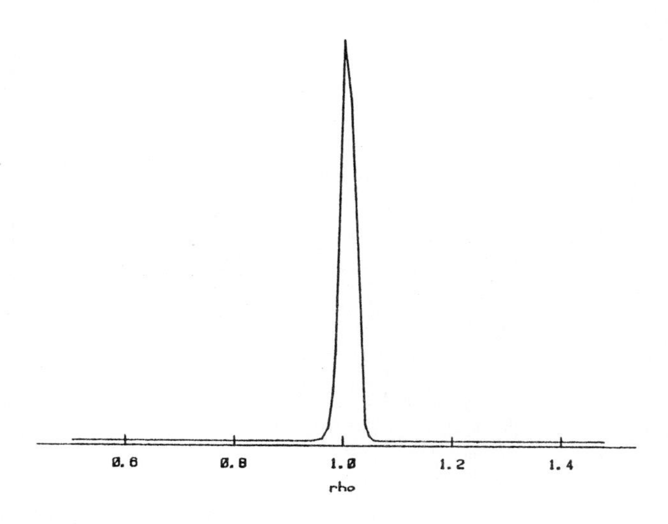

Figure 4.4. *Based on the model (2.1)-(2.2).*

ACKNOWLEDGEMENTS

We wish to thank Joyce Bernstein for helping with the computing and the various graphs. We also wish to thank Prof. A. Zellner for helpful suggestions that led to a substantial improvement over the first draft.

REFERENCES

Bagchi, P. and Guttman, I. (1987). *Bayesian Regression Analysis with Non-Normal Errors. Tech. Rep.* 09, University of Toronto, Canada.

Broemeling, L. (1985). *Bayesian Analysis of Linear Models, Chapter 5*. New York: Marcel Dekker, Inc.

Zellner, A. (1971). *An Introduction to Bayesian Inference in Econometrics.* Toronto: John Wiley & Sons.

Zellner, A. (1976). Bayesian and Non-Bayesian Analysis of the Regression Model with Multivariate Student-*t* Error Terms. *J. Amer. Statist. Assoc.* 400–405.

Zellner, A. and Tiao, G. C. (1964). Bayesian Analysis of the Regression Model with Autocorrelated Errors. *J. Amer. Statist. Assoc.* **59**, 763–778.

BAYESIAN STATISTICS 3, pp. 553–556
J. M. Bernardo, M. H. DeGroot, D. V. Lindley and A. F. M. Smith, (Eds.)

On Kolmogorov's Partial Sufficiency

J. A. CANO SÁNCHEZ, A. HERNÁNDEZ BASTIDA and E. MORENO BAS
Universidad de Murcia, and *Universidad de Granada*

SUMMARY

In the case of positive likelihood for all sample points, and for discrete parametric space, K-sufficiency notion of partial sufficiency is equivalent to sufficiency. If $f(x/\theta)$ has some values equal zero, to ensure the equivalence of both concepts an additional condition is needed.This condition has been given *ad hoc* for discrete parametric space and so in the case of continuous parametric space additional assumptions to obtain the same conclusion are needed.

This paper is devoted to the study of these conditions, which are related to the behaviour of the likelihood function as well as to the underlying σ-fields.

Keywords: SUFFICIENCY; K-SUFFICIENCY.

1. INTRODUCTION

In a Bayesian setup, the concepts of partial sufficiency —or sufficiency in the presence of a nuisance parameter— and sufficiency have been proposed by Kolmogorov (1942). Whereas the notion of sufficiency does not promote many disputes, Kolmogorov's partial sufficiency called K-sufficiency, has been controversial. Hàjek (1967) gave a theorem proving the equivalence of the notions. However, the proof was given in the case of a discrete parametric space avoiding the situation in which the likelihood function had some zeros. Later, Martin, Petit and Littaye (1973) and Mouchart and Rolin (1984), observing this fact, gave necessary and sufficient conditions for Hàjek's theorem in the case of a discrete parametric space.

Finding conditions under which Hàjek's theorem holds in the case of a continuous parametric space with non-equivalent sampling probabilities is an open problem. The solution to this creates the need for a new concept of sufficiency in the presence of a nuisance parameter from a Bayesian viewpoint, at least in those cases in which K-sufficiency and sufficiency coincide.

This paper offers some answers concerning the problems stated above for dominated cases with non-constant support. Section 2 recalls the K-sufficiency notion and Section 3 deals with the comparison of sufficiency and K-sufficiency.

2. K-SUFFICIENT

Let $(\mathcal{X}, \mathcal{S}, P_\theta)$ be a sample probability space, where P_θ is a probability measure for each $\theta \in \Theta$. We assume that $P_\theta \ll \mu$, where μ is σ-finite on $\mathcal{S}$, and let $f(x|\theta)$ represent the density function for each $\theta \in \Theta$. We also consider the prior measurable space $(\Theta, \mathcal{A})$, $\mathcal{A}$ being the smallest σ-field with respect to which the likelihood is a measurable function.

Let $b(\theta)$ be a non-trivial transformation.

Definition 1. *(Kolmogorov, 1942)*
The statistic $T(X)$ is K-sufficient (or sufficient for the subparameter $b(\theta)$), if, for every prior probability ξ on $\mathcal{A}$, the equality

$$\xi(b|x) = \xi(b|T(x))$$

holds, for each $x \in \mathcal{X}$.

If $\mathcal{B}$ denotes the sub-σ-field induced by $b(\theta)$, it can be easily shown that Definition 1 is equivalent to the one following.

Definition 1′. *The statistic $T(X)$ is K-sufficient, (for the subparameter $b(\theta)$), if*

$$E_\xi 1_B(x_1|\theta) f(x_1|\theta) = K_\xi(x_1, x_2) E_\xi 1_B(\theta) f(x_2|\theta)$$

whatever prior ξ on $\mathcal{A}$ is used and every $B \in \mathcal{B}$ such that $\xi(B) > 0$, where x_1, x_2, are two sample points belonging to the coset of $T(x) = t$, i.e. $T(x_1) = T(x_2) = t$, and $K_\xi(x_1, x_2)$ does not depend upon the subset B.

3. RELATIONS BETWEEN SUFFICIENCY AND K-SUFFICIENCY

Firstly, we are going to summarise conditions for Hàjek's theorem in the case of discrete parametric space (Th.1, Th. 2). Hence, consider $\Theta = \{\theta_n; n \geq 1\}$ and $\mathcal{A} = \mathcal{P}(\Theta)$, $\mathcal{P}(\Theta)$ containing all the subsets of Θ. Theorems 1 and 3 below are particular cases of more general results in Martin Petit and Littaye (1973) for the case of equivalent sampling measures. However, we give them because of the simplicity of the proofs.

Theorem 1. *If the likelihood $f(x|\theta_n)$; $n \geq 1$, is positive for all $x \in \mathcal{X}$, then K-sufficiency for a non-trivial subparameter is equivalent to sufficiency.*

Proof. Let $T(X)$ be K-sufficient for $b(\theta)$. Then, under the above conditions it is clear that

$$f(x_1|\theta_i) \propto f(x_2|\theta_i), i \geq 1,$$

for all x_1, x_2, satisfying $T(x_1) = T(x_2)$. The converse is obvious. This completes the proof.

Theorem 2. *(Mouchart and Rolin, 1984)*
Let $f(x|\theta_n)$ be the likelihood, not necessarily positive for all $\theta_n \in \Theta$. Then, condition (H), given below, is necessary and sufficient to ensure that any statistic K-sufficent for the non-trivial subparameter $b(\theta)$ is also sufficient.

Condition (*H*). *From $f(x|\theta_i) = f(y|\theta_i) = 0$, for some i, it follows that $f(x|\theta_i) \propto f(y|\theta_i)$ for all i.*

Proof. See Mouchart and Rolin, 1984.

Secondly, let us consider a parametric space Θ, not necessarily discrete, which is endowed with a topology.

Theorem 3. *Let $(\Theta, \mathcal{A})$ be a measurable space where Θ is separable and $\mathcal{A}$ is the Borel σ-field on Θ. Let us also assume that the likelihood $f(x|\theta)$ satisfies the following conditions:*

(i) $f(x|\theta) > 0$, *for all* $(x, \theta) \in \mathcal{X} \times \Theta$,

(ii) $f(x|\cdot)$ *is a continuous function for all* $x \in \mathcal{X}$.

In this situation, a statistic $T(X)$ is sufficient if and only if it is K-sufficcient for the subparameter

$$b(\theta) = \begin{cases} \theta_n, & \text{if } \theta = \theta_n, \\ \theta^* & \text{otherwise,} \end{cases}$$

where $\theta^* \notin D$ and $D = \{\theta_n; n \geq 1\}$ denotes a subset dense in Θ.

Proof. Let $T(X)$ be a K-sufficient statistic for $b(\theta)$. Selecting the prior probability defined by

$$\xi(\theta_i) = \begin{cases} p_i, & \text{if } \theta_i \in D, \\ 0, & \text{otherwise,} \end{cases}$$

where $p_i > 0$, $i \geq 1$, $\Sigma p_i = 1$, it follows that

$$f(x_1|\theta_i) = K_\xi(x_1, x_2) f(x_2|\theta_i), i \geq 1,$$

for x_1, x_2 satisfying $T(x_1) = T(x_2)$. That is to say $f(x_1|\theta_i) \propto f(x_2|\theta_i)$, $i \geq 1$. The application of condition (ii) yields

$$f(x_1|\theta) \propto f(x_2|\theta), \theta \in \Theta.$$

Therefore $T(X)$ is sufficient. The converse is straighforward and hence the proof is completed.

Under some conditions on the sampling model it is possible to find in the general case a necessary and sufficient condition to ensure that any K-sufficient statistic is also sufficient. This condition, called Condition (H'), is close to Condition (H) given above.

Condition (H'). *The non-trivial sub-σ-field $\mathcal{B}(\mathcal{B} \subset \mathcal{A})$ (or the subparameter generating it $b(\theta)$), the family of likelihoods $\{f(x|\cdot), x \in \mathcal{X}\}$ and the statistic $T(X)$ satisfy Condition (H') if, from*

$$f(x|\theta)1_B(\theta) = f(y|\theta)1_B(\theta) = 0$$

for some $B \neq \phi$, $B \in \mathcal{B}$ where $T(x) = T(y)$, it follows that

$$f(x|\theta) \propto f(y|\theta)$$

for $\theta \in \Theta - B$.

Theorem 4. *Let $(\theta, \mathcal{A})$ be the parametric mesurable space where the sets $\{\theta\}$ belong to $\mathcal{A}$.*

Then, Condition (H') is a necessary and sufficient condition to ensure that the K- sufficient statistic $T(X)$ is sufficient.

Proof. Necessity of (H')

This can be easily shown from the fact that $\{\theta\} \in \mathcal{A}$,

Sufficiency of (H')

Assume Condition (H') holds. *Ab contrario*, it suffices to prove that for a statistic $T(X)$ which is not sufficient it follows that $T(X)$ is not K-sufficient for an arbitrary non-trivial subparameter.

Indeed, if $T(X)$ is not sufficient there exist two sample points x_1, x_2, and two parametric points θ_1, θ_2 such that

$$f(x_1|\theta_1)/f(x_2|\theta_1) \neq f(x_1|\theta_2)/f(x_2|\theta_2) \tag{1}$$

where $T(x_1) = T(x_2)$.

On the other hand, given $b(\theta)$ the following two situations may occur:

(a) $b(\theta_1) \neq b(\theta_2)$

(b) $b(\theta_1) = b(\theta_2)$.

Under situation (a), let $B_i = b^{-1}(b(\theta_i))$, $i = 1, 2$. Obviously, $B_1 \cap B_2 = \phi$. Taking the prior measure defined by $\xi(\cdot) = \frac{1}{2}(\delta_{\theta_1}(\cdot) + \delta_{\theta_2}(\cdot))$, it follows that $\xi(B_i|x_1) \neq \xi(B_i|x_2)$ for $i = 1, 2$. Hence, $T(X)$ is not K-sufficient and the proof is finalized.

Under situation (b), let $B = \{\theta : b(\theta) \neq b(\theta_1) = b(\theta_2)\}$. Since $b(\cdot)$ is not constant then $B \neq \phi$. Furthermore, there exists $\theta_3 \in B$ such that one of the two values $f(x_1|\theta_3)$, $f(x_2|\theta_3)$ is different from zero. For, if $f(x_1|\theta)1_B(\theta) = f(x_2|\theta)1_B(\theta) = 0$ then the application of Condition (H') contradicts expression (1). Therefore, without loss of generality, assume that $(f(x_1|\theta_3)/f(x_2|\theta_3)) \neq (f(x_1|\theta_1)/f(x_2|\theta_1))$. Then, taking the prior measure $\xi(\cdot) = \frac{1}{2}(\delta_{\theta_1}(\cdot) + \delta_{\theta_3}(\cdot))$ it follows that $\xi(B_i|x_1) \neq \xi(B_i|x_2)$ for $i = 1, 3$, where $B_3 = \{\theta : b(\theta) = b(\theta_3)\}$. Consequently, $T(X)$ is not K-sufficient for the subparameter $b(\theta)$ and the proof is completed.

Remark. It should be noted that Condition (H'), and consequently Theorem 4, is in line with the comment in Mouchart and Rolin (1984) that one should look for a property linking $T(X)$ and the subparameter $\mathcal{B}$, instead of considering restrictions on the model only.

CONCLUSION

The conclusion that can be derived from this paper is that not only is Kolmogorov's concept of partial sufficiency not void but also some stringent conditions are needed in the case of continuous parametric space to make it equivalent to sufficiency.

ACKNOWLEDGEMENTS

We are indebted to Prof. Mouchart for very valuable comments and suggestions.

REFERENCES

Hàjek, J. (1967). On Basic Concepts of Statistics. *Proc. Fifth Berkeley Symp. Math. Statist. Prob.* **1**, 139–162.

Kolmogorov, A. N. (1942). Determination of the Center of Dispersion and Degree of Accuracy for a Limited Number of Observations, (in Russian), *Izvestija Akademi Nank SSSR, Ser. Mat.* **6**, 3–32.

Martin, F., Petit, J. L. and Littaye, M. (1973). Independance conditionnelle dans le modèle statistique Bayesien. *Ann. Inst. Henri Poincarè, B* **9-1**, 17–40.

Mouchart, M. and Rolin, J. M. (1984). On K-sufficiency. *Statistica* **3**, 367–372.

BAYESIAN STATISTICS 3, pp. 557–567
J. M. Bernardo, M. H. DeGroot, D. V. Lindley and A. F. M. Smith, (Eds.)

Invariance and Bayesian Predictive Analysis with an Application to the Diallel Cross Design

G. CONSONNI and P. VERONESE
Bocconi University, Milan

SUMMARY

The predictive approach to Bayesian inference is enriched with invariance considerations which naturally occur in experimental designs. Invariance is used extensively both to compute predictive distributions and to derive predictive models together with the posterior parameter distribution. Under the usual assumption of normality, a simple and direct relationship between predictive and posterior distributions is shown to exist. General results are then illustrated in detail with reference to the diallel cross experimental design.

Keywords: INVARIANCE; PREDICTIVE SUFFICIENCY; EXPERIMENTAL DESIGNS.

1. INTRODUCTION

The predictive approach to Bayesian inference, see e.g. Geisser (1986), constitutes the main motivation of this paper. The basic novelty, however, is represented by the explicit use of *invariance* ideas which naturally occur in experimental designs.

Invariance is used extensively both to compute predictive distributions and to derive predictive models together with the posterior parameter distribution. In particular, under the usual assumption of normality we show that there exists a simple and direct relationship between predictive and posterior distributions; as a consequence, the role of the model in this case turns out to be somewhat redundant for predictive purposes.

The paper is divided into four sections. The first two deal with general issues of invariance, predictive distributions and model building; these are applied to the diallel cross experimental design in Section 3. In particular, under normality, we derive the predictive distributions of natural types of predictands, construct a corresponding predictive statistical model and then compute the posterior parameter distribution. Finally, Section 4 provides a brief discussion.

2. IMPLICATIONS OF INVARIANCE ON PREDICTIVE DISTRIBUTIONS AND MODELS

2.1. *Invariance, predictive distributions and sufficient statistics*

We start with a sequence of observables $\{X_u\}$ and with a group acting on this sequence. More precisely we assume that a repetitive structure —see Lauritzen (1984) and Dawid (1982)— is assigned, and that the nature of G is specified through its action on the sequence of observable spaces belonging to the repetitive structure so that natural compatibility conditions are satisfied. Informally we have a sequence of structured sets of observables partially ordered, so that each such set, which we shall call an array, can be seen as the projection from a "larger" array.

Typically, in experimental designs, G is a permutation group which relabels units without destroying the structure of the experiment.

For any given predictive problem, we shall have a finite set of predictands, Y, and data Z. The vector Z will be assumed to be an array of the repetitive structure, i.e. all types of actions of G are allowable on Z, so that in particular "unbalanced" data are not considered. This is necessary if invariance of the repetitive structure is to be fully exploited.

Now consider the smallest array containing both Y and Z and the corresponding group acting on it which, by an abuse of notation, we shall still write as G. Let $\mathcal{Y}$ and $\mathcal{Z}$ be, respectively, the sample space of Y and Z and consider all those $g \in G$ such that

$$g(Y, Z) = (\bar{g}Y, g^* Z, \qquad \text{where } \bar{g}Y \in \mathcal{Y}, g^* Z \in \mathcal{Z},$$

i.e. those transformations under which both $\mathcal{Y}$ and $\mathcal{Z}$ are closed. In this case if $p_{YZ}(y, z)$ represents the (generalized) density of (Y, Z) evaluated at $Y = y$ and $Z = z$, because of invariance we have

$$p_{YZ}(y, z) = p_{\bar{g}Y, g^* Z}(y, z) = p_{YZ}(\bar{g}y, g^* z) = p_{\bar{g}Y, Z}(y, g^* z).$$

In particular it follows that $p_Z(z) = p_Z(g^* z)$. We thus have, in terms of conditional density:

$$p_{Y|Z=z}(y) = p_{YZ}(y, z)/p_Z(z) = p_{\bar{g}Y, Z}(y, g^* z)/p_Z(g^* z) = p_{\bar{g}Y|Z=g^* z}(y), \tag{2.1}$$

in other words the conditional distribution of Y given Z is equivariant. As a corollary suppose that $g(Y, Z) = (Y, \hat{g}Z)$, i.e. that $\hat{g}$ is the identity, then

$$p_{Y|Z=z}(y) = p_{Y|Z=\hat{g}z}(y). \tag{2.2}$$

Now let $T = T(Z)$ be a function such that, for all y, $p_{Y|Z}(y) = p_{Y|T}(y)$. In this case T is said to be a predictive sufficient statistic (p.s.s.) of Z for Y, see Cifarelli and Regazzini (1982). We now investigate the consequences of (2.2) upon the structure of T. Before doing this, however, we need some prerequisities. We start by remarking that the conditional distribution of Y given Z naturally induces a partition of $\mathcal{Z}$, i.e. $\mathcal{Z} = \cup_s \mathcal{Z}_s$, where z_1 and $z_2 \in \mathcal{Z}_s$ iff $p_{Y|Z=z_1}(y) = p_{Y|Z=z_2}(y)$ for all y. Now if T is a p.s.s. of Z for Y then it induces its own partition of $\mathcal{Z}$, i.e. z_1 and z_2 belong to the same equivalence class iff $T(z_1) = T(z_2)$. A reasonable requirement at this stage is that the two partitions should coincide. This can be shown to be equivalent to either of the following conditions (see Consonni and Veronese, 1987):

(i) T is identified, i.e.

$$p_{Y|T(z_1)}(y) = p_{Y|T=T(z_2)}(y) \Rightarrow T(z_1) = T(z_2) \tag{2.3}$$

(ii) T is minimal, i.e. T is a p.s.s. function of any other p.s.s..

Let us now go back to formula (2.2). If T is a p.s.s., then $p_{Y|T=T(z)}(y) = p_{Y|T=T(\hat{g}z)}(y)$, for all $\hat{g}$ such that $g(Y, Z) = (Y, \hat{g}Z)$; and if (2.3) is satisfied $t(\hat{g}Z) = T(Z)$, i.e. T is invariant with respect to all such $\hat{g}$. For this reason we shall henceforth consider only minimal p.s.s.'s. Notice that (2.2) and its main consequence $T(\hat{g}Z) = T(Z)$ hold independently of the distribution of the observables under consideration. For specific choices of this distribution, however, T will exhibit more symmetries than those described by $\hat{g}$. This becomes transparent, for example, when the observables are normally distributed for, in this case, T will appear only in the expected value of Y. As a consequence it is possible to apply (2.2) to each scalar predictand Y_q, thus reducing the constraints to which $\hat{g}$ must satisfy. The whole p.s.s. for Y will then be obtained by collecting together the individual p.s.s.'s for Y_q.

2.1.1. The Normal Case.

As a special but very important case we consider the situation in which the sequence $\{X_u\}$ is normal, so that the predictive distribution is also normal. In this case one needs to consider only predictive expectation and dispersion.

(i) It is well known that T appears only in the expression of the conditional expectation which is a linear transformation of Z. Hence T is a vector of sums of elements of Z. The structure is actually derived by considering the conditional expectation of a single element Y_q of Y. If T_q is the p.s.s. of Z for Y_q, then $T = (T_q)$ is also a functionally minimal p.s.s. for Y, although T will generally not satisfy the property of being dimensionally minimal. This will typically be achieved by removing redundant components, functions of the remaining ones.

In practice a further simplification does arise whenever there exists a $\bar{g}$ such that $Y_r = \bar{g}Y_q$. In this case, in fact, it follows from (2.1) that $Y_q|T(Z)$ is distributed as $Y_r|T(g^*Z)$, so that, letting $T_q = T(Z)$ be the p.s.s. for Y_q, the corresponding p.s.s. for Y_r, T_r, is seen to be $T_r = T(g^*Z)$. Consequently only a p.s.s. for each single $\bar{g}$-orbit need be computed. Obviously we also have that $\mathrm{E}(Y_q|Z = z) = a + b'T_q(z)$ (where b' is a row-vector) and it follows, on applying (2.1), that $\mathrm{E}(Y_r|Z = z) = \mathrm{E}(\bar{g}Y_q|Z = z) = a + b'T_q(g^*z) = a + b'T_r(z)$, i.e. the coefficient vectors remain unchanged. To find T_q explicitly we apply (2.2) to the conditional expectation of Y_q and find, having set $Z = (Z_u)$, that

$$\Sigma a_{uq} Z_u = \Sigma a_{uq}(\hat{g}Z_u). \tag{2.4}$$

Equation (2.4) imposes equality constraints on the coefficients a_{uq} which completely identify T_q. More precisely if there exists a $\hat{g}$ such that $\hat{g}Z_u = Z_s$, then $a_{uq} = a_{sq}$. See also, although in a different context, Dawid (1977).

Since under normality the nature of G is characterized by the covariance matrix, a quick and easy method of finding out which coefficients in (2.4) are equal is to look at the covariance vector between Y_q and Z. If $\mathrm{Cov}(Y_q, Z_u) = \mathrm{Cov}(Y_q, Z_s)$ then $a_{uq} = a_{sq}$, and so we recognize immediately which elements of Z must be summed, or equivalently averaged, together.

(ii) With respect to the predictive dispersion structure, we remark that, by (2.1), $\mathrm{Cov}(Y|T(Z)) = \mathrm{Cov}(\bar{g}Y|T(g^*Z))$; but since the dispersion matrix does not depend on the specific sample realization, we can take $Z = 0$, so that $Z = g^*Z$, and hence $\mathrm{Cov}(Y|T(Z)) = \mathrm{Cov}(\bar{g}Y|T(Z))$, i.e. the predictive dispersion is invariant with respect to all $\bar{g}$. This clearly simplifies computations, since only a limited number of pairs of predictands need be considered.

2.2. *Invariance, Predictive Models and Posterior Distributions*

In a completely predictive approach to inference, one can dispense with the use of parameters and hence no statistical model is introduced.

One can however employ a model for prediction purposes by introducing a parameter, θ say, such that, conditionally on θ, predictands and data are independent. A way to construct such models has been suggested by Cifarelli and Regazzini (1980-81, 1982), where the parameter appears as a limit of a p.s.s., so that a prior distribution on θ becomes more meaningful. Clearly since different types of prediction generally imply different p.s.s.'s, the statistical model will change according to the prediction problem considered.

If a p.s.s. T converges almost surely, as the number of observations increases in a suitable way, to a random quantity, θ say, and $p_{Y|T}$ —where Y is the predictand— converges almost surely to $p_{Y|\theta}$, then this distribution is also used to model the data and θ is taken to be the parameter. For this to be possible of course Y must be such that, through $p_{Y|\theta}$ and p_θ, the joint distribution of any array of observables may be recovered.

If in the above discussion T is not a dimensionally minimal p.s.s. then it is easy to check that it will give rise to a model that is not identified. Informally, if T and T^* are two (functionally minimal) p.s.s.'s with T^* not dimensionally minimal —i.e. $T^* = (T, U)$— then

$p_{Y|(T,U)} = p_{Y|T}$ and so, if T converges to θ and U to φ, we shall that have $p_{Y|\theta,\varphi} = p_{Y|\theta}$, i.e. θ is a sufficient parameter and φ is redundant so that, in particular, the model given θ and φ is not identified, see Dawid (1979). When making inference about (θ, φ) only the posterior distribution of θ is necessary. Indeed, recalling that we model the data as Y given θ, we have $\varphi \perp Z|\theta$, i.e. $p(\varphi|z,\theta) = p(\varphi|\theta)$. Thus to obtain $p(\varphi|z)$ we have simply to take the expectation of $p(\varphi|\theta)$ with respect to the posterior distribution of θ.

We now consider invariance aspects of model and parameter distribution and to fully exploit invariance properties we shall assume that (Y, Z) is an array.

Consider (Y, Z) and let T be the p.s.s. of Z for Y. Then the action of $g*$ on Z —see Subsection 2.1— induces an action g_T on the T-space $\mathcal{T}$, in the sense that $g_T T = T'$ whenever $T' = T(g^* Z)$. Notice that such an action is well defined because of the identifiability of T. Clearly g_T reduces to the identity when g^* is of type $\hat{g}$.

Recalling that (Y, Z) is an array, then, by a property of invariant repetitive structures, all the original transformations g on the array Z may be seen as the union of transformations of type $\hat{g}$ and g^*. In this case the total invariance of Z is induced on T through g_T. Finally since θ is the limit of the p.s.s. T, invariance of the distribution of T reflects on the prior distribution of θ too.

Now if θ and θ' are, respectively, the limit, of T and $T' = g_T T$, recalling that g_T is a permutation, it follows that θ and θ' have the same distribution, i.e. that θ is invariant w.r.t. g_θ where $\theta' = g_\theta \theta$. Typically, however, the limit process may generate extra symmetries, so that the prior distribution of θ turns out to be invariant w.r.t. transformations other than g_θ.

Consider now the predictive distribution of Y given Z described in (2.1). Its equivariance can be reformulated in terms of Y given T, namely

$$p_{Y|T=t}(y) = p_{\bar{g}Y|T=g_T t}(y) \qquad \text{where } g_T T = T(g^* Z).$$

If we pass to the limit in this distribution we shall obtain

$$p_{Y|\theta}(y) = p_{\bar{g}Y|g_\theta \theta}(y). \tag{2.5}$$

where (2.5) stems from equality in distribution of (Y, T) and $(\bar{g}Y, g_T T)$ so that: if $(Y, T) \rightarrow (Y, \theta)$ and $(g\bar{Y}, g_T T) \rightarrow (\bar{g}Y, \theta')$, then $\theta' = \tilde{g}_\theta$. Notice that the group of $\tilde{g}_\theta$'s is not bigger than that of g_θ's because of the extra constraints imposed by the joint consideration of Y and θ.

Since in our approach the statistical model is derived from $p_{Y|\theta}$ by taking the distribution of Y given θ to coincide, for a suitable Y, with that of Y given θ, it follows that $p_{Z|\theta}$ will be equivariant according to (2.5). Hence, applying a result by Dawid (1977), we have $p_{\theta|Z=z}(\theta) = p_{\tilde{g}_\theta \theta|Z=g^* z}(\theta)$ or, equivalently,

$$p_{\theta|T=t}(\theta) = p_{\tilde{g}_\theta \theta|T=g_T t}(\theta), \tag{2.6}$$

i.e. the posterior distribution of θ is equivariant.

2.2.1. The normal case

When the joint distribution of the observables is normal, one can make some further remarks:

(i) T is a vector of averages over suitable blocks of data, so that, if the variables appearing in each block belong to a stationary process, then T converges almost surely.

(ii) Since T and $Y|T$ are normal, the limit distributions are still normal (as can be seen by inspecting the characteristic functions), so that it is enough to derive the limit expectations and dispersions.

(iii) Formula (2.6) can be used to derive the posterior expectation and dispersion of θ in a simplified way, exactly as formula (2.1) was used to derive the predictive expectation and dispersion of Y.

(iv) There exists a simple relationship between the structure of the predictive distribution of a linear function of Y, $h_1(Y) = W$ say, and that of the posterior distribution of $\tau = h_2(\theta) = E(W|\theta)$, so that in practice one can be immediately recovered from the other. Indeed τ is normal since h_2 is linear, and

$$E(\tau|Z) = E(E(W|\theta)|Z) = E(E(W|\theta, Z)|Z) = E(W|Z) = E(W|T), \qquad (2.7)$$

where the second equality follows from the independence of Y and Z given θ. Notice that, as is known under more general conditions, T is sufficient for τ, so we can write $E(\tau|T)$ instead of given Z.

Now let Σ_W, Σ_T, Σ_{WT} be the dispersion matrices of W, T and (W, T) and similarly for Σ_τ and $\Sigma_{\tau T}$. Then $E(W|T) = E(W) + \Sigma_{WT}\Sigma_T^{-1}(T - E(T))$ and $E(\tau|T) = E(\tau) + \Sigma_{\tau T}\Sigma_T^{-1}(T - E(T))$. Now, $E(W) = E(\tau)$ and from (2.7) it follows that $\Sigma_{WT} = \Sigma_{\tau T}$.

Letting now $\Sigma_{W|T}$ and $\Sigma_{\tau|T}$ be, respectively, the predictive and posterior dispersion matrices we have $\Sigma_{W|T} = \Sigma_W - \Sigma_{WT}\Sigma_T^{-1}\Sigma_{TW}$ and $\Sigma_{\tau T} = \Sigma_\tau - \Sigma_{\tau T}\Sigma_T^{-1}\Sigma_{T\tau} = \Sigma_\tau - \Sigma_{WT}\Sigma_T^{-1}\Sigma_{TW}$, so that

$$\Sigma_{\tau|T} = \Sigma_{W|T} + (\Sigma_\tau - \Sigma_W). \qquad (2.8)$$

3. AN APPLICATION TO THE DIALLEL-CROSS EXPERIMENTAL DESIGN

3.1. Description of the Design

The diallel cross is an experimental design used in genetics which consists in making all possible crosses between several genotypes (e.g. plants or animals) in order to study and compare performances of lines in hybrid combination; see e.g. Yates (1947), and Griffing (1956).

If there are k lines and we arrange male parents on rows and female parents on columns then the results of this experiment are typically summarized in a $k \times k$ table with entries $(X_{ij} : 1 \leq i, j \leq k)$. Actually there are several configurations of this experiment since crosses of type X_{ii} or the reciprocal cross of X_{ij}, namely X_{ji}, may or may not be included; we shall deal with the case in which only X_{ii} is not included, so that we have altogether $k(k-1)$ observables $(X_{ij} : i \neq j)$.

The natural repetitive structure for this experiment is represented by the sample spaces corresponding to the elements of the squares without diagonal entries $(X_{ij} : n+1 \leq i, j \leq n+r, i \neq j, n \geq 0, r \geq 1) = A_{nr}$. An array A_{ms} will be regarded as a projection of A_{nr} whenever $n \leq m$ and $r \geq s$ with at least one strict inequality.

As far as the invariance of the single array A_{nr} is concerned, we shall follow Speed (1982) and assume that the distribution of A_{nr} is invariant with respect to permutations of lines, namely $X_{ij} \rightarrow X_{\pi(i)\pi(j)}$ (where π is a permutation of the integers $\{n+1, \ldots, n+r\}$) and to permutation of males and females for all lines, i.e. $X_{ij} \rightarrow X_{\lambda(i,j)}$ for all i, j, where λ is a permutation of $\{i, j\}$. Therefore the action of G on the array A_{nr} is generated by $\pi \times \lambda$.

Under the assumption of normality of the joint distribution of the observables, the previous invariance hypotheses amount to $E(X_{ij}) = m$ (say) assuming, as usual, here and in the following $i \neq j$, and for the dispersion structure:

$$\begin{aligned}
&\text{Cov}(X_{ij}, X_{i^*j^*}) = u(i \neq i^*, j \neq j^*, i \neq j^*, j \neq i^*), \text{Cov}(X_{ij}, X_{ij^*}) = \\
&= \text{Cov}(X_{ij}, X_{i^*j}) = s(i \neq i^*, j \neq j^*), \text{Cov}(X_{ij}, X_{i^*i}) = \text{Cov}(X_{ij}, X_{jj^*}) = \\
&= a(i \neq i^*, j \neq j^*, i^* \neq j, i \neq j^*), \text{Cov}(X_{ij}, X_{ji}) = r, \text{Var}(X_{ij}) = v.
\end{aligned}$$

The constants u, s, a, r and v will be such that the dispersion structure of any array is positive definite.

3.2. *Predictive Sufficient Statistics*

Suppose that data Z are represented by $(X_{ij} : n+1 \leq i, j \leq n+k, i \neq j)$. Then there are three single basic predictands to consider: $X_{iq}; X_{qi}$ and X_{qq^*} where: $n+1 \leq i \leq n+k$ and $q, q^* \notin \{n+1, \ldots, n+k\}$. Typically a predictand Y will be a set of elements of these three types.

In order to find the p.s.s. for Y, as mentioned in Subsection 2.1.1. (i) we may consider the p.s.s. of each scalar predictand,and in particular only those for X_{iq} and X_{qq^*}. Indeed because of invariance, one can derive from the p.s.s. of X_{iq} all the p.s.s.'s for predictands of type $X_{i^*q^*}$ and $X_{q^*i^*}(n+1 \leq i^* \leq n+k)$.

Consider first prediction of X_{iq}. By inspection of the dispersion matrix $\Sigma_{X_{iq}Z}$ we derive, following the arguments of Subsection 2.2., that the p.s.s. of Z for X_{iq} is given by

$$T_i = (X_{i\cdot}, X_{\cdot i}, {}_iX_{\cdot\cdot}), \quad \text{where } X_{i\cdot} = (k-1)^{-1}\sum_{j \neq i} X_{ij}; X_{\cdot i} = (k-1)^{-1}\sum_{i^* \neq i} X_{i^*i};$$

$${}_iX_{\cdot\cdot} = ((k-1)(k-2))^{-1} \sum_{\substack{i^* \neq i \\ i^* \neq j}} \sum_{j \neq i} X_{i^*j}.$$

It is easy to verify that T_i is also the p.s.s. of Z for X_{qi}. Finally, since $\Sigma_{X_{qq^*}Z} = u1_{k(k-1)}$, where 1_k is the unit vector of size k, the p.s.s. for X_{qq^*} is given by $X_{\cdot\cdot} = (k(k-1))^{-1}\Sigma\Sigma_{i \neq j} X_{ij}$.

3.3. *Predictive Expectation*

To obtain the predictive distribution of Y we need to compute its expectation and dispersion given Z.

Here and in the following we will omit most of the repetitive calculations and we will outline only the general procedure. Details may be found in Consonni and Veronese (1987).

We consider predictive expectation first. As already recalled, it is enough to consider separately the posterior expectations of X_{iq}, X_{qi} and X_{qq^*}. Now, since $\Sigma_{X_{iq}T_i} = (s, a, u)$, and setting for simplicity Var$(X_{i\cdot})$ =Var$(X_{\cdot i}) = A$, Var$({}_iX_{\cdot\cdot}) = B$, Cov$(X_{i\cdot}, X_{\cdot i}) = C$, Cov$(X_{i\cdot}, {}_iX_{\cdot\cdot}) =$ Cov$(X_{\cdot i}, {}_iX_{\cdot\cdot}) = D$, $E = AB - D^2$, $F = BC - D^2$, $G = (A - C)$, $H = A + C$, we have

$$\begin{aligned} E(X_{iq}|Z) &= E(X_{iq}|T_i) = m + \Sigma_{X_{iq}T_i}\Sigma_{T_i}^{-1}(T_i' - m1_3) \\ &= m + G^{-1}(E+F)^{-1}\{(sE - aF - uGD)(X_{i\cdot} - m) \\ &+ (aE - sF - uGD)(X_{\cdot i} - m) + G(uH - (s+a)D)({}_iX_{\cdot\cdot} - m)\}. \end{aligned} \tag{3.1}$$

Now, applying the λ-transformation and recalling (2.1) we have

$$E(X_{qi}|(X_{i\cdot}, X_{\cdot i}, {}_iX_{\cdot\cdot})) = E(X_{iq}|(X_{\cdot i}, X_{i\cdot}, {}_iX_{\cdot\cdot})),$$

so that, in practice, the predictive expectation of X_{qi} can be derived from (3.1) by interchanging the coefficients of $X_{i\cdot}$ and $X_{\cdot i}$.

Finally the predictive expectation of X_{qq^*} is directly established to be

$$E(X_{qq^*}|Z) = E(X_{qq^*}|X_{\cdot\cdot}) = m + (\text{Var}X_{\cdot\cdot})^{-1}\text{Cov}(X_{qq^*}, X_{\cdot\cdot})(X_{\cdot\cdot} - m). \tag{3.2}$$

Next we consider the dispersion structure of Y given Z. As described in (ii) of Subsection 2.1.1., the predictive dispersion structure of Y is invariant with respect to all $\bar{g}$, so that we shall compute the predictive covariances only of those pairs which can generate, through $\bar{g}$, all other pairs. It is instructive to remark that one can dispense altogether with matrix manipulations —which in some cases, e.g. to derive $\text{Cov}(X_{iq}, X_{i^*q}|Z)$ $i \neq i^*$, may involve the inversion of the covariance matrix of a five-dimensional p.s.s.— by means of the following identity which holds under normality, namely:

$$\text{Cov}(Y_1, Y_2|Z) = \text{Cov}(Y_1, Y_2) - \text{Cov}(E(Y_1|Z), E(Y_2|Z)), \tag{3.3}$$

where Y_1, Y_2 are any scalar predictands belonging to Y.

3.4. Predictive Dispersion

We now consider the predictive dispersion of Y. In order to obtain a complete description of its structure, it is expedient to derive the predictive dispersion structure of all predictands belonging to the array $(\bar{Y}, Z)$ where $\bar{Y}$ contains Y. Actually it is useful to partition $\bar{Y}$ as $S_1 \cup S_2 \cup S_3$ as in the following table

S_1	S_3
S_2	Z

where $S_1 = (X_{qq^*})$, $S_2 = (X_{iq})$, $S_3 = (X_{qi})$, with $n+1 \leq i \leq n+k$, $n_1 \leq q$, $q^* \leq n$, and $n - n_1 \geq 3$. This last condition and the choice of working with $\bar{Y}$ instead of Y allows one to derive predictive covariance types for all possible predictands (and hence for Y) by means of invariance alone.

We now discuss separately the predictive dispersion structure for each block S_t and for pairwise combinations therefrom.

(i) Predictive dispersion within S_1. Since the group of transformations $\bar{g}$ induced by $g(Y, Z) = (\bar{g}Y, g^*Z)$ —see Subsection 2.1.— has the same nature as the full group G, the predictive dispersion structure is analogous to the marginal one and so there will be at most four distinct types of covariances, besides the variance.

(ii) Predictive dispersion within $S_2(S_3)$. In this case the constraint that both Y and Z be closed under permutation transformations means that there are fewer possible transformations of type $\bar{g}$ than under (i). Consequently all variances are equal, and there exist at most three distinct covariances within S_2, namely those identified by the pairs:

$$(X_{iq}, X_{i^*q})i \neq i^*, (X_{iq}, X_{iq^*})q \neq q^*, (X_{iq}, X_{i^*q^*})i \neq i^*, q \neq q^*.$$

Similar remarks apply to S_3 by obvious symmetry.

We consider now covariances which arise from mixing two basic predictive sets.

(iii) Predictive dispersion between S_1 and $S_2(S_3)$. In this case the action of $\bar{g}$ is such that there are three distinct covariances conditionally on data Z (and indeed unconditionally), identified by the pairs: (X_{qq^*}, X_{iq}), (X_{q^*q}, X_{iq}) and $(X_{q^*\tilde{q}}, X_{iq})(q \neq q*$ and $\tilde{q} \neq q)$. Similarly when S_2 is replaced by S_3.

(iv) Predictive dispersion between S_2 and S_3. Here again, as in case (ii), the predictive covariance is less structured than the marginal one and is characterized by the pairs: (X_{iq}, X_{qi}), (X_{iq}, X_{qi^*}), (X_{iq}, X_{q^*i}), $(X_{iq}, X_{q^*i^*})i \neq i^*$.

Finally, we have further equality relationships between covariance values across different predictive sets due to the permutation λ which interchanges indexes: thus $\text{Cov}(X_{iq}, X_{i^*q^*}|\cdot$

$Z) = \text{Cov}(X_{qi}, X_{q^*i^*}|\ Z)$ and $\text{Cov}(X_{iq}, X_{q^*\tilde{q}}|Z) = \text{Cov}(X_{qi}, X_{\tilde{q}q^*}|Z)$ for all i, q and all choices of i^*, q^* and $\tilde{q}$.

Because of previous considerations, it follows that we need to compute two predictive variances and fourteen distinct covariances, which can be evaluated as indicated in (3.3).

Thus, for example, to compute the variance of X_{iq} given Z —which is equal to that of X_{qi} given Z— we may proceed as follows:

$$\text{Var}(X_{iq}|Z) = \text{Var}(X_{iq}|T_i) = \text{Var}(X_{iq}) - \text{Var}(E(X_{iq}|T_i)),$$

where $\text{Var}(X_{iq}) = v$ and, from (3.1),

$$\text{Var}(E(X_{iq}|T_i)) = G^{-1}(E+F)^{-1}\{(s^2+a^2)E + u^2GH - 2asF - 2u(a+s)GD\}.$$

3.5. *Predictive Models*

In this section we shall apply to the diallel cross experiment the methodology suggested in Subsection 2.2. to construct a statistical model for predictive purposes.

Consider, as usual, data $Z = (X_{ij} : n+1 \leq i, j \leq n+k)$ and a predictand $Y = ((X_{qq^*} : 1 \leq q, q^* \leq n), (X_{iq} : n+1 \leq i \leq n+k, 1 \leq q \leq n), (X_{qi} : 1 \leq q \leq n, n+1 \leq i \leq n+k), n \geq 4)$.

From previous arguments a p.s.s. for Y, T^*, is obtained by collecting together all p.s.s.'s for each single component of Y, i.e. $T^* = (X_{..}, (X_{i.}), (X_{.i})({}_iX_{..}))$. Such a p.s.s., is not dimensionally minimal. A first reduction may be obtained by omitting $({}_iX_{..})$; this leaves us with $(X_{..}, (X_{i.}), (X_{.i}))$, or equivalently $T = (X_{..}, (X_{i.} - X_{..}), (X_{.i} - X_{..}))$. Since $\sum(X_{i.} - X_{..}) = \sum(X_{.i} - X_{..}) = 0$, the dimension of T can be further reduced by omitting two components such as $(X_{1.} - X_{..})$ and $(X_{.1} - X_{..})$. This, however, would generate a loss of symmetry in the induced parametric model. For this reason we shall prefer T even though the corresponding model will not be identified, see Subsection 2.2.

We now derive the predictive model for Y corresponding to such a choice of T.

To obtain the parameter we need to compute the limit of T as the number of elements within each average extends to infinity. In order to do so it is useful to distinguish between the actual number of lines involved under Z —namely k— and the conceptual number of lines, N say, over which averaging is performed. We may thus rewrite each component of T using N so that, for example, $X_{i.}$ becomes:

$$(N-1)^{-1}\sum_{n+1}^{n+N} X_{ij} \qquad \text{for } i = n+1, \ldots, n+k.$$

From points (i) and (ii) of Subsection 2.2.1. we derive directly that T converges almost surely $(N \to \infty)$ to $\theta = (\mu, (\alpha_i), (\beta_i))$, $i = n+1, \ldots, n+k$ where θ is normal and:

$$\begin{aligned} &E(\mu) = m, E(\alpha_i) = E(\beta_i) = 0, \text{Var}(\mu) = u, \text{Var}(\alpha_i) = \text{Var}(\beta_i) = s - u, \\ &\text{Cov}(\mu, \alpha_i) = \text{Cov}(\mu, \beta_i) = 0, \text{Cov}(\alpha_i, \alpha_{i^*}) = \text{Cov}(\beta_i, \beta_{i^*}) = 0, \\ &\text{Cov}(\alpha_i, \beta_i) = a - u, \text{Cov}(\alpha_i, \beta_{i^*}) = 0, \qquad \text{where } i \neq i^*. \end{aligned}$$

In order to derive the distribution of Y given θ we make use of the equivariance of $p_{Y|\theta}$; see (2.5). In particular, only the conditional expectation on each $\bar{g}$-orbit, and the conditional covariances of those pairs which characterize the $\bar{g}$-invariant dispersion structure need be considered.

We consider conditional expectation first. It is easy to see that there are two $\bar{g}$-orbits, namely, one for all predictands of type X_{qq^*} and one for the remaining predictands.

Having computed all the possible covariances between predictands and the components of the p.s.s. it is possible to derive $E(X_{qq^*}|\theta)$ and $E(X_{iq}|\theta)$. Indeed we have

$$\begin{aligned} E(X_{qq^*}|\theta) &= \lim_{N\to\infty} E(X_{qq^*}|T) = \lim_{N\to\infty} E(X_{qq^*}|X_{..}) \\ &= \lim_{N\to\infty} \left[E(X_{qq^*}) + \text{Var}(X_{..})^{-1}\text{Cov}(X_{qq^*}, X_{..})(X_{..} - E(X_{..}))\right] \\ &= m + u^{-1}u(\mu - m) = \mu \end{aligned}$$

and setting $T_i' = (X_{..}, X_{i.} - X_{..}, X_{.i} - X_{..})$

$$\begin{aligned} E(X_{iq}|\theta) &= \lim_{N\to\infty} E(X_{iq}|T) = \lim_{N\to\infty} E(X_{iq}|T_i) \\ &= m + \left(\lim_{N\to\infty} \Sigma_{X_{iq}T_i}\right)\left(\lim_{N\to\infty} \Sigma_{T_i}\right)^{-1} \begin{pmatrix} \mu - m \\ \alpha_i \\ \beta_i \end{pmatrix}, \end{aligned}$$

where we have used the fact that $\lim \Sigma_{T_i}^{-1} = (\lim \Sigma_{t_i})^{-1}$, whenever the operations involved are allowable. We thus obtain

$$E(X_{iq}|\theta) = \mu + \alpha_i,$$

and, recalling the equivariance of $p_{Y|\theta}$,

$$p_{X_{iq}|\theta}(y) = p_{X_{qi}|\tilde{\lambda}\theta}(y), \qquad \text{where } \tilde{\lambda}(\theta) = \tilde{\lambda}(\mu, (\alpha_i), (\beta_i)) = (\mu, (\beta_i), (\alpha_i)),$$

we have also

$$E(X_{qi}|\theta) = \mu + \beta_i.$$

Next consider the conditional dispersion of Y. Since such a dispersion is $\bar{g}$-invariant, only a limited number of covariances, as usual, need be computed as in Subsection 3.4., so, for example, we have:

$$\begin{aligned} \text{Cov}(X_{iq}, X_{i^*q}|\theta) &= \text{Cov}(X_{iq}, X_{i^*q}) - \text{Cov}(E(X_{iq}, |\theta), E(X_{i^*q}|\theta)) \\ &= a - \text{Cov}(\mu + \alpha_i, \mu + \alpha_{i^*}) = a - u. \end{aligned}$$

It follows from the previous discussion that the model for the predictand Y is given by:

$$X_{qq^*} = \mu + \delta_{qq^*} \tag{3.4.1}$$

$$X_{iq} = \mu + \alpha_i + \varepsilon_{iq} \tag{3.4.2}$$

$$X_{qi} = \mu + \beta_i + \tau_{qi} \tag{3.4.3}$$

with $n+1 \le i \le n+k$, $1 \le q$, $q^* \le n (q \ne q^*)$, where (δ_{qq^*}), (ε_{iq}) and (τ_{qi}) are suitable error terms.

We now extend representations (3.4) to the data $X_{ij} : n+1 \le i, j \le n+k$, $i \ne j$, in order to perform a full parametric Bayesian inference.

Since (3.4.2) and (3.4.3) are sufficient to model the data and involve the full parameter set $\theta = (\mu, (\alpha_i), (\beta_i))$, we shall take as our statistical model

$$\begin{aligned} X_{qq^*} &= \mu + \delta_{qq^*} \\ X_{iJ} &= \mu + \alpha_i + \varepsilon_{iJ} \qquad i > J \\ X_{Ij} &= \mu + \beta_j + \tau_{Ij} \qquad I < j \end{aligned} \tag{3.4.4}$$

where $1 \leq I,\ J \leq n+k,\ 1 \leq q,\ q^* \leq n$ and $n+1 \leq i,\ j \leq n+k$.

Notice however that model (3.4.4) can be further simplified if the predictands involve only new lines. In this case, in fact, the model for predictands and data may be written as

$$X_{IJ} = \mu + \delta_{IJ}.$$

The joint distribution of parameters and errors will be normal and is completely specified in Consonni and Veronese (1987). We simply remark that the covariance structure between parameters and errors may be obtained by first expressing errors as functions of observables and parameters, and then computing the covariances between observables and parameters using the fact that the latter are limits of observables.

3.6. Posterior Distributions and Predictive Distribution

Having obtained a predictive model, we perform traditional prior-to-posterior Bayesian analysis. This can be done either directly or exploiting the relationship between posterior and predictive distributions —see point (iv) of Subsection 2.2.1.

As usual, because of normality, we simply compute $E(\theta|Z) = E(\theta|T)$ and the posterior dispersion matrix $\Sigma_{\theta|T}$.

Because of equivariance of the posterior distribution of θ and recalling the action of g_T which induces $\tilde{g}_\theta$ since (Y, Z) is an array, it is readily seen that the basic $\tilde{g}_\theta$ actions are: (i) $(\mu, (\alpha_i), (\beta_i)) \rightarrow (\mu, (\alpha_{\pi(i)}), (\beta_{\pi(i)}))$ and (ii) $(\mu, (\alpha_i), (\beta_i)) \rightarrow (\mu, (\beta_i), (\alpha_i))$. It therefore follows that, in order to evaluate the posterior expectation of θ, it is enough to compute $E(\mu|T)$ and $E(\alpha_i|T)$ using arguments similar to those employed when deriving the predictive distribution of Y.

Furthermore, the posterior dispersion of θ is identified by Var $(\mu|T)$, Var $(\alpha_i|T) =$ Var $(\beta_i|\ T)$; Cov $(\mu, \alpha_i|T)$= Cov $(\mu, \beta_i|T)$,Cov $(\alpha_i, \beta_i|T)$, Cov $(\alpha_i, \alpha_{i^*}|T) =$ Cov $(\beta_i, \beta_{i^*}|T)$, Cov $(\alpha_i, \beta_{i^*}|T)$, $n+1 \leq i, i^* \leq n+k, i \neq i^*$, which can be computed using a formula similar to (3.3).

Notice that the lack of identification of the model is reflected in the posterior distribution of θ through symmetric constraints on posterior expectations and dispersions such as $E(\sum \alpha_i|T) = E(\sum \beta_i|T) = 0$.

Clearly, if the predictive distribution is already available, then, because of remark (iv) of Subsection 2.2.1., we can directly derive the posterior distribution of θ, noting that

$$E(\mu|T) = E(X_{qq^*}|T), E(\alpha_i|T) = E(X_{iq}|T) - E(X_{qq^*}|T).$$

Furthermore, the relevant distinct posterior covariances may be evaluated according to (2.8) having set $\tau = \theta$ and $W = (X_{qq^*}, (X_{iq} - X_{qq^*})), (X_{qi} - X_{qq^*})$.

4. DISCUSSION

We have seen in this paper how the joint consideration of invariance and predictive ideas may be fruitfully put to use. In particular, the notion of predictive sufficient statistic (p.s.s.) has proved useful both as a computational aid when deriving the predictive distribution, and as a conceptual tool in constructing and interpreting a suitable predictive model. Furthermore invariance ideas help appreciate structural properties of the underlying design and considerably reduce the actual dimensionality of the problem, which is especially useful from a computational viewpoint.

Our analysis reinforces the concept of duality between parameters and observables and the allied duality between predictive and posterior distributions. This point is especially transparent under normality, where an easy relationship allows one to move from posterior to predictive distribution and *vice versa.*

ACKNOWLEDGEMENTS

Research partially financed by C.N.R. - contract. n. 85.01143. 10-115.16992 and by Ministero Pubblica Istruzione (60% grants).

REFERENCES

Cifarelli, D. M. and Regazzini, E. (1980–81). Sul ruolo dei riassunti esaustivi ai fini della previsione in contesto bayesiano. *Part I, Riv. Mat. Sci. Econom. Sociali* **3**, 109–125; *Part II, Riv. Mat. Sci. Econom. Sociali* **4**, 3–11.

Cifarelli, D. M. and Regazzini, E. (1982). Some considerations about mathematical statistics teaching methodology suggested by the concept of exchangeability. *In Exchangeability in Probability and Statistics*, 185–205, (G. Koch and F. Spizzichino, eds.). Amsterdam: North-Holland.

Consonni, G. and Veronese, P. (1987). Invariance and bayesian predictive analysis with an application to the diallel cross design. Research Report - Studi Statistici n. 16, Universita L. Bocconi, Milan.

Dawid, A. P. (1977). Invariant distributions and analysis of variance models. *Biometrika* **64**, 291–297.

Dawid, A. P. (1979). Conditional independence in statistical theory (with discusion). *J. Roy. Statist. Soc. B* **41**, 1–31.

Dawid, A. P. (1982). Intersubjective statistical models. *In Exchangeability in Probability and Statistics*, 217–232, (G. Koch and F. Spizzichino, eds.). Amsterdam: North-Holland.

Geisser, S. (1986). Predictive analysis. *Encyclopedya of Statistical Science*, (S. Kotz, N.L. Johnson and B. Read, eds.). New York: John Wiley.

Griffing B. (1956). Concept of general and specific combining ability in relation to diallel crossing systems. *Australian J. Biological Sciences* **9**, 463–493.

James, A. T. (1982). Analysis of variance determined by symmetry and combinatorial properties of zonal polynomials. *In Statistics and Probability: Essays in Honor of C. R. Rao*, 329–341, (G. Kallianpur, P. R. Krishnaiah and J. K. Ghosh, eds.). Amsterdam: North-Holland.

Lauritzen, S. L. (1984). Extreme point models in statistics. *Scand. J. Statist.* **11**, 65–91.

Speed, T. P. (1982). The analysis of variance. (To appear).

Yates F. (1947). The analysis of data from all possible reciprocal crosses between a set of parental lines. *Heredity* **1**, 287–301.

BAYESIAN STATISTICS 3, pp. 569–577
J. M. Bernardo, M. H. DeGroot, D. V. Lindley and A. F. M. Smith, (Eds.)

On Differentiability Properties of Bayes Operators

A. CUEVAS and P. SANZ
Universidad Complutense, Madrid

SUMMARY

Bayes formula can be viewed as an operator B depending on three "variables": the sample x, the likelihood f and the prior P. We study differentiability properties of this operator, with respect to P and f. Specifically, we obtain two theorems on differentiability (in the Fréchet sense) of B with respect to P. The statistical meaning of these results is discussed in connection with the theory of influence functions.We also prove a result on differentiability of B with respect to f. A general discussion is included in the last section.

Keywords: INFLUENCE CURVE; FRÉCHET DIFFERENTIAL; BAYESIAN ROBUSTNESS.

1. INTRODUCTION

Let $\Omega \subset \Re^k$ be a Borel set (parameter space) and let $\{f(\cdot|\theta), \theta \in \Omega\}$ be a family of μ-density functions on $\Re^p$, for some σ-finite measure μ on $\mathcal{B}(\Re^p)$ ($\equiv$ Borel σ-algebra on $\Re^p$). In the sequel $x = (x_1, x_2, \ldots, x_n)$ will denote a (fixed) random sample and $f(x|\theta)$ will represent the likelihood $\prod_{i=1}^n f(x_i|\theta)$.

The space of probability measures on $\mathcal{B}(\Omega)$ will be denoted by $\mathcal{P}$. Unless otherwise stated, $\mathcal{P}$ is endowed with the Prohorov metric

$$d(P,Q) = \inf\{\varepsilon > 0 : P(A) \leq Q(A^\varepsilon) + \varepsilon \quad \forall A \in \mathcal{B}(\Omega)\} \quad P, Q \in \mathcal{P}.$$

For each "prior distribution" $P \in \mathcal{P}$ the notation $P(\cdot|x)$ will represent the "posterior distribution" associated with the P-density $p(\theta|x) = f(x|\theta)/\int_\Omega f(x|\theta)\, dP(\theta)$ (the integration space Ω will be omitted henceforth).

The elements $P, Q, \ldots$ of $\mathcal{P}$ will be also denoted, when convenient, in terms of their distributions functions $F, G, \ldots$.

Loosely speaking, the posterior distribution can be viewed as an operator depending on three "variables": the sample x, the prior distribution P and the likelihood function f. This simple observation suggests, as a natural plan of work, the idea of studying the analytical properties (measurability, continuity, differentiability, ...) of this operator with respect to its variables (jointly or separately).

In our opinion, the interest of such a study in Bayesian theory is motivated for two main reasons:

(i) Measurability, continuity and differentiability are fundamental properties of primary interest in any mathematical model. Every result about them provides qualitative information on the structure of the model. In our case the element to be analyzed (the posterior distribution) is of central interest in the theory under study.

(ii) The above mentioned properties can be interpreted in statistical terms. For instance, measurability of the transformation $\boldsymbol{x} \to P(\cdot|\boldsymbol{x})$ would allow us to consider the posterior distribution as a "generalized estimate" taking values in $\mathcal{P}$. Thus, consistency, sampling distributions and other relevant concepts can be defined for the posterior distribution as a generalization of the corresponding concepts for parametric estimates. This idea has been used by Cuevas (1988) in proposing a new approach to Bayesian robustness, based upon an adaptation of Hampel's theory of qualitative robustness.

Continuity properties are closely related to the notion of robustness. For example, the continuity of the posterior distribution with respect to P, can be interpreted as an assessment of stability against infinitesimal pertubations in the prior distribution.

On the other hand, differentiability properties of the Bayes operator can be used to translate to a Bayesian context the theory of influence curves and its applications. A partial discussion of this topic is included below.

Finally, it is known that differential methods are useful tools in the study of asymptotic results; however, this application is not considered here.

The present paper does not attempt an exhaustive development of the above schedule of work. In fact, we just obtain some results on differentiability and (secondarily) continuity of Bayes operator. Several open problems arise in this context (some of which will be discussed below).

As far as we know, the first result on the differentiability of the Bayes operator (with respect to P) is due to Diaconis and Freedman (1986).

The aim of our work is to provide a further element in the elucidation of the mathematical structure involved in Bayes methodology.

The paper is organized as follows: in Section 2 we obtain two results on differentiability of the Bayes operator with respect to the prior. In Section 3 we suggest some ideas on influence functions in the Bayesian framework. Section 4 includes a theorem on Fréchet differentiability of the Bayes operator considered as a function of the likelihood. Finally, Section 5 is devoted to discussion and comments.

2. BAYES OPERATOR AS A FUNCTION OF THE PRIOR

With the notation introduced in Section 1, let us fix $\boldsymbol{x}$ and f such that

$$\int f(\boldsymbol{x}|\theta)d\theta < \infty \tag{2.1}$$

and

$$0 < \int f(\boldsymbol{x}|\theta)dP(\theta) < \infty, \quad \text{for all } P \in \mathcal{P} \tag{2.2}$$

We will require the following supplementary notation in this Section:

$\mathcal{L}^1 \equiv$ space of real Lebesgue-integrable functions on Ω, endowed with the L_1-norm: $||g||_1 = \int |g(\theta)|d\theta$

$B \equiv$ Bayes operator, $B : \mathcal{P} \to \mathcal{L}^1$, defined by $B(P) = p(\cdot|\boldsymbol{x})$.

$\mathcal{G} \equiv$ vector space of finite signed measures on $\mathcal{B}(\Omega)$.

The parameter space will be assumed to be a closed (possibly unbounded) interval in $\Re$. Let us observe that, in this case, the elements of $\mathcal{G}$ can be represented in terms of distribution functions, in the form $\Delta = F - G$, where F, $G : \Re \to \Re$ are bounded, right-continuous and non-decreasing functions with $F(-\infty) = G(-\infty) = 0$. The space $\mathcal{G}$ endowed with the norm

$$||\Delta||_\infty = \sup_\theta |\Delta(\theta)| \tag{2.3}$$

will be denoted by $\mathcal{G}_\infty$.

The choice of $\mathcal{L}^1$ as the final space of the operator B is discussed in Section 4.

It is straightforward to establish the continuity of B, provided that $f(\boldsymbol{x}|\theta)$ is a bounded continuous function of θ.

Since $\mathcal{P}$ is not a vector space, the usual concept of the Fréchet differential cannot be directly employed in studying differentiability of B. In order to avoid this difficulty, two alternative methods may be adopted:

(a) We may consider a "weakened" version of the Fréchet differential which retains the essential meaning of the concept without imposing a linear structure in the space of definition. To be concrete, we will consider the following definition, which has been previously used by Clarke (1983) for the study of statistical real-valued functionals defined on $\mathcal{P}$:

Definition 1. *We will say that a functional $T : \mathcal{P} \to \mathcal{L}^1$ is differentiable at $F \in \mathcal{P}$, with respect to the metric D defined in $\mathcal{P}$, iff there exists a linear functional $T'_F : \mathcal{G} \to \mathcal{L}^1$ such that*

$$\frac{||T(F) - T(G) - T'_F(F - G)||_1}{D(F, G)} \to 0 \tag{2.4}$$

as $D(F, G) \to 0$.

(b) The operator B can be extended to the normed space $\mathcal{G}_\infty$. Thus, we define $\hat{B} : \mathcal{G}_\infty \to \mathcal{L}^1$ by

$$\hat{B}(\Delta) = \begin{cases} f(\boldsymbol{x}|\cdot)/\int f(\boldsymbol{x}|\theta)d\Delta(\theta) & \text{if } \int f(\boldsymbol{x}|\theta)d\Delta(\theta) \neq 0 \\ g & \text{otherwise} \end{cases} \tag{2.5}$$

where g is a density function on Ω (not depending on Δ).

The standard definition of the Fréchet differential (see. e.g., Kolmogorov-Fomin (1975)) can now be applied to this operator.

Next, we obtain theorems on differentiability for B and $\hat{B}$, according to both approaches (a) and (b).

Theorem 1. *Let us assume that $f(\boldsymbol{x}|\cdot)$ is a continuous function of bounded variation on Ω.*

Then, the operator B is differentiable (in the sence of Def. 1) at F, for all $F \in \mathcal{P}$ with respect to the Kolmogorov metric (that is, for $D = d_k$, with $d_k(F, G) = \sup_\theta |F(\theta) - G(\theta)|$).

Moreover, if $F \in \mathcal{P}$ is absolutely continuous with bounded density function, then B is also differentiable at F, with respect to the Prohorov and Lévy metrics (the latter defined by $d_L(F, G) = \inf\{\delta > 0 : F(\theta - \delta) - \delta \leq G(\theta) \leq F(\theta + \delta) + \delta, \forall\theta \in \Omega\}$).

Proof. We assume $\Omega = \Re$ (the proof would be analogous for $\Omega = [a, \infty)$ or $\Omega = [a, b]$). The natural candidate for B'_F is given by the "directional derivative"

$$B'_F(\Delta) = \frac{d}{dt}\hat{B}(F + t\Delta)\bigg|_{t=0}$$

where $\hat{B}(F + t\Delta) = f(\boldsymbol{x}|\cdot)/[\int f(\boldsymbol{x}|\theta)dF(\theta) + t\int f(\boldsymbol{x}|\theta)d\Delta(\theta)]$. (This expression is well defined for t small enough, since $\int f(\boldsymbol{x}|\theta)dF(\theta) > 0$.)

A direct calculation gives

$$B'_F(\Delta) = \frac{-f(\boldsymbol{x}|\cdot)\int f(\boldsymbol{x}|\theta)d\Delta(\theta)}{[\int f(\boldsymbol{x}|\theta)dF(\theta)]^2} = -B(F)\cdot\frac{f(\boldsymbol{x}|\theta)d\Delta(\theta)}{\int f(\boldsymbol{x}|\theta)dF(\theta)}$$

Now, in accordance with Definition 1, we must prove

$$C(F,G) \to 0, \text{as } D(F,G) \to 0$$

where $C(F,G)$ is the quotient appearing in (2.4); in our case, by substituting the expressions of B and B'_F in $C(F,G)$, we have

$$C(F,G) = \frac{1}{D(F,G)}\|B(F)\|_1 \cdot \frac{\left[\int f(\boldsymbol{x}|\theta)d(F-G)(\theta)\right]^2}{\int f(\boldsymbol{x}|\theta)dF(\theta)\cdot\int f(\boldsymbol{x}|\theta)dG(\theta)}$$

By integration by parts, the term between square brackets becomes

$$\int f(\boldsymbol{x}|\theta)d(G-F)(\theta) = (G-F)(\infty)\cdot f(\boldsymbol{x}|\infty) - (G-F)(-\infty)\cdot f(\boldsymbol{x}|-\infty) -$$

$$-\int (G-F)(\theta)df(\boldsymbol{x}|\theta) = -\int (G-F)(\theta)df(\boldsymbol{x}|\theta)$$

since $G(\infty) = F(\infty) = 1$, $G(-\infty) = F(-\infty) = 0$ and $f(\boldsymbol{x}|\cdot)$ is bounded. Thus

$$\left|\int f(\boldsymbol{x}|\theta)d(G-F)(\theta)\right| = \left|\int (G-F)(\theta)df(\boldsymbol{x}|\theta)\right| \le V\cdot\sup_\theta |G(\theta)-F(\theta)|$$

where V denotes the total variation of $f(\boldsymbol{x}|\cdot)$.

So, we have, for $D = d_k$

$$C(F,G) \le \|B(F)\|_1 \cdot \frac{V^2\cdot d_k(F,G)}{\int f(\boldsymbol{x}|\theta)dF(\theta)\cdot\int f(\boldsymbol{x}|\theta)dG(\theta)} \to 0, \qquad (2.6)$$

as $d_k(F,G) \to 0$ (observe that $\int f(\boldsymbol{x}|\theta)dG(\theta) \to \int f(\boldsymbol{x}|\theta)dF(\theta)$). This proves differentiability of B with respect to d_k.

To prove the second part of the theorem, let us take an absolutely continuous $F \in \mathcal{P}$, with $|F'| \le c$. Then, as Clarke (1983) points out, we have

$$d_k(F,G) \le (c+1)d_L(F,G) \qquad (2.7)$$

Indeed, for all $\delta \ge d_L(F,G)$,

$$|F(\theta)-G(\theta)| \le |F(\theta)-F(\theta+\delta)| + |F(\theta+\delta)-G(\theta)| \le c\delta+\delta,$$

which proves (2.7).

Differentiability of B with respect to d_L is now obtained easily by combining (2.6) and (2.7).

Finally, differentiability with respect to Prohorov metric d follows from $d_L \le d$. ◁

The next theorem provides another result on differentiability of B in the setting of approach (b) above mentioned.

Theorem 2. *If $f(x|\cdot)$ is a continuous function of bounded variation on Ω then the "extended Bayes operator" $\hat{B}$, defined by (2.5), is Fréchet differentiable at F, for all $F \in \mathcal{P}$.*

Proof. Given $F \in \mathcal{P}$, we calculate, for every $\Delta \in \mathcal{G}_\infty$, the so-called "weak differential"

$$\begin{aligned} \hat{B}'_F(\Delta) &= \frac{d}{dt}\hat{B}(F+t\Delta)\Big|_{t=0} \\ &= \frac{-f(x|\cdot)\int f(x|\theta)d\Delta(\theta)}{\left[\int f(x|\theta)dF(\theta)\right]^2} = -\hat{B}(F)\cdot\frac{\int f(x|\theta)d\Delta(\theta)}{\int f(x|\theta)dF(\theta)} \end{aligned} \tag{2.8}$$

Clearly, $\hat{B}'_F$ is a member of the space $\mathcal{L}(\mathcal{G}_\infty, \mathcal{L}^1)$ of linear continuous operators mapping $\mathcal{G}_\infty$ on $\mathcal{L}^1$. Moreover, under the conditions imposed on $f(x|\cdot)$, we have (by a similar reasoning to that employed in the proof of Theorem 1)

$$\int f(x|\theta)dH_r(\theta) \underset{r\to\infty}{\longrightarrow} \int f(x|\theta)dF(\theta) > 0 \tag{2.9}$$

for every sequence $\{H_r\} \subset \mathcal{G}_\infty$ such that $||H_r - F||_\infty \to 0$. Thus, there exists a neighbourhood $\mathcal{U} = \mathcal{U}(F)$ such that we can define the map

$$\begin{aligned} W : \mathcal{U} &\longrightarrow \mathcal{L}(\mathcal{G}_\infty, \mathcal{L}^1) \\ H &\longrightarrow \hat{B}'_H \end{aligned}$$

where $\hat{B}'_H$ is obtained from (2.8), by substituting F by H. If $\mathcal{L}(\mathcal{G}_\infty, \mathcal{L}^1)$ is endowed with the usual norm

$$||L|| = \sup_{||H||_\infty \neq 0} [||L(H)||_1/||H||_\infty]$$

a standard result of Fréchet differential theory (Kolmogorov-Fomin, 1975, p. 270) states that continuity of W at F is a sufficient condition for differentiability of $\hat{B}$ at F. Hence, let $\{H_r\} \subset \mathcal{U}$ be such that $||H_r - F||_\infty \to 0$. We need to prove $||W(H_r) - W(F)|| \to 0$ as $r \to \infty$. Indeed, we have

$$||W(H_r) - W(F)|| = K_1 \cdot K_2 \cdot \left| \left[\int f(x|\theta)dH_r(\theta)\right]^{-2} - \left[\int (f(x|\theta)dF(\theta)\right]^{-2} \right| \tag{2.10}$$

where $K_1 = \int f(x|\theta)d\theta < \infty$, $K_2 = \sup_{||H||_\infty \neq 0}[|\int f(x|\theta)\ dH(\theta)|/||H||_\infty]$. Observe that K_2 is finite because it can be interpreted as the norm of the linear functional $T : \mathcal{G}_\infty \to \Re$ defined by $T(H) = \int f(x|dH(\theta)$. So, by the argument used in (2.9), the right-hand side of (2.10) converges to zero, and the proof is concluded. ◁

Remark 1. A similar result is obtained if we consider the space $\mathcal{G}$ endowed with the total variation norm. In this case, continuity and boundedness of $f(x|\cdot)$ are sufficient conditions for ensuring differentiability of $\hat{B}$.

3. SOME SUGGESTIONS ON INFLUENCE FUNCTIONS IN BAYESIAN METHODOLOGY

The influence function $IC_F(x)$ is known to be a useful tool in robust statistics. A comprehensive study of this topic can be found in Hampel *et al.* (1986). The development of Section 2 suggests that an analogous technique could be adapted to the Bayesian framework.

Next, we outline some ideas in this respect.

(1) By analogy with the standard concept of influence curve, we could define the "Bayesian influence function" in terms of the transformation

$$\theta_0 \longrightarrow \hat{B}'_F(\delta_{\theta_0} - F) = B(F) \cdot \left[1 - f(\boldsymbol{x}|\theta_0) / \int f(\boldsymbol{x}|\theta) dF(\theta)\right] \in \mathcal{L}^1,$$

where δ_θ denotes the degenerate distribution at θ. Roughly speaking, $\hat{B}'_F(\delta_{\theta_0} - F)$ represents the effect caused in the posterior distribution (corresponding to the prior F) by an infinitesimal contamination in F of the form $(1-t)F + t\delta_{\theta_0}$. The values θ_0, where $\hat{B}'_F(\delta_{\theta_0} - F) \equiv 0$ are interesting points in a first analysis.

(2) Since $\hat{B}'_F(\delta_{\theta_0} - F)$ is actually a function for each θ_0, it could be interesting to summarize the information provided by $\hat{B}'_F(\delta_{\theta_0} - F)$ in a single numerical value. Thus, in the theory of Robust Estimation the "gross-error sensitivity" is defined as $\sup_x |IC_F(x)|$. In our case, we could consider.

$$\begin{aligned} I_1(F;\theta_0,\theta) &= |\hat{B}'_F(\delta_{\theta_0} - F)(\theta)| \\ I_2(F;\theta) &= \sup_{\theta_0} |\hat{B}'_F(\delta_{\theta_0} - F)(\theta)| \\ I_3(F;\theta) &= ||\hat{B}'_F(\delta_{\theta_0} - F)||_1. \end{aligned}$$

Informally, the values of θ where these functions are maximized (resp. minimized) could be interpreted in different senses as points of maximum (reps. minimum) sensitivity of the posterior distribution against local changes in the prior. However, these aspects would require a more detailed study.

(3) Some examples:

(a) Let $f(\boldsymbol{x}|\theta)$ be the likelihood function corresponding to the normal distribution with mean θ and standard deviation 1 ($f(\cdot|\theta) \sim N(\theta, 1)$). Let the prior be $F \sim N(0,1)$. It can be shown that $\hat{B}'_F(\delta_{\theta_0} - F) \equiv 0$ for

$$\theta_0 = n\bar{x}/(n+1) \pm [\log(2+n)/(n+1)] + [(n\bar{x})^2/((n+2)(n+1)^2)]^{1/2}$$

($\bar{x}$ denotes the sample mean).

For $n > 2$, $I_2(F;\theta)$ is maximized at $\theta_1 = n\bar{x}/(n+1)$ and minimized for $\theta \longrightarrow \pm\infty$. Moreover, the supremum defining $I_2(F;\theta)$ is also attained at θ_1, (observe that this value coincides with the posterior mean).

(b) If $f(\boldsymbol{x}|\theta)$ corresponds to a Poisson distribution with parameter $\theta > 0$ and the prior is a Gamma distribution with parameters 1, 1, then we also obtain that $I_2(F;\theta)$ is maximized at $\theta_0 = n\bar{x}/(n+1)$ and minimized when $\theta \longrightarrow 0$ or ∞. The supremum defining $I_2(F;\theta)$ is also attained at θ_0 (observe that this value is close to the posterior mean which is $(n\bar{x}+1)/(n+1)$).

The explicit calculation of the values θ_0 where $\hat{B}_F(\delta_{\theta_0} - F) \equiv 0$ is more difficult than in Example (a).

(4) In the standard theory of influence functions, $IC_F(x)$ is defined in terms of a simple "directional derivative". No further assumptions of differentiability are required. However, if the functional under study is differentiable in the Fréchet (or Hadamard, or Von Mises)

sense, then the influence function presents a deeper meaning, because it is closely related to asymptotic properties.

We believe that this situation could be partially extended to the Bayesian context: in addition to the results of Fréchet differentiability obtained in Section 2, it should not be difficult to establish a sort of Von Mises differentiability of the type

$$B'_F(F+t(G-F)) \equiv \int B'_F(F+t(\delta_\theta - F))dG(\theta), \quad \text{for all } G \in \mathcal{P}.$$

(5) In the Bayesian theory, point estimates are obtained as functionals depending on posterior distributions; Bayes estimates (obtained by minimizing the expected posterior loss) are a typical example. If $f(\cdot|\theta)$ and x remain fixed, a Bayesian point estimate can be viewed as a functional $T = T(P)$ depending on the prior distribution P. Results on differentiability of T can be proved either for a direct reasoning on $T(P)$, or from above Theorems 1 and 2, by using a suitable version of Chain Rule. The required assumptions would be similar to those used in these theorems.

On the other hand, this approach allows us to define, in a natural way, an influence function for T

$$T'(P;\theta) = \lim_{t \downarrow 0}[T(P+t(\delta_\theta - P)) - T(P)]/t. \tag{3.1}$$

The analogy between this definition and that of the ordinary influence function is obvious. Of course the difference lies in the fact that the latter refers to the underlying distribution whereas the expression (3.1) defines an influence function with respect to the prior.

For example, if $T(P)$ is the posterior mean, we obtain

$$T'(P;\theta) = \frac{f(x|\theta)}{\int f(x|\theta)dP(\theta)}[\theta - T(P)].$$

Thus, the minimum sensitivity (in absolute value) corresponds to $\theta_0 = T(P)$. If, in addition, $f(x|\theta) \sim N(\theta, 1)$, we have $\lim_{\theta\to\pm\infty} T'(P;\theta) = 0$, which represents an assessment of robustness against contaminations in the tails of P.

4. BAYES OPERATOR AS A FUNCTION OF THE LIKELIHOOD

If the prior distribution $P \in \mathcal{P}$ remains fixed, the natural space of likehood functions is given by the set $\mathcal{L}_P$ of P-integrable functions $g : \Omega \to \Re$, endowed with the norm

$$||g||_P = \int |g|dP.$$

This leads us to consider the operator $\bar{B} : \mathcal{L}_P \to \mathcal{L}_P$, defined by

$$\bar{B}(g) = \begin{cases} g/\int g dP & \text{if } \int g dP \neq 0 \\ 1 & \text{otherwise} \end{cases} \tag{4.1}$$

Obviously, $\bar{B}$ is continuous at g, for all $g \in \mathcal{L}_P$ such that $\int g dP \neq 0$. With respect to the Fréchet differentiability, we have the following result

Theorem 3. *The operator $\bar{B}$ defined by (4.1) is Fréchet differentiable at g, for all $g \in \mathcal{L}_P$ such that $\int g dP \neq 0$.*

Proof. Given $g \in \mathcal{L}_P$, with $\int g dP \neq 0$, the "weak differential" is given by

$$\bar{B}'_g(h) = \frac{d}{dt}\bar{B}(g+th)\bigg|_{t=0} = h/\int g dP - \bar{B}(g) \cdot \int h dP/\int g dP \qquad (4.2)$$

for every $h \in \mathcal{L}_P$. Since $\bar{B}'_g$ is linear and continuous in h, the Fréchet differentiability of $\bar{B}$ will be stated if we prove

$$||\bar{B}(g+h) - \bar{B}(g) - \bar{B}'_g(h)||_P/||h||_P \longrightarrow 0, \quad \text{as } h \to 0.$$

Indeed, from (4.1) and (4.2) this quotient becomes

$$\left|\frac{\int h dP}{\int g dP}\right| \cdot \frac{||\bar{B}(g) - \bar{B}(g+h)||_P}{||h||_P}$$
$$\leq \left|1/\int g dP\right| \cdot ||\bar{B}(g) - \bar{B}(g+h)||_P \underset{h\to 0}{\longrightarrow} 0. \qquad \triangleleft$$

5. DISCUSSION AND COMMENTS

(1) Diaconis and Freedman (1986) have computed the Fréchet differential of the operator taking priors to posteriors. Specifically, they consider the "extended Bayes operator", defined on the space $\mathcal{G}_V$ of finite signed measures (with the variation norm) and taking values also in $\mathcal{G}_V$. As these authors point out, the Fréchet derivative "helps to identify data sets x where small changes in the prior cause large changes in the posterior".

The present work attempts to offer a complementary perspective by providing alternative results of differentiability and suggesting other aspects in which they could be meaningful.

(2) We have considered the Bayes operator in terms of posterior densities (instead of posterior measures). We believe that this does not imply any serious limitation; in fact, posterior distributions are mostly used in terms of densities. Of course, when the prior P is variable, the functions $f(x|\theta)/\int f(x|\theta)dP(\theta)$ are not densities with respect to a common measure, but their use can also be justified as we indicate in the next point.

(3) In Section 2, the assumption (2.1), $\int f(x|\theta)d\theta < \infty$, is imposed in order to embed the posterior densities in the space $\mathcal{L}^1$. Observe that, in this setting, the convergence $B(P_r) \longrightarrow_{r\to\infty} B(P)$ is equivalent to convergence (at x) of the corresponding predictive densities. This provides a statistical interpretation for the mathematical framework that we have considered. If we impose alternative assumptions on $f(x|\cdot)$, the posterior densities can conceivably be included in other spaces where the above interpretation remains valid. So, the use of L^1-norm is not really essential: the important fact is the proximity between predictive densities.

(4) The regularity assumption $0 < \int f(x|\theta)dP(\theta) < \infty$ is clearly not relevant. It can be dropped by introducing some technical modifications.

(5) The concepts of differentiability (Fréchet, Hadamard, ...) borrowed from mathematical analysis, are often too rigid for statistical applications: this could be the case of the present study. Theorem 1 provides an example of the methodology based upon "weakened" concepts of differential. Other examples of this methodology can be found in the statistical literature; perhaps some of them could be adapted to our problem (see, e.g., Boos (1977), who also suggests a useful class of norms for the space $\mathcal{G}$).

(6) It should be noted that the hypothesis concerning bounded variation in Theorems 1 and 2 is not actually too restrictive. For example, if there exists a maximum likelihood estimate $\hat{\theta}$, such that $f(x|\cdot)$ increases for $\theta < \hat{\theta}$ and decreasese for $\theta > \hat{\theta}$, then it is clear that the likelihood can be expressed as a difference of two bounded increasing functions and, consequently, is of bounded variation.

(7) Ramsay and Novick (1980) have proposed, in the Bayesian context, three definitions of influence functions which are given in terms of partial derivatives of the likelihood and the utility function. Similarly to the non-Bayesian case, the boundedness of these influence functions may be interpreted as a property of robustness.

Of course, this approach is different to that we have followed here, although it could be partially related to the general plan of work outlined in Introduction.

REFERENCES

Boos, D. D. (1977). The differential approach in statistical theory and robust inference. Ph. D. Thesis. Florida State University.

Clarke, B. R. (1983). Uniqueness and Fréchet differentiability of functional solutions to maximum likelihood type equations. *Ann. Statist.* **11**, 1196–1205.

Cuevas, A. (1988). Qualitative robustness in abstract inference. *J. Statist. Plann. and Inference.* **18**, 277-289.

Diaconis, P. and Freedman, D. (1986). On the consistency of Bayes estimates. *Ann. Statist.* **14**, 1–26.

Hampel, F. R., Ronchetti, E. M., Rousseeuw, P. J. and Stahel, W. A. (1986). *Robust Statistics: The Approach Based On Influence Functions.* New York: Wiley.

Kolmogorov, A. N. and Fomin, S. V. (1975). *Elementos de la Teoría de Funciones y del Análisis Funcional.* Moscow: Mir.

Ramsay, J. O. and Novick, M. R. (1980). PLU robust Bayesian decision theory: point estimation. *J. Amer. Statist. Assoc.* **75**, 901–907.

BAYESIAN STATISTICS 3, pp. 579–583
J. M. Bernardo, M. H. DeGroot, D. V. Lindley and A. F. M. Smith, (Eds.)

Parametric Estimation With L^1 Distance

J. DE LA HORRA
Universidad Autónoma de Madrid

SUMMARY

The value of a parameter $\theta \in \Theta$ is estimated by means of a decision problem, where the loss function is the L^1 distance between the estimate and a density over a small interval around the actual value of θ. Two cases are considered: $\Theta = \mathcal{R}$ and $\Theta = (0, \infty)$; in both, the Bayes rule is a probability density over a small interval around the posterior median, when sufficient sample information has been obtained. These results are applied to location parameters and scale parameters.

Keywords: PARAMETRIC ESTIMATION; L^1 DISTANCE; BAYES RULE; LOCATION PARAMETER; SCALE PARAMETER; POSTERIOR MEDIAN.

1. INTRODUCTION

Let us consider a sample x from the random variable X, with distribution given by the probability measure P_θ, $\theta \in \Theta \subset \mathcal{R}$. Our problem is to find a "good" estimate of θ, in the sense of being close to the "actual" value of the parameter. Gatsonis (1984) proposed obtain the best density for a location parameter by means of a decision problem, where the loss function was the squared L^2 distance between the posterior density and the estimate.

The idea developed in this paper is to consider a decision problem, where the loss function is the L^1 distance between a density over a small interval around the actual value of the parameter and the estimate. This same idea was used in De la Horra (1987), where the best generalized estimator or inference (Eaton (1982)) was obtained (for a finite parameter space), by means of a decision problem (difference distances were taken as loss functions). Some modifications will, of course, be needed for adapting the decision problem to the continuous case.

In Section 2, the case $\Theta = \mathcal{R}$ is studied, and the Bayes rule is obtained. Results are applied to the problem of finding the best invariant rule for a location parameter. In Section 3, a similar study is carried out for the case $\Theta = (0, \infty)$, with application to scale parameters.

2. THE CASE $\Theta = \mathcal{R}$

Let $x = (x_1, \ldots, x_n)$ be a sample, where $x_1, \ldots, x_n$ are independent and identically distributed observations from a random variable X (taking values in a sample space $\mathcal{X}$), with distribution P_θ, where $\theta \in \Theta = \mathcal{R}$. We want to choose a "good" distribution over Θ, that is to say, a density concentrating its mass around the "true" value of θ. To find such a "good" distribution we next define a decision problem, which is a natural generalization of that used in De la Horra (1987) for finite Θ. The decision problem $(\Theta, \mathcal{A}, L_k)$ is defined as follows:

$$\mathcal{A} = \{f \in L^1(\mathcal{R}) : f(\theta) \geq 0 \text{ for all } \theta \in \Theta \text{ and } \int_\Theta f(\theta) d\theta \leq 1\}$$

and, for $\omega \in \Theta$, $f \in \mathcal{A}$:

$$L_k(\omega, f) = \int_{\mathcal{R}} |f(\theta) - \frac{1}{2k} I_{(\omega-k,\omega+k)}(\theta)| d\theta$$

where $k > 0$ is fixed, $I_{(\omega-k,\omega+k)}$ is the usual indicator function, and integrals are calculated with respect to the Lebesgue measure.

Some remarks are needed for understanding why we have chosen $\mathcal{A}$ and L_k in this way:

1) We are interested in finding a probability density, and nevertheless, $\mathcal{A}$ contains subprobability densities. The reason for considering subprobability densities is that the optimization problem is easier to solve for this action space. Morever, we shall see that a probability density is obtained, if the number of observations is sufficiently large. Gatsonis (1984) defined the action space in an analogous way.

2) The idea behind the definition of the loss function is very simple: if ω is the "actual" and unknown value of the parameter and we take the density f as an "estimate" of ω, the error can be measured by some notion of distance between the distribution given by f and the distribution giving mass one at ω. A reasonable way for carrying out this idea is to replace the last distribution by the uniform distribution over a small interval $(\omega - k, \omega + k)$ (we must choose a convenient $k > 0$). In practice, this replacement means that it is impossible to measure the value of the parameter, in an exact way; we shall therefore be satisfied with a good approximation.

3) L^1 distance is a very natural one when we consider densities; see, Devroye and Györfi (1985; chapter 1).

4) We are considering densities with respect to Lebesgue measure. But it is interesting to remark that L^1 distance is independent of the particular choice of the dominating measure as is easily proved. The proof is very similar to that in Pitman (1979; p. 7), for Hellinger's distance.

We next determine the Bayes rule for this decision problem.

Theorem 2.1. *The Bayes rule for the decision problem* $(\Theta, \mathcal{A}, L_k)$ *defined above, with respect to a prior probability density* π, *is the function* $d^*: \begin{smallmatrix} \mathcal{X} \longrightarrow \mathcal{A} \\ x \longrightarrow f_x \end{smallmatrix}$, *where*

$$f_x(\theta) = \begin{cases} \frac{1}{2k} & \text{if } \int_{\theta-k}^{\theta+k} \pi(\omega|x) d\omega > 1/2 \\ 0 & \text{otherwise} \end{cases}$$

$\pi(.|x)$ *is the posterior probability density.*

Proof. We must find that $f_x \in \mathcal{A}$ which minimizes the posterior expected loss:

$$\int_{\mathcal{R}} L_k(\omega, f_x) \pi(\omega|x) d\omega = \int_{\mathcal{R}} \left[\int_{\mathcal{R}} |f_x(\theta) - \frac{1}{2k} I_{(\omega-k,\omega+k)}(\theta)| d\theta \right] \pi(\omega|x) d\omega$$
$$= \int_{\mathcal{R}} \left[\int_{\mathcal{R}} |f_x(\theta) - \frac{1}{2k} I_{(\theta-k,\theta+k)}(\omega)| \pi(\omega|x) d\omega \right] d\theta$$

For each fixed θ, the expression inside the brackets is minimized by taking $f_x(\theta)$ to be the median of the distribution induced by the function $\frac{1}{2k} I_{(\theta-k,\theta+k)}$, from $\pi(.|x)$. This function takes the values:

$$\begin{cases} \frac{1}{2k} & \text{with probability } \int_{\theta-k}^{\theta+k} \pi(\omega|x) d\omega \\ 0 & \text{with probability } 1 - \int_{\theta-k}^{\theta+k} \pi(\omega|x) d\omega \end{cases}$$

The function f_x is now easily obtained. We must prove that $\int_{\mathcal{R}} f_x(\theta)d\theta \leq 1$. Let M_x be the median of the posterior distribution $\pi(.|x)$. If $\theta \leq M_x - k$ or $\theta \geq M_x + k$, then $\int_{\theta-k}^{\theta+k} \pi(\omega|x)\, d\omega \leq 1/2$. So, f_x takes the value $\frac{1}{2k}$, only on some subset of $(M_x - k, M_x + k)$ and $\int_{\mathcal{R}} f_x(\theta)d\theta \leq 1$. $\triangleleft$

Next, we make some remarks, which will show that f_x is a probability density, when sufficient sample information has been obtained. Let us assume that $\pi(\omega|x)$ decreases strictly as ω moves away from the modal value in either direction. When n increases, $\pi(.|x)$ becomes more and more concentrated around its median M_x; thus, when we have obtained a sufficiently large number of observations, we can expect that $\int_{M_x-2k}^{M_x+2k} \pi(\omega|x)\, d\omega \simeq 1$, and so:

$$f_x(\theta) = \begin{cases} \frac{1}{2k} & \text{if} \quad \theta \in (M_x - k, M_x + k) \\ 0 & \text{otherwise.} \end{cases}$$

Now, $\int_{\mathcal{R}} f_x(\theta)d\theta = 1$.

Next, we shall apply Theorem 2.1 to the problem of finding the best invariant rule for a location parameter. Therefore, let us assume that x is an observation from a random variable X (taking values in $\mathcal{X} = \mathcal{R}$), with distribution given by a density $h_\theta(x) = h(x - \theta)$, $\theta \in \Theta = \mathcal{R}$. The next theorem derives the best invariant rule, under the group of translations $G = \{g_c : g_c(x) = x + c, c \in \mathcal{R}\}$.

Theorem 2.2. *The best invariant rule (under G) for the location parameter θ, when we consider the decision problem defined above, is the function $d^*: \begin{smallmatrix}\mathcal{X} \longrightarrow \mathcal{A} \\ x \longrightarrow f_x\end{smallmatrix}$, where:*

$$f_x(\theta) = \begin{cases} \frac{1}{2k} & \text{if} \quad \int_{x-\theta-k}^{x-\theta+k} h(y)dy > 1/2 \\ 0 & \text{otherwise} \end{cases}$$

Proof. First, we must verify that the decision problem is invariant under G. It is known that the family of distributions is invariant under G. The parameter space Θ consists of a single orbit ($\bar{G}$ is transitive). The loss function is also invariant under G; in fact

$$\begin{aligned} L_k(\omega, f) &= \int_{\mathcal{R}} |f(\theta) - \frac{1}{2k} I_{(\omega-k,\omega+k)}(\theta)|d\theta \\ &= \int_{\mathcal{R}} |f(\theta' - c) - \frac{1}{2k} I_{(\omega-k,\omega+k)}(\theta' - c)|d\theta' \\ &= \int_{\mathcal{R}} |f(\theta' - c) - \frac{1}{2k} I_{(\omega+c-k,\omega+c+k)}(\theta')|d\theta' \\ &= \int_{\mathcal{R}} |f_c(\theta') - \frac{1}{2k} I_{(\omega+c-k,\omega+c+k)}(\theta')|d\theta' \quad \text{(where } f_c(\theta) = f(\theta - c)) \\ &= L_k(\omega + c, f_c) = L_k(\bar{g}_c(\omega), f_c) \end{aligned}$$

Therefore, L is invariant under G, with $\tilde{g}_c(f) = f_c$ and $f_c(\theta) = f(\theta - c)$. But the best invariant rule is the same as the Bayes rule with respect to the right invariant Haar density (see for instance, Berger (1985; p. 410)). In this problem, the right invariant Haar density is $\pi^r(\omega) = 1$, for all $\omega \in \mathcal{R}$, and $\pi^r(\omega|x) = h(x - \omega)$. Therefore, if we apply Theorem 2.1, the best invariant rule is:

$$f_x(\theta) = \begin{cases} \frac{1}{2k} & \text{if} \quad \int_{\theta-k}^{\theta+k} h(x - \omega)d\omega > 1/2 \\ 0 & \text{otherwise.} \end{cases}$$

But, defining $y = x - \omega$,

$$\int_{\theta-k}^{\theta+k} h(x-\omega)d\omega = \int_{x-\theta-k}^{x-\theta+k} h(y)dy. \quad \triangleleft$$

3. THE CASE $\Theta' = (0,\infty)$

Let $x = (x_1, \ldots, x_n)$ be a sample, where $x_1, \ldots, x_n$ are independent and identically distributed observations, from a random variable X (taking values in sample space $\mathcal{X}$), with distribution P_θ, $\theta \in \Theta' = (0,\infty)$. Also, a decision problem will be considered for choosing a "good" distribution over Θ'. Now, we define:

$$\mathcal{A}' = \{f \in L^1(\mathcal{R}) : f(\theta) \geq 0 \text{ for all } \theta \in \Theta' \text{ and } \int_{\Theta'} f(\theta)d\theta \leq 1\}$$

$$L_k'(\omega, f) = \int_0^\infty |f(\theta) - \frac{-1}{2\theta \log k} I_{(\omega k, \omega/k)}(\theta)| d\theta$$

where $k \in (0,1)$ is fixed.

There is a slight difference with respect to the loss employed in Section 2. But this difference is easy to justify: if ζ is a random variable with density $-1/(2\theta \log k) I_{\omega k, \omega/k)}$, $\log \zeta$ is uniformly distributed over the interval $(\log \omega + \log k, \log \omega - \log k)$, and this was the structure used when $\Theta = \mathcal{R}$. Moreover, $(\Theta', \mathcal{A}', L_k')$ will be well behaved under the group of scale changes, when θ is a scale parameter. We remark that $(\omega k, \omega/k)$ is always contained in Θ'.

Theorem 3.1. *The Bayes rule for the decision problem $(\theta', \mathcal{A}', L_k')$, with respect to a prior probability density π, is the function $d^*: \begin{smallmatrix} \mathcal{X} \longrightarrow \mathcal{A}' \\ x \longrightarrow f_x \end{smallmatrix}$, where:*

$$f_x(\theta) = \begin{cases} \frac{-1}{2\theta \log k} & \text{if } \int_{\theta k}^{\theta/k} \pi(\omega|x)d\omega > 1/2 \\ 0 & \text{otherwise} \end{cases}$$

Proof. This is similar to the proof of Theorem 2.1. $\triangleleft$

When we have obtained a sufficiently large number of observations, the "estimate" of θ is the distribution given by the probability density:

$$f_x(\theta) = \begin{cases} \frac{-1}{2\theta \log k} & \text{if } \theta \in (M_x k, M_x/k) \\ 0 & \text{otherwise} \end{cases}$$

Now, we shall apply Theorem 3.1 to the problem of finding the best invariant rule for a scale parameter. Therefore, let us assume that x is an observation from a random variable X (taking values in $\mathcal{X} = (0,\infty)$), with distribution given by a density $h_\theta(x) = (1/\theta)h(x/\theta)$, $\theta \in \Theta' = (0,\infty)$. The next theorem derives the best invariant rule, under the group $G' = \{g_c : g_c(x) = x/c, c > 0\}$.

Theorem 3.2. *The best invariant rule (under G') for the scale parameter θ, when we consider the decision problem $(\theta', \mathcal{A}', L_k')$ is the function $d^*: \begin{smallmatrix} \mathcal{X} \longrightarrow \mathcal{A}' \\ x \longrightarrow f_x \end{smallmatrix}$, where:*

$$h_x(\theta) = \begin{cases} \frac{-1}{2\theta \log k} & \text{if } \int_{(xk)/\theta}^{x/(\theta k)} h(y)dy > 1/2 \\ 0 & \text{otherwise} \end{cases}$$

Proof. This is similar to the proof of Theorem 2.2, with $g_c(f) = f_c$, $f_c(\theta) = c\, f(c\theta)$, $\pi^r(\omega) = 1/\omega$ and $\pi^r(\omega|x) = (x/\omega^2)h(x/\omega)$. $\triangleleft$

REFERENCES

Berger, J. O. (1985). *Statistical Decision Theory and Bayesian Analysis*. New-York: Springer-Verlag.

De la Horra, J. (1987). Generalized estimators: A Bayesian decision-theoretic view. *Statist. & Dec* **5**, 347–352.

Devroye, L. and Györfi, L. (1985). *Nonparametric Density Estimation: The L^1 View*. New York: Wiley.

Eaton, M. L. (1982). A method for evaluating improper prior distributions. *Statistical Decision Theory and Related Topics III*, **1**, (S.S. Gupta and J.O. Berger eds.) New York: Academic Press.

Gatsonis, C. A. (1984). Deriving posterior distributions for a location parameter: A decision theoretic approach. *Ann. Statist.* **12**, 958–970.

Pitman, E. J. G. (1979). *Some Basic Theory for Statistical Inference*. London: Chapman and Hall.

BAYESIAN STATISTICS 3, pp. 585–592
J. M. Bernardo, M. H. DeGroot, D. V. Lindley and A. F. M. Smith, (Eds.)

Choosing a Quality Supplier A Bayesian Approach

J. J. DEELY and W. J. ZIMMER
Univ. of Canterbury and *Univ. of New Mexico*

SUMMARY

This paper treats the problem of selecting the "best" of several competing suppliers, each of which has supplied a series of lots from which an assessment of the true unknown quality of that particular supplier can be obtained. Each lot has associated with it a parameter θ and a sample statistic x with distribution $f(x|\theta)$ which in this paper is assumed to be normal with mean θ. The "quality" of a supplier is determined by its true unknown distribution of θ over lots, say $g(\theta|\gamma)$, again in this paper assumed to be normal with unknown mean γ. For this simple situation the "best" supplier, i.e. the one with highest quality, can easily be defined as a function of γ; usually largest or smallest γ is desired. It is shown that this "mixture" model reduces to a model treated by Berger and Deely (1988) and thus their methodology of a hierarchical Bayesian model can be used to select the best supplier. As is to be expected however different forms of prior information particular to quality data are available and this can influence the choice of certain "priors" in the hierarchical model. To illustrate the methodology, data from two suppliers of electronic components is analyzed and the relevance of specific prior information is discussed. The paper concludes with remarks about other possible models and definitions of "best" supplier.

Keywords: QUALITY SUPPLIERS; NORMAL DISTRIBUTIONS; HIERARCHICAL BAYESIAN MODEL; PROCEDURES; MIXTURES; PRIOR INFORMATION.

1. INTRODUCTION

Suppose one were faced with the task of chosing a supplier whose product would be purchased for a lenhgt of time. It seems appropriate that such a choice should be made, not on the basis of an individual lot, but on the basis of several lots in an attempt to describe a supplier's overall quality distribution. The selection is then made by comparing the quality distributions of the competing suppliers.

The problem as stated can be seen to be a selection or ranking problem where the selection process seeks to find the "best" mixing distribution (a supplier's quality distribution) $g_i(\theta|\gamma_i)$ using data from the mixture $f_i(x|\gamma_i)$ where

$$f_i(x|\gamma_i) = \int f_i(x|\theta)g_i(\theta|\gamma_i)d\theta.$$

The θ's represent true lot means and γ_i is a ranking parameter for each of the k suppliers, $i = 1, 2, \ldots, k$. It is assumed that knowledge of $\gamma = (\gamma_1, \ldots, \gamma_k)$ allows a perfect ranking of the suppliers. The data from each supplier consist of observations from r_i lots where $x_{ij} = (x_{i1}, \ldots, X_{ir_i})$ represents the observations from the jth lot from the ith supplier, $j = 1, \ldots, r_i$. The distribution $f(x|\theta_i)$ of the observations within a lot is dependent on a lot parameter θ_i which is an unobserved random variable from the quality distribution g_i, i.e., $\theta_{i1}, \theta_{i2}, \ldots, \theta_{ir_i}$ are i.i.d from g_i.

It should be noted that there are two different problems that might be addressed in this situation: (1) if one is interested in comparing suppliers on the basis of individual lots then a ranking of the θ_i's is required where θ_i is the true mean from a lot from the ith supplier; (2) if however one is interested in comparing suppliers on the basis of the long-term process quality, then a ranking of the g_i's is required. The first problem has been investigated often and many approaches and results are available. The second problem has received limited attention and will be addressed in this paper. A Bayesian approach which provides effective and intuitively appealing solutions is proposed.

This problem of choosing a long-term high-quality supplier is similar in model and assumptions to an analysis of variance (AOV) problem relating to nested designs and described as pure subampling in three stages, here denoted as: supplier, lots, observations (see Neter, Wasserman, Kutner (1985)). The solution proposed in this paper is based on a Bayesian approach which can give straightforward probabilistic answers to both the question of the truth of the hypothesis and to the question of which supplier is best if the hypothesis of equality is false. In fact, in this paper the posterior distribution of supplier differences is obtained. The Bayesian solution proposed is based on modelling exchangeability through the use of a hierarchical Bayesian analysis. The technique of using such a model to both test the hypothesis of equality and then to select a desired treatment was developed in Berger and Deely (1988) and is extended here to treat this application of a mixture problem. It is seen from results presented herein that one can add another level of hierarchy without further complicating the desired calculation. This leads to the conjecture that the number of levels of hierarchy colud be extended indefinitely and then, with appropriate information available, collapsed so that the calculations would remain at the level of a simple Bayesian analysis. Furthermore since this mixture problem is a form of a multifactor randomized AOV, it seems that the Berger and Deely results can be extended to all types of AOV models.

2. GENERAL MODEL AND THE PRIOR DISTRIBUTION

It is assumed that $f_i(x|\theta_{ij})$ is a normal distribution with mean θ_{ij} and known variance σ_{ij}^2 and that the quality distribution of the ith supplier, $g_i(\theta|\gamma_i), i = 1, 2, \ldots, k$, is also a normal distribution with mean γ_i and known variance τ_i^2. Throughout, the best supplier will be taken as the one with smallest γ_i but the case for largest γ_i can also be treated easily.

To describe the uncertainty about γ, we will use the well-known hierarchical Bayesian model which is described for example in Berger (1985) (Sections 3.6 and 4.6) and more explicity for an AOV type selection problem in Berger and Deely (1988). In fact it will be seen that the above model can be reduced to a special case of the model treated in that paper and hence their results can be readily applied to the problem at hand. Emphasis should be placed on the versatility of the hierarchical Bayesian model and hence its appeal. For example it will be seen that both exchangeability and independence amongst the γ_i's can be incorporated in the one model as well as any trends amongst the γ_i's. In addition prior beliefs as to equality amongst the γ_i's can easily be accommodated.

Specifically, we will assume the prior distribution on γ is obtained as a mixture of a distribution of γ given hyperparameters (β, σ_π^2) with a distribution of (β, σ_π^2) written $\pi_1(\gamma|\beta, \sigma_\pi^2)$ and $\pi_2(\beta, \sigma_\pi^2)$ respectively. The components of the vector γ conditional on β, σ_π^2 are assumed to be i.i.d. and thus $\pi_1(\gamma|\beta, \sigma_\pi^2)$ will be the k-product of the univarate distribution of each γ_i. In particular let y be a $k \times \ell$ known matrix and d a k-vector of known constants, the use of which will be explained later. Let

$$\pi_1(\gamma_i|\beta, \sigma_\pi^2) = N_k(y\beta + d, \sigma_\pi^2 I) \qquad \text{and} \qquad \pi_2(\beta, \sigma_\pi^2) = \pi_{2,1}(\beta) \cdot \pi_{2,2}(\beta, \sigma_\pi^2)$$

where

$$\pi_{2,1}(\beta) = N_\ell(\beta^0, A) \qquad \text{and} \qquad \pi_{2,2}(\sigma_\pi^2) \text{ arbitrary.}$$

Then the prior distribution $\pi(\gamma)$ is given by

$$\pi(\gamma) = \int \pi_1(\gamma|\beta, \sigma_\pi^2)\pi_2(\beta, \sigma_\pi^2)d\beta d\sigma_\pi^2,$$

where

$$\pi_1(\gamma|\beta, \sigma_\pi^2) = \prod_{i=1}^{k} \pi_1(\gamma_i|\beta, \sigma_\pi^2).$$

Some mention should be made of the advantages of this seemingly complicated model. Firstly, the matrix y and the vector d allow the suppliers' ranking parameters $\gamma_1, \ldots, \gamma_k$ to be given as *different* regression functions of the *same* variables $\beta_1, \ldots, \beta_\ell$. For example, β_1 might be time and β_2 cost; then each γ_i could be a different function of these two variables with interest still being on the smallest γ_i. Of course, simple exchangeability of the γ_i's can be modelled by taking $\ell = 1$ and $y = 1$.

Secondly, the hyperpriors $\pi_{2,1}$ and $\pi_{2,1}$ allow considerable flexibility for dealing with prior information. For example, in the important case of exchangeability amongst the γ_i's, β^0 represents what the best prior elicitation can obtain about an estimate of the mean of the lot means. That is, the quality engineer would assess (from experience and other prior inputs) what the average of the true unobserved lot means might be. The variance A in the distribution of β represents, in the same spirit, what range the quality engineer might expect in this opinion on the mean quality. When this opinion is completely vague we take $A = \infty$. Calculations for A finite and $A = \infty$ are given in the numerical example in Section 4.

Similarly the distribution $\pi_{2,2}(\sigma_\pi^2)$ is a description of the variability of the quality between suppliers. Under vague prior opinion we take $\pi_{2,2}(\sigma_\pi^2)$ as some reasonable noninformative prior (reasonable in the sense that posterior distributions are defined). The numerical example treated in Section 4 should help to clarify these concepts.

Finally, it should be noted that we will not, in fact, require the computation of the prior $\pi(\gamma)$ from this complicated model. Rather we will need to compute only the posterior probabilities that each supplier is best. We now develop this concept.

3. SELECTING THE "BEST" SUPPLIER

To select the "best" supplier, we will compute p_j, the posterior probability that γ_j is smallest, and $p_j(b)$ the posterior probability that $\gamma_j - \gamma_i$ is no larger than b for all $i \neq j$. There are a number of important variations in the model which affect these computations. Hence this section will be broken into the following parts: (i) showing that the mixture model can use the methodology presented in Berger and Deely (1988); (ii) giving the general formulas for the relevant probabilities; (iii) treating the important special cases of exchangeability and independence; (iv) discussing the priors $\pi_{2,1}$ and $\pi_{2,2}$.

(i) We have that there are k suppliers, the i-th supplier having r_i lots with the j-th lot from the i-th supplier having a sample of size n_{ij}. Thus the data consititutes and array x with k columns (suppliers), the i-th column having r_i numbers being the sample means from the lots from that supplier. The array of true unknown means for this data will be denoted by θ. Specifically we have

$$x|\theta \text{ has distribution } f(x|\theta, \gamma, \beta, \sigma_\pi^2) = \prod_{i=1}^{k}\prod_{j=1}^{r_i} N\left(\theta_{ij}, \frac{\sigma_{ij}^2}{n_{ij}}\right)$$

and

$$\theta|\gamma \text{ has distribution } g(\theta|\gamma, \beta, \sigma_\pi^2) = \prod_{i=1}^{k}\prod_{j=1}^{r_i} N(\gamma_i, \tau_i^2)$$

where $\gamma = (\gamma_1, \ldots, \gamma_k)$ are the k quality distribution means which in turn will have a prior distribution with parameters β and σ_π^2 as indicated in Section 2. The two distributions above do not actually depend on β and σ_π^2 and the first one does not depend on γ. But by writing them in that fashion one can see that the posterior distribution of γ given the data can be written as

$$\pi(\gamma|x) = [f(x)]^{-1} \int_0^\infty \int_\beta \int_\theta f(x|\theta, \gamma, \beta, \sigma_\pi^2) g(\theta|\gamma, \beta, \sigma_\pi^2) \pi_1(\gamma|\beta, \sigma_\pi^2) \pi_2(\beta, \sigma_\pi^2) d\theta\, d\beta\, d\sigma_\pi^2 \tag{3.1}$$

where $f(x)$ is the unconditional marginal distribution of x. The conditional distribution of x given γ is obtained by

$$\begin{aligned} f(x|\gamma) &= \int_\theta f(x|\theta, \gamma, \beta, \sigma_\pi^2) g(\theta|\gamma, \beta, \sigma_\pi^2) d\theta \\ &= \prod_{i=1}^{k} \prod_{j=1}^{r_i} N\left(\gamma_i, \frac{\sigma_{ij}^2}{n_{ij}} + \tau_i^2\right). \end{aligned} \tag{3.2}$$

Substituting (3.2) into (3.1), we obtain

$$\pi(\gamma|x) = [f(x)]^{-1} \int_0^\infty \int_\beta f(x|\gamma) \pi(\gamma|\beta, \sigma_\pi^2) \pi_2(\beta, \sigma_\pi^2) d\beta\, d\sigma_\pi^2 \tag{3.3}$$

Hence the problem can then be viewed as having an observation array x with k columns, the i-th column of which has r_i observations from r_i independent (conditional on γ_i) normal random variables with mean γ_i but possibly unequal variances. This is intuitively clear since the samples from the i-th supplier come from a mixture and hence the mixing parameter θ can be eliminated by integrating over the θ space. In addition it can be seen from (3.2) that $\bar{x}_i = \frac{1}{r_i} \sum_{j=1}^{r_i} x_{ij}$ is sufficient for γ_i so that $f(x|\gamma)$ can be replaced by

$$f(\bar{x}|\gamma) = \prod_{i=1}^{k} N(\gamma_i, \nu_i^2)$$

where $\bar{x} = (\bar{x}_1, \ldots, \bar{x}_k)$ and $\nu_i^2 = \frac{1}{r_i^2} \sum_{j=1}^{r_i} \frac{\sigma_{ij}^2}{n_{ij}} + \frac{\tau_i^2}{r_i}$. It is now easy to see that this problem satisfies the model treated in Berger and Deely and therefore their results can be readily applied.

(ii) For our purposes in this paper the most general situation will be to compute, for each j, posterior probabilities of the form

$$p_j(b) = P(\gamma_j \le \gamma_i + b \text{ for all } i \ne j|\text{data})$$

for any real number b. This posterior probability is a Bayesian analog to multiple comparisons since it gives a probability to the simultaneous events $\gamma_j - \gamma_i \le b$ for *all* $i = 1, 2, \ldots, k$, $(i \ne j)$. In addition, the posterior probability that supplier j is best is obtained by setting $b = 0$. That is,

$$p_j = p_j(0) = P(\text{supplier } j \text{ is best}|\text{data}).$$

In the distributions given in Section 2 take $\ell = 1$, $y = 1$ then

$$\begin{aligned} p_j(b) &= \int_0^\infty \int_{-\infty}^\infty P(\gamma_j \le \gamma_i + b \text{ for all } i \ne j|\text{data}) \pi_{2,1}(\beta|\bar{x}) \pi_{2,2}(\sigma_\pi^2|\bar{x}) d\beta\, d\sigma_\pi^2 \\ &= \int_0^\infty \int_{-\infty}^\infty \int_\infty \\ &\prod_{i \ne j} \left(1 - \Phi\left(\frac{\gamma_j - u_i - b}{\sqrt{V_i}}\right)\right) \times \pi_1(\gamma_i|\bar{x}, \beta, \sigma_\pi^2) \pi_{2,1}(\beta|\bar{x}) \pi_{2,2}(\sigma_\pi^2|\bar{x}) d\gamma_i\, d\beta\, d\sigma_\pi^2 \end{aligned} \tag{3.4}$$

where $\pi_{2,1}$ and $\pi_{2,2}$ are the posterior distributions of β and σ_π^2, respectively, and noting that $\gamma_1, \ldots, \gamma_k$ conditional on $\bar{x}, \beta, \sigma_\pi^2$ are independent normal random variables, i.e. $\pi(\gamma_i|\bar{x}, \beta, \sigma_\pi^2)$ is normal with mean and variance given respectively by

$$u_i = \frac{\sigma_\pi^2 \bar{x}_i + \nu_i^2(\beta + d_i)}{\nu_i^2 + \sigma_\pi^2}, \qquad V_i = \frac{\nu_i^2 \sigma_\pi^2}{\nu_i^2 + \sigma_\pi^2}.$$

Expressions for $\pi_{2,1}$ and $\pi_{2,2}$ can be found in Berger and Deely (1988) (see Equation 3.2) and are amenable to computer calculations of (3.4) which involves only a three-dimensional integral regardless of the choice of $\pi_{2,2}(\sigma_\pi^2)$.

(iii) There are two important cases which reflect the extremes of prior knowledge about the quality parameters $\gamma_1, \ldots, \gamma_k$. Firstly, in the absence of any information concering differences between individual suppliers, it is reasonable to assume that the γ_i's are exchangeable. At the other extreme, it may be that the assumption of independent $\gamma_i's$ is appropriate. We are not advocating this approach in this paper but merely pointing out that this case is covered; in particular, if γ_i has a prior distribution which is normal with mean μ_i and variance ξ_i^2, then taking $\sigma_\pi^2 = 0$, $\ell = 1$, $y = (\xi_1^2, \ldots, \xi_k^2)$, $d = (\mu_1, \ldots, \mu_k)$, $\beta^0 = 0$ and $A = 1$ will model this case. Hence, $\gamma_i = y_i\beta + d_i$ with probability one and $\beta \sim N(0,1)$ and thus γ_i has the desired prior distribution. The posterior distribution of γ_i given $\bar{x}_i$ is thus $N(\tilde{\mu}_i, \tilde{\xi}_i^2)$ where

$$\tilde{\mu}_i = \frac{\xi_1^2 \bar{x}_i + \nu_i^2 \mu_i}{\xi_1^2 + \nu_i^2}, \tilde{\xi}_1^2 = \frac{\xi_1^2 \nu_i^2}{\xi_1^2 + \nu_i^2}$$

and a direct computation gives

$$p_j(b) = \int_{-\infty}^{\infty} \prod_{i \neq j} [1 - \Phi(b_{ij} z + \delta_{ij})]\, \phi(z) dz \qquad (3.5)$$

where $b_{ij} = \tilde{\xi}_j / \tilde{\xi}_i$ and $\delta_{ij} = \tilde{\xi}_i^{-1}(\tilde{\mu}_j - \tilde{\mu}_i - b)$. Evaluation of (3.5) can be accomphished with straightforward numerical integration techniques.

When prior information about differences between suppliers is quite vague, the assumption that the γ_i's are exchangeable can be used to model this situation. This case is obtained by setting $\ell = 1$, $y = 1$, $d = 0$ and arbitrary choices for β^0, A and $\pi_{2,2}(\sigma_\pi^2)$ corresponding to the type of available prior information about suppliers in general.

Sometimes, in practice, it may be possible to assume that the various suppliers have equal variances $\nu_i^2 = \nu^2$. This assumption allows considerable simplication and reduction of (3.4) to a two-dimensional integral as given below:

$$p_j = \int_0^{\infty} \int_{-\infty}^{\infty} \prod_{i \neq j} \left[1 - \Phi\left(\frac{\sigma_\pi/\nu}{\sqrt{\sigma_\pi^2 + \nu^2}}(\bar{x}_j - \bar{x}_i) + z\right)\right] \phi(z) \pi_{2,2}(\sigma_\pi^2|\bar{x}) dz\, d\sigma_\pi^2. \qquad (3.6)$$

Further simplifications and derivations can be found in Berger and Deely. In Section 4 a numerical example will be given to illustrate these and other simplifications.

(iv) It is clear that considerable latitude is available in selecting the second stage priors, $\pi_{2,1}(\beta)$ and $\pi_{2,2}(\sigma_{pi}^2)$. We have assumed normality for $\pi_{2,1}$ although other functional forms are possible. Under normality the mean β^0 and the variance A must be chosen. The noninformative case would be $\beta^0 = 0$ and $A = \infty$ but when prior information is available other values may be elicited. For example, if the quality engineer were asked: "Where do you expect the mean quality over all lots and all suppliers to be?", the answer could easily be an interval (a, b) in which case a reasonable choice of β^0 and A would be $\beta^0 = (a + b)/2$ and $A = (b - a)^2/16$.

For $\pi_{2,2}(\sigma_\pi^2)$, there are several noninformative choices. Their form and discussion of such can be found in Berger and Deely. But when prior information is available then again elicitation would result in an informative prior. Thus if the quality engineer were asked a second question: "How much variation do you expect over the lots and within what range?", then a member of the parametric "Shoe" family below could be determined. The "Shoe" family is given by

$$\pi_{2,2}(\sigma_\pi^2) = \begin{cases} m/[(m+1)c], & 0 < \sigma_\pi^2 \le c \\ \frac{mc^m}{(m+1)(\sigma_\pi^2)^{m+1}}, & c < \sigma_\pi^2 \end{cases} \tag{3.7}$$

and has the property that its mean is $mc/[2(m-1)]$ and its $100q$-th percentile $\left(\frac{m}{m+1} < q < 1\right)$, $\sigma_\pi^2(q)$, is given by

$$\sigma_\pi^2(q) = \frac{c}{1 - [q(m+1) - m]^{1/m}}.$$

Generally $m = 2$ or 3 is adequate to approximate the elicited probabilities. An example is given in Section 4. Other informative choices for $\pi_{2,2}(\sigma_\pi^2)$ are possible.

4. NUMERICAL EXAMPLE

Data from two suppliers of a Zoner diode are given in the table below:

Supplier 1				Supplier 2			
Lot	x_{1j}	s_{1j}	n_{1j}	Lot	x_{2j}	s_{2j}	n_{2j}
1	19.1	4.2	26	1	13.3	17.7	30
2	8.2	16.4	29	2	15.9	13.1	29
3	9.1	8.5	30	3	10.1	16.8	30
4	17.4	17.5	30	4	24.9	17.3	30
5	12.1	7.7	29	5	27.8	14.2	30
6	9.0	5.6	29	6	15.7	11.5	30
$\bar{x}_1 = 10.8$		$s_{\bar{x}_1} = 1.30$		$\bar{x}_2 = 17.9$		$s_{\bar{x}_2} = 2.57$	

Firstly, since $k = 2$, there will be considerable simplification of the general formula (3.4). The expression for $p_j(b)$ becomes

$$\begin{aligned} p_2(b) &= P(\gamma_2 < \gamma_1 + b|\bar{x}) \\ &= \int_0^\infty \Phi(U(\sigma_\pi^2))\pi_{2,2}(\sigma_\pi^2|\bar{x})d\sigma_\pi^2 \end{aligned} \tag{4.1}$$

where U and $\pi_{2,2}$ are given in Berger and Deely (1988) (see Equation 5.1).

We now evaluate (4.1) for several different exchangeable models, i.e., all cases assume $\ell = 1$, $y = 1$, $d = 0$. Also we will use the sample variance for ν_i^2. A comment will be made in conclusion concerning known variances.

Case 1: (noninformative) $\beta = 0$, $A = \infty$, $b = 0$, $\nu_i = s_{\bar{x}_i}$ and $\pi_{2,2}$ given by

$$\pi_{2,2}(\sigma_\pi^2) = \left[\left(\frac{c}{\nu_1^2 + \sigma_\pi^2}\right)\left(\frac{c}{\nu_2^2 + \sigma_\pi^2}\right)\right]^{\frac{3}{4}}.$$

Numerical results give $p_2 = 0.039$, $p_1 = 0.961$, overwhelming evidence that Supplier 1 is best.

Case 2: (informative) $\beta^0 = 9$, $A = 9$, $b = 0$, $\nu_i = s_{\bar{x}_i}$ and $\pi_{2,2}$ taken as a "Shoe" distribution with $c = 10$, $m = 2$. Numerical results give $p_2 = 0.047$, $p_1 = 0.953$, still overwhelming evidence that Supplier 1 is best.

Case 3: Same as Case 1 except assume variances are equal.
Take $\nu_1^2 = \nu_2^2 = \frac{(1.3)^2+(2.57)^2}{2} = 4.15 = \nu^2$. In this case, a closed form expression for p_2 can be obtained and is given by (see Berger and Deely)

$$p_2 = \frac{\Phi(\zeta) - e^{-\upsilon}\left(\frac{1}{2} + \frac{\zeta}{\sqrt{2\pi}}\right)}{1 - e^{\upsilon}}$$

where $\zeta = \frac{\bar{x}_1-\bar{x}_1}{\nu\sqrt{2}} = \frac{10.8-17.9}{2.04\sqrt{2}} = -2.46$ and $\upsilon = \frac{\zeta^2}{2} = 303$. Thus $p_2 = 0.032$ and again there is overwhemingly evidence that Supplier 1 is better. Finally, if we change the choice of $\pi_{2,2}(\sigma_\pi^2)$ to

$$\pi_{2,2}(\sigma_\pi^2) = \frac{5\nu^5}{2(\nu^2+\sigma_\pi^2)^{7/2}},$$

then

$$p_2 = \frac{\Phi(\zeta) - e^{-\upsilon}\sum_{i=0}^{2}\left\{\frac{\upsilon^i}{i!}\left[\frac{1}{2}\frac{\zeta e_i}{\sqrt{2\pi}}\right]\right\}}{1 - e^{-\upsilon}\sum_{i=0}^{2}\frac{\upsilon^i}{i!}} = .0985$$

where $c_0 = 1$, $c_1 = \frac{2}{3}$,$c_2 = \frac{8}{15}$. Again Supplier 1 is seen to be the better of the two.
Case 4: Same as Case 2 except that b is arbitrary so we can compute $p_j(b)$, which gives the distribution of the difference $\gamma_2 - \gamma_1$.

b	–3	0	3	6	9	12
$p_2(b)$	.0008	.043	.360	.787	.973	.999

This table gives a much clearer picture of just how γ_1 and γ_2 compare after observing this particular data set.

5. CONCLUDING REMARKS

(i) It clear that the approach taken in this paper is a simple one, yet applicable. The mean of the quality distribution of a supplier is an important parameter. But so is the variance. A next step would be to incorporate both parameters in the model and put prior distributions on both via the hierarchical approach. Perhaps some function of the mean and variance would be a better indicator of the quality of a supplier. For example, suppose the supplier with largest proportion of unknown lot means below a specified critical value is deemed "best", i.e. $p_i = P(\theta_{ij} < c)$. Then p_i is largest iff $(c - \gamma_i)/\tau_i$ is largest. When both parameters are unknown, a prior distribution on both is required, or at least a prior on the quantity $(c - \gamma_i)/\tau_i$ or even p_i itself.

(ii) For the model treated in this paper, only the mixed variance ν_i^2 was required for any of the computations. Knowledge of σ_{ij}^2 and τ_i^2 is sufficient for ν_i^2 but we can also obtain a good estimate of ν_i^2 if the number of lots is large simply by using the sample variance as was indicated in the numerical examples. Of course, in the data presented there were only six lots and this may not provide an accurate estimate of ν_i^2. The model can be extended to put another prior on ν_i^2 and this will simply add another dimension to the integration. This can be seen by viewing all of the results presented conditional on ν_i^2's; then integrating over their common prior, assuming the ν_i^2's are i.i.d., and taking the k-th product will give the desired result. The effect of this change has yet to be studied.

(iii) Another intuitively appealing aspect of the problem is to somehow take into the selection procedure the history of lots; that is, the individual sample lot means may be showing an important trend which would be helpful in the selection of a "best" supplier. Also, it is suggested in this model that a trade off between sample size within a lot and number of lots can be done in an optimal way, i.e. to maximize the probability of getting the "best" supplier. Since there is a "within" lot variance and a "between" lots variance, considerations of how many items from a lot and how many lots are required would add to the applicability of the model suggested herein. Hopefully some or all these ideas can be developed in the future.

(iv) Finally, we want to point out that incorporating a prior probability that all suppliers have the same mean quality can easily be done. The details are contained in Berger and Deely (1987). Essentially, it simply means that a posterior probability, p_0 of the null hypothesis is computed using a prior probability that $\sigma_\pi^2 = 0$. Then all of the p_j's computed in this paper would be multiplied by $(1 - p_0)$.

REFERENCES

Berger, J. O. (1985). *Statistical Decision Theory and Bayesian Analysis*, New Yok: Springer-Verlag.

Berger, J. O. and Deely, J. J. (1988). A Bayesian Approach to Ranking and Selection of Related Means with Alternatives to AOV Methodology. *J. Amer. Statist. Assoc.* **82**,

Neter, J., Wasserman, W. and Kutner, M. H. (1985). *Applied Linear Statistical Models*, Homewood, Illinois: Richard D. Irwin, Inc.

AYESIAN STATISTICS 3, pp. 593–599
. M. Bernardo, M. H. DeGroot, D. V. Lindley and A. F. M. Smith, (Eds.)
) Oxford University Press, 1988

Global v Local Screening

I. R. DUNSMORE and R. J. BOYS
University of Sheffield and *University of Newcastle*

SUMMARY

Screening problems in binary response models correspond to a form of diagnosis or discrimination problem in which restrictions are placed on the probability of "success". We develop here an approach suggested in Dunsmore & Boys (1987) which concentrates on the predictive probability of success for a particular future individual.

eywords: BINARY REGRESSION; DIAGNOSTIC APPROACH; PREDICTIVE PROBABILITY; SAMPLING APPROACH; SCREENING.

1. INTRODUCTION

ndividuals can be categorized as "success" ($t = 1$) or "failure" ($t = 0$), but such a categoriza-ion may be difficult. By using some related feature variable $X = (X_1, X_2, \ldots, X_q)$, which s much easier to obtain, we attempt to screen a future individual on the basis of the value $X = x$, so that we can confidently assert whether such an individual should be categorized as success.

We assume that a data set $(x_1, t_1), (x_2, t_2), \ldots, (x_n, t_n)$ is available from n individuals vhose categories $t_1, t_2, \ldots, t_n$ are known with certainty. We wish to assess the category t of future individual for whom we have observed the value x of the feature variable. We do his by use of the predictive diagnostic probability function $p(t|x, \text{data})$, and in particular seek specification region C_X^* such that

$$p(t = 1|x, \text{data}) \begin{cases} \geq \delta_p^* & \text{for} \quad x \in C_X^*, \\ < \delta_p^* & \text{for} \quad x \notin C_X^*. \end{cases}$$

Such an approach may be termed a *local* approach as distinct from the *global* approach leveloped in Boys and Dunsmore (1987) and Dunsmore and Boys (1987). There, a region C_X vas sought such that $p(t = 1|x \in C_X, \text{data}) = \delta_p$. The region was considered to be of specific orm $C_X = \{x : a'x \geq \omega\}$, and an additional criterion, namely minimization of the error robability $\varepsilon_p = p(t = 1|x \notin C_X, \text{data})$, was necessary to avoid indeterminacy. Such a global pproach averages $p(t = 1|x, \text{data})$ over the conditional distribution given by $p(x|x \in C_X)$. Ve use the notation C_X^* and δ_p^* in the local approach to emphasize the distinction from C_X nd δ_p in the global approach. Notice that if we set $\delta_p^* = \delta_p$ we are unlikely to find that $C_X^* = C_X$.

Similar related global problems were discussed in Boys and Dunsmore (1986), where neasurements (Y, X) were obtained and success (i.e. $t = 1$) corresponded to $Y \in C_Y$, a pecification region for Y. A normal model for (Y, X) for the case $q = 1$ was developed. A ocal Bayesian approach in this situation has been investigated by Wong, Meeker and Selwyn (1985).

A pure predictive approach to screening should place paramount importance on the di-agnostic probability level of the individual. This is what the local approach developed here aims to do. In this paper we restrict attention to the case of $q = 1$. Further research and development is under way for the multivariate situation.

2. DERIVATION OF C_X^*

The keystone of the Bayesian approach is the predictive probability $p(t = 1|x, \text{data})$. Two parametric pathways present themselves, namely through the diagnostic and sampling models suggested by Dawid (1976) and Aitchison and Begg (1976).

2.1 Diagnostic approach

The more direct modelling procedure is the diagnostic approach through a binary regression model. We consider here the linear logistic form

$$p(t = 1|x, \xi) = \frac{e^{\xi_0+\xi_1 x}}{1 + e^{\xi_0+\xi_1 x}}.$$

Similar analyses follow with the probit and complementary log-log models.

Boys and Dunsmore (1987) discuss the problems involved in the updating to

$$p(t = 1|x, \text{data}) = \int p(t = 1|x, \xi)p(\xi|\text{data})d\xi, \tag{1}$$

and present several approximations. We use here three of these alternatives. In the first two the approximate normality of (ξ_0, ξ_1) *a posteriori* is assumed, so that $\zeta = \xi_0 + \xi_1 x$ is approximately $N(b_x, d_x^2)$, where

$$b_x = \hat{\xi}_0 + \hat{\xi}_1 x, \quad d_x^2 = (1\ x)(-H)^{-1}\begin{pmatrix}1\\ x\end{pmatrix},$$

$\hat{\xi}_0$, $\hat{\xi}_1$ are the maximum likelihood estimates of ξ_0, ξ_1 and H is the Hessian of the log likelihood evaluated at $\hat{\xi}_0$, $\hat{\xi}_1$. We can then approximate (1) through

$$\int \frac{e^{\zeta}}{1 + e^{\zeta}}\phi(\zeta|b_x, d_x^2)d\zeta,$$

either by I: quadrature or II: the normal logistic approximation

$$\Phi\left\{\frac{b_x}{(k^2 + d_x^2)^{\frac{1}{2}}}\right\};$$

(see Boys and Dunsmore (1987) for more details and Lauder (1978) for discussion on the choice of constant k).

Approximation III follows Bernardo's (1988) suggestion that $p(t = 1|x, \text{data})$ be forced to logistic form

$$\frac{e^{\tilde{\xi}_0+\tilde{\xi}_1 x}}{1 + e^{\tilde{\xi}_0+\tilde{\xi}_1 x}}$$

for suitably chosen $\tilde{\xi}_0$, $\tilde{\xi}_1$ obtained by maximizing the expected utility function based on a logarithmic scoring rule, or equivalently minimizing the directed divergence between the true and the approximated logistic probabilities.

For the local screening problem we then proceed as follows.

Approximation I: Evaluate $p(t = 1|x, \text{data})$ by quadrature as a function of x. From a plot of $p(t = 1|x, \text{data})$ against x read off the region C_X^* such that $p(t = 1|x, \text{data}) \geq \delta_p^*$. In many situations $p(t = 1|x, \text{data})$ will be a monotonic function of x in a neighbourhood of the solution, and correspondingly C_X^* will have a very simple form.

Approximation II: For a suitable value of k, solve for x the inequality

$$\Phi\left\{\frac{b_x}{(k^2+d_x^2)^{\frac{1}{2}}}\right\} \geq \delta_p^*.$$

Approximation III: Derive estimates $\tilde{\xi}_0, \tilde{\xi}_1$. Then

$$C_X^* = \left\{x : \tilde{\xi}_0 + \tilde{\xi}_1 x > \ell n\left(\frac{\delta_p^*}{1-\delta_p^*}\right)\right\}.$$

2.2 *Sampling approach*

The specification of the model here is through a different parameterization, namely

$$p(t, x|\theta) = p(x|t, \eta)p(t|\psi).$$

Suppose that X follows a normal distribution in each group, that is

$$p(x|t=i, \eta) \sim N(\mu_i, \sigma_i^2) \quad (i = 0, 1).$$

With vague prior assumptions on $\eta = (\mu_0, \mu_1, \sigma_0, \sigma_1)$ the predictive densities are of Student form, namely

$$p(x|t=i, \text{data}) \propto \left\{1 + \frac{n_i}{n_i^2-1}\frac{(x-\bar{x}_i)^2}{s_i^2}\right\}^{-\frac{1}{2}n_i} \qquad (i = 0, 1),$$

where there are n_i individuals with $t = i$, and $\bar{x}_i$, s_i^2 are the group sample means and variances; see Boys and Dunsmore (1987).

Similarly with a vague prior on ψ, where $p(t = 1|\psi) = \psi$, the predictive probability function for t is

$$p(t = i|\text{data}) = \frac{n_i}{n_0+n_1} \quad (i = 0, 1).$$

The local screening problem requires that we use the specification region

$$C_X^* = \left\{x : \frac{p(x|t=1, \text{data})}{p(x|t=0, \text{data})} > \frac{n_0}{n_1}\left(\frac{\delta_p^*}{1-\delta_p^*}\right)\right\}.$$

This reduces to a region of the form

$$C_X^* = \begin{cases} \{x : x > \text{constant}\} & \text{if} \quad \bar{x}_0 \leq \bar{x}_1, \\ \{x : x < \text{constant}\} & \text{if} \quad \bar{x}_0 \geq \bar{x}_1 \end{cases}.$$

2.3 *Global approach*

Details for the derivation of C_X for a specified δ_p for the global approach are provided in Boys and Dunsmore (1987).

3. EXAMPLE

Hodgkin's disease involves a malignant neoplasm that usually arises in a group of lymph nodes. Patients are treated with radiotherapy (RT) but additional chemotherapy (CT) can be given if necessary, although the effects of this are very unpleasant. Neither treatment guarantees survival however.

Data is available for 42 patients who have been treated with RT and who then either survived for three months after treatment ($t = 1$) or died ($t = 0$). (The survival times would perhaps be more appropriate data to analyse and we are developing suitable models to incorporate such extra information.) Consider the following screening problem. We seek to identify those patients we believe will survive for at least three months after undergoing RT, and we want to guarantee that for such a patient the predictive probability of survival is at least 0.95. Such patients will be treated with RT and all other patients will be given the more drastic CT as well.

Clearly other formulations of the screening problem could be appropriate here, and other information may well be available. For example we may have data available from patients who received CT as well. We simply present the above screening problem as an illustration of our methods.

Four screening variables are available, namely age, erythrocyte sedimentation rate (ESR), lymphocyte count (LC) and haemoglobin level (HB). However for illustration here we use a discriminant from a linear logistic model based on the logarithms of the variables, namely

$$x = -9.711 + 0.082 \log(\text{age}) - 0.722 \log(\text{ESR}) \\ + 0.080 \log(\text{LC}) + 5.104 \log(\text{HB}).$$

Logarithms of the variables are used because the discriminant then satisfies (reasonably well) the different model assumptions, that is, normality within the two groups $t = 1$ and $t = 0$ and also overall, properties which are required in the sampling framework and the global model. A linear discriminant on the untransformed variables is less satisfactory in this respect. The values of x are given in Table 1.

	x				
t = 1 (n_1 = 33)	-0.49	-0.35	-0.03	0.63	0.78
	1.01	1.18	1.21	1.27	1.62
	1.65	1.86	2.04	2.07	2.31
	2.33	2.39	2.54	2.73	2.74
	2.86	2.91	3.01	3.17	3.33
	3.61	3.74	3.86	3.95	4.42
	4.94	5.16	5.27		
t = 0 (n_0 = 9)	-2.37	-1.17	-0.19	-0.04	0.07
	0.50	0.80	1.11	3.53	

Table 1. *Values of discriminant x for survivors ($t = 1$) and dead ($t = 0$)*

For this screening variable x the specification regions are of the form $C_X^* = \{x : x > \omega^*\}$

and $C_X = \{x : x > \omega\}$, and we set both δ_p^* and δ_p equal to 0.95. Table 2 shows the values of the cut-off points ω^* and ω for the local and global approaches under the different assumptions. The agreement within the local approach between the different models is reasonably good. The outlying value in the global approach for the sampling model is probably due to the normality assumption for X within the $t = 0$ group being slightly suspect. Figure 1 shows the predictive success probability $p(t = 1|x, \text{data})$ under approximation I as a function x.

		Local	Global
		ω^*	ω
Diagnostic	I	3.40	2.41
	II (k = 1.714)	3.30	2.27
	III	3.51	2.53
Sampling		3.77	3.75

Table 2. *Cut-off values in the local and global approaches*

In the global approach a marginal distribution on X overall is required; for this data a normal distribution seems reasonable. Following the methods in Boys and Dunsmore (1987) we can then derive such quantities as $\gamma_p = p(t = 1|\text{data})$, $\beta_p = P(X \in C_X|\text{data})$ and $\varepsilon_p = p(t = 1|x \notin C_X, \text{data})$. These are shown in Table 3. One distinct advantage of the local approach within the diagnostic framework is that the specification of a distribution on X is not required.

		γ_p	β_p	ε_p
Diagnostic	I	0.77	0.40	0.66
	II (k = 1.714)	0.77	0.43	0.64
	III	0.79	0.37	0.69
Sampling		0.79	0.16	0.75

Table 3. *Summary of global approach*

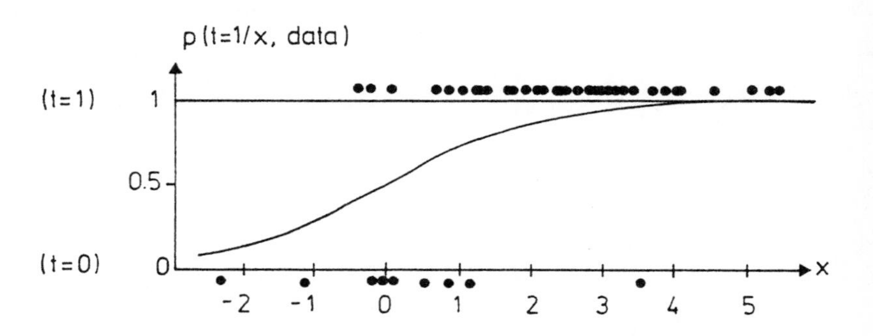

Figure 1. *Predictive success probability $p(t = 1|x, data)$ under approximation I as a function of x*

We see that prior to screening the proportion of survivors is about 0.77. The screening procedure retains about 40% of individuals; this is the proportion whom we believe will survive and to whom RT only will be given. The relatively large values of ε_p imply that a relatively large proportion of patients given CT also survive.

It is of interest to determine the value of δ_p^* which would result if we used the cut-off ω from the global approach as the ω^* in the local approach. When using approximation I we find that, in the neighbourhood of $\delta_p^* = 0.95, \omega^*$ may be very well expressed as a quadratic in δ_p^*, namely

$$\omega^* = 95.009 - 220.041\ \delta_p^* + 130.116\ \delta_p^{*2}.$$

(The coefficient of determination R^2 is 99.8%.) Hence if we use a cut-off of 2.41 in the local approach we would obtain a δ_p^* of 0.90 only. The corresponding value from use of approximation III is 0.89.

Similarly we can determine the value of δ_p which would result if we used the cut-off ω^* from the local approach as the ω in the global approach. Thus, for example, with approximation I if we use $\omega = 3.40$ in the global approach we would attain a δ_p of 0.97.

4. COMMENT

The local approach provides a personal view, that is, the predictive probability assessment $p(t = 1|x, \text{data})$ is specific to the particular individual (or x value). We know for this individual whether the probability exceeds δ_p^* — indeed we have the actual value (perhaps approximately). In the global approach an overall view is provided. For a given x in C_X there is no guarantee that $p(t = 1|x, \text{data})$ will be greater than δ_p. For some x the predictive probability will be less that δ_p; for other x it will be greater. It was for this reason that we distinguished δ_p and δ_p^*.

A major advantage of using the local approach with a diagnostic model is that there is no need to specify a model for X. Only if comparisons are required between, for example, δ_p and δ_p^* or ε_p and $\varepsilon_p^* = p(t = 1|x \notin C_x^*, \text{data})$ would such a model be required.

REFERENCES

Aitchison, J. and Begg, C. B. (1976). Statistical diagnosis when basic cases are not classified with certainty. *Biometrika* **63**, 1–12.

Bernardo, J. M. (1988). Bayesian linear probabilistic classification. *Statistical Decision Theory and Related Topics* **4**, 151–161. (S. S. Gupta and J. O. Berger, eds.) New York: Springer.

Boys, R. J. and Dunsmore, I. R. (1986). Screening in a normal model. *J. Roy. Statist. Soc. B* **48**, 60–69.

Boys, R. J.and Dunsmore, I. R. (1987). Diagnostic and sampling models in screening. *Biometrika* **74**, 365–374.

Dawid, A. P. (1976). Properties of diagnostic data distributions. *Biometrics* **32**, 647–658.

Dunsmore, I. R. and Boys, R. J.(1987). Predictive screening methods in binary response models. *Probability and Bayesian Statistics* (R. Viertl ed.). New York: Plenum, 151–158.

Lauder, I. J. (1978). Computational problems in predictive diagnosis. *Compstat1978* (L. C. A. Corsten and J. Hermans, eds.) Vienna: Physika Verlag. 186–192

Wong, A., Meeker, J. B. and Selwyn, M. R. (1985). Screening on correlated variables: a Bayesian approach. *Technometrics* **27**, 423–431.

BAYESIAN STATISTICS 3, pp. 601–607
J. M. Bernardo, M. H. DeGroot, D. V. Lindley and A. F. M. Smith, (Eds.)

Estimated Public Welfare Quality Control Error Rates and Penalties

W. B. FAIRLEY and D. FAIRLEY
Philadelphia, Analysis & Inference and *Ohio State Univ.*

SUMMARY

In the Food Stamp Program of the U.S. federal government, state governments which administer this program can be penalized by loss of federal benefits. The penalty amount is based on the difference between the estimated rate of dollar errors that state caseworkers make in giving out Food Stamp benefits to low income recipients and an estimated target rate for such errors. The largest positive estimated rates and penalty amounts tend to be classically upwardly biased estimates of true rates and penalty amounts. We refer to such bias as extreme value selection bias. James-Stein estimates for 53 states and territories will have lower aggregate mean square error than the observed state values. With exchangeable true state period rates or penalties and symmetric losses no basis exists for preferring observed state values over James-Stein estimates for any individual state. We describe three alternative sets of Bayesian models for period rates for each state: 7 models, one for each period, for all 53 states; 53 models, one for each state, for 7 periods; and a single model for all 53 states in 7 periods in which we use a two-way ANOVA model for the prior mean of the state period means. Empirical Bayes estimates are obtained as estimates of the posterior means for the state period means.

Keywords: BAYESIAN MODELS; EXCHANGEABILITY; JAMES-STEIN ESTIMATES; EMPIRICAL BAYES ESTIMATES; QUALITY CONTROL; EXTREME VALUE SELECTION BIAS; MEAN SQUARE ERROR; TWO-WAY ANOVA; FOOD STAMP PROGRAM.

1. INTRODUCTION

The Food Stamp Program of the U.S. federal government issues some $11 billion in food benefits to lower income persons. Benefits are in the form of "food stamps" entitling the bearer to purchase food at grocery stores for the benefit amount. State agencies issue these benefits to eligible households in every state, but the dollar costs of the benefits are paid entirely by the federal goverment. In 1977 the Congress of the Unites States passed legislation ("Food Stamp Act of 1977") that mandated a "quality control program to reduce the rate of errors made in determining eligibility for benefits or, given eligibility, in determining the correct dollar amount of benefits to which eligible recipients are entitled. Eligibility and entitlement amounts are based primarily upon income level, wealth, and household status. The requirements are complex and are spelled out in legislation and in regulations issued by the U.S. Department of Agriculture's Food and Nutrition Service, which administers the program at the federal level. Similar quality control programs have been mandated by Congress for then even larger public welfare programs of assistance to families with dependent children (AFDC) and medical benefits for the needy (Medicaid).

The Food Stamp Act of 1977 (pages 35–36) describes the quality control program as follows:

(d) Effective October 1, 1981, and annually thereafter, each State ... shall ... submit ... a quality control plan for the State which shall specify actions such State proposes to take in order to reduce:

(1) the incidence of error rates in and the value of

(A) food stamp allotments for households which fail to meet basic program eligibility requirements;

(B) food stamp allotments overissued to eligible households, and

(C) food stamp allotments underissued to eligible households...

(e) As used in the section "quality control" means monitoring and reducing the rate of errors in determining basic eligibility and benefit levels.

The Act goes on the specify that the states are responsible for keeping their rate of dollars in error —the "payment error rate", defined as the ratio of dollars in error to total dollars paid out— below specified target levels in every six month period. A state's target for a given period is the larger of (a) the national average payment error rate for all states in a prior base period (one to one and a half years previous), or (b) the state's own payment error rate in the base period decreased by 10 percent towards a final goal of 5%. A state that fails to meet its target rate must return to the federal government a percentage of the dollars provided to it equal to the difference between its estimated payment error rate and its target payment error rate. The Act has been amended subsequently, but for purposes of the present paper the changes are not important.

For the fiscal year 1985 dollar "penalties" for failure to meet targets amounted to $20 million dollars. States may request a "good cause" waiver of the penalty by showing that extraordinary circumstances beyond their control led to the excess their estimated payment error rate over their target. Failing a good cause waiver, states may appeal the penalty levied upon them to a federal Appeals Board. Finally, if the Board does not drop the penalty, states may take the federal government to court and seek to have the penalty removed.

To implement the quality control program for Food Stamps the Food and Nutrition Service required each state to take for each 6 month period a probability sample of households participating in the Food Stamp Program. For each household in the sample, a state auditor reviews the food stamp allotment made to the household to determine if any errors were made by the state worker who initially made the allotment. The total of dollar errors so determined, as a ratio to the total dollars issued in the 6 month period, is the state's estimated payment error rate. Subsequently, the federal agency auditors (Food and Nutrition Service) take a random sub-sample of the households in the state's sample and perform their own review of the cases in the sub-sample.

The agency then determines its own estimated payment error rate by using a regression estimator. The regression estimator uses the regression of the federally-determined errors in the sub-sample on the state-determined error in the sub-sample to obtain an estimate that adjusts the federal sample mean payment error rate by the calibrated difference between the sample mean for the large state sample and the state mean for the federal sub-sample. Cochran (1977) discusses such regression estimates. The federally-determined regression estimates of states' payment error rates are the ones ultimately used to determine the target rates, the penalty period rates, and form these the estimated excess of states' error rates over the target rates.

The authors of the present paper have prepared, on behalf of states receiving penalties, critiques of the federal quality control programs for public welfare programs. (See Fairley (1985, 1986)). The purpose of the present paper is to discuss important extreme value bias in federal statistical estimation of states' payment error rates and of states' penalties based upon these, and empirical Bayes estimates that reduce the mean square error of estimation. Section 2 discusses the bias in the use of either state sample means or federal regression estimates as estimates of penalties for the states. Section 3 discusses alternative methods of determining error rates and penalties that reduce the biases and have desirable statistical properties. We give Bayesian models for payment error rates and penalties and empirical Bayes estimates of these.

2. A STATISTICAL BIAS IN ERROR RATE AND PENALTY INFERENCE

The point of view toward the objectives of statistical inference adopted in the present paper is deliberately eclectic. While Bayesian models are most helpful in structuring the inference issues, they face a hurdle of acceptance in courts, at least in the United States.We believe that this hurdle can and will be overcome (Fienberg and Kadane (1980)). Meanwhile, it is important to emphasize that statistical problems can be discussed from a variety of inference points of view. In particular, the problems of statistical bias that we discuss below can be understood from both Bayesian and frequentist perspectives. Important practical conclusions from these perspectives are not in conflict here.

Extreme value selection bias in the quality control estimates is a bias in both the estimated error rates and estimated penalties. For each period the federal agency makes 53 state and territory error rate estimates and 53 corresponding penalty estimates. Those states whose difference between estimated state penalty period rate and estimated target rate is positive are penalized. Thus the penalty estimates that count are selected conditional on being positive. They are order statistics from the 53 populations. But the largest order statistics are upwardly biased estimates of their respective population means, while the smallest tend to be downwardly biased. We refer to the difference between the expected value of these estimates and their respective population means as extreme value bias. The biases depend on the distributions of the estimates. Inferences in this situation are discussed, for example, by Gibbons, Olkin and Sobel (1977) and Miller (1981). A major feature of empirical Bayes estimates, which we discuss below, is to correct for extreme value bias by shrinking observed values towards a models estimate that is less affected by this kind of bias.

Even if the estimated difference between the penalty period rate and the target were unbiased, the estimated penalty itself would be biased. This arises from the definition of the penalty as a function of the estimated difference. If this difference is negative, so that the period rate is below target, the state is not penalized but it gets no credit, whereas if this difference is positive the penalty is proportional. In mathematical terms the penalty is a convex function of the difference and therefore, by Jensen's inequality, the expected value of the penalty exceeds the same function of the expected value of the difference.

Federal agencies implementing quality control systems for public welfare programs have justified the use of the state estimated rates as the "best" single point estimates for each state. "Best" has generally been understood as lowest mean squared error conditional on the true rates. For a single state taken in isolation, this might be argued. However, when simultaneous estimates of 53 state rates are made, the average state mean square error conditional on the true rates is not minimized by using 53 single point estimates, even if they are admissible individually —as James and Stein (1960) demonstrated. In multivariate estimation, superior estimates for the ensemble are obtained by using shrinkage estimators, no matter what the underlying distribution (Brown (1966)). Furthermore, provided that the true rates or the deviations of the true rates from a linear model are exchangeable, there is no basis for preferring the so-called single "best" point estimate for a given state over the shrinkage estimate for that state. Under the assumption of exchangeability, a better single point estimate for any given state can be expected to be the shrinkage estimate for that state.

Elimination of bias in and of itself is not a suitable criterion for inference. A state is interested in the entire (posterior) distribution of penalties given its prior distribution on the true difference. However, for a state with a prior distribution on the difference symmetric about 0, the posterior distribution of penalties will have a positive mean. This state therefore has a positive expected cost each period even though its true penalty is zero. Were there features of the penalty estimates that offset the bias so that the mean squared error (the risk) of the estimate or some other symmetric expected loss of the error of estimate were reduced, then the state would probably be indifferent to the bias, at least from a purely numerical point of view. (Biases arising from specific known sources have other problems.) But this is not the case here. The bias causes an increased expected value of any reasonable loss function of interest.

3. ALTERNATIVE PROCEDURES

A. *Bayesian Models*

In will be useful as an expository device to build up to the best estimated model for error rates by describing assumptions and estimates for three models of increasing complexity. The first is a model for error rates of 53 states in a given period, the second a model for error rates over 7 periods for a given state, and the third a model for all 53 states over all 7 periods. We call these the Single Period model, the Single State model, and the Multi-Period-Multi-State model, respectively, We generally follow the notation of Morris (1983).

For each of these models, we use cross-validation to compare the empirical Bayes (EB) estimators to the naive estimator. We fit the models excluding the "eighth" period and compare use of the observed estimates for the seventh period to estimate the eighth. Here we assume that a given state's rate is approximately constant over a period as short as one year. In actuality, the "eighth" period is the fifth period chronologically in a consecutive set of eight 6-month periods, so we are predicting the fifth period using as the naive estimator the state estimated error rates for the 53 states in the previous or the fourth period chronologically. The seven periods used in fitting the ANOVA model used by EB estimates are actually the periods 1 through 4 and 6 through 8 chronologically.

Single Period Model. Suppose that we had error rates available for only a single period for all 53 states and territories. The estimated error rates, $R_i, i = 1, 2, \ldots, 53$, conditional on the population values, $\theta_i = 1, 2, \ldots, 53$, in the given period are essentially sample means of large samples. Their variability reflects finite random sampling error and measurement error. Ignoring measurement error bias, the R_i can be assumed to be distributed normally with means θ_i and variances V_i. The V_i will differ primarily because sample sizes differ, and to a lesser extent because the size distributions of errors differ. For the moment we assume V_i are constant ($V_i = V$), though the ratio of maximum to minimum sample sizes and variances is about 8 to 1. Because samples in the various states are drawn independently and error determinations are made independently, the R_i given θ_i can be assumed independent. Thus:

$$R_i|\theta_i \overset{\text{ind}}{\sim} N(\theta_i, V), \qquad i = 1, 2, \ldots, 53 \tag{1}$$

In the first prior for which error rates were estimated, it might have been reasonable to model the true state rates as normally and independently distributed with mean μ and variance A:

$$\theta_i \overset{\text{ind}}{\sim} N(\mu, A), \qquad i = 1, 2, \ldots, 53 \tag{2}$$

In a personal personal probability interpretation, the θ_i's might have been exchangeable at that point in time.

The marginal distribution of R_i, is then

$$R_i \overset{\text{ind}}{\sim} N(\mu, V + A), \qquad i = 1, 2, \ldots, 53 \tag{3}$$

Regarding V as known and using standard results (Morris (1983, p. 48), the least squares estimate R of μ (the overall sample mean of R_i's) and the sum of squares of deviations of the error rates from this estimate, S, the between-state variance, lead to estimates of μ and A. Here:

$$S = \sum_{i=1}^{53} (R_i - \bar{R})^2$$

The posterior distribution of θ_i, conditional on the observed R_i's is then:

$$\theta_i | R_i \stackrel{\text{ind}}{\sim} N(\theta_i^*, V(1-B)) \tag{4}$$

with mean

$$\theta_i^* = B\bar{R} + (1-B)R_i \tag{5}$$

and shrinkage factor

$$B = V/(V+A). \tag{6}$$

With B estimated as

$$\hat{B} = V/(S/51), \tag{7}$$

Stein's estimator for θ_i becomes

$$\hat{\theta}_i = \hat{B}\bar{R} + (1-\hat{B})R_i. \tag{8}$$

Empirical Bayes (unweighted) estimates of the θ_i under these assumptions, using an average within-state variance of 4, a between-state variance of 13.26 and consequently an estimated shrinkage factor B of 0.3 towards the arithmetic mean rate of all 53 states produces an interestingly lower mean square error of prediction for error rates in other periods compared to prediction using the observed error rates R_i themselves. The results given here were obtained for regressed rates. State rate results are similar. However, we pass over these results to go on to the more realistic models.

Single-State Models. The Single-State Model take single state's error rates for 7 periods as normally and independently distributed with a common variance, V_i. Normality was discussed above. Independence follows from the independent drawings of successive random samples for each period. The common variance is believed to be a good approximation since sample sizes generally do not change radically and the size distribution of errors for a given state is believed stable —though unequal variances for each period could be easily modelled. The error rate model for the i-th state conditional on the true period rates, is then:

$$R_{ij} \stackrel{\text{ind}}{\sim} N(\theta_i, V_i), \qquad j = 1, 2, \ldots, 7. \tag{9}$$

Were data on no other state available, it might be reasonable to take the state's rate θ_i as being drawn from a normal distribution with mean μ_i and variance A:

$$\theta_{ij} \stackrel{\text{ind}}{\sim} N(\mu_i, A), \qquad j = 1, 2, \ldots, 7. \tag{10}$$

This model of course simply assumes no time variation in the mean of the state's rate (μ_i is constant over periods). The model can be extented to allow time variation, for example, by a) modelling average time period parameters in a two-way ANOVA as is done below, or b) fitting individual time trends to each of the state rates. See, for example, the model suggested by Reinsel (1985). Empirical Bayes estimates of the 53 θ_i's for the "eighth" period from 53 Single-State Models, estimating the conditional variances by second differences of the R_j's and the variance of the marginal distribution by the usual sample variance, produces a substantially improved average mean square error of prediction and number of predictions closer over the 53 states and territories as compared to use of the state error rates themselves in the period prior to the predicted period. Again, we pass over the details for these simplified Single-State Models.

Multi-Period-Multi-State Model. By taking error rate data for every state in every period we get a model that incorporates the distinct realistic features of the two simplified models discussed above. Combining the assumptions in previous models for the estimated error rates R_{ij} conditional on the true i-th state rate in a given j-th period, θ_{ij}, gives

$$R_{ij}|\theta_{ij}, \stackrel{\text{ind}}{\sim} N(\theta_{ij}, V_i), \quad \begin{array}{l} i = 1, 2, \ldots, 53 \\ j = 1, 2, \ldots, 7. \end{array} \tag{11}$$

Here, we allow the conditional variance V_i to vary across states but, for simplicity and to a reasonable approximation, not across time. The true rates θ_{ij} are drawn from normal distributions whose means are modelled by a linear two-way ANOVA without interactions:

$$\theta_{ij}|\mu, \alpha_i, \beta_j, A_i \stackrel{\text{ind}}{\sim} N(\mu + \alpha_i + \beta_j, A_i), \quad \begin{array}{l} i = 1, 2, \ldots, 53 \\ j = 1, 2, \ldots, 7. \end{array} \tag{12}$$

The exchangeability of the mean rates, θ_{ij} under this model is plausible, unlike the case under the two simplified models discussed above—in each of which available information was withheld.

Empirical Bayes estimates were obtained under the assumption, first, that the state θ_{ij}'s had equal variances ($A_i = A$) and, second, that they were unequal and proportional to the conditional variances of the error rates ($A_i \propto V_i$). The second assumption is convenient as it leads to simple, closed form solutions for estimates, but it is also plausible that smaller states, having fewer cases and fewer social workers, would have larger variances than larger states. Note that in this case B will be independent of i since B is $V_i/(V_i + A_i)$ which equals $1/(1+\alpha)$ if A_i equals αV_i. Results were similar under the two models, so we describe the proportional variance case, which was marginally superior in prediction. The marginal distribution of the R_{ij}'s is then:

$$R_{ij}|\mu, \alpha_i, \beta_j A_i, V_i \stackrel{\text{ind}}{\sim} N(\mu + \alpha_j + \beta_j, V_i + A_i). \tag{13}$$

Weighted least squares estimates, $\hat{\mu}^*, \hat{\alpha}_i^*$ and $\hat{\beta}_j^*$ of the two-way ANOVA parameters are used in the empirical Bayes estimates:

$$\hat{\theta}_{ij} = \hat{B}(\hat{\mu}^* + \hat{\alpha}_i^* + \hat{\beta}_i^*) + (1 - \hat{B})R_{ij}. \tag{14}$$

The weights are the inverses of estimates of $V_i + A_i$ where $V_i + A_i$ is estimated as

$$S_i = (kn/(kn - k - n - l)) \sum_{j=1}^{7} (R_{ij} - (\hat{\mu} + \hat{\alpha}_i + \hat{\beta}_j))^2/7$$

where $\hat{\mu}$, $\hat{\alpha}_i$ and $\hat{\beta}_j$ are from an unweighted two-way ANOVA. The value of the shrinkage factor B_i for the i-th state, the value of $V_i/(V_i + A_i)$ is independent of i, so $B_i = B$, since by assumption $A_i \propto V_i$. The value of $V = 2.41$ was determined as the average value of V_i estimated from second differences of the R_{ij} for given i. The value of $V + A = 4.07$ was determined as the sum of squares of residuals from the weighted least squares estimates of $\mu + \alpha_i + \beta_j$. The above method follows Morris except that, because the V_i's are poorly estimated, we chose the more conservative $B = 0.5$ as an approximate optimal value for the mean square error criterion.

The proportional variance model estimated by empirical Bayes gave an average estimated mean square error of prediction for the 8-th period of 1.14 percent $(60.60/53)$ compared to a value of 2.96 percent (156.9/53) for predictions of the same period from the prior period's error rates, or a reduction of 60.11%. Out of 53 states the empirical Bayes estimates were closer to the predicted error rates than were the prior period's error rates in 36 cases of 53, or 68% $(36/53 = .68)$. The observed percentage reduction in mean square error is to be compared to a theoretically expected reduction of about 50% for a true B of 0.5. The empirical Bayes estimates were therefore, by either criterion, substantially better predictors of the 8-th period's error rates.

ACKNOWLEDGEMENTS

The authors thank Arvid Roach, Mark Berliner, Murray Selwyn, and Anne Edwards for helpful ideas and assistance in the work that led to the present paper. These individuals, however, are not responsible for the views expressed.

REFERENCES

Berger, J. O. (1985). *Statistical Decision Theory and Bayesian Analysis*, 2nd Ed. Springer-Verlag.

Brown, L. D. (1966). On the admissibility of invariant estimators of one or more location parameters. *Annals* **37**, 1007–1136.

Cochran, W. G. (1977). *Sampling Techniques*, 3rd Ed. New York: Wiley.

Fairley, W. B. (1985). Affidavit in the Matter of Connecticut Department of Income Maintenance. Before the State Food Stamp Appeals Board U.S. Department of Agriculture. Washington, D. C.: Covington and Burling.

Fairley, W. B. (1986). Affidavit in regard to fiscal year 1981 Medicaid quality control disallowance. Before the Departmental Grant Appeals Board, U.S. Department of Health and Human Services. Washington, D. C.: Covington and Burling.

Fienberg, S. and Kadane, J. (1983). The Presentation of Bayesian Statistical Analyses in Legal Proceeding. *The Statistician* **32**, 88–98.

James, W. and Stein, C. (1960). Estimation with quadratic loss. *Proceedings Fourth Berkeley Symposium on Mathematical Statistics and Probability*, University of California Press.

Gibbons, Olkin and Sobel (1977). *Sele:.ng and Ordering Populations: A New Statistical Methodology*. New York: Wiley.

Miller, R. (1981). *Simultaneous Statistical Inference*. Springer-Verlag.

Morris, C. (1983). Parametric empirical Bayes inference: theory and applications, *J. Amer. Statist. Assoc.* **78**, 47–55.

Pratt, J. W. (1965). Bayesian interpretation of standard inference statements. *J. Roy. Statist. Soc. B* **27**, 169–203. (with discussion).

Reinsel, G. C. (1985). Mean squared error properties of empirical Bayes estimators in a multivariate random effects general linear model. *J. Amer. Statist. Assoc.* **80**, 642–650.

Rubin, D. B. (1980). Using empirical Bayes techniques in the law school validity studies. *J. Amer. Statist. Assoc.* **75**, 801–827. (with discussion).

BAYESIAN STATISTICS 3, pp. 609–613
J. M. Bernardo, M. H. DeGroot, D. V. Lindley and A. F. M. Smith, (Eds.)

Iterative Procedures for Continuous Bayesian Designs

K. FELSENSTEIN
Technische Universität Wien

SUMMARY

The search for optimal experimental designs in the linear model is regarded as a Bayesian decision problem. Under quadratic loss the Bayesian-information matrix describes a design. The equivalence theorem for optimal designs serves as advice for constructing optimal continuous designs. The standard procedure is to add points according to the equivalence theorem until it converges. Some modifications of that method are given.

Keywords: LINEAR REGRESSION MODEL; CONTINUOUS DESIGNS; OPTIMAL BAYESIAN DESIGNS, EQUIVALENCE THEOREM; BAYESIAN INFORMATION MATRIX; CONVERGING PROCEDURES.

1. INTRODUCTION

The optimal allocation of regressors has been discussed for a long period and especially the classical linear regression model has been inquired into the dependence of statistical properties and the choice of the controlled variables. The observations Y_i are assumed to depend upon the control-variable x through

$$Y_i = \eta(x_i, \theta) + \varepsilon_i, \quad i = 1, \ldots, n$$

where ε_i are uncorrelated errors with $\mathcal{E}\varepsilon_i = 0$ and same variance σ^2. The response $\eta(x, \theta)$ is a known function usually linear in the parameter $\theta \varepsilon \Re^r$.

$$\eta(x, \theta) = f(x)'\theta.$$

The function $f(x)$ is defined on the experimental region χ, the set of points x where Y is observable.

The design problem consists of the choice of the regressors and the estimator $\hat{\eta}$ of the response η. A decision theoretical formulation demands the specification of a loss function $L(\theta, v, \hat{\eta})$ evaluating a strategy $(v, \hat{\eta})$ where $v = v_n$ denotes a discrete design $v_n = \{(x_1, p_1), \ldots, (x_m, p_m)\}$ with weights $p_i = \frac{n_i}{n}$, n_i being the number of observations taken at x_i. The estimator $\hat{\eta}$ depends upon the design v_n.

In the Bayesian framework the resulting observations and the prior information about θ are combined to obtain a decision for the response η. The problem of finding an optimal strategy can be divided into a Bayesian estimation problem when v_n is fixed and the design problem. The optimization concerning v_n can be carried out separately if the Bayes estimator is found. Assuming quadratic loss

$$L(\theta, v_n, \tilde{\theta}) = (\theta - \tilde{\theta})'U(\theta - \tilde{\theta})$$

with $U \varepsilon \Re^{r \times r}$ and positive definite, where U might depend upon the design v_n and the experimental point x in $\eta(x, \theta)$, the Bayes-estimator of θ is

$$\theta^* = (P + F'F)^{-1}(F'\mathbf{Y} + P\mu)$$

where $F = (f(x_1), \ldots, f(x_n))$ is the design matrix of v_n, P is the prior precision matrix and μ the prior mean and $\mathbf{Y}$ the vector of observations. θ^* is optimal under a conditional normal prior distribution of θ given the precision $\tau = \frac{1}{\sigma^2}$ of the errors which are expected to be normally distributed. θ^* remains the Bayes estimate if the assumptions about the distributions are dropped but the search is confined to linear estimators only. The corresponding Bayes risk turns out to be

$$tr(UM^{-1})\mathcal{E}\frac{1}{\tau} \tag{1.1}$$

with the (conditional) posterior precision τM. The matrix M is defined as the Bayesian information matrix

$$M_B = M = (P + F'F).$$

An appropriate choice of the loss matrix U leads to the usual optimality criteria as $A - U = I-$, $L - U = U_L-$, $C - U = cc'-$, $D - U = \P\lambda_i t_{\max} t'_{\max}$, λ_i, are the ordered eigenvalues and $t_{\max}$ the orthonormal eigenvector of λ_r-, $E - U = t_{\max}t'_{\max}$ etc. The Bayesian design minimizes $tr(UM^{-1})$ and analogous to the classical approach an extension of the definition of a design facilitates that task of finding the optimal design. $v_n = \{(x_i, p_i)\}$ can be interpreted as a distribution on χ and therefore any measure ξ on the experimental region χ could be regarded as a design. In order to complete the embedding of discrete designs the information matrix is normalized by $\frac{1}{n}$ and has the form

$$M_B(\xi) = \int f(x)f(x)'d\xi(x) + \frac{1}{n}P.$$

Contrary to the classical approach the Bayesian information matrix involves the number of observations. The new problem of minimizing $tr(UM_B(\xi)^{-1})$ according to any measure ξ only gives an approximat optimal design except in some special cases where the continuous optimal design happens to be a discrete one. After rounding the weights p_i of the continuous design the resulting discrete design $v_n = \{(x_1, p_1), \ldots, (x_m, p_m)\}$ with weights $\tilde{p}_i = \frac{n_i}{n}$, $n_i \varepsilon \mathcal{N}$, differs from the optimal discrete design v_n^* through

$$\frac{tr(UM_B(v_n)^{-1})}{tr(UM_B(v_n^*)^{-1})} \leq 1 + \frac{\Sigma|p_i - \tilde{p}_i|}{\min(n_i)}$$

(See Chaloner (1984).) Many problems of Bayesian designs are treated in detail in the comprehensive monograph of Pilz (1983).

2. EQUIVALENCE THEOREM

The search for optimal designs is substantially facilitated by the analysis of the structure of optimal designs, whose properties are often revealed by equivalence theorems. The first result in this context was Kiefer's equivalence theorem between D-optimal and mini-max designs. Whittle gave a more general theorem for convex functions Ψ of designs. The idea is geometric in its nature: Ψ is supposed to have directional derivatives and at any design the direction is selected that causes the steepest gradient. An optimal design is reached when no further descent can be achieved. The optimal design is equivalent to "no descent in any direction". Using the convex functional $tr(UM_B(\xi)^{-1})$ the equivalence theorem for fixed loss matrix U reads

Theorem 1. *The following assertions a), b) and c) are equivalent:*

a) ξ^* *is a Bayesian design, meaning that* $tr(UM_B(\xi^*)^{-1})$ *is minimal.*

b) $tr(UM_B(\xi^*)^{-1}) - tr(UM_B(\xi^*)^{-1}M_B(\xi)M_B(\xi^*)^{-1}) \geq 0 for\ all\ \xi.$

c) $tr(UM_B(\xi^*)^{-1}) - \frac{1}{n}tr(UM_B(\xi^*)^{-1}PM_B(\xi^*)^{-1})$

$$= \sup_x f(x)'M_B(\xi^*)^{-1}UM_B(\xi^*)^{-1}f(x)$$

The supremum in c) is attained in the points of the spectrum of ξ^x (points x with positive weigt). If x is out of the spectrum of ξ^* then

$$f(x)'M_B(\xi^*)^{-1}UM_B(\xi^*)^{-1}f(x)$$
$$= tr(UM_B(\xi^*)^{-1}) - \frac{1}{n}tr(UM_B(\xi^*)^{-1}PM_B(\xi^*)^{-1}).$$

This condition is necessary only. Convexity and Caratheodory's theorem imply that it is possible to construct an optimal design containing at most $\frac{r(r+1)}{2}$ points in the spectrum. If the loss matrix is singular this upper bound can be lowered (Chaloner (1984)). Since the function $tr(U(.)^{-1}$ is strictly convex on the set of non-negative definite matrices the information matrix of the optimal design is unique. These properties yield methods for the construction of optimal designs. Some special cases (like polynomial regression) can be analysed with the equivalence theorem. A general method for the construction of designs arises.

3. ITERATIVE PROCEDURES

The structure of optimal designs explained in the equivalence theorem offers useful method for constructing the points of the spectrum. Starting with an arbitrary design ξ_s (attained by direct minimization of $tr(UM_B(.)^1)$ among a small number of designs) new point is added to the initial design:

$$\xi_{s+1} := (1-\alpha_s)\xi_s + \alpha_s\xi_{x_{s+1}},$$

where ξ_x denotes a design which consists of one points x only, s indicates the number of supporting points and x_{s+1} is a solution of the equation

$$f(x_{s+1})'M_B(\xi_s)^{-1}UM_B(\xi_s)^{-1}f(x_{s+1}) = \max_x f(x)'M_B(\xi_s)^{-1}UM_B(\xi_s)^{-1}f(x). \quad (3.1)$$

The weight α_s of x_{s+1} is selected according to the minimum of

$$tr(U\Phi^{-1}) - \frac{1}{1+f(x_{s+1})\Phi f(x_{s+1})}f(x_{s+1})'\Phi^{-1}U\Phi^{-1}f(x_{s+1}) \quad (3.2)$$

where

$$\Phi = (1-\alpha_s)\int f(x)f(x)'d\xi_s(x) + \frac{1}{n}P.$$

Fedorov (1972) suggested this algorithm in principle for D-optimal classical designs. The choice of α in (3.2) provides fast progress towards the optimal design. The convergence of ξ_s holds if α_s is an arbitrary sequence converging to 0 slowly.

Theorem 2. *Let* ξ^* *be a Bayesian design. The sequence of designs* ξ_s *approaches* ξ^**in the sense*

$$tr(UM_B(\xi_s)^{-1}) \to tr(UM_B(\xi^*)^{-1}), \qquad s \to \infty,$$

if x_s *is taken as in (3.1) and the sequence of corresponding weights* α_s *fulfils*

$$\alpha_s \to 0 \text{ and } \Sigma\alpha_s = \infty.$$

Proof. Assume ξ_s not to be optimal. Then

$$tr(UM_B(\xi_s)^{-1} \leq \sup_x f(x)'M_B(\xi_s)^{-1}UM_B(\xi_s)^{-1}f(x)$$
$$+ \frac{1}{n}tr(UM_B(\xi_s)^{-1}PM_B(\xi_s)^{-1}).$$

Writing $M_s = M_B(\xi_s)$, the difference of the information matrices at s and $s+1$ is

$$\begin{aligned} &tr(UM_s^{-1}) - tr(UM_{s+1}^{-1}) \geq tr(UM_{s+1}^{-1}) \\ &- tr(UM_{s+1}^{-1})((1-\alpha)^{-1}M_{s+1} - \frac{\alpha}{1-\alpha}M_B(\xi_{x_{s+1}}))M_{s+1}^{-1} \\ &= \frac{\alpha}{1-\alpha}(tr(UM_{s+1}^{-1}M_B(\xi_{x_{s+1}})M_{s+1}^{-1}) - tr(UM_{s+1}^{-1})). \end{aligned} \tag{3.3}$$

Because of the choice of x_{s+1} the last term is non negative. $tr(UM_s^{-1})$ is monotone and bounded by $tr(UM_B(\xi^*)^{-1})$ and therefore convergent. If this limit does not equal $tr(UM_B(\xi^*)^{-1})$ then $\delta < 0$ exists and

$$tr(UM_s^{-1}M_B(\xi_x)M_s^{-1}) > tr(UM_s^{-1}) - \delta.$$

This implies that (3.3) is greater than $-\frac{\delta\alpha}{1-\alpha}$ for all s. Due to the property of α_s this contradicts the convergence of $tr(UM_s^{-1})$.

The stopping rule of this iterative algorithm might have the following form. The calculations are terminated if the gain of information is less than some given constant ε

$$|tr(UM_s^{-1}) - tr(UM_{s+1}^{-1})| < \varepsilon$$

or if

$$|f(x_{s+1})'M_s^{-1}UM_s^{-1}f(x_{s+1}) + \frac{1}{n}tr(UM_s^{-1}PM_s^{-1}) - tr(UM_s^{-1})| < \varepsilon$$

which says that condition c) in the equivalence theorem is nearly satisfied. The stopping rule might involve the weights and the iteration is continued untill α_s is less than a given constant ε. If the weights are $\alpha_s = \frac{1}{s}$ which are used if a solution of (3.2) is out of reach then the calculation is terminated after a fixed number of iterations.

The number of points in the spectrum of an optimal design can be limited by $\frac{r(r+1)}{2}$. In practice it is necessary to reduce the spectrum. If some points lie close to each other, which happens in practical calculation, these points are merged with some central point and the weights are added up in order to avoid designs with large spectrum. Also, isolated points with extremly small weights should be deleted from the design.

A method of keeping the number of supporting points constant provides the "forward - backward" procedures suggested by Fedorov (1972). In the forward part the new point x_{s+1} is choosen according to (3.1). Now the point out of the spectrum of ξ_{s+1} which minimizes

$$f(x)'M_B(\xi_{s+1})^{-1}UM_B(\xi_{s+1})^{-1}f(x)$$

is deleted from the design and the weights are adapted properly.

The number of forward and backward steps can vary due to the number of solutions of (3.1). Let ξ_m be a design with m supporting points which are solutions of (3.1). ξ_s changes to

$$\xi_{s+m} = (1-\alpha)\xi_s + \alpha\xi_m, \qquad \text{with } \alpha = \frac{m}{s+m}$$

Hereafter the supporting points of the new design ξ_{s+m} are examined according to $d(x_i) > d$ where

$$d(x) = |tr(UM_B(\xi_{s+m})^{-1}) - tr(UM_B(\xi_{s+m})^{-1}M(\xi_x)M_B(\xi_{s+m})^{-1})|.$$

The spectrum points with $d(x_i) > d$, where d is some positive constant, are deleted from ξ_{s+m} and the remaining weights are adjusted. The procedure stops if no point fulfils $d(x_i) > d$ or a given number of iterations is attained.

In general the modifications of the iterative procedure mentioned above entail the loss of convergence. Especially the change of the conditions at every step causes possibly non-converging procedures which nevertheless are important in practical situations.

The initial design might be found by gradient method to get a rapidly converging iteration. In many situations it seems to be reasonable to start with points on the boundary of the operation region with equal weights and maximal distance between the points.

In case of non-linear response η local optimal designs are investigated. The linearization of η leads to

$$f(x,\theta) = \text{grad } \eta(x,\theta)$$

To some extent it is justified to work with local information matrices

$$M(\theta) = \int f(x,\theta)f(x,\theta)'d\xi$$

if η is smooth in the parameter. In the Bayesian approach the local information is combined with the prior information and

$$\int \mathcal{E}_\theta f(x,\theta)f(x,\theta)'d\xi(x) + \frac{1}{n}P$$

replaces the Bayesian information matrix. Essential properties of information matrices remain valid. In principle the iterations work in this concept. Of course this conception serves only as approximation since the Bayes estimator and the information matrix were developed under concrete assumptions.

REFERENCES

Bandemer, H., Näther, W., and Pilz, J. (1987). Once more: optimal experimental designs for regression models, (with discussion) *Statistics* **18**, 2 171–217.

Chaloner, K. (1984). Optimal Bayesian Experimental Design for Linear Models, *Ann. Stat.* **12**, 283–300.

Fedorov, V. (1972). *Theory of Optimal Experiments*, New York: Academic Press.

Fedorov, V. (1980). Convex Design Theory, Math. Operationsforsch. *Stat. Ser. Stat.* **11**, 403–413.

O'Hagan, A. (1978). Curve fitting and optimal design for prediction (with discussion), *J. Roy. Statist. Soc. B* **40**, 1–42.

Pilz, J. (1979). Konstruktion von optimalen diskreten Versuchsplänen für eine Bayes-Schätzung im linearen Regressionsmodell, *Freiberger-Forschungshefte* **117**, 123–152.

Pilz, J. (1983). *Bayes estimation and experimental design in linear regression models.* Teubner-Texte zur Mathematik, Band 55, Leipzig: Teubner-Verlag.

BAYESIAN STATISTICS 3, pp. 615–616
J. M. Bernardo, M. H. DeGroot, D. V. Lindley and A. F. M. Smith, (Eds.)

A Report on Continuity of Uncertainty Functions

PILAR GARCÍA–CARRASCO
Universidad Complutense, Madrid

Pilar García-Carrasco died prematurely, in a mountain accident, only two months after the Conference took place. This is an extended abstract of her contribution, as prepared by an anonymous referee. She was a respected colleague and a beautiful person. She will be missed; she *is* being missed.

Keywords: UNCERTAINTY; INFORMATION; EXPERIMENT; CONTINUITY; PRIOR DISTRIBUTION; POSTERIOR DISTRIBUTION.

1. INTRODUCTION

Most work related to information theory involves logarithmic functions. However, as DeGroot (1962) pointed out, there are no satisfactory motives for this restriction. DeGroot proposed working with the more general concept of uncertainty function. It is possible to define an information measure for each uncertainty function. We have studied some applications of these generalized measures of information to statistics (García-Carrasco, 1986b and García-Carrasco and De la Horra, 1988). In order to obtain most of these results it was necessary to assume continuity of uncertainty and information functions. In this paper, we examine the relation between both hypotheses.

2. RESULTS

Let $W = \{w_1, \ldots, w_m\}$ be the set of all possible values of some parameter w, with $m \geq 2$. Let $P = \{p = (p_1, \ldots, p_m) : \sum_{i=1}^m p_i = 1,\ p_i \geq 0,\ i = 1, \ldots, m\}$ be the set of probability vectors over W. Let P^0 be its interior.

An uncertainty function is a concave and nonnegative function $u : P \to R$ (DeGroot, 1962). An experiment X is a random variable, defined on some probability space, with specified conditional probability density functions $f_i(x)$, $i = 1, \ldots, m$, with respect to a σ-finite measure μ.

If the prior distribution over W is $p = (p_1, \ldots, p_m)$, then the marginal density function of X with respect to μ is $f(x) = \sum_{i=1}^m p_i f_i(x)$. Moreover, after having performed the experiment X and observing the value $X = x$, the posterior distribution over W, $p(x) = (p_1(x), \ldots, p_m(x))$, is, from Bayes' Theorem $p_i(x) = p_i f_i(x)/f(x)$, $i = 1, \ldots, m$.

Let X be an experiment and p the prior distribution. Also, let $A_i = \{x : f_i(x) > 0\}$, $i = 1, \ldots, m$, and $B = \{x : f(x) > 0\}$.

The information $I_u(X; p)$ relative to the uncertainty function u, in an experiment X, when the prior distribution over W is $p = (p_1, \ldots, p_m)$, is defined as $I_u(X; p) = u(p) - E_X[u(p(x))]$, where expectation is computed under the marginal distribution $f(x)$. (DeGroot, 1962). Some interesting results concerning continuity of uncertainty and information functions are given below.

Proposition 1. *Let $u : P \to R$ be a continuous uncertainty function on P. Then, for every experiment X, the information function $I_u : P \to R$ is continuous on P.*

Lemma. *Let $m = 2$ and $u : P \to R$ be an uncertainty function. Let $\{p^n\}$ and $\{q^n\}$ be two sequences with $p^n \in P^0$ and $q^n \in P^0$ for every $n \in N$ and $\lim_{n\to\infty} p^n = \lim_{n\to\infty} q^n$. Then, $\lim_{n\to\infty} u(p^n) = \lim_{n\to\infty} u(q^n)$.*

Theorem 1. *Let $m = 2$ and let u be an uncertainty function with a single discontinuity on $(0,1)[(1,0)]$. Then, $I_u(X;p)$ is continuous on P if $P_2(\overline{A}_1 \cap A_2) = 0$ $[P_1(A_1 \cap \overline{A}_2) = 0]$.*

Proof. The "only if" part follows from an application of the above Lemma. The "if" part is fairly easy to prove. ◁

Theorem 2. *Let $m = 2$. Then, u is continuous on P if $I_u(X;p)$ is continuous on P for all experiments X.*

Proof. This result is obtained from Theorem 1 and Proposition 1. ◁

3. DISCUSSION

The above results indicate that for every continuous uncertainty function, the associated information measure is also continuous. However, if the uncertainty function is discontinuous then the associated information measure is either continuous or discontinuous, depending on each experiment.

Note that, if the support set for each density $f_i(x)$ is the same, then the condition in Theorem 1 is automatically satisfied. Thus, we only need to worry about experiments with different support sets over w_1 and w_2.

REFERENCES

DeGroot, M. H. (1962). Uncertainty, information and sequential experiments. *Ann. Math. Statist.* **33**, 404–419.

García-Carrasco, M. P. (1986a). Distribuciones mínimo-informativas; caso de espacio paramétrico finito. *Qüestiió* **10**, 7–12.

García-Carrasco, M. P. (1986b). Algunas propiedades y casos particulares de la incertidumbre generalizada. *Estadística Española* **28**, 59–79.

García-Carrasco, M. P. and De la Horra, J. (1988). Maximizing uncertainty functions under constraints on quantiles. *Statistics and Decisions* (to appear).

BAYESIAN STATISTICS 3, pp. 617–630
J. M. Bernardo, M. H. DeGroot, D. V. Lindley and A. F. M. Smith, (Eds.)

A Bayesian Approach to the Analysis of LD50 Experiments

A. P. GRIEVE
Horsham, Ciba-Geigy

SUMMARY

The determination of the *LD50* from acute toxicity tests has recently been criticized on the grounds that such tests often provide insignificant results. In this paper we propose a Bayesian approach in which point estimation and confidence limits are not the primary aim. Instead interest centers on predefined toxicity classes giving a range of *LD50* values. In our approach the posterior probabilities that the *LD50* lies within each of the given ranges is calculated. We develop approximations to the posterior distributions of interest which should be readily programmable on any small computer. In addition we consider three different methods for the elicitation of prior belief in the parameters of the probit model.

Keywords: ACUTE TOXICITY; *LD50*; PROBIT MODEL; BAYESIAN ANALYSIS; TOXICITY CLASSES; APPROXIMATION; ASSESSMENT OF PRIOR BELIEF.

1. INTRODUCTION

Recently, the controversy surrounding the validity and usefulness of the acute toxicity test has markedly increased. On the one hand animal protection groups question the biological relevance of such tests citing examples in which limited, or insignificant, information is obtained.Toxicologists, on the other hand, emphasize the need to quantify the toxic potential of a chemical while at the same time they encourage procedures designed to limit the number of animals required to give an assessment of lethality (Bass *et al.*, 1982; Dayan *et al.*, 1984). This desire to limit the number of experimental animals has given rise to a number of recent suggestions for modifying the standard practice in acute toxicity testing (see Müller and Kley, 1982; Schütz and Fuchs, 1982; Lorke, 1983).

In such toxicity tests the response of an animal to the test substance is dichotomous; alive/dead or no response/response. The design of such a test consists of k dose levels on an appropriate scale. The experiments may be characterized by the triplets $d_i, n_i, r_i (i = 1, \dots, k)$ where d_i is the dose administered to n_i animals of which r_i respond in the i-th dose group. A mathematical dose-response function relating the probability of response to the dose, usually the probit or logit model, is specified. Based on the above triplets, the parameters of either model are traditionally estimated by maximum likelihood, weighted least-squares or minimum chi-square (Finney, 1971). In this type of experiment the *LD50* is of main interest, being defined as the dose, or quantity, of the substance which kills 50% of the animals exposed to it.

It is well known that under certain conditions the traditional methods of analysis give rise to inadequate results, in that although they provide a point estimate of the *LD50*, the fiducial limits, at some specified level of confidence, will consist of the whole real line (Fieller, 1954; Finney, 1971, Section 4.7). In this paper the view is taken that the object of estimating the *LD50* is to determine an index of the toxicity of a subtance by means of some predefined toxicity classes. For instance, the European Economic Community defines the following toxicity classes for classifying the lethality of substances based on the *LD50* values from oral studies in rats (Annex VI of the Council Directive 67/548/EEC - Sixth Amendment):

Toxicity Class	Description	Range of *LD50* (mg./kg.)
1	very toxic	<25
2	toxic	25–200
3	harmful	200–2000
4	practically non-toxic	>2000

A second example comes from the recent, 1983, Swiss poison regulation again using oral *LD50* values in rats:

Toxicity Class	Range of *LD50* (mg./kg.)
1	<5
2	5–50
3	50–500
4	500–2000
5	2000–5000

The motivation for the present work arose from the need to classify a substance which in an acute toxicity test gave the data in Table 1.

Dose (mg./kg.)	Number of Animals Exposed	Number of Animals Dying
500	5	1
1000	5	2
2500	5	3
5000	5	2

Table 1. *Results from an Acute Toxicity Experiment*

Using maximum likelihood to estimate the parameters of the probit model (using a log dose scale) gives 4049 mg./kg. as a point estimate for the *LD50*. However, this is one of the above examples for which the 95% fiducial limits comprise the whole positive real axis, a result practically useless for classifying the substance. We would argue that classical methods cannot answer the question of interest —which toxicity class does the substance belong to?— whereas a Bayesian approach can. However, even if the regulatory authorities require a point estimate and a confidence interval, a Bayesian approach is preferable since a highest posterior density (H.P.D.) interval will always exist, if a proper prior is used; and will exist in all but pathological cases if an improper prior is used (Tsutakawa, 1975). This is not the case for fiducial limits.

In the present paper, part of whose results were reported in Racine *et al.*(1986), a Bayesian analysis is developed in which emphasis is placed on calculating the posterior probabilities of a substance belonging to predetermined toxicity classes. Two cases are considered: (i) an improper prior distribution for the parameters of the model; (ii) a normal prior distribution for the parameters of the model. Methods of determining from toxicologists their prior beliefs in the parameters of the model are considered. Approximations to the various posterior distributions are developed.

It can be argued (Zbinden and Flury-Roversi, 1981; Kimber, 1986) that it is incorrect to judge a substance's toxicity on the basis of the *LD50.* We do not dispute this point of view but agree with Finney (1985) that "as long as the *LD50* is used, there is no excuse (scientific or economic) for not estimating according to some accepted criteria of optimality".

2. A BAYESIAN ANALYSIS USING AN IMPROPER PRIOR

In this paper we choose to use the probit model, although the methods have also been successfully implemented using the logit. Let k doses of a substance $d_i(i = 1, \ldots, k)$ be administered to $n_i(i = 1, \ldots, k)$ animals of which $r_i(i = 1, \ldots, k)$ respond, then the likelihood $L(\alpha, \beta|Y)$, where α and β are the parameters of the probit model and Y denotes the data, is given by,

$$L(\alpha, \beta|Y) = \prod_{i=1}^{k} \Phi(\alpha + \beta x_i)^{r_i} [1 - \Phi(\alpha + \beta x_i)]^{n_i - r_i} \tag{2.1}$$

where $\Phi(\cdot)$ is the standard normal distribution function and $x_i = \ln(d_i)$. Assuming an improper prior for α and β, that is

$$p_c(\alpha, \beta) = \text{constant} \quad -\infty < \alpha < \infty, \quad 0 < \beta < \infty \tag{2.2}$$

(c denotes the constraint $\beta > 0$) use of Bayes' theorem gives,

$$p_c(\alpha, \beta|Y) = \frac{L(\alpha, \beta|Y) p_c(\alpha, \beta)}{p_c(Y)} \tag{2.3}$$

where,

$$p_c(Y) = \int_{-\infty}^{\infty} \int_{0}^{\infty} L(\alpha, \beta|Y) p_c(\alpha, \beta) d\beta d\alpha. \tag{2.4}$$

Letting $w = -\alpha/\beta$ be the ln(*LD50*) we have,

$$p_c(w|Y) = \int_{0}^{\infty} \beta p_c(-w\beta, \beta|Y) d\beta \quad -\infty < w < \infty \tag{2.5}$$

and supposing a toxicity class, on the log-scale, to be defined by w_L and w_U, we have,

$$P[w_L < w < w_U|Y] = \int_{w_L}^{w_U} p_c(w|Y) dw. \tag{2.6}$$

Equations (2.1) to (2.6) provide all the necessary information to made inferences concerning the *LD50.* However because of the non-linearity of the probit model the integrations in (2.4), (2.5) and (2.6) cannot be performed analytically. Thus either numerical integration methods need to be resorted to, or approximations sought.

The numerical integration problem may be simplified by redefining the range of β in (2.2). Thus if β is not constrained to be greater than zero, and writing $p_U(\alpha, \beta)$ for this unconstrained prior distribution then,

$$p_U(\alpha, \beta|Y) = \frac{L(\alpha, \beta|Y) p_U(\alpha, \beta)}{p_U(Y)} \tag{2.7}$$

where,

$$p_U(Y) = \int_{-\infty}^{\infty}\int_{-\infty}^{\infty} L(\alpha,\beta|Y)p_U(\alpha,\beta)d\beta d\alpha. \tag{2.8}$$

Using the results in Box and Tiao (1973, Section 1.5) we have,

$$p_c(\alpha,\beta|Y) = \frac{p_U(\alpha,\beta|Y)P\{\beta>0|\alpha,\beta,Y\}}{P\{\beta>0|Y\}} = \frac{p_U(\alpha,\beta\wedge\beta>0|Y)}{P\{\beta>0|Y\}} \tag{2.9}$$

and

$$p_c(w|Y) = \frac{\int_0^\infty \beta p_U(-w\beta,\beta\wedge\beta>0|Y)d\beta}{P\{\beta>0|Y\}}, \quad -\infty<w<\infty. \tag{2.10}$$

The methods described by Naylor and Smith (1982) could be used to perform the integrations in (2.8), (2.9), (2.10) and (2.6). However this is not recommended since it would involve using indicator functions for calculating $P(\beta > 0|Y)$ and for calculating (2.6), which practice has shown can seriously underestimate or overestimate the required probabilities. Alternatively a modification of the quadrature rules developed by Galant (1969) and Steen *et al.* (1969) may be used to integrate over β. These rules were developed for integrals of the form,

$$\int_0^b \exp(-x^2)f(x)dx$$

but may be simply modified for integrals of the form,

$$\int_b^\infty \exp(-x^2)f(x)dx.$$

If the doses have been chosen on a true log-scale, or if they are not far from it, instead of using numerical integration an approximation can be developed which may be used even for small sample sizes. We illustrate how this may be achieved using the hypothetical example shown in Table 2.

Dose (mg./kg.)	Number of Animals Exposed	Number of Animals Dying
100	3	1
1000	3	2

Table 2. *Results from a Hypothetical Experiment*

Using the Gauss-Hermite quadrature described by Naylor and Smith (1982) the double integration in (2.8) is efficiently performed from which (2.7) is simply obtained. Figure 1 shows the bivariate 50% and 95% H.P.D. regions for α and β for the data in Table 2.

The contours in Figure 1 are very nearly elliptical, suggesting that a bivariate normal (BN) approximation to $p_U(\alpha,\beta|Y)$ may be reasonable. The parameters of the *BN* approximation may be obtained as a by-product of the Naylor and Smith approach, or by a second approximation.

Denoting by $\hat{\alpha}$, $\hat{\beta}$, $\hat{\sigma}^2_\alpha$, $\hat{\sigma}^2_\beta$, $\hat{\sigma}_{\alpha\beta}$ the maximum likelihood estimates of α and β and their asymptotic variances and covariance, and by $\tilde{\alpha}$, $\tilde{\beta}$, $\tilde{\sigma}^2_\alpha$, $\tilde{\sigma}^2_\beta$, $\tilde{\sigma}_{\alpha\beta}$, the posterior means, variances and covariance, then from Lindley (1980) the following results are obtained.

$$\tilde{\alpha} = \hat{\alpha} + \frac{1}{2}L_{30}\hat{\sigma}^4_\alpha + \frac{3}{2}L_{21}\hat{\sigma}^2_\alpha\hat{\sigma}^2_{\alpha\beta} + \frac{1}{2}L_{12}(\hat{\alpha}^2_\alpha\hat{\sigma}^2_\beta + 2\hat{\sigma}^2_{\alpha\beta}) + \frac{1}{2}L_{03}\hat{\sigma}^2_\beta\hat{\sigma}_{\alpha\beta} + O(N^{-1}) \tag{2.11}$$

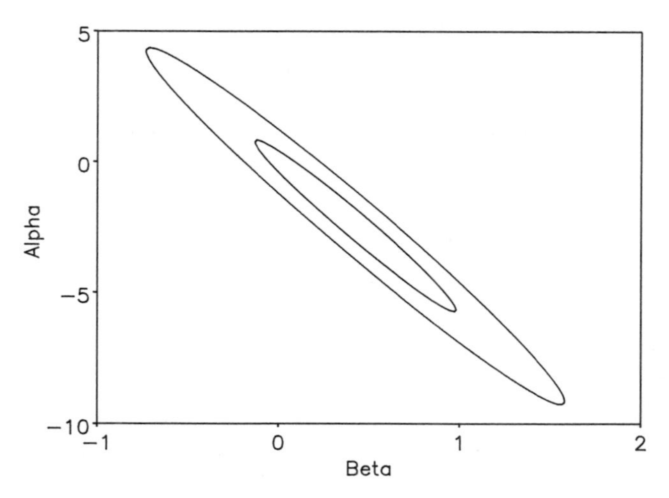

Figure 1. *Joint Posterior Distribution of α and β Data from Table 2.*

$$\tilde{\beta} = \hat{\beta} + \frac{1}{2}L_{03}\hat{\sigma}_{\beta}^{4} + \frac{3}{2}L_{12}\hat{\sigma}_{\beta}^{2}\hat{\sigma}_{\alpha\beta}^{2} + \frac{1}{2}L_{21}(\hat{\alpha}_{\alpha}^{2}\hat{\sigma}_{\beta}^{2} + 2\hat{\sigma}_{\alpha\beta}^{2}) + \frac{1}{2}L_{30}\hat{\sigma}_{\alpha}^{2}\hat{\sigma}_{\alpha\beta} + O(N^{-1}) \quad (2.12)$$

where

$$L_{30} = \left.\frac{\partial^3\{\ln[L(\alpha,\beta|Y)]\}}{\partial\alpha^3}\right|_{\alpha=\hat{\alpha}} \quad \text{and} \quad N = \sum_{i=1}^{k} n_i.$$

As Lindley notes that similar corrections to $O(N^{-1})$ are not available for the variances and covariance. However, although it has not been possible to prove the following result, in a large number of cases over a wide range of total sample sizes and different values of k, the number of groups, it has been found to be very accurate. Define $\delta = (\tilde{\alpha} - \hat{\alpha})/\tilde{\alpha} \approx (\tilde{\beta} - \hat{\beta})/\tilde{\beta}$ then take,

$$\frac{\tilde{\sigma}_{\alpha}^{2} - \hat{\sigma}_{\alpha}^{2}}{\tilde{\sigma}_{\alpha}^{2}} \approx \frac{\tilde{\sigma}_{\beta}^{2} - \hat{\sigma}_{\beta}^{2}}{\tilde{\sigma}_{\beta}^{2}} \approx \frac{\tilde{\sigma}_{\alpha\beta} - \hat{\sigma}_{\alpha\beta}}{\tilde{\sigma}_{\alpha}} \approx \delta/2$$

To ilustrate these corrections they have been applied to the data of Table 2, the results being shown in Table 3.

Parameter	Maximum Likelihood Estimates	Posterior Moments	Approximate Posterior Moments
α	–2.154	–2.443	–2.436
β	0.374	0.424	0.423
σ_{α}^{2}	7.284	7.730	7.706
σ_{β}^{2}	0.211	0.224	0.224
$\sigma_{\alpha\beta}$	–1.217	–1.291	–1.287

Table 3. *Approximations to Posterior Moments for Hypothetical Example*

The results in Table 3 are satisfactory in that they correspond, in the case of approximate posterior means to a relative error of less than 0.3%, while the corresponding relative errors for approximate second moments are less than 0.4%. (An alternative approach would be to use the approximation developed by Tierney and Kadane (1986) which has the advantage that it also provides corrections for second order moments.)

Suppose now that $p_U(\alpha, \beta|Y)$ may be approximated by a BN density with means $\tilde{\alpha}, \tilde{\beta}$, variances $\tilde{\sigma}_\alpha^2$, $\tilde{\sigma}_\beta^2$; and covariance $\tilde{\sigma}_{\alpha\beta}$ denoted by $BN(\mu, \Sigma)$ where,

$$\mu = \begin{pmatrix} \tilde{\alpha} \\ \tilde{\beta} \end{pmatrix} \quad \text{and} \quad \Sigma = \begin{pmatrix} \tilde{\sigma}_\alpha^2 & \tilde{\sigma}_{\alpha\beta} \\ \tilde{\sigma}_{\alpha\beta} & \tilde{\sigma}_\beta^2 \end{pmatrix}. \tag{2.13}$$

then

$$p_c(\alpha, \beta|Y) = \frac{BN(\mu, \Sigma)}{\Phi\left(\frac{\tilde{\beta}}{\tilde{\sigma}_\beta}\right)}, \quad \beta > 0.$$

Using the results in the Appendix the posterior distribution of $w = \ln(LD50)$ given the constraint $\beta > 0$ may be calculated from (A.2), while inferences of the form (2.6) may be derived from (A.4).

Application of the double-fold approximation, the BN distribution for $p_U(\alpha, \beta|Y)$ and the approximate means, variances and covariance is shown in Table 4 using the Swiss toxicity classes, from which it can be seen that the approximations are satisfactory. The exact probabilities in this table were calculated using subroutine $DBLIN$ from the $IMSL$ library of subroutines.

Toxicity Classes

	1	2	3	4	5	>5000 mg/kg
Exact Probabilities	0.041	0.062	0.570	0.224	0.038	0.065
Probabilities based on Normal Approx. (exact Moments)	0.042	0.061	0.576	0.219	0.037	0.065
Probabilities based on Normal Approx. (approx. Moments)	0.042	0.061	0.576	0.219	0.037	0.066

Table 4. *Exact Posterior Probabilities of Toxicity Classes and Approximate Probabilities for the Data in Table 2.*

Returning to the example in Table 1, which was the motivation behind this work, the probabilities that the substance belongs to the various toxicity classes are shown in Table 5. From these results it may be seen that, although we may not definitely decide into which class the subtance should be placed, it is extremely unlikely that it belongs to classes 1, 2, or 3, since their total probability is 0.03. Further experimentation would be necessary to determine which of classes 3 or 4 it belongs to or whether the *LD50* is greater than 5000 mg./kg.

Toxicity Classes

	1	2	3	4	5	>5000 mg/kg
Probabilities	0.005	0.004	0.021	0.232	0.402	0.336

Table 5. *Posterior Probabilities of Toxicity Classes for the Data in Table 1.*

We need not be restricted to calculating the posterior probabilities of the toxicity classes. The results in the Appendix allow us to simply calculate the posterior distribution of ln(*LD50*) and its cumulative distribution function, or to calculate H.P.D. limits for ln(*LD50*). To illustrate, Figure 2 and 3 show the posterior distribution and cumulative posterior distribution respectively, for the data of Table 1 whose 95% H.P.D. limits for ln(*LD50*) are 4.71 and 15.67 corresponding to 111 and 6.38×10^6 mg./kg.

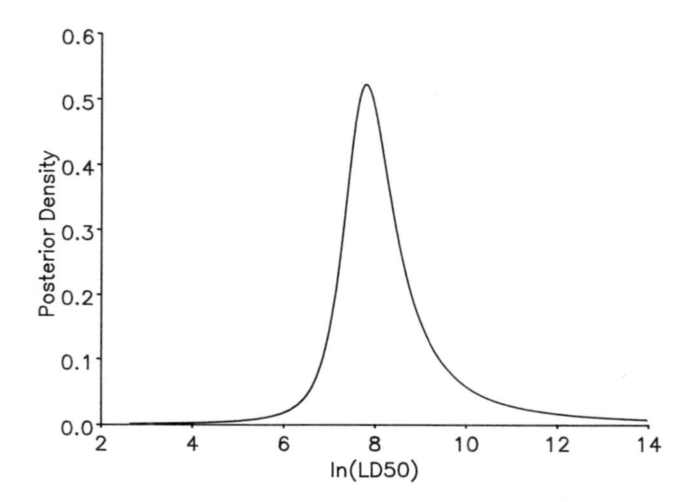

Figure 2. *Posterior Density of ln (LD50) – Data from Table 1.*

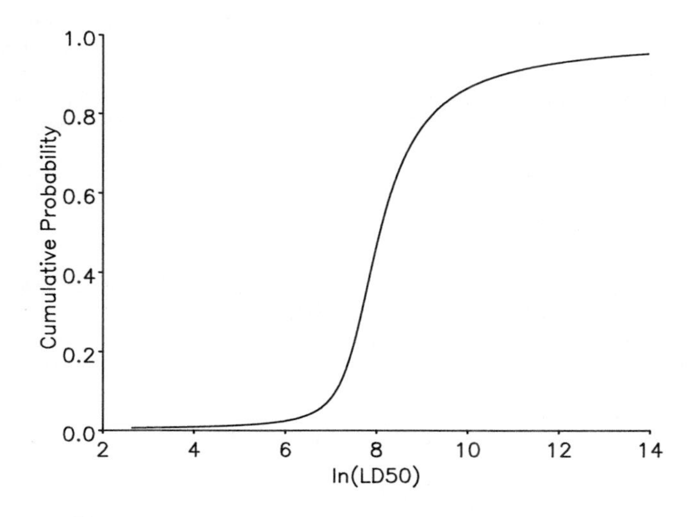

Figure 3. *Cumulative Posterior Distribution – Data from Table 1.*

3. A BAYESIAN ANALYSIS USING AN INFORMATIVE PRIOR FOR α AND β

To begin we suppose that prior to performing the current experiment a previous experiment has been performed yielding data Y_0. Assuming further that prior to the previous experiment our joint prior distribution for α and β was improper, it follows from standard Bayesian arguments that we may pool the data as if they came from a single experiment. The analysis in Section 2 may then be carried out.

Suppose now, however, that we can approximate our unconstrained prior distribution for α and β by a BN distribution; an assumption which will be justified in Section 4. Denoting this unconstrained prior by $BN(\mu_0, \Sigma_0)$, then, since as we have seen in the previous Section that for doses on a log-scale the likelihood may be approximated by $BN(\mu, \Sigma)$ {see (2.13)}, standard Bayesian calculations give,

$$p_U(\alpha, \beta|Y) = BN(\mu^*, \Sigma^*) \tag{3.1}$$

where,

$$\Sigma^* = (\Sigma_0^{-1} + \Sigma^{-1})^{-1} \quad \text{and} \quad \mu^* = \Sigma^* \left(\Sigma_0^{-1}\mu_0 + \Sigma^{-1}\mu\right)$$

(see for instance Lindley and Smith, 1972, Section 2). Since $p_U(\alpha, \beta|Y)$ in (3.1) is approximated by a BN distribution the results in the Appendix may be used to make posterior inferences concering $w = \ln(LD50)$ exactly as in Section 2.

If the normal approximation (2.13) does not hold, which can be checked by calculating the third and fourth moments of the posterior marginal distributions of α and β, Lindley's or Tierney and Kadane's approximations for marginal distributions may be directly applied to the product of the prior and likelihood in the parametrization w and β.

4. DETERMINING A PRIOR DISTRIBUTION FOR α AND β

In this section we consider ways in which one can determine an experimenter's prior belief in the parameters of the probit model. Each method which is considered leads to a normal prior distribution for α and β, so that the methods in the previous section may then be used.

For the logit model Tsutakawa (1975) suggests that a parametrization of the model which is familiar to the experimenter should be chosen. He considers two methods using an experimenter's prior beliefs in the probabilities of response P_1 and P_2 at two dose levels d_1 and d_2. First we investigate the implications of Tsutakawa's methods for the probit model, and second consider a method based on eliciting the experimenter's prior beliefs in the toxicity class to which the test substance belongs.

4.1. A Semi-Uninformative Prior Distribution for α and β Determined using Probabilities of Response

Following Tsutakawa (1975) suppose that P_1 and P_2 are uniformly distributed over the region $0 < P_1 < P_2 < 1$. Tsutakawa shows this to imply that the *LD50* lies between d_1 and d_2 with probability $1/2$, d_1 and d_2 are respectively the lower and upper prior quantiles for the *LD50*. Further he shows that the prior distribution in terms of α and β belongs to the natural conjugate family of distributions. We now show that the above construction leads, for the probit model, to a BN prior for α and β.

Suppose that P_1 and P_2 are *a priori* uniformly distributed such that $0 < P_1 < P_2 < 1$, then,

$$p(P_1, P_2) = 2 \quad 0 < P_1 < P_2 < 1.$$

Make the transformation,

$$P_i = \Phi(\alpha + \beta x_i) \quad \{x_i = \ln(d_i), i = 1, 2\} \tag{4.1}$$

vith Jacobian,

$$(x_2 - x_1)\phi(\alpha + \beta x_1)\phi(\alpha + \beta x_2) \tag{4.2}$$

o that,

$$\begin{aligned} p_1(\alpha, \beta) &= \frac{2(x_2 - x_1)}{2\pi} \exp\left[\frac{-1}{2}\{(\alpha + \beta x_1)^2 + (\alpha + \beta x_2)^2)\}\right] \\ &= 2BN(\mu_1, \Sigma_1) - \infty < \alpha < \infty, 0 < \beta < \infty \end{aligned} \tag{4.3}$$

vhere,

$$\mu_1 = \begin{pmatrix} 0 \\ 0 \end{pmatrix} \quad \text{and} \quad \Sigma_1 = \begin{pmatrix} \frac{x_1^2 + x_2^2}{(x_2 - x_1)^2} & \frac{-(x_1 + x_2)}{(x_2 - x_1)^2} \\ \frac{-(x_1 + x_2)}{(x_2 - x_1)^2} & \frac{2}{(x_2 - x_1)^2} \end{pmatrix}$$

'or practical application (4.3) is replaced by $BN(\mu_1, \Sigma_1)$ and the analysis given in Section 3 s carried out.

It is clear from the form of Σ_1 that this method should not be used for cases in which x_1 and x_2 are chosen such that $x_2 - x_1$ is very small or very large. In the former cases the 'ariances of α and β tend to infinity and ρ tends to -1, while in the latter cases the variance of β tends to 0. A second disadvantage is the zero prior modal values for α and β.

4.2. An Alternative Determination of $p(\alpha, \beta)$ Using Probabilities of Response

A second suggestion of Tsutakawa for the logit model, when moderate amounts of prior nformation are available is that the experimenter should specify the modal probabilities of esponse, $\hat{P}_1$ and $\hat{P}_2$ corresponding to doses d_1 and d_2. Supposing that the prior distribution of P_1 and P_2 is a member of the family of natural conjugate prior densities, that is

$$p(P_1, P_2) \propto P_1^{l_1 - 1}(1 - P_1)^{m_1 - l_1 - 1} P_2^{l_2 - 1}(1 - P_2)^{m_2 - l_2 - 1} \tag{4.4}$$

hen,

$$l_i = 1 + \hat{P}_i(m_i - 2) \quad (l_i > 1, m_i > 2). \tag{4.5}$$

The values of m_i, and hence from (4.5) l_i, should be chosen to reflect the weight to be given o the two dose levels.

Supposing that the m_i's and l_i's have chosen combining (4.1), (4.2) and (4.4) gives,

$$\begin{aligned} p_2(\alpha, \beta) \propto\ & [(x_2 - x_1)\phi(\alpha + \beta x_1)\phi(\alpha + \beta x_2)] \\ & \times \Big[\Phi(\alpha + \beta x_1)^{l_1 - 1}\{1 - \Phi(\alpha + \beta x_1)\}^{m_1 - l_1 - 1} \\ & \quad \Phi(\alpha + \beta x_2)^{l_2 - 1}\{1 - \Phi(\alpha + \beta x_2)\}^{m_2 - l_2 - 1}\Big]. \end{aligned} \tag{4.6}$$

The expression in the first square bracket in (4.6) is the same as (4.2) so that it may be vritten as $BN(\mu_1, \Sigma_1)$, c.f. (4.3), while the expression in the second square bracket may be pproximated by $BN(\mu, \Sigma)$ as in Section 2. Thus using (3.1) gives,

$$p_2(\alpha, \beta) = BN(\mu^*, \Sigma^*) \tag{4.7}$$

vhere,

$$\Sigma^* = (\Sigma_1^{-1} + \Sigma^{-1})^{-1} \quad \text{and} \quad \mu^* = \Sigma^*(\Sigma_1^{-1}\mu_1 + \Sigma^{-1}\mu).$$

n practice it is recommended that the experimenter is given information as to the consequences of his choice of d_1, d_2, m_1 and m_2. Thus (4.7) could be used to show the implied *a priori* probabilities of the test substance being in the toxicity classes of interest. Using (4.7) the analysis in Section 3 may be carried out.

4.3. Determining α and β using Prior Probabilities of the Toxicity Classes

Suppose that the experimenter is prepared to supply the following information:

(i) prior information concerning the *LD50* in terms of a discrete probability distribution,

(ii) the most likely value for the slope parameter β (modal value).

If an experimenter is prepared to choose the dose levels in an experiment it is necessary for him to have some idea, albeit subconscious, of the likely values of the *LD50* and the slope since he will not choose dose levels for which he is *a priori* sure he will get no response or 100% response.

To illustrate how the above information may be used, suppose that prior to the experiment in Table 1 the experimenter specifies the probabilities of the substance being in each of the Swiss toxicity classes, as shown in Table 6.

	Toxicity Classes					
	1	2	3	4	5	>5000 mg/kg
Probabilities	0.01	0.04	0.10	0.35	0.40	0.10

Table 6. *Prior Probabilities of Toxicity Classes.*

Suppose further that the experimenter's unconstrained prior distribution for α and β is $BN(\mu, \Sigma)$, where μ and Σ are defined in the Appendix. By equating the cumulative distribution given in Table 6. to the theoretical cumulative distribution defined by (A.4) and (A.5) it might be hoped that μ and Σ can be determined. However, in the Appendix, h and γ may be written as,

$$h = \frac{w\frac{x_0}{\sigma_x}\frac{\sigma_x}{\sigma_y} - \frac{y_0}{\sigma_y}}{\sqrt{\left(w^2\frac{\sigma_x^2}{\sigma_y^2} - 2\rho w\frac{\sigma_x}{\sigma_y} + 1\right)}}, \quad \gamma = \frac{w\frac{\sigma_x}{\sigma_y} - 1}{\sqrt{\left(w^2\frac{\sigma_x^2}{\sigma_y^2} - 2\rho w\frac{\sigma_x}{\sigma_y} + 1\right)}}$$

so that (A.5) depends only on the four parameters,

$$c_1 = \frac{x_0}{\sigma_x}, \quad c_2 = \frac{y_0}{\sigma_y}, \quad c_3 = \frac{\sigma_x}{\sigma_y}, c_4 = \rho \tag{4.8}$$

{c.f. Hinkley, 1970}. This result implies that any four of the prior probabilities in Table 6 are sufficient to determine c_1, c_2, c_3, and c_4, but additional information is required to determine μ and Σ. We choose to do this through the specification of x_0, that is the slope parameter.

4.4. Numerical Examples of the Determination of $p(\alpha, \beta)$

Each of the three methods given above for determining a prior distribution $p(\alpha, \beta)$ lead to a BN prior. Thus the methods in Section 3 may be used to make inferences. In this section we compare the inferences which are made when these methods are applied to the experiment in Table 1. In order to use these methods a number of subjective assessments need to be made. These are as follows:

(i) In order to use the method of Section 4.1, two doses, d_1 and d_2, need to be chosen within which *a priori* the *LD50* lies with probability 1/2. These were chosen to be 1000 mg./kg. and 3000 mg./kg.

(ii) For the method of Section 4.2, in addition to the doses d_1 and d_2 which were taken to be as above, the modal responses $\hat{P}_1$ and $\hat{P}_2$ at these doses and the weights m_1 and m_2 need to be chosen. $\hat{P}_1$ and $\hat{P}_2$ were chosen to be 1/4 and 3/4 while the influence of the weights was investigated by choosing $m_i = 3, 4$ and 5.

(iii) For the method based on the prior probabilities of the toxicity classes, in addition to the probabilities, given in Table 6, the model value of β needs to be chosen. In this case it was set to 0.5.

In Table 7 are shown the prior distributions which are given by (i), (ii) and (iii) together with their corresponding inferences; for completeness the inferences for the improper prior are given again.

			Prior Moments			
Parameters of Prior	Improper	Sec. 4.1.	Sec. 4.2. $m_i = 3$	Sec. 4.2. $m_i = 4$	Sec. 4.2. $m_i = 5$	Sec. 4.3.
α	–	0.000	–4.227	0.224	–6.350	–3.786
β	–	0.000	0.567	0.748	0.852	0.500
σ_α^2	–	92.646	63.197	46.645	36.817	5.282
σ_β^2	–	1.657	1.130	0.834	0.659	0.070
ρ	–	–0.997	–0.997	–0.997	–0.997	–0.987
Toxicity Class			Probability			
1	0.005	0.000	0.005	0.005	0.004	0.002
2	0.004	0.000	0.004	0.003	0.003	0.002
3	0.021	0.000	0.020	0.019	0.018	0.017
4	0.232	0.008	0.271	0.289	0.306	0.323
5	0.402	0.206	0.428	0.441	0.453	0.561
>5000 mg/kg	0.336	0.794	0.272	0.243	0.216	0.094

Table 7. *Prior Distributions and Posterior Inferences*

The results in Table 7 are worthy of comment for a number of reasons:

(i) The prior based on the results in Section 4.1 is not recommended. Although the choice of d_1 and d_2 does not lead to either a very small or a very large variance, the *a priori* modal value of β, i.e. 0, tends to have a relatively extreme effect on the modal posterior value of β thus increasing the most likely value of the *LD50*.

(ii) For fixed values of the modal responses $\hat{P}_1$ and $\hat{P}_2$, increasing the variable parameters m_i has a smooth effect on the posterior probabilities of the toxicity classes; $m_i = 3$ may be considered as semi-uninformative.

(iii) The method of Section 4.3, based on prior probabilities of the toxicity classes allows a considerable amount of prior information to be incorporated.

(iv) For the present data set an analysis based on the improper prior, or on any of the informative priors, shows that there is a very small probability that the *LD50* is less than 500 mg./kg (toxicity classes 1, 2 and 3), the maximum posterior probability being 0.03.

5. DISCUSSION

In this paper we have considered a Bayesian approach to the analysis of *LD50* experiments. We have provided operational techniques for providing the inferences which the toxicologists and regulatory authorities require, and have seen that such techniques, even when the traditional method of maximum likelihood gives results which are not useful for practical purposes, give a meaningful classification of the lethality (toxicity) of a substance. Ross (1986), Bailey and Gower (1986) and Cox (1986) in discussion of Racine *et al.* (1986) all suggest that the failure of the traditional method to give an adequate result is due to the invalidity of the normal approximation to the binomial for such small samples. They conclude that use of likelihood intervals solves the problem (Aitkin, 1986 and Williams, 1986 show how these may be calculated for the logit model using GLIM). While it is certainly true that for the example in Racine *et al.* this method produces a 95% interval for the ln(*LD50*), it does not for the data in Table 1. In fact it is possible to show that there always exists a "confidence level" for which this likelihood method fails to produce an interval exactly as with Fieller's theorem, the only difference being that the critical "confidence level" for the likelihood approach is greater than that of Fieller's

The ideas which have been presented will almost certainly be new to toxicologists; in particular they may find it strange to specify a prior distribution for α and β, or one for the *LD50*, which has a subjective content. We argue that such a subjective assessment is inherent in the toxicologist's choice of design, since he is hardly likely to choose doses which he considers to have practically no chance of being lethal, or doses which he is practically sure will be lethal.

It has been suggested that the computer has a role to play in toxicity testing as an alternative to laboratory animals. Although, in some cases, the approximations which we have developed reduce the amount of numerical work involved, a computer is necessary. Thus the computer certainly has a role to play; however, we suggest that this role be confined to the implementation of adequate statistical methods, which can produce answers to the questions of interest.

REFERENCES

Aitkin, M. (1986). Statistical modelling: the likelihood approach. *The Statistician* **35**, 102–113.

Bass, R., Guenzel, P., Henschler, D., Koenig, J., Lorke, D., Neubert, D., Schütz, E., Shuppan, D. and Zbinden, G. (1982). *LD50* versus acute toxicity - critical assessment of the methodology currently in use. *Arch. Toxicol.* **51**, 183–186.

Box, G. E. P. and Tiao, G. C. (1973). *Bayesian Inference in Statistical Inference.* Reading, Massachusetts: Addison-Wesley.

Cooper, B. E. (1968). Algorithm AS 4: An auxiliary function for distribution integrals. *Appl. Statist.* **17**, 190–192, and Correction in *Appl. Statist.* **19**, 204 (1970).

Cox, D. R. (1986). In discussion of Racine *et al.*

Dayan, A. D., Clark, B., Jackson, M., Morgan, H. and Charlesworth., F. A. (1984). Role of the *LD50* test in the pharmaceutical industry. *The Lancet*, 555–556.

Galant, D. (1969). Gauss quadrature rules for the evaluation of $2\pi^{-1/2}\int_0^\infty \exp(-x^2)f(x)dx$. *Math. Comp.* **23**, 674.

Fieller, E. C. (1954). Some problems in interval estimation. *J. Roy. Statist. Soc. B* **16**, 175–186.

Finney, D. J. (1971). *Probit Analysis*, 3rd. Ed. Cambridge, U.K.: Cambridge Universtity Press.

Finney, D. J. (1985). The median lethal dose and its estimation. *Arch. Toxicol.* **56**, 215–218.

Bailey, R. A. and Gower, J. C. (1986). In discussion of Racine *et al.*

Hinkley, D. V. (1969). On the ratio of two correlated normal variables. *Biometrika* **56**, 635–639, and Correction in *Biometrika* **57**, 683 (1970).

Kimber, G. R. (1986). In discussion of Racine *et al.*

Lindley, D. V. (1980). Approximate Bayesian methods. *Bayesian Statistics.* (J. M. Bernardo, M. H. DeGroot, D. V. Lindley and A. F. M. Smith, eds.). Valencia: University Press

Lindley, D. V. and Smith, A.F.M. (1972). Bayes estimates for the linear model, (with discussion). *J. Roy. Statist. Soc. B* **34**, 1–41.

Lorke, D. (1983). A new approach to practical acute toxicity testing. *Arch. Toxicol.* **54**, 275–287.

Müller, H. and Kley, H.-P. (1982). Retrospective study on the reliability of an "approximate LD50" determined with a small number of animals. *Arch. Toxicol.* **51**, 189–196.

Naylor, J. C. and Smith, A. F. M. (1982). Applications of a method for the efficient computation of posterior distributions. *Appl. Statist.* **31**, 214–225.

Owen, D. M. (1956). Tables for computing bivariate normal probabilities. *Ann. Math. Statist.* **27**, 1075–1090.

Racine, A., Grieve, A. P., Flühler, H. and Smith, A. F. M. (1986). Bayesian methods in practice: experiences in the pharmaceutical industry (with discussion). *Appl. Statist.* **35**, 93–150.

Ross, G. J. S. (1986). In discussion of Racine *et al.*

Schütz, E. and Fuchs, H. (1982). A new approach to minimizing number of animals used in acute toxicity testing and optimizing the information of test results. *Arch. Toxicol.* **31**, 197–220.

Steen, N. M., Byrne, G. D. and Gelbard, E. M. (1969). Gaussian quadrature for the integrals $\int_0^\infty \exp(-x^2) f(x)dx$ and $\int_0^b \exp(-x^2) f(x)dx$. *Math. Comp.* **23**, 661–671.

Tierney, L. and Kadane, J. B. (1986). Accurate approximations for posterior moments and marginal densities. *J. Amer. Statist. Assoc.* **81**, 82–86.

Tsutakawa, R. K. (1975). Bayesian inference for biossay. *Tech. Rep.* **52**, Mathematical Sciences: University of Missouri, Columbia.

Williams, D. A. (1986). Interval estimation of the median lethal dose. *Biometrics* **42**, 641–645.

Young, J. C. and Minder, Ch. E. (1974). Algorithm AS 75: An integral useful in calculating non-central t and bivariate normal probabilities. *Appl. Statist.* **23**, 455–457, and Correction in *Appl. Statist.* **28**, (1979).

APPENDIX

Suppose that the posterior distribution of two variables y and x is bivariate normal with means y_0 and x_0 variances σ_y^2 and σ_x^2 and correlation ρ so that,

$$p(y,x) = BN(\mu, \Sigma)$$

where,

$$\mu = \begin{pmatrix} y_0 \\ x_0 \end{pmatrix} \quad \text{and} \quad \Sigma = \begin{pmatrix} \sigma_y^2 & \rho\sigma_y\sigma_x \\ \rho\sigma_y\sigma_x & \sigma_x^2. \end{pmatrix}$$

Then,

$$p(w = y/x | x > 0) = \frac{\int_0^\infty x p(wx, x)dx}{\int_0^\infty \int_{-\infty}^\infty p(x,y)dydx} = \frac{A}{B}. \tag{A.1}$$

Following Hinkley (1969) it may be shown that,

$$A = \frac{b(w)d(w)}{\sqrt{(2\pi)}\sigma_y\sigma_x a^3(w)}\Phi\left(\frac{b(w)}{\sqrt{(1-\rho^2)}a(w)}\right) + \frac{\sqrt{(1-\rho^2)}}{2\pi\sigma_y\sigma_x a^2(w)}\exp\left(\frac{-c}{2(1-\rho^2)}\right) \tag{A.2}$$

where,

$$a^2(w) = \frac{w^2}{\sigma_y^2} - \frac{2\rho w}{\sigma_y\sigma_x} + \frac{1}{\sigma_x^2}, \quad b(w) = \frac{wy_0}{\sigma_y^2} - \frac{\rho(y_0 + wx_0)}{\sigma_y\sigma_x} + \frac{x_0}{\sigma_x^2},$$

$$c = \frac{y_0^2}{\sigma_y^2} - \frac{2\rho y_0 x_0}{\sigma_y\sigma_x} + \frac{x_0}{\sigma_x^2}, \quad d(w) = \exp\left[\frac{b^2(w) - ca^2(w)}{2(1-\rho^2)a^2(w)}\right].$$

It is simply shown that,

$$B = \Phi\left(\frac{x_0}{\sigma_x}\right). \tag{A.3}$$

(A.1), (A.2) and (A.3) define the posterior distribution of w.

The cumulative distribution, $F(w)$, can be written as,

$$F(w) = \frac{\int_0^\infty \int_{-\infty}^{wx} p(y,x)dydx}{\sigma\left(\frac{x_0}{\sigma_x}\right)} = \frac{B(h,k,\gamma)}{\Phi(k)} \tag{A.4}$$

where,

$$B(h,k,\gamma) = \frac{\int_{-\infty}^h \int_{-\infty}^k \exp\left[\frac{-(u^2 - 2\gamma uv + v^2)}{2(1-\gamma^2)}\right] dudv}{2\pi\sqrt{(1-\rho^2)}} \tag{A.5}$$

$$h = \frac{w x_0 - y_0}{\sigma_y \sigma_x a(w)}, \quad k = \frac{x_0}{\sigma_x}, \quad \gamma = \frac{w \sigma_x - \rho \sigma_y}{\sigma_y \sigma_x a(w)}.$$

Numerical evaluation of (A.5) may be carried out using equation (2.1) of Owen (1956) together with a program for evaluating the integral,

$$T(h, a) = (2\pi)^{-1} \int_0^a \frac{\exp\left[\frac{-h^2(1+x^2)}{2}\right]}{1 + x^2} dx.$$

Two such programs are given by Cooper (1968) and Young and Minder (1974).

BAYESIAN STATISTICS 3, pp. 631–640
J. M. Bernardo, M. H. DeGroot, D. V. Lindley and A. F. M. Smith, (Eds.)

Outliers and Influence: Evaluation by Posteriors of Parameters in the Linear Model

I. GUTTMAN and D. PEÑA
University of Toronto and Universidad Politécnica, Madrid

SUMMARY

This paper shows how the posterior distribution of the parameters involved in the normal linear model may be used to convey information about outliers and influential observations. Using Kullback-Leibler measures to evaluate changes in distributions, we show that the changes in the posterior distribution of the regression parameter vector when a certain set of observations is dropped measures the influence of that set of observations, while the changes in the posterior distribution of the scale parameter provides information about whether the set of observations can be considered to be outlying, in that the change is a function of the standard outlier test.

Keywords: BAYESIAN INFERENCE; OUTLIERS; KULLBACK-LEIBLER DISTANCE.

1. INTRODUCTION

Outliers have been for a long time an important topic for statistical research (Barnett and Lewis, 1978), but the concept of influential observations has appeared in the statistical literature only recently (Cook and Weisberg, 1982; Belsley, Kuhn and Welsh, 1980).

From the Bayesian point of view, Johnson and Geisser (1983) showed how to use the symmetric Kullback-Leibler distance to measure the change in the predictive distribution when a row of the design matrix and its corresponding observation are deleted. They proved that their predictive Bayesian measures are asymptotically equivalent to the sum of Cook's influence statistic and a convex function of the studentized residual. Related work is also given in Geisser (1983), and Johnson and Geisser (1985).

Moulton and Zellner (1983) developed a generalization of the standard frequentist influential measures from the linear model using non-informative priors. Pettit and Smith (1983) suggested the use of the Kullback-Leibler distance to monitor the change on the posterior distribution and applied this idea to linear models with known σ.

In this paper we discuss the question of how to deal in a unified framework with both outliers and influential observations, using the symmetric Kullback-Leibler distance for posterior distributions. It is shown that the change on the marginal posterior for the regression parameter vector, β, offers a straightforward diagnostic for influence that is a function of those previously suggested. On the other hand, the change in the marginal posterior for σ^2 is a measure of "outlyingness", whereas the change in the joint posterior directs attention to those observations that really matter in the problem.

This paper is organized as follows. Section 2 presents the notation and some general background for the problem. In Section 3 a measure of change in the marginal of β is obtained, and in Section 4 the same procedure is applied to measure the change in the marginal for σ^2. The change in the joint posterior is developed in Section 5. The performances of these measures are presented in an example in Section 6.

2. DROPPING OBSERVATIONS IN THE BAYESIAN LINEAR MODEL

Suppose $y_j, j = 1, \ldots, n$, independent, are generated in accordance with the model

$$y = X\beta + \varepsilon \tag{2.1}$$

where $y = (y_1, \ldots, y_n)$, and $\varepsilon = N(0, \sigma^2 I_n)$. We assume that X, $(n \times p)$, is of full rank. Suppose too that to good approximation, use of the non-informative prior for (β, σ^2) is appropriate, that is, we will employ

$$p(\beta, \sigma^2) \propto (\sigma^2)^{-1}. \tag{2.2}$$

Then as, is well known, the posterior of (β, σ^2), given the data $(y; X)$ is such that

$$(i) \quad p_{\beta}(\beta|y; X) = K_{\beta}[1 + Q(\beta; \hat{\beta}; s^2(X'X)^{-1})/n-p]^{-n/2} \tag{2.3}$$

where

$$\hat{\beta} = (X'X)^{-1}X'y \tag{2.3a}$$

$$\begin{aligned} Q(\theta; \theta_0; M) &= (\theta - \theta_0)'M^{-1}(\theta - \theta_0) \\ (n-p)s^2 &= y'[I - X(X'X)^{-1}X']y, \end{aligned} \tag{2.3b}$$

and

$$K_{\beta} = \frac{|X'X|^{1/2}\Gamma\left(\frac{n}{2}\right)}{(s^2)^{p/2}(\pi)^{p/2}\Gamma\left(\frac{n-p}{2}\right)(n-p)^{p/2}}. \tag{2.3d}$$

That is, in summary, $\beta|y \sim MT_p(\hat{\beta}; s^2(X'X)^{-1}; n-p)$. Also

$$(ii) \quad p_{\sigma^2}(\sigma^2|y; X) = K_{\sigma^2}(\sigma^2)^{-[(n-p)/2+1]}\exp\{-(n-p)s^2/2\sigma^2\} \tag{2.4}$$

with

$$K_{\sigma^2} = [(n-p)s^2]^{(n-p)/2}/[2^{(n-p)/2}\Gamma(n-p)/2)]. \tag{2.4a}$$

That is, in summary, $\sigma^2|y \sim (n-p)s^2/\chi^2_{n-p}$.

We now introduce an indexing system to monitor the dropping or deletion of certain subsets of observations from $(y_1, \ldots, y_n) = y'$.. Let the sets i and $\bar{i}$ be a partition of the set $\{1, \ldots, n\}$,, that is,

$$i = (i_1, \ldots, i_k), \bar{i} = (j_1, \ldots, j_{n-k})$$

where

$$\begin{aligned} i \cup \bar{i} &= \{1, \ldots, n\} \\ i \cap \bar{i} &= \phi. \end{aligned} \tag{2.5}$$

Also, we will denote the deletion of observations $y_{i_1}, \ldots, y_{i_k}$ from y (which yields the vector $(y_{j_1}, \ldots, y_{j_{n-k}})'$ by "(i)", so that

$$y_{(i)} = (y_{j_1}, \ldots, y_{j_{n-k}})' \tag{2.5a}$$

and of course

$$y_{(\bar{i})} = (y_{i_1}, \ldots, y_{i_k})'.$$

Similarly, $X_{(i)}$ denotes the $(n-k) \times p$ matrix formed from X by deleting rows $(i_1, \ldots, i_k)$ from X.

So suppose now that a set of observations, $y_{(\bar{i})}$, are deleted from consideration, and that the posterior of the parameters based on the remaining data, $y_{(i)}$, is denoted by $p_{\beta|(i)}$ and $p_{\sigma^2|(i)}$ for β and σ^2 respectively. Letting $\hat{\beta}_{(i)}$ and $S^2_{(i)}$ be defined by

$$\begin{aligned} \hat{\beta}_{(i)} &= (X'_{(i)}X_{(i)})^{-1}X'_{(i)}y_{(i)} \\ (n-k-p)s^2_{(i)} &= y'_{(i)}[I - X_{(i)}(X'_{(i)}X_{(i)})^{-1}X'_{(i)}]y_{(i)}, \end{aligned} \tag{2.6}$$

we have that $p_{\beta|(i)}$ and $p_{\sigma^2|(i)}$ are of the same form as in (2.3) and (2.4) respectively, but with n replaced by $n-k$, $\hat{\beta}$ replaced by $\hat{\beta}_{(i)}$, s^2 replaced by $s^2_{(i)}, X$, replaced by $X(i)$, and relabelling K_β and K_{σ^2} by $K_{\beta|(i)}$, and $K_{\sigma^2|(i)}$, respectively.

Now, in general, given two *pdf's* f_1 and f_2 the average information of proceeding from f_1 to f_2, as defined by Kullback and Leibler (1951), is

$$I(f_1, f_2) = E_{f_1}[\ell n f_1/f_2] = \int f_1 \ell n \frac{f_1}{f_2} dx. \tag{2.7}$$

A discussion of this measure is given in Kullback (1959).

The directed divergency, or mean discrimination, (2.7), is not symmetric. A more natural measure of distance is the symmetric divergence defined as

$$J(f_1, f_2) = I(f_1, f_2) + I(f_2, f_1). \tag{2.8}$$

By taking f_1 to be the posterior of some parameters, given the full data set, and f_2 to be the posterior of the same parameters based on a data set which omits from the original (full) data set, k observations, we may monitor the discrepancy between the posteriors using (2.7) or (2.8), etc.

We proceed to do this for β (Section 3) and σ^2 (Section 4). We mention in passing that if $f_j (j = 1, 2)$ is the *pdf* of a p-order multivariate-normal, mean μ_j, variance covariance matrix Σ_j assumed positive definite, then, as is easily verified,

$$J(f_1, f_2) = \frac{1}{2}(\mu_1 - \mu_2)'(\Sigma_1^{-1} + \Sigma_2^{-1})(\mu_1 - \mu_2) + \frac{1}{2}tr(\Sigma_1 \Sigma_2^{-1} + \Sigma_2 \Sigma_1^{-1}) - p$$

where, in general, $tr(M)$ stands for the trace of the matrix M. We will use this measure to monitor the change on the distributions.

3. A MEASURE OF CHANGE IN THE MARGINAL FOR β

As discussed in Section 2, the marginal posterior for β is multivariate-t (see (2.3)), and approximating the *pdf* of a t of order p by a p order multivariate normal, we have

$$\beta|y \dot{\sim} MN_p(\hat{\beta}; s^2(X'X)^{-1}). \tag{3.1}$$

[The symbol "$\dot{\sim}$" means "approximately distributed as".] We are also concerned with the "influence" of $y_{(\bar{i})}$ on inference about β, and hence we will want to compare the posterior of $\beta|y$ with that of $\beta|y_{(i)}$. Of course

$$\beta|y_{(i)} \dot{\sim} MN_p(\hat{\beta}_{(i)}; s^2_{(i)}(X'_{(i)}X_{(i)})^{-1}). \tag{3.2}$$

Now on applying (2.7), the Kullback-Leibler distance from $p_{\beta|(i)}$ to p_{β}, which measures the change due to basing the posterior of β on $y_{(i)}$ rather than y, is, for large n, using (3.1) and (3.2), such that

$$\begin{aligned}2I(p_{\beta|(i)}, p_{\beta}) &= (\hat{\beta}_{(i)} - \hat{\beta})'[s^2(X'X)^{-1}]^{-1}(\hat{\beta}_{(i)} - \hat{\beta}) \\ &+ tr\{[s^2_{(i)}(X'_{(i)}X_{(i)})^{-1}][s^2(X'X)^{-1}]^{-1}\} + \ell n|(s^2(X'X)^{-1})(s^2_{(i)}(X'_{(i)}X_{(i)})^{-1})^{-1}| \\ &- p.\end{aligned} \tag{3.3}$$

The first term on the "right-hand side" of (3.3) has been labelled as pD^2 by Cook (1977) and D^2 has been used by Cook and many others as a measure of influence of $y_{(i)}$. Doing some algebra (see Appendix I for the definition of $H_{(i)}$ at (A.6)), we may write (3.3) as

$$\begin{aligned}2I(p_{\beta|(i)}, p_{\beta}) &= p[D^2 + \ell n(s^2/s_{(i)}) + (s^2_{(i)}/s^2) - 1] + \ell n|I_k - H_{(\bar{i})}| \\ &+ (s^2_{(i)}/s^2)tr\{H_{(\bar{i})}[I_k - H_{(\bar{i})}]^{-1}\}\end{aligned} \tag{3.4}$$

For the case $k = 1$, we have that $(i) = (i_1)$ denotes the deletion of one observation, say y_i, so that (3.4) takes the from

$$\begin{aligned}2I(p_{\beta|(i)}, p_{\beta}|k = 1) &= p[D^2 + \ell n\ s^2/s^2_{(i)} + s^2_{(i)}/s^2 - 1] \\ &+ \ell n(1 - h_i) + (s^2_{(i)}/s^2)[h_i/(1 - h_i)].\end{aligned} \tag{3.5}$$

Now in a similar manner it is easy to see that for general k

$$\begin{aligned}2I(p_{\beta}, p_{\beta|(i)}) &= pD^2_{(i)} + [s^2/s^2_{(i)}]tr[(X'X)^{-1}(X'_{(i)}X_{(i)})] \\ &- p\ell n(s^2/s^2_{(i)}) - \ell n|(X'_{(i)}X_{(i)})(X'X)^{-1}| - p\end{aligned} \tag{3.6}$$

with

$$pD^2_{(i)} = (\hat{\beta} - \hat{\beta}_{(i)})'X'_{(i)}X_{(i)}(\hat{\beta} - \hat{\beta}_{(i)})/s^2_{(i)}. \tag{3.7}$$

Combining (3.6) with (A.1) of Appendix I, we find that the measure of influence of $y_{(i)}$, say $J(p_{(i)}, p)$, is such that

$$\begin{aligned}2J(p_{(i)}, p) &= p(D^2 + D^2_{(i)}) + (s^2_{(i)}/s^2)tr[(X'X)(X'_{(i)}X_{(i)})^{-1}] \\ &+ (s^2/s^2_{(i)})tr[(X'X)^{-1}(X'_{(i)}X_{(i)})] - 2p\end{aligned} \tag{3.8}$$

or

$$\begin{aligned}2J(p_{(i)}, p) &= p(D^2 + D^2_{(i)}) + (s^2_{(i)}/s^2)\{p + trH_{(\bar{i})}[I_k - H_{(\bar{i})}]^{-1}\} \\ &+ (s^2/s^2_{(i)})(p - trH_{(\bar{i})}) - 2p,\end{aligned} \tag{3.9}$$

since $tr[(X'X)^{-1}(X'_{(i)}, X_{(i)})] = tr[I - (X'X)^{-1}X'_{(\bar{i})}X_{(\bar{i})}]p - trH_{(\bar{i})}$. For the case $k = 1$ we obtain

$$2J(p_{(i)}, p) = p(D^2 + D^2_{(i)}) + \frac{s^2_{(i)}}{s^2}(p + \frac{h_i}{1 - h_i}) + \frac{s^2}{s^2_{(i)}}(p - h_i) - 2p \tag{3.10}$$

where h_i is the i-th diagonal term of H. We set

$$J(p_{(i)}, p) = M(\beta) \tag{3.11}$$

in Table 6.1.

4. MEASURE OF CHANGE IN THE POSTERIOR OF σ^2

The posterior of σ^2 given the full data set is given by (2.4) so that if we delete $(y_{i_1}, \ldots, y_{i_k})$, we have that

$$p_{\sigma^2|(i)}(\sigma^2|y_{(i)}, X_{(i)}) = K_{\sigma^2|(i)}(\sigma^2)^{-[(n-k-p)/2]-1} \exp -(n-k-p)s^2_{(i)}/2\sigma^2 \tag{4.1}$$

where

$$K_{\sigma^2|(i)} = (n-k-p)^{(n-k-p)/2}[2^{(n-k-p)/2}\Gamma(n-k-p)/2)]^{-1}(s^2_{(i)})^{(n-k-p)/2}.$$

Hence we have that

$$\begin{aligned} I(p_{\sigma^2|(i)}, p_{\sigma^2}) &= \ell n \frac{K_{\sigma^2|(i)}}{K_{\sigma^2}} - \frac{k}{2} E_{\sigma^2|(i)} \ell n \frac{1}{\sigma^2} \\ &+ \frac{(n-p)s^2 - (n-k-p)s^2_{(i)}}{2} E_{\sigma^2|(i)}[(\sigma^2)^{-1}|(i)]. \end{aligned} \tag{4.2}$$

Here, $E_{\sigma^2|(i)}$ stands for expectation with respect to the distribution of (4.1), which is such that

$$(1/\sigma^2) \simeq \chi^2_{n-k-p}/(n-k-p)s^2_{(i)}. \tag{4.3}$$

Using the fact that

$$E(\ell n\ \chi^2_m) \simeq \ell n\ m \tag{4.4}$$

we have from (4.2) that

$$\begin{aligned} I(p_{\sigma^2|(i)}, p_{\sigma^2}) &\simeq \ell n \frac{K_{\sigma^2(i)}}{K_{\sigma^2}} + \frac{k}{2} \ell n s^2_{(i)} \\ &+ \frac{(n-p)s^2 - (n-k-p)s^2_{(i)}}{2} \times \frac{1}{s^2_{(i)}}. \end{aligned} \tag{4.5}$$

Similarly, it is easy to see that

$$\begin{aligned} I(p_{\sigma^2}, p_{\sigma^2|(i)}) &\simeq -\ell n \frac{K_{\sigma^2|(i)}}{K_{\sigma^2}} - \frac{k}{2} \ell n s^2 \\ &- \frac{[(n-p)s^2 - (n-k-p)s^2_{(i)}]}{2} \frac{1}{s^2}. \end{aligned} \tag{4.6}$$

The addition of (4.5) and (4.6) gives the divergence between p_{σ^2} and $p_{\sigma^2|(i)}$, that is, the influence of $(y_{i_1}, \ldots, y_{i_k})$, as

$$\begin{aligned} J(p_{\sigma^2}, p_{\sigma^2|(i)}) &= \frac{k}{2} \ell n \frac{\sigma^2_{(i)}}{s^2} \\ &+ \left[\frac{(n-p)s^2 - (n-k-p)s^2_i}{2}\right] \left[\frac{1}{s^2_{(i)}} - \frac{1}{s^2}\right]. \end{aligned} \tag{4.7}$$

For the case $k = 1$, it is well known that

$$(n-p)s^2 - (n-p-1)s^2_{(i)} = e^2_{(i)}/(1-h_i), \tag{4.8}$$

where e_i is the ordinary residual of y_i, so that we have, approximately

$$\ell n \frac{s^2_{(i)}}{s^2} = \ell n \left(1 - \frac{r_i^2}{n-p}\right) \cong \frac{-r_i^2}{n-p} \tag{4.9}$$

where r_i^2 is the standardized residual, given by

$$r_i^2 = \frac{e_i^2}{s^2(1-h_i)}. \tag{4.10}$$

Now $r_i^2/(n-p)$ follows a Beta distribution (see Cook and Weisberg, 1982, p. 19) and is expected to be small if n is large. Assuming that this is the case, (4.7) can be written, for $k = 1$, to good approximation.

$$J(p_{\sigma^2}, p_{\sigma^2|(i)}) = \frac{1}{2}(t_i^2 - r_i^2) \tag{4.11}$$

where t_i is the studentized residual defined by

$$t_i^2 = \frac{e_i^2}{s^2_{(i)}(1-h_i)}. \tag{4.12}$$

Clearly, use of this measure is equivalent to the use of an outlier test statistic. We label (4.11) as $M(\sigma^2)$, and use it for the example of Section 6 in Table 6.1.

5. MEASURE OF CHANGE IN THE JOINT POSTERIOR OF β, σ^2

We first wish to establish $I(p_{\beta,\sigma^2|(i)}, p_{\beta,\sigma^2})$, which of course is the expectation of

$$\log[p_{\beta,\sigma^2|(i)}(\beta, \sigma^2|y_{(i)})/p_{\beta}, \sigma^2(\beta, \sigma^2|y)], \tag{5.1}$$

with respect to $p_{\beta,\sigma^2|(i)}$. Now we may write

$$p_{\beta,\sigma^2|(i)}(\beta, \sigma^2|y_{(i)}) = p(\sigma^2|y_{(i)})p(\beta|\sigma^2; y_{(i)}) \tag{5.2}$$

with a similar expression for p_{β,σ^2}. Hence we wish to find the expection of

$$\log[p(\sigma^2|y_{(i)})/p(\sigma^2|y)] + \log[p(\beta|\sigma^2; y_{(i)})/(p(\beta|\sigma^2; y)]. \tag{5.2a}$$

The expectation of the first term in (5.2a) leads, of course, to (4.5). To find the expectation of the second term, we may proceed by first evaluating the expectation of the second term with respect to the distribution of $\beta|\sigma^2; y_{(i)}$, and then taking the expectation of the latter with respect to the distribution of $\sigma^2|y_{(i)}$. The result of this prescription and adding (4.5) yields

$$\begin{aligned} I(p_{\beta,\sigma^2|(i)}, p_{\beta,\sigma^2}) &= \ell n \frac{K_{\sigma^2|(i)}}{K_{\sigma^2}} + \frac{k}{2} \ell n s^2_{(i)} \\ &+ \frac{1}{2}[(n-p)s^2/s^2_{(i)} - (n-k-p)] + \frac{p}{2}\frac{s^2}{s^2_{(i)}} D^2 + \frac{1}{2} tr H_{(\bar{i})}[I - H_{(\bar{i})}]^{-1} \\ &+ \frac{1}{2} \ell n |I_k - H_{(\bar{i})}|. \end{aligned} \tag{5.3}$$

Similarly, it is straightforward to verify that

$$I(p_{\beta,\sigma^2}, p_{\beta,\sigma^2|(i)}) = -\ell n \frac{K_{\sigma^2|(i)}}{K_{\sigma^2}} - \frac{k}{2}\ell n s^2 - \frac{1}{2}[(n-p)-(n-p-k)s^2_{(i)}/s^2] + \frac{1}{2}\frac{s^2_{(i)}}{s^2}pD^2_{(i)} - \frac{1}{2}trH_{(i)} - \frac{1}{2}\ell n|I_k - H_{(\bar{i})}|. \tag{5.4}$$

Adding (5.3) and (5.4) gives, after some algebra, that the divergence between p_{β,σ^2} and $p_{\beta,\sigma^2|(i)}$ is

$$J = \frac{k}{2}\ell n\frac{s^2_{(i)}}{s^2} + \frac{(n-p)s^2-(n-k-p)s^2_{(i)}}{2}\left(\frac{1}{s^2_{(i)}} - \frac{1}{s^2}\right) + \frac{k}{2}[s^2_{(i)}D^2 + \frac{s^2_{(i)}}{s^2}D^2_{(i)}] + \frac{1}{2}trH_{(\bar{i})}[I-H_{(i)}]^{-1}H_{(\bar{i})} \tag{5.5}$$

and we may use J to measure the influence of $y_{(\bar{i})} = (y_{i_1}, \ldots, y_{i_k})'$. For $k = 1$, using previous results, we have to good approximation

$$J = \frac{1}{2}(t_i^2 - r_i^2) + \frac{p}{2}[\frac{s^2}{s^2_{(i)}}D^2 + \frac{s^2_{(i)}}{s^2}D^2(i)] + \frac{1}{2}h_i^2/[1-h_i] \tag{5.6}$$

and this is a mixture of the influence measures for β and outlier test statistics. We label (5.6) by $WP(\beta, \sigma^2)$, and use it for the example of Section 6 in Table 6.1.

A referee has suggested the use of the "expected change in divergence", say E. D., given by

$$\text{E.D.} = WP(\beta, \sigma^2) - M(\sigma^2), \tag{5.7}$$

instead of $M(\beta)$, given by (3.11), since the latter was derived using the approximation (3.2). In the example of Section 6, we use (5.7), entering the results in Table 6.1, and find that E. D. takes on values that are very similar to $M(\beta)$, often matching to two places and sometimes 3 places of accuracy.

6. AN EXAMPLE

To illustrate the behaviour of the suggested measures we will reconsider the data by Mickey, Dunn and Clark (1967) which have been previously used to compare regression diagnostics by Cook and Weisberg (1982), Draper and John (1981), and Pettit and Smith (1983) among others. This set of data is plotted in Figure 1.

Table 1 gives a summary of some relevant statistics for these data. Looking first at the change on the whole posterior (WP) it is obvious that there are two important points, observations 18 and 19. The analysis of the marginals indicates perfectly the difference between them: observation number 18 is very influential but is not an outlier, and observation 19 is an outlier, but is not influential.

It is interesting to notice that the Bayesian measures differentiate between both types of data better than either the statistic $D^2_{(i)}$ (column 4) or the DFITS statistic, defined as $pS^2D^2/S^2_{(i)}$, and are clearly able to point out the outlier.

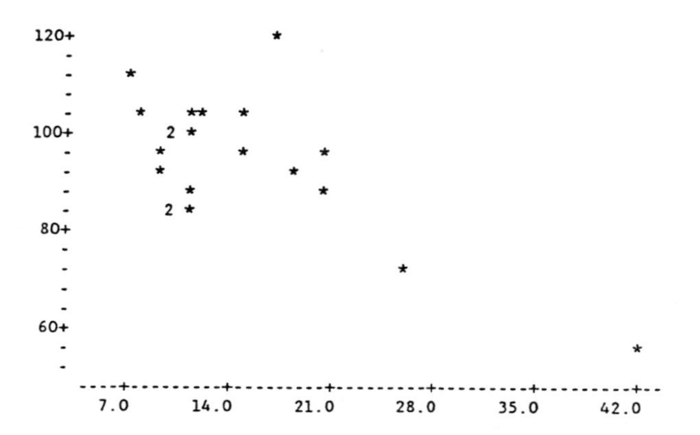

Figure 6.1. *Plot of the Mickey, Dunn and Clark Data.*
(the "2" 's appearing on the graph indicate that two observations fell at that point.)

No.	h_i	t_i	r_i	D_i^2	DFITS	$M(\beta)$	$M(\sigma^2)$	$WP(\beta,\sigma^2)$	E.D.*
1	0.05	0.18	0.19	0.00	0.04	.01	.00	.00	.00
2	0.15	-0.94	-0.94	0.08	-0.40	.15	.00	.15	.15
3	0.06	-1.51	-1.46	0.07	-0.39	.12	.04	.16	.12
4	0.07	-0.81	-0.82	0.03	-0.22	.20	.12	.32	.20
5	0.05	0.83	0.84	0.02	0.19	.04	.00	.03	.03
6	0.07	-0.03	-0.03	0.00	-0.01	.01	.00	.00	.00
7	0.06	0.31	0.32	0.00	0.08	.01	.00	.00	.00
8	0.06	0.23	0.24	0.00	0.06	.01	.00	.00	.00
9	0.08	0.29	0.30	0.00	0.09	.01	.00	.00	.00
10	0.07	0.62	0.63	0.02	0.17	.01	.00	.00	.00
11	0.09	1.05	1.05	0.06	0.33	.11	.00	.11	.11
12	0.07	-0.34	-0.35	0.01	-0.09	.01	.00	.00	.00
13	0.06	-1.51	-1.46	0.07	-0.39	.12	.04	.16	.12
14	0.06	-1.28	-1.26	0.05	-0.31	.08	.01	.09	.08
15	0.06	0.41	0.42	0.01	0.10	.02	.00	.01	.01
16	0.06	0.13	0.13	0.00	0.03	.01	.00	.00	.00
17	0.05	0.80	0.81	0.02	0.19	.04	.00	.03	.03
18	0.65	-0.84	-0.85	0.68	-1.16	1.66	.00	1.64	1.64
19	0.05	3.61	2.82	0.22	0.85	1.67	1.98	2.49	.51
20	0.06	-1.08	-1.07	0.04	-0.26	.05	.00	.05	.05
21	0.06	0.13	0.13	0.00	0.03	.01	.00	.00	.00

*E.D. = $WP(\beta,\sigma^2) - M(\sigma^2)$

Table 6.1. *Relevant Statistics for the Mickey, Dunn and Clark data.*

APPENDIX I

In Section 3, we are involved with the determination of $I(f_1, f_2)$, given by (2.7), with the role of f_1 played by (3.2), and the role of f_2, played by (3.1). With these $f_j's$, I is easily found to be

$$2I(p_{|\beta(i)}, p_\beta) = pD^2 + [s^2_{(i)}/s^2]tr[(X'X)(X'_{(i)})^{-1}] + p\ell n(s^2/s^2_{(i)}) \\ + \ell n|(X'X)^{-1}(X'_{(i)}X_{(i)})| - p. \tag{A.1}$$

To further simplify (A.1), let us recall that we are considering

$$y = \begin{Bmatrix} y_{(i)} \\ y_{(\bar{i})} \end{Bmatrix} = \begin{Bmatrix} X_{(i)} \\ \cdots \\ X_{(\bar{i})} \end{Bmatrix} \beta + \varepsilon \tag{A.2}$$

where $X(\bar{i})$ is the $(k \times p)$ matrix obtained by "deleting $(\bar{i})$" from X, that is, by deleting rows $j_1, \ldots, j_{n-k}$ from X. It is now easily verified, using the well known identity

$$(I - AB)^{-1} = I + A(I - BA)^{-1}B, \tag{A.3}$$

that

$$(X'_{(i)}X_{(i)})^{-1}X'X = I_p + (X'X)^{-1}X'_{(\bar{i})}[I_k - H_{(\bar{i})}]^{-1}X_{(i)} \tag{A.4}$$

where H is the so-called "hat" matrix

$$H = X(X'X)^{-1}X'. \tag{A.5}$$

In (A.4), $H_{(\bar{i})}$ is the $(k \times k)$ matrix found by deleting rows $j_1, \ldots, j_{n-k}$ from H, or

$$H_{(\bar{i})} = X_{(\bar{i})}(X'X)^{-1}X'(\bar{i}) \tag{A.6}$$

Hence, from (A.4),

$$tr[(X'X)(X'_{(i)}X_{(i)})^{-1}] = p + trH_{(\bar{i})}[I_k - H_{(\bar{i})}]^{-1}, \tag{A.7}$$

and we may use this to help compute the second term of (A.1). It is also easily verified that since

$$X'_{(i)}X_{(i)} = X'X(I_P - (X'X)^{-1}X'_{(\bar{i})}X_{(\bar{i})} \tag{A.8}$$

and because $|I - AB| = |I - BA|$, $|CD| = |DC|$, etc., we may write

$$|(X'X)^{-1}X'_{(i)}X_{(i)}| = |I_k - H_{(\bar{i})}|, \tag{A.9}$$

which may be used to compute term 4 of (A.1).

Using (A.7) and (A.9) in (A.1) yields (3.4).

ACKNOWLEDGEMENTS

This research was supported in part by NSERC (Canada) through grant No. A8743, and CAYCIT (Spain) through grant No PB86-0538. We also wish to thank Prof. A. F. M. Smith and an anonymous referee for very helpful comments that have improved an earlier version of this paper.

REFERENCES

Barnett, V. and Lewis, T. (1978). *Outliers in Statistical Data.* Wiley.

Belsley, D. A., Kuh, E. and Welsch, R. E. (1980). *Regression Diagnostics.* Wiley.

Cook, R. D. (1979). Influential Observations in Linear Regression. *J. Amer. Statist. Assoc.* **74**, 169–174.

Cook, D. R. and Weisberg, S. (1982). *Residuals and Influence in Regression.* Chapman and Hall.

Draper, N. R. and John J. A. (1981). Influential Observations and Outliers in Regression. *Technometrics* **23**, 21–26.

Geisser, S. (1983). On the prediction of Observables: A Selective Update. *Bayesian Statistics 2*, 203–230. (J. M. Bernardo, M. H. DeGroot, D. V. Lindley and A. F. M. Smith, eds.). Elsevier Science Publishers.

Johnson, W. and Geisser, S. (1983). A Predictive View of the Detection and Characterization of Influential Observations in Regression Analysis. *J. Amer. Statist. Assoc.* **78, 381**, 137–144.

Johnson, W. and Geisser, S. (1985). Estimative Influence Measures for the Multivariate General Linear Model. *Journal of Statistical Planning and Inference* **11**, 33–56.

Kullback, S. and Leibler, R. A. (1951). On Information and Sufficiency. *Ann. Math. Statist.* **22**, 79–86.

Kullback, S. (1959). *Information Theory and Statistics.* Wiley.

Mickey, M. R., Dunn, O. J. and Clark. V. (1967). Note on Use of Stepwise Regression in Detecting Outliers. *Computers and Biomed. Res.* **1**, 105–111.

Moulton, B. R. and Zellner, A. (1983). Bayesian Regression Diagnostics, (to appear).

Pettit , L. I. and Smith, A. F. M. (1983). Outliers and Influential Observations in Linear Models. *Bayesian Statistics 2*, 473–474 (J. M. Bernardo, M. H. DeGroot, D. V. Lindley and A. F. M. Smith, eds.). Elsevier Science Publishers.

BAYESIAN STATISTICS 3, pp. 641–651
J. M. Bernardo, M. H. DeGroot, D. V. Lindley and A. F. M. Smith, (Eds.)

Combining Opinions in a Predictive Case

A. B. HUSEBY
University of Oslo

SUMMARY

In this paper the well-known problem of combining experts' opinions about some uncertain quantity is considered. When the uncertain quantity is some unknown distribution parameter, μ, a useful method for doing this is to model the information available to the experts, as a set of observations $X_1, \ldots, X_n$, reconstruct these observations as far as possible from the information provided by the experts, and use the derived statistics as a basis for the combined judgement. In other cases, however, it is more natural to think of the uncertain quantity as a future observation, X, from the same distribution as the X_i's, and thus the above model has to be modified in order to serve the purpose. The paper provides a methodology for solving the problem in this case.

Keywords: CONSENSUS DISTRIBUTION; COMBINING OPINIONS; RETROSPECTION; PREDICTION; BAYESIAN METHODS.

1. INTRODUCTION

A decision maker (DM) searching for relevant information concerning uncertain quantities, is frequently faced with the problem of combining information from many different sources. In the present paper we consider the case where the main source of information is some uncertainty judgements provided by a panel of "experts". For more background on this problem and a bibliography, we refer to Genest and Zidek (1986).

Clearly, the Bayesian solution to this problem would be to consider the judgements provided by the experts as data being entered into the model through a likelihood function, and then calculate the posterior distribution using Bayes'theorem. The main problem now is how to find a suitable likelihood function.

In Huseby (1986) we approached this problem by establishing a model for how the experts arrived at their judgements. The precise assumption we made was that all the available information about the uncertain quantity of interest, could be modelled as a sequence of (imaginary) observations, $X_1, \ldots, X_n$. Moreover, we assumed that each of the experts possessed partial knowledge about this sequence.

If μ was the uncertain quantity, it was typically assumed that for given μ the X_i's were independent and identically distributed, with common density $q(x|\mu)$. That is, μ was an uncertain parameter of a distribution, and the X_i's were simply a random sample from this distribution.

For more details we refer to Huseby (1986). See also Winkler (1968) and Clemen (1987) for similar approaches.

In this paper we consider the case where the uncertain parameter more naturally can be described as a future observation from the same distribution as the available (imaginary) data. Moreover, we assume that each of the experts provides a set of fractiles from a probability distribution reflecting their state of uncertainty about X. The aim of the paper is to describe a complete procedure for combining such information, including the model fitting.

2. THE GENERAL MODEL

In order to get information about some uncertain quantity, X, we assume that k experts are interviewed by the DM and that each expert provides a set of fractiles from a probability distribution reflecting his state of uncertainty about X.

Instead of using these fractiles directly, the DM starts out by modelling how the experts arrived at their fractiles. In order to do this, the DM postulates that the information about X available to the experts, may be modelled as a sequence of (imaginary) observations, $X_1, \ldots, X_n$.

Having done so the next step is to model how the experts have used these observations to calculate their uncertainty distributions for X. In general different persons may use different likelihood functions for the same set of data and thus arrive at different uncertainty statements. However, in this case the DM simply assesses that the same likelihood is used by all experts. Moreover, we assume that the DM himself agrees on this likelihood. Specifically, this likelihood model, as assessed by the DM, is formulated as follows:

Model 2.1. Given a certain hyperparameter, θ, the uncertain quantity X, and the (imaginary) data $X_1, \ldots, X_n$, are independent and identically distributed with density $q(x|\theta)$. The density function $q(x|\theta)$, is completely specified except for the parameter θ.

The prior state of uncertainty about θ (before observing any of the X_i's) is assumed to be expressed equally by the DM and the experts in terms of a common vague prior density g_0.

The DM next assesses (as in Huseby (1986)) that the k experts possess partial knowledge about the sequence $X_1, \ldots, X_n$. Specifically, let $A_1, \ldots, A_k$ be subsets of the index set $\{1, \ldots, n\}$. Then the DM assesses that $\{X_i : i \in A_j\}$ are the observations available to the j-th expert, $j = 1, \ldots, k$. We stress that the sets $A_1, \ldots, A_k$ have to be assessed subjectively by the DM. Finally, for simplicity we assume that none of the observations $X_1, \ldots, X_n$ are available to the DM himself.

As mentioned, each expert provides a set of fractiles from a probability distribution reflecting their uncertainty about X. If the above probability model were "true" in the sense that it perfectly described how the experts had worked out their uncertainty statements, the fractiles provided by the j-th expert ($j = 1, \ldots, k$) should correspond to a distribution having a density of the following form:

$$q_j(x|\boldsymbol{x}(A_j)) = \int_{\Theta} q(x|\theta) g_j(\theta|\boldsymbol{x}(A_j)) d\theta \tag{2.1}$$

where $\boldsymbol{x}(A_j)$ denotes the subvector of the vector $\boldsymbol{x} = (x_1, \ldots x_n)$ corresponding to the indices in A_j, Θ is the set of all possible value of the hyperparameter θ, and $g_j(\theta|\boldsymbol{x}(A_j))$ denotes the conditional density of θ given $\boldsymbol{x}(A_j)$. That is, for $j = 1, \ldots, k, g_j(\theta|\boldsymbol{x}(A_j))$ is given by:

$$g_j(\theta|\boldsymbol{x}(A_j)) = \frac{g_0(\theta) \prod_{i \in A_j} q(x_i|\theta)}{\int_{\Theta} g_0(\theta) \prod_{i \in A_j} q(x_i|\theta) d\theta}, \quad \theta \in \Theta. \tag{2.2}$$

Thus, it seems reasonable to fit a distribution with density, $q_j(x|\boldsymbol{x}(A_j))$, to the given fractiles. Since the shape of the density (2.1) typically depends on $\boldsymbol{x}(A_j)$, knowledge about this density, obtained by fitting fractiles, may be used retrospectively to derive information about $\boldsymbol{x}(A_j)$. Usually, however, $q_j(x|\boldsymbol{x}(A_j))$ depends on $\boldsymbol{x}(A_j)$ only through some sufficient statistic $\boldsymbol{t}_j = \boldsymbol{t}_j(\boldsymbol{x}(A_j)), j = 1, \ldots, k$. Thus, it is only possible to derive information about these statistics.

Due to the fact that Model 2.1. is just an aproximation to reality, it is in most cases unlikely that the given fractiles will fit perfectly to a distribution of the form (2.1). As a result it is typically not possible to reconstruct the underlying data entirely. The precise values of the $\boldsymbol{t}_j$'s will remain uncertain.

From a Bayesian point of view the correct way to handle this problem, is to calculate a posterior distribution for the t_j's given the experts' fractiles, and incorporate this distribution in the further calculations. However, especially in cases with many experts, introducing so many uncertain quantities into the model soon makes it impossible to compute the resulting distribution of X.

If the fitting is not too poor though, (indicating that the t_j's may be well estimated from the fractiles), a reasonable approximation can be obtained by neglecting the uncertainty about the t_j's and just using point estimates as if they were the true values of the t_j's. In the examples considered in this paper, we have used this type of approach. Specifically, in Section 3, we have used the mean values of the distribution of the t_j's rather than the complete distribution. In the more complex example considered in Section 4, the estimates are obtained based on a "least squares" approach. We, however, feel that in a more thorough discussion of the model this problem should be investigated further. At least, there is a need for a careful study where different estimates for the t_j's are compared.

The t_j's estimated from the fractiles play the same role in this approach as in approaches of Winkler (1981) and Lindley (1983), i.e. the they are the information extracted from the experts judgements. However, in our setting the t_j's are interpreted as functions of the underlying imaginary observations, $\boldsymbol{x}$.

Having done all this, the DM may calculate his posterior distribution for X given the information extracted from the experts' judgement, i.e. the estimated t_j's (considered henceforth as the true values of these statistics), by applying the following formula:

$$f(x|t_1,\ldots,t_k) = \int_{\Theta} q(x|\theta)g(\theta|t_1,\ldots,t_k)d\theta \tag{2.3}$$

where $g(\theta|t_1,\ldots,t_k)$ denotes the $DM's$ posterior distribution for θ given $t_1,\ldots,t_k$, and can be found by computing the likelihood of θ for $t_1,\ldots,t_k$ (using the standard transformation formula) and then using Bayes' theorem to combine this likelihood with the vague prior g_0

Alternatively, the posterior distribution for X given $t_1,\ldots,t_k$ may be computed using the following formula:

$$f(x|t_1,\ldots,t_k) = \int_{\chi}\int_{\theta} q(x|\theta)g(\theta|\boldsymbol{x})p(\boldsymbol{x}|t_1,\ldots,t_k)d\theta d\boldsymbol{x} \tag{2.4}$$

where $g(\theta|\boldsymbol{x})$ denotes the distribution of θ given the whole vector, $\boldsymbol{x}$, $p(\boldsymbol{x}|t_1,\ldots,t_k)$ denotes the distributions of $\boldsymbol{x}$ gives $t_1,\ldots,t_k$, and χ denotes the set of possible values of $\boldsymbol{x}$.

[We can write $g(\theta|\boldsymbol{x})$ instead of $g(\theta\ |\boldsymbol{x},\ t_1,\ldots,\ t_k)$ and $q(x|\theta)$ instead of $q(x|\theta,\ \boldsymbol{x}, t_1,$ $\ldots,\ t_k)$ in the integrand of (2.4) since the t_j's are deterministic functions of $\boldsymbol{x}$, and since $X, X_1,\ldots,X_n$ are independent given θ.]

If $t = t(\boldsymbol{x})$ is a sufficient statistic for $\boldsymbol{x}$, then $g(\theta|\boldsymbol{x}) = g(\theta|t)$. In this case (2.4) may be replaced by the following usually simpler expression:

$$f(x|t_1,\ldots,t_k) = \int_{T}\int_{\Theta} q(x|\theta)g(\theta|t)p(t|t_1,\ldots,t_k)d\theta dt \tag{2.5}$$

where T denotes the set of possible values of t, and $p(t|t_1,\ldots,t_k)$ is the conditional distribution of t given $t_1,\ldots,t_k$.

When modelling information as an imaginary sequence of observations, one of the main advantages is that it is fairly easy to represent the dependence between the experts. According to the model this is reflected by the degree of overlap between the A_j-sets. Thus, instead of specifying some complicated covariance matrix characterizing the dependence, the DM specifies these sets.

Computationally, it is easiest to handle the model in cases where $X, X_1, \ldots, X_n$ are normally distributed given θ. Still, at least in principle, it is possible to use the same approach in other situations as well. In such cases, however, it is often more difficult to obtain closed form solutions, and typically it is necessary to use numerical integration to calculate the posterior distribution.

Although models yielding closed form solutions in a certain sense are more appealing, they should not be used unless they are sufficiently flexible to describe the vital aspects of the problem. Moreover, it is important that the parameters of the model have intuitive interpretations. If this is not so, they soon becomes too difficult to assess.

In Section 3 we shall investigate a case where the distribution $q(x|\theta)$ is exponential. The reason for including this example is partly that we consider this application interesting in itself, and partly that we wanted to show that it is not too complicated to obtain a solution in a non-normal case.

In Section 4 we discuss how other parameters of the model may be derived. In the model considered in Section 3 this can be done right away, while in a situation where $q(x|\theta)$ is normal, it is shown that this can be done by requiring that the experts provide some additional information.

3. AN EXAMPLE FROM RELIABILITY THEORY

In this section we let X be the lifetime of a component in a reliability system. The DM asks k experts to express their opinions on X. The j-th expert describes his uncertainty about X by specifying a vector of r numbers, $\boldsymbol{z}_j = (z_{j1}, \ldots, z_{jr})$ such that:

$$\Pr(X > z_{js}) = \pi_s, \quad s = 1, \ldots, r \tag{3.1}$$

where $\pi_1, \ldots, \pi_r \in (0, 1)$ are r probabilities specified by the DM. The DM then faces the problem of computing his posterior distribution for X given the data $\boldsymbol{z}_1, \ldots, \boldsymbol{z}_k$.

We assume that the DM assesses that given the hyperparameter, θ, X and the imaginary observations $X_1, \ldots, X_n$ are independent and exponentially distributed with density $q(x|\theta) = \theta \exp(-\theta x)$, and as the vague distributions for θ, g_0 a gamma-density with parameters close to zero is used. That is:

$$g_0(\theta) \propto \theta^{-1}, \qquad \theta > 0. \tag{3.2}$$

We now turn to the problem of assessing the sets $A_1, \ldots, A_k$. Typically, the resulting posterior distribution will depend on the number of elements in each of the A_j's as well as on the number of elements in all possible intersections of these sets. In general this implies that unless one applies some sort of simplified model, one has to assess $(2^k - 1)$ numbers. Even for moderate sizes of k this is of course a hopeless task. We have therefore simplified the model by assuming that every imaginary observation is either accessable to all the experts or to just one.

That is, let $B_0, B_1 \ldots, B_k$ be disjoint subsets of the index set $\{1, \ldots, n\}$ such that $B_0 \cup B_1 \cup \ldots \cup B_k = \{1, \ldots, n\}$. It is then assessed that the j-th expert has access to the X_i's with indices in the set $A_j = (B_0 \cup B_j), = 1, \ldots, k$. Thus, B_0 is the set of common observations, while the B_j's are the set of individual observations.

We also introduce the following notation:

$$n_j = \text{The number of elements in } A_j, \quad j = 1, \ldots, k \tag{3.3}$$

$$t_j = \sum_{i \in A_j} x_i, j = 1, \ldots, k \text{ and } t = \sum_{i=1}^{n} x_i$$

The parameters $n_1, \ldots, n_k$ and n are assessed subjectively by the DM, while information about $t_1, \ldots, t_k$ and t will be derived by fitting distributions to the z_j's as follows.

By standard calculations it is easily seen that:

$$\Pr(X > x | x(A_j)) = \left(\frac{t_j}{t_j + x}\right)^{n_j}, \quad j = 1, \ldots, k \tag{3.4}$$

and:

$$\Pr(X > x | x) = \left(\frac{t}{t + x}\right)^{n}.$$

Inserting the z_{js} into (3.4) yields:

$$\left(\frac{t_j}{t_j + z_{js}}\right)^{n_j} = \pi_s, s = 1, \ldots, r \text{ and } \; j = 1, \ldots, k. \tag{3.5}$$

Hence, for each j we have r equations determining t_j. However, unless the model fits data perfectly, these solutions will not be equal. Denoting by t_{js} the estimate of t_j obtained from the s-th equation, $s = 1, \ldots, r$ and $j = 1, \ldots, k$, we get:

$$t_{js} = \phi_{js} z_{js}, s = 1, \ldots, r \text{ and } j = 1, \ldots, k, \text{ where } \phi_{js} = \frac{\pi_s^{1/n_j}}{1 - \pi_s^{1/n_j}}, s = 1, \ldots, r. \tag{3.6}$$

As mentioned in Section 2, the correct way to proceed now, would be to calculate a posterior distribution for the t_j's given the z_{js}'s (or equivalently the t_{js}'s), and use this further on. As an approximation we just specify the means of this distribution, and use these as if they were the true values of the t_{js}-s. Specifically, we estimate the t_{js} as follows:

$$t_j = \sum_{s=1}^{r} \alpha_s t_{js} = \sum_{s=1}^{r} \alpha_s \phi_{js} z_{js}, \quad j = 1, \ldots, k, \tag{3.7}$$

where $\alpha_1 + \ldots + \alpha_r = 1$ and $\alpha_s \geq 0, s = 1, \ldots, r$, and the weights $\alpha_1, \ldots, \alpha_r$ reflect how precise the corresponding z_{js} are considered to be. [In practical situations it is often felt to be easier for the experts to specify 50%-fractiles, than for example the 10%- or 90%-fractiles. Hence, it appears to be reasonable to assess a greater weight on a 50%-fractile than the 10%- or 90%-fractiles.]

A weakness with the above procedure is that the t_j's are estimated separately. Because of the positive dependence between the t_j's, a simultaneous approach could yield a better result. We, however, leave this problem to a future study.

Having derived the statistics, $t_1, \ldots, t_k$ from the experts' judgements, it remains to compute the posterior distribution of X given $t_1, \ldots, t_k$, By (3.4) it is clear that $t = t(x)$, as defined in (3.3), is a sufficient statistic for x. Thus, $\Pr(X > x | t(x)) = \Pr(X > | x)$.

Hence, using (2.5) and (3.4) it follows easily that the posterior reliability of the component at time x is given by:

$$\begin{aligned} \Pr(X > x | t_1, \ldots, t_k) &= \int_T \Pr(X > x | t(x)) p(t | t_1, \ldots, t_k) dt \\ &= \int_T \left(\frac{t}{t + x}\right)^n p(t | t_1, \ldots, t_k) dt \end{aligned} \tag{3.8}$$

where T denotes the set of possible values of t and $p(t | t_1, \ldots, t_k)$ as usually denotes the conditional distribution of t given $t_1, \ldots, t_k$.

If $k = 1$, then clearly $t = t_1$, and thus the distribution of t given t_1 degenerates to the point $t = t_1$.

We next consider the nontrivial case where $k \geq 2$.

By (3.3) it can easily be seen that T is the set of all t such that:

$$\sum_{j=1}^{k} t_j - (k-1) \min_{1 \leq j \leq k} t_j \leq t \leq \sum_{j=1}^{k} t_j. \tag{3.9}$$

In order to determine $p(t|t_1, \ldots, t_k)$, we start out by introducing:

$$s_j = \sum_{i \in B_j} x_i, \quad j = 0, 1, \ldots, k. \tag{3.10}$$

By standard results it is clear that given $\theta, s_0, s_1, \ldots, s_k$ are independent and that for $j = 0, 1, \ldots, k$, $s_j|\theta \sim$ Gamma (m_j, θ), where $m_0, m_1, \ldots, m_k$ are the number of elements in the sets $B_0, B_1, \ldots, B_k$ respectively.

Integrating out θ with respect to its marginal density g_0, as defined in (3.2), we obtain the following expression for the simultaneous density of $s_0, s_1, \ldots, s_k$:

$$u(s_0, s_1, \ldots, s_k) \propto \left[\prod_{j=0}^{k} s_j^{m_j - 1}\right] \Big/ \left[\sum_{j=0}^{k} s_j\right]^n. \tag{3.11}$$

Clearly, the relations between $s_0, s_1, \ldots, s_k$ and $t, t_1, \ldots, t_k$ are as follows:

$$t_j = s_0 + s_j, \quad j = 1 \ldots, k \quad \text{and} \quad t = s_0 + s_1 + \ldots + s_k \qquad ;(3.12)$$

or equivalently,

$$s_0 = t^* - \frac{1}{k-1} t, \quad s_j = t_j - t^* + \frac{1}{k-1} t, \quad j = 1 \ldots, k, \quad \text{where } t^* = \frac{1}{k-1} \sum_{j=1}^{k} t_j. \tag{3.13}$$

Since the transformation (3.13) is linear, the Jacobian will just be a constant. Hence it follows that the simultaneous density of $t, t_1, \ldots, t_k$, and thus also the conditional density of t given $t_1, \ldots, t_k$, are proportional to (3.11) where $s_0, s_1, \ldots, s_k$ are replaced by the expressions (3.13). In other words we have:

$$p(t|t_1, \ldots, t_k) \propto t^{-n} \left(t^* - \frac{t}{k-1}\right)^{m_0 - 1} \prod_{j=1}^{k} \left[(t_j - t^*) + \frac{t}{k-1}\right]^{m_j - 1}. \tag{3.14}$$

Inserting (3.14) into (3.8) yields:

$$\Pr(X > x|t_1, \ldots, t_k) = C \int_a^b (t + x)^{-n} \left(t^* - \frac{t}{k-1}\right)^{m_0 - 1} \prod_{j=1}^{k} \left[(t_j - t^*) + \frac{t}{k-1}\right]^{m_j - 1} dt \tag{3.15}$$

where a and b are respectively the lower and upper bounds of T given in (3.11), and C is a normalizing constant determined so that $\Pr(X > 0|t_1, \ldots, t_k) = 1$.

Assume specifically that 5 experts are interviewed by the DM, and that the j-th expert provides three numbers, z_{j1}, z_{j2}, z_{j3}, corresponding respectively to the probabilities $\pi_1 = 0.90, \pi_2 = 0.50$, and $\pi_3 = 0.10$, and that DM assesses that $n_1 = \ldots = n_5 = 5$, while

$n = 13$. Moreover, the DM assesses the weights used in (3.7) to be $\alpha_1 = 0.2, \alpha_2 = 0.6$ and $\alpha_3 = 0.2$. Table 3.1. lists the actual values of these fractiles, the t_j's as well as the estimates t_j-values (the numbers are fictitions):

Expert	z_{j1}	z_{j2}	z_{j3}	t_{j1}	t_{j2}	t_{j3}	Estimated t_j
1	0.6	4.3	15.2	28.2	28.9	26.0	28.2
2	0.5	3.8	14.4	23.5	25.6	24.6	25.0
3	0.9	6.4	25.3	42.3	43.0	43.3	42.9
4	0.7	5.3	20.8	32.9	35.6	35.6	35.1
5	0.3	2.0	8.8	14.1	13.5	15.0	13.9

Table 3.1. *Fractiles (in years) given by 5 experts.*

By (3.11) we get that in this case $T = [89.5, 145.1]$ and $t^* = 36.3$. Moreover, $m_0 = 3$ while $m_1 = \ldots = m_5 = 2$. Finally, numerical integration yields that the normalizing constant in (3.14) $C = 9.1992 \cdot 10^{16}$. In Figure 3.1. we have plotted the reliability of the component as a function of time.

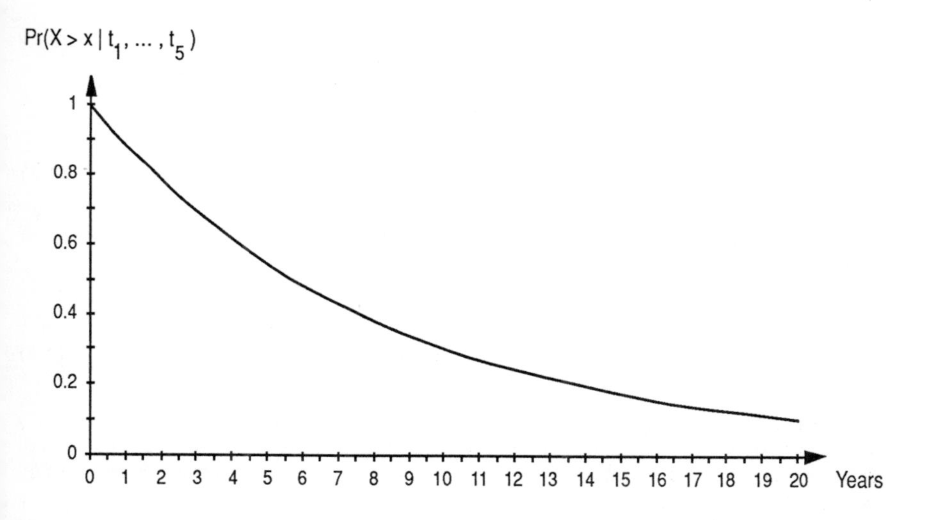

Figure 3.1. *The reliability of the component as a function of time.*

Note that the calculated t_{j1}, t_{j2} and t_{j3} for expert j are quite close to each other, $j = 1, \ldots, 5$ indicating that the model fits the data well. As a consequence, unless one suspects some sort of systematic bias in the experts' judgements, the estimated t_j's should be close to the "true" values. Thus neglecting the uncertainty about the t_j's appears to be a reasonable approximation in this case.

4. ESTIMATION OF OTHER MODEL PARAMETERS

We now turn to a situation where the distribution $q(x|\theta) = \text{norm}(x|\theta, \sigma)$, the normal density with expectation θ and standard deviation σ. As the vague distribution for θ, one uses $g_0(\theta) \equiv 1$.

For simplicity we assume that the sets $A_1, \ldots A_k$ are disjoint. [Generalizations to the more interesting cases where the A_j's overlap are in principle straight forward but more involved computationally. In order to focus on the main point in this section: estimation of model parameters, we have chosen to work with the simpler situation.]

Following the procedure from the previous sections, we see that in this case the DM has to assess subjectively the cardinalities of the sets $A_1, \ldots A_k$, as well as parameter σ. In this section, however, we shall see how these parameters can be estimated from the information gathered from the experts.

Under the above assumptions it is easily seen that the fractiles assessed by the j-th expert should be fitted to a density of the following form:

$$q_j(x|\boldsymbol{x}(A_j)) = \text{norm}\left(x|t_j, \sigma\sqrt{\frac{n_j+1}{n_j}}\right), \quad j = 1, \ldots, k \tag{4.1}$$

where n_j is the number of elements in the set A_j and the statistic t_j is given by:

$$t_j = t_j(\boldsymbol{x}(A_j)) = \frac{1}{n_j} \sum_{i \in A_j} x_i. \tag{4.2}$$

Moreover, the distribution of x given all the imaginary observations, $\boldsymbol{x}$, is given by:

$$f(x|\boldsymbol{x}) = \text{norm}\left(x|t, \sigma\sqrt{\frac{n+1}{n}}\right) \tag{4.3}$$

where as usually n is the total number of imaginar observations, and t is the mean value of these observations.

Since the A_j's in the case are assumed to be disjoint, it follows that:

$$t = \frac{1}{n} \sum_{j=1}^{k} n_j t_j, \text{ and } n = \sum_{j=1}^{k} n_j \tag{4.4}$$

Assume first that the DM has assessed the n_j's (and hence n as well) and σ subjectively. He then tries to derive information about the t_j's (and hence t as well) by fitting distributions of the form (4.1) to the given fractiles. As in Section 3 we assume that the fractiles provided by the experts, i.e. the $\boldsymbol{z}_j$'s satisfy the equation (3.1) for a suitable set of probabilities $\pi_1, \ldots \pi_r$ specified by the DM. Moreover, let $u_1, \ldots u_r$ be the corresponding fractiles of the standard normal distribution. That is, if U is normally distributed with expectation 0 and standard deviation 1, the u_s's satisfy:

$$\Pr(U > u_s) = \pi_s, \quad s = 1, \ldots, r. \tag{4.5}$$

For convenience we assume that the π_s's are chosen such that the mean value of the u_s's is zero. [Indeed it is very natural to do so anyway.]

Since we have that:

$$\Pr(X > x|\boldsymbol{x}(A_j)) = \Pr\left(U > \frac{z_{js} - t_j}{\sigma\sqrt{(n_j+1)/n_j}}\right) = \pi_s, \quad s = 1, \ldots, r, \quad j = 1, \ldots, k. \tag{4.6}$$

it is reasonable to determine $t_1, \ldots, t_k$ such that:

$$\frac{z_{js} - t_j}{\sigma\sqrt{(n_j + 1)/n_j}} \approx u_s, \quad s = 1, \ldots, r \text{ and } j = 1, \ldots, k. \tag{4.7}$$

By solving (4.7) with respect to t_j we get that:

$$t_j \approx z_{js} - u_s\sigma\sqrt{(n_j + 1)/n_j} \quad s = 1, \ldots, r \text{ and } j = 1, \ldots, k. \tag{4.8}$$

More precisely, the t_j are determined such that Q_j, given by:

$$Q_j = \sum_{s=1}^{r}\left(t_j - \left(z_{js} - u_s\sigma\sqrt{(n_j + 1)/n_j}\right)\right)^2 \tag{4.9}$$

is minimized, Hence, (using the fact that the u_s's have zero mean) we get that:

$$t_j = \frac{1}{r}\sum_{s=1}^{r} z_{js}, \quad j = 1, \ldots, k. \tag{4.10}$$

Having derived the t_j's, the sufficient statistic, t for x may be calculated using (4.4). Thus, the posterior distribution for X given the t_j's is equal to the distribution of X given x, i.e. the distribution given in (4.3).

Now assume that the DM does not want to assess the parameters $n_1, \ldots, n_k$ and σ subjectively. Instead the DM wishes to derive as much as possible information about these parameters from the experts' judgements. However, since the fitted distributions depened on these parameters only through $\sigma[(n_j + 1)/n_j]^{1/2}, j = 1, \ldots, k$, it is clear that it is impossible to derive both σ and the n_j's from the given information. Thus, the DM has to ask the experts to provide some more information. The question now is: What kind of information should he ask for? It will obviously not help to get more fractiles of the distributions on X, since the problem of deriving σ and the n_j's is independent of s, the number fractiles from each of the experts. Furthermore, the DM can hardly ask the experts to assess σ and the n_j's directly, since these parameters only are parts of the $DM's$ model, and hence probably not very meaningful quantities to the experts.

A possible solution to this problem, however, is to ask each expert the following question:

"Suppose that You (Expert j) are given an observation $Y_j = y_j$ of a quantity "similar" to the uncertain quantity X. How would You update your uncertainty about X given this information?".

The precise mathematical interpretation of the statement that X an Y_j are "similar" is of course that X and Y_j exchangable. However, unless the expert is familiar with such technical terms, it is probably wiser to use a more common expression.

If a question like the above is used in an interview of the experts, the quantity y_j should be replaced by a specific (but fictitions) number. A simple choise is for each expert to let y_j be equal to the 50%-fractile of his first uncertainty judgements. Then the expert just needs to update the spread and not the location of his previous assessments.

Assume now that each expert is asked the above question, and let $w_1, \ldots, w_k$ be the updated fractiles for expert 1 up to expert k respectively. In order to model this updating, we assume that given θ, $X_1, \ldots, X_n, Y_1, \ldots, Y_k$ and X are independent and normally distributed with expectation θ and standard deviation σ. Supposing that the experts' updated judgements can be reasonably well described by this model, the w_j's should be fitted to a density of the following form:

$$q_j^*(x|\boldsymbol{x}(A_j), y_j) = \text{norm}\left(x|\frac{n_j t_j}{n_j+1} + \frac{y_j}{n_j+1}, \sigma\sqrt{\frac{n_j+2}{n_j+1}}\right), \quad j = 1, \ldots k \tag{4.11}$$

Given both the z_j's and the w_j's, it is now possible to derive all the parameters of the model as follows:

Assume that the DM specifies k sets of integers $N_1, \ldots, N_k$, reflecting the areas he believes that respectively $n_1, \ldots, n_k$ belongs to. For each $\boldsymbol{n} = (n_1, \ldots, n_k) \in N_1 \times \ldots \times N_k$, the DM determines $t_1, \ldots, t_k$ and σ such that $Q(\boldsymbol{n}) = Q_1(\boldsymbol{n}) + Q_2(\boldsymbol{n})$ is minimized, where $Q_1(\boldsymbol{n})$ and $Q_2(\boldsymbol{n})$ are given by:

$$\begin{aligned} Q_1(\boldsymbol{n}) &= \sum_{j=1}^{k}\sum_{s=1}^{r}\left(\sigma\sqrt{\frac{n_j+1}{n_j}}u_s + t_j - z_{js}\right)^2 \\ Q_2(\boldsymbol{n}) &= \sum_{j=1}^{k}\sum_{s=1}^{r}\left(\sigma\sqrt{\frac{n_j+2}{n_j+1}}u_s + \frac{n_j t_j}{n_j+1} + \frac{y_j}{n_j+1} - w_{js}\right)^2. \end{aligned} \tag{4.12}$$

[The expressions $Q_1(\boldsymbol{n})$ and $Q_2(\boldsymbol{n})$ can be motivated in a similiar fashion as the Q_j-s given in (4.9)]

The minimizing values of $t_1, \ldots, t_k$ and σ are given by:

$$\begin{aligned} t_j &= \frac{\frac{1}{r}\sum_{s=1}^{r}(z_{js} + w_{js}) - \frac{y_j}{n_j+1}}{\left(1 + \frac{n_j}{n_j+1}\right)}, \quad j = 1, \ldots, k, \text{ and} \\ \sigma &= \frac{\sum_{s=1}^{r}\sum_{j=1}^{k} u_s\left(z_{js}\sqrt{\frac{n_j+1}{n_j}} + w_{js}\sqrt{\frac{n_j+2}{n_j+1}}\right)}{\left(\sum_{s=1}^{r} u_s^2\right)\sum_{j=1}^{k}\left(\frac{n_j+1}{n_j} + \frac{n_j+2}{n_j+1}\right)} \end{aligned} \tag{4.13}$$

By inserting the optimal values of $t_1, \ldots, t_2$ and σ into $Q(\boldsymbol{n})$, the minimal $Q(\boldsymbol{n})$-value is calculated for this particular value of $\boldsymbol{n}$. An estimate of $\boldsymbol{n}$ is then obtained as the vector minimizing this minimal $Q(\boldsymbol{n})$-value, while estimates of $t_1, \ldots, t_2$ and σ are calculated by inserting this vector into (4.13).

5. CONCLUDING REMARKS

In the present paper we have investigated a model for combining experts' judgements about some uncertain quantity X. The model is specially designed for cases where the available information about X, can be interpreted as a set of observations $X_1, \ldots, X_n$ such that it is reasonable to assess that $X_1, X_1, \ldots, X_n$ are exchangeable quantities.

In two particular cases we have suggested some approximate methods for deriving the necessary parameters of the model from the given judgements. Athough these methods appear to produce reasonable results, we still feel that they should be used with caution. Especially, in a practical application of these models, one should investigate carefully how well models fit the given data.

In a more thorough study it is perhaps desirable to introduce some sort of error structure on the given data in order to get a better impression of the performance of the estimates. If an error structure is established, it is also possible to incorporate uncertainty about the fitted parameters into the calculations. However, this typically increases the computational problems considerably.

In Section 4 we showed that in a normal case it is possible to estimate certain parameters of the model by questioning the experts about certain additional information. Practical experiences with this method shows that the question one has to ask to get this information, very easily is misunderstood by the experts. Hence, when using such a method, one should spend quite a lot of time explaining the exact meaning of the question.

REFERENCES

Clemen, R.T. (1987). Combining overlapping Information, *Management Sci.* **33**, 373–380.

Genest, C. and Zidek. J. (1986). Combining Probability Distributions: A Critique and an Annotated Bibliography, *Statistical Sci* **1**, 114–48.

Huseby, A. B. (1986). Combining Experts' Opinions, A Retrospective Approach. *Tech Rep.*, Center for Industrial Research.

Lindley, D. V. (1983). Reconciliation of Probability Distributions. *Operations Res.* **31**, 866–880.

Winkler, R. L. (1968). The Consensus of Subjective Probability Distributions. *Management Sci.* **15**, B61–B75.

Winkler, R. L. (1981). Combining Probability Distributions from Dependent Information Sources. *Management Sci.* **27**, 479–488.

BAYESIAN STATISTICS 3, pp. 653–656
J. M. Bernardo, M. H. DeGroot, D. V. Lindley and A. F. M. Smith, (Eds.)

Bayesian Estimation of the Variance Components in a General Random Linear Model

H. JELENKOWSKA
Lublin University

SUMMARY

The general linear model, with random effects, is considered. Using a non-informative prior for the variance components and a normal distribution with mean $\mu = (\mu_1, \ldots, \mu_c)$ and covariance matrix $\Sigma = \text{diag}\,(\sigma_1^2 P_{m_1}, \ldots, \sigma_c^2 P_{m_c})$ for the random effects, the conditional and joint marginal posterior distributions of the random effects, are obtained. The joint posterior distribution for the following vector of ratios of variance components

$$\left(\frac{\sigma^2}{\sigma_1^2}, \frac{\sigma_1^2}{\sigma_2^2}, \ldots, \frac{\sigma_{c-1}^2}{\sigma_c^2}\right)$$

is given as an inverted t-dimensional Dirichlet distribution.

Keywords: GENERAL LINEAR MODEL; RANDOM EFFECTS; VARIANCE COMPONENTS; MULTIPLE t DISTRIBUTION; INVERTED t-DIMENSIONAL DIRICHLET DISTRIBUTION; POSTERIOR DISTRIBUTION.

1. MODEL AND ASSUMPTIONS

Consider a general random linear model

$$Y = \sum_{i=1}^{c} U_i b_i + e = Ub + e, \tag{1.1}$$

where Y is a $(n \times 1)$ vector of observations,

U_i is a $(n \times m_i)$ known design matrix of rank m_i, $i = 1, \ldots, c$,

b_i is a $(m_i \times 1)$ vector of unobservable random effects,

e is a $n \times 1$ vector of error terms,

$b = (b_1', \ldots, b_c')_{m \times 1}$, $U = (U_1, \ldots, U_c)_{n \times m}$, and

$m = m_1 + m_2 + \cdots + m_c$.

Assume, that the vectors $b_1, b_2, \ldots, b_c$, and e are independent and vector e has a normal distribution with mean vector 0 and covariance matrix $\sigma^2 I_n$.

Prior information for the parameters of the model is introduced in two stages.

First, the conditional prior distribution for the random effects b, given the variance components σ_i^2 is a normal distribution with mean vector $\mu = (\mu, \ldots, \mu_c)$ and covariance matrix $\Sigma = \text{diag}\,(\sigma_1^2 P_{m_1}, \ldots, \sigma_c^2 P_{m_c})$, where the μ_i are known m_i-dimensional vectors and P_{m_i} is a symmetric and full rank matrix.

At the second stage, prior information for the variance components $\sigma^2, \sigma_1^2, \ldots, \sigma_c^2$ is provided. Suppose σ^2, σ_i^2 are independent and they have a non-informative reference prior of the following form

$$p(\sigma^2, \sigma_1^2, \ldots, \sigma_c^2) \propto \sigma^{-2} \prod_{i=1}^{c} \sigma_i^{-2}. \tag{1.2}$$

The joint prior density of the parameters is

$$\begin{aligned} p(b, \sigma^2, \sigma_1^2, \ldots, \sigma_c^2) &\propto \prod_{i=1}^{c} (\sigma_i^2)^{-(\frac{m_i}{2}+1)} \\ &\times \exp -\frac{\sigma_i^{-2}}{2} \left[(b_i - \mu_i)' p_{m_i}^{-1} (b_i - \mu_i) \right], \end{aligned} \tag{1.3}$$

when $\sigma_i^2 > 0$, $\sigma^2 > 0$, and $b_i \in R^{m_i}$, $i = 1, 2, \ldots, c$.

From (1.1), the distribution of Y is normal with mean Ub and covariance matrix $\sigma^2 I_n$, thus the likelihood function is

$$L(b, \sigma^2/Y) \propto (\sigma^2)^{-\frac{n}{2}} \exp -\frac{\sigma^{-2}}{2} (Y - Ub)'(Y - Ub) \tag{1.4}$$

when $b \in R^m$, and $\sigma^2 > 0$.

Combining the likelihood function with the prior density and using Bayes' theorem the joint posterior density of the parameters is obtained as

$$\begin{aligned} p(b, \sigma^2, \sigma_1^2, \ldots, \sigma_c^2/Y) &\propto (\sigma^2)^{-(\frac{n}{2}+1)} \\ &\times \exp -\frac{1}{2\sigma^2} \left[(Y - Ub)'(Y - Ub) \right] \prod_{i=1}^{c} (\sigma_i^2)^{-(\frac{m_i}{2}+1)} \\ &\times \exp -\frac{1}{2\sigma_i^2} \left[(b_i - \mu_i)' P_{m_i}^{-1} (b_i - \mu_i) \right], \end{aligned} \tag{1.5}$$

when $Y \in R^n$, $b \in R^m$, σ^2, $\sigma_1^2, \ldots, \sigma_c^2 > 0$.

2. POSTERIOR DISTRIBUTIONS OF THE VECTOR b

The function (1.5) can be rewritten as

$$\begin{aligned} p(b, \sigma^2, \sigma_1^2, \ldots, \sigma_c^2/Y) &\propto (\sigma^2)^{-(\frac{n}{2}+1)} \\ &\times \exp -\frac{1}{2\sigma^2} \left[Y'Y - \hat{b}'U' \cdot U\hat{b} + (b - \hat{b})'U' \cdot U(b - \hat{b}) \right] \\ &\times \prod_{i=1}^{c} (\sigma_i^2)^{-(\frac{m_i}{2}+1)} \exp -\frac{1}{2\sigma_i^2} \left[(b_i - \mu_i)' \cdot P_{m_i}^{-1} (b_i - \mu_i) \right], \end{aligned} \tag{2.1}$$

where $\hat{b} = (U'U)^- U'Y$ and $(U'U)^-$ is the Moore-Penrose generalized inverse of $U'U$.

Integrating (2.1) with respect to $\sigma^2, \sigma_1^2, \ldots, \sigma_c^2$, the marginal density function of the a posteriori distribuition of b is given by

$$\begin{aligned} p(b/Y) &\propto \left[Y'Y - \hat{b}'U'U\hat{b} + (b - \hat{b})'U'U(b - \hat{b}) \right]^{\frac{n}{2}} \\ &\times \prod_{i=1}^{2} \left[(b_i - \mu_i)' P_{m_i}^{-1} (b_i - \mu_i) \right]^{-\frac{m_i}{2}}, b \in R^m. \end{aligned} \tag{2.2}$$

This proves the following.

Theorem 1. *The marginal posterior density function of* b *is a poly-t density, given by following form*

$$p(b/Y) \propto \left[1+(b-\hat{b})'A(b-\hat{b})\right]^{-\frac{n}{2}} \prod_{i=1}^{c} \left[1+(b_i-\hat{\hat{b}}_i)'A_i(b_1-\hat{\hat{b}}_i)\right]^{-\frac{m_i}{2}} \tag{2.3}$$

where

$$A = \frac{U'U}{Y'Y - \hat{b}'U'U\hat{b}}, \qquad A_i = \frac{P_{m_i}^{-1}}{\mu_i \cdot \mu_i - \hat{\hat{b}}_i \cdot P_{m_i}^{-1}\hat{\hat{b}}_i}, \qquad \hat{\hat{b}}_i = P_{m_i}\mu_i.$$

Since the poly-t distribution is in general multimodal and asymmetric, it may be approximated by a multivariate normal with the same first two moments as the multivariate t-distribution. Using this approximation with each of the c factors (2.3), one may show the density of b may be approximated by

$$p(b/Y) \propto \exp -\frac{1}{2}\left[(b-b^*)'A^*(b-b^*)\right], \quad b \in R^m, \tag{2.4}$$

where

$$\begin{aligned} A^* &= A_1 + A_2, \\ b^* &= (A^*)^{-1}A_1\hat{b}, \\ A_1 &= \frac{(n-m-2)U'U}{Y'Y - \hat{b}U'U\hat{b}} \\ A_2 &= \{\text{DIAG}\,(A_i)\}. \end{aligned}$$

The matrix A_2 is $m \times m$ block diagonal with i-th diagonal matrix A_1.

Thus the posterior distribution of b is approximately normal with mean vector b^* and precision matrix A^*. The mean of this distribution b^* is a point estimator of the vector b.

3. POSTERIOR DISTRIBUTION OF THE VARIANCE COMPONENTS

For an exact Bayesian analysis of the variance components, the marginal posterior distribution of the variance components is needed. This can be obtained from (2.1) by integration with respect to b but in this case we have to use approximations, because the joint posterior density of $\sigma^2, \sigma_1^2, \ldots, \sigma_c^2$ is very hard to deal with. Thus, using the results of Tiao and Box (1973) we can derive the posterior distribution of variance components as

$$p(\sigma^2, \sigma_1^2, \ldots, \sigma_c^2/Y) \propto \frac{1}{\gamma} \prod_{i=1}^{c} f_{i\chi^{-2}}\left(\frac{\sigma_i^2}{S_i}\right) f_{\chi^{-2}}\left(\frac{\sigma^2}{S}\right), \tag{3.1}$$

where $f_{\chi^{-2}}$ is the inverted chi square density function and

$$S = Y'Y \quad S_i = \hat{b}_i U'U\hat{b}_i \quad \text{and} \quad \hat{b}_i = (U_i'U_i)^{-}U_i'Y,$$

where $\gamma = S\prod_{i=1}^{c} S_i P_c\left(\frac{S_1}{S}, \ldots, \frac{S_c}{S}\right)$ and $P_c\left(\frac{S_1}{S}, \ldots, \frac{S_c}{S}\right)$ is the cumulative probability function of an inverted c-dimensional Dirichlet distribution. The function (3.1) is not analytically intractable.

The posterior distribution of the following variables

$$X_1 = \frac{\sigma^2}{\sigma_1^2}, \quad X_2 = \frac{\sigma_1^2}{\sigma_2^2}, \ldots, \quad X_c = \frac{\sigma_{c-1}^2}{\sigma_c^2} \tag{3.2}$$

can be obtained from (3.1) by making a transformation from $(\sigma^2, \sigma_1^2, \ldots, \sigma_c^2)$ to $\left(\frac{\sigma^2}{\sigma_1^2}, \frac{\sigma_1^2}{\sigma_2^2}, \ldots, \frac{\sigma_{c-1}^2}{\sigma_c^2}, \sigma^2\right)$ and integrating out σ^2, as

$$
\begin{aligned}
p(X_1, X_2, \ldots, X_c/\boldsymbol{Y} &\propto \frac{\Gamma\left(\frac{1}{2}\sum_{i=1}^c m_i\right)}{\prod_{i=1}^c \Gamma\left(\frac{m_i}{2}\right)} \frac{S_c}{S} \left(\frac{X_1 S_1}{S}\right)^{\frac{m_1}{2}} \\
&\times \left(\frac{X_1 X_2 S_2}{S}\right)^{\frac{m_2}{2}} \cdots \left(\frac{X_1 \ldots X_c S_c}{S}\right)^{\frac{m_c}{2}-1} \\
&\times \left(1 + \frac{X_1 S_1}{S} + \frac{X_1 X_2 S_2}{S} + \cdots + \frac{X_1 \ldots X_c S_c}{S}\right)^{-\frac{1}{2}} \sum_{i=1}^c m_i
\end{aligned}
\tag{3.3}
$$

Thus

$$
W_1 = \frac{X_1 S_1}{S}, \quad W_2 = \frac{X_1 X_2 S_2}{S}, \ldots, \quad W_c = \frac{X_1 X_2 \ldots X_c S_c}{S},
$$

have a joint inverted Dirichlet distribution truncated at $\frac{S_1}{S}$ for w_1, $\frac{S_2}{S}$ for w_2 and $\frac{S_c}{S}$ for w_c and given by

$$
\begin{aligned}
f(w_1, w_2, \ldots, w_c/\boldsymbol{Y}) &\propto \frac{\Gamma\left(\frac{1}{2}\sum_{i=1}^c m_i\right)}{\prod_{i=1}^c \Gamma\left(\frac{m_i}{2}\right)} w_1^{\frac{m_1}{2}-1} w_2^{\frac{m_2}{2}-1} \ldots w_c^{\frac{m_c}{2}-1} \\
&\times \left(1 + \sum_{i=1}^c w_i\right)^{-\frac{1}{2}\sum_{i=1}^2 m_i}, \\
0 < w_1 < \frac{S_1}{S}, &\quad 0 < w_2 < \frac{S_2}{S}, \ldots, \quad 0 < w_c < \frac{S_c}{S}
\end{aligned}
\tag{3.4}
$$

Using (3.3) it follows for

$$
\begin{aligned}
&k_1 < \frac{m_1}{2}, \; k_2 < \frac{m_2}{2} + k_1, \ldots, k_c < \frac{m_{c-1}}{2} + k_{c-1} \\
&\mu_{k_1}, \ldots, k_c = E\left(X_1^{k_1} X_2^{k_2} \ldots X_c^{k_c}/\boldsymbol{Y}\right) \\
&= \left(\frac{S}{S_1}\right)^{k_1} \left(\frac{S_1}{S_2}\right)^{k_1} \cdots \left(\frac{S_{c-1}}{S_c}\right)^{k_c} \Gamma\left(\frac{m_1}{2} - k_1\right) \\
&\times \Gamma\left(\frac{m_1}{2} + k_1 - k_2\right) \cdots \Gamma\left(\frac{m_{c-1}}{2} + k_{c-1} - k_c\right) \frac{\Gamma\left(\frac{m_c}{2} + k_c\right) P_c^{k_1 \ldots k_c}}{\prod_{i=1}^c \Gamma\left(\frac{m_i}{2}\right) P_c^{0 \ldots 0}}
\end{aligned}
\tag{3.5}
$$

where $P_c^{k_1 \ldots k_c} = P_c\left(\frac{S_1}{S}, \frac{S_2}{S}, \ldots, \frac{S_c}{S}; \frac{m_1}{2} - k_1, \ldots, \frac{m_c}{2} + k_2\right)$.

REFERENCES

Box, G. E. P. and G. C. Tiao (1973). *Bayesian Inference in Statistical Analysis.* Reading, Mass.: Addison-Wesley.

Broemeling, L. D. (1985). *Bayesian Analysis of Linear Models.* New York: Dekker.

Hartley, H. O. and Rao J. N. K. (1967). Maximum likelihood estimation for the mixed analysis of variance model. *Biometrica* **54**, 93–108.

Henderson, C. R. (1953). Estimation of variance components. *Biometrica* **9**, 226–252.

Kleffe, J. (1977). Optimal estimation of variance components. *Sankhya B.* **39**, 211–244.

Lindley, D. V. and Smith, A. F. M. (1972). Bayes estimates for the linear model. *J. Roy. Statist. Soc. B* **34**, 1–18.

Searle, S. R. (1971). *Linear Models.* New York: Wiley.

Searle, S. R. (1978). A summary of recently developed methods of estimating variance components. *Tech. Rep.* BU-338. Biometrics Unit, Cornell University, New York.

Smith, A. F. M. (1973). Bayes estimates in one-way and two-way models. *Biometrika* **60**, 319–329.

BAYESIAN STATISTICS 3, pp. 657–663
J. M. Bernardo, M. H. DeGroot, D. V. Lindley and A. F. M. Smith, (Eds.)

A Heteroscedastic Hierarchical Model

W. S. JEWELL
University of California at Berkeley

SUMMARY

Hierarchical models are important in Bayesian prediction because they enable the use of collateral data from related risks with exchangeable parameters. The Lindley and Smith normal-normal-normal model with random means show clearly how the linear predictive mean for a single risk is improved by the availability of cohort data. However, this model has the disadvantage that the predictive density is homoscedastic, that is, the posterior variance depends only on the design (number of risks and number of samples). In most applications, one would assume that the variance also depended upon the data values.

One can change the variances at each level into random parameters, but this modifies the predictive mean formulae and leads to messy results in general. This paper describes an extended normal model with variances that are quadratic in the data, and with the additional advantage that the linear mean formulae are unchanged.

Keywords: HIERARCHICAL MODELS; COLLATERAL DATA; BAYESIAN PREDICTION; HETEROSCEDASTIC VARIANCES; CREDIBILITY FORMULAE.

1. INTRODUCTION

Consider an *individual risk* (#1), characterized by an unknown *risk parameter*, $\tilde{\theta}_1$, from which n_1 i.i.d. *sample observations*, $\mathcal{D}_1 = \{x_{1t}; (t = 1, 2, \ldots, n_1)\}$, are available; the problem is to predict a *future observation*, say $\tilde{w}_1 = \tilde{x}_{1,n_1+1}$. Given the *model density*, $p(x_{1t}|\theta_1)$, and *prior density*, $p(\theta_1)$, finding the *forecast density*, $p(w_1|\mathcal{D})$, is then a simple exercise in Bayes' law.

For a variety of simple likelihoods and priors (Jewell (1974) (1975a)), the forecast mean turns out to be a linear function of the data:

$$\mathcal{E}\{\tilde{w}_1|\mathcal{D}_1\} = f_1(\mathcal{D}_1) = (1 - z_1)m + z_1(\Sigma x_{1t}/n_1), \tag{1.1}$$

where the mixing coefficient,

$$z_1 = n_1/(n_1 + (e/d)), \tag{1.2}$$

is called the *credibility factor*, and the three required marginal moments are:

$$m = \mathcal{E}\mathcal{E}\{\tilde{x}_{1t}|\tilde{\theta}_1\}; \quad e = \mathcal{E}\mathcal{V}\{\tilde{x}_{1t}|\tilde{\theta}_1\}; \quad d = \mathcal{V}\mathcal{E}\{\tilde{x}_{1t}|\tilde{\theta}_1\}. \tag{1.3}$$

The *credibility formula*, $f_1(\mathcal{D}_1)$, has an obvious interpretation as a mixture of the prior mean with sample mean, according to a learning curve, z_1, which tends towards unity with increasing sample size at a rate governed by a *time constant*, (e/d). $f_1(\mathcal{D}_1)$ is a robust formula in the sense that it is also the *best least-squares fit* to the true $\mathcal{E}\{\tilde{w}_1|\mathcal{D}_1\}$ for arbitrary $p(x_{1t}|\theta_1)$ and $p(\theta_1)$ (Bühlmann (1967)).

In many applications, the number of samples from risk #1 will be small, but there may be additional information available from *related risks*, that is, $(\tilde{x}_{it})$ charecterized by a *different* risk parameter, $\tilde{\theta}_i$, but by the *same* form of model density, $p(x|\theta)$ $(i = 2, 3, \ldots, r)$. For example, in insurance we may have a portfolio of risks which, a priori, are similar in nature, as measured by some risk classification scheme. Or, in health statistics we may have a cohort of apparently similar lives, with the same ages, heritage, diet, etc. Of course, if these risk parameters were independent, with the same density $p(\theta)$, then the *collateral data* would have no predictive value.

However, if we assume that $(\tilde{\theta}_1, \tilde{\theta}_2, \ldots, \tilde{\theta}_r)$ are *exchangeable*, we keep the assumed similarity between risks, yet introduce correlation between the risks in a natural way; this adds another unknown parameter, $\tilde{\varphi}$, which characterizes additional uncertainty at the portfolio or cohort level. In other words, the prior $p(\theta)$ is changed to a *conditional prior* $p(\theta|\varphi)$, and we assume a *hyperprior density*, $p(\varphi)$, is known, from which we can calculate the exchangeable *joint prior*:

$$p(\theta_1, \theta_2, \ldots, \theta_r) = \int \Pi p(\theta_i|\varphi) p(\varphi) d\varphi. \tag{1.4}$$

The resulting three-level structure is called a *hierarchical model*.

We can also use credibility theory to find the best linear predictive mean in the least-squares sense for a hierarchical model with arbitrary $p(x|\theta)$, $p(\theta|\varphi)$ and $p(\varphi)$, using the *total cohort data*, $\mathcal{D} = \{x_{it}; (i = 1, 2, \ldots, r)\ (t = 1, 2, \ldots, n_i)\}$ Jewell (1975b) shows that (1.1) becomes a combination of two credibility-like forecasts:

$$\begin{aligned} \mathcal{E}\{\tilde{w}_1|\mathcal{D}\} \approx f_1(\mathcal{D}) &= (1 - z_1) f_0(\mathcal{D}) + z_1 y_1 \\ f_0(\mathcal{D}) &= (1 - z_0) m + z_0 Y_0; \end{aligned} \tag{1.5}$$

with $r + 1$ linear *sufficient statistics*:

$$y_i = (\Sigma x_{it}/n_i); \quad Y_0 = (\Sigma z_i y_i / \Sigma z_j); \tag{1.6}$$

and $r + 1$ credibility factors:

$$z_i = \frac{n_i}{n_i + (f/g)}; \quad z_0 = \frac{\Sigma z_i}{\Sigma z_j + (g/h)}; \tag{1.7}$$

for each risk i and for the portfolio as a whole. The four required marginal moments are:

$$\begin{aligned} m &= \mathcal{E}\mathcal{E}\mathcal{E}\{\tilde{x}_{it}|\tilde{\theta}_i|\tilde{\varphi}\}; \quad & f &= \mathcal{E}\mathcal{E}\mathcal{V}\{\tilde{x}_{it}|\tilde{\theta}_i|\tilde{\varphi}\}; \\ g &= \mathcal{E}\mathcal{V}\mathcal{E}\{\tilde{x}_{it}|\tilde{\theta}_i|\tilde{\varphi}\}; \quad & h &= \mathcal{V}\mathcal{E}\mathcal{E}\{\tilde{x}_{it}|\tilde{\theta}_i|\tilde{\varphi}\}; \end{aligned} \tag{1.8}$$

where multiple operators and conditioning are to be paired "inside out". $f_0(\mathcal{D})$ is the credibility approximation to the portfolio mean, $\mathcal{E}\{\tilde{\varphi}|\mathcal{D}\}$.

Unfortunately, the number of cases in which explicit analytic results can be obtained from hierarchical models is quite limited. As far as we know, the only case in which the predictive individual mean is linear in the individual and collateral data is in a simplified form of the random-mean normal-normal-normal linear model of Lindley and Smith (1972).

Although full-distributional results are available from this model, and (1.5) is now an exact result, their approach has the major drawback that all of the posterior and predictive (co)variances are *homoscedastic* in the data, by which we mean that they depend only upon the $r + 1$ *sampling design parameters* (n_i, r), and not upon the actual data values. (This can be seen in (2.4) (2.9) (2.11) and (2.13) below by holding the precisions constant.) This is, we believe, a serious limitation, since we would expect in a general setting to be able to learn about any unknown variances from sufficient portfolio data. In what follows, we generalize the normal-normal-normal model to obtain heteroscedastic forecast formulae, while retaining the simplicity of the credible mean forecast (1.5). The formulation also clarifies the numerical integration that would be necessary to extend the generalization further.

2. A GENERALIZED HIERARCHICAL MODEL

To obtain heteroscedastical results from the Lindley and Smith model, we permit the variances at each level to be random quantities. Specifically, for risk i at time t, we assume:

$$
\begin{aligned}
(\tilde{x}_{it}|\mu_i,\varphi,\omega,\gamma,\eta) &\sim (\tilde{x}_{it}|\mu_i,\omega) \sim No(\mu_i;\omega^{-1}); \\
&\qquad (i = 1,2,\ldots,r) \\
(\tilde{\mu}_i|\varphi\gamma,\eta) &\sim (\tilde{\mu}_i|\varphi\gamma) \sim No(\varphi;\gamma^{-1}); \\
&\qquad (i = 1,2,\ldots,R) \\
(\tilde{\varphi}|\eta) &\sim \tilde{\varphi} \sim No(m;\eta^{-1});
\end{aligned}
\tag{2.1}
$$

where $\tilde{\mu}$ and $\tilde{\varphi}$ are the usual $r+1$ random mean parameters, and $\tilde{\omega}$, $\tilde{\gamma}$, and $\tilde{\eta}$ are new random precisions, corresponding to f^{-1}, g^{-1}, and h^{-1} in (1.8), respectively. The variable length observed data, $\mathcal{D} = \{x_{it}\}$, is now to be used to estimate *all* of the unknown parameters, $\tilde{\Theta} = (\tilde{\mu}_i,\tilde{\varphi};\tilde{\omega},\tilde{\gamma},\tilde{\eta})$, plus make predictions of future $\tilde{x}_{it}$. We assume that the three precision parameters, $\tilde{\Omega} = (\tilde{\omega},\tilde{\gamma},\tilde{\eta})$, are statistically independent of the mean parameters, and temporarily governed by some general prior joint density, $p(\Omega)$. Box and Tiao (1973 Chapter 5) refer to this structure as a three-component random effects model with a single batch sample.

Using the statistics $y_i = (\Sigma x_{it}/n_i)$ and $y_{ii} = (\Sigma x_{it}^2/n_i)$, we find the joint density of the data and the mean parameters, conditional on the precisions, as:

$$
\begin{aligned}
p(\mathcal{D},\mu,\varphi|\Omega) = &\left(\frac{\omega}{2\pi}\right)^{1/2\Sigma n_i}\left(\frac{\gamma}{2\pi}\right)^{1/2r}\left(\frac{\eta}{2\pi}\right)^{1/2}\times \\
&\exp\{-1/2\omega\sum_i[n_i y_{ii} - 2\mu_i n_i y_i + n_i\mu_i^2]\}\times \\
&\exp\{-1/2\gamma\sum_i[\mu_i-\varphi]^2 - 1/2\eta[\varphi-m]^2\}.
\end{aligned}
\tag{2.2}
$$

The various conditional densities can then be extracted in the usual tedious way. We summarize the main results, without proof.

(1). The $(\tilde{\mu}_i|\varphi,\Omega,\mathcal{D})$ are independent and normally distributed, with:

$$\mathcal{E}\{\tilde{\mu}_i|\varphi,\Omega,\mathcal{D}\} = f_i(\varphi,\Omega,\mathcal{D}) = [1-z_i(\omega,\gamma)]\varphi + z_i(\omega,\gamma)y_i; \tag{2.3}$$

$$\mathcal{V}\{\tilde{\mu}_i|\varphi,\Omega,\mathcal{D}\} = (\gamma+\omega n_i)^{-1} = \frac{1}{\gamma}[1-z_i(\omega,\gamma)]; \tag{2.4}$$

where we have defined (conditional) *individual credibilities*:

$$z_i(\varphi,\Omega) = z_i(\omega,\gamma) = \left(\frac{\omega n_i}{\gamma+n_i}\right). \tag{2.5}$$

(2). $(\tilde{\varphi}|\Omega,\mathcal{D})$ is also normally distributed, with mean:

$$\mathcal{E}\{\tilde{\varphi}|\Omega,\mathcal{D}\} = f_0(\Omega,\mathcal{D}) = [1-z_0(\Omega)]m + z_0(\Omega)Y_0(\Omega); \tag{2.6}$$

where we define a (conditional) *cohort credibility factor*:

$$z_0(\Omega) = \frac{\gamma\Sigma z_i(\omega,\gamma)}{\eta+\gamma\Sigma z_j(\omega,\gamma)}; \tag{2.7}$$

and the (conditional) *credibility-weighted average observation*:

$$Y_0(\Omega) = Y_0(\omega, \gamma) = \frac{\Sigma z_i(\omega, \gamma) y_i}{\Sigma z_j(\omega, \gamma)}. \tag{2.8}$$

The variance is:

$$\mathcal{V}\{\tilde{\varphi}|\Omega, \mathcal{D}\} = \mathcal{V}\{\tilde{\varphi}|\Omega\}(\eta + \gamma\Sigma z_i(\omega, \gamma))^{-1} = \frac{1}{\eta}[1 - z_0(\Omega)]. \tag{2.9}$$

(3). Combining the above, we find that $(\tilde{\mu}|\Omega, \mathcal{D})$ is *multinormal*, with moments, say, $\mathcal{E}\{\tilde{\mu}|\Omega, \mathcal{D}\} = f(\Omega, \mathcal{D})$ and $\mathcal{V}\{\tilde{\mu}|\Omega, \mathcal{D}\} = \Sigma(\Omega, \mathcal{D})$ and:

$$[f(\Omega, \mathcal{D})]_i = f_i(\Omega, \mathcal{D}) = [1 - z_i(\omega, \gamma)] f_0(\Omega, \mathcal{D}) + z_i(\omega, \gamma) y_i; \tag{2.10}$$

$$\begin{aligned}[\Sigma(\Omega, \mathcal{D})]_{i,j} &= \sigma_{ij}(\Omega, \mathcal{D}) = \\ &\gamma^{-1}[1 - z_i(\omega, \gamma)]\delta_{ij} + \eta^{-1}[1 - z_i(\omega, \gamma)][1 - z_j(\omega, \gamma)][1 - z_0(\Omega)];\end{aligned} \tag{2.11}$$

where $\delta_{ij} = 1$ if $i = j$, zero otherwise.

(4). Predictive densities are important in applications. Let $\tilde{w}_{it} = \tilde{x}_{it}$ for any $t > n_i$. Then it can be shown that $p(w|\Omega, \mathcal{D})$ is also *multinormal*, with:

$$\mathcal{E}\{\tilde{w}_{it}|\Omega, \mathcal{D}\} = f_i(\Omega, \mathcal{D}); \tag{2.12}$$

$$\mathcal{C}\{\tilde{w}_{it}; \tilde{w}_{ju}|\Omega, \mathcal{D}\} = \omega^{-1}\delta_{ij}\delta_{tu} + \sigma_{ij}(\Omega, \mathcal{D}); \tag{2.13}$$

for all $t > n_i$ and $t > n_j$.

The remainder of (2.2) can now be matched with the still-general prior $p(\Omega)$ to give the conditional posterior joint density of the precision parameters as:

$$p(\Omega|\mathcal{D}) \propto p(\Omega)\omega^{1/2\Sigma n_i}\{\prod_i [1 - z_i(\omega, \gamma)]^{1/2}\}\{[1 - z_0(\omega, \gamma)]^{1/2}\}\exp\{-\tfrac{1}{2}A(\Omega, \mathcal{D})\}, \tag{2.14}$$

where the function A is:

$$A(\Omega, \mathcal{D}) = \omega\Sigma_i[n_i y_{ii} - n_i z_i(\omega, \gamma) y_i^2 + \eta\{m^2 - [1 - z_0(\Omega)]^{-1} f_0^2(\Omega, \mathcal{D})\}. \tag{2.15a}$$

By expansion, and the use of the new definitions:

$$\begin{aligned} N &= \Sigma n_i; \bar{z}(\omega, \gamma) = N^{-1}\Sigma n_i z_i(\omega, \gamma); \\ Z(\omega, \gamma) &= \Sigma z_i(\omega, \gamma); Y_{00} = N^{-1}\Sigma n_i y_{ii};\end{aligned} \tag{2.16}$$

the exponent can be put into several equivalent forms, such as:

$$\begin{aligned} A(\Omega, \mathcal{D}) = \omega[NY_{00} - \Sigma n_i z_i(\omega, \gamma) y_i^2] + \eta z_0(\Omega)[m - Y_0(\omega, \gamma)]^2 \\ - \gamma Z(\omega, \gamma) Y_0^2(\omega, \gamma),\end{aligned} \tag{2.15b}$$

or

$$\begin{aligned} A(\Omega, \mathcal{D}) = N\omega\Big\{ [Y_{00} - Y_0^2(\omega, \gamma)] + N^{-1}\Sigma n_i z_i(\omega, \gamma)\,[Y_0^2(\omega, \gamma) - Y_i^2] \\ + [1 - \bar{z}(\omega, \gamma)][1 - z_0(\Omega)][m - Y_0(\omega, \gamma)]^2 \Big\}.\end{aligned} \tag{2.15c}$$

We apologize for this heavy notation, but in developing special models it is important to know explicitly where the various unknown precisions occur. Certainly, we have now achieved our objective of a heteroscedastic model, since the final results desired can now be obtained by a triple integration using (2.14). But it is equally clear from the above formulae that there is no "magic" general prior $p(\Omega)$ that will give a tractable, closed-form posterior density, $p(\Omega|\mathcal{D})$!

3. A GENERAL HIERARCHICAL MODEL WITH ALL VARIANCES LINKED

The motivation for this paper arose from unpublished work with H. Bühlmann on credibility approximations for the hierarchical (co)variances, in which second-order portfolio statistics augment a second-moment forecast using individual data (Jewell and Schnieper (1985) have the non-hierarchical version). In order to validate these approximations, an analytic test case was sought, with some variation on the Lindley and Smith model as a natural candidate. To make our heteroscedastic generalization useful, we must now specialize $p(\Omega)$ to give more tractable results. Furthermore, to keep the link with the credibility formula (1.5), it desirable not to disturb the linear data dependence of the predictive means (2.3), (26), (2.10) and (2.12).

By examining these formulae, we see that only the ratios of the precisions are involved in the credibility factors. This suggests that an appropriate simplification would be to link all three precisions together by assuming:

$$\tilde{\gamma} = n_0\tilde{\omega}; \tilde{\eta} = r_0\tilde{\gamma} = r_0 n_0 \tilde{\omega}, \tag{3.1}$$

for appropriate positive n_0 and r_0. This is equivalent to assuming that we have very tight knowledge about the division of total variance among the three levels of the model, but are uncertain about the value of the total variance. This assumption not only simplifies the various credibility factors:

$$z_i = n_i/(n_0 + n_i); Z = \Sigma z_i; z_0 = Z/(r_0 + Z); \bar{z} = \Sigma n_i z_i/N;, \tag{3.2}$$

but also makes the (f_i), f_0, and Y_0 independent of $\tilde{\omega}$!

More importantly, (2.14) now simplifies to:

$$p(\omega|\mathcal{D})\alpha p(\omega)\omega^{1/2N}\exp(-1/2A(\omega, \mathcal{D})),$$

with $A(\omega, \mathcal{D}) = \omega N B(\mathcal{D})$, say. Thus, the unconditioning on ω required in the general formulae of Section 2 reduces to simple expectations over $(\tilde{\omega}|\mathcal{D})$, and we can use the natural conjugate *Gamma* prior for $p(\omega)$! For completeness, we now display all of the final results for the linked-variance, Gamma prior model, using notation developed previously.

3.1 Posterior Parameter Results

If $p(\omega)$ is $Ga(\alpha, \beta)$, then from (2.15c) and (3.3) we see that $p(\omega|\mathcal{D})$ is $G(\alpha', \beta')$, with updated parameters:

$$\begin{aligned} \alpha' &= \alpha + 1/2N; \quad \beta' = \beta + 1/2NB(\mathcal{D}); \\ B(\mathcal{D}) &= [Y_{00} - Y_0^2] + (1 - \bar{z})(1 - z_0[m - Y_0]^2) + N^{-1}\Sigma n_i z_i [Y_0^2 - y_i^2]. \end{aligned} \tag{3.4}$$

The most important moment formula for our purposes is $\mathcal{E}\{\tilde{\omega}^{-1}|\mathcal{D}\} = \alpha'/(\beta' - 1)$, which can be written:

$$\mathcal{E}\{\tilde{\omega}^{-1}|\mathcal{D}\} = (1 - z_*)E\{\tilde{\omega}^{-1}\} + z_* B(\mathcal{D}), \tag{3.5}$$

where β is eliminated in favor of the prior average variance, and a *variance credibility factor* has been defined:

$$z_* = N/(N + 2(\alpha - 1)). \tag{3.6}$$

Posterior-to-data moments of the portfolio mean are:

$$\mathcal{E}\{\tilde{\varphi}|\mathcal{D}\} = f_0(\mathcal{D}) = (1 - z_0)m + z_0 Y_0; \tag{3.7}$$

$$\mathcal{V}\{\tilde{\varphi}|\mathcal{D}\} = \left[\frac{1 - z_0}{r_0 n_0}\right]\mathcal{E}\{\tilde{\omega}^{-1}|\mathcal{D}\}. \tag{3.8}$$

In fact, because of the position of ω in the simplified version of $p(\varphi|\Omega,\mathcal{D})$, we see that $p(\varphi|\mathcal{D})$ must be a one-dimensional Student-t density, with $2\alpha' = 2\alpha + N$ degress of freedom, and the above moments.

Progressing to the individual risk means, we find:

$$\mathcal{E}\{\tilde{\mu}_i|\mathcal{D}\} = f_i(\mathcal{D}) = (1-z_i)f_0(\mathcal{D}) + z_i y_i; \tag{3.9}$$

$$\mathcal{C}\{\tilde{\mu}_i;\tilde{\mu}_j|\mathcal{D}\} = \sigma_{ij}(\mathcal{D}) = \left[\frac{(1-z_i)}{n_0}\sigma_{ij} + \frac{(1-z_0)(1-z_i)(1-z_j)}{r_0 n_0}\right]\mathcal{E}\{\tilde{\omega}^{-1}|\mathcal{D}\}. \tag{3.10}$$

Again, one can argue that $p(\mu|\mathcal{D})$ is an r-dimensional Student-t density, with $2\alpha + N$ degrees of freedom.

3.2. *Posterior Predictive Results*

Passing to the forecasts of future values, we find simply:

$$\mathcal{E}\{\tilde{w}_{it}|\mathcal{D}\} = f_i(\mathcal{D}) = (1-z_i)m + z_i y_i, \tag{3.11}$$

$$\mathcal{C}\{\tilde{w}_{it};\tilde{w}_{ju}|\mathcal{D}\} = \sigma_{ij}\sigma_{tu}\mathcal{E}\{\tilde{\omega}^{-1}|\mathcal{D}\} + \sigma_{ij}(\mathcal{D}). \tag{3.12}$$

From these, various multivariate *Student-t* densities can be generated "over future times" for a single risk, "over risks" at one future epoch, or for various mixtures of risks and epochs.

4. EQUAL DATA LENGTHS

Notation simplifies dramatically if all data record lengths are the same, i.e., $n_i = n$ with $i = 1, 2, \ldots, r$. Then:

$$N = nr; z_i = z = \bar{z} = n/(n_0+n); Z = rz; z_0 = rz(r_0+rz); z_* = nr/(nr+2(\alpha-1)), \tag{4.1}$$

and the overall statistics Y_0 and Y_{00} can be replaced by the simpler versions:

$$y_0 = \Sigma y_i/r = \Sigma\Sigma x_{it}/nr; y_{00} = \Sigma y_{ii}/r = \Sigma\Sigma x_{it}^2/nr. \tag{4.2}$$

Then (3.4) simplifies to:

$$B(\mathcal{D}) = y_{00} - (1-z)y_0^2 - z(\Sigma y_i^2/r) + (1-z)(1-z_0)(m-y_0)^2. \tag{4.3}$$

5. MORE GENERAL PRECISION PRIORS

To illustrate the difficulties caused by general priors, consider the single random variance model analyzed by Berger (1985, 4.6), in which it is assumed (in our notation) that $\omega = f^{-1}$ and $\eta = h^{-1}$ are known with certainty, so that the unknown parameters are $(\tilde{\mu}, \tilde{\varphi}, \tilde{\gamma})$. For simplicity, assume all data lengths are equal to n (Berger has $n = 1$).

There are now two conditional credibility factors to consider:

$$z_1(\gamma) = \frac{n}{f\gamma + n}; \quad z_0 = \frac{(rnh)\gamma}{(rnh+f)\gamma + n}; \tag{5.1}$$

plus the individual statistics (y_i, y_{ii}) and the simplified portfolio statistics (y_0, y_{00}). Other factors in (2.16) become: $N = nr$, $\bar{z}(\gamma) = z_1(\gamma)$, and $Z(\gamma) = rz_1(\gamma)$.

The posterior precision density (2.14) is now a one dimensional formula that can be "simplified" to reveal the structure on γ:

$$p(\gamma|\mathcal{D}) \propto p(\gamma)\left[\frac{f\gamma}{f\gamma+n}\right]^{1/2}\left[\frac{f\gamma+n}{(rnh+f)\gamma+n}\right]^{1/2}\exp(-1/2C(\gamma)); \tag{5.2}$$

$$C(\gamma) = (rnf^{-1})\left[\left[\frac{n}{f\gamma+n}\right][(y_0^2-\Sigma y_i^2/r)] + \left[\frac{f\gamma}{(rnh+f)\gamma+n}\right](m-y_0)^2\right]. \tag{5.3}$$

It is clear that no analytical prior will lead to a tractable posterior density, so that numerical methods will have to be used in this case. Our notation reveals that the essential work in finding the posterior and predictive first and second moments will be in taking expectations of various powers of the two credibility factors, plus finding $\mathcal{E}\{\tilde{\gamma}^{-1}|\mathcal{D})$. The situation is similar if either $\tilde{\omega}$ or $\tilde{\eta}$ is the unknown variance, except that, in the latter case, only z_0 will be modified by the data.

Box and Tiao (1973 Chapter 5) tackle the general problem (2.1) by expressing the likelihood in terms of three unknown variances: $\tilde{\sigma}_1^2 = \tilde{\omega}^{-1}$; $\tilde{\sigma}_{12}^2 = \tilde{\sigma}_1^2 + n\tilde{\gamma}^{-1}$; and $\tilde{\sigma}_{123}^2 = \tilde{\sigma}_{12}^2 + nr\tilde{\eta}^{-1}$; and in terms of $\tilde{\theta} = \tilde{m}$, also considered to be a random quantity. A non-informative prior on $\tilde{\theta}$ and the log-variances is then assumed. The computations, which must be carried out over a space constrained by the relations $\tilde{\sigma}_{123}^2 > \tilde{\sigma}_{12}^2 > \tilde{\sigma}_{12}^2$, are extremely complex. We believe that our notation, using credibility factors, would give a better foundation for setting up numerical approximations for more general $p(\Omega)$. It should also be noted that many of the issues raised by Box and Tiao, such as the ratios of the variances, are trivial in our specialization.

6. FINAL REMARKS

The results just presented can be interpreted in a variety of different ways. For instance, the time constants, n_0 and r_0 introduced in (3.1) have the interpretation that the prior information is equivalent to r_0 risks each having n_0 data samples with common value, m. The credibility factors themselves reveal the "learning curves" appropriate to the various statistics. Further interpretations and explanations are in the original report, available from the authour.

The formulae can easily be extended to the case where additional data is available at the highest level, from different portfolios or "batches". The model can also be generalized, with only an increase in notational detail, to the general linear structure of Lindley and Smith (1972). Finally, it is clear that the linkage introduced in (3.1) will also simplify heteroscedastic hierarchical models with more than three layers.

I would be interested in hearing from others who have studied second-moment hierarchies, especially in non-normal models.

REFERENCES

Berger, J. O. (1985). *Statistical Decision Theory and Bayesian Analysis.* New York: Springer-Verlag, 2nd ed.

Box, G. E. P. and Tiao, G. C. (1973). *Bayesian Inference in Statistical Analysis.* Reading: Addison-Wesley.

Bühlmann, H. (1967). Experience rating credibility. *ASTIN Bulletin* **4**, 199–207.

Bühlmann, H. and Jewell, W. S.(1987). Hierarchical credibility revisited. *Bulletin Association of Swiss Actuaries* **1**, 35–54.

Jewell, W. S. (1974). Credible means are exact Bayesian for simple exponential families. *ASTIN Bulletin* **8**, 77–90.

Jewell, W. S. (1975a). Regularity conditions for exact credibility. *ASTIN Bulletin* **8**, 336–341.

Jewell, W. S. (1975b). The use of collateral data in credibility theory: a hierarchical model. *Giornale dell'Istituto Italiano degli Attuari* **38**, 1–16.

Jewell, W. S. and Schnieper, R. (1985). Credibility approximations for Bayesian prediction of second moments. *ASTIN Bulletin* **15**, 103–121.

Lindley, D. V. and Smith A. F. M. (1972). Bayes estimates for the linear model. *J. Roy. Statist. Soc. B* , 1–41.

BAYESIAN STATISTICS 3, pp. 665–668
J. M. Bernardo, M. H. DeGroot, D. V. Lindley and A. F. M. Smith, (Eds.)

Algorithmic Diagnosis of Appendicitis Using Bayes' Theorem and Logistic Regression

G. LINDBERG and G. FENYÖ
Huddinge Univ. Hospital and *Nacka Hospital, Stockolm*

SUMMARY

An algorithm for the diagnosis of *acute appendicitis* was constructed using data from medical history taking, physical examination, and laboratory tests of 746 patients with acute abdominal pain. The diagnostic value of symptoms and signs was expressed as a score based on the logarithm of the likelihood ratio $p(S+|D+)/p(S+|D-)$. The scores were then entered into a logistic regression model with presence of acute appendicitis as the dependent variable. The resulting β-values were used to adjust the likelihood ratio scores.

Reclassification showed the final algorithm to be well calibrated and propective testing showed that the algorithm is a reliable tool for identifying both patients in need of an operation and patients with little risk for acute appendicitis.

Keywords: ACUTE APPENDICITIS; DIAGNOSIS; DIAGNOSTIC VALUE; LOGISTIC REGRESSION; BAYES' THEOREM; COMPUTER AIDED DIAGNOSIS.

1. INTRODUCTION

Early recognition of acute appendicitis is important in order to avoid potentially dangerous complications of this disease if left untreated. This has led many surgeons to adopt a strategy which means accepting a high proportion of negative laparotomies in order to minimize the number of missed cases with acute appendicitis. A large number of studies have been performed to assess the value of symptoms and signs for identifying patients with acute appendicitis and other diseases causing acute abdominal pain and several attempts have been made to utilize the digital computer for carrying out the tedious calculations necessary. The most widely applied system for computer-aided diagnosis of acute abdominal pain is one developed by deDombal and his co-workers in Leeds (deDombal *et al.*, 1972).

Most systems developed in this area have been built on Bayes' theorem assuming independence between variables. As has been shown previously this assumption is not valid for much medical data (Spiegelhalter, 1982). One way to get around this problem is to express Bayes' theorem as an additive model using logarithms of likelihood ratios and to subject this model to a logistic regression analysis. Recent work from Denmark has succeeded in implementing this combination for the differential diagnosis of jaundice (Matzen *et al.*, 1984).

The present work examines the possibility of developing a simple scoring system for the diagnosis of acute appendicitis.

2. PATIENTS

The study material consisted of 830 consecutive patients presenting with acute abdominal pain (abdominal pain with a duration of less than one week) to Nacka Hospital in Stockholm. Data were registered on admission by the attending surgeon who filled in a specially constructed questionnaire with 22 indicants from medical history, physical examination and laboratory tests. In 84 patients, one or more findings had not been registered and these were therefore excluded from the analysis. The remaining 746 patients formed the construction sample.

The evaluation sample consisted of 96 patients with acute abdominal pain in a new consecutive series from the same hospital.

VARIABLE	INDICANT	LOG LIKELIHOOD	BETA	SCORE
SEX	MALE	0.628	1.21	+8
	FEMALE	–0.598		–7
WBC	<9000	–1.576	0.83	–14
	9000–13900	0.201		+2
	≥14000	1.142		+10
DURATION OF PAIN	<12 HRS	0.359	0.85	+3
	12–48 HRS	–0.054		±0
	>48 HRS	–1.390		–12
PROGRESSION OF PAIN	PRESENT	0.329	0.86	+3
	ABSENT	–0.559		–5
RELOCATION OF PAIN	PRESENT	0.818	0.88	+7
	ABSENT	–0.995		–9
VOMITING	PRESENT	0.720	0.91	+7
	ABSENT	–0.500		–5
AGGRAVATION WITH COUGHING	PRESENT	0.527	0.67	+4
	ABSENT	–1.590		–11
REBOUND TENDERNESS	PRESENT	0.752	0.66	+5
	ABSENT	–1.562		–10
RIGIDITY	PRESENT	1.756	0.74	+13
	ABSENT	–0.544		–4
PAIN OUTSIDE RLQ	PRESENT	–0.537	0.92	–5
	ABSENT	0.391		+4

Table 1. *Likelihood ratios, beta values from logistic regression analysis, and final scores for the ten variables included in the algorithm for the appendicitis. RLQ = right lower quadrant. WBC = white blood cell count*

3. METHODS

The construction sample was divided into two groups according to the presence or absence of acute appendicitis. For each symptom and sign a SCORE reflecting its weight of evidence for the diagnosis of acute appendicitis was calculated from the likelihood ratio:

$$\text{SCORE} = 10 \times \ln[p(S+|D+)/p(S+|D-)]$$

$S+$ means symptom or sign present, $D+$ disease present, and $D-$ disease absent. These "raw" scores were then subjected to a stepwise logistic regression analysis with acute appendicitis as the dependent variable. The resulting β-values served to adjust the raw scores.

The adjusted scores were first tested with reclassification of the patients included in construction sample. Performance of the model was tested using the quadratic score of Hilden *et al.* (1978). In a second experiment the model was tested on a separate test sample. For this purpose the scores were tabulated in form of a simple pocket-chart (Figure 1).

Constant			-10
Sex	Male	+6	
	Female		-8
W.B.C.	≤ 8.9		-14
	9.0-13.9	+2	
	≥ 14.0	+10	
Pain duration	<24 hrs	+3	
	24-48 hrs	0	0
	>48 hrs		-12
Progression of pain	Yes	+3	
	No		-5
Relocation of pain	Yes	+7	
	No		-9
Vomiting	Yes	+7	
	No		-5
Aggravation w. cough	Yes	+4	
	No		-11
Rebound tenderness	Yes	+5	
	No		-10
Rigidity	Yes	+13	
	No		-4
Pain outside R.L.Q.	Yes		-6
	No	+4	
Total			
	SUM SCORE		

Date:....................
Time:....................
Sign:....................

Figure 1. *A pocket chart for the probabilistic diagnosis of acute appendicitis using scores for representing the diagnostic weight of evidence of indicants.*

4. RESULTS

Acute appendicitis was diagnosed in 219/746 patients. The Chi-squared test identified 18/22 symptoms, signs and test results with a significantly different incidence in the two groups.

Logistic regression analysis showed that in combination only ten of these were significant at the 0.01 level. The raw scores, betas, and adjusted scores for the ten indicants are shown in Table 1. The adjusted scores served to construct a simple pocket-chart (Figure 1).

Reclassification of the construction sample using the scores of the pocket-chart showed these to be well calibrated (Figure 2). The quadratic score was 0.92, slightly greater than its expected value of 0.91.

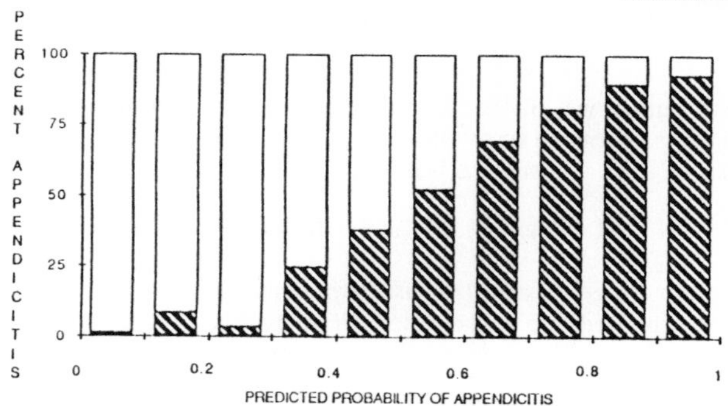

Figure 2. *Reclassification of the construction sample of 746 patients using an algorithm with 10 indicants.*

In the evaluation sample three groups of patients were identified on the basis of calculated scores. A *high risk* group with scores for acute appendicitis of ± 0 or more ($p(App) \geq 0.50$, a *low risk* group with scores of –26 ($p(App) < 0.07$) or less and an intermediate or *grey zone* group. The results in these groups are summarized in Table 2.

	PREDICTED RISK FOR APPENDICITIS		
	LOW RISK	GREY ZONE	HIGH RISK
TRUE DIAGNOSIS	SCORE $\leq$–26	–26$<$ SCORE $<$ 0	SCORE $\geq$ 0
ACUTE APPENDICITIS	0	3	24
OTHER	47	15	5

Table 2. *Results of applying the algorithm to the evaluation sample.*

5. DISCUSSION

An easy-to-use algorithm for probabilistic diagnosis of acute appendicitis has been developed. Contrary to previous work about probabilistic diagnosis in acute abdominal pain, the proposed algorithm takes into account only the diagnosis of acute appendicitis.

Previous work with the so called Independence Bayes' model has shown that interdependency of diagnostic data is common and that this leads to over-confident models. The quadratic score indicates that the proposed algorithm is highly discriminatory and the closeness of this score to its expected value indicates that the probabilities predicted from the algorithm are reliable.

The reliability of the predicted probabilities makes it possible to do a formal analysis of the diagnostic decision making in suspected acute appendicitis. A possible strategy would be to select for laparotomy all patients in the high risk group. In our evaluation sample this led to a negative laparotomy rate of only 11% in this group. Patients in the low risk group can probably be sent home, and observation reserved for those in the intermediate "grey zone".

REFERENCES

DeDombal, F. T., Leaper D. J., Staniland, J. R., McCann, A. P. and Horrocks, J. C. (1972). Computer-aided diagnosis of abdominal pain. *Br. Med. J.* **2**, 9–13.

Spiegelhalter, D. J. (1982). Statistical aids in clinical decision making. *Statistician* **31**, 19–36.

Matzen, P., Malchow-Moller, A., Hilden, J., Thomsen C., Svendsen, L.B., Gammelgaard, J. and Juhl, E. (1984). And the Copenhagen Computer Icterus Group. Differential diagnosis of jaundice: a pocket diagnostic chart. *Liver* **4**, 360–371.

Hilden, J., Habbema, J. D. F. and Bjerregaard, B. (1978). The measurement of performance in probabilistic diagnosis. III. Methods based on continuous functions of the diagnostic probabilities. *Methods Inf. Med.* **17**, 238–246.

BAYESIAN STATISTICS 3, pp. 669–675
J. M. Bernardo, M. H. DeGroot, D. V. Lindley and A. F. M. Smith, (Eds.)

Prior and Posterior Tail Comparisons

P. MAIN
Universidad Complutense, Madrid

SUMMARY

We consider the classification of distribution functions depending on their tail behaviour. Some comparisons between the prior and posterior distributions, for large values, are studied. Results that agree with intuitive considerations about the forms arising in different circumstances are established.

Keywords: CLASSIFICATION OF DISTRIBUTIONS; TAIL BEHAVIOUR; BAYESIAN INFERENCE.

1. INTRODUCTION

There are many Bayesian methods for outlier study. See, for example, the general review by Barnett and Lewis (1984), Ch. 12. The aspect considered in this paper is the analysis of tail-behaviour for the different distributions in the Bayesian approach.

Many authors, among them Dawid (1973) and Berger (1980), have pointed out the importance in the choice of the tail in Bayesian analysis. The effect of extreme observations is taken up by O'Hagan (1979) who proposes new definitions of outlier-proneness and outlier-resistance related to the notions introduced earlier by Green (1976). The basic distribution is said to be outlier-prone or outlier-resistant depending on the behaviour of the posterior distribution as the observation increases. He considers Bayesian estimation of a location parameter, so that an outlier is accommodated or ignored automatically depending on whether the basic distribution is outlier-resistant or outlier-prone, respectively. See also Goldstein (1982, 1983).

We now extend the classification of distributions considered by Green (1976), changing the outlier-proneness concept and proposing the new topic of outlier-neutrality.

Let X be a random variable with distribution function $F(\cdot)$ and density function $f(\cdot)$, having infinite endpoint. We can consider its right tail behaviour depending on the limit of the difference or the ratio of the two upper extremes $X_{n-1:n}$ and $X_{n:n}$, in random samples of size n. The notation will be $X_{k:n}$ for the k-th order statistic and

$$S_n = X_{n:n} - X_{n-1:n}$$
$$R_n = X_{n:n}/X_{n-1:n}.$$

Definition 1.1.

(i) *F is SR (outlier-**resistant** with the **sum**) if $S_n \xrightarrow{P} 0$ as $n \to \infty$.*

(ii) *F is SN (outlier-**neutral** with the **sum**) if*

$$\left.\begin{array}{l} S_n \equiv O_P(1) \\ S_n \not\equiv o_P(1) \end{array}\right\} \quad \textit{as } n \to \infty$$

(iii) *F is SP (outlier-**prone** with the **sum**) if $S_n \xrightarrow{P} \infty$ as $n \to \infty$.*

Definition 1.2.

(i) *F is RR (outlier-resistant with the ratio) if* $R_n \xrightarrow{P} 1$ *as* $n \to \infty$.

(ii) *F is RN (outlier-neutral with the ratio) if*

$$\left.\begin{array}{r} R_n \equiv O_P(1) \\ R_n - 1 \not\equiv o_P(1) \end{array}\right\} \quad \text{as } n \to \infty.$$

(iii) *F is RP (outlier-prone with the ratio) if* $R_n \xrightarrow{P} \infty$ *as* $n \to \infty$.

Left tail behaviour may be handled similarly replacing upper by lower extremes.

In the following we will use some helpful characterizations. The first follows as a corollary from a result of Schuster (1984).

Theorem 1.1. *Suppose the limit* $h = \lim_{x\to\infty} \frac{f(x)}{1-F(x)}$ *exists and is possibly infinite. Then*

(i) F *is* $SR \Leftrightarrow h = \infty$.

(ii) F *is* $SN \Leftrightarrow h = c$, *with* $0 < c < \infty$.

(iii) F *is* $SP \Leftrightarrow h = 0$.

We can now obtain a similar result but with the ratio of the two upper extremes just working with the logarithm of the basic random variable.

Theorem 1.2. *Suppose the limit* $k = \lim_{x\to\infty} \frac{xf(x)}{1-F(x)}$ *exists and is possibly infinite. Then*

(i) F *is* $RR \Leftrightarrow k = \infty$.

(ii) F *is* $RN \Leftrightarrow k = c$, *with* $0 < c < \infty$.

(iii) F *is* $RP \Leftrightarrow k = 0$.

Section 2 uses the definitions relative to the difference S_n in location parameter distributions. Prior and posterior tails are considered for comparisons. Some results that are in good agreement with intuitive ideas are provided. Illustrative examples are given at the end of that section.

Similar results are available in terms of ratio definitions for scale parameter distributions. In section 3 some indicative results in this line are stated briefly.

2. LOCATION PARAMETER DISTRIBUTIONS

Denote by X a random variable with density function $f_\theta(x) = f(x-\theta)$, $\theta \in \Theta \subset \Re$. If we define a prior density function $\pi(\theta)$, then the posterior density function

$$\pi(\theta|x) = \frac{f(x-\theta)\pi(\theta)}{p(x)},$$

where $p(x) = \int_{-\infty}^{+\infty} f(x-\theta)\pi(\theta)d\theta$, is the predictive density for x. The posterior right tail, expressed in terms of the prior one, is then

$$1 - \Pi(\theta|x) = (1-\Pi(\theta))\frac{\int_{-\infty}^{+\infty} f(x-u)\pi_\theta^R(u)du}{p(x)},$$

where $\pi_\theta^R(u)$ is the right truncated prior density in θ

$$\pi_\theta^R(u) = \begin{cases} 0 & u < \theta \\ \frac{\pi(u)}{1-\Pi(\theta)} & u \geq \theta \end{cases}$$

and $\Pi(\cdot)$, $\Pi(\cdot|x)$ the corresponding prior and posterior distribution functions.

Therefore

$$\frac{1-\Pi(\theta|x)}{1-\Pi(\theta)} = \frac{p_\theta^R(x)}{p(x)},$$

where $p_\theta^R(x) = \int_{-\infty}^{+\infty} f(x-u)\pi_\theta^R(u)du$ is the corresponding right truncated predictive density in θ for x.

Also,

$$\frac{\pi(\theta|x)}{1-\Pi(\theta|x)} = \frac{\pi(\theta)}{1-\Pi(\theta)}\frac{f(x-\theta)}{p_\theta^R(x)}$$

Since Theorem 1.1 refers to limits of the form $\lim_{x\to\infty}\frac{f(x)}{1-F(x)}$, if we want to compare the prior and posterior tail behaviour we have to study the factor $\frac{f(x-\theta)}{p_\theta^R(x)}$, for θ sufficiently large.

Suppose that $f(\cdot)$ is symmetric with infinite endpoint and monotone tails. Then

$$f(x-u) < f(x-\theta)$$

for every $u > \theta$ and θ sufficiently large.

If follows that

$$\int_\theta^\infty \frac{f(x-u)}{f(x-\theta)}\pi(u)du < (1-\Pi(\theta)).$$

Thus, for a prior distribution with an infinite endpoint, we have

$$p_\theta^R(x) = \int_\theta^\infty f(x-u)\frac{\pi(u)}{1-\Pi(\theta)}du < f(x-\theta),$$

for θ sufficiently large.

So that

$$\lim_{\theta\to\infty}\frac{f(x-\theta)}{p_\theta^R(x)} \geq 1,$$

Hence

$$\lim_{\theta\to\infty}\frac{\pi(\theta|x)}{1-\Pi(\theta|x)} \geq \lim_{\theta\to\infty}\frac{\pi(\theta)}{1-\Pi(\theta)} \quad (1)$$

Under the previous assumptions we can establish the following results.

Proposition 2.1.

(i) *If we have a SP posterior distribution, then the prior distribution has to be, necessarily, a SP distribution.*

(ii) *If we have a SR prior distribution, then we also get a SR posterior distribution.*

Proof. In both cases inequality (1) is used to obtain the implication.

(i) $\lim_{\theta\to\infty}\frac{\pi(\theta|x)}{1-\Pi(\theta|x)} = 0 \Rightarrow \lim_{\theta\to\infty}\frac{\pi(\theta)}{1-\Pi(\theta)} = 0.$

(ii) $\lim_{\theta\to\infty}\frac{\pi(\theta)}{1-\Pi(\theta)} = \infty \Rightarrow \lim_{\theta\to\infty}\frac{\pi(\theta|x)}{1-\Pi(\theta|x)} = \infty.$

Proposition 2.2.

(i) *If the prior distribution is SN, then the posterior distribution cannot be SP.*

(ii) *If the posterior distribution is SN, then the prior distribution cannot be SR.*

Proof. The proof is similar to that of Proposition 2.1.

Remark. It is easy to see how these results follow for the context of a random sample $x_1, \ldots, x_n$. Notice that the factor used for comparisons is now

$$\frac{p_\theta^R(x_1, \ldots, x_n)}{\prod_{i=1}^n f(x_i - \theta)} = \int_\theta^\infty \prod_{i=1}^n \frac{f(x_i - u)}{f(x_i - \theta)} \frac{\pi(u)}{1 - \Pi(\theta)} du.$$

However, the following results need some changes to be applied in the general case.

There are also some relations between the likelihood function and the posterior density behaviour in the right tail. Let us supposse a prior density with a monotone right tail and an infinite endpoint. We need also to assume that the likelihood function is symmetric.

Proposition 2.3.

(i) *With a SR likelihood we get a SR posterior distribution.*

(ii) *If the posterior distribution is SP, then the likelihood must be SP.*

(iii) *If the posterior distribution is SN, then the likelihood cannot be SR.*

(iv) *With a SN likelihood we cannot get a SR posterior distribution.*

Proof. If the prior density has a monotone right tail we have

$$\pi(u) < \pi(\theta),$$

for every $u > \theta$ and θ sufficiently large. Then

$$\frac{f(x-\theta)}{p_\theta^R(x)} = \frac{f(x-\theta)}{\int_\theta^\infty f(x-u)\frac{\pi(u)}{1-\Pi(\theta)}du} > \frac{1-\Pi(\theta)}{\pi(\theta)} \frac{f(x-\theta)}{\int_\theta^\infty f(x-u)du}$$

and so

$$\frac{\pi(\theta|x)}{1-\Pi(\theta|x)} > \frac{f(x-\theta)}{F(x-\theta)} = \frac{f(\theta-x)}{1-F(\theta-x)}.$$

Hence, the result follows by taking limits as θ increases to infinity.

Remark. These results lead us to confirm our intuitive considerations. If we obtain a heavy-tailed posterior distribution we can neither have started from a concentrated prior distribution nor have used a likelihood function with light tails. Similarly, if we take a prior density or a likelihood function with very light tails we obtain a posterior density with the same characteristics.

Notice that if we have a SP likelihood function and a SP prior distribution we always get a SP posterior distribution. However, for SN or SR prior distributions, even using a SP likelihood function, the posterior tail behaviour can be different to the prior one.

As explained above, some assumptions about SP likelihood functions will have to be required so that the prior and posterior tail behaviour are the same for SN and SR distributions. We consider prior distributions with infinite endpoint.

Theorem 2.1. *Let $f(\cdot)$ be a positive, symmetric about zero and continuous likelihood function with monotone tails. If for every $\varepsilon > 0$*

$$\lim_{x\to\infty} \frac{f(x+\varepsilon)}{f(x)} = 1,$$

then, the prior and posterior distributions have the same tail behaviour.

Proof. We just need to demonstrate it for SN and SR prior distributions.

As we stated before,

$$\frac{\pi(\theta|x)}{1-\Pi(\theta|x)} = \frac{\pi(\theta)}{1-\Pi(\theta)} \cdot \frac{f(x-\theta)}{p_\theta^R(x)}$$

To make comparisons we have to calculate

$$\lim_{\theta\to\infty} \frac{f(x-\theta)}{p_\theta^R(x)},$$

but

$$\frac{p_\theta^R(x)}{f(x-\theta)} = \frac{\int_\theta^\infty \frac{f(x-u)}{f(x-\theta)}\pi(u)du}{1-\Pi(\theta)},$$

and, by symmetry,

$$\frac{p_\theta^R(x)}{f(x-\theta)} = \int_0^\infty \frac{f(\theta-x+v)}{f(\theta-x)}\frac{\pi(\theta+v)}{1-\Pi(\theta)}dv.$$

On the other hand, for every $\varepsilon > 0$

$$\lim_{x\to\infty} \frac{f(x+\varepsilon)}{f(x)} = 1.$$

Then, this relation holds uniformly for ε in any fixed finite closed interval (Seneta (1976)).

Hence we have for $\varepsilon_0 > 0$

$$\lim_{\theta\to\infty} \frac{\int_0^{\varepsilon_0} \frac{f(\theta+v)}{f(\theta)}\pi(\theta+x+v)dv}{\Pi(\theta+x+\varepsilon_0)-\Pi(\theta+x)} = 1,$$

because of the uniform convergence. Furthermore,

$$\int_{\varepsilon_0}^\infty \frac{f(\theta+x)}{f(\theta)}\pi(\theta+x+v)dv < 1-\Pi(\theta+x+\varepsilon_0),$$

because of the tail monotony of $f(\cdot)$, for θ sufficiently large, so that it converges to zero as ε_0 tends to infinity.

Summing up, given $\varepsilon_0 > 0$

$$\int_0^{\varepsilon_0} \frac{f(\theta+v)}{f(\theta)}\frac{\pi(\theta+x+v)}{1-\Pi(\theta+x)}dv \sim 1 - \frac{1-\Pi(\theta+x+\varepsilon_0)}{1-\Pi(\theta+x)},$$

as θ tends to infinity.

But the prior distribution function has to satisfy one of the next conditions (Main (1986)).

For SN prior distributions,

$$\lim_{\theta\to\infty} \frac{1-\Pi(\theta+x+\varepsilon_0)}{1-\Pi(\theta+x)} = e^{-c\varepsilon_0} \quad (0 < c < \infty)$$

For SR prior distributions,

$$\lim_{\theta\to\infty} \frac{1-\Pi(\theta+x+\varepsilon_0)}{1-\Pi(\theta+x)} = 0.$$

In both cases, rewriting $\frac{p_\theta^R(x)}{f(x-\theta)}$ it follows that,

$$\lim_{\theta\to\infty}\frac{p_\theta^R(x)}{f(x-\theta)}=\lim_{\theta\to\infty}\int_0^\infty\frac{f(\theta+v)}{f(\theta)}\frac{\pi(\theta+x+v)}{1-\Pi(\theta+x)}dv=1.$$

Remark. Using L'Hopital's rule we have for every $\varepsilon>0$,

$$\lim_{x\to\infty}\frac{f(x+\varepsilon)}{f(x)}=1\Rightarrow\lim_{x\to\infty}\frac{1-F(x+\varepsilon)}{1-F(x)}=1,$$

and the last condition is equivalent to $F(\cdot)$ being a SP distribution (Main (1986)).

Example 1. Let $f(x)$ be the density function of a Student-t_m distribution

$$f(x)=K(m+x^2)^{-\frac{(m+1)}{2}}$$

This is known to be symmetric about the origin, positive and with monotone tails as required and also satisfies

$$\lim_{x\to\infty}\frac{f(x+\varepsilon)}{f(x)}=1\qquad\text{for every }\varepsilon>0$$

Now, let the prior distribution of the location parameter have an extreme value type distribution function.

$$\Pi(\theta)=\exp\{-e^{-\theta}\},\quad\theta\in\Re$$

and density function

$$\pi(\theta)=\exp\{-\theta-e^{-\theta}\},\quad\theta\in\Re.$$

Therefore it corresponds to a SN distribution

$$\lim_{\theta\to\infty}\frac{\pi(\theta)}{1-\Pi(\theta)}=1.$$

If we determine the posterior density function

$$\pi(\theta/x)\alpha e^{\theta-e^{-\theta}}[m+(x-\theta)^2]^{-\frac{m+1}{2}}$$

it follows that

$$\lim_{\theta\to\infty}\frac{\pi(\theta+\varepsilon/x)}{\pi(\theta/x)}=e^{-\varepsilon}\qquad\text{for every }\varepsilon>0.$$

Hence we have also a SN posterior distribution.

Example 2. Let us consider a normal distribution $N(\theta,1)$, therefore a SR distribution,

$$f(x-\theta)=\frac{1}{\sqrt{2\pi}}\exp\left\{-\frac{1}{2}(x-\theta)^2\right\},\quad x\in\Re.$$

If we have also the prior density

$$\pi(\theta)=\exp\{-\theta-e^{-\theta}\},\quad\theta\in\Re,$$

we get

$$\pi(\theta/x)\alpha\exp\left\{-\frac{\theta^2}{2}+\theta(x-1)-e^{-\theta}\right\},$$

and it is easily verified that this corresponds to a SR distribution in agreement with the results stated earlier.

3. SCALE PARAMETER DISTRIBUTIONS

Let $f_\theta(x) = \theta f(\theta x)$ be a density function with scale parameter $\theta > 0$ having a prior density $\pi(\theta)$. Then the posterior density has the form

$$\pi(\theta|x) = \frac{\theta f(\theta x)\pi(\theta)}{p(x)}.$$

Similarly to location parameter distributions we get

$$\frac{\theta\pi(\theta|x)}{1-\Pi(\theta|x)} = \frac{\theta f(\theta x)}{p_\theta^R(x)}\frac{\theta\pi(\theta)}{1-\Pi(\theta)},$$

so that the comparisons of the prior and posterior tail behaviour depend on the factor $\frac{\theta f(\theta x)}{p_\theta^R(x)}$ for θ sufficiently large.

Also,

$$\frac{p_\theta^R(x)}{\theta f(\theta x)} = \int_\theta^\infty \frac{uf(ux)}{\theta f(\theta x)}\frac{\pi(u)}{1-\Pi(\theta)}du$$

Here we have to use the ratio definitions and in a similar way to the above some relations between the prior and posterior tail behaviour can be established.

ACKNOWLEDGEMENTS

I would like to express my appreciation to Prof. M. A. Gómez Villegas who introduced me to this problem; to Prof. Lindley for helpful comments; and to Prof. H. Navarro for several useful conversations on this topic.

REFERENCES

Barnett, V. and Lewis, T. (1984). *Outliers in Statistical Data.* New York: Wiley.

Berger , J. O. (1980). *Statistical Decision Theory.* Springer-Verlag.

Dawid, A. P. (1973). Posterior expectations for large observations. *Biometrika* **60**, 664–667.

Goldstein, M. (1982). Contamination distributions. *Ann. Statist.* **10**, 174–183.

Goldstein, M. (1983). Outlier resistant distributions: where does the probability go?. *J. Roy. Statist. Soc. B* **45**, 355–357.

Green, R. F. (1976). Outlier-prone and outlier-resistant distributions. *J. Amer. Statist. Assoc.* **71**, 502–505.

Main, P. (1986). Clasificación de distribuciones y datos atípicos. *Ph. D. Thesis*, Univ. Complutense of Madrid.

O'Hagan, A. (1979). On outlier rejection phenomena in Bayes inference. *J. Roy. Statist. Soc. B* **41**, 358–367.

Seneta, E. (1976). Regularly varying functions. *Lecture Notes in Mathematics* **508**. Springer-Verlag.

Schuster, E. F. (1984). Classification of probability laws by tail behaviour. *J. Amer. Statist. Assoc.* **79**, 936–939.

BAYESIAN STATISTICS 3, pp. 677–696
J. M. Bernardo, M. H. DeGroot, D. V. Lindley and A. F. M. Smith, (Eds.)

Coherent Priors for Ordered Regressions

T. MÄKELÄINEN and P. J. BROWN
Univ. of Helsinki and *Univ. of Liverpool*

SUMMARY

A class of inverse Wishart priors is developed for a finite or countably infinite dimensional normal model with unknown covariance matrix. The point of view is that of regression of one variable on the others, and the model provides a setting for coherent estimation of all submodels containing less than the full regressor set. The beliefs of the researcher are supposed to imply that most of the explanation is due to small subsets of regressors. The beliefs are considered in the framework of the marginal multivariate T distribution.

The prior incorporates expectations concerning the performance of finite subsets of variables in prediction in two ways: (i) There is a best regressor set among those of any given size and the best sets are linearly ordered by inclusion, (ii) all conditional distributions arising on conditioning on a best regressor set are, up to change of scale and to second order, identical. The class of priors satisfying a condition of type (ii) is given a characterization, and a number of subclasses satisfying (i) is found. Among these one is put forward on the basis of its regression properties which are thought to be reasonable. The paper concludes with a discussion of the newly discovered determinism property of the inverse Wishart prior as well as its consequences for the application of our priors.

Keywords: NORMAL-INVERSE WHISHART MODELS; COHERENT SEQUENCES; REGRESSOR ORDERINGS; DETERMINISTIC PROCESSES; LATENT VARIABLES.

1. INTRODUCTION

We consider the situation of a researcher whose statistical data consists of observations on a variable of special interest, Y_0, and of a number, q, possibly infinite, of further variables Y_1, $Y_2, \ldots$ which can be used to fit a model to predict Y_0. In particular, assuming joint normality of $Y_0, Y_1, \ldots$ conditional on their covariance matrix, Y_0 will have linear regressions on any subset of $Y_1, Y_2, \ldots$. Such subsets correspond to different submodels having parameters of different dimensions. In the past the assignment of prior distributions to submodels has occasionally taken place without proper attention to their interconnections. For an example, in a context different from ours, showing that priors should be made to cohere over submodels see Lindley (1978) and Brown (1980).

Lindley (1978), and subsequently Dickey, Lindley and Press (1985), achieved the coherence with the jointly normal model by assigning consistent exchangeable inverse Wishart priors to the finite dimensional covariance matrices, and Dawid (1987) analyzed the general inverse Wishart priors having no such special structure. An important finding by Dawid is that, due to the inverse Wishart form of the prior, the process Y_k is implicitly deterministic, cf. Section 3 below. We follow these authors in using the normal-inverse Wishart model introduce a structural assumption reflecting beliefs different from those behind an exchangeable prior.

The present paper is an attempt to formulate a simple class of priors reflecting the beliefs that small subsets of the variables $Y_1, Y_2, \ldots$ can explain most of the explainable variation of Y_0. Throughout the main development the hyperparameters of the inverse Wishart distribution are examined via the joint marginal (or prior predictive) distribution of the variables Y_k.

Turning to a general discussion of our approach, assume

1. The variables are linearly ordered in such a way that for each p the first p variables $Y_1, Y_2, \ldots, Y_p$ constitute the best regressor set of size p.

The foremost merit of this assumption is its simplicity. For a few remarks concerning those problems in which this assumption would not be appropriate see Sections 5 and 6.

Suppose then that the researcher sets out to assess the reduction of the error of prediction to be brought by the addition of one variable, say Y_{p+1}, to the regressor set already consisting of the previous variables. To this end he must consider the joint conditional distribution of Y_0 and Y_{p+1} given $Y_1, \ldots, Y_p$. Suppose he is going to make such an assessment for each p. In order to utilize coherence he may as well expand the set Y_0, Y_{p+1} to include all the later variables, which will anyway be involved in the later assessments. We may call this stage of assessment the researcher's $(p+1)$-th prediction situation (along with Y_0 he is now predicting $Y_{p+1}, Y_{p+2}, \ldots$). We make the assumptions that

2. Any two prediction situations are assessed to be roughly similar up to an adjustment of their scales (cf. Section 6).
3. Successive additions to the regressor set are assessed to reduce the error variance for Y_0 at a constant rate.

These assumptions essentially concern the form of the marginal covariance matrix of $Y_0, Y_1, \ldots$ (cf. Section 6). It is the last assumption which makes it possible to have small sets of regressors accounting for most of the explanation for Y_0. The class of reduction schemes envisaged is simple and the only one we know of that leads in conjunction with the second assumption to a simple representation theorem (Section 7), which is important for the sequel.

Consider the prediction of Y_0 in the conditional distribution of $Y_0, Y_{p+1}, Y_{p+2}, \ldots$ given $Y_1, \ldots, Y_p$, a distribution in which Y_0 would be regressed on $Y_{p+1}, Y_{p+2}, \ldots$. Subject to the second assumption this problem is similar (up to scale) to the original unconditional prediction problem of using $Y_1, Y_2 \ldots$ to predict Y_0, the former problem being the same problem for the $(p+1)$-th prediction situation.

The second and third assumptions in particular imply that no expectations of a strong effect are admitted which are specific to Y_p and by which that variable would distinguish itself in the various prediction situations. Conversely, any variable which is expected to dramatically influence the prediction should be considered separately. In general all explanatory variables should initially be divided into two groups depending on whether their inclusion is, or is not, in doubt. The present work is then solely concerned with the marginal prior of the residual covariance matrix of the variables in doubt given those not in doubt.

Firstly, Section 5 gives, in passing, a simple partially exchangeable model recognizing the special role of Y_0 but without an ordering of the other variables. Next, neglecting only an additive random component of Y_0, the main class of priors incorporating the ordering are specified in Section 6. Section 7 begins the exploration of its implications by developing an equivalent latent variable representation. Particular simple structures involve the specification of just 3 hyperparameters and the implications of their specification are detailed in Section 8. Special attention is given to the theoretical regression coefficients, and their general behaviour, both in a fixed model and with increasing number of regressors, is found to be satisfactory. The implicit determinism is discussed here as well as in the concluding Section 9.

Though our priors imply the belief that just a few variables can explain most of the explainable variation of Y_0 we have not attempted to discuss the use of a small subset of regression variables. For ideas in that direction in the context of polynomial regression see Young (1977). Issues of model choice, see for example Jeffreys (1983), Lindley (1968) and Box (1980), are approached with our prior distribution in Mäkeläinen and Brown (1987). In that paper predictive analysis is provided using data from the model. The predictive distribution now depends on the posterior distribution, also of the normal-inverse Wishart form, in the same way in which it depends on the prior in the present study. Being a compromise between the prior distribution and the data the posterior distribution may in general lead to a modification of the original linear ordering of the regressors.

2. MATRIX-VARIATE DISTRIBUTIONS

We shall be following the notation introduced by Dawid (1981) for matrix-variate distributions. Throughout the paper bold face letters are used for matrices or vectors, and capitals for matrices or random variables. With $\boldsymbol{U}$ a matrix having independent standard normal entries, $\boldsymbol{K} + \mathcal{N}(\boldsymbol{\Lambda}, \boldsymbol{M})$ will stand for the matrix-variate normal distribution of $\boldsymbol{V} = \boldsymbol{K} + \boldsymbol{A}'\boldsymbol{U}\boldsymbol{B}$ where $\boldsymbol{K}$, $\boldsymbol{A}$, and $\boldsymbol{B}$ are fixed matrices satisfying $\boldsymbol{A}'\boldsymbol{A} = \boldsymbol{\Lambda}$ and $\boldsymbol{B}'\boldsymbol{B} = \boldsymbol{M}$. Thus $\boldsymbol{K}$ is the matrix mean of $\boldsymbol{V}$, and $\lambda_{ii}\boldsymbol{M}$ and $m_{jj}\boldsymbol{\Lambda}$ are the covariance matrices of the i-th row and the j-th column, respectively, of $\boldsymbol{V}$. If $\boldsymbol{U}$, furthermore, is of order $n \times p$ with $n \geq p$, the symbol $IW(\delta, \boldsymbol{M})$ with $\delta = n - p + 1$, will stand for the distribution of $\boldsymbol{B}'(\boldsymbol{U}'\boldsymbol{U})^{-1}\boldsymbol{B}$, an inverse Wishart distribution. The degrees of freedom parameter δ may be generalized, using the density function, to take on any positive real values. The matrix-variate T distribution $\boldsymbol{K} + T(\delta; \boldsymbol{\Lambda}, \boldsymbol{M})$ is the distribution of $\boldsymbol{T}$ where, conditional on Σ, say, and with $\Sigma \sim IW(\delta, \boldsymbol{M})$, $\boldsymbol{T}$ follows the $\boldsymbol{K} + \mathcal{N}(\boldsymbol{\Lambda}, \Sigma)$ distribution. Subscripts on symbols or distributions may be used to indicate the order of the variate.

3. THE MODEL

We shall denote the data matrix of n observations by $\boldsymbol{X}^q = (\boldsymbol{X}_0, \boldsymbol{X}_{(q)})$. Here $\boldsymbol{X}_0$ is the vector containing the observations on the regressand and $\boldsymbol{X}_{(q)}$ is the regressor matrix of q regressors. Following Lindley (1978) we shall consider random regressions and make the assumption that $\boldsymbol{X}^q$ is a sample from a $(q+1)$-variate normal distribution having, without loss of generality, mean zero. The assumption of random regressions is a necessary prerequisite if considerations of coherence are to be made over different sets of regressors. An exception are any regressors which are part of all the regressions involved. These regressors can always be taken as fixed.

The researcher also contemplates a future independent observation $\boldsymbol{Y}^q = (Y_0, \boldsymbol{Y}_{(q)})$ from the same distribution. Finally the covariance matrix of the distribution is taken to follow an inverse Wishart distribution with positive definite expectation $\boldsymbol{M}^q$, an assumption we shall shortly comment on.

To sum up, the assumptions made are, in the notation introduced by Dawid (1981), that

$$\boldsymbol{X}^q|\Sigma \sim \mathcal{N}_{n,q+1}(\boldsymbol{I}_n, \Sigma), \tag{3.1}$$

$$\Sigma \sim IW_{q+1}(\delta, (\delta - 2)\boldsymbol{M}^q), \qquad \delta > 2, \boldsymbol{M}^q > 0, \tag{3.2}$$

$$\boldsymbol{Y}^q|\Sigma \sim \mathcal{N}_{1,q+1}(1, \Sigma), \tag{3.3}$$

$$\boldsymbol{X}^q \textit{ and } \boldsymbol{Y}^q \textit{ are independent, given } \Sigma. \tag{3.4}$$

The assumption that the number of degrees of freedom δ exceeds 2 is there only to ensure that Σ has finite expectation. By leaving out the factor $\delta - 2$ of $\boldsymbol{M}^q$ any positive degrees of freedom could be accommodated.

In some contexts it may be natural to consider an infinite sequence of regressors. (This is the point of view of Dawid 1981, 1988.) For $q = 1, 2, \ldots$ the formulae (3.1)-(3.4) provide a model for this situation, of course with the understanding that for $p < q$ each of $\boldsymbol{X}^p, \boldsymbol{Y}^p$, and $\boldsymbol{M}^p$ is the appropriate submatrix of $\boldsymbol{X}^q, \boldsymbol{Y}^q$, and $\boldsymbol{M}^q$, respectively. This applies, in particular, to the subclass of models which is the principal object of our study in Section 8.

As is well-known the kind of beliefs which are expressible by means of model (3.1)-(3.2) are rather special (Press 1972, Lindley 1978). Thus the marginal distribution of $\boldsymbol{X}^q$ is

$$T_{n,q+1}(\delta; \boldsymbol{I}_n, (\delta-2)\boldsymbol{M}^q). \tag{3.5}$$

The matrix parameters of this distribution can be obtained via the variances

$$\text{Var}(\boldsymbol{s}'\boldsymbol{X}^q\boldsymbol{t}) = \boldsymbol{s}'\boldsymbol{s}\cdot\boldsymbol{t}'\boldsymbol{M}^q\boldsymbol{t}, \tag{3.6}$$

for instance. [In dimension 1×1 the distribution (3.5) becomes $\{(\delta-2)m_{00}/\delta\}^{\frac{1}{2}} \cdot t_\delta$, where t_δ denotes the Student(δ) distribution.] Formula (3.6) shows how there is essentially just one covariance matrix for the rows, $\boldsymbol{I}_n$, and one for the columns, $\boldsymbol{M}^q$. (This limitation is similar to a property of the Wishart distribution itself discussed by Press 1972.) Any other rotatable prior is also burdened by the same implication (Dawid, 1981). Moreover, for the inverse Wishart there is the disadvantage emphasized by Lindley (1978) that a single shape parameter, δ is to account for the uncertainty concerning each element of Σ [or for the tail behaviour of (3.5)]. Finally, Dawid (1988) has discovered that the infinite-dimensional Wishart prior gives probability 1 to the set of those covariance matrices of the infinite-dimensional normal distribution which permit exact prediction of Y_0 from the sequence $Y_1, Y_2, \ldots$, a phenomenon he calls the *determinism* of the Wishart distribution or the *implicit* determinism of the model. See Sections 8 and 9 and Subsection 8.3.

Noting these limitations our primary concern in this paper is the specification of special forms of $\boldsymbol{M}^q$ (cf. Section 6), also the marginal(or prior predictive) covariance matrix of $\boldsymbol{Y}^q$, The marginal distribution itself is given by

$$\boldsymbol{Y}^q \sim T_{1,q+1}(\delta; 1, (\delta-2)\boldsymbol{M}^q). \tag{3.7}$$

An alternative and much more restrictive interpretation of our work which strips the model of its parameters is to say that the main development is only concerned with the second order properties of the future data. In the infinite-dimensional model to be studied in Section 8 a distinction will also be made between the deterministic part of the *marginal* model, where most of the work is needed in this paper, and the complete marginal model (cf. Subsection 8.3). Both have finite marginal T distributions as in (3.7) and are called (infinite-dimensional) T distributions (Dawid 1981). Also both are marginal distributions of an implicity deterministic normal-inverse Wishart model.

4. SUBMATRICES

We shall be considering the subvectors of

$$\boldsymbol{Y}_{(q)} = (Y_1, Y_2, \ldots, Y_q) \tag{4.1}$$

which obtain on deletion of some regressors. Let I be a subsequence of $(1, 2, \ldots, q)$, I^c its complement, and $|I|$ the number of its elements. This italic use of I should be distinguished from boldface $\boldsymbol{I}$ denoting the identity matrix. For the sequence $(1, 2, \ldots, p)$ we write (p). Deletion of the regressors with indices in I^c leaves the future observation vector of $|I| + 1$ entries

$$\boldsymbol{Y}^I = (Y_0, \boldsymbol{Y}_I) \tag{4.2}$$

containing the regressor vector

$$\boldsymbol{Y}_I = (Y_i)_{i \in I}. \tag{4.3}$$

Note that according to (4.2) the notation $\boldsymbol{Y}^q$ now emerges as a slight simplification of $\boldsymbol{Y}^{(q)}$. We shall write $\boldsymbol{Y}_{(q)} = (\boldsymbol{Y}_I, \boldsymbol{Y}_{I^c})_I$ or, dropping the last subscript,

$$\boldsymbol{Y}_{(q)} = (\boldsymbol{Y}_I, \boldsymbol{Y}_{I^c});$$

also, by an obvious extension,

$$\boldsymbol{M}^I = \begin{bmatrix} m_{00} & \boldsymbol{m}_{0I} \\ \boldsymbol{m}_{I0} & \boldsymbol{M}_{II} \end{bmatrix},$$

and similarly for Σ^I. Finally, we denote

$$\begin{aligned} \boldsymbol{b}_{0I} &= \boldsymbol{m}_{0I}\boldsymbol{M}_{II}^{-1}, & \boldsymbol{\beta}_{0I} &= \boldsymbol{\sigma}_{0I}\boldsymbol{\Sigma}_{II}^{-1}, \\ m_{00\cdot I} &= m_{00} - \boldsymbol{b}_{0I}\boldsymbol{m}_{I0}, & \sigma_{00\cdot I} &= \sigma_{00} - \boldsymbol{\beta}_{0I}\boldsymbol{\sigma}_{I0}, \end{aligned}$$

and similarly when other subscripts outside I appear in lieu of 0. On the left hand side we have the regression coefficients and the residual variance, respectively, of Y_0 given Y_I, coming from the marginal distribution (3.7). On the right hand side are the same quantities of the conditional distribution given Σ, which is the distribution (3.3).

5. A PARTIALLY EXCHANGEABLE MODEL

In his search coherent priors of the inverse Wishart type (3.2) Lindley (1978) studied for the case in which $\boldsymbol{Y}^q$ is exchangeable given Σ (hence marginally also).

A slightly more general prior is obtained with marginal covariance matrices of the form

$$\boldsymbol{M}^q = c^2 \begin{bmatrix} 1 & b & b & \dots & b & b \\ b & 1 & a & \dots & a & a \\ b & a & 1 & \dots & a & a \\ \vdots & \vdots & \vdots & & \vdots & \vdots \\ b & a & a & \dots & 1 & a \\ b & a & a & \dots & a & 1 \end{bmatrix}, \qquad a \geq b^2. \tag{5.1}$$

While the whole model is still exchangeable on the part of the regressors, which have mutual correlations a, it also recognizes the special role of Y_0 by allowing a different correlation b between it and the regressors. Making this choice for every q yields an infinite dimensional distribution with coefficient of determination

$$r^2_{0\cdot(\infty)} = b^2/a. \tag{5.2}$$

The model is thus marginally nondeterministic. Also in Dawid's (1988) terminology the model is marginally tail-dominated. In the completely exchangeable case, $b = a$, (5.2) equals a. One immediate and often undesirable conclusion for the exchangeable model is that expectations of a high degree of explanatory capability on one hand and of moderate mutual correlations among the regressors on the other hand are incompatible. The extension (5.1) is free from this particular limitation.

The marginal residual variances become

$$m_{00\cdot I} = c^2\left(1 - \frac{|I|b^2}{1 + (|I| - 1)a}\right)$$

yielding

$$m_{00\cdot I} - m_{00\cdot(\infty)} = \frac{b^2(1-a)}{a[1+a(|I|-1)]} = O\left(\frac{1}{|I|}\right). \tag{5.3}$$

In placing all subsequences I having a common number of elements on an equal footing the present partially exchangeable prior attains a rate of decrease for $m_{00\cdot I}$ which makes the information of an added regressor comparable with that of only one single observation in a sample. It is clear that if a prior is to represent opinions which hold that a small number of regressors will suffice, the rate of decrease must be more rapid.

It is also clear that such a prior cannot treat the regressors as equal in explanatory capability. Among the distinguishing features of the priors to be considered below are for one that there is a linear ordering of the regressors according to such capability and for another that the rate of improvement brought about by the p first of them is exponential in p. No doubt the former requirement can in practice be relaxed to some degree by using mixtures of a small number of priors based on alternative orderings.

6. BASIC ASSUMPTIONS ON PRIORS

As indicated above conditions will be placed on the prior distribution via the marginal distribution of the observation $\boldsymbol{Y}^q$. The components Y_i are indexed in an order of decreasing efficiency in the sense that for each p the best set of regressors with p elements, as judged from the residual variances, is $\boldsymbol{Y}_{(p)}$.

We remind the reader that those regressors which are a part of every regression involved in the considerations of coherence should in the first place be thought of as having been eliminated from discussion by conditioning on them. In other words the discussion is only concerned with the conditional covariance matrix of those regressors which are present in some, but not all, of the models of interest.

The situation in which there are two good sets of regressors neither of which contains the other is, according to the stated assumption, not likely to get a high prior density and therefore does not suggest a direct application of any of the priors to be developed below.

From the next assumption on we shall proceed as far as until Subsection 8.3 by ignoring that part of Y_0 which cannot be predicted from $Y_1, Y_2, \ldots$. Thus for the time being we shall be concerned with a marginally deterministic model, the deterministic component of the final model.

The second assumption implies that the researcher's ideas about the way that the regressors work are rather vague in a sense to be explained now. We consider the conditional prediction problem of predicting Y_0 by means of $Y_2, Y_3, \ldots, Y_q$ given that the best predictor Y_1 is already in use. The assumption is that the researcher's expectations about this problem are similar to those concerning the prediction of Y_0 unconditionally by means of $Y_1, \ldots, Y_{q-1}$. What must change due to the conditioning is the scale. In general we assume that, with $\boldsymbol{Y}_{(p)}$ given, prediction of Y_0 with the help of the rest of the regressors $Y_{(q)\setminus(p)}$ is similar to the unconditional prediction of Y_0 using $\boldsymbol{Y}_{(q-p)}$. However, to utilize the coherence more fully we shall extend the discussion from mere prediction to the whole of the conditional distribution. We assume that given $Y_{(p)}$ all of the conditional distribution of $Y_0, Y_{p+1}, Y_{p+2}, \ldots, Y_q$ is similar to the original unconditional distribution of $Y_0, Y_1, Y_2, \ldots, Y_{q-p}$.

Finally we require that each addition to the regressor set diminishes the scale of the error by a fixed fraction, an assumption which reflects a rather definite belief that relatively few regressors are needed.

The notion of vague ideas about prediction is best formulated in the limiting case $\delta \to \infty$ of the distribution (3.7), i.e. when

$$\boldsymbol{Y}^q \sim \mathcal{N}_{1,q+1}(1, \boldsymbol{M}^q). \tag{6.1}$$

The covariance matrix Σ is here treated as known, or as equal to its expection M^q. By the given motivation we require the existence of a number d such that

$$\begin{bmatrix} Y_0 \\ Y_{p+1} \\ Y_{p+2} \\ \vdots \\ Y_q \end{bmatrix} - E\left(\begin{bmatrix} Y_0 \\ Y_{p+1} \\ Y_{p+2} \\ \vdots \\ Y_q \end{bmatrix} \middle| \boldsymbol{Y}_{(p)}\right) \Bigg| \boldsymbol{Y}_{(p)} = y \sim d^p \begin{bmatrix} Y_0 \\ Y_1 \\ Y_2 \\ \vdots \\ Y_{q-p} \end{bmatrix} \tag{6.2}$$

for every p $(1 \leq p < q)$ and all y in R^p.

Note that for normal distributions the outer conditioning is actually redundant whilst if (3.7) were assumed (6.2) would be impossible because (a) the left side would become dependent on y and (b) conditioning changes the degrees of freedom in a T distribution.

With (6.1) the condition is simply a condition on second moments,

$$\boldsymbol{M}_{(0,p+1,\ldots,q)(0,p+1,\ldots,q)\cdot(p)} = d^{2p}\boldsymbol{M}^{q-p}, \qquad 1 \leq p < q, \tag{A.1}$$

where the left side symbol stands for the residual covariance matrix in the regression of $Y_0, Y_{p+1}, \ldots, Y_q$ on $\boldsymbol{Y}_{(p)}$. As a second order condition (A.1) is equally feasible for the T distribution. Its import is that the conditional prediction problems are still similar to the initial unconditional prediction problems of the same dimension, *to second order.* Subject to (A.1) differences between the distributions of the two members of (6.2) in the T case are furthermore of no concern to a researcher who is content with measuring errors of prediction quadratically, because by the linearity of the regressions the point predictors from the two members will be identical linear functions.

Moreover, on account of the linearity of the regressions in the T distribution and of the closedness of the $T(\delta)$ family under linear transformations, Condition (A.1) is again expressible in terms of residuals as

$$\begin{bmatrix} Y_0 \\ Y_{p+1} \\ Y_{p+2} \\ \vdots \\ Y_q \end{bmatrix} - E\left(\begin{bmatrix} Y_0 \\ Y_{p+1} \\ Y_{p+2} \\ \vdots \\ Y_q \end{bmatrix} \middle| \boldsymbol{Y}_{(p)}\right) \sim d^p \cdot \begin{bmatrix} Y_0 \\ Y_1 \\ Y_2 \\ \vdots \\ Y_{q-p} \end{bmatrix}, \quad 1 \leq p < q. \tag{A.1'}$$

Let

$$\boldsymbol{R}^q = [r_{ik}] \tag{6.3}$$

be the correlation matrix deriving from the marginal covariance matrix M^q. We shall assume that

$$r_{01} > 0, \tag{A.2}$$

which is to state a mere convention, and finally that

$$r_{01} \geq \max_{1<k\leq q} |r_{0k}|. \tag{A.3}$$

Condition (A.3) states that Y_1 is initially as good as any other regressor.

We shall turn to some simple consequence of (A.1-3). The $(0,0)$ entry of the matrix (A.1) yields the condition

$$m_{0,0\cdot(p)} = d^{2p} m_{00}, \quad 1 \leq p < q. \tag{6.4}$$

For $p = 1$ it gives

$$d^2 = 1 - r_{01}^2. \tag{6.5}$$

It can be shown that (6.4) combined with case $p = 1$ of (A.1) alone is enough to imply all of (A.1). Also, with $q \to \infty$, (6.4) implies that Y_0 can be exactly predicted from $Y_1, Y_2 \ldots$, i.e. that the marginal model is deterministic.

A simple way of turning the model into a nondeterministic one is to add to Y_0 another component which is uncorrelated with all the Y_is. Condition (A.1) must now be restricted to the deterministic component. However, the inequalities for the correlations, (A.2-3), are the same for the new Y_0.

The first form of (A.1), Condition (6.2), despite being valid only for the limiting normal form, is nevertheless useful for deducing various second order properties. For instance combining with (6.2) the fact that partial correlations are ordinary correlations for the pertinent conditional distributions (this is true for the T distributions as well) immediately yields the equations

$$r_{0k\cdot(p)} = r_{0,k-p}, \quad 1 \le p < k \le q. \tag{6.6}$$

Together with (A.3) this in turn implies that

$$r_{0,p+1\cdot(p)} \ge \max_{p+1<k\le q} |r_{0k\cdot(p)}|, \quad 1 \le p \le q-2.$$

That is, given that the p best regressors are already in use there will be no better choice for the next regressor than Y_{p+1} among the rest of the variables. This demonstrates that, due to (A.1) and (A.3) we do have the linear ordering of the regressors mentioned in the beginning of this section.

7. REPRESENTING VAGUE IDEAS ABOUT PREDICTION

We shall develop here a representation for a vector $\boldsymbol{Y}^q$ whose covariance matrix satisfies Condition (A.1). This condition expresses vague ideas about prediction in the sense explained in the previous section. The representation will be in terms of the residuals

$$\tilde{Y}_{k\cdot I} = Y_k - \boldsymbol{b}_{kI}\boldsymbol{Y}_I' \tag{7.1}$$

of the variables Y_k from their best linear predictors given $\boldsymbol{Y}_I$. We shall write

$$\boldsymbol{b}_{k(i)} = (\boldsymbol{b}_{k(i-1)\cdot i}, b_{ki\cdot(i-1)}), \tag{7.2}$$

where $\boldsymbol{b}_{k(i-1)\cdot i}$ and $b_{ki\cdot(i-1)}$ are the partial regression coefficients of Y_k on $\boldsymbol{Y}_{(i-1)}$ given Y_i and on Y_i given $\boldsymbol{Y}_{(i-1)}$, respectively. The best linear predictor of Y_k in terms of $\boldsymbol{Y}_{(i)}$ is

$$\begin{aligned}
\boldsymbol{b}_{k(i)}\boldsymbol{Y}_{(i)}' &= \boldsymbol{b}_{k(i-1)\cdot i}\boldsymbol{Y}_{(i-1)}' + b_{ki\cdot(i-1)}Y_i \\
&= \boldsymbol{b}_{k(i-1)\cdot i}\boldsymbol{Y}_{(i-1)}' + b_{ki\cdot(i-1)}\left[\boldsymbol{b}_{i(i-1)}\boldsymbol{Y}_{(i-1)}' + \tilde{Y}_{i\cdot(i-1)}\right] \\
&= \left[\boldsymbol{b}_{k(i-1)\cdot i} + b_{ki\cdot(i-1)}\boldsymbol{b}_{i(i-1)}\right]\boldsymbol{Y}_{(i-1)}' + b_{ki\cdot(i-1)}\tilde{Y}_{i\cdot(i-1)} \\
&= \boldsymbol{b}_{k(i-1)}\boldsymbol{Y}_{(i-1)}' + b_{ki\cdot(i-1)}\tilde{Y}_{i\cdot(i-1)},
\end{aligned} \tag{7.3}$$

where the last equation may be proved by checking that it produces $\mathrm{Cov}(\boldsymbol{Y}_{(i-1)}, \tilde{Y}_{k\cdot(i-1)}) = 0$ correctly. The recursion (7.3) yields the representations

$$\begin{aligned}
\boldsymbol{b}_{k(i)}\boldsymbol{Y}_{(i)}' &= b_{k1}Y_1 + \sum_{j=2}^{i} b_{kj\cdot(j-1)}\tilde{Y}_{j\cdot(j-1)}, \\
Y_k &= b_{k1}Y_1 + \sum_{j=2}^{i} b_{kj\cdot(j-1)}\tilde{Y}_{j\cdot(j-1)} + \tilde{Y}_{k\cdot(i)}
\end{aligned} \tag{7.4}$$

with $Y_1, \tilde{Y}_{2\cdot 1}, \ldots, \tilde{Y}_{i\cdot(i-1)}, \tilde{Y}_{k\cdot(i)}$ uncorrelated. The latter sequence is simply a Gram-Schmidt orthogonalization (but not normalization) of the sequence $Y_1, Y_2 \ldots, Y_i, Y_k$, and the representation (the Gram-Schmidt orthogonalization) is valid for an arbitrary $k \geq 0$, of course with $\tilde{Y}_{k\cdot(i)} = 0$ if $1 \leq k \leq i$.

The coefficient $b_{ki\cdot(i-1)}$ is the regression coefficient of Y_i in the regression of Y_k on Y_i given $\boldsymbol{Y}_{(i-1)}$. Now suppose $k > i$. Subject to (A.1) or, more vividly, to (6.2) with normality, the regression is the same as that of $d^{i-1}Y_{k-i+1}$ on $d^{i-1}Y_1$ unconditionally. Consequently,

$$b_{ki\cdot(i-1)} = b_{k-i+1,1}, \quad 1 \leq i < k \leq q. \tag{7.5}$$

For $k = 0$ the same argument yields

$$b_{0i\cdot(i-1)} = b_{01}, \quad 1 \leq i \leq q. \tag{7.6}$$

For the index $i = 1$ the equations (7.5-6) are to be understood as the definitions of the left side symbols. The representations (7.4) simplify, on account of (7.5-6), to

$$\begin{aligned} Y_0 &= b_{01} \sum_{i=1}^{q} \tilde{Y}_{i\cdot(i-1)} + \tilde{Y}_{0\cdot(q)}, \\ Y_k &= \sum_{i=1}^{k-1} b_{k-i+1,1} \tilde{Y}_{i\cdot(i-1)} + \tilde{Y}_{k\cdot(k-1)}, \quad 1 \leq k \leq q. \end{aligned} \tag{7.7}$$

Here $\tilde{Y}_{1\cdot(0)}$ means Y_1. Turning to an investigation of the distribution of

$$\begin{aligned} \tilde{\boldsymbol{Y}}^q &= (\tilde{Y}_{0\cdot(q)}, Y_1, \tilde{Y}_{2\cdot 1}, \ldots, \tilde{Y}_{q\cdot(q-1)}) \\ &= (\tilde{Y}_{0\cdot(q)}, \tilde{\boldsymbol{Y}}_{(q)}) \end{aligned} \tag{7.8}$$

we note that $\mathrm{Var}(\tilde{Y}_{i\cdot I}) = m_{ii\cdot I}$, hence

$$\mathrm{Var}\left(\tilde{Y}_{i\cdot(i-1)}\right) = m_{ii\cdot(i-1)} = d^{2(i-1)} m_{11}, \quad 1 \leq i \leq q \tag{7.9}$$

from (A.1). In the case of $\tilde{Y}_{0\cdot(q)}$, using the uncorrelatedness of the components of (7.8) in the first equation of (7.7) it is easily seen that

$$\mathrm{Var}\left(\tilde{Y}_{0\cdot(q)}\right) = d^{2q} m_{00}. \tag{7.10}$$

Since the components of (7.8) are uncorrelated linear functions of $\boldsymbol{Y}^q$, the results can be summarized by

$$\tilde{\boldsymbol{Y}}^q \sim T_{1,q+1}\left(\delta; 1, (\delta - 2) \cdot \mathrm{diag}\left(d^{2q} m_{00}, m_{11}, d^2 m_{11}, \ldots, d^{2(q-1)} m_{11}\right)\right). \tag{7.11}$$

If the distribution of (Y_0, Y_1) is fixed (by fixing m_{11}, b_{01} and d, e.g.), there remain $q - 2$ free hyperparameters $b_{21}, \ldots, b_{q-1,1}$. The formulae (7.7) and (7.11) show that these parameters completely determine the distribution of $\boldsymbol{Y}^q$. We shall next show that (7.7) makes full use of the condition (A.1), i.e., that any $\boldsymbol{Y}^q$ with a representation of form (7.7) satisfies (A.1). To this end we employ the form (A.1′) of the condition.

Note that, for $1 \le p < q$ the transformation $\tilde{\boldsymbol{Y}}_{(p)} \mapsto \boldsymbol{Y}_{(p)}$ is invertible so that

$$E\left(\begin{bmatrix} Y_0 \\ Y_{p+1} \\ \vdots \\ Y_q \end{bmatrix} \middle| \boldsymbol{Y}_{(p)}\right) = E\left(\begin{bmatrix} Y_0 \\ Y_{p+1} \\ \vdots \\ Y_q \end{bmatrix} \middle| \tilde{\boldsymbol{Y}}_{(p)}\right).$$

Making use of this fact, the representations (7.7), the linearity of the conditional expectations, and the uncorrelatedness of the components of $\tilde{Y}^q$ in the first step below, and the distribution (7.11) directly in the second, we find that

$$\begin{aligned}
&\begin{bmatrix} Y_0 \\ Y_{p+1} \\ \vdots \\ Y_q \end{bmatrix} - E\left(\begin{bmatrix} Y_0 \\ Y_{p+1} \\ \vdots \\ Y_q \end{bmatrix} \middle| \boldsymbol{Y}_{(p)}\right) \\
&= \begin{bmatrix} b_{01} \sum_{i=p+1}^{q} \tilde{Y}_{i\cdot(i-1)} & +\tilde{Y}_{0\cdot(q)} \\ & \tilde{Y}_{p+1\cdot(p)} \\ & \vdots \\ \sum_{i=p+1}^{q-1} b_{q-i+1,1} \tilde{Y}_{i\cdot(i-1)} & +\tilde{Y}_{q\cdot(q-1)} \end{bmatrix} \\
&\sim d^p \begin{bmatrix} b_{01} \sum_{i=p+1}^{q} \tilde{Y}_{i-p\cdot(i-p-1)} & +\tilde{Y}_{0\cdot(q-p)} \\ & Y_1 \\ & \vdots \\ \sum_{i=p+1}^{q-1} b_{q-i+1,1} \tilde{Y}_{i-p\cdot(i-p-1)} & +\tilde{Y}_{q-p\cdot(q-p-1)} \end{bmatrix} \\
&= d^p \begin{bmatrix} b_{01} \sum_{i=1}^{q-p} \tilde{Y}_{i-p\cdot(i-p-1)} & +\tilde{Y}_{0\cdot(q-p)} \\ & Y_1 \\ & \vdots \\ \sum_{i=1}^{q-p-1} b_{q-p-i+1,1} \tilde{Y}_{i\cdot\cdot(i-1)} & +\tilde{Y}_{q-p\cdot(q-p-1)} \end{bmatrix} \\
&= d^p \begin{bmatrix} Y_0 \\ Y_1 \\ \vdots \\ Y_{q-p} \end{bmatrix}.
\end{aligned}$$

This is (A.1′). The following is a second order formulation of what has been proved.

Theorem. *A necessary and sufficient condition for* $\mathrm{Cov}(\boldsymbol{Y}^q) = \boldsymbol{M}^q$ *to satisfy* (A.1) *is that* $\boldsymbol{Y}^q$ *have a representation of the form* (7.7) *where* $\tilde{Y}_{0\cdot(q)}$ *and* $\tilde{Y}_{i\cdot(i-1)}$, $1 \le i \le q$, *are uncorrelated and have variances* (7.9-10).

To recapitulate we have shown that (A.1) is equivalent to the existence of a linear latent variable representation in terms of uncorrelated innovations (a Gram-Schmidt orthogonalization of $Y_1, Y_2, \ldots, Y_q, Y_0$, in that order). That representation displays the effect of (A.1) in that it only involves $q-1$ different coefficients $b_{01}, b_{21}, b_{31}, \ldots, b_{q-1,1}$. Here the b_{j1}'s provide the coefficients for a representation of Y_k as a linear process for $k \ge 1$. The representation (7.7) will be used as a tool for further exploration of (A.1). An inverted (or autoregressive) version of (7.7) arising in a special case in Subsection 8.3 will illuminate the nature of the prior beliefs concerning the way in which new information is incorporated with the addition of new regressors.

8. GEOMETRIC COEFFICIENT SEQUENCES

The latent variable representation (7.7) of an expectation structure with vague ideas about prediction involves $q-1$ free regression coefficients, those of the simple regressions of $Y_0, Y_2, Y_3, \ldots, Y_{q-1}$ on Y_1. The rest of the paper will be concerned with the description of the subclass of these structures defined by the geometric form

$$b_{k1} = bc^{k-2}, \quad 1 < k \leq q \tag{8.1}$$

of the regression coefficient sequence. Inevitably, if q is large, the reduction of the number of linear hyperparameters to just two further narrows the scope of the structure and can make it unattractive to the user. Nevertheless some realistic features are retained with (8.1) and study of it sheds light on the other possibilities in (7.7).

The representation (7.7) admits immediate calculation of moments. Their principal use below is in the investigation of the correlation structure. Condition (A.3), which has a complementary role to (A.1) or (7.7) is central to this analysis. Subsections 8.1 and 8.2 are devoted to these matters.

A feature of the marginal distribution of more direct interest than correlations are the regressions of Y_0 on various sets of variables. These are the relationships which would be employed if predictions were called for prior to obtaining the statistical data.

In (7.7) the regressor process $(Y_k)_{k=1,2,\ldots}$ is given as a linear process. We also give an autoregressive representation which exhibits the prior predictive beliefs about a new regressor Y_k given all the preceding regressors.

The form (8.1) is precisely the one for which we have expressions in closed form for the various regression relationships of interest. The regressions will be studied in Subsection 8.3.

Besides the description of the marginal distribution such regression coefficients also provide information which is directly relevant to the estimation of the regression coefficients arising conditionally given Σ. These were introducted in Section 4. To discuss these matters we need the distributions of the latter.

If $I \subset (q)$, we have that

$$\beta_{0I}|\sigma_{00\cdot I} \sim \boldsymbol{b}_{0I} + \mathcal{N}(\sigma_{00\cdot I}, \frac{1}{\delta - 2}\boldsymbol{M}_{II}^{-1}), \tag{8.2}$$

$$\sigma_{00\cdot I} \sim \frac{(\delta - 2)m_{00\cdot I}}{\chi^2_{\delta+|I|}} \tag{8.3}$$

yielding

$$\beta_{0I} \sim \boldsymbol{b}_{0I} + T_{1,|I|}(\delta + |I|; m_{00\cdot I}, \boldsymbol{M}_{II}^{-1}). \tag{8.4}$$

In Subsection 8.3 we shall take $I = (p)$ and exhibit the locations $\boldsymbol{b}_{0(p)}$ and the variances of the individual coefficients $\beta_{0k\cdot(p)\backslash\{k\}}$ in the geometric case (8.1). Some general considerations concerning the variances will now be taken up.

In his exchangeable model Lindley (1978) studied, instead of a marginal distribution like (8.4), the conditinal prior distribution (8.2). The purpose was to try and see if the judgments for different I implied by the prior were reasonably consistent. He found that augmentation of the regressor set decreases the common expectation of the individual regression coefficients, which are exchangeable, and increases their uncertainty. This is what one might expect when too many regressors are included. However, the role of $\sigma_{00\cdot I}$ needs examination, and this also brings us to the notion of implicit determinism mentioned in Section 3.

By (8.3) we have that

$$E\sigma_{00\cdot I} = \frac{\delta - 2}{\delta + |I| - 2} \cdot m_{00\cdot I}. \tag{8.5}$$

As I increases towards an infinite set of natural numbers I_0, (8.5) converges to zero, hence also $\sigma_{00\cdot I} \to 0$ in probability. Here it is assumed that $\sigma_{00\cdot I}$ is derived from an infinite-dimensional covariance matrix Σ so that it is decreasing in I and must converge in any case. The limit is must then be zero. As shown by Dawid (1988) the limit is the variance of the prediction error in the infinite-dimensional distribution. This implies that exact prediction is possible with $\{Y_i : i \in I_0\}$ *if* Σ is known. Here we have a statement about the public part of the model which would often be inappropriate. The result holds irrespective of the particular choice of the M's (provided they are consistent). —The fact that the marginal distribution strongly depends on $|I|$ suggests to us using the marginal distribution (8.4) in coherence considerations, rather than the conditional distribution (8.2).

In fact, the above conclusion about the uncertainty of regression coefficients in Lindley (1978) is reversed with the marginal distribution. The general form of the marginal variance of the k-th coefficient is

$$\begin{aligned} \mathrm{Var}(\beta_{0k\cdot\backslash\{k\}}) &= \frac{1}{\delta + |I| - 2} \cdot m_{00\cdot I}[M_{II}]^{kk} \\ &= \frac{1}{\delta + |I| - 2} \cdot \frac{m_{00\cdot I}}{m_{kk\cdot I\backslash\{k\}}}, \end{aligned} \tag{8.6}$$

the double superscripts above referring to an element in the inverse of the bracketed matrix. As compared to (8.2) $\sigma_{00\cdot I}$ is replaced by $m_{00\cdot I}$ and the new factor $1/(\delta + |I| - 2)$ emerges. Even though $m_{00\cdot I}$ already is decreasing in I it is the latter factor which makes the difference to the exchangeable model. Yet the fact that the same factor is at the root of the determinism of the Wishart distribution makes one less inclined to fault the particular exchageable form here. But it turns out that in our examples the same factor is not critical to the behaviour of (8.6), neither one way nor the other. In this sense our models show a stronger resistance against the effects of implicit determinism, the usually undesirable property of a model. Below we shall discuss (8.6) with $I = (p)$.

In general the notion that (8.6) should increase with $p = |I|$ means tying the beliefs about the internal relationships between the regressors rather closely to those concerning the accuracy of prediction, as is seen by inspecting the last quotient in (8.6). Thus, however well the regressors are thought to contribute to the prediction of the quantity of interest, Y_0, with increasing variances (8.6) they are believed to be far more given to predicting each one of the old regressors, with ratios of residual variances growing faster than linearly in $p = |I|$. In Subsection 8.3 we give an example of a *deterministic* marginal covariance structure of this type which satisfies Condition (A.1) and has the geometric coefficients (8.1). However, there is a conflict with (A.3) and we have not found any case in which an ultimate increase of (8.6) could be accommodated with (A.3). We therefore conclude that, if taken seriously, these considerations alone suggest looking for modifications of the model as developed so far.

Another, and we think, more compelling consideration which calls for modification is the very determinism of the marginal model. As already mentioned, taking Y_0 to include an additional random component, uncorrelated with all the old variables, removes the determinism. Interestingly, it also makes it possible to accommodate the increasingness of (8.6) from some p_0 on with the model (A.1-3).

8.1. Correlation structures: Non-decreasing correlation

The present subsection explores an extreme case under Condition (A.3) and will contribute relatively little to subsequent development. Condition (A.3),

$$r_{01} \geq \max_{1<k\leq q} |r_{0k}|$$

delimits the correlation structures which we may consider acceptable. In fact, the entertainment of a correlation structure violating (A.3) would imply that the predictive residual variance could already at the first step be reduced by a higher factor than the assumed d^2 by suitably deviating from the supposedly best order of inclusion of the variables. The term "order of inclusion" provides us with a convenient means of speaking about our basic ordering of the regressors but we do not here have in mind any actual process of choosing them.

It is quite straightforward to prove the following facts, which are completely independent of special assumptions concerning the sequence b_{k1}.

Lemma. *Subject to Conditions (A.1-2) and for $1 < p \leq q$ the following statements are equivalent:*

(a) $|r_{01}| = |r_{02}| = \cdots = |r_{0,p-1}| \geq |r_{0p}|$,

(b) $r_{01} = r_{02} = \cdots = r_{0,p-1} \geq |r_{0p}|$,

(c) $b_{k1} = \frac{1}{2}r_{01}^2\left(1 - \frac{1}{2}r_{01}^2\right)^{k-2}$ *for* $1 < k < p$, *and* $b_{p1} \leq \frac{1}{2}r_{01}^2\left(1 - \frac{1}{2}r_{01}^2\right)^{p-2}$.

These results lead directly to the first of our special cases of (8.1). This is also covered as a special case in Case D of Subsection 8.2.

Case A: $b = \frac{1}{2}r_{01}^2$, $c = 1 - \frac{1}{2}r_{01}^2$. By the Lemma this is equivalent to all r_{0k}s being equal. The other simple correlations are found by means of the formulae (7.9-10) and (7.7). We have that

$$r_{ik} = 1 - \frac{(1 - r_{01}^2)^i}{\left(1 - \frac{1}{2}r_{01}^2\right)^{2i-1}}, \quad 1 \leq i < k \leq q. \tag{8.7}$$

Because of the inequality $1 - r_{01}^2 < \left(1 - \frac{1}{2}r_{01}^2\right)^2$ all correlations in (8.7) are positive and we have

$$r_{1k} < r_{2k} < \cdots < r_{k-1,k} \tag{8.8}$$

with $r_{ik} \to 1$ as i, k, and q increase without bound. With increasing index the message conveyed by a variable is expected to become more and more repetitive, apparently owing to the relatively slow rate of decrease of b_{k1}.

Since the r_{0k}s are all equal, as a single regressor any variable Y_p is as good as any other. However, *if the constant rate of reduction of the predictive residual variance, d^2 is to be maintained, for no p can Y_p be included later than as the $(p+1)$-th regressor.*

For suppose $\boldsymbol{Y}_{(p-1)}$ and Y_i with $i > p$ have been included. We assert that only Y_p should now be given consideration. If $p = 1$, we know that no loss of the rate of reduction has taken place. For an arbitrary $p > 1$ consider first the case in which Y_i is the last addition to the regressor set. By (6.6) and the equality of the r_{0k}'s this is as good an addition as any other so that a reduction by factor d^2 has been achieved. But the total reduction achieved by a set of regressors only depends on the whole set and not on the order. We conclude that even for $p > 1$ no loss of the rate of reduction has taken place.

The same will continue to be true if Y_p is included as the $(p+1)$-th regressor. Suppose instead that variable Y_j (perhaps other than Y_p) is included. The reduction of the predictive residual variance is by the factor $1 - r_{0j\cdot(p-1)\cup\{i\}}^2$. Here the partial coefficient of correlation reduces, on using (6.6) and an analogue with index $p-1$, to

$$r_{0,j-p+1\cdot i-p+1} = \frac{r_{0,j-p+1} - r_{0i-p+1}r_{i-p+1,j-p+1}}{\sqrt{(1 - r_{0i-p+1}^2)(1 - r_{i-p+1,j-p+1}^2)}} = \frac{r_{01}(1 - r_{i-p+1,j-p+1})}{\sqrt{(1 - r_{01}^2)(1 - r_{i-p+1,j-p+1}^2)}}.$$

Simple manipulation shows that $r_{0,j-p+1\cdot i-p+1} \leq r_{01}$ is equivalent to

$$\frac{\frac{1}{2}r_{01}^2}{1 - \frac{1}{2}r_{01}^2} \leq r_{i-p+1,j-p+1} < 1$$

the equalities occurring simultaneously in the two inequalities. By (8.7) the first member equals r_{i-p+1} for $j < i$ and the middle one, $r_{i-p+1,j-p+1}$, is strictly increasing in j until it reaches the value i, and equals thereafter $r_{i-p+1,i-p+2}$. Thus only $j = p$ achieves the full reduction of the predictive residual variance.

To summarize, the present case corresponds to the intersection of all the boundary planes $r_{0k} = r_{01}$ of the set of the correlation matrices satisfying (A.3). As regards the order of inclusion of the regressors, Case A conveys the least definite beliefs permitted under (A.3). In fact the way in which (A.3) has been motivated suggests that one should stand by a definite ordering and exclude the boundary of (A.3) from consideration.

8.2. Correlation structures: Further cases

Before turning to a more general investigation of the geometric coefficient sequences (8.1) we shall do away with two further cases which, although very special, provide examples of correlation matrices belonging to the interior of (A.3).

Case B: $b = 0$. For $1 \leq i \leq q$ we have

$$Y_i = \tilde{Y}_{i\cdot(i-1)}$$

and beliefs center on the independence of the regressors. The only nonzero elements outside the diagonal of $\boldsymbol{R}$ are

$$r_{0i} = r_{01}(1 - r_{01}^2)^{\frac{i-1}{2}}, \quad 1 \leq i \leq q.$$

We conclude that (A.3) is satisfied with strict inequality and that this will continue to be true in the general form (8.1) for all $|b|$ and $|c|$ small enough.

Case C: $c = 0$. The correlations

$$r_{0k} = \frac{r_{01}(b + d^2)}{\sqrt{b^2 + d^2}} \cdot d^{k-2}, \quad 1 < k \leq q,$$

decrease in absolute value whenever b does not exceed $\frac{1}{2}r_{01}^2$ and so satisfy (A.3). Strict inequality obtains if $b < \frac{1}{2}r_{01}^2$. For the general from (8.1), and any $b < \frac{1}{2}r_{01}^2$, taking $|c|$ small enough ensures (A.3) with strict inequality. The special choice $b = -d^2$ makes all the r_{0k}s with $k > 1$ vanish.

The other nonzero correlations are

$$r_{i,i+1} = \frac{bd}{b^2 + d^2}, \quad 1 \leq i < q.$$

In the following study of (A.3) we shall restrict ourselves to certain cases of (8.1) which exhibit one these simplifying features: (a) r_{0k} is monotone and decreasing in absolute value, (b) the denominator of r_{0k} is positive and increasing and the numerator is monotone with a useful upper bound to its absolute value, (c) complementary subsequences of r_{0k} are each of type (a) or (b). These cases make a list of simple sufficient conditions for (A.3). The conditions can often be sharpened and still others can be produced, but at the expense of simplicity. The presentation follows the lexicographically decreasing ordering of the pairs (c, b). Assuming that $c \neq 0$ a general expression for r_{0k} is

$$r_{0k} = r_{01} \cdot \frac{\frac{b}{c}\sum_{i=1}^{k-1}\left(\frac{d^2}{c}\right)^{i-1} + \left(\frac{d^2}{c}\right)^{k-1}}{\sqrt{\frac{b^2}{c^2}\sum_{i=1}^{k-1}\left(\frac{d^2}{c^2}\right)^{i-1} + \left(\frac{d^2}{c^2}\right)^{k-1}}}, \quad 1 \leq k \leq q.$$

Other simple correlations than these will not be considered any further.

Case D: $d < c \leq \frac{1}{2}(1+d^2)$, $\sqrt{c^2-d^2} \leq b \leq c-d^2$. The second inequality is a necessary and sufficient condition for the nonemptiness of the interval given for b. Note that $\frac{1}{2}(1+d^2) = 1 - \frac{1}{2}r_{01}^2$. A convenient expression for r_{0k} is

$$r_{0k} = r_{01} \cdot \frac{\sqrt{c^2-d^2}}{c-d^2} \cdot \frac{b-(b-c+d^2)(d^2/c)^{k-1}}{\sqrt{b^2-(b^2-c^2+d^2)(d^2/c^2)^{k-1}}}. \tag{8.9}$$

The last factor on the right has a positive decreasing numerator and an increasing denominator. Thus r_{0k} is decreasing with the positive limit

$$r_{01} \cdot \frac{\sqrt{c^2-d^2}}{c-d^2}$$

which can made to take on any value in the interval $(0, r_{01}]$. The situation is of type (a) and (A.3) is satisfied. Except in one case r_{0k} is even strictly decreasing and (A.3) is strict. When $c = \frac{1}{2}(1+d^2)$ and $b = \sqrt{c^2-d^2}$, r_{0k} is constant and we are thus back to Case A.

Case E: $d < c < \frac{1}{2}(1+d^2)$, $d^2 - c \leq b \leq -\sqrt{c^2-d^2}$. Consider for $k > 1$ the last factor on the right hand side of (8.9). The denominator has the lower bound $\sqrt{c^2-d^2}$ (corresponding to $k=1$), and the absolute value of the numerator has the upper bound $\max\{-b, c-d^2\}$, which is $c-d^2$ by assumption. Thus (A.3) follows, in fact as a strict inequality.

Case F: $c = d$, $0 \leq b \leq d(1-d)$. Here

$$r_{0k} = \frac{r_{01}}{1-d} \cdot \frac{b-[b-d(1-d)]d^{k-1}}{\sqrt{d^2+b^2(k-1)}}$$

is strictly decreasing with limit zero.

Case G: $c < d$, $b = c-d^2$. For $c \neq -d$ we have

$$r_{0k} = r_{01}\left\{(c^2-d^2)/(b^2-(b^2-c^2+d^2)(d^2/c^2)^{k-1})\right\}^{1/2}$$

and this is strictly decreasing. If $|c| < d$ the limit is zero and if $c < -d$ it is

$$r_{01} \cdot \frac{\sqrt{c^2-d^2}}{d^2-c}$$

with range $(0, r_{01})$. For $c = -d$

$$r_{0k} = \frac{r_{01}}{\sqrt{1+(k-1)(1+d)^2}}$$

strictly decreases to zero. Thus in both cases (A.3) holds as a strict inequality.

Case H: $d^2 < c < d$, $0 \leq b < c-d^2$. Taking into account the changes of sign from (8.9) we write

$$r_{0k} = r_{01} \cdot \frac{\sqrt{d^2-c^2}}{c-d^2} \cdot \frac{b-(b-c+d^2)(d^2/c)^{k-1}}{\sqrt{(b^2-c^2+d^2)(d^2/c^2)^{k-1}-b^2}}.$$

This is strictly decreasing with limit zero.

Case I: $c = d^2$, $0 \leq b \leq d(1-d)$. We have

$$r_{0k} = r_{01} \cdot \frac{[d^2 + b(k-1)]d^{k-1}}{\sqrt{d^2[d^2(1-d^2)+b^2] - b^2 d^{2k}}}.$$

The numerator is decreasing. It follows that r_{0k} is strictly decreasing with limit zero. Again (A.3) holds with strict inequality.

Case J: $-d^2 < c < 0, (c-d^2)(\log d)/\log(-c) \leq b < 0$. The numerator is strictly decreasing, separately for even and odd indices, and with limit zero in both cases. The denominator is increasing. By Lemma 1, $r_{01} > r_{02}$. Thus (A.3) holds with strict inequality.

Case K: $c = -d, -d(1+d) \leq b \leq -d^2$. The subsequence $r_{0,2k-1}$ strictly decreases to 0. For $r_{0,2k}$ the situation is of type (b). Condition (A.3) holds strictly.

Case L: $c < -d$, $\sqrt{c^2 - d^2} \leq b \leq d^2 - c$. Here $|r_{0,2k}|$ is strictly decreasing, and $r_{01} > r_{02}$ by Lemma 1. The sequence $r_{0,2k+1}$ is of type (b). Condition (A.3) holds strictly.

8.3. Marginally nondeterministic models

We shall now make the promised extension to a marginally nondeterministic (but still *implicitly* deterministic) model. We write

$$Y_0 = Y_0^{\infty} + \tilde{Y}_{0\cdot(\infty)}, \tag{8.10}$$

where Y_0^{∞} is the deterministic component of Y_0 and $\tilde{Y}_{0\cdot(\infty)}$ the component which is uncorrelated with all of $Y_1, Y_2, \ldots$. The elaborate notation is hoped to be suitably suggestive. When an old symbol like Y_0 acquires a new meaning the superscript ∞ is used to indicate the earlier quantity. For (8.10) the variance and the residual variance given $\boldsymbol{Y}_{(p)}$ thus become

$$\begin{aligned} m_{00} &= m_{00}^{\infty} + m_{00\cdot(\infty)}, \\ m_{00\cdot(p)} &= m_{00\cdot(p)}^{\infty} + m_{00\cdot(\infty)}. \end{aligned} \tag{8.11}$$

By making the above modification of the $(0,0)$ entry (with $m_{00\cdot(\infty)} > 0$) of each $\boldsymbol{M}^p$ we have arrived at new marginal distributions $T_{1,p+1}(\delta; 1, (\delta-2)\boldsymbol{M}^p)$ defining a nondeterministic infinite-dimensional marginal T distribution.

8.3.1. Geometric coefficient sequences: regresion structures

We shall in this subsection be concerned with judging the reasonableness or otherwise of those descriptive features of both the marginal and the conditional distributions (given Σ) of $(Y_0, Y_1, Y_2, \ldots)$ which are revealed by a few particular regressions.

Regressions of interest can be obtained by inverting the triangular equation system (7.7). This can be done recursively. For the coefficient sequence (8.1) it results in regression equations of closed form. Remembering to add the nondeterministic component these become

$$Y_k = b \sum_{i=1}^{k-1} (c-b)^{k-i-1} Y_i + \tilde{Y}_{k\cdot(k-1)}, \quad 1 \leq k \leq q, \tag{8.12}$$

and

$$Y_0 = \frac{b_{01}}{1-c+b} \sum_{i=1}^{q} [1 - c + b(c-b)^{q-i}] Y_i + \tilde{Y}_{0\cdot(q)}^{\infty} + \tilde{Y}_{0\cdot(\infty)} \quad \text{for } c \neq b+1, \tag{8.13}$$

$$Y_0 = b_{01} \sum_{i=1}^{q} [1 - b(q-i)] Y_i + \tilde{Y}_{0\cdot(q)}^{\infty} + \tilde{Y}_{0\cdot(\infty)} \quad \text{for } c = b+1. \tag{8.14}$$

Rather than again produce a list of cases we shall focus on a single case from Subsection 8.2, is the one that is suggested by the previous discussion of (8.6). We need the diagonal entries of the inverse of

$$\begin{aligned} \boldsymbol{M}_{(p)(p)} &= \text{Cov}(\boldsymbol{Y}'_{(p)}) \\ &= \text{Cov}(\boldsymbol{T}_p \tilde{\boldsymbol{Y}}'_{(p)}) \\ &= \boldsymbol{T}_p \text{Cov}(\tilde{\boldsymbol{Y}}'_{(p)}) \boldsymbol{T}'_p, \end{aligned}$$

say, where $\boldsymbol{T}_p$ is the triangular matrix which transforms $\tilde{\boldsymbol{Y}}'_{(p)}$ into $\boldsymbol{Y}'_{(p)}$ in (7.7) and the inverse of which is displayed in (8.12) for the geometric coefficients (8.1). We find for the k-th diagonal element

$$\begin{aligned} m^{-1}_{kk\cdot(p)\setminus\{k\}} &= \sum_{i=1}^{p} \{[\boldsymbol{T}'_p]^{ki}\}^2 \{\text{Var}\tilde{Y}_{i\cdot(i-1)}\}^{-1} \\ &= m_{11}^{-1} d^{-2(k-1)} + \sum_{i=k+1}^{p} b^2 (c-b)^{2(i-k-1)} m_{11}^{-1} d^{2(i-1)}. \end{aligned}$$

We shall take $|c-b| \neq d$. For (8.6) we find, using (8.11),

$$\begin{aligned} \text{Var}(\beta_{0k\cdot(p)\setminus\{k\}}) =& \frac{1}{\delta + p - 2} \cdot \frac{m_{00}^{\infty} d^{2p} + m_{00\cdot(\infty)}}{m_{11} d^{2k} [(c-b)^2/d^2 - 1]} \\ & \cdot \left\{ b^2 \left(\frac{c-b}{d} \right)^{p-k} - b^2 + (c-b)^2 - d^2 \right\}. \end{aligned} \tag{8.15}$$

Case E of Subsection 8.2 assumes that $d < c < \frac{1}{2}(1+d^2)$ and $d^2 - c \leq b \leq -\sqrt{c^2 - d^2}$. We know that Condition (A.3) is satisfied as a strict inequality. Note that since b is negative, $d < c < c - b$, so that we do have $|c - b| \neq d$. Also, $\text{Var}(\beta_{0k\cdot(p)\setminus\{k\}})$ must be strictly increasing in p from some p_0 on if $m_{00\cdot(\infty)} > 0$. It is this qualitative behaviour which we wanted to single out.

However, suppose $m_{00\cdot(\infty)} = 0$ so that we are back to the deterministic marginal distribution. Then the 3rd and 2nd inequalities of Case E yield

$$c - b \leq c - (d^2 - c) = 2c - d^2 < 1.$$

(We note in passing that case $c = b + 1$, or (8.14), is now exluded.) But for (8.15) to be ultimately increasing it is required that $c - b > 1/d$ in contradiction with the assumptions of Case E and also with (A.3). As mentioned before, we know of no case of the deterministic distribution in which both (A.3) and the ultimate growth of the absolute value of (8.15) could occur together. But this does not seem surprising to us.

We shall now turn to the inspection of the marginal regression coefficients themselves continuing to work with Case E.

In order to get an idea of the magnitudes of the coefficients in the regressions (8.12-13) those coefficients still need to be scaled by the standard deviation of the

$$\sqrt{m_{ii}} = m_{11}^{1/2} [b^2 c^{2(i-1)} - (b^2 - c^2 + d^2) d^{2(i-1)}]^{1/2} / (c^2 - d^2)^{1/2}.$$

From the autoregressive representation (8.12) we find

From the autoregressive representation (8.12) we find

$$\sqrt{m_{ii}} \cdot b_{ki\cdot(k-1)\backslash\{i\}} = m_{11}^{1/2} b(c-b)^{k-2} \cdot \left(\frac{c}{c-b}\right)^{i-1} \cdot [b^2 - (b^2 - c^2 + d^2)(d/c)^{2(i-1)}]^{1/2}/(c^2 - d^2)^{1/2}. \tag{8.16}$$

This is negative (since $b < 0$) and of the two factors which depend on i the first is decreasing and the second increasing (since $d < c < c - b$ and $b^2 - c^2 + d^2 \geq 0$). By considering the ratios of consecutive scaled coefficients one can show that the net effect is to make (8.16) decreasing in absolute value. [Here and in (8.16) we correct Mäkeläinen and Brown (1987).] That is, in the expectations of a researcher who entertains beliefs consistent with a prior within Case E, a contemplated new regressor Y_k will most closely be linked with the best of the old regressors.

Subject to the assumptions of Case E the regression coefficients $b_{0i\cdot(p)\backslash\{i\}}$, which are given for $p = q$ in (8.13), are positive and decreasing in i with minimum value b_{01} at $i = p$. The behaviour of the more relevant scaled coefficients $\sqrt{m_{ii}} b_{0i\cdot(p)\backslash\{i\}}$ may be studied by considering their consecutive ratios. By writing the ratios from (8.16) as factors in them one can easily show that even the scaled coefficients are strictly decreasing. This is, of course, what one expects in view of the built-in order of importance of the regressors.

Now consider a fixed i. The coefficient $b_{0i\cdot(p)\backslash\{i\}}$ will grow with p. The asymptotic value

$$b_{01} \cdot \frac{1-c}{1-c+b}$$

could be large. The expectations concerning the influence of the later regressors are seemingly inflated, a phenomenon which is apparently the clearest manifestation of the implict determinism of the model we have seen in the marginal distribution under Case E. However, the asymptotically exponential growth of the variance (8.15) is likely to rapidly nullify the effect for the marginal distribution of $\beta_{0k\cdot(p)\backslash\{k\}}$.

To sum up we have here confined ourselves to Case E of Subsection 8.2, showing first that the variance of any single theoretical regression coefficient in the predictor of Y_0 will eventually exceed any given bound (the rate of growth being asymptotically exponential) if the number of regressors is allowed to become large enough, a result which depends on the model being *marginally* nondeterministic. The expectation of the same coefficient, which is positive, will at the same time grow towards a limit, but this effect will be outweighed by the rapid increase in the uncertainty. In a fixed model these expectations, appropriately scaled, form a decreasing sequence thus reflecting the order of importance of the regressors. When a new regressor is predicted instead of Y_0 the ordering of the regressors continues to have a similar role. To us the general behaviour with regard to the introduction of many regressors seems reasonable.

9. DISCUSSION

We have constructed a family of marginal T distributions with three essential shape parameters and three other parameters, including a random variance component. The existence of cases in which the basic assumptions of our model combine with not unreasonable regression and correlation properties (Case E) has been demonstrated. These considerations include looking at the qualitative consequences of changing the number of regressors in the model. Although only a very limited selection of regressions have been considered those studied are believed to be the most relevant for our purposes.

Aside from the very limited scope that the inverse Wishart distribution offers for parametrizing uncertainty, one of the properties of consequence of this prior is the way in which the expectations of the theoretical residual variances change with the number of regression coefficients. The formulae (8.5), (6.4), and (8.11) yield

$$E\sigma_{00\cdot(p)} = \frac{\delta - 2}{\delta + p - 2} \cdot \left(m_{00}^{\infty} d^{2p} + m_{00\cdot(\infty)} \right), \tag{9.1}$$

an expression depending on four constants which are to be chosen by the researcher. The expectation is of order $O(1/p)$ if $m_{00\cdot(\infty)}$ is positive, that is if the model is marginally nondeterministic. This behaviour is common to all marginally nondeterministic normal-inverse Wishart models, not just our model. Otherwise the specific form (9.1) only contributes by a speeding up of the initial decrease. For a rough assessment of the speed of convergence p is to be measured in units of $\delta - 2$. —It may be recalled that the decrease of (9.1) to zero is directly responsible for the implicit determinism of the model.

The researcher's choices of p being restricted to $1 \leq p \leq q$ considerations of implicit determinism only come into play when q is large, "large" being understood in the sense that $\sigma_{00\cdot(q)}/\sigma_{00}$ has a high probability of being small. (With the inverse Wishart priors those probabities can be easily evaluated.) In order to be able to accept an inverse Wishart distribution as a representation of his beliefs the researcher must under these circumstances practically accept the notion of implict determinism for his problem. Moreover, he must be believing that the true rate of decrease in the range $1 \leq p \leq q$ is not much slower than that implied by the prior. For instance, suppose it were thought that some (continuous!) meteorological phenomenon could be exactly predicted if only the present state of the atmospheric system were completely known. Then a model of marginally nondeterministic type also states that prescribed accuracy of prediction can be attained with relatively few measurements on the system, an implication a meteorologist may find unacceptable.

We conclude that the kind of prior judgments concerning the accuracy of prediction which are required when q is large must in practice be based on actual experience in the particular field of application. Specifically, there is no justification for letting q grow by way of trying haphazard sets of regressors. Instead the model demands the opposite attitude: the larger the q the more thought must be given to the assumptions. By emphasizing in this way the use of common sense in model building the phenomenon of implicit determinism thus also conveys a constructive message.

REFERENCES

Box, G. E. P. (1980). Sampling and Bayes' inference in scientific modelling and robustness (with discussion). *J. Roy. Statist. Soc. A* **143**, 383–430.

Brown, P. J. (1980). Coherence and complexity in classification problems. *Scand. J. Statist.* **7**, 95–98.

Dawid, A. P. (1981). Some matrix-variate distribution theory: Notational considerations and a Bayesian application. *Biometrika* **68**, 265–274.

Dawid, A. P. (1988). The infinite regress and its conjugate analysis. (In this volume).

Dickey, J. M., Lindley, D. V. and Press, S. J. (1985). Bayesian estimation of a dispersion matrix of a multivariate normal distribution. *Commun. Statist. -Theor. Meth.* **14**, 1019–1034.

Jeffreys, H. (1983). *Theory of Probability.* 3rd Ed., Hong Kong: Clarendon Press.

Lindley, D. V. (1968). The choice of variables in multiple regression, (with discussion). *J. Roy. Statist. Soc. B* **30**, 31-66.

Lindley, D. V. (1978). The Bayesian approach (with Discussion). *Scand. J. Statist.* **5**, 1–26.

Mäkeläinen, T. and Brown, P. J. (1987). Priors for choice of regressors. Reports of the Department of Math., Univ. of Helsinki.

Press, S. J. (1972). *Applied Multivariate Analysis.* New York: Holt, Rinehart and Winston.

Young, A. S. (1977). A Bayesian approach to prediction using polynomials. *Biometrika* **64**, 309–317.

BAYESIAN STATISTICS 3, pp. 697–699
J. M. Bernardo, M. H. DeGroot, D. V. Lindley and A. F. M. Smith, (Eds.)

On Stochastic Approximation and Bayes Linear Estimators

U. E. MAKOV
University of Haifa

SUMMARY

Stochastic approximation is defined and is shown to be related to a broad class of linear Bayes estimators. This motivates the idea of the Bayesianification of stochastic approximation, which is demonstrated in the case of estimating the mean of a contaminated normal distribution.

Keywords: BAYESIANIFICATION; CONTAMINATED NORMAL DISTRIBUTIONS.

1. STOCHASTIC APPROXIMATION

The method of stochastic approximation, originally proposed by Robbins and Monro (1951), is concerned with the problem of finding the root of a regression function which is neither known nor directly observable. In particular let $\mu(t)$ be a fixed function having a unique θ such that, $\mu(\theta) = 0$. It assumed that to each value of t corresponds an observable random variable $\rho(t)$ and a probability density function $f(\rho|t)$ such that

$$\mu(t) = E[\rho(t)] = \int \rho(t) f(\rho|t) d\rho.$$

$\mu(t)$, the regression function of ρ conditioned on t, and θ are assumed unknown and, further, $f(\rho|t)$ need not be known.

Stochastic approximation theory examines the random sequence $t_1, t_2, \ldots$, in which t_1 is an arbitrary number and $t_2, t_3, \ldots$, are found in terms of the observed $\rho(t_i)$ by

$$t_{n+1} = t_n - a_n \rho(t_n), \tag{1}$$

where a_n is a gain sequence of positive numbers. According to Gladyshev (1965), who sharpened the results of Robbins and Monro (1951), the following conditions guarantee convergence of t_n to θ with probability 1:

(i) $\sum_1^\infty a_n = \infty$

(ii) $\sum_1^\infty a_n^2 < \infty$

(iii) $\text{Inf}(t - \theta)\mu(t) > 0$ for each $\varepsilon > 0$ such that

$$\varepsilon < |t - \theta| < \varepsilon^{-1}$$

(iv) $E\rho^2(t) \leq d(1 + t^2)$, where d a positive number.

Sacks (1958) studies the asymptotic normality of (1) and showed that under regularity conditions and for a gain sequence of the type $a_n = a/n$, $n^{-1/2}(t_n - \theta)$ is asymptotically normal with mean zero and variance $\sigma^2 a^2/(2a\beta - 1)$, where $\text{Var}[\rho(t_n)] = \sigma^2$ and $\beta = \beta(\theta) = \frac{d}{dt}\mu(t)|_{t=\theta}$.

It is straightforward to verify that the asymptotic variance is minimal for $a = 1/\beta$. We shall therefore refer to $a_n = 1/\beta n$ as the optimal gain sequence, and note that since β is not necessarily known, the optimal gain function is not always available.

2. LINEAR BAYES ESTIMATORS AND STOCHASTIC APPROXIMATION

In this section we discuss a family of linear Bayes estimators and show that the recursive form of these estimators follow a stochastic approximation recursion. This recursion is characterized by a gain sequence, the parameters of which reflect some of the prior belief on θ. We further show that the Bayes estimators are optimal in the sense described in the previous section.

Let $x_1, x_2, \ldots, x_n$, be independent identically distributed random variables having an unknown mean θ and a known finite variance φ^2. Let the prior mean and variance of θ be $E'(\theta)$ and $V'(\theta)$ respectively. Ericson showed that if the posterior mean takes the linear form $E''(\theta|x_1, \ldots, x_n) = \alpha\bar{x} + (1-\alpha)E'(\theta)$, where $\bar{x} = \sum_{i=1}^{n} x/n$, and α does not depend on the data, then $\alpha = V'(\theta)/[V'(\theta) + E_\theta n^{-1}\varphi^2$. Diaconis and Ylvisaker (1979) characterized conjugate prior distributions with respect to this property when θ is a natural parameter in the exponential family.

Assuming the linearity of the posterior mean in Ericson's sense and adopting a squared error loss function, the linear Bayes estimator takes the recursive from

$$\hat{\theta}_{n+1} = \hat{\theta}_n - a_n^\star[\hat{\theta}_n - x_{n+1}], \tag{2}$$

where $\hat{\theta}_0 E'(\theta)$ and $a_n^\star$, the Bayesian gain sequence, is given by

$$a_n^\star = \frac{1}{n + 1 + \frac{E'_\theta \varphi^2}{V'(\theta)}}$$

Equation (2) clearly has a stochastic approximation structure, (1), and it is straightforward to check that conditions for convergence of $\hat{\theta}$ to θ in probability 1 are satisfied.

We note that $\eta = \frac{E'_\theta\varphi^2}{V'(\theta)}$ is a discounting factor "representing all that is known prior to sampling". This, perhaps, justifies referring to η as the sample size equivalent of the prior.

We further note that for large n, $a_n^\star \approx \frac{1}{n} = \frac{1}{n\beta}$, since $\beta = \frac{d}{d\theta}E_{x|\theta}(\hat{\theta}_n - x_{n+1})|_{\hat{\theta}=\theta} = 1$ and thus the Bayesian gain sequence is optimal in terms of minimum asymptotic variance.

Chien and Fu (1967) studied the optimal gain sequence in (1) such that the mean square error is minimized at each and every stage. Assuming X to be normally distributed, they proved that the optimal sequence is identical with the Bayesian gain sequence when the prior on θ is normal too. This is not surprising and is true in general since the posterior expected loss is known to be minimized by this choice of estimator. This criterion of optimality will be used in the next section.

3. BAYESIANIFICATION OF THE STOCHASTIC APPROXIMATION RECURSION

There are problems for which a coherent and tractable Bayesian solution is unavailable while a convergent stochastic approximation recursion exists. For these problems an approximate Bayesian solution can be devised by modifying the stochastic approximation recursion in such a way that it retains the Bayesian flavour. Such an idea led to the development of the Quasi-Bayes estimator (Makov (1980)) in the context of unsupervised learning. We now propose a different approach discussed in the context of sequentially estimating the mean of an ε-contaminated normal distribution.

Let $x_1, x_2, \ldots, x_n$ be i.i.d. generated from an ε-contaminated normal density $(1-\varepsilon)f_N(\theta;\sigma^2) + \varepsilon f_N(\theta; k\sigma^2)$, where ε, k and σ^2 are assumed known and $f_N(a;b)$ denotes a normal density with mean a and variance b. The Bayesian solution for estimating θ becomes, as n increases, intractable due to a combinatorial explosion, and this remains the case even if a conjugate prior is chosen. A simple stochastic approximation estimator, based on a moment estimator, is

$$\hat{\theta}_{n+1} = \hat{\theta}_n - \frac{1}{n}(\hat{\theta}_n - x_{n+1}) \tag{3}$$

Bayesianification of (3) is carried in three stages:
- In the first stage we suppose $E'(\theta)$ and $V'(\theta) = \tau^2$ known.
- In the second stage we choose $\hat{\theta}_0 = E'(\theta)$, contrary to the arbitrary choice mentioned in Section 1.
- In the last stage we replace the simple gain sequence in (3), $a_n = 1/n$, by a sequence which minimizes the mean square error (m.s.e) at every stage. From (3), where α_n replaces $1/n$ as a gain sequence, we obtain an expression for the m.s.e:

$$E(\hat{\theta}_{n+1} - \theta)^2 = (1 - \alpha_n)^2 E(\hat{\theta}_n - \hat{\theta})^2 + \alpha_n^2 E(x_{n+1} - \theta)^2,$$

or

$$V_{n+1}^2 = (1 - \alpha_n)^2 V_n^2 + \alpha_n^2 \mathrm{Var}(X),$$

where V_n^2 is the m.s.e at stage n and $\mathrm{Var}(X)$ is the variance of the ε-contaminated normal density defined above. By setting the first derivative of V_{n+1}^2 with respect to α_n equal to zero and solving for α_n, we minimize the m.s.e at stage $n+1$ and obtain a locally optimal gain sequence $a_n^+ = V_n^2/(\mathrm{Var}(X) + V_n^2)$. Iterating V_n^2 and a_n^+ alternately and noting that $V_0^2 = V'(\theta)$ and $\mathrm{Var}(X) = \sigma^2[1 + (k-1)\varepsilon]$, we obtain a gain sequence of the type $a_n^0 = \frac{1}{1+\frac{\sigma^2}{\tau^2}C+n}$, where $\tau^2 = V'(\theta)$ and $C = 1 + (k-1)\varepsilon$. The gain sequence discounts the effect a new observation has on the evaluation of the new estimate, relative to the previous estimate. The degree of discount is a function of n, the number of observations available at the last stage and of C. Clearly, the more prevalent (ε) and the more noisy (k) is the contaminant, the higher is the discount. The role of σ^2/τ^2 is identical to a similar ratio found in the Bayesian solution where no contamination is in evidence ($\varepsilon = 0$).

REFERENCES

Robbins, H. and Monro, S. (1951). A stochastic approximation method. *Ann. Math. Statist.* **29**, 373–386.

Gladyshev, E. G. (1965). On stochastic approximation. *Theory of Prob. and its Appl.* **10**, 275–278.

Sacks, J. (1958). Asymptotic distribution of stochastic approximation procedures. *Ann. Math. Statist.* **29**, 373–386.

Ericson, W. A. A note on the posterior mean of a population mean. *J. Roy. Statist. Soc. B* **31**, 332–334.

Diaconis, P. and Ylvisaker, D. (1979). Conjugate priors for exponential families. *Ann. Statist.* **7**, 269–281.

Chien, Y. T. and Fu, K. S. (1967). On Bayesian learning and stochastic approximation. *IEEE Trans. on Sys. Science and Cyber.*, SSC-**3**, 28–38.

Makov, U. E. (1980). Approximations of unsupervised Bayes learning procedures. *Bayesian Statistics.* (J. M. Bernardo, M. H. DeGroot, D. V. Lindley and A. F. M. Smith, eds.) Valencia: University Press.

BAYESIAN STATISTICS 3, pp. 701–704
J. M. Bernardo, M. H. DeGroot, D. V. Lindley and A. F. M. Smith, (Eds.)

Reparameterisation for Bayesian Inference in ARMA Time Series

J. M. MARRIOTT
Trent Polytechnic, Nottingham

SUMMARY

In this paper a method of parameter transformation to facilitate Bayesian inference for ARMA time series models is proposed and illustrated, together with graphical displays of posterior densities.

Keywords: AUTOREGRESSIVE; MOVING AVERAGE; ADMISSIBLE REGION; REGULARISING TRANSFORMATIONS.

1. INTRODUCTION

Modern methods for the implementation of Bayesian analysis have been motivated by the need to perform the necessary integration in high dimensions. However, the key aspect of the approach proposed by Smith *et al.* (1987) is the transformation of the parameters to the real line in order to make the Gauss-Hermite and importance sampling approaches feasible. In this paper a reparameterisation strategy is suggested for ARMA time series that ensures the admissibility of the autoregressive-moving average parameters and is easily interpreted in terms of the untransformed model.

2. BAYESIAN ANALYSIS OF ARMA MODELS

The process $\{y_t\}$ is said to be an autoregressive-moving average process of order (p, q), ARMA(p, q), if it is generated by the difference equation

$$\phi(B)y_t = \theta(B)\varepsilon_t \tag{1}$$

where B is the backward shift operator, $B^k y_t = y_{t-k}$, $\{\varepsilon_t\}$ is a white noise process,

$$\phi(B) = 1 - \phi_1 B - \phi_2 B^2 - \cdots - \phi_p B^p, \theta(B) = 1 - \theta_1 B - \theta_2 B^2 - \cdots - \theta_q B^q.$$

For the process $\{y_t\}$ to be stationary the roots of $\phi(B) = 0$ must lie outside the unit circle and if the roots of $\theta(B) = 0$ also lie outside the unit circle the process is said be invertible, in which case there will be a unique model corresponding to the likelihood function [Box and Jenkins (1970)].

These conditions restrict the parameter vectors ϕ and θ to the regions C_p and C_q, say, so that admissible values of (ϕ, θ) must lie in $C_p \times C_q$.

For first and second order models the regions are simple, for example for the ARMA$(1, 1)$ process the region is the square, $C_1 \times C_1$ defined by $-1 < \phi < 1, -1 < \theta < 1$ and for the ARMA$(2, 0)$ process C_2 is the triangle defined by $-1 < \phi_2 < 1, |\phi_1| < 1 - \phi_2$. However for $k > 2, C_k$ becomes complicated.

In likelihood analysis estimation must proceed either by constrained optimisation to ensure stationarity or by checking the estimates at each iteration. In practice the most widely used methods do not enforce the invertibility condition and use the approximation to the likelihood function suggested by Box and Jenkins (1970). For a fuller discussion see Granger and Newbold (1977) and Harvey (1981). In a Bayesian analysis $C_p \times C_q$ is the region for integration when obtaining joint and marginal posterior distributions of the parameters or for evaluating posterior expected values of the form

$$E[g(\psi)|y] = \int g(\psi)p(\psi|y)d\psi, \tag{2}$$

where $\psi' = [\phi', \theta']$ and $p(\psi|y)$ is the posterior density.

A special case of (2) is the posterior predictive density

$$p(y_F|y) = \int p(y_F|\psi)p(\psi|y)d\psi, \tag{3}$$

where $p(y_F|\psi)$ is the density of future data.

The problem is to find a regularising transformation from $C_p \times C_q$ to the plane in order to use the Nottingham University BAYES 4 integration techniques [see Smith *et al.* (1987) and the references therein].

Monahan (1983) gives a transformation of the parameters that permits integration over $C_p \times C_q$ as part of a fully Bayesian approach to ARMA time series. However the approach suffers from several drawbacks; it is not suitable for maximum likelihood estimation, it involves checking each integration grid point for admissibility and, as shown by Marriott (1987) the integration rules used can be greatly improved on. Monahan (1984) proposed a different transformation that could be used to overcome the first two of these objections but it lacks an intuitive appeal in that the transformed parameters have no clear interpretation when they are obtained from θ.

3. A REGULARISING TRANSFORMATION

The approach proposed here (as did those of Monahan (1983) and Osborn (1976)) starts by recognising that the polynomial in x, with real coefficients

$$1 - \tau_1 x - \tau_2 x^2 - \cdots - \tau_m x^m \tag{4}$$

can be factored as

$$\begin{aligned} &\prod_{j=1}^{m/2}(1 - a_{1,j}x - a_{2,j}x^2) && \text{for } m \text{ even} \\ \text{and } (1 - bx) &\prod_{j=1}^{(m-1)/2}(1 - a_{1,j}x - a_{2,j}x^2) && \text{for } m \text{ odd} \end{aligned} \tag{5}$$

where $a_{i,j}$ and b are also real.

While the transformation from the $a_{i,j}, b$ to the τ_j is not one to one, for any given $a_{i,j}, b$ the τ_j can be obtained uniquely by expanding (5) and, when the polynomial at (4) is $\phi(B)$ or $\theta(B)$, the condition that (ϕ, θ) lies in $C_p \times C_q$ is satisfied by ensuring that each pair $(a_{1,j}, a_{2,j})$ lies in the triangle C_2 defined for the ARMA$(2,0)$ process above, and, in the case of m odd, $|b| < 1$.

For integration purposes this transformation and an algorithm for the Jacobian are presented in Marriott and Smith (1988).

In order to then use the Nottingham University BAYES 4 integration techniques Marriott (1987) has already shown that regularising transformations from the admissible region to the plane will be provided by

$$a^{\star}_{1,j} = \ln\left\{\frac{1 - a_{2,j} + a_{1,j}}{1 - a_{2,j} - a_{1,j}}\right\}, \quad a^{\star}_{2,j} = \ln\left\{\frac{1 + a_{2,j}}{1 - a_{2,j}}\right\} \tag{6}$$

for any pair $(a_{1,j}, a_{2,j})$ and

$$b^{\star} = \ln\left\{\frac{1 + b}{1 - b}\right\} \tag{7}$$

for any b when p or q are odd.

Numerical integration is then performed over R^n, the space of the $a^{\star}$ (and $b^{\star}$) parameters. For each integration point the tranformations at (6) and (7) uniquely determine a set of a (and b) parameters and for any given set of a (and b) there will be a unique set of τ obtained by expanding the expression at (5), these τ values are used to compute the likelihood using Ansley's (1979) algorithm.

There then remains the problem of finding starting values for the BAYES 4 integration. First stage consistent estimates for (ϕ, θ) are obtained (see Saikkonen (1986)) and the roots of the estimated polynomials $\hat{\phi}(B)$ and $\hat{\theta}(B)$ are obtained numerically.

A well defined factorisation of the type (5) must now be constructed taking into account the fact that the transformation from the τ's to the a's and b's is not one-to-one.

One straightforward way to achieve this is to take complex conjugate pairs to obtain quadratic factors with real coefficients and, where more than one real root exists, to rank the real roots in terms of their absolute values, $r_1, r_2, \ldots, r_k$, say, and then take them in pairs (r_1, r_k), $(r_2, r_{k-1}), \ldots$ to provide the quadratic factors.

This rule serves both to alleviate the potential many-to-one problem arising in the choice of quadratic factors and also helps to encourage the stability of subsequent numerical search procedures.

The parameters are then transformed to the plane using (6) and (7) and maximum likelihood estimates of the transformed parameters are used as the basis of the start of the BAYES 4 integration.

This approach has been successfully tested on problems involving up to six parameters and Figure 1 and Figure 2 illustrate the effect of the transformations.

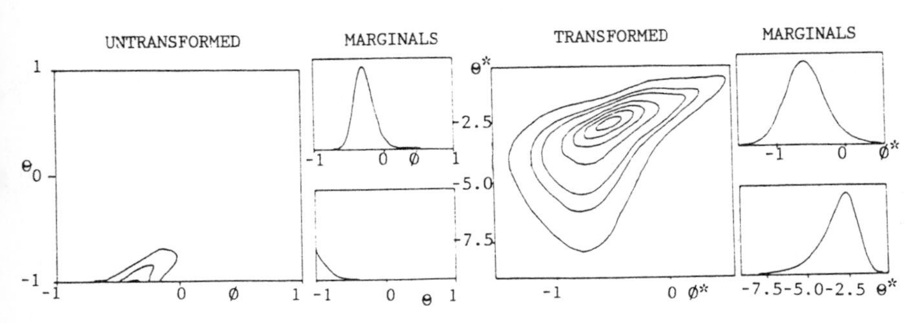

Figure 1. *Special Case of ARMA*$(1, 1)$.

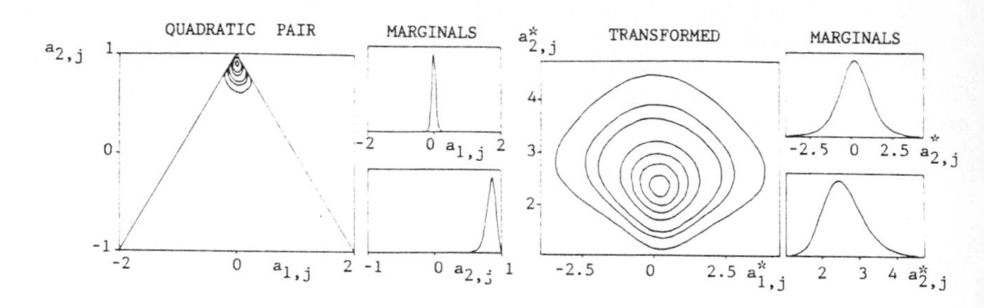

Figure 2. *The effect of the transformation from a quadratic pair.*

4. CONCLUSION

In conclusion it can be said that by using this approach to fitting ARMA models to time series data

1. the admissibility of the ARMA parameters is ensured whether likelihood or Bayesian analysis is involved.
2. by concentrating on a quadratic factorisation the results can be readily interpreted in terms of the roots of the original ARMA polynomials without the need for any numerical approximation.
3. the regularising transformations enable the use of the powerful BAYES 4 procedure to its best advantage.

ACKNOWLEDGEMENTS

I would like to thank the referee for a number of useful comments.

REFERENCES

Ansley, C. F. (1979). An algorithm for the exact likelihood of a mixed autoregressive-moving average process, *Biometrika* **66**, 59–65.

Box, G. E. P. and Jenkins, G. M. (1970). *Time series analysis forecasting and control.* San Francisco: Holden Day.

Granger, C. W. J. and Newbold, P. (1977), *Forecasting Economic time series.* Academic Press.

Harvey, A. C. (1981). *Time series Models.* Oxford: Philip Allan.

Marriott, J. M. (1987). Bayesian numerical and graphical methods for Box-Jenkins time series. *The Statistician,* special issue: *Practical Bayesian Statistics.*

Marriott, J. M. and Smith A. F. M. (1988). Aspects of numerical Bayesian methodology for ARMA models, *Tech. Rep.* **88-04**, Department of Mathematics, Nottingham University.

Monahan, J. F. (1983). Fully Bayesian analysis of ARMA time series models, *Journal of Econometrics* **21**, 307–331.

Monahan, J. F. (1984). A note on enforcing stationarity in autoregressive-moving average models. *Biometrika* **71**, 403–404.

Osborn, D. R. (1976). Maximum likelihood estimation of moving average processes, *Annals of Economic and Social measurement* **5/1**, 75–87.

Saikkonen, P. (1986). Asymptotic properties of some preliminary estimators for autoregressive-moving average time series models. *Journal of time series analysis* **7**, 133–155.

Smith, A. F. M., Skene, A. M., Shaw, J. E. H. and Naylor, J. C. (1987). Progress with numerical and graphical methods for practical Bayesian Statistics. *The Statistician,* special issue: *Practical Bayesian Statistics.*

BAYESIAN STATISTICS 3, pp. 705–711
J. M. Bernardo, M. H. DeGroot, D. V. Lindley and A. F. M. Smith, (Eds.)

Inferences About the Ratio of Linear Combinations of the Coefficients in a Multiple Regression Model

M. MENDOZA
Dept. de Matemáticas, Fac. Ciencias UNAM
and
Dept. de Estadística, Universidad de Valencia.

SUMMARY

In some applications it is required to produce inferences about a ratio of normal means. In many other cases the parameter of interest is again of this ratio-type. The available procedures for this kind of situation are always related to specific problems and it is usually difficult to determine their general properties analytically. In this paper, a general structure which includes all the usual problems as particular cases is analyzed. A class of prior distributions leading to an easily tractable posterior is proposed. The results obtained generalize previous analysis for the ratio of normal means and the slope ratio assay.

Keywords: BIOASSAY; LINEAR MODELS; MEANS RATIO; NON-INFORMATIVE PRIORS; SLOPE RATIO.

INTRODUCTION

The problem of estimating a ratio of normal means has been investigated within the Bayesian framework, by among others, Kappenman, Geisser and Antle (1970), Bernardo (1977) and Sendra (1982). Darby (1980) and more recently, Mendoza (1987) have dealt with some closely related problems concerning the estimation of the horizontal distance between two parallel lines and the estimation of the slope ratio for two straight lines with a common intercept.

The situation where a statistical analysis is required for a parameter of this ratio-type arises in a natural way in the context of *Bioassay*. One of the most usual problems in bioassay is that of assessing the relative potency of two different stimuli. If a direct assay is considered, the parameter describing the relative potency is a *ratio of means*. The case of an indirect assay leads, when the most common models (*parallel lines, slope ratio*) are used, to the estimation of a parameter directly related to the functions studied by Darby (1980) and Mendoza (1987). It must be stressed however, that the basic elements of a bioassay may be identified with the constituents of some other research problems (e.g. quality control, marketing) where the appropriate statistical procedures might have a useful application.

An inspection of the literature reveals that the available results are always related to very particular problems and typically provide analytical conclusions only for the case where a *non-informative prior* is utilized. Here, the analysis is extended to cover a more general situation, namely that of making inferences about the ratio of linear combinations of the coefficients in a multiple regression model, which includes as particular cases all the problems mentioned before. Moreover, we shall propose a family of prior distributions which not only leads to a number of interesting, analytical properties for the posterior distribution of the parameter of interest, but also allows the scientist to either express his subjective prior knowledge about the parameters involved or to use an appropriately defined non-informative prior.

2. THE MODEL

Suppose that $\boldsymbol{Y} \in \boldsymbol{R}^n$ is a random vector following a multivariate normal distribution with mean vector $\boldsymbol{X}\theta$ and covariance matrix $\sigma^2 \boldsymbol{I}$ where $\boldsymbol{X}$ is a fixed $(N \times k)$, full rank design matrix, $\theta \in \boldsymbol{R}^k (k < N)$ is a vector of unknown coefficients and $\sigma > 0$ is also unknown. Suppose, in addition, that the parameter of interest is given by

$$\rho = \lambda_1^t \theta / \lambda_2^t \theta$$

where λ_1 and λ_2 are two fixed, linearly independent vectors in $\boldsymbol{R}^k (\lambda_2^t \theta \neq 0)$. Using a common reparametrization technique we may write

$$\begin{aligned} \boldsymbol{X}\theta &= \boldsymbol{X}(\boldsymbol{L}^{-1}\boldsymbol{L})\theta \\ &= \boldsymbol{X}(\boldsymbol{L}^{-1})(\boldsymbol{L}\theta) \\ &= \boldsymbol{Z}\beta \end{aligned}$$

where $\boldsymbol{Z} = \boldsymbol{X}\boldsymbol{L}^{-1}, \beta = \boldsymbol{L}\theta$ and $\boldsymbol{L}$ is a $(k \times k)$ full rank matrix such that $\beta_1 = \lambda_1^t \theta$ and $\beta_2 = \lambda_2^t \theta$. Therefore, our model can be written as

$$\boldsymbol{Y} \sim N(\boldsymbol{Z}\beta, \sigma^2 \boldsymbol{I}) \tag{1}$$

and the parameter of interest is simply

$$\rho = \beta_1 / \beta_2. \tag{2}$$

An additional transformation of the parameter space may be applied to express the likelihood function explicitly in terms of ρ. Let $\gamma \in \boldsymbol{R}^k$ be a vector such that

$$\gamma_1 = \rho, \gamma_i = \beta_i; \quad i = 2, \ldots, k$$

This is a one-to-one transformation so that β can be expressed in terms of γ $(\beta = \beta(\gamma))$. The likelihood function may now be written as

$$L(\gamma, \sigma | \boldsymbol{Y}) = (2\pi\sigma^2)^{-N/2} \exp\{-(\boldsymbol{Y} - \boldsymbol{Z}\beta(\gamma))^t (\boldsymbol{Y} - \boldsymbol{Z}\beta(\gamma))/(2\sigma^2)\} \tag{3}$$

in order to be combined with the prior $p(\gamma, \sigma)$.

3. PRIOR - POSTERIOR ANALYSIS

We shall firstly recall that our model is parametrized by the vector

$$(\gamma^t, \sigma) = (\rho, \beta_2, \ldots, \beta_k, \sigma)$$

and not only by the parameter of interest ρ. Under such circumstances, $\beta_2, \ldots, \beta_k$ and σ may be considered as nuisance or incidental parameters that should be, in a sense, "eliminated" from the analysis. As it has been stated, most of the available results on this topic have been obtained using some form of a *joint* non-informative prior for (γ, σ) or equivalently for (β, σ). Typically, these results are analytically involved and require some numerical analysis. Here, we shall distinguish the role of ρ from that of the nuisance parameters. Let us write

$$p(\rho, \beta_2, \ldots, \beta_k, \sigma) = p(\rho) p(\beta_2, \ldots, \beta_k, \sigma | \rho)$$

Even if some subjective prior information is available about the nuisance parameters, it may be difficult to describe the knowledge about $(\beta_2, \ldots, \beta_k, \sigma | \rho)$, specially if k is large.

As an alternative we propose to "eliminate" the nuisance parameters using a conditional non-informative prior. For this purpose is useful to note that for a fixed value of ρ, the mean vector $Z\beta$ where

$$Z = (Z_1, Z_2, \ldots, Z_k) \qquad (N \times k)$$

and

$$\beta^t = (\beta_1, \beta_2, \ldots, \beta_k)$$

can be replaced by $W\Psi$ where

$$W = (\rho Z_1 + Z_2, Z_3, \ldots, Z_k)(N \times (k-1))$$

and

$$\Psi^t = (\beta_2, \beta_3, \ldots, \beta_k).$$

Therefore, conditional on ρ, the model (1) can be written as

$$Y \sim N(W\Psi, \sigma^2 I) \tag{4}$$

and a conditional non-informative prior for the nuisance parameters can be derived as the limit of the respective conjugate family (DeGroot 1970, chapter 9), or using any other method as, for example, Jeffrey's rule (Box and Tiao, 1973 section 1.3) or reference analysis (Bernardo, 1979). The most important result is that for all those methods the prior for $\beta_2, \ldots, \beta_k$ and σ is *independent* of ρ and has the form

$$\pi(\beta_2, \beta_3, \ldots, \beta_k, \sigma | \rho) \propto \sigma^{-r}$$

for some $r > 0$. Hence, we claim that a reasonable approximation to prior beliefs about (γ, σ) may be

$$p(\gamma, \sigma) \propto p(\rho)\sigma^{-r} \tag{5}$$

where $p(\rho)$ is the marginal prior density of ρ describing the initial knowledge that the scientist has about ρ. If we use this prior, the joint posterior density is given by

$$\begin{aligned} p(\gamma, \sigma | Y) &\propto p(\rho)\sigma^{-r}(2\pi\sigma^2)^{-N/2} \exp\{-(Y - Z\beta(\gamma))^t (Y - Z\beta(\gamma))/(2\sigma^2)\} \\ &\propto p(\rho)\sigma^{-(N+r)} \exp\{-(Y - Z\beta(\gamma))^t (Y - Z\beta(\gamma))/(2\sigma^2)\} \end{aligned} \tag{6}$$

and then

$$p(\gamma | Y) \propto p(\rho) \int \sigma^{-(N+r)} \exp\{-(Y - Z\beta(\gamma))^t (Y - Z\beta(\gamma))/(2\sigma^2)\} d\sigma.$$

Using some results related to the inverted Gamma distribution is easy to show that

$$p(\gamma | Y) \propto p(\rho)[(Y - Z\beta(\gamma))^t (Y - Z\beta(\gamma))]^{-(N+r-1)/2} \tag{7}$$

Finally, the posterior density for the parameter of interest ρ can be obtained from the expression

$$\begin{aligned} p(\rho|\boldsymbol{Y}) &\propto p(\rho) \int [(\boldsymbol{Y} - \boldsymbol{Z}\beta(\gamma))^t (\boldsymbol{Y} - \boldsymbol{Z}\beta(\gamma))]^{-(N+r-1)/2} d\beta_2 d\beta_3 \dots d\beta_k \\ &= p(\rho) \int [(\boldsymbol{Y} - \boldsymbol{W}\boldsymbol{\Psi})^t (\boldsymbol{Y} - \boldsymbol{W}\boldsymbol{\Psi})]^{-(N+r-1)/2} d\Psi. \end{aligned} \tag{8}$$

If we let $\hat{\psi}$ be the usual least square estimate of ψ and recall some elementary results of the theory of linear models, then we may write

$$\begin{aligned} p(\rho|\boldsymbol{Y}) &\propto p(\rho)\{H(\rho)\}^{-(N+r-1)/2}. \\ &\cdot \int [1 + (\hat{\psi} - \boldsymbol{\Psi})^t (\boldsymbol{W}^t \boldsymbol{W})(\hat{\psi} - \boldsymbol{\Psi})/H(\rho)]^{-(N+r-1)/2} d\Psi \end{aligned} \tag{9}$$

where $H(\rho) \equiv (\boldsymbol{Y} - \boldsymbol{W}\hat{\psi})^t (\boldsymbol{Y} - \boldsymbol{W}\hat{\psi})$. The kernel of a multivariate Student's t distribution can be easily recognized so that

$$\begin{aligned} p(\rho|\boldsymbol{Y}) &\propto p(\rho)\{H(\rho)\}^{-(N+r-1)/2} |(\boldsymbol{W}^t \boldsymbol{W})/H(\rho)|^{-1/2} \\ &= p(\rho)\{H(\rho)\}^{-(N+r-1)/2} |\boldsymbol{W}^t \boldsymbol{W}|^{-1/2} \{H(\rho)\}^{(k-1)/2} \\ &= p(\rho)\{H(\rho)\}^{-(N+r-k)/2} |\boldsymbol{W}^t \boldsymbol{W}|^{-1/2}. \end{aligned} \tag{10}$$

Some additional calculations are necessary to show that, as a function of ρ,

$$|\boldsymbol{W}^t \boldsymbol{W}| \propto Q(\rho),$$

where $Q(\rho) = c_2\rho^2 + c_1\rho + c_0$ is such that

$$c_2 = v_{22}/\Delta, c_1 = -2\nu_{12}/\Delta, c_0 = v_{11}/\Delta,$$

and

$$\begin{aligned} &v_{22} = \text{Var}(\hat{\beta}_2)/\sigma^2, \nu_{12} = \text{Cov}(\hat{\beta}_1, \hat{\beta}_2)/\sigma^2, \nu_{11} = \text{Var}(\hat{\beta}_1)/\sigma^2, \\ &\Delta = \nu_{22}\nu_{11} - \nu_{12}^2, \end{aligned}$$

$\hat{\beta}_1$ and $\hat{\beta}_2$ being the usual least squares estimators of β_1 and β_2 for the reparametrized model (1). As a consequence, we have that

$$p(\rho|\boldsymbol{Y}) \propto p(\rho)\{H(\rho)\}^{-(N+r-k)/2}\{Q(\rho)\}^{-1/2}. \tag{11}$$

Finally, and using some results from the classical theory of linear models, it can be shown that

$$p(\rho|\boldsymbol{Y}) \propto p(\rho)\{Q(\rho)\}^{-1/2}[\eta + h(\rho)]^{-(N+r-k)/2} \tag{12}$$

where

$$\begin{aligned} h(\rho) &= (\hat{\beta}_1 - \rho\hat{\beta}_2)^2/[S^2\{\nu_{22}\rho^2 - 2\nu_{12}\rho + \nu_{11}\}], \\ S^2 &= (\boldsymbol{Y} - \boldsymbol{Z}\hat{\beta})^t (\boldsymbol{Y} - \boldsymbol{Z}\hat{\beta})/\eta, \\ \eta &= N - k. \end{aligned}$$

As a first general result, we have that since the positive function $Q(\rho)$ is a second degree polynomial with no real roots, then $p(\rho|\boldsymbol{Y})$ is a proper density for any proper prior $p(\rho)$.

Even if the marginal prior is improper the fact that $\{Q(\rho)\}^{-1/2}$ is an integrable function (Gradshteyn and Ryzhik 1965, p. 81) guarantees a proper posterior for any *bounded* prior $p(\rho)$. As a second general result, it can be recognized that the function $h(\rho)$ in (12) is precisely the pivotal quantity used in the classical approach to produce, via the celebrated Fieller theorem (Fieller 1954), the so-called *confidence* regions for ρ.

Other, more specific, properties of the posterior density $p(\rho|\boldsymbol{Y})$ depend on the choice of $p(\rho)$. In any case, the scientist could try to describe his initial state of knowledge by means of a function leading to an analytically tractable posterior. In a situation where only relatively vague prior information about ρ is available, the result obtained using the reference analysis is particularly interesting. Applying the formulae developed by Bernardo (1979) to obtain the reference prior (i.e. the prior which maximizes the missing information about ρ and the missing residual information about the nuisance parameters given ρ) we get

$$\pi(\gamma,\sigma) \propto \{Q(\rho)\}^{-1/2}\sigma^{-k}. \tag{13}$$

It follows that the reference prior $\pi(\gamma,\sigma)$ is a member of the family of prior densities proposed above. In fact, it suffices to take $r = k$ and $p(\rho) = \{Q(\rho)\}^{-1/2}$. An interesting property of this reference prior is that the marginal density $p(\rho)$ is *proper*. The posterior reference density for ρ is given by

$$\pi(\rho|\boldsymbol{Y}) \propto \{Q(\rho)\}^{-1}[\eta + h(\rho)]^{-N/2}. \tag{14}$$

Some additional remarks can be made for this specific posterior. In particular, $\pi(\rho|Y)$ may have one or two modes; they can be explicitly determined as the roots of the third degree polynomial which is obtained from equating the first derivative of $\pi(\rho|\boldsymbol{Y})$ to zero. In any case, it can be verified that this posterior generalizes previous analysis for the ratio of normal means (Bernardo, 1977) and for the slope ratio assay (Mendoza, 1987).

4. A NUMERICAL EXAMPLE

In order to obtain a graphical idea of some of the results that may be produced if a prior of the type proposed in this paper is adopted, we have analyzed a set of data simulated using an *slope ratio* bioassay model. Specifically, we have simulated an experiment where four doses $(X_{11}, X_{12}, X_{13}, X_{14})$ of a first stimulus and four doses $(X_{21}, X_{22}, X_{23}, X_{24})$ of a second stimulus are assayed three times so that a set $\{Y_{ijk};\ i = 1,2; j = 1,\ldots,4;\ k = 1,\ldots,3\}$ of $N = 24$ conditionally independent Normal observations is obtained with common variance $\sigma^2 = 1$ and such that

$$\begin{aligned} E(Y_{1jk}) &= \alpha + \delta_1 X_{1j}; \quad j = 1,\ldots,4; \quad k = 1,\ldots,3 \\ E(Y_{2jk}) &= \alpha + \delta_2 X_{2j}; \quad j = 1,\ldots,4; \quad k = 1,\ldots,3 \end{aligned} \tag{15}$$

where $\alpha = 0, \delta_1 = 10$ and $\delta_2 = 20$. A study of the relative potency of the stimuli can be accomplished by estimating the ratio $\rho = \delta_2/\delta_1 (= 2)$. It is easy to identify this structure with a special case of the model discussed in section 1. The data simulated using this model is given in Table 1. The posterior marginal density obtained when the reference prior

$$\pi(\gamma,\sigma) \propto \{Q(\rho)\}^{-1/2}\sigma^{-3} \tag{16}$$

is used has two modes located at $\rho_1 = -14.721$ and $\rho_2 = 2.009$, respectively. Since this posterior satisfies the condition $\pi(\rho < 0|\boldsymbol{Y}) < 10^{-10}$, we can ignore the first mode and analyze the behaviour of $\pi(\rho|\boldsymbol{Y})$ in a neighbourhood of the second mode which is, in fact, very close to the true value of the parameter. The posterior probability is clearly accumulated around ρ_2. As an example, it can be verified that $\pi(1.91 < \rho < 2.09|\boldsymbol{Y}) = 0.954$. This

Stimulus	1				2			
Doses	0.5	1.5	2.5	3.5	1.0	2.0	3.0	4.0
	5.784	16.470	25.948	33.770	21.361	39.818	60.333	81.391
	3.608	15.623	24.347	37.347	21.508	40.989	57.723	80.396
	5.508	14.594	25.228	33.228	21.047	38.550	59.430	81.443

Table 1. *Simulated data.*

fact and many others are illustrated in Figure 1 where the main part of $\pi(\rho|\boldsymbol{Y})$ is exhibited. Figure 1 also shows the posterior density $p(\rho|\boldsymbol{Y})$ associated with the prior

$$p(\gamma, \sigma) \propto \sigma^{-3} \tag{17}$$

which could be considered as another type of non-informative prior where the marginal for ρ is locally uniform.

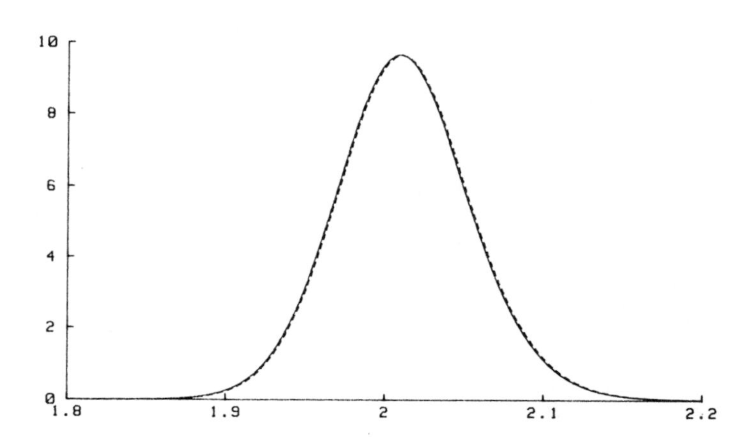

Figure 1. *Posterior densities* ($\pi(\rho|\boldsymbol{Y})$: —; $p(\rho|\boldsymbol{Y})$: – –)

Both posterior densities, $\pi(\rho|\boldsymbol{Y})$ and $p(\rho|\boldsymbol{Y})$, lead to practically the same numerical results as shown in the Figure. In fact, it can be verified that for a large range of positive values of m a prior of the form

$$p(\gamma, \sigma) \propto \{Q(\rho)\}^{-m}\sigma^{-3} \tag{18}$$

leads to a very similar posterior.

5. CONCLUDING REMARKS

There are many models, not necessarily arising from biological applications that may be analyzed using the results of this paper. For example, the problem of estimating the intersection point of two non-parallel straight lines, which has relevance in the study of switching regression models (Smith and Cook, 1980), can be easily identified as a particular case of the structure analyzed in this paper. Some results concerning this specific problem have been obtained and will be published elsewhere.

An important remark is that the full rank condition assumed for the matrix $\boldsymbol{X}$ can be removed since only the inverse of the matrix $(\boldsymbol{W}^t\boldsymbol{W})$ is required for the calculations. In this sense, the results can be extended to cover some non-full rank models. A more general conclusion might be obtained if $(\boldsymbol{W}^t\boldsymbol{W})^{-1}$ could be replaced by some generalized inverse.

ACKNOWLEDGEMENTS

This work was partly supported by the Sistema Nacional de Investigadores of México. The author wishes to express his deepest gratitude to the staff of the Departamento de Estadística, Univ. of Valencia, where he has spent a splendid time.

REFERENCES

Bernardo, J. M. (1977). Inferences about the ratio of normal means: A Bayesian approach to the Fieller-Creasy problem in *Recent Developments in Statistics.* (Barra et al. eds.). Amsterdam: North Holland, 345–50.

Bernardo, J. M. (1979). Reference posterior distributions for Bayesian inference (with discussion). *J. Roy. Statist. Soc. B* **16**, 186–94.

Box, G. E. P. and Tiao, G. C. (1973). *Bayesian Inference in Statistical Analysis.* Reading, Mass: Addison-Wesley.

Darby, S. C. (1980). A Bayesian approach to parallel line bioassay. *Biometrika* **67**, 3, 607–12.

DeGroot, M. H. (1970). *Optimal Statistical Decision.* New York: Mc Graw-Hill.

Fieller, E. C. (1954). Some problems in interval estimation. *J. Amer. Statist. Assoc.* **16**, 175–85.

Gradshteyn, I. S. and Ryzhik, I. M. (1965). *Tables of Integrals, Series and Products.* New York: Academic Press.

Kappenman, R. F., Geisser, S. and Antle, C. E. (1970). Bayesian and fiducial solutions for the Fieller-Creasy problem. *Sankhya B* **32**, 331–40.

Mendoza, M. (1987). A Bayesian analysis of the slope ratio bioassay. *Tech. Rep.* Univ. Valencia

Sendra, M. (1982). Distribuci6n final de referencia para el problema de Fieller-Creasy. *Trabajos de Estadística e Investigaci6n Operativa* **33**, 55–72.

Smith, A. F. M. and Cook, D.G. (1980). Straight lines with a change point: a Bayesian analysis of some renal transplant data. *Applied Statistics* **2**, 180–9

BAYESIAN STATISTICS 3, pp. 713–724
J. M. Bernardo, M. H. DeGroot, D. V. Lindley and A. F. M. Smith, (Eds.)

Optimal Design for Threshold Attainment

R. J. OWEN and M. CAIN
University College of Wales, Aberystwyth

SUMMARY

The predictive distribution is given for the linear model with possibly correlated errors and an informative prior distribution. Two types of problem are considered: the minimisation of cost (risk) subject to constraints on the probability of threshold attainment and the optimisation of a balanced combination of the probability of threshold attainment and cost. Restricted and unrestricted designs are considered and the procedures determine an optimal design in each case.

Keywords: DESIGN; LINEAR MODEL; MULTIPLE REGRESSION; THRESHOLD ATTAINMENT; PREDICTIVE DISTRIBUTION; PREDICTOR LEVELS; PRIOR INFORMATION.

1. INTRODUCTION

It is often the case that explanatory variables are to be chosen so that the corresponding response is to exceed a given hurdle h. Often, cost or resource constraints or randomness preclude overcoming the hurdle with certainty so a stochastic model is appropriate. Moreover, both prior information and previous data may be relevant to the choice of values of the explanatory or control variables. The class of problems considered here has all of these ingredients.

In the sequel $N(\boldsymbol{\mu}, \boldsymbol{D})$ denotes a multivariate normal distribution with mean vector $\boldsymbol{\mu}$ and dispersion matrix $\boldsymbol{D}$. Unless otherwise stated it is understood that the distribution is non-singular so that $\boldsymbol{D}$ is positive definite.

Let $\boldsymbol{Y}_1(n_1 \times 1)$ and $\boldsymbol{Y}_2(n_2 \times 1)$ denote the response vectors where $\boldsymbol{Y}_1$ is thought of as occurring before $\boldsymbol{Y}_2$ in time. A model is assumed where the combined vector $\boldsymbol{Y}' = (\boldsymbol{Y}_1', \boldsymbol{Y}_2')$ has a multivariate normal distribution:

$$(\boldsymbol{Y}|\boldsymbol{\beta}, \rho) \sim N(\boldsymbol{X}\boldsymbol{\beta}, \rho^{-1}\boldsymbol{V}) \tag{1.1}$$

with $\rho > 0$ and $p \times 1$ vector $\boldsymbol{\beta}$ both unknown but $n \times p$ matrix $\boldsymbol{X}$ and $n \times n$ matrix $\boldsymbol{V}$ known. The matrix $\boldsymbol{X}$ of explanatory variables and dispersion matrix $\boldsymbol{V}$ are partitioned:

$$\boldsymbol{X} = \begin{bmatrix} \boldsymbol{X}_1 \\ \boldsymbol{X}_2 \end{bmatrix} \qquad \boldsymbol{V} = \begin{bmatrix} \boldsymbol{V}_{11} & \boldsymbol{V}_{12} \\ \boldsymbol{V}_{21} & \boldsymbol{V}_{22} \end{bmatrix}$$

where $n_1 \times p$ matrix $\boldsymbol{X}_1$ is given and $n_2 \times p$ matrix $\boldsymbol{X}_2$ is controllable. The matrix $\boldsymbol{V}_{ij}$ is $n_i \times n_j (i, j = 1, 2)$. Here typically $n_1 > p$ although none of $\boldsymbol{X}_1$, $\boldsymbol{X}_2$ or $\boldsymbol{X}$ need be of full column rank for the analysis to be valid. In particular the expressions for the predictive distribution hold for the analysis of variance case as well as that of regression.

Various correlation structures could arise for instance when $\boldsymbol{Y}$ is obtained as a linear transformation of uncorrelated responses or where components of $\boldsymbol{Y}$ are subject to common random fluctuations.

It is assumed that prior information may be adequately represented for β, ρ by a normal-gamma distribution

$$(\beta|\rho) \sim N(\boldsymbol{m}_0, \rho^{-1}\boldsymbol{D}_0), \qquad \rho \sim \gamma\left(\frac{1}{2}d_0, \frac{1}{2}g_0\right), \tag{1.2}$$

where $\boldsymbol{m}_0$, $\boldsymbol{D}_0$, $d_0 > 2$ and $g_0 > 0$ are known. In the elicitation process it may be better to elicit $\bar{\boldsymbol{D}}_0$, the unconditional dispersion matrix of β, and then $\boldsymbol{D}_0$ would be given by

$$\boldsymbol{D}_0 = [(d_0 - 2)/g_0]\bar{\boldsymbol{D}}_0.$$

Some useful suggestions for elicitation are made in Zellner (1985), Kadane *et al.* (1980) and references therein.

Interest centres on choice of $\boldsymbol{X}_2$ after observing $\boldsymbol{Y}_1$ so as to make it most likely that the components of $\boldsymbol{Y}_2$ surpass their specified hurdles $h_1, h_2, \ldots, h_{n_2}$. It is therefore required to determine the predictive distribution of $\boldsymbol{Y}_2$ (Section 2). Section 3 considers optimisation with probabilistic constraints and Section 4 deals with optimisation of a threshold probability and cost combination.

Notation.
$\Re^p$ denotes the set of p-tuples of reals and ϕ the empty set.
$\mathcal{S}^c$ denotes the complement of set $\mathcal{S}$ and $\bar{\mathcal{S}}$ the convex closure of $\mathcal{S}$.

2. THE PREDICTIVE DISTRIBUTION

For the case of homoscedastic and independent responses the predictive distribution of $\boldsymbol{Y}_2$ given $\boldsymbol{Y}_1$ is given by Zellner (1971) and Press (1982) for the case of a diffuse prior distribution and by Broemeling (1985) for an informative prior distribution of the form (1.2).

In the more general case considered here it may be shown that the unconditional distribution of $\boldsymbol{Y}$ is multivariate t with

$$\left.\begin{array}{l}\text{degrees of freedom } = d_0, \text{ location vector } \mathbb{E}\,(\boldsymbol{Y}) = \boldsymbol{X}\boldsymbol{m}_0 \\ \text{and dispersion matrix Var } (\boldsymbol{Y}) = [g_0/(d_0-2)]\boldsymbol{K} \\ \text{where } \boldsymbol{K} = \boldsymbol{V} + \boldsymbol{X}\boldsymbol{D}_0\boldsymbol{X}'.\end{array}\right\} \tag{2.1}$$

In the sequel it is convenient to write

$$\boldsymbol{K} = \begin{bmatrix} \boldsymbol{K}_{11} & \boldsymbol{K}_{12} \\ \boldsymbol{K}_{21} & \boldsymbol{K}_{22} \end{bmatrix}$$

where $\boldsymbol{K}_{ij}$ is $n_i \times n_j (i, j = 1, 2)$ so that

$$\boldsymbol{K}_{ij} = \boldsymbol{V}_{ij} + \boldsymbol{X}_i\boldsymbol{D}_0\boldsymbol{X}_j' \quad i, j = 1, 2. \tag{2.2}$$

If follows from the properties of the multivariate t distribution that the predictive distribution:

$$(\boldsymbol{Y}_2|\boldsymbol{Y}_1) \sim \text{ multivariate } t \tag{2.3}$$

with

(i) Degrees of freedom $= d_0 + n_1 = d$;
(ii) $\mathbb{E}\,(\boldsymbol{Y}_2|\boldsymbol{Y}_1) = \boldsymbol{a} + \boldsymbol{X}_2\boldsymbol{b}$;
(iii) $\text{Var}\,(\boldsymbol{Y}_2|\boldsymbol{Y}_1) = [g/(d-2)](\boldsymbol{X}_2'\boldsymbol{A}\boldsymbol{X}_2 - 2\boldsymbol{B}\boldsymbol{X}_2' + \boldsymbol{F})$;

where

$$\begin{aligned}
a &= V_{21}K_{11}^{-1}(Y_1 - X_1 m_0)\\
b &= m_0 + D_0 X_1' K_{11}^{-1}(Y_1 - X_1 m_0)\\
A &= (D_0^{-1} + X_1' V_{11}^{-1} X_1)^{-1} = D_0 X_1' K_{11}^{-1} X_1 D_0\\
B &= V_{21}K_{11}^{-1} X_1 D_0\\
F &= V_{22} - V_{21}K_{11}^{-1}V_{12} \quad \text{and}\\
g &= g_0 + (Y_1 - X_1 m_0)' K_{11}^{-1}(Y_1 - X_1 m_0).
\end{aligned}$$

Further notation

In subsequent sections,

(i) $k = \sqrt{d/g}$; (2.4)

and for $i = 1, 2, \ldots, n_2$:

(ii) Y_{2i} denotes the i-th component of Y_2;

(iii) x_i' denotes the i-th row of X_2;

(iv) f_i denotes the i-th diagonal element of F;

(v) B_i denotes the i-th row of B;

(vi) $G(x_i) = \sqrt{x_i' A x_i - 2B_i x_i + f_i}$ where it is noted that the quadratic expression in x_i is positive definite [c.f. (2.3iii)] so $f_i > B_i A^{-1} B_i'$.

In this notation, for given Y_1',

$$k(Y_{2i} - a_i - b' x_i)/G(x_i) \tag{2.5}$$

has Student's (standard) t distribution on d degrees of freedom.

3. OPTIMISATION WITH PROBABILISTIC CONSTRAINTS

Let the cost of X_2 be

$$c(X_2) \tag{3.1}$$

where the scalar valued function $c(\cdot)$ is convex on its domain. More generally (3.1) could have arisen as a predictive risk incorporating cost. Here and in the next section it is assumed that each element of the matrix X_2 may be chosen from the reals without restriction unless otherwise stated. It is required to choose X_2, at least cost, to satisfy the conditions

$$\mathcal{P}(Y_{2i} > h_i | Y_1) \geq 1 - \varepsilon_i \tag{3.2}$$

($i = 1, \ldots, n_2$) where $0 < 1 - \varepsilon_i < 1$ is a specified probabilistic threshold for given hurdle h_i. This problem occurs, for intance, in quality control, where the response variables are quality measurements of successive items, the hurdles are specification limits and the explanatory variables are inputs for the manufacturing process. Costs of the input variables would depend on returns to scale of supply and so $c(\cdot)$ is not necessarily separable with respect to rows.

Denoting by t_i the upper ε_i point of Student's (standard) t-distribution on d degrees of freedom, (3.2) is equivalent to

$$k^{-1} t_i G(x_i) \leq a_i - h_i + b' x_i \tag{3.3}$$

and X_2 is chosen to minimise (3.1) subject to (3.3) ($i = 1, \ldots, n_2$). Since the square root function is increasing and concave, it follows that

$$G(x_i) \text{ is strictly convex } (i = 1, \ldots, n_2). \tag{3.4}$$

Case: each $0 < \varepsilon_i \leq \frac{1}{2}$
Here, for each $i, t_i \geq 0$ and, in view of (3.4) the set of $\boldsymbol{x}_i$ satisfying (3.3) is convex. It follows that, for each i, the set of $\boldsymbol{X}_2$ satisfying (3.3) is also convex and so the intersection of these sets ($i = 1, \ldots, n_2$) is convex. Hence the problem is infeasible or there is an optimal $\boldsymbol{X}_2$ which can be found by convex programming [Fletcher (1987), Gill *et al.* (1981)]. This solution naturally modifies to cover the case where there is a restriction $\boldsymbol{x}_i \geq \mathbf{0}$ on some rows of $\boldsymbol{X}_2$, indeed to allow the $\boldsymbol{x}_i$ to be restricted to arbitrary convex sets.

Case: $\frac{1}{2} < \varepsilon_i < 1$ *for at least one* i
Define the index sets I_0, I_1 as:

$$I_0 = \left\{i = 1, \ldots, n_2 : 0 < \varepsilon_i \leq \frac{1}{2}\right\}, I_1 = \left\{i = 1, \ldots, n_2 : \frac{1}{2} < \varepsilon_i < 1\right\},$$

and for $i = 1, \ldots, n_2$ let $\mathcal{S}_i$ denote the set of matrices $\boldsymbol{X}_2$ satisfying (3.3). It follows that

$$\mathcal{S}_i \text{ is convex for } i \in I_0 \text{ and } \mathcal{S}_i^c \text{ is convex for } i \in I_1.$$

In this notation the set of $\boldsymbol{X}_2$ constrained by (3.3) $i = 1, \ldots, n_2$ is $\mathcal{X}_0 \cap \mathcal{X}_1$ where

$$\mathcal{X}_0 = \cap_{i \in I_0} \mathcal{S}_i \text{ and } \mathcal{X}_1^c = \cup_{i \in I_1} \mathcal{S}_i^c.$$

It is noted that $\mathcal{X}_1^c$ is expressible as the union of disjoint sets each of which is the union of one or more $\mathcal{S}_i^c$, $i \in I_1$. Let $c(\boldsymbol{X}_2)$ attain its minimum over $\mathcal{X}_0$ at $\boldsymbol{X}_2^*$. If $\boldsymbol{X}_2^* \in \mathcal{X}_1$ then it is clearly optimal. Otherwise $\boldsymbol{X}_2^*$ belongs to one of the disjoint sets of $\mathcal{X}_1^c$ in which case $c(\boldsymbol{X}_2)$ attains its optimum (over $\mathcal{X}_0 \cap \mathcal{X}_1$) on the boundary of the intersection of this set with $\mathcal{X}_0$. This point is therefore specified algebraically by equality in one or more of the constraints (3.3) $i \in I_1$, and can be determined by Lagrangean methods. Again this solution modifies naturally if the $\boldsymbol{x}_i$ are restricted to convex sets.

The above solutions can incorporate additional constraints which further restrict $\boldsymbol{X}_2$ to a convex set. A case of some interest is where the rows of $\boldsymbol{X}_2$ are all the same so that $\boldsymbol{X}_2$ takes the form

$$\boldsymbol{X}_2 = \mathbf{1}\boldsymbol{x}'$$

with $\mathbf{1}$ an $n_2 \times 1$ vector of ones. The above argument goes through essentially with $\boldsymbol{X}_2'$ and each $\boldsymbol{x}_i$ replaced by $\boldsymbol{x}$.

Feasibility of the constraints
It is required to determine the minimal value of ε_i for which (3.2) is feasible or equivalently the maximal value of t_i for which (3.3) is feasible; hence define:

$$\bar{t}_i = k \sup_{\boldsymbol{x}_i \in Z_i} \{(a_i - h_i + \boldsymbol{b}'\boldsymbol{x}_i)/G(\boldsymbol{x}_i)\} \tag{3.5}$$

where Z_i denotes the set of possible choices of $\boldsymbol{x}_i$.

In the case where the model (1.1) includes a constant term the first component of each $\boldsymbol{x}_i$ is fixed at 1. However $\bar{t}_i$ may be expressed in the same form as in (3.5) but with $\boldsymbol{x}_i$ having one component fewer. Specifically, writing

$$\boldsymbol{x}_i' = (1, \boldsymbol{\omega}_i'), \quad \boldsymbol{b}' = (b_0, \boldsymbol{b}_1'),$$

$$\boldsymbol{A} = \begin{bmatrix} \alpha_0 & \boldsymbol{\alpha}_1' \\ \boldsymbol{\alpha}_1 & \boldsymbol{A}_1 \end{bmatrix}, \quad \boldsymbol{B}_i = (\gamma_{0i}, \boldsymbol{\gamma}_{1i}')$$

(3.5) would be modified according to

$$\boldsymbol{x}_i \to \boldsymbol{\omega}_i, \quad a_i - h_i \to a_i - h_i + b_0, \quad \boldsymbol{b} \to \boldsymbol{b}_1$$

$$\boldsymbol{A} \to \boldsymbol{A}_1, \quad \boldsymbol{B}_i \to (\boldsymbol{\gamma}_{1i} - \boldsymbol{\alpha}_1)', \quad f_i \to f_i + \alpha_0 - 2\gamma_{0i}.$$

Hence for models both with or without a constant term it suffices to consider a problem of the form (3.5) where the first component of $\boldsymbol{x}_i$ is not restricted to unity.

(a) *Unconstrained case*
Here $Z_i = \Re^p$ and the optimisation (3.5) is carried out, firstly maximising on the plane $\boldsymbol{b}'\boldsymbol{x}_i = \theta$ (θ constant) and then with respect to θ. The following further notation is used ($i = 1, \ldots, n_2$):

$$q_i = f_i - \boldsymbol{B}_i\boldsymbol{A}^{-1}\boldsymbol{B}_i', \quad r = \boldsymbol{b}'\boldsymbol{A}^{-1}\boldsymbol{b}, \quad s_i = a_i - h_i + \boldsymbol{b}'\boldsymbol{A}^{-1}\boldsymbol{B}_i' \tag{3.6}$$

and it is noted [cf. (2.4vi)] that $q_i > 0$.

Case $\boldsymbol{b} = \boldsymbol{0}$

$$\text{if } a_i < h_i \text{ then } \bar{t}_i = 0 \text{ when } \boldsymbol{x}_i'\boldsymbol{A}\boldsymbol{x}_i = \pm\infty \tag{3.7a}$$

(and in particular, when any component of $\boldsymbol{x}_i$ equals $\pm\infty$);

$$\text{if } a_i = h_i \text{ then } \bar{t}_i = 0 \text{ for all } \boldsymbol{x}_i \varepsilon \Re^p; \tag{3.7b}$$

$$\text{if } a_i > h_i \text{ then } \bar{t}_i = k(a_i - h_i)/\sqrt{q_i} \text{ at } \boldsymbol{x}_i = \boldsymbol{A}^{-1}\boldsymbol{B}_i'. \tag{3.7c}$$

Case $\boldsymbol{b} \neq \boldsymbol{0}$

$$\text{if } s_i \leq 0 \text{ then } \bar{t}_i = k\sqrt{r} \text{ at } \boldsymbol{x}_i = \boldsymbol{A}^{-1}\boldsymbol{B}_i' + \lim_{u\to\infty}(u\boldsymbol{A}^{-1}\boldsymbol{b}); \tag{3.7d}$$

$$\text{if } s_i > 0 \text{ then } \bar{t}_i = k\sqrt{r + s_i^2/q_i} \text{ at } \boldsymbol{x}_i = \boldsymbol{A}^{-1}\boldsymbol{B}_i' + (q_i/s_i)\boldsymbol{A}^{-1}\boldsymbol{b}. \tag{3.7e}$$

It is noted that $\bar{t}_i$ is a continuous function of $\boldsymbol{b}$ and so is the optimal $\boldsymbol{x}_i$ when finite ($s_i > 0$).

This gives a solution for the case where the rows of $\boldsymbol{X}_2$ may be chosen separately and unrestrictedly and therefore provides an absolute upper bound on the probability in (3.2).

(b) *Constrained case* $\boldsymbol{x}_i \geq \boldsymbol{0}$
Here Z_i is the positive orthant of $\Re^p$. If the unconstrained solution (3.7) has non-negative components then it is the required point. Otherwise the solution hinges on which of the sets

$$\begin{aligned} M_i &= \{\boldsymbol{x}_i : \boldsymbol{x}_i \geq \boldsymbol{0}, \quad \boldsymbol{b}'\boldsymbol{x}_i + a_i > h_i\}, \\ E_i &= \{\boldsymbol{x}_i : \boldsymbol{x}_i \geq \boldsymbol{0}, \quad \boldsymbol{b}'\boldsymbol{x}_i + a_i = h_i\}, \\ L_i &= \{\boldsymbol{x}_i : \boldsymbol{x}_i \geq \boldsymbol{0}, \quad \boldsymbol{b}'\boldsymbol{x}_i + a_i < h_i\} \end{aligned}$$

are empty. This may be established by standard linear programming procedures.

If $M_i \neq \phi$ then $\bar{t}_i > 0$ and

$$\bar{t}_i = k \sup_{\boldsymbol{x}_i \in M_i} \{(a_i - h_i + \boldsymbol{b}'\boldsymbol{x}_i)/G(\boldsymbol{x}_i)\}. \tag{3.8}$$

To evaluate this consider the one to one transformation

$$u_0 = 1/(\boldsymbol{b}'\boldsymbol{x}_i + a_i - h_i), \quad \boldsymbol{u} = u_0\boldsymbol{x}_i, \quad \boldsymbol{x}_i \in M_i \tag{3.9}$$

which has inverse

$$\boldsymbol{x}_i = \boldsymbol{u}/u_0 \text{ for } (u_0, \boldsymbol{u}) \in U$$

where

$$U = \{(u_0, \boldsymbol{u}) : \quad u_0 > 0, \quad \boldsymbol{u} \geq \mathbf{0}, \quad (a_i - h_i)u_0 + \boldsymbol{b}'\boldsymbol{u} = 1\}.$$

It follows that the expression in (3.8) may be evaluated via

$$\bar{t}_i^{-2} = k^{-2} \min_{(u_0, \boldsymbol{u}) \in \bar{U}} (\boldsymbol{u}'\boldsymbol{A}\boldsymbol{u} - 2u_0\boldsymbol{B}_i\boldsymbol{u} + f_i u_0^2). \tag{3.10}$$

This can be computed by a convex quadratic programming procedure [see Goldfarb and Idnani (1983) or Gill and Murray (1978)].

If $M_i = \phi$ then $\bar{t}_i \leq 0$ and if in addition $E_i \neq \phi$ then $\bar{t}_i = 0$. Moreover if $M_i = \phi$ and $E_i = \phi$ so that $a_i < h_i$ and $\boldsymbol{b} \leq \mathbf{0}$ then $Z_i = L_i$ and

$$\bar{t}_i^{-2} = k^{-2} \sup_{\boldsymbol{x}_i \in L_i} \{[G(\boldsymbol{x}_i)]^2/(h_i - a_i - \boldsymbol{b}'\boldsymbol{x}_i)^2\}.$$

This is evaluated by means of the one to one transformation

$$\omega_0 = 1/(h_i - a_i - \boldsymbol{b}'\boldsymbol{x}_i), \quad \boldsymbol{\omega} = \omega_0\boldsymbol{x}_i, \quad \boldsymbol{x}_i \in L_i \tag{3.11}$$

with inverse map

$$\boldsymbol{x}_i = \boldsymbol{\omega}/\omega_0 \text{ for } (\omega_0, \boldsymbol{\omega}) \in W$$

where

$$W = \{(\omega_0, \boldsymbol{\omega}) : \omega_0 > 0, \quad \boldsymbol{\omega} \geq \mathbf{0}, \quad \boldsymbol{b}'\boldsymbol{\omega} + (a_i - h_i)\omega_0 = -1\}.$$

Thus

$$\bar{t}_i^{-2} = k^{-2} \max_{(\omega_0, \boldsymbol{\omega}) \in \bar{W}} \{\boldsymbol{\omega}'\boldsymbol{A}\boldsymbol{\omega} - 2\omega_0\boldsymbol{B}_i\boldsymbol{\omega} + f_i\omega_0^2\}.$$

Here the function to be maximised is strictly convex and it follows that one of the $p + 1$ extreme points of $\bar{W}$ is optimal so mere enumeration is all that is required to evaluate $\bar{t}_i$.

This completes the determination of (3.5) when $\boldsymbol{x}_i \geq \mathbf{0}$.

For given hurdle h_i and probability ε_i the feasibility of constraint (3.2), $t_i \leq \bar{t}_i$, is therefore determined for the case $\boldsymbol{x}_i$ unrestricted and for the case where the restriction $\boldsymbol{x}_i \geq \mathbf{0}$ is imposed.

4. STOCHASTIC OBJECTIVES

Here attention is restricted to a single future response, Y_2, so that $n_2 = 1$. The notation of previous sections obtains with the subscript i deleted and $\boldsymbol{X}_2'$ and $\boldsymbol{x}_i$ replaced by $\boldsymbol{x}$.

A. *Probability Maximisation*

Consider firstly the problem of maximising the probability of surpassing a hurdle without cost considerations. Here it is required to choose $\boldsymbol{x}$ so as to

$$\text{maximise } \mathcal{P}(Y_2 > h|\boldsymbol{Y}_1). \tag{4.1}$$

This may be regarded as choice of $\boldsymbol{x}$ according to a zero-one predictive loss function for surpassing-failing the hurdle. It follows from (2.5) that

$$\mathcal{P}(Y_2 > h|\boldsymbol{Y}_1) = \mathcal{P}\{S \geq k(h - a - \boldsymbol{b}'\boldsymbol{x})/G(\boldsymbol{x})\} \tag{4.2}$$

where S has a Student's (standard) t distribution on d degrees of freedom. Hence the problem is to

$$\underset{\boldsymbol{x}\in Z}{\text{maximise}}\ [(a-h+\boldsymbol{b}'\boldsymbol{x})/G(\boldsymbol{x})] \tag{4.3}$$

where Z denotes the set of possible choices of $\boldsymbol{x}$. It follows at once that the maximal probability in (4.1) is

$$\mathcal{P}(S > -\bar{t}) \tag{4.4}$$

where $\bar{t}$ and the associated optimal choice of $\boldsymbol{x}$ is given in the previous section for both the case where $\boldsymbol{x}$ is unrestricted and also the case where $\boldsymbol{x}$ is restricted to have non-negative components. Moreover, this maximal probability provides an upper bound for any more restrictive situations.

The above development may be modified to incorporate additional constraints. For instance in the otherwise unconstrained case (where $Z = \mathcal{R}^p$) consider the imposition of the constraint

$$\boldsymbol{j}'\boldsymbol{x} = J$$

when $\boldsymbol{b} \neq \boldsymbol{0}$ and $\boldsymbol{j} \neq \boldsymbol{0}$ are not proportional. Let

$$\boldsymbol{\Delta} \equiv \begin{bmatrix} \delta & \delta_{12} \\ \delta_{12} & \delta_{22} \end{bmatrix} \text{ denote the inverse of } \begin{bmatrix} \boldsymbol{b}' \\ \boldsymbol{j}' \end{bmatrix} \boldsymbol{A}^{-1}(\boldsymbol{b},\boldsymbol{j});$$

$$\omega = J - \boldsymbol{j}'\boldsymbol{A}^{-1}\boldsymbol{B}' \text{ and } \lambda = \{\delta(\delta_{22}\omega^2 + q) - (\delta_{12}\omega)^2\}/(\delta s - \delta_{12}\omega)^2$$

(when the denominator $\neq 0$).

The following solution may be established by an argument similar to that which gave (3.7).

Case $s > \delta_{12}\omega/\delta$

$$\bar{t} = k\sqrt{(1+1/\lambda)/\delta} \text{ at } \boldsymbol{x} = \boldsymbol{A}^{-1}\boldsymbol{B}' + \boldsymbol{A}^{-1}(\boldsymbol{b},\boldsymbol{j})\boldsymbol{\Delta}\begin{pmatrix} \{s\lambda - \omega\delta_{12}(1+\lambda)\}/\delta \\ \omega \end{pmatrix}.$$

Case $s \leq \delta_{12}\omega/\delta$

$$\bar{t} = k/\sqrt{\delta} \text{ at } \boldsymbol{x} = \boldsymbol{A}^{-1}\boldsymbol{B}' + \lim_{u\to\infty}\{u\boldsymbol{A}^{-1}(\delta\boldsymbol{b} + \delta_{12}\boldsymbol{j})\}.$$

In the constrained case ($\boldsymbol{x} \geq \boldsymbol{0}$) also, an additional linear constraint may be imposed as in the development below.

B. *Incorporation of Cost*

In the case $\boldsymbol{x} \geq \boldsymbol{0}$, assume a predictive loss function of the form:

$$L(y,\boldsymbol{x}) = \begin{cases} \boldsymbol{c}'\boldsymbol{x} + 1 & \text{if } y \leq h, \\ \boldsymbol{c}'\boldsymbol{x} & \text{if } y > h \end{cases}$$

where $c_1 > 0, \ldots, c_p > 0$ are suitably scaled cost components of $\boldsymbol{c}$. The problem is therefore to choose $\boldsymbol{x} \geq \boldsymbol{0}$ so as to minimise the associated predictive risk:

$$\rho(\boldsymbol{x}) = \mathbb{E}\{L(Y_2,\boldsymbol{x})|\boldsymbol{Y}_1\} = \boldsymbol{c}'\boldsymbol{x} + \mathcal{P}(Y_2 \leq h|\boldsymbol{Y}_1). \tag{4.5}$$

However, as an intermediate step consider the problem:

$$\underset{\boldsymbol{x}\in Z(C)}{\text{maximise}}\ \mathcal{P}(Y_2 > h|\boldsymbol{Y}_1) \tag{4.6}$$

where $Z(C) = Z \cap \Xi$ and $\Xi = \{\boldsymbol{x} : \boldsymbol{c}'\boldsymbol{x} \leq C\}, C > 0$.

If $M \cap \Xi = \{\boldsymbol{x} : \boldsymbol{x} \geq \boldsymbol{0}, \boldsymbol{b}'\boldsymbol{x} + a > h, \boldsymbol{c}'\boldsymbol{x} \leq C\} \neq \phi$, the probability in (4.6) can exceed 1/2 (see (3.8)) and the problem posed in (4.6) (see (4.3)) is equivalent to

$$\underset{\boldsymbol{x} \in M \cap \Xi}{\text{minimise}} \ \{[G(\boldsymbol{x})]^2/(\boldsymbol{b}'\boldsymbol{x} + a - h)^2\}. \tag{4.7}$$

Making the one-to-one transformation

$$u_0 = 1/(\boldsymbol{b}'\boldsymbol{x} + a - h), \boldsymbol{u} = u_0\boldsymbol{x}, \boldsymbol{x} \in M \cap \Xi$$

with inverse map

$$\boldsymbol{x} = \boldsymbol{u}/u_0, (u_0, \boldsymbol{u}) \in \Gamma$$

where

$$\Gamma = \{(u_0, \boldsymbol{u}) : u_0 > 0, \boldsymbol{u} \geq \boldsymbol{0}, \boldsymbol{c}'\boldsymbol{u} \leq Cu_0, \boldsymbol{b}'\boldsymbol{u} - 1 = (h - a)u_0\},$$

problem (4.7) becomes

$$\underset{(u_0, \boldsymbol{u}) \in \bar{\Gamma}}{\text{minimise}} \ \{\boldsymbol{u}'\boldsymbol{A}\boldsymbol{u} - 2\boldsymbol{B}u_0\boldsymbol{u} + fu_0^2]. \tag{4.8}$$

The quadratic form in (4.8) is positive definite and hence the problem can be solved as a quadratic programme [see Goldfarb and Idnani (1983) or Gill and Murray (1978)] to yield a unique solution.

If $M \cap \Xi = \phi$ then $\bar{t} \leq 0$ and if in addition $E \cap \Xi \neq \phi$ then $\bar{t} = 0$. Moreover, if $M \cap \Xi = \phi$ and $E \cap \Xi = \phi$ so that $a < h$ and $\boldsymbol{b} \leq \boldsymbol{0}$ then $Z(C) = L \cap \Xi$ and problem (4.6) becomes:

$$\underset{\boldsymbol{x} \in L \cap \Xi}{\text{maximise}} \ \{[G(\boldsymbol{x})]^2/(h - a - \boldsymbol{b}'\boldsymbol{x})^2\} \tag{4.9}$$

where $L \cap \Xi = \{\boldsymbol{x} : \boldsymbol{x} \geq \boldsymbol{0}, \boldsymbol{b}'\boldsymbol{x} + a < h, \boldsymbol{c}'\boldsymbol{x} \leq C\}$. Making the one-to-one transformation

$$\omega_0 = 1/(h - a - \boldsymbol{b}'\boldsymbol{x}), \quad \boldsymbol{\omega} = \omega_0\boldsymbol{x}, \quad \boldsymbol{x} \in L \cap \Xi$$

with inverse map

$$\boldsymbol{x} = \boldsymbol{\omega}/\omega_0 \text{ for } (\omega_0, \boldsymbol{\omega}) \in \Pi$$

where $\Pi = \{(\omega_0, \boldsymbol{\omega}) : \omega_0 > 0, \boldsymbol{\omega} \geq \boldsymbol{0}, \boldsymbol{c}'\boldsymbol{\omega} \leq C\omega_0, \boldsymbol{b}'\boldsymbol{\omega} + (a - h)\omega_0 = -1\}$, the problem may be stated as

$$\underset{(\omega_0, \boldsymbol{\omega}) \in \bar{\Pi}}{\text{maximise}} \ \{\boldsymbol{\omega}'\boldsymbol{A}\omega - 2\omega_0\boldsymbol{B}\boldsymbol{\omega} + f\omega_0^2\}.$$

The objective function to be maximised here is strictly convex and hence one of the extreme points of $\bar{\Pi}$, of which there are at most $\frac{1}{2}(p + 1)(p + 2)$, is optimal. Explicit enumeration provides the solution.

This completes the maximisation (4.6) and denoting its solution by $x(C)$ the optimum of (4.5) is determined by iterative choice of $C > 0$ to minimise

$$\boldsymbol{c}'\boldsymbol{x}(C) - \max_{\boldsymbol{x} \in Z(C)} \mathcal{P}(Y_2 > h|\boldsymbol{Y}_1) + 1.$$

5. CONCLUDING REMARKS AND ILLUSTRATIVE EXAMPLES

The solutions of the problems in Sections 3 and 4 hinge on the relationship between the predictive means and variances [(3.3), (4.3), (2.3ii), (2.3iii)]. For illustration the case $p = 1 = n_2$ and $Z = \{x : x \geq 0\}$ is considered so the optimal value x^* of a single non-negative explanatory variable x is sought. Referring to (3.5) and (4.3) it is seen that both feasibility in Section 3 and the maximal probability in Section 4 are determined from the behaviour of the function

$$t(x) = k(a - h + bx)/\sqrt{Ax^2 - 2Bx + f} \qquad (x \geq 0).$$

Now in the range $-\infty < x < \infty$ this function has a turning point at

$$\hat{x} = \{bf + B(a - h)\}/\{bB + A(a - h)\}$$

if this exists. Moreover, this gives a maximum iff $bB + A(a - h) > 0$ so

$$x^* = \hat{x} \text{ iff } bf + B(a - h) \geq 0 \text{ and } bB + A(a - h) > 0.$$

This agrees with (3.7e).

Example 1.

$$a = 2, b = -1, f = 3/2, h = 1, A = 3, B = 2.$$

This illustrates the case $M \neq \phi$. Here (see Figure 1) $\hat{x} = \frac{1}{2} = x^*, \bar{t} = k$.

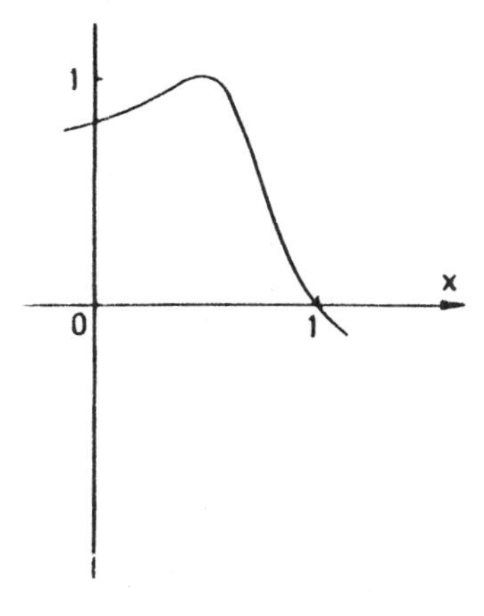

Figure 1. *Graph of* $t(x)/k$

For this example the appropriate transformation (3.9) is

$$u_0 = 1/(1 - x), u = u_0 x$$

so the solution may be obtained by solving

$$\underset{(u_0,u)\in\bar{U}}{\text{minimise}} \; (3u^2 - 4u_0 u + 3u_o^2/2)$$

where

$$\bar{U} = \{(u_0, u) : u_0 \geq 0, u \geq 0, u_0 - u = 1\}.$$

This is equivalent to

$$\underset{(u \geq 0)}{\text{minimise}} \left(\frac{1}{2}u^2 - u + 3/2 \right)$$

hence $u^* = 1, u_0^* = 2$ and $x^* = u^*/u_0^* = \frac{1}{2}$ as determined above.

Example 2.

$$a = 1, b = -1, f = 5, h = 2, A = 1, B = -2.$$

This illustrates the case $M = \phi$ and $E = \phi$. Here $t(x)$ is decreasing on Z and so $x^* = 0$ and $\bar{t} = -k/\sqrt{5}$ (see Figure 2).

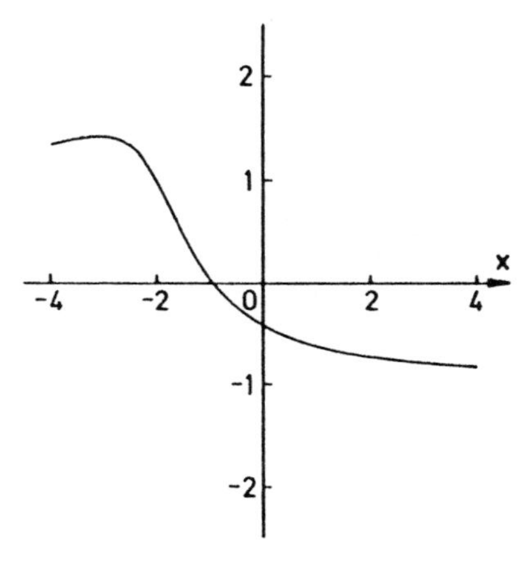

Figure 2. *Graph of* $t(x)/k$

For this example the appropriate transformation (3.11) is

$$\omega_0 = 1/(1+x), \qquad \omega = \omega_0 x$$

and the problem is equivalent to

$$\underset{(\omega_0, \omega) \in \bar{W}}{\text{maximise}} \ (\omega^2 + 4\omega_0\omega + 5\omega_0^2)$$

where

$$\bar{W} = \{(\omega_0, \omega) : \omega_0 \geq 0, \omega \geq 0, \omega_0 + \omega = 1\}.$$

This objective function may be written as $2(\omega - 3/2)^2 + \frac{1}{2}$ where $0 \leq \omega \leq 1$ and hence $\omega^* = 0$, $\omega_0^* = 1$ and $x^* = 0$, in agreement with the value obtained above. Observe that this value of x maximises the predictive mean $= 1 - x$ and minimises the predictive variance $= [g/(d-2)](x^2 + 4x + 5)$.

Example 3.

$$a = 2, b = -1, f = 2, h = 5, A = 1, B = -1.$$

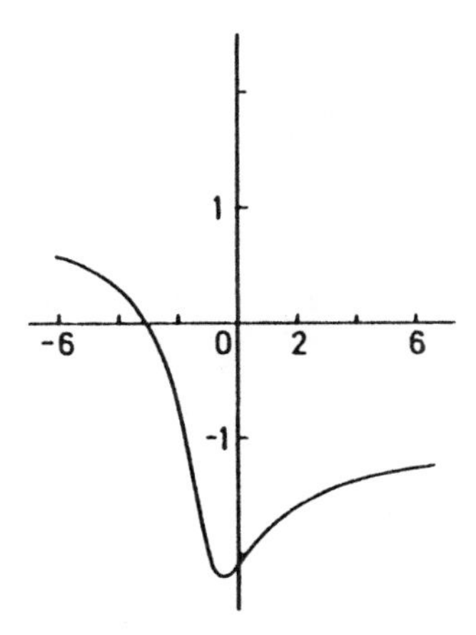

Figure 3. *Graph of* $t(x)/k$

In this example also $M = \phi = E$ so $\bar{t} \leq 0$. Here $t(x)$ is increasing on Z so $x^* = \infty$ and $\bar{t} = -k$ (see Figure 3).

In this case the appropriate transformation (3.11) is

$$\omega_0 = 1/(3 + x), \quad \omega = \omega_0 x, \quad x \geq 0$$

and hence

$$\bar{t}^{-2} = (5/9k^2) \max_{(0 \leq \omega \leq 1)} [(\omega + 1/5)^2 + 9/25]$$

giving $\omega^* = 1$, $\omega_0^* = 0$ and so agreement with the solution above.

In contrast to the previous example it is optimal to expand the predictive variance at the expense of the predictive mean, "successes" being achieved by virtue of the greater scatter.

In general (when $n_2 \geq 1$ and $p \geq 1$) the form of the unrestricted optimum depends on the sign of s_i; observe that here $\bar{t}_i$ is a continuous function of s_i and, if $\boldsymbol{b} \neq \boldsymbol{0}$, so is the optimal choice of $\boldsymbol{x}_i$. Note that the predictive variance of Y_{2i} is least, with value $gq_i/(d-2)$, when $\boldsymbol{x}_i = \boldsymbol{A}^{-1}\boldsymbol{B}_i'$ and s_i may be expressed in terms of the predictive mean at this point:

$$s_i = \mathbb{E}\,(Y_{2i}|\boldsymbol{Y}_1, \boldsymbol{x}_i = \boldsymbol{A}^{-1}\boldsymbol{B}_i') - h_i.$$

In the case $\boldsymbol{b} = \boldsymbol{0}$ the results (3.7a, b, c) are in accord with one's intuition, the predictive variance being maximised, irrelevant or minimised according respectively as the predictive mean a_i $<$, $=$ or $>$ h_i. When $\boldsymbol{b} \neq \boldsymbol{0}$ and $s_i \leq 0$ the optimal solution asserts that both the predictive mean and the predictive variance tend to $+\infty$. Thus an infinite expansion of the predictive variance is optimal if any expansion (form its minimum) is needed to obtain a better than evens chance of success. Unless $\boldsymbol{x}_i$ is restricted, only in the case $s_i > 0$ is there a unique finite optimal point (3.7e) and this gives a balance between maximising the predictive mean and minimising the predictive variance.

REFERENCES

Broemeling, L. D. (1985). *Bayesian Analysis of Linear Models.* New York and Basel: Marcel Dekker.
Fletcher, R. (1987). *Practical Methods of Optimisation.* New York: Wiley.
Gill, P. E. and Murray, W. (1978). Numerically stable methods for quadratic programming. *Mathematical Programming* **14**, 349–372.

Gill, P. E., Murray, W. and Wright, M. H. (1981). *Practical Optimization*. London and New York: Academic Press.

Goldfarb, D. and Idnani, A. (1983). A numerically stable dual method for solving strictly convex quadratic programming problems. *Mathematical Programming* **27**, 1–33.

Kadane, J. B., Dickey, J. M., Winkler, R. L., Smith, W.S. and Peters, S. C. (1980). Interactive Elicitation of Opinion for a Normal Linear Model. *J. Amer. Statist. Assoc.* **75**, 845–854.

Press, J. S. (1982). *Applied Multivariate Analysis*, (2nd Ed.). Malabar, Florida: Robert E. Krieger Publishing Company.

Zellner, A. (1971). *An Introduction to Bayesian Inference in Econometrics*. New York: Wiley.

Zellner, A. (1985). Bayesian statistics in econometrics. *Bayesian Statistics* **2**. (J. M. Bernardo, M. H. DeGroot, D. V. Lindley and A. F. M. Smith, eds.). Amsterdam: North-Holland .

BAYESIAN STATISTICS 3, pp. 725–732
J. M. Bernardo, M. H. DeGroot, D. V. Lindley and A. F. M. Smith, (Eds.)

Bayesian Diagnostic Testing in the General Linear Normal Regression Model

D. J. POIRIER
University of Toronto

SUMMARY

This paper considers the linear normal regression model with a general covariance structure of the form $\sigma^2\Psi(\alpha)$, where $\Psi(\cdot)$ is of a known, but arbitrary, form depending on the unknown parameter vector α. Posterior odds tests of the maintained hypothesis $H_\star : \Psi(\alpha) = I_T$ are considered, and a particularly simple diagnostic Bayesian analog of a score (LM) test is suggested.

Keywords: DIAGNOSTIC TEST; *LM* TEST; NATURAL CONJUGATE PRIORS; POSTERIOR ODDS; SCORE TEST; TIERNEY-KADANE APPROXIMATION.

1. INTRODUCTION

In this author's view [see Poirier (1988)] both the prior and the likelihood are subjective entities which facilitate expression of researchers' uncertainties about the world. The principal distinction between the two is that the likelihood is chosen so that a relatively large group of researchers are willing to agree to disagree principally in terms of the prior to assign to the parameters of the likelihood. The researcher engaging in public research must engage in a sensitivity analysis in which various statistical activivies involving prediction and inference are performed in such a way as to convey to readers their sensitivity over a fairly rich class of priors.

The choice of the likelihood amounts to choosing a parametric window through which to view the world. In reality the researcher is rarely in a position to assign probability of unity to the adequacy of a particular choice. Recognition of this fact can lead to a potential infinite regress in which a likelihood is continually nested in a richer family and priors are assigned to the new parameters to make the original likelihood "highly probable" but not dogmatic. Pragmatism must assert itself at some point, however, in order for the analysis to begin. At the point at which the regress halts, a particular likelihood is entertained, and implicitly, if not explicitly sharp hypotheses defining the chosen likelihood within still broader families are maintained.

At this point it seems attractive for the researcher to have diagnostic tests available to perform on the maintaned likelihood in order to determine the adequacy of the breadth of window chosen in light of the data. Such tests are particularly attractive if they largely involve computation under the maintained hypothesis since often the pragmatic termination of the preceding parametric expansion is determined by computational considerations.

This paper discusses examples of such tests in the context of the general linear normal regression model

$$y = X\beta + u, \tag{1}$$

where y is $T \times 1$, X is a fixed $T \times k$ matrix with rank k, $\beta \varepsilon R^k$ and $u \sim N_T(0_T, \sigma^2\Psi(\alpha))$, where $\sigma^2 > 0$ and $\Psi = \Psi(\alpha)$ is a $T \times T$ positive definite matrix of known form depending on the unknown identified vector $\alpha \varepsilon R^n$. Given a proper prior probability density function (p.d.f.) $p(\beta, \sigma^{-2}, \alpha)$, Bayesian analysis is in principle straightforward[1]. The observed behaviour of many researchers, however, suggests that the maintained likelihood defined by $H_\star : \Psi(\alpha) = I_T$ is a preliminary window at which the researcher may wish to check to see whether further elaboration is warranted. It would be nice to have a rationalization for this observation, say, as a possible outcome of testing $H_\star : \alpha = \alpha_\star$ *vs* $H : \alpha \neq \alpha_\star$, where $\Psi(\alpha) \neq I_T$ for $\alpha \neq \alpha_\star$. This paper considers a conventional Bayesian posterior odds analysis of H *vs* $H_\star$, and suggests an approximation not requiring computation under H. For obvious reasons, such a test can be thought of as a Bayesian analog of an LM or score test. The motivation, however, is purely Bayesian.

2. ARBITRARY INFORMATIVE PRIORS

The log-likelihood corresponding to (1) can be written as $L(\theta|y) = \phi_T(y|X\beta, \sigma^2\Psi(\alpha))$, where $\theta \equiv [\beta', \sigma^{-2}, \alpha'\}'$ and $\phi_T(y|X\beta, \sigma^2\Psi(\alpha))$ denotes the T-dimensional multivariate normal density for y with mean $X\beta$ and covariance matrix $\sigma^2\Psi(\alpha)$. Suppose prior beliefs are independent between $[\beta', \sigma^{-2}]'$ and α. Also suppose the prior odds in favour of H *vs* $H_\star$ are $(1 - \underline{\pi}_\star)/\underline{\pi}_\star$. Then the posterior odds are

$$\frac{1 - \overline{\pi}_\star}{\overline{\pi}_\star} = B\left[\frac{1 - \underline{\pi}_\star}{\underline{\pi}_\star}\right], \tag{2}$$

where

$$B \equiv \frac{p(y|H)}{p(y|H_\star)}$$
$$= \frac{\int_{R^n}\left[\int_0^\infty \int_{R^k} \phi_T(y|X\beta, \sigma^2\Psi(\alpha))p(\beta, \sigma^{-2})d\beta d\sigma^{-2}\right]p(\alpha)d\alpha}{\int_0^\infty \int_{R^k} \phi_T(y|X\beta, \sigma^2 I_T)p(\beta, \sigma^{-2})d\beta d\sigma^{-2}} \tag{3a}$$
$$= \int_{R^n} b(y|\alpha)p(\alpha)d\alpha \tag{3b}$$

is the Bayes factor for H *vs* $H_\star$, and

$$b(y|\alpha) = \frac{p(y|H, \alpha)}{p(y|H_\star)} \tag{4}$$

is the conditional Bayes factor evaluated at a point $\alpha \neq \alpha_*$.

AQUI

Evaluation of Bayes factor (3) can be computationally demanding, especially in cases where a closed form expression is not available for conditional Bayes factor (4). In the latter case the approximation technique suggested by Tierney and Kadane (1986) is most attractive. Let $\theta_\star \equiv [\beta', \sigma^{-2}, \alpha_\star']'$ and consider

$$f_\star(\theta_\star) \equiv T^{-1}\{\ln[L(\theta_\star|y)] + \ln[p(\beta, \sigma^{-2})]\}, \tag{5}$$

$$f(\theta) \equiv T^{-1}\{\ln[L(\theta|y)] + \ln[p(\beta, \sigma^{-2})] + \ln[p(\alpha)]\}. \tag{6}$$

Define $\tilde{\theta}_\star \equiv [\tilde{\beta}_\star', \tilde{\sigma}_\star^{-2}, \alpha_\star']' \equiv$ argmax $\{f_\star(\theta_\star)\}(\alpha_\star$ fixed), $\tilde{\theta} \equiv [\tilde{\beta}', \tilde{\sigma}^{-2}, \tilde{\alpha}'] =$ argmax $\{f(\theta)\}$ and let $F_\star(\cdot)$ and $F(\cdot)$ denote minus the Hessians of $f_\star(\cdot)$ and $f(\cdot)$, respectively. Then

[1] The notation "$p(\cdot)$" is used throughout this paper in a generic fashion in which the arguments identify the p.d.f. in question in a particular context.

the Tierney-Kadane approximation of (3) corresponds to the ratio of Laplace approximations to the numerator and denominator of (3a):

$$\tilde{B} = \frac{|F(\tilde{\theta})|^{-1/2}}{|F(\tilde{\theta}_\star)|^{-1/2}} \exp\{T[f(\tilde{\theta}) - f_\star(\tilde{\theta}_*)]\} \tag{7a}$$

$$= \left[\frac{|F(\tilde{\theta})|^{-1}}{|F(\tilde{\theta}_\star)|^{-1}}\right]^{1/2} \left[\frac{L(\tilde{\theta}|y)}{L(\tilde{\theta}_\star|y)}\right] \left[\frac{p(\tilde{\beta}, \tilde{\sigma}^{-2})}{p(\tilde{\beta}_\star, \tilde{\sigma}_\star^{-2})}\right] p(\tilde{\alpha}). \tag{7b}$$

Note that (7b) provides three corrections to the classical generalized likelihood ratio $L(\tilde{\theta}|y)/L(\tilde{\theta}_\star| \, y)$: (i) a correction $[|F(L(\tilde{\theta}) \, |^{-1}/|F_\star(\tilde{\theta}_\star)|^{-1}]^{1/2}$ reflecting relative generalized variances under H and $H_\star$, (ii) a correction $p(\tilde{\beta}, \tilde{\sigma}^{-2})/p(\tilde{\beta}_\star, \tilde{\sigma}_\star^{-2})$ reflecting prior beliefs on the nuisance parameters, and (iii) a correction $p(\tilde{\alpha})$ reflecting prior beliefs concerning the parameters involved in the comparison of $H vs H_\star$. While the computational requirements of (7) are fairly minimal (essentially the same level of difficulty as maximum likelihood (ML) estimation under $H_\star$ and H), the accuracy is quite high: the relative error of approximation being $O(T^{-2})$.

An obvious place from which to start an iterative search for the posterior mode $\tilde{\theta}$ under H is the posterior mode $\tilde{\theta}_\star$ under $H_\star$. One Newton step away from $\tilde{\theta}_\star$ corresponds to step $j = 1$ in the iterative process

$$\tilde{\theta}^{(j)} = \tilde{\theta}^{(j-1)} - \left[F\left(\tilde{\theta}^{(j-1)}\right)\right]^{-1} \left[\frac{\partial f\left(\tilde{\theta}^{(j-1)}\right)}{\partial \theta}\right] \qquad (j = 1, 2, \ldots) \tag{8}$$

with $\tilde{\theta}^{(0)} \equiv \tilde{\theta}_\star$. The direction of the first step away from $\tilde{\theta}_\star$ depends on the score

$$\frac{\partial f(\tilde{\theta}_\star)}{\partial \theta} = \frac{\partial[\ln L(\tilde{\theta}_\star|y)]}{\partial \theta} + \frac{\partial[\ln p(\tilde{\beta}_\star, \tilde{\sigma}_\star^{-1})]}{\partial \theta} + \frac{\partial \ln p(\alpha_\star)}{\partial \theta}. \tag{9}$$

The first term in (9) corresponds to the score appearing in an LM test; the latter two terms in (9) provide adjustments based on prior beliefs. If $\alpha = \alpha_\star$ corresponds to an interior mode of $p(\alpha)$, as will often be the case, then the third term in (9) vanishes. $\tilde{B}$ evaluated at $\tilde{\theta} = \tilde{\theta}^{(1)}$ is a Bayesian analog of an LM test since it involves only quantities readily available under $H_\star$. The length of the first step, in the metric $F(\tilde{\theta}_*)$, is

$$\left[\frac{\partial f(\tilde{\theta}_\star)}{\partial \theta}\right]' \left[F(\tilde{\theta}_\star)\right]^{-1} \left[\frac{\partial f(\tilde{\theta}_\star)}{\partial \theta}\right] \tag{10}$$

and is also closely related to the classical LM test statistic.

In cases where a closed from expression is available for conditional Bayes factor (4) then the preceding analysis can be greatly simplified. The next two sections consider specific cases: conjugate and non-informative priors for β and σ^{-2}.

3. CONJUGATE PRIORS

Consider the natural conjugate normal-gamma prior

$$p(\beta,\sigma^{-2}) = \phi_k(\beta|\underline{\beta},\sigma^2\underline{Q})\gamma(\sigma^{-2})|\underline{s}^2,\underline{\nu}), \tag{11}$$

where $\underline{s}^2 > 0$, $\underline{\nu} > 0$, $\underline{Q}$ is a $k \times k$ positive definite matrix, and

$$\gamma(\sigma^{-2}|\underline{s}^2,\underline{\nu}) = [\Gamma(\underline{\nu}/2)]^{-1}[\underline{\nu}\underline{s}^2/2]^{\underline{\nu}/2}\sigma^{2-\underline{\nu}}\exp(-\underline{\nu}\underline{s}^2\sigma^{-2}/2) \tag{12}$$

denotes the gamma p.d.f.. Then the conditional posterior p.d.f. $p(\beta,\sigma^{-2}|y,H,\alpha)$ is of the normal-gamma form with hyperparameters

$$\begin{aligned}
\overline{\beta} &\equiv \overline{Q}[\underline{Q}^{-1}\beta + (X'\Psi^{-1}X)\hat{\beta}], \hat{\beta} = (X'\Psi^{-1}X)^{-1}X'\Psi^{-1}y, && (13)\\
\overline{Q} &\equiv [\underline{Q}^{-1} + X'\Psi^{-1}X]^{-1}, && (14)\\
\overline{\nu} &\equiv \underline{\nu} + T, && (15)\\
\overline{\nu}\overline{s}^2 &\equiv \underline{\nu}\underline{s}^2 + (y - X\overline{\beta})'(\Psi + X\underline{Q}X')^{-1}(y - X\overline{\beta}) && (16a)\\
&= \underline{\nu}\underline{s}^2 + (y - X\hat{\beta}_\star)'\Psi^{-1}(y - X\hat{\beta}_\star) + (\hat{\beta}_\star - \underline{\beta})'\underline{Q}^{-1}(\hat{\beta}_* - \underline{\beta})\\
&\quad -(\hat{\beta}_\star - \overline{\beta})'\overline{Q}^{-1}(\hat{\beta}_* - \overline{\beta}). && (16b)
\end{aligned}$$

Furthermore, the conditional p.d.f. $p(y|H,\alpha)$ is the t-distribution p.d.f.:

$$t(y|X\underline{\beta},\underline{s}^2(\Psi + X\underline{Q}X'),\underline{\nu}) = \left[\frac{\Gamma(\overline{\nu}/2)(\underline{\nu}\underline{s}^2)^{\underline{\nu}/2}}{\pi^{T/2}\Gamma(\underline{\nu}/2)}\right](\overline{\nu}\overline{s}^2)^{-\overline{\nu}/2} \tag{17}$$

The analysis under $H_\star$ can be obtained by simply replacing α by $\alpha_\star$ in (13)-(16) to obtain the hyperparameters

$$\begin{aligned}
\overline{\beta}_\star &\equiv \overline{Q}_\star[\underline{Q}^{-1}\underline{\beta} + (X'X)\hat{\beta}_*], \hat{\beta}_* = (X'X)^{-1}X'y, && (18)\\
\overline{Q}_\star &\equiv [\underline{Q}^{-1} + X'X]^{-1}, && (19)\\
\overline{\nu}_\star &\equiv \underline{\nu} + T = \overline{\nu}, && (20)\\
\overline{\nu}_\star\overline{s}_\star^2 &\equiv \underline{\nu}\underline{s}^2 + (y - X\underline{\beta})'(I_T + X\underline{Q}X')^{-1}(y - X\underline{\beta}) && (21a)\\
&= \underline{\nu}\underline{s}^2 + (y - X\hat{\beta}_\star)'(y - X\hat{\beta}_\star) + (\hat{\beta}_\star - \beta)'Q^{-1}(\hat{\beta}_\star - \beta)\\
&\quad -(\hat{\beta}_\star - \overline{\beta})'\overline{Q}_\star^{-1}(\hat{\beta}_\star - \overline{\beta}). && (21b)
\end{aligned}$$

The conditional Bayes factor (4) equals

$$b(y|\alpha) = \left[\frac{|\overline{Q}|}{|\Psi||\overline{Q}_\star|}\right]^{1/2}\left[\frac{\overline{s}^2}{\overline{s}_\star^2}\right]^{-\overline{\nu}/2}. \tag{22}$$

Given the availability of (22) in the conjugate case, it seems wise to dispense with Tierney-Kadane approximation (7), and simply proceed directly to the Laplace approximation of (3b). This amounts to replacing the log integrand of (3b) by a second-order Taylor series around the posterior mode $\hat{\alpha} \equiv \text{argmax}\ \{h(\alpha)\}$, where[2]

[2] Support for expanding around the posterior mode, rather than the *ML* estimate, can be found in Lindley (1980).

$$h(\alpha) = T^{-1}\{\ln[b(y|\alpha)] + \ln[p(\alpha)]\}. \tag{23}$$

Letting

$$H(\alpha) \equiv -\frac{\partial^2 h(\alpha)}{\partial\alpha\partial\alpha'}, \tag{24}$$

integral (3b) can be approximated by

$$\begin{aligned}\hat{B} &\equiv \int_A \exp[Th(\hat{\alpha}) - \frac{1}{2} - (\alpha - \hat{\alpha})'H(\hat{\alpha})(\alpha - \hat{\alpha})]d\alpha \\ &= (2\pi)^{n/2}|H(\hat{\alpha})|^{-1/2}p(\hat{\alpha})b(y|\hat{\alpha}).\end{aligned} \tag{25}$$

Tierney and Kadane (1986) note that the relative error in (25) is of order $O(T^{-1})$.

An obvious place from which to start an iterative search for the posterior mode $\hat{\alpha}$ is the maintained hypothesis $H_\star : \alpha = \alpha_\star$. One Newton step away from $\alpha_\star$ corresponds to step $j = 1$ in the iterative process

$$\hat{\alpha}^{(j)} = \hat{\alpha}^{(j-1)} - \left[H(\hat{\alpha}^{(j-1)})\right]^{-1}\left[\frac{\partial(\hat{\alpha}^{(j-1)})}{\partial\alpha}\right] \quad (j = 1, 2, \ldots) \tag{26}$$

with $\hat{\alpha}^{(0)} = \alpha_\star$,

$$\frac{\partial h(\alpha)}{\partial\alpha'} = T^{-1}\left[\frac{\partial\ln[b(y|\alpha)]}{\partial\alpha'} + \frac{\partial\ln[p(\alpha)]}{\partial\alpha'}\right], \tag{27}$$

$$\begin{aligned}H(\alpha) = {}& -T^{-1}\left[\{\text{vec}\,(\Psi - X\overline{Q}X' - \overline{s}^{-2}\overline{uu}')\}' \otimes I_n\right]\left[\frac{\partial\Psi^{-1}}{\partial\alpha\partial\alpha'}\right] \\ & - \left[\frac{\partial\,\text{vec}\,(\Psi^{-1})}{\partial\alpha}\right]'\left[(\Psi\otimes\Psi) - (X\overline{Q}X' + \overline{s}^{-2}\overline{uu}')\otimes(X\overline{Q}X' + \overline{s}^{-2}\overline{uu}')\right] \\ & \left[\frac{\partial\,\text{vec}\,(\Psi^{-1})}{\partial\alpha}\right] + \left[\frac{\partial^2\ln p(\alpha)}{\partial\alpha\partial\alpha'}\right],\end{aligned} \tag{28}$$

and

$$\frac{\partial\Psi^{-1}}{\partial\alpha\partial\alpha'} \equiv \frac{\partial}{\partial\alpha}\left\{\text{vec}\left[\frac{\partial\,\text{vec}\,(\Psi^{-1})}{\partial\alpha'}\right]\right\}. \tag{29}$$

The direction of steps (26) [assuming $\alpha = \alpha_\star$ is the mode of $p(\alpha)$] is determined by the score of conditional Bayes' factor (22):

$$\frac{\partial\ln[(y|\alpha)]}{\partial\alpha'} = \frac{1}{2}\left[\frac{\partial\,\text{vec}\,(\Psi^{-1})}{\partial\alpha}\right]'\text{vec}\,(\Psi - X\overline{Q}X' - \overline{s}^{-2}\overline{uu}') \tag{30}$$

where

$$\overline{u} \equiv y - X\overline{\beta} = \Psi V\underline{u} \tag{31}$$

with $V \equiv V(\alpha) = [\Psi(\alpha) + X\underline{Q}X']^{-1}$ and $\underline{u} \equiv y - X\underline{\beta}$. Score (30) differs sharply from the conventional likelihood score

$$\frac{\partial\ln[(\theta|y)]}{\partial\alpha'} = -\frac{1}{2}\left[\frac{\partial\,\text{vec}\,(\Psi^{-1})}{\partial\alpha}\right]'\text{vec}\,(\Psi - \sigma^2 uu'). \tag{32}$$

Unlike $b(y|\alpha)$, from which β and σ^{-2} have been marginalized-out, score (32) depends on these nuisance parameters. Standard practice is to evaluate these parameters in (32) at their conditional maximum likelihood (ML) estimates $\hat{\beta}$ in (13) and

$$\hat{\sigma}^{-2} \equiv T(\hat{u}'\Psi^{-1}\hat{u})^{-1}, \tag{33}$$

where

$$\hat{u} \equiv y - X\hat{\beta}. \tag{34}$$

Then given α, (32) can be evaluated at $\hat{\beta}$ and $\hat{\sigma}^{-2}$ to yield

$$\frac{\partial \ln[\phi_T(y|X\hat{\beta}, \hat{\sigma}^2\Psi(\alpha))]}{\partial \alpha'} = \frac{1}{2}\left[\frac{\partial \text{ vec } (\Psi^{-1})}{\partial \alpha}\right] \text{vec } (\Psi - \hat{\sigma}^{-2}\hat{u}\hat{u}'). \tag{35}$$

Scores (30) and (35) differ importantly in the presence of the term $-X\overline{Q}X'$ in (30). For the frequentist omission of this term from (35) is somewhat embarrassing. Although $E_{y|\beta,\sigma^{-2},\alpha}(\hat{\sigma}^{-2}\hat{u}\hat{u}'|\beta = \hat{\beta}, \sigma^{-2} = \hat{\sigma}^{-2}, \alpha) = \Psi$, a more important quantity is the expectation which takes into account the sampling uncertainty of $\hat{\beta}$ and $\hat{\sigma}^{-2}$, namely,

$$\begin{aligned} E_{y|\alpha}(\hat{\sigma}^{-1}\hat{u}\hat{u}'|\alpha) &= E_{\hat{\sigma}^{-2}|\alpha}\left[E_{y|\sigma^2,\alpha}(\hat{\sigma}^{-2}\hat{u}\hat{u}'|\alpha, \sigma^{-2} = \hat{\sigma}^{-2})\right] \\ &= E_{\hat{\sigma}^{-2}|\alpha}\left[\Psi - X\overline{Q}X'\right] \\ &= \Psi - X\overline{Q}X'. \end{aligned} \tag{36}$$

Thus Bayesian score (30) naturally includes (even in the non-informative case!) the "corrected" covariance matrix the frequentist seeks[3].

In the case of the first step of (26), (27) and (28) reduce [again assuming $\alpha = \alpha_\star$ is the mode of $p(\alpha)$] to

$$\frac{\partial \ln h(\alpha)}{\partial \alpha} = -\frac{1}{2}\left[\frac{\partial \text{ vec } \{[\Psi(\alpha_\star)]^{-1}\}}{\partial \alpha}\right]' \text{vec } (I_T - X\overline{Q}_\star X' - \overline{s}_\star^{-2}\overline{u}_\star\overline{u}_\star'). \tag{37}$$

$$\begin{aligned} H(\alpha_\star) &= T^{-1}\left[\{\text{vec } (I_T - X\overline{Q}_\star X' - \overline{s}_\star^{-2}\overline{u}_\star\overline{u}_\star)\}' \otimes I_n\right]\left[\frac{\partial[\Psi(\alpha_\star)]^{-1}}{\partial\alpha\partial\alpha'}\right] \\ &- \left[\frac{\partial \text{ vec } [\Psi(\alpha_\star)]^{-1}}{\partial \alpha}\right]' \left[(I_T \otimes I_T) - (X\overline{Q}_\star X' + \overline{s}_\star^{-2}\overline{u}_\star\overline{u}_\star') \otimes (X\overline{Q}_\star X' + \overline{s}_\star^{-2}\overline{u}_\star\overline{u}_\star')\right] \\ &\left[\frac{\partial \text{ vec } [\Psi(\alpha_\star)]^{-1}}{\partial \alpha}\right] - \frac{\partial^2[\ln p(\alpha)]}{\partial\alpha\partial\alpha'}, \end{aligned} \tag{38}$$

where $\overline{Q}_\star$ and $\overline{s}_\star^2$ are given by (19) and (21), respectively, and

$$\overline{u}_\star = y - X\overline{\beta}_\star \tag{39}$$

At $\hat{\alpha}^{(0)} = \alpha_\star, b(y|\hat{\alpha}) = 1$ and approximation (25) yields

$$\hat{B}^{(0)} = (2\pi)^{n/2}|H(\alpha_\star)|^{-1/2}p(\alpha_\star). \tag{40}$$

Using (37) and (38) in (26) to compute the first update $\hat{\alpha}^{(1)}$, approximation (25) yields

$$\hat{B}^{(1)} = (2\pi)^{n/2}|H\left(\hat{\alpha}^{(1)}\right)|^{-1/2}p\left(\hat{\alpha}^{(1)}\right)b\left(y|\hat{\alpha}^{(1)}\right). \tag{41}$$

Both (40) and (41) are Bayesian analogues of a LM test in that they involve computations readily available under $H_\star$.

[3] Frequentists sometimes recommend the inclusion of the asymptotically negligible term $-X\bar{Q}X'$ in a second stage attempt to estimate α from a regression with transformed OLS residuals as a dependent variable. For example, see Judge *et al.* (1985, pp. 431-441).

4. NON-INFORMATIVE PRIORS

Care must be utilized in deriving approximations in the non-informative case. Under most definitions of non-informativeness, the numerator and denominator of (3a) are not strictly defined. However, since the same non-informative prior appears in both the numerator and denominator of (3a), the ratio may be defined. In the conjugate non-informative case

$$\underline{\nu} \to 0, \tag{42}$$

$$\underline{Q}^{-1} \to 0_{k \times k}, \tag{43}$$

conditional Bayes factor (4) approaches

$$b(y|\alpha) = \left[\frac{|X'\Psi^{-1}X|}{|\Psi|\;|X'X|}\right]^{1/2} \left[\frac{\hat{u}'\Psi^{-1}\hat{u}}{\hat{u}_\star'\hat{u}_\star}\right]^{-T/2}. \tag{44}$$

In the case the non-informative prior $p(\beta, \sigma^{-2}) \propto \sigma^2$, the limiting ratio is slightly different:

$$b(y|\alpha) = \left[\frac{|X'\Psi^{-1}X|}{|\Psi|\;|X'X|}\right]^{1/2} \left[\frac{\hat{u}'\Psi^{-1}\hat{u}}{\hat{u}_\star'\hat{u}_\star}\right]^{-(T-k)/2}. \tag{45}$$

Both (44) and (45) are closely related to the generalized likelihood ratio test statistic

$$\lambda = |\Psi|^{-1/2} \left[\frac{\hat{u}'\Psi^{-1}\hat{u}}{\hat{u}_\star'\hat{u}_\star}\right]^{-T/2}. \tag{46}$$

for comparing the point $\alpha \neq \alpha_\star$ to $\alpha = \alpha_\star$. From (44) it is seen that the Bayesian approach amounts to an "averaging" of (46) using the prior $p(\alpha)$ and the adjustment factor $|X'\Psi^{-1}X|^{1/2}/|X'X|^{1/2}$.

In the non-informative case described by (42) and (43), evaluation under $H_\star$ implies $\overline{Q}_\star \to (X'X)^{-1}$, $\overline{u}_\star \to \hat{u}_\star \equiv y - X\hat{\beta}_\star$ and $\overline{s}_\star^2 \to \hat{\sigma}_\star^2 \equiv \hat{u}_\star'\hat{u}_\star/T$ which implies a first step away from $\alpha_\star$ in the direction

$$\frac{\partial \ln[b(y|\alpha_\star)]}{\partial \alpha} = -\frac{1}{2}\left[\frac{\partial \text{ vec } \{[\Psi(\alpha_\star)]^{-1}\}}{\partial \alpha}\right]' \text{vec}\,(I_T - X(X'X)^{-1}X' - \hat{\sigma}_\star^{-2}\hat{u}_\star\hat{u}_\star'). \tag{47}$$

Again, the inclusion of the term $-X(X'X)^{-1}X$ in (47) distinguishes score (47) from that used in standard LM tests.

5. CLOSING REMARKS

This paper has suggested some computationally simply approximations to the Bayes factor for testing whether the disturbance covariance matrix in the general linear normal regression model is proportional to the identity matrix. The alternative hypothesis is presumed well-specified, but the analysis herein treats it in a completely general fashion. The appendix provides all log-likelihood derivatives needed in computing the Tierney-Kadane approximation (7), in terms of the derivatives of $\Psi(\alpha)$ with respect to α. The necessary derivatives in the case of conjugate or noninformative priors are provided in the text. An important topic for future research is to investigate the accuracy of approximations (7) and (25) in a variety of settings, as well as the accuracy of their corresponding values at or one-step away from the maintained hypothesis.

ACKNOWLEDGEMENTS

The research support of the Social Sciences and Humanities Research Council of Canada under Grant No. 410-86-0319, is gratefully acknowledged.

REFERENCES

Judge, G. G., Griffiths, W. E., Hill, R. C., Lütkepohh, H. and Lee, T. C. (1985). *The Theory and Practice of Econometrics.* (2nd ed.) New York: Wiley.

Lindley, D. V. (1980). Approximate Bayesian methods (with discussion). *Bayesian Statistics* **1**. (J. M. Bernardo, M. H. DeGroot, D. V. Lindley and A. F. M. Smith, eds.). Valencia: University Press.

Poirier, D. J. (1988). Frequentist and subjectivist perspectives on the problems of model building in economics (with discussion). *Journal of Economic Perspectives* **2**, 121–170.

Tierney, L. and Kadane, J. B. (1986). Accurate approximations for posterior moments and marginal densities. *J. Amer. Statist. Assoc.* **81**, 82–86.

APPENDIX

This appendix provides the first and second derivatives of the log-likelihood $L(\theta)$.

$$\frac{\partial L(\theta)}{\partial \beta} = \sigma^{-2} X' \Psi^{-1} u, \qquad \frac{\partial L(\theta)}{\partial \sigma^{-2}} = \frac{T\sigma^2}{2},$$

$$\frac{\partial L(\theta)}{\partial \alpha} = \frac{1}{2} \left[\frac{\partial \text{ vec } (\Psi^{-1})}{\partial \alpha} \right] \text{vec } (\Psi - \sigma^2 u u'),$$

$$\frac{\partial^2 L(\theta)}{\partial \beta \partial \beta'} = -\sigma^{-2} X' \Psi^{-1} X, \qquad \frac{\partial^2 L(\theta)}{\partial \beta \partial \sigma^{-2}} = X' \Psi^{-1} (y - X\beta),$$

$$\frac{\partial^2 L(\theta)}{\partial \beta \partial \alpha'} = \sigma^{-2} \left[\frac{\partial \text{ vec } (\Psi^{-1})}{\partial \alpha} \right]' (u \otimes X), \qquad \frac{\partial^2 L(\theta)}{\partial (\sigma^{-2})^2} = -\frac{T}{2\sigma^4},$$

$$\frac{\partial^2 L(\theta)}{\partial \sigma^{-2} \partial \alpha} = \frac{1}{2} [\text{vec } (uu')]' \left[\frac{\partial \text{ vec } (\Psi^{-1})}{\partial \alpha'} \right],$$

$$\frac{\partial^2 L(\theta)}{\partial \alpha \partial \alpha'} = \frac{1}{2} \sigma^{-2} \sum_{i=1}^{T} \sum_{j=1}^{T} (\Psi_{ij} - u_i u_j) \frac{\partial (\Psi^{-1})_{ij}}{\partial \alpha \partial \alpha'} - \frac{1}{2} \left[\frac{\partial \text{ vec } (\Psi^{-1})}{\partial \alpha} \right]' (\Psi \otimes \Psi) \left[\frac{\partial \text{ vec } (\Psi^{-1})}{\partial \alpha} \right],$$

Note that vec$(\cdot)$ denotes a column vector formed by stacking the columns of its matrix argument, and that the subscripts "i, j" denote the corresponding element in the matrix to which they are attached.

All of the above derivatives for α are expressed in terms of derivatives of Ψ^{-1}. In some cases it may be more convenient to compute derivatives for Ψ and then use

$$\frac{\partial \text{ vec } (\Psi^{-1})}{\partial \alpha} = -(\Psi^{-1} \otimes \Psi^{-1}) \frac{\partial \text{ vec } (\Psi)}{\partial \alpha}.$$

BAYESIAN STATISTICS 3, pp. 733–745
J. M. Bernardo, M. H. DeGroot, D. V. Lindley and A. F. M. Smith, (Eds.)

Transfer Response Models: a Numerical Approach

A. POLE
University of Warwick

SUMMARY

An efficient numerical approach for the Bayesian posterior analysis and forecasting of state-space time series models which contain an unknown transfer response component is described. The basic tools used are Gauss Hermite integration and a time evolving parameter grid. Also discussed are methods for subjective intervention and variance discounting within this framework.

Keywords: TRANSFER RESPONSE; DYNAMIC BAYESIAN ANALYSIS; STATE-SPACE MODELS; DISCOUNTING.

1. INTRODUCTION

This paper describes a numerical approach for the Bayesian posterior analysis and forecasting of state space time series models which contain an unknown transfer response component. These models are non-linear in the effect (or transfer response) parameter(s) and thus exact analytic methods are generally unavailable. Previous work on models of this kind has been concerned with various analytic approximations in order to achieve tractable posterior/forecast distributions.

In mixture models (Sorenson and Alspach, 1971; Willsky 1976) where nonlinearities are approximated with linear conbinations of a discrete set of conditionally Gaussian structures, the prior to posterior computation involves an exponential increase in the number of possible models. In the context of a sequential analysis this process quickly gets out of hand and various approximations have been suggested to overcome the problem. The basic technique is one of mixture collapsing (Sorenson and Alspach, 1971; Harrison and Stevens 1976) which, in essence, allows a single representative model to replace a group of models in the aforementioned posterior analysis. In this way the set of representative model definitions carried forward to each new observation remains 'small' (with appropriately adjusted probabilistic characteristics) and the numerical problems are obviated.

An alternative approach to analysing certain kinds of non-linear models was proposed in West, Harrison and Migon, 1986, and used the notion of a "guide-relationship" —a modeler's guide to reality. In this scheme the non-linear form of the model is used only as a means to determine distributional moments, the form of the actual distribution being determined separately— and conventionally a member of some closed from conjugate family.

In this paper, in contrast to these analytic kinds of approximation. we describe a feasible approach to implementing a numerical analysis of transfer response models on a small microcomputer. The core of the problem here is to define a procedure which will constrain the numerical requirements to a realistic level while allowing accurate information on the non-linear model components in addition to any conditionally linear structures to be ascertained. A direct approach which throws many points at the problem, that is, defines a "fine" grid of points in the non-linear parameter space at which to perform numerical prior/posterior and forecast calculations, is not at all satisfactory — the resulting demands on processor resources required for even modest amounts of data in quite simple models is prohibitive for all but large mainframe computer systems, which would put the techniques out of reach of many practitioners who could benefit from these kinds of models.

In order to reduce these computational overheads a system of time varying grids is proposed in Section 2. In Section 3 a modification of the discounting technique, (Ameen and Harrison, 1985) which is used to specify the system noise covariance at each stage is described. Section 4 discusses intervention strategies which allow the introduction of subjective information concerning possible structural changes at any time. These ideas are applied to some illustrative examples in Section 5. Finally, in Section 6 some conclusions are drawn from the analysis presented.

2. MODEL AND ANALYSIS

2.1. *Model*

The basic form of the first order transfer response model is

Observation Equation	$y_t = E_t + v_t,$	$v_t \sim N(0, V)$	(2.1)
System Equation	$E_t = \lambda E_{t-1} + \rho X_t + w_t$	$w_t \sim N(0, w_t),$	(2.2)

where y_t is the observed value of the time series under investigation, X_t is an input, E_t is the effect of the input sequence and λ is the non-linear transfer response parameter whose domain is usually a subset of the unit interval $[0, 1]$. Extending the model in (2.1.) and (2.2.) to include additional linear components such as a trend, seasonal factors or exogenous regressors is straightforward (see, for example, Harrison and West, 1986). However since these elements do not alter the fundamental model structure, their inclusion here will simply cloud the presentation and so they will be omitted in this paper. Furthermore, we will, for the most part, make no reference beyond the simplest first order models. In practical applications second order models are also extremly important, particularly in the case of complex (conjugate) roots which admits the possibility of cyclical transfer response behaviour. The techniques for this case are simple extensions of those to be described—involving a cartesian product in two dimensions for the parameters grid and corresponding integration rule.

For any of the above kinds of models, if the non-linear parameter has a known value then the models reduce to familiar normal dynamic linear models (NDLM). Technical details of the analysis of these models and extensions are given in Harrison and Stevens, 1976. However, in the interesting case where λ is unknown the analysis is not directly tractable and some form of approximation is required. Our solution is basically a dynamic variation of the Gaussian quadrature based numerical integration strategy of Smith *et al.* 1985, and we begin with a brief review of some of the underlying concepts, based on Naylor and Smith, 1982.

2.2. *Gauss-Hermite Integration*

A well known formula for the approximate evaluation of univariate integrals of the form

$$\int_{-\infty}^{\infty} e^{-t^2} f(t)dt, \tag{2.3}$$

is the Gauss-Hermite (G-H) quadrature formula

$$\sum_{i=1}^{n} w_i f(t_i), \tag{2.4}$$

where $w_i = \frac{2^{n-1}n!\sqrt{\pi}}{n^2[H_{n-1}(t_i)]^2}$ and t_i is the i-th zero of the Hermite polynomial $H_n(t)$ —see, for example, Davis and Rabinowitz, 1967. Indeed, the form of the remainder function,

$$R_n = \frac{n!\sqrt{\pi}}{2^n(2n)!}f^{(2n)}(\xi), \tag{2.5}$$

for some ξ, means that approximation (2.4) is exact for (2.3) whenever $f(t)$ is a polynomial of degree not exceeding $(2n-1)$. More generally, if $h(t)$ is a suitably regular function and

$$g(t) = h(t)(2\pi\sigma^2)^{\frac{1}{2}} \exp\left\{-\frac{1}{2}\left(\frac{t-\mu}{\sigma}\right)^2\right\}, \tag{2.6}$$

then it is straightforward to see that

$$\int_{-\infty}^{\infty} g(t)dt \approx \sum_{i=1}^{n} m_i g(z_i), \tag{2.7}$$

where $m_i = w_i \exp(t_i^2)\sqrt{2}\sigma$, $z_i = \mu + \sqrt{2}\sigma t_i$. Tables of t_i, w_i, and $w_i \exp(t_i^2)$ are available for $n = 1, (1), 20$ (Salzer, Zucker and Capuano, 1952) and the error will be small if $h(z)$ is well approximated by a polynomial over the region for which $g(z)$ is appreciable (see Smith, Sloan and Opie, 1983). For the present paper let $t = \eta$ where $\eta = \ln\left(\frac{\lambda - a}{b - \lambda}\right)$ and $0 < a < \lambda < b \leq 1$, then (2.7) can be applied to integrals of the form

$$E(\underline{\theta}_t) = \int_{-\infty}^{\infty} E(\underline{\theta}_t|\eta, \text{Data})p(\eta|\text{Data})d\eta, \tag{2.8}$$

provided that we can find a normal density which, when multiplied by a polynomial in η given an adequate approximation to $p(\eta|\text{Data})$ Here, $\underline{\theta}_t$ is the state parameter vector ($\equiv (E_t, \rho)^T$ for the model in (2.1) and (2.2)).

2.3. *The Dynamic Grid*

At time $(t-1)$ assume we have available the posterior density $p(\lambda|D_{t-1})$ where λ is the non-linear parameter, defined at points on a grid, L_0 say, and D_{t-1} denotes all the information available up to and icluding $(t-1)$; together with associated conditional quantities such as $\underline{m}_{t-1}(\lambda)$, $C_{t-1}(\lambda)$—the state vector posterior conditional means and covariances respectively—also with respect to L_0. We also then have estimates of $E(\lambda|D_{t-1})$, $V(\lambda|D_{t-1})$ and the unconditional moments $\underline{m}_{t-1}$, C_{t-1}.

Now proceed to time t. Using $E(\lambda|D_{t-1})$ and $V(\lambda|D_{t-1})$ as initial estimates of the posterior moments for λ after time t we construct a new λ-grid—call it L_1—using the G-H formula of Section 2.2. Based on this new grid we require the time t prior specification for λ, which is precisely the time $(t-1)$ posterior (assuming no additional information, methods for including such are discussed below). Since $p(\lambda|D_{t-1})$ is not well defined by a closed form mathematical function, but available only numerical on a grid L_0, we require a procedure to obtain the implied density values on the new gird L_1. This is accomplished in the following manner.

1. fit "not-a-knot" cubic splines (De Boor, 1978) to the log density on grid L_0;
2. interpolate the log density values at the new abcissae L_1;
3. convert the interpolated log density values to the density scale.

We now have a prior on λ for time period t. In a similar manner we also require the time t prior specification for the linear (or state) parameters. These are fully determined in the usual normal/Student-t manner by the conditional moments $\underline{m}_{t-1}(\lambda)$, $C_{t-1}(\lambda)$ and, as for λ, we now require these on the new grid L_1. Numerical interpolation is again applied but the nature of the present quantities imposes certain additional constraints which must be satisfied by the interpolated values. That is, the conditional covariance matrices must remain positive definite. At present we meet this condtion by using simple linear interpolation for $C(\lambda)$—and also for the conditional state mean vector, $\underline{m}(\lambda)$, for compatibility. Although there are undoubtedly more sophisticated techniques for achieving this goal (using spectral decomposition for example—and suggestions would be welcome) the simple approach used here does have the considerable advantages of computational ease, and it works!

On completion of this new-grid/interpolation stage we are readly to perform the time t updates. The conditional posterior state moments at time $(t-1)$ evolve to their time t values via familiar recurrance relationships see Harrison and West, 1986 for details, with all the usual quantities being defined conditionally upon λ. That is, at this stage we are simply updating in parallel a number of familiar NDLMs having idenditical structure and distinguishable only by the value (not the form) of one component, namely the state equation matrix G which, for the model in (2.2) is defined as

$$G_t = \begin{bmatrix} \lambda & X_t \\ 0 & 1 \end{bmatrix}$$

With this updating process complete we are able to compute the time t posterior on λ over the grid L_1. This is done from a direct implementation of Bayes rule and the posterior moments are obtanied by the G-H integration procedure. Furthermore, all other unconditional quantities-$\underline{m}_t$, C_t etc.—also use the G-H quadrature weights.

2.4. Analysis Summary

The procedure described above can be summarised in the following scheme.

1. Specify initial prior moments—usually conditional moments of the state vector will be identical;
2. On the basis of the prior moments for λ construct an eight point (say) grid;
3. Usual updates and posterior calculations given first data point.

For all additional time periods t the following steps are repeated.

A. Using time $(t-1)$ posterior moments for λ construct the new grid;

B. Interpolate the time $(t-1)$ posterior quantities $E(\underline{\theta}_{t-1}|D_{t-1},\lambda)$, $\text{Cov}(\underline{\theta}_{t-1}|D_{t-1},\lambda)$, $p(\lambda|D_{t-1})$ on the new grid;

C. Perform the usual evolution for each conditional model to obtain the conditional prior moments $E(\underline{\theta}_t|D_{t-1},\lambda)$, $\text{Cov}(\underline{\theta}_t|D_{t-1},\lambda)$ etc.;

D. Calculate $P(\lambda|D_t)$ and construct quadrature weights;

E. Calculate any required marginal posterior quantities, for example $E(\underline{\theta}_t|D_t)$, using G-H integration (i.e. the quadrature weights).

3. DISCOUNTING

The concept of discounting, Ameen and Harrison, 1985, for dynamic Bayesian models to obviate the need for practitioners to specify evolution covariance matrices is also applied to the present non-linear models. However, the precise nature of its use requires some additional consideration. For a linear model the simplest discounted form of the evolution disturbance covariance matrix is a single discount factor applied to the posterior covariance of the state parameter. A direct translation of this procedure to the conditionally linear models of this paper will be inappropriate since it would imply different amounts of uncertainty in the state evolution for each conditionally linear model (or, equivalently, for each value of λ). A procedure which we feel is more suitable and which we have adopted here is to apply the discount method to the unconditional posterior covariance of the state parameter and use this single term as the evolutionary uncertainty for each conditional model,

$$W_t(\lambda) = W_t = \left(1 - \frac{1}{\sigma}\right) \mathrm{Cov}(\underline{\theta}_{t-1} | D_{t-1})$$

More complex discounting schemes using multiple discount factors are also applied in this way.

4. INTERVENTION

4.1. State Parameters

In a typical forecasting situation there will be occasions when additional information becomes available to the forecaster which he would like to include in an analysis at some point. For example, if the demand for a company's product is being forecast then knowledge that a competitor's product is to be withdrawn (say) is clearly an important source of prior information. In this situation the forecaster would want to modify the prior mean and uncertainty associated with some component(s) of his model—a level component almost surely. For concreteness suppose that the fortecaster decides that as a result of the removal of his competitor's product, demand for his own product might be expected to increase by 20%, but consideration of other factors causes him to be more uncertain than before. What is the best way of including this information—mean level increase of 20%, level variance increase of 50% (say)—into his existing priors?

There are two fundamental levels at which such changes can be made. We can think either in terms of the individual conditional models (micro level) or at the unconditional (macro) level. We believe that latter is the correct way to proceed for two reasons. Firistly, the micro model is really composed of a continuum of conditional models (λ is a continuous parameter) and so thinking in terms of the current set of representative conditional models is rather artifical in this context. Secondly, when considering the market for his product the forecaster would implicitly be thinking in terms of a single overall model. Thinking about global components such as level, growth, seasonality, transfer responses etc., is certainly likely, but going beyond this to the micro level of conditional models seems to be unreal. Indeed, in such a situaton why stop at conditioning on λ? For these reasons it would appear that the most appropriate intervention strategies for the state parameter of these models are

1. increased variation only: use suitable alternative discount factors;
2. component mean changes: reset all conditional moments to the suitably adjusted macro beliefs.
3. more complex structural changes: on rare occasions the entire form of prior distributions and/or model components may require adjustment.

4.2. *The Effect Parameter*

Formally, we can deal with intervention on λ in a similar way to that just described for $\underline{\theta}$. Operationally, however, this involves additional considerations in the present context. It is not enough simply to reset $p(\lambda|D_{t-1})$ here because of the dicrete approximation scheme which is employed, that is, we must also consider the implications of such changes on the conditional linear sub-models. Suppose that we have $p(\lambda|D_{t-1})$ defined on a grid L_0. Now, a simple method for increasing the uncertainty on λ while retaining the overall from of the distribution is to redefine the grid points L_0 by spreading them out more widely in the domain of λ—see Figure 4.1. To achieve this simply reset the time $t-1$ posterior mean and variance for λ so that the calculation of the grid for time t is automatically adjusted. The prior probability values need not be altered since appropriate normalisations are handled by the program.

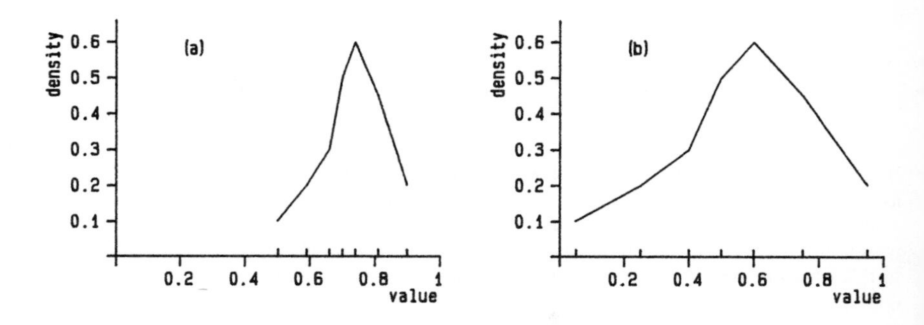

Figure 4.1. *Increasing uncertainty on the effect parameter (a) pre-intervention, (b) post-intervention*

The problem which now arises is what to do about the conditional moments of $\underline{\theta}|D_{t-1}$ which are currently defined on grid L_0. In the scheme outlined in Section 2 these values would simply be interpolated/extrapolated on the new grid L_0 and the analysis taken from there. However, in the present situation this is not regarded as satisfactory since there is very likely to be a greater extent to the extrapolation, in other words, the grid will be widened condiderably more than in routine running. An approach which we feel is better suited to this case is to reset all conditional state prior moments to their unconditional values obtained from the G-H integration rule over grid L_0. This takes our thinking back to the macro model where the prior information on the state vector after the intervention on λ is unchanged. Hence at this level we have decoupled the external information on λ from beliefs about θ which is generally desirable—see the discussion of component-wise model construction and intervention strategies in the linear case in (Harrison and West, 1986). It may be noted that this decoupling procedure parallels the method of information discounting discussed in Section 3.

5. EXAMPLES

A simple first order model of the form (2.1) and (2.2) was applied to the advertising data discussed in Migon and Harrison, 1983. Although such a model is a vast simplification it does retain one of the most important aspects of the model constructed in Migon and Harrison, 1983, and that is the memory decay effect which is obtained with a value of $\lambda \in (0, 1)$. In fact, the nature of the data suggested that a value of λ near to the top end of this range and this led to the prior domain specification $0.8 < \lambda < 0.9999$. A four point grid was selected and initial prior moments on λ set to 0.9 for the mean and 0.025 for the variance (set to produce a grid approximately uniform on the entire interval) and prior probabilities of 0.25 were assigned to each. The discount factor for E_t was set to 0.95. Intervention was performed on λ at times 35, 95, and 155—immediately before the start of each new advertising campaign—with the prior on λ being reset to the values used at $t = 0$. Figure 5.1 illustrates the one step ahead foreacats from this analysis and it can be seen that the major qualitative nature of the data has been captured by the model. Figure 5.2 is a plot of the dynamic grid—the effect of the interventions can clearly be seen as can the natural way in which knowledge of λ accumulates through each advertising campaign.

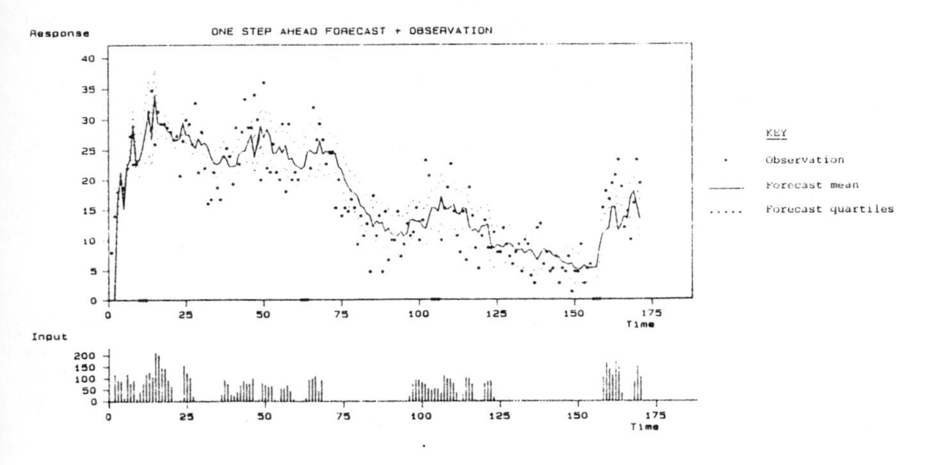

Figure 5.1. *Advertising example*

Next, two sets of simulated data were analysed in order to see how well these models can adapt to structural changes. Data was generated using a first order model with a level component.

$$
\begin{aligned}
y_t &= \mu_t + E_t + v_t & w_{1t} &\sim N(0,1), \\
\mu_t &= \mu_{t-1} + w_{1t} & w_{1t} &\sim N(0,0.1), \\
E_t &= \lambda E_{t-1} + \rho X_t + w_{2t} & w_{2t} &\sim N(0,0.1).
\end{aligned}
\tag{5.1}
$$

Fifty observations were generated with $\mu_0 = 100$, $E_0 = 0$, $\rho = 25$ and $X_t \sim N(0,1)$ at predesignated times, and 0 elsewhere. In the first data set a level change was imposed at time $t = 25 : \mu_{25}$ was increased by 20% over what it would otherwise would have been. The initial priors have accurate means with "large" variances (vague prior information) and a single discount factor of 0.95 was chosen for μ_t, E_t. The level change at $t = 25$ is anticipated by greatly increasing the prior variance on $\mu_{25}|D_{24}$: the results are very encouraging. Forecast uncertainty for $t = 25$ is increased, y_{25} provides the information necessary to learn about μ_{25} and for $t = 26$ and beyond forecasts are about as good as before the level change. These is no discernible effect on λ of the intervention, as can be seen from Figure 5.3. Inference remains extremely tight (not surprising given the almost noiseless data) around the true value of 0.8.

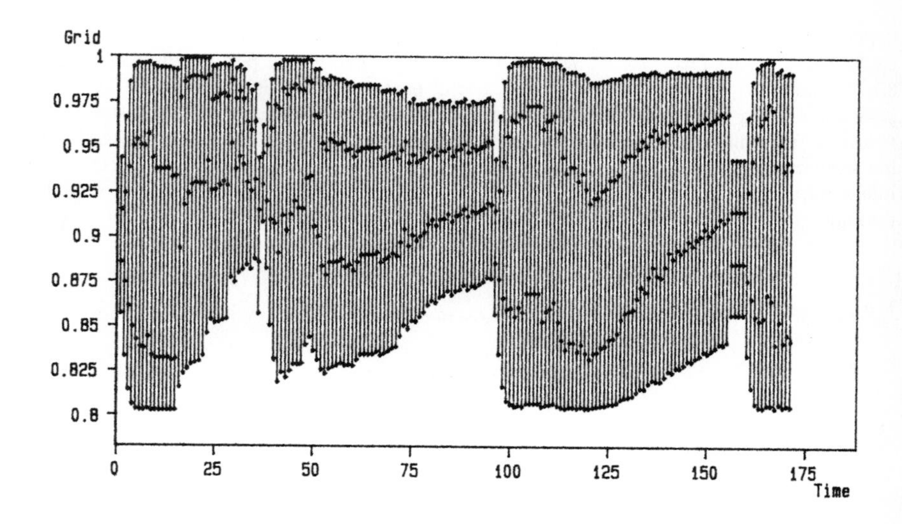

Figure 5.2. *Advertising data: the dynamic grid*

Technical Aside. It might be thought that the intervention on the prior variance of μ_{25} ought to feed through in some manner to the dynamic grid—that is, to inference on λ. Clearly any such mechanism must work through $P(\lambda|D_{25})$ since the intervention is applied after the time $t = 24$ updates have been calculated. Thus, the question here is, how does the increased variation in $(\mu_{25}|D_{24})$ feed through to $P(\lambda|D_{25})$? Consider the form of the posterior distribution on λ,

$$p(\lambda|D_{25}) \propto p(\lambda|D_{24})|S_{24}(\lambda)Q_{25}(\lambda)|^{-\frac{1}{2}} \left[1 + \frac{e_{25}(\lambda)^2}{n_{24}S_{24}(\lambda)Q_{25}(\lambda)}\right]^{-\frac{1}{2}(n_{24}+1)} \tag{5.2}$$

where S and Q are the usual quantities defined for NDLMs—see Migon and Harrison, 1983:

- $e_t(\lambda)$ is the one step-ahead forecast error at time t;
- $Q_t(\lambda)$ is the one step-ahead forecast variance conditional on known observation disturbance variance V;
- $S_{t-1}(\lambda)$ is a scale parameter which multiplies Q_t to give the one step-ahead forecast variance in the case of unknown V;
- n_t is the degress of freedom for the case of unknown V.

This shows that the only quantities affected by the intervention which are reflected in inference on λ are Q_{25} and e_{25}. Clearly $e_{25}(\lambda)$ will be "large"—in this example aproximately 20 (the size of the shift) for all λ—so that practically speaking $e_{25}(\lambda)$ is constant over λ. $Q_{25}(\lambda)$ is also approximately constant because of the intervention: in fact, variation over λ was less than 1% compared with 100% before intervention. Together, these results mean that posterior inference on λ at time $t = 25$ is approximately unchanged from the prior (or, equivalently, the time $t = 24$ posterior) and hence the dynamic grid will be little changed. Now, by time $t = 26$, much is learned about the level and the conditional variances $V(\mu_{26}|D_{26}, \lambda)$ are greatly reduced, which means that the situation is restored to its general pre-intervention/level change state. Hence the intervention on the level prior variance does not lead to any (noticeable) effect on λ.

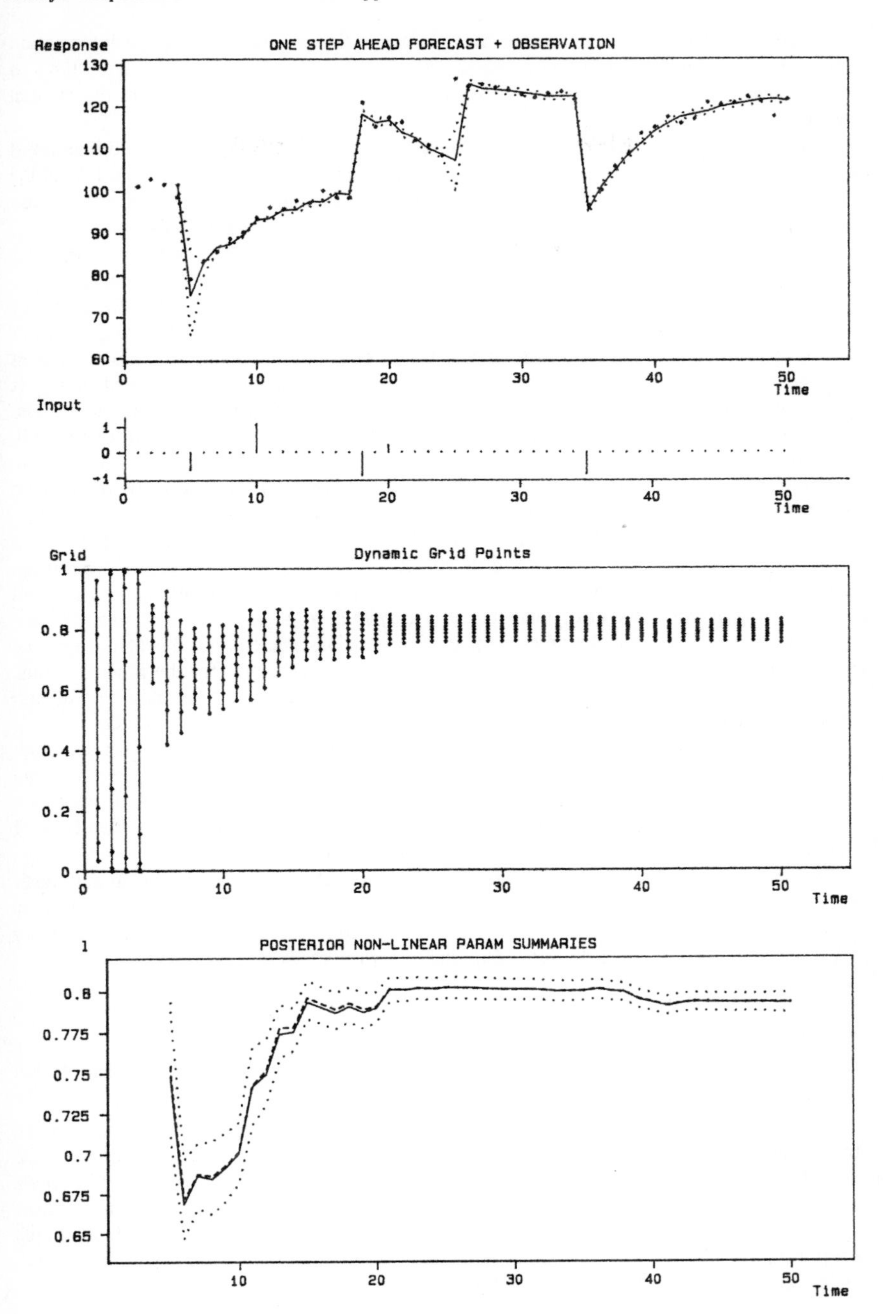

Figure 5.3. *Simulation 1: Intervention analysis of level change*

In the second simulation of model (5.1) similar initial conditions and prior information for the analysis were used but this time the value of λ was reduced by half from 0.85 to 0.425 at time $t = 25$. Two analysis were performed: (a) fully automatic, and (b) manual intervention.

In (a) initial learning about λ follows a familiar pattern. "Nothing" happens until an input occurs ($X_t \neq 0$) then rapid reaction produces immediate response in inference on λ (aided in this case by the size of the input, its coefficient and the small observation noise variance). Inference is later improved and sharpened alter additional inputs are processed.

Reacting to the λ-shift: clearly, as expected, nothing happens until an input ocurs after the shift and then the automatic reaction is immediate. The input at $t = 35$ leads to a clear under-forecast (the input is negative) at $t = 36$ which immediately increases the posterior variance on λ and widens the grid for $t = 37$. This process continues until the effect of the input dies out about three periods later. By this time the inference on λ has a reduced mean and lower precision but clearly there has been insuficient data in the form of inputs and related responses to pull the posterior on λ fully into line with the actual situation. The forecast plot in Figure 5.4 reflects these comments on the learning about the λ-shift very well. Indeed, monitoring such a plot on line should alert the investigator to the potential problem, suggesting manual intervention before the next input. Alternatively, an automatic monitoring and adaptation system can be used—one such scheme is described in West and Harrison, 1986.

It is interesting to note that the under forecasting after the λ-shift causes the posterior estimate of the level to increase, and also its variance. The resulting over-forecasts when the effect of the input has died out subsequently leads to a lowering of the level again. This kind of confounding between the effect and the level is common in models of this sort. In fact, when λ is close to 1 the model suffers from an identification problem, but this can usually be overcome by specifying different rates of evolution (discount factors) for these components. Effects of the λ-shift can also be seen in the posterior development of the input coefficient, ρ. Alter $t = 35$ the point estimate of ρ drops slightly, with precision decreasing too.

These results are quite encouraging. The dynamic grid reacts as we might wish, although we note that a single input of the given magnitude is insufficient to complete the learning process. The post-shift value of λ means that there are only about three periods from which to adapt to the new conditions. This is too short a time as we have seen since the effect of the shift is confounded in both the level and input parameters at first.

In (b) an analysis supposing prior information about a change in the non-linear parameter circumstances was attempted. Here it was assumed that it was known that something of an unspecified nature might occur in λ at $t = 25$. In accordance with the intervention strategy for λ described in Section 4.2 the following changes were made.

(1) prior mean of λ at $t = 25$ set to 0.5;

(2) prior variance of λ at $t = 25$ set to 0.6;

(3) conditional prior means and covariances of the state vector, (μ_t, E_t, ρ), all set to their time $t = 24$ unconditional posterior values.

The results of this analysis, illustrated in Figure 5.5, are excellent. The altered prior on λ has mean in the middle of the unit interval domain which happens to be quite close to the post-shift value of 0.425. This accounts for much of the improved forecast for $t = 36$ so we should not be misled to over optimism here. More importantly, it is encouraging to see how much the uncertainty in λ is reduced by just the two additional observations which supply any information on the post shift value. Also, the point estimate (posterior mean) of λ moves quite close to its true value—but do not forget that this data is almost noiseless.

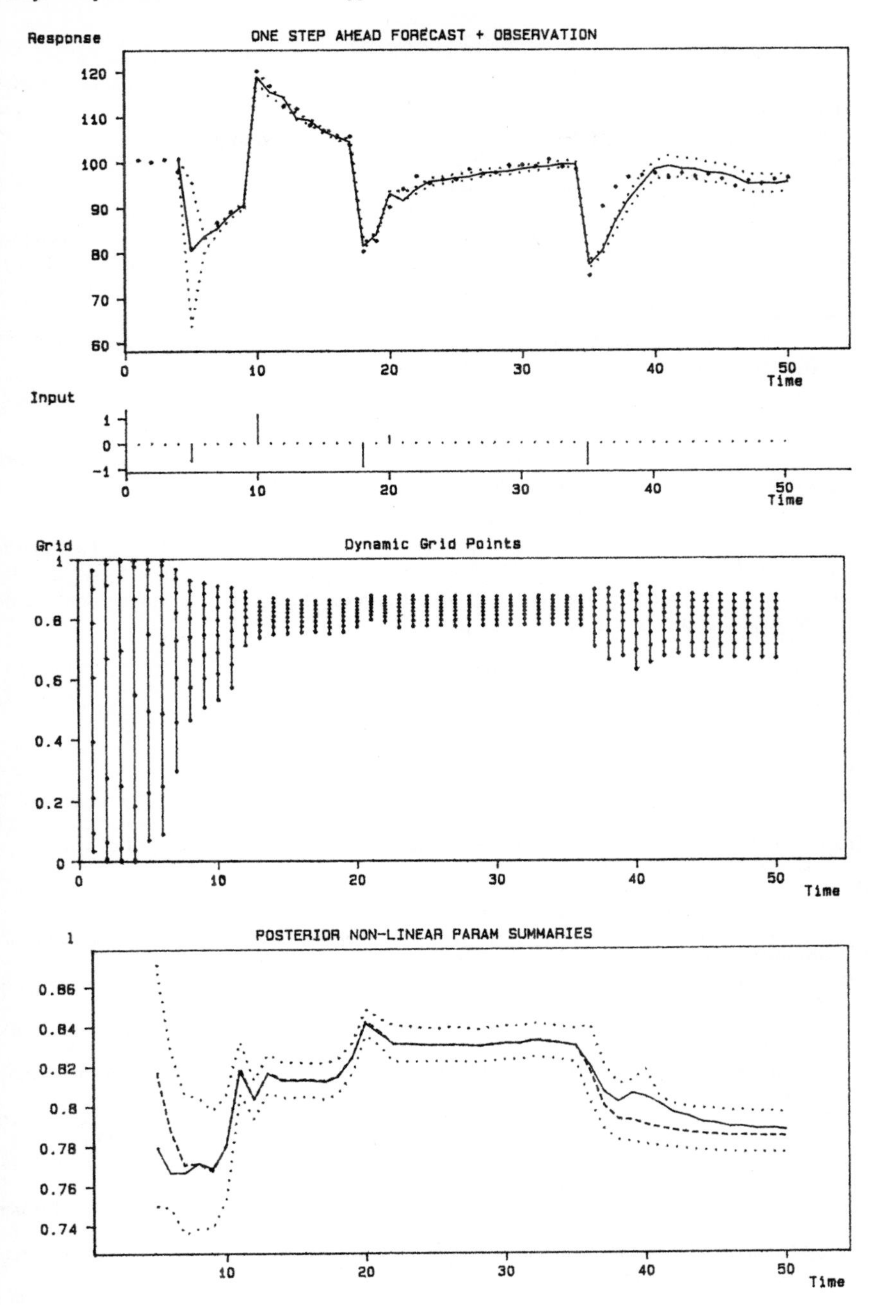

Figure 5.4. *Simulation 2: Automatic analysis of λ-shift*

6. CONCLUSION

This paper has presented a feasible approach for the numerical analysis of dynamic transfer response models. Although only first order models were explicitly considered the technique readily generalises to the practically important class of second order models, and also to higher order models if necessary. The method involves a dynamic generalisation of the efficient numerical integration technique in Smith *et al.* 1985 and certain numerical interpolation strategies. Methods for intervening in an analysis at any stage in order to admit external information including the case of structural change were discussed and three illustrative examples were analysed. These analyses, although of over-simplified models, clearly demonstrate certain desirable features of the methodology. In particular, the ability of the dynamic grid to react quickly to underlying structural change is seen.

Modifications to the basic techinque presented in this paper which are currently being investigated include

(1) iteration of the Gauss-Hermite procedure at each time point in the manner of (the static) BAYES 4 (Smith *et al.* 1985);

(2) iterate on a monitoring basis: if $p(\lambda|D_t)$ changes from $p(\lambda|D_{t-1})$ by more than some predetermined amount, iterate until the inference stabilizes;

(3) update the grid, with iterate, only periodically.

In both (2) and (3) we would also iteration after any external intervention. This should improve analyses of the kind reported above, for example. Investigation of these schemes with particular reference to the improved accuracy weighed against the greater demand on computational resources will be reported in a future paper.

REFERENCES

Ameen, J. R. M. and Harrison, P. J. (1985). Normal discount Bayesian models. *Bayesian Statistics 2*. Amsterdam: North Holland.

Davis, P. J. and Rabinowitz, P. (1967). *Numerical Integration.* Waltham, Mass.

De Boor, C. (1978). *A Practical Guide to Splines.* New York: Springer Verlag.

Harrison, P. J. and Stevens, C. F. (1976). Bayesian forecasting, (with discussion). *J. Roy. Statist. Soc. B* **38**, 205–247.

Harrison P. J. and West, M. (1986). Practical Bayesian forecasting. *Proc. of the 2nd Conference on Practical Bayesian Statistics.*

Migon, H. S. (1985). An approach to non-linear Bayesian forecasting problems with applications. *Unpublished Ph. D Thesis,* Warwick University.

Migon, H. S. and Harrison, P. J. (1983). An approach to non-linear Bayesian forecasting to television advertising. *Bayesian Statistics 2*. Amsterdam, North Holland. 681–696.

Naylor, J. C. and Smith, A. F. M. (1982). Applications of a method for the efficient computation of posterior distributions. *Ann. Statist.* **31**, 214–225.

Salzer H. E., Zucker, R. and Capuano, R. (1952). Tables of the zeroes and weight factors of the first twenty Hermite polynomials. *J. Research of the National Bureau of Standards* **48**, 111–116.

Smith, A. F. M., Skene, A. M., Shaw, J. E. H., Naylor, J. C. and Dransfield, M. (1985). The implementation of the Bayesian paradigm. *Commun.Statist.- Theor Meth.* **14**, 1076–1102.

Smith, W. E., Sloan, I. H. and Opie, A. H. (1983). Product integration over infinite integrals I. Rules based on the zeroes of Hermite polynomials. *Mathematics of Computation* **40**, 519–535.

Sorenson, H. W. and Alpach, D. L. (1971). Recursive Bayesian estimation using gaussian sums. *Automatica* **7**, 465–479.

West, M. and Harrison, P. J. (1986). Monitoring and adaptation in Bayesian forecasting models. *J. Amer. Statist. Assoc.* **81**, 741–750.

West, M., Harrison, P. J. and Migon, H. S. (1986). Dynamic generalised linear models and Bayesian forecasting (with discussion). *J. Amer. Statist. Assoc.* **80**, 73–97.

Willsky, A. S. (1976). A survey of design methods for failure detection in dynamic systems. *Automatica* **12**, 601–611.

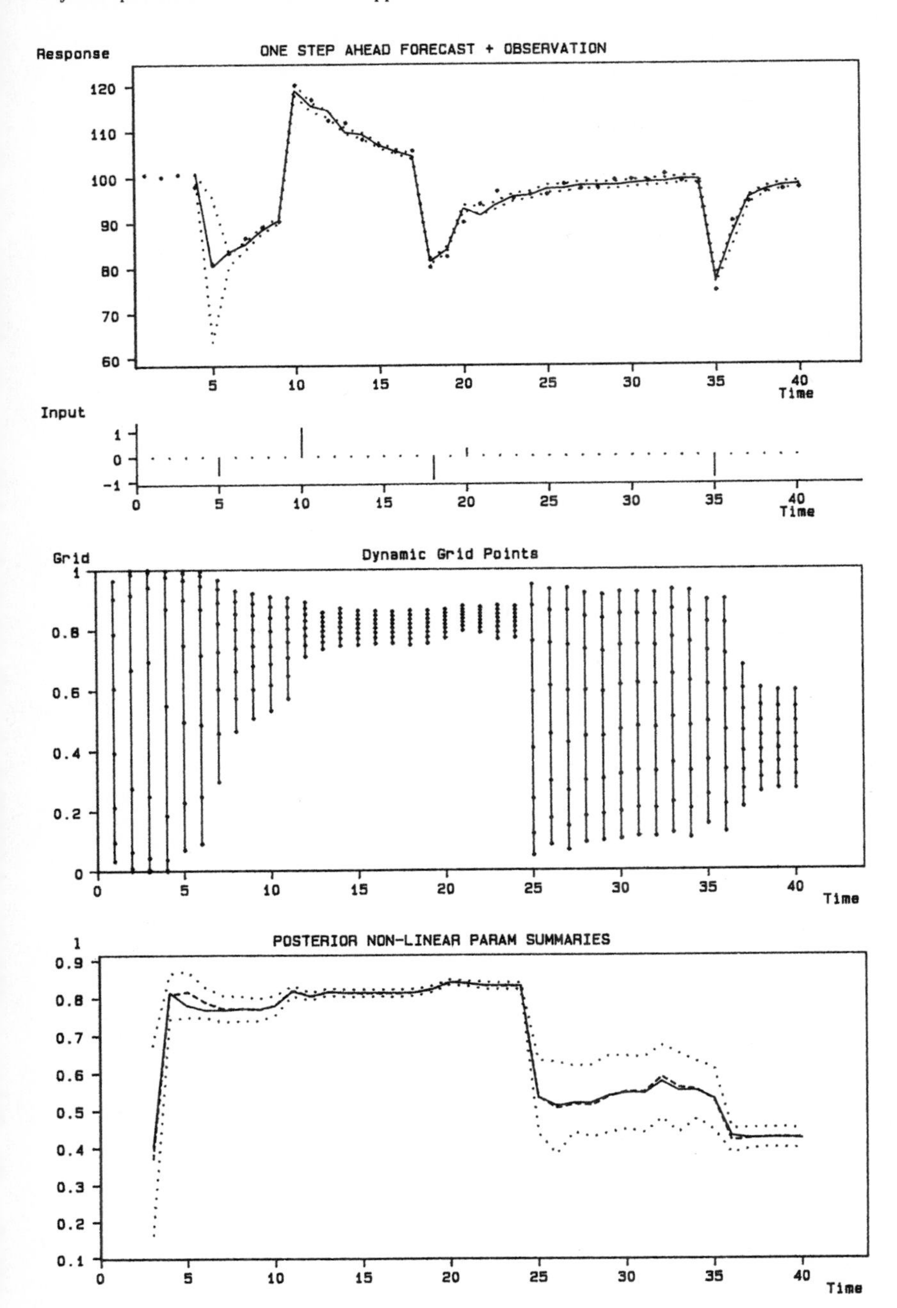

Figure 5.5. *Simulation 2: Intervention analysis of λ-shift*

BAYESIAN STATISTICS 3, pp. 747–756
J. M. Bernardo, M. H. DeGroot, D. V. Lindley and A. F. M. Smith, (Eds.)

Time Series Analysis of Compositional Data

J. M. QUINTANA* and M. WEST*
University of Warwick

SUMMARY

We discuss the use of a class of logistic normal models for analysis of time series of compositional data. The approach embeds log/logistic transforms of multivariate data within dynamic, multivariate regression models that allow for learning about covariance matrices of several series. The primary use of these models is in retrospective analysis of the joint structure of several series, and, in particular, in the assessment of the nature of changes in covariance structure over time.

Keywords: COMPOSITIONAL DATA, DYNAMIC REGRESSION, LOGISTIC-NORMAL DISTRIBUTION, MULTIVARIATE TIME SERIES.

1. INTRODUCTION

Consider a multivariate time series consisting of vectors of positive quantities. Often the focus of attention is on the relative rather than the absolute values, leading to multivariate time series of vectors of proportions. Typical examples are opinion polls, market shares, composition of populations, composition of economic quantities such as imports/exports, gross national product, etc.

This paper concerns compositional data time series, the modelling method being as follows. The log ratio tranformation is applied to the time series, the result being modelled via a dynamic multivariate regression (Quintana, 1985, 1987; Quintana and West, 1987). This procedure enables us to apply the power and versatility of dynamic Bayesian models for time series together with the elegant and rich interpretation of the multivariate log ratio/logistic transformation for compositional data. A key use of the models is the smoothing of multivariate series for the retrospective analysis of patterns of change over time in correlation structure across series. This is illustrated in an application to compositional imports series.

2. DYNAMIC MULTIVARIATE REGRESSION

2.1 Basic model structure.

A class of dynamic, multivariate regression (DMR) models has recently been developed for modelling multi-, and matrix-, variate time series (Quintana, 1985, 1987; Quintana and West, 1987). For a column vector time series y_t, $(t = 1, \ldots,)$, the basic model structure is as follows.

$$\text{Observation Equation}: \quad y_t' = x_t'\Theta_t + e_t', \qquad e_t \sim N(0, v_t\Sigma) \tag{1a}$$

$$\text{Evolution Equation}: \quad \Theta_t = G_t\Theta_{t-1} + F_t, \qquad F_t \sim N(0, W_t, \Sigma) \tag{1b}$$

$$\text{Prior Information}: \quad (\Theta_{t-1}|\Sigma) \sim N(M_{t-1}, C_{t-1}, \Sigma), \quad \Sigma \sim W^{-1}(S_{t-1}, d_{t-1}). \tag{1c}$$

* J.M. Quintana is now at AT&T, USA; M. West is now at Duke University, USA.

In the above equations $\boldsymbol{y}_t$ is a $(q \times 1)$ vector of observations made at time $t, \boldsymbol{x}_t$,is a $(p \times 1)$ vector of independent variables, Θ_t is an unknown $(p \times q)$ matrix of system (regression) parameters, $\boldsymbol{e}_t$ is a $(q \times 1)$ observation error vector, v_t is a scalar variance associated with $\boldsymbol{e}_t$, and Σ is an unknown $(q \times q)$ system scale variance matrix providing cross-sectional correlation structure for the components of $\boldsymbol{y}_t$. $N(\boldsymbol{M}, \boldsymbol{C}, \Sigma)$ and $W^{-1}(\boldsymbol{S}, d)$ denote the general matrix normal and inverted-Wishart distributions as described in the Appendix. These lead, from (1c), to multivariate $\boldsymbol{T}$ distributions for elements of Θ_{t-1} for inference about model parameters unconditional on Σ.

The equation (1a) defines the distribution of the observations given Θ_t and Σ. The system evolution is determined by (1b) which specifies the distribution of Θ_t in terms of Θ_{t-1} given Σ. The system prior information (1c) represents the joint distribution of Θ_1 and Σ at time $t-1$. It is also assumed that $\boldsymbol{e}_t, \boldsymbol{F}_t$ and Θ_{t-1} are independent, with $\boldsymbol{e}_t, \boldsymbol{F}_t$, independent over time given Σ. All distributions appearing in (1) are implicitly assumed conditional on the relevant information available at time $t-1$ including (if any) previous observations $\boldsymbol{y}_{t-1}, \boldsymbol{y}_{t-2}, \ldots$. The distributions (1c) sufficiently summarise this historical information.

The DMR (1) is a multivariate generalisation of the univariate dynamic linear model (DLM) as formulated in Harrison and West (1986, 1987). The relationship between the DMR and the univariate DLM is the dynamic counterpart of the relationship between the multivariate regression and the univariate multiple regression models. It is important to notice that the DMR is not equivalent to the usual vector-variate DLM, in fact, both models can be embedded in a general matrix-variate DLM (Quintana, 1985, 1987).

The nature of the model component series can be seen as follows. For $j = 1, \ldots, q$,, let y_{tj} be the observation on the jth series, simply the jth element of $\boldsymbol{y}_t; e_{tj}$ the corresponding element of $\boldsymbol{e}_t; \boldsymbol{\theta}_{tj}$ the jth column of $\Theta_t; \boldsymbol{f}_{tj}$ the jth column of $\boldsymbol{F}_t; \boldsymbol{m}_{tj}$ the jth column of $\boldsymbol{M}_t$; and σ_j^2 the jth diagonal element of Σ. Then y_{tj} *marginally* follows the DLM

$$\begin{aligned}
&\text{Observation Equation:} && y_{tj} = \boldsymbol{x}_t'\boldsymbol{\theta}_{tj} + e_{tj}, && e_{tj} \sim N(0, v_t\sigma_j^2), && \text{(2a)}\\
&\text{Evolution Equation:} && \boldsymbol{\theta}_{tj} = \boldsymbol{G}_t\boldsymbol{\theta}_{t-1,j} + \boldsymbol{f}_{tj}, && \boldsymbol{f}_{tj} \sim N(\boldsymbol{0}, \boldsymbol{W}_t\sigma_j^2), && \text{(2b)}\\
&\text{Prior Information:} && (\boldsymbol{\theta}_{t-1,j}|\Sigma) \sim N(\boldsymbol{m}_{t-1,j}, \boldsymbol{C}_t\sigma_j^2). && && \text{(2c)}
\end{aligned}$$

The joint structure comes in via the covariances, conditional upon Σ,

$$\begin{aligned}
\text{Cov}(e_{ti}, e_{tj}) &= v_t\sigma_{ij}, && \text{(3a)}\\
\text{Cov}(\boldsymbol{f}_{ti}, \boldsymbol{f}_{tj}) &= \boldsymbol{W}_t\sigma_{ij}, && \text{(3b)}\\
\text{Cov}(\boldsymbol{\theta}_{ti}, \boldsymbol{\theta}_{tj}) &= \boldsymbol{C}_{t-1}\sigma_{ij}, && \text{(3c)}
\end{aligned}$$

for $i \neq j$, where σ_{ij} is the ij off-diagonal element of Σ.

Thus the individual series $y_{tj}, (j = 1 \ldots, q)$, follow the same *form* of DLM, the model parameters $\boldsymbol{\theta}_{tj}$ being different across series. Correlation structure induced by Σ affects both the observational errors through (3a) and the evolution errors $\boldsymbol{f}_{tj}$ through (3b). Thus if, for example, σ_{ij} is large and positive, series i and j will tend to follows similar patterns of behaviour, although their scales σ_i and σ_j may differ. This common structure carries over to the parameter vectors in (3c). Thus, although generally the estimates $\boldsymbol{m}_{tj}$ will differ, the variance matrix $\boldsymbol{C}_{t-1}$ is common to each of the q models.

2.2 Sequential updating equations

As time evolves the distributions in (1c) are revised via a set of matrix-variate updating equations analogous those standard in DLMs. Details are found in Quintana (1985) and Quintana and West (1987). Briefly, define $\boldsymbol{R}_t = \boldsymbol{G}_t\boldsymbol{C}_{t-1}\boldsymbol{G}_t' + \boldsymbol{W}_t, Q_t = \boldsymbol{x}_t'\boldsymbol{R}_t\boldsymbol{x}_t + v_t, \boldsymbol{a}_t = \boldsymbol{R}_t\boldsymbol{x}_t/Q_t$ and $\boldsymbol{r}_t' = \boldsymbol{y}_t' - \boldsymbol{x}_t'\boldsymbol{G}_t\boldsymbol{M}_{t-1}$. Here $\boldsymbol{r}_t$ is the $(q \times 1)$ vector of one-step ahead, point forecast errors, and $\boldsymbol{a}_t$ the $(q \times 1)$ vector of adaptive coefficients. The updating equations are then

$$\boldsymbol{M}_t = \boldsymbol{G}_t\boldsymbol{M}_{t-1} + \boldsymbol{a}_t\boldsymbol{r}_t', \tag{4a}$$

$$\boldsymbol{C}_t = \boldsymbol{R}_t - \boldsymbol{a}_t\boldsymbol{a}_t'Q_t, \tag{4b}$$

$$d_t = d_{t-1} + 1, \tag{4c}$$

$$\boldsymbol{S}_t = \boldsymbol{S}_{t-1} + \boldsymbol{r}_t\boldsymbol{r}_t'/Q_t. \tag{4d}$$

Note that the matrix updating equation (4a) can be decomposed column by column, giving q separate updates, one for each series. These individual equations are essentially the usual *Kalman Filter* equations for the models (2). However, the associated variance matrix updates are identical, being defined by the single $(p \times p)$ matrix update (4b), leading to a reduction in the necessary calculations, particularly when q is large. Finally, from the updates (4c) and (4d) we obtain the posterior estimate $\boldsymbol{V}_t = \boldsymbol{S}_t/d_t$ for Σ.

2.3 Discount factors

In the DMR, the use of discount factors to assign suitable structure and magnitude to the sequence of evolution variance matrices $\boldsymbol{W}_t$ is just as in univariate DLMs (Harrison and West, 1986, 1987). Details can be found in these references, particularly concerning the use of multiple discount factors for indvidual components of the model θ_t, and derives from the fact that, given Σ, the model can be viewed as q individual DMLs for the scalar component series y_{tj}. Here we restrict ourselves to the use of a single discount factor for the entire θ_t vector, this simplifying structure being appropriate for the model used in the application of Section 4 below. Thus, for a given discount factor $\delta, 0 < \delta \leq 1$, the matrix $\boldsymbol{W}_t$ in (1b) is defined as

$$\boldsymbol{W}_t = (\delta^{-1} - 1)\boldsymbol{G}_t\boldsymbol{C}_{t-1}\boldsymbol{G}_t' \tag{5}$$

Notice that $\delta = 1$ implies $\boldsymbol{W}_t = \boldsymbol{0}$ and a purely deterministic evolution for Θ_t, or a *static* model, whereas smaller values model greater random variation in Θ_t.

2.4 Dynamic scale variance matrix

In the DMR formulation (1) the system scale variance is static. It is often more realistic to assume a steady, random evolution for Σ, allowing for time changes in the variance-covariance structure. This effect can be obtained by introducing a further discount factor $\beta, (0 < \beta \leq 1)$ associated with Σ, as in Quintana and West (1987), for example. The net effect is simply to increase the uncertainty about Σ over the time intervals $(t - 1, t)$, leading to (4c) and (4d) being modified as

$$d_t = \beta d_{t-1} + 1 \quad \text{and} \quad \boldsymbol{S}_t = \beta\boldsymbol{S}_{t-1} + \boldsymbol{r}_t\boldsymbol{r}_t'/Q_t.$$

Here β discounts the prior quantities d_{t-1} and $\boldsymbol{S}_{t-1}$ before updating. Clearly $\beta = 1$ corresponds to the original model with Σ static.

2.5 Choice of discount factors

For a model with $\boldsymbol{x}_t$ and $\boldsymbol{G}_t$ specified, the effects of different values of δ and β can be explored by fitting several such models and assessing their relative merits using visual examination of forecasts, forecast errors, simple numerical summaries of model performance, etc. Experience strongly indicates *a priori* plausible ranges of values, typically $0.75 \leq \delta \leq 1$ and $0.95 \leq \beta \leq 1$ (see also Harrison and West, 1986, 1987). One numerical measure of the relative support from the data for different values is provided by the observed predictive densities of the observations under each of the models. This quantity may be easily sequentially calculated (West, 1986), and, as δ and β are varied between models, it provides a likelihood function for the discount factors. This likelihood may, of course, be formally used to update a specified prior for discount factors. The practical benefits of this are really rather doubtful, however, particularly in view of the use of interventions on discount factors etc. that are necessary to cater for discontinuities routinely encountered with real data (Harrison and West, 1986, 1987; West and Harrison, 1986, 1987). Thus, as a rough guide, large values of this likelihood function indicate data-supported values for the discount factors that guide model choice so long as these values do not wildly conflict with the *a priori* plausible ranges.

2.6 Smoothing for retrospective analysis

At time t, the quantities $\boldsymbol{M}_t$, $\boldsymbol{C}_t$, d_t and $\boldsymbol{S}_t$ sufficiently summarise the historical information $\boldsymbol{y}_t, \boldsymbol{y}_{t-1}, \ldots$, providing current estimates $\boldsymbol{M}_t$ and $\boldsymbol{V}_t$ of Θ_t and Σ respectively. However, having processed a full series of n observations, these quantities may be revised to be based on *all* the data, including observations obtained after time t and up to time n. The corresponding *smoothed* estimates of Θ_t and Σ are denoted by $\boldsymbol{M}_t^*$ and $\boldsymbol{V}_t^*$. For *retrospective* time series analysis these smoothed, estimates are required, being calculated using backwards filtering equations similar to those in standard DLMs (Harrison and West, 1986). Full details are lengthy and therefore omitted, the key calculations are given by the following backwards recursions. Recall the matrices $\boldsymbol{R}_t$ from Section 2.2, and define the $(p \times p)$ matrices $\boldsymbol{B}_t = \boldsymbol{C}_t \boldsymbol{G}_{t+1} \boldsymbol{R}_{t+1}^{-1}$. Then, for $t = n-1, n-2, \ldots, 1$,

$$\begin{aligned}
\boldsymbol{M}_t^* &= \boldsymbol{M}_t + \boldsymbol{B}_t(\boldsymbol{M}_{t+1}^* - \boldsymbol{G}_{t+1}\boldsymbol{M}_t), \\
\boldsymbol{C}_t^* &= \boldsymbol{C}_t - \boldsymbol{B}_t(\boldsymbol{R}_{t+1} - \boldsymbol{C}_{t+1}^*)\boldsymbol{B}_t', \\
\boldsymbol{V}_t^{*-1} &= \boldsymbol{V}_t^{-1} + \beta(\boldsymbol{V}_{t+1}^{*-1} - \boldsymbol{V}_t^{-1}),
\end{aligned}$$

with starting values $\boldsymbol{M}_n^* = \boldsymbol{M}_n$, $\boldsymbol{C}_n^* = \boldsymbol{C}_n$ and $\boldsymbol{V}_n^* = \boldsymbol{V}_n$. Clearly $\beta = 1$ implies $\boldsymbol{V}_t^* = \boldsymbol{V}_n$ for all t.

3. LOGISTIC/LOG RATIO TRANSFORMATION

The logistic/log ratio transformation (LLT) has found wide use in analysis of proportions (Aitchison, 1986) and is used here in the time series context. Let $\boldsymbol{z}_t' = (z_{t1}, \ldots, z_{tq})$, $t = 1, 2 \ldots$, be a multivariate time series such that $z_{ti} > 0$ for all i and t. Suppose that we are concerned only with the proportions

$$\boldsymbol{p}_t = \left(\sum_{i=1}^{q} z_{ti} \right)^{-1} \boldsymbol{z}_t. \tag{6}$$

Log ratio transformations map the vector of proportions $\boldsymbol{p}_t$ into a vector of real-valued quantities $\boldsymbol{y}_t$. A particular, symmetric log ratio transformation is given by

$$y_{tj} = \log\left(\frac{p_{tj}}{\hat{p}_t}\right) = \log p_{tj} - \log \hat{p}_t, \quad j = 1, \ldots, q, \tag{7}$$

where $\hat{p}_t$ is the geometric mean of the p_{tj}. The inverse of this log ratio transformation is the logistic transformation

$$p_{ti} = \frac{\exp(y_{ti})}{\sum_{j=1}^{q} \exp(y_{tj})}, \quad i = 1, \ldots, q. \tag{8}$$

Modelling $\boldsymbol{y}_t$ with the DMR of Section 2 implies a conditional multivariate normal structure. Thus the observational distribution of the proportions $\boldsymbol{p}_t$ is the multivariate logistic-normal distribution as defined in Aitchison and Shen (1980). Clearly a sufficient condition for the vector $\boldsymbol{p}_t$ to have a logistic-normal distribution is that the series $\boldsymbol{z}_t$ is distributed as a multivariate log-normal, although the converse does not hold.

Suppose that $\boldsymbol{y}_t$ in (7) follows the model (1). A complication arising is the singularity of the model due to the zero-sum constraint $\boldsymbol{y}_t'\boldsymbol{1} = 0$, for all t, where $\boldsymbol{1}' = (1, \ldots, 1)$. This follows directly from the definition (7) and leads to singularity of the matrices Σ, $\boldsymbol{V}_t$, $\boldsymbol{V}_t^*$ etc. This, and other, singularities can be handled as follows. Suppose that we model $\boldsymbol{y}_t$ ignoring the constraint, assuming, as in the case throughout Section 2, Σ, $\boldsymbol{V}_t$, $\boldsymbol{V}_t^*$, etc., to be non-singular. Then the constraint can be imposed directly by transforming the series to $\boldsymbol{y}_t'\boldsymbol{K}$ where

$$\boldsymbol{K} = \boldsymbol{I} - q^{-1}\boldsymbol{1}\boldsymbol{1}', \quad \boldsymbol{1} = [1, \ldots, 1]'.$$

Then, from equations (1),

$$\boldsymbol{y}_t'\boldsymbol{K} = \boldsymbol{x}_t'\boldsymbol{\Psi}_t + (\boldsymbol{K}\boldsymbol{e}_t)', \qquad \boldsymbol{K}\boldsymbol{e}_t \sim N(\boldsymbol{0}, v_t\Xi), \tag{9a}$$

$$\boldsymbol{\Psi}_t = \boldsymbol{G}_t\boldsymbol{\Psi}_{t-1} + \boldsymbol{F}_t\boldsymbol{K}, \qquad \boldsymbol{F}_t\boldsymbol{K} \sim N(\boldsymbol{0}, \boldsymbol{W}_t, \Xi) \tag{9b}$$

$$\boldsymbol{\Psi}_{t-1} \sim N(\boldsymbol{M}_{t-1}\boldsymbol{K}, \boldsymbol{C}_{t-1}, \Xi), \qquad \Xi \sim W^{-1}(\boldsymbol{K}'\boldsymbol{S}_{t-1}\boldsymbol{K}, d_{t-1}) \tag{9c}$$

where $\boldsymbol{\Psi}_t = \boldsymbol{\Theta}_t\boldsymbol{K}$ and $\Xi = \boldsymbol{K}'\Sigma\boldsymbol{K}$. Thus the constrained series follows a DMR. For a derivation of this, and related, results see Quintana (1987). Note that the quantities $\boldsymbol{x}_t'$, v_t, $\boldsymbol{G}_t$, $\boldsymbol{W}_t$ and $\boldsymbol{C}_t$ are unaffected by such linear transforms, as is the use of discount factors to define $\boldsymbol{W}_t$.

With this in mind, initial priors for $\boldsymbol{\Theta}_1$ and Σ at time $t = 1$ can be specified in non-singular forms ignoring effects of the constraint, and then transformed as in (9d). This procedure is similar to that for setting a prior for a form-free seasonal DLM component (Harrison and West, 1986). Moreover, the unconstrained model is simply a DMR for the logarithms of $\boldsymbol{z}_t$, rather than the log-ratios, so that the unconstrained priors are those of meaningful quantities.

4. EXAMPLE: MEXICAN IMPORTS

The data in Table 1 are monthly Mexican imports from January 1980 through December 1983 according to use: Consumer, Intermediate and Capital (this order being used in the data vectors for each month). The time period includes the drop in world oil prices, the effects on imports being visible during 1981 and 1982. The log-ratio transformed series are plotted in Figure 1, where the compositional features are clearly apparent. Note, in particular the relative increase in imports of intermediate use goods at the expense of consumer goods during the economic crisis. An important feature of the transformation is that it removes some of the common variation over time evident on the original scale, such as inflation effects and national economic cycles.

The model chosen used for the transformed series is a linear growth DMR, defined by

$$\boldsymbol{x}_t = [1, 0], \quad v_t = 1, \quad \boldsymbol{G}_t = \begin{bmatrix} 1 & 1 \\ 0 & 1 \end{bmatrix}.$$

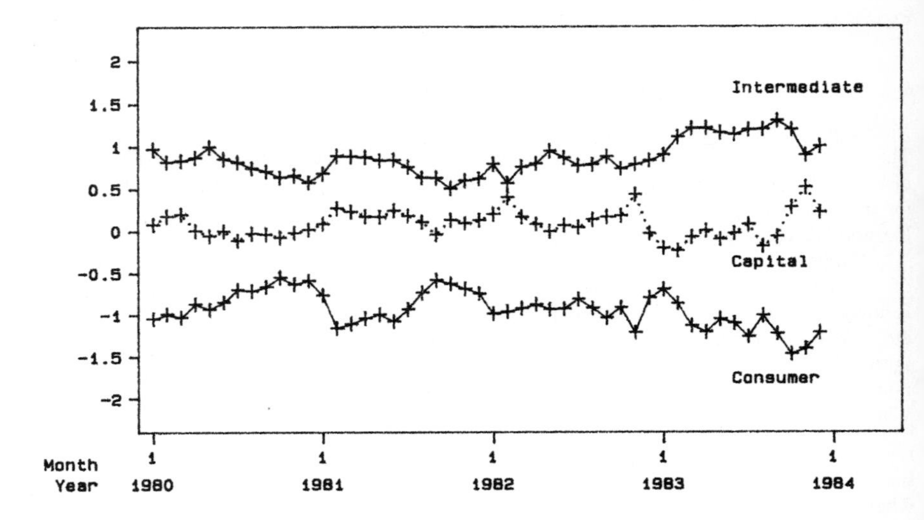

Figure 1. *Log-ratio of Mexican imports composition by use*

This is *not* a model to be seriously entertained for *forecasting* the series other than in the very short-term (eg. 1 step-ahead). However, as in many other applications, it serves here as an extremely flexible *local smoothing* model, that can track the changes in trend in the data and provide excellent retropective estimates of the time varying trend and cross-sectional correlation structure. Thus, for our purposes here, this simple, local trend description is adequate, the basic, *locally linear* form of the model being supported by the graphs in Figure 1.

A relatively uninformative initial prior is used for illustration (the problems involved in specifying an informative prior for $\boldsymbol{\Sigma}$ are well discussed by Dickey, Dawid and Kadane, 1986). For the unconstrained series.

$$\boldsymbol{M}_1 = \begin{pmatrix} -1 & 1 & 0 \\ 0 & 0 & 0 \end{pmatrix}, \quad \boldsymbol{C}_1 = \begin{pmatrix} 1 & 0 \\ 0 & 1 \end{pmatrix}.$$

This implies that the initial growth in each of the series has prior mean 0, the levels having means –1 (Consumer), 1 (Intermediate) and 0 (Capital). The prior for $\boldsymbol{\Sigma}$ has $\boldsymbol{S}_1 = 10^{-5}\boldsymbol{I}$ and $d_1 = 10^{-3}$, a very vague specification. These values are used to initialise the analysis, the corresponding posterior and smoothed values for each time t then being transformed as in equations (9) to impose the linear constraint.

The analyisis summarised in Figure 2 to 5 is based on this model with discount factors taking the typical values $\delta = 0.9$ and $\beta = 0.98$. Of several models with values in the *a priori* plausible ranges $0.8 \leq \delta \leq 1$ and $0.95 \leq \beta \leq 1$, that illustrated has essentially the highest likelihood, as discussed in Section 2.5. In particular, the static $\boldsymbol{\Sigma}$ models with $\beta = 1$ are very poor by comparison, indicating support for time variation in $\boldsymbol{\Sigma}$. The inferences illustrated are representative of inferences from models with discount factors near the chosen values.

In addition to the above specifications, one further modification is made to the model (and to each of those explored). It is clear from Figure 1 that the trends in the data change rather abruptly at one or two places, essentially responding to major events in the international markets, such as the oil price crash in 1981. Such abrupt changes, and other exceptional events, can be catered for through intervention, either subjective or automatic (Harrison and West, 1986; West and Harrison, 1987). Here three interventions are made to allow for changes in model parameters that are greater than those anticipated through the standard discount factors δ and β. The points of intervention are January 1981, September 1981 and January 1983. At these points only, the standard discount factors are replaced by lower values $\delta = 0.1$ and $\beta = 0.9$ to model greater *random* variation in both Θ_t and Σ though not anticipating the direction of such changes.

Figures 2, 3 and 4 display the smoothed estimates of the trends (the *fitted* trends) in three imports series, with approximate 90% posterior intervals (These should not be confused with the associated *predictive* intervals that, though not displayed, are roughly twice as wide). The times of intervention are indicated by arrows on the time axes, and the response of the model is apparent at points of abrupt change. Though the model could be refined, improvements are likely to be very minor, the fitted trends *smooth* the series rather well. The main interest is in correlation structure across series, and Figure 5 clearly identifies the time variation in Σ. The graph is of the correlations taken from the smoothed estimates V_t^* of Σ. Slow, random changes are apparent, accounting for changing economic conditions that affect the three import areas in different ways. A more marked change occurs at intervention prior to the onset of recession in early 1981. Cross-sectional inferences made in early 1981 based on this model would reflect these changes, which are quite marked, whereas those based on a static Σ model would be much more heavily based on the pre-1981 estimates of Σ. Thus, for example, this model would suggest a correlation near –0.25 between the Intermediate and Capital series, whereas the pre-1981 based value is near –0.75.

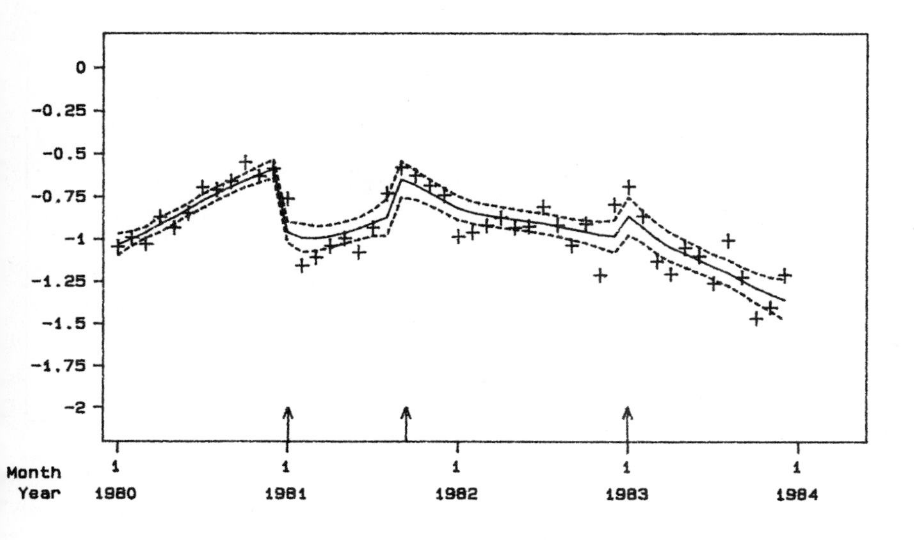

Figure 2. *Smoothed trend in Consumer series*

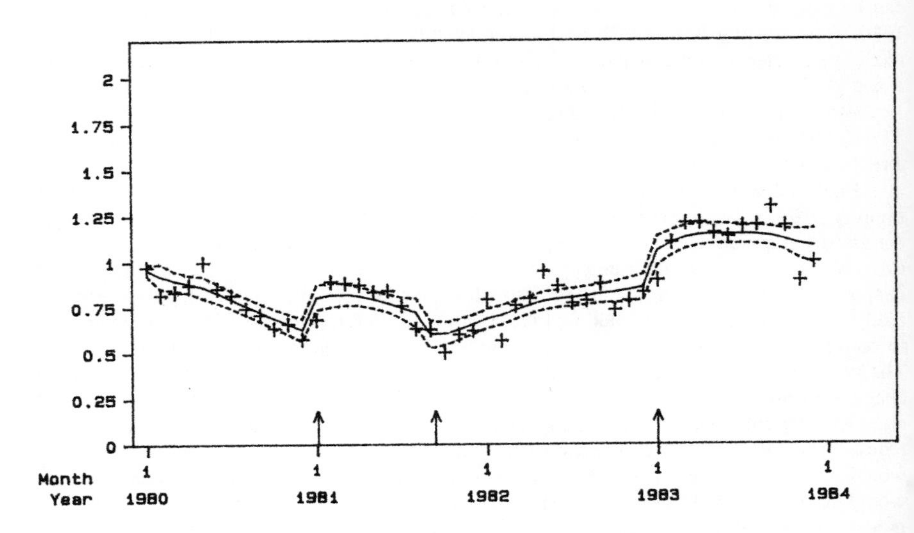

Figure 3. *Smoothed trend in Intermediate series*

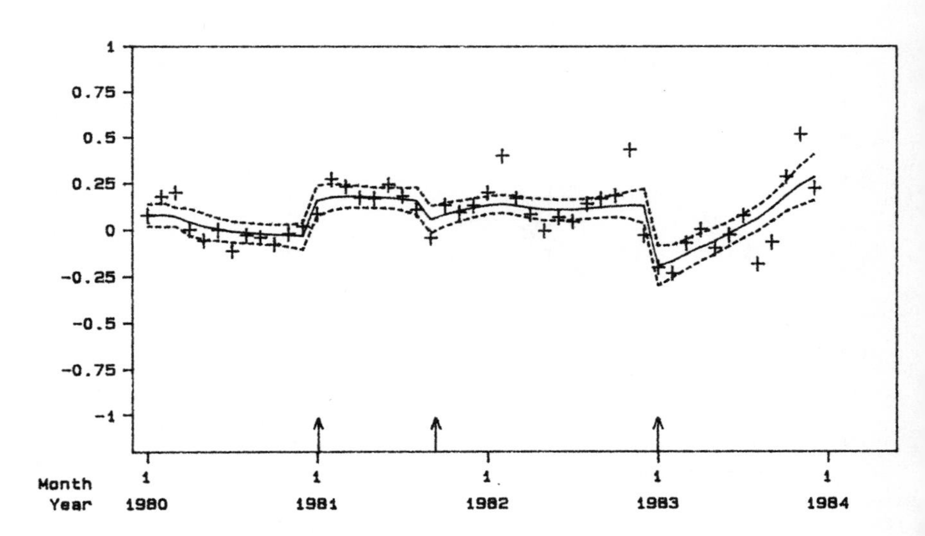

Figure 4. *Smoothed trend in Capital series*

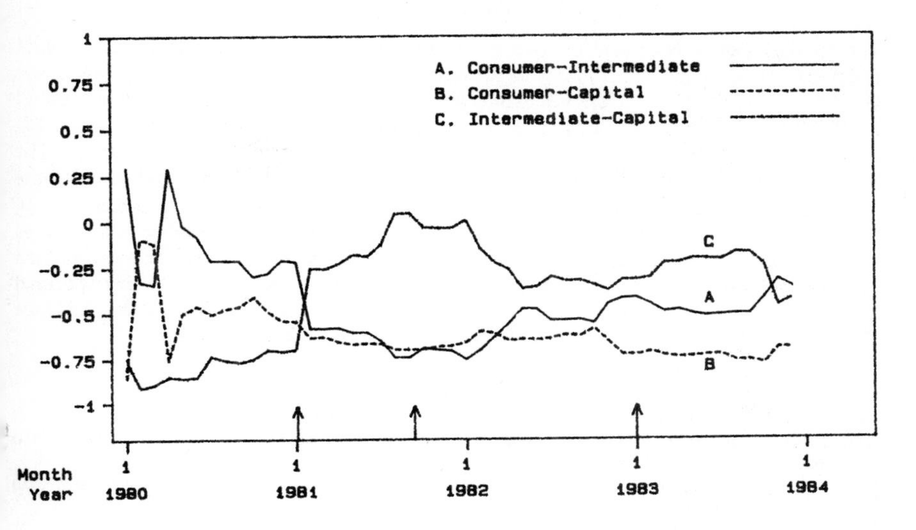

Figure 5. *Smoothed estimates of correlations*

		J	*F*	*M*	*A*	*M*	*J*	*J*	*A*	*S*	*O*	*N*	*D*
1980	*A*	98.1	112.1	123.1	150.2	148.0	180.2	244.9	225.0	253.1	321.5	262.4	331.1
	B	736.3	684.1	793.9	856.2	1020.9	986.9	1111.8	962.9	994.0	1049.2	954.0	1058.6
	C	301.2	360.7	4290.9	358.7	354.6	419.8	438.4	445.2	470.3	514.8	483.4	606.1
1981	*A*	246.5	135.4	183.4	178.2	194.4	189.2	226.6	254.8	298.9	362.2	270.2	273.0
	B	1046.9	1051.6	1342.2	1208.0	1213.6	1291.1	1229.6	994.4	999.3	1120.4	977.1	1067.2
	C	575.3	565.4	699.2	598.3	622.6	709.7	689.6	588.8	510.9	773.9	588.3	653.3
1982	*A*	157.2	175.4	191.5	167.9	123.5	143.9	156.8	115.0	70.0	90.6	42.6	82.3
	B	933.4	808.9	1029.2	894.3	811.8	866.1	762.0	631.0	476.6	470.0	314.9	419.5
	C	514.5	684.7	568.4	436.0	312.6	388.7	367.1	330.1	232.9	270.1	220.8	176.5
1983	*A*	52.3	44.5	42.9	39.2	55.7	56.6	41.9	60.5	40.2	33.1	39.5	47.4
	B	257.0	318.6	457.2	440.3	506.4	529.6	488.1	548.3	502.2	474.9	392.4	431.7
	C	85.2	83.2	126.7	131.3	144.4	165.5	159.5	137.4	128.2	190.7	268.8	198.1

A = Consumer; B = Intermediate; C = Capital

Table 1. *MONTHLY MEXICAN IMPORT DATA* ($\$10^6$)
Source: Bank of México

APPENDIX

THE GENERAL MATRIX-NORMAL AND INVERTED-WISHART DISTRIBUTIONS

The $(p \times q)$ matrix $\mathbf{\Theta}$ is said to have the matrix-normal distribution with $(p \times q)$ matrix mean $\boldsymbol{M}$, $(p \times p)$ left and $(q \times q)$ right variance matrices $\boldsymbol{C}$ and $\mathbf{\Sigma}$, if and only if vec $\mathbf{\Theta} \sim N(\text{vec}\,\boldsymbol{M}, \mathbf{\Sigma} \otimes \boldsymbol{C})$. Here vec stands for the column-vectorisation of a matrix and $\otimes$ for the Kronecker direct product. The distribution of $\mathbf{\Theta}$ is denoted by $\mathbf{\Theta} \sim N(\boldsymbol{M}, \boldsymbol{C}, \mathbf{\Sigma})$. It is of note that $\boldsymbol{C}$ and/or $\mathbf{\Sigma}$, and consequently $\mathbf{\Theta}$, may or may be not be singular. The inverted-Wishart distribution may be defined as follows. Consider a $(n \times n)$ positive definite symmetric matrix $\mathbf{\Omega}$ with density proportional to $|\mathbf{\Omega}|^{(\frac{1}{2}d+n)} \exp[-\frac{1}{2}tr(\mathbf{\Omega}^{-1})]$ where $d > 0$. The $(q \times q)$ variance matrix $\mathbf{\Sigma}$ is said to have an inverted-Wishart distribution with $(q \times q)$ scale matrix parameter $\boldsymbol{S}$ and shape parameter d if, and only if, it has the same distribution as that of $\boldsymbol{A}'\mathbf{\Omega}\boldsymbol{A}$ and $\boldsymbol{S} = \boldsymbol{A}'\boldsymbol{A}$. The $(n \times q)$ matrix $\boldsymbol{A}$ and n need not be further specified because if a $(N \times q)$ matrix $\boldsymbol{B}$ is such that $\boldsymbol{B}'\boldsymbol{B} = \boldsymbol{A}'\boldsymbol{A}$ then $\boldsymbol{A}'\mathbf{\Omega}_{(n,d)}A \sim \boldsymbol{B}'\mathbf{\Omega}_{(N,d)}\boldsymbol{B}$; see Dawid (1981) for details. The distribution of $\mathbf{\Sigma}$ is denoted by $\Sigma \sim W^{-1}(\boldsymbol{S}, d)$

REFERENCES

Aitchison J. (1986). *The Statistical Analysis of Compositional Data.* London: Chapman-Hall.

Aitchison J. and Shen, S. M. (1980). Logistic-normal distribution: some properties and uses, *Biometrika* **2** (67), 261–272.

Dawid, A. P. (1981). Some matrix-variate distribution theory : notational considerations and a Bayesian application, *Biometrika* **1** (68), 265–274.

Dickey, J. M., Dawid A. P. and Kadane J. B. (1986). Subjective-probability assessment methods for multivariate- t and matrix- t models, in *Bayesian inference and decision techniques: Essays in honor of Bruno de Finetti*, Goel, P. K. and Zellner, A. (editors), Amsterdam: North-Holland.

Harrison, P. J. and West, M. (1986). Bayesian forecasting in practice, Warwick Bayesian Statistics Study Year. *Tech. Rep.* **13**.

Harrison, P. J. and West M. (1987). Practical Bayesian forecasting, *The Statistician* **36**, 115–125.

Quintana, J. M. (1985). A Dynamic Linear Matrix-variate Regression Model, Warwick, *Tech. Rep* **83**.

Quintana, J. M. (1987). *Multivariate Bayesian forecasting models*, PhD Thesis, Warwick.

Quintana, J. M. and West, M. (1987). Multivariate time series analysis: new techniques applied to international exchange rate data, *The Statistician* **36**, 275–281.

West, M. (1986). Bayesian model monitoring, *R. Statist. Soc.*, **B, 48** 1, 70–78.

West, M. and Harrison P. J. (1986). Monitoring and adaptation in Bayesian forecasting models, *J. Amer. Statist. Ass.*, **8** 395, 741–750.

West, M. and Harrison P. J. (1988). Subjective intervention in formal models, J. Forecasting, to appear.

BAYESIAN STATISTICS 3, pp. 757–764
J. M. Bernardo, M. H. DeGroot, D. V. Lindley and A. F. M. Smith, (Eds.)

Bayesian Forecasting of Time Series by Gaussian Sum Approximation

S. SCHNATTER
Technische Universität Wien

SUMMARY

The problem of analytic approximation of Bayesian predictive distributions derived from an autoregressive process is reviewed and a Gaussian sum approximation is proposed and illustrated.

Keywords: AUTOREGRESSIVE PROCESS; BAYESIAN PREDICTIVE DISTRIBUTIONS; GAUSSIAN SUM APPROXIMATION.

1. INTRODUCTION

Many real world processes can be described by a model of the autoregressive (AR) type. In this case the process $\{y_t, t = -p+1, \ldots, -1, 0, 1, \ldots\}$ follows the equation

$$y_t = \sum_{j=1}^{p} \alpha_j y_{t-j} + \varepsilon_t, \quad t = 1, 2, \ldots \tag{1.1}$$

where the ε_t's are normally and independently distributed with zero mean and variance $1/\tau$. The starting values $y_{-p+1}, \ldots, y_0$ are assumed to be known.

Given an observed time series $\{y_t, t = -p+1, \ldots, n\}$ two questions will arise:

(a) How to use the given data to make inference about the unknown parameters $\alpha_1, \ldots, \alpha_p$ and the unknown precision τ?

(b) How to produce forecasts $\{y_t, t > n\}$?

Both problems have been discussed from a Bayesian viewpoint, where $\alpha_1, \ldots, \alpha_p$ and τ are treated as random variables. Bayesian techniques for estimating the parameters of an AR model are described by Zellner (1971) or Broemeling (1985). These techniques yield posterior probability density functions (p.d.f.) of the parameters which are of well known types.

Problems arise with Bayesian techniques if one intends to predict more than one step ahead. The predictive p.d.f. of future observations is no more of well known type. Different ways of approximating the predictive p.d.f. have been suggested (see Section 3.1). In this paper an approximation by Gaussian sums will be proposed.

2. ESTIMATION OF TIME SERIES PARAMETERS FROM A BAYESIAN VIEWPOINT

The problem of estimating the parameters of an $AR(p)$ process has been extensively discussed in the literature (e.g. Zellner (1971), Broemeling (1985)). I will restrict myself to a brief review.

Given n observations from an $AR(p)$ process the likelihood function of the observations is calculated by

$$l(y_1,\ldots,y_n|\alpha_1,\ldots,\alpha_p,\tau) \sim \tau^{n/2}\exp\left\{-\tau/2\sum_{k=1}^{n}\left(y_k-\sum_{j=1}^{p}\alpha_j y_{k-j}\right)^2\right\} \tag{2.1}$$

The conjugate class of prior p.d.f.s is the normal-gamma family.

Using Bayes' theorem the posterior p.d.f. of the parameters given the observations $y^n = (y_1,\ldots,y_n)$ can be shown to be also of the normal-gamma family (see Broemeling (1985)):

$$p(\alpha,\tau|y^n) \sim \tau^{(n+2a+p)/2}\exp\{-\tau/2\cdot(\alpha-A^{-1}C)'A(\alpha-A^{-1}C)+D\} \tag{2.2}$$

By integrating the joint density (2.2) with respect to τ one obtains the marginal posterior p.d.f. of α as a t-density:

$$p(\alpha|y^n) \sim \{(\alpha-A^{-1}C)'A(\alpha-A^{-1}C)+D\}^{-(n+2a+p)/2} \tag{2.3}$$

The marginal posterior p.d.f. of τ is gamma p.d.f.:

$$p(\tau|y^n) \sim \tau^{(n+2a)/2-1}\exp\{-\tau/2\cdot D\} \tag{2.4}$$

Thus the Bayesian estimation of the parameters of an $AR(p)$ model is clear and straightforward.

3. BAYESIAN FORECASTING OF AN $AR(P)$ MODEL

3.1. The Predictive Distribution

Since the famous work of Box/Jenkins (1970) forecasting of time series concentrated mainly on producing point forecasts of future observations. Forecasting from a Bayesian viewpoint requires the calculation of the joint p.d.f. of future observations conditioned on observations up to n:

$$p(y_{n+1},y_{n+2},\ldots,y_{n+l}|y^n), \quad l=1,\ldots,L \tag{3.1}$$

Zellner (1971) has shown that the one step ahead predictive distribution for $AR(1)$ and $AR(2)$ models is a t-density. This result can be generalized to $AR(p)$ models (Broemeling (1985)).

Problems arise when multiperiod forecasts are required. Land (1981) writes the l-step joint predictive distribution as product of l univariate conditional t-densities. As such a product is generally not a multivariate t-density the predictive p.d.f. is usually not of a well known type.

Rather than trying to obtain its analytical form Thompson (1984) proposed to simulate the predictive p.d.f. Choosing τ from the posterior p.d.f. (2.4) and $\alpha_1,\ldots,\alpha_p$ from the posterior p.d.f. (2.3) a single future path of the time series can be geneated from (1.1). By repeating this procedure about 10000 times Thompson (1984) obtains a bundle of all possible future developments of the time series.

If Bayesian forecasting is performed in practice one may be interested in marginal predictive p.d.f.s

$$p(y_{k+l}|y^n), \quad l=1,\ldots,L \tag{3.2}$$

The mentioned result of Land (1981) does not yield a closed form for (3.2). The method of Thompson (1984) described above can be used to approximate the marginal predictive p.d.f.s by means of histograms.

3.2. Gaussian Sum Approximation of the Marginal Predictive Distributions

Instead of simulating the marginal predictive p.d.f. an analytical approximation can be looked for. Gaussian sum approximations of the posterior p.d.f. have been successfully applied to nonlinear estimation problems (Alspach (1970), Alspach/Sorenson (1972)). In this paper we approximate the marginal predictive p.d.f. (3.2) by a Gaussian sum:

$$p(y_{n+l}|y^n) \sim \sum_{j=1}^{m} w_j p\left(y_{n+l}; \mu_{n+l}^j, (\sigma^2)_{n+l}^j\right) \quad l = 1, \ldots, L \tag{3.3}$$

where $p\left(y_{n+l}; \mu_{n+l}^j, (\sigma^2)_{n+l}^j\right)$ is a normal p.d.f. with mean μ_{n+l}^j and variance $(\sigma^2)_{n+l}^j$. This approximation requires the determination of the number m of members, the weights and the means and variances of the m Gaussian densities. To obtain these values we combine results from numerical integration with the fact that the marginal predictive p.d.f. of y_{n+l} is normal for fixed parameters $\alpha_1, \ldots, \alpha_p$ and τ.

First we show how the weights in (3.3) are obtained. Considering the rules for conditional densities $p(y_{n+l}|y^n)$ can be represented by the following integral over a τ-conditional p.d.f.:

$$p(y_{n+l}|y^n) = \int p(y_{n+l}|y^n, \tau) p(\tau|y^n) d\tau$$

This is a weighted integration over $p(y_{n+l}|y^n, \tau)$ with weight function $p(\tau|y^n)$, which is a gamma p.d.f. (see (2.4)). Such an integration can be performed numerically by a "Generalized Laguerre Integration":

$$p(y_{n+l}|y^n) \sim \sum_{j_0=1}^{m_0} v_0^{(j_0)} p\left(y_{n+l}|y^n, \tau_{j_0}\right) \tag{3.5}$$

where for given m_0 the abscissas $(\tau_{j_0}, j_0 = 1, \ldots, m_0)$ are chosen as the zeros of a generalized Laguerre polynomial $L_{m_0}(\tau)$ of order m_0 (see Appendix A).

Any of the m_0 p.d.f.s $p\left(y_{n+l}|y^n, \tau_{j_0}\right)$ in (3.5) again can be represented by a weighted integration:

$$p\left(y_{n+l}|y^n, \tau_{j_0}\right) = \int p\left(y_{n+l}|y^n, \tau_{j_0}, \alpha_1\right) p\left(\alpha_1|y^n, \tau_{j_0}\right) d\alpha_1 \tag{3.6}$$

with weight function $p\left(\alpha_1|y^n, \tau_{j_0}\right)$ which is Gaussian (from (2.3) it can be seen that $p(\alpha_1, \ldots, \alpha_p|\tau, y^n)$ is normal; therefore also any marginal p.d.f. $p(\alpha_j|\tau, y^n)$ is normal). Integration with this type of weight function is usually performed numerically by Hermite Integration:

$$p\left(y_{n+l}|y^n, \tau_{j_0}\right) \sim \sum_{j_1=1}^{m_1} v_1^{(j_1)} p\left(y_{n+l}|y^n, \tau_{j_0}, \alpha_1^{(j_1)}\right) \tag{3.7}$$

where for given m_1 the abscissas $\left(\alpha_1^{(j_1)}, j_1 = 1, \ldots, m_1\right)$ are chosen as the zeros of a Hermite polynomial $H_{m_1}(\alpha_1)$ of order m_1 (Abramowitz/Stegun (1970)).

This Hermite integration can be repeated with any of the m_1 p.d.f.s $p(y_{n+l}|y^n, \tau_{j_0}, \alpha_1^{(j_1)})$ in (3.7), each of which yields a weighted sum of $p(y_{n+l}|y^n, \tau_{j_0}, \alpha_1^{(j_1)}, \alpha_2^{(j_2)})$. In the same way Hermite integration is continued with any of these p.d.f.s until integration over α_p is performed.

Substituting the last sum

$$\sum_{j_p=1}^{m_p} v_p^{(j_p)} p\left(y_{n+l}|y^n, \tau_{j_0}, \alpha_1^{(j_1)}, \ldots, \alpha_p^{(j_p)}\right)$$

$$\boxed{p(y_{n+l}|y^n)}$$

$\downarrow$ generalized Laguerre integration

$$\sum_{j_0=1}^{m_0} v_0^{(j_0)} \boxed{p(y_{n+l}|y^n, \tau_{j_0})}$$

$\downarrow$ Hermite integration

$$\sum_{j_1=1}^{m_1} v_1^{(j_1)} \boxed{p(y_{n+l}|y^n, \tau_{j_0}, \alpha_1^{(j_1)})}$$

$\downarrow$ Hermite integration

$$\sum_{j_2=1}^{m_2} v_2^{(j_2)} \boxed{p(y_{n+l}|y^n, \tau_{j_0}, \alpha_1^{(j_1)})\alpha_2^{(j_2)})}$$

$\downarrow$

....

$$\sum_{j_{p-1}=1}^{m_{p-1}} v_{p-1}^{(j_{p-1})} \boxed{p(y_{n+l}|y^n, \tau_{j_0}, \alpha_1^{(j_1)}, \ldots, \alpha_{p-1}^{(j_{p-1})})}$$

$\downarrow$ Hermite integration

$$\sum_{j_p=1}^{m_p} v_p^{(j_p)} \boxed{p(y_{n+l}|y^n, \tau_{j_0}, \alpha_1^{(j_1)}, \ldots, \alpha_p^{(j_p)})}$$

Figure 1. *Approximation of* $p(y_{n+l}|y^n)$

into the preceding and so on until we reach sum (3.5) we obtain an approximation of $p(y_{n+l}|y^n)$:

$$p(y_{n+l}|y^n) \sim \sum_{j=1}^{m} w_j p(y_{n+l}|y^n, (\alpha_1, \ldots, \alpha_p, \tau)_j) \tag{3.8}$$

with

$$m = \prod_{k=0}^{p} m_k, \quad w_j = \prod_{k=0}^{p} v_k^{(j_k)}$$

and

$$(\alpha_1, \alpha_2, \ldots, \alpha_p, \tau)_j = \left(\alpha_1^{(j_1)}, \alpha_2^{(j_2)}, \ldots, \alpha_p^{(j_p)}, \tau_{j_0}\right)$$

where

$$j_k = \begin{cases} \text{int } \left((j-1)/\prod_{l=k+1}^{p}\right) \bmod m_k + 1 & 0 \le k < p \\ (j-1) \bmod m_p + 1 & k = p \end{cases}$$

(see Figure 1).

As $p(y_{n+l}|y^n, (\alpha_1, \ldots, \alpha_p, \tau)_j)$ in (3.8) is obviously is a normal p.d.f. the approximation derived is a Gaussian sum.

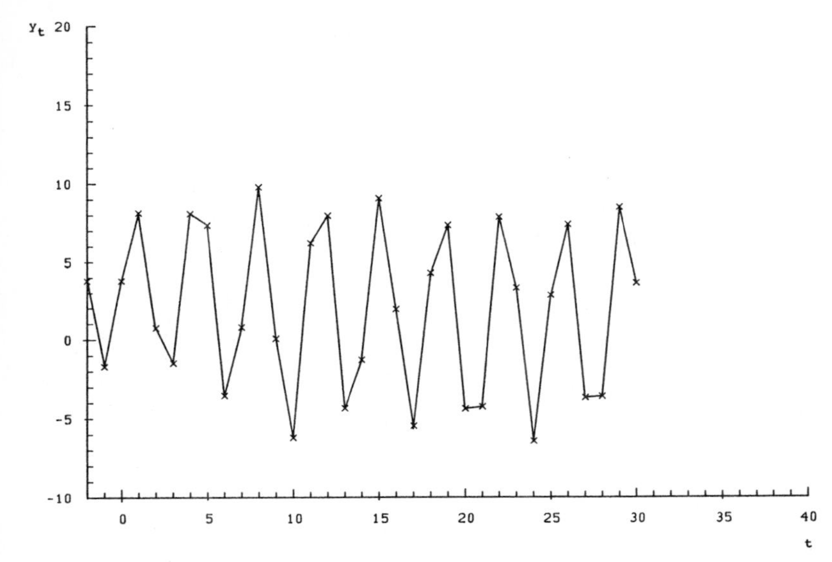

Figure 2. *Time series to be forecasted ($AR(3)$ process)*

4. EXAMPLE

The approximation of the marginal predictive p.d.f of a simulated time series will give an idea of the practical use of the proposed algorithm. The "observed" time series ($y_k, k \leq 30$) is part of a simulated path of an $AR(3)$ process (see Figure 2) with $\alpha_1 = 0.5454$ $\alpha_2 = -0.6363$, $\alpha_3 = 0.9091$, $\tau = 1$, $y_{-2} = 3.792$, $y_{-1} = -1.694$ and $y_0 = 3.796$.

Starting with the priors

$$p(\alpha_1, \ldots, \alpha_3|\tau) \sim \tau^{3/2} \qquad \text{and} \qquad p(\tau) \sim e^{-\tau}$$

the Bayesian estimation of the parameters yields (see (2.3), (2.4)):

$$p(\tau|y^{30}) \sim \tau^{\nu-1} \exp\{-\tau\beta\}, \nu = 16, \beta = 10.627$$

$$p(\alpha_1, \ldots, \alpha_3|\tau, y^{30}) \sim \tau^{3/2} \exp\{-\tau/2(\alpha - \mu)' A(\alpha - \mu)\}$$

with $\mu = (.56985, -.64275, .98469)'$

$$A^{-1} = \begin{bmatrix} .172{*}10^{-2} & .1020{*}10^{-3} & .1135{*}10^{-2} \\ & .1112{*}10^{-2} & .8889{*}10^{-4} \\ & & .1851{*}10^{-2} \end{bmatrix}$$

Approximation of the marginal predictive p.d.f.s $p(y_{31}|y^{30})$, $p(y_{32}|y^{30}), \ldots, p(y_{36}|y^{30})$ were performed for different degrees $m_0 = m_1 = m_2 = m_3 = M$. The approximations were compared with the "true" marginal predictive p.d.f. found by simulation (see Section 3.1) If $M = 1$ all predictive p.d.f. are approximated by a single Gaussian p.d.f. (see Figure 3). Because of the skewness and kurtosis of the predictive p.d.f. the Gaussian p.d.f. fits less the more periods ahead forecasting is required.

If $M = 2$ all predictive distributions are approximated by a sum of $m = M^4 = 16$ Gaussian densities (see Table 1). The approximation gives good results especially for multi-period forecasts (see Figure 3). Skewness and kurtosis are well reproduced. For this example approximations of degrees higher than $M = 2$ hardly improved the result.

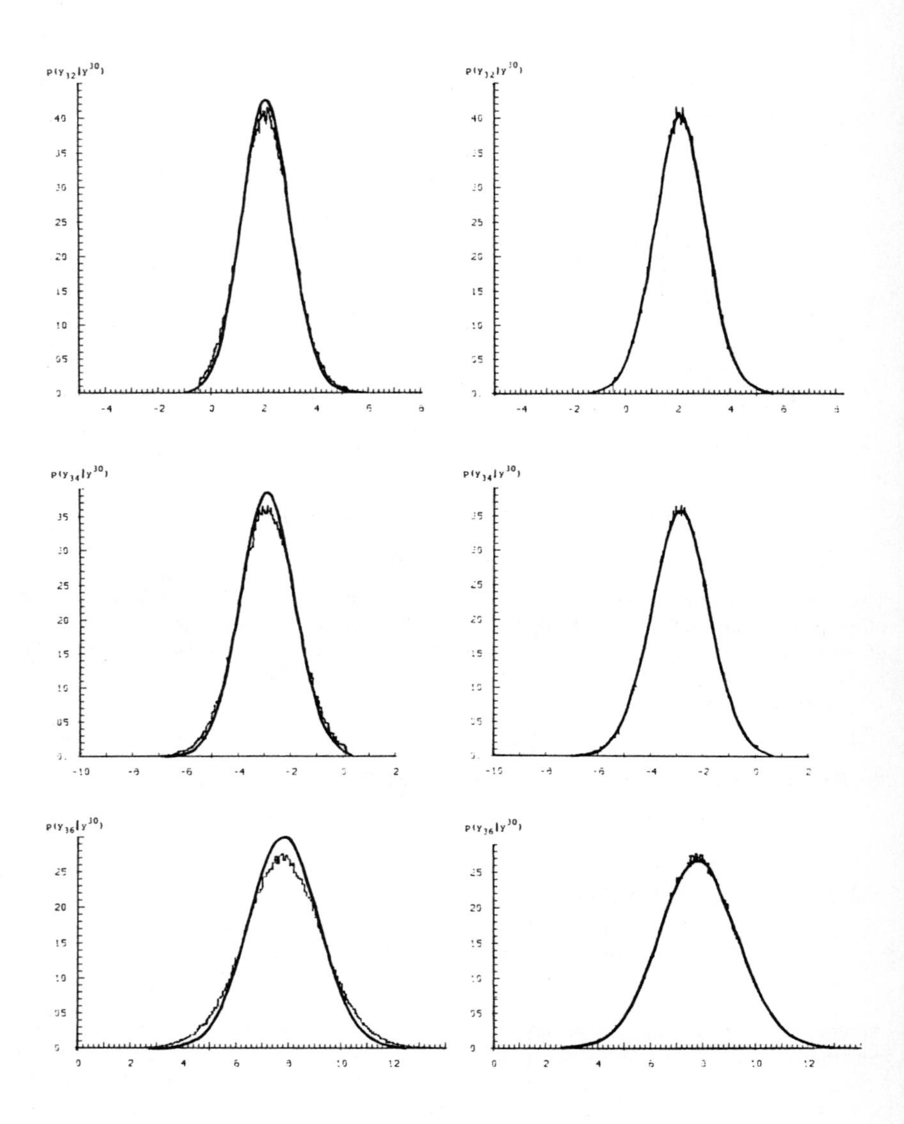

Figure 3. *Gaussian sum approximation of the marginal predictive p.d.f.s* $p(y_{30+l}|y^{30})$, $l = 2, 4, 6$, *compared with the "true" p.d.f.s found by simulation.*
Left side: $m_0 = m_1 = m_2 = m_3 = 1$, *right side:* $m_0 = m_1 = m_2 = m_3 = 2$.

i	τ	α_1	α_2	α_3	w_i	i	τ	α_1	α_2	α_3	w_i
1	1.212	.532	–.675	.929	.07766	9	1.988	.540	–.668	.941	0.4734
2	1.212	.532	–.675	.989	.07766	10	1.988	.540	–.668	.988	0.4734
3	1.212	.532	–.615	.930	.07766	11	1.988	.540	–.621	.942	0.4734
4	1.212	.532	–.615	.991	.07766	12	1.988	.540	–.621	.989	0.4734
5	1.212	.608	–.671	.979	.07766	13	1.988	.599	–.665	1.027	0.4734
6	1.212	.608	–.671	1.039	.07766	14	1.988	.599	–.665	.980	0.4734
7	1.212	.608	–.610	.980	.07766	15	1.988	.599	–.617	.981	0.4734
8	1.212	.608	–.610	1.040	.07766	16	1.988	.599	–.617	1.028	0.4734

Table 1. *Chosen paramater combination for* $M = 2$

5. CONCLUSION

The proposed algorithm is fast and gives in practice good results (see example in Section 4). However the theoretical properties are unknown and many questions remain unanswered for the moment:

(a) Is the approximation optimal in any sense?

(b) What about error bounds? Abramowitz/Stegun (1971) give error bounds for one dimensional orthogonal integration, which are difficult to generalize to multi-dimensional integration.

(c) How should one choose the degrees $m_0, m_1, \ldots, m_p$? It seems possible that m_i will depend on the "peakedness" of the marginal p.d.f. of the parameters, but no theoretically based choise can be provided.

APPENDIX A. GENERALIZED LAGUERRE INTEGRATION

Generalized Laguerre integration can be applied if the weight function $w(t)$ in the integral is of the form $t^a \cdot e^{-t}$:

$$\int f(t)t^a e^{-t} dt = \sum_{j=1}^{m} w_j \cdot f(t_j) + R_m \tag{A.1}$$

The abscissas $t_1, \ldots, t_m$ are chosen as the zeros of the generalized Laguerre polynomial $L_m^a(t)$ (see Abramowitz/Stegun (1970)). These polynomials are w-orthogonal, i.e.

$$\int L_m^a(t) L_n^a(t) w(t) dt = \frac{\Gamma(a+n+1)}{n!} \cdot \delta_n^m \qquad a > -1 \tag{A.2}$$

and can be defined recursively:

$$\begin{aligned} L_{m+1}^a(t) &= \frac{1}{m+1} \cdot ((2m+a+1-t) \cdot L_m^a(t) - (m+a) \cdot L_{m-1}^a(t)) \qquad m \geq 1 \\ L_0^a(t) &= 1, L_1^a(t) = -t + a + 1 \end{aligned} \tag{A.3}$$

For $m = 1$ and $m = 2$ the zeros can be calculated analytically:

$$\begin{aligned} &t_1 = a + 1, \quad m = 1 \\ &t_1 = a + 2 - \sqrt{a+2}, t_2 = a + 2 + \sqrt{a+2}, \quad m = 2 \end{aligned}$$

For $m > 2$ the zeros have to be calculated numerically. As the zeros of $L_m^a(t)$ lie between the zeros of $L_{m+1}^a(t)$ the regula falsi can applied to determine them. As any integration formula using m abscissas integrates polynomials of degree less or equal $m-1$ exactly the weights can be calculated from the following equations:

$$\int t^i t^a e^{-t} dt = \sum_{j=1}^{m} w_j (t_j)^i = \Gamma(a+i+1) \quad i = 0, 1, \ldots, m-1 \qquad (A.4)$$

If we integrate over a non standard gamma p.d.f.

$$\gamma(x; \nu, \beta) = \beta^\nu / \Gamma(\nu) x^{\nu-1} e^{-x\beta}$$

the abscissas $(t_1, \ldots, t_m)$ and the weights $w_1, \ldots, w_m$ need to be transformed:

$$\int f(x) \cdot \gamma(x; \nu, \beta) dx = \sum_{j=1}^{m} w_j^* f(x_j)$$

with $x_j = t_j / \beta$ and $w_j^* = w_j / \Gamma(\nu)$.

REFERENCES

Abramowitz, M. and Stegun, I.(1970). *Handbook of Mathematical Functions*. New York.

Alspach, D. L. (1970). A Bayesian approximation technique for estimation and control of time-discrete stochastic systems. Ph. D. dissertation, University of California, San Diego.

Alspach, D. L. and Sorenson, H. W. (1972). Nonlinear Bayesian Estimation using Gaussian Sum Approximation. *IEEE Trans. Autom. Control* **17**, 439–448.

Box, G. E. P. and Jenkins, G. M. (1970). *Time Series Analysis. Forecasting and Control*. Holden-Day.

Broemeling, L. (1985). *Bayesian Analysis of Linear Models*. New York: Marcel Dekker.

Jazwinski, A. H. (1979). *Stochastic Processes and Filtering Theory*. New York: Academic Press.

Kalman, R. E. (1960). A new Approach to Linear Filtering and Prediction Problems. *Trans Asme, Journal of Basic Eng.* **82**, 35–44.

Land, M. (1981). Bayesian Forecasting for Switching Regression and Autoregressive Processes. Ph. D. dissertation, Oklahoma State University, Stillwater, OK.

Stoer, J. (1972). *Einführung in die numerische Mathematik*. Berlin/New York: Springer.

Thompson, P. A. (1984). Bayesian Multiperiod Prediction: Forecasting with Graphics. Ph. D. dissertation, University of Wisconsin, Madison, WI.

Zellner A. (1971). *An Introduction to Bayesian Inference in Econometrics*. New York: Wiley.

BAYESIAN STATISTICS 3, pp. 765–776
J. M. Bernardo, M. H. DeGroot, D. V. Lindley and A. F. M. Smith, (Eds.)

Models, Optimal Decisions and Influence Diagrams

J. Q. SMITH
University of Warwick

SUMMARY

Generalised conditional independence is defined. Its properties are satisfied by a variety of subjective information measures including probability. Directed graphs called influence diagrams are then defined in terms of these properties. Influence diagrams are a useful representation of a decision problem or model. We will discuss how conditional independence statements can be manipulated within an influence diagram to provide the decision analyst with valuable information about the consequences of using a given model or description of a problem. Some of these results are known, some are new. Two new examples will then be given, one illustrating how influence diagrams can provide a rigorous and insightful proof of a theorem, the other illustrating how the structure of different models can be critically appraised by comparing their corresponding diagrams.

Keywords: BAYES GAME THEORY, BELIEF NETS; CONDITIONED INDEPENDENCE; EXPERT OPINION; INFLUENCE DIAGRAMS; SUFFICIENCY.

1. INTRODUCTION

Influence diagrams are directed graphs which represent sets of conditional independence statements. Like most good statistical tools they were first developed as a practical modelling aid —enabling the modeller to discuss a problem and derive its basic modelling structure by asking the client a set of simple questions. One of their uses is that they do not require a client to give numerical values for his probabilities and utilities early in the modelling process (see Smith, 1987a).

Examples of the use of influence diagrams in modelling can be found in Agogino (1985), Barlow and Zhang (1986), Barlow *et al.* (1986), Howard and Matheson (1981), Holtzman (1985), Miller *et al.* (1981), Oliver (1986), Rege and Agogino (1986a). More recently the theoretical foundations of influence diagrams have been addressed; Olmsted (1984), Pearl (1986), Shachter (1986a,b), Smith (1986, 1987a, 1987b, 1987c).

Influence diagrams can be drawn whenever it is possible to define some form of conditional independence in a problem. *Generalised conditional independence* is defined in Smith (1986) as any tertiary operator $\cdot \amalg \cdot | \cdot$ defined on all subsets X of a set A of uncertain quantities where $X \amalg Y | Z$ reads "Given Z, X is uniformative about Y. This operator must satisfy the following three properties

$P1$ for all subsets $X, Y, Z,$ $\quad X \amalg Y | Y \cup Z$ which reads —
Once Y (and anything else, Z) is known, X conveys no further information about Y

$P2$ for all subsets $X, Y, Z,$ $\quad X \amalg Y | Z \Leftrightarrow Y \amalg X | Z$ which reads —
Once Z is known, if X is uninformative about Y, then Y is uninformative about X

$P3$ for all subsets W, X, Y, Z

$\boldsymbol{X} \amalg \boldsymbol{Y}|\boldsymbol{Z} \cup \boldsymbol{W}$

together imply and are implied by $\boldsymbol{X} \amalg \boldsymbol{Y} \cup \boldsymbol{Z}|\boldsymbol{W}$.

$\boldsymbol{X} \amalg \boldsymbol{Z}|\boldsymbol{W}$

Once $\boldsymbol{W}$ is known, "$\boldsymbol{X}$ uninformative about $\boldsymbol{Y}$ and $\boldsymbol{Z}$" in equivalent to the two statements "$\boldsymbol{X}$ is uninformative about $\boldsymbol{Z}$" and "$\boldsymbol{Z}$ is uninformative about $\boldsymbol{Y}$ once $\boldsymbol{Z}$ is also known".

It is easily checked that these 3 properties hold when $\{X_1, \ldots, X_m\}$ are random vectors and $\cdot \amalg \cdot|\cdot$ is, in Dawid's (1979) notation, conventional c.i. However Smith (1986) shows that $\cdot \amalg \cdot|\cdot$ can be defined on many other structures. For example he shows that if $\boldsymbol{X} \amalg \boldsymbol{Y}|\boldsymbol{Z}$ reads "a best linear estimate (under quadratic loss) of random vector $\boldsymbol{X}$ based on random vectors $\boldsymbol{X}$ and $\boldsymbol{Z}$ need only include components of the $\boldsymbol{Z}$ vector", then this operator satisfies P1, P2 and P3. In fact I would postulate that any sensible concept of informedness between variables should exhibit such a c.i. structure.

2. WHAT INFLUENCE DIAGRAMS ARE AND HOW THEY CAN BE USED

An *influence diagram* $(I.D.)$ I over the m subsets $\boldsymbol{X}_1, \ldots, \boldsymbol{X}_m$ of a set A is any directed graph with nodes labelled $\boldsymbol{X}_1, \boldsymbol{X}_2, \ldots, \boldsymbol{X}_m$ satisfying the property

$$\boldsymbol{X}_r \amalg \boldsymbol{X}^{r-1}|P(\boldsymbol{X}_r) \qquad r = 2, 3, 4, \ldots, m \tag{2.1}$$

where $\boldsymbol{X}^{r-1} = \{\boldsymbol{X}_1, \boldsymbol{X}_2, \ldots, \boldsymbol{X}_{r-1}\}$ and $P(\boldsymbol{X}_r) \subseteq \boldsymbol{X}^{r-1}$ are all the nodes in I connected by an arc to the node $\boldsymbol{X}_r$.

By property P1 and P3, note that an equivalent set of defining equations for a $I.D.$ is

$$\boldsymbol{X}_r \amalg \overline{P}(\boldsymbol{X}_r)|P(\boldsymbol{X}_r) \qquad r = 2, 3, \ldots, m \tag{2.3}$$

where $\overline{P}(\boldsymbol{X}_r) = \boldsymbol{X}^{r-1} \backslash P(\boldsymbol{X}_r)$ is the complement of $P(\boldsymbol{X}_r)$ in $\boldsymbol{X}^{r-1}$. In words, for each node X_r in I, X_r is independent of $\overline{P}(\boldsymbol{X}r)$ conditional on the uncertain quantities in $P(\boldsymbol{X}_r)$ whose nodes are attached to $\boldsymbol{X}_r$ by arcs.

$I.D.$'s are defined only in terms of properties P1, P2 and P3 and use no other type of property associated with joint probability distributions. Therefore not only can they be applied in a variety of forms of inference but it is also often easier to prove theorems about how they can be manipulated than for competing graphical representations of relationships between variables (see e.g. Spiegelhalter 1986, 1987, Wermuth and Lauritzen, 1983). In particular no awkward positivity conditions are usually required. Furthermore, as described later, they can be used as the basis for representing a full decision problem.

Since the c.i. relationships between variables as expressed in the $I.D.$ with $P(\boldsymbol{X}_r) = \boldsymbol{X}^{r-1}$ are vacuous, there always exists at least one $I.D.$ over variables $\boldsymbol{X}_1, \boldsymbol{X}_2, \ldots \boldsymbol{X}_m$. In fact there is usually no unique $I.D.$ representation of a problem, although we will see later that some graphs are more useful than others. Note from the definition above that $\boldsymbol{X}_i$ is attached to $\boldsymbol{X}_r$, that is $\boldsymbol{X}_i \in P(\boldsymbol{X}_r)$, implies $i < r$, so all $I.D.$'s must be acyclic.

Why should we choose to represent the set of c.i. statements of equation (2.1) in terms of a graph? Well, one good reason (and there are many others) is as follows. Suppose we are given a directed graph with nodes representing uncertain quantities and suppose these nodes are not indexed. Then Smith (1986) proves, using P1, P2 and P3, that *any* indexing of the nodes consistent with $X_i \in P(X_r) \Rightarrow i < r$, to form an $I.D.I_1$ implies and is implied by any other such indexing of nodes to form another $I.D.I_2$. The acyclic directed graph therefore expresses the important *partial order* deduced from the c.i. structure across the variables without introducing any extraneous information about the total order in which variables were introduced into a problem. For example if a variable $\boldsymbol{X}$ has no arc out of it in I we can immediately state

$$\boldsymbol{X} \amalg \boldsymbol{X}^m \backslash \boldsymbol{X}|P(\boldsymbol{X})$$

where $P(\boldsymbol{X})$ are the nodes connected into $\boldsymbol{X}$. This is because there is an ordering of the variables consistent with I introducing the variable $\boldsymbol{X}$ last. This fact can be read from the $I.D.$ with no calcultion needing to be made. It is shown in Smith (1986) that the property given above is *not* shared by another popular directed graphical representation of uncertainty (Wermuth and Lauritzen, 1983).

If $\boldsymbol{X}$ is connected by an arc to $\boldsymbol{X}_r$ so that $\boldsymbol{X} \in P(\boldsymbol{X}_r)$, it is called a *direct predecessor* (d.p.) of $\boldsymbol{X}_r$ and $\boldsymbol{X}_r$ is called a *direct successor* (d.s.) of $\boldsymbol{X}$. The set of all direct successors of X is denoted by $S(X)$. $\boldsymbol{X}$ is called a *predecessor* of $\boldsymbol{Y}$ and $\boldsymbol{Y}$ called a *successor* of $\boldsymbol{X}$ if there exists a directed path from $\boldsymbol{X}$ to $\boldsymbol{Y}$ in a $I.D.I$. The set of all successors of $\boldsymbol{Y}$ is denoted by $\sigma(Y)$ and $\overline{\sigma}(\boldsymbol{Y})$ denotes the complement of $\sigma(Y)$ in $\boldsymbol{X}^m \backslash \boldsymbol{X}$.

Because they just represent a set of c.i. statements $I.D.$'s have their own algebra. For the remainder of this section I shall list some rules, all which can be proved rigorously, by which you can manipulate c.i. statements directly on graphs. In the final section I give two short examples where such graphical manipulation is useful.

I have already given one result which in the notation given above, can be written:

Rule 1. *[Partial order characterisation]*
In any $I.D.I$, for all nodes $\boldsymbol{X}$

$$\boldsymbol{X} \amalg \overline{\sigma}(\boldsymbol{X}) | P(\boldsymbol{X})$$

i.e. given its d.p.s $\boldsymbol{X}$ is independent of the set of all variables which are not is successors.

Proof. See Smith (1986). $\triangleleft$

An $I.D.I_1$ is said to be *implied by* an $I.D.I_2$ (written $I_2 \Rightarrow I_1$) if all c.i. statements in I_1 can be deduced from those in I_2. If $I_1 \Rightarrow I_2$ and $I_2 \Rightarrow I_1$, $I_1 \& I_2$ are said to be *equivalent* (written $I_1 \equiv I_2$). Unless otherwise stated, $I.D.$'s will be assumed to be on the same set of m nodes and the d.p.s and d.s.s. of Z in I_i will be denoted by $P_i(Z)$ and $S_i(Z)$ respectively, $i = 1, 2, 3, \ldots$

A rule which is the direct consequence of P3 is the following:

Rule 2. *[Addition of arcs]*
If, for two $I.D.$'s I_1 and I_2,

$$P_1(X_i) \subseteq P_2(X_i) \qquad 1 \le i \le m$$

then

$$I_1 \Rightarrow I_2.$$

Rule 3. *[Node Elaboration]*
Let $I.D.I_1$ have nodes $\{\boldsymbol{X}_1, \ldots, \boldsymbol{X}_m\}$ introduced in that order.

Let $I.D.I_2$ have nodes $\{\boldsymbol{X}_1, \ldots, \boldsymbol{X}_{k-1}, \boldsymbol{X}_k(1), \ldots \boldsymbol{X}_k(n)\}$, introduced in that order and let $I.D.I_3$ have nodes $\{\boldsymbol{X}_1, \ldots, \boldsymbol{X}_{k-1}, \boldsymbol{X}_k(1), \ldots, \boldsymbol{X}_k(n), \boldsymbol{X}_{k+1}, \ldots, \boldsymbol{X}_m\}$, introduced in that order, where $P_1(\boldsymbol{X}_i) = P_2(\boldsymbol{X}_i)$, $1 \le i \le k-1$ and

$$\boldsymbol{X}_k = \cup_{i=1}^{n} \boldsymbol{X}_k(i). \tag{2.3}$$

If

(a) $P_3(\boldsymbol{X}_k(r)) = P_2(\boldsymbol{X}_k(r)) \qquad 1 \le r \le n$

(b) $P_3(\boldsymbol{X}_i) = \begin{cases} P_1(\boldsymbol{X}_i) \cup \{\boldsymbol{X}_k(1), \ldots, \boldsymbol{X}_k(n)\} \backslash \boldsymbol{X}_k & \text{if } \boldsymbol{X}_k \in P(\boldsymbol{X}_i) \\ P_1(\boldsymbol{X}_i) & \text{otherwise, } 1 \le i \le m \end{cases} \qquad i \ne k.$

then I_1 and I_2 together are equivalent to I_3.

Proof.

(a) For these nodes the c.i. statements (2.1) are identical for both I_2 and I_3.

(b) By (2.3) and P3, for any variables $\boldsymbol{Z}$ and $\boldsymbol{W}$

$$\boldsymbol{Z} \amalg \boldsymbol{X}_k | \boldsymbol{W} \Leftrightarrow \boldsymbol{Z} \amalg \{\boldsymbol{X}_k(1), \ldots, \boldsymbol{X}_k(u)\} | \boldsymbol{W}$$

So, in particular, from the c.i. statements (2.1)

$$\boldsymbol{X}_k \notin P_1(\boldsymbol{X}_i) \Leftrightarrow \{\boldsymbol{X}_k(1), \ldots, \boldsymbol{X}_k(n)\} \notin P_3(\boldsymbol{X}_i)$$

The result now follows. ◁

This rule is used in the two examples given at the end of this paper. It allows a large $I.D.I_3$ to be simplified by a smaller one I_1. The c.i. statements contained in I_3 can then be reintroduced by elaborating I_1 at some later stage. Notice that by Rule 3 we can add arcs until

$$P_3(\boldsymbol{X}_k(i)) = P_3(\boldsymbol{X}_k(j))$$

$$1 \leq i, j \leq n$$

$$S_3(\boldsymbol{X}_k(i)) = S_3(\boldsymbol{X}_k(j))$$

when I_3 is in the form of the theorem. So in particular $I_3 \Rightarrow I_1$ where I_1 is defined in its statement.

Rule 4. *[The Howard-Matheson Theorem]*
If $I.D.I_1$ has $X_j \in P_1(X_k)$ and $I.D.I_2$ has its d.p.s defined by

$$\begin{aligned} P_2(X_i) &= P_1(X_i) & i \neq j, k \\ P_2(X_j) &= \{P_1(X_j) \cup P_1(X_k) \cup X_k\} \backslash X_j & \\ P_2(X_k) &= P_1(X_k) \cup P_1(X_j) & \end{aligned}$$

then $I_1 \Rightarrow I_2$.

Proof. See Olmstead (1983), Shachter (1986) or more generally Smith (1986). ◁

Since I_2 has the arc (X_j, X_k) reversed from I_1, the theorem above is an analogue of Bayes Theorem under general c.i. Note that often the $I.D.I_2$ will have more arcs than I_1 in which case some c.i. information will be lost through performing this "arc reversal". There is however one important class of $I.D.$s where reversing arcs in I_1 does not lose information.

An $I.D.$ is said to be *decomposible* if the d.p.s $P(X)$ of any node X in I have the propery that X_1, $X_2 \in P(X)$ then $X_1 \in P(X_2)$ or $X_2 \in P(X_1)$. Two $I.D.$s I_1 and I_2 defined in the same set of nodes are called *similar* (written $I_1 \sim I_2$) if the undirected graph obtained by deleting arrows from the directed graph of I_1 is identical that obtained by deleting arrows from I_2.

Rule 5. *[The Decomposition Theorem]*
If I_1 and I_2 are both decomposible $I.D.$s and $I_1 \sim I_2$ then $I_1 \equiv I_2$.

Proof. See Smith (1986). ◁

This result is analogous to the theorem proved by Wermuth and Lauritzen (1983) from which it takes its name. Although this rule is not used in this paper it is very useful especially in multivariate problems (see Smith, 1987c).

Sometimes inherent in a problem are additional pieces of information about c.i. not in the $I.D.$ A node $\boldsymbol{W}$ is said to be identified by a set of nodes $(\boldsymbol{X}_1, \ldots, \boldsymbol{X}_r)$ if

$$\boldsymbol{W} \amalg B|(\boldsymbol{X}_1, \ldots, \boldsymbol{X}_r)$$

where B is any set of nodes not containing $\boldsymbol{W}$. For example if $\boldsymbol{W}$ is a function of $\boldsymbol{X}_1, \ldots, \boldsymbol{X}_r$ it is identified by these variables.

Rule 6. *[Reduction by identification]*
If $\boldsymbol{W}$ is identified by $(\boldsymbol{X}_1, \ldots, \boldsymbol{X}_r)$ and $(\boldsymbol{W}, \boldsymbol{X}_1, \ldots, \boldsymbol{X}_r) \subseteq P_1(\boldsymbol{Y})$ for some $\boldsymbol{Y} \in I_1$ then $I_1 \Rightarrow I_2$ where I_2 has all nodes with the same d.p.s as I_1 except the d.p.s of $\boldsymbol{Y}$ which are given by $P_2(\boldsymbol{Y}) = P_1(\boldsymbol{Y}) \backslash \boldsymbol{W}$.

Proof. See Smith (1987a). ◁

This is a useful way of cutting down the number of extra arcs added between a set of nodes by employing Rule 4.

It is often convenient to represent a full decision problem by an $I.D.$ To do this simply let your probabilistic $I.D.$ represent your (i.e. a decision analyst's) beliefs about how your client will act provided that he choses to act wisely. Since such a diagram represents only quantities which are uncertain to you, you are able now to legitimately define c.i. over your client's utility, decisions and uncertain quantities.

Traditionally (see Howard and Matheson, 1981) nodes representing utility, decision variables and your client's uncertain quantities are called respectively value nodes (diamond shapes), decision nodes (a square shape) and chance (circular shapes). For you to specify appropriate c.i. statements across these variables it is easiest if you can introduce the nodes in the problem in a useful order.

A *proper influence diagram* ($P.I.D.$) is a probabilitic $I.D.$ with a single value node, introduced last; decision nodes introduced consistent with the order in which their respective decisions need to be taken and where, if a chance node X will have been observed when decision D is taken then D is introduced after X. Furthermore a decision node is *only* attached to nodes whose variables are known at the time D is taken.

Clearly if your client has no information about $Q(\boldsymbol{X}_r) \subseteq \boldsymbol{X}^{r-1}$ and has observed or knowns $\overline{Q}(\boldsymbol{X}_r) = \boldsymbol{X}^{r-1} \backslash Q(\boldsymbol{X}_r)$ at the time he takes decision $\boldsymbol{X}_r$, then a logical consequence needs to be

$$\boldsymbol{X}_r \amalg Q(\boldsymbol{X}_r)|\overline{Q}(\boldsymbol{X}_r) \tag{2.4}$$

It is therefore always possible to find an $I.D.$ which satisfies condition (2.4) above —e.g. set $Q(\boldsymbol{X}_r) = \overline{P}(\boldsymbol{X}_r)$.

If you believe your client's c.i. statements, your $P.I.D.$ representation of a decision problem need attach U to those variables of which it is an explicit function [See Rule 6], you will attach a decision node D to any nodes representing information he has and needs to know in order to find an optimal policy [by the comment above] and include in $\overline{P}(X)$ of a chance node X those previously listed chance nodes which are independent of X given $P(X)$ and those decision nodes which do not affect the distribution of X given $P(X)$. A rigorous justification of these statements can be found in Smith (1987a).

Once your client has given you his probabilities in a probabilistic $P.I.D.$ you can use the $P.I.D.$ to help you evaluate an optimal policy [see Olmsted (1984), Shachter (1986),

Smith (1987(a), (b))]. It is helpful for this evaluation to use Rule 3 to find a new $P.I.D.$ in "extensive form").

A $P.I.D.$ is said to be in *extensive form* (called an $E.F.I.D.$) If the only node to have no successor is the value node and for every decision node $D, Q(D) = \phi$ (the empty set) where Q is defined above.

Several rules, specifically for the manipulation of probabilistic $P.I.D.$s are given below and will be illustrated later. It should be noted however that analogous results can be proved for less conventional forms of decision analysis (e.i. Goldstein's inference, based on linear structures).

Rule 7. *[Addition of nodes in a P.I.D.]*

(a) *Let $I.D.I_2$ be formed from I_1 by adding to I_1 a node W which is connected by arcs from each member in a set of nodes $\{X_1, \ldots, X_r\} \in I_1$. Then the identification of W by $\{X_1, \ldots, X_r\}$ and I_1 imply the c.i. statements in I_2.*

(b) *Let W above be a chance node in a $P.I.D.I_1$ and a $P.I.D.I_3$ be formed from the $P.I.D.I_2$ by attaching from W arcs to each member of a set of decision nodes $\{D_1, \ldots, D_k\}$ which have*

$$\{X_1, \ldots, X_r\} \subseteq P(D_i) \qquad\qquad 1 \leq i \leq k$$

Then I_1 and W being a function of $\{X_1, \ldots, X_r\}$ together imply the statements contained in $P.I.D.I_3$.

Proofs. Follow directly from the definitions of a $P.I.D.$ (Smith 1987a). ◁

Rules 8. *[Barren Node Reduction]*
If X in $P.I.D.I_1$ is a chance or decision node with no successor and I_2 is formed from I_1, by deleting X from I_1 together with all its connecting arcs, then an optimal policy for the problem depicted by I_1 will also be an optimal policy for the problem depicted by I_2.

Proof. See Olmsted (1983), Shachter (1986) or Smith (1987a). ◁

Rule 9. *[Arc Reduction through Sufficiency]*
Let I be an $E.F.I.D.$ introducing decision nodes in the order $\{D_1, D_2, \ldots, D_m\}$, let C_i denote the set of chance nodes (possibly empty) introduced between decision nodes D_{i-1} and D_i and let U denote the value node. Then, for the purposes of evaluating an optimal policy, the influence diagram I_1 which deletes arcs from $X \in P(D_i)$ whenever $x \notin T(D_i)$ is equivalent to I where $T(D_i) = \{\cup_{j>i}^{m} P(C_j) \cup P(D_j)\} \cup P(C_{m+1}) \cup P(U) \setminus [\{\cup_{j>i}^{m} C_j \cup D_j\} \cup C_{m+1}]$ and $P(C_j)$ is the union of the direct predecessor of nodes in C_j.

Proof. See Smith (1987a). ◁

Rule 10. *[On unnecessary elicitation]*
If a chance node W is a d.p. of all decision nodes in a $P.I.D.I_1$ then its distribution need not be elicited.

Proof. By the definition of a $P.I.D.$, W will be known *before* any decision needs to be taken. So its marginal distribution will not affect an optimal policy. ◁

3. TWO PROBLEMS MANIPULATED USING INFLUENCE DIAGRAMS

Example 1 An Influence Diagram proof in Bayesian Game Theory.

A game is played repeatedly between two players P_1 and P_2, P_1 being a Bayesian. Assume that the repeated game has continued to time point t. At the next time point $t+1$ either the repeated game will terminate or continue. The time of termination of the game will be independent of all information available to the players. If the game continues then P_1 and P_2 simultaneously choose a move $m_i(t+1)$ from a set M_i, $i = 1, 2$. At time $t+1$ the move pair $(m_1(t+1), m_2(t+1))$ will give P_1 a monetary pay-off $x_1(m_1(t+1), m_2(t+1))$ which is a known function of this move pair only. At the time of their $(t+1)$-st move pair, the only information available to either player will be the information D_t from the t-th time point and before.

Being a Bayesian, P_1 needs to find a strategy which will maximise his expected utility given his beliefs about P_2 which are represented as a distribution on the types of strategies he believes P_2 will play. A *strategy* S_i for P_i is a decision rule which specifies the probability of the move $m_i(t+1)$ for P_i as a function of the past $i = 1, 2$, and $t = 1, 2, 3 \ldots$. We shall assume that P_i's utility function U is increasing in his pay-off aggregated over the whole of the game. This utility would be appropriate for many types of experimental games (see Smith (1985)). Let P_1s pay-off aggregate to time t be denoted by Σ_t. In P_1's model he assumes that P_2's move $m_2(t)$ depends only on the past through the statistic vector $\tau(t)$. P_1 knowns that the game will terminate at or before time T.

Some of the statements given above are summarised in the influence diagram I_1 in Figure 3.1. Notice that I_1 is in extensive form. The dependencies depicted in I_1 hold, if only vacuously, even if the game has terminated before T, since the time the game ends is independent of what happened in the past. [In fact U is independent of the hypothetical "moves" played after the game has terminated.]

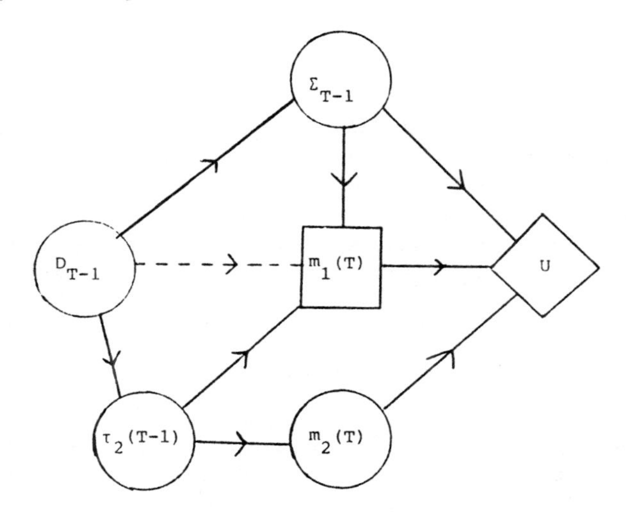

Figure 3.1. *The I.D.I_1. Delete $(D_{T-1}, m_1(T))$ arc.*

Since I_1 is an *E.F.I.D.* by using Rule 9 we can, without loss, delete the $(D_{T-1}, m_1(T))$ arc. This then reads "At the last stage of the game there is an optimal policy for P_1, (if any exists) which depends only on the value $\tau(T-1)$ and P_1's aggregate pay-off Σ_{T-1}, accrued before he moved.

If we decompose node D_{T-1} into the nodes (D_{T-2}, Σ_{T-2}, $m_1(T-1)$, $\tau(T-2)$, $m_2(T-1)$) using Rule 3 and our stated assumptions we obtain the influence diagram I_2 given in Figure 3.2. We can again now use Rule 9 to delete the $(D_{t-2}, m_1(T-1))$ arc. Decomposing $D_{T-k}, k = 1, 2, \ldots, T-1$ and successively deleting the $(D_{T-k-1}, m_1(T-1))$ arcs we have proved the following theorem.

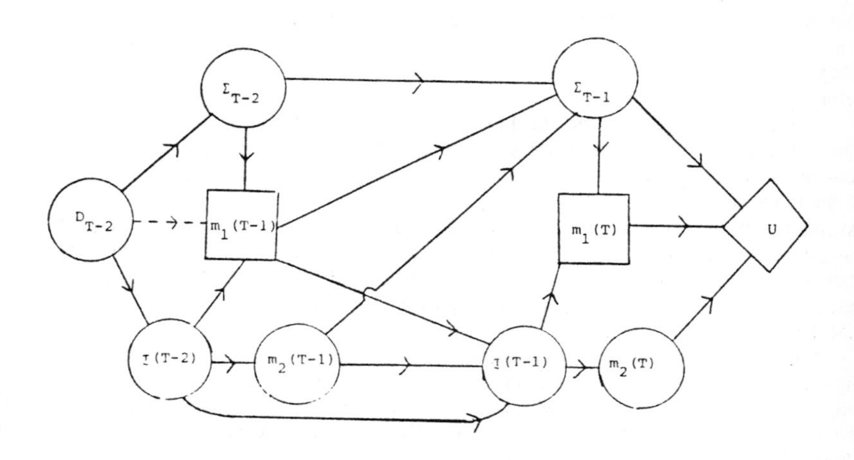

Figure 3.2. *The I.D.I_2. Delete $(D_{T-2}, m_1(T-1))$ arc.*

Theorem. *If P_1's model is as described above, then if an optimal policy exists, there will be an optimal policy in which P_1's t-th move $m_1(t)$ will depend on the past only through $\tau(t-1)$ and his pay-off Σ_{t-1} aggregated up to that time.*

The proof given above is essentially rigorous and does not embroil us in inelegant and opaque algebraic equations. Furthermore, by using such a proof it is often easy to see how the result might be extended. Here if Σ_t is any function of only $(\Sigma_{t-1}, m_1(t), m_2(t))$ $t = 1, 2, \ldots$ where $\Sigma_0 = 0$, our proof is still appropriate and unchanged. So, for example, our result would still hold if P_1 had a utility which was any function of any discounted aggregate pay-off.

Example 2. Making probability forecasts using data.

You need to specify your probability, *d*, it will rain tomorrow (an event $X = 1$) or not (an event $X = 0$) and your decision will be judged using a scoring rule U (for example U may be the Brier Score $(x - d)^2$). You are allowed to make your foercast after hearing a weatherman's forecast of rain, q. You also have available a history of his forecasts q_i and whether or not it rained on that day x_i, $1 \leq i \leq n$. So your information at the time you make your own forecast d is (q, y) where $y = \{(x_i, q_i) : 1 \leq i \leq n\}$.

This decision problem can be expressed by the $I.D.I_3$ given in Figure 3.3 which I will call the calibration-form model. Lindley's helpful insight was to point out that q should be treated as data in this problem. He went further, however. He maintained (see for example Lindley (1985)) that to specify your joint distribution on the triple (Y, Q, X) you should start by specifying your prior margin on X and then the distribution of (Y, Q) conditional on X.

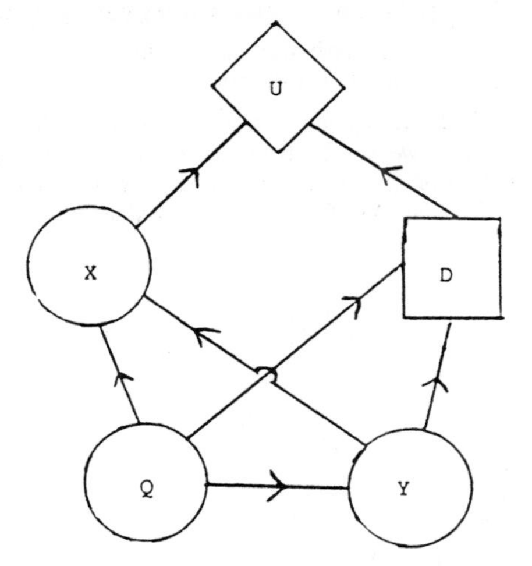

Figure 3.3. *The calibration-form model I_3.*

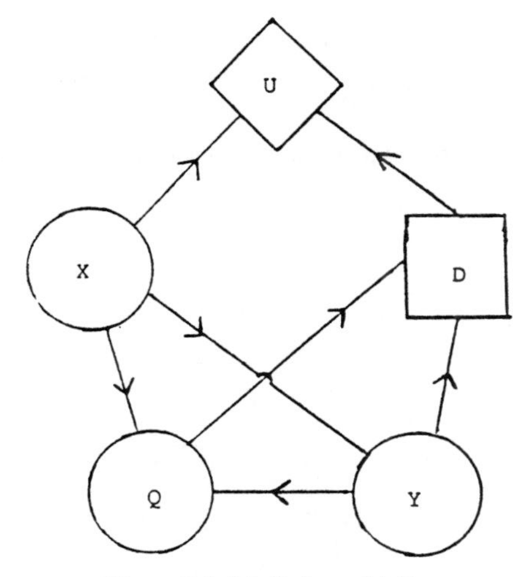

Figure 3.4. *Lindley's model, I_4.*

In terms of $I.D.$s he suggests that the problem should be posed by an $I.D.I_4$ (See Figure 3.4) equivalent to I_3 with all the arcs between (X, Y, Q) reversed. [It is easy to see this is an equivalent description of the problem by using Rule 4 on arcs (Q, Y) (Q, X) and then (Y, X) and notice we need add no new arcs to the diagram.] You can then obtain your distribution of X given q and using Bayes Rule.

Although such a methodology has other advocates (e.g. See West (1988)) personally I find it unattractive for three reasons:

(a) it is not the usual form of Bayesian model. Normally, to exploit latent conditional independence, we condition on *parameters* not future observables, The predictive distribution of variables like X is then obtained by marginalising out these parameters.

(b) as Morris (1986) points out,it is often very difficult to elicit directly the distribution of a past observable conditional on a future observable.

(c) it is inefficient (as pointed out by McConway in his discussion of Lindley's paper). If we use the description I_3 we see by Rule 10 that it is unnecessary for us to specify our marginal distribution of Q. So by using description I_4 you choose to specify, albeit indirectly, a considerable amount of information ancillary to the problem.

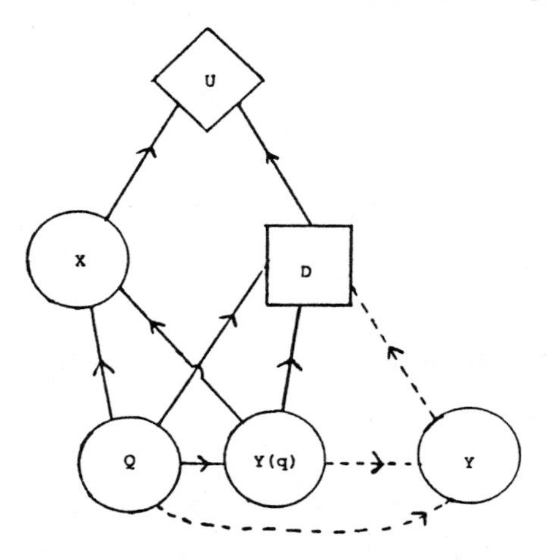

Figure 3.5. *A model incorporating a conditionl independence statement,* I_5.

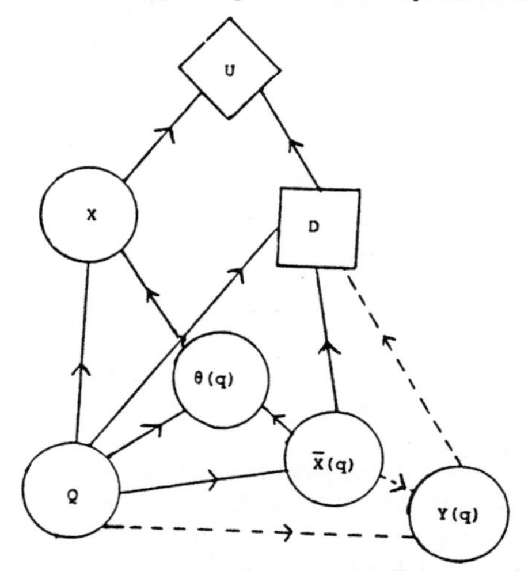

Figure 3.6. *A model incorporating exchangeable beliefs.*

How does one set about an optimal poicy using I_3? Well, the answer is to start introducing some conditional independence statements; in my opinion the starting point of all good Bayesian modelling.

Let $Y(q) = \{(X_i, Q_i) : Q_i = q\}$. The simplest assumption to make is

$$X \amalg Y|Y(q) \tag{3.1}$$

i.e. the only data informative about X are those days when the forecaster has predicted the probability of rain as q, his forecast today. The random vector $Y(q)$ can be added to I_3 using Rule 7 to give I_5. The node Y can then be removed together with the dotted arcs (Y, D), $(Y(q), Y)(Q, Y)$ using Rule 9 and then Rule 8. Another assumption you might choose to make is that given each value of q the event of rain on these days is exchangeable to you. Let

$$\overline{X}(q) = \{(\Sigma X_i, \#(i)) : Q_i = q\}$$

By De Finetti's Theorem, I_5 can now be written in the form I_6 and then, by Rules 9 and 8 $Y(q)$, $(Y(q)D)$, $(\overline{X}(q), Y(q))$, and $(Q, Y(q))$ deleted from the diagram where $\theta(q)$ is your probability of rain on *any* day (not just tomorrow) given a forecast q by the weatherman.

Now explicit modelling is easy. For example set $\theta(q)$ to have a prior Beta distribution with parameters $(\alpha(q), \beta(q))$ set represent your prior beliefs about the calibration of the weatherman (e.g. $E(\theta(q)) = q, P\{q - \epsilon < \theta(q) < q + \epsilon\} = \alpha$.) The usual Beta-Binomial analysis gives a Beta posterior for $\theta(q)$ in terms of $\alpha(q)$, $\beta(q)$ and $\overline{X}(q)$ above.

There has been a considerable amount of discussion about the role of calibration in probability forecasting models. Under this particular Bayesian model, the empirical distribution of forecasts by our weatherman $\{\#(i) : Q_i = q, 1 \leq i \leq n\}$ and his empirical calibration curve $\{(\#(i))^{-1}\Sigma X_i : Q_i = q\}$ are *sufficient* for $\{\theta(q) : q\epsilon[0, 1]\}$ and hence sufficient for making your predictions.

Of course, the model above can be improved upon. In particular assumption (3.1) looks naïve. Surely the forecaster's performance when stating probabilities close to but not equal to q should be informative about X. However the $I.D.$s above suggest ways in which our simple model might be modified to take such effects into account. Essentially we could require

$$X \amalg Y|Q, \overline{\boldsymbol{X}}$$

where $\overline{\boldsymbol{X}}\{\overline{\boldsymbol{X}}(q) : q$ over all stated values$\}$.

This will maintain the sufficiently of empirical calibration property above. Using Goldstein's $\cdot \amalg^W \cdot|\cdot$ best linear estimates on terms in $\overline{\boldsymbol{X}}$ of $\boldsymbol{\theta}$ can then be derived and hence best linear estimates D of X found. Details of this approach will be given in a later paper.

4. CONCLUSIONS

Influence diagrams are a useful pictorial representation of inference and decision problems which make it much easier for the statistician or decision analyst and his client to understand the implications of using a particular model before explicit calculations are made. Now that influence diagrams have been sshown to possess a formal algebra they can also be used to prove theorems that would otherwise be algebraically messy.

REFERENCES

Agogino, A. M. (1985). Use of probabilistic inference in diagnostic expert systems. *Proceedings of the 1985 ASME International Computers in Engineering Conference* 2, 305–310. Boston, Mass.

Barlow, R. and Zhang, X. (1986). Bayesian analysis of acceptance sampling procedures discussed by Deming. Operations Research Center Report No. 86-2, University of California, Berkeley, California. To appear in *Journal of Statistical Planning and Inference.*

Barlow, R., Mensing, R. W. and Smiriga, N. G. (1986). Using influence diagrams to solve a calibration problem. *Proceedings of International Symposium on Probability and Bayesian Statistics,* 23-36. Austria.

Dawid, A. P. (1979). Conditional indepence in statistical theory. (with discussion). *J. Roy. Statist. Soc. B* **43**, 105–130.

Holtzman, S. (1985). Intelligent Decision Systems. Ph. D. Thesis, Engineering-Economic Systems Dept., Stanford University, Stanford, California.

Howard, R. A. and Matheson, J. E. (1981). Influence diagrams, Chapter 8. *The Principles and Applications of Decision Analysis* **2**. Menlo Park, California: Strategic Decisions Group.

Lindley, D. V. (1985). Reconciliation of discrete probability distributions. *Bayesian Statistics* **2**. 375–390. (J. M. Bernardo, M. H. DeGroot, D. V. Lindley and A. F. M. Smith, eds.). Amsterdam: North-Holland.

Miller, A. C., Merkhofer, M. M. and Howard, R. A. (1976). Development of Automated Aids for Decision Analysis. Stanford Research Institute, Menlo Park, California.

Morris, P. A. (1986). Observations on expert aggregation. *Man. Sci.* **32**, 321–328.

Olmsted, S. M. (1984). On representing and solving decision problems. Ph. D. Thesis, Engineering-Economic Systems Dept., Stanford University, Stanford California.

Oliver, R. M. (1986). Influence diagrams in forecasting. To appear in *Journal of Forecasting*.

Owen, D. (1978). The use of influence diagrams in structuring complex decision problems. *Proceeding of the Second Lawrence Symposium on Systems and Decision Sciences*. Berkeley, California.

Pearl, J. (1986). Markov and Bayes Networks: a comparison of two graphical representions of probabilistic knowledge. *Tech. Rep.* **CSD 860024** Cognitive Systems Lab., Comp. Sci. Dept., University of California, Los Angeles.

Rege, A. and Agogino, A. M. (1986b). Sensor integrated expert system for manufacturing and process diagnostics, ORC Report No. 86-15, University of California, Berkeley.

Shachter, R. D. (1986). Evaluating influence diagrams, *Reliability and Qulity Control*, 321–344. (A. P. Basu, ed.). Elsevier, North-Holland.

Shachter, R. D. (1986). *Intelligent Probabilistic Inference in Uncertainty and Artificial Intelligence*, 371–382. (L. N. Kanal and J. Lemmer, eds.). Amsterdam, North Holland.

Smith, J. Q. (1985). A statistical approach to the analysis of reapeated two player games, *Tech. Rep.* **62**. Dept. of statistics, University of Warwick.

Smith, J. Q. (1986). Diagrams of influence in statistical models, *Tech. Rep.* **99**. Dept. of statistics, University of Warwick.

Smith, J. Q. (1987a). Influence diagrams for Bayesian decision analysis. *European Journal of Operations Research*. (To appear).

Smith, J. Q. (1987b). *Decision analysis: A Bayesian approach*, Chapman Hall Chapter 5.

Smith, J. Q. (1987c). Statistical priciples on Graphs. *Tech. Rep.* **138**. Dept. of Statistics, University of Warwick. (To appear).

Spiegelhalter, D. J. (1986). Probabilistic reasoning in predictive expert systems. To appear in *Uncertainty in Artificial Intelligence*, (L.N. Kanal and J. Lemmer, eds.). Amsterdam, North-Holland.

Spiegelhalter, D. J. (1987). Coherent evidence propagation in expert systems, *The Statistician*. (To appear).

Wermuth, H. and Lauritzen, S. L. (1983). Graphical and recursive models for contingency tables. *Biometrika* **70**, 537–557.

West, M. (1988). Probability forecasting and modelling expert opinion. (In this volume).

BAYESIAN STATISTICS 3, pp. 777–783
J. M. Bernardo, M. H. DeGroot, D. V. Lindley and A. F. M. Smith, (Eds.)

Nonsequential Designs for Model Discrimination and Parameter Estimation

F. SPEZZAFERRI
University of Rome

SUMMARY

A quadratic utility function for a probability distribution is used to obtain an optimality criterion for non sequential designs. Optimal designs can be derived for: a) estimation b) model discrimination and c) the dual problem of parameter estimation and model choice. Nested normal linear models are considered. With vague prior information the criterion gives the usual D-optimal designs in cases a) and b), while in c) it generates new designs which depend on the model's prior probability.

Keywords: BAYESIAN DESIGNS; D-OPTIMAL DESIGNS; MODEL DISCRIMINATION; NESTED LINEAR MODELS; UTILITY FUNCTION.

1. INTRODUCTION

Let $\mathcal{E}$ be an experiment consisting in observing a random vector at the points of an experimental region. Let $\boldsymbol{X}$ and $\boldsymbol{y}$ denote, respectively, the design matrix, to be fixed by the experimenter, and the vector of observations.

We shall consider the problem of choice of $\boldsymbol{X}$ when $\mathcal{E}$ is performed in order to make inference on an unknown random vector $\boldsymbol{\beta}$. Some of the components of $\boldsymbol{\beta}$ may have discrete probability distributions and the others absolutely continuous distributions. Let $p(\boldsymbol{\beta})$ denote the generalized probability density function (g.p.d.f.) of $\boldsymbol{\beta}$ prior to performing the experiment $\mathcal{E}$ and let $p(\boldsymbol{\beta}|\boldsymbol{y}, \boldsymbol{X})$ be the g.p.d.f. of $\boldsymbol{\beta}$ after $\mathcal{E}$ is performed.

If $\boldsymbol{\beta}$ is the true value, we assume that the utility associated with the choice of $f(\cdot)$ as g.p.d.f. of $\boldsymbol{\beta}$ is the quadratic utility function

$$U\{\boldsymbol{\beta}, f(\cdot)\} = 2f(\boldsymbol{\beta}) - \int \cdots \int f^2(\boldsymbol{\beta})d\boldsymbol{\beta}$$

where the integral becomes a sum for the discrete components of $\boldsymbol{\beta}$. The quadratic utility function is a proper scoring rule and was introduced by De Finetti (1962) in the discrete case. The usefulness of the experiment $\mathcal{E}$ can be measured by the expected relative increase of the utility before and after $\mathcal{E}$ is performed:

$$\begin{aligned} U(\mathcal{E}) &= \frac{\int \cdots \int [U\{\boldsymbol{\beta}, p(\boldsymbol{\beta}|\boldsymbol{y}, \boldsymbol{X})\} - U\{\boldsymbol{\beta}, p(\boldsymbol{\beta})\}]\, p(\boldsymbol{\beta}|\boldsymbol{y}, \boldsymbol{X})d\boldsymbol{\beta}}{\int \cdots \int U\{\boldsymbol{\beta}, p(\boldsymbol{\beta})\}\, p(\boldsymbol{\beta})d\boldsymbol{\beta}} \\ &= \frac{\int \cdots \int \{p(\boldsymbol{\beta}|\boldsymbol{y}, \boldsymbol{X}) - p(\boldsymbol{\beta})\}^2\, d\boldsymbol{\beta}}{\int \cdots \int p^2(\boldsymbol{\beta})d\boldsymbol{\beta}} \end{aligned}$$

The optimal design $\boldsymbol{X}^*$ is obtained as

$$\begin{aligned}\boldsymbol{X}^* &= \arg\max \int U(\mathcal{E})p(y|\boldsymbol{X})dy \\ &= \arg\max \int \frac{\int p^2(\beta|y,\boldsymbol{X})d\beta}{\int p^2(\beta)d\beta} p(y|\boldsymbol{X})dy\end{aligned}$$

where $p(y|\boldsymbol{X})$ is the conditional distribution of y given $\boldsymbol{X}$.

In this paper, the above procedure is applied to obtain nonsequential designs for the case of normal linear models. Depending on the purpose of the experiment, that is on the choice of β as the parameter of interest, different designs can be obtained. Designs can be derived either for the parameter estimation problem or for model discrimination or for the dual problem of model choice and parameter estimation.

A review of the literature on the design of experiments is given by Atkinson (1982). Hill (1978) reviewed the experimental procedures for discriminating between models and also stressed the important need for simultaneous consideration of discrimination and parameter estimation. The present paper is related to that of Borth (1975), who developed sequential designs for the simultaneous problem using a criterion based on the logarithmic utility function.

2. PARAMETER ESTIMATION

Suppose that the conditional joint distribution of y is a multivariate normal distribution with mean vector $\boldsymbol{X}\boldsymbol{\theta}$ and covariance matrix $\sigma^2\boldsymbol{I}$. Assume σ^2 known and let $\beta = \boldsymbol{\theta}$ be the unknown d-vector of interest. With the usual conjugate normal prior distribution for $\boldsymbol{\theta}$ with mean vector $\boldsymbol{\mu}$ and covariance matrix $\sigma^2\boldsymbol{W}$ we obtain

$$\int \cdots \int p^2(\theta|y,\boldsymbol{Y})d\theta = \frac{|\boldsymbol{X}^T\boldsymbol{X}+\boldsymbol{W}^{-1}|^{\frac{1}{2}}}{(2\sqrt{\pi})^d(\sigma^2)^{d/2}}.$$

In this case, $\boldsymbol{X}^*$ turns out to be the Bayesian D-optimal design. If σ^2 is unknown, let us consider a gamma prior distribution for σ^{-2} with scale parameter g. If $\boldsymbol{X}$ is a full rank $n \times d$ matrix, then, under the assumption $\boldsymbol{X}^T\boldsymbol{X}+\boldsymbol{W}^{-1} \approx \boldsymbol{X}^T\boldsymbol{X}$, the classical D-optimal design is obtained whether when $\beta = \boldsymbol{\theta}$ or when $\beta = (\boldsymbol{\theta}, \sigma^2)$.

It can be seen that the integrals of the squares of the marginal posterior distribution of $\boldsymbol{\theta}$ and of the joint posterior distribution of $\boldsymbol{\theta}$ and σ^2 are respectively proportional to

$$|\boldsymbol{X}^T\boldsymbol{X}|^{\frac{1}{2}}(g_1)^{-d/2} \quad \text{and to} \quad |\boldsymbol{X}^T\boldsymbol{X}|^{\frac{1}{2}}(g_1)^{-(1+d/2)}$$

where

$$\begin{aligned} g_1 &= g + \frac{\nu}{2}\hat{s}^2 \\ \nu\hat{s}^2 &= y^T(\boldsymbol{I} - \boldsymbol{X}(\boldsymbol{X}^T\boldsymbol{X})^{-1}\boldsymbol{X}^T)y \\ \nu &= n - d. \end{aligned}$$

The D-optimal design is obtained in both cases by observing that the following integral is independent of $\boldsymbol{X}$.

$$\begin{aligned}\int (g_1)^h p(y|\boldsymbol{X}) &= \int \left\{ \int (g_1)^h p(y|\boldsymbol{X},\sigma^2)dy \right\} p(\sigma^2)d\sigma^2 \\ &= \int \left\{ \int (g + \sigma^2\chi^2)^h p(\chi^2)d\chi^2 \right\} p(\sigma^2)d\sigma^2,\end{aligned}$$

where $p(y|\boldsymbol{X},\sigma^2)$ is a normal distribution with mean vector $\boldsymbol{X}\boldsymbol{\mu}$ and covariance matrix $\sigma^2(\boldsymbol{I}+\boldsymbol{X}\boldsymbol{W}\boldsymbol{X}^T)$, $p(\chi^2)$ is a central chi-square distribution with ν d.o.f. and $p(\sigma^2)$ is the prior distribution of σ^2.

3. MODEL DISCRIMINATION

Suppose that two models M_1, M_2 are available and that the purpose of the experiment is to discriminate between M_1 and M_2.

Let $p(y|\boldsymbol{X}, M_h)$ be the density function for y given $\boldsymbol{X}$ according to the h-th model, $h = 1, 2$. Define the parameter of interest β as 1 if M_1 is true and 2 if M_2 is true and denote by π_h and p_h, respectively, the prior and the posterior probability that $\beta = h$, $h = 1, 2$. Straightforward application of Bayes theorem gives $p_1 = K/(1+K)$ and $p_2 = 1/(1+K)$ where

$$K = \frac{[\pi_1 p(y|\boldsymbol{X}, M_1)]}{[\pi_2 p(y|\boldsymbol{X}, M_2)]}$$

is the posterior odds ratio for the two models. The design criterion in Section 1 is to maximize with respect to $\boldsymbol{X}$:

$$\int (p_1^2 + p_2^2) p(y|\boldsymbol{X}) dy$$

where $p(y|\boldsymbol{X}) = \pi_1 p(y|\boldsymbol{X}, M_1) + \pi_2 p(y|\boldsymbol{X}, M_2)$.

Equivalently, we can minimize w.r.t. $\boldsymbol{X}$ $\int p_1 p_2 p(y|\boldsymbol{X}) dy$ and, after some algebra, it can be seen that this corresponds to minimizing either $\int p_2 p(y|\boldsymbol{X}, M_1) dy$ or $\int p_1 p(y|\boldsymbol{X}, M_2) dy$. So the design criterion of Section 1 leads to the intuitively sensible criterion of minimizing the expectation of the posterior probability of one model, when the other is assumed to be true.

As an example, consider two nested normal linear models, $M_1 \subset M_2$, defined by

$$\begin{aligned} \boldsymbol{Y}|\boldsymbol{A}_h, \boldsymbol{\theta}_h, \sigma^2, M_h &\sim N(\boldsymbol{A}_h \boldsymbol{\theta}_h, \sigma^2 \boldsymbol{I}), \qquad h = 1, 2, \\ \boldsymbol{A}_2 &= (\boldsymbol{A}_1 \vdots \boldsymbol{A}), \boldsymbol{\theta}_2^T = (\boldsymbol{\theta}_1^T \vdots \boldsymbol{\theta}^T), \end{aligned} \tag{1}$$

where $\boldsymbol{A}_h$ is a full rank matrix, $\boldsymbol{\theta}_h$ is a d_h-vector of unknown parameters, σ^2 is known and $\boldsymbol{A}_2$ is the design matrix. Let $p(\boldsymbol{\theta}_2|\sigma^2)$ and $p(\boldsymbol{\theta}_1|\sigma^2)$ be normal densities with zero means and covariance matrices $\sigma^2 \boldsymbol{W}_1$, $\sigma^2 \boldsymbol{W}$ respectively, such that

$$\boldsymbol{W}_2 = \begin{pmatrix} \boldsymbol{W}_1 & \vdots & 0 \\ \cdots & \cdots & \cdots \\ 0 & \vdots & \boldsymbol{W} \end{pmatrix}, \boldsymbol{W}_1^{-1} + \boldsymbol{A}_1^T \boldsymbol{A}_1 \approx \boldsymbol{A}_1^T \boldsymbol{A}_1 \text{ and } \boldsymbol{W}^{-1} + \boldsymbol{A}^T \boldsymbol{A} \approx \boldsymbol{A}^T \boldsymbol{A}. \tag{2}$$

The posterior odds ratio for the two models is given by

$$K = \frac{\pi_1}{\pi_2} \frac{|\boldsymbol{I} + \boldsymbol{A}_2^T \boldsymbol{W}_2 \boldsymbol{A}_2|^{\frac{1}{2}} \exp\{-\frac{1}{2}(Q_1 - Q_2)\}}{|\boldsymbol{I} + \boldsymbol{A}_1^T \boldsymbol{W}_1 \boldsymbol{A}_1|^{\frac{1}{2}}}$$

where

$$\sigma^2 Q_h = \boldsymbol{y}^T \left[\boldsymbol{I} + \boldsymbol{A}_h^T \boldsymbol{W}_h \boldsymbol{A}_h\right]^{-1} \boldsymbol{y}, \quad h = 1, 2.$$

Using (2) and well known identities for the determinant and inverse of matrices of the form $\boldsymbol{D} + \boldsymbol{E}\boldsymbol{F}\boldsymbol{E}^T$, (see, e.g. Smith, 1973) we obtain

$$K = \frac{\pi_1}{\pi_2} |\boldsymbol{W}|^{\frac{1}{2}} |\boldsymbol{S}|^{\frac{1}{2}} \exp\left\{-\frac{1}{2} Q\right\} \tag{3}$$

where

$$\boldsymbol{S} = \boldsymbol{A}^T \boldsymbol{A} - \boldsymbol{A}^T \boldsymbol{A}_1 (\boldsymbol{A}_1^T \boldsymbol{A}_1)^{-1} \boldsymbol{A}_1^T \boldsymbol{A} \text{ and}$$

$$\sigma^2 Q \approx y^T\{A_2(A_2^T A_2)^{-1}A_2^T - A_1(A_1^T A_1)^{-1}A_1^T\}y.$$

we are looking for the design which minimizes $\int p_2 p(y|A_1, M_1)dy$. The odds ratio, K, and, consequently, p_2, depends on y only through the quadratic form Q. Since, from the Fisher-Cochran theorem, Q, under M_1, has a central chi-square distribution $p(\chi^2)$ with $d_2 - d_1$ d.o.f., then

$$\int p_2 p(y|A_1, M-1)dy = \int \frac{p(\chi^2)}{1 + \frac{\pi_1}{\pi_2}|W|^{\frac{1}{2}}|S|^{\frac{1}{2}}\exp\{-\frac{1}{2}\chi^2\}}d\chi^2 \tag{4}$$

and the design criterion leads us to maximize $|S|$. Note that this is the D-optimal design for testing the hypothesis $\theta = 0$ (see e.g. Atkinson 1972).

4. DISCRIMINATION AND ESTIMATION

Consider now the case of two normal linear models M_1, M_2 when there is joint interest in discrimination and estimation. To deal with this problem, it is convenient to assume the same sampling distribution for the two models and discriminate between them through different prior distributions on the common unknown parameter. More specifically, assume the distribution in (1) when $h = 2$ as the sampling distribution $p(y|A_2, \theta_2, \sigma^2)$ for both models M_1 and M_2 and let

$$P(\theta_2|M = M_h, \sigma^2) = N(0, \sigma^2 \begin{pmatrix} W_1 \vdots & 0 \\ \cdots\cdots\cdots \\ 0 & \vdots \overline{W}_h \end{pmatrix}, \qquad h = 1, 2$$

be the prior distributions of the vector θ_2 under M_h. Let us also assume that (2) holds, σ^2 is known and $P_r(M = M_1) = \pi_1$, $P_r(M = M_2) = \pi_2 = 1 - \pi_1$.

The vector of interest for this joint problem is $\beta = (\theta_2, M)$ and the design criterion is to maximize with respect to A_2

$$\int \frac{\sum_{h=1,2}\int p^2(\theta_2, M_h|y, A_2, \sigma^2)d\theta_2}{\sum_{h=1,2}\int p^2(\theta_2, M_h|\sigma^2)d\theta_2} p(y|A_2)dy \tag{5}$$

where:

$$p(y|A_2) = \pi_1 p(y|A_2, M_1) + \pi_2 p(y|A_2, M_2),$$

$$p(y|A_2, M_h) = \int p(y|A_2, \theta_2, \sigma^2)p(\theta_2|M = M_h)d\theta_2,$$

$$p(\theta_2, M_h|y, A_2, \sigma^2) = \frac{p(y|A_2, \theta_2, \sigma^2)p(\theta_2, M_h|\sigma^2)}{p(y|A_2)} \quad \text{and}$$

$$p(\theta_2, M_h|\sigma^2) = p(\theta_2|M = M_h, \sigma^2)\pi_h, \qquad h = 1, 2.$$

Note that if $\overline{W}_1 \to 0$ then

$$\int p(y|A_2, \theta_2, |\sigma^2)p(\theta_2|M = M_1, \sigma^2)d\theta \to N(A_1\theta_1, \sigma^2 I)$$

that is the sampling distribution under M_1 assumed in (1). Therefore to deal with the case M_1 nested in M_2 we shall consider the limit case of (5) when $\overline{W}_1 \to 0$.

Expression (5) can be rearranged to obtain:

$$\frac{\sum_{h=1,2}\pi_h \int \overline{p}_h \left\{\int p^2(\theta_2|y, A_2, M_h, \sigma^2)d\theta_2\right\} p(y|A_2, M_h)dy}{\sum_{h=1,2}\pi_h^2 \int p^2(\theta_2|M = M_h, \sigma^2)d\theta_2} \tag{6}$$

where

$$p(\theta_2|y, A_2, M_h, \sigma^2) = \frac{p(y|A_2, \theta_2, \sigma^2)p(\theta_2|M = M_h, \sigma^2)}{p(y|A_2, M_h)}$$

and $\overline{p}_h = \frac{\pi_h p(y|A_2, M_h)}{p(y|A_2)}$ is the posterior probability of M_h, $h = 1, 2$.

The integrals between the curly brackets and in the denominator of expression (6) can be easily evaluated with the above distributional assumptions. We observe that the integrals corresponding to $h = 2$ are independent on $\overline{W}_1$, while those corresponding to $h = 1$ diverge as $\overline{W}_1 \to 0$. After some simple matrix algebra we have

$$\int p^2(\theta_2|y, A_2, M_1, \sigma^2)d\theta_2 \approx \frac{|A_1^T A_1 + W_1^{-1}|^{\frac{1}{2}}}{|\overline{W}_1|^{\frac{1}{2}}} \left(\frac{1}{4\pi\sigma^2}\right)^{\frac{d_2}{2}}$$

$$\int p^2(\theta_2|M = M_1, \sigma^2)d\theta_2 \approx \frac{1}{|\overline{W}_1|^{\frac{1}{2}}|W_1|^{\frac{1}{2}}} \left(\frac{1}{4\pi\sigma^2}\right)^{\frac{d_2}{2}}$$

as $\overline{W}_1 \to 0$ and, identifying $\overline{W}_2$ with W appearing in (2),

$$\lim_{\overline{W}_1 \to 0} \frac{\overline{p}_1}{\overline{p}_2} = \frac{p_1}{p_2} = K$$

where K is defined in (3).

The limit case of (6) is therefore equivalent to

$$\left|A_1^T A_1\right|^{\frac{1}{2}} \int \frac{p_1}{\pi_1} p(y|A_2, M_1)dy \tag{7}$$

or, applying (4), to

$$\int \frac{\left|A_2^T A_2\right|^{\frac{1}{2}}}{\pi_2 \exp\left\{\frac{1}{2}\chi^2\right\} + \pi_1 \left|\overline{W}_2\right|^{\frac{1}{2}} |S|^{\frac{1}{2}}} p(\chi^2)d\chi^2. \tag{8}$$

The optimal design for the dual problem of estimation and model discrimination is obtained by maximizing (7) or (8) w.r.t. A_2. Note that expression (7) is the product of two factors and that D-optimal designs to estimate θ_1 and to discriminate between M_1 and M_2 are found by maximizing respectively the first and the second factors in (7) (see Section 3). Expression (8) shows the following consistent result: D-optimal designs for parameter estimation in M_1 or M_2 are obtained when either $\pi_1 = 1$ or $\pi_1 = 0$.

5. EXAMPLES

In this section we apply the results of Section 4 to derive optimal designs in two special cases of polynomial regression. As a first application, we analyze the case of a first and a second degree polynomial. In this case the i-th row of A_2 is given by $(1 x_i x_i^2)$, where $x_1, x_2, \ldots, x_n$ assumed to take the values $-1, 0, 1$, are the design points corresponding to the components $y_1, y_2, \ldots, y_n$ of the n-vector y of observations. Considering the usual symmetric designs, placing $\frac{1}{2}np$ at $x = \pm 1$ and $n(1-p)$ at $x = 0$, applying (8), we see that the optimal design for model discrimination and parameter estimation is given by

$$p^* = \arg\max \int_0^\infty \frac{(p^2 - p^3)^{\frac{1}{2}} z^{-\frac{1}{2}} \exp\left\{-\frac{1}{2}z\right\}}{\pi_2 \exp\left\{\frac{1}{2}z\right\} + \pi_1 w^{\frac{1}{2}} [np(1-p)]^{\frac{1}{2}}} dz$$

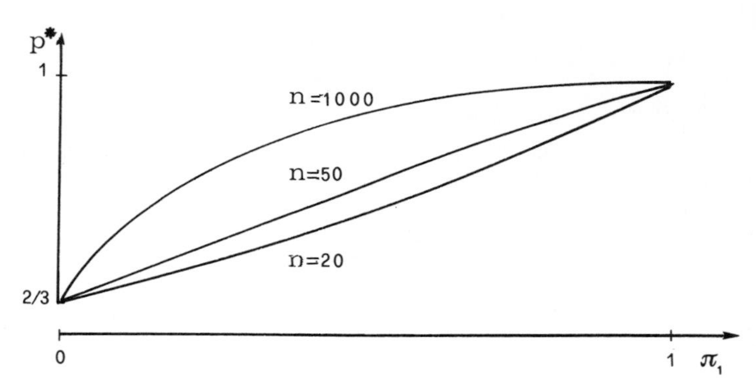

Figure 1.

In Figure 1, p^* is graphed as function of π_1 for selected values of n, while w has been fixed at, the prior variance of the extra term, 3/2 using the imaginary training sample device of Spiegelhalter and Smith (1982).

Note that choosing n in a reasonable range of values does not have a strong effect on p^*. The extreme case $n = 1000$ shows how a large n favours values of p^* close to 1 meaning that a small proportion of the observations is required on the design middle point.

As a second application consider a quadratic and a cubic model, for which the i-th row of $\boldsymbol{A}_2$ is given by $(1 x_i x_i^2 x_i^3)$.

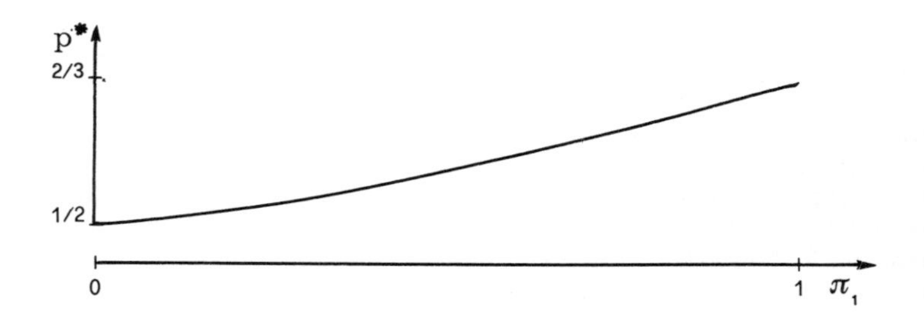

Figure 2.

If we restrict attention to designs intermediate between the forms of classical D-optimal designs for parameter estimation in quadratic and cubic models, such designs have proportions $\frac{1}{2}np$ at $x = \pm 1$ and $\frac{1}{2}n(1-p)$ at $x = \pm a$, for some a.

Note that the D-optimal design for the smaller (larger) model is obtained if $a = 0$ and $p = 2/3$ ($a = 1/5$ and $p = 1/2$). Applying (8), the optimal design $((p^*, a^*)$ is obtained by maximizing w.r.t. (p, a)

$$\int_0^\infty \frac{p(1-p)a(1-a^2)^2 z^{-\frac{1}{2}} \exp\left\{-\frac{1}{2}z\right\}}{\pi_2 \exp\left\{\frac{1}{2}z\right\} + \pi_1 a(1-a^2)\left[\frac{np(1-p)w}{p+a^2(1-p)}\right]^{\frac{1}{2}}} dz$$

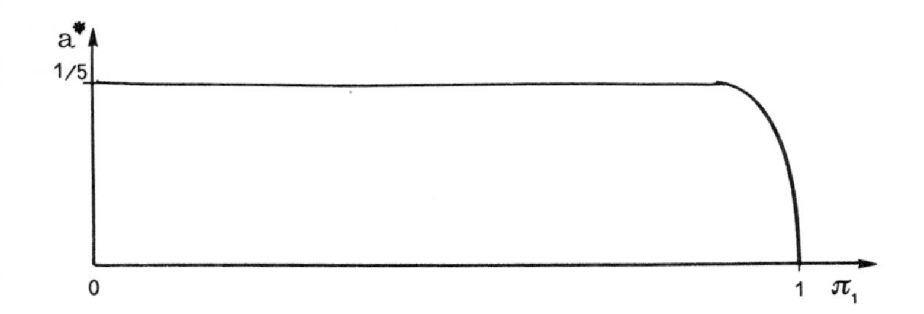

Figure 3.

In Figure 2 and Figure 3, p^* and a^* are graphed as function of π_1, when $n = 20$ and $w = 1.1$ has been found by using the imaginary training sample device.

Figure 2 and 3 show that p^* increases monotonically from $1/2$ to $2/3$, while a^* is steadly equal to 1/5 and decreases sharply to zero when π_1 is close to 1.

REFERENCES

Atkinson, A. C. (1972). Planning experiments to detect inadequate regression models. *Biometrika* **59**, 275–293.

Atkinson, A. C. (1982). Developments in the design of experiments. *Intern. Statist. Review* **50** 161–177.

Borth, D. M. (1974). A total entropy criterion for the dual problem of model discrimination and parameter estimation. *J. Roy. Statist. Soc. B* **37**, 77–87.

De Finetti, B. (1962). Does it makes sense to speak of "good probability appraisers"? *The scientist speculates.* (I. J. Good, ed.). New York.

Hill, D. H. (1978). A review of experimental design procedures for regression model discriminition. *Technometrics* **20**, 15–21.

Smith, A. F. M. (1973). A general bayesian linear model. *J. Roy. Statist. Soc. B* **35**, 67–75.

Spiegelhalter, D. J. and Smith, A. F. M. (1982). Bayes factors for linear and log-linear models with vague prior information. *J. Roy. Statist. Soc. B* **44**, 377–387.

BAYESIAN STATISTICS 3, pp. 785–790
J. M. Bernardo, M. H. DeGroot, D. V. Lindley and A. F. M. Smith, (Eds.)

The 2 x 2 Table in Observational Studies

R. A. SUGDEN
University of London, Goldsmiths' College

SUMMARY

A "face-value" analysis of data from a 2×2 contingency table in an observational study to compare two treatments is shown to be valid when an assumption of non-informativeness or strong ignorability holds for the treatment assignment mechanism. Inference is investigated when this condition fails.

Keywords: OBSERVATIONAL STUDY; 2×2 TABLE; IGNORABILITY; NON-INFORMATIVE; CAUSAL EFFECTS; TREATMENT ASSIGNMENT MECHANISM.

1. INTRODUCTION

The analysis of data from observational studies in order to make causal inferences has recently been of interest e.g. Rosenbaum and Ruben (1983, 1984). Such influences use covariance adjustments, matching, etc. to provide unbiased estimates of treatment effects under an assumption of strong ignorability of the treatment assignment mechanism, which in the study is uncontrolled by the experimenter. Moreover, the definition of causal effect (one treatment versus another) has been the subject of lively debate, Holland (1986).

In the absence of covariate information and with no external data, however, there is little a Bayesian can do when faced by data from an observational study other than to model his beliefs about the assignment mechanism and examine the effects of varying these assumptions on the final causal inference.

2. THE PROBLEM

Consider a simple situation with a binary response 0 (failure) or 1 (success) and two treatments T_1 and T_2. Suppose that there is no sampling, so that the entire population of N individuals is partitioned by the treatment assignment mechanism into two groups of sizes N_1 and N_2 respectively. Then the data forms a 2×2 table:

		Response 0	Response 1	
Treatment	T_1	$N_1 - M_1$	M_1	N_1
	T_2	$N_2 - M_2$	M_2	N_2
		$N - M$	M	N

Table 1. *Observed data.*

Taking this table at face-value one is led to inference based on the *apparent causal effect* (Smith and Sugden, 1985, 1988) or equivalently prima facie causal effect (Holland, 1986) given by

$$M_1/N_1 - M_2/N_2 \tag{1}$$

Under Rubin's model for causal inference (Rubin, 1978), attached to each unit is pair of (binary) responses, one for each treatment and the causal effect is the simple difference of these responses for the ith unit:

$$Y_i^{(1)} - Y_i^{(2)}, \quad i = 1, \ldots, N \tag{2}$$

Using this notation implicitly makes the stable treatment unit value assumption (Stuva) (Rubin, 1980).

The population can also be represented as 2×2 table of counts with overall total N, but of rather different structure from that of Table 1

		Response under T_2 0	1	
Response under T_1	0	a_1	a_2	$a_1 + a_2$
	1	a_3	a_4	$a_3 + a_4$
		$a_1 + a_3$	$a_2 + a_4$	N

Table 2. *True population*

Note that none of these frequencies save the overall total N is observable except if all units are assigned to one treatment, when the corresponding margin is observed.

The true mean causal effect is the finite population mean of the unit effects.

$$\overline{Y}^{(1)} - \overline{Y}^{(2)} = (a_3 - a_4)/N \tag{3}$$

It is important to realise that the apparent and true causal effects may differ widely. Even with the simple model of treatment assignment considered below and disregarding chance fluctuation in actual numbers assigned (i.e. considering expected numbers only), there may be wide differences unless assignment is non-informative, that is the probability of assigning treatment 1 to a unit is constant over all units in the population. Smith and Sugden (1985, 1988) give a general framework following Rubin (1978) and including the effects of unit selection but also give a numerical example which shows under informative treatment assignment that apparent causal effects can be reversed in the true population, even in extreme cases such as the comparison of deaths due to lung cancer under "treatments" of smoking and non-smoking. Even when the mechanism is non-informative, chance fluctuations may produce differences though are these likely to be small if N_1, N_2, are large.

The model for treatment assignment is to assume that each unit has independently a probability of being assigned to T_1 which depends on the potential responses

		Response under T_2 0	1
Response under T_1	0	p_1	p_2
	1	p_3	p_4

Table 3. [*prob receive* T_1].

The asumption of non-informative treatment assignment or strong ignorability here is that

$$p_1 = p_2 = p_3 = p_4 = p \quad (4)$$

but otherwise these probabilities may take any values between 0 and 1.

Conditional on the true population (Table 2) we have four independent binomially distributed random variables representing units assigned to treatment 1

		Response under T_2		
		0	1	
Response under T_1	0	X_1	X_2	$X_1 + X_2 = N_1 - M_1$
	1	X_3	X_4	$X_3 + X_4 = M_1$
		$X_1 + X_3$	$X_2 + X_4$	N_1

Table 4. *Units assigned to* T_1.

where $X_i \sim$ Bin $(a_i; p_i)$ $\quad i = 1, 2, 3, 4$.

Furthermore, let $Z_i = a_i - X_i (i = 1, 2, 3, 4)$ be the corresponding Binomial complements so that we have

		Response under T_2		
		0	1	
Response under T_1	0	Z_1	Z_2	$Z_1 + Z_2$
	1	Z_3	Z_4	$Z_3 + Z_4$
		$Z_1 + Z_3$ $= N_2 - M_2$	$Z_2 + Z_4$ $= M_2$	N_2

Table 5. *Units assigned to* T_2.

Note that in each table the cell frequencies are unobservable but in Table 4, the row margins and in Table 5, the column margins are observed using Table 1. From these relations and the overall total N given in Table 2 we may deduce easily the following inequalities in complementary pairs

$$\begin{aligned} N - M_1 \geq a_1 + a_2 \geq N_1 - M_1 \\ M_1 + N_2 \geq a_3 + a_4 \geq M_1 \end{aligned} \quad (5)$$

and

$$\begin{aligned} N - M_2 \geq a_1 + a_3 \geq N_2 - M_2 \\ N_1 + M_2 \geq a_2 + a_4 \geq M_2 \end{aligned}$$

These bounds can be attained by simply considering extreme cases e.g. if every unit "succeeding" under T_2 is assigned T_1 and every unit "failing" under T_2 is assigned T_2 then in Table 1, $N_1 - M_1 = a_2$, $M_1 = a_4, N_2 - M_2 = a_1 + a_3$, $M_2 = 0$. This will happen if $p_1 = 0$, $p_2 = 1$, $p_3 = 0$, $p_4 = 1$.

The bounds for the true causal effect (3) are thus very wide, in fact of width unity:

$$(M_1 - M_2) - N_1 \leq a_3 - a_4 \leq (M_1 - M_2) + N_2 \tag{6}$$

Clearly the data considered in isolation provides very little (if any) information on which to base a causal inference and we must rely on prior knowledge. If, for example, our prior opinions lead us to believe that non-informativeness (4) is at least approximately true then the observed data in Table 1 in conjunction with prior opinion of the efficiency of the two treatments should lead to sensible inferences and with vague prior knowledge our posterior opinions of (3) in this case should be centred around the apparent effect (1).

It is mathematically difficult to obtain a tractable expression for the likelihood with "parameters" in Table 2 even using well known results such that the distribution of X_1 in Table 4 is hypergeometric conditional on the margins. A more elegant likelihood results from a "superpopulation" prior for the responses considered at unit level, each assumed to be IID.

		Response under T_2		
		0	1	
Response under T_1	0	$r_1 =$ $1 - \pi_1 - \pi_2 + \pi_{12}$	$r_2 =$ $\pi_2 - \pi_{12}$	$1 - \pi_1$
	1	$r_3 = \pi_1 - \pi_{12}$	$r_4 = \pi_{12}$	π_1
		$1 - \pi_2$	π_2	1

Table 6. *Joint prior at unit level.*

The hyperparameters π_1, π_2 measure the efficiency of the treatments and for large N, $\pi_1 - \pi_2$ is approximately the true causal effect (3). Note that the precise meaning of the hyperparameter π_{12} should be considered carefully. The implied prior for the "parameters" in Table 2 is therefore multinomial with the above cell probabilities.

A generating function argument then shows that the joint distribution of $X_i, Y_i \quad i = 1, 2, 3, 4$, is also multinomial with probabilities (7)

$$\xi_1 = p_i r_i, \quad \eta_i = (1 - p_i) r_i \quad i = 1, 2, 3, 4 \quad \text{respectively} \tag{7}$$

Thus the likelihood of the data in Table 1 in terms of the parameters $p_i (i = 1, 2, 3, 4)$; π_1, π_2, π_{12} is proportional to

$$(\xi_1 + \xi_2)^{N_1 - M_1} (\xi_3 + \xi_4)^{M_1} (\eta_1 + \eta_3)^{N_2 - M_2} (\eta_2 + \eta_4)^{M_2} \tag{8}$$

The seven parameters are not individually estimable from the data but if we assume that (4) non-informativeness of treatment assignment holds, then (8) becomes

$$p^{N_1} (1 - p)^{N_2} \times (1 - \pi_1)^{N_1 - M_1} \pi_1^{M_1} \times (1 - \pi_2)^{N_2 - M_2} \pi_2^{M_2} \tag{9}$$

a product of three "face-value" likelihoods (Dawid and Dickey, 1977) which do not depend on π_{12}, this parameter being non-estimable. Thus under a priori independence, inference on π_1, π_2, and hence predictive inference on (3), proceeds separately using the observed proportions M_1 out of N_1, M_2 out of N_2, as for a designed experiment in which the treatment had been assigned by the experimenter at random.

Under the more restrictive assumption that the assigment probabilities do not depend on the potential responose under T_2, so that

$$p_1 = p_2 = q \text{ and } p_3 = p_4 = s \tag{10}$$

the likelihood becomes

$$q^{N_1-M_1} \times (1-\pi_1)^{N_1-M_1}\pi_1^{M_1} \times [(1-q)(1-\pi_2)+(\pi_1-\pi_{12})(q-s)]^{N_2-M_2} \times [\pi_2(1-q)+\pi_{12}(q-s)]^{M_2} \tag{11}$$

which does not give simple face-value inferences on either π_1 or π_2.

If, however q is close to s we may neglect the second term in each bracket and face value inferences are again approximately valid for both π_1 and π_2 (and hence predictive inference on (3)). Under the additional condition that the potential responses are independent for each unit

$$\text{i.e.}\quad \pi_{12} = \pi_1\pi_2 \tag{12}$$

then (11) shows that face value inference on π_2 is exact whereas that on π_1 is modified by a correction term

$$[(1-q)] + \pi_1(q-s)]^{N_2} \tag{13}$$

No such simplification occurs when the assumption of prior independence holds without any condition on the assignment mechanism, but a similar condition to (10) regarding T_1 produces analogous conclusions.

In the general case the posterior predictive distribution for the parameters in Table 2 is such each entry is *a posteriori* the sum of 2 independent binomially distributed random variables with probabilities determined by (7) but we can also express this at the unit level by the following table:

Class in which unit observed in Table 1	Value of (2) -1	0	1
$N_1 - M_1$	$\xi_2/(\xi_1+\xi_2)$	$\xi_1/(\xi_1+\xi_2)$	0
M_1	0	$\xi_4/(\xi_3+\xi_4)$	$\xi_3/(\xi_3+\xi_4)$
$N_2 - M_2$	0	$\eta_1/(\eta_1+\eta_3)$	$\eta_3/(\eta_1+\eta_3)$
M_2	$\eta_2/(\eta_2+\eta_4)$	$\eta_4/(\eta_2+\eta_4)$	0

Table 7. *Posterior probabilities of the unit causal effect (2) (all units a posteriori independent.)*

From this table we can obtain immediate expressions for the posterior mean and variance, say, of the true mean causal effect (3) in terms of the parameters $p_i, i = 1, 2, 3, 4$ and π_1, π_2, π_{12} (If these are themselves unknown then such expressions can be used in a further stage of averaging). Under the twin assumption of non-informativeness (4) and *a priori* independence (12) the expressions reduce to the intuitive forms

$$\text{E}[N(\overline{Y}^{(1)} - \overline{Y}^{(2)})|\text{data}] = (M_1 + N_2\pi_1) - (M_2 + N_1\pi_2)$$
$$\text{V}[N(\overline{Y}^{(1)} - \overline{Y}^{(2)})|\text{data}] = N_1\pi_2(1-\pi_2) + N_2\pi_1(1-\pi_1)$$

When π_1 and π_2 are given independent priors proportional to $[\pi_i(1-\pi_i)]^{-1}$, $0 < \pi_i < 1$, $i = 1, 2$, then their posterior means using the face-value likelihood (9) are just M_1/N_1 and M_2/N_2 respectively and the posterior mean of the true causal effect (3) reduces to the apparent effect (1).

Rubin (1978) considers a similar example but his arguments are asymptotic. His results, on the posterior distribution of $\pi_1 - \pi_2$, do not depend on any explicit characterisation of the joint distribution of potential responses embodied in our parameter π_{12}.

An alternative approach would be to regard the data as an incomplete three way classification and use an extension of the methods of Albert (1985). The third dimension is the assigned treatment and only two sets of marginal totals are observed: see Tables 4 and 5. This approach may be the subject of further investigation.

REFERENCES

Albert, J. H. (1985). Bayesian estimation methods for incomplete two-way contingency tables using prior beliefs of association. 589–602, *Bayesian Statistics* **2.** (J. M. Bernardo, M. H. DeGroot, D. V. Lindley and A. F. M. Smith, eds.). Amsterdam: North-Holland .

David, A. P. and Dickey, J. M. (1977). Likelihood and Bayesian inference from selectively reported data. *J. Amer. Statist. Assoc.* **72**, 845–850.

Holland, P. W. (1986). Statistics and causal inference. *J. Amer. Statist. Assoc.* **81**, 945-970.

Rosenbaum, P. R. and Rubin, D. B. (1983). The central role of the propensity score in observational studies for causal effects. *Biometrika* **70**, 41–55.

Rosenbaum, P. R. and Rubin, D. B. (1984). Reducing bias in observational studies using subclassification on the propensity score. *J. Amer. Statist. Assoc.* **79**, 516–524.

Rubin D. B. (1978). Bayesian inference for causal effects: The role of randomisation. *Ann. Statist.* **6**, 34–58.

Rubin D. B. (1980). Discussion of Randomisation analysis of experimental data: The Fisher randomisation test by D. Basu, *J. Amer. Statist. Assoc.* **75**, 591–593.

Smith, T. M. F. and Sugden, R. A. (1985). Inference and the ignorability of selection for experiments and surveys. *Bull. Int. Statist. Inst. 45th Session*, Amsterdam.

Smith, T. M. F. and Sugden, R. A. (1988). Selection and allocation in experiments, surveys and observational studies. *Int. Statist. Rev.* **56.** (To appear)

BAYESIAN STATISTICS 3, pp. 791–799
J. M. Bernardo, M. H. DeGroot, D. V. Lindley and A. F. M. Smith, (Eds.)

Approximate Posterior Distributions in Censored Regression Models

T. J. SWEETING
University of Surrey

SUMMARY

Location-scale regression models for possibly right-censored observations are widely used in the analysis of failure time and survival data. It has recently been shown that good posterior approximations based on standard χ^2 and multivariate t distributions are to be expected in such models. These approximations turn out to be particularly accurate in the case of loglinear Weibull models. In the present paper we describe some simple data-dependent transformations which improve these approximations, especially for other error models. The attempt to find easily-handled, yet accurate approximations is regarded as being complementary to a full numerical integration approach. The work is motivated partly by asymptotic considerations, and partly by the structure of the information function. Some clinical survival data are used to illustrate the approximations.

Keywords: CENSORED DATA; LOCATION-SCALE REGRESSION MODELS; BAYESIAN ANALYSIS; APPROXIMATE INFERENCE.

1. INTRODUCTION

In Sweeting (1987b), the possibility of using simple approximations based on χ^2 and multivariate t distributions to posterior distributions arising from location-scale regression models for censored survival, or failure time data was explored. If successful, the obvious appeal of these approximations is that otherwise complex analyses can be performed with relative ease, interpretation aided, and a unified approach to the analysis of such data made available. Such approximations are regarded here as complementing full numerical integration.

Normal approximations based on maximum likelihood (ML) estimates are of course the simplest general approximations to posterior distributions, but they can be poor in small samples; in particular, they do not give any degrees of freedom reduction in the error estimation, leading to over-precise statements regarding the regression parameters. In the present context, they can also be poor in moderately large samples; a large number of regressor variables, censored observations, and a nonnormal error distribution all combine to reduce the effective degrees of freedom available (cf. Sweeting (1987b), Section 4). The approximations described in Sweeting (1984, 1987b) are essentially adaptations of ML which take into account the particular model structure. The overall conclusion is that such approximations generally perform very well indeed, although accuracy is of course dependent on the underlying error model. In particular, they are found to be very good in loglinear Weibull models.

In the present paper we explore the reasons for the high accuracy in Weibull models, and obtain refinements appropriate for other error models. In addition, simple transformations to correct skewness in the marginal posterior distributions of the regression coefficients are obtained. The approximations are applied to some clinical survival data considered previously.

Similar ideas, but motivated from a classical point of view, have been discussed by Sprott; see for example Sprott (1980). The transformations used here to eliminate skewness are based on the same considerations as those in Sprott (1980), except that we use data-dependent versions. The treatment of the scale parameter here, however, is quite different.

As is well known (Lawless (1982) for example), in the uncensored case the Bayesian analysis of location-scale regression models with the usual vague prior is formally equivalent to exact frequency-based conditional inference, as well as fiducial inference and other similar approaches. Sprott (1980) also comments that unconditional inference based on ML estimation and observed information is very close in practice to exact conditional inference in these models (which is another instance of the observation that most of the relevant ancillary information is usually contained in the observed information). For censored data, one would expect similar conclusions to hold for an appropriate set of approximate ancillary statistics, as discussed in Sweeting (1987b).

2. REGRESSION MODELS FOR CENSORED DATA

In this section we define the class of location-scale regression models we will be dealing with, and study the form of the observed information. Let $y_1, \ldots, y_n$ be independent observations from the location-scale regression model

$$y_i = x_i^T \theta + \sigma z_i$$

where θ is an unknown k-vector, $\sigma > 0$ is the unknown scale parameter, $X = (x_1, \ldots, x_n)^T$ is an $n \times k$ matrix of full column rank and $x_i^T = (1, x_i^{*T})$, so that the model contains a constant term θ_1. The z_i are i.i.d. with known density f, assumed twice differentiable throughout $\Re$. In addition, we suppose that $n - m$ of the observations are right-censored values, and let C, U be the set of censored and uncensored values respectively. This is the set-up in Sweeting (1987b).

Let $g(z) = \log f(z)$, $S(z)$ be the survivor function and $h(z)$ the hazard function corresponding to $f(z)$ and let $\phi = \sigma^{-2}$ be the precision parameter. Let $z_i(\theta, \phi) = \phi^{\frac{1}{2}}(y_i - x_i^T \theta)$, $q_i(z) = -g(z)$ if $i \in U$ and $q_i(z) = -\log S(z)$, $i \in C$ and $\hat{\theta}(\phi)$ be the ML estimate of θ for fixed ϕ. The second-order partial derivatives evaluated at $\theta = \hat{\theta}(\phi)$, given in Sweeting (1987b), are

$$\begin{aligned}
\ell_{\theta\theta}(\hat{\theta}(\phi), \phi) &= -\phi \Sigma q_i''(z_i) x_i x_i^T = -\phi D^{-1}(\phi) \\
\ell_{\theta\phi}(\hat{\theta}(\phi), \phi) &= \left(2\phi^{\frac{1}{2}}\right)^{-1} \Sigma z_i q_i''(z_i) x_i = \left(2\phi^{\frac{1}{2}}\right)^{-1} v(\phi) \\
\ell_{\phi\phi}(\hat{\theta}(\phi), \phi) &= -(2\phi^2)^{-1} \left[m + \frac{1}{2} \Sigma \left\{ z_i^2 q_i''(z_i) - z_i q_i'(z_i) \right\}\right] \\
&= -(2\phi^2)^{-1} N(\phi)
\end{aligned}$$

where here $z_i = z_i(\hat{\theta}(\phi), \phi)$.

The basic thesis is that good χ^2 and multivariate t approximations are to be expected provided that, relative to n, (i) $D(\phi)$ is approximately constant in ϕ, (ii) $v(\phi)$ is approximately zero, and (iii) $N(\phi)$ is approximately constant in ϕ. Conditions (i)-(iii) are of course exact in the case of a normal error model with no censored values. For approximation (ii) to hold, it is necessary to parametrize the model so that either $E(zg''(z))$ or $\Sigma \hat{z}_i q_i''(\hat{z}_i)$ is zero, where $\hat{z}_i = z_i(\hat{\theta}, \hat{\phi})$, $\hat{\theta} = \hat{\theta}(\hat{\phi})$, and $\hat{\phi}$ is the estimate of ϕ obtained in Section 3. The latter data-dependent parametrization is particularly appropriate in the case of censored data, as discussed in Sweeting (1987b), and we assume throughout this paper that this has been carried out.

The accuracy of the final approximations depends in part on sample size, and in part on the assumed error distribution. They perform especially well in the case of Gumbel, or more generally log Gamma, error models. One reason for this is that, as we shall see, the basic approximation (i) above is particularly good for these distributions. Also, in very small samples the approximations may be adversely affected by skewness of the likelihood function in θ_j for fixed ϕ. We show here that simple data-dependent parameter transformations lead to a wider range of applicability of the approximations.

3. APPROXIMATE MARGINAL POSTERIOR DISTRIBUTION OF ϕ

It was shown in Sweeting (1984, 1987b) that if the posterior distribution of θ given ϕ is approximately normal, then the structure of the observed information discussed in Section 2 leads to an approximate χ^2 posterior distribution for ϕ. Although we are generally thinking of cases where such approximate normality is going to hold good, at least for some simple transformation of θ, there is actually no need at this stage to make such an assumption in order to investigate the posterior distribution of ϕ if we appeal to asymptotic considerations. In the sequel the order symbol $0(\cdot)$ will mean $0_p(\cdot)$ under the true value (θ_0, ϕ_0); also all expectations are calculated under (θ_0, ϕ_0).

Let $p(\theta, \phi)$ be the joint prior density of θ, ϕ. Then, to $0(n^{-1})$ uniformly on bounded neighbourhoods of $\hat{\phi}$, marginal posterior density of ϕ is

$$p(\phi|y) \propto p(\hat{\theta}(\phi), \phi)|\phi D^{-1}(\phi)|^{-\frac{1}{2}k} p(y|\hat{\theta}(\phi), \phi) \tag{1}$$

This follows from the general second-order approximation given by Kass, Kadane and Tierney (1989), which is similar to that in Tierney and Kadane (1986) where $\hat{\theta}(\phi)$ is taken to be the conditional posterior mode. The formula is also given in Sweeting (1987a) in discussion of Cox and Reid (1987); omitting the prior density factor, this expression is related to a modified or conditional profile likelihood.

If $D^{-1}(\phi)$ is approximately constant, then

$$p(\phi|y) \propto p(\hat{\theta}(\phi), \phi)\phi^{-\frac{1}{2}k} p(y|\hat{\theta}(\phi), \phi) \tag{2}$$

approximately. It is of interest to investigate the order of accuracy of the final approximation (2). Write $G(\phi) = D^{-1}(\phi)$; ignoring for the moment errors of approximation, the argument is briefly as follows. First, if approximation (ii) of Section 2 holds, then $\theta'(\hat{\phi}) \approx 0$, which implies that $\hat{\theta}(\phi) \approx \hat{\theta}$. Then $z_i(\hat{\theta}(\phi), \phi) \approx z_i(\hat{\theta}, \phi)$, and

$$\begin{aligned} G'(\phi) &\approx \frac{1}{2}\phi^{-1}\Sigma z_i(\hat{\theta}_i, \phi) q_i'''(z_i(\hat{\theta}, \phi)) x_i x_i^T \\ &= \frac{1}{2}\hat{\phi}^{-1}\Sigma z_i q_i'''(\hat{z}_i) x_i x_i^T \end{aligned}$$

at $\phi = \hat{\phi}$. Next consider the case of a log Gamma distribution with

$$f(z) \propto \exp\left\{\alpha(z + a) - e^{z+a}\right\} \tag{3}$$

where a is the orthogonalizing constant defined in Sweeting (1987b). Here we find that $g''(z) = g'''(z)$. But the orthogonal parametrization has $\Sigma \hat{z}_i q_i''(\hat{z}_i) = 0$, and so in the uncensored case $n^{-1}G'(\hat{\phi})$ is likely to be small. Therefore $G(\phi)$ is likely to vary slowly near $\hat{\phi}$.

We show that this behaviour can be obtained for other error models, provided we carry out a simple reparametrization. We formalize the above argument in the uncensored case. Assume that information increases linearly with n; specifically assume that $\lambda_{\max}(X^T X) \leq Cn$,

$\lambda_{\min}(X^TX) \geq cn$ for all n where $\lambda_{\max}(A)$, $\lambda_{\min}(A)$ are the maximum and minimum eigenvalues of A. This condition wiil justify the various order calculations to be made below, although full details are not given for reasons of space.

First we show that the approximate observed orthogonality of θ and ϕ implies that

$$\hat{\theta}(\phi) = \hat{\theta} + 0(n^{-1}) \tag{4}$$

when $\phi - \hat{\phi} = 0(n^{-\frac{1}{2}})$ (cf. Cox and Reid (1987) for the case of "expected" orthogonality). We have

$$\hat{\theta}(\phi) - \hat{\theta} = (\phi - \hat{\phi})\hat{\theta}'(\hat{\phi}) + 0(n^{-1})$$

and $\hat{\theta}'(\hat{\phi}) = \frac{1}{2}\hat{\phi}^{-\frac{1}{2}}D(\hat{\phi})v(\hat{\phi})$ (on implicit differentiation of $\ell_\theta(\hat{\theta}(\phi), \phi) = 0$. We show that $v(\hat{\phi}) = 0(n^{\frac{1}{2}})$ from which (4) will follow, since $D(\hat{\phi}) = 0(n^{-1})$. The orthogonal parametrization of Section 2 has $\Sigma\hat{z}_i q_i''(\hat{z}_i) = 0$ and, writing $z_{i0} = z_i(\theta_0, \phi_0)$, it follows that $\Sigma z_{i0} q_i''(z_{i0}) = 0(n^{\frac{1}{2}})$ since $\hat{z}_i = z_{i0} + 0(n^{-\frac{1}{2}})$. Now $n^{-1}\Sigma z_{i0} q_i''(z_{i0}) = E(z_{i0} q_i''(z_{i0})) + 0(n^{-\frac{1}{2}})$, and so $E(z_{i0} q_i''(z_{i0})) = 0(n^{-\frac{1}{2}})$. Therefore $n^{-1}\Sigma z_{i0} q_i''(z_{i0}) x_i = 0(n^{-\frac{1}{2}})$, which implies that $v(\hat{\phi}) = 0(n^{\frac{1}{2}})$ as required.

Let $\gamma = \phi^{1+p}$ where

$$p = \frac{\Sigma\hat{z}_i q_i''(\hat{z}_i)}{2\Sigma q_i''(\hat{z}_i)}$$

Define $H(\theta, \gamma) = \phi^{-p} G(\theta, \phi)$ where $G(\theta, \phi) = -\phi^{-1}\ell_{\theta\theta}(\theta, \phi)$. If $\phi = \hat{\phi} + 0(n^{-\frac{1}{2}})$, we have

$$n^{-1}(H(\hat{\theta}, \gamma) - H(\hat{\theta}, \hat{\gamma})) = n^{-1}(\gamma - \hat{\gamma})H'(\hat{\theta}, \hat{\gamma}) + 0(n^{-1}). \tag{5}$$

Now

$$H'(\theta, \gamma) = (1 + p)^{-1}\phi^{-(1+2p)}[\phi G'(\theta, \phi) - pG(\theta, \phi)].$$

Let $p_0 = \{2E(q''(z_0))\}^{-1}E(z_0 q'''(z_0))$; then

$$E(\phi_0 G'(\theta_0, \phi_0) - p_0 G(\theta_0, \phi_0)) = \Sigma x_i x_i^T \left[\frac{1}{2}E(z_0 q'''(z_0)) - p_0 E(q''(z_0))\right] = 0.$$

Since $p_0^{-1}p = 1 + 0(n^{-\frac{1}{2}})$, it follows that $H'(\theta_0, \gamma_0) = 0(n^{\frac{1}{2}})$ and so also $H'(\hat{\theta}, \hat{\gamma}) = 0(n^{\frac{1}{2}})$ since $\hat{\theta} = \theta_0 + 0(n^{-\frac{1}{2}})$, $\hat{\gamma} = \gamma_0 + 0(n^{-\frac{1}{2}})$. The r.h.s. of (5) is therefore $0(n^{-1})$. Finally, write $H(\gamma) = H(\hat{\theta}(\gamma), \gamma)$; from (4) we have $n^{-1}(H(\gamma) - H(\hat{\theta}, \gamma)) = 0(n^{-1})$, and so we have shown that

$$n^{-1}(H(\gamma) - H(\hat{\theta}, \hat{\gamma})) = 0(n^{-1}). \tag{6}$$

It now follows from (6) that, with relative error $0(n^{-1})$,

$$p(\gamma|y) \propto p(\hat{\theta}(\gamma), \gamma)\gamma^{-\frac{1}{2}k}p(y|\hat{\theta}(\gamma), \gamma) \tag{7}$$

when $\gamma - \hat{\gamma} = 0(n^{-\frac{1}{2}})$.

Although the above reparametrization is motivated by second-order asymptotic considerations in the uncensored case, when condition (ii) of Section 2 holds the near-constancy of $H(\gamma)$ is to be expected with a moderate amount censoring. We therefore use and investigate it as a general refinement in the case of censored observations. For computation note that when $i \in C$, $q_i''' = h(\{h + g'\}\{2h + g'\} + g'')$. Returning to the case of the Gumbel model with $\alpha = 1$ in (3), note that here $h(z) = g'(z)$ and it follows that $q_i''(z) = q_i'''(z)$ even for the *censored* values! Thus no adjustment is necessary in the case of censored data from loglinear Weibull models.

For simplicity, we treat the case of vague prior knowledge about θ and ϕ, and assume that $p(\theta, \phi) \propto \phi^{-1}$. The case of substantial prior knowledge may be treated as described in Sweeting (1984), if appropriate. The approximate posterior mode $\hat{\gamma}$ of γ is obtained from the equation

$$2\ell_\gamma(\hat{\theta}(\gamma), \gamma) - (k+2)\gamma^{-1} = 0$$

which on differentiation of (7) w.r.t. γ is seen to be equivalent to the equation

$$\Sigma z_i q_i'(z_i) = m - (1+p)(k+2).$$

After some algebra, the second derivative of the logarithm of the approximate posterior density (7) is found to be

$$-(2\gamma^2)^{-1}(1+p)^{-2}\left[N(\phi) - (1+p)(k+2) - \frac{1}{2}v^T(\phi)D(\phi)v(\phi)\right]$$

$$= (-2\gamma^2)^{-1}[\nu(\gamma) - 2]$$

say. As discussed in Sweeting (1984), provided $\nu(\gamma)$ is approximately constant over the main range of (7), the posterior distribution of $d\gamma$ is approximately $\chi^2(\nu)$, where $\nu = \nu(\hat{\gamma})$ and $d = (\nu - 2)\hat{\gamma}^{-1}$ is the deviance. This amounts to a generalized Gamma distribution for ϕ, but it is preferable to work in terms of the transformed precision γ. Note that this final approximation cannot be justified entirely from these asymptotic considerations, since generally $\nu(\phi)/\nu(\hat{\phi}) = 1 + 0(n^{-\frac{1}{2}})$ only. It is possible to investigate further refinements at this stage, but the simple $\chi^2 - t$ analysis is then lost. Empirical work suggests however that the error here is generally small.

4. APPROXIMATE POSTERIOR DISTRIBUTION OF REGRESSION COEFFICIENTS

If for fixed γ the observed information $-\ell_{\theta\theta}(\theta, \gamma)$ is approximately constant over a sufficiently large region about $\hat{\theta}(\gamma)$, then the posterior distribution of θ given γ will be approximately $N_k(\hat{\theta}(\gamma), \gamma^{-1}\Sigma(\gamma))$, where $\Sigma(\gamma) = H^{-1}(\gamma) = \phi^{p-1}D(\phi)$. If in addition approximation (ii) of Section 2 holds, then $\hat{\theta}(\gamma) \simeq \hat{\theta}$ and $\Sigma(\gamma) \simeq \hat{\Sigma} \equiv \Sigma(\hat{\gamma})$ as discussed in the previous section, so that $\theta|\gamma, y \sim N(\hat{\theta}, \gamma^{-1}\hat{\Sigma})$ approximately. The χ^2 approximation for the marginal distribution of γ in the previous section now gives rise to a multivariate t approximation for the marginal distribution of θ, so reparametrization to γ causes no change to the argument in Sweeting (1984, 1987b).

In the uncensored case, $\hat{\theta}(\gamma)$ and $\Sigma(\gamma)$ are constant to $0(n^{-1})$ from (4) and (6) respectively, which leads us to explore possible transformations of θ to achieve an error of normal approximation of $0(n^{-1})$. From a more general point of view, we would wish to remove any significant skewness in the distribution of θ given γ before attempting to use a t approximation. Numerical experience suggests that skewness is often not too serious a problem. However, simple and easily automated transformations will often remove skewness, thus widening the range of applicability of the approximations. From a purely asymptotic point of view, it appears that a t approximation is not necessary, being essentially an $0(n^{-1})$ correction. However, in cases where there are many regressor variables and a fair degree of censoring, the improvement has been found to be clearly worthwhile.

Quite generally, if θ is a *single* real parameter, then on transformation to $\beta = \beta(\theta)$, the third derivative w.r.t. β of the loglikelihood function $\ell(\theta)$ evaluated at $\theta = \hat{\theta}$, the ML estimate, is found to be

$$[\beta'(\hat{\theta})]^{-3}[\ell'''(\hat{\theta}) - 3\{\beta'(\hat{\theta})\}^{-1}\beta''(\hat{\theta})\ell''(\hat{\theta})]$$

where a' indicates differentiation w.r.t. θ. There are many choices of $\beta(\theta)$ which will make this quantity zero. If we allow data-dependent parametrizations, then one possibility is to choose $\beta(\theta)$ so that the above expression is zero for *all* θ on replacing $\beta'(\hat{\theta})$, $\beta''(\hat{\theta})$ by $\beta'(\theta)$, $\beta''(\theta)$ respectively. It is easily seen that the solution is $\beta(\theta) = \lambda^{-1}(e^{\lambda\theta} - 1)$ where

$$\lambda = \{3\ell''(\hat{\theta})\}^{-1}\ell'''(\hat{\theta}). \tag{8}$$

Returning to the present problem, the marginal posterior density of a regression coefficient θ_j for fixed γ is, to $0(n^{-1})$, proportional to

$$p(y|\theta_j, \hat{\eta}_j(\theta_j), \gamma)|\ell_{\eta\eta}(\theta_j, \hat{\eta}_j(\theta_j), \gamma)|^{-\frac{1}{2}} \tag{9}$$

(cf. Tierney and Kadane (1986)), where η_j is the vector of regression coefficients omitting θ_j. The value λ_j for the above transformation $\theta_j \to \beta_j$ can then be determined from (8), giving rise to a t approximation for the posterior distribution of β_j. The resulting expression however is complicated, involving up to fifth order partial derivatives of ℓ. Alternatively, assuming the first factor in (9) (i.e. the profile likelihood) is the dominant one, a simpler method is to determine λ_j from the profile likelihood alone. In practice this appears to successfully account for any significant skewness present. The more difficult problem of finding a suitable multivariate transformation $\theta \to \beta$ for approximate normality is not pursued here; we consider only univariate transformations $\theta_j \to \beta_j$ for removing skewness in the marginal distributions.

Write $k_j(\theta_j, \gamma) = \ell(\theta_j, \hat{\eta}_j(\theta_j), \gamma)$; we therefore reparametrize to $\beta_j = \lambda_j^{-1}(e^{\lambda_j\theta_j} - 1)$ where $\lambda_j = (3\hat{k}_j'')^{-1}\hat{k}_j'''$, and a' indicates differentiation w.r.t. θ_j. It can be shown that $\hat{k}_j'' = \sigma_{jj}^{-1}\gamma$ where σ_{jj} is the j-th diagonal element of $\Sigma \equiv \Sigma(\gamma)$. (cf. Sprott (1980)), and $\hat{k}_j'''$ may be obtained by a further differentiation. In our case, however, there is a simpler way of obtaining $\hat{k}_j''$ and $\hat{k}_j'''$. First orthogonalize to θ_j, in an observed sense, by defining $\eta_j^* = \eta_j - d_j\theta_j$ where

$$d_j = -\hat{\ell}_{\eta_j\eta_j}^{-1}\hat{\ell}_{\eta_j\theta_j} = \eta_j'(\hat{\theta}_j) = \sigma_{jj}^{-1}\Sigma_{\eta_j\theta_j}$$

where $\Sigma_{\eta_j\theta_j}$ is the j-th column of Σ omitting θ_j. Clearly $\eta_j^{*'}(\hat{\theta}_j) = 0$. Now let w_{ij} be i-th vector of regressor variables omitting x_{ij}, and note that

$$x_i^T\theta = x_{ij}\theta_j + w_{ij}^T\eta_j = x_{ij}^*\theta_j + w_{ij}^T\eta_j^*$$

where $x_{ij}^* = x_{ij} + w_{ij}^T d_j = \sigma_{jj}^{-1}\Sigma_j^T x_i$, and Σ_j is the j-th column of Σ. It now follows that for fixed ϕ we have $\hat{k}_j'' = -\phi A_j, \hat{k}_j''' = \phi^{\frac{3}{2}}B_j$ where $A_j = \Sigma q_i''(z_i)x_{ij}^{*2}$, $B_j = \Sigma q_i'''(z_i)x_{ij}^{*3}$, and hence $\lambda_j = -\phi^{\frac{1}{2}}(3A_j)^{-1}B_j$. Finally, γ is evaluated at the approximate posterior mode for a single transformation index for θ_j.

If a joint normal approximation is to be used, it would seem sensible to work with the β-parametrization, rather than with θ, The posterior approximation then becomes $\beta|\gamma, y \sim N(\hat{\beta}, \gamma^{-1}\hat{\Sigma}_1)$ where $\hat{\Sigma}_1 = \Lambda\hat{\Sigma}\Lambda$, $\Lambda = \text{diag}(e^{\lambda_j\theta_j})$, and the marginal posterior distribution of β is taken to be multivariate t. For fitting reduced models, note that $\beta_j = 0$ corresponds to $\theta_j = 0$. These transformations also have an invariance property: if $x_{ij}' = cx_{ij}$ then the induced parameter transformation is $\theta_j' = c^{-1}\theta_j$, and it is readily verified that the corresponding transformation index is $\lambda_j' = c\lambda_j$. Hence $e^{\lambda_j'\theta_j'} = e^{\lambda_j\theta_j}$, and so the same transformation would be made, however the x's are coded.

5. COMPUTATIONAL ASPECTS

Computation of $\hat{\theta}, \hat{\gamma}$ is achieved by simultaneously solving the modal equations for $\hat{\theta}(\gamma)$ and $\hat{\gamma}$, while also updating the transformation index p for ϕ. It is convenient to compute the orthogonality constant at the end of the iterative cycle as discussed in Sweeting (1987b), in which case p may be obtained during the iteration from the formula

$$p = \frac{1}{2}[\Sigma q_i''(z_i)]^{-2}[(\Sigma q_i''(z_i))(\Sigma z_i q_i'''(z_i)) - (\Sigma z_i q_i''(z_i))(\Sigma q_i'''(z_i))].$$

Some increase in accuracy is achieved if the final estimates for θ and the matrix $\hat{\Sigma}$ are obtained using the approximate posterior mean $\overline{\gamma} = \nu/d$, rather than the modal value.

The θ-reparametrization can be made quite simply after the main iterative cycle, and the formulae given in Section 4 require no comment. When λ_j is small, one has the option of retaining θ_j i.e. setting $\lambda_j = 0$. A third-order Taylor expansion of the profile log likelihood $k_j(\theta_j)$ of θ_j shows that the appropriate quantity to consider for this is $[-k_j''(\hat{\theta}_j)]^{-\frac{1}{2}}|\lambda_j|$; when this is less than some specified accuracy, then reparametrization of θ_j will not be necessary.

As described in Sweeting (1984), an augmented matrix may be set up for Sweep operations, in the usual manner for a normal linear model. Here this matrix is taken to be

$$\begin{pmatrix} \hat{\Sigma} & \hat{\beta} \\ \hat{\beta}^T & d \end{pmatrix}$$

and appropriate submodels assessed by a comparison of deviances using the F distribution. As already noted, $\beta_j = 0$ is equivalent to $\theta_j = 0$.

Finally, we note that the approximations here may be used to obtain exact posterior densities and moments by numerical integration, as described in Sweeting (1987b). We regard the above computational approach as complementing full numerical integration; posterior quantities of interest should also be obtained by integration, at least at the final stage of an analysis.

6. AN EXAMPLE: CARCINOGENIC DATA

The data to be considered form a small part a large body of data from a clinical trial carried out by the Radiation Therapy Oncology Group in the U.S. These data consist of all the data from Institution 3, and are describe in Kalbfleisch and Prentice (1980). There are 41 observations, including 14 right-censored values, and models using 4 regressor variables were fitted in Sweeting (1987b). In that paper, the approximate and exact posterior densities of ϕ and the treatment effect θ_2 were compared using a Gumbel error model, along with lognormal and normal approximations for ϕ and θ_2 respectively based on ML. It was seen that the approximations worked extremely well in this case. As mentioned in Section 3, no adjustment to the χ^2 distribution for ϕ is necessary for the Gumbel model.

Corresponding comparisons were also made in the case of a logistic error model, where again the approximations do well, but the distribution of ϕ is less accurate then the Gumbel case. Here the transformation index $(1 + p)$ for ϕ is introduced, and Figure 1 shows the approximate, exact and "ML" posterior distributions for ϕ. The accuracy here is of a similar order to the Gumbel case, as anticipated. The value of p here is -0.142.

There is very little skewness present in the posterior distribution of θ_2 and also of θ_3, the effect of patient condition, and there is no need for a skewness tranformation in either case. Some slight skewness is present however in the distributions of θ_4 and θ_5, which represent the effect of tumour size ("T-staging"), at three levels here. For the Gumbel error model the skewness transformation indexes λ_4 and λ_5 are 0.176, 0.194 respectively. For the purpose of illustration, Figure 2 shows the posterior distribution of θ_4 along with the approximate distribution, which manages to capture the skewness present here reasonably well.

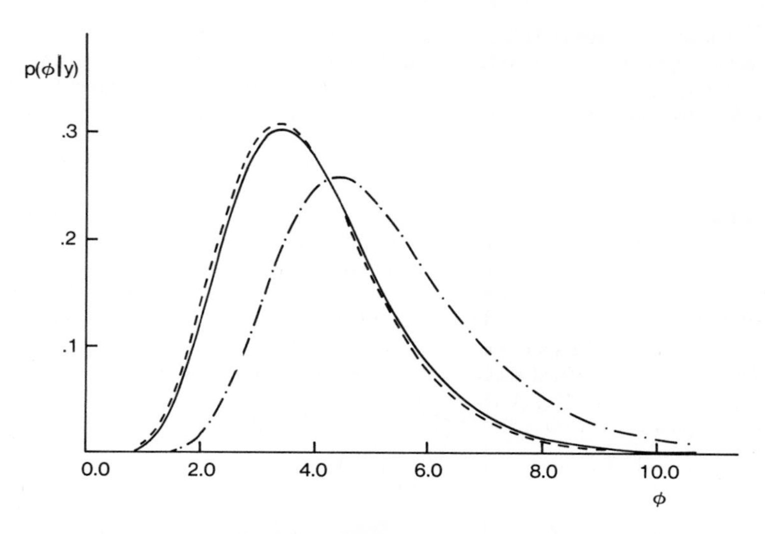

Figure 1. *Exact marginal posterior density of ϕ(———), generalized Gamma (– – –) and lognormal (. – .–) approximations for the R.T.O. group data, assuming a logistic error law.*

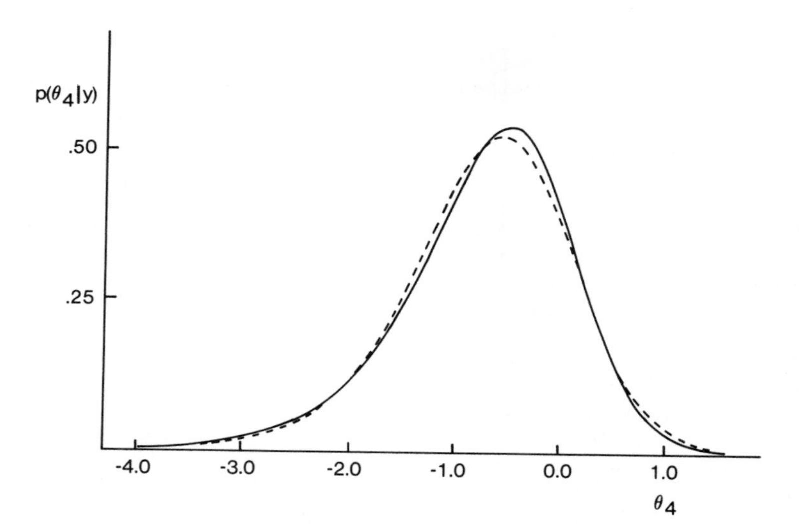

Figure 2. *Exact marginal posterior density of θ_4(———) and t approximation with skewness correction (– – –) for the R.T.O. group data, assuming a Gumbel error law.*

7. FURTHER DEVELOPMENTS

The straight location-scale regression model discussed in the present paper can of course be parametrically enriched: for example, by replacing y_i by a transformation function $g_\lambda(y_i)$, or by including unknown parameters α in the error specification. The approximations described here applied for fixed values of these parameters, will give rise to approximations to the marginal posterior density of these parameters, in the form of a modified-type profile likelihood. In the former case, suitable forms of prior specification for $(\lambda, \theta, \sigma)$ are discussed in Sweeting (1985). Regarding the latter case, one would attempt to use a parametrization which has α near-orthogonal to (θ, σ). In a similar way, the approximations given here can be used to yield approximate predictive distributions.

REFERENCES

Cox, D. R. and Reid, N. (1987). Orthogonal parameters and approximate conditional inference (with discussion). *J. Roy. Statist. Soc. B* **49**, 1–39.

Kalbfleisch, J. D. and Prentice, R. L. (1980). *The Statistical Analysis of Failure Time Data.* New York: Wiley.

Kass, R. E., Kadane, J. B. and Tierney, L. (1989). Asymptotics in Bayesian computation. (In this volume).

Lawless, J. F. (1982). *Statistical Models and Methods for Lifetime Data.* New York: Wiley.

Sprott, D. A. (1980). Maximum likelihood in small samples: estimation in the presence of nuisance parameters. *Biometrika* **67**, 515–23.

Sweeting, T.J. (1984). Approximate inference in location-scale regression models. *J. Amer. Statist. Assoc.* **69**, 246–9.

Sweeting, T.J. (1985). Consistent prior distributions for transformed models. 755–762. *Bayesian Statistics* **2**. (J. M. Bernardo, M. H. DeGroot, D. V. Lindley and A. F. M. Smith, eds.). Amsterdam: North-Holland .

Sweeting, T.J. (1987a). Discussion of Cox and Reid. *J. Roy. Statist. Soc. B* **49**, 20–21.

Sweeting, T. J. (1987b). Approximate Bayesian analysis of censored survival data. *Biometrika* 74, 809–916.

Tierney, L. and Kadane, J. B. (1986). Accurate approximations for posterior moments and marginal densities. *J. Amer. Statist. Assoc.* **81**, 82–86.

BAYESIAN STATISTICS 3, pp. 801–805
J. M. Bernardo, M. H. DeGroot, D. V. Lindley and A. F. M. Smith, (Eds.)

Information Contained in Nuisance Parameters

R. WILLING
Veterinärmedizinischen Universität Wien

SUMMARY

Statisticians often simplify an analysis by using the concept of sufficiency and arguing that no information is lost. This may be misleading in the context of nuisance parameters. If a parameter λ has been identified as uninteresting and an analysis can be made by transforming the data $D \longrightarrow t(D)$ so that λ is not involved, information about the relevant parameter α can be lost, even if the posterior marginal density of $\tilde{\alpha}$ in the two parameter model depends on the data via $t(D)$ only. An example in reliability theory is given which provides insight into this problem.

Keywords: SUFFICIENCY; NUISANCE PARAMETERS; INFORMATION; BAYESIAN INFERENCE.

1. INTRODUCTION

In practice, we often meet the problem where we are interested in a quantity α, but experiments only provide data D, the distribution of which is governed not only by α but also by additional unknown quantity λ (a nuisance parameter). If the likelihood factors in the form

$$f(D|\alpha, \lambda) = g(t_1(D)|\alpha) \cdot h(t_2(D)|\lambda),$$

where $t_1(D)$ is a statistic sufficient for α and ancillary for λ, with $t_2(D)$ having the reverse property (see Lindley, 1965), we need not consider λ, because we can use the marginal model for α alone (Dawid, 1980). If this factorization is not possible, we have to take λ into account to avoid losing valuable information.

Transforming the data $D \longrightarrow t(D)$, so that the distribution of $t(D)$ depends on α alone (which Basu, 1977 calls an α-oriented statistic), is in general throwing away information.

2. THE PROBLEM

Consider the following problem. The lifetime of an object is tested under two different levels of stress S_1 and S_2. Under both levels of stress, the distributions belong to the same family, which, for simplicity, we assume to be the exponential distribution. The model of the stochastic behaviour is defined by their density functions with common parameter λ, which is the hazard rate for the lower stress S_1, together with the acceleration factor α (the parameter of interest) relating the upper to the lower stress level,

$$f(t|S_1) = \lambda e^{-\lambda t}$$
$$f(t|S_2) = \alpha\lambda e^{-\alpha\lambda t}.$$

n independent observations $t_{11}, t_{12}, \ldots, t_{1n}$ on stress level S_1 are available, together with n independent observations $t_{21}, t_{22}, \ldots, t_{2n}$ on stress level S_2. One easy way of analysing α is to combine the observations into n pairs (t_{1i}, t_{2i}), $i = 1, \ldots, n$, and calculate the quotients $u_i = t_{1i}/t_{2i}$. The probability distribution of these latter quantities is a transformed β-distribution of the second order with density function

$$g(u_i|\alpha) = \frac{\alpha}{(u_i + \alpha)^2}.$$

In Basu's terminology, $U_i = T_{1i}/T_{2i}$ is α-oriented. But this formation of pairs, as described in Viertl 1985, is arbitrary. Another possibility is to use the α-oriented statistic $V = T_1/T_2$ with $T_1 = \sum_{i=1}^n T_{1i}$, $T_2 = \sum_{i=1}^n T_{2i}$ and density function

$$h(v|\alpha) = \frac{1}{\beta(n,n)} \frac{\alpha^n v^{n-1}}{(v + \alpha)^{2n}},$$

which is again derived from a β-distribution of the second order. This method has the advantage of using more of the data. These distributions do not depend on the nuisance parameter λ, so the information contained in the parameter λ is not used. Calculating the posterior distributions of α for a non-informative prior distribution $\pi(\alpha) \propto \alpha^{-1}$ we can draw further conclusions.

To show that it is important to use all the observations and not only the quotients or the quotient of the sums alone, a simulation study has been carried out. Figure 1 shows the posterior densities of α with different prior information about λ used. In all the four cases, the non-informative prior distribution $\pi(\alpha) \propto \alpha^{-1}$, for the scale parameter α was used. The data has been simulated with $\alpha = 1.5$ and $\lambda = 2.0$.

a) the first line *(full)* is the posterior density of α using the quotients $u_i, i = 1, \ldots, n$, so that no information about λ is used

$$\pi_1(\alpha|D) \propto \frac{\alpha^{n-1}}{\prod_{i=1}^n (u_i + \alpha)^2}$$

b) the second line *(dash)* is the marginal posterior density of α using all the information $D = (t_{11}, t_{12}, \ldots, t_{1n}, t_{21}, t_{22}, \ldots, t_{2n})$ and a non-informative prior information $\pi(\lambda) \propto \lambda^{-1}$ for the nuisance parameter

$$\pi_2(\alpha|D) \propto \frac{\alpha^{n-1}}{[\alpha t_2 + t_1]^{2n}}$$

where $t_1 = \sum_{i=1}^n t_{1i}$ and $t_2 = \sum_{i=1}^n t_{2i}$ are sufficient statistics for α.

c) the third line *(dot)* is the marginal posterior density of α using all observations D and an informative prior information for λ, the natural conjugate prior $\gamma(m, s)$ with the same prior mean m/s as the simulated value of λ and a small variance m/s^2

$$\pi_3(\alpha|D) \propto \frac{\alpha^{n-1}}{[s + \alpha t_2 + t_1]^{2n+m}}$$

d) the fourth line *(dash dot)* is the posterior density of $\tilde{\alpha}$ using all observations D and perfect prior information about λ (i.e.; in this case the observations on stress level S_1 do not provide any information about α, the value of λ as in the simulation of the data)

$$\pi_4(\alpha|D) \propto \alpha^{n-1} e^{-\lambda(\alpha t_2 + t_1)} \propto \alpha^{n-1} e^{-(\lambda t_2)\alpha}$$

40 pairs of pseudorandom numbers were generated. For the informative prior for λ, a $\gamma(10, 5)$ was used. From the first line to the fourth height of the density at the mode rises. This picture has been found in all the simulation runs. Sometimes due to the stochastic nature of simulation the peak of the posterior density with perfect information about the nuisance parameter was

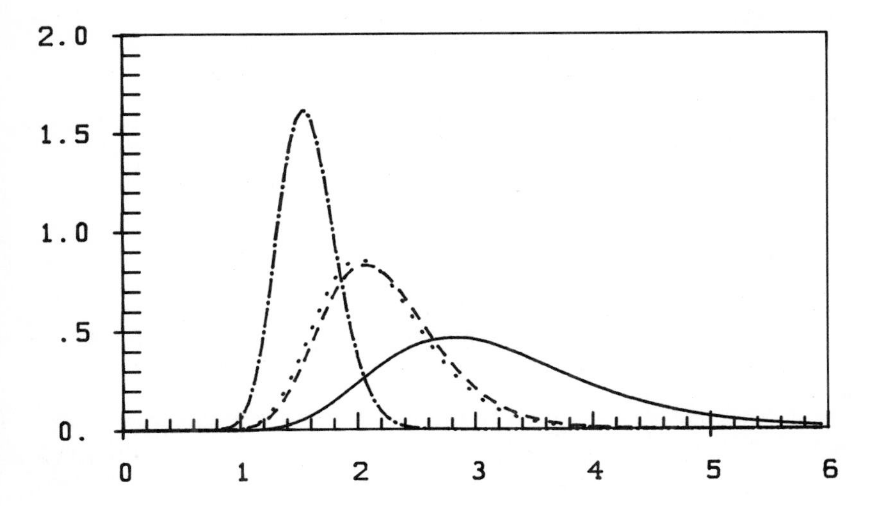

Figure 1. *Posterior densities of the acceleration factor* α

farther away from the simulated value than the other peaks with less information used. But in most of the cases the situation was the other way round.

When we calculate the posterior expectations of α for the different information levels we find (for π_1 an analytical expression is not available)

$$\begin{aligned} \boldsymbol{E}_{\pi_2}\tilde{\alpha} &= \frac{n}{n-1}\cdot\frac{t_1}{t_2} \\ \boldsymbol{E}_{\pi_3}\tilde{\alpha} &= \frac{n}{n+m-1}\cdot\frac{t_1+s}{t_2} \\ \boldsymbol{E}_{\pi_4}\tilde{\alpha} &= \frac{n}{\lambda}\cdot\frac{1}{t_2} \end{aligned}$$

depending on the data $D=(t_1,t_2)$. To compare these values, we can use preposterior analysis and calculate the expected values of the posterior means

$$\begin{aligned} \boldsymbol{E}_{\alpha,\lambda}\boldsymbol{E}_{\pi_2}\tilde{\alpha} &= \frac{n}{n-1}\cdot\frac{n\alpha}{n-1} = \left(\frac{n}{n-1}\right)^2\alpha \\ \boldsymbol{E}_{\alpha,\lambda}\boldsymbol{E}_{\pi_3}\tilde{\alpha} &= \frac{n}{n+m-1}\cdot\frac{n+\lambda s}{n-1}\alpha = \left(\frac{n}{n-1}\right)\left(\frac{n+m}{n+m-1}\right)\alpha \\ \boldsymbol{E}_{\alpha}\boldsymbol{E}_{\pi_4}\tilde{\alpha} &= \frac{n}{\lambda}\cdot\frac{\alpha\lambda}{n-1} = \left(\frac{n}{n-1}\right)\alpha. \end{aligned}$$

They are getting nearer to the value of α. The same can be done for the posterior variances

$$\mathrm{Var}_{\pi_2}\tilde{\alpha} = \frac{n(2n-1)}{(n-1)^2(n-2)}\left(\frac{t_1}{t_2}\right)^2$$
$$\mathrm{Var}_{\pi_3}\tilde{\alpha} = \frac{n(2n+m-1)}{(n+m-1)^2(n+m-2)}\left(\frac{t_1+s}{t_2}\right)^2$$
$$\mathrm{Var}_{\pi_4}\tilde{\alpha} = \frac{n}{\lambda^2}\cdot\frac{1}{t_2^2}$$

and the expected posterior variances

$$\boldsymbol{E}_{\alpha,\lambda}\mathrm{Var}_{\pi_2}\tilde{\alpha} = \frac{n^2(n+1)(2n-1)}{(n-1)^3(n-2)}\alpha^2$$
$$\boldsymbol{E}_{\alpha,\lambda}\mathrm{Var}_{\pi_3}\tilde{\alpha} = \frac{n[(n+m)^2+n)](2n+m-1)}{(n-1)(n+m-1)^2(n-2)(n+m-2)}\alpha^2$$
$$\boldsymbol{E}_{\alpha}\mathrm{Var}_{\pi_4}\tilde{\alpha} = \frac{n}{(n-1)(n-2)}\alpha^2.$$

The posterior variance is the *Bayes Risk*, when using quadratic loss. The expected variance is about twice the size if perfect information is not available for the nuisance parameter.

3. SEMI-SUFFICIENCY

The distributions of the lifetimes D form a *semi-exponential family* as described in Willing 1987

$$f(D|\alpha,\lambda) = \alpha^n\lambda^{2n}e^{-\lambda(\alpha t_2+t_1)}$$

with the semi-sufficient statistic $S(\lambda|\alpha) = (n, \alpha t_2 + t_1)$. One can use the natural conjugate family of the γ-distributions as prior for $\lambda|\alpha$. Reparametrization can transform it into an exponential family, but then the likelihood l of $\phi = \alpha\lambda$ and $\psi = \lambda$ factors

$$l(\phi,\psi;D) = \phi^n e^{-\phi t_2}\cdot\psi^n e^{\psi t_1}$$

in a way which enables one to discard t_1 in an analysis of ϕ. And reparametrization does not give one the opportunity to concentrate on α. Exponential families narrow the choice of conjugate priors, if you want to exploit its advantages.

But using it as a semi-exponential family simplifies the calculations of the marginal posterior distribution of α. Taking

$$\pi(\alpha,\lambda) \propto \alpha^{m_1-1}e^{-s_1\alpha}\cdot\lambda^{m_2-1}e^{-s_2\lambda}$$

as a prior for (α,λ) we obtain the joint posterior density

$$\pi(\alpha,\lambda|D) \propto \alpha^{n+m_1-1}e^{-s_1\alpha}\cdot\lambda^{2n+m_2-1}e^{-(\alpha t_2+t_1+s_2)\lambda}$$

with the marginal posterior density for α

$$\pi(\alpha|D) \propto \frac{\alpha^{n+m_1-1}e^{-s_1\alpha}}{[s_2+\alpha t_2+t_1]^{2n+m_2}}.$$

Instead of the prior hyperparameter s_2, one can use an arbitrary function of α to model prior judgement. This would not complicate the analysis. This density is computationally even easier to handle than the density obtained by using the quotients of the arbitrary pairs. The normalizing factor can be calculated, together with the posterior mean and variance.

4. CONCLUSIONS

Seeking to eliminate nuisance parameters from a statistical analysis is only legitimate when they contain no information about the parameters of interest. It may be easier to use a marginal (reduced) model, but in general information is lost. If it is necessary to simplify the model, use semi-exponential families with their natural conjugates. These allow one to make use of the special form of the likelihood function, and nuisance parameters can be included without much extra work.

REFERENCES

Basu, D. (1977). On the elemination of nuisance parameters. *J. Amer. Statist. Assoc.* **72**, 355–366.

Dawid, A. P. (1980). A Bayesian look at nuisance parameters. *Bayesian Statistics.* (J. M. Bernardo, M. H. DeGroot, D. V. Lindley and A. F. M. Smith, eds.). Valencia: University Press 167–184.

Lindley, D. V. (1965). *Introduction to Probability and Statistics from a Bayesian Viewpoint, Part 2: Inference.* Cambrigde: University Press.

Viertl, R. (1985). A Note on Bayesian Methods in Semiparametric Accelerated Life Testing. Proceedings of the 6-th Symposium Reliability Electronics, Budapest.

Willing, R. (1987). Semi-sufficiency in Accelerated Life Testing. *Probability and Bayesian Statistics,* (R. Viertl, ed.). New York: Plenum.